History of Western Nebraska and Its People.
General History. Cheyenne, Box Butte, Deuel,
Garden, Sioux, Kimball, Morrill, Sheridan,
Scotts Bluff, Banner, and Dawes Counties. A
Group Often Called the Panhandle of
Nebraska

History of Western Nebraska and its People

HISTORY OF
WESTERN NEBRASKA
AND ITS PEOPLE

BANNER, BOX BUTTE, CHEYENNE, DAWES, DEUEL, GARDEN,
KIMBALL, MORRILL, SCOTTS BLUFF, SHERIDAN, AND
SIOUX COUNTIES.   A GROUP OFTEN CALLED
THE PANHANDLE OF NEBRASKA

GRANT L. SHUMWAY, SCOTTSBLUFF, NEBRASKA

EDITOR-IN-CHIEF

V. 3

ISSUED IN THREE ROYAL OCTAVO VOLUMES

VOLUME III.

ILLUSTRATED

THE WESTERN PUBLISHING & ENGRAVING COMPANY
LINCOLN, NEBRASKA
1 9 2 1

THE TORCH PRESS
CEDAR RAPIDS
IOWA

H. M. Sender — $125.00 (3 vols)

1192387

B. K. Bushnell.

BIOGRAPHIES OF SOME MEN PROMINENT IN THE DEVELOPMENT OF WESTERN NEBRASKA

BERTON KENYON BUSHEE, banker and statesman, has had a career significantly marked by courage, self reliance, marked initiative and executive ability, which bring normally in their train a full measure of success He has begotten the popular confidence and esteem that are important along the line of enterprise in which he has engaged and led to his election to high office in political life and he thus has secured status as a representative figure in the financial and political life of western Nebraska and the Panhandle No further voucher for him is needed than the statement that he is president of the Citizens State Bank, of Kimball It has been through the effective policies inaugurated by Mr Bushee that the Citizens Bank has increased its deposits and business within the late years and materially assisted in the development of Kimball county

Berton Bushee was born at Dartford Wisconsin, May 3, 1871, the son of Ezra Kenyon and Alzina Spooner Bushee He was reared in the beautiful little town surrounded by its encircling hills, attended the public schools until the family came to Nebraska in 1888 The father came to the Panhandle with his family when this section of the state was a veritable wilderness, settlers were few and great stretches of unbroken prairie stretched for miles Ezra Bushee filed on a homestead in what is now Kimball county and at once began the arduous task of breaking his land and establishing a home for the family Young Berton assumed his share of the burdens of a frontier farm and became sturdy and self-reliant As soon as the young man attained his majority he filed on a homestead of his own in 1892, proved up on it and engaged in ranching and frontier farming for several years At the same time he was offered and accepted a position to teach school in Kimball county, thus earling a livelihood during the lean years of farm life In 1898 Mr Bushee engaged in merchandising in Kimball, met with success in his enterprise, became recognized as one of the leading business men of Kimball, not disposing of his interests in this line until 1915

From first coming to this section of the state Mr Bushee entered into the civic life of his community and the Panhandle He was elected superintendent of schools of the county, serving three terms from 1896 to 1900 and from 1902 to 1903, an office which he filled with great efficiency to his own credit and the benefit of the educational interests of this section

From time to time as he could buy to advantage Mr Bushee increased his land holdings around Kimball and is one of the large landed proprietors of the southwest today His business life was but a start in a rising commercial career for Mr Bushee became interested financially in the Citizens State Bank, bought a controlling interest of its stock and became the executive head of the institution which is regarded as one of the safest and soundest banks in Nebraska Interested in the welfare of his community both as a landholder and banker it was but natural that Mr Bushee should enter public life to take care of and improve such interests He entered politics more than twelve years ago as a member of the Nebraska House of Representatives, serving from 1908 to 1912, then was elected to the State Senate, has proved such an able statesman and materially assisted in placing so many excellent laws upon the statute books that he has been reelected and is still serving During the session of the legislature in 1919-1920 Mr Bushee had the honor of being elected president of the senate, and as presiding officer of that body won a wide reputation as a legislator and leader of men

In politics Mr Bushee has been a consistent member of the Republican party He is a member of the Modern Woodmen of America, of the Knights of Pythias and the Masonic order With his wife Mr Bushee is a member of the Methodist church

April 4, 1894, occurred the marriage of Berton Bushee and Miss Ruth Cunningham, the event taking place at Sidney Draw Mrs Bushee was the daughter of George H and Martha Cunningham, the father was a Missourian, while the mother was a native of

Maine They came to western Nebraska at an early day and were well known pioneers of the Panhandle Two children have been born to this union, Helen Bernice and Elizabeth Ruth

As one of the representative business men and legislators, and progressive, public-spirited citizens of Kimball county, Mr Bushee merits special recognition in the annals of the Panhandle and this section in the opening up and development of which he has taken such an energetic and active part

BENJAMIN F GENTRY — Commonwealths have great need of capable men of broad vision and conscientious purpose, who will take time to study various problems of public necessity and serve faithfully for the general welfare In Benjamin F Gentry, western Nebraska has such a man He is prominently identified with the state's vast irrigation projects, and is proud of the fact that he was one of the two men who plowed the first furrow for irrigation purposes in his part of the state He is known all over Scottsbluff county through serving in public capacities, and is active in the business life of Gering in the line of abstracts and real estate

Benjamin Franklin Gentry was born in Nodaway county, Missouri, March 24, 1861, the son of William E and Rebecca (Wiles) Gentry, the former of whom was probably born in Kentucky and the latter in Indiana The father of Mr Gentry served in the Civil War for a short time, returning then to his home and succumbing to an attack of sickness Benjamin F was then nine months old, the youngest of his parents' three children The others survive Milton, who is in the teaming business at Weeping Water, Nebraska, and Rachel Catherine, the widow of J W Hostetter She resides at Omaha and has property in Cass county, Nebraska After the death of the father, the mother moved with her children to Mills county, Iowa, and subsequently to Cass county, Nebraska, where she died in 1917 Her second marriage was to Mattis Akeson, and they have had two children Thor W , a farmer near Weeping Water, and Emma, the wife of James Breckenbridge, a farmer living near Manley, Nebraska She was an admirable woman in every way and was a devoted member of the Christian church

Mr Gentry remained at home and attended the public schools in Cass county until he was fourteen years old, when he went to live with an uncle, Captain Isaac Wells on a farm one mile from Plattsmouth, Nebraska, where he later attended high school After taking a commercial course in the college at Valparaiso, Indiana, he worked for six months as a deputy collector in the office of the county treasurer of Cass county Mr Gentry then went to Hamilton county, Nebraska, where he handled grain for W H Newell and Company of Plattsmouth, until 1886, when he came to what is now Scottsbluff county, which at that time was Cheyenne county, as Scottsbluff county was not yet organized, and homesteaded, continuing to live on his farm until he was elected county clerk, in the fall of 1888, being the first man elected to that office, when he came to Gering, where he has resided ever since He served two terms as county clerk, and while in office became interested in the abstract business, which led to his purchase in 1909 of the O W Gardner Scottsbluff Abstract Company Mr Gentry has since given close attention to this important business, also handling a large amount of real estate in city and county, nevertheless he has found the time to accept responsibility in connection with the great irrigation projects that are interesting progressive men all over the state He has been a director of a number of the ditch commissions, and is serving as such at present in reference to the Minatare ditch in Scottsbluff county During a long directorate, he assisted in the building of what is known as the "nine-mile ditch " He is a member of the school board, on which he has given careful, honest service for thirty years

On November 30, 1890, Mr Gentry was united in marriage to Miss Cora E Johnson, who was born in Cass county, Nebraska, near Weeping Water, a daughter of Daniel D and Elizabeth A (Lathrop) Johnson Mrs Gentry's mother was born in Ohio and died near Weeping Water, Nebraska The father was born in Pennsylvania After marriage he moved to Iowa and during the Civil War served three years as a member of the Twenty-ninth Iowa volunteer infantry, suffering wounds at Helena, Arkansas After the close of the war he came with his family to Nebraska and homesteaded on the present site of Wabash, Nebraska He survives and makes his home at Scottsbluff, Scottsbluff county Mr. and Mrs Gentry have four children Harold E , who was educated in the State University at Lincoln, is chief chemist for the Great Western Sugar Company, Willard Max, a graduate of the Wesleyan University at Lincoln, will enter the medical profession; and Elizabeth, a student in both universities at Lincoln Mrs Gentry is a member of the where he remained until he was elected county Methodist Episcopal church Mr Gentry belongs to the Masonic fraternity and to the Eastern Star as does Mrs Gentry Politically he is a Republican

AMON R DOWNER, who is serving in his second consecutive term as treasurer of Scottsbluff county, belongs to old pioneer stock, and his business interests have always been centered here Although young in years for the heavy responsibilities of his office, he was not without official experience when first elected, and the efficient, careful, methodical performance of his public duties has afforded universal satisfaction

Amon R Downer was born in Hitchcock county, Nebraska, October 8, 1891, the elder of two sons born to Marion R and Jennie (Ball) Downer Mr Downer's brother, Marvin T, entered military service in September, 1917, and was sent to Europe as an army truck driver, with the American Expeditionary Forces The parents of Mr Downer were born in Iowa and came from there to Nebraska, where his father homesteaded in the eighties He was a Republican in politics and he belonged to the Methodist Episcopal church The mother of Mr Downer, who is now the wife of W R Wolffenden, a merchant, resides at Gering Her father, James Ball, came to Nebraska soon after the close of the Civil War, in which he had participated as a soldier and suffered from wounds He homesteaded in this county and lived here the rest of his life, his death occurring in 1916, one of the old veterans of the Grand Army

Amon R Downer was educated in the public schools of Gering and in the State University at Lincoln, where he was a student two years Afterward he was employed for some months in the construction department of the Union Pacific Railroad, and then went into the office of the county treasurer as clerk and deputy In 1917 he was elected treasurer and re-elected in 1918 and is still serving

In 1915 Mr Downer was united in marriage o Miss Bertie Margaret Lackey, who was born at Elmwood, Nebraska She is a daughter of Andrew and Eliza (Campbell) Lackey, natives of Toronto, Canada, who came to Nebraska in 1879 and homesteaded The father died in 1904 but the mother still lives at Gering. They had fourteen children of whom eight survive Mr and Mrs Downer had one child, Virginia Bess, who died in infancy They belong to the Methodist Episcopal church He s a Scottish Rite Mason and has been secretary of his lodge Politically he is a Republican

VALLE B KIRKHAM, one of the popular officials of Scottsbluff county, now serving in his second term as county clerk was born at Orrick, in Ray county, Missouri, March 14, 1883 When he came to Nebraska, he brought with him no capital except an excellent education, even a technical one He has made his own way in the world and in such a manner as to command the respect and confidence of his fellow-citizens, and in large measure he has their friendly esteem

The Kirkhams, in the person of David R Kirkham grandfather of Valle B, came many years ago from Virginia to Missouri and became a tobacco manufacturer there On the maternal side, the Blythes, were of Tennessee, and from that state the grandfather, Riley Blythe, came to western Missouri, where he acquired 1500 acres of land He became a man of political significance, served two terms in the state legislature and afterward was elected to the state senate Mr Kirkham's parents are C R and Elizabeth (Blythe) Kirkham, the former of whom was born seventy years ago at St Louis, Missouri, and the latter near Orrick, where they yet reside The mother is a member of the Christian church The father is a Democrat in politics and for many years has been an Odd Fellow Prior to 1874 when he moved to Ray county he was associated with his father in the tobacco manufacturing business at St Louis Of his family of eight sons and daughters, the following are living W H, county surveyor and a civil engineer, who lives at Richmond, Missouri, A J, a farmer near Orrick Valle B, who resides at Gering, Nebraska, Pattie the wife of W P Wolfe, living near Orrick, Dallas, married Claud Heather, who is a farmer in Ray county, Ross, who is a farmer near Orrick and Lillian, is a teacher at Orrick

After completing the high school course at Richmond, Missouri, Valle B Kirkham spent three years in the state normal school at Warrensburg, six months in the state university, and then took a special course in pharmacy, at Highland Park, Des Moines, Iowa For some years afterward he divided his attention between farm labor and railroad work with a civil engineering outfit, but in none of these activities did he accumulate a perceptible fortune It was in 1908 that he came to Scottsbluff county, accepting a job in the Irrigation Bank, an association which continued for seven and a half years, going from there to the Great Western Sugar Company's office clerk in 1916. His administration of the office was so satisfactory that he was reëlected in the fall of 1918, his popularity being demonstrated by the fact that he was elected in a Republican county with a majority of two hundred and seventy votes in 1916, and of four hundred and seventy in the second campaign.

On June 19, 1913, Mr. Kirkham was united in marriage to Miss Willie Gordon. She was born at Fort Smith, Arkansas, and is a daughter of Richmond and Lillie Gordon, retired residents of that place. Mr. Gordon has been an educator during the greater part of his life. Mr. and Mrs. Kirkham had one child, Vivian Lucile, who died when aged six months. They are members of the Christian church. He belongs to the Odd Fellows, both subordinate lodge and Encampment, and has a Grand Lodge degree. During the continuance of the World War, he served freely on the local draft board and contributed to its maintenance in many patriotic ways.

ROBERT G. NEELEY. — In the younger generation of business men in Nebraska will be found those who, like Robert G. Neeley, register of deeds for Scottsbluff county, early take an understanding interest in public affairs, cultivating serious political convictions, thereby raising the standard of true citizenship, and inevitably become useful and influential in their communities. The broadening effect of this wider vision may be seen in what this younger generation is accomplishing.

Robert G. Neeley was born in the Mitchell valley, Scottsbluff county, Nebraska, December 12, 1894, the son of Robert F. and Jennie (Yates) Neeley, natives of Missouri, who now reside at Gering, where Mr. Neely is engaged in the real estate business. He was the founder of the Gering National Bank and for some years was president of the institution. Prior to that he was in the cattle business and dealt in real estate in Mitchell valley. To his first marriage, with a Miss Burgess, two sons were born: Franklin E., cashier of the Gering National Bank, and A. Raymond, a druggist at Gering. A son and a daughter have been born to his present marriage: Robert G., register of deeds, who lives at Gering, and Mildred, the wife of E. S. Slafter, who conducts a garage business at Dubois, Wyoming.

Robert G. Neeley obtained his education in the public schools of Gering. For two years he was employed in the Gering National Bank, for the next two years was in the office of the present county judge, for one year served as deputy county clerk, and for more than a year filled office as deputy register of deeds. With this thorough experience in county offices, he was particularly well equipped when he was elected register of deeds in 1918. He is one of the most popular of the county officials, those doing business with his office always finding exact knowledge and courteous treatment.

On April 14, 1917, Mr. Neeley was united in marriage to Miss Eunice M. Barton, who was born near Council Bluffs, Iowa, a daughter of Samuel and Mary Elberta (Heft) Barton. The father was born in England and the mother in Illinois. They came to Nebraska in 1902 and Mr. Barton is carrying on extensive farming enterprises near Gering. Mr. and Mrs. Neeley are members of the Christian church, in which they are somewhat active, and they take part in the pleasant social life of the city. Mr. Neeley is a Republican in his political views but is not illiberal, hence he has many political as well as personal friends.

WILLIAM H. LAMM. — Public service carries with it the supposition of business efficiency, as well as sterling character, and the progressive little city of Gering has no more trustworthy public official than William H. Lamm, who has been postmaster since 1915. Mr. Lamm followed agricultural pursuits during a large portion of his life, and school teaching also in early manhood, and in every line of endeavor in which he has been engaged, has commanded the respect and confidence of his fellow citizens. The Gering post office is a busy place, but under Mr. Lamm's administration, the work is expedited, and the service entirely satisfactory because of his practical ideas and careful, methodical oversight.

William H. Lamm was born at Thayer, in Union county, Iowa, April 22, 1877, the eldest of nine children born to William and Jane (Knotts) Lamm, both born in 1854 in Iowa, the former near Madison and the latter near Burlington. Her death occurred November 4, 1918. In addition to William H., their children are: Ernest F., a farmer near Glendo, Wyoming; Bert, a farmer near Meridian, Idaho; Bertha, the wife of John M. Gross, farmer and stockman, near Glendo; Bess, the wife of Joseph E. Nisley, farmer, near Gering; Carmie, the wife of Harry C. Barton, assistant cashier of the First National Bank of Gering; Lauretta, the wife of Ivor C. Davies, a druggist at Gering; True R., a farmer and stockraiser near Glendo; and Grace, the wife of Peter B. Schmidt, employed in the First National Bank of Scottsbluff. He entered the

R. M. Hampton

National army in 1918, and was in training at Camp Funston when he received his honorable discharge The father of Mr Lamm is one of the capitalists of Scottsbluff county, to which he came in 1904 He purchased a section of irrigated land and still owns a part of it together with other valuable properties He has been one of the sound disciples of Democracy in the county, but has never accepted a public office He is a member of the Christian church

William H Lamm first attended the country schools near his father's farm in Iowa, then Palmer College at Marshalltown, and Capital City Commercial College at Des Moines, Iowa He, as the eldest of the family, early took on responsibility, assisted his father in his agricultural industries, and for several years engaged also in teaching school in Iowa, during one year teaching at Thayer In April, 1904, he came to Scottsbluff county, and from that time until 1915 was mainly occupied with farm activities, although, during three years of this period he served as deputy sheriff From early manhood he had interested himself intelligently in public affairs, believing good citizenship demanded such a course He has always been identified with the Democratic party and is fully in accord with the present administration at Washington In 1915 he was called from his farm to become postmaster of Gering and, as indicated above, has fulfilled every expectation

On May 13, 1911, Mr Lamm was united in marriage to Miss Maude L Abbott, who was born in Indiana They have two children Thelma Maxine, who attends school, and Lyman Abbott, who celebrates his fifth birthday in May, 1919 Mr and Mrs Lamm are members of the Christian church Fraternally he is a Mason and belongs also to the Woodmen of the World

RODOLPHUS M HAMPTON, president of the First National Bank of Alliance, is one of the early settlers of Box Butte county No history of the county would be complete without the name of R M Hampton, for he has been a resident of this section for thirty-five years He has lived to see the wide open prairies developed into a smiling green countryside dotted with prosperous towns that are the barometer of prosperity and success In all movements for opening up the country, the building of railroads, villages and cities he has taken an active and aggressive part and it is such men who have made history in the Panhandle

Mr Hampton was born in New Lexington, Ohio, November 14, 1861, the son of William and Mary (Young) Hampton, the fifth of seven children born to his parents The father owned an eighty acre farm in the Buckeye state and there the boy was reared He attended the country schools near his home and being assigned the usual tasks to a small boy on the farm grew up sturdy and self-reliant While still a small lad of twelve he began to earn money for himself by digging coal at Moxahala, Ohio, which was not far from his home, but as he was paid by the bushel he did not make more than sixty-five cents a day After completing his education Mr Hampton, at the age of twenty followed the pedagogic profession for five years, in both the country and city schools He was ambitious to succeed in the world, and after reading of the many openings for a young, vigorous man in the new country in the west came to Nebraska in 1885 to learn what fortune might have in store for him on the plains There were scarcely more than fifty families in what is now Box Butte county when Mr Hampton arrived, so that he settled down in a locality where habitations were few, comfort and conveniences scarce and the elements of civilization at their lowest, but the tide of immigration was setting toward the upper Platte valley in the middle eighties, and within a few months after his arrival the population had more than doubled Soon after reaching the present Box Butte county, Mr Hampton selected a pre-emption and timber claim and broke out ten acres on each and putting up a "soddy" he kept "bachelor hall" as he expresses it He realized that he was not cut out for a frontier farmer, so sought a professional life, forming a partnership in a law firm with James H Danskin, opening an office at Hemingford The new firm was kept busy with the many land cases arising over confused titles, where contests had been filed Many land sharks tried to secure title to land that had been filed on by bone fide settlers previously and used every pretext to obtain possession Because of this many lively legal contests ensued, but the firm of Hampton and Danskin did their best for the honest settlers and as a result had a fine practice They tried cases at Hay Springs, Rushville, Chadron Hemingford and Nonpariel The present territory of Box Butte county was at that time included in Dawes county with the seat of justice at Chadron, but in 1887, Box Butte was erected as a separate county and the seat of justice located at Nonpariel, so the young lawyers moved their office there   They continued in business until February, 1889, when Mr Hampton resigned from the firm to devote his time and energies to the organization of the American Bank of Alliance, as he believed that there was a great future for banking business in the newly developed country   Associated with him in this enterprise were O M Carter of Omaha, as president, A S Reed of Alliance, vice-president, and Mr Hampton assumed office as cashier.   The board of directors consisted of these same officers, J H Danskin and I E Tash   The new bank was established with a capital stock of $25,000 and operated one year when it was merged into the First National Bank of Alliance, which had a capital stock of $50,000   It was conducted under the same board of directors and the same personnel as to officers   For thirty years Mr. Hampton has taken a leading part in the financial life of the county and the Panhandle   Three years ago he assumed the office of president of this sound, prosperous and progressive house   His high integrity, steady purpose and business foresight have begotten that popular confidence which is so essential in the furtherance of the important enterprise along which he had directed his attention and energies for a quarter of a century, and through which he has gained secure status as one of the representative figures in the financial circles of northwestern Nebraska Mr Hampton is also interested in the Lake Side State Bank, of which he is a stockholder and director   From 1900 to 1911, in addition to his financial affairs, Mr Hampton operated a ten thousand acre ranch located southwest of Alliance which he sold to Hall and Graham

Today the First National Bank is the leading financial institution of Alliance and Box Butte county, it has a capital stock of $50,-000, surplus of $50,000 and deposits of $1,-250,000   The personnel of the banking house in 1919 was as follows   R M Hampton, president, C E Ford, vice-president; F Abegg, cashier   The board of directors consists of R M Hampton, M Hampton, C E Ford, F Abegg and J M Kimberling

In October, 1888, Mr Hampton was married at Logan, Ohio, to Miss Minnie Fickell, a native of that place, a daughter of Joseph and Hannah (O'Hara) Fickell   They have but one child living, Dorothy who is attending the Alliance high school   The Hampton family are all members of the Methodist church of which they are liberal supporters   Both Mr Hampton and his wife are broad gauged liberal minded people who keep abreast of the trend of events and are interested in the development and progress of their commuity and are willing to support with time and money every laudable enterprise that tends to civic and communal welfare   Mr Hampton is a Republican in politics and though he takes an active interest in political affairs has ever been too busy to accept public office

ADA M HALDEMAN. — The fact that a woman can hold important public office and has the capacity to direct affairs with executive energy, can no longer be denied or be considered a subject for criticism even by those who once were frankly incredulous   The truth, however, may be acknowledged, that there are comparatively few women in any community who are qualified for such service   In many fields the sex has undoubtedly won laurels, but men have, as a rule, been a little backward in assisting women to positions of great responsibility   Naturally then it may be assumed that unusual personal qualities and marked scholarship pertain to a woman who has three times been elected to the exacting office of superintendent of schools, which testimonial has been given and honor paid to Miss Ada M Haldeman, in Scottsbluff county

Superintendent Haldeman was born at Avoca, in Pottawattamie county, Iowa, one of a family of five children born to Francis Wayland and Martha E (Lewis) Haldeman   The father of Miss Haldeman was born in 1846, at Marion, Ohio, and died in Iowa, in 1886   The mother was born at West Liberty, Iowa, and resides at Gering   Miss Haldman has one brother and one sister, namely   Henry, who gives attention to the homestead in Scottsbluff county, formerly traveled for the Remington Typewriter Company, but now occupies his leisure in writing for magazines, and Virginia, who is the wife of Nyle Jones, of New Orleans, Louisiana

Francis Wayland Haldeman accompanied his people to Iowa in boyhood and was reared and educated there   During the Civil War he served two years as a bugler   He came early to Nebraska, went through some pioneer experiences here and did some hunting while looking over land in different parts of the state   He was a nurseryman and understood horticulture and as such was able to treat and preserve many of the early orchards from Grand Island westward   His death occurred in his old home just after he had reserved his homestead in Nebraska   He was a Republican in his political views

Miss Haldeman completed the high school course before leaving Avoca, Iowa, later securing her B A degree in the University of Colorado, and was quite young when she entered the educational field, teaching in the schools of Iowa, Wyoming, and Nebraska After coming to Nebraska she taught one year in the city of Lincoln and later for one year in the Scottsbluff high school In 1914 she was elected county superintendent of Scottsbluff county, and after a faithful service was reelected in 1916 and 1918 She is devoted in her work, conscientiously giving the best that is in her to maintain the high standards she has set for the county's educational progress She resides with her mother at Gering, the latter of whom still owns the old homestead, on which a feature was made last year of growing sugar beets The venture proved very satisfactory and the 100 acres in beets yielded a large income Miss Haldeman owns a homestead in Scottsbluff county, situated one mile north of Toohey Both she and her mother belong to the Congregational church

WILLIAM HENRY HARDING — The stable character of Gering's commercial life is shown in the many solid, well-financed industries that are prospering here There are many industrial concerns that have a wide market for their products and thus carry the name and fame of Gering to other sections, while, at the same time, they promote local prosperity by paying high wage scales to expert workmen One of these to which attention may be called is the large blacksmithing and wagonmaking business that was founded and is carried on here by William H Harding

William Henry Harding was born in Decatur county, Kansas, August 5, 1884, the eldest of a family of eight children born to William T and Mary (Nehls) Harding The father of Mr Harding was born in Wisconsin, and the mother was a native of Iowa They were married in Kansas, and her death occurred in 1900 William T Harding went to Kansas in early manhood and well remembers seeing great herds of buffalo in the section of the state where he settled He bought a relinquishment claim in Decatur county and lived on his farm there until 1889, when he came to Nebraska and bought another farm near Gering, and also secured a Kinkaid claim that he has recently sold He now lives retired at Morrill, in Scottsbluff county

William H Harding had public school advantages and was graduated from the Gering high school in 1899 After that he went to work on a ranch and in 1905 took a Kinkaid claim, proved up on it and resided there for five years and then sold In 1912 he came to

Gering and started his present plant and has developed a large business In addition to manufacturing, he handles farm machinery As a business man he is held in high esteem, his methods being fair and honorable

In 1911 Mr Harding was united in marriage to Miss Bessie Cole, who was born at Miller, Nebraska, and they have four children, namely Henry, James, Daniel, and Ella Mr Harding has never been very active in politics but nevertheless is an intensely active citizen where Gering interests are at stake He belongs to the Odd Fellows Modern Woodmen, Woodmen of the World, and B P O E

JAMES P WESTERVELT, whose numerous business interests have made him well acquainted with different sections of the western country, has been a resident of Gering since May 15, 1887, when this was Cheyenne county, before Scottsbluff was organized, and is the able manager of the Farmers Mercantile Company of this city Despite the handicap of meager educational advantages in youth, Mr Westervelt has not only been a successful man in several business lines, but in at least two counties in Nebraska has served for years in public offices of trust and responsibility with extreme efficiency

James P Westervelt was born in Ionia county, Michigan, March 12, 1869 His parents were James H and Lorena A (Day) Westervelt, the former of whom was born in New Jersey in 1840 and died in 1908, and the latter in Vermont in 1848, and died in 1912 They married in Vermont and five children were born to them, James P being the second of the four surviving, all of whom live in this county Eugene the eldest conducts the Scottsbluff Republican, Claude who carries on a blacksmith business and Goldie, the wife of P Gilbert a commercial traveler The parents were members of the Baptist church They moved from Vermont to Michigan in 1868, where James H Westervelt carried on work as a blacksmith until 1879, when they left Ionia county for Custer county Nebraska The family lived on the homestead until 1887 and then moved to Gering, where Mr Westervelt started a general store which he conducted until 1900, when he returned to work at his trade He voted with the Republican party

James P Westervelt was so circumstanced in boyhood that work on the farm was considered more necessary than that he should gain a good education He remained in the country until 1888 and then went to Banner county, Nebraska, and started a general store in the village of Freeport where he remained until

1891, during which time he was postmaster. From there he removed to Sheridan, Wyoming, where he followed ranching until 1905, the year he came to Gering. In the meanwhile, for twenty-five years he had engaged in the practice of dentistry, not continuously, but as occasion seemed to demand, having a natural skill in the use of delicate tools and a fair knowledge of the profession through reading and experience. In 1913 Mr. Westervelt assisted in the organization of the Farmers Mercantile Company at Gering, incorporated and capitalized at $20,000, since which time he has been general manager, dividing his time between the store and a valuable farm he owns in the environs of Gering.

In 1908 Mr. Westervelt was united in marriage to Miss Edith W. Sayer, who was born in Iowa. Her father, Reverend E. H. Sayer, came to Gering in 1897 as pastor of the Presbyterian church. He now lives retired. Mr. and Mrs. Westervelt have one son, Leon, who assists in his father's store. Mr. and Mrs. Westervelt are church members and active in many benevolent organizations. He is a Scottish Rite Mason, has held all the offices in the Blue lodge and for four years was master of his lodge. In politics he has always been a Republican and has served in public office in this county, for some years being on the school and town boards and for four years, from 1904 to 1909, was sheriff of Scottsbluff county.

Mr. Westervelt's brother, Claude, drove into Scottsbluff county, May 15, 1887, in true pioneer style with a yoke of oxen, coming across the prairies from Custer county, where he had been living. The next year the father came and opened a small store about the time Mr. Westervelt opened his first mercantile establishment at Freeport, Banner county, so the family has a true pioneer history and has become well and favorably known in the upper Platte valley.

MATTHEW H. McHENRY, clerk of the District Court and for many years a resident of Gering, was born in Harrison county, Iowa, November 4, 1869. His parents were Oliver O. and Mary Jane (Hall) McHenry. His father was born in Missouri, near the Iowa line, in 1844, and died in Scottsbluff county, Nebraska, in 1917. His mother was born in 1849, near London, England, and died in Nebraska, December 29, 1915.

The parents of Mr. McHenry came to Scottsbluff county in February, 1889, homesteaded and spent the rest of their lives here. During life he was a merchant and farmer and also operated an elevator. During the Civil war he belonged to an organization of state guards. In politics he was a Republican. The mother of Mr. McHenry was a member of the Baptist church, but the father belonged to the Latter Day Saints. Of their five children Matthew H. was the second in order of birth, the others being as follows: Elizabeth Ann, who is the wife of John A. Burton, a retired citizen of Upland, California; Lucy Jane, who is the wife of John Springer, a farmer in California; Harry H., who lives on his ranch near Torrington, Wyoming; and Lewellyn O., who is a druggist at Morrill, Nebraska.

Matthew H. McHenry was educated in the public schools and a business college at Woodbine, Iowa, after which he worked on a ranch, still later operating a ranch of his own. He still owns a fine ranch in Sioux county, Nebraska, and a valuable farm situated south of Morrill, Nebraska. Mr. McHenry has always deemed an interest in politics a necessary part of good citizenship. In November, 1911, he was made clerk of the District Court, but had served for four years already as county clerk, attending to the duties ex-officio of the district clerk before the latter office was established. Mr. McHenry has been continued in office ever since. He is one of the county's best informed and most courteous officials.

In December, 1895, Mr. McHenry was united in marriage to Miss Mary Belle Weeks, who was born in Missouri and died in Nebraska, January 18, 1910. She was a member of the Baptist church. She was the mother of three children: Winifred, Wesley O., and Coral, the two younger children being in school. Winifred is the wife of Marvin Downar, who entered military service in the United States on September 22, 1917, accompanied the American Expeditionary Force to Europe and at the time of this writing was with the Army of Occupation in Germany. As a member of company D in ammunition train 314 he went to the front in September, 1918, and was under fire for forty-two days. Mrs. Downar is a deputy clerk under her father. On November 11, 1911, Mr. McHenry was married to Miss Amanda Sappington, who was born in Keith county, Nebraska, and they have one son, John Roger McHenry, who was born in November, 1916. Mrs. McHenry is a member of the Episcopal church. Mr. McHenry is identified with the Elks at Alliance, and the Odd Fellows at Mitchell, Nebraska.

JOSEPH L. GRIMM, county attorney, has justified the confidence that his fellow citizens of Scottsbluff county reposed in him, when they elected him to this important office in

November, 1918 A native of Nebraska, all her interests are dear to him, and from the beginning of his professional career until the present, it has been his aim to defend her citizens and maintain their rights in the face of the world

Joseph L Grimm was born in Saline county, Nebraska, December 23, 1883, one of a family of eight children born to Joseph H and Esther E (Hess) Grimm The late Joseph H Grimm was a man of both professional and political distinction He was born in Licking county, Ohio, in 1848, and died January 15, 1911 In 1875 he came to Nebraska and located first at Pleasant Hill, later becoming prominent in public affairs in Saline county and serving two terms in the state legislature from that county He was an able member of the bar and twice was elected county attorney on the Republican ticket He married Esther E Hess, at Mount Vernon, Iowa, who was born in Linn county, Iowa, in 1854, and died June 25, 1907 Of their six surviving children, Joseph L is the fourth in order of birth, the others being Mabel, who is a teacher in the schools of Wilber, Nebraska, James J, who is county judge in Saline county, May A, who is the wife of Ralph Woods, a lawyer, of Tacoma, Washington, Clarence, who has been in military service since September, 1917, is a first lieutenant in a regiment of American troops sent to keep order in Siberia, and Hazel, who is the wife of E H Shary, of Chester, Pennsylvania The mother of the above family was a member of the Lutheran church

Joseph L Grimm completed the public school course at Wilber, after which he entered the law department of the University of Nebraska, from which he was graduated in 1908 In June of that year he entered into practice at Wilber and continued there until May, 1918, when he came to Gering and opened an office in the Gering National Bank building His legal talent soon became known and on September 2, 1918, he was made deputy county attorney, on October 8 following, was appointed county attorney and his election followed in November Mr Grimm has charge of the bond issue for the proposed new courthouse

On May 12, 1909, Mr Grimm was united in marriage to Miss Sady E Kimport, who was born at Garrison, Iowa, and they have two children Benjamin Hayes, born May 6, 1910, and Esther Rosalee, born June 10, 1915 Mrs Grimm is a member of the Methodist Episcopal church It has been some years since Mr Grimm became a Mason and he has continued in good standing ever since and has reached the Royal Arch degree He is past master of Blue Valley lodge No 64, F & A M, at Wilber, which he served three years and during that entire period missed but two meetings Politically he has always been affiliated with the Republican party

LEWIS L RAYMOND, whose name carries with it the high regard that comes of honorable achievement, is a leader of the bar at Scottsbluff, and a citizen of the county who has served in numerous important official capacities with marked efficiency and great public spirit A native son of Nebraska, he is a representative of an old pioneer family that settled within its borders almost a half century ago Mr Raymond was born October 19 1871, in Butler county, the son of Seth and Elizabeth (Lovelace) Raymond

Seth Raymond was born at Millersburg, Ohio, September 9, 1835, but was a resident of Wisconsin when the Civil War came on He enlisted April 3 1861, in Company G, Third Wisconsin volunteer infantry and served faithfully as a soldier until he was honorably discharged at Beaufort, North Carolina, in February, 1865 Until the day of his death, June 10, 1910, he bore the marks of the wounds he received at the battle of Winchester, Virginia On January 16, 1864, he was united in marriage, at Janesville, Wisconsin, to Elizabeth Lovelace, who was born October 4, 1843, at Erie Pennsylvania She resides at Scottsbluff, where she is active in the Methodist Episcopal church to which her husband also belonged Of their family of nine children Lewis L was the fifth in order of birth In October, 1870, Seth Raymond and his family came to Nebraska and he homesteaded in Butler county, remaining on his land there until August, 1884, when he moved to Dawson county, from there coming to Scottsbluff county in March, 1887 He took up land five miles southwest of Gering and remained on that farm until March, 1892, when he moved into Gering, where he lived a somewhat retired life until 1905 In the spring of that year he came to Scottsbluff, where his remaining years were passed He was a man of sterling character, was somewhat active in the Republican party and a Mason in good standing in his lodge

Lewis L Raymond had public school advantages in early youth and later spent four years in study in the normal school at Fremont Like many another intellectual young man, he began business life in the schoolroom and more or less continuously taught school for the following fourteen years in Scottsbluff county, during a part of this time devoting himself to the study of law, F. A. Wright being his preceptor. Mr. Raymond was admitted to the bar, November 17, 1902, and soon afterward started practice in this county, where since then he has been identified with many of the most important cases that have come before the courts. He has not, however, been able to devote his entire time to his profession for his fellow citizens have often called him into public life. He served one term as deputy county clerk, two terms as county superintendent, two terms as county attorney, and one term as county judge, while in 1909 he was elected to the state senate. Since retiring from the political field his law practice has absorbed his attention to a great extent and his high standing at the bar is unquestioned.

On September 1, 1897, Mr. Raymond was united in marriage to Miss Mable Shumway, a member of the prominent Shumway family of this section of the state, and they have three children, two sons and one daughter: Charles R., Jack L., and Evelyn. The daughter is still in school. Both sons entered military service as volunteers in 1918, neither of them being of military age, but loyal and patriotic American youths to the core. Charles R. is a sergeant in the Four Hundred Forty-seventh Labor battalion, at Camp Humphrey, Virginia, and Jack L. is a member of the United States Marines. Mr. Raymond and his family are members of the Methodist Episcopal church. In his political convictions he is a Republican.

WILLIAM H. LYMAN. — The banking interests that go so far to substantiate the importance of Gering, are fortunately in the hands of able financiers and trustworthy business men. With sound, reliable banking institutions in the background, a community is helped in many ways, this possession giving confidence to investing capitalists who largely base their opinion on the showing of the banks. One of these substantial banks at Gering that has been doing a large and safe business here for many years, is the State Bank of Gering, of which William H. Lyman is vice president and active in its management.

William H. Lyman was born at Kearney, Nebraska, August 12, 1886, the youngest of a family of nine children born to William H. and Maria J. (Van Cleek) Lyman, the latter a native of Canada, who now lives at Weeping Water, in Cass county, Nebraska. Mr. Lyman's father was born at Spencer, Massachusetts, came to Nebraska as a pioneer and died here in November, 1917, having conducted an insurance and real estate business at

Weeping Water for a number of years. In politics he was a Republican, belonged fraternally to the A. O. U. W., and was a faithful member of the Congregational church.

William Henry Lyman was educated at Weeping Water. After his graduation from the high school in 1906 his studies were continued in the academy. When prepared to enter business he chose the jewelry line and Holyoke, Colorado, as his business field. He remained at Holyoke for ten years and during that time became prominent in public affairs of the town, serving on the town board and also as mayor. In 1917 he came to Gering and has been identified with the State Bank of Gering ever since, first as assistant cashier. In August, 1918, Mr. Lyman, together with Mr. Denslow, bought the controlling interest in the bank at the same time becoming vice president, with Lloyd Denslow as president. The latest bank statement of 1919, gives the following: Capital stock, $25,000; undivided profits and surplus, $23,000; deposits, $400,000. Mr. Lyman is interested in Scottsbluff county irrigated and ranch lands and is a persistent booster for what he says is the best county on earth.

In 1908 Mr. Lyman was united in marriage to Miss Grace Mowry, who was born on a farm near Marysville, Missouri. Her father, the late Charles M. Mowry, was in the hardware business at Holyoke, Colorado, for twenty-five years preceding his death. Mr. and Mrs. Lyman have one daughter, Anna Mae, an attractive little maiden of ten years. They are members of the Baptist church and willingly helpful in its many avenues of benevolence. Mr. Lyman is somewhat prominent in the order of Knights of Pythias, serving for a time as chancellor commander.

Mr. Lyman has been an independent voter and has at all times felt free to give his political support to those whose measures his own judgment approved.

ROBERT G. SIMMONS, an able member of the Scottsbluff county bar and formerly county attorney, is widely and favorably known, not only as a young man of brilliant promise in his profession, but as a patriotic soldier when his country needed defenders. Mr. Simmons was in the aviation service.

Robert G. Simmons was born at Scottsbluff, in Scottsbluff county, Nebraska, December 25, 1891, fifth in a family of seven children born to Charles H. and Alice M. (Sheldon) Simmons. Both parents were born in the state of New York, and the mother died in Nebraska in 1918. The father of Mr. Simmons came to Scottsbluff county and homesteaded in 1886 and his family joined him in the following year. He remained on his farm until 1898 when he came to Gering where he conducted a grocery store for a time. He moved then to Scottsbluff, of which place he has been a continuous resident and important citizen ever since, serving for ten years as postmaster of the town and subsequently accepting a place on the board of water commissioners, which he still fills. In politics he is somewhat active in Republican councils, and fraternally he is identified with the order of Modern Woodmen. He is a pillar of the Presbyterian church. He takes satisfaction in the fact that not only his son Robert G. has chosen Scottsbluff county as his permanent home when choice might be made of any other section, but the other members of his family have done likewise, as follows: William L., who is in a contracting business; Otis W., who is also a contractor; Charles S., who is a sign painter by trade; Edith, who is the wife of Lee Harrison; and Ada and Ida, who attend school.

Robert G. Simmons was afforded educational advantages and was graduated from the high school in 1909, following which came two years in Hastings College, and a course in law in the Nebraska State University, from which he was graduated in 1915. He immediately opened a law office at Gering and in 1916 was elected county attorney, continuing to serve in that capacity until October 29, 1917, when he entered the aviation department of the National army and was sent to Fort Omaha for training. Passing every test, and they are many and exacting, he made rapid progress and won the rank of second lieutenant in the air service, for five months having command of a company, with jurisdiction over four under officers. When relieved from service he returned home and resumed practice at Gering where his friends and admirers are many, although for family reasons he is considering the transfer of his office to Scottsbluff. He possesses every requisite for professional advancement.

Mr. Simmons was united in arriage to Miss Gladys Weil, on June 23, 1917. She also is a native of Nebraska and was born at Harvard, her people being old settlers of that section. Mr. and Mrs. Simmons have one son, Robert G. They are members of the Presbyterian church. Politically Mr. Simmons is a sound Republican and has the reputation of being loyal to his political friends. He is a Scottish Rite Mason and a Shriner.

LLOYD DENSLOW. — Because of his success in managing the affairs of the Gering State Bank, of which he is president, it might be inferred that Lloyd Denslow has been in the banking business all his life. This, however, is not the case, although, undoubtedly the business ability which he has shown here, has been a leading factor in other enterprises in which he has been equally successful. He is an example of Nebraska's native born, thoroughly educated, steady, ambitious and enterprising young citizens who gives great promise for the future of this commonwealth.

Lloyd Denslow was born at Hooper, in Dodge county, Nebraska, November 7, 1884, the youngest of four children born to Jeremiah and Anna (Sutton) Denslow. The father was born in the state of New York, in 1843, came to Nebraska in 1854, and died in Dodge county, April 22, 1907. The mother was a native of Illinois, born there in 1843, came to Nebraska in 1857, married Jeremiah Denslow at Fontanelle, this state, and died October 21, 1907. Lloyd Denslow has one brother and two sisters: J. H., who owns a large irrigated farm near Denver, Colorado; May, the wife of Charles H. Lyman, who is in the real estate business at Polson, Montana, and Nina, the wife of Dr. T. Wiglesworth, a practicing physician at Twin Falls, Idaho. Jeremiah Denslow was a freighter in early days. He became a man of wealth and prominence, at one time was put forward by the Prohibitionist party as its candidate for lieutenant-governor of the state. Yet, he was in very humble circumstances when he came to Nebraska and has been heard to declare that his sole capital was represented by thirty-five cents and no expectations. He possessed, however, capital of another kind, and in the honorable acquisition of property, and in the building up of a stable reputation, he proved that industry, prudence and personal integrity have high value. Mr. Denslow homesteaded in Dodge county and the family still have this land now grown very valuable. For twenty-five years before he retired from business he was president of a bank at Hooper. He was a Knight Templar Mason and always an ardent supporter of the cause of temperance and a firm believer in its final triumph although he was not permitted to see his judgment confirmed.

Lloyd Denslow was graduated from the Hooper high school in 1901, and from the Fremont high school in 1903. His graduation from the Nebraska State University followed in 1908 and after that came one year of post graduate work at Columbia University, New York City. He then entered business and spent two years in Old Mexico, Idaho and Washington, being twenty-six years old when he returned to Hooper, where he embarked in the real estate business, in which he continued for three years. Mr. Denslow owned a tract of forty acres of land and to that he added forty acres, and was engaged there in farming and hog-raising, together with dealing in real estate up to 1916, when he came to Gering. Here he was identified for a while with the Great Western Sugar Company. Later he sold his eighty acres of farm land, though retaining some holdings in Wyoming, and in association with William H. Lyman bought the controlling stock in the Gering State Bank, of which he was made president. This banking institution is considered one of the most solvent in the state, is amply financed and carefully and conservatively directed.

Mr. Denslow was united in marriage to Miss Alda Gray, November 8, 1913, who is a native daughter of Nebraska, her birthplace being Pleasant Dale. She is a member of the Methodist Episcopal church and the devoted mother of their little son and daughter, Jerry and Dorothy, the former of whom was born in 1914 and the latter in 1919. An independent voter from early manhood, Mr. Denslow has felt free to give his political support to men and measures approved by his own judgment. During the World War in which the country was engaged, he never failed in any patriotic duty and served as chief clerk on the draft board for almost a year. He is a Scottish Rite Mason and a Shriner, and is past senior warden of his lodge. In religious belief the faith of the Unitarians attracts him.

JOHN S. PECKHAM, at the head of a prosperous general automobile business operated under the style of the Peckham Motor Company, at Gering, Nebraska, is a practical and experienced automobile man, having had special training in this line of mechanics. He has been in business for himself since 1911 and has been located at Gering since 1917.

John S. Peckham was born at Kearney, Nebraska, February 8, 1888, one of a family of eight children born to George F. and Roselle (Lyons) Peckham. The father was born in 1843 in Wisconsin, and died in Nebraska in 1914. The mother was also born in Wisconsin, seventy years ago, and still survives. Of their seven surviving children, John S. was the seventh in order of birth, the others being: Mina, the widow of Charles Esley, formerly with Booth & Co., Denver, Colorado; Nellie, the wife of H. W. Van Meter, of Lincoln, Nebraska; George, who resided at Kearney; Eva, the wife of B. P. Cutting, a traveling salesman in Nebraska for the Nebraska Buick

Automobile Company; Cornelia, the wife of Roy Flemming, a contracting painter at Scottsbluff, and Ralph, a conductor on the Burlington railroad. The father of the above family served four years in the Civil War, as a wagoner in the Eleventh Wisconsin infantry, escaping permanent injury although often in great danger. He came to Kearney, Nebraska, in the eighties and for a number of years afterward was in the pump and windmill business. He was a member of the Baptist church.

John S. Peckham attended the public schools of Kearney until the tenth grade, and had further advantages at Lincoln. In that city he went to work for the Cushman Motor Company, in the gas engine department, and remained employed at Lincoln for the next ten years remaining three years with the Cushman people and the rest of the time was with the Nebraska Buick Automobile Company and the E. E. Mockett Auto Company. He had the best possible mechanical training with these representative concerns and by 1911 was ready to embark in the same business on his own account. He located at Douglas, Nebraska, where he cotinued until 1917, when he came to Gering where he has done well. He is sales agent for the Buick cars, which has added to his business popularity because of the general confidence in these motors.

In 1913 Mr. Peckham was united in marriage to Miss Madge Allison, who was born at Sterling, Nebraska, and they have two children, namely: Ruth and Ray. Although not unduly active in politics, Mr. Peckham is intelligently watchful of public events as well as local affairs, and casts a Republican vote. He has been a Mason for a number of years and is in good standing in his lodge.

GEORGE B. PECKHAM, well known both in Kearney and Gering in the automobile industry, being an expert mechanician, belongs to an old pioneer family of Nebraska, his father coming here early in the eighties. Mr. Peckham was born in 1877, in Wisconsin. His parents, George F. and Roselle (Lyons) Peckham, were also born in Wisconsin, his father in 1843 and his mother in 1849. The latter survives. The father served four years in the Civil War as a wagoner in a Wisconsin regiment. He was in the well-digging business and after coming to Kearney, Nebraska, he engaged in that business and expanded it and for many years was the leading man in Buffalo county in the pump and windmill business. His death occurred in 1914. George B. Peck-

Yours truly
V. Anderson

ham was the fourth in a family of eight children, the others being Mina, the wife of Charles Esley, who died at Denver, Colorado, in 1918, William, who died in 1912 when aged forty-two years, Nellie, the wife of H W Van Meter, of Lincoln, Nebraska, Cornelia, the wife of Roy Flemming, of Scottsbluff, Eva, the wife of B P Cutting, of Lincoln, John S, in the automobile business at Gering, and Ralph, who resides at Lincoln

George B Peckham accompanied his parents when a boy from Wisconsin to Kearney, Nebraska, where he attended school His first business experience was as an employee of a lumber company Later he went into the mechanical department of an automobile company and has been interested in this business ever since When his brother, John S Peckham, came to Gering in 1917 and established the Peckham Motor Company, he accompanied him and has ever since been identified with this business, of which he will soon become a co-partner

On December 14, 1898, George B Peckham was united in marriage to Miss Mable Hodges, and they have one daughter, Edna, who is the wife of J M Branson of Gering Mrs Peckham is a member of the Methodist Episcopal church, but Mr Peckham was reared in the Baptist faith

VICTOR ANDERSON, M D, proprietor of the leading drug store at Bridgeport and a general medical practitioner of wide experience, has spent the greater part of a very useful life in the United States but his birthplace was in another country He was born in Sweden, March 21, 1867, came to this country when twelve years old and through his own efforts not only acquired a literary but a professional education

The parents of Dr Anderson were Andrew and Carrie (Magnuson) Anderson, both of whom died in Sweden Of their four children Dr Anderson was the third in order of birth, the others being Charles, Alice, who is a resident of St Joseph, Missouri and Carrie, who is Mrs Sangren, lives near Salina, Kansas The parents were members of the Lutheran church The father was a well read man but not professional, being a farmer all his life

Victor Anderson attended the public schools in his native land In 1880 he came to the United States and went to work on a farm in Republic county, Kansas, but soon found an opportunity to work in a drug store at Scandia, Kansas, and during his three years in that position applied himself so diligently to the study of drugs and early medical reading, that he was able to enter the University Medical College of Kansas City, Missouri, from which he was graduated in 1887 He began practice in the Wabash Railway hospital service at Peru, Indiana, moving later to Springfield Illinois When he retired from hospital service he located at Butler, Indiana, where he engaged in a general practice for three years, removing then to Deweese, Nebraska, where he remained for eight years and then established himself at Bridgeport Here he has a well equipped drug store and his fellow citizens know that all prescriptions are put up under his own supervision In connection with handling drugs, Dr Anderson has many of the other features which the public has learned to expect in a modern establishment of this kind

In 1894 Dr Anderson was united in marriage to Elsie Spanogle, a sister of Mark and Clyde Spanogle, bankers, at Bridgeport They have two children Howard Lloyd, who is associated with his father in the drug store, and Helen M, a graduate of the Kearney normal school Dr Anderson's family belong to the Episcopal church In politics he is a Democrat and elected on that party ticket he has served as coroner of Morrill county For many years he has been identified with Masonry and has passed the Consistory degrees

WILLIAM F FRENCH — An interesting example of business enterprise at Gering is the success attending the firm of French & Hanks, who established their general repair automobile business here in 1916, being also selling agents for a number of the best known cars now manufactured Mr French has had experience in other lines of business, but seems particularly well qualified for the automobile trade, and since coming to Gering has built up a sound business reputation and made many personal friends

William F French was born at Jamestown, Kansas, January 25, 1885 His parents are B C and Anna (Duffey) French, the former of whom was born in Canada and the latter in Wisconsin their marriage taking place at Jamestown Kansas He went to that state in 1872 and followed farming there until 1893 when he moved to Oklahoma, bought a farm there and continued to operate it as long as he remained in active life He now lives retired on his property in Grant county In his political views he is a Democrat The mother of Mr French is a member of the Roman Catholic Church Of their five children, William F was the first born, the others being Andrew, who entered military service in the National army, August 15, 1918, is now in Germany, a member of a hospital corps of evacuation camp No 37, Stella, the wife of R L Thompson, a farmer near Pond Creek, Oklahoma, May, the wife of Glenn R. Ratcliff, a business man of Mankota, Kansas, and Gladys, who resides with her parents

William F French obtained his education in Oklahoma and was graduated from the high school of Pond Creek, in 1904 For several years afterward he assisted his father and followed agricultural pursuits, then embarked in the insurance business Later he learned telegraphy and for four and a half years was a telegraph operator for the Union Pacific railroad He then became interested in automobiling for pleasure and this led to practical results In 1916 he came to Gering and in partnership with R M Hanks, established his motor repair works and the enterprise has prospered The firm also displays and sells Maxwell, Haynes, Reo, and Mitchell cars, their territory covering Scottsbluff, Boxbutte, Banner, Morrill, and Sioux counties, Nebraska, and Goshen county, Wyoming In the present favorable condition of the automobile trade, a great future may be in store for honest, intelligent young business men, who devote themselves to its best interests and identify themselves only with sterling cars

In October, 1910, Mr French was united in marriage to Miss Lottie Brown, who was born at Clyde, Kansas She is a member of the Christian church Mr French is a Scottish Rite Mason Politically he is identified with the Democratic party but has never desired public office, his preference being for the business rather than the political field

ROBERT M HANKS, an automobile dealer at Gering in association with William F French, has spent many years in Scottsbluff county where he owns valuable property, is widely known and has been prominently identified with the great work of irrigation in this section He divides his time between the management of his ranch on which he carries on extensive cattle feeding, and his automobile business in the city

Robert M Hanks was born in Montgomery county, Illinois, in July, 1859, the fourth in a family of eight children born to James and Armina (Witherspoon) Hanks, the former a native of Mississippi and the latter of Kentucky. They moved to Illinois in the early forties, and both died in that state They were members of the Presbyterian church, and the father was a Republican from the time of the organization of that party He was a farmer all his active life His surviving children besides Robert M are one son and three daughters Ritta, the widow of Robert Hughes, of Hillsboro, Illinois, James, an agent for the Illinois Central railroad, at Martinsville, Addie, a teacher in the public schools of Hillsboro, and Alma, who also resides at Hillsboro

Robert M. Hanks attended school at Hillsboro, Illinois, grew up on his father's farm and later engaged in farming for himself until 1886, when he came to Nebraska and homesteaded in Scottsbluff county After residing on his land for four and a half years, Mr Hanks went to Kearney, where he worked in a brickyard the summer of 1891, then returned to Scottsbluff county and became interested here in farming and ditch building He built the Gering irrigation ditch and served three years on the board of directors For twenty seasons Mr Hanks operated a threshing machine and with his outfit visited many parts of the county in earlier years In 1916 in association with William F French he established an automobile repair business at Gering, which has proved a profitable undertaking, for the use of motor cars of some kind is almost universal through this section The demand for the best of well built modern cars induced Mr Hanks and Mr French to become selling agents for the Maxwell, Haynes, Reo and Mitchell cars, and they have a wide territory and are doing a large amount of business

In 1884 Mr Hanks was united in marriage to Miss Nettie Kern, who was born in Ohio but was reared in Illinois They have two daughters Fay, a teacher, resides at home, and Alta, the wife of Miller Cooper, who manages the Hanks ranch near Gering Mr Hanks and his family are members of the Methodist Episcopal church He has always been interested in politics, not as a politician, but as a citizen, and casts his vote with the Republican party He has been and still is, one of the strong men of the county, honorable and upright both in business and private life

EVAN G DAVIES, who conducts a grain, feed and draying business at Gering, is not an old-time resident of Scottsbluff county, but has become well-known and esteemed as a business man since he established himself in the above city Mr Davies had experience in handling wheat and other precious grain products of American farms, on which the eyes of a hungry world are centered at the present time, before he became a grain buyer, for he was a grower on the old homestead farm in Hitchcock county for many pears previously

Evan G Davies is proud of his sturdy Welsh ancestry He was born in Hitchcock county, Nebraska, June 24, 1886 His parents were Samuel and Margaret (Morris) Davies, both of whom were born and reared in Wales Shortly after their marriage in their native land they came to the United States and settled at first in Iowa and then came to Hitchcock, Nebraska, where the father homesteaded The mother died on the farm but the father died at Trenton, where he had lived retired for some time Of their eleven children, ten are living, Evan G being the tenth in order of birth The parents were members of the Congregational church

Evan G Davies obtained his education in the Trenton schools He was reared on the home farm and for a number of years confined his attention to agricultural pursuits In October, 1915, he came to Gering and began to buy grain and feed He opened a grain and feed store in a good business section and subsequently added draying, exercising business prudence in all of his ventures He is doing a fine business and a safe one, and is the leading merchant in his line in the city

In December, 1908, Mr Davies was united in marriage to Miss Amy Houser, who was born in Seward county, Nebraska, and is a daughter of William W and Alma (Coover) Houser The father of Mrs Davies was born in Hardin county, Ohio, and was taken to Iowa when young and grew up there In February, 1884, he went to Hastings, Nebraska, where he became interested in farming but later moved to Seward county and still later to Lincoln, his present home being at Cambridge, Nebraska Mr and Mrs Davies have two children Heath and Inez The family belongs to the Christian church In politics he is a Republican

OTTO J PROHS — An important business enterprise at Gering is the Prohs Brothers Hardware Company, a name that covers several lines of merchandising, several industries and also undertakings This business is a growing concern It was founded in this city early in 1914, on a small margin of capital, which has been greatly increased with the rapid development of five years of legitimate dealing, until now a stock value of $30 000 is carried The firm is made up of three brothers, William, Otto J and Edward S Prohs

Otto J Prohs was born October 14, 1887,

at Juniata, Adams county, Nebraska His parents are Louis and Emma (Doll) Prohs, the former of whom was born at Stuttgart, Germany in 1857, and the latter at St Louis, Missouri, in 1861 They were married at Hollowayville, in Bureau county, Illinois, and eleven children were born to them, eight of whom are living They are members of the Evangelical Lutheran church In 1884 the parents of Mr Prohs came to Nebraska and the father rented land in Adams county, but in 1888 the family returned to Illinois and the parents have resided at Peru ever since

Otto J Prohs attended the public schools in Peru until he finished tenth grade work, after which he was a student for a time in Brown's Business college and also took a business course with the International Correspondence school He went to work first in a printing office, but soon afterward found a position with the Big Ben Clock works, starting to work for $3 a week That Mr Prohs remained with this company for the next fourteen years speaks well for his industry and efficiency and when the further fact is mentioned that when he severed his long relation he was receiving a large salary as assistant stock manager, it may be inferred that he had proved faithful to every responsibility he had assumed On March 9, 1914 he came to Gering and with his brothers established the present business The firm handles a general line of hardware furniture and carpets, do tinning and plumbing and also are undertakers In all their business transactions they have been fair and honorable and they enjoy the respect and confidence of the public

In 1913 Otto J Prohs was united in marriage to Miss Flora West, who was born at Peru, Illinois They have one son, Wesley Richard, born at Gering, Nebraska, in September, 1914 Mr Prohs, like his father, is a Republican in politics, and, while never particularly active, feels a citizen's responsibility and keeps well informed on all public questions He belongs to the Masonic fraternity and also to the Knights of Pythias

WILLIAM LAMM, Sr, one of the extensive landowners of Scottsbluff county, and for many years one of the heaviest cattle feeders, has been a resident of Gering since 1909, where his judgment on business matters and public affairs is considered of great value Although Mr Lamm has been a capitalist for many years, he began his business career a penniless boy thrown upon his own resources, hence his advice from lessons learned through experience, may well be treasured He was born April 23, 1854, in Des Moines county, Iowa His father was Benedict Lamm, and both parents were born in Germany They came to the United States in 1853 and settled in Iowa, where Benedict Lamm bought a farm and on that place both he and wife died Of their four children, two besides William are living Oscar, a stonecutter by trade, lives at Burlington, Iowa, and Mary, the wife of Charles Bassett, lives in New York The parents of Mr Lamm were members of the Roman Catholic church

William Lamm had comparatively few educational advantages in boyhood He worked on his father's farm as long as he remained at home, and continued farm work after he was thrown at an early age on his own responsibility By nature industrious and through necessity frugal and saving, he gradually advanced himself and in the course of time became the owner of farm land in Union county, Iowa, where he resided for many years. When he came to Scottsbluff county, Nebraska, in 1904, his business sense led to his investing largely in irrigated land and at the present time he owns many valuable acres, where his operations in cattle and sheep feeding have proved exceedingly profitable

In 1875 Mr Lamm was united in marriage to Miss Jane Knotts, who died December 3, 1918 They became the parents of the following children William H, postmaster at Gering, Ernest, a farmer in Wyoming, Bert, a farmer in Idaho, Bertha, the wife of John Gross, a farmer in Wyoming; Bess, who married Joseph Nisley, in the monument business at Gering, Carrie, the wife of Harry Barton, a farmer near Gering, Cloreta, the wife of E Davies, True, who resides at home, having been honorably discharged from the military camp where he had been in training since October, 1918, and Grace, the wife of P E Schmidt, who entered the National army in April, 1918, was honorably discharged in March, 1919 and is now connected with the First National Bank of Scottsbluff Mr Lamm and his family belong to the Christian church Politically he is a sound Democrat He has never cared for public office as his time and energies are occupied in the management of six hundred and forty acres of valuable land which he owns near Gering, all under cultivation Mr Lamm is essentially the architect of his own fortunes and is self made He was one of the organizers and heavy stockholders of the Gering National Bank, being a member of the board of directors of that institution

HOWARD O JONES, D D S — The different professions are well represented at Gering, and a leading Practitioner of dentistry is Dr Howard O Jones, who established himself in this city in 1916 Although not in continuous practice since then, nevertheless he has succeeded in building up a large clientele made up of those who demand scientific treatment and appreciate thorough dental knowledge Dr. Jones has but recently resumed his practice after an absence of four months in military training at Camp Logan, Houston, Texas He entered the National army September 16, 1918, was assigned to the dental corps, was ranked a first lieutenant, and was honorably discharged February 22, 1919

Howard O Jones was born August 29, 1894, at Granger, Scotland county, Missouri His parents are Henry Harrison and Ida (Sullinger) Jones, the former of whom was born at Granger and the latter at Glasgow, Missouri Of their three children, Howard O is the second born, the others being Lois, the wife of B O Reeves, county attorney of Scotland county, and Paul, attending school The parents of Dr Jones are members of the Methodist Episcopal church The father has been very active in Republican politics for many years and ex-county clerk of Scotland county He is a lawyer by profession, and for twenty-five years has been in the abstract business at Memphis, Missouri He is also president of the Mutual Fire Insurance Company, and he is identified with various Masonic bodies

After completing his public school course at Memphis, Howard O Jones entered a dental college at Denver, Colorado, from which he was graduated in 1916, immediately afterward coming to Gering Dr Jones keeps fully abreast of the times in relation to the remarkable progress being made in his profession, devoting much study to oral and plastic surgery, the wonders of which have more or less revolutionized every branch of the healing art By inheritance and inclination he is a Republican and belongs to the religious body in which he was reared by careful home influences He is unmarried

ROBERT G MILLER — The wisdom shown by large corporations in placing experienced and practical men in charge of their industries in which they have invested immense capital is to be commended, and a case

in point is the Great Western Sugar Company, at Gering, of which Robert G Miller, construction superintendent, has been made superintendent Mr Miller not only has had thorough training in construction work, having been identified with machinery and mechanics all his business life, but he is an able man in other directions He possesses great executive ability, has proved himself able to cope with changing conditions in the industrial world, and has had much to do with the present prosperity of the plant

Robert G Miller was born at Burlington, Iowa, April 13, 1871 His father, Peter Miller, was born in Switzerland, and his mother was born in Germany They came to the United States when young and met and were married at Milwaukee, Wisconsin Nine children were born to them, Robert G being the fourth, and seven still survive The mother was a member of the Roman Catholic church The father of Mr Miller was a machinist by trade and he followed the same for a number of years at Burlington, Iowa, and later in California He was an American citizen and in politics was affiliated with the Republican party

After securing a good, common school education in the city of his birth, Mr Miller learned the machinist trade and familiarity with every line of mechanics followed While living in California he embarked in construction work and there and in other places he has erected two mills and like structures In 1916 he came to Gering and as superintendent of construction erected the mill for the Great Western Sugar Company, and in 1917 went to Bayard and built a mill there In 1918 he was called back to Gering to become superintendent of the mill he had erected here and has so continued

In 1897 Mr Miller was united in marriage to Miss Katie Sawyer, who is a daughter of Jacob Sawyer, a well-known real estate dealer of Los Angeles, California They have one daughter, Helen, who resides at home In politics Mr Miller has always voted with the Republican party

AUBURN W ATKINS, who for many years has been a man of prominence in Nebraska came to Cheyenne county as a cow puncher in 1880 Well educated and with comfortable home environment in the East, when he reached his majority, he chose the freedom and adventure of the West, where he has achieved no small measure of distinction and has accumulated a fortune Colonel At-

kins was born in Ashtabula county, Ohio, April 16, 1859, and received his title when serving as a member of Governor Neville's staff

The Colonel's parents were Levi and Persis Amanda (Clarke) Atkins natives of Ohio The father was a soldier in the Civil war, a member of the Eighth Ohio Volunteer, and died from disease caused from exposure, while in the service Of his four children, Auburn W is the second of the three survivors, the others being Angelo, a noted teacher of music at Bowling Green Ohio, and Frances Genevieve, a widow, who resides at David City, Nebraska The mother was a member of the Baptist church

Auburn W Atkins was young when he lost his father He obtained his first schooling at Sullivan, Ohio, and Greenville, Mississippi, and later attended the high schools at Tabor and Hamburg Iowa He worked on a cattle ranch for a number of years after coming to Nebraska, in the meanwhile homesteading and pre-empting land as opportunity presented, and at the present owns about 4,000 acres and has eight hundred acres under irrigation, this farm being under rental For many years he has been a heavy raiser of cattle, his activities including buying and selling, and he has prospered exceedingly Colonel Atkins has been closely identified with irrigation projects and also with railroad construction in this section of the state, as well as Montana, where he built eight miles of road at one time He also constructed ten miles of the Burlington Railroad south of Bridgeport He is president of the Alliance Ditch Company and was one of its organizers

On July 16, 1893, Col Atkins was united in marriage to Miss Luha Barnhart, who was born in Bedford county, Pennsylvania, a daughter of David A and Malinda (Moore) Barnhart, who settled at Kimball, Nebraska, in 1878 He was a successful cattle man during his active years and his death occurred in 1911 at Sidney The mother of Mrs Atkins lives at Cheyenne Four children have been born to Colonel and Mrs Atkins Clarke W, who enlisted in the aviation department for service in the World War, was in France for seven months and did his full duty, Allan B, is working on his father's ranch Auburn H, a member of the Naval Reserve corps during the war, has reached home, and Luha Virginia, is yet in school All the members of the family have been confirmed in the Episcopal church

Colonel Atkins has been very prominent in

Democratic politics for many years and was his party's choice for the General Assembly, being defeated by less than one hundred votes He has served as county commissioner in Cheyenne county and has been a member of the town board and the school board of Bridgeport many times In 1916-17 he served, as mentioned above, on Governor Neville's staff He is a Scottish Rite Mason, an Odd Fellow, an Elk and Knight of Pythias At present he devotes his time mainly to looking after his land and irrigation interests, and maintains a beautiful and hospitable home in Bridgeport

WILLIAM G BROWN, D D S, who has been engaged in the practice of dentistry at Gering for a number of years, occupies a prominent place in professional circles here and is a member of the medical advisory board of Scottsbluff county Since locating at Gering he has taken a commendable interest in civic affairs in general and is now serving in the office of city clerk He was born in Ralls county, Missouri, May 4, 1886 His parents are George and Virginia (Elzea) Brown, natives of Virginia, who located in Missouri prior to the Civil War, and now live retired at New London, Missouri, where the father was in the hardware business for a number of years Of their five children William Guy was the third born, the others being Ernest, in the employ of the United States government, lives at Muscogee, Oklahoma, Clifford, a dentist at Ashton, Idaho, Elizabeth, a teacher in the public schools at Lincoln, and Deskin, a sailor on the United States ship Mayflower, having entered military service in 1913

William G Brown completed his public school course at New London, Missouri, when he was graduated from the high school in 1903 He then went to work for the Portland Cement Company at Hannibal, Missouri, where he remained four years, being manager of the empty bag department From early youth, however, he had taken an interest in dentistry and when prepared to study the art scientifically, he entered Creighton Dental college, from which he was graduated three years later, in 1910 He located immediately at Emmerson, Nebraska, where he remained in practice until the fall of 1914, when he came to Gering, where he has had much professional success

In 1910 Dr Brown was united in marriage to Miss Matie Gaeth, who was born in Nebraska She is a member of the Episcopal church, is interested in charitable movements

and is well-known in social life Dr Brown was reared in the Presbyterian church Politically he is identified with the Democratic party and fraternally he is an Odd fellow and has passed through all the chairs of the local lodge Also B P O E

TED L IRELAND — While the luxuries of life may be desirable, they can be dispensed with in the interests perhaps of patriotism, or health or economy, but there are certain basic commodities, represented by the general name of groceries, that are absolutely necessary for consumption in every household, in order to keep the balance that means nutrition or ill health While they may never pose as philanthropists, nevertheless the honest and wide awake grocer in of beneficial influence in a community The reliable grocer insures his customers receiving full weight and standard goods and his business alertness protects them from unwholesome products that may be put forward under the lure of cheaper price A leading grocery house of the better class at Gering and one that handles dependable goods only, is that conducted by Ted L Ireland, in association with his brother Roy M Ireland

Ted L Ireland was born at Arapahoe, in Furnas county, Nebraska, April 13, 1888, the eighth in a family of ten children born to George M and Mary E (Sexon) Ireland They were married in Nebraska, but the father was born in West Virginia and the mother in Iowa They came to Furnas county in the eighties, where the father homesteaded and lived on his farm for thirty years, retiring then to Mitchell, in Scottsbluff county, where he died The mother still resides at Mitchell She is a member of the Methodist Episcopal church, as was the father He served in the Civil War from 1861 until 1865, with rank of first sergeant, and during that time spent six months as prisoner at Andersonville, Georgia. Ted L Ireland has six brothers and one sister Wilbur J, in the grocery business at Scotsbluff, William B, an instructor in the university at Lincoln, Charles C, in business at Mitchell, Nebraska, George M, conducts a general merchandise store at Mitchell, Cecil H, in business at Mitchell, Roy M, associated with his younger brother in business at Gering, and Anna, the wife of William Cockle, who is in business with George M. Ireland at Mitchell

From home on the farm and the country schools, Ted L Ireland went first to Kearney, where he attended the normal school, (1907), and afterward was a student in the Wesleyan

University, (1915) He was seventeen years old when he went to work for the Mitchell Mercantile Company, with which concern he continued eight years In 1916 he started a store at Scottsbluff in partnership with a brother, which was proving a profitable investment when Mr Ireland's plans were disarranged by the call of the government He entered military service September 6, 1918, at Omaha, going into training for the balloon branch, but the signing of the armistice in November hastened his discharge and on December 23 following he was released He came then to Gering and embarked in his present grocery enterprise, to which he devotes himself with every indication of unusual business success

On March 18, 1918, Mr Ireland was united in marriage to Miss Ethel G Long, who was born at Holdridge, Nebraska, and they have one daughter, Ruth Elenor Mr and Mrs Ireland are members of the Methodist Episcopal church His political convictions make him a Republican, and in fraternal life he is an Odd Fellow, and a member of the American Legion

FRANK B YOUNG, M D, a physician and surgeon of wide experience, who has been established at Gering since the fall of 1916, occupies a foremost place in medical circles here as he has done elsewhere Dr Young was born in Sherman county, Kansas, August 11, 1878, the son of John and Sophia (Franklin) Young, who were married at St Louis, Missouri, and settled in Kansas in 1877 Dr Young's father, also a physician of eminence was born in Tennessee, in 1836, while his mother was a native of Canada The father was a graduate of the Missouri Medical college after which he practised there until 1879 when he moved to Arkansas and continued active in his profession there until his death in 1914 He served as captain in the Third Missouri volunteer infantry in the Confederate army during the war between the states, and was several times wounded After the Civil War Dr John Young was a citizen of Nebraska for several years, being connected with the freighting work of Majors Russell and Waddell, and the Wells Fargo Company His father died and is buried at Weeping Water, Nebraska He was a Mason and Odd Fellow, a Democrat in politics and with his family belonged to the Methodist Episcopal church The mother of Dr Frank B Young resides at Springdale, Arkansas The two other children of the family are Daisy, the wife of

Bruce Holcomb, a banker at Fayetteville, Arkansas, and John, who owns a ranch in New Mexico

Following his graduation from the University of Arkansas, Frank B Young began the study of medicine and was graduated in 1900 from the Kansas City Medical college He entered into practice at Springdale, Arkansas, in partnership with his father until 1913 when he was appointed State Health Officer and spent one year in that position at Little Rock, then became superintendent of the State Insane Asylum there He continued at the head of that institution until January 1, 1916, when he resumed private practice in the capital, but in the fall of the year came to Gering, where he has built up a gratifying clientele In 1913-1914 he was president of the Arkansas Medical society was a member of the Arkansas State Board of Medical Examiners from 1907 to 1913, and was president of the first Board of Health in Arkansas

In 1912 Dr Young was united in marriage with Mrs Jessie Keefer, who was born at Denver, Colorado Mrs Young has two children of her first marriage, Charlotte and Hamilton, and they reside with Dr and Mrs Young He is a Scottish Rite Mason of the 14th degree, is Past Grand in the Odd Fellow fraternity, is Past Chancellor in the Knights of Pythias, and belongs to the Elks, and for many years has been a member of the American Medical association Dr Young is held in the highest esteem in this city both professionally and personally

LUTHER F HAMILTON — To all citizens proud of the acknowledged general intelligence of the United States, the published fact that an examining government board in recent years, found so large a proportion of the individuals coming before it illiterate, brought a feeling of astonishment, less, perhaps, to the country's educators than to others Scholarly men like Luther F Hamilton superintendent of the public schools of Gering and widely known in the state, who have devoted their lives to educational effort possibly understand more completely than others, the lamentable lack in modern days of that consuming thirst for real knowledge that will lead youth to scale mountains of difficulty in order to obtain knowledge For many years it has been Superintendent Hamilton's conscientious task to inspire this love of learning in the young by whom he has been continuously surrounded and his highest aim has been the opening of doors of opportunity for future usefulness through awakened and enlightened minds. He came to Gering from other educational fields where he had been highly appreciated, and the influence he has exerted has been marked by constant progress in the city schools.

Luther F. Hamilton was born March 19, 1872, in Macoupin county, Illinois. His parents are William and Mary (Stephens) Hamilton, the former of whom was born in Scotland and the latter in England. They now live retired at Eddyville, Nebraska, having come to this state in 1889 from Illinois. The father purchased land in Otoe county and engaged in farming and raising cattle. In politics he is a Democrat, and both father and mother belong to the United Brethren church. Of their five children Luther F. was the second born, the others being: C. R., who conducts a goat ranch in New Mexico; Ida, the wife of John Johnston, of Seattle, Washington; Minnie, the wife of Ruford Williams, a farmer near Arcadia, Nebraska; and Maggie, the wife of Victor Wall, a farmer and cattleman near Eddyville.

Luther F. Hamilton attended school in Otoe county, in 1888 being graduated from the Pelmyra high school, immediately following which he began to teach school. In 1890 he entered the Nebraska State University, where he continued his studies for three years, in 1906 securing his A. B. degree and his B. A. degree in 1912, in 1914 winning his M. A. degree. In the meanwhile he continued teaching, first at York college where he was science instructor, and at other points. In the course of years he was made superintendent of the schools of Panama, in Lancaster county, where he remained five years, going then to Douglas in the same capacity for six years, after which he was superintendent of schools at Cook for two years. From there he came to Gering in 1916, and his services here have been of great value.

In 1896 Luther F. Hamilton was united in marriage to Miss Estella Weston, who w s born in Wisconsin, and is a daughter of Perry Weston, who located at Panama, Nebraska, in 1889. They have three children, two sons and one daughter, namely: Cecil C., who enlisted in the American army in December, 1917, for service in the aviation department, is yet in France; Keith, who is fourteen years old; and Genevieve, who is eleven years old, both of whom attend school. Mr. Hamilton and his family are members of the Methodist Episcopal church. In politics he is a Republican

and fraternally is a Knight Templar Mason and past master of the blue lodge.

FREMONT SCOTT. — The men who may most confidently be depended upon to build up the substantial structures of business in any community are those who have a varied experience to fall back on. By the light of their experience, often painfully gained, they are able to plan successfully for the future. One of the experienced and representative businss men of Gering is Fremont Scott, who has the real estate situation well in hand in the Panhandle of Nebraska. He has been a resident of the state since his fifteenth year and to the unusual opportunities offered to those seeking them, in both past and present Nebraska, he attributes much of his success in life, although his friends are not slow in calling to mind his personal efforts that made these opportunities fruitful.

Fremont Scott was born January 15, 1857, in Shenaugo county, New York, the son of Ezckiel G. and Ruth (Wilcox) Scott, both of whom were born and reared in the Empire state. They came to Wisconsin in 1857 and from that state Ezekiel Scott enlisted for service in the Union army during the Civil War. Shortly after becoming a soldier he was taken sick and was so seriously ill that he had to be brought home on a stretcher. After recovery he resumed his former pursuits, being a man of education, and continued to live in Wisconsin until 1872, when, accompanied by his family with one team, he came to Nebraska, driving across country in pioneer style, and homesteaded in Hamilton county. While living there he is credited with killing the last buffalo that was slain in Nebraska. Later he moved to Phillips county, Colorado, where he remained eight years. In March, 1894, he returned to Nebraska and settled in Scottsbluff county and here both he and his wife lived the rest of their lives. The latter was a member of the Presbyterian church. Of their seven children but two survive, Washington and Fremont, both of whom reside in Scottsbluff county.

Fremont Scott completed his public school course after coming to Nebraska. His boyhood and early youth were spent on a farm. Forced by circumstances to depend upon his own efforts, he developed sturdy qualities which have been useful to him ever since, undergoing as have other self-made men a discipline not altogether to be deplored. His first purchase of land was a tract held at $10 an acre, and he earned the money to pay for it

P. Magennis

by working at ditching. That was the nucleus of his present valuable farm of four hundred and twenty acres, which lies in Scottsbluff county. To farming and allied pursuits he devoted himself until 1913, in which year he came to Gering. Here he embarked in the real estate business and his interests now cover a wide territory, particular attention being given to lands in Scottsbluff county and eastern Wyoming.

In 1878 Mr. Scott was united in marriage with Miss Adelia Florence Moore, who was born in Almakee, Iowa, and they have the following children: Rosie, the wife of Alvee Leonard, residing on Mr. Scott's farm; Daisey, the wife of Emerson Ewing, of Carter Canyon, Scottsbluff county; Ruby Lillian who married Bert Scott, a farmer near Mitchell, Nebraska; Violet, the wife of Zonoua Yates, a farmer south of Gering; Pansy, the wife of Charles Gering, of Gering; Emery G., a farmer and stockman on a ranch in Banner county; and Pearl and Harold, both of whom are at home. All the children have had excellent educational advantages. Mr. and Mrs. Scott are members of the First Baptist church at Gering. Fraternally he is a member of the Modern Woodmen order. Like his father before him, Mr. Scott has always been a sound Republican and at different times has served with great public satisfaction in important county offices. For four years he was in charge of the county jail and also was deputy sheriff for some time.

PATRICK MAGINNIS, pioneer, frontiersman and early settler, today capitalist, landowner, banker and successful business man, has had a career of varied and interesting experiences, from hunting buffalo on these western plains when western Nebraska was a veritable wilderness with settlements few and far between, to the civilized existence of modern days, and few men twenty years his junior show so few of the scars of life. A resident of Nebraska for nearly forty years, Mr. Maginnis knew this country when most of the houses in the central and western section were of sod and has watched with the eye of proprietor the various changes that have been wrought with the passage of the years and the sturdy progressive work of the settlers. He has borne a full share of the labor of development from the earliest years and since irrigation was first attempted along the Platte river, has been one of the most prominent figures and important factors in making what was known as the "Stakes Plains," of

the middle west, blossom like the rose and today the rich valley lands of the Panhandle are the most productive in this wide country. It is said that the Irish-Americans always succeed, whether on the public rostum, where they are possessed of golden speech; behind the counter, where business acumen counts for capital; on the farm and ranch, where energy and thrift are in demand and in commercial life of wide range; Mr. Maginnis has proved this to be true and his personal success is so bound up with the development and success of the south west section and all the people who are living, prospering and thriving there that he should be given special mention in the annals of the Panhandle.

Patrick Maginnis is a son of the Emerald Isle, born in County Down, near Belfast, January 6, 1867, the son of Hugh and Alice Maginnis, who were married in their native country in February, 1864, and became the parents of seven children, four girls and three boys: Patrick, of this review; Arthur, who came to the United States, spent his life here and died February 20, 1920, at Lawrence, Massachusetts; Mary Maginnis McAlinden, who has nine children and lives at Airdrie, Scotland; Charles H., who came to America when young and now lives in San Francisco, California, where he is in the government service as a member of the staff of the pure food department, and has one son; Elizabeth, the wife of John Morgan, of Airdrie, Scotland, who is connected with one of the large rolling mills of that city where they are rearing five children; Allen, who married a Mr. Kelly, and Susan who now lives with the mother in Ireland. Mrs. Maginnis is a remarkable old lady of seventy-five years, who still retains much of the vigor of her youth and all her mental faculties. Hugh Maginnis died in his native land in 1905.

Patrick Maginnis attended the public schools in his native country until he was thirteen years of age but being an ambitious lad he had paid attention to the stories told by returning Irishmen from America of the many advantages and opportunities open for a youth willing to work, and August 11, 1880, broke all the dear home ties and sailed for the United States. After landing on our shores he came west to Illinois, locating in Brown county, remained there two years then came to Nebraska to take advantage of whatever business openings he might find in the frontier country. Mr. Maginnis had worked in a blacksmith shop in Illinois and after settling in his new home at Aurora, Hamilton county, followed that vocation until he moved to Sweetwater county, Wyoming where he was employed on a ranch, part of his time being devoted to the necessary blacksmithing of such an enterprise. In 1885, a true pioneer, Mr. Maginnis came to the Panhandle, one of the early residents of the present Kimball county, making his home in what was then known as the town of Antelope, now Kimball. He opened the first blacksmith shop when there were but a few houses and has seen this little frontier settlement grow into one of the flourishing communities of the state. With a successful and growing business, Mr. Maginnis was not too busy to let his inventive genius mature and ripen and during the years from 1885 to 1910 was ever alive to the demands and necessities of the new country opening and developing under his eyes. He was one of the first men of the section to see that the first attempts at irrigation were crude and expensive; he studied over the question of betterment and invented a flume which greatly helped in the infant reclamation projects along the Platte. Within a short period he began the manufacture of the flumes extensively, applied for patents which were granted in 1902. The state used the flume on all its projects in the Panhandle; it was introduced into Porto Rico and Mexico but during the time it has been in use the patents were infringed on which caused Mr. Maginnis long and expensive litigation and it was necessary for him to obtain a restraining order from the Federal Court which held for years until a decision in his favor was handed down and the question settled for all time by Judge Lewis, of Denver. Since then Mr. Maginnis has increased the output of the flume and in the last year sold more than a quarter of a million dollars' worth. As his many and varied interests had grown to such proportions he was induced to sell his patent rights in the flume at an enormous profit and now devotes his time to his landed and commercial interests in Kimball county and the state of Oregon.

Mr. Maginnis' sons have been running a fine hardware store in Kimball for years, and when the father decided to build a large commercial block, the store was not disturbed, the new building was erected around the one doing business and is now housed in the well-known Maginnis block which consists of store on the first floor with office rooms above, one of which Mr. Maginnis keeps for his own use.

Believing in the future of this section, Mr. Maginnis began buying land in Kimball county, when his purchases were the raw prairie. He continued to increase his holdings until he was known as one of the largest land owners in this section of the state. From time to time in recent years he has sold or traded off the largest part but still owns three hundred and twenty acres under water rights and eight hundred and sixty acres of rich farming and grazing land. Not confining this business to Nebraska alone, Mr. Maginnis has purchased valuable land in Oregon, near the railroad station of Redmond, and this land is also under water.

Some time ago Mr. Maginnis purchased a block of stock in the American Bank, of Kimball and was elected one of the executive heads of that institution and is vice president. This is one of the progessive banking houses of Nebraska, having today a paid up capital of a hundred thousand dollars.

In 1888 Mr. Maginnis married Miss Margaret A. Marshall, the daughter of Holmes H. and Isabella Marshall. Mrs. Marshall's first husband was a Mr. Weir, who was the father of the two small children left fatherless when he was accidentally killed. Mrs. Weir later married Mr. Marshall and her children took the Marshall name. The Marshalls were old and respected residents of Kimball, locating here at an early date. Ten children have been born to Mr. and Mrs. Maginnis; Arthur F., Alice Isabella, Robert J., Edward Dewey, Hugh Marshall, Lizzie Margaret, William P., Mary Ellen and a son Charles, who died in 1899.

Mr. Maginnis has taken an active part in the life of Kimball county and the city of his adoption since first coming here to live and may be said to be its foremost citizen in years of residence and worldly goods. His standing with his fellow citizens and in the county is testified to by his election as sheriff of the county which he served for years before being elected county treasurer in 1902 and reëlected, serving until 1905, when he refused another nomination as his growing business interests demanded all his time and energy. The Maginnis family is one of the best known and prominent in the southern Panhandle where every member has taken and is now taking an important part in the upbuilding of the southwest region. They are one of the oldest families in years of residence, the boys and girls are all popular and well liked, taking after their parents, who are among the most genial and best liked people of the county, ever ready to help their friends, open handed in giving to any cause to build up and develop Kimball and the county.

ARTHUR M FAUGHT, M D, formerly mayor of Scottsbluff, and a physician and surgeon whose professional reputation extends over the state, for a decade has been one of this city's most virile and constructive citizens. Illustrative of his public spirit and civic interest, he has been the means of establishing here a umber of worthy enterprises included in which is the Mid-West Hospital, an institution deserving high praise as to its aim and accomplishments

Arthur M Faught was born July 27, 1884, at Plattsmouth, Cass county, Nebraska. His parents were John and Martha (Root) Faught, both of whom were born also at Plattsmouth. The mother of Dr Faught died in 1915, but the father survives and resides at Lincoln, living retired after an active business life covering many years, eighteen of which he spent at Phillips, Nebraska, where he was interested in lumber. In his political views he is a Democrat. Arthur M Faught is the eldest of the family of children born to his parents, the others being Mrs Ralph Murphy, of Hastings, Nebraska, Claude J, in charge of the L C Smith Typewriter interests at Sioux City, Iowa, Mrs Grace Busby, of Minneapolis, Justice L, connected with the Bell Telephone Company at Rochester, Minnesota, and Ruth, residing with her father at Lincoln. The family was reared in the Christian church

Arthur M Faught is a graduate of the Nebraska State University. In 1905 he was graduated from the medical department of Colton University, immediately afterward engaging in medical practice in Seward county, Nebraska, where he remained four years. In 1909 he came to Scottsbluff and many are the ties that now bind him to this city, where a friendly greeting meets him on every side, and where his devotion to his professional work is deeply appreciated. In 1911 he established here the Mid-West Hospital, which he owns, a thoroughly modern structure with thirty beds. He has been exceedingly successful in his surgical cases, to which he devotes the most of his time many patients availing themselves of his skill, some from a long distance but many nearer home as indicated by a record that shows that he has preformed over 2,000 major operations. He has taken post-graduate courses in operative surgery and watched many major operations in the clinics of noted institutions in Chicago

For six years Dr Faught was city physician of Scottsbluff, is chairman of the examining board of Scottsbluff county, and is ex-

aminer for civil service positions in government offices. He is a member of the American Medical association, the Nebraska State Medical association, the Scottsbluff county Medical society, and the National Electic Medical association. Politically he is a Republican, but largely because of his general popularity he was elected mayor of this city in 1917. His administration was an admirable one but taxed his strength because of his professional duties, hence he declined to again become a candidate

On July 25, 1906 Dr Faught was united in marriage to Miss Rosa Hartman, of Staplehurst, Seward county, Nebraska, and they have two children. Ardon M and Audry H Dr and Mrs Faught are members of the Episcopal church. Like his father, he is a Thirty-second degree Mason, and belongs also to the Odd Fellows the Knights of Pythias and the Elks

NELSON H RASMUSSEN, M D who is meeting with success in the practice of medicine and surgery at Scottsbluff, came to this city in 1917 and already has built up a satisfactory practice and has made many personal friends. Although not born in America, almost his entire life has been spent here

Dr Rasmussen was born September 18, 1881, in Denmark one of seven children born to J S and Carrie (Jensen) Rasmussen, who were born, reared and married in Denmark. They came to the United States in 1882 and established a home at Winona Minnesota, where the father secured employment in a big business plant. He was a steady, dependable workman and not only secured the confidence of his employers, but also of his neighbors. He and family belonged to the Methodist Episcopal church, in which he was an earnest worker, and he was equally active in the cause of temperance. Dr Rasmussen is the oldest of their children, the three other survivors being. John, who owns a ranch in North Dakota, Mary, the wife of Paul Nelson, of Oregon and Margaret, the wife of Rudolph Offerman who operates a hardware store and lumber yard at Cobden Minnesota

The public schools of Winona, in which he remained a student until he was graduated from the high school, gave Nelson H Rasmussen a fair preparation for a future career, but at first it helped him very little in the matter of securing a medical education which was the goal of his ambition. His father could give him but little assistance and the youth realized that he must depend on his own efforts

In no way discouraged and never giving up the hope of ultimate success, he went to the Klondyke region and worked four years in the gold fields there, meeting with some reward. Prior to this, however, he had worked on a Minnesota farm, had tried railroading, and had endeavored to learn the blacksmith trade. When finally he found himself an enrolled student of medicine in the John A. Creighton Medical College at Omaha, he took upon himself various duties in order to pay his way. Possibly it was not always agreeable to act as janitor in the church edifice, as a waiter at table or to spend his Saturday afternoons selling goods in a clothing store while others of his age were on holiday pleasures bent, but, to his credit be it said, he persisted and in these and other ways earned sufficient to not only enable himself a fine medical education, but to also enable his sister to take a course in nursing at Omaha. After his graduation, Dr. Rasmussen served a year as an interne in the Creighton and St. Joseph hospitals, and afterwards was associated with Dr. T. J. Butler, at Omaha, and later as assistant to Dr. J. E Conn, a prominent surgeon at Ida Grove, Iowa. Since coming to Scottsbluff Dr. Rasmussen has given special attention to surgical cases, making specialties of obsterities and pediatrics.

In 1917 Dr. Rasmussen was united in marriage to Miss Theresa C. Lzingle, who was born at Ashton, Nebraska, and was graduated as a nurse from St. Catherine Hospital, Omaha. They have one daughter, Betty. Mrs. Rasmussen is a member of the Roman Catholic church, while the Doctor belongs to the Christian church. He is an independent voter in politics, but is identified with leading organizations, belonging to various Masonic bodies including the Mystic Shrine, and a member also of the Yoemen and the Danish Brotherhood.

RALPH W. HOBART, judge of the Seventeenth Judicial District of Nebraska, is eminently qualified for the high position to which he has twice been called by the votes of his fellow citizens, and in which he has served with conspicuous judicial ability since April, 1911. Through a long and successful career as a lawyer, Judge Hobart won distinction at the bar, and when the Seventeenth District neaded a broad-minded, well balanced, firm and resolute judge, it was fortunate that he was elected to the bench.

Judge Hobart was born at Calais, Washington county, Maine, March 24, 1865, of English extraction and old colonial stock. The first of the Hobart family on record in this country bore the name of Edmund. He came from England in 1633 and assisted in the settlement of Charlestown, Massachusetts. Judge Hobart's parents were Daniel Kilby and Amy Elizabeth (Whidden) Hobart. His father was born April 15, 1823, in Maine, and died March 15, 1891. In civil life he was in the lumber and commission business, and for sixteen years he held a consular office in relation to the Dominion of Canada. He was married in Maine to Amy Elizabeth Whidden, who was born in New Brunswick, May 21, 1827, and died June 9, 1896. Her father, Reudol Whidden, was a native of New Hampshire. Of the seven children born to the parents of Judge Hobart, he is the second of the three survivors, having two brothers Charles E. and Harry K.

Ralph Whidden Hobart had collegiate training in Kings College, Nova Scotia. Subsequently he came to the United States, entered the University of Michigan where he was graduated in law at Ann, Arbor, in 1888, and the following year located for practice in Dell Rapids, Minnehaha county, South Dakota, where he remained eleven years. In 1900 he removed to Columbus, Nebraska, where he practiced until 1906 when he came to Mitchell, and through appointment was called from the bar to the bench in April, 1911. Twice since then he has been elected judge of the Seventeenth Judicial District, his jurisdiction extending over the counties of Scottsbluff, Banner, Morrill, Garden and Arthur. Both in public and in private life, Judge Hobart stands as an example of useful, high-minded, exemplary American citizenship.

In 1898 Judge Hobart was united in marriage to Miss Anna Maldrum, who was born in Ontario, Canada, and the have one son, Edmund Maldrum. Mrs. Hobart is a member of the Congregational church. Judge Hobart is a Republican. For many years he has been a Mason and Odd Fellow.

THOMAS M. MORROW, the subject of this sketch and the second son of Thomas and Mary (McDonald) Morrow, was born in Lewis county, New York, on the 25th day of October, 1868. His parents were born in Ireland and came to America when quite young. They were united in marriage in Lewis county, New York. Their children are as follows: John, who for the past five years has been rtceiver of public moneys in the United States land office at Alliance, Nebraska; Thomas M.

Morrow; Lavinia, a teacher in the public schools of Bayard, Nebraska; William, law partner of Thomas M. Morrow; Anna, wife of George G. Cronkleton, cashier of The First National Bank of Bayard, Nebraska; Frank, physician and surgeon at Columbus, Nebraska, and Mary, the wife of J. P. Golden, a real estate dealer of O'Neill, Nebraska, all of whom were born in Lewis county, New York, except Mary, who was born in Holt county, Nebraska. In 1879, Thomas and Mary McDonald Morrow moved with their family from Lewis county, New York, to Holt county, Nebraska, where they settled upon a homestead and continued to reside until 1906 when they sold the homestead. They then went to Denver where they remained for about one year and moved from there to Scottsbluff, Nebraska, where they now reside.

Thomas M. Morrow received his early educational training in the public schools of Holt county, Nebraska, and in 1892 graduated from the Fremont normal school. During the following year he was principal of the public schools of Oakdale, Antelope county, Nebraska, and during the next year superintendent of the public schools of O'Neill, Holt county, Nebraska. While engaged in educational work, he also pursued the study of law and was admitted to the bar in February, 1894. In September of the same year he began the practice of law at Gering, Nebraska, where he remained until 1899. In 1896, he was united in marriage to Miss Lizzie Carlon. On October 26, 1897, William and Mary Morrow, twins, were born of this marriage at Gering, Nebraska.

While located at Gering, Nebraska, Mr. Morrow acquired a reputation as a lawyer which was far above the average practitioner in western Nebraska. In 1899, Mr. Morrow desiring to enter a broader field for the practice of his profession, moved to Omaha with his family with the intention of making that city his permanent home. Shortly after establishing himself in the practice of law in Omaha, his wife's health commenced to fail and he was compelled to move to Denver in the following year for the benefit of her health. Hoping that the climate of Denver would soon restore her health, he abandoned all idea of returning to Omaha and opened a law office in Denver, but her condition gradually grew worse and she died in April, 1901. On June 13, 1907, he was again married to Margaret L. Rank and they have two daughters, Lettie and Catherine, aged respectively eleven and seven years. Owing to the reputation acquired by him while located at Gering,

Nebraska, he was from time to time, while living in Denver, called back to western Nebraska for the trial of important cases.

In 1903, his brother, William Morrow, graduated from the law school of the University of Nebraska, and for three years thereafter was associated with him in the practice of law in Denver. The two brothers realized the great possibilities of the North Platte Valley and in 1906 shortly after the construction of irrigation ditches in the valley was begun on a large scale, it was decided that William Morrow should open a law office in Scottsbluff and conduct the business under the firm name of Morrow & Morrow. Although it was then thoroughly understood that the connection of Thomas M. Morrow with the firm was limited to such cases as he actually and personally assisted in trying, yet all of his former friends and clients brought their business to William, and in a few years his business grew beyond the capacity of any one man to handle it. In 1915, Thomas M. Morrow came to Scottsbluff and entered into an equal partnership with his brother in the practice of law. They continued the firm name of Morrow & Morrow. This firm handles a large proportion of the important law business in Scottsbluff and surrounding counties. Both members of the firm are considered among the best lawyers of the state and have participated in nearly all of the important litigation of western Nebraska. They are held in the highest esteem by the general public, and their business is constantly growing.

Thomas M. Morrow belongs to the Roman Catholic church, is a member of the Denver council of Knights of Columbus, the Modern Woodmen of America, and other fraternal organizations. He has a wide circle of friends here and in all other places where he resided.

CHARLES F. COFFEE. — Few men in Nebraska are better known in the cattle business than Charles F. Coffee, an honored resident of Chadron, and his interests in this line connect him with this great industry throughout the entire country. Mr. Coffee has been closely identified with the development of western Nebraska for many years and his spirit of public service, marked even in boyhood, and his gift of business organization, have been vitally helpful over a long extended period. He is an important factor in political life and a dominant one in the state in the financial field.

Charles F. Coffee was born March 22, 1847, at Greenfield, Dade county, Missouri, a son of John T. and Harriet L. (Ware) Coffee, the latter of whom died in 1863, in Dade county. Of the six children born to his parents, Charles F. is the only one living in Dawes county at the present time. One brother was accidentally killed in Wyoming, in 1879, and another, Samuel B. Coffee, died at Harrison, in Sioux county, after which his family moved to Chadron.

John T. Coffee, father of C. F. Coffee, was born in Tennessee and educated for the law and in 1855, through the good offices of Honorable John S. Phelps of Missouri, was awarded a commission as captain in the regular army and in 1856 or 1857 was elected a member of the Missouri Legislature and chosen speaker of the house. In 1861, he raised a very efficient regiment of soldiers for the confederacy, serving under General Price and General Sheby of Missouri, and was in all the battle of any note during the war. He distinguished himself on many occasions by unusual bravery and military tactics. On one occasion at the Battle of Lone Jack, capturing a body of Federal troops, almost unaided and was promoted to colonel. However, after the war between the states was over, being tired of military life he resumed the practice of law and drifted to Georgetown, Texas, and continued to follow this vocation till his death in 1893, at the age of seventy-five years.

During boyhood Charles F. Coffee had but indifferent educational opportunities in reference to school training, but the practical lessons he learned while earning his own living from the age of twelve years, were doubtless, of far more value to him in after days than any knowledge he could have absorbed from textbooks. When the Civil War came on he succeeded in being accepted as a soldier in the Confederate army although only thirteen years of age. He was mainly engaged in his father's regiment and after the war closed went to Texas and tried several lines of industry, with very indifferent results. He farmed some, clerked in a store awhile, then owned a store and went broke. Mr. Coffee tried raising cotton, but the prices went down and the young financier went with it, and in 1871, he hired as a "cowboy" to help drive a herd of about eighteen hundred head of longhorns from Texas to Cheyenne, Wyoming, for Snyder Brothers. Young Charles showed such ability for his work that in one month he was promoted to foreman and his pay advanced twenty dollars. He worked for this company two years trailing cattle from Texas to Wyoming and remembers a circumstance which happened on the first drive. The first white man they struck in Wyoming was the station

agent at Pine Bluff. This was the entire population at that time, and the little box depot the only building. Mr. Coffee entered the door and the agent was facing the other way and did not deign to look around. In the ticket window facing the cow puncher was a human skull and neatly printed on the forehead were these words, *This man was talked to death by immigrants.* Mr. Coffee after surveying this gruesome object for a short time mustered up courage to say, "Mister, I may be committing a rash act by disturbing you sir, but I am strictly in it. I am driving a large bunch of cattle to Cheyenne. I understand from here on water is scarce, can you tell me where the next watering place is located?" The agent proved to be a very pleasant man after all, but knew nothing about watering places, all he knew was to board the train and go through to water. The herd was driven all that day and a dry camp made, with no water for the cattle; they drove the next day till about one o'clock and the cattle were beginning to get pretty well fagged, when they fortunately struck a creek with a sandy bottom but no water, but found a place that still showed dampness. Mr. Coffee held his herd on this wet sand and milled the cattle around and packed the sand until the water raised sufficiently to water them. The next day they came out in sight of a beautiful lake of water and could see Cheyenne in the distance — the sight was a glorious one and the cowboys shouted with delight and the cattle scenting the water were bellowing as they made a wild stampede for the water, and were soon up to their sides enjoying the first good drink they had had since leaving Pine Bluffs. Looking down towards the town they saw a cloud of dust rapidly nearing them and discovered it was caused by a horseman coming toward them. They did not have long to wait to find what the trouble was, as a very red faced and angry man reined up in front of them and shouted, "Get your cattle out of here, I am the mayor of Cheyenne." One of the boys said, "The H— you are, we thought you was the butcher and wanted to buy some beef." This enraged the mayor to such an extent that he was in danger of having apoplexy. "Don't you know," he shouted, "this is the reservoir from which Cheyenne draws her drinking water?" Mr. Coffee tried to reason and conciliate him for nearly twenty-five minutes; he getting madder every minute, but by this time the cattle had satisfied their thirst and did not object to be again on the move to Cheyenne. Mr. Snyder met them; he had sold the cattle to a rancher by the name of J. H. Durban and Mr. Coffee drove the herd to Pole Creek about thirteen miles away where the nearest grass and water could be found, and tallied the cattle out. He was then ordered to deliver the cattle to Mr. Durban's ranch about twenty miles distance and all his instructions consisted of was to follow a line of small cedar trees which Mr. Durban had cut and set in a line to mark the way to his ranch. He was told to line his cattle out single file and beat a road to the ranch which he did and that same cattle trail is the present road which Mr. Coffee started with his cattle nearly fifty years ago.

In 1879 Mr. Coffee homesteaded and pre-empted land in Sioux county, Nebraska, this land never since having gone out of his possession, its location being in Hat Creek Basin. In 1873, when he established his first ranch, in Goshen Hole, Wyoming, sixty-five miles north of Cheyenne, his nearest neighbor in one direction was eighteen miles distant and twenty-seven in the other. At that time the Platte river was the dividing line between the white settlers and the Indians, but the latter notably evaded every law, making the carrying and often the use of a gun an actual necessity and Mr. Coffee for six years never went to the spring for water without carrying his trusty rifle. With other settlers Mr. Coffee lost heavily in stock at times, and once, in 1877, while he was away on a trip to the nearest town, Indians stole every horse on the ranch, all he had left being the four animals he had been driving. In an interesting way he tells how the horses were taught to recognize danger when they heard shooting, and seemingly with almost human intelligence, would gallop to the corral for safety, led by a favorite horse which the Indians killed to demoralize the herd so they could drive them away, but the plan was not a success, as the horses scattered and he rounded them up the next day. Those early ranch days were hard on both man and beast and Mr. Coffee relates that often he would be out searching for his horses and cattle for three weeks without removing his clothing. Mr. Coffee ran cattle for about thirty years before quitting the range.

In partnership with his eldest son, John T. Coffee, Mr. Coffee owns twenty-one thousand acres of land, all being operated, the son being foreman. It is known as the Square 3-Bar ranch, brand Z, and there about six hundred calves are branded yearly. Mr. Coffee owns also a ranch of about fifteen thousand acres, near Lusk, Wyoming, in partnership with a Mr. Tinnan, where they brand fifteen hundred calves annually, the brand being the 0-10, this ranch being known as the 0-10 Bar. A part of this land is irrigated. Mr. Coffee in the beginning stock his ranches with registered cattle, and has kept his line of White Face cattle thoroughbred but has not continued registration. He probably owns ten thousand head of White Face cattle.

In April, 1879, Mr. Coffee was married at Camden, Arkansas, to Miss Jennie A. Toney, who died in November, 1906. Her parents were James R. and Jane (McClain) Toney, her father being a merchant and a former slaveholder. Mr. and Mrs. Coffee had four children, the three survivors being as follows: John T., who is associated with his father as above mentioned; Blanche M., who resides with her father at Chadron; and Charles F., who is vice-president of the First National Bank of Chadron.

Mr. Coffee first became interested in banking in 1888, when he became president of the Commercial Bank of Harrison, Nebraska, which he converted into a National bank and afterward sold his interest. In 1900, he became connected as vice-president, with the First National Bank of Chadron, of which he subsequently became president. In 1900, he bought the bank at Gordon, Nebraska, serving first as its president and still is a director. In 1912, he still further added to his financial interest by the purchase of the First National Bank of Hay Springs, in Sheridan county, becoming its president, and at the same time bought stock in the Stockyards National Bank of South Omaha, of which institution he continues to be a director. In 1911, Mr. Coffee and F. W. Clark bought the Nebraska National of Omaha, of which he is vice-president, and in 1915, Mr. Coffee bought the First National Bank, of Douglas, Wyoming, of which he is president. He owns considerable valuable real estate at Chadron, both residential and business, and erected the Coffee-Pitman building, a modern garage and other structures.

In political life Mr. Coffee has always been a Democrat. In 1900, he was the fusion candidate for state representative from the 53d District, was nominated on the Democratic ticket and endorsed by the Populists and served in 1901, so acceptably that he received the party vote for United States senator, but declined to accept. Personally he is esteemed and in all business relations bears an unimpeachable character.

WILLIAM MORROW, who is recognized as one of the ablest members of the Scottsbluff bar, established the business now conducted under the style of Morrow & Morrow, when he came to this city in May, 1906 He belongs to a pioneer family of the state, and with the exception of three years spent in Colorado, has lived here since 1879

William Morrow was born in Lewis county, New York, April 10, 1875, a son of Thomas and Mary (McDonald) Morrow, both of whom were born in Ireland They came to the United States at an early age and lived in the state of New York until 1879, when they came to Nebraska The father homesteaded in Holt county and continued there until 1906, when he sold the homestead and in the following year he and wife retired to Scottsbluff, where they yet reside They are well kown in this city and are highly esteemed and respected by all who know them They have children as follows John, receiver of public funds of the United States land office at Alliance, Nebraska, Thomas M, a member of the law firm of Morrow & Morrow, at Scottsbluff, of whom personal mention is found on other pages of this work, Lavinia, a teacher in the public schools at Bayard, Nebraska, William, who resides and owns property at Scottsbluff, Anna, the wife of George Cronkleton, cashier of the First National Bank at Bayard, Frank, a prominent surgeon in active practice at Columbus, Nebraska, and Mary, the wife of J P Golden, who is in the real estate and abstract business at O'Neil, Nebraska

William Morrow attended the public schools and was graduated from the high school of Atkinson, Holt county In 1897 he entered the Nebraska State University, and in 1903 was graduated from the law department of that institution The same year he established himself in practice at Denver, where he remained until May, 1906, when he came to Scottsbluff Here he engaged in the practice of law, continuing alone until 1916 when he former a partnership with his brother, Thomas M Morrow, under the firm name of Morrow & Morrow This firm has a wide and sound reputation and individually its members are accounted astute lawyers and honorable men The business of the firm covers a wide territory

On September 18, 1912, Mr Morrow was united in marriage to Miss Philomena Congdon, who was born at Flint, Michigan, and they have had three children, namely Helen, who is five years old, William, who died at the age of fourteen months, and John Philip, who is not yet a year and a half old Mr and Mrs Morrow are members of the Roman Catholic church He is active in Democratic politics in the county, formerly was city attorney of Scottsbluff and was county attorney from 1907 to 1911 Fraternally he is identified with the Knights of Columbus, the Elks, the Royal Highlanders and the Woodmen of the World

During the World War Mr Morrow almost abandoned the practice of his profession and devoted himself almost entirely to the various war activities He was the first financial chairman of the Red Cross of Scottsbluff county, was a member of the County Council of Defense, a member of the Liberty Loan Committee, town chairman of the War Savings campaign, besides serving on various other committees of less importance

ARTHUR R HONNOLD, LL B, an able member of the Scottsbluff bar, is devoting his entire attention to private practice He is a Nebraska man, born, reared and educated in the state, and has all the local pride in the resources, development and wonderful progress of his native commonwealth, that marks the true Nebraskan Mr Honnold was born March 7, 1876, at Ord, Nebraska

The parents of Mr Honnold were Richard and Eceneith (McMichael) Honnold, both of whom were born in Ohio His father died on his farm in Valley county, Nebraska, April 6, 1906, at the age of fifty-five years, and his mother now resides at Denver, Colorado Of the three survivors of their family of four children, Arthur Rankin is the eldest, the two others being Irving T, who is secretary of an oil company at Louisville, Kentucky, and Chester W, who returned to private life, and his former occupation as a druggist, when honorably discharged from military service in the American Expeditionary Forces, in March, 1919 His sister, Nora (Honnold) Cromwell, died at Thomas, Oklahoma, in 1919

The father of the above family came to Nebraska, and homesteaded near Ord, in Valley county, in 1874, and engaged in farming there until the close of his life In politics he was a Republican

Arthur R. Honnold graduated from the high school at Ord, in 1898, and from the Grand Island business college in 1902 His first official work was as a state accountant at Lincoln, Nebraska, where he remained two years He entered the University of Nebraska, graduating from the law department in 1904, with an LL B degree The same year, he entered into the practice of his profession at Ord, but three

Charles E. Lockwood.

years later moved to Denver, Colorado, for family health reasons. He continued his law practice in his new home until 1909, when he was appointed district counsel of the United States Reclamation Service, which legal division includes the states of Nebraska, Wyoming, South Dakota, and Oklahoma, and the onerous duties of this position he performed for ten years.

Desiring to resume private practice of law, Mr. Honnold resigned his federal office on April 1, 1919, the government thereby losing a faithful and tireless official. In connection with his general practice, he now gives special attention to irrigation and water law, and to oil and corporation practice.

In 1904, Mr. Honnold was united in marriage with Miss Julia Christianson, who was born at Le Seuer, Minnesota, she dying in Denver, in 1909. Mr. Honnold's second marriage took place in 1912, to Miss Marguerita E. Van Horn, who was born in Belle Bourche, South Dakota. They have one son, Arthur Rankin, Jr. The family attends the Episcopal church at Scottsbluff. In his political attitude, Mr. Honnold is a Republican. Fraternally, he belongs to the Masons and Eastern Star organizations, and also to the Modern Brotherhood of America, and the Modern Woodmen. A man of intellectual strength and wide reading, historical facts have always engaged his interest, and he has been a member of the Nebraska State Historical Society for a number of years.

CHARLES ELMER LOCKWOOD. — In noting the representative men of the Panhandle who qualify as early settlers, there are few who came here with more determined purpose to secure a permannt home in what was then a veritable wilderness than the man whose name heads this review. Pioneer, farmer, stock-raiser and real estate dealer, he has played an important part in the development of Boone and Kimball counties and it is to such men that the opening up and settlement of the Panhandle has been due, as he came here in the early days, had courage to hold out during the hard years of drought, winter blizzards and other hardships, for he had faith in the country and this has been justified, for today there are few more prosperous men in this section and not many of them have achieved such a fortune with so small a start.

Mr. Lockwood was born in Odessa, Iowa, June 9, 1866, the son of Alfred O., and Mary Vesta Lockwood, the former born in Delaware, December 21, 1841. He was reared and educated in his native state, then came west as did so many young men of the period, to engage in farming and stock-raising in Iowa. During the Civil War Alfred Lockwood enlisted in the Union Army as a member of the Iowa Volunteers, serving till peace was established, when he returned to his farm and soon afterward married. His wife was the daughter of Aaron and Mary Brown. Eight children were born to this union: Ella, became the wife of Frank Phillips, both now deceased; Charles Elmer, of this review; Birt O., who lives in Roseburg, Oregon; Maggie M., the wife of Fred Kinney, of Ellenburg, Washington; Emily S., the wife of James Garner, of Rathdrum, Idaho; Alfred J. D., a farmer of this county; Carrie Pearl, the wife of Mr. Ring, Falls City, Oregon.

In 1878 the Lockwood family left Iowa as the father was desirous to secure some of the good cheap land in Nebraska. Coming to this state he located on the prairies of Boone county, took up a large tract of land and so became one of the earliest settlers of this section. He worked hard to place as much land as possible under cultivation but passed away January 5, 1885 and was followed by his wife almost a year later, leaving the family of children alone. Mr. and Mrs. Lockwood were members of the Methodist Episcopal church and its staunch supporters in the new territory in Boone county.

With the death of the father, the care of the family fell upon the shoulders of Charles Lockwood, then a young man of nineteen years. He had been reared on the ranch, early learned the practical side of farm industry and cattle raising and so assumed entire charge of the business. In 1887, Mr. Lockwood bought the ranch when it was sold to settle his father's estate, he borrowed every dollar it cost as he had nothing of his own to start with and the old homestead became the start of his extensive ranching properties, for he continued to buy more land as he had the money and saw where he could buy advantageously, until he was the owner of 5,400 acres of grazing and farm property in a solid body. Most of this lay in the famous Beaver Valley and nearby. Starting with little but his determination to succeed, and his known ability, which gave him excellent credit, Mr. Lockwood began to handle from 1,000 to 1,500 head of sheep and from three to six hundred head of cattle, about 500 head of hogs and in addition pastured many hundred head of cattle and horses. N. P. Dodge, a distant relative, believed in the young man and it was through him that credit was obtained in Council Bluffs, Iowa. This

Dodge was a brother of the well-known Gen. G. M. Dodge, of Union Pacific fame.

In 1909 Mr. Lockwood sold his ranch in Boone county and came to Kimball county, locating his home in the town of Kimball in 1910. He at once bought several sections of land here and at the same time conducted a real estate office. During the short time he was engaged in this business, Mr. Lockwood sold a large amount of land in the county. From time to time he has purchased other ranch property and today holds some 8,000 acres of valuable Kimball county land. He has invested in property in the states of Oregon and Florida, owns valuable holdings in the city of Kimball and has a fine home at Long Beach, California and today is regarded as one of the successful and substantial citizens of Nebraska, where he has resided for more than forty years.

January 8, 1890, Mr. Lockwood married Mrs. Abbie Derbyshire, the daughter of Mr. and Mrs. Joseph St. Louis, of Boone county, Nebraska. They were of French Canadian ancestry, though born near Oswego, New York. Three children were born to this union: Myrtle, the wife of Roy H. Kennedy, a merchant of Grand Island, Nebraska, and they have two fine children; Joseph Alfred, associated with his father in business is now farming in Kimball county, this son entered the army during the World War and was sent to Manhattan, Kansas, for his training and received his honorable discharge at the close of the war; and Lloyd Lincoln, who married Miss Nellie Rose, the daughter of Mr. and Mrs. George Rose, of Kimball county. This son entered the army during the World War and was sent to Camp Funston for his training. At the signing of the Armistice he received his honorable discharge and returned home and has been associated with his father in the real estate business. Mrs. Lockwood died Dec. 21, 1899, in Boone county. She was a member of the Christian church and the Royal Neighbors. January 2, 1901, Mr. Lockwood was married a second time to Mrs. Anna R. Sams, the daughter of Mr. and Mrs. Martin Brooks, of Carthage, Missouri, where they had settled at an early day and became well and favorably known in the southwestern part of the state. Three children have been born to this union: Charles Oliver Martin, now in high school in Kimball and a well-known athlete of the western part of Nebraska; Nadine Onetta, also in the high school, and Odetta Vesta attending the grade schools.

For over thirty years Mr. Lockwood has been a member of the Modern Woodmen of America, he is also a member of the Ancient Order of United Workmen and the Royal Highlanders, while the family are members of the Presbyterian church in which the children take an active part, as Charles has served as delegate to the Christian Endeavor Society conventions of the state.

Mr. Lockwood is one of the progressive men of his community, stands for progress and advocates those measures which tend to the upbuilding of his county and city; for years he has been a leading factor in every important public-spirited movement promulgated and his high standing in business circles makes his influence a valued and valuable one.

JOHN B. COOK, one of the younger generation of business men at Scottsbluff, is setting an example in energy and enterprise that may well be imitated. Within the comparatively short period that he has resided here, he has displayed business ability of a high order, and has exhibited both business and social qualities that reflect credit on his upbringing. Mr. Cook was born in his parents' beautiful home at Beatrice, Nebraska, March 24, 1897, the youngest of four children born to Daniel Wolford and Elizabeth (Case) Cook.

The late Daniel Wolford Cook, was a man of large affairs in Gage county, Nebraska, where his death occurred in March, 1916. He was born March 27, 1860, at Hillsdale, Michigan, a son of John P. and Martha (Wolford) Cook, and a descendant in a direct line from William Bradford, who came to the shores of America in 1620, in the Mayflower, and who for thirty years was governor of Plymouth colony. His great-great-granddaughter, Mary Bradford, who married Captain David Cook, who distinguished himself in the Revolutionary War. But Daniel Wolford Cook needed no long line of illustrious ancestors to establish his place in the history of his country or the hearts of his fellow men. In his thirty years of active business life at Beatrice, he was largely, although not exclusively, interested in the Beatrice National Bank and was president of its board of directors from 1905 until his death. He devoted a part of his time to agricultural pursuits, and took much interest in the breeding of fine stock. In business, however, especially banking, Mr. Cook was best known. From 1891 until his demise, he was vice president of the Bankers Life Insurance Company of Lincoln, Nebraska, and in this enterprise was associated with large financiers in the state. Always interested in the growth and development of Beatrice from the time he located there in 1884, he contributed generously to many public enterprises, notably to the establishment of the numerous beautiful parks of the city. His marriage to Miss Elizabeth Case was celebrated December 22, 1883, and the following children were born to them: Daniel Wolford, cashier of the Beatrice National Bank; Mary E., the wife of William C. Ramsey, of Omaha; William W., who was accidentally drowned in the Big Blue river, in August, 1905, and John Bradford, who is now a resident of Scottsbluff. A more extended memoir of Mr. Cook may be found in the History of Gage county, recently issued by the Western Publishing & Engraving Company.

John Bradford Cook was graduated from the Beatrice high school in 1914, and from the State University in 1918, in March of the latter year coming to Scottsbluff. Here he was bookkeeper in the First National Bank for four months, at the end of which period he entered the naval service on the United States Steamer Philadelphia and belongs yet to the reserves. On his return to Scottsbluff, he went into the real estate and farm loan business and sold $70,000 worth of real estate during his first month of effort. He proposes to continue in this line and also to utilize his 1080 acres of irrigated land in cattle feeding, going into this as a profitable prospect and as a patriotic measure.

At Chadron, Nebraska, Mr. Cook was united in marriage to Miss Edna Coffee, who is a daughter of Buffington Coffee. She is a highly accomplished lady, was educated in the Chadron schools and the State University, and is a member of the Methodist Episcopal church. Mr. Cook retains his interest and membership in his college fraternity, the Phi Kappa Psi.

FRANK A. McCREARY, mayor of Scottsbluff, has been an active business man of this city for a number of years. Because of his sterling character and upstanding American citizenship, he has been a man of influence in the community, and his circle of friendly acquaintance reaches all over the state. He was born in Ashtabula county, Ohio, March 23, 1868, a son of James and Catherine (Craig) McCreary.

Mayor McCreary's parents came to Nebraska from Illinois. The father was born in Lawrence county, Pennsylvania, September 26, 1838, and the mother in 1843. She is a much esteemed resident of Scottsbluff, but the father died here in March, 1919. In politics he was a Republican, and he belonged to the order of Modern Woodmen. He was one of the pillars of the Methodist Episcopal church,

in which organization the mother of Mayor McCreary continues to be active. Of their family of five children three survive: Craig, who is associated with his brother in business; Frank A., who is mayor of Scottsbluff, and Lula, the wife of William Bentley, who is a merchant at Morrill, Nebraska. After their marriage in Pennsylvania, James McCreary and wife moved to Ohio, from there to Illinois, and in 1873 to Nebraska. Mr. McCreary homesteaded in Buffalo county and lived on his land there until 1890, when he moved to Shelton, and from there came to Scottsbluff in 1915. **1192387**

Frank A. McCreary spent his early life on a farm and attended school at Shelton, where he later embarked in the mercantile business and remained so occupied for five years. In 1899 he came to Scottsbluff county and remained one year at Gering, in business as a general merchant, then came to Scottsbluff and formed a partership with George B. Lift. Within a year he bought his partner's interest and continued the business alone for another year, when his brother Craig also came to Scottsbluff. Since then the brothers have been associated under the firm style of McCreary Bros. The business has been expanded to include undertaking, while in the handling of general hardware, furniture, rugs, queeswars and musical instruments, no other house in the county approaches them in complete lines or value of stocks.

In 1901 Mr. McCreary was united in marriage to Miss Madalaide Robb, who was born in Texas, daughter of Seymour Robb, formerly sheriff of Cheyenne, Wyoming. They have one child, Lorraine, who is attending school. Mrs. McCreary is a member of the Presbyterian church. Mayor McCreary is a Scottish Rite Mason and a Shriner and belongs also to the Odd Fellows and the Modern Woodmen. Since early manhood he has been zealous in the interests of the Republican party, and on numerous occasions has been selected as its candidate for public office. He was a member of the first village board of Scottsbluff, and served one term as coronor of this county. On April 1, 1919, he was elected mayor, and being a thorough business man as well as public-spirited citizen, commanding the support of the best element of the public, Mayor McCreary will undoubtedly give the city an admirable administration. With the added pressure of public responsibility to his business cares, he decided to part with his fine farm of one hundred and sixty acres of irrigated land, and a satisfactory sale has recently been effected.

FRANKLIN E. NEELEY, cashier of the Gering National Bank, came to the institution in this capacity, in May, 1910, at which time he had the distinction of being the youngest bank cashier in the state of Nebraska. Mr. Neeley has continued with the bank ever since, an obliging yet careful, conservative official who holds its interests paramount, although necessarily giving some attention to other enterprises in which he is individually concerned, and to the duties that several public offices impose.

Franklin E. Neely was born at Fremont, Nebraska, August 21, 1890, a son of Robert F. Neeley, one of the old and substantial residents of Scottsbluff county. Mr. Neeley is indebted to Fremont, Gering, and Omaha for thorough educational training. Following his graduation from the Gering high school in 1907, he entered Creighton College, where he took a business and a law course. His banking experience began in 1909, at Sheridan, Wyoming, shortly afterward transferring to the Scottsbluff National Bank, where he remained a year and then came to the Gering National Bank as an executive. His early life had been spent on a farm, but his talents so unmistakably indicated a business career that it was the part of good judgment to educationally prepare for commercial life. For eight years Mr. Neeley has been in charge of the finances of Gering, being both school treasurer and city treasurer. He also supervises the management of three farms.

In 1916 Mr. Neeley maried Miss Ruth Carroll, who was born at Butte, Montana, moved later to Michigan, but was educated in the University of Nebraska. She is a member of the Episcopal church. Mr. Neeley is a Scottish Rite Mason and a Shriner. In his political views he is a Democrat with independent tendencies.

SEVERIN SORENSEN, who is one of the best known men in the brick industry at Gering, Nebraska, has built up a fine business as a brick manufacturer and contractor since he came to this city in 1908. He has supplied brick for many off the finest structures here and has a solid reputation as a business man. Mr. Sorenson was born in Denmark and lived there until he was thirteen years old.

The parents of Mr. Sorenson were Jens P. and Christiana (Jensen) Sorenson, natives of Denmark. They had ten children and Severin, who was born November 21, 1855, was the fourth in order of birth. They came to America and settled at Avoca, Iowa, June 22, 1869, where the father worked at brickmaking, moving later to Harlan, in Shelby county, where both parents of Mr. Sorenson died. They were members of the Baptist church.

Severin Sorenson attended school in Denmark and after accompanying his parents to the United States, worked for three years on an Iowa farm. He knew that his father's trade was a good one and chose the same for himself, learning brickmaking at Council Bluffs, where he worked four years. He then located at Harlan, Iowa, where he began contracting and remained until 1882, when he transferred his business to Minden, Nebraska, where he made brick and engaged in brick contracting until 1889 and then moved to Denver, Colorado. At that time business prospects in his line were very promising at Denver and Mr. Sorenson accepted many large contracts, on the most of which he lost heavily when a business panic paralyzed all industries. Hence, when he came to Gering in 1908, Mr. Sorenson practically had to begin all over again. He has much more than retrieved his fortunes since coming here and is in comfartable circumstances.

On August 2, 1881, Mr. Sorenson was united in marriage to Miss Anna Markusen, who was born also in Denmark, and they have the following children, a large family, the members of which are respected wherever known: Carl, who is a carpenter and bricklayer, at Gering; Herman, who is in partnership with his father; May, who is the wife of Bernee Knudson, of Denver, Colorado; Emma, who is the wife of Dr. Warrick, in the garage business at Scottsbluff; Louis, who has but recently been discharged from military service, entered the National army in December, 1917, was first assigned to duty in Texas, later in New York and still later in England, where he was in the air service; Peter, who entered military service in the fall of 1917, remained in the training camp at Fort Funston until he was honorably discharged in January, 1917; Anna, who is the wife of R. W. Smith, now a farmer northwest of Morrill, but previously the contractor who built the Fraternity building at Gering; and Otto, Martin, Raymond, Walter and Helen, all of whom live with their parents. Mr. Sorenson has never been an office seeker, but he is intelligently interested in public affairs and gives ihs political support to the Democratic party. He is a member in good standing of the Gering lodge of Odd Fellows.

CAPTAIN ALBERT M PETITE, one of Nebraska's most gallant sons who twice has responded to his country's call and offered the greatest gift a man possesses, his life, as a soldier of the United States, is one of the best known and most popular of the business men of Scottsbluff county and the city of Scottsbluff itself where he has been in the real estate business for many years The captain is a true American patriot in whose veins flows the blood of a long line of French ancestors who played an important part in France and later in the French settlements of America and the representative of the present generation but lives up to the high standard attained by the forebears of his race and as Theodore Roosevelt so often has said "blood will tell "

Albert May Petite was born at Fon du Lac, Wisconsin, June 13, 1868, just at the close of the Civil War and it may be that some of the iron resolution, indomitable courage and determination, a spirit that permeated the north during that memorable conflict may have entered into his mental make up for he has proved himself a veritable son of Mars His ancestors were among the French settlers who came to Wisconsin at an early day when the thoroughfare to the West from Quebec and Montreal lay up the great lakes, down Green Bay, then known by the French name of "Le Baye," up the Fox river to ' the portage" now Portage and thence down the Wisconsin to the Mississippi The descendants of these fine old French families are still to be found along this old route and a fine race they have proved to be Albert Petite received his excellent educational advantages in the schools of Iowa, and it was from that state that he enlisted when President McKinley called for volunteers at the outbreak of the war with Spain After entering the service in 1898 he was assigned to the Second Regiment, United States Engineers for service in Cuba, as first lieutenant of his company Following the close of the Cuban campaign he took part in the reconstruction work accomplished by the United States before turning over the island to the Cuban government, and for some time was in charge of the old fortress Moro Castle and also of Cabanas, which guard the entrance to Havana harbor Captain Petite has many interesting stories to tell of the greusome discoveries made by him while in charge of the work of cleaning up and putting in a sanitary condition, these old fortresses which for hundreds of years under the Spanish regime had been landmarks of terror and dread to the inhabitants of the Island During the Philip-

pine Insurrection, Captain Petite served in the islands under Colonel—now General—Bullard as first lieutenant of the infantry, Thirty Ninth regiment and was twice wounded in a battle near Manila When peace was finally established in the Philippines, the captain resigned from the service to return to peaceful pursuits After returning to the United States he returned to his home in Iowa where he engaged in handling real estate until 1910 leaving in that state his son William C Petite of Des Moines who has two children, William C, Jr, and Mary Louise, and a daughter, Grace Celia, the wife of Donald McGiffen of Fairfield, and they have one son, Donald, Jr

Coming to the Panhandle in 1910, Captain Petite located in the city of Scottsbluff, opened a real estate office and was engaged in business here alone until he formed an association with the Payne Investment Company after which he handled the land and water right end of the business for the firm In politics Captain Petite has been a member of the Republican party since he cast his first vote, has taken a somewhat active part in local political circles but has never been willing to accept public office himself, but ever throwing his influence to the man he believed best fitted to serve the people

On November 23 1910, Captain Petite married Miss Ruby L Wildy, who was born at Lenzburg, Illinois, March 23, 1887, the daughter of Albert and Carrie W (Dueker) Wildy, early settlers of Scottsbluff where they now reside Mrs Petite's father built the first two-story building in the town and thus is numbered among the honored pioneers of this section When he first came to the Panhandle Mr Wildy took up a homestead in Box Butte county where he operated a frontier hotel for the accommodation of travelers as towns were few and far apart and people could make the trip from one to the next in a day This first land was homesteaded in 1887 and it was but recently that Mr Wildy disposed of it at a most satisfactory figure Later a postoffice named Melinda was established on his ranch, his wife being the first postmistress He and his wife are charter members of the Methodist Episcopal church, to which both Captain and Mrs Petite also belong Mrs Petite has one brother, Clinton D Wildy, cashier of the American State Bank of Scottsbluff

Captain Petite did not entirely give up military life upon his discharge from the army and upon returning home he became captain of a company in Fifty-fifth Regiment, Iowa National Guard, thus a third time entering the service of his country and of the state. When the United States declared war against Germany he again placed himself and his services at the disposal of his country and volunteered for any branch of the army where he would do the most valuable work in prosecuting the war. While in Cuba and later in the Philippines he had much and valuable experience in the quartermaster's department as to make him valuable to the government. He passed the physical requirements for this branch of the service and was commissioned captain of the quartermaster's corps being detailed as assistant to the general superintendent United States Army Transport Service for the Port of New York, February 14, 1918. As one of the prominent and patriotic men of Nebraska and the Panhandle it is but just to the citizens of this section that they should know what an important part his service has played and in a history of the Panhandle the work of a man of this section should be told. For this reason we give a brief resume. There has been so much waste — wanton waste and extravagance—in many departments of the army since war was declared that it should be known that one department at least has not only been paying its own way, but which, up to date, has earned hundreds of thousands of dollars for the administration's coffers. The thrifty group of workers who have accomplished this are a group of governmental workers — mostly officers — in the army machine, a department that has received scant credit for the tremendous work they have done because of wastefulness during the war. It is the labor employment branch of the army, and when the record of its service is written, though it may be garnished with silver chevrons, denoting exclusively at home service—the public will be proud to acclaim it. Captain Cox is in charge of the bureau and under him in charge of a special department is Captain Petite. No better summing up of his work can be made than that of a New York newspaper which we take the privilege of quoting. "The amount of money can not be figured to the dollar—but it is certain that it has totaled nearly $1,000,000 in the employment department alone.—All this money is saved by the insurance and compensation department, under the direction of Captain Petite, a veteran officer who has proved himself adept in his new calling as he was in the numerous campaigns in which he participated. Captain Cox supervises the establishment, which has three floors in the Dey Street Building — 54 Dey Street. — One hudnred per cent efficient himself, he has with him a staff as

capable." Captain Petite is still in the service at the office in New York, while his wife has remained at Scottsbluff looking after their property. She is much better equipped for this work than the ordinary woman as she was reared on a pioneer Nebraska homestead where by circumstances she was forced to grow up self-reliant, to be quick of thought and action. She attended a "soddy" school house while her parent lived on the ranch before coming to take advantage of the educational facilities of the town of Scottsbluff and had early learned of avenues in which to direct her energies as well as resourcefulness and thrift. Even before her marriage she displayed unusual business abilities for she became a successful dealer in horses, having learned their qualities and value on the home farm, and by this business made enough capital to build a fine ten-room house, which she conducted as an European hotel. Since Captain Petite has been in the army she has had charge of their joint interests and during the past year has managed them so well that she is now operating three large apartment houses in the same manner, always having a waiting list of tenants. Her entire family is well known in county and both she and the captain hold an estimable place in the community where they are regarded as two of the most patriotic, substantial and progressive citizens as they support most liberally all movements for the civic and communal welfare.

PETER O'SHEA, who has the reputation of having developed a larger acreage of land than any other man in Scottsbluff county, has been engaged in the real estate business with offices at Scottsbluff, since 1907, but his many interests have made his name well-known through the valley. Mr. O'Shea was born in Pike county, Missouri, January 23, 1864, the son of Patrick and Anna (Nolan) O'Shea, notable names in Ireland, where the father was born on the shores of Lake Killarney, and the mother in County Tipperary. They came to the United States in 1847, in one of the old slow-moving sailing vessels, but were landed safely in New Orleans, Louisiana. From there they came up the great Mississippi as far as St. Louis, where the father secured work with a construction company building levees on the river, remaining in St. Louis for seven years. Afterward Mr. O'Shea worked at Clarksville, Missouri, and from there on down into Louisiana. In 1874 the family came by wagon to Madison county, Nebraska, where the father bought land in the hope of comfortably rearing his family of nine children. During the early years in Madison county the struggle was hard and the first crops were devoured by the grasshoppers. Better times came, however, and at the time of his death, he left an estate worth $60,000. Both he and wife died on the Madison county homestead, his life being prolonged to ninety-three years. He was a man of strong political as well as religious convictions, being identified with the Democratic party, and faithful to every observance of the Catholic church. Of his surviving children, Peter is the fourth in order of birth, the others being Thomas, in the banking business at Madison, Nebraska; Edward, identified with the Home Savings Bank at Madison; Ella, who resides with her brother Edward at Madison; and John J., of Newman Grove, Nebraska, retired banker and real estate man.

Peter O'Shea was ten years old when his parents located in Madison county and there he received his schooling. In that section and at that time, no one took any particular pains to interest and amuse youths that were strong and sturdy, but no doubt Peter, with lads of his acquaintance did not work on the farm all the time even under the strictest discipline, but found occasional means of recreation. Work, however, was the order of the day, and while yet young Peter started to labor as a miner and continued in that line for six years. Afterwards, for seven years he was in a grain business at Humphreys, Nebraska, and from there, in 1907, came to Scottsbluff. Here he embarked in the real estate business, also invested in a ranch and went into the cattle business, and in all his undertakings has done remarkably well. He possesses what is called business foresight and this natural faculty has ruled his judgment in his large land investments. At one time he bought 1,700 acres of land and has developed every acre of it.

In 1900 Mr. O'Shea married Miss Matilda Fricke, who was born and reared in Nebraska, and they have three children. Helen, John and Frank, the two younger being yet in school. Mr. O'Shea and his family are members of the Roman Catholic church. Immersed in his business, Mr. O'Shea entertains no desire for public office, but he is too enterprising a citizen not to recognize the value of political convictions and heartily supports the Democratic party.

THEODORE D. DEUTSCH, has been practically identified with all the great irrigation projects that have been of so much importance to the people of Scottsbluff and adjacent counties. He began to build ditches in 1891 for the Tri-State Company, and continued until 1909, although prior to coming to Scottsbluff county, in 1886, he had been interested in different sections of the country along similar lines in connection with railroading. Mr. Deutsch is widely known for his enterprise, his public usefulness and his extensive ownership of valuable lands.

Theodore D. Deutsch was born February 28, 1861, in Richland county, Wisconsin, the son of Daniel and Catherine (Lewis) Deutsch, the former born in Manheim Germany, February 28, 1821, and died in 1896. Both came to the United States with their parents who settled first in Ohio and then moved to Wisconsin, where the grandparents died. Daniel Deutsch was a cooper by trade. In Wisconsin he was employed for some years by the government, to operate boats used in clearing the channels of Wisconsin rivers. In religious faith he belonged to the Mennonite sect, while his wife was a member of the Catholic church. In 1872 they moved to Iowa, where he bought land and both died there. Of the five children three are living. Theodore D., whose home is at Scottsbluff, Anna, the wife of Eli Swihart, of West Newton, Iowa, and Albert, who lives on the old home place in Iowa.

Theodore D. Deutsch obtained his education in Iowa and remained a farmer until he was about twenty years old. In 1880 he began to work at railroad construction and helped build the grade for the old Diagonal road from Waterloo to Des Moines. In 1884 he went to Washington and remained on the Pacific coast for two years, engaged in teaming at Walla Walla for several months. From there he went to Yakima and built grade on the Northern Pacific road, and when that job was finished returned to Iowa. Finding no business opening to please him in the old neighborhood, he remained only one month, before locating at Elk Point, South Dakota, where he went into the cattle feeding business. In the meanwhile he homesteaded in Banner county Nebraska, having the honor of naming that county, but later sold his homestead there for $1 per acre. In March 1886, he came to what is now Scottsbluff county, Cheyenne at that time, and was one of the county commissioners when Scottsbluff county was organized. From the beginning of the plans for the building of the great irrigation ditches to their completion, Mr Deutsch was active in the work He has been identified with all the ditch building in this section and additionally built five miles of the grade for the Burlington railroad Mr Deutsch has been engaged in the real estate business since 1909, has a large loan business to which he gives close attention, and not only owns valuable city realty but has eight hundred acres of fine irrigated land in the valley. In 1888 Mr Deutsch married Miss Laura Ammerman, who was born in Pennsylvania, and they have two daughters Blanche, the wife of Joseph Kottall, who died April 21, 1919, aod Edna, who resides with her parents Mrs Deutsch and Mis Emma are members of the Christian church In politics Mr Deutsch is a Democrat He was one of the first county commissioners of Scottsbluff and continued in that office for thirteen years He is one of the older members of the Masonic fraternity at Scittsbluff, and belongs also to the Modern Woodmen

Without a sense of humor, the trials and tribulations of the pioneers might often have weighed heavier than was the fact Few of them in recalling events now passed fail to remember amusing occurence that are worth the telling and none are more appreciative of a joke, even upon themselves This is the case with Mr Deutsch when he refers to early hardships, when even getting married entailed considerable thought and inconvenience no was to be contrasted with the easy methods of the present After receiving the consent of the lady he wished to wed, he started of on a hundred and fifty mile horseback ride to secure the license, and on the way home stopped at a town emporium and invested in tow white shirts, unusual possessions, from which he promised himself much satisfaction He had yet another horseback ride to take, one of a hundred miles, to secure a preacher When the latted arrived, in order to do the occasion honor, Mr Deutsch lent one of his precious shirts to the minister for the ceremony, who was held in a dugout Possibly, Mr Deutsch reminisces, the latter thought the shirt a gift as he never saw it again

HARRY S FIESBACH, president of one of the largest mercantile firms of Scittsbluff, has been actively identified with the business for the past ten years He is a man of marked business ability and his large enterprise is conducted along the lines of personal and public service that in any undertaking will assure worth while success Mr Fliesbach is a ative of Nebraska, and was born at Seward, January, 28, 1884, the son of Otto and Nina Louise (Senter) Fliesbach, both of whom have passed away The father was born in Illinois and died in Scottsbluff, in 1916, while Mrs Fliesbach was born at Nashua, New Hampshire, and died at Scottsbluff, April 11, 1919 Of their five children, Harry S is the eldest, the others being Chester, the secretary and treasure of the above mentioned mercantile business, Glenn, a merchant in Montana, Amelia R, the wife of Ralph W Smith, a mining engineer at Denver, but now associated with the firm at Scottsbluff, and Laura G, the wife of C L Howey, residing at Dallas, Texas twenty yeats Otto Fliesbach carried on a mercantile business at Imperial, Nebraska In 1909 he sold out and the family moved to Denver, residing there until the early part of 1916, when Mr Fliesbach came to Scottsbluff, where he had business investments His death occurred shortly afterward He was a man of impeachable character, and a member of the Christian Science church

Harry S Fliesbach's mother was widely known, not only as a Christian Science practitioner, but as an inspiring personality She was affectionately known as "Mother" and at the time of her passing a local newspaper wrote of her as follows "Mrs Fliesbach was best known to all as 'Mother' and that is one of the best tributes that can ever be paid to her, for she represented all the pure, loving, unselfish and exalted thoughts that 'Mother' brings to mind, not nly to her family, but to many others whom she helped She became interested in the Christian Science movement about thirty years ago, and for some years has given practically her whole attention to it She was a practitioner and also First Reader of the Christian Science Society of Scottsbluff" She was a daughter of Addison and Roxana (Cutler) Senter, and spent her childhood in New Hampshire When about fourteen years old she came to Omaha, Nebraska, to attend school, several years later becoming a teacher at Osceola, Nebraska, where she met and was subsequently married to Otto Fliesbach After his death she made short visits to her children, then went to California and still later visited Boston, and after her return to Scottsbluff in the early part of 1919, began to make preparation for a permanent home in this city where she was so sincerely admired adn so much beloved

Eng by E.G. Williams & Bro. N.Y.

Western Pub & Eng Co

Harry S Fliesbach was educated in the public schools at Imperial and a business college at Lincoln He obtained his early business training in his father's store and afterward was in the wholesale line for ten years at St Louis, Missouri In 1909 he stablished a department store in Scottsbluff, in partnership with his brother Chester At first they made dry goods the main feature, but subsequently added one department after another until their stock now covers all the commodities usually found in modern establishments of this kind The business was incorporated for $100 000, the two owners being the officials and managers They are honorable, upright, conscientious business men, who have the confidence and esteem of every one

In 1916, Harry S Fliesbach was united in marriage to Miss Grace B Weybright, who was born in Nebraska They are members of the Christian Science church In politics Mr Fliesbach is a Democrat, as was his father, but he is inclined to be somewhat independent

ANDREW T CRAWFORD, proprietor of the A T Crawford Garage, is one of the old established business men of Scottsbluff He stands high in public esteem both as an upright, honest man of affairs and as an active, useful, dependable citizen He is a native of Nebraska, born at Omaha, March 31, 1885, the son of Dr Andrew and Anna (Hall) Crawford, the former of whom was born in County Donegal, Ireland, and the latter in Canada Dr Crawford is widely known in this state and is a graduate of the old Omaha Medical College He has been one of the leading members of his profession at Scottsbluff since he came here in 1902 He is a member of the Scottsbluff Medical association and of the Theosophical society, and has been president of that able body, the Scottsbluff County Medical society A review of Dr Crawford will be found on another page of this history

Andrew T Crawford attended school at Omaha and then started to learn the harnessmaking trade, but not being sufficiently interested in that line, gave it up and went to Colorado, where he worked as a cowboy for seven years Mr Crawford returned to Nebraska, and in 1911 established the Central Garage at Scottsbluff, an early venture here in this business, later he changed the name to the A T Crawford Garage He bought a lot of ground favorably situated and had a suitable building erected, but later bought the commodious building in which he is now es-

tablished, to which however, he has since been obliged to make two additions to accommodate the expansion of his business He has more than 16,000 feet of ground space He handles the Hudson and Essex cars and sells all through the Platte Valley and in South Dakota and Wyoming The business belongs to Mr and Mrs Crawford and together they have made every dollar that is invested This is the oldest firm in Scottsbluff to be actively engaged in their line of business, and with the exception of a period of ten months, Mr Crawford has had no other partner than his wife

In 1907 Mr Crawford married Miss Blanche Pearl Jones, who was born at Wahoo, Nebraska Her parents were John J and Diana (Mattison) Jones, the former born in Wales and the latter in Wisconsin They came to Nebraska in 1885, where the father engaged in farming and stockraising Mrs Crawford is the only survivor of their family of three children She is a member of the Presbyterian church In politics Mr Crawford is a staunch supporter of the principles of the Republican party He belongs to the order of Modern Woodmen and to the Elks, being identified with lodge No 961 at Scottsbluff

ANDREW CRAWFORD, M D, one of the leading members of the medical fraternity of Scottsbluff, where he is loved by the people and honored by the profession, came to Nebraska in 1883 and in 1902 to this city During the World War, now happily ended, he was a member of the volunteer medical reserve board, and on other occasions he has freely placed his professional skill at the service of the public without thought of remuneration

Dr Crawford was born in County Donegal, Ireland, April 19, 1851, as son of William and Margaret (Crawford) Crawford, both natives of the Emerald Isle, honest, worthy, working people The mother was a member of the Episcopal church The father was a thoughtful, reasoning man and voted with the Conservative party in Canada In Ireland he was a farmer, but after emigrating to Canada in 1852, found it more profitable to work as a laborer In 1854 his family joined him in Canada and that remained the family home Dr Crawford has one sister, Ellen, the wife of William Barlow, a railroad engineer in the Dominion

Andrew Crawford attended the common schools in Canada After finishing the elementary schools he studied medicine but did not complete his medical course until after coming to Nebraska. Soon after locating in this commonwealth he matriculated at the Medical School of the State University at Omaha, then known as the Omaha Medical College, graduating with the class of 1888. From 1888 to 1900 he was engaged in the practice of his profession in the city of Omaha, then moved to Harrisburg, Nebraska in August of the latter year and from there came to Scottsbluff on January 29, 1902.

In 1874 Dr. Crawford married at Mamilton, Canada, Miss Anna Hall, who was born in Canada, and they have four children: Anna Grace, the wife of Charles Hamer, of Scottsbluff; Mrs. Helen R. Eastman, who lives at Scottsbluff; Mary, the wife of L. M. Kinney, of this city, and Andrew T., in the automobile business at Scottsbluff. Dr. Crawford has been president of the Scottsbluff County Medical society; is a member of the Scottsbluff Medical association, and of the Theosophical society, and belongs to the fraternal order of Modern Woodmen.

CHESTER FLIESBACH, a promiennt business man of Scottsbluff, has been identified with merchandising ever since he left college. He is secretary and treasurer of the mercantile firm operating the largest department store in this city, a business that is capitalized at $100,000.

Chester Fliesbach was born at Imperial, Nebraska, May 3, 1888. His parents were Otto and Nina Louise (Senter) Fliesbach, people widely and favorably known in this state for many years. His father was born in Illinois and his mother in New Hampshire. Both died at Scottsbluff, the father in 1916 and the mother in 1919. They had six children and Chester is the second of the five survivors, the others being: Harry S., who is president of of the mercantile firm referred to above; Glenn, a merchant in Montana. Amelia R., the wife of Halph W. Smith, a mining engineer, but now associated with the firm at Scottsblff; and Laura G., the wife of L. G. Howey, who is in the banking business near Dallas, Texas. For twenty years the father of the above family was in the mercantile business at Imperial, Nebraska. After disposing of his interests there, in 1909 he moved to Denver, and early in 1916 came from there to Scottsbluff. Both parents were members of the Christian Science church, in which the mother had been prominent for many years. She was a successful practitioner and at the time of her deeply lamented death, was First Reader in the church at Scottsbluff.

Following his graduation in 1902 from the high school of Imperial, Nebraska, Chester Fliesbach took a course in the Gem City Business College, at Quincy, Illinois. He the entered his father's store at Imperial, where he had invaluable business training and remained until 1909, then came to Scottsbluff and in association with his brother Harry S. Fliesbach, organized the department store which has been expanded until it is one of the largest in Scottsbluff county.

In 1912 Chester Fliesbach married Miss Rhea Matheny, who was born at Carthage, Illinois, and they have two children: Gordon, who is six years old; and Chester, aged fifteen months. Mr. and Mrs. Fliesbach are members of the Christian Science society. Politically he casts an independent vote.

JAMES C. McCREARY, on of the prominent men in numerous business enterprises at Scottsbluff as well as other points, was born in Ashtabula county, Ohio, and was brought to Nebraska by his parents in boyhood. He grew up on his father's farm in Buffalo county and obtained his education in the country schools and at Shelton.

When nineteen years old, Mr. McCreary went to work for M. A. Hostetter, with whom he remained for seven years and after that engaged in a general mercantile business with F. A. McCreary and E. T. Peck. For two years before coming to Scottsbluff, in 1901, when he embarked in business with F. A. McCreary, though he still lived on a farm, never having lost his interest in agricultural pursuits and surroundings, but since then has devoted himself rather closely to the enterprises in which he has made large investments. He is secretary of the Scottsbluff Investment Company, is president of the McCreary Brothers Company, is president of the Scottsbluff Creamery Company, and owns an interest in a store at Morrill, Nebraska, with unusual business acumen being able to direct all these undertakings profitably.

In 1896 Mr. McCreary married Miss Belle Bently, who was born in the state of New York and died at Scottsbluff, January 19, 1915, having been the mother of five sons: Victor, who died March 10, 1918, aged twenty-one years and one day; Pearson, born November 16, 1899; Harold J., born in December, 1900; J. Curtis, born in May, 1909; and Willis H., born in January, 1911. Mr. McCreary's second marriage took place on March 9, 1916, to Miss Jane B Polk, a native of Kentucky, a highly cultured woman who has all the gracious hospitality and charm of the daughters of the Blue Grass State They are members of the Methodist Episcopal church In politics he is a Republican of no uncertain stamp, and fraternally is a Mason and a Shriner

JONAS ZOELLNER, one of the representative business men of Scottsbluff, to which city he came in 1905, has built up a large and dependable mercantile establishment which is a leading business house of the upper valley He was born in Germany, July 6, 1851, one of seven children born to Ephraim and Fredericka (Kaufman) Zoellner Both parents died in Germany, where the father was a merchant ailor Jonas is one of the two children to come to America, a brother, Charles Zoellner, being a merchant at Deadwood, South Dakota

Jonas Zoellner was twenty years old when he left his native land for the United States, reaching these hospitable shores in February, 1871, a stranger His objective point was Memphis, Tennessee, where he was employed by one of his countrymen as a clerk in a large store, and he remained there from 1871 until 1877 Then he went to Deadwood, South Dakota, where he was associated in a mercantile business with his brother for thirty-nine years It has been Mr Zoellner's policy to expand his business interests where he has seen opportunity, hence, in 1905 he came to Scottsbluff and here opened a mercantile house under the name of Zoellner Brothers In 1908 he bought his brother's interest and admitted his only son, Charles, to partnership, and since then the firm name has been Zoellner & Son

In 1882 Mr Coellner was united in marriage with Miss Anna Goldbloom, who was born at Indianapolis, Indiana, and they have one son, Charles This young man was educated at Deadwood, South Dakota, and at Highland Park, Illinois, later taking a business course in the Bryant & Stratton Commercial college, Chicago He entered upon his business career as a traveling representative of a wholesale house of St Joseph, Missouri, continuing wit house at St Joseph, Missouri, continuing with the same firm for seven years, since severing his connection with that firm he has been associated with his father In addition to the Scottsbluff establishment, the Zoellners own and conduct a large, up-to-date store at Gering, in both cities handling shoes, clothing and

men's furnishings Charles Zoellner, the younger member of the firm, married Miss Mamie O'Connor, who was born at Wisner, Nebraska, and they have one daughter Dorothea

In politics both father and son are Republicans Both have been too actively engaged in business to feel able to give attention to public office to any extent, although Jonas Zoellner did serve for six years as treasurer of Spearfish, South Dakota He is of Jewish extraction and faith and belongs to the Hebrew congregation at Scottsbluff The high regard in which both he and son are held in this city and elsewhere is indicated in their Masonic connections and both are Shriners, are member of the Modern Woodmen and of the Elks They belong to many benevolent organizations and in all charitable movement are among the foremost to contribute and also support all worthy movements of the community

JEROME H SMITH, on the leading real estate men of the younger generation, who has built up a gratifying business, largely handled in Scottsbluff county, belongs to a well known pioneer family of Hamilton county, Nebraska, the member of which have had much to do with the development of the state Mr Smith was born at Aurora, Nebraska, October 17, 1888 He received his early academic training in Arizona and graduated from the high school before entering the University of Nebraska, where he completed his course in 1911 Soon after leaving college he accepted a position with the Lincoln Traction Company, Lincoln, Nebraska, of which concern he became cashier Afterward he went to Washington and subsequently became advertising man for N K Fairbank, with Washington and Oregon as his territory In 1918 Mr Smith came to Scottsbluff and embarked independently in the real estate business, which he continued again after an interruption that began with his entering military service in July, 1918 and ended in December of the same year He was attached to the signal corps and was in the special training camp at College Station, Texas

In March 1915, Mr Smith was united in marriage to Miss Lenora Frances Stadler who was born at Fort Wayne Indiana They are members of the Episcopal church While he is deeply interested in public affairs as all good citizens must be, Mr Smith declines political party affiliation and when he casts his vote it is according to the dictates of his own good judgement

JESSE B LANE, postmaster of Scottsbluff and a leader in Democratic political circles in this section of Nebraska, formerly was engaged in the real estate business here and owns a large amount of valuable realty He has been a resident of Nebraska for a number of years, but his birth took place at Lancaster, Ohio, December 22, 1854

Mr Lane's parents were Jesse D and Matilda (Loofborough) Lane, both natives of Ohio, and both of English extraction The grandfathers, John Lane and William Loofborough, were born in Pennsylvania, and their parents were born in England Mr and Mrs Jesse Lane were reared in Ohio and married there but later moved to Illinois, where the father became a very successful farmer Both died in that state and six of their children are yet living, Jesse B being the only one to come to Nebraska

Jesse B Lane attended school both in Ohio and Illinois, and after his education was finished remained on the home farm until his marriage when he moved to Herrick, Illinois, where he conducted a hardware business and for three years served as postmaster Mr Lane established a general mercantile business in Edgar county, Illinois, in partnership with his brother, L F Lane, an association which continued for a year In 1888 he came to Nebraska, and after settling in Cumming county engaged in handling real estate for a number of years and continued in that line after coming to Scottsbluff in the fall of 1905, carrying it on very successfully until 1915, when he assumed the duties of postmaster His management of the office here has been very satisfactory, notwithstanding a great increase in office business since he took charge

In 1879 Mr Lane married Miss Martha Strohl, who was born in Ohio They have had nine children, of whom the following are living Mrs George Elquist, who lives on a ranch near Torrington, Wyoming, J Ray, in the real estate business at Scottsbluff, Mable, a bookkeeper and cashier for a large business firm at Scottsbluff, Guy, associated with his brother in the real estate business, and Mildred, a bookkeeper for the firm of McCreary Brothers The family belongs to the Presbyterian church Mr Lane is a member of the Fraternal Union and the Modern Woodmen

SILAS G ALLEN, M D, one of the able and experienced medical men engaged in the practice of their profession at Scottsbluff, for any years has also been interested in farm production and now owns over seven hundred acres of fine irrigated land in Scottsbluff county Dr Allen was born in Shelby county, Iowa, April 6, 1874, the son of Daniel and Mary (Bothwell) Allen, the former was born in the state of New York and the latter in Jones county, Iowa Their marriage took place in Illinois and from there they moved to Iowa early in the seventies In earlier years the father was an engineer but later in life was a farmer in Iowa where he died in 1901 He was a Republican in politics and was an advanced Mason and a Shriner Dr Allen's mother resides at Harlan, Iowa, and is a member of the Methodist Episcopal church Of the seven children of the family Dr Allen was the second born, the others being as follows Cora, the wife of Herbert Wilcox, a farmer near Tilden, Nebraska, George, a specialist in diseases of the eye, ear, nose and throat, in practice at Topeka, Kansas, Sadie, the wife of Thomas J Newby, county treasurer, residing at Harlan, Iowa, Daisie, who lives with her mother at Harlan, Mamie, who lives in California, and Cleo, the wife of Dr E F Zoerb, a physician and surgeon at Genoa, Nebraska

Silas G Allen attended the public schools in Shelby county, Iowa and the Woodbine normal school For some time after completing his education he engaged in teaching school, in the meanwhile preparing for a medical career, and in 1901 was graduated from the Nebraska State Medical college, at Omaha, at which time he had the honor of being class president He remained at Omaha for one year as physician in the Methodist Episcopal Hospital, then settled at Clarkson, where he continued in successful practice for seventeen years He owned land in that locality which he sold before coming to Scottsbluff in 1918 Since coming to the Panhandle the Doctor has invested heavily in irrigated land in this district, and now owns three farms

In 1904 Dr Allen married Miss Louise Beran, a native daughter of Nebraska, and they have one daughter, Viola, who is in school In politics Dr Allen was reared in the Republican party and still adheres to its principles of patriotism and Americanism While living at Clarkson he served on the

Ruben Thomas Reeves

town board and as county coroner He is a Scottish Rite Mason and a Shriner

CHARLES C McELROY, whose able management of his several interests places him with the successful business and professional men of Scottsbluff, was born in South Dakota, December 26, 1886, the only child of Charles and Mattie (Arbuckle) McElroy

The McElroy family is of Irish extraction but has been American for generations The grandfather of Charles McElroy, John P McElroy, was born in the state of New York, moving later to Illinois and still later to South Dakota The father of Mr McElroy was a rancher and farmer in South Dakota and died at Rapid City in 1892 The mother was born and reared in Iowa, receiving her education in O'Brien county She now lives at Lincoln, Nebraska, where she is an active member of the Congregational church

Charles C McElroy attended school at Rapid City and in 1902 completed his high school course at Wisner, Nebraska, after which he spent one year in a business college at Omaha In 1908 he completed his law course in the University of Nebraska, and after eighteen months of practice at Lincoln, came to Scottsbluff in March, 1910 Here he has not only won a definite place at the bar but has broadened his interests and does a large business in insurance and loans He is also the representative of the R G Dun agency in the Platte Valley He has been very active in Masonry and has received the Thirty-second degree, belonging to the consistory at Lincoln, and he attended the dedication of the Masonic Temple in that city In politics he is affiliated with the Democratic party

WILLIAM REEVES was actively associated with his father in railroad contract work in Nebraska during the period of his early youth, and thus gained at first hand a definite familiarity with the conditions that prevailed in central and western Nebraska in the pioneer days With his father he came to Cheyenne county before Scottsbluff county was segregated therefrom, and both took up and perfected title to land about five miles southeast of where Scottsbluff is now located The subject of this sketch eventually became the owner of both of these tracts and he improved the same into one of the valuable farm estates of the county This property he still owns and to its management he continued to give his active attention until 1918, when he rented the farm and removed with his family to Scottsbluff where he purchased the attractive residence in which the family now makes its home and

where the children are afforded the advantages of the excellent public schools Mr Reeves has been closely identified with progressive movements that have conserved the civic and industrial advancement of Scottsbluff county, and is a citizen who has a secure place in popular confidence and good will His father was one of the builders of the Winter creek irrigation ditch and served as the first president of the company that constructed the same, while the son William was a director for thirteen years, as the owner of eighteen shares of the stock In politics he gives his allegiance to the Democratic party and he and his wife hold to the faith of the Christian church

William Reeves was born in Mercer county, Missouri, August 28, 1861, and is a son of Ruben Thomas Reeves, who was born in Christian county, Kentucky, March 10, 1826, and who was about eighty-one years of age at the time of his death His wife was born in Ohio She died in Illinois Their marriage was solemnized in Missouri William Reeves has gained his education almost entirely in the school of practical experience and through self-discipline, as he early became associated with his father in railroad construction work in remote localities and was thus denied the customary school privileges He was but ten years old when he thus began work with his father, who was engaged in 1875 in construction work on the levee along the Mississippi river from Hannibal to Hamburg Bay, later taking a contract for the building of one mile of the roadbed of the Union Pacific railroad near Callaway, Custer county, Nebraska, where he utilized in this work an average of about twenty-five teams Later he constructed under contract two miles for the Burlington & Missouri River railroad, near Central City, and in both of these enterprises his son gave valuable cooperation It was in 1886 that the father and son came to what is now Scottsbluff county and entered claim to the land which is now owned by the latter In the intervening years William Reeves has stood exponent of the most progressive citizenship the while he has worked for and won distinctive prosperity

In 1898 Mr Reeves wedded Miss Susan V Lacey, who was born in Texas in July 1874 They have two children both of whom were born in the primitive sod house which still stands on the old home farm in Scottsbluff county Shelley was born January 22, 1900 and Nellie July 18, 1906 both now being students in the public schools of Scottsbluff

J RAY LANE who has been established in the real estate business at Scottsbluff since 1908, has been the means of bringing a large amount of capital to the Platte Valley, owning a large acreage of valuable land and having control of vast properties on the commission basis Mr Lane has become as favorably known in the business as he farmerly was in the educational field He was born at Herrick, in Shelby county, Illinois, June 1, 1884

The parents of Mr Lane were Jesse B. and Martha (Strohl) Lane, both of whom were born in Ohio and when young accompanied their parents to Illinois For three years the father was postmaster of Herrick, Illinois, an office he has filled at Scottsbluff since 1915 In 1888 Mr Lane's parents came to Nebraska and settled in Cuming county where the father engaged in the real estate business until 1905, when he moved to Scottsbluff, where he continued his former activities until he was appointed postmaster Of his nine children J Ray is the second of the five survivors

In 1902 Mr Lane was graduated from the Wisner high school, and in 1904 from the Nebraska Normal college at Wayne, with the B C degree, being president of his class During the three following years he taught school, for one year in the country near Wayne, for one year being principal of the schools of Wolbach, in Greeley county, and one year superintendent of schools at Franklin He then joined his father at Scottsbluff and is now associated with his brother Guy in the same business

In 1915 Mr Lane was united in marriage to Miss Dora J Carter, a musician of note, who is a graduate of the New England Conservatory of Music, Boston, Massachusetts, and prior to the World War, was a student in Germany for two years Mr and Mrs Lane have no children In his political views he is a Democrat like his father and grandfather He is active in Masonic circles and was secretary of the first Masonic lodge installed at Scottsbluff He is a member of the Episcopal church

GUY LANE. whose aggressive yet well planned business activities have given him high standing in commercial circles, is identified with a brother in the real estate line at Scottsbluff Mr Lane was born at Wisner, Cuming county, Nebraska, March 28, 1889 The family history appears in this work as it is an old and important one in the state

Guy Lane enjoyed educational advantages at Wisner, and after completing the high school course, went into the telephone business, and during the following eight years he gained so broad a knowledge of electricity, that he might qualify for a number of positions where such knowledge is indispensable In the meanwhile, however, the family moved to Scottsbluff and he joined them here, had some experience in the real estate line with his father, then became associated with his brother as the firm of Lane Brothers, and the partners carry on an extensive business, to the extension of which Mr Lane gives close attention

In 1908 Mr Lane was united in marriage to Miss Ann Konkle, and they have two children, namely Helen Louise and Audrey Lee As an intelligent, upstanding, expectant citizen, with ambition to not only forward his own fortunes but to also advance the best interests of county and state, Mr Lane takes a hearty interest in politics and proves the sincerity of his convictions when he gives support to the Democratic party

CLYDE N MOORE, M D , president of the Scottsbluff County Medical society, is a leading member of his profession at Scottsbluff, where he has built up a large practice and become thoroughly identified with the best interests of this section Dr Moore was born at Macomb, in McDonough county, Illinois, March 13, 1882, the son of H N and Anna (Cooper) Moore, the former born in Ihio and the latter in McDonough county, Illinois They had two sons born to them Roscoe P , who is manager of the Ogallala Lumber Company, at Ogallala, Nebraska , and Clyde N , who was an infant when his parents came to Nebraska It was in 1882 that they left their cultivated land in Illinois and came to a sparsely settled section of Seward county, where the father invested in school land for which he paid $7 an acre He became wealthy as a farmer and stock feeder and remained on his homestead in Seward county until the close of his life, his death occuring in August, 1908 Dr Moore's mother survives and resides at Scottsbluff She is a member of the Presbyterian church and is interested in numerous benevolent enterprises

Clyde N Moore completed his high school course at Seward in 1900, but before taking up a scientific cause, devoted some time to the study of human nature by spending a short period on a ranch near Buffalo, Wyoming, and conducting a hotel In the meanwhile he had done enough preparatory medical reading to enable him to enter Lincoln Medical college, in 1907, graduating in 1911 with the degree of M D , and immediately entered into practice at Gering, where he continued one and a half years before coming to Scottsbluff Dr Moore is engaged in medical and surgical practice, meeting with the success that not only is a source of gratification to every conscientious medical man, but that has proved his worth to his fellow citizens He is active in all the leading medical organizations, is president of the county society, is a member of the Nebraska State Medical association, of the American Medical association and belongs also to the Volunteer Medical Service Corps of the United States Dr Moore's personal standing is as high as is his professional He has long been identified with the Masonic fraternity and still retains membership in his college Greek letter society Dr Moore owns two valuable farms but his practice demands too much of his time and attention to permit his being much of a practical agriculturist

On October 11, 1911, Dr Moore married Miss Udoris M Wilmeth, a native of Salem, Iowa She had a liberal education there and later took advanced courses in other schools Dr and Mrs Moore have one son, bearing his father's name, who was born May 9, 1917 In politics Dr Moore follows the example set by his venerated father and gives support to the principles of the Republican party and upholds its vindication of true Americanism

HORACE E BROWN — It is surprising how many interesting stories come to light when real lovers of Nebraska get together and exchange reminiscences, and could the readers of the history of the Panhandle have these stories at first hand, few would ever afterward relish more romances of courage, endurance, persistency, of neighborhood brotherliness or exemplification of sincere Christianity Question where you will, among the stable, representative people of this great state and you will pass on with a deeper respect for the primitive qualities that have helped build up so great a commonwealth The family history of Horace E Brown, the leading druggist at Scottsbluff, goes back to Illinois and Indiana, then to Iowa, and after four years of fighting in the Civil War, reaches Nebraska, where different but almost as fatal enemies were found, and finally were overcome

Horace E Brown was born at Mount Pleasant, in Henry county, Iowa, May 19, 1867 the eldest of six children born to Richard T and Catherine (Allen) Brown Richard T Brown was born at Bedford, Indiana, in 1840 and in early life accompanied his father, John Brown,

to Iowa During his growing period he worked with a railroad company and was the first agent at Pacific Junction, in Mills county When the Civil War was precipitated, he enlisted and served four years as a member of the Fourth Iowa cavalry All his life he was a man of good standing in the community, was a member of the Odd Fellows and a pillar in the Methodist Episcopal church After the war he was married in Iowa to Catherine Allen, who was born in 1843 and died in 1915 Her father, John Allen, came early to Iowa, where he was a merchant, his death occurring while on a business trip on the Mississippi river between New Orleans and Burlington The Browns settled in Johnson county when they came to Nebraska They were not prepared to endure the climatic changes, nor could the father of Mr Brown prevail against the grasshoppers that devastated his fields, so return was made to Iowa, but it was too late, the charm of the wide, open prairies, the deep blue skies, the freshening winds and the fruitful land, had made living in any other section impossible, and in 1880 the Johnson county residents were once more increased by the Brown family, who settled at Tecumseh The father died in January, 1917

Horace E Brown had excellent school advantages at Tecumseh While attending school in the winters, he worked on a farm in the summers and learned to punch cattle on a ranch near Tecumseh, Beatrice and Nebraska City An agricultural life, however, did not appeal to him, and after spending two years in the drug business, in Idaho, he went to Louisville, Nebraska, where he carried on a drug business for ten years In 1905 he came to Scottsbluff and opened his drug store here which he has conducted ever since Both graduates Nebraska State University and registered druggists

In Idaho, in 1890, Mr Brown was united in marriage to Miss Mary Lindsey, who was born in Boise City, and they have two children, Richard and Raymond Richard is an American soldier with the Army of Occupation in Germany, entering military service in September 1917, as a member of the Eighty-ninth Division He was educated in the University of Nebraska He married Beatrice McIntosh Raymond was also educated in the State University He married Zona Cline

In politics Mr Brown is a sound Republican as was his late father He has served at times, as a member of the city council and when his party brought him forward as a candidate for mayor he lost the election by but seven votes Mr Brown is the most advanced

Mason in the county, a member of K. C. C. H., and a shriner, and belongs also to the Modern Woodmen. Mrs. Brown is active in social circles to some extent, is interested in all charitable enterprises and is a member of the Episcopal church.

DANIEL R. SCHENCK, justice of the peace and police judge, of Scottsbluff, who has served continuously and with the greatest efficiency since 1911, is well and favorably known in different sections of the state. He was born in Parke county, Indiana, September 8, 1849, the second of eight children born to Cyrenius and Mildred H. (Reeder) Schenck.

Judge Schenck's father was born in 1827, in Butler county, Ohio, of Holland ancestry. In 1846 he was married in Parke county, Indiana, to Mildred H. Reeder, who was born in Virginia, in 1827, and died in 1913, surviving her husband one year. They were faithful members of the Methodist Episcopal church. In 1856 removal was made to Iowa, where the father engaged in the practice of medicine. At the outbreak of the Civil War he entered the Union army and served on hospital duty for four years, being quartered at Jefferson barracks during the greater part of the time. In 1876 Dr. Schenck came to Webster county, Nebraska, where he later served in the offices of coroner and justice of the peace. He was a Republican in his political views and belonged to the Masonic fraternity.

Daniel R. Schenck began life on a farm after obtaining a country school education, in Davis, Decatur and Warren counties, Iowa. In 1872, while in Warren county, he met with the serious accident that cost him his hand, it having been caught in a circular saw. When able once more to resume active life, he taught one term of school in Warren county and then went to Decatur to reënter a mill and completed the miller's trade. Through work at this trade, he came to Republican Valley, in 1876, and engaged in the milling business there until 1909, when he came to Scottsbluff and took charge of a mill for his brother-in-law, O. R. Brown, which he operated for a year, when it was destroyed by fire. In 1907, in Republican Valley, he had been elected justice of the peace and servtd with much general satisfaction. In 1911 he was appointed both justice of the peace and police judge of Scottsbluff. No one could perform the duties of these offices with more discrimination on the side of justice than Judge Schenck and there is never any danger but that the dignity of his courtroom will be upheld. In his political affiliation he has always been a Republican.

On March 31, 1881, Judge Schenck was united in marriage with Miss Alice L. Brown, who was born in Illinois, and they have three children: Albert O., Lloyd C. and Emma E. Albert O. went to Europe with the American Expeditionary Force in June, 1918, entered military service in December, 1917, and has proved himself a brave and gallant soldier; Lloyd C., a soldier with the Army of Occupation in Germany, was sent, after enlistment, to Jefferson barracks, where his grandfather had been stationed during his military service, many years ago. At the time of entering service, Judge Schenck's sons had just been graduated from Kansas City Business College, Kansas City, Missouri. Emma E., the only daughter, is trying to keep up the home atmosphere for her father, as Mrs. Schenck was called away in 1915. Judge Schenck and his children all belong to the Methodist Episcopal church.

WINFIELD EVANS, who is serving his second term as water commissioner of Scottsbluff, has been identified with Scottsbluff county since 1886. He is widely known, for through his scientific agricultural efforts much has been done to bring this section of the Panhandle into a "place in the sun." While modest in regard to his achievements, he naturally takes pleasure in his success, and there are few representative agricultural bodies in the state that have not taken a deep interest in the methods which have produced the remarkable exhibits of vegetables and fruits that for some years have carried off medals and premiums at various state fairs.

Winfield Evans was born at Knoxville, Illinois, May 17, 1864, the son of Charles and Jane Margaret (Wilber) Evans, the former born in Hartford county, Connecticut, in June, 1819, and died in 1888, and the latter in Schoharie county, New York, in 1830, and died in April, 1886, in Illinois. Of their seven children the following, besides Winfield, survive: Ada, the wife of Frank Hardesty, a druggist at Rigby, Idaho; Harry, a traveling salesman, of Milwaukee, Wisconsin; Ralph, a printer of Milwaukee, and Grace, the wife of Walter Reiter, of Indiana. The mother of the family was a member of the Episcopal church. The father was a carpenter and cabinetmaker by trade and lived in several states. He came to Scottsbluff county, Nebraska, and homesteaded in 1887 and died here. He was a Republican in politics and a member of the Masonic fraternity for many years.

In the public schools of Avoca, Iowa, Mr. Evans secured some educational training but

George E. Mason

he was only nine years old when he began farm work and continued interested in that line until 1894, at the same time acquiring a working knowledge of the building trades He came to Scottsbluff county in 1886 and assisted in building the first houses in Scottsbluff village, to which place he moved in 1900 He carried on building and contracting until 1915 when he erected his own comfortable residence and since taking possession of it has applied himself entirely to intensive gardening, his main object being to grow exceptionally fine vegetables to exhibit at state fairs During the five years his products have appeared on exhibition he has won first premiums for four years, and won first premium for the best county display in the world at the International Soil Produce Exhibition, at Kansas City in 1918 This is a notable distinction and reflects great credit on Mr Evans

On July 28, 1886, Mr Evans married Miss Minnie J Coakes, at Council Bluffs, Iowa, and they had two children Charles I, who is employed with a sugar company at Bayard, Nebraska, and Ada Appoline, the wife of Ernest Parmenter, of San Diego, California The mother of these children died July 26, 1893, on her twenty-fifth birthday On January 20, 1901, Mr Evans married Miss Henrietta E Hughes, who was born at Eldora, Hardin county, Iowa, and they have four children Donald, Allen, Dorothy, and Winfield James, who took first premium when two years old at the 1916 county baby show, also at the Lincoln State Fair, when twenty-eight months old Mrs Evans is a member of the Presbyterian church In politics Mr Evans is staunch in his adherence to Republican principles He is a Scottish Rite Mason and has passed through the chairs of the local lodge

GEORGE E MASON is a sterling citizen who contributes no negligible quota to the business prestige of the village of Bayard Morrill county where he is successfully conducting a well-equipped general wood-working shop Further interest attaches to his career by reason of the fact that he is distinctively one of the pioneers of this favored section of the state

Mr Mason was born in New York city on the 6th of February 1852, and is a scion of the staunchest of American ancestry of German origin, his parents and his paternal grandfather having likewise been natives of the national metropolis and his paternal great-grandfather having been born in Hessen Germany whence he was sent by his sovereign to the American colonies This sturdy patriot joined the Continental forces and served with utmost valor as

a soldier in the war of the Revolution Frederick E Mason, father of the subject of this sketch, upheld the military prestige of the family name by his service in the defense of the Union when the Civil War was precipitated He became a member of Company C, Sixtyninth New York Volunteer Infantry in which he rose to the rank of lieutenant He was killed at the battle of Antietam in August, 1863 He was by trade a wood-carver and pattern-maker His widow, whose maiden name was Hattie Wemerger, eventually contracted a second marriage when she became the wife of Ferdinand Dippel In 1874 they removed to Indianapolis, Indiana, where the devoted mother is still living (1919) at the venerable age of eighty-seven years Her parents were natives of Germany and were residents of New York at the time of their death

George E Mason passed the period of his childhood and youth in New York city, where he was afforded good educational advantages, including those of a leading academy of music, an institution in which he developed his exceptional musical ability In the national metropolis he served a four-years apprenticeship to the trade of pattern-maker, and thereafter he obtained from the government a position as chief musician and instructor in organizing and instructing a government band in the city of Chicago In his official capacity he was later sent to various other localities, and in 1877 he was assigned to duty at Fort Laramie, where he remained six weeks He returned to this frontier post in the following year and there served as chief musician, with the rank of lieutenant until 1879, when he resigned his governmental post and engaged in the work of his trade, in New York city There he continued his activities until 1884, when he removed to Indianapolis, Indiana, where he continued the work of his trade until the autumn of the following year when he heard and responded to the call of the progressive west It was thus in the fall of 1885 that Mr Mason came to western Nebraska, where he located a homestead in what is now Scottsbluff county his pioneer home being situated four miles east of the present village of Minatare To the developing and improving of his claim Mr Mason continued to give his attention until 1901, when he sold the property and removed to Bayard, Morrill county and established himself as a carpenter and builder He continued to be thus engaged for a period of about three years, within which he erected some of the first of the more substantial and permanent buildings of the new town Since that time he has successfully conducted his general wood-working shop, gaining high reputation as a skilled artisan, as well as a reliable and substantial citizen to whom is accorded pioneer honors   Mr Mason has never abated his interest in music He had the distinction of organizing the first band at Gering, Scottsbluff county, as well as the first at Bayard   While residing on his claim he drove a distance of fifteen miles to instruct the band at Gering, and as a skilled musician he has otherwise done much to develop general musical interests in this section of the state

As a pioneer Mr Mason bore his full share of responsibilities in connection with civic development and progress   He has never wavered in his allegiance to the Republican party He served as justice of the peace in both Scottsbluff and Morrill counties, his services in this office covering a period of fully a quarter of a century, besides which he was a member of the first school board organized in what is now Scottsbluff county

October 15, 1884, recorded the marriage of Mr Mason to Miss Christina Ruehl, who was born and reared in the city of Cincinnati, Ohio, her parents having been natives of Germany and her father having served as a gallant soldier of the Union in the Civil War   In 1886 Mrs Mason joined her husband on the pioneer farm in western Nebraska, and she had the distinction of being the third white woman to become a resident of this now favored section of the Platte Valley, where she bravely bore her share of pioneer trials and vicissitudes   Of the eight children of Mr and Mrs Mason three died in early childhood   George E conducts a barber shop at Bayard, and is also the leader of the Bayard band and head of a well-trained orchestra in this village, Edith is the wife of Nelson Wysong, of Harrison, Arkansas, Maude Emily is the wife of Lloyd Staples, of Los Angeles, California, Lydia L is the wife of Frederick Young, of Bayard, Hazel E remains at the parental home

Mr Mason is one of the well known and highly esteemed citizens and business men of Morriss county, and in connection with his wood-working shop, which is fifty by sixty-four feet in dimensions, he conducts a blacksmith shop, so that he is prepared to handle diversified work with expedition and ability

JOHNSON H GRAVES, for many years identified with lumber interests in different states of the Union, has been connected with Scottsbluff enterprises more or less continuously since 1908   He is vice president and treasurer of L W Cox & Co, of this city Mr Graves is a native of Nebraska, born at

Palmyra, August 7, 1873, the son of James A and Eva T (Quick) Graves, the former born in Illinois and the latter in Pennsylvania James Graves came to Nebraska in 1868 and was married at Nebraska City   There were five children in the family, three of these survive   Johnson H, of Scottsbluff, May, the wife of Charles Young, of Freeport, Illinois, and Carroll, a farmer near Fort Lupton, Colorado   The parents were members of the Baptist church   The father was identified with the Populist party during its political ascendancy, and he belonged to the orders of United Workmen and Woodmen of the World   The paternal grandfather, John Graves, spent his last years in the state of Washington and died there in his ninety-fourth year

Johnson Graves was reared on his father's homestead in Otoe county, attended the local schools and the State University for three years   He then accepted a position in the state land commissioner's office at Lincoln, which he filled four years, and in 1896-1897 was a clerk in the state legislature   For five months he was associated with the Barnett Lumber Company at McCook, Nebraska, and from that time may be dated his interest in the lumber industry, in which he has since been an important factor   In different capacities he has been connected with the lumber trade in Nebraska, Colorado, Montana and Idaho   With the intention of locating permanently   Mr. Graves came to Scottsbluff in 1908 and bought out the Pathfinder Lumber Company, afterward he had interests at other points for five years, then returned here and bought an interest in the large enterprise conducted under the name of L W Cox & Co, incorporated, of which he is vice president, treasurer and manager In business circles he stands high

On August 24, 1899, Mr Graves married Miss Jennie Holland, who was born in Otoe county, Nebraska, and is a daughter of L J and Sidney E Holland, the former was a prominent farmer in Red Willow county, from which he was elected a member of the state legislature in 1900   Mr and Mrs Graves have two children Jackson, who is in school, and Elizabeth May, who has just passed her second birthday   Mr Graves has settled convictions in regard to politics and has always been affiliated with the Democratic party

HENRY W NEFF, an enterprising business man of Scottsbluff, is a member of the firm doing business here under the name of the Carr-Neff Lumber Company, which has the distinction of being the oldest business firm in this city   Mr Neff was born in Pennsylvania, came to Nebraska in 1890 and to Scottsbluff in 1900

His parents were Benjamin Landis and Mary (McMurtry) Neff, the former born in the Keystone state, a descendant of fine old Pennsylvania stock, while the mother was of Scotch-Irish stock They were married in Pennsylvania and the father died in that state Their eldest son came to Lincoln, Nebraska, in 1880, later moving to Sidney and still later to Lexington, Nebraska, and in 1890 the other members of the family joined him there and all still reside there except Henry W They are as follows Maggie, the widow of J E Robb, Ada, the wife of J D Eger, John, in the lumber business, and Benjamin Landis, in the real estate business The family belongs to the Methodist Episcopal church, in which the father was an exhorter

Henry W. Neff was graduated from the Lexington high school in 1897, then attended the university at Denver, after which he returned to Lexington and remained six months Desiring a business career Mr Neff took this time to look about for a promising opening and in 1900 he associated himself with J M Carr at Gering, and they organized the Carr & Neff Lumber Company, which, now incorporated, is the Carr-Neff Lumber Company, capitalized at $60,000, with an investment of $200,000 They maintained a plant at Gering and at Scottsbluff until 1903, when they moved the main plant to the latter city The business has prospered, although both partners started the enterprise on borrowed capital They have lumber yards at Mitchell, Bridgeport and Northport, and they do a general lumber and coal business and handle paints, oils and other commodities Mrs Neff is treasurer of the company

In February, 1903, Mr Neff married Miss Libbie Johnston, of Lexington, Nebraska, who died May 13, 1910, leaving one son, Kenneth Landis, who was born January 7, 1905 Mr Neff was married a second time in August, 1913, to Miss Anna Burnham, and they have one daughter, Margaret Ann, who was born in July, 1914 Mr and Mrs Neff are members of the Presbyterian church In politics he is a Republican and formerly was very active in village affairs While serving on the village board, of which he was chairman, he brought about the installation of the electric light plant and the city water works, these public utilities doing as much as anything else to bring population and capital here He has given encouragement to many of the stable enterprises which are rapidly making this beautiful little city known far and wide

LEE E LEWIS, one of the progressive business men of the younger generation who are making financial history in the Panhandle, resides at Scottsbluff and is the owner of a stock ranch in this county He attributes his business success to the opportunities he found awaiting when he decided to make Nebraska his permanent home, as he came to the state in 1897 and to Scottsbluff in 1911

Mr Lewis was born in Rice county, Minnesota, April 12, 1870, the son of Richard D and Adelia (Wales) Lewis, the former born in the state of New York and the latter in Wisconsin, in which state they were married Of their five children but two survive Lee E and Incy D The father came to Wisconsin with his parents in childhood He worked at the carpenter trade when he reached manhood When the war between the North and South was precipitated, Richard D Lewis enlisted in the Union army and served three years and three months as a member of the Twentieth Wisconsin volunteer infantry After the war closed he moved to Minnesota, where he homesteaded He was a Republican in his political views and both he and wife were members of the Methodist Episcopal church

Lee E Lewis had school advantages at Faribault, Minnesota, after which he was a clerk in a store for a time, then was a farmer for six years in northern Minnesota In 1897, with sixty cents in his pocket as capital, Mr Lewis came to Nebraska and settled in Valley county Gradually he became independent in the stock business, not through any great good luck, but through the old reliable method of hard work and a saving sense of thrift In 1911 Mr Lewis came to the Panhandle, locating in Scottsbluff, and has been a vitalizing force here ever since He became associated in the furniture business with G L Wilcox and also was an auctioneer until 1918 In the meanwhile he had acquired one of the finest cattle ranches in this county, which lies eighteen miles north of Scottsbluff, where he feeds and ships right off the grass Mr Lewis is very appreciative of what Nebraska has done for him, but his friends call attention also to his capacity for hard work and the business integrity which has backed all his ventures

In 1900 Mr Lewis married Miss Ida L Sheldon, who was born in Greeley county, Nebraska, and they have two children, Irma May and Donald D, both attending school Mr Lewis and his family belong to the Methodist Episcopal church While never unduly active in politics, he has firm political convictions and has always been affiliated with the Republican party He has belonged to the order of Odd

Fellows for many years and also is a member of the Modern Woodmen

FRANK C MAGRUDER, civil engineer by profession, came to Scottsbluff in the spring of 1915 and took charge of the Farmer Irrigation District that, under able management, is making Nebraska one of the garden spots of the country   Mr. Magruder was born at Webb City, Missouri, January 16, 1879

The parents of Mr. Magruder, William Edward and Mary Alice (Randall) Magruder, now reside near Appleton City, Missouri   The father was born at Kirksville, Missouri, a son of John Henry Magruder, who was born near Baltimore, Maryland   The grandfather came to Missouri at an early day and went to California in 1849   After he returned to Missouri, he was a stock buyer and conducted a meat business   William Edward Magruder is a blacksmith by trade. For a number of years he was a miner, but now is a farmer near Appleton City   In politics he is a Democrat and fraternally is a Mason   He married Mary Alice Randall, who was born at Macomb, Illinois, and of their eight children the following are living   Claude, a blacksmith at Lamar, Missouri, Harry Edward, a blacksmith and miner, at Milford, Utah, Frank Cecil, who resides at Scottsbluff, Ralph E, who lives in South Dakota; Alfred and Raymond J, both of whom are farmers near Appleton City The parents are members of the Christian church

Frank C Magruder was educated at the Missouri State University, from which he was graduated as a civil engineer in 1903   He soon attracted attention in his profession and was sent to Fort Laramie, Wyoming, on government work, later was transferred to South Dakota, and in the spring of 1915 was appointed to his present responsible position and came to Scottsbluff. He has inspired confidence and the thorough manner in which he attends to the small details as well as the great ones, gives promise of still more marvelous results than those already brought about

In 1908 Mr Magruder was united in marriage to Miss Martha Driver, of Hill City, South Dakota, and they have two children   Lida Jane and William Henry   They are members of the Episcopal church   He is a Mason and both he and wife belong to the Eastern Star, of which he was worthy patron at Bellefourche, South Dakota   He is a Republican in politics

GUY CARLSON — The twentieth century is notable for the important commercial interests established and ably managed by men young in years but old in their business visions   An able representative of this class in the upper valley is Mr Carlson of Scottsbluff, who came to the Platte valley in 1910 and to this city in 1915, where he has since been in business, and is now senior partner in the Carlson-Scott Implement Company   Mr Carlson is a native son of Nebraska, born in Kearney county, October 25, 1886, his parents being C J and Anna V (Gustafson) Carlson, who now live comfortably retired at Axtell, Nebraska   They were born in Sweden   The father came to the United States at the age of nineteen years and took a homestead in Kearney county, Nebraska, in 1881   The mother accompanied her parents on the journey to the United States when she was a small girl of six   Besides Guy they have two other children   Elmer, who carries on the home farm near Axtell, Kearney county, and Lawrence, a farmer near Twin Falls, Idaho   The parents are members of the Presbyterian church   In politics the father and sons are all Republicans

Guy Carlson attended the public schools of Axtell, Nebraska, after which he spent nine months taking a business course in a commercial college at Hastings   After his studies were finished he spent some years on the homestead in Kearney county as a practical farmer   In 1910 he came to the Platte valley and for four years bought grain for the Central Granaries Company, of Minatare   In 1915 he located in Scottsbluff and engaged in the implement business with a Mr Bennett, whom he subsequently bought out, and in 1917 sold a half interest in the concern to Ambrose E Scott, since which time the firm carries on business under the name of the Carlson-Scott Implement Company   The trade territory of the firm is largely the Platte valley, and their stock is complete, including modern threshers and farm tractors   Both partners give personal attention to the business which is one of the largest at Scottsbluff

In 1916 Mr Carlson married Miss May Lane, who was born in Iowa   June 9, 1919, was born a daughter, Bonney Elane   Mr Carlson is a member of the Modern Woodmen and the Knights of Pythias, of which order he is vice chancellor   He is interested in all that concerns the welfare of the city and at present is serving in the office of fire chief, much to the satisfaction of his fellow citizens.

JAMES R MURPHY, who occupies an exceedingly important position as general superintendent of the Intermountain Railway Light & Power Company, has made his head-

MR. AND MRS. J. T. THOELICKE

quarters at Scottsbluff since July, 1918, has identified himself with local interests and has made many personal friends

James R Murphy was born at Elkhart, Illinois, in 1878, the ninth in a family of ten children born to Patrick and Ann E (Barron) Murphy. Both parents were born in Ireland but passed the greater part of their lives in the United States, to which the father of Mr Murphy came at the age of seventeen years, and the mother when a babe of six months They were married at Freeport, Illinois The father died in 1909, at the age of eighty-seven years, and the mother in 1911, at the age of eighty-three They were members of the Roman Catholic church. In earlier years Patrick Murphy was a superintendent of construction work for the Baltimore & Ohio Railroad, but spent his final years as a farmer in Illinois

James R Murphy was educated in the public schools, including the high school at Williamsville, after which he worked for the Callahan-Kratz Construction Company of Omaha, on the Illinois Drainage & Mississippi canal Then he spent a year in the state normal school at Normal, Illinois, following which he continued his studies for three years in the University of Illinois Thus well equipped for both professional and business life, he chose the latter and soon became identified with the Commonwealth Edison Company, Chicago, six months later, in 1907, transferring to the Western Electric Company of the same city, where he continued until 1909, when he was called to take charge of the Hoisington Light & Ice Company, of Hoisington, Barton county, Kansas, where he remained until 1917 In the meanwhile, in 1912 he had taken charge of the Great Bend Water & Electric Company, of Great Bend, Kansas, and served as vice president and general manager of both plants Mr Murphy then went to Wellington, Kansas, as city engineer and superintendent of public utilities, and from there came to Scottsbluff in July 1918 The Intermountain Railway Light & Power Company distributes power to Gering, Minatare, Meibeta, Bayard and Scottsbluff, furnishing heat to Scottsbluff, and ice throughout the entire valley, Mr Murphy being the alert, capable general superintendent of all the plants The importance of this work largely claims his time, but he is not indifferent as a citizen In every way possible he has shown an interest in Scottsbluff, contributing to local movements and encouraging worthy enterprises, with the expectation of making this city his permanent home

In 1915 Mr Murphy was united in marriage to Miss Elinor Lewis, who was born at Great Bend Kansas, and they have two little daughters, namely Margaret Ann and Genivieve Mrs Murphy was reared in the Methodist Episcopal faith but Mr Murphy belongs to the Catholic church He is a member of the Knights of Columbus and of the Elks at Scottsbluff, Neb In his political views he is a Democrat with a tendency toward independence

JULIUS THEO THOELECKE, who was one of the pioneer merchants of Sidney Cheyenne county Nebraska, is now a resident of Pocatello, Idaho He was born in Stade province of Hanover, Germany, April 12, 1854, and when given regular courses in education, he learned the jeweler's trade by working four years as an apprentice

In 1872 on the 26th day of June, he landed in New York city and went at once to Iowa City, where he landed on July 3d Here he took up his trade working for his brother until the summer of 1874 when he went overland to Omaha, where he again took up his trade, and worked for about three months, after which he went to Plattsmouth At this place he remained for about a year, falling ill with typhoid fever Then he went to Saint Joseph, Missouri, and remained until October 7 1875 From there he returned to Germany for a year's visit, after which he returned to Omaha, and resumed work at his trade October 7, 1877, he bought a jewelry store at Tekamah, Nebraska, which he operated until April, 1879, when he sold out In July of the same year, he located at Sidney Nebraska, then in the wild frontier, and opened a jewelry store, which he operated continuously until December 9 1894 He then went to Pocatello, Idaho, and engaged in the jewelry business successfully until the present time He is now closing out with the intention to retire permanently from business activity

On November 15, 1878 he was married to Miss Lyda E Ringland, at Iowa City, Iowa She was born reared and educated at Iowa City and died at Pocatello, Idaho March 16, 1918 In her younger years she was a vigorous woman of exceptional strength and courage, and none of the frontier dangers had any terrors for her

In August 1885 Mr Thoelecke took a homestead in that part of Cheyenne county, Nebraska that later became Banner county He made final proof in due time and still owns the land (1919) He is well acquainted with Grant L Shumway the historian of this work, and unconsciously contributed to the unknown or rather obscure history of this county and vicinity years ago

Upon the Thoelecke homestead was employed Francois Jourdain, from whom the historian first learned of the story of Mallet brothers and their journey through this country, and from his friend Tommy Chaunavierre (Shunover) came the connecting link between the past and the present This story is told in full elsewhere in this history

Mr Thoelecke was an active Republican, and was a delegate to a Republican state convention held in Omaha He never wanted office for himself, but was active for his friends and party, taking part in all the town, county, and state elections

At present he is an active member of the B P O E No 674 at Pocatello, Idaho He has been prominent in both the Knights of Pythias and Odd Fellows, being a charter member of both at Sidney, but he has since dropped out of them, retaining his fraternal allegiance only to an active interest in the Elks This lodge has made him its representative to the Grand Lodge

Mr and Mrs Thoelecke were never blessed with any children of their own, but they have an adopted son, Stanley H Thoelecke, who for some time has been associated in business with his foster father

Stanley was a twin child from a distant relative of the Thoeleckes At the age of six months, when he was taken to care for by them, he lacked the vitality to hold up his head Dr Stewart, then of Sidney, said that his lungs were gone, or never had been sound Mrs Thoelecke, fondly called "Lyde" by her friends, then took the child in hand, and in her own vigorous way she treated it In a year the boy was a strong, healthy child He has charge of the acetalene welding and repairing department in the traction company at Pocatello, and is a splendid entertainer, with high ambitions, in addition They are now talking him for mayor of his city

WILLIAM A McCAIN, who through thrift and good management has, in a comparatively short time, built up a fine garage and automobile business, came to Scottsbluff in another line in 1905 He was born in Bradford county, Pennsylvania, April 22, 1882

The parents of Mr McCain are William W and Orpha A (Granger) McCain, who reside at Stevensville, Pennsylvania, where the father has been a merchant for thirty years During the Civil War period he was a captain in the Home Guards Of his three children the two survivors, William A and Mildred, both re-

side at Scottsbluff The latter is the wife of Clarence L Morris of this city In politics the father of Mr McCain is a Republican, and both parents are members of the Presbyterian church

William A McCain completed the public school course at Stevensville and then assisted his father in his store In 1905 he came to Scottsbluff and was a clerk in the store of J A Smith for three years before engaging in business for himself In 1908, in partnership with C O Harris, and with a capital of $600, they opened a garage and continued together for five years, when James D Shaw became Mr McCain's partner This firm has made great headway and now has an investment of $50,000. A general automobile business is done and the firm handles the Overland and the Willys-Knight cars Through wise investments the firm has accumulated valuable city property and two farms

In June, 1910, Mr McCain was united in marriage to Miss Leda A Ross, who was born in Iowa, a daughter of William Ross, who is a produce merchant at Maitland, Missouri They have three sons, namely William Ross, Jack L. and James A , their ages ranging from eight to three years Mrs McCain is a member of the Presbyterian church Mr McCain is a solid Republican, not a seeker for any office, but a man of sincere political convictions who conscientiously maintains them

ARTHUR L SELZER, city engineer of Scottsbluff, worthily represents that admirable class of American modern young men, who leave the schoolroom with fixed ideas of usefulness and seldom fail in reaching satisfactory results In a disordered world this fact has saving grace Mr Selzer was born at Carroll, Iowa, April 9, 1887

The parents of Mr Selzer are Michael and Munzen (Maier) Selzer, the former of whom was born in Baden, Germany, October 26, 1858, and the latter at Des Moines, Iowa The Selzer family was founded in the United States by the grandparents of City Engineer Selzer, George and Mary (Marz) Selzer, who came here from Germany in the spring of 1868 and spent the rest of their lives on their homestead in the state of Iowa The father of Mr Selzer came to Nebraska City, Nebraska, in 1884, the year of his marriage, and resided there until 1913, when he came to Scottsbluff, where he is a substantial business man

Arthur L Selzer had the best of educational advantages afforded him, and in 1911 was graduated as a civil engineer from the Nebraska State University He immediately went to work for the Tri-State Irrigation Ditch Commission, resuming his activities with the Tri-State after his return and continuing until he entered upon the duties of city engineer, to which he was elected in 1914, and has remained in office ever since because of marked efficiency. He devotes his whole time to the duties of his office.

In 1912 Mr. Selzer was united in marriage to Miss Willa J. Wallace, who is a daughter of Wilbur Wallace, who is in the banking business at Henry, Nebraska, and they have one child, Bettie, an engaging little daughter of three years. Mr. and Mrs. Selzer are members of the Presbyterian church. He is a Scottish Rite Mason and a Shriner.

MICHAEL SELZER, who is a leading business man of Scottsbluff, is identified with the Scottsbluff Creamery Company, of which he is a stockholder and treasurer and also general manager. Mr. Selzer was born in Baden, Germany, October 26, 1858. His parents were George and Mary (Marz) Selzer, who came to the United States in April, 1868, and spent the rest of their lives on their homestead in Iowa. They had the following children: George, in the automobile business at Carroll, Iowa, Marie, a widow, lives at Denver, Michael, a valued resident of Scottsbluff, Kate, who lives on the old homestead in Iowa, Jack, who farms the old homestead, Barbara, who lives with her sister and brother on the homestead, and John, in the ice cream business at Carroll, Iowa. The parents were members of the Lutheran church.

Michael Selzer remained on the old homestead until he was twenty-three years old, then worked for a time in a bottling factory before engaging in the business for himself in 1882 at Carroll. In 1884 he came to Nebraska and bought out a bottling plant at Nebraska City, where he carried on the business until 1913, when he came to Scottsbluff and became identified with an important concern here as above mentioned. Through thrift and business alertness he has accumulated much valuable property which is represented in farming lands in both Morrill and Scottsbluff counties. The Scottsbluff Creamery Company has an authorized capital of $50,000. The manufacture of butter and ice cream is carried on and bottling of soft beverages is also a feature.

At Des Moines, Iowa, in 1884, Mr. Selzer was united in marriage to Miss Munzen Maier who was born in that city, and they have three children, Arthur L., city engineer of Scottsbluff, Caroline, the wife of D. C. Leach who is cashier of a bank at Beard, Nebraska, and

Milton R., now at home after almost two years of service in the aviation corps of the United States, which he entered in June, 1917. Mr. Selzer and his family attend the Christian church. He belongs to the Elks at Nebraska City and to other organizations. In politics he chooses to be independent and while at Nebraska City accepted no political office except on the school board.

ASA F. MIDDAUGH, who has been identified with the business interests of Scottsbluff since 1913, exemplifies the progressive spirit that so signally marks the young American business man. Honest, active, enterprising, well educated and of courteous demeanor, the country's commercial interests seem to be safe in such hands. Mr. Middaugh is president and general manager of the Scottsbluff Motor Company.

Mr. Middaugh was born at Denver, Colorado August 26, 1890, one of five children born to Asa F. and Amelia (Siever) Middaugh, who were married at Cimarron, Colfax county, New Mexico. The father of Mr. Middaugh was born at Erie, Pennsylvania, a son of William Middaugh, who was also born in Pennsylvania. In 1860 he came to Colorado and was elected the first sheriff of Denver county. The father of Mr. Middaugh accompanied his father to Colorado but after his marriage and the birth of his children established himself in the mercantile business at Del Norte where he was a merchant and banker until he retired when he returned to Denver, which remains the family home. The mother of Mr. Middaugh was born at St. Louis, Missouri. Of their three surviving children A. F. is the youngest, the others being Nettie the wife of A. M. Johnson, of Chicago and Florence, who resides with her parents.

A. F. Middaugh was graduated from the Denver high school in 1909 after which he spent two years in the Colorado State University. He had two years of business experience with his father at Del Norte following which he came to Scottsbluff and in October, 1913, in partnership with Ray Smith opened a garage and handled the Ford automobiles. In 1916 he sold his Ford interest and organized the Scottsbluff Motor Company which is capitalized at $15,000. The business is in a highly prosperous condition, the company handling the Dodge and Cadillac cars.

On October 30, 1916 Mr. Middaugh was united in marriage to Miss Mable Maxon who is a daughter of F. E. Maxon. Mrs. Middaugh is a member of the Presbyterian church

Mr. Middaugh belongs to the Elks and retains his membership in his old college fraternity, Phi Delta Theta.

ALEXANDER MESTON, who owns and operates one of the best equipped and most modern laundry plants in the Platte valley, has been established at Scottsbluff since 1912, but has had many years of experience in this business at other points. Mr. Meston was born in April, 1870, in Black Hawk county, Iowa.

The parents of Mr. Meston were Alexander and Agnes (Hutchinson) Meston, both of whom were born in Scotland. They came to the United States in 1867 and before coming to Nebraska lived in Iowa and in Wisconsin, where the father worked as a blacksmith. For five years after coming to this state he was in the lumber business at Harvard, in Clay county, and then moved to Spring Ranch, where he was engaged in milling until his death, in 1890. Of his children, Alexander was the third in order of birth, the others being as follows: Sarah Ann, the wife of Dr. F. W. Dean, of Council Bluffs, Iowa; Agnes, a teacher in the public schools of Hastings, Nebraska; John James, in the hardware business at Bradford, Illinois, and Helen, who resides with her mother at Hastings. Miss Meston is a highly educated lady and for six years was dean of the Women's department at Doane College. The family attends the Congregational church.

Alexander Meston attended school at Spring Ranch and spent two years in Doane College, after which he assisted his father in the milling business. In 1890 he first embarked in the laundry business and continued in that line at Hastings until 1900, removing then to North Platte, where he owned a laundry which he conducted until October, 1912, when he came to Scottsbluff. Up to that time laundry facilities here were indifferent, but Mr. Meston bought a plant that he could remodel and started into business. By 1916 he found it necessary to enlarge his quarters and erected a suitable building of brick construction in which he has continued ever since. He has introduced modern machinery and laundry equipments of the best class and his trade has continued to expand until now he ships laundry all through the valley. While his success has been marked it was brought about entirely through his own efforts. He is a large employer of labor, keeping thirty regular laundry workers throughout the year.

In 1901 Mr. Meston was united in marriage to Miss Maude Mable Martin, who was born in Adams county, Nebraska. Her father, S. Lewis Martin, an old pioneer of Adams county, was one of the first sheriffs and afterward was chief of police at Hastings. He arrested the famous Olive robber gang that operated in Custer county. Mr. and Mrs. Meston have three children, namely: Alexander, Margaret and Dorothy, their ages ranging from three to eight years. Following in the political footsteps of his father, Mr. Meston is a Republican. He belongs to no fraternal organization except the Elks.

NATHANIEL M. SNYDER, electrical engineer, is a member of the firm of C. D. Snyder & Son, in the battery business at Scottsbluff. He was born at Weeping Water, Nebraska, April 4, 1880, and is a son of Cecil D. and Florence M. (Hizart) Snyder.

The father of Mr. Snyder was born in the village of Tripps Corners, near Oshkosh, Wisconsin, in 1855. In 1872 he came to Weeping Water, Nebraska, and went into the milling business, removing to Alliance in 1896 and from there came to Scottsbluff in 1897. At one time he had an extensive milling business and his special brand of flour took the first prize at the state fair in Lincoln. For about fourteen years Mr. Snyder was a miller at Scottsbluff, using alfalfa for fuel during the time that Kimball was the nearest railroad shipping point. He then went into the feed business, and subsequently, with his son, embarked in the battery business under the style of C. D. Snyder & Son. They have operated a Willard Storage Battery station for a number of years.

Nathaniel Marion Snyder was graduated from the Alliance high school in 1898, then entered the state university and was graduated from the electrical mechanical course in 1901. He began the battery business with the Studebaker firm at South Bend, Indiana, where he remained eight years and had entire charge of the electrical automobile department. He is identified with several professional bodies, including the Institute of Electric Engineers of London, the Institute of Radio Engineers and others.

In 1907 Mr. Snyder was united in marriage to Miss Mable Grace Earnest, who was born in Bureau county, Illinois, a daughter of Hamilton and Emma (Charlton) Earnest. Both parents died while she was young, her father in Illinois and her mother in South Dakota. She was adopted in her infancy by an uncle and aunt, John R. and Anna Elizabeth (Charlton) Earnest, the former of whom was born in Pennsylvania and the latter at Philadelphia in the same state. In his younger years Mr. Earnest was a mining operator in Missouri,

GUSTAV ADOLPH THOMAS   CARL THOMAS
GOTTFRIED THOMAS   CHRISTIAN HENRY THOMAS

later in the express business in Illinois, and now lives retired at Joplin, Missouri Mr and Mrs Snyder attend the Methodist Episcopal church He is a Knight Templar Mason and belongs to other organizations

VALENTINE THOMAS, a resident of Sioux county, belongs to that class of men who have not only been eye witnesses of the wonderful changes that have taken place but have contributed in large measure to the development and upbuilding of this part of Nebraska No finer body of land can be found in this section of the country than "Dutch Flats," a name given to this fertile valley by the subject of this record, who was the first settler to locate here

Valentine Thomas was born in Rhine Province, Germany, June 5, 1856 He was reared and educated in his native land and was married there in 1885 to Miss Elizabeth Kamann who was born in the same locality In 1887 they bade adieu to home and friends and sailed for America, landing in Baltimore, Maryland, in June From there they made their way to Nebraska, where Mr Thomas had a half-brother living in Saunders county A month later they came to what was then Cheyenne county and took a preemption of one hundred and sixty acres in what is now Scottsbluff county Their first home was a very primitive one, a dug-out, and here the family lived until they were entitled to receive a deed to their land from the government Mr Thomas then took a tree claim one mile north, in what is now Sioux county, and established another home After proving up on this he went five miles further north and took a homestead where he engaged in the sheep business for fifteen years Here they endured all the hardships and privations incident to the settling up of a new country, but they were filled with that determination characteristic of their race meeting and overcoming all obstacles, played their part well and as the years have gone by they have prospered, and Mr Thomas is one of the wealthy land owners of the Panhandle He returned to the tree claim where he has erected modern improvements and where he now makes his home, being the owner of three hundred and thirty-nine acres of valuable land, due to the extensive irrigation system that has been inaugurated, of which Mr Thomas has always been an enthusiastic advocate and to which he is a liberal contributor Their first home a soddy," still stands, though it has been moved a mile from its original location His judgment has been good, and seeing an opportunity to increase his fortune, he invested in three and one-half sections of land in Arkansas and Prairie

counties, Arkansas, devoted to rice culture, which is now under the management of two of his sons As proof of the value of this investment we may mention that in the year 1919 one hundred thousand dollars of rice was raised and marketed from this plantation

Mr Thomas has been public-spirited to a high degree No movement for the good of his community ever seeks his aid in vain He was instrumental in getting others to come to this country and all are loud in their praises of having been induced to cast their lot in a community that is excelled by no other portion of Nebraska

The home of Mr and Mrs Thomas has been blessed with the birth of five children Anna Katrina, who was born in Germany, died in childhood, Christian Henry, the first white child born in Dutch Flats who is now operating a sheep ranch in Wyoming, Gottfried and Gustav Adolph, who are managing their father's rice plantation in Arkansas and Carl, who is a successful farmer near the home place

Mr Thomas is independent in politics and has served his district as school director and road overseer The family are members of the Presbyterian church

While he has been successful and acquired a large amount of this world's goods, he has not been remiss in any duty of citizenship and wherever known has a host of friends

Mr and Mrs Thomas relate many interesting experiences of the early days The second year they had twenty-five acres of wheat and the market price was forty cents a bushel at Alliance, sixty-five miles away They brought a supply of money with them from Germany and it looked odd to them to see people pick up bones on the prairie and haul them to Alliance and sell them for $6 to $10 per ton But Mr Thomas was glad to do that when his money had been invested and he needed cash Mrs Thomas has pillows made from feathers picked from wild geese more than thirty years ago They ground wheat in a coffee mill and made bread Once when vieing with a neighbor to see who could make a pound of coffee last the longer Mrs Thomas made a pound last six weeks, but it was not very good coffee

CHARLES H IRION — Among the prospering citizens of Scottsbluff are many men of high personal standing and wide business experience and one of these, whose life story is very interesting to follow, is Charles H Irion, who for a number of years has been extensively engaged in handling choice real estate here and all over the country Mr Irion was born in McLean county, Illinois, May 8, 1860, the son of John and Susan (Osborn) Irion, who still survive ad live at Miles City, Montana. The father was born in Germany and the mother in Kentucky. They were married at Jacksonville, Illinois, and eleven children were born to them, Charles H. being the eldest of the family. The others were: Edward, a stockman in Montana; William, in the horse business in Montana; John, operates a ranch in Montana; Lewis, in the stock business in the same state; Sadie, the wife of Jack Mettlin, a retired farmer at Alliance, Nebraska; Maggie, the wife of Mr. Kelley, a sheepman in Montana; Albert J., who has been a government horse buyer, has a ranch in Montana; Ray and Farber, both of whom are in the stock business in Montana, and one child deceased. The parents are members of the Christian church. In politics the father is a Republican. He is a man of education and is particularly well posted in history. In 1877 he moved from Illinois to Arkansas, where he remained but a short time, returning then to Illinois but shortly afterward he moved to Iowa and then to Missouri and Nebraska from there to Montana in 1899. He has been a farmer and stockman all his life.

Although Charles H. Irion was an unusually intelligent boy, he had very little encouragement in the way of education after the family moved to Iowa. His first work away from home was a season spent as a harvest hand in Missouri. He then found an employer in Minnesota, who consented that he could attend school and work for his board and clothes, and it was in this way that Mr. Irion secured a teacher's certificate and taught his first school at New Richland, Minnesota, in the meantime he put in a crop, on some rented land, that turned out well financially. He was able to take some money back home with him when he joined his parents at Oregon, Missouri, to which place the family had moved in 1883. He bought an interest in his father's team, put in a crop with his father and after it matured and he had paid his debt, he yet had $180 in cash. On March 16, 1885, the entire family started westward, with three teams but all of the horses were old and worn out animals, however they managed to haul the wagons into Nebraska, and on April 15, 1885, the family camped near running water and Mr. Irion took a claim on land in Box Butte county, three of his brothers and his father also taking claims. They had nothing, however with which to carry on either farm or domestic life. Mr. Irion tells of how he started for Camp Clark in order to get flour

having about $80 by that time to buy necessities with. A blizzard set in, through which he drove all one day and had to pay $2 to cross the river on the bridge. On the home trip, when within a half mile of the cabin, the horses gave out and he turned them loose and walked the rest of the way, having been absent two days.

For a number of years Mr. Irion broke prairie for other settlers for a living, also did freighting and has seen great herds of deer, antelope and buffalo on the then, open prairies. His father made the necessary improvements on the different claims while his son was away. On his pre-emption land he had to pay $1.25 an acre, then borrowed $500 on the place and with a small capital he had, bought cattle and afterward started a little store at Lawn, Nebraska, which he conducted until 1895, in the meanwhile securing a postoffice under the name of Belle, of which he was postmaster for four years. He then sold his interests there and moved to Marsland, Nebraska, where he bought a store building for $100 and a residence for $150 and went into business. He prospered there and remained until 1902 whon he sold his property, bought three hundred head of steers and a ranch in Sioux county, later purchased more cattle and the whole investment has proved very profitable. In 1903 he came to Scottsbluff and rented the Emery hotel, which he operated advantageously for three years. In 1911 he embarked in the real estate business and today has an extensive business all over the country, making a specialty of ranch properties.

In 1893 Mr. Irion married Miss Ada M. Lane, who was born at Hale, Iowa, a daughter of L. F. Lane. Four children were born to Mr. and Mrs. Irion: Lettie R., the wife of J. Newton Hughes, of Scottsbluff; Archie R., born April 25, 1897; and Charles and Donald, both of whom are in school. The eldest son, Archie R. Irion, brought the dreaded yet precious "gold star" into the family, for he met a soldier's death on the soil of France. He entered the service of his country in April, 1917, left home in June for Omaha, on June 16, went to Deming, New Mexico, where he completed his military training and by July 17 had reached France as a member of the American Expeditionary Force. He belonged to Battery B One Hundred and nineteenth artillery, in which he was a sergeant. He was wounded September 29, 1918, and his brave spirit passed away November 11, 1918. His name belongs on Nebraska's Roll of Honor.

Mr Irion and his family are members of the Presbyterian church He has always been a Republican and has known many of the party leaders who have maintained its principles through stormy times, but has been unwilling to accept political office, believing he could be more effective as a loyal, law-abiding private citizen

JAMES D SHAW, who is a reputable business man of Scottsbluff, whose experience has been gained in several lines of effort has made his home at Scottsbluff for a number of years and since April, 1915 has been in the automobile and garage business He has a wide acquaintance and a host of business as well as personal friends

James D Shaw was born at Baresville, Ohio, January 18, 1882, and is a son of Richard and Elizabeth Shaw He has one sister and one brother, namely. Delilah Ann, who is the wife of A F Petersen, a farmer and rancher of Buffalo Gap, South Dakota, and William M, who is a farmer and feeder near Seward, Nebraska The father of Mr Shaw served four years in an Ohio regiment in the Civil War, during which time he was thrice captured by the enemy, made two escapes and once was exchanged He died in Ohio in 1882 In 1889 the mother of Mr Shaw removed with her children to Omaha, Nebraska, and still lives there She is a member of the First Christian church of that city

In the graded schools of Omaha, James D Shaw received educational training The first money he earned was by working on a farm near Omaha Afterward he entered the employ of M C Peters Mill Company, for whom he traveled for seven years selling alfalfa feeds, visiting Iowa, Missouri, Illinois, Indiana, Idaho and Wyoming He then located at Scottsbluff and went into the hay business, buying hay all through the Platte valley In the meanwhile he became interested in the automobile business and embarked in the same with Mr McCain, in April, 1915 The firm handles the Overland and Willys-Knight cars and the Republic trucks and does a large business

On June 24, 1908, Mr Shaw was united in marriage to Miss Cynthia Ellen Raymond, who was born at Florence, Nebraska, and is a daughter of H S Raymond, who is a fruit-grower near Omaha Mrs Shaw is a member of the Episcopal church at Scottsbluff, while Mr Shaw belongs to the First Christian church at Omaha Like his father before him, Mr Shaw is a Republican in politics As a citizen and as a business man he stands high in public regard

CLARENCE E BOGGS, who has led an active business life ever since completing his education, is a young man of business dependability, social standing and personal uprightness He is one of the younger circle of business men of Scottsbluff, and is president and general manager of the Scottsbluff Milling Company

Mr Boggs is a native of Illinois, born at Havana, in Mason county, August 8, 1877, a son of James W and Elizabeth C (Caldwell) Boggs who had three other children, namely James W, who was in the first draft for service in the great war, was with an engineering corps in the American Expeditionary Force that went to France in October, 1917, now resides at Lincoln, Nebraska, Charlotte Rose, who resides with her father at Lincoln, and Allen M, who is now at home, was in a soldiers' training camp at Fort Worth, when the great war closed The father was born in Ohio and the mother in Illinois, and they were married at Crete, Nebraska Her death occurred at Lincoln in 1909 In politics the father is a Republican and for thirty-four years was deputy county treasurer of Lancaster county He came to Lincoln, Nebraska, in 1879 and for a number of years was in the insurance business He is a member of the Unitarian church, and belongs to the Odd Fellows

After completing the high school course at Lincoln and being graduated in 1898, Clarence E Boggs spent two years in the state university Immediately afterward he went into the towel supply business at Lincoln, in which he continued for eighteen years and then engaged in the milling business In October, 1917, Mr Boggs came to Scottsbluff and organized the Scottsbluff Milling Company, which is an incorporated concern, capitalized at $25,000, and since then has given his main attention to the development of his business The selling territory is all through the Platte Valley and the business is very prosperous

In 1902 Mr Boggs was united in marriage to Miss Cora M McGrew, who was born at Lincoln, and they have three children Alice, Barbara and Robert Mr and Mrs Boggs are members of the Presbyterian church Politically he is a Republican but in no sense is he a politician just a good, reliable earnest and law supporting citizen

GEORGE W STOCKWELL, who has charge of the battery and electrical business for himself at Scottsbluff, has had considerable electrical experience and is considered an expert in his line of work   Mr Stockwell was born in Dawson county, Nebraska, November 22, 1889

The parents of Mr Stockwell are Frank E and Emily Kate (Adams) Stockwell, who now are esteemed residents of Wilder, Idaho   The father was born in Iowa and the mother in Missouri   They were married at Loup City, Nebraska, and their children are Edna, the wife of Leo Rengler, a merchant at Overton, Nebraska , George W , resides at Scottsbluff, Ray, foreman of the H Gilchrist ranch in Montana , James, lives at Bayard, Nebraska, where he is assistant cashier of a bank, and Herman, lives at Wilder, Idaho   Frank E Stockwell came to Sweetwater, Nebraska, in 1876, where he followed farming for a time, then homesteaded in Dawson county and remained for twenty-three years   He has always been a foresighted business man and that took him to Grand Island, where he profitably engaged in the horse business for four years and then returned to Dawson county, but soon afterward bought a store at Paxton, in Keith county, which he operated for four years   Mr Stockwell also conducted a store at Beard for a while, then moved to Wheatland, Wyoming, and from there to Wilder, Idaho, where he owns a productive fruit farm   He belongs to the Odd Fellows and the United Workmen

George W Stockwell was educated at Overton, Nebraska, where he was graduated from the high school, then learned the telephone business and followed that for ten years, being engaged at different points   In 1916 he came to Scottsbluff to work in the battery department of the automobile business of McCain & Shaw, and now has full charge as mentioned above

In January, 1915, Mr Stockwell was united in marriage to Miss Pauline Dilla, who was born in Missouri, and they have two children, Elaine and Wayne   Mrs Stockwell is a member of the Catholic church, but Mr Stockwell was reared in the Methodist Episcopal church by his mother   In party politics he maintains an independent attitude

JAMES M CARR, who has been identified with the lumber industry at Scottsbluff for almost twenty years, is a native of Nebraska, born in 1875 at Lexington, in Dawson county, and has practically spent his entire life in the state   Mr Carr bears a name that has long been held in high repute in business circles, his father having been active and successful in this section for many years   Mr Carr is secretary and outside manager of the Carr & Neff Lumber Company of Scottsbluff

The parents of Mr Carr are James P and Ada M (Martin) Carr, the former of whom was born in Pennsylvania and the latter in Ohio   They were married at Lexington, Nebraska, where they now reside   They have had two children James M and J C   The latter is in the stock business at Lexington   James P Carr came to Nebraska in 1872 and homesteaded in Dawson county and still owns the old place   Later he engaged in the mercantile business, in which he continued until 1893, when he sold out and since then has devoted himself to looking after numerous business interests in which he has investments, one of these being the Carr & Neff Lumber Company of Scottsbluff, of which he is president   Since coming to Nebraska he has built up his entire fortune, natural business capacity combining with generous opportunity, and he now is one of the substantial men of this section

James M Carr attended the Lexington public schools and was graduated from the high school in 1893, after which he spent one year in the Lincoln normal school   Mr Carr entered business as a clerk in a general store and continued with his first employer for seven years   In 1900 he came to Scottsbluff and embarked in the lumber business with a partner under the style of Carr & Neff, which has since been changed to the Carr & Neff Lumber Company   In addition to acting as secretary of the company, Mr Carr attends to the outside yards and business details

In 1903 Mr Carr was united in marriage to Miss Ada Johnston, a daughter of G S Johnston, a farmer near Lexington, and they have one daughter, Dorothy, attending school   Mr and Mrs Carr belong to the Presbyterian church   He is a good citizen but is identified with no particular political party

CHARLES M MATHENY, who is entitled to affix a number of letters to his name, indicating high scholarship, has practically spent his life in the school room and has high standing as an educator in Nebraska as well as in his native Ohio   For seven years he has been superintendent of the Scottsbluff schools

Charles M Matheny was born at Athens, Ohio, January 6, 1874   His parents were Rev L G and Hannah (Martin) Matheny, the former of whom was born in Ohio and the latter in New Jersey   The mother of Professor Matheny died in April, 1914   Her father, William Martin, was born in Ireland,

MR. AND MRS. DICK PICKETT

came from there to the United States and settled first in New Jersey but later moved to Athens, Ohio, where he died On the paternal side the ancestry is French The paternal grandfather, Isaac Matheny, was born in Ohio His parents came from France, as Huguenot refugees, and settled at first in Virginia but later moved to Ohio and established the home in which L G Matheny was born and reared He entered military service in the beginning of the Civil War, in which he served four years, was lieutenant of company I in an Ohio regiment that took part in the memorable struggles at Memphis, Shiloh and Vicksburg Later he became a minister in the Methodist Episcopal church and since retiring from active church work has carried on a fire insurance business at Nelsonville, Ohio Of his seven children, Charles M is the eldest of the survivors, the others being William, in the employ of the General Electric Company at St Louis, Missouri, Harry, employed as an inspector by the Hupmobile Company, at Detroit, Michigan, Gertrude, the wife of W A Pride, a dental practitioner at Gloscester, Ohio, Luella, head saleswoman in a wholesale millinery house at Detroit, and Marie, private secretary for an attorney at Cleveland, Ohio

Following his graduation in 1889, from the Beverly, Ohio, high school, Charles M Matheny began to teach school and thereby earned his way through college, his method being to teach during the winter season and enter school in the spring Thus he paid his way through the Ohio University at Athens, between 1894 and his graduation in 1900, with the degree of B Ped For two years he was superintendent of the schools of Coolville, Ohio, for three years was principal of the public schools of Athens, for three years afterward taught mathematics at Circleville, and in 1908 was offered a fellowship in American history and political science, at Columbus, Ohio, receiving his Master's degree in 1909 For three years before coming to Nebraska, he was principal of the schools of Defiance, Ohio, and afterward, for two years was school superintendent at Emerson, in Dixon county, Nebraska He came then to Scottsbluff and took over the superintendence of the city schools At the present time he has heavy responsibilities, having charge of nine school buildings, 51 teachers and 1,562 pupils Supt Matheny is a man of progressive ideas and many modern innovations have been planned and accepted by him for the benefit of the school service He has a capable trained nurse inspect the pupils twice each week He

has done much to raise the standard of scholarship and gives encouragement to various school movements designed to arouse ambition and emulation

In 1898, Professor Matheny was united in marriage to Miss Lolo Wiley who was born at Guysville, Ohio, her father, A P Wiley, being a substantial farmer and stockman and a veteran of the Civil War Mrs Matheny is a highly educated lady, a teacher, and much interested in higher education They have one son, H Claire, who was born August 1, 1901 At present he is attending the University of Colorado The family belong to the Methodist Episcopal church In politics, like his honored father, Mr Matheny is a Republican He belongs to the Masonic fraternity and is a member of the Eastern Star order

DICK PICKETT, one of the leading and progressive business men of Scottsbluff, who has been and is playing an important part in the development of this western section of the state is the exponent of what ability and determination may do There are numerous instances in western Nebraska where men have arrived in the Panhandle without acquaintances or friends and have worked their way to affluence and position, but there are few which equal the record of the man whose name heads this review He has been the architect of his comfortable fortune and has the pride of knowing that it has been by his own unaided efforts that his present competency has been accumulated through honest business methods and his own hard work His ability, given the opportunity finally to evince itself has placed him in an enviable position, for today Mr Pickett is accounted one of the leading citizens of Scottsbluff and the surrounding commercial district

Dick Pickett was born in Perry county, Indiana January 3 1859, the son of James H and Maryanna (Evett) Pickett The father was a Hoosier by birth was reared and educated in his native state where he received his educational advantages in the public schools, and after attaining manhood's estate, engaged in the business with which he had become familiar in his early youth agricultural industry, and was accounted one of the best farmers and successful stockmen of that section and time Maryanna (Evett) Pickett was born in Ireland, and accompanied her parents to America when she was a young girl of thirteen years After reaching the United States the family located in Indiana where she grew to womanhood was educated and there met and married her future husband She was a loving wife and devoted mother and lived to see all her children well started in life before she was called by the Grim Reaper to her last rest in her forty-fifth year. In 1893, James Pickett left his old home in Indiana and came west, locating at Ravenna, Nebraska, about three years, then to Springfield, Missouri, where he died. He lived to be eighty years old. He enlisted in 1861 in Company H, Twenty-third Indiana Regiment and served three years, also three of his sons, all in same regiment and company.

Dick Pickett was reared on his father's farm in Indiana; attended the public school nearest his home and grew up enured to the invigorating but strict discipline of farm life. He early learned all the practical methods of farming and such stock raising as was conducted in Indiana, where he engaged in business when old enough to conduct his own affairs. He was an ambitious youth, read of the many advantages afforded a man willing to hazard his fortunes in the newer country west of the Missouri river, and while still a young man determined to strike out from the old and more thickly settled districts for the west, which has ever had a lure for the youth of this broad land. In 1883, he came to Nebraska and settled in Buffalo county on a farm which he cultivated for seventeen years, bringing the soil up to a high state of fertility, making permanent improvements in the way of buildings, and becoming one of the well-to-do farmers of that section. In 1900 Mr. Pickett sold his first farm in Nebraska and purchased a better location in the vicinity of Hershey, but he was wide-awake, kept abreast of all agricultural questions of the day, and with a keen, far vision soon realized that the great future of the agriculturist lay in that section where a man was not dependent upon the rainfall which in this semi-arid country made farming rather a gamble than an assured commercial enterprise, and selling his holdings he came to Scottsbluff county to take advantage of the irrigation projects, both private and government, for he knew that the soil was fertile enough provided water could be had in proper quantity and at just the proper growing season. Mr. Pickett purchased twenty acres of land just east of the city of Scottsbluff, but within the corporation limits, and an eighty-acre tract a mile north of town. He has raised feed and engaged extensively in buying western cattle, feeding them to fatten for the market and then shipped to the big packing centers of Kansas and Nebraska, and along this line has met with gratifying success as he is a skilled buyer, a good manager and hard worker, a combination that

must bring good results in business when a man devotes his energies and abilities to a desired end. In politics Mr. Pickett is a staunch adherent of the Democratic party, though he does not draw strict party lines in mere local elections, as he is broad-minded enough and has the affairs of his community so at heart that he desires to throw his influence to the best man fitted to serve the people. Fraternally he is allied with the Independent Order of Odd Fellows and with his family is a member of the Presbyterian church. Mr. Pickett has not been remiss in any duty of citizenship and is today regarded as one of the progressive and influential men of Scottsbluff, and his success is well merited.

On November 3, 1886, Mr. Pickett married Miss Lizzie Herbaugh, in Buffalo county, Nebraska. She was born in Indiana but came to Nebraska with her parents when a child and was reared and educated in this state, so may almost be regarded as a native daughter. Her father, John Herbaugh, was also a Hoosier by birth, reared and educated in that state, and after attaining manhood, established himself independently in business as a farmer. Indiana was well settled up at that time and land was high, so he decided to take advantage of the fine offers of land given by the homestead plan in the west, and in 1873 he came to Nebraska, locating on a claim in Buffalo county, and thus became one of the hardy, sturdy, brave pioneers of the middle west. The family passed through all the hardships and privations that settlers had to contend with, but they were not discouraged by blizzards or droughts, and lived here to see their faith in this great, wide, open country justified. Mr. Herbaugh, invigorated by his strenuous life, was a hearty old man who lived to be seventy-three years old. He served three years in the Rebellion. Rachel Ann Crawford Herbaugh was born and bred in Indiana, where she received her education and after her schooling was over, met and married John Herbaugh, accompanied him to the new home in the west, and was a loving wife and faithful helpmate during all the trying years they spent on the frontier, establishing a home and winning a comfortable fortune before the sunset years of life overtook them. Mrs. Herbaugh passed away in her sixty-sixth year. They had a family of ten robust children and lived to see them become capable, upstanding, honorable men and women.

Mr. and Mrs. Pickett have been blessed with ten children, of whom eight survive: James M., of Glendo, Wyoming, a farmer owning his own homestead, who during the World War served in the Coast Defense Artillery at San

Francisco and also in the Fortieth Coast Artillery, receiving his honorable discharge December 23, 1918; Mrs. Verdie R. Roseman, of Torrington, Wyoming, has two children: William O., of Glendo, Wyoming; John C., of Scottsbluff, who also served in the Coast Defense Artillery in San Francisco, being later transferred to the Forty-first Coast Artillery at Fortress Monroe, is now a student at the University of Nebraska, at Lincoln; Willard, deceased; Dorsey M., Theodore, Ivedell, Richard, and Raymond, all of whom are still at home with their parents, who intend to give each child every educational advantage afforded in the city and state that they may be well equipped to start out in life.

CLYDE L. HARRISON, a representative business man of Scottsbluff, who is doing a large business as contractor and builder, was born at Greenfield, Iowa, February 10, 1880, and has been a resident of Scottsbluff for seventeen years.

The parents of Mr. Harrison are John J. and Clara E. (Rice) Harrison, both of whom survive. The father came to Iowa in young manhood and served three years in the Civil War as a member of company C Twenty-third Iowa infantry. He was a daring soldier and in one of the big battles was so seriously wounded that he had to be placed in a hospital and later was honorably discharged. He was a carpenter and contractor before the war. He now resides in the Soldiers' Home at Leavenworth, Kansas, while the mother is a resident of Ainsley, Nebraska. Of their nine children seven survive and four of these live in Nebraska, three other sons besides Clyde L., namely: Worley M., who is a resident of Gordon; Orien A., who lives at Ainsley; and E. Lee, who is in the contracting business at Scottsbluff. In politics the father is a Republican, and both parents are members of the Presbyterian church.

Clyde L. Harrison obtained his education in the public schools in Nebraska. He learned the carpenter trade and after locating at Scottsbluff in 1902, was, for years, engaged in more housebuilding than any other builder and contractor in this city. In 1918 he opened his garage, where a general automobile repair business was carried on, and he handled the King and Oldsmobile cars. This business he sold out in the spring of 1919.

In 1903 Mr. Harrison was united in marriage to Miss Amy B. Fink, who was born at Seward, Nebraska, and they have the following named children: Velma Gertrude, Ivan Ray, Clyde, Helen Ruth and Howard Sheldon, the older children being in school. Mr. and Mrs. Harrison are members of the Presbyterian church, in which he is an elder and member of the board of trustees. Men of Mr. Harrison sturdy character are not apt to be unduly active in politics with a view to securing public office, but he is a faithful, ernest citizen and conscientiously supports the principles of the Republican party.

JOHN W. MONTZ, whose business enterprise and natural adaptability have placed him among the successful men in the automobile industry at Scottsbluff, is one of Nebraska's own sons, born at Harrisburg, August 5, 1891. With him in conducting the garage is his brother, Martin R. Montz, and an extensive business is done.

The parents of Mr. Montz are Martin and Gertrude (Simon) Montz, the former of whom was born in Steuben county, New York, July 20, 1858, and the latter in DeKalb county, Missouri. They were married December 30, 1880, at Cameron, Missouri, and six children have been born to them: Elizabeth Matilda, the wife of J. R. Naird, a farmer and stockman in Sioux county, Nebraska; Lebanna and Martie R., twins, the former of whom lives at Alberta, nine miles north of Scottsbluff, and the latter of whom is in the garage business at Scottsbluff; Gertrude Malissa, the wife of John Burnstock, a railroad man of Bridgeport, Nebraska; John William, of Scottsbluff; and Verna Ruth, an accomplished stenographer. The parents are members of the Christian church. Politically the father is an independent voter, and he belongs to the order of United Workmen. He came to Nebraska in the spring of 1884 and in 1886 homesteaded in Banner county, where he engaged in farming for several years. Later he came to Scottsbluff and for a number of years was in the meat market business. He now assists in the garage owned by his sons. He is well known and much respected.

John W. Montz remained at Harrisburg, where he attended school until 1900, then worked on a farm near Scottsbluff and as a stockman for a while. About 1910 he embarked in an automobile livery business at Scottsbluff, which he conducted for four years, then worked as a mechanic in a garage until he had learned the business in every detail, including the mechanism of every type of automobile. In 1918 he opened his own garage and since then has devoted himself closely to his business with satisfactory results. On

December 4, 1913, he was united in marriage to Miss Bertha May Klingman, and they have one daughter, Loraine Genevieve He is independent in his political views but no one doubts his good citizenship

MARTIE R MONTZ, who is well-known in the garage business at Scottsbluff, in association with his brother, John W Montz, was born in Missouri, February 7, 1884 He is one of a family of six children born to Martin and Gertrude (Simon) Montz, the former of whom was born in Steuben county, New York, and the latter in Missouri, in which latter state they were married

The parents of Mr Montz came to Nebraska in the spring of 1884, when he was but an infant His father homesteaded in Banner county in 1886 and his early years were spent on the home farm but he attended school at Harrisburg Mr Montz then went farther west and for fifteen years rode range in Montana, South Dakota and Idaho as well as Nebraska, meeting with many thrilling adventures during that time In 1917 he located at Scottsbluff and went into business with his brother, and a large business connection has been built up The partners are practical men, both have had solid experience and the public has confidence in them

Mr Montz was married on December 25, 1918, to Miss Sylvia Folden While interested and well posted on public affairs, Mr Montz like his father has always preferred to be an independent voter.

GEORGE F KIMBROUGH — There are few men in the automobile business at Scottsbluff who have advanced to the front in this line more rapidly or substantially than George F Kimbrough, and not always do men of collegiate training and professional prestige, find equal success in the practical field of business Mr Kimbrough is the owner of the Scottsbluff plant of the Platte Valley Motor Company, and also owns the Bayard Motor Company

George F Kimbrough was born at Denver, Colorado, October 18, 1887, the second in a family of three children born to James W and Norah (White) Kimbrough The other members of the family are James T , who is a railroad man, of Denver, and Corinne, who is the wife of Stephen M Hall, a stockman of Denver county The mother of the above family was born at Bellefontaine, Ohio, April 18, 1859, and the father at Carthage, Illinois, December 7, 1849 He came to Denver in

1878, was married at Denver, and for years has been a railroad man, at present being one of the older conductors on the Colorado & Southern line He is a Democrat in his political views and belongs to the Masonic fraternity.

With his graduation from the high school in 1907, George F Kimbrough completed the entire public school course at Denver, and in 1912 graduated from the law department of the Colorado State University with his LL B degree, during his college life being a member of the Phi Delta Theta and the Phi Delta Phi Greek letter fraternities He was admitted to the bar in the same year and engaged in the practice of law with the firm of Macbeth & May, of Denver, for four years Mr Kimbrough then became interested in the automobile business and accepted the office of secretary of the Sharman Automobile Company, with which concern he remained eighteen months and then took charge of the Scottsbluff branch of the Platte Valley Motor Company, in which he bought a one-half interest in April, 1918, and the remaining interest in January, 1919, and also became owner of the Bayard Motor Company as mentioned above Mr Kimbrough handles Ford cars and Fordson tractors exclusively His sale field is all through the Platte Valley where these cars and tractors are very satisfactory

On June 24, 1914, Mr Kimbrough was united in marriage to Miss Helen Ryals, who was born at Macon, Georgia, and they have a daughter, who was born April 19, 1919 He belongs to Union Lodge No 7, A F & A M , Denver Chapter No 2 R A M and Colorado Commandery No 1, Knights Templar and El Jebel Temple Shriners He is not affiliated with any political party but is a watchful citizen nevertheless and casts a careful, well considered vote according to his own free judgment

FRANK B DE CONLY — One of the interesting men of Scottsbluff is found in Frank B De Conly, vice president of the Scottsbluff Live-Stock Commission Company, and of other important business enterprises, not only because of his pleasing personality, but on account of the fact that he has built up a substantial fortune, entirely through his own efforts in the comparatively short time since he reached manhood He was born in Custer county, Nebraska, in 1888

The parents of Mr De Conly, Frank and Mary E (Ellington) De Conly, reside at Hastings, Nebraska The father was born in Penn-

MRS. ELIJAH McCLENAHAN

sylvania and the mother in Virginia and their marriage took place at Plum Creek, Nebraska Of their five children the following are living Emma, who is the wife of Charles Godby, of McCook, Nebraska, Edwin, who is a printer, lives at Scottsbluff, Florence, who is the wife of Harry Fiest, lives in Colorado, and Frank B, who is so widely and favorably known in the Panhandle The father, a printer by trade, came to Nebraska in the seventies and settled first in Dawson county, where he worked in a mill and on a ranch before moving to Custer county where he homesteaded He is a Democrat in politics, is a Knight of Pythias, and both he and wife belong to the Episcopal church

Frank B De Conly attended the Callaway public schools and the Lexington high school, after which he went to work for the Union Pacific Railway at Callaway, where he remained fourteen months He then spent five months at Hastings in the paint shop of Haines Brothers, and seven months for the Burlington Railroad as checker He then went into the incubator factory of the M M Johnson Company, later becoming an office man there and remaining eight years In the meanwhile, from being an enthusiastic baseball player for recreation, he became an expert in the national game, and for seven years played professional baseball as third baseman in the State League and the Tri-State League In this connection he is remembered admiringly all over the country In 1912 Mr De Conly came to Scottsbluff and embarked in the real estate business and in the fall of that year went into the stock business He now has three large farms, his main activities being feeding cattle and sheep, his record showing that in one year alone he fed 10,400 head of sheep, and each year ranges from 3,000 to 10,000, and from 500 to 1,000 cattle He has demonstrated great business capacity, has invested wisely and at present is identified with a number of prospering business concerns. In addition to being vice president of the Scottsbluff Live-Stock Company, he is vice president of the Fisher Grocery Company, and owns one-third of the company stock and one-fourth of the livestock, in the former organization He owns three hundred and forty acres of fine land

In 1910 Mr De Conly was united in marriage to Miss Neva Wyman Palmer, who was born in Seward county, Nebraska, and is a daughter of David B Palmer, a heavy stockman and leading citizen of Seward county Mr. and Mrs De Conly have one son who has

about reached the engaging age of two years and bears his maternal grandfather's honored name Mr De Conly is a vestryman in the Episcopal church at Scottsbluff He is an independent in his political opinions but is a very active and influential citizen in all matters pertaining to the progress of Scottsbluff During the two years of his service as a member of the city council, he was president of that body the entire time He is a Scottish Rite Mason and a Shriner He has never lost interest in manly sports and is a member of the Athletic Club at Omaha

ELIJAH McCLENAHAN, pioneer in irrigation, farmer and financier, who is now numbered among the substantial business men of Scottsbluff, has been the architect of his own fortune, and having based his life's structure on firm, substantial foundations, has builded soundly and well When he entered upon his career he was possessed of little save inherent ability great ambition and the determination to succeed, and these have been sufficient, through their development to enable him to become a large landholder, progressive farmer, and man of finance in a well-to-do community that does not lack for able and successful men of enterprise and progress

Elijah McClenahan was born in Keokuk county, Iowa October 26 1866, the son of Elijah and Elizabeth (Wilson) McClenahan, the former a native of the famous state of Kentucky, who settled in Illinois at a very early date at the time when the government was having difficulties with the Indians over their refusal to give up the lands they had ceded to the United States under a promise of removing west of the Mississippi river Mr McClenahan (senior) was one of the men who helped build a log fort in Stark county when Black Hawk and his band went on the warpath with the idea of driving the whites out of their territory and forts were necessary in various localities where the whites could gather for protection against their Indian foes, who crept stealthily upon the outlying settlements and murdered the unsuspecting women and children when the men were away or out in the fields After remaining in Illinois for some years Mr McClenahan removed still farther toward the frontier and settled in Keokuk county Iowa, where he engaged in agricultural pursuits passing away there in his seventy-fifth year Elizabeth Wilson McClenahan was born in Ohio, where she spent her early childhood, receiving an excellent practical education in the public schools of that state, when a young girl her parents removed to Iowa and she accompanied them to the new home in the west. In 1887, accompanied by her children, she became a pioneer settler of the Panhandle, settling on a homestead where the city of Scottsbluff now stands. She was a hale, hearty woman, enured to the hardships and privations through which she passed during the hard and trying years of frontier life, but was thrifty, willing to work, and of many good deeds, that stood in number as the years of her life counted by days. She was a devoted mother and it was for the advantages that her children might have that she located in this section at a time when habitations were few, civilization in its dawn on the high prairies, and privations many, but she lived to see that her faith in this section was justified, as she was over eighty years of age before summoned to her last long rest. She was a member of the Christian church. Elijah was the second child in the family and one of the seven who accompanied his mother to western Nebraska. Andy J., who lives in Utah, was the oldest but Elijah and Mrs. John Emery are the only members of the family in Scottsbluff.

Elijah McClenahan spent his youthful years in Iowa, where he received the educational advantages afforded by the excellent school system of that state. He helped his parents on the farm, thus at an early age becoming well acquainted with the practical side of farm industry, as he early assumed what duties he was capable of carrying on a pair of young shoulders and what his growing strength permitted, for there is always plenty for a boy to do on a farm, from herding cattle to feeding stock and driving team and plow. When his mother came west Elijah was a young man just past his twenty-first birthday and he determined to establish himself independently and took up a homestead two miles west of the present site of Scottsbluff. He proved up on the land, broke the sod of the prairie for his early crops, and when his capital allowed, made good and permanent improvements on the place in the way of a house and farm buildings. He engaged in diversified farming and stock-raising and during slack periods of farm work or when he could get some other member of the family to care for his stock, rode the range as a cowboy, as that was the period when the great cattle companies had vast herds on the plains and required great cattle camps for their many men who guarded and directed the manner in which the cattle ranged in feeding. It was in this way that he materially aided his financial resources and at the same time gained an invaluable knowledge of the cattle business which was of great use to him in his own business

enterprises when the cattle barons and their monopolies were a thing of the past in the Panhandle and where once was range is now a smiling countryside where the green crops wave in the breeze with many a flourishing village and town which are fine indications of the prosperity of this section once known and called the "Great American Desert." Mr. McClenahan from his first coming to this section had great faith in its agricultural possibilities; he was determined that not only himself but others should have the most that their lands could produce. He was a man who kept abreast of the times, the improvements in farm methods and any project that would give a great yield from the soil, so that it is not surprising that he was one of the first to believe in and advocate irrigation for the Platte valley. The soil was fertile, the sunshine unfailing in the high prairie country, all that was needed to make this a garden spot was assured water, and there was plenty of it in the river. The problem lay in getting a sufficient quantity onto the land. He was one of the projectors and the first superintendent of the Winter Creek irrigation ditch, the pioneer project in this section. He helped not only materially but financially in the building of the ditch, being the man who removed the first shovel of earth on the construction work, it might be said he laid the foundation stone for it. For fourteen years he devoted a large part of his time and much of his energy to the great and paramount question of the Platte valley, will irrigation pay? He kept a careful record of the amount of water used in the Winter Creek district, the number of acres it watered and the greater yield per acre under ditch, and it was from his careful and painstaking work, a report of which was filed with the government that the Reclamation Service decided to place Scottsbluff county under government reclamation, which has been the making of the small landholder along the Platte. Later, Mr. McClenahan was instrumental in the work of building the Mitchell ditch and the Enterprise project, which have so materially changed conditions of farming and settlement in this vicinity, and have developed a semi-arid region into one of the most beautiful and productive regions of the great commonwealth of Nebraska. In truth, "the desert now blossoms like the rose."

Inherited from his Blue Grass father, Mr. McClenahan has had a fine taste for horses all his life and when his capital permitted he invested in some fine blooded stock, raising polo ponies for the eastern market and high grade riding horses. He also owned one of the fastest quarter milers of western Nebraska, "Tenpins," who won many a race and was a source of pride to his owner. However, this was but a side-line of the extensive business in which Mr. McClenahan engaged, for, after the railroads were built through Scottsbluff county he began to be one of the heavy and extensive cattle feeders of this section. Buying in the west he shipped here, fattened his stock and then shipped to the big packing centers of eastern Nebraska and Kansas, a business which proved most successful, due to his early experiences in range cattle and his keen ability as a buyer. Up to the time of Mrs. McClenahan's death, Elijah lived with her, managing her landed interests. He is now a partner with Charles Beatty in a four hundred acre ranch southeast of Minatare, which they are devoting to diversified farming. Both men are progressive in their ideas, have introduced modern methods and use modern machinery and are reaping the reward which justly comes to men who devote time, brains and effort to the business in hand. Mr. Clenahan owns ten acres of land in the southeastern part of Scottsbluff which he is arranging for an extensive cattle-feeding yard, a project which has long been needed in this section, of which Scottsbluff is the center. In politics Mr. McClenahan is a member of the Republican party, and though he takes no active interest in the politics of the state, is intensely interested in the men who run for local office, believing that only good, conscientious men should fill public positions. On December 26, 1912, Mr. McClenahan married Miss Nellie Boone, a Hoosier by birth, being reared and educated in her native state of Indiana. She was the daughter of John and Martha (Southerlin) Boone, both of whom were born and reared on the Wabash river. Mr. and Mrs. McClenahan have five children, Pearl A., Merle E., Joseph A. and twins, Nellie and Ellen, who have a bright future, as their parents are determined that they all shall have every social and educational advantage afforded by the schools of the town and state for the equipment of life's battle which is strenuous at best. Mr. and Mrs. McClenahan are estimable people, who believe advocate and support every movement for the betterment of civic and communal life, and are held in high esteem by their neighbors, fellow-townsmen, and a large circle of friends.

WILLIAM P. HODNETT, M. D., who has been engaged in medical practice at Scottsbluff for some years, is highly esteemed professionally and is equally valued personally. Dr. Hodnett was born at Danville, Virginia, September 2, 1883. His parents, William P.

and Belle (Price) Hodnett, are natives of Virginia and still live in the old home at Danville. The father of Dr. Hodnett is a man of ample fortune, now practically retired. When the Civil War closed he, like many other residents of the South, found it necessary to entirely rebuild his fortunes and was entirely successful. He owns valuable business property at Danville. He is of high personal standing there, has served in the city council for many years, is a sturdy supporter of the Democratic party and is a consistory Mason and a Knight of Pythias. Both parents of Dr. Hodnett are members of the Methodist Episcopal church. He is the only one of their seven children to establish a home in Nebraska.

Dr. Hodnett was liberally educated. After attending private schools he spent one year in Randolph-Macon college, Ashland, Virginia, two years in the Virginia State University, and in 1912 was graduated from the medical department of the University of Colorado. After graduation he practiced for one year in St. Luke's and Mercy hospitals, Denver, then in the city of Denver and the mining camps near Telluride, in San Miguel county, Colorado. In the fall of 1916 Dr. Hodnett came to Scottsbluff, finding a ready welcome for a man of his professional ability, and continued alone until March, 1918, when he formed a partnership with Dr. F. W. Plehn, the firm being recognized as one of the ablest in the city.

In 1912 Dr. Hodnett was united in marriage to Miss Eleanor Finley, of Denver, Colorado, and they have two children, William Finley and Virginia Belle. Dr. Hodnett and wife are members of the Presbyterian church and take active part in social affairs. In politics he is a sound Democrat and for many years he has been identified with the Masonic fraternity. Dr. Hodnett belongs also to representative medical organizations and occasionally contributes to their literature.

D. J. POLLOCK, who is well-known through the Platte Valley as a cattleman and judge of stock, has been a resident of Scottsbluff for some years and is interested in dealing in stock and also real estate. Mr. Pollock was born in Union county, Iowa, December 29, 1860, and is a son of James P. and Eliza (McVay) Pollock.

Mr. Pollock's father was born in Knox county, Ohio, a son of Samuel Pollock, a native of Scotland, and the mother in Greene county, Pennsylvania, a daughter of Vincent

McVay, who was also born in Scotland and came in early manhood to the United States. Both parents died in Iowa, the father when aged eighty-seven and the mother sixty-eight years. They had children as follows: W. V., who resides at Gering, Nebraska, is a retired farmer; D. J., who resides at Scottsbluff; J. L., a resident and officeholder at Des Moines, Iowa; R. M., of Larned, Kansas, is a traveling salesman; and one deceased. The father was a farmer all his life. Both parents were members of the Scotch Presbyterian church.

After his school period ended, Mr. Pollock began to assist his father on the farm and has been identified more or less with farm activities all his life. In 1916 he came to Scottsbluff county and settled on land near Scottsbluff that he had previously bought, and has made raising thoroughbred stock the main feature of his business since coming to the upper valley. He raises Duroc hogs extensively and has paid as high as $400 for a thoroughbred boar.

On October 21, 1891, Mr. Pollock married Miss Lillie B. Stalcup, who was born in Iowa and died in that state February 26, 1913, the mother of four children: Etha, connected with a business house at Scottsbluff; Zaida, at home; Dorothy, a student in Doane college, Crete, Nebraska; and Howe, a mechanic for the Page Motor Company. The family belongs to the Presbyterian church. In politics Mr. Pollock is a Democrat, and while living in Iowa, was the first of his political party to be elected to the office of assessor of his district. Mr. Pollock is held in high esteem as a man of sterling character.

MATHEW J. HIGGINS, general merchant at Scottsbluff, and an active, interested, public spirited citizen, is a business man of long experience. He came to this city in 1913 and founded the Golden Rule store, through honorable methods and business integrity making the name significant. He was born at Camden New Jersey, November 4, 1879.

The parents of Mr. Higgins were M. J. and Esther (Rodgers) Higgins, the latter of whom was born and married in the city of Philadelphia, and now resides in Iowa. The father of Mr. Higgins was born in Wilmington, Delaware, and from that state enlisted for service in the Civil War, entering company C, Fifty-first Delaware infantry, in which he served during the closing months of the war, during that time contracting disease which finally caused his death. After his marriage he engaged in the hotel business at Philadelphia, in 1876 removing to Iowa, where he was a merchant. He was a Republican in politics and was a member of the Presbyterian church. Of his seven children M. J. was the third in order of birth, the others being: Frank, in the grocery business at Malvern, Iowa; William, in the employ of the Standard Oil Company, at Malvern; John, a commercial traveler for a San Francisco business house; Edward C., manager of the Penny store, at Blackwell, Oklahoma; and Charles, in the grocery business at Malvern. The mother of the above family is a member of the Episcopal church.

Mr. Higgins attended the public schools in his native state and later the Chicago University. He began business life as clerk in a store and had fine training as an employe of the great house of Marshall Field & Co., first in the Chicago establishment and later as one of the firm's highly regarded traveling salesmen. He then embarked in business for himself at Las Animas, Bent county, Colorado, where he confined himself to handling dry goods and shoes, and remained in business there for seven years. In 1913 he came to Scottsbluff, invested in property and started the Golden Rule store which has proved an exceedingly successful enterprise, his amount of business having doubled each year. He has been obliged to enlarge his quarters to accomodate his large stock of dry goods, shoes and clothing. As a merchant here he stands in the first rank.

In September, 1904, Mr. Higgins was united in marriage to Miss Eva K. Knox, who was born at Grand Island, Nebraska, and is a member of the Christian church. They have an interesting family of four children, namely: Frank, Harold, Chester and Paul. Mr. Higgins is interested in all that concerns Scottsbluff, its schools, its business, its social advantages, and as a member of the city council, in which he is serving his second term, he carefully considers such matters and lends his influence accordingly. In the political field, Republican principles and candidates have always been his choice. He has long been identified with the Odd Fellows.

FRANK R. BECKER, who is well-known in business circles at Scottsbluff, is part owner and general manager of Diers Bros. & Company store, with which important commercial house he has been identified for a period approaching twenty-one years. He was born in Dearborn county, Indiana, in 1878.

The parents of Mr. Becker, J. P. and Mary T. (McCracken) Becker are deceased. The father was born in one of the Rhine provinces, Germany, and the mother in a New England state. The father came to Indiana when young and was married in that state and followed the carpenter trade and was an auctioneer. He came with his family to Butler county, Nebraska, in 1883, and during his later years engaged in market gardening. His family consisted of four sons and two daughters. He was a Democrat in politics, a member of the order of Odd Fellows and he belonged to the German Lutheran church.

Frank R. Becker attended school at David City, Nebraska, but left when he reached the eighth grade, in order to become self supporting. For eight years he was a clerk with the Diers Bros firm at Fullerton, Nebraska. In the spring of 1905 he came to Scottsbluff and went to work for the same people operating here under the firm name of Luft & Diers Bros. After the death of Mr Luft, Mr Becker continued with the other partners for three years and then resigned and went to the Mitchell Mercantile Company, where he had charge of the clothing department for three years. He then homesteaded six miles from Mitchell, on Dutch Flats, where he now owns eighty acres of irrigated land. On January 1, 1913, he came back to Scottsbluff to become manager of Diers Bros Company store, in which he purchased stock, which he increased to a one-third interest on June 8, 1914. Mr Becker has demonstrated great business capacity, having built up a comfortable fortune entirely through his own efforts.

On August 28, 1912, Mr Becker was united in marriage to Miss Lacy Bryan, and they have one son, Frank M, who was born August 23, 1916. Mrs Becker is a member of the Methodist Episcopal church, to which Mr Becker's mother also belonged. In politics he is a Democrat but no seeker for office. He is identified with the Knights of Pythias and the Scottsbluff Country Club.

JOHN R. KELLY, who is one of Banner county's progressive agriculturists and leading citizens, has lived in this county many years, homesteading in 1888 and never parting with his original purchase, which now comprises some of the most valuable land in the county. He was born in Worth county, Missouri December 7, 1867.

The parents of Mr Kelly were John and Jerusha (Millican) Kelly, the former of whom was born in Ohio in 1838 and the latter in Illinois, March 17, 1841. Her death occurred in January, 1907. She was a faithful member

of the Baptist church from girlhood. Of their six children, the two sons in Nebraska are John R and Samuel. In boyhood the father of Mr Kelley went to Illinois and lived there seven years as a farmer, married there, and then moved to Missouri where he died in 1872.

John R Kelley was only five years old when his father died. He started to go to school in Missouri, later went to school for a short time in Page county, Iowa, and when ten years old went to work on a farm in Kansas. He remained one year and then went back to Missouri, where he followed farm life for six years and then went again to Iowa for two years. After another year in Missouri, on March 23, 1887, he came to old Cheyenne county, now Banner, and in July following secured his homestead. At that time $100 would purchase 160 acres of land that now would bring $50 an acre. Mr Kelly had to depend entirely on his own efforts, and after securing his claim, found it a serious undertaking to make enough money to make his payments. In those days real money was scarce in Nebraska and remuneration for any kind of labor was small, while farm produce brought but inadequate returns in the market. Mr Kelly relates that in 1892 he and his brother raised wheat and hauled it a distance of twenty-five miles to Kimball and sold it for twenty-four cents a bushel, and pork, at the present time one of the world's luxuries commanded so small a price that it became a question whether the raising of hogs was worth while. The interest on money at that time had risen to thirty-nine per cent. In the fall of 1889 Mr Kelly went to Hall county and husked corn in the vicinity of Wood river for a cent and a half a bushel, working for Fremont Dodge. The latter advised Mr Kelly to keep his Banner county land at all hazards, and the taking of this advice proved very advantageous to Mr Kelly, although it necessitated much hard work to follow it. During those early years he worked for $1 a day, then acceptable by workers and employers alike, and to secure this had to travel as far as Greely, Colorado, and Cheyenne, Wyoming.

However, those times have long since passed away. Starting with 160 acres, Mr Kelly acquired more land as his improving circumstances permitted until at present he is the owner of 3,200 acres. It is mainly ranch land and his stock interests are very important. He believes Hereford cattle and Percheron horses the most profitable and feeds about ten head of horses a year and about 200 head of cattle, and raises annually sixty fine cows for breeding purposes. It was on Mr Kelly's land that the Prairie Oil & Gas Company sunk a shaft that struck an extra good grade of oil but at that time and with the company's facilities, did not seem to indicate oil in paying quantity. Further investigation has convinced Mr. Kelly, however, that some day he will have a well here with a profitable flow of oil.

On December 25, 1900, Mr. Kelly was united in marriage to Miss Anna McKinnon, the ceremony taking place at Harrisburg, Nebraska. She is a daughter of Hugh and Elizabeth (Mickle) McKinnon, natives of Scotland, who settled in Banner county in 1889. They died in Harrisburg, the father in 1904 and the mother in Scottsbluff, May 3, 1918. Three of their children live in Scottsbluff county, Nebraska, and two others, Mrs. Kelly and Edward McKinnon, in Banner county. Mr. and Mrs. Kelly have had two children, the one survivor, Allison, living at home.

Since early manhood Mr. Kelly has taken an active part in public matters that he has believed come within the scope of good citizenship. Politically he is a Democrat and has wide influence in county politics but has never accepted any public office except that of sheriff, serving one year (1896) by appointment, and two years by election, his term expiring in January, 1899. He is a member of the Farmers Union, and is financially interested in the Scottsbluff Creamery and the Independent Lumber Company of Scottsbluff.

ERNEST H. KLINGMAN, a representative business man of Scottsbluff, proprietor of a grocery house and a storage business, was born in Clayton county, Iowa, September 6, 1864, the son of Lewis and Elizabeth (Lowe) Klingman, the former born in Germany and the latter in Connecticut. The father was a blacksmith by trade and owned his own shop in Iowa, in which state he married and both he and wife died in Iowa. Of their seven children Ernest H. is the only one who lives in Nebraska.

Ernest Klingman had only country school advantages, and after his school days ended he remained at home and worked as a farmer until twenty-one years old. In 1888 he came to Nebraska and settled on the Middle Loup river in Custer county, where he remained two years, then lived one year in Holt county, being a farmer in both places. Mr. Klingman then went to Oklahoma and from there to Kansas, in which latter state he remained six years working for farmers, then came back to Nebraska and accepted employment with Charles Richardson, who conducted a livery business at Broken Bow. In 1901 he came

to Scottsbluff county and engaged in a draying business after which he invested in property at Scottsbluff and opened a confectionery store. In 1917 he erected a fine store building on his home lot and put in a stock of fancy and staple groceries from which he has received gratifying remuneration for the money invested as well as the thought and labor he has expended. His storage business is also a profitable source of income. This is but the merest outline of Mr. Klingman's career but it gives convincing proof that persistent industry and honest effort, will bring reward.

In 1889 Mr. Klingman married Miss Matilda Predmore, who was born in Hardin county, Iowa, a daughter of John and Nancy Jane (Peters) Predmore, natives of Ohio. Mrs. Klingman was the fourth born in her parent's family of fourteen children, twelve of whom are living. Mr. and Mrs. Klingman have four children: Charles, Roy, May and Lloyd. The one daughter is the wife of John Montz of Scottsbluff. All three sons of Mr. Klingman have been in military service and attached to the heavy artillery and all are safe at home again after overseas service. Mrs. Klingman is a member of the Methodist Episcopal church. In politics Mr. Klingman is a Republican.

JOHN M. MARTIN, who is an enterprising and progressive business man of Scottsbluff, extensively interested in the handling of real estate, is a Nebraska product and is proud of the fact. He was born at Hastings, in 1888, the sixth in a family of ten children born to John and Mary (Rose) Martin.

The father of Mr. Martin was born in the state of New York, one of a large and important family. His father, Solomon Martin, a native of New York, came to Nebraska in 1874 with his son John and family, being then aged ninety-five years. He had 131 grandchildren and great-grandchildren. John Martin drove a team and covered wagon the entire distance from Illinois and when he reached Adams county where he intended to homestead, he camped and tethered his horses on the present site of the courthouse at Hastings. He was a farmer all the rest of his life, his death occurring in April, 1918. He was married in Nebraska to Mary Rose, who was born in Ohio and now resides at Mullen, Nebraska. Her father, Peter Rose, was a veteran of the Civil War. She is a member of the Methodist Episcopal church.

John M. Martin attended school for five

years at Guide Rock, Nebraska, and after his school period was over he became an auctioneer, having the gift of ready speech and continued in that line at Minatare and Mullen, for eight years. He was so successful that his services were engaged by C H Irion for the selling of real estate after he came to Scottsbluff in 1916, and since the latter part of 1917 they have been equal partners in the business. The operate all through the Platte Valley, doing a large business in farm property.

On May 28, 1913, Mr Martin was united in marriage to Miss Sylvia Hendrickson, who was born in Harlan county, Nebraska, a daughter of James Hendrickson, a prominent farmer in Harlan county. Mrs Martin is a member of the Methodist Episcopal church. Mr Martin is much more interested in business than in politics, although he is a thoughtful and careful citizen, but he has not identified himself with any particular party, voting according to his own judgment. He belongs to the Odd Fellows. His acquaintance is wide and his personal friends are everywhere.

SAMUEL WILLARD RIPLEY, a well-known resident of Scottsbluff and an active, useful citizen, came to this city in 1900, from his homestead in what was then Cheyenne county, but which has since been organized as Scottsbluff county, Nebraska, where he had pre-empted and taken a tree claim in 1886. He has been greatly interested in the success of the irrigation projects, and perhaps only a few of his neighbors are aware that he, with B F Gentry established the first irrigation project in Nebraska, when they ran water by ditch onto a tract of millet, for D D Johnson. Later Mr Ripley was superintendent of the Enterprise Ditch for one year. He was born in Fremont county, Iowa, August 28, 1861, and was a crowing, happy infant when his father marched away to take part in the Civil War.

Mr Ripley's parents were Samuel A and Nina E (Barger) Ripley, the former born in the state of New York and the latter in Iowa. The paternal grandfather was S W Ripley, a native of New England, who practiced medicine first in Ohio and later in Iowa and died in Fremont county. The family settled in Iowa before the Civil War and from that state Samuel A Ripley enlisted for service in company E twenty-ninth Iowa infantry, and did his full duty as a private soldier for over three years, never being either wounded or captured. Early in life he was a farmer but later a butcher. He was a fine, honest man whom many

mourned when he died in 1889, but, because of his generous instincts never was a success financially. He was a Republican in politics, a member of the Odd Fellows and throughout life was influenced by the Christian training he had received in a good home. He was married in Iowa to Nina E Barger, who is also deceased, their burials taking place at Weeping Water, Nebraska. She was reared in the Methodist Episcopal faith. Of their eight children Samuel Willard is the oldest survivor, the others being William Jasper, a farmer and carpenter living in Wyoming, Guy Douglas, in the electrical business in California, Feribey, the wife of Rev James G Clark, a Presbyterian minister at Beaver City, Nebraska, and Loy E, the wife of Charles C Spencer, of Wyoming.

Samuel W Ripley learned the trade of a butcher and was a successful farmer for two years in Nebraska. In 1886 he located in Cheyenne county, now Scottsbluff county, four miles northeast of the city of Scottsbluff, and lived there until 1900, passing through many of the hardships that made pioneering in the state a difficult and trying process. After coming to Scottsbluff he operated a hotel and a meat business, then was appointed superintendent of the Enterprise Ditch. In August, 1905, Mr Ripley accepted a position with the Standard Oil Company as local manager, with headquarters at Scottsbluff, and has since continued with this corporation. He has taken part in civic affairs quite actively, has accepted the responsibilities of office when called on and has and still is assisting in the substantial development of this place.

In 1889 Mr Ripley married Miss Anna M Johnson, of Missouri Valley, Iowa, a daughter of D D Johnson, a wounded veteran of the Civil War, who makes his home with Mr Ripley. Mr and Mrs Ripley have an adopted daughter, Clara Lois, a schoolgirl of fourteen years. The family belongs to the Presbyterian church. Mr Ripley is an Odd Fellow and is a Republican in politics. During his service of four years on the town board, he was chairman a part of the time. Mr Ripley can relate many interesting facts concerning early days here when he was engaged in freighting between Alliance and Gering and Kimball and Gering when the actual necessities of life were hard to secure and had to be hauled by teams from those railroad towns. Scottsbluff was only a year old when Mr Ripley came here and took charge of the hotel with, probably, not over one hundred inhabitants. Since that early day he has been a continuous resident of this now flourishing and prosperous town.

ASA E. CHILES, who represents one of the leading piano and music houses of the country at Scottsbluff, the A. Hospe Company of Omaha, Nebraska, has been located in this city since 1917, and has assisted in developing a fine musical taste here. In addition to being an excellent business man, Mr. Chiles has shown a hearty interest in everything pertaining to this city and has made many personal friends.

Asa E. Chiles was born at Riverside, Washington county, Iowa, May 23, 1880. His parents are Jacob S. and Susan E. (Armagost) Chiles, the former of whom was born in Maryland and the latter in Pennsylvania. In 1872 the father went to Iowa and was married in 1876, and they have three children: Asa E., who is of Scottsbluff; George S., who is chief draftsman for the American Steel Foundries Company, Chicago; and Amy, who is the wife of Lewis E. Schmidt, who is in the telephone business at Council Bluffs. In politics the father is a Republican and belongs to the order of Knights of Pythias. Both parents are members of the Methodist Episcopal church at Clarinda, Iowa, where they live retired. The father engaged in farming for many years in Iowa and owns a body of land in Canada, on which he spends a part of his time.

Asa E. Chiles was educated in Page county, Iowa, and attended the high school at Clarinda. He has been the builder of his own fortune, beginning with Swift & Company, packers, at the age of fourteen years and continuing with that company until he was twenty-two. Musically inclined and possessing musical gifts, he then accepted the opportunity to go into the piano business at Bushnell, Illinois, where he remained two years. During the next six years he was on the road in special sale work for different piano houses, following which, for four years he was with E. L. Benedict & Sons at Clarinda. In 1916 he became associated with the A. Hospe Piano Company of Omaha, and on May 3, 1917, came to Scottsbluff and took charge of their piano and music business here and future prospects are all that could be desired. Mr. Chiles has six employes, five in this city and one at Alliance.

In 1899, at Clarinda, Iowa, Mr. Chiles was united in marriage to Miss Sudie I. Leffler, who was born at Des Moines, Iowa, and is a daughter of George W. Leffler, who is in the book and music business at Butte, Montana. Two children have been born to Mr. and Mrs. Chiles: Aileen, aged fourteen years, and Warren, aged twelve years. She is a member of the Methodist Episcopal church. Mr. Chiles is not particularly active in politics but has always been identified with the Republican party. He belongs to the order of Knights of Pythias and to several musical organizations.

LOU SCHWANER, who is in the jewelry and optical business at Scottsbluff, is associated with his brother, Charles H. Schwaner, and they operate under the firm name of Schwaner Brothers. They were born in Valley county, Nebraska, Lou Schwaner on March 6, 1883, and Charles H. Schwaner on May 10, 1885. They have practically spent their entire lives in the jewelry business, the older partner beginning at the age of twenty-one and the younger when fourteen years old.

The parents of the Schwaner brothers are H. J. and Margaret (Reese) Schwaner, the former of whom was born in Wisconsin and the latter in Indiana. They came to Iowa when young and were married in Polk county. In 1882 the father homesteaded in Valley county, Nebraska, and his children have heard him tell of the hardships that faced the pioneers of that time when the nearest neighbors were four miles distant over a trackless prairie covered with high-growing, wild, red-topped grass. Fortunately easier times succeeded and it is a great satisfaction to their sons that the parents are now enjoying all the comforts of life at Ord, where they live retired. Besides the two sons mentioned they have two daughters, namely: Lydia, who is the wife of R. E. Mickelwait, a banker at Richfield, Idaho; and Minnie, who is the wife of H. Snedeker, a farmer near Thompson, Iowa.

Lou Schwaner obtained his public school education at Ord, Nebraska. His first venture in the jewelry business was at Greeley Center, Nebraska, where he remained from 1903 to 1904, when he returned to Ord and in partnership with his brother, bought the business of the jeweler with whom they had learned the trade. They continued together until 1909, when they sold, and both moved to Gooding, Idaho, where they engaged in farming for four months and then C. H. returned to Ord, and Lou bought a store at Loup City. In 1913 Charles H. sold his store at Ord and during that winter engaged in the real estate business in southern Texas, after which he was in the jewelry business for three years at Burwell, Nebraska, then traveled in the same

HOPE BROWN AND FAMILY

line out of Loup City for a while In April, 1917, the brothers came to Scottsbluff and opened their present store and they have prospered They carry a fine line of jewelry, and optical goods, making a feature of the latter and manufacturing and grinding their own lenses

In 1907, Lou Schwaner was united in marriage to Miss Mayme Auble, at Ord, Nebraska, and they have two children Charles and Martha He and family belong to the Presbyterian church Charles H Schwaner married Miss Bessie Rawles, of Ord, Nebraska, and they have one daughter, Georgia The brothers are Republicans in politics and members of the Knights of Pythias They are enterprising citizens and honorable business men and command respect and enjoy the confidence of everyone

HOPE BROWN, who is a prominent and highly respected citizen of Banner county, is owner and proprietor of Big Horn ranch, consisting of 5,000 acres of range and farming land, and that this property has been acquired through his own unassisted efforts, speaks well for his industry, good judgment and business foresight He was born in the city of Glasgow, Scotland, October 10, 1867, and is a son of Robert and Agnes (Boyd) Brown

From Scotland the parents of Mr Brown came to the United States in 1868 For several years they lived at Madison, Wisconsin, where the father followed his trade of stonecutter, then moved to Omaha, and one year later to Colfax county, Nebraska The father homesteaded near Schuyler and lived on his land there until his death, which occurred in 1878 As opportunity offered he worked at his trade always frugal and industrious, a man of sterling integrity The mother of Mr Brown still lives in Colfax county and owns the original homestead Both parents belonged to the Presbyterian church Of their nine children, six survive, and of these Hope and David live in Banner county

Hope Brown went to school until twelve years old and then began to be self-supporting After working for farmers, both in Nebraska and Iowa, he turned to the home farm and was engaged there for five years, coming then to Banner county In 1889 he bought a relinquishment and homesteaded, later bought additional land in the county and finally the property on which he has resided for seventeen years Formerly this place was known as Big Horn postoffice and for seven years Mrs Brown was postmistress Mr Brown raises 100 head of White Face cattle and several car-

loads of hogs yearly, while 1,200 acres are devoted to general farming

On April 5, 1893, Mr Brown was united in marriage to Miss Maggie E Maynard, who is a daughter of Alexander G and Eva (Vincent) Maynard, who now live retired in Minatare To Mr and Mrs Brown the following children have been born William O, who married Pinkie, a daughter of Mervin Snyder Edith, who married Rolland Sickles and now lives in Maxwell county, Agnes, who resides at home, Alice, who is the wife of Frederick Haskell, a farmer near Reddington, and Lillie, Hope, Jr, Byron, Eunice, Eva, and Lois, all of whom reside at home, an intelligent and happy family prominent in the social life of the neighborhood

In national matters Mr Brown is a Republican, but sometimes issues come up in local affairs that cause him to cast an independent vote He has served in a trustworthy manner in public office at times was a county commissioner from 1902 to 1908 and was one whose judgment was consulted about consolidating three school districts with district No 8 He is well known in fraternal life, belonging to the Masons, Knights of Pythias, Modern Woodmen of America, and United Workmen Mr Brown is now a man of ample fortune He went a few years ago to Cheyenne, Wyoming, and there worked in a brickyard, and helped in the removal of Camp Carlin to Fort Russell Although he yet oversees the operation of his large property for the past five years ill health has somewhat reduced his activity and he has shifted some of his responsibility to younger shoulders

ALBERT B KERNS, D D S, who is engaged in the practice of dentistry at Scottsbluff, came to this city in 1917, and with the exception of a period of military training, has been in continuous practice here ever since Dr Kerns has thoroughly demonstrated his knowledge of modern dentistry and has won the confidence of the public in a professional way, and at the same time has gained respect and esteem as a young man of high personal character

Albert B Kerns was born at Auburn, Nemaha county, Nebraska, in 1891, and is a son of James W and Alice J (Crowley) Kerns, the former of whom was born in Ireland and the latter in the state of Illinois Their marriage took place in Illinois, and in 1878 they came to Nebraska The father was in the lumber business almost all his life prior to retirement first embarking in the same at Omaha, but later removing to Phelps, Missouri. At the latter place the town was submerged when unexpected rises took place in the river, entailing great loss of property. He then moved to Auburn, Nebraska, which was but a little hamlet at that time, and resumed his operations in lumber and become one of the substantial men of the place and still resides there. For many years he has been a prominent factor in Republican politics and served one term in the state legislature. Both parents of Dr. Kerns are members of the Roman Catholic church. Of their eleven children Albert B. was the fourth in order of birth.

Albert B. Kerns was graduated from the Auburn parochial school in 1907, after which he spent two years in Creighton University, Omaha, and in 1912 was graduated from Creighton Dental college, Omaha, and immediately entered into practice at Elgin, Nebraska, where he continued until 1916, spending the rest of the year in Fremont and then establishing himself at Scottsbluff. He entered military service in the National army, September 7, 1918, taking a medical officers training course at Fort Oglethorpe, Georgia, where he remained on duty until his honorable discharge, December 22, 1918. In February following the Doctor was found ready for professional work in his office, through unusual experience better qualified than before for the problems continually being presented to a dental surgeon.

In 1917 Dr. Kerns was united in marriage to Miss Helen Celia Toillion, who was born at North Platte, Nebraska, a daughter of Xavier Toillion, who was born in France and is now retired and lives at Sterling, Colorado. Dr. and Mrs. Kerns are members of the St. Agnes Catholic church, and he is very active in the Knights of Columbus, also the local B. P. O. E. In his political views he is a Republican.

THOMAS F. KENNEDY, one of the energetic, progressive men of Scottsbluff, has been prominent in business circles and in civic affairs here ever since he chose this place for his home in 1905. At present he is one of the city officials and is also secretary and treasurer of the Tri-State Land Company, and in addition manages an extensive produce business. Mr. Kennedy was born at St. Joseph, Missouri, March 26, 1873, the son of Thomas H. and Mary H. (Furman) Kennedy, the former born at Drogheda, County Meath, Ireland, September 15, 1835, and died December 8, 1908, while the mother was a native of

the Empire State, born July 25, 1839, and died November 2, 1902, The parents were married at Florence, Massachusetts, and two of their three children survive, Thomas F. and Ruth D., the eldest, Philip H., died at the age of forty years. Mr. Kennedy's sister is the widow of Dante Barton, who died August 6, 1917. For a number of years he was an editorial writer for the *Kansas City Star*. Mrs. Barton resides in Washington, where she is connected with the National War Labor Board. The life of Mr. Kennedy's father had many elements of romance in it, his whole career being well worth repeating. He was a runaway from home at the age of thirteen years, reached the United States as a stowaway, picked up a fair education in his adopted country as best he could, and was forty years old when he was graduated from the law school of the Kansas University. In 1868 he located at St. Joseph, Missouri, in 1875 removed to Lawrence, Kansas, and in 1880 to Kansas City, where his death occurred. He became well known in his profession and for some years made a specialty of pension cases.

Thomas F. Kennedy obtained his school training at Kansas City. His first business experience was with a firm of building contractors, after which he was engaged in the produce business at Kansas City for a number of years. In May, 1905 he came to Scottsbluff and for a year was cashier and office manager for the Tri-State Land Company, with which organization he has continued as secretary and treasurer. To some extent he has engaged in the produce business, making a specialty of buying potatoes.

On December 19, 1900, Mr. Kennedy married Miss Alice Beesley, who was born at Mossy Creek, Tennessee, and they have one daughter, Alice Alberta, a student in the Scottsbluff high school. Mr. Kennedy and family are members of the Presbyterian church. His parents belonged to the Congregational church. He was reared in the Republican party and has loyally supported its principles all his life. Since coming to Scottsbluff he has been interested in the city's progress in every way, has served as village clerk and almost continuously on the school board, of which he is the present secretary.

FRED M. BRYAN, who is a prominent and reliable business man of Scottsbluff, is at the head of one of the largest jewelry establishments in Western Nebraska, is widely and favorably known to the trade, and is vice president of the Nebraska Retail Jewelers association Mr Bryan came to this city in 1913 and has identified himself with her best interests, and his public spirit and usefulness are being utilized in his earnest civic efforts as alderman of his city ward

Fred M Bryan was born at Mason City, Nebraska, in 1887, and is a son of Millard C and Mary A (Boden) Bryan They were married in Illinois, came to Nebraska in 1879 and settled in Seward county, then the father embarked in a mercantile business at Mason City, where he continued until 1915, when he disposed of his interests there and came to Scottsbluff Of their five children, Fred M was the third in order of birth, the others being Mrs Becker, a resident of Scottsbluff, Mrs R C Smith, who lives on a farm in Butler county, Paul Franklin, in business with his brother Fred, and William Lloyd, who died November 7, 1918, was also a member of the firm of Bryan Bros

Fred M. Ryan attended the public schools of Ulysses, Nebraska, and afterward spent two years at Omaha, attending a trade school where he learned watchmaking Afterward for four years he was in the jewelry business at Ulysses, then sold out and in 1913 came to Scottsbluff Mr Bryan has a beautiful store well stocked in his line, his goods being carefully selected to suit the most critical taste and of great value He has patrons all through the Platte Valley and is prepared to supply jewels or jewelry designs for all occasions

In 1909 Mr Bryan was united in marriage to Miss Lillian Peterson, who was born at Fremont, Nebraska, and educated in the high school there Her father, David Peterson, carries on a plumbing business at Fremont She is a member of the Methodist Episcopal church They have two children, Maude and Rex Politically he is a Republican and fraternally an Odd Fellow, a Scottish Rite Mason and Shriner

PAUL F BRYAN, who is a member of the jewelry firm of Bryan Bros, at Scottsbluff, is well known in this city, to which he came in 1911 and was identified with a prominent firm here before his present one was organized Mr Bryan was born at Ulysses, Nebraska, in 1892, and is a son of Millard C and Mary A (Boden) Bryan, who now live retired at Scottsbluff

The parents of Mr Bryan came from Illinois to Nebraska in 1879 The father located first in Seward county, later moved to Ulysses in Butler county and then embarked in the mercantile business at Mason City in Custer county, where he continued until 1915, when he retired to Scottsbluff He is a Republican in his political views

Paul Franklin Bryan is one of a family of five children, two daughters and three sons All three sons engaged in business together under the style of Bryan Bros, at Scottsbluff, in 1913, but only two survive, Fred and Paul F, the other, William Lloyd, having died November 7, 1918 Paul F Bryan was educated at Ulysses and after completing the high school course, learned the jewelry trade under his brother In 1911 he came to Scottsbluff and was connected with the firm of Diers Bros, until he went into partnership with his brother as Bryan Bros

Mr Bryan is one of the returned soldiers from overseas service in the Great War, his experiences while in France for six months, including the terrific fighting in Argonne Forest Many of his brave comrades fell there and he was so exhausted that he had to be sent to a hospital in Bordeaux He was a member of Company A Three hundred fifty-fifth infantry, Eighty-ninth division, a sergeant in rank He suffered first from a gas attack, August 8, 1918, went back to the front lines on September 15, left the hospital December 8 and sailed for home, and with duty well done, was honorably discharged January 11, 1919 He earned a place on the record that Nebraska will cherish of her best and bravest sons

GUS W LAWTON, who owns an attractive jewelry store at Scottsbluff, is a man of marked executive ability, a good citizen and quite active in civic affairs He is one of the younger business men of the this city and came here in 1915, but he has shown business ability and a recognition of the highest standard of commercial integrity Mr Lawton was born at Fairhope, Alabama, June 2, 1889, the son of John and Clara (Cranuto) Lawton, both born at Leeds, in Yorkshire England The father traveled for a number of years in different parts of the world to secure specimens for the British Museum, and he was a hunter of wild animals in Africa and Australia and shipped them to England He came to the United States and was married at Chicago Illinois, and in 1885 settled in Alabama Mr Lawton is now of venerable age, being in his ninetieth year the mother being aged seventy-five years They are highly respected and esteemed residents of Greeley, Colorado Of their three surviving children, Gus W is the eldest, the others being daughters: Mrs. E. J. Preston, of Kansas City, and Madeline, a teacher near Greeley.

During Mr. Lawton's school period, the family lived in Texas, and in 1904 he was graduated from the Dennison high school and shortly afterward went to Denver, Colorado, where he served an apprenticeship at the jeweler's trade. For one year he was in a jewelry house at Chamita, New Mexico, then went to Salt Lake, Utah, and from there to Galveston, Texas. Later he located in Chicago, Illinois, in all these cities working in the manufacturing departments of large business houses in his line, and before he came to Scottsbluff, Buffalo, Kansas City and Denver had been added to the list of cities where he had resided and been associated with jewelry concerns. He has a well arranged store with a complete stock and has built up an excellent business.

In politics Mr. Lawton is a Republican but aside from partisan activities, has shown much public spirited interest concerning the progress of Scottsbluff. He is a member of the Knights of Pythias and has served through all the chairs except vice chancellor. He has a wide circle of social acquaintances and is a member of the Country Club and its secretary.

CLARENCE G. STEEN, D. D. S., who is well known professionally all through the Platte Valley, enjoys the distinction of being the oldest dental practitioner in point of time at Scottsbluff. Dr. Steen is a native of Nebraska and was born at Wahoo, in Saunders county, November 1, 1883.

The parents of Dr. Steen, John and Mary Louise (Hought) Steen, natives of Norway, came to Iowa when young and were married at Decorah in that state. Of their four children Dr. Steen is the third in order of birth, the others being: Mrs. A. C. Killian, whose husband is a clothing merchant; Theron H., who is in the stock business in South Dakota; and Mona, who looks after the domestic affairs of her brother at Scottsbluff. The parents still reside at Wahoo, where they are active in the Methodist Episcopal church. The father is a member of the Masonic fraternity, in politics he is a Republican and in earlier years was quite prominent in public life, at one time being city treasurer of Omaha and state land commissioner. He is now engaged in the real estate line at Wahoo.

After his public school course, C. G. Steen spent three student years at the University of Nebraska and in 1908 was graduated from the school of dentistry of Creighton University, Omaha. He began the practice of his profession at Scottsbluff and continued until 1915 when he went to Omaha and practiced there for two years and then returned to Scottsbluff and resumed practice. Dr. Steen has a reputation for skill in his profession that places him in the front rank of dental surgeons.

On June 6, 1908, Dr. Steen was united in marriage to Miss Mable Mellinger, who was born at Burlington, Iowa, and died December 13, 1914, survived by three children, namely: Jane, John M. and Virginia Louise, their ages ranging from nine to six years. Dr. Steen has been prominent in Republican political circles and active in public affairs here, at one time serving as mayor of the city, 1914-1915. He is a Consistory Mason and has been an official of the Blue Lodge.

JAMES T. ANDERSON. — It is a fact of modern medicine that when mysterious diseases attack, in many cases the dentist is called in to diagnose and successful treatment follows his advice. An experienced dental surgeon of the modern school at Scottsbluff, is found in Dr. James T. Anderson, who has been established in this city since August, 1917. Dr. Anderson was born in 1875, at Red Wing, Minnesota.

The parents of Dr. Anderson were John A. and Elizabeth (Johnson) Anderson, the former of whom was born in Sweden and the latter in Pennsylvania. Both came to Minnesota as young people and were married there. Of their seven children, James T. is the youngest of the five survivors, the others being: Minnie, the wife of John Fryer, a resident of Minneapolis, Minnesota; Charlotte L., an artist, lives at Minneapolis; .... the wife of William Richards, formerly an educator but now in the real estate business; and Louise, a teacher of physical culture and dancing. The parents were members of the Methodist Episcopal church. The father was a farmer all his life when not engaged in serving his country (1887) in the state legislature, and in 1861-65 as a soldier in the Civil War. As a member of company D Third Minnesota infantry, he participated in many of the serious battles of that struggle.

James T. Anderson was graduated in the agricultural course from the University of Minnesota in 1898. He then taught school for two years in North Dakota and for one year afterward was clerk in a store. In 1904 he completed his course in dentistry at the Indiana Dental college, following which he located at Axtell, in Kearney county, Nebraska,

CHRIS KRONBERG AND FAMILY

where he continued in active practice until he came to Scottsbluff. Dr Anderson has a well earned reputation for professional skill, and is in the enjoyment of a large and lucrative practice.

In June, 1906, Dr Anderson was united in marriage to Miss Anna Halberg, born at Greenville, Illinois, and they have four children Loretta, Hobson, Francis and Benjamin John. In politics Dr Anderson is a Republican and he and wife belong to the Pres-Presbyterian church.

CHRIS KRONBERG — Over a quarter of a century of connection with the agricultural interests of Scottsbluff county has made Kris Kronberg one of the substantial and well-known men of this vicinity. A native of Germany of Danish descent, when he came to the United States in 1882, he brought with him many of the admirable traits of the people of both those countries, and the success that has come to him has been won by legitimate participation in the enterprises of this section. Mr Kronberg says that next to the pride he takes in the fact that his sons did their full duty to the United States during the war with Germany, is that in his good farm and his record when he served his community and the county as assessor and deputy sheriff. He is progressive in his ideas and methods, takes an active part in all questions for the upbuilding of this section, as well as state and national affairs.

Chris Kronberg was born in North Schleswig, Germany, January 18, 1862 the son of A Kronberg, a native of Denmark, and Lena (Andersen) Kronberg, who was born in Germany. The father was an innkeeper in the old country where he and his wife passed their lives. They had three children Georgia, who died in Germany, Martin located in Sidney, Nebeaska, and Chris. The brother is now dead. As a youth the boy received an excellent education in the public schools of Germany which are supervised by the government but he saw little future for a man without money in the old country and determined that he would go to America and in the new country secure a foothold from which to climb the ladder of fortune, and he set sail for the United States arriving in 1882. He had little knowledge of language, conditions or methods however he was quick to familiarize himself with both the tongue and customs of his adopted country. Soon after landing on our shores he came west as the idea of every man from European countries is to possess land, but as he had little money he began punching cows for a cattle

outfit near Ogallala, Nebraska. Afterward he removed to Sidney and still with his original determination in mind, to Scottsbluff county in 1888, where he preempted one hundred and sixty acres of land, proved up on it, made some improvements, and then was able to dispose of it to advantage. He then came to his present place, section 5 township 22-55, where he bought a hundred and sixty acre tract, homesteaded twenty-seven additional acres and on this land developed a fine farm, and there established a home.

In 1888 Mr Kronberg married Miss Betty Smith, a native of Illinois, and to this happy couple were born ten children. Mary, the wife of Roy Konkle, lives on a Scottsbluff farm, Bertha, the wife of Sam Perkins, lives on a farm north of Mitchell, Charles has recently returned home after thirteen months service in France, during which time he took part in some of the most important battles of the war and won promotion to the rank of sergeant, having been a member of the One Hundred and Sixteenth Machine Gun Battalion, William is still in France at this writing, being a member of the One Hundred and Ninth Engineers, Jesse was in the army, but was discharged for disability, Sophie, Roy, Grace, Ruth, and Gladys are still members of the family circle. Mr Kronberg is an active member of the Modern Woodmen, and in politics votes independently, believing the best man should be elected to office in local affairs regardless of party lines. Mr Kronberg served one term as assessor and two years as deputy sheriff. He helped organize the company that built the Enterprise ditch and has assisted in the management of same for more than thirty years. His service for the county was highly satisfactory and a host of friends are proud of the record he made while in office. Mr Kronberg believes that a public official owes a real duty to the people who elect him and he did his best to demonstrate in a practical way the ideas he advocates to the satisfaction of his adherents and his own conscience.

HARLIN I BROWN, M D chiropractor, has been a resident of Scottsbluff since the fall of 1911 and has built up a large and lucrative practice here and has an established reputation all through the valley for unusual success. Dr Brown was born April 24, 1873, at Canton, Missouri.

The parents of Dr Brown were Abner D and Matilda (Mullen) Brown, the former of whom was born at Indianapolis Indiana and died in Custer county, Nebraska, in 1910, when aged fifty-six years. The latter was born at Streator, Illinois, where their marriage took place, and she now lives in Custer county. Of the family of ten children Dr. Brown is the second in the list of eight survivors, the others being: C. L., a farmer in Arkansas; L. A., a chiropractic at Kearney; Stella, the wife of William Halliday, a farmer in Montana; F. C., a farmer in Custer county; Earl W., also a substantial farmer in Custer county; Lila, who is the wife of William Phifer, who is in the draying business at Arnold, Nebraska; and Oma, who lives at home. The father came to Custer county, Nebraska, in 1881 and homesteaded, later becoming active in Republican political circles and serving as a county commissioner. He belonged to the Christian church and was both a Mason and an Odd Fellow..

Harlin J. Brown attended the public schools in Custer county. He was graduated in his school of medicine at Universal college, Davenport, Iowa, in 1910, immediately afterward beginning practice at Calloway, Nebraska, but in November, 1911, established himself at Scottsbluff. He has some remarkable cures to his credit and his patients come from all ranks in life.

In 1897 Dr. Brown was united in marriage to Miss Elizabeth Holliday, a daughter of C. T. Holliday, an early settler in Custer county. Dr. and Mrs. Brown have two children, Fay S. and Fonda, the latter of whom, now twelve years old, is yet in school. The former entered the aviation department of the National army, March 4, 1918, and was in training for thirteen months at Ebertsfield, Arkansas, making many flights. He was honorably discharged and reached home in April, 1919. Dr. Brown is one of the city's sterling citizens but is not active in politics, his profession making such heavy demands that added official service, if his desires were in that direction, would be almost impossible. For many years he has been an Odd Fellow, passed through all the chairs at Arnold, Nebraska, then entered the Encampment, and has served two terms as district deputy grand master.

ROBERT E. GILLETTE, who operates a first class blacksmith and carriage shop at Scottsbluff, does a large business because the public has learned that he is a competent workman and reliable business man. He has been a resident of Scottsbluff since the spring of 1911 and is numbered with the town's useful and representative citizens.

Robert E. Gillette was born in the southern part of Wisconsin, May 4, 1869, and is a son of Hamilton and Margaret (Downs) Gillette, the latter of whom was born in Ireland and the former in New York. He was twenty-five years old when he located in Wisconsin, where he married, and some years afterward moved to Gage county, Nebraska. He was a carriage-maker by trade and worked at the same in New York, Wisconsin and Nebraska, conducting his own shops at Beatrice and Adams in Gage county, his death occurring at Adams. Of his six children the following survive: Elizabeth, the wife of John Frederick, a retired farmer of Beatrice; Emily, who resides at Adams; Robert E., who lives at Scottsbluff; and Minnie, the wife of Harry Smith, a farmer in Michigan. The father of Mr. Gillette was a Republican in politics and he belonged to the Masonic fraternity.

Robert E. Gillette attended the public schools at Adams, Nebraska, after which he worked as a farmer until he was twenty-two years old, at which time he learned the blacksmith trade. He conducted his own shop at Adams until he was burned out, in 1910, and in the spring of the following year came to Scottsbluff. Here he has a good business location with modern tools and equipments, and has no fault to find with the large volume of business coming his way.

In 1899 Mr. Gillette was united in marriage with Miss Sadie E. Annabell, who was born near Adams, Gage county, Nebraska, and they have one daughter, Gladys. Mr. Gillette and his family belong to the Methodist Episcopal church. In politics he is a Republican.

JESSE C. COOMES. — There are many lines of business carried on in every modern community that are rightly deemed important but, considering the relation that meat products bear to the sustaining of life, it would seem that the meat industry in all its branches, is among the foremost of all. A leading butcher and meat dealer at Scottsbluff is found in Jesse C. Coomes, who, in a short time here has built up a fine business.

Jesse C. Coomes was born in Illinois, February 27, 1884, and is a son of John W. and Sarah (McDonald) Coomes, the latter of whom was born in Illinois and the former in Iowa, in which state they were married. In 1892 they came to Nebraska and the father bought a farm nead Wood River, in Hall county. The mother died there but the father survives and now lives retired. Of the family of five children Jesse C. is the second of three survivors, his two sisters being as follows: Pearl, the wife of William Mankin, a hardware dealer at Glisco, Nebraska, and Edna, the wife of John Mankin, a merchant at Oshkosh, Nebraska. The parents were members of the Christian church. In politics the father is a Democrat, and he belongs to the Knights of Pythias

In the excellent schools at Wood River, and later in a military school at Kearney, Jesse C Coomes was prepared educationally for the future. After his graduation at Kearney in 1904, he went to Green River, Wyoming, where he learned the butcher's trade and remained there until 1912, when he came to Mitchell, Nebraska, and worked in the butcher shop of Harry Naylor until February 1, 1919, when he came to Scottsbluff. Here, in partnership with Mr Naylor, he bought the shop of Charles Deulen. The firm, although a comparatively new one in this city, is doing well. Both partners being experienced in the business, they are able to offer the best meat products, carefully selected and prepared, have commodious quarters and do business according to honorable methods

Mr Coomes was married in 1905 to Miss Ida Mansfield, who was born at Salt Lake Utah, and they have one daughter, Anna who is attending school. Mr Coomes is a Democrat in politics and an intelligent, enterprising man in business

WILBUR J IRELAND, who is prominent in the grocery trade in Scottsbluff county, interested in three cities in this line and manager for the firm of Ireland Bros at Scottsbluff, is widely known in this section both in business and public affairs

Wilbur J Ireland was born at Saling's Grove, Nebraska, October 1, 1872. His parents were George M and Mary E (Sexson) Ireland, the former of whom was born in West Virginia, and the latter in Iowa. They were married near Omaha, Nebraska. In 1878 they came to Furnas county, Nebraska, where the father homesteaded. In 1907 he removed to Mitchell, in Scottsbluff county, and his death occurred there November 18, 1915. The mother of Mr Ireland still resides at Mitchell. He grew up on the homestead and attended the country schools. Until 1911 he continued work as a farmer, then entered the employ of the Carr-Neff Lumber Company at Mitchell and remained so connected for five years. In April, 1916, with his brothers he established the grocery store at Scottsbluff and also a store at Gering, under the same firm name, and a third store at Mitchell which is operated under the firm style of Ireland & Cockle

These are all high class business houses and are conducted carefully and systematically

On May 28, 1902, Mr Ireland was united in marriage to Miss Lola Whitten, who was born in Michigan and is a daughter of Lorenzo D and Martha Whitten, who moved to Saline county, Nebraska, in 1882. The mother of Mrs Ireland died in 1884, and the father died December 9, 1914, residing at that time with Mr and Mrs Ireland. They have two children Raymond, born January 5, 1906, and Eunice, born December 5, 1909. Mr and Mrs Ireland are members of the Methodist Episcopal church. He belongs to the order of Odd Fellows. In politics he is a Republican and while a resident of Mitchell served on the town council. Mr Ireland has built up his fortune through his own efforts, in earlier years teaching school, farming, working as a section hand and with a threshing outfit, all of which reflects credit upon him and offers an example that might well be profitably emulated

RUBY P DORAN, who has business interests of importance at Scottsbluff and other points, is a native of Nebraska, born in Seward county, February 23, 1877. Mr Doran has been the builder of his own fortunes, circumstances making such a course necessary in his boyhood

The parents of Mr Doran were Barney W and Chrissie (Dobson) Doran, the latter of whom was born in County Leitrim, Ireland, and the former at Toronto, Canada, of Irish ancestry and of the Catholic faith. He was a college bred man, educated for the priesthood, but never was ordained. By trade he was a cabinetmaker. After his marriage in Canada, he came, early in the seventies, to Nebraska and homesteaded in Butler county. His death was the result of an accident at Sheridan, Wyoming. Of his seven children Ruby P was the third in order of birth, the others being William John Henry, an importer and broker in the coffee trade, and a wholesale coffee roaster, at Denver, Colorado, Ada May, an artist in china painting, resides at Omaha. Claude James, a stockman at Grand Island, Collins the fifth in order of birth, Fred employed in a shoe factory at St Louis, Missouri and Nellie the wife of Lewis Davis, a farmer near Valley Falls, Kansas. The mother of the above family belonged to the Presbyterian church

Ruby P Doran attended school at Ulysses, Nebraska. His business connections before the state of Nebraska became prohibition territory, were with the retail liquor trade. In 1915, he came to Scottsbluff and established himself in the bakery and confectionery business, erecting a substantial one-story brick building with dimensions of 25x100 feet. He has prospered greatly in this enterprise which has expanded to large proportions.

In June, 1908 Mr. Doran was united in marriage to Miss Myrtle Coleman, who was born at Ulysses, Nebraska, a daughter of George and Katie Coleman, residents of Ulysses, Mr. Coleman being a farmer. Mr. and Mrs. Doran have two sons, namely: Richard Peter, who was born June 25, 1914; and William Elmer, who was born in November, 1916. Mr. and Mrs. Doran are members of the Episcopal church. In politics Mr. Doran is independent but not indifferent, the best interests of his country being very dear to him. He is identified with the Masonic fraternity and as a prominent man is active in many worthy organizations.

ROBERT L. COSNER, who for many years was prominent in the dental profession, and was the first dental practitioner at Scottsbluff, belonged to one of the old families that had come from Illinois to Nebraska in pioneer days. Dr. Cosner was born in Illinois, April 12, 1869, and passed out of life at Scottsbluff, in the beautiful home he had just completed, December 30, 1917.

Dr. Cosner's parents were William and Rosetta (Epperson) Cosner, the former died at Clayton, Nebraska, but the latter now resides at Scottsbluff. With the death of her son Robert L., Mrs. Cosner has but three living children: Harry, in the real estate business at Malta, Montana; Mrs. Edith Patterson, a widow, who lives with her mother; and Mrs. Harry Johnson, who also resides at Scottsbluff. Mrs. William Cosner is a member of the Presbyterian church.

Robert L. Cosner attended school in Nebraska through boyhood and then entered the dental school of Northwestern University, Chicago, Illinois, from which he was graduated. For a while he practiced in Chicago, then in Wayne and Schuyler, Nebraska, and also in Montana, so when he came to Scottsbluff in January, 1919, it was as an experienced dental practitioner. He homesteaded in Scottsbluff county and his widow still owns the property. Dr. Cosner was skilled in his profession and built up a wide reputation and a large practice, of such extent that he was required to hire an assistant during the last three years of his life. He was a man of high personal charac-

ter, a member and liberal supporter of the Presbyterian church, a faithful Mason and Knight of Pythias, and an earnest, public-spirited citizen. He was liberal in his benefactions to charity and conscientious in his support of movements for the public good.

In June, 1914, Dr. Cosner was united in marriage with Miss Carrie Young, who was born in Scottsbluff county, a daughter of William and Mary (Schumacher) Young, people of importance and wide acquaintance in this section of Nebraska. Dr. and Mrs. Cosner had one daughter, Florence May, a most engaging child who is a great comfort to her bereaved mother. Mrs. Cosner is active in the Presbyterian church.

William Young, father of Mrs. Cosner, was born in Iowa and her mother was born in Wisconsin. They came to Nebraska and homesteaded in Scottsbluff county in 1885, Mrs. Young being the first woman to live in the Gering valley. Her two nearest neighbors were miles distant and even the smoke from their cabins could not be seen across the pathless prairie covered with red topped, swaying grass. The Youngs went through many harrowing pioneer experiences but bravely survived them all, reared and educated a fine family, and survive with vigor left to carefully and efficiently look after their numerous interests, Mrs. Young remaining for this purpose in Scottsbluff county, while Mr. Young is engaged in attending to a profitable fruit farm in Florida. Of their seven children the following survive: George, Leonard F., Mrs. Cosner, Ernest S., Minnie R. and Florence E. The eldest son, George, is in the lumber business at Marsland, Nebraska. Leonard F. is a civil engineer and consultant on construction work. He has worked on many of the irrigation projects in Nebraska, and for four years was on the Tri-State Ditch. For the past six years he has been associated with one of the largest concerns in New York, Sanderson & Porter, builders of some of the most extensive plants of all kinds in the world. Ernest S., who entered military training at the Presidio, California, sailed for France in September, 1917, and was on the Tuscania when it was torpedoed by the enemy. Formerly he was attached to an artillery division, but now is in a civilian division and is port commander at St. Denis, France. Minnie R., who has chosen the noble calling of a trained nurse, is in the Northwestern Hospital, Minneapolis, Minnesota. Florence E., is a bookkeeper and stenographer in the First National Bank of Scottsbluff. The mother of

MR. AND MRS. WALTER E. JENNINGS

the above family is a member of the Catholic church, but the father was reared a Lutheran They are widely known and universally esteemed

WALTER E JENNINGS was an infant pioneer of this great commonwealth who remembers that his first home here was a sod house half in and half out, but a rather good warm home at that and while the family prosperity was so great that he has no distinct recollections of that warm sod house, he believes it must have had an excellent influence upon his infant character for it must have played its part in making him the upstanding fearless, progressive citizen of today Before his eyes have passed the kaleidoscopic panorama of change that has worked silently but unceasingly since territorial days to change the silent rolling prairies, the "Great American Desert," as it was known for so many years, into a great agricultural state, one of the richest in the Union, now covered with thriving farms, populous towns and cities knit together with threads of steel He has watched from year to year and even today as his eyes travel across the wide fields of Scottsbluff, asks himself, 'Is it real ?" For today he is a prosperous, well-to-do farmer on land that even the Indians held of little value save for the wild game they killed upon it

Walter Jennings was born in Iowa in 1873, the son of William A and Mary E (Whipple) Jennings, the former born in Illinois in 1848, while the mother was of fine old New England ancestry, born in Connecticut in 1841 and died in 1916 The father was a farmer in Ohio who emigrated to Nebraska soon after the admission of this state to the Union He located in Valley county in 1873, took up a homestead on which he proved up and made some improvements, then disposed of it to profit He seems to have been a pioneer by nature and when settlements began to be marked, moved on to more virgin country After leaving Valley county the family made a home in Boone county on land purchased by the father, but the lure of the west was in his blood and before long they went to Midland Montana, but later returned to Scottsbluff county, where at last the goal of his desire was reached for he still resides on his farm in the vicinity of Mitchell, a hale, hearty old man of seventy-three years, who can recount many thrilling and interesting experiences of the early days in this state He is a Republican in politics The family were members of the Episcopal church Eight children constituted the younger members of the family Gustavus, a farmer near Mitchell,

James W, on a farm in Montana, Walter, Mary E, the wife of Oscar Collins, a farmer of Valley county, John Elbert a farmer in Boone county, Edward M is located on a farm near Bayard. Nebraska, Frederick, also on a farm near Bayard, and Charles, who has a farm not far from his two brothers there

Though he has not advanced far beyond the psalmist's span of three score years and ten and still possesses to the full amount his physical and mental vigor, Mr Jennings has the distinction of having lived in Nebraska nearly a half century, and it is gratifying to him to know that he has been able to play a part in the civic and industrial progress that have taken place since his parents first brought him here as an infant in arms He spent his boyhood days on his father's farm in Valley county, acquiring his early education in the public schools afforded in the new country at that period He made the various changes with the other members of the family in Nebraska, working during his youth for his father and later independently for himself His taste was for rural life and in 1906 he came to Scottsbluff to establish a permanent home He took up a homestead of eighty acres on which he proved up and at once engaged in general farming and stock-raising Mr Jennings has a high grade of stock on his farm and specializes in Duroc Jersey hogs He has a beautiful home, well built and kept farm buildings, and no better cultivated land is to be found in the Mitchell district Times have changed, but so has the subject of this sketch He is up-to-date in methods, buys the latest farm machinery, and thus today enjoys the well-earned fruits of a well-spent, profitable life, standing high in local circles for his honesty and kind-heartedness He now owns 240 acres He is a Republican in politics, while with the family he is a member of the Methodist Episcopal church

In 1896 was solemnized the marriage of Mr Jennings and Miss Ethel D Weare, the daughter of Burney and Sarah E (Coffin) Weare, who live in Mitchell Mrs Jennings is a woman of high intellectual attainments being a graduate of the high school at Ord, who for some years before her marriage taught school She is a charming, gracious woman who has aided her husband in every way to attain his present success and comfortable fortune The following children belong to the family Cecil May, the wife of Henry S Sullivan who was a soldier during the World War, being a member of a supply company in the Eighty-ninth Division, Geneva L the wife of Luther Stiver, who lives on a farm north of

Mitchell, and William, Walter, and Evelyn, all at home. Mr Jennings is enjoying a well-earned success though he is only a man of middle age, for in every relation of life he has measured up to the full standard of manhood and loyal citizenship

WILLIAM E KENT, who is president and general manager of the Scottsbluff Potato Growers' Association, is a well-known business man in several states other than Nebraska, for he long was an important factor in the lumber industry and was financially interested in all the numerous plants operated by the Walrath & Sherwood Lumber Company in Nebraska. He has long been recognized as an able, dependable business man, whose natural sagacity has been invaluable in the large enterprises in which he has engaged. Mr Kent however, had little assistance in building up reputation and fortune, early beginning to depend on his own efforts, and his entire career has been marked by persevering industry assisted by intelligent judgment. He was born in Portage county, Wisconsin, in 1860

The parents of Mr Kent were Edward L. and Sarah L. (McGuire) Kent, the latter of whom was born in Scotland in 1833, and died in 1917. The father of Mr Kent was born in England in 1830, and died in 1917. He came to Detroit, Michigan, and from there went to Wisconsin, where he was married in Milwaukee, in 1851. In Michigan he was a farmer and buyer of logs and in Wisconsin was in the lumber business. He served three years and three months in the Civil War as a member of the Nineteenth Wisconsin volunteer infantry, and suffered both capture and slight wounding. Of his five sons and two daughters, two sons, William E and Frank J, and two daughters, Jennie and Cora, are living. Frank J Kent is a wheat grower near Walla Walla, Washington. Jennie is the widow of James McIcroe, a large rancher and state trustee of prisons, and Cora is the wife of Frank Hammil and they own and live on the old family homestead. The parents were members of the Methodist Episcopal church. The father was active in the Republican party and prominent in the Order of Odd Fellows

William E Kent had high school advantages at Almond, Wisconsin. In 1879 he began working in the northern woods of Wisconsin, and for a number of years spent much time in the great timber country. After working seven months for a logging firm J J Kennedy & Co, of Spencer, Wisconsin, he used the money he earned to complete his education

In 1882 he came to Nebraska and homesteaded in Antelope county, but soon sold his claim and with his brother went to work with the construction gang on the Oregon Short Line Railroad, from March to September, 1882. He came then to Platte Center, Platte county, and for eleven years was manager of the Chicago Lumber Company of Omaha, retiring when the business was sold to Walrath & Sherwood. Subsequently, however, he became financially interested with this firm when they bought his plant at Monroe, and was in business at North Bend as a member of the firm of the Walworth, Sherwood & Kent Lumber Company, acquiring interests in every plant operated by the firm. He became auditor of the company and handled all the business in Nebraska, North and South Dakota and Iowa. He founded the Platte Valley Cement & Tile Company of Fremont, Nebraska, and was president of that concern until 1916, when he sold his lumber and other interests and moved to Sioux county. He owns a quarter section of irrigated land and lived in Sioux county on his farm until the spring of 1919, when he came to Scottsbluff to assume the duties of president and general manager of the Scottsbluff Potato Growers' Association. He has greatly improved the business outlook of this organization, which is a mutual body that expects to have warehouses erected in a dozen towns throughout the valley

On December 16, 1886, Mr Kent was united in marriage to Miss Anna Bucknell, of Waupaca, Wisconsin, and they have two children. Pearl, who is the wife of Fred Young, a farmer near Mitchell and they have two children, William Andrew and Andrew Kent, and John Edward, who married Hester Collins, of Dodge county, Nebraska, has one child, Helen Marie, and is with the Union Pacific Railroad. Mr Kent is a member of the Federated church at Mitchell. He is a Republican in politics and is a Consistory Mason

WILLIAM E CALHOUN, who is proprietor of the Star Moving Picture house at Scottsbluff, has been identified with this industry since 1913, and through excellent judgment and careful management, provides much enjoyable entertainment to his patrons. Mr Calhoun was born in Adair county, Iowa, in 1881

The parents of Mr Calhoun, William and Margaret (Emmons) Calhoun, were born in Pennsylvania and accompanied their parents early to Iowa. The father was a farmer in that state, near Greenfield in Adair county, but retired from active life in 1900, when he moved to Nebraska. In politics he is a Democrat and both he and wife belong to the Methodist Episcopal church. Of their eight children, William E. is the youngest of the survivors, the others being Jennie, widow of Samuel Miller, lives in Idaho; Myrtle, the wife of C. T. Jackman, a real estate dealer in Idaho; Hattie, the wife of James Pence, a railroad master mechanic, at Deadwood, South Dakota; Frank, in the furniture business at Cambridge, Nebraska, where the parents yet reside.

William E. Calhoun obtained his education in the public schools, had some farm experience and then learned the carpenter trade and after coming to Scottsbluff in 1910 was engaged as a carpenter and contractor until 1913, when he became interested in his present enterprise. The Star, in size and equipment, compares favorably with like places of entertainment in other cities, and there is much evidence to show that Mr. Calhoun's efforts are appreciated.

In 1904 Mr. Calhoun was united in marriage to Miss Maude Allen, who was born in Nebraska. He takes no very active part in politics, voting independently, but is very much interested in the further development of Scottsbluff and the welfare of its people, for here he has been able to lay the foundation of what promises to be an ample fortune.

SANFORD STARK, a member of Scottsbluff's retired colony, and for years a prominent citizen, belongs to an old New England family of military distinction and of Scotch descent. The records of this family in Connecticut date back to 1658. Mr. Stark was born in New London county, Connecticut, December 3, 1849, the son of Henry S. and Mary E. (Rathbun) Stark, who spent their entire lives in Connecticut. The father was born in 1822 and died in 1857, the mother, born in 1826, died in 1909. They had four children, Elizabeth, Charles R. and Sanford yet surviving. Elizabeth is the widow of John F. Randall, who left Yale college to enter the Union army in the Civil War in which he served as a commissioned officer and afterward was prominent in the insurance field at St. Louis. Charles R., has been treasurer of the Rhode Island Horse Shoe Company for many years. His son, Charles R., Jr. has just returned home from honorable service in the World War. The parents of the above family were members of the Baptist church. The father followed the sea all his life, was captain of many vessels and was widely known in seafaring circles. His parents were Sanford and Nancy (Park) Stark, of Connecticut, where they lived and died, the former serving a short time during the War of 1812, and his ancestors were members of the Colonial army under General George Washington, and thus their names occur in the history of Revolutionary days. Elisha Rathbun, the maternal grandfather of Sanford Stark of Scottsbluff married into the old Connecticut family of Parker. Both he and wife lived to advanced old age as did the paternal grandparents. Grandmother Stark being ninety-six years old at the time of her death.

Sanford Stark was educated in an academy at Mystic, Connecticut, the Civil War breaking into his academic studies, however. On account of his father being a seafaring man, ships were familiar and interesting to him in boyhood, and during the last year of the war he succeeded in being the captain's helper on a supply vessel running to Key West and Pensacola. He returned then to his studies and afterward became a clerk in a store, but the sea called him once more and he took passage on a vessel from New York to San Francisco by way of Cape Horn, and from the western city sailed for Europe and by the time he reached New York again, thirteen months had elapsed. He recalls that experience with pleasurable emotions but his life since then has been passed on land. Business affairs have mainly engaged his attention and prior to coming to Scottsbluff, in August, 1909, he was cashier for the Great Western Sugar factory, at Longmont, Boulder county, Colorado. When the company began the construction of its plant at Scottsbluff, Mr. Stark was transferred to this city and continued as cashier until he resigned in November, 1918, at which time he retired from business. He has continued an active citizen, however, and during the late war assisted very materially in the war loan drives and the Red Cross work. Mr. Stark is well and favorably known at Denver where, from 1879 to 1893, he conducted a wholesale boot and shoe business.

On November 4, 1872 Mr. Stark married Miss Lucy Latham Ransom, who was born at New London, Connecticut a daughter of Nathaniel and Catherine (Latham) Ransom, lifelong residents of that state. Mrs. Stark's father left a prosperous lumber business to become a soldier in the Union army during the Civil War, as a member of the Twenty-first Connecticut infantry. He suffered wounds that required hospital care. Mrs. Stark has one sister, Kittie, the wife of Edwin H Tift, a lumber merchant of Boston, Massachusetts. Mr and Mrs Stark have the following children: Catherine, the wife of A K Sage, proprietor of a large plumbing and steamfitting plant in Brooklyn, New York, Harry S, vice-president of the First National bank of Scottsbluff, Frederick B, a farmer near Scottsbluff, and Helen, who married J B Badgley, a bookkeeper with the sugar factory in this city

Mr Stark and his family belong to the Baptist church   In politics he is identified with the Republican party   He is a member of the Sons of the American Revolution

JOHN F RAYMOND, for many years profitably interested in agricultural pursuits and still owning valuable farm properties, came to Scottsbluff in 1901, but has been a resident of Nebraska for more than forty years   Of New England birth and ancestry, he possesses many characteristics that have made that section notable, business foresight being included

John F Raymond was born at Hartford, Connecticut, in 1852, and is a son of Josiah and Fannie A (Hurlbut) Raymond   His father was born in Connecticut in 1815, a son of Joshua Raymond, who spent his life in that state   Josiah Raymond was a man of brilliant parts, a prominent lawyer at Hartford and also a farmer near that city, and for some years served in the state legislature   He died in Connecticut in 1862   He was married there to Fannie A Hurlbut, who was born in the same house as was Noah Webster, the lexicographer, in which house her father, Samuel Hurlbut, died   She came to Otoe county, Nebraska, with her family, in 1879, bought railroad land and died in 1889   Of the family of seven children, the survivors are as follows   Robert O, a farmer near Gurley! John F, an esteemed resident of Scottsbluff, Fannie E, who lives at Scottsbluff, Charlotte H., who also resides in this city, and Henry J, a farmer in Cheyenne county   Both parents were members of the Presbyterian church

John F Raymond was educated in his native city and as a young man came to Nebraska in 1878 and bought land in Otoe county, removing in 1885 to Cheyenne county, where he pre-empted land on which he continued to live for many years   He engaged in general farming and raised a large amount of stock, becoming a well-known shipper   Mr Raymond was active in his farm industries until he came to Scottsbluff county and retains full ownership of his land, which is some of the finest in Cheyenne county, but his investment in a tree claim on the edge of Scottsbluff he subsequently sold to the sugar company of this city for $28,000   He owns considerable realty in the city that he has under favorable rental

In November, 1914, Mr. Raymond was united in marriage with Mrs Adelaide During, who was born at Milton, Illinois, a daughter of Charles and Mary (Davis) Chaplin   Mrs Raymond's mother is deceased, but the father survives and resides at Pittsfield, Illinois   He is a veteran of the Civil War, having been wounded in the service of his country   Of his eight children, there are but three survivors, Mrs Raymond and her two sisters Mrs Charles Johnson, of Pittsfield, and Miss Nellie Chaplin, who resides with her father   Mr and Mrs Raymond are members of the Presbyterian church   He has never had any political ambitions, but, like his father before him, has always believed in the sound principles upon which the Republican party was founded and has supported this organization

REV FRANK A WOTEN, pastor of the Christian church at Gering, Nebraska, is probably as well-known as any citizen of Scottsbluff county   He is a young man of versatile gifts, of sound philosophy and vigorous personality   While in no sense a crusader, he carries his religion into the most practical things of life, through example as well as precept, proving the saving grace that follows honorable industry and strict adherence to the principles of law and justice   He is a native of Nebraska, born in Gage county, December 5, 1883

The parents of Dr Woten were William I and Susan (Swaner) Woten, the former of whom was born in Jay county, Indiana, December 5, 1857, and the latter April 4, 1856   The mother died in January, 1917, but the father still resides on his Gage county homestead which he secured in 1881   Of his family of nine children, Frank A was the second in order of birth, and four others survive Claude, who lives at Fresno, California, is a National bank examiner, Goldie, who resides with her father, Sylvia, who also lives at home, and Grace, the wife of Howard Hall, a farmer near Wellfleet, in Lincoln county, Nebraska   The father has followed agricultural pursuits all his life   In politics he is affiliated with the Democratic party   The Christian church holds his membership

Frank A Woten grew up on the family

FRANK L. FOREMAN AND FAMILY

homestead near Adams, in Gage county, in 1903 being graduated from the Adams high school In 1911 he was graduated from Cotner University with the degree of A B, later took special work in astronomy at the State University of Missouri, and completed his theological course at Cotner His first ministerial charge was Palmer, Nebraska, where he remained two years, then went to Alliance and during his term of two years there built up the congregation and erected the first stucco church edifice in Western Nebraska He then came to Scottsbluff as pastor of the First Christian church, which charge he subsequently resigned and went to the southern part of Sioux county, where he took up a homestead, and while proving up, supplied the church at Gering, and accepted the regular pastorate of this church in the fall of 1918 As a minister Reverend Woten exemplifies his Christian faith in every possible way, but he is a liberal-minded man and a strong advocate of practical Christianity The needs of his congregation spiritually are well looked after without encroaching too much on his time, and he gives attention to a transportation business and operates an omnibus line between Gering and Scottsbluff, which carries the mail between the two points

In 1912 Mr Woten was united in marriage to Miss Lena Colborn, who was born at Palmer, Merrick county, Nebraska They have three little daughters, namely Arlene, Pauline and Frances, their ages ranging from five to two years Mr Woten is a Democrat in his political opinions He belongs to the order of Odd Fellows, and has served as chaplain of the local lodge, and has also served in the highest office of the local organization W O W The Woten name is of German origin, but the ancestors of Reverend Woten have belonged to Great Britain since the Fifteenth century

FRANK L FOREMAN who is one of the substantial and representative farmers of the Mitchell valley, has been a resident of this great commonwealth for more than three decades, so that his personal experience covers virtually the entire period marking the development and progress of this now favored section of Nebraska He is a man born to the soil who deserted it, but with the passing years found no satisfaction in the turmoil of cosmopolitan life and returned to a farm where kind mother earth has given him a bountiful reward for his labors

Frank Foreman was born in McDonough county, Illinois, March 5, 1868, the son of James and Hettie (Lamb) Foreman The father was a native of the Buckeye state, born in Bellmont county, in 1838, who died in Gering at the age of sixty-eight years Hettie Lamb was born in Cincinnati, Ohio, in 1848 and passed away in Nebraska in 1916, a woman of great honor and warm heart There were seven children in the Foreman family, of whom Frank was the eldest, the others were William, a freighter at Thermopolis Wyoming, Elmer, a freighter at Big Trail, Wyoming, Zella, the wife of Leonard Early, lives on the old homestead in Scottsbluff county, Bessie, the deceased wife of Thomas Bracken; one who died in infancy, and Charles, the second boy, who died in Missouri

James Foreman was a barber by trade but a farmer by vocation He also had the honor of being a member of the Union army during the Civil War, enlisting in Illinois under Colonel Bob Ingersoll in the Eleventh Illinois Cavalry He served four years and two months during that memorable conflict, taking part in many of the hardest fought battles of the war Twice his mount was shot from under him but he lived to return home after the close of hostilities

Like so many men who had been in the army, Mr Foreman was not contented with the conditions he had known before his service and determined to avail himself of the opportunities afforded farther west With his family he came to Cheyenne county, Nebraska, in 1886, when that country was still unbroken prairie, took up a homestead, proved up on it, established a home, made good improvements on his farm and there engaged in general agriculture and stock-raising for a number of years Later he retired and located in Gering where he took an active part in communal affairs He was a member of the Grand Army of the Republic, helping the other members in the direction of the affairs of the local post, while in politics he was a staunch supporter of the Democratic party The family were members of the Christian church

Frank Foreman received his educational advantages in the public schools of Illinois and as usual with a boy on a farm assisted with such work as his years and strength permitted during the vacations He grew to manhood sturdy, resourceful and self-reliant, all qualities which stood him in good stead when he accompanied the family to the new settlement in Cheyenne county, where he also took up a homestead on which he proved up, made many improvements necessary in a frontier community, established a home and soon was engaged in general farming and stock-raising He knew and overcame many of the hardships and trials of a frontiersman, such as drought, insect pests and lack of adequate machinery for agricultural work, but none daunted him. After some years Mr. Foreman was able to dispose of his property to great advantage and left the farm to locate in Gering, where he opened and operated a barber shop, but the call of the land was in his blood and he responded by purchasing a farm in section 28-23-56, where he is the proprietor of forty acres of highly cultivated irrigated land on which he raises beets and conducts a general truck farm. The country looks very different today with its green cover than did the prairies when Mr. Foreman first located in the state, and he often speaks of the great happiness that comes to the farmer today with his insured crop no matter what the weather conditions may be.

Mr. Moreman has ever been a man of active mind, he takes interest in all questions of the day entering actively into the civic life of the community and had the honor of being a delegate to the first county convention which located the seat of justice of Scottsbluff county at Gering. Independent all his life, it is but natural that this man should be independent in politics and he draws no party line in casting his vote directing his influence to the best man. Fraternally he is connected with the Independent Order of Odd Fellows, the Rebeccas, and the Woodmen of the World. There are seven children in the happy Foreman family: Glen, who has a ranch in Sioux county and a farm in the Scottsbluff locality; Loren, a teamster in Mitchell; Ray, also living in Mitchell; Zeta, Gwelda, and Wayne all at home. The family are members of the Methodist church.

Mr. Foreman married at Hull, Nebraska, Grace Beck, who was born in Indiana, but reared in Nebraska from the age of ten years.

HENRY A. SCHMODE, who, as superintendent of the plant of the Great Western Beet Sugar Company at Scottsbluff, fills a position of responsibility as he should, for he is a highly trained man in this particular industry in which he has had much practical experience.

Henry A. Schmode was born in Silesia, Germany, in 1870. His parents were Constantine and Ernestina (Bleich) Schmode, natives of Posen, Germany. Of their eight children five survive, but only two of these live in the United States, Henry A. and Frederick, the latter being a machinist at Denver, Colorado. The father owned a woolen factory and employed fifty-five men in producing broadcloth. The parents were members of the Lutheran church.

After being graduated from the high school in his home town, Henry A. Schmode served two years in the German army, as was the law. In 1893 he came to the United States and located at Norfolk, Nebraska, and there started the first Steffin process in the United States, and was superintendent of construction for one year, then went to California and in the following year started the second Steffin process plant in the United States, at Chino, in San Bernardino county. He remained there five years with the American Beet Sugar Company. From there he came to Grand Island, Nebraska, as master mechanic of the factory of the same firm and continued two years, then went to Ames, Nebraska, where he was associated with the Standard Beet Sugar Company for one year as master mechanic and for six years as superintendent of the factory. Mr. Schmode's services were then secured by the Great Western Sugar Company and he was so connected at Fort Collins for six months, then was master mechanic of the factory at Windsor, Colorado, from which plant he came to Scottsbluff and after superintending the construction of the Great Western's plant here, became superintendent and has continued his efficient service in that capacity ever since.

At Norfolk, Nebraska, in 1897, Mr. Schmode was united in marriage to Hulda Mittelstadt, and they have three children, namely: Mart C., who is employed in the sugar factory; Edwin H. and Dorothy Irma, both of whom are attending school.

Mr. Schmode and his family are members of the Presbyterian church, of which he has been a trustee for the past five years and during 1918 was president of the board. He is a Scottish Rite Mason and both he and wife belong to the Eastern Star. In addition to his scientific knowledge and executive efficiency, Mr. Schmode is a broad-minded, intelligent man and a valuable citizen. He gives his political support to the Republican party.

FRED ANSEN. — There were many residents of Scottsbluff county as well as other sections in this state and in Colorado, who knew, respected and esteemed the late Fred Ansen, whose family is a valued one in the county. He was an honest, upright, industrious man, fulfilling every duty of life to the best of his ability, injuring none and helping many.

Fred Ansen was born in Alsace-Lorraine, France, August 15, 1856. In 1881 he came to the United States and for a year worked in one of the big packing plants in Chicago, Illinois, but he was not satisfied there as he had come to America in the hope of owning a farm This hope he fulfilled in 1882 by locating in Buffalo county, Nebraska, where he secured a tract of land on which he lived three years He then had a chance to sell it to advantage and went to Colorado, but on the way through Cheyenne county, Nebraska, stopped long enough to take up a homestead He continued on his trip to Colorado where he was employed as a cook in mining camps until 1887, then returned to settle on his homestead and remained seven years In the meanwhile he bought a farm in Mitchell valley, Nebraska, and moved there in 1894, and that farm remained the family home until 1905, when he came to Scottsbluff Here Mrs Ansen and the children remained while he once more returned to work in Colorado His death occurred March 27, 1909 He was a Republican in politics, and both he and wife were members of the Lutheran church

In 1885 Fred Ansen was united in marriage to Miss Theresa Siebke, who was born in Germany, a daughter of John and Caroline (Haase) Siebke, who spent their lives in Germany Mrs Ansen came to the United States in 1881 and was married in Buffalo county, Nebraska All of their six children are living Mable, Margaret, Charles, Maude and May, twins, and Gladys Maude is the wife of Steward Rice, a farmer in Scottsbluff county The other daughters reside with their mother at Scottsbluff Charles Ansen the only son, entered military service in September, 1917, and crossed to Europe as a member of the American Expeditionary Force and is with the Army of Occupation yet in Germany He is a young man of fine qualities, and has done his duty as a soldier

WILLIAM S CLINE, one of the retired residents of Scottsbluff, has been one of the substantial farmers and highly esteemed citizens of Scottsbluff county for many years When he came first to this section he made wise investments and now owns some of the best farm land in the upper valley He has taken active and useful part in all civic and commercial movements for the betterment of the county, since making the Panhandle his home

William S Cline was born in Hendricks county, Indiana, February 27 1857 the eldest of a family of six children born to John F and Mary Jane (Goben) Cline Both parents were born in Indiana from which state they moved to Clark county, Iowa, in 1866, where

the father bought a farm on which he and his wife passed the remainder of their days They were members of the Christian church, most worthy people in every relation of life William S Cline has two brothers and two sisters Francis Marion, lives retired at Scottsbluff, Jesse Bennett, is a farmer in Iowa, Margaret, is the wife of Andrew Adams, a farmer in Iowa, and Rose, is the wife of Perl King, an Iowa farmer

William S Cline attended the public schools and was reared to farm pursuits In 1905 he came to Scottsbluff county and purchased one hundred and sixty acres of land east of the town and eighty acres where the sugar factory now stands He owns eighty acres of fine irrigated land north of Scottsbluff that would command a high price in the market if it were for sale It may not be out of place to say that when Mr Cline first came to Nebraska he was practically without capital, but he had the good judgment to take advantage of the opportunities offered here to a young man of energy and industry, with the result that by the time he had reached middle age he was able to retire with a competency

On November 11, 1883, Mr Cline married Miss Margaret Bevins, who was born in Iowa, a daughter of Asher and Anna Bevins, the former born in Highland county, Ohio, and the latter in Delaware They died in Iowa Mr and Mrs Cline have the following children Rose, the wife of C C Terhune, a railroad man at Omaha, Nellie, the wife of Otis Simmons, a carpenter at Scottsbluff, Millie, the wife of Charles Bisel, of Scottsbluff, Walter, is a farmer but lives in Scottsbluff, Homer, operates a garage in Scottsbluff, and Lola the wife of George Brown, who is employed in the sugar factory here Mr Cline and his family are members of the Presbyterian church He is a Democrat in politics and has always been a loyal party man but has not been willing to accept public office Since 1910 he has been a resident of Scottsbluff, a welcome addition to the town's most reputable citizenship

EDWARD C DUNHAM, who is a prominent citizen of Scottsbluff county and for many years active in the agricultural field, since 1917 when he retired and moved into Scottsbluff has given much attention to the Farmers Mutual Fire Insurance Company, of which he is president Mr Dunham was born at St Louis Missouri, December 13, 1856

The parents of Mr Dunham were Cornelius L and Mary (Buswell) Dunham, the former of whom was born at New Haven,

Connecticut, and the latter in the state of Vermont He was a student at Jacksonville, Illinois, and she, a graduate of Knox College, was a teacher in the School for the Blind, when they met and they were married in that city Later they moved to St Louis, Missouri, where the father of Mr Dunham taught school for four years, then moved to Bureau county, Illinois, where he engaged in farming until 1862 In that year he entered the Union army for service in the Civil War as a member of Company H Ninety-third Illinois Volunteer Infantry His service covered three and a half years, during which he was promoted to a sergeancy He participated in the siege of Vicksburg and in the campaign around Chattanooga and Memphis, Tennessee After the war closed he returned to his Illinois farm and lived there until 1876 when he removed to Grinnell, Iowa He accumulated a competency and later he and wife gave themselves the pleasure of extensive travel They were members of the Congregational church In his political views the father believed the principles of the Republican party the most advantageous for the country and supported this organization until the end of his life

Of the seven children born to Cornelius L Dunham and his wife, Edward C was the eldest, the others being Ida G, the wife of S H Blackwell, a farmer near Longmont, Colorado, Cornelius L, an orange grower in Florida, Ralph W, a farmer in southwestern Missouri, E H, a farmer near Grinnell, Iowa, Alice C, the wife of J. R Hannay, a farmer near Grinnell, and Mary Cornelia, who still lives in the old home at Grinnell, Iowa

Edward C Dunham was educated in Iowa and carefully trained by his intellectual father From 1876 until 1896 he had entire charge of his father's farm and afterward bought a part of the old home place In 1896 he moved to Arkansas and was interested there until 18 . in growing strawberries and apples, after which he came to Scottsbluff county and bought land in Pleasant Valley In 1917 the family left the farm and came into the city, where they have a wide social circle and Mr Dunham gives time, as indicated above, to the affairs of the Farmers Mutual Fire Insurance Company, that does business in five counties in the western part of the state He is also city assessor Mr Dunham located in this county on a government unit as did his daughter, so the family really owns two units The farm on which the family lived is now devoted to grain and sugar beets and Mr Dunham also owns 173 acres, all of irrigated land, and an additional tract near the city limits

In 1881 Mr Dunham was united in marriage to Miss Hannah M Mann, who was born in Oswego county, New York, a daughter of John H and Susan (Willis) Mann, natives of New York, who moved to a farm in Iowa in 1871, later to Iowa City for three years and still later to Grinnell Mrs Dunham was the fourth born in a family of six children, five of whom are living Mr and Mrs Dunham have two sons and one daughter, namely Dwight Mann, who homesteaded in South Dakota, Robert E, who homesteaded in Nebraska, has been rural mail carrier out of Scottsbluff for seven years, and Mary Florence, who lives at home Mr Dunham and his family belong to the Presbyterian church At one time while living in Iowa, Mr Dunham was quite active in the Populist party but now he casts an independent vote He is a broad-minded, thoughtful man and has always had the best interests of his country at heart

WILLIAM W EMICK, who is secretary and treasurer of the Farmers Mutual Fire Insurance Company, has been identified with this important business enterprise since January, 1915, since which time he has devoted himself largely to its concerns, although he has many additional personal interests He came to Nebraska in early manhood, invested in land in Scottsbluff county in 1909, and has been a resident of Scottsbluff since 1918

William W Emick was born in Wayne county, Ohio, April 22, 1872, and is a son of Adam and Catherine (Sweigert) Emick, the former of whom was born in Germany and the latter in Pennsylvania Adam Emick was brought to the United States when two years old, was reared in Ohio and married there Of his twelve children seven are living, two of these being in Nebraska, William W and Charles, the latter of whom is a merchant at Creighton Adam Emick was a hard-working man, was a farmer and a carpenter, whereby he accumulated a competency After his first wife died he married Alice Okehauf and they had three children, two of whom live in Ohio and one at Chadron Nebraska In politics Adam Emick was a Democrat, and all his life he was a member of the German Lutheran church He lived to be ninety-four years old

William W Emick obtained his education in the public schools in his native state and remained assisting his father on the farm until he was twenty years of age He then came to Knox county, Nebraska, and for three years was a clerk in a store at Bazile Mills and afterward at other points, also traveled as a

ALVA A. SMITH AND FAMILY

canvasser and commercial agent For nine years he lived at Deadwood, South Dakota, during which time he was engaged as collector, clerk and general manager of stores there and at Lead City For one year he was on the road selling groceries for the firm of Shankberg, of Sioux City, through the Big Horn basin, when a railroad wreck in which he was a victim, kept him off the road for some time He then accepted his old position at Lead City, following which he became traveling representative of Raymond Bros & Clark, through Western Nebraska, for five years Mr Emick then took charge of his brother's store in South Dakota, for some eight months In 1909 he came to Scottsbluff county and bought an irrigated farm, on which he lived until 1918, when he sold that property and moved into Scottsbluff, where he has valuable realty He also owns a farm near the city

In 1911 Mr Emick was united in marriage to Miss Myrtle Fry, who was born in Fall River county, South Dakota She died in 1914 leaving one infant daughter, Myrtle Josephine In the fall of 1917 Mr Emick was married to Miss Julia Coony who was born in Custer county, Nebraska, and they have one daughter, Willemetia Mrs Emick is a member of the Christian church In politics he is a Democrat and fraternally belongs to the order of Elks In addition to being secretary and treasurer of the Farmers Mutual Fire Insurance Company, Mr Emick has been general manager of the work in Scottsbluff, Banner, Morrill and Sioux counties, the company's charter covering eleven counties in the Panhandle A man of wide and varied experience, Mr Emick is particularly well qualified for the responsible position he fills so well in the business world

ALVA A SMITH — It has often been said that Smith is a name hard to distinguish yet it remains that the possessor of the name at the head of this review has succeeded at least in a modest way, in distinguishing his cognomen in the realm of ordinary citizenship and practical, profitable farming This is an ordinary story that has been duplicated perhaps a thousand times in western Nebraska but it ever becomes interesting when narrowed down to an individual whose achievements are worthy of being published to the world Mr Smith is one of the homesteaders of Scottsbluff county who passed through many privations and hardships, courageously persevering in the face of discouraging situations, overcoming seemingly insurmountable obstacles, and eventually winning a way to well deserved success

This man is a Wolverine by birth, born in McComb county, Michigan, February 12, 1862, amidst the throes of our great Civil War and it may be that some of the dogged determination that was imbued in the citizens of the north to preserve the Union at all cost entered into his mentality for nothing has daunted his spirit He is the son of Andrew and Esther (Arnold) Smith, the former born in the Empire state in 1838 died in Michigan in 1899, while the mother like her son, was a native of Michigan, born there in 1841, who lived until 1896 The father was a successful Michigan farmer who reared his family in great comfort, giving them all the advantages afforded in their community excellent educations and such practical knowledge as could be attained under his careful guidance on the farm during the vacations and after leaving school

There were eight children in the family Iowa, who died at eighteen years of age Alva, Alma his twin who became the wife of Joe Burgess, lived on a farm in the state of Michigan, and later removed to Gering, and now lives in Oregon, Eugene, a farmer in Michigan, Florence, the wife of William Drinkwater of Michigan, Minnie, deceased, Lila, the wife of Fred Drinkwater, also lives on a farm in Michigan, and Frances, the wife of George McVittie, a government mail clerk resides in Detroit Both the parents were members and supporters of the Methodist Episcopal church, while Mr Smith took an active and prominent part in the councils of the Republican party

Mr Smith worked on a Michigan farm upon reaching manhood but he heard of the great opportunities afforded on the prairies of the middle west and determined to put his fortune to the hazard, and breaking all the home ties and intimate associations, started for the Dakotas He took plenty of time to look the country over as he had determined that wherever he located was to be a permanent home and Dakota did not measure up to his standard so he came to Cheyenne county in 1887, where he homesteaded 160 acres of land, preempted another tract of equal acreage, proved up on it and at the same time was engaged in making permanent and efficient improvements His first home, like that of nearly all the pioneer settlers, was a sod house but Mr Smith met with success in his chosen vocation and before long the sod structure gave way to a comfortable farm home Later Mr Smith removed somewhat west and north of his first claim, locating in section 32-23-56 Scottsbluff county, in what is locally known as the Mitchell valley is today one of the garden spots of the great state of Nebraska, that under modern methods and intensive farming is producing more to the acre than ever was dreamed in the pioneer days when the Smith family located here. Mr. Smith is one of the men of the county who has made good use of his opportunities, and his life record illustrates what may be accomplished by one who is industrious, far-sighted, and has an ambition to succeed. His harvests have been cut short by drought, his crops ruined by hail and insect pests, but he was never discouraged to the extent of giving up, and the succeeding years brought prosperity, and today he is the owner of 160 acres of highly improved land, all under irrigation, so that he never worries about the weather as he is insured a crop with water a-plenty and the never failing sunshine of this section. He has substantial and practical farm buildings, a good home and latest farm equipment, being engaged in general agriculture and stock-raising.

In 1891 Mr. Smith married Miss Alma Tappan at Broken Bow, Nebraska, the daughter of Bradford Tappan, both she and her father being natives of Michigan. Five children have become members of the Smith family: Floyd, who died in infancy; Kem, on a farm in Wyoming; Eunice, the wife of Claud Godbey, is at home, as are also Emmet and Craig.

Mr. Smith is one of the progressive business men of Scottsbluff county, is public-spirited, advocating every movement for the advancement of the community, which is attested by the fact that he is a school director and chairman of the irrigation board. A man of high ideals in life and commercial affairs, he is held in esteem by all his friends and associates. In politics he is a Democrat, but has never had the time or desire to hold public office, while his fraternal associations are with the Odd Fellows and Modern Woodmen.

JOHN SCHUMACHER. — To set down a true history of Nebraska in all its counties, mention must be made of those who came into the state without capital, and through hard work and great self denial finally became of independent fortune because of ownership of valuable lands. Some of these early settlers, it is true, had not the courage to endure inevitable hardships and gave up before their battles over storm, drought and loss of crops and stock had been won, but there were others, like the late John Schumacher, who held on, worked harder, hoped for the best, and were well rewarded.

John Schumacher was born in Roxbury, Dane county, Wisconsin, January 21, 1862. He had school advantages near his father's farm and worked as a farmer until he determined to start out for himself. That he was a young man not easily discouraged may be assumed from the fact that with practically empty pockets, he walked the entire distance from his old home in Wisconsin to Cheyenne county, Nebraska. There he homesteaded in what is now Scottsbluff county, six miles southeast of the present town, and remained on his farm, developing and improving it, until the end of his life, his death occurring November 8, 1915. At that time he owned a section of irrigated land. He made a specialty of stock-raising and under his care this industry proved very profitable.

In 1893 John Schumacher was united in marriage with Miss Katie Gaugler, a schoolmate, who was born in Dane county, Wisconsin, a daughter of Joseph and Mary (Retsler) Gaugler, Mrs. Schumacher being the youngest of their family of fifteen children, nine of whom are living. Two daughters were born to Mr. and Mrs. Schumacher: Helena, the wife of Philo Tillson, a farmer north of Minatare, Nebraska, and Elsie, who resides with her mother and is attending school. Mrs. Schumacher is, as was her husband, a member of the Catholic church. On April 11, 1916, she moved to Scottsbluff and purchased a beautiful residence on Fourth avenue, but found it too great a care to keep it up, hence sold and now resides in great comfort at No. 1814 Fifth avenue. Mr. Schumacher was a Republican in politics and served on the school board in his township. He was an honest, upright man and was very highly regarded by all who knew him.

FRANK B. MORGAN, who has had quite a great deal to do with the material development of Scottsbluff, is a leading builder and contractor here, owner of valuable city realty both unimproved and built upon. Since coming here in 1914, he has shown personal and public-spirited interest in the city, has invested judiciously and has been an encourager of a number of worthy enterprises.

Frank B. Morgan was born in Caldwell county, Missouri, June 22, 1869, a son of Joseph and Tabitha (Hobbs) Morgan, the latter of whom is now in Illinois and is now deceased. The father of Mr. Morgan was born at Indianapolis, Indiana, a son of George Morgan, who was a native of Virginia. Joseph Morgan was a soldier in the Civil War, enlisting in the Fourteenth Missouri Infantry, served three years and was wounded at Shiloh. After a long period in a hospital, he reënlisted but was soon afterward discharged on account of disability. In 1883 he came to Nebraska and bought a section of land in Furnas county, afterward selling the same and buying land in Oklahoma His death occurred at Beaver City, Nebraska He was a Democrat in politics Both he and wife were members of the Christian church They had seven children, Frank B being the fourth of the five survivors, who are George R, a farmer near Hendley, Nebraska, Delilah, the widow of ———— Whiteman, of Hendley, Thomas, a butcher in business at Hendley, and Mary, the wife of Roy Goebel, a farmer in Furnas county

Frank B Morgan obtained his education in the public schools and afterward followed farming until he was thirty years old He had always been deft in the use of tools and then began to work at the carpenter trade, going to Denver in 1901, and worked as a carpenter there and at Fort Collins until 1905, when he came to Morrill county, Nebraska He secured a homestead there, which he late sold, then bought land in Wyoming and subsequently sold that In 1914 he came to Scottsbluff and has proved a valuable citizen He invested in vacant property here and through improving it with attractive residences, has added greatly to the appearance of every section in which he owns lots He has found ready sale for his houses, for the people of Scottsbluff are homemakers, in the main, and his enterprise has been appreciated Mr Morgan has assisted in the organization of the Commercal Bank at Scottsbluff

In 1890 Mr Morgan was united in marriage to Miss Mary E Martin, who was born in Illinois, and is a daughter of John and Matilda Martin The mother of Mrs Morgan died at Stamford, in Harland county, Nebraska The father died at the home of a daughter in Iowa Mr and Mrs Morgan have had children as follows Mahlon C, who is a farmer in Scottsbluff county, Merlin O, who was honorably discharged in February, 1919, from military service in the World War, entered the navy, was first located on the Pacific coast near Seattle, Washington, then spent one year in the Canal Zone, later was sent to New York and was discharged at Brooklyn, as chief carpenter, Mable, who is a student in the high school, and John and Audry, who are doing well in their several grades in school Mr Morgan and his family are members of the Christian church Fraternally he is a Mason and belongs also to the Modern Woodmen He is not active politically and votes independently

CHARLES S SIMMONS, who finds his time fully occupied with the work of his profession, sign painting, can show a large amount of fine, artistic work from his brush at Scottsbluff, in which city he painted his first sign, in April 1900

Charles Sheldon Simmons belongs to an old county family and was born near Scottsbluff, Nebraska May 20, 1887 He is a son of Charles H Simmons, extended mention of whom will be found in this work Mr Simmons attended the public schools Very early he displayed talent with pencil and brush and after learning the painting trade, decided to specialize on sign painting This branch of the business requires not only practical knowledge but real skill and Mr Simmons went to Chicago and entered a class in the Art Institute, where he remained a year, securing necessary technical training as well as artistic inspiration His work is very much admired and his services are in constant demand

In 1897 Mr Simmons was united in marriage to Miss Estella M Snyder, who was born at Garrison, Iowa, a daughter of Edward H and Belinda (Hilka) Snyder They were born in Pennsylvania The father of Mrs Simmons is engaged in truck farming near Sterling, Nebraska Mr and Mrs Simmons have two children, namely Cleo who is seven years old, and Charles, who is a babe of eight months Mr and Mrs Simmons are active members of the Presbyterian church In politics he is a zealous Republican, and fraternally he is identified with the Knights of Pythias, the Modern Woodmen, and the A F & A M

AUGUST DORMANN, who is a well known business man of Scottsbluff, where his beautiful residence and other property are located, for a number of years has been identified with commercial enterprises of large importance here and elsewhere

August Dormann was born at Wisner, Nebraska, in 1877 He is a son of August Dormann who was born in Germany, came to the United States, has been a merchant all his life, and now resides at Denver, Colorado At Omaha, Nebraska, he was married to Frederick A Kemenbley, who was born in New Jersey, February 9 1851, and died February 28, 1910 Of their five children August is the fourth in order of birth, the others being Agnes the wife of William Reisendorfor a lumber merchant in California, George W, foreman of the Skinner & Eddy shipbuilding plant at Seattle Washington, Anna, resides with her father, and Fred, a consulting engineer at Denver The family is of the Lutheran faith In politics the father is a Republican He was one of the first three men to take out policies in the New York Life Insurance Company, in Nebraska, and is the only survivor of the three

August Dormann obtained his educational training in the public schools of Wisner, after which he was associated with his father in the mercantile business until 1906 when he came to Scottsbluff county He bought a farm on which he resided two years, then took charge of the Zoellner clothing store at Scottsbluff, which he managed for five years On retiring from that connection he went into business of buying and selling mercantile stocks, in which he continued until 1916, when he organized the August Dormann Company, for the purpose of buying farms and ranches for sale or trade Early in 1919 he bought all the company's interests with the result that he owns many acres of fine land in Michigan, South Dakota, and western Nebraska He now devotes himself to his large farming interests

In 1899 Mr Dormann was united in marriage to Miss Katherine O'Connor, who was born at Wisner, Nebraska, and they have had the following children born to them Charles August, who was born December 23, 1900, entered military service in January, 1917, and is a member of a hospital corps now stationed in the Philippine Islands, George Eugene, who was born October 4, 1902, Genevieve Ruth, who was born May 3, 1904, Jerome Wilbur, who was born August 11, 1908, and Katherine Virginia, who was born July 7, 1911, and two deceased Victor Hugo, and Herald Mrs Dormann and the children belong to the Catholic church Mr Dormann is a Republican in politics and is a Mason in good standing

JOHN W BROSHAR for many years was well known in Nebraska, and his honorable name is preserved by a surviving family of Scottsbluff He was a native of Ripley county, Indiana, born May 7, 1845 His parents moved to Champaign, Illinois, in 1852, and Mr Broshar was educated there and from that town enlisted and served during the thirteen closing months of the Civil War After leaving the army he began business independently as a farmer in Illinois and continued agricultural pursuits until 1888, when he came with his family to the Panhandle, took up a homestead in Box Butte county, and lived on this farm for a number of years, then took a Kinkaid claim near the line of Sioux county His death occurred at Canton, Nebraska, February 6, 1913

At Paris, Illinois, in 1875, John W Broshar was united in marriage with Miss Jennie Waggoner, who was born in Fayette county, Ohio, and is a daughter of E D and Elizabeth F (Bush) Waggoner Mrs Broshar's father was born in Virginia and the mother in Ohio They moved to Illinois in 1864 and both died there Mrs Broshar was the eldest of their four children To Mr and Mrs Broshar three daughters were born Pearl, the wife of Arthur Barr, a farmer near Melbeta, in Scottsbluff county; Myrtle, the widow of Henry Safford, resides at Scottsbluff, and Edith, who resides with her mother, is connected with the Irrigation Bank

Following her husband's death, Mrs Broshar displayed business capacity by securing a homestead for herself and resided on her property until 1915, when she came to Scottsbluff and is now enjoying the comforts of a beautiful home at No 1601 Fourth avenue The family belongs to the Baptist church Mr Broshar was a man of sterling character He was successful as a farmer and stockraiser, and took considerable interest in public matters in Republican political circles in Box Butte county, although he never consented to accept public office

ALBERT W PETERSON, who is one of the quiet, industrious, useful business men of Scottsbluff, has been a resident of this city since the spring of 1915, but has lived in Nebraska since he was five years old He is a carpenter and contractor who has built up a large business and has the reputation here and elsewhere of business integrity and dependability

Albert W Peterson was born at Princeton, Illinois, April 14, 1880, and is the oldest of seven children born to Nels W and Anna C (Swanson) Peterson, both of whom were born in Sweden and came young to the United States They were married at Princeton, Illinois, where he was foreman for the Bryant Nursery Company for nine years In 1885 Nels W Peterson brought his family to Nebraska and bought land near Aurora, in Hamilton county, on which he has lived ever since, at the present time owning 320 acres of fine land He is an example of the citizenship of Nebraska that has prospered within her welcoming borders through faithful and law-abiding industry, for he came here from Illinois not only without capital but burdened with debt He has worked hard but feels well repaid He has never been active in politics, has always favored prohibition legislation, and in casting his vote as a citizen, gives his support to candidates he believes will unselfishly do their best, irrespective of party policy, for the country He and wife are members of the Swedish Mission church

Ranch of John Engstrom

John Engstrom and Family

After his school days were over in Hamilton county, Albert W Peterson worked as a farmer and also as a carpenter, early showing skill with tools, and continued to live in Hamilton county until March, 1915, when he came to this city, the rapid settlement of which and expansion of industries, offered abundant opportunity for his line of work He has done a great deal of substantial building here and gives all his time to carpentering and contracting Formerly he owned a farm in the eastern part of Scottsbluff county, but this he sold in the spring of 1919 He has property in the city which includes a comfortable and extremely attractive residence at No 2008 Avenue A, a beautiful home

Mr Peterson was married February 24, 1906, to Miss Mina Hanson, who lived at Chicago, Illinois, and is a daughter of Hans Hanson and his wife, who spent their entire lives in Sweden Mr and Mrs Peterson are members of the Swedish Mission church and greatly interested in its various avenues of benevolence Like his father, Mr Peterson prefers to be independent in politics

JOHN ENGSTROM — This representative agriculturist and stock-grower of Scottsbluff county has been a resident of Nebraska for nearly twenty-seven years and by making use of the advantages here offered he has made his way forward to the goal of independence and marked prosperity Today he is the owner of 160 acres of the finest property in the western part of the state Mr Engstrom is one of the sterling citizens here who has had the prescience and energy to make the most of the opportunities offered in connection with civic and material development and progress He is a native of Sweden, that land which has furnished this great country so many of its earnest and progressive men of affairs His natal day was May 20, 1859, being the son of Swan and Louise (Carlson) Engstrom both of whom were Scandinavians, born in Sweden Both parents were vigorous and sturdy, the father living to the advanced age of seventy-four years and the mother to seventy-six years of age John's father was a farmer but as land allotments are not large in Sweden he learned the shoe-maker's trade, in which he was engaged a part of the time There were six children in the family but only three brothers broke the home ties to begin a career in America Gustav, Emil, and John, who landed in the United States in 1882, Gustav subsequently returned to the mother country but John remained, determined to win a way in the great west Emil died in Kansas He has been in

the greatest sense the architect of his own fortunes and few men have played a more sturdy part in the development of the communities in which they lived Both civic and industrial lines have been benefitted by the interest displayed in them by this young Swede, who soon after coming to this great land went to St Paul, Minnesota, where so many of his countrymen had established homes He soon found employment in Minnesota but later went north, being employed by a railroad in Canada for a considerable period before deciding to become an owner of land the great desire of nearly all men who came to America from Europe With this idea in mind he came to Nebraska and soon had filed on a claim in Sidney Valley, Cheyenne county he proved up on this land, living there six years before removing to Scottsbluff county, purchasing 160 acres of land in section 15, forty-five acres being under ditch Mr Engstrom has proved himself one of the world's constructive workers and in the furtherance of his own prosperity has aided in the civic and material development and progress of the country and state of his adoption and as one of the prosperous representatives of the western section of this great commonwealth deserves recognition in this history

Mr Engstrom first married Anna Carlson in Sweden, to this union five children were born Carl J, who lives at home, Gustav A, deceased, Anna, the wife of Otto Swanson, a farmer in Scottsbluff county, and Betty Louise, now living in Chicago For his second wife Mr Engstrom married Ada Carlson, also a native of Sweden, and they have one daughter, Hilda Carlson who is at home The family are members of the Swedish Mission church Mr Engstrom is a man of sterling character and ability so he has been called upon to serve as a school director, an office in which he has made a record for liberality and progressiveness as he takes great interest in the welfare of his community and the prosperity and happiness of the rising generation Politically he is an independent, believing that the man best fitted to hold office should be elected

Mr Engstrom sold his farm in 1919 and moved to Scottsbluff where he has bought property He worked for fifteen years to make his farm one of the best improved properties in the county and deserves much credit for what he has accomplished

ALBERT PAXTON, for several years one of the active and energetic men engaged in handling real estate at Scottsbluff, is well known as a handler of realty and other activities in the middle west, and also owns some very valuable farm land near Henry, in Goshen county, Wyoming. Mr. Paxton was born at Rensellaer, Jasper county, Indiana, August 27, 1867, the son of William F. and Isabella (Sharpe) Paxton, the former born at Bedford and the latter at Johnstone, Pennsylvania. They were married in that state, then moved to Ohio and later settled in Indiana, where they lived out the allotted span of life. Of their family of eight children two reside in the West, Ralph and Albert, the former of whom owns and operates the Paxton hotel at Torrington, Wyoming, and also owns a farm in that vicinity. The father was a farmer in Jasper county and was active in the Democratic party of that section. Both parents were members of the Methodist Episcopal church.

Albert Paxton was given educational opportunities in his youth, his father being a man of education himself. The young man began his business life as a clerk in a store at Montpelier, Indiana, and later, for a number of years was manager of the New York Store Company in that city. He became prominent in Democratic party circles, was recognized as a man of civic influence, was elected to the city council of Montpelier and served as president of that body. In 1908 Mr. Paxton came west to Wyoming, locating at Torrington, where for a number of years he engaged in the live-stock business, being an extensive buyer of horses, on one occasion buying seven carloads in a month, at Henry, Nebraska. In 1917 he embarked in the real estate business at Scottsbluff and still owns his handsome residence here, although recently he has transferred his real estate business to Torrington, Wyoming, where he is associated with G. E. Gannon. They are doing an extensive business in general real estate, farm loans and insurance.

In 1894 Mr. Paxton married Miss Anna Bebout, who was born in Indiana, and they have two children: Albert E. and Melva, aged respectively seventeen and thirteen years. The family belongs to the Methodist Episcopal church. He is identified with the Knights of Pythias.

LIGGETT FURNITURE CO. — A recent business enterprise at Scottsbluff that may confidently be expected to be of substantial importance to the city, is the furniture and house furnishing goods store established here May 24, 1919, by Clarence D. and Dwight W. Liggett, under the firm name of Liggett Furniture Co. Both members of the firm are men of business experience and of the highest possible personal character.

Clarence D. Liggett was born in 1884 and Dwight W. Liggett in 1890, both in Union county, Ohio. Their parents are John W. and Mary (Hardy) Liggett, both natives of Ohio, the father born in 1852 and the mother in 1853. In addition to the two sons mentioned, they have two others, namely: Raymond H., who is connected with the Mid-west Construction Company, and James Bruce, who, since his military service overseas ended, has been associated with his father in business at Fort Morgan, Colorado. He was in the Thirty-sixth Division in France and was wounded in the battle before Compiegne, but fortunately not fatally and has been recently welcomed home. The paternal grandfather was John Liggett, who was born in Virginia, moved to Ohio and spent the rest of his life there. The maternal grandfather, W. D. Hardy, was born in Scotland, came to the United States and died on his farm in Greene county, Ohio. When John W. Liggett left Ohio, he was ready to invest and enter into business at some favorable location in a western state and he selected Fort Morgan, Colorado, where he went into the furniture business and has continued there ever since. He is a Republican in politics and both he and wife are members of the United Presbyterian church.

Clarence D. Liggett was educated at Cedarville College, in Ohio. He began his business career as proprietor of a bicycle shop at Fort Morgan, and later became associated with his father at that point in the furniture business and still owns a half interest in the store there. He then entered into partnership with his brother at Scottsbluff, in 1919, and already the firm has an established place and firm standing in the city's commercial life. In 1912 he was united in marriage to Miss Blanche Coulter, a native of Iowa, but at that time a resident of Colorado, and their little daughter Helen is four years old, and a son, Howard Dean, born November 21, 1919.

Dwight W. Liggett was educated at Cedarville College in Ohio. In 1913 he came to Scottsbluff, where he was interested in the Mid-West Concrete Construction Company, but sold his interest to his brother in the spring of 1918 established the Liggett Furniutre Co., at Antioch, Nebraska, and Mr. Liggett came back to Scottsbluff and the present business was founded here May 24, 1919. In 1915 Mr. Liggett was married to Miss Lona Smith, who was a school teacher at Lodgepole, Cheyenne county, Nebraska, but a native of Iowa, and they have a little daughter of eighteen months whom they have named Dorothy Dell Dwight W Liggett is a member of the Presbyterian church at Scottsbluff, and Clarence D Liggett belongs to the United Presbyterian church at Fort Morgan The brothers are Republicans in their political affiliation and both are men of sterling worth

MILTON E HARRIS — One of the leading business men of the Platte valley is found in Milton E Harris, rancher and cattle feeder, and also proprietor of the most extensive meat business in the section west of Lincoln Mr Harris came to Scottsbluff in the spring of 1907 and his immense business is the result of his energy and good business judgment Mr Harris is a self-made man, starting out for himself at the age of eleven years and has fought his way steadily upward with the old watchwords of industry and perseverance ever in mind

Milton Evan Harris was born in Hancock county, Illinois, August 12, 1879, and is a son of John G and Jane (Latham) Harris and the youngest of eight children Both parents were born in Ohio, later were residents of Illinois, and the mother died subsequently at Cedar Rapids, Iowa The family was established at Brush, Colorado, in 1901, where the father served in some town offices, and died at that place in February, 1902 Three of his mother's brothers were soldiers in the Civil War He was a Republican in politics and both he and wife belonged to the Methodist Episcopal church All through life his business was farming but his efforts did not bring great financial independence

When eleven years old Milton E Harris started to work for Dr Martin for his board and attended school at La Harpe, Illinois and two terms after 1893 at Ray, Colorado At the latter place he worked in a butcher shop for two years, then went on a ranch and was employed near Ray on a ranch for about two years He then worked in a brother's meat shop at Brush, Colorado for two years, following which, in May, 1907, he came to Scottsbluff and bought a meat shop here Mr Harris prospered from the first and has continued to prosper, as indicated by the report of business for the year of 1907 showing its amount as $9,862 15 and the acknowledgment that since then it has amounted to $125 000 per year To provide facilities for such expansion, Mr Harris erected a handsome building of brick construction, on Broadway, with dimensions 26 x 140 feet with fine basement 7 feet high and 124 feet long in which every modern improvement is installed and all devices for the sanitary handling and preparation of meats provided Wherever the "T H S" trademark is seen, representing the Harris Sanitary market, his patrons, and people at large, feel confident as to the quality of meat and produce appearing under this style As manager Mr Harris has his brother-in-law, George Hillerege, an old experienced meat man, entire efficiency marking every detail of the business Mr Harris owns a fine ranch near Scottsbluff and feeds from 500 to 800 cattle and hogs

In politics Mr Harris votes the Republican ticket but gives the greater part of his time to his business affairs rather than public matters He is identified with the order of Odd Fellows

He married, April 30, 1901, Miss Helen Dow, a native of Jo Davies county, Illinois They have three children Beulah, Chas L , and Emmett G

OTIS W SIMMONS who is a member of the contracting firm of Simmons Brothers, at Scottsbluff is exceedingly well known in the construction line here, and is numbered with the city's active and representative citizens in many ways Mr Simmons was born at North Bend, Dodge county, Nebraska, September 16, 1885, and is a son of Charles H Simmons, extended mention of whom will be found in this work

Otis W Simmons attended his first school in Scottsbluff, held in one of the primitive sod houses very numerous in his boyhood days all through the newly settled sections of the West, later had advantages at Gering, and in 1903 was graduated from the high school of Scottsbluff He then learned the carpenter trade and about that time became greatly interested in amateur photography, which he still sometimes engages in as a recreation owning many pictures of artistic value taken all through the beautiful Platte valley In   Mr Simmons became associated with his brother, W——— L Simmons in the contracting business and they have done a large amount of substantial residence building here at the present time having ten residences in course of construction and giving steady employment to five skilled men The firm enjoys the reputation of perfect reliability and their work is constantly increasing in volume

In 1907 Mr Simmons was united in marriage to Miss Nellie G Cline who was born at Osceola Iowa, and is a daughter of W S Cline, extended mention of whom will be found on another page of this work Mr and Mrs Simmons have two children namely Helen who was born February 25, 1909, and

Harold, who was born August 11, 1911. Mr. and Mrs. Simmons are members of the Presbyterian church. Mr. Simmons is in active membership with the Knights of Pythias and the Modern Woodmen, has passed the chairs and has represented the local lodge in the Grand Lodge on three occasions, and has served almost continuously since 1905 as keeper of records and seal. Mr. Simmons is an earnest and straightforward citizen, ever ready to do his part in bringing about the best of conditions. Politically he is a Republican.

DANIEL D. DAVIS, who is one of Scottsbluff's most esteemed retired citizens, has long been identified with the substantial development and material progress of this city and county. He is widely known, as he came to Nebraska in 1884 and homesteaded in Scottsbluff county in 1886. This section has been his chosen home ever since, where many marks of public confidence have been shown him, and where mutual and sincere regard has followed the acquaintanceships of years. Mr. Davis was born on Catawba Island, Ottawa county, Ohio, March 7, 1859, the son of Captain Daniel N. and Sarah (Prentiss) Davis. The father was born on Long Island, New York, and followed a seafaring life, being captain of a sailing vessel on Lake Erie at the time he was attacked by a highway robber and murdered for his money, while on a visit on land, in November, 1868. The Davis family was reared in the Baptist church by a good mother. Captain Davis was active in politics long ago and was a Democrat until the party accepted the candidacy of C. L. Vallandigham, who had favored the cause of the Confederacy during the war in which his own son had suffered, when he changed his political views entirely and became a Republican.

Daniel D. Davis, on account of the early loss of his father, had fewer educational advantages than otherwise, and when twelve years old spent one hard winter working in the northern woods. Following that season he assisted his brothers in their fishing industries, which they carried on at Willoughby and Wickliffe, Ohio, remaining with them about five years. In 1884 he came to Nebraska and went to work for a brick manufacturing company near Lincoln, then opened a little store in the town, and struggled on as a youth does who has but little capital and is entirely dependent upon his own resources. On March 1, 1886, Mr. Davis came to Scottsbluff county and immediately filed on a homestead claim, building his own little dugout on his land and making himself as comfortable as possible. In

the meanwhile he had impressed his neighbors so favorably that in 1889 they had him appointed deputy county clerk when the county organized and he served in that office for a year, and having read law, he was admitted to the bar. Mr. Davis then returned to his homestead, which he had been improving and developing. In later years he became an important factor in Republican politics in the county, forwarding by his influence many movements of importance and accepting responsibilities when he believed such a course would be beneficial for the community. He served four years as county assessor.

In January, 1891, Mr. Davis married Miss Frances E. Brown, a lady of unusual intellect and educational prominence. Mrs. Davis was born at Sidney, Iowa, a daughter of James N. and Lois (Clark) Brown, the former born in Canada and the latter in Michigan. They had seven children and Mrs. Davis is the youngest of the five survivors. She attended school at Sidney and Hamburg, Iowa, and after being graduated from the high school at Tabor, in 1889 came to Nebraska and engaged in teaching school at Madison, returning then to Iowa. In 1890 she was elected the second superintendent of the schools of Scottsbluff county and served ably in this responsible position for two years. Mr. and Mrs. Davis have two children, Edwin P. and Alice E. Edwin P. Davis was born April 22, 1897, attended school at Minatare, Nebraska, and was graduated at Ames, Iowa. He volunteered June 5, 1917, in Company E, Fifth Nebraska National Guards, was in training at Camp Cody, was transferred to the One Hundred Thirty-fourth Infantry, and accompanied the American Expeditionary Forces to France in October, 1918, and remained in Europe in the Army of Occupation for twenty-five months. During this time Mr. Davis served in Company K, Forty-seventh Infantry, Fourth Division, during the World War, marched into Germany with the American Army of Occupation, and after the armistice was stationed at Addrian, Remogen and Coblenz, Germany until relieved and returned home, being honorably discharged at Camp Dodge, August 4, 1919. Alice E., the one daughter of the family, is a graduate of the Scottsbluff high school and is attending college at Ames, Iowa. Mrs. Davis is a member of the Presbyterian church and is actively interested in social questions of the day and in various charities that appeal to her benevolent impulses. Mr. and Mrs. Davis suffered a severe bereavement in the death of his sister, who was the wife of the late Edwin F. Moulton, a noted educator, superintendent of

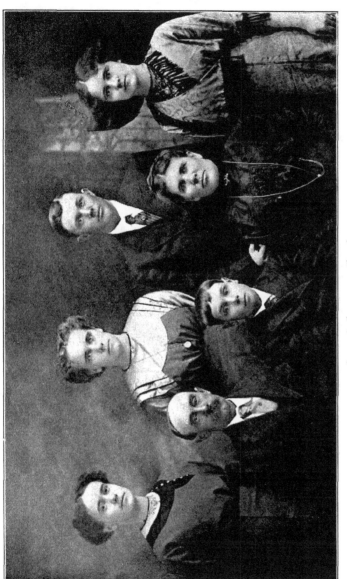

John A. Jones and Family

schools of Cleveland, Ohio, and at time of death was president of the board of trustees of the Kent State Normal School Mrs Moulton was very prominent in club life and at the time of her death, in 1911, was chairman of the civics department of the Federation of Women's Clubs

From the earliest date of agitation looking toward the great irrigation project, Mr Davis has been actively and intelligently interested and served as secretary of the first meeting called to consider ditch construction, and for some time was secretary of the Beard ditch, now the nine mile ditch. He was equally alert concerning other worthy enterprises, was a member of the North Platte Water-users Association for seven years and was secretary of the first Beet Growers Association. In 1916 Mr Davis sold his farm and bought a comfortable and attractive residence in Scottsbluff. He is identified with the Odd Fellows and the Knights of Pythias, and serves on many committees looking to the public good. Men like Daniel D Davis never find a retired life useless or lonesome

There is an interesting historical record of the first Davis in this country. He was Capt Dolor Davis, who came from England in 1634. He had land granted him in 1659, near Concord, was prominent in matters affecting the Plymouth colony, and followed the useful trades of carpenter and surveyor

JOHN A JONES, a native son of the Sunflower state, while he is a comparative newcomer in Scottsbluff county, where he took up his residence in 1900, has succeeded in firmly establishing himself in a position of prominence in agricultural circles, as well as in the confidence of his many friends and associates. When he entered upon independent commercial life he was possessed of little save inherent ability and the determination to succeed and these have been sufficient to enable him to gain a comfortable fortune and fine landed property

John Jones was born in Kansas, January 15 1860, the son of D B and Margaret (Cowen) Jones, the former a sturdy son of New England, who came west at an early day to take advantage of the opportunity of acquiring land in the new territory opened for settlement during slave days. The children in the family numbered fourteen, eleven of whom are still living. Laura, married Herman Gofferth and now lives in Emerson, Iowa, Ella the wife of Roswell Crossett of Spokane, Washington, Anna, married Henry Hawk, of Oregon. Arthur, a resident of Burns, Oregon. Florrie and

William live in Scottsbluff county, Fred lives in Buffalo county, Alice, the wife of George Veal, of Buffalo county, Addie, the wife of Thomas Wells, lives in Stapleton, Nebraska. Mary, Hettie and Frank are dead. The Jones family removed from Kansas to Illinois, then to Iowa, and from there to Buffalo county, Nebraska, and it was there that John received the education afforded by the public schools, leading the life of a boy on a frontier farm, gaining experience in agricultural life from direct association with it, and many a good lesson is thus learned in the hard but sure school of experience, lessons never to be forgotten in later life. Mr Jones was a mature man of forty when he determined to take advantage of new country and located in Scottsbluff county in 1900, buying eighty acres of land on which he has since made his home. He is a sincere advocate of intensive farming with irrigation and his marked success in this line may well be followed by those of less experience. When he arrived in the western section of the state the only improvement on the land was a sod and log house of some early settler, this was entirely inadequate to his use and for his family and within a short time he had a large comfortable residence erected, to which he may point with pride

All of his land is under irrigation which insures crops no matter what the weather conditions. Mr Jones is engaged in general farming and dairying but devotes special time and attention to his bees, making such a success and specialty of this side-line that one year he shipped three tons of honey, which at present prices means a comfortable income. Because of his apiary, Mr Jones has become known far and near as "Honey" Jones a title in which he takes no small pride. He is a tireless worker and engages in any industry that promises a worthy recompense, as eleven years he has supplied Scottsbluff with ice and another year in addition to his regular farm undertakings, milked seventeen cows. He has a keen, shrewd brain, is far-sighted and these qualities added to hard work are safe and sure landmarks along the highway to fortune. On June 1, 1887 Mr Jones married Ida M Tottersman, a native of Indiana and to this union five children have been born. Mabel, married Earl Enes of Bayard Nebraska. Roy A who lives next to the home place, Irna the wife of James Chris Lynch, living at Lingle Wyoming, Hazel, the wife of J H Cassidy also a resident of Lingle, Wyoming, and B W, who is at home

When the Jones family first came to Nebraska they made the trip in a prairie schooner to

Buffalo county, encountering all the dangers and sharing all the hardships incident to settlers in the new country, but they had been reared to some privations and soon were too busy establishing the new home to think of mere comforts that they knew soon would be theirs. When Mr. Jones removed to Scottsbluff county he followed the same method his father had and drove overland in a covered wagon, though it was not a necessity, as at the time of the first trip. Since then he has seen all the wonderful progress and material developments that have taken place in this great commonwealth in over a half century and today may sometimes be induced to recount some of the early experiences he and the family had when first arrived west of the great Muddy, as the Missouri river was known. Having a keen, preceptive, and retentive mind these stories of pioneer days are not only interesting but instructive. Mr. Jones is an independent thinker on all subjects pertaining to the civic and political life of his section and the nation and it but follows that he is an independent in politics; is a public-spirited citizen and an all-American.

ROBERT HAXBY, who is a highly respected resident of Scottsbluff, a retired farmer of ample means and possessed of the good judgment and experience that make him a wise and prudent advisor on many questions of civic importance, is a native of Iowa, born in Dubuque county, in 1852, came to Nebraska in 1871 and to this pleasant, healthful little city in 1916.

The parents of Mr. Haxby were William and Dorothy (Bradley) Haxby, both of whom were born in the same neighborhood in Yorkshire, England. Shortly after their marriage they came to the United States, settled in Iowa and there he followed his trade of wagonmaker and wheelwright until the end of his life. After his death, the mother of Mr. Haxby came to Nebraska and died in this state. Of their seven children Robert is one of the three survivors, the others being: John, a retired farmer living at Fremont, Nebraska, and Mary, the wife of Joseph Smith, who is a retired farmer living at Cedar Bluffs, Nebraska. The parents were most worthy people in every relation of life, and they were faithful members of the Methodist Episcopal church.

Robert Haxby began life on a farm and continued to be interested in agricultural pursuits during his entire active life. In 1871 he came to Saunders county, Nebraska, bought land that he has never parted with, and at the present time has 154 acres, which command $300

an acre. Mr. Haxby has been unusually successful in all his undertakings and has been numbered with the solid men of the Platte valley. In 1916 he moved to Scottsbluff with the hope that this city's even climate might be beneficial to Mrs. Haxby, a hope that has been realized.

In 1886 Mr. Haxby was united in marriage to Miss Amelia Rasch, who was born in Saunders county, Nebraska, and is a daughter of William and Philopena (Tillman) Rasch, both of whom were born in Germany. They came to the United States in 1858, lived for a time in New York, came then to Omaha, Nebraska, and subsequently homesteaded in Saunders county. The first home that Mrs. Haxby remembers in Saunders county, was a sod house, with outbuildings also of sod construction, this kind of house prevailing over wide areas in early days here when the transportation of lumber was difficult. To Mr. and Mrs. Haxby the following children were born: William C., who lives on his wheat ranch in Cheyenne county, Nebraska; Benjamin Robert, who manages his father's property in Saunders county; Esther Fay, who is an independent young business woman at Scottsbluff; Myrtle Ruth, who is the wife of Frank Anderson, formerly in the plumbing business at Scottsbluff, returned June, 1919, an overseas soldier in the American army in France; and Everett Lyle, who is attending school. Mr. Haxby and his family are members of the Methodist Episcopal church. In politics he is a Republican as was his father before him.

JAMES BAXTER, who came to what is now Scottsbluff county, Nebraska, in 1888, has lived in this county ever since and is one of its substantial and well thought of citizens. Mr. Baxter has been a good farmer, an honorable and useful citizen, and on his own land and underneath his own care has reared a worthy and creditable family. Since 1917 he has lived retired at Scottsbluff.

James Baxter was born in Fermanaugh county, Ireland, September 16, 1854. His parents were James and Margaret (Montgomery) Baxter, both of whom died in Ireland, both Scotch. Of their eight children four came to America, namely: Alexander, Mrs. Martha Kennedy, Mrs. Maria Beatty, and James. Alexander came to the United States and engaged in the draying business at Pittsburg, Pennsylvania, from 1864 until his death in 1905. Maria is the wife of George Beatty, a farmer in Madison county, Nebraska. The parents were members of the Presbyterian church.

James Baxter attended school in his native land and made himself useful to his father, who was a farmer, until 1874, when he came to the United States During one year he remained with his brother Alexander at Pittsburg, then went to Delaware county, Iowa, where he worked for five years on different farms, for about $200 a year, then rented a farm in Iowa for four years In 1888 he came to Nebraska and pre-empted his present farm in what was then Cheyenne but is now Scottsbluff county, and for years afterward devoted himself closely to its improvement Mr Baxter now owns 120 acres of well improved, irrigated land, as a just reward for his perseverance and industry In 1917, he decided the time had come when he could take a rest, therefore he bought a nice property on a pleasant avenue in Scottsbluff, having turned his farm over to the management of his eldest son-in-law

Mr Baxter was married while living in Iowa, to Miss Anna Crothers, who was born October 15, 1853, in Ireland, and came to the United States with her parents in 1864, who settled in Delaware county She was a daughter of William and Margaret (Ramsey) Crothers, both natives of Ireland He died July 4, 1865, one year after coming to the United States, leaving a large family of children The mother died in Iowa at the age of sixty-seven, they were both members of the Methodist church and of Scotch descent They have had children as follows Mary E who is the wife of Howard H Elsabach, a lumber man at Henry, Nebraska, James, who operates his own farm near Scottsbluff Sarah, who is the wife of W O Powell, a railroad man in Montana, Mattie, who is the wife of Mike I Helen, a farmer on Mr Baxter s old homestead, Etta, who is the wife of C G Nicholas, a farmer near Mr Baxter, John Alexander, who was in military service at Camp Funston, during the great war for seventeen months, and Alma Lillian, who was a victim of influenza, in November, 1918 Mr Baxter and family belong to the Methodist Episcopal church In politics he is a Republican

MILLARD F CLUCK — Perhaps no American story writers have a wider audience than those who write understandingly, or otherwise, of western ranch life Their unconventional tales are usually most interesting, for a spirit of freedom and adventure lies dormant in every one, and more emphasis is laid on this feature than on the sordid details that are necessary in the practical conduct of a great ranch, in which thousands of dollars are pretty sure to have been invested In many ways life on a ranch in Nebraska undoubtedly pleases and satisfies, but it must also profit or such experienced business men as Millard F Cluck, a well-known resident of Scottsbluff, would not devote his valuable time to operating a a ranch Mr Cluck owns four thousand acres of land in Banner county

Millard F Cluck was born at Newport, Perry county, Pennsylvania, September 25, 1878 the son of William and Barbara (Creek) Cluck, both born in Perry county In 1879 they moved to Iowa, where the father followed the blacksmith trade for ten years, then came to Nebraska and homesteaded in Banner county The family lived on the ranch until 1899, then moved to Scottsbluff, in which place Mr Cluck died some years later He was a man of sturdy character, upright in every act, and a leading member of the Evangelical Lutheran church Of his children the following survive in addition to Millard F Curtis M, a farmer in Morrill county, Nebraska, Catherine, the wife of Allen Chamberlain, who is pastor of the Methodist Episcopal church at Ord, Nebraska, Alice, who married Timothy M Granshaw, of Council Bluff, Iowa, she died September 10, 1919, and Anna L, the wife of Dee England, a farmer near Orient, Iowa The paternal grandfather of this family was Jacob Cluck, who spent his life in Pennsylvania

Millard F Cluck was well educated in the public schools of Gering, and his first business experience was in the Irrigators Bank, an association which continued for seven years and when he retired in order to give attention to his own affairs, had become cashier of the institution In 1907 he moved to Scottsbluff, purchased a desirable property and has since maintained his home here, and taking much interest in the town's progress His large ranch in Banner county he uses for feeding cattle and is one of the big shippers of this section

November 28, 1906, Mr Cluck was united in marriage with Miss Tina Barrett, who was born in Cass county, Nebraska, not far from Weeping Water She is a daughter of George W and Ollie (Wolcott) Barrett, the former a native of New York and the latter of Illinois After their marriage near Elizabeth, Illinois, the parents of Mrs Cluck came to Cass county, Nebraska, where the father homesteaded During their first years they lived in a barn, but prosperity came to them and Mr Barrett is retired from business life and resides on his estate in Florida. Mrs. Barrett died in 1913, all her life a faithful member of the Christian church. Mrs. Cluck has three brothers and one sister: Lynn, a farmer and rancher in Canada; Tillie, the wife of John Todd, a rancher in Canada; Loren, a farmer in Kansas; and Ray, a farmer near Burlington, Kansas. Mr. and Mrs. Cluck have two daughters, Velma and Mildred, aged respectively, eleven and five years. Mrs. Cluck was educated in the high school at Elmwood and in 1901 was graduated from the training department of the Peru normal school. She is a member of the Christian church. In his political views Mr. Cluck is a Republican.

WILLIAM H. GABLE, has for many years been identified with important business interests, both in Nebraska and Wyoming, was born in the southern part of Germany, February 5, 1864, the youngest son of Henry Gable. Both parents spent their lives in Germany, where the father kept a hotel. They were members of the Lutheran church.

William H. Gable was left an orphan when yet young. He attended school and followed such pursuits in his native land as were open to him, and June 1, 1878 came to the United States by himself . After landing in the United States Mr. Gable located first in St. Louis, where many of his countrymen lived. As he was learning the English language he accepted any honest employment for a living during the first two years. In 1880 he went to Denver, Colorado, but a year later moved to Wyoming where he became a cowboy on a ranch, an occupation he followed for five years. Not such a very long time elapsed however, before his industry had brought him some capital, and for some years he kept working on farms, but later secured land of his own and went into the sheep business, in which he prospered. He continued to operate his Wyoming ranch until 1906, when he sold and came to Nebraska. He bought land within two miles of Scottsbluff and now has three hundred acres of irrigated land under rental. He has important interests at Scottsbluff, being concerned in an independent lumber yard and also in the Scottsbluff creamery.

In 1900 Mr. Gable married Miss Lina Danulat, who was born September 28, 1876 in Alsace-Lorraine, and they have three children: Theodore, who is on the home farm; Martin, employed in the Scottsbluff creamery during vacations; and Lilly, who is attending school. Mr. Gable reared two nephews, Arnold and Fred Pistorius, both in military service dur-ing the World War, the latter being a wireless operator. Mr. Gable and his family are members of the Christian Science church. In politics he has been identified continuously with the Republican party. He is valued highly as a citizen of Scottsbluff, where he has served on the school board and has furthered many worthy enterprises of different kinds.

JOHN W. BLY, who is a member of the Weller Company, at Scottsbluff, is a practical hardware man, having been identified with this line of work ever since leaving his school books. He has been a resident of Scottsbluff since 1912, and in April, 1918, was elected city clerk.

John W. Bly was born at Big Bend, Kansas, December 27, 1887, and is a son of Lucian G. and Catherine (McDonald) Bly, the former of who was born in Illinois and the latter in Indiana. In 1893 the father located in Colorado, being a traveling representative of a wholesale hardware house, and ever since then the family home has been at Greeley. Of the family of five children, John W. was the second in order of birth, the others being: Winnie, the wife of H. J. Guise, in charge of the poultry department work in connection with the agricultural college at Davis, California; Hazel, the wife of Allen Straight, a farmer near Loveland, Colorado; Lucian, who conducts a tin shop at Scottsbluff; and Helen, attending school. The parents of the above family are members of the Presbyterian church. In politics the father is a Republican, and for many years he has been identified with the Masonic fraternity.

John W. Bly enjoyed excellent educational advantages and in 1906 was graduated from the Teachers College, at Greeley. With the intention of thoroughly learning the hardware business, he started in at the bottom, beginning as a stove polisher. Mr. Bly has advanced to his present position through industry and close attention and now owns a block of stock in the Weller Company in addition to being its credit manager.

On June 1, 1913, Mr. Bly was united in marriage to Miss Sarah E. Lowery, who was born in Iowa, and they have one son, Robert Walter, who has not yet reached his second birthday. Mr. and Mrs. Bly are members of the Presbyterian church. He is a Scottish Rite Mason and has always lived up to the moral standards such a connection makes necessary. His political affiliation has always been with the Republican party. He has been an earnest and active citizen ever since coming here and individually and in business is held dependable and trustworthy

EDWIN A CURRIE who is now numbered among the substantial agriculturists of Scottsbluff county, has been the architect of his own fortune, and having based his life's structure on substantial foundations, has builded soundly and well When he entered upon his career he was possessed of little save inherent ability and a determination to succeed, and these have proved ample, through their development, to enable him to become a well-to-do farmer, stockman, and banker in a community which does not lack for able men

Edwin A Currie is of staunch Scotch ancestry on both the paternal and maternal sides as the respective names fully indicate He was born in northeastern Ohio March 6 1858, the son of James and Marion (Hamilton) Currie both natives of the vicinity of the famous Scotch city of Glasgow To them were born four children Lucille M, who married Jason Beach, is deceased, Edwin, James R, who lives in Ohio, and Maggie H, the wife of Fred Simpson, also lives in Ohio

The parents were some of the fine Scotch settlers who came to the United States during the nineteenth century, as they emigrated from their native land about 1849 and soon after reaching America located in Ohio where the father followed farming as a life work, being eminently successful in this chosen vocation The family were members of the Methodist Episcopal church in Ohio where the father died in 1898, being survived by his wife who lived to a hearty old age, passing away in 1903

Edwin Currie attended the district schools in Ohio, receiving an excellent rudimentary education supplemented by the instruction of his parents and the reading they induced him to do by himself He began working by the month for a short time, then bought a team and began running a huxter wagon and dealing in stock The Ohio valley was thickly settled at this period and the young man determined to take advantage of the opportunities of securing government land in the newer state west of the Mississippi, and the Great Muddy, as the Missouri was known, for here on the rolling prairies was land, and room enough for all who desired to come and take it On April 6 1886 the start for the new home in the west was made Mr Currie was a bachelor so it was not necessary to take as much household goods as though more members were to make the trip In Missouri he and his uncle, John H Currie, purchased a team and wagon for the long journey overland Leaving there May

18th it was the 1st of July before Mr Currie reached Scottsbluff county where he took a tree claim, preempted 160 acres and homesteaded another 160 acre tract in 1887 He at once began to make improvements upon his land engaged in general farming and stock-raising The early struggles taxed the young man's strength to the full but he possessed determination and persistence and in the end they triumphed over all obstacles Mr Currie had the utmost confidence in the community where he had selected to make his home, and during the drought years of 1890 and 1894, when other settlers were discouraged and were leaving for their former homes in the east, he bought more land, and has lived to reap the reward of this confidence He still owns the original homestead and claim but has added to them until today he has a rural estate of 6,000 acres dry grazing land, about 600 acres of which are under irrigation From 1886 to the present Mr Currie has been actively engaged in agricultural pursuits, his talents seem naturally adapted to these lines and today he is the owner of a splendid enterprise which is but a just reward for a man of industry and energy, enterprise and spirit, which was so well demonstrated during the trying years when crops failed He is noted for his integrity and the manner in which he lives up to his business obligations

Mr Currie was married in 1906 to Miss Jennie G Richards of New England extraction, as she was born in Vermont and came to western Nebraska while her father at the time of settlement in this section Mr Currie is a staunch Republican in politics, he and his wife are members of the Federated Congregational church, while he is fraternally a Scottish Rite Mason He now rents his irrigated land and keeps his pasture land He was one of the organizers and the first president of the American Bank of Mitchell

ADELBERT A MILLER, widely known in Western Nebraska as special agent for the Occidental Building & Loan Association of Gering, for a number of years was prominent in the educational field and has also been connected with business enterprises of some magnitude Mr Miller was born at Tekonsha, Calhoun county, Michigan, July 14, 1873, and has been a resident of Gering, Nebraska, and a leader in many of its affairs of moment, for the past eighteen years

Mr Miller's parents were Daniel S and Elizabeth Ann (Harsh) Miller, both born near Canton, Ohio The grandparents were Peter Miller and Adam Harsh, both natives of

Pennsylvania who moved to Ohio, where the latter died but the father passed away in Michigan. Daniel S. Miller served in the Civil War as a member of the Ninety-eighth Ohio infantry and accompanied his regiment with General Sherman on the memorable march to the sea. In 1866 he moved to Michigan, bought land in Calhoun county and there both he and wife died. They had the following children: Maggie the wife of William Creore, of Battle Creek, Michigan; Adelbert A., who resides at Gering; Lawrence L., a retired merchant at Gering; and two who are deceased. The father was reared in the Lutheran faith and never changed his church relationship, while the mother was a faithful member of the Methodist Episcopal church. Mr. Miller's father continued his interest in the Grand Army of the Republic up to the time of his death, belonging to the post at Tekonsha, Michigan.

Adelbert A. Miller attended the country schools near his father's farm and afterward the normal school at Ypsilanti with the intention of making teaching a part of his life work and for a number of years he was very prominent in the educational field, first in Michigan, later in North Dakota, where he was superintendent of schools of Milnor, in Sargent county, for four years, and afterward at Gering, where he filled the same office. Mr. Miller then embarked in the lumber business, which he followed for nine years, retiring from that line to enter the mercantile business with his brother. Six years later he sold his store interest to accept the position of special agent for the Occidental Building & Loan Company, a business concern of large importance, and Mr. Miller now has charge of all the loans in Western Nebraska. He devotes all his time to furthering the interests of this corporation, but during the progress of the World War he put aside most of his personal interests in order to work for the public weal, serving early and late as a member of the Council of Defense and as food administrator.

In 1898 Mr. Miller was united in marriage with Miss Elsie Johnson, who was born in Southern Michigan, a daughter of Homer Johnson, who was a substantial farmer. To Mr. and Mrs. Miller the following children have been born: Margaret, educated at Gering, occupied a position of chemist in the sugar factory here for one year, and is now ticket and express agent for the Union Pacific Railroad at this point; and Murray, Dorothy, Stanley, Esther, Adelia, Jack, Elizabeth and Catherine. Mr. Miller and family belong to the Methodist Episcopal church. Politically he is a staunch Republican and frequently he has served in public offices at Gering. He was the first city treasurer and has served on the school board for twelve years. He is prominent in the order of Odd Fellows and is a member of the grand lodge, having passed through all the chairs. Mr. Miller is recognized as one of Gering's representative citizens.

HENRY EBERHARDT, who is engaged in the mercantile business at Scottsbluff, has, in a few years, built up a large business enterprise here, on a foundation of business honesty and courtesy to everyone. Mr. Eberhardt came to the United States from a far distant country, but soon adapted himself to American ways and is able to count his acquaintances as friends.

Henry Eberhardt was born in Russia, April 12, 1891, the youngest of a family of six children born to Jacob and Mary (Milburger) Eberhardt. The other members of the family are: George and Jacob, who are farmers in Russia; Mary, who lives with her mother in Russia; Fred, who came to the United States and is in the creamery business in Kansas; and Lizzie, who lives in Russia. The father died on his farm in Russia when Henry was but six months old. He attended school in his native land and thus was well informed when he came to the United States and settled in Kansas in 1910. There he worked in a store and also learned the language of the country with the quick intelligence for which his countrymen are noted. In 1914 he came to Scottsbluff and started a small store, stocking it with reliable and seasonable goods, and from that modest beginning has built up a large trade and now has a commodius general store. Having sold this out he and Dr. C. N. Moore bought the Harris market on Broadway and have one of the most modern and up-to-date markets in Western Nebraska.

In 1915 Mr. Eberhardt was united in marriage to Miss Mary Kuxhousen, and they have two children, namely: Leo and Ruth. They are members of the Lutheran church.

WILLIAM L. SIMMONS, who is one of the leading contractors at Scottsbluff, is a member of a very important old family that settled first in Dodge and later became known in other couties of the state of Nebraska. He was born at North Bend, Dodge county, June 7, 1882, and is a son of Charles H. Simmons, extended mention of whom will be found in this work.

William L Simmons did not have the educational advantages that he is able to afford his own children, but he remembers when school was held in a sod house within walking distance over the prairie from his father's homestead He helped on the farm but his inclination was toward mechanics and when eighteen years old he learned the carpenter trade and has spent the greater part of his time at Scottsbluff ever since as a carpenter and contractor He is now associated with his brother, O W Simmons, and they do a large volume of business, running two crews of men all the time The business reputation of the firm is above par

In 1904 Mr Simmons was united in marriage to Miss Alpha McCartney, who was born at Sibley, Iowa, a daughter of James S and Alice (Darling) McCartney, natives of Illinois who were married in Iowa to which state they were taken when young The father of Mrs Simmons was a farmer He died January 15, 1913 The mother resides at Scottsbluff Mrs Simmons has two sisters Mary, the wife of W G Munser, a farmer in Wyoming, and Alice, the wife of Arthur Marley, a farmer near Lingle, Wyoming Mr and Mrs Simmons have a son and daughter, Harry and Harriett, now fourteen years old and attending school, who have the distinction of being the first twins born in the city, and Fred, who is ten years old and also in school Mrs Simmons is a member of the Presbyterian church Mr Simmons is a Republican in politics but he has never desired any political office He is somewhat prominent in the order of Odd Fellows, belonging to the grand lodge and the Encampment

DAVID W HILL, who is a highly respected citizen of Scottsbluff, has been one of the county's extensive cattle feeders for many years, and since moving into this city in 1914 still devotes his Scottsbluff irrigated land to this industry Mr Hill has been a resident of Nebraska for thirty-three years

David W Hill was born at Lockport New York, December 15, 1864, one of a family of fourteen children born to Minard and Almira (June) Hill, the former of whom was born in England, and the latter in New York, in which state they were married In 1865 removal was made to Michigan where the father bought land and both parents died there They were quiet, Christian people and both belonged to the Methodist Episcopal church

Of the seven surviving members of his parents' family, David W Hill is the only one

living in Nebraska He was afforded a thorough public school education, being graduated from the high school, in Van Buren county, Michigan, after which he intelligently took up work on the home farm, paying considerable attention to stockraising In January, 1886 he moved to Buffalo county, Nebraska and shortly afterward took a pre-emption in Banner county, and for a number of years following lived in that section For fifteen years he was in partnership with T C Eggleston in the cattle business, their operations being extensive in raising and handling high grade and registered White Face cattle Mr Hill then came to Scottsbluff county and bought irrigated land and went into the cattle feeding and shipping business, in which he continued actively engaged until 1914, when he came to Scottsbluff and took possession of his comfortable residence on Avenue A and identified himself with the best interests of the place

In the spring of 1894 Mr Hill was united in marriage to Miss Gertrude Grafiuse, who was born in Pennsylvania and is a daughter of Thomas and Jennie (De Remer) Grafiuse The parents of Mrs Hill moved to Buffalo county Nebraska, in 1878, homesteaded and resided on their land until the father's death in 1910 Mrs Hill has one brother, Charles, who is a hardware merchant at Kearney, Nebraska Mr and Mrs Hill have the following children Bernice, who is a student in the state university, Jennie, who is also a university student and Ivis, Charles and Dorothea, all of whom are in school with higher educational advantages in prospect Mrs Hill is a member of the Baptist church Mr Hill was quite active in Republican politics while living in Banner county and for six years served as a member of the board of county commissioners He belongs to the Odd Fellows and the Knights of Pythias and is past chancellor in the latter organization

FRANK E COWEN, who is a representative business man of Scottsbluff, has been engaged here for a number of years as a cement contractor and has built up an unquestioned reputation for reliability Mr Cowen was born in Marshall county Iowa, July 2 1873, and is a son of Elisha M and Elvira (Triplett) Cowen extended mention of whom will be found in this work

Frank E Cowen attended the public schools in Chicago Illinois in boyhood, then came to Cheyenne county Nebraska, worked at farming for some years and also at times on the "round ups" in Wyoming. In 1904 Mr. Cowen came to Scottsbluff and has been interested in the cement business and identified with the Cowen Construction Company, ever since, with the exception of three years which he spent as a farmer in Arkansas.

On December 31, 1894, at Harrisburg, Banner county, Nebraska, Mr. Cowen was united in marriage to Miss Maude Dennison, who is a daughter of Edward and Mary (Urban) Dennison, the former of whom was born in Illinois, and the latter in Bohemia. The parents of Mrs. Cowen live at Scottsbluff and her father is interested in the cement industry. Mr. and Mrs. Cowen have had children as follows: Grace, who died when aged eighteen years; Luretta, who is employed by the local telephone company; Vera, who is also a telephone operator; and Lovella, Nellie and Edward Mason, who are yet in school. Mr. Cowen has taken much interest in civic matters ever since coming to Scottsbluff and his attitude on many public questions has won him the confidence of his fellow citizens, which they have evidenced by electing him a member of the city council for the third time. He belongs to the Masonic fraternity and to the Modern Woodmen.

ELISHA M. COWEN, who is at the head of the Cowen Construction Company, Scottsbluff, for the past fourteen years has been identified with building interests here, and to him the city is largely indebted for the substantial character of the larger number of its residences and business houses. The Cowen stamp on a building marks material and workmanship the best that can be secured.

Elisha M. Cowen was born at Cummington, Berkshire county, Massachusetts, March 28, 1848. His parents were James M. and Julia M (Mason) Cowen, the former of whom was born at Glasgow, Scotland, and the latter in Massachusetts. In early youth James M. Cowen was bound out to an uncle, in whose cotton-spinning mill at Preston, England, he learned the trade of spinner and before leaving the mill had become a foreman. In 1840 he crossed the Atlantic ocean to the United States in a sailing vessel and found work in one of the great mill districts of Massachusetts. Of his two children Elisha M. is the only survivor.

Elisha M. Cowen was educated at Albany, New York, passing through the high school and the normal college, and in 1863 was graduated from the Bryant & Stratton Commercial college of that city. While attending school in Albany, he was occupied during a part of the time with the duties of a page in the House of Representatives there. Although opportunities were afforded him for a professional or possibly a political career, for he made many influential friends at the capital, his inclinations were in an entirely different direction. He learned the bricklaying trade and worked at the same in Albany until 1866, when he went to Chicago, Illinois, worked there at his trade, then to Iowa for three years, and then to Banner county, Nebraska. Mr. Cowen then homesteaded and remained on his land for seventeen years, removing then to Colorado Springs, and from there, in 1905 came to Scottsbluff. Mr. Cowen has been very successful in his business undertakings and is ranked with the leading men in his line in this part of Nebraska.

On January 25, 1870, Mr. Cowen was united in marriage to Miss Elvira Triplett, who was born at Princeton, Illinois and is a daughter of Edward and Lucinda Triplett, natives of Ohio. They moved on a farm in Illinois and both died there. Mr. and Mrs. Cowen had two children born to them, namely: Nellie and Frank E. The latter is prominent in public affairs at Scottsbluff and is serving in his third term as a member of the city council. He is a cement contractor and is connected with the Cowen Construction company.

The only daughter of Mr. and Mrs. Cowen, Nellie, was the widow of W. M. J. Brozee, and had two children: Stanley and Etola, the latter of who is the wife of Frank H. Burbank, who is a railroad man. Stanley Brozee was an employe of the Mid-Continent Oil Company at Bartlesville, Oklahoma, prior to entering military service in the World War, in May, 1918. He was trained in ambulance service at Tarrytown, New York. His mother died in 1900 at the early age of twenty-two years. Mrs. Cowen is a member of the Baptist church. Since 1870 Mr. Cowen has belonged to the order of Odd Fellows and for many years has been a Mason and twice has been master of his lodge.

PHILO J. McSWEEN, chief of Police Department at Scottsbluff, occupies a position that requires personal courage, together with a large measure of discriminative judgment and understanding of human nature. Since entering upon the duties of this office, Chief McSween has enforced the law intelligently and without fear or favor and to the fullest extent enjoys the confidence of the law-abiding public.

Andrew J. Faulk, M. D.

Philo J McSween was born in Burnet county, Texas, September 11, 1872, and is a son of Dr John and Elizabeth (Wright) McSween The father was born in Tennessee and the mother in Mississippi, in which latter state they were married and the father obtained his medical degree from a Mississippi college After some years of practice there he moved to Texas, in which state he also practiced his profession for some years and then went into the cattle business in which he continued until he retired from active life His death occurred in Texas and that of the mother of Chief McSween in Colorado Of their nine children Philo J was the youngest and is the only one of the five survivors living in Nebraska Before the war between the states, Dr McSween was a man of large fortune, but like many others he lost heavily through circumstances over which he had no control As long as he lived he was a conscientious supporter of the priciples of the Democratic party He was a faithful Mason and strict in his adherence to the tenets of the Presbyterian church

Philo J McSween obtained his education in the schools of Burnet, Texas When eighteen years old he went to Colorado and for sixteen years he was concerned with farm industries there, having had experience in his native state From Colorado he came to Nebraska and in 1907 embarked in the meat business at Scottsbluff in partnership with M E Harris, but two years later bought a farm in the county, which he conducted until March, 1919, when he sold it and moved back into town, accepting the appointment of chief of police

At Brush, Colorado, January 10, 1899, Philo J McSween was united in marriage to Miss Lureada Lee, who was born in Iowa and is a daughter of Joseph and Rosaline Lee, natives of Kentucky, who moved to Iowa and became farming people there Chief and Mrs McSween have had five children, namely Myrtle, who was graduated from the high school at Scottsbluff in the class of 1919, Raymond L, who fell a victim of influenza in the epidemic of 1918, a promising and talented youth of sixteen years, and Merle, Mildred and Fred, aged respectively thirteen, seven and three years A staunch Democrat in politics a loyal member of the Knights of Pythias a public-spirited citizen and an efficient and reliable official, all these may be truthfully cited of Chief McSween

ANDREW J FAULK, M D, is one of the favored mortals whom nature launches into the world with the heritage of sturdy ancestry, a splendid physique, a masterful mind and energy enough for many men Added to these attributes are exceptional intellectual and professional attainments and useful lessons of a wide and varied experience stored away He is a type of the true gentleman and representative of the best in communal life, dignified, yet possessed of an affability and abiding human sympathy that have won him warm friends among all classes and conditions of men Of sturdy pre-Revolutionary stock, he was born strong of decision, with judgment and pronounced independence If a man comes of a good family he ought to be proud of it and he performs an immeasurable duty when he employs the best means to preserve the family record in enduring form, that future generations may receive instruction through the principles and influences, the personality and career of the forbears The subject of this biography can trace his lineage to colonial days, as two of his great-grandfathers came to this country and located in the Keystone state before the Revolutionary war and aided in reclaiming Pennsylvania from the virgin forest and possession of the Indians

Andrew J Faulk was born in Pittsburgh, Pennsylvania, June 13, 1858 the son of Thomas B and Sarah (Reed) Faulk, both natives of that great commonwealth Thomas Faulk received an excellent education and served on the editorial staff of one of the Pittsburgh daily papers for several years He was an ardent supporter of the Republican party and as a young man eagerly entered into political life, taking a leading part in party policy and administration, holding one office for twenty-seven years being reelected term after term Andrew's grandfather removed to Dakota Territory in 1862, locating in Yankton, where he entered prominently into the communal life of the city and surrounding country, and was territorial governor He was a member of the Republican national convention that nominated John C Fremont The grandfather died in 1898 passing away a man of honor in his eighty-fourth year Thomas B Faulk died at the age of fifty-nine years at Kittanning, Pennsylvania in 1898

Reared in such a family with its many traditions and high ideals it was but natural that the boy should receive an excellent elementary education afforded by the public schools of Yankton followed by broader and more comprehensive courses that developed his fine mentality and prepared him for the career to which he intended to devote his life's labors At the close of his academic career the young man entered upon the study of law, being admitted to practice in South Dakota in 1881, passing a brilliant examination before the bar For some years he followed the profession of law and his name became well known in the territory and state, but the career of a lawyer did not entirely satisfy him for he wished his life to be significantly one of service, for he is a man of unwavering optimism and abiding human sympathy and to satisfy these qualities he entered the Sioux City College of Medicine, receiving his degree of M D in 1901 For a short period the doctor was engaged in the practice of his profession in westeren Iowa, but in 1903, he removed to western Nebraska, and on the 25th day of September opened an office in Mitchell On coming here his ability soon gained him recognition, with the result that great success has attended his earnest efforts in his chosen calling In the work of his humane mission Dr Faulk spares himself neither mental nor physical effort, and carries relief and solace to those in affliction and distress His practice has grown to immense proportions for he has gained a reputation throughout the entire Panhandle as physician and surgeon The doctor is a namesake of his illustrious grandfather, while his ancestor, General Daniel Brodhead, was an officer in the Revolutionary War, and while the doctor has never sought military honors he is a worthy representative of his family, as he volunteered among the first on the declaration of war in 1917 He is a leading figure in all patriotic movements and takes an active part in all civic and national affairs that tend toward the betterment of living conditions in state and county His aunts have been at various times delegates of the Daughters of the American Revolution and one also served as president of this organization in both South Dakota and Nebraska Dr Faulk does not neglect his duties as a citizen of the city in which he makes his home but enters actively in the political life of Scottsbluff county as a staunch Republican, having served several terms as chairman of the Republican Central Committee and as Congressional committeeman He is a member of the Masonic fraternity and a Shriner, having taken his thirty-second degree, is also a member of the B P O E and the Independent Order of Odd Fellows and a generous supporter of the Episcopal church, of which the family are members He has served as surgeon general of the Patriarchs Militant for Nebraska, was the organizer and first president of the Scottsbluff County Med-

ical Society, is a member of the American Medical Society and state representative to the Nebraska State Medical Society, and was three times elected delegate thereto At the present time he is serving in important local offices, being president of the school board, and city physician, a position he is well qualified to fill as he is not only a highly educated man but is one of broad outlook who keeps abreast of the times and well up on all questions of the day, and at all times advocates the latest equipment and most advanced methods in the schools for the benefit of the rising generation

Dr Faulk was first married in 1881 to Mina L Fletcher, a native of the Empire state, who became the mother of two children Carl F, who chose law for his career and is now practicing in Alaska, and Mina Lucille, who is deceased Mrs Faulk was a highly educated woman of wide attainments, who for some time previous to her marriage taught "Methods" in New York State Normal She died in April, 1902 Two years later the doctor married Miss Maude E Baldwin of Minnesota, who is a woman of splendid talents and utmost sincerity, taking a very active part in all benevolent, charitable, and war work, and with her husband enjoys great popularity

E FRANK KELLEY, a man of ripened school experience and high scholarship, efficiently fills the office of county superintendent of schools in Morrill county His name carries weight in representative educational circles all over the state Mr Kelley was born in Illinois, October 16, 1876, but received most of his educational training in Nebraska

He is the son of James Dallas and Eugenia (Smith) Kelley, the former of whom was born in Pennsylvania and the latter in Iowa, the marriage taking place at Fort Madison, that state The mother of Superintendent Kelley died in 1914, but his father still resides at Portland, Oregon, where he was a mechanic in the railway shops for ten years, following similar trade employment in Illinois, Iowa and Nebraska He has always been considered a man of good judgment, has ever been faithful to his trade contracts, is a staunch advocate of the principles of the Democratic party and a member of the Presbyterian church He belongs to the order of Modern Woodmen Of his five children E Frank is the only one living in Nebraska

E Frank Kelley attended the Osceola high school following his graduation he matriculated at Fremont college, graduating after four year with his Bachelor's degree in 1909 and immediately began teaching school in Polk county, Nebraska Finding work in this profession congenial, he continued his pedagogical work in Polk county for five years, then served as principal of the Lodgepole schools for one year, and subsequently for three years was principal of the schools of Bayard In the meanwhile he became connected with the First National Bank of Bayard, an association which lasted two years before he became an official of the Bank of Bayard an institution with which he was connected five years until the fall of 1916 when he was elected county superintendent of Morrill county and assumed office in 1917, being re-elected in the fall of 1918 Mr Kelley has been pleasantly associated with the county teachers, who have found in him not only a competent educational leader, but also a wise and helpful friend this condition working beneficially for the schools all over the county

On June 24, 1903, Mr Kelley was united in marriage to Miss Clara Goldsmith, who was born at Ashland, Nebraska, in 1881, a daughter of David G and Helen Goldsmith, both of whom survive, the father being a retired farmer, with a home at North Platte, Nebraska Superintendent and Mrs Kelley have two children Helen, born May 14, 1905, and Dallas, born January 15, 1916 The family belong to the Episcopal church In politics a Democrat like his father, Mr Kelley also belongs to the order of Modern Woodmen

JOHN H STEUTEVILLE, who has most ably exercised judicial powers in Morrill county as County Judge for the last decade, stands in foremost rank with the substantial and loyal citizens of Bridgeport Judge Steuteville was born in Grayson county, Kentucky, December 1, 1875, the son of Richard Foggatt and Narcissa E (Haynes) Steuteville, who moved from Grayson Springs, Kentucky, to Brownville, Nebraska, in 1880, and still reside there Both parents were born in Kentucky, the father a son of Richard and Mary (Phillips) Steuteville, natives of Louisiana and Kentucky respectively, and the mother a daughter of Henry and Jane (Stith) Haynes, natives of Virginia and Kentucky These old names are yet familiar and honored in different sections of the South Judge Steuteville has two brothers and two sisters Earl, the postmaster at Bridgeport, Nebraska, William V, an attorney at Sioux City, Iowa, Jessie E Berlin, who resides at Brownville, Nebraska, and Mary, a teacher of mathematics in the high school at Sioux City, Iowa

Following his graduation from the Brown-

ville high school, John Henry Steuteville entered upon the study of law and in 1899 was graduated from the University of Nebraska College of Law For some years he was active in the educational field, at first teaching country schools in Nemaha county, Nebraska, and afterward served as principal of the city schools of Howe Johnson and Brownville, Nebraska, and of Belle Fourche in South Dakota He then engaged in the practice of his profession, first at Gering and later at Bridgeport When the county was divided, at the first county election, he was elected county judge, in which office he has continued ever since, being re-elected five times Not only on the bench has Judge Steuteville been a prominent and representative citizen of state and county, but in other relations and movements contributive to the general welfare, he has been a valuable co-operating force He is a 32d degree Scottish Rite Mason, and has filled all the chairs in the Blue Lodge at Bridgeport During the World War he served as secretary of the Council of Defense, was county Food Administrator, a member of the Home Guards, of the Four Minute Men, and was active on committees in the Y M C A movements and in other war preparations Judge Steuteville owns several farms in Morrill county, and today is accounted one of the substantial and representative professional men of the Panhandle

MABEL J JOHNSON, the county treasurer of Morrill county, Nebraska, has earned the reputation of being one of the most capable, energetic, efficient and likable officials that have been elected and re-elected to responsible office here for many years The spirit of progress that marks Nebraska in so many ways, is no more notably manifested than in opening doors of equal opportunity to both sexes and the calling of women as well as men who have the confidence of the public to positions of trust Miss Johnson, after one term of difficult duty meritoriously performed was re-elected county treasurer in 1918 and is still serving

Mabel Johnson was born at Omaha, Nebraska Her parents were Charles and Josephine (Palmquist) Johnson, both of whom were natives of Sweden They came to the United States in 1879 and resided at first at Minneapolis, Minnesota In 1899 they located at Omaha but subsequently the father homesteaded in Morrill county where he was engaged in farm industry until the time of his death in 1909 He was a member of the

Lutheran church, was a Republican in politics and belonged to the Independent Order of Odd Fellows   The mother of Miss Johnson survived her husband and now lives at Bridgeport   Of the family of ten children eight survive and two of the sons were in military service during the World War, August W and David G   The former saw seven months of hard service in France and participated in the memorable battle of the Argonne Forest, where he was severely wounded   This young hero has not yet recovered from his injuries and, although once more on American soil, is yet a sufferer in a military hospital at Des Moines, Iowa   David was yet in a training camp at the time the armistice was signed with Germany   The other brother and sisters of Miss Johnson are as follows   A C, who is a Broadwater ranchman and farmer, Anna V, who is the wife of L C Curtis, engaged in the sand business at Fremont, Mary, who is deputy treasurer of Morrill county, and Helen and Alice, who are yet in school

Miss Johnson was educated in the Omaha public schools   Her first public position was as an employe of the post office for four years, after which she served as deputy county treasurer for six years and was first elected treasurer in 1916 and re-elected in 1918   She votes the Republican ticket   Miss Johnson is a member of the Presbyterian church at Bridgeport

Z HAROLD JONES, district and county clerk of Morrill county, has been identified with county offices since 1914, entering public life from the educational field, in which he had been favorably known for some years   Mr Jones was born at Gretna, Sarpy county, Nebraska, March 28 1891, and his interests have always been centered in this state

Mr Jones' parents were Ziba and Mary I (Stansberry) Jones, who were born, reared and married in Iowa   In 1879 they came to Nebraska and settled in Sarpy county but later moved to Dawson county where the father bought land   This farm he subsequently sold   It was before the settlers had commenced to benefit by the irrigation project that later has brought such plenteousness into even the most arid territories   Mr Jones and his family returned to Sarpy county and located on a farm twenty-five miles southwest of Omaha, on which the family lived for twenty years   The father retired from active work at that time and moved to Gretna, where his death occurred in 1900, when fifty-

two years old.   He had been a man of considerable importance in Sarp county, was active in the Republican party and was a member of the Congregational church   For a number of years he had been a member of the order of Modern Woodmen of America, under the auspices of which he was buried, and in which organization he carried insurance to the amount of $3,000   Of his eight children three besides Z Harold survive: Ella J, the widow of John Hickey, lives at Marsland, Nebraska and owns two large ranches in Sioux county, Nebraska, George P, a miller at Hemingford, in Boxbutte county, and Augusta who is the wife of Arthur E Simonds, of Bellevue, Nebraska, agent for the Burlington Railroad   The mother of the family lives with her son at Bridgeport and belongs to the Presbyterian church of this city

Following his graduation from the Gretna high school in 1907, Mr Jones for six years alternated teaching and attending school to carry on higher and extended studies and thus qualify for better positions   He taught one year in the Bridgeport high school, being re-elected at the close of his contract   The offer he did not accept, however, and in 1914 entered the county clerk's office as deputy, which position he served until January 1, 1917, when he was elected clerk for the two-year term, and in November, 1918, was re-elected   His duties include those of both district and county clerk and complete efficiency marks their administration

On January 1, 1919, Mr Jones was united in marriage to Miss Nell Jeffords, who was born at St Paul, Nebraska, a daughter of John F and Rose (Cordell) Jeffords, who were born, reared and married in Illinois   Some thirty years ago they came to Nebraska, locating first at Loup City but moving later to St Paul and afterward coming to Bridgeport, where the father engaged in the jewelry business, a vocation in which he built up an excellent clientele and which he conducted until his death here   Mrs Jeffords survives her husband and yet resides here

Mr Jones is affiliated with the Democratic party, takes an active part in civic affairs in Bridgeport in many ways, and is secretary of the school board and treasurer of the Home Guards   He is present chancellor commander of the Knights of Pythias, and belongs to the Bridgeport Progressive Club   Mr and Mrs Jones are members of the Presbyterian church, and are prominent in all social activities of the city

BENAJAH A ROSEBROUGH, a prosperous business man and thoroughly respected citizen of Mitchell, Nebraska, who is connected in a business way with the Mitchell Mercantile company, lays no claim to being a pioneer of this section, though he has undergone many of the pioneer conditions He is one of those who, having spent a period on a farm, deserted the soil to enter commercial pursuits and has found success and prosperity therein, for today he enjoys great popularity due to courteous treatment, absolute fidelity to engagements reasonable prices and expeditious service All these qualities have served to attract to the store trade that extends over a wide stretch of the surrounding countryside His standing in business circles is excellent, and rests upon more than a decade of honorable and straightforward dealing

Mr Rosebrough is a native of Illinois born at Havana, August 20, 1868, the son of Ben A and Matilda (Tomlin) Rosebrough the former a native of Ohio and the latter of New Jersey To them were born five children Elizabeth, the wife of Joseph P Fisher of Mitchell, deceased, Cora, married George Drake Coon, Pecos, Texas, and is deceased, Benajah A, Frank, who lives in Rockport, New York, and Bertha, the wife of Gilbert Carey a resident of Dewitt, Nebraska, deceased For many years the father of the family was a carpenter and contractor in Illinois but later became a farmer, a vocation he followed until his death which occurred April 8, 1907, his wife having passed away in 1876

Benajah received an excellent education in the public schools of Illinois and upon graduating from the high school entered Lincoln University, Lincoln, Illinois where he finished a course of study before graduation Soon after the close of his college career the young man was engaged in Y M C A work for about a year, but this impaired his health to such an extent that he was forced to seek less continuous occupations and accepted a position with the Hoosier Furniture Company of Lincoln, Illinois Thirteen months later he returned to New Holland for a vacation but left to accept the position of manager of the Ryan furniture store, of New Holland, Illinois Mr Rosebrough heard the call of the west however and after looking up various localities decided to come to Nebraska, which he did in 1904, the country looked good to him as he says today and he determined to make this great commonwealth his future home For a year he lived much as did some of the pioneers of the earlier days, but in 1905 he came to Mitchell to accept a position with the Mitchell Mercantile Company, is head of the undertaking, furniture, and hardware departments He at once began the study of embalming and received a license to practice in February, 1908, the following June he passed second in the class at Omaha, receiving his Nebraska license in 1910 Not satisfied with his preparation for this important profession Mr Rosebrough took a post graduate course in embalming under Professor Howard Eckles, being one of eleven men in a class of thirty-eight members to pass the examination in dermo surgery He has recently embalmed the largest known man in this section of the world as he was six feet and two inches tall, and weighed seven hundred and twenty pounds

On April 26, 1899, was solemnized the marriage of Benajah Rosebrough and Nellie Derr, who was born and reared in Illinois, and to them have been born five children Mary, at home, LaVerne, in Scottsbluff Paul, at home, Immogene and Dorothy The family are members of the Presbyterian church while Mr Rosebrough's fraternal affiliations are with the Independent Order of Odd Fellows and the Knights of Pythias while he exercises his privilege of the ballot as an independent, voting for the best man to fill office.

CHARLES D CASPER editor and proprietor of the Bridgeport Herald, has been known and appreciated in journalism in Nebraska for many years, and through the medium of his own facile pen might disclose much that is interesting in relation to newspaper work and political movements during that time For Mr Casper is equally well-known in public affairs and as a member of both houses of the state legislature, has been influential in placing some very important laws on the statute book He is a self made man and struggled up from a boyhood environment of orphanage and limited opportunity

Charles D Casper was born at Red Lion, near New Castle, Delaware, being one of two children born to Richard and Margaret (Reed) Casper His sister, Emma, is the widow of Richard Dilmore and resides in the city of Philadelphia Both parents spent their lives in Delaware the mother of Mr Casper dying in his childhood The father married Mary Reed sister of his first wife and they had two children, both of whom are now deceased The father never accumulated property Mr Casper's birth took place December 10, 1845, and his school privileges were limited, as he practically looked after himself until he enlisted for service as a soldier in a cavalry regiment in the Civil War, with which

he served two years and one month   In 1866
he came west and for three years was a member of the regular army of the United States,
receiving his honorable discharge in Dakota

Mr Casper then went to Iowa, where there
was need of harvest hands and after the season was over accepted work as a section hand
on the railroad   It was in 1872, at Victor,
Iowa, that he started in a printing office to
learn the trade and continued there and even
owned a newspaper in that town for a short
time before locating at David City, Nebraska,
where he established his first permanent residence   Mr Casper became a prominent factor in Democratic politics, was elected the
first county clerk of Morrill county and served three years   In 1885 he was elected to the
lower house of the state legislature, in 1886
was sent to the upper house from Polk and
Butler counties, and in 1893 was returned to
the house and served two terms   He returned
to David City and resided there until 1905,
when he came to Morrill county and homesteaded and in 1906 came to Bridgeport   In
the meanwhile, Mr Casper had conducted the
Bridgeport *Blade* for one year, and the Bayard
Transcript for eighteen months   On March 1,
1911, Mr Casper founded the Bridgeport
Herald, a weekly journal, which has built up
a wide circulation and fills a long felt want
It is ably edited and its columns give both the
news of the outside world and of local happenings that interest subscribers   In connection
with his newspaper, Mr Casper owns and
operates a fine job printing office

Mr Casper was married December 21, 1880,
to Nancy M Brownsett, who was born in the
Province of Quebec, Canada, and they have
three daughters   Emma M, the widow of
Earl M Duncan, is her father's able assistant
in the newspaper office, Grace A, the wife of
F J Hansen, a railroad agent at Shelton, Nebraska, and Ruby L B, the wife of A T
Bjoraas, a brick contractor at Torrington,
Wyoming   Mr Casper is a member of the
Presbyterian church   For many years he has
been a Mason and at the present time is serving as master of his lodge at Bridgeport

CLYDE SPANOGLE, who is one of the
three owners of the Bridgeport Bank, the pioneer banking institution here, is prominent in
other fields than banking, public affairs having
engaged his attention for some years, although
at present he gives the most of his attention to
the rapidly growing business of the bank
Mr Spanogle enjoys the distinction of having
been elected the first mayor of Bridgeport

Clyde Spanogle was born in Hamilton county, Nebraska, May 10, 1880   His parents were
Andrew J and Catherine (Stover) Spanogle,
who were born and married in Pennsylvania
They came from there in 1879 to Nebraska,
and the father bought two sections of land in
Hamilton county, in association with his brother, and latter established the first bank at Phillips, which he conducted for a number of
years, then sold, retiring from business, and
his death occurred in 1892   The mother of
Mr Spanogle died in 1902   In youth the
father and mother belonged to the Dunkard
church but lated united with the Baptist
church   The father was a man of sterling
character and in 1883 was honored in Hamilton county by election to the state legislature,
in which body he served with steadfast adherence to what he believed to be right

Clyde Spanogle attended the public schools
and completed his education in the Williamson School of Mechanical Trades at Philadelphia, Pennsylvania, graduating in 1900   Since
1903 he has been connected with the Bridgeport Bank, in the ownership of which he is
associated with his brother Mark Spanogle,
and Fred R Lindberg   Fred R Lindberg is
president of the bank, Mark Spanogle is cashier and Clyde Spanogle is assistant cashier

In 1909 Clyde Spanogle was united in mar-
riage to Miss Martha Sheffel, who was born at
St Louis, Missouri, and they have one son,
Andrew John, born in 1915

Mr Spanogle stands deservedly high in public esteem at Bridgeport, where for years he
has been an earnest citizen and a worker for
civic betterment   For five years he was chairman of the village board, elected on the Republican ticket, at different times has been
city clerk, and in 1918 was elected mayor of
Bridgeport   He has given encouragement to
many worthy business enterprises here as an
aid to commercial development and has been
liberal in his support of patriotic and charitable movements affecting the whole community   He attends the Episcopal church

FRANK H PUTNAM, who has been interested in the lumber business at Bridgeport
since 1905, has been active in the public affairs
of city and county and is well and favorably
known in Western Nebraska   Mr Putnam is
a native of Iowa, born in Davis county, September 13, 1855, a son of Green M and Mary
M (Kelsey) Putnam, the former of whom
was born in Illinois and the latter in Indiana
The paternal grandfather, Elijah Putnam, was
born in Virginia, moved from there to Illinois

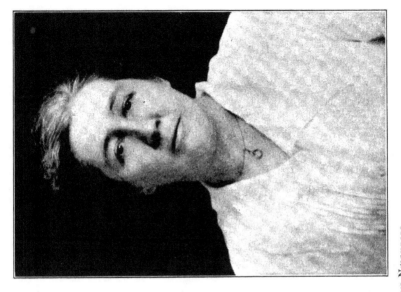

Mr. and Mrs. Joseph Neighbors

and later to Iowa where he engaged in farming during the rest of his life. The maternal grandfather, George Kelsey, died in Missouri but removed from Indiana to Iowa when the mother of Mr Putnam was a child. Both parents were reared in Iowa, were married there and both died in that state. Of their ten children six are living, Frank H., the only one in Nebraska, being the first born. The parents were members of the Methodist Episcopal church.

After his period of school attendance was over Mr Putnam assisted his father on the home farm until he was twenty-one years of age and for one year afterward engaged in agricultural pursuits on his own account. In 1878 he came to western Nebraska and for several years worked for a cow outfit, but in 1884 homesteaded in Morril county. Later he traded his homestead for Sand Hill ranch, and there was engaged in the cattle business until 1905, when he came to Bridgeport and bought the Bridgeport Lumber yard. The business at that time was incorporated for $25,000, but under his able management was increased to $75,000. He served as president of this concern until 1917, when he sold his interest but still fills the office of manager.

Mr Putnam was married in 1890, to Miss Emma C Hutchinson, who was born in Wisconsin, and they have two children Glenn G., formerly a farmer in Morrill county, and Hazel M., the wife of Chester Carter, who has returned to Bridgeport after two years of overseas service with the American Expeditionary Force in France. Mrs Putnam is a member of the Episcopal church. A zealous Republican, Mr Putnam has been honored by party choice for responsible public positions and has served on the city council and also as county commissioner. He is somewhat prominent in the order of Odd Fellows and has passed all the chairs in the local lodge.

THOMAS F NEIGHBORS, one of the younger members of the bar at Bridgeport where his friends and well wishers are many has been in practice here since 1915 not continuously, however, as he spent almost two years as a soldier in training during the World War. Mr Neighbors is a native of Nebraska, born at McGrew in Scottsbluff county in 1891.

Mr Neighbors comes of military ancestry as his paternal grandfather Joseph Neighbors was a soldier in the Civil War and fell at the battle of Nashville. His maternal grandfather, Dr Thomas Franklin served with the

rank of captain in the Civil War, under the command of General Grant. Afterward he became a physician at Gering, Nebraska and continued in practice there until his death.

The parents of Mr Neighbors are Joseph G and Carrie A (Franklin) Neighbors, who were born, reared and married in the state of Missouri. They came to Nebraska in 1885, settling first in Custer county, but in 1887 the father homesteaded in Scottsbluff county and the family home has been near McGrew ever since. The father has always been affiliated with the Democratic party but has never accepted public office. He is a member of the Baptist church and was one of the founders of the lodge of Odd Fellows at Bayard. The mother was reared in the Methodist Episcopal faith. Of their five children three survive. Grace, the wife of Samuel Shove, a merchant at Glenrock, Wyoming, Thomas F, of Bridgeport, and Melvin, who resides on a farm near McGrew.

Thomas F Neighbors attended the country schools in early boyhood, in 1908 was graduated from the high school at Bayard, from the Wesleyan Academy at Lincoln, in 1912, and in 1915 completed his course in law at the University of Nebraska. He immediately entered into practice with F E Williams, the partnership being dissolved when both answered the call to arms. Mr Neighbors entering service May 12, 1917. For three months afterward he was in the training camp at Fort Snelling, Minnesota and afterward until his discharge in February, 1919, was at Camp Dodge. Upon his return to private life Mr Neighbors immediately picked up the broken threads of his personal business and re-established his law practice at Bridgeport where he has found his professional efforts appreciated. He has served as city attorney both at Bayard and Bridgeport.

On September 4, 1918, Mr Neighbors was united in marriage to Miss Irene Welsher, who was born at Knoxville, Iowa a daughter of B R Welsher. Mrs Neighbors grew up in the Methodist Episcopal church, but Mr Neighbors is an Episcopalian. In politics he is active in his support of Republican doctrine and fraternally is identified with the Knights of Pythias. He is a young man of stable, well poised character, able in his profession and earnest and public spirited as a citizen.

ROBERT E BARRETT — The purchaser of land who is careless about securing a clear title to the same, often finds himself involved in serious legal difficulties as to real ownership Hence the careful, patient abstractor is best called in, his accurate and attested documents making investments sound and safe In Robert E Barrett, city clerk of Bridgeport, Morrill county has one who has had long experience in the abstract business Mr Barret is a native of Nebraska, born at North Platte, October 23, 1872

His parents were Harry and Jane (Barchard) Barrett, the former born in Ireland and the latter in England Both came to the United States as young people and were married in the state of Missouri Of their twelve children Robert E was the seventh born and six still survive The father was connected with railroad construction work all his life, a vigorous, hardy, dependable man While yet young, in Missouri, he was foreman of section gangs and after coming to Nebraska in 1867, was continued in the same position and for years was employed in construction work in the vicinity of Lodgepole, where both he. and wife died They were faithful members of the Roman Catholic church In his earlier years the father was a Democrat but the issues brought forward in the campaign of 1884 when Hon James G Blaine was a candidate, caused him to change his party allegiance and ever afterward was a Republican

Robert E Barrett attended school at Lodgepole and Chappell, Nebraska, and the first work he ever did was as a laborer on the railroad In the course of years he was interested along other lines, and in 1904 he was elected county clerk in old Cheyenne county, serving four years It was while acting in this public capacity that he did his first abstract work and was the only abstractor in the county Later he moved to Julesburg, Coloroda, where he engaged in the lumber business for seven years before coming to Bridgeport to open an abstract office, and has developed this business into one of great importance In politics he is a Republican, was city clerk of Bridgeport, was census enumerator in 1900, and is secretary of the Northport Irrigation District In 1897 Mr Barrett was united in marriage to Miss Grace Durkee, who was born in the state of New York, a daughter of David Cook Durkee, a homesteader in Nebraska Later he and wife removed to Julesburg, Colorado, and still live there Mr and Mrs Barrett have three children : Maude, Barchard and Leander The family belongs to the Presbyterian church It is not always that men immersed in business cares find leisure for literary expression even if they have talent, and it must be conceded that Mr Bar-

rett has had a fairly busy life Nevertheless, he has found time to add to the world's contribution of enjoyable literature, has published one book, "Treading the Narrow Way," and has written poetry of high literary quality

WILLIAM E GUTHRIE, whose extensive business activities and public efforts have made him prominent for years in Wyoming and Nebraska, has been a resident of Bridgeport since 1904, and he is now secretary of the board of irrigation in this district Mr Guthrie was born at Rue, in Marion county, Ohio, July 26, 1849, the son of Isaac F and Rachel (Fredrick) Guthrie The father was born in Ohio, a son of Joseph Guthrie, and a grandson of Colonel John Guthrie, an officer in the Revolutionary War, who was born in Pennsylvania and settled at an early day in Pike county, Ohio The mother was born in Virginia, a daughter of John Fredrick, an early settler of Ohio Mr Guthrie's parents were married in Ohio and he was the second born of their twelve children, the other survivors being as follows· S A , in the sheep business in Wyoming, a sister, who is the wife of County Clerk Clelland, of Converse county, Wyoming, P. E , in the cattle business, lives at Broken Bow, Nebraska , and another sister, the wife of J B Russell, a capitalist of Savannah, Missouri The father of this family was very prominent in Marion county, Ohio, for many years He was a successful farmer there and owned his Ohio farm until the time of his death, although, in 1885 he came to Merrick County, Nebraska, bought land near Clarks, and died on that place In politics he was a Democrat For twelve years he was county commissioner of Marion county and for fifteen years was a justice of the peace He belonged to the Masonic fraternity and lived up to every rule of the order The mother of Mr Guthrie was a member of the Methodist Episcopal church and the father was a liberal contributor They were people of solid worth and their descendants recall them with emotions of pride and veneration

William E Guthrie enjoyed educational advantages in the district schools in boyhood and later in the Wesleyan University at Delaware, Ohio From college he returned home to give his father assistance and remained until 1878, when he came to Wyoming and there, for twenty-five years prospered in the cattle business In 1895 he located in Omaha and shortly afterward bought a farm and feedyard at Clarks, in Merrick county, where he continued to handle cattle for the next twenty years

ELMER Z. JENKINS AND WIFE

In the meanwhile he had become active in the political field and in 1890 was elected to the Wyoming state legislature on the Republican ticket and took part in bringing about some very important legislation In 1904, when Mr Guthrie came first to Morrill county he became deeply interested in the irrigation projects and bought land along the Belmont Irrigation Canal, has continued his active interest and, as mentioned above is secretary of the board that is expending $75 000 in putting in drains and headgate in the Morrill county irrigation district Mr Guthrie owns four irrigated farms and gives much of his time to their development

In 1885 Mr Guthrie was united in marriage to Miss Margaret Hewitt, who was born at Zanesville, Ohio, but was reared in Des Moines, Iowa They have one daughter, Margaret, the wife of I P Hewitt who is connected with the Puget Sound Navy Yard, at Everett, Washington They have two children William Guthrie Hewitt and Helen Hewitt Mr Guthrie is a York Rite Mason and a Shriner and belongs also to the Knights of Pythias and the Elks

ELMER Z JENKINS — In section 1, township 23-56, near the thriving town of Mitchell, in the north central part of Scottsbluff county, will be found the attractive and admirably improved farm home of him whose name initiates this paragraph, who is successfully engaged in general farming and stockgrowing and who is known and valued as one of the influential and representative citizens of the community

Mr Jenkins is a contribution made to Nebraska by the fine old Buckeye state, but there is no faltering in his appreciation of and loyalty to the great state in which he has achieved prosperity through his own well ordered endeavors He was born in Jackson county, Ohio, September 30, 1872 and is a son of Andrew J and Charlotte (Moore) Jenkins he a native of Ohio, and she of Missouri, the father being eighty years of age and the mother above seventy years at the time of this writing, in the winter of 1919 Andrew J Jenkins was a farmer in Ohio but he went as a pioneer into Kansas, where he took up and perfected title to a tree claim, a property upon which he made good improvements and upon which he continued to reside several years He was a member of a gallant Ohio regiment that did valiant service in defense of the Union during the Civil War, and in his venerable years he finds deep satisfaction in his affiliation with the Grand Army of the Republic He is a stalwart Re-

publican in politics, and his wife is a devoted member of the Methodist Episcopal church Of their seven children, Elmer Z, of this review, is one of the two eldest, his twin sister, Bertina now a resident of Lincoln, Nebraska, being the widow of Lafayette Sherrow, Mary is married and resides at Shenandoah, Iowa, William is a painter and decorator by vocation and resides at Kansas City, Missouri, Albert is a street-car conductor in the city of Lincoln, Nebraska, Roy resides in Kansas City, and Lottie is deceased

Elmer Z Jenkins gained his youthful education in the public schools of Ohio and Kansas, and he has been a resident of Scottsbluff county since 1908, when he entered claim to the homestead upon which he has since maintained his residence, the same comprising eighty acres, and the entire tract having excellent irrigation facilities He has erected good buildings and made other modern improvements on the place, and is making definite success in connection with his vigorous enterprise as an agriculturist and stock-raiser He has added eighty acres to his holdings by purchase He is influential in community affairs, has served nine years as a member of the board of directors of the consolidated schools of which his district is a part, and he is treasurer of the Farmers Union at Mitchell In politics he maintains an independent attitude and votes in consonance with the dictates of his judgment He and his wife are zealous members of the Methodist Episcopal church at Mitchell and he is a valued and popular teacher in its Sunday school as well as liberal in the support of all departments of its work

In 1898 was solemnized the marriage of Mr Jenkins to Miss Gertrude E Hoy, daughter of Daniel Hoy, who came from Virginia to Nebraska and who is now a prosperous farmer near Saltillo, Lancaster county this state To Mr and Mrs Jenkins have been born four children Arthur married Helen Lukens and will reside in this neighborhood, Carl and Clarence are at the parental home, and Inez died at the age of nine years In a fraternal way Mr Jenkins is actively identified with the Mitchell Camp of the Modern Woodmen of America

ALBERT F FISHER, who owns the controlling interest in the Nebraska State Bank at Bridgeport, of which he is cashier, is not only widely known in financial circles but for many years was one of the foremost educators in the state Mr Fisher was born at Wyanet in Bureau county, Illinois, November 5, 1871 and was brought by his parents to Nebraska in infancy He has spent his life in this state and some years ago homesteaded in Dawes county, south of Chadron

The parents of Mr Fisher, Eugene K and Hulda S (Smith) Fisher, were born, reared and married in Illinois Of their six children Albert E is one of the four survivors, the others being Henry L, a retired ranchman who lives at Chadron, Nellie M, the wife of Morgan H Nichols, a merchant at Chadron, and Ralph W, a traveling salesman out of San Francisco, lives at Oakland, California In 1872 the parents came to Fillmore county, Nebraska The father has been a farmer all his life and owns a section of land in Dawes county, but now lives retired at Chadron, where the mother died She was a faithful member of the Presbyterian church and a woman of beautiful Christian character The father was reared in the Baptist faith He has always given his political support to the Republican party and for many years has been identified with the Independent Order of Odd Fellows

Albert E Fisher entered Chadron Academy, from which he was graduated in 1898 He had taught school in the meantime, for six years, then entered the university at Omaha, from which he was graduated in 1905 He found teaching a congenial vocation and continued in the educational field until 1917, during this time serving as superintendent of schools at Beemer, Nebraska, from 1905 to 1908, at Neligh, from 1908 to 1910, and was superintendent of the Aurora schools from 1910 to 1917, when he retired from educational work He then embarked in the banking business at Bayard where he remained one year as president of the Farmers State Bank of Bayard, then came to Bridgeport, taking over the controlling interest in the Nebraska State Bank of this city This institution is a sound, reliable, prosperous bank, with a capital of $25,000, surplus and profits $5,000, and average deposits $70,000 In 1918 Mr Fisher was elected president of the Bankers Association of Nebraska He has been honored many times by the Nebraska State Teachers Association and has served in all the higher offices of that body and for five years was a member of the state examining board

On December 26 1905 Mr Fisher was united in marriage to Miss Katherine C Clark who was born at Craig, Nebraska, and they have three children Katherine, John A and Helen C, all attending school Mr Fisher and family belong to the Presbyterian church He is active in the Masonic lodge at Bridge-

port, is past master at Aurora and Bayard, belongs to the Royal Arch at Aurora, and is a charter member and master under dispensation at Bayard and now a member of Camp Clark lodge at Bridgeport He belongs also to the order of Highlanders In his political views he is a Republican

CHARLES E STEUTEVILLE, postmaster at Bridgeport, from the nature of his office is one of the city's best known citizens, and because of his efficient administration of the same, is one of the most popular He belongs to an old and most worthy Nebraska family and was born in Nemaha county in 1885, being a son of Richard F and a brother of Judge J H Steuteville.

Charles E. Steuteville completed the high school course at Brownville and then spent one year in the normal school at Peru, Nebraska He embarked in business as manager of a hardware store and lumber yard and for a number of years was identified with the lumber industry For eight years he was an employe of the Edwards-Bradford Lumber Company of Sioux City, Iowa, and for three years filled an important position with the C N Dietz Lumber Company of Omaha In 1908 he came to Morrill county and homesteaded, later worked for the Bridgeport Mercantile Company, and for two years acted as assistant postmaster On May 15, 1918, he was appointed postmaster and took charge of the office, with the duties of which he was already familiar This post office is continually growing in importance and Mr Steuteville has taken advantage of every opportunity afforded him to improve the local service

In 1911 Mr Steuteville was united in marriage to Miss Eva L Todd, who was born in Missouri They have one son, John Richard, born July 3, 1913 Mrs Steuteville is a member of the Presbyterian church Postmaster Steuteville's political affiliation has always been with the Democratic party and he is a member of high standing in the Independent Order of Odd Fellows at Bridgeport

MARTIN HANNAWALD — When that great artery of transportation, the Union Pacific Railroad, had been completed across Nebraska and the territory had become a state of the Union, many substantial and far-seeing men of the states farther eastward, began to take an interest in the prairie lands of the new state, and one of these was Martin Hannawald, then a farmer in Michigan and now a retired resident of Bridgeport For forty-five years this family has belonged to Nebraska, and for almost thirty years Mr Hannawald was a representative farmer and stockman in Hamilton county

Martin Hannawald was born in the state of New York, November 11, 1848, and obtained his schooling there He was yet a young man when he came as far west as Chicago, Illinois, where he lived for three years, during that time driving an express wagon for a livelihood, and then went to Van Buren county, Michigan, as a farm worker It was while so engaged that he met and was subsequently married to Miss Elizabeth Mather, who was born in Van Buren county and is one of the two survivors of three children born to Reuben and Celia (Caveney) Mather They were natives of New York, where Mr Mather was a well-to-do farmer They drove from New York in a covered wagon to Michigan, where the father of Mrs Hannawald became a prominent man He was a Republican in politics and served as township treasurer for seventeen years Both parents of Mrs Hannawald were members of the Baptist church Her one brother, Wright Mather, is a produce merchant at Saginaw Michigan

In 1874 Mr and Mrs Hannawald come to Nebraska and bought land in Hamilton county and for many years lived on that property and then sold and came to Morrill county Here Mr Hannawald purchased a large ranch and was in the stock business during the rest of his active life, in 1911 retiring to Bridgeport Of the family of eight children born to Mr and Mrs Hannawald the following survive Hattie, the wife of C D James, a farmer near Ericson, Nebraska, Celia, who married M J Cass, a retired farmer near Long Beach, California, Thomas J, who lives at Aurora, Nebraska, N L, who homesteaded as also did his wife and live near Bridgeport and Blondena, the wife of M Beerline, a hardware merchant at Bridgeport Mr Hannawald has always voted the Republican ticket but has never been willing to serve in public office though often urged, as a man of high standing and sound judgment, to accept such responsibility He belongs to the Masonic fraternity and both he and wife are members of the Eastern Star, in which Mrs Hannawald has been active and prominent She served as the first Worthy Matron of Bridgeport Chapter No 260 Mr and Mrs Hannawald can recall many interesting events of pioneer life in Nebraska, and they cherish many kind thoughts of those who, like themselves, had the courage and endurance to bear the inevitable hardships and helped to bring about better conditions

RALPH O CANADAY, one of the younger members of the Bridgeport bar, came to this city to establish himself in his profession in March, 1919, after his return from military service during the World War Lieutenant Canaday was born at Minden, Nebraska April 4 1891, the elder of two sons born to Joseph S and Mary Jane (Winters) Canaday His brother, Walter A Canaday, is in the real estate business at Bridgeport, and his sister, Mary Golda, is a senior in the State University

Senator Canaday, father of Ralph O Canaday, was born in Sullivan county, Indiana, a son of John Canaday, who was born in Kentucky, lived subsequently in Indiana, Illinois and Nebraska and died in the last named state in 1900 The Canadays probably settled in Kentucky contemporary with Daniel Boone and the grandfather of Senator Canaday was the the only member of his family that escaped during an Indian attack on the unprotected settlements Joseph S Canaday was married in Illinois to Mary Jane Winters, who was born in Crawford county, that state, and in 1887 they came to Nebraska He bought land in Kearney county and still lives at Minden He has been very prominent in Democratic politics in the county, served in the state senate, was county superintendent of schools and also county treasurer and has frequently been suggested for other public positions of responsibility He was the organizer of the Co-operative Elevator Association found all over the state and is president of the same With his family he belongs to the Christian Science church

Ralph O Canaday was graduated from the Minden high school in 1909 and spent six years in the State University, in 1915 being graduated with the degree of A B and in 1918 received his LL B degree He was admitted to the bar in 1917 and practiced at Minden until May 17, 1918, when he entered the National army going to the officers' training school at Camp Dodge and was commissioned second lieutenant of Company D Eighty-eighth infantry on August 26 1918 The end of hostilities came before his regiment left Camp Dodge and he received his discharge January 31, 1919 In March following he came to Bridgeport, formed a partnership with William Ritchie Jr, and has been engaged in the practice of law here ever since with encouraging success He has charge of the Central States Investments Company's business in Morrill county In politics Mr. Canaday is a Democrat and fraternally a Mason, belonging to Lodge No 19 A F & A M at Lincoln He belongs to the Christian Science church

WALTER A CANADAY, second son of Hon, Joseph S Canaday and Mary Jane (Winters) Canaday, was born at Minden, Nebraska, March 22, 1893 He was graduated from the high School of Minden in 1913, after which he took a commercial course in Boyle's Business college at Omaha He then went on his father's farm in Kearney county and remained interested there until in August, 1917, when he joined a medical corps for service in the World War, accompanied the American Expeditionary Force to France, where he served from August, 1918, until May, 1919, when he was discharged He returned home and visited one week, then came to Bridgeport and embarked in the real estate business in partnership with R C Neumann Mr Canaday's business future looks bright Like his brother he belongs to the Christian Science church Both are held in the highest possible esteem

RAYMOND C NEUMANN, a leader in the real estate business at Bridgeport in partnership with Walter A Canaday, is widely and favorably known For many years he was identified with agricultural interests in the state and later with business enterprises in this city Mr Neumann is a native of Nebraska and was born at Sidney, July 19, 1875, a son of Henry and Fidelia (McMurray) Neumann, the former of whom was born in Hanover, Germany, and the latter in Iowa They were married at Sidney, Nebraska, and three of their four children survive, Hank R , Raymond C and Rosebud The father came to the United States when fourteen years of age and shortly afterward enlisted at New York in the United States army, came to the western country as a soldier, took part in Indian warfare and assisted in guarding the railroad workers when the Union Pacific Railroad was being built into Sidney Later he became a stockman in Cheyenne county, Nebraska, and his death occurred at Denver, in 1910, where the widow yet survives Mr Neumann was a prominent factor in Republican politics in Cheyenne county and served as county commissioner

Raymond C Neumann obtained his education in the public schools and later taught school for two years, during 1896-97 He be-

gan life on a farm and early became interested in stock and particularly cattle, and there is a story told in the family that he was but six years old when he attended a round up and surprised the other cow punchers with his skill as a rope thrower In later years he substantiated this reputation Sometime later he rented his father's ranch for five years and went into the stock business, raising many horses and some cattle and making a success of his enterprise Afterward he engaged in the oil business at Denver for a time In 1905 he came to Bridgeport and was one of the first business men to go into the restaurant business here and two years later widened his business by opening a hotel, which he conducted until August, 1917, when he sold out and retired from that line, although he still owns the building Since then Mr Neumann has been interested in the real estate and insurance lines of business in which he has demonstrated his usual enterprise and good judgment

On November 20, 1897, Mr Neumann was united in marriage to Miss Callie Capron, who was born in Ohio, and they have three daughters, namely Violet, who fills an important position in the Bridgeport Bank , and Opal and Callie, both of whom are attending school While never unduly active , Mr Neumann has always been faithful to the principles of the Republican party, believing them safest for a real foundation upon which true Americanism can build For many years he has been identified with the Masonic fraternity He is numbered with the useful and representative citizens of Bridgeport

FRANK N HUNT, whose successful operations in real estate in Morrill county have resulted in a change of ownership of large and valuable tracts of land, and thus brought considerable outside capital to this section, has won a place among the leading business men of even much riper experience Mr Hunt belongs to Nebraska, having been born at Omaha, October 19, 1887, the son of George J and Margaret (Bouldin) Hunt, both descended from ancestors who settled in Maryland at an early day His father was born in the city of Baltimore, in 1856, where the Hunt family for generations has been prominent in financial and political affairs Immediately after his graduation from the University of Maryland, in 1876, he came to Nebraska and the impression he received during a year spent at Omaha, was so favorable that after his admission to the bar two years later, he came back to that city and became prominent as a member of the

William D. Linden

law firm of Condon, Clarkston & Hunt In the spring of 1893 he came to Morrill county and prior to locating at Bridgeport in 1904 gave his professional attention to the affairs of the Belmont Ditch, in which he was heavily interested He stands at the head of the Bridgeport bar His marriage took place in 1883 to Miss Margaret Bouldin, of Belair, Maryland, and Frank N is the youngest of their three children, of whom personal mention and a steel portrait appears on other pages of this work

After his preliminary educational training in the public schools, Frank N Hunt entered a military school in Missouri, afterward attending an academy at Macon, Missouri, for two years, then became a student at Lehigh University, South Bethlehem, Pennsylvania, from which he was graduated with the class of 1910 The same year he came to Bridgeport and spent one season on his father's ranch in Morrill county, and the following two years found him at work in the Bridgeport Bank He then took up Kinkaid land, of which he yet owns six hundred and forty acres, lived on his property for three years and then came to Bridgeport to enter the real estate business, opening his office in 1916 The firm does a general land and abstract business, has valuable clients all through the valley and takes pride in its reputation for business integrity

In 1912 Mr Hunt was united in marriage to Miss Sybil Ball, who was born in the city of London, England They have two children Lesa Mary and Frank Mr and Mrs Hunt belong to the Episcopal church Mr Hunt is a Democrat in politics and a loyal party worker, but personally is more interested in business than in politics He is devoted to the welfare of Bridgeport and is ever ready to cooperate with other good citizens for the city's benefit, and has had influence in bringing about improvements in many directions For some years he has been an Odd Fellow and at present is serving as secretary of the lodge at Bridgeport

WILLIAM D LINDEN —The Panhandle district of Nebraska is indebted to the neighboring state of Iowa for many of its representative citizens, and among the prominent figures in the industrial life of Scottsbluff county who is thus to be designated as a native of the Hawkeye commonwealth is Mr Linden He is one of the most loyal, progressive and valued citizens of Mitchell, where he has developed an important and successful enterprise in the conducting of a mill for the grinding of alfalfa and the manufacturing of a valuable

product, and where also he has been specially vigorous and enthusiastic in furthering the advancement of the town and the county, it having been his privilege to serve as the first mayor of Mitchell

Mr Linden was born at Mount Pleasant, Iowa, on the 14th of September 1872, the son of Andrew G and Elizabeth (Hakanson) Linden, both natives of Sweden, where they were reared and educated Andrew G Linden emigrated from his native land to the United States in 1864 and first established his residence at Galesburg, Illinois, where he became a teamster Eventually he became a pioneer settler near Fort Dodge, Iowa, where he became the owner of 160 acres of land, which he reclaimed and developed into a productive farm In 1873 he disposed of this property and came to Clay county, Nebraska, where he took up and perfected title to a homestead of 160 acres and where he became one of the representative agriculturists and stock-growers of the county He made excellent improvements on his homestead and there continued to reside until he was well advanced in years and was justified in retiring He and his wife now reside at Hershey, Lincoln county and are enjoying the gracious rewards of former years of earnest and honest endeavor, the former being seventy-four and the latter sixty-four years of age at the time of this writing in the winter of 1919 They are earnest communicants of the Lutheran church and politically the Republican party enlists the loyal support of Mr Linden Of the fine family of children the subject of this sketch is the eldest, Martin is a prosperous farmer near Funk, Phelps county Reka is the wife of Alvin Johnson, who likewise is a prosperous Nebraska farmer Anna is the wife of Morey Johnson a farmer near Hershey, Lincoln county Lillian is a popular teacher in the public schools of the state of Montana, Gustavus is a prominent ranchman and merchant in Tripp county South Dakota where he is serving as postmaster at Linden an office named in his honor, Esther died at the age of seven years

William D Linden was an infant at the time of the family removal to Nebraska and is the only one of the children born in Iowa He was reared on the old home farm in Clay county, there received the advantages of the public schools, and there he continued his active alliance with farm industry until he had attained to the age of twenty-seven years In the meanwhile he had the distinction of doing the first plowing by steam power in that county In 1900 he removed to Phelps county, where for three years he was engaged in farming and the raising of pure-blood Duroc-Jersey hogs and Red Polled cattle He then removed to Lincoln county, where he developed and irrigated a farm for his father, near Hershey, in which village he likewise established and put into successful operation an alfalfa mill In 1909 he removed to Mitchell, Scottsbluff county, and assumed control and active management of a well-equipped alfalfa mill which had been established by others, but which had proved a failing venture His energy, executive ability and technical knowledge proved potent in the redemption of the enterprise and placing the same on a most substantial working basis, so that it now proves a definite adjunct to the industrial activities of the village and the county.

Mr Linden has been a leader in movements and enterprises tending to advance the interests of Mitchell, and he was chosen the first mayor of the town, in which position he gave a most able and progressive administration, while he also rendered equally effective service during the five years that he was a member of the city council He is now president of the Mitchell Alfalfa Milling Company and the Mitchell Electric Light Company, and is one of the foremost boosters of the fine little city, in which he has established his home His political support is given to the Democratic party, he is affiliated with the Independent Order of Odd Fellows, and he is a member of the Congregational church his wife holding membership in the Christian church

The year 1902 made record of the marriage of Mr Linden to Miss Elizabeth S Sullivan, who was born in Illinois and who is a daughter of John Sullivan Mr and Mrs Linden have a pleasant home in Mitchell and the same is notable for its generous hospitality and good cheer, the while it is brightened by the presence of their five children — Raymond, Devona, Kenneth, Dorothy, and Doris

MIKE BEERLINE — The senior member of the firm of Beerline & Scott, hardware merchants of Bridgeport, Mike Beerline, belongs to that class of men who have fought their own way to position and independence through the exercise of qualities which have been developed under the stimulating influence of their own necessities When he entered upon his career he had little save his ambition and his yet undeveloped native ability to assist him, but so ably has he directed his activities that he has elevated himself to a place of commercial prestige and has assisted in making the enterprise with which he is connected one of Bridgeport s necessary business adjuncts

Mr Beerline was born October 16, 1857, in Auglaize county, Ohio, a son of Henry and Christina (Elsass) Beerline, the former a native of Kentucky and the latter of Ohio They were married in the Buckeye state, and in 1865 moved to Nebraska, where the father was engaged in agricultural pursuits until 1867, when the family removed to Missouri, where Mr Beerline met with a measure of success and remained until his death In 1871, the widow and children returned to Nebraska, locating in Sarpy county The father was a Democrat in politics, and his religious faith was that of the Lutheran church To this denomination belongs Mrs Beerline, who survives him as a resident of Papilion, Nebraska They were the parents of five children Mike, Mrs Eaton, the wife of an automobile garage proprietor at Papilion, George, who is carrying on agricultural operations in Morrill county, this state, Henry, successfully engaged in the implement business at Papilion, and William, a farmer in the vicinity of Broadwater, Nebraska

Mike Beerline was but sixteen years of age when he accompanied his widowed mother to Papilion, Nebraska, where he attended the public school, residing there until 1887, in which year he located in Morrill county and took up a tree claim On this property he resided for some eighteen years, continuing to till the soil and make improvements until 1905, the year of his advent in Bridgeport He had carefully saved his earnings, and upon his arrival in this enterprising and promising community invested his capital in a hardware business, in partnership with Elbert Scott, under the style of Beerline & Scott As a result of the splendid work and honorable methods of the partners the business has grown to large proportions, and now commands an excellent trade in Bridgeport and throughout the surrounding country Not only is Mr Beerline one of the thoroughly capable business men of Bridgeport, whose standing in the confidence of his associates is of the highest order, but he has also taken an active and constructive part in civic affairs He has supported ably and generously all movements tending to make for higher education, cleaner morals and better citizenship, and as a member of the town council for ten years was able to apply his inherent gifts to the securing of needed legislation and the obtaining of improvements for the place of his adoption He is a Democrat in his political allegiance and is accounted an influential factor in the ranks of his party in this community His only fraternal affiliation is with the local lodge of the Knights of Pythias, in which he has numerous friends

Mr Beerline was married in 1906 to Miss Blondina Hanewald, daughter of Martin Hanewald They have no children

ELBERT SCOTT — With the coming of Elbert Scott to Bridgeport, in 1894, there was added to this community the services of a young man who had both the ambition and the ability to become a factor of great general usefulness Since that time he has steadily advanced in position and prosperity, and at the present day, as a member of the hardware firm of Beerline & Scott, is accounted one of his community's substantial business men, occupying a position which presents him with opportunities for the commercial and civic advancement of the locality

Mr Scott, like numerous other residents of Morrill county, is a native of the Buckeye state He was born at Palmyra, Ohio, March 1, 1869, his parents being Alexander and Harriet (McKensie) Scott His father, born in Pennsylvania, was a young man when he migrated to Ohio, and there married a native of that state and entered upon his career as the operator of a farm In later life, when his agricultural ventures had proven successful, he branched out into other pursuits and for some years carried on a successful contracting business Both he and his wife were faithful members of the Latter Day Saints, in the faith of which denomination they passed away Mr Scott was a Democrat in politics, was elected to township offices on several occasions, and was a man of considerable influence in his community, where he was universally respected There were eight children in the family, of whom seven are living, but only two reside in Nebraska Rosel P, who is engaged in farming near Bridgeport, and Elbert

Elbert Scott enjoyed the benefits to be acquired through attendance at the public schools of Ohio, and remained in his native state as an associate of his father in farming until June, 1894, when he came to Nebraska and located at Bridgeport Here he secured employment with the Belmont Canal Company, and later conducted a ranch for several years He received his introduction to the hardware business with the concern of Elter & Company an association which continued from 1904 until 1906, the latter year Mr Scott embarked in an enterprise on his own account, in partnership with Mike Beerline, under the firm style of Beerline & Scott From a modest beginning the partners have built up an excellent business and their affairs are in a flourishing condition,

while their standing in business circles is of the best Mr Scott is a man of sound and practical ideas, possessed of clear judgment and good business sense, and has made a thorough study of the hardware trade as well as of the needs of the community at large and his patrons in particular His reputation for integrity has been honestly won through years of honorable dealing, while his good citizenship has been evidenced by his cooperation in movements of a progressive and beneficial character

Mr Scott was married in 1910 to Miss Mabel Pool who was born in Illinois They have no children of their own, but have an adopted daughter, Sylvia, who was born in April, 1918 Fraternally, Mr Scott is affiliated with the local lodge of the Independent Order of Odd Fellows, in which he has passed through the chairs and both he and Mrs Scott are members of the Rebekahs His political tendencies cause him to support the candidates and principles of the Democratic party

LLOYD WIGGINS — Beginning as a clerk in a country store, Lloyd Wiggins, cashier of the First National Bank of Bridgeport, Nebraska, probably learned lessons in patience, accuracy and courtesy that have remained with him through many years of business life and have been helpful from the beginning to the present A bank cashier very often has need of all these qualities in his dealings with his fellow men and responding to the many and varied demands of the public Mr Wiggins has been identified with the First National since 1915 and is one of its most popular officials He was born May 8, 1881, in Coshocton county, Ohio, the son of Warren and Ruth (Pigman) Wiggins, both of whom were born in Coshocton county, where the father still lives The mother died there in January, 1914 Of their six children Lloyd is the youngest of the three survivors, the others being Mrs Etta Russell, of Martinsburg, Knox county Ohio and D M a blacksmith at Bladensburg Ohio The family was reared in the Christian church The father is a retired carpenter His father, Kinsey Wiggins, was born in Ohio and died there and he also was a carpenter The maternal grandparents of Mr Wiggins were James and Mary (Hooker) Pigman both of whom came to Ohio from Maryland in 1910 He was a preacher in the Methodist Episcopal church, a circuit-rider, and continued to travel back and forth performing his religious duties as long as he lived

Lloyd Wiggins attended the public schools and then accepted a position as clerk in a country store, but this did not offer advantages for the future and Mr Wiggins soon made his way to Zanesville There, for three years, he saw life from the platform of a street car The next change was to the West and for one year he served as a clerk in a grocery store at Victory, Colorado, but this also was but a stepping-stone, for Mr Wiggins had the ambition and knew he had the ability, to satisfactorily fill a much more important position In December, 1906, he came to Mitchell, Nebraska, to enter the employ of Carr & Neff, lumber dealers, and subsequently bought an interest in the business, an association which lasted for eight years, as he attended to the company's interests at Mitchell, Scottsbluff and Bridgeport, in this way becoming well known to solid men of business who were not slow in recognizing his business capacity and strict integrity In 1915 he came to the First National Bank as assistant cashier and in the following year became cashier This flourishing institution is capitalized at $25,000, has a surplus of $5,000, and average deposits of $180,000

On June 8, 1910, Mr Wiggins was united in marriage to Miss Augusta M Mack, who was born in Germany She is a member of the Presbyterian church Mr Wiggins has never been unduly interested in politics but is a sound Republican and freely expresses his reasons for being of that political faith, but has never been willing to accept any public office He is devoted to his business, his home and friends, and finds many of the latter among the Masons and Odd Fellows, to which fraternities he has belonged many years, at present being senior warden of the Blue Lodge, F & A M, and past grand in the latter organization

ALBERT T SEYBOLT, whose name stands for business integrity at Bridgeport, established himself here in the real estate line, in a small way, in September, 1910 The business has expanded into a large enterprise, in which Mr Seybolt requires the help of three assistants Good business judgment, careful attention to clients strict honesty and prompt fulfilling of obligations have brought this progress about, and perhaps no dealer in real estate, insurance and abstracts in this section enjoys, in greater degree, the confidence of the public Mr Seybolt is a native of Nebraska, born at Plattsmouth, February 26, 1875, the son of George A and Mary J (Thorne) Seybolt, both born in 1843, in Orange county, New York They were reared, educated and married in the Empire state, and came from there to Nebraska late in 1874, but unfavorable conditions in the new country caused them to return to New York in 1877 Ten years later, in 1887, they again came to Nebraska and Mr Seybolt preempted land in Custer county and took a tree claim In later years he resided at Lincoln, engaging there in the real estate business, and to some extent, was interested in Democratic politics The mother died in 1904 and the father now lives at San Diego, California _ Of their five children, Albert T was the third born, the others being as follows · Anna, who resides at Ashland, Nebraska, Floyd, who now lives retired at Lincoln, was formerly a Federal bank examiner, Andrew D, whose home is at Ashland, Nebraska, and Sarah, the wife of James A Kurk, who is in the real estate business at Broken Bow, Nebraska

After Albert T Seybolt had completed his course in the public schools at Lincoln, he went to work on his father's farm in Custer county where he remained nine years, then removed to Douglas county and continued agricultural pursuits, for a time enjoying the same, and still takes an interest in looking after his fine irrigated farms After some experience on the road for a well known insurance company, Mr Seybolt came to Bridgeport and embarked in the insurance business himself, in connection with handling real estate, and has greatly prospered as noted above His is a conspicuous example of close attention to business, bringing commensurate results

On December 31, 1901, Mr Seybolt was united in marriage to Miss Bessie I. Milmine, who was born at Kenney, Dewitt county, Illinois, the only daughter of Murray M and Delia (Kent) Milmine The father was born at Hamilton, Canada, September 23, 1843, and now resides at Lincoln, Nebraska The mother was born at Lansingville, New York, September 14, 1845, and died October 11, 1896 They were married at Maroa, Illinois, December 31, 1867, lived for a time in Illinois but later came to Nebraska Mrs Seybolt has one brother, Edward K, who takes care of the abstract department of Mr Seybolt's office Mr and Mrs Seybolt have one daughter, Marian, now attending school The family belongs to the Presbyterian church and Mr. Seybolt has served on its board of trustees at Bridgeport Ever since locating here he has been active in all movements promising substantial benefit to city or county From principle, he votes with the Republican party, but political office has no attraction for him equal to that of business

ZADOCK GOODWIN, a resident of Mitchell, Scottsbluff county, who takes pride in claiming the fine old Hoosier state as the place of his nativity, was a young man of twenty-three when he came with his father to Nebraska and established his residence in Box Butte county, and thus to whom may be justly accorded pioneer honors in the famous Panhandle of the state, to which this history is dedicated. He has been a prominent and influential force in connection with the development of farm industry in this section of the state and he and his wife are now the owners of a large and well-improved landed estate in Scottsbluff county, where their attractive home is situated about twenty-five miles distant from the thriving little city of Mitchell, which is their postoffice address.

Mr. Goodwin was born in the vicinity of Greencastle, Putnam county, Indiana, on the 9th day of January, 1867, a son of Zadock Goodwin, who was born in Ohio and who was one of the most venerable citizens of Box Butte county, Nebraska, at the time of his death, in 1912, when he was ninety years of age. The maiden name of the mother of the subject of this sketch was Nancy Sigler, and she likewise was born near Greencastle, Indiana, her death having occurred when she was about forty years of age. Zadock Goodwin, Sr., was a farmer in Indiana and from that state he finally removed to Iowa, where he became a pioneer in the realm of agricultural and live-stock enterprise and where he remained until 1886 when he came to Nebraska and took up a homestead and a tree claim in Box Butte county. To this tract of 320 acres he later made very appreciable additions, and he became largely and prominently identified with the raising of cattle in this section of the state, where he became well known as a man of sterling character and marked business ability.

He whose name initiates this article was a boy at the time of the family removal to Pottawatamie county, Iowa, where he received the advantages of the public schools and where also he gained his initial experience in connection with agricultural enterprise and the stock business. In 1886 he accompanied his father to Box Butte county, Nebraska, where he took a preemption claim, to which he perfected his title and which he finally sold. In 1895 he became a pioneer farmer in Scottsbluff county. He has wisely made investment in land in this county and now has a valuable estate of 2000 acres, of which 160 acres are supplied with irrigation and given over to the raising of diversified crops. On his extensive ranch Mr. Goodwin likewise gives special attention to the raising of cattle and other livestock of excellent types and he has erected good buildings and made other modern improvements on the place. Though he still gives a general supervision to this fine estate, which he now rents, he is living virtually retired, in the enjoyment of the ample rewards for former years of intense and well-ordered activity as one of the world's productive workers. He is a staunch advocate of the principles for which the Democratic party stands sponsor in a basic sense and is a liberal and loyal citizen. His wife and daughters hold membership in the Presbyterian church.

In 1894 was recorded the marriage of Mr. Goodwin to Miss Lennie Shull, who was born in Monroe county, Iowa. Her father, Isaac Shull, was born in Indiana, as a representative of one of the very early pioneer families of the state. Mr. and Mrs. Goodwin have two children. Hildred, who remains at the parental home, was graduated in the University of Nebraska with the degree of Bachelor of Arts, and Mary, who likewise remains at the parental home, was graduated in the Mitchell high school, as a member of the class of 1919.

CHARLES F. CLAWGES, who has been identified with Bridgeport interests since the town's earliest days, serving as its first postmaster and in other important capacities, is a native of Missouri, born at Trenton February 23, 1865. He has lived in at least four states in the Union but has long claimed Nebraska as his home. He came to Cheyenne county in 1900.

The parents of Mr. Clawges were Dr. J. W. F. and Charlotte (Galander) Clawges, the former of whom was born in Kentucky and died in Missouri in 1869. The mother of Mr. Clawges was born at Gottenberg, Sweden, eighty-five years ago. She vividly recalls the long voyage from Sweden to the United States made in her youth in a sailing vessel that was on the water for three months before landing its passengers at New Orleans. In her long life she has witnessed many wonderful things come to pass, but the marvels of rapid transportation perhaps interest her most. She was united in marriage to Dr. J. W. F. Clawges at Annawan, Henry county, Illinois. During the Civil War he was regimental surgeon of the Seventh Missouri Cavalry, and afterward engaged in the practice of his profession in Missouri until his death. He belonged to the Masonic fraternity. There were six children born to Dr. and Mrs. Clawges as follows: Una, who is the wife of J. W. Cartwright, a carpenter and contractor at Bridgeport; Lottie, who is the wife of William Forrest, an attorney at Peoria, Illinois; Laura, who is the wife of W. A. Shellheimer, a farmer near Chillicothe, Missouri; Charles F. and Jack, who are twins, both of whom live at Bridgeport, and Daniel F., who is assistant postmaster at Kansas City, Missouri. The mother, who resides with her son Charles F., is a member of the Seventh Day Adventist church.

Charles F. Clawges completed his high school course at Annawan, Illinois, after which he spent some time in the Northwest Normal School at Geneseo, Illinois. For two years afterward he taught school in Illinois, then went to Kansas, where for three years he was employed in a railroad office, when he was transferred to St. Louis, Missouri, and remained in the same capacity there for three years, following which he spent one year at Spokane Falls, Washington. In 1900 Mr. Clawges came to Cheyenne county and went on a ranch with his brother Jack, the latter at the present time being superintendent of the boiler room in the Burlington shops at Bridgeport.

In 1895 Mr. Clawges was united in marriage to Miss Mary Leaf, who was born in Boone county, Iowa. She was the first wife and mother in the Bridgeport settlement and the first child born here was the son of Mr. and Mrs. Clawges, Dan, whose bright young life went out during the influenza epidemic, November 24, 1918, at the age of seventeen years. Mr. and Mrs. Clawges have a daughter, Una, who is attending school. Mrs. Clawges is a member of the Adventist church.

In 1901 Mr. Clawges was appointed postmaster of the new town of Bridgeport and continued in office for four years, and in many ways, as an intelligent and reputable citizen, was useful in bringing about stable conditions. For some years he conducted a barber shop and was influential in bringing other business concerns to the place. He invested in land as his good judgment recognized the opportunity, and now owns a valuable farm of 200 acres all irrigated. Since retiring from active business life at Bridgeport he has been a very successful salesman of automobiles for the Mitchell Car Company. In politics he is a Republican, and he has long been identified with the order of Knights of Pythias.

EDGAR C. PORTER. — Among Bridgeport's retired farmers are found some of the most substantial citizens of Morrill county. They are more than that, for they usually are men of such good business judgment and stable personal character, as to be a valuable controlling element in the regulation of civic affairs and a check on unwise expenditures. They have had experience. Not many of them had wealth when they came to Nebraska, and the ample fortunes they now enjoy, have only been secured through hard work, self denial, and close economy. They are examples of the value of the above qualities that, in times of national extravagance and distress, may well be listed as virtues. A prominent retired farmer of Bridgeport is found in Edgar C. Porter, who came to Nebraska in 1894. He was born in Madison county, Iowa, February 8, 1859.

The parents of Mr. Porter were John and Nancy Ellen (Crager) Porter, the former of whom was born in West Virginia and the latter in Ohio, in which state they were married. They were pioneers in Madison county, Iowa, where the father secured government land at $1.25 an acre, to the development and improvement of which he devoted the rest of his active life. Ten children were born to them and eight still live, Edgar C. being the fifth born. He has one brother, Samuel, living in Dakota, but the other brothers and sisters have remained in Iowa. The father supported with vigor the principles of the Republican party, and both parents were faithful members of the Methodist Episcopal church, with which the mother united when fourteen years old. Edgar C. Porter attended the country schools near his father's farm in Madison county, and early learned to be useful. He remained at home until about twenty-six years old and then started out for himself, pioneering after the manner of his father by coming westward, reaching Denver, Colorado, in 1885. He homesteaded in that vicinity and lived on his place for five years and then disposed of it. After eight years in Colorado, he came to Nebraska and in 1894 settled on North river in Cheyenne county, renting land for several years. In 1903 he purchased a tract of school land, to which he has added from time to time, until he now owns an entire section, 100 acres of which are irrigated and the rest is operated under a dry farming system. It may be remarked that the only property Mr. Porter owned when he came to Nebraska was a team and wagon, cow and calf. Now, in addition to his land above mentioned, Mr. Porter owns valuable town property as does Mrs. Porter, who also owns a section in Morrill county. They have a beautiful residence at Bridgeport into which they moved in 1913.

In 1894 Mr. Porter was married to Miss Hattie Mount, who is a daughter of William and Sarah (Stumpff) Mount, the former of whom was born in Shelby county, Illinois, May 2, 1849, and the latter in Fairfield county, Ohio In 1884 William Mount came to Buffalo county, Nebraska, subsequently lived in Logan, Weld and Sedgwick counties, and in 1894 came to western Nebraska He bought land along the Belmont Ditch and followed farming there for six years, then bought farms in Morrill county Later he sold his homestead, retired to Bridgeport in 1910, and looks after several acres of land adjacent to other town property In 1871 Mr Mount was married to Sarah Stumpff, and Mrs Porter is the eldest of their nine children Mr Mount is a Republican and has long taken an active part in political affairs, believing good citizenship demands it

Mr and Mrs Porter have three children Claudia, a popular teacher and well known in social circles at Bridgeport, Florence, who resides at home, and Marjorie, who is yet in school Mr Porter and his family are members of the Presbyterian church He has always been affiliated with the Republican party but has never consented to hold office He occupies his leisure time pleasantly with looking over his farm and stock within a short distance of Bridgeport

CHARLES O MORRISON — Foremost among the citizens of Morrill county, whose business success and high personal character entitle them to prominence, is Charles O Morrison, vice president of the First National Bank of Bayard, and the owner also of a large acreage of valuable land Although a native of another state, Mr Morrison has passed the greater portion of his life in Nebraska, lived on a farm until he was twenty-three years old and started out for himself on a limited capital He was born at Dixon, Illinois, August 10, 1862

The parents of Mr Morrison, William F and Virginia (Lichtenberger) Morrison, were born, reared and married in Pennsylvania Of their ten children eight are living, two of them being in Morrill county, namely Charles O and E W, the latter a retired resident of Bayard, Nebraska The parents located at Dixon, Illinois, in 1861 and the father engaged in farming in Lee county until 1870, when he decided to seek better opportunities in the West He brought his family to York county, Nebraska, the journey being made in a covered wagon after the fashion of the old Conestoga, dear to the pioneers, and shortly after reaching here he homesteaded and both parents of Mr Morrison spent the rest of their lives in York county, passing away at Bradshaw The father

became a man of consequence, serving in the early organization of the county, later as county commissioner and in other offices of responsibility He was one of the early Masons in York county and assisted in establishing the Christian church

Charles O Morrison attended the country schools in York county and remained on the home farm until 1884 when he embarked in the mercantile business For twenty-six years he was a merchant, first at Bradshaw, later at Phillips and then at Bayard, being in this line at Bayard for seventeen years He disposed of his mercantile interests in December, 1916 Mr Morrison assisted in the organization of the First National Bank at Bayard, in 1910, and has served ever since in the office of vice president The latest bank statement gives the following facts concerning this reliable financial institution Capital, $50 000, profits and surplus, $20,000, average deposits, $450,000 The accommodations rendered by this bank and the courtesy accorded patrons have been greatly appreciated

On September 1, 1896, Mr Morrison was united in marriage to Miss Catherine Miller, who was born at Toledo, Ohio Her father, James C Miller, came to Phillips, Nebraska, in 1889 and subsequently died there Mr Morrison is a member of the Episcopal church He is a Knight Templar Mason and belongs also to the Royal Highlanders and the Modern Woodmen His political affiliation has always been with the Republican party and at times he has served very usefully in town offices and for thirteen years has been a member of the town board Mr Morrison has invested extensively in land in Morrill county, presumably with the foresight of a keen and experienced business man, and now owns 1040 acres, 400 of which is irrigated

WILLIAM T McKELVEY, who is one of Bayard's respected retired citizens, can look back over thirty-three busy years in Nebraska, during which he built up an ample fortune, from a very small beginning Mr McKelvey was born in Clark county, Illinois in 1857

The parents of Mr McKelvey were Patrick and Mary (Campbell) McKelvey, the former of whom was born in County Donegal, Ireland, and the latter in Kentucky They both came to Clark county, Illinois, in early life and were married there The father of Mr McKelvey served in the Mexican War and afterward was a farmer and merchant in Clark county He held a number of township offices in the gift of the Democratic party He died in Clark county, but the mother of Mr Mc-

Kelvey died in Nebraska, in 1918, having lived with her son William T. for over twenty years.

Of their eight children the following are living, in addition to William T.: Lavona, who is the widow of Lafayette Beard, of Topeka, Kansas; Lydia May, who is the wife of James Beacham, a retired farmer of O'Neill, Nebraska; and Horatio A., who is a farmer in Minnesota. The parents were members of the Methodist Episcopal church.

William T. McKelvey obtained his education in the district schools in Clark county and worked on his father's farm. He was not robust as a boy and remained at home until he was twenty-seven years old, then came west, and in the active outdoor life and strenuosity of existence on the range, found health as did one of America's greatest statesman, the late Theodore Roosevelt. Mr. McKelvey located in Cheyenne, now Morrill county, in 1886, where he homesteaded and for a number of years rode range as a cowboy. He owns a large acreage in the county yet, for years being interested extensively as a stockman, and has property at Bayard, where, for one year before he retired, he was engaged in a real estate business. Mr. McKelvey passed through the hardships that attended the pioneers in the early eighties in this section of Nebraska, but he never became discouraged and now enjoys the fruits of his endurance and toil. He has been active in the Republican party both before and since locating at Bayard, has served in public office and was an exceedingly useful member of the first county board of commissioners. In every way, for years, he has done much to further the interests of Morrill county.

In 1892 Mr. McKelvey was united in marriage to Miss Jennie Webb, who was born in Clark county, Illinois. During the great war, Mr. McKelvey was foremost in patriotic work and was particularly active and interested in the Y. M. C. A. activities and was chairman of the local board.

HENRY E. RANDALL. — To the pleasant town of Bayard have come many men of ample fortune after many years of toil and financial struggle, finding here well earned ease with agreeable surroundings and pleasant companionship. These retired farmers and stockmen are desirable citizens in every respect and form a solid, dependable body that adds to the community's resources and gives assistance in the maintenance of law and order. One of the highly respected retired residents of Bayard is Henry E. Randall, who is well known all over Morrill county. Mr. Randall was born in Trempealeau county, Wisconsin, April 25, 1869.

The parents of Mr. Randall were James M. and Lucy (Hasson) Randall, the former of whom was born in Michigan, and the latter in New York. They were married in Wisconsin, this being the father's second union. Three children of his first marriage survive, namely: Charles, who is a miner in Nevada; Otis, who lives near Bridgeport, and Mrs. Elmer Hathaway, who is a resident of Morrill. Three children were born to the father's second marriage: Henry E., Dean and Arthur. Dean is a farmer near Melbeta in Scottsbluff county, Nebraska, and Arthur has been a mail carrier at Gering for a number of years. The father survives and resides at Gering. He is a member of the G. A. R. post there, having served in the Civil War as a member of Company I, Thirty-sixth Wisconsin Infantry. The family came to Nebraska in 1886 and the mother died here.

Henry E. Randall obtained his education in the public schools of Minnesota, where his parents lived for a time before coming to Nebraska. In the spring of 1886 the family reached what now is Morrill county, Cheyenne at that time, and in 1886 he homesteaded and kept the property until quite recently when he sold to advantage. As opportunity offered he bought other land and at one time owned 800 acres, his farms, four in number, being situated at different points. On one of these farms Mr. Randall lived for twenty-five years and during that time was an extensive raiser of cattle and stock. He retired to Bayard in March, 1916, and is a stockholder and one of the directors of the Farmers State Bank.

In 1891 Mr. Randall was united in marriage to Miss Melissa Belden, who was born in Kansas, and they have three children: Gerald, Gladys and Mack, the youngest son being yet in school. The one daughter is the wife of Merl Garwood, of Morrill county. The eldest son of Mr. Randall has an honorable military record. He was born July 24, 1895, was educated in Morrill county, and entered military service for action in the World War, 1917. He was attached to six different training camps, namely: Waco, Kelly, Field, and others, and at the time the armistice was signed with the enemy, was at Fort Sill and at Lee Hall, Virginia, just ready to sail for France. Since his discharge and return home, he has gladly resumed peaceful pursuits and has a place with the working force in the sugar factory.

Mr. Randall was quite active in Republican politics for many years and as a member of

Mr. and Mrs. Denver N. Plummer

the board of county commisioners, on which he served six years, when he resigned on account of ill health, he was able to greatly further the best interests of the county in many ways He belongs to the Odd Fellows and also to the Modern Woodmen

DENVER NEWTON PLUMMER is another of the progressive citizens who has shown the skill and enterprise that make for success in connection with industrial enterprise in the favored section of Nebraska to which this history is dedicated In section 7, township 23-56, about four and one-half miles distant from Morrill, Scottsbluff county, is to be found the well improved and ably managed farm of Mr Plummer He came to the county in 1910 and here purchased land in the north central part of this progressive county To his original domain he has since added until he now has a valuable landed property of 240 acres, all under effective irrigation and having the intrinsic richness of soil that makes irrigation farming so remarkably profitable in this locality He has made excellent improvements on his land, including the erection of good buildings, and is one of the resourceful and representative agriculturists and stock-growers of his county with a high personal standing that indicates fully the estimate placed upon him by his fellow men Loyal to all civic duties and responsibilities, Mr Plummer has given his support to measures and enterprises that have been projected for the benefit of his community and county, and in local politics he is independent, though in a basic way he advocates and upholds the principles for which the Republican party stands sponsor Both he and his wife hold membership in the Methodist Episcopal church in Dutch Flats

Denver Plummer was born near the city of Des Moines, Iowa, on the 29th of January, 1871, a son of James and Sarah Lavena (Garrett) Plummer The parents were born and reared in the old Buckeye state and the father was one of the early settlers of Iowa, where he accompanied his parents at an early age and where he reclaimed a pioneer farm The family later removed to Colorado and Ezra Plummer and his wife passed the closing years of their lives in the state of Colorado, both having been earnest members of the Methodist Episcopal church and his political faith having been that of the Republican party Concerning their children adequate mention is made on other pages, in the sketch of the career of John W Plummer, an elder brother of the subject of this review

Denver Plummer was about eleven years of age at the time of the family removal from

Iowa to Colorado, in which latter state he was reared to maturity and received the advantages of the public schools Prior to coming to Nebraska he had given his attention principally to farming and was owner of land in Larimer county, and the maximum success that has marked his career has been that gained since he established his home in Scottsbluff county and assumed the labors and responsibilities incidental to the development of a productive farm He has not waited for success but has won it through his own efforts, though he gives full credit to Scottsbluff county for the splendid opportunities it has afforded him

In 1905 Mr Plummer was united in marriage to Miss Cora Drummond, who was born in the state of Missouri, and their marriage has been blessed by three children — Veda Virginia, Ezra Allen, and Vivian Margaret — who lend brightness and cheer to the pleasant family home

WALTER J ERICSON, who is one of Bayard's representative citizens and substantial business men, is president of the Farmers State Bank and is also the head of the Ericson Hardware Company In many ways he has been active in the development of Bayard and stands deservedly high in public esteem He was born at Bertrand, in Phelps county, Nebraska, in 1885

The parents of Mr Ericson, John and Mary (Peterson) Ericson, were born in Sweden They came to the United States in the early eighties, acquired land in Phelps county, Nebraska, that is still in the possession of the family, and the father died on the homestead in the spring of 1919 The family lived at first in a sod house, as did many of their pioneer neighbors, and the father cultivated his land with oxen Of his family of nine children, Walter J was the fourth in order of birth The other survivors are Charles, in the drug business at Loomis, Nebraska, Frank, a general merchant at Hillrose, Morgan county, Colorado, Axel, a druggist at Bayard, Harry, also in the drug business at Bayard; Esther, the wife of Victor J Johnson, operating the old Ericson homestead, and Emil, associated with his brother, Walter J, in the hardware business He was born at Bertrand in 1890 and came to Bayard in August, 1915 On November 21 1917, Emil Ericson was married to Miss Alta Durnal, who is a daughter of R F Durnal

Walter J Ericson was reared on his father's farm near Bertrand, where he attended school, and remained in Phelps county until 1911, when he came to Bayard, where his brother

Frank was conducting a hardware business. He purchased his brother's hardware store and reorganized the business and it is now widely known as the Ericson Hardware Company, his brother Emil, as mentioned above, being associated with him. As president of the Farmers State Bank, he is additionally well known in commercial circles. While business claims much of his time, Mr. Ericson, as an earnest citizen, has concerned himself with civic development and betterment, and is an active, fearless and useful member of the city council, in which body he is serving his second term. In politics Mr. Ericson is a Republican, as was his father. He belongs to the Masons and the Odd Fellows.

GEORGE G. CRONKLETON, who is cashier of the First National Bank of Bayard, has been identified with the banking business during many years of his life and has been connected with the above institution since the spring of 1911. Mr. Cronkleton is held in high esteem at Bayard, where he has, on numerous occasions, been called to public office, in which he has served faithfully and conscientiously, and he has furthered many desirable public enterprises by the influence of his active interest. He is a native of Iowa, and was born March 14, 1876, at Dunlap, in Harrison county.

The parents of Mr. Cronkleton were Ezra J. and Julia (O'Hare) Cronkleton, the former of whom was born in Ohio, and the latter in Ireland. The father went to Iowa when a young man and served in the Civil War as a member of Company C, Second Iowa Cavalry, until captured by the enemy. He was a prisoner of war for ten months in Alabama. After returning from his exhausting experiences in the war, he traveled about for a time and then was married at Boone, Iowa, located soon afterward at Dunlap, and for many years was engaged in business there as a contractor and builder. His death occurred at Dunlap on August 17, 1913. In his earlier years he belonged to the Christian church but later became a Catholic, of which church his wife was a devoted member. Of their five children, George G. is the third of the survivors, the others being: Charles J., who is a resident of Council Bluffs, Iowa; Mary J., and Eugenia, both of whom live at Council Bluffs.

After completing the high school course at Dunlap, Mr. Cronkleton became deputy postmaster and subsequently deputy county auditor of Harrison county, his financial talents being thus early recognized and called into play. After retiring from office he accepted the position of assistant cashier in the First National Bank of Dunlap and remained with that institution for six years. He then made a visit to the Pacific coast and during his sojourn there served as cashier of the First National Bank of Ritzville, Washington. In the meanwhile his brother had engaged in the grocery trade at Council Bluffs, and when Mr. Cronkleton returned to Iowa he entered his brother's store and remained there three years. He then visited Wyoming and once more became identified with a large financial institution, serving for one year as assistant cashier of the bank of Noble, Lane & Noble, at Lander, Wyoming. He then accepted the office of cashier of the bank at Henry, Nebraska, and continued as such for eighteen months. In 1909 he came to Bayard and in May, 1911, became cashier of the First National Bank, an institution of which city and county are proud. It operates with a capital of $50,000; surplus, $10,000; average deposits, $450,000.

In November, 1913, Mr. Cronkleton was united in marriage to Miss Anna Morrow, a daughter of Thomas Morrow, extended mention of whom will be found in this work. Mr. and Mrs. Cronkleton are members of the Catholic church and he belongs to the Knights of Columbus. In his political affiliation he has always been a Republican. He has served as a United States commissioner, and since coming to Bayard has been town clerk and also a member of the school board.

THOMAS F. WATKINS, who, probably is as well known in Morrill county as any other individual, unless Mrs. Watkins, his admirable wife be excepted, came to Bayard in 1910. Since then Mr. and Mrs. Watkins have owned the Commercial hotel and have had much to do with the development of what was then a village into the close semblance of a city. Mr. Watkins was born at Swansea, Wales, May 27, 1848.

The parents of Mr. Watkins were Thomas and Mary (Davis) Watkins, the former of whom was born in Wales, December 11, 1814, and the latter February 27, 1822. They came to the United States and located on 160 acres of land in Monroe county, Iowa, when Thomas F. was an infant. There were two older children in the family, nine more were born in America, and besides Thomas F., the following are living: W. D., who resides at Long Beach, California; D. M., who owns the old family homestead in Monroe county, Iowa; Mary M., who is the wife of Thomas Lewis, of Long Beach, California, and Mittie, who is the wife of Martin Haller, a farmer near Springfield, Missouri The parents were members of the Baptist church

In his boyhood Thomas F Watkins had but meager educational advantages He early learned to perform farm work and labor in the coal mines In 1890 he came to Alliance, Nebraska, where he carried on a meat business for four years, after which he worked on a ranch in Box Butte county for four years In 1898 he came to Morrill county and after marriage took charge of his wife's homestead and put the property in fine shape Later he bought a tract of land near the homestead and subsequently his wife secured a second homestead In 1910 they moved into Bayard, as mentioned above, and went into the hotel business The Commercial hotel is known all through this section and patronage never fails

At Hemmingford, Nebraska, Mr Watkins was married to Mary Nebraska (Joice) Dual, who is affectionately known by man in Nebraska as she bears the distinction of having been the first white child born in Nebraska City, Neb In this connection, by her kind permission, part of a private letter is here quoted that will prove interesting to every reader

"My father, Jacob H Joice, of Dayton, Indiana, emigrated to eastern Nebraska in the year 1854 and settled in what is now the thriving city of Nebraska City, building the fifth house which was built in this city, obtaining the material from the natural forest surrounding the little place The population of the surrounding country consisted chiefly of Indians from the Otoe tribe which, as a general rule, were very friendly to the white settlers J H Joice was of Irish descent while his wife, formerly Angeline Blacklidge, was of Scotch descent I was born on the 17th day of December, 1854, and by the request of a prominent man of the community at that time, I was named Mary Nebraska, he promising if the child was so named he would deed her a quarter section of land when she reached her majority, but this promise was never fulfilled At the age of three years I moved with my parents to Iowa, locating where at one time Eastport stood From Eastport we moved to Hamburg, Iowa, where I grew to young womanhood and was married It was only a few years until I was left a widow, during which time I lived in Savannah, Missouri After the death of my parents the call of my native state appealed to me so strongly, that I emigrated to Bayard, Nebraska, where I met and married Thomas F Watkins The first year of our married life was spent in the city of Alliance, Nebraska, after which we located on a homestead three miles due east of Bayard, Nebraska, where we underwent the hardships and privations of the early pioneers of that day We lived the life of the pioneer ranchman for about sixteen years, when we moved to Bayard in the year 1910, purchasing and operating the Commercial hotel of that place Bayard at that time consisted of a population of about 200 men, women and children At this time the settlers coming from eastern Nebraska and adjoining states, began locating on the lands adjacent to Bayard, where they organized a successful irrigating project which transformed a former desert into a veritable Garden of Eden" Mrs Watkins concludes with expressions of pleasant anticipation concerning the annual homecoming celebration of the early pioneers of Nebraska City to which she had been invited and in which she has found herself a highly honored guest whenever she has been able to attend

Mr and Mrs Watkins have had no children of their own but they raised two children, Stella Slausen and Richard Dual, the former of whom is deceased, the latter being a resident of Bayard They also raised a nephew of Mrs Watkins, Frank Joice, deceased For forty-nine years Mr Watkins has belonged to the order of Odd Fellows and both he and Mrs Watkins are old members of the auxiliary order of Rebekah, and both have represented their local body in the Grand Lodge Mr Watkins has never been a politician and at present he maintains an independent attitude on public questions and votes according to his own judgment that has been ripened by many years of thought and mingling with his fellowmen Mrs Watkins has one brother and one sister H A Joice and Mrs Hattie White, both of whom reside at Bigelow, Holt county, Missouri

JOHN L LOEWENSTEIN who is the able manager of the L W Cox & Company lumber business at Bayard, is not only an alert and enterprising business man, but is also an earnest and useful citizen in a public capacity During the seven years of his residence at Bayard he has so firmly established himself in the confidence of his fellow citizens that in April 1918 he was elected a member of the city council and has faithfully performed every duty pertaining to this office

John L Loewenstein was born at Keokuk, Iowa, in 1882, the only son of Christopher and Caroline (Schultz) Loewenstein both of whom were born at Keokuk, of German parentage The mother died there in 1916, but the father survives and carries on his business of hardware merchant in that city. In politics the father is a Republican, fraternally is an Odd Fellow, and all his mature life has been a member of the German Evangelical church. Mr. Loewenstein has one sister, Mildred, who is the wife of Frank Wiseman, who is a salesman in the electrical line, at Oakland, California.

After his public school course at Keokuk, John L. Loewenstein attended a commercial school. The first business opening he found was in a shoe factory and for several years he remained there and learned the trade. From there he went into railroad work as an employe in the main office of the Chicago, Burlington & Quincy. One year later he accepted a position with the Iowa State Insurance Company, with which concern he continued for two years and then embarked in the drug business on his own account, at Cantril, Iowa, and remained so connected for four years. In the meanwhile Mr. Lowenstein kept alert as to other business opportunities, and when he found a congenial opening in the lumber trade, with E. G. Caine, at Indianola, Nebraska, took advantage of it and continued there until 1912, when he came to Bayard and accepted his present position. He has substantial knowledge along several lines of activity and a very wide acquaintance, has a genial manner that wins friendly attention and an upright character that in the business world means trustworthiness.

In 1904 Mr. Loewenstein was united in marriage to Miss Edna Frances Caine, who was born at Keokuk, Iowa. They have three children: Madeline, Lillian, and Josephine. Mr. and Mrs. Loewenstein are members of the Presbyterian church. He is a Republican in his political affiliation, and he belongs fraternally to both the Masons and Odd Fellows. Bayard has made wonderful progress within the last few years and credit is due those men of business foresight and true public spirit who have in every possible way furthered her interests and it is but just to say that Mr. Loewenstein is one of these.

CHARLES H. HARPOLE. — There are few lines of reputable business that do not have adequate representation at Bayard, in fact the little city can claim same progressive concerns that would be creditable anywhere. Reference may be made to the Burke & Harpole Company, dealers in general hardware and furniture and undertakers, the founder of the business being Charles H. Harpole, who came to Bayard in 1900. He was born in Warrick county, Indiana, December 15, 1863.

The parents of Mr. Harpole were W. S. and Elizabeth (Griffith) Harpole, the former of whom was born in Virginia and the latter in the city of New Orleans, Louisiana. Their people settled in Indiana when they were young and they were married there and remained until 1881, when they moved to southwestern Missouri. The father bought land there and both died on the home farm. Of their ten children seven are living, Charles H. being the only one residing in Nebraska. His educational opportunities were somewhat meager, confined to a little country school near his father's farm in southern Indiana. He grew up on the home farm, accompanied his parents to Missouri and afterward followed an agricultural life there until 1900, when he came to Bayard, Nebraska. Here he saw a business opening in the hardware line and started in a small way, in partnership with D. J. Burke. Immediate success followed as the village grew into a town and then a city, and Mr. Harpole and Mr. Burke proved equal to the occasion. At first they increased their stock but later found themselves needing more room and purchased the large brick building they now occupy. The business is now incorporated as the Burke-Harpole Company, which is capitalized at $30,000. There is hardly an instance in the city where a business enterprise has developed more rapidly or substantially. In addition, Mr. Harpole is interested to some extent in farming, owning valuable land in Morrill county.

In August, 1893, Mr. Harpole was united in marriage to Miss Elneta Mingus, who was born in Ohio. She died without issue, in June, 1914. Mr. Harpole was married second in 1916 to Miss Emma De Vault, who was born at St. Louis, Missouri. They are members of the Methodist Episcopal church and Mrs. Harpole is active in the various avenues of beneficence carried on by the church. Mr. Harpole belongs to no fraternal body except the Modern Woodmen. He is a staunch Republican in politics and has served with great efficiency on the school board and in other town offices. He is held in high esteem at Bayard and is numbered with the representative business men of the place, one ever ready to encourage worthy enterprises and generous in his support of charitable movements.

WILLIAM WEBER. — The year 1887 marked the arrival of Mr. Weber in that part of old Cheyenne county that is now comprised in Scottsbluff county, and he became a pioneer homesteader in the vicinity of the present

GOTTFRIED KAMANN

county seat, Gering Of this original homestead of 160 acres he later disposed, after
having made good improvements on the place,
and he then invested in other land, his valuable holdings now comprising 240 acres, devoted to diversified agriculture and the raising of excellent types of live stock, and the
greater part of the tract having been supplied
with good irrigation facilities This admirable
ranch property, accumulated through the
earnest and honorable endeavors of the owner, is situated in section 1, township 15, and
is eligibly situated about one-half mile east of
Gering

Mr Weber was born in Germany, on the
10th of March, 1863, a son of Anton and
Gertrude (Petz) Weber, both of whom were
born in Cologne, Germany, and both of whom
passed their entire lives in the fatherland of
their nativity

William Weber acquired his youthful education in the excellent schools of his native
land, and he was an ambitious young man of
eighteen years when he severed the home
ties and immigrated to America, in 1881 He
arrived in April of that year and soon made
his way to Illinois, where he continued to be
employed—principally at farm work—until
1887, when he came to Nebraska and numbered himself among the pioneers of what is
now Scottsbluff county Here he has kept
pace with the splendid march of development
and progress, and at all times he has stood
exemplar of the most loyal and public-spirited
citizenship, so that he has a secure place in
popular esteem He has assisted in the furtherance of those movements that have conserved the best interests of the community and
was for some time president of the Central
Ditch Company, controlling one of the important irrigation projects of the county His political allegiance is given to the Republican
party, he and his family are communicants of
the Catholic church and he is affiliated with
the Modern Woodmen of America

In 1889 Mr Weber wedded Miss Minnie
Brown, of Utica, New York, and her death
occurred at the home in Scottsbluff county
Of this union were born four children William H, is a prosperous farmer in this county, Bert R, likewise is identified with farm
enterprise in the same county and he is individually mentioned on other pages, Anna
is the wife of John Fohland, a farmer south
of Melbeta, this county, and Harry died at the
age of five years In 1904 Mr Weber was
united in marriage to Miss Winnie Newby, a

native of Missouri, and she likewise is deceased, the two children of this union having
died in infancy Mr Weber contracted a
third marriage, when Miss Ida Davis became
his wife, she having been born and reared in
Nebraska

GOTTFRIED KAMANN is one of the
sturdy, hardy pioneers of Nebraska, who has
known the hardships and privations of early
settlers in the west and who has contributed
his share to the upbuilding of the county It
is to this citizen that recognition is here accorded

Gottfried Kamann was born in Rhine province of the German Empire in 1862 being the
son of Heinrich and Gertrude (Bovenschen)
Kamann, both natives of Germany, where they
were reared and educated The father was a
blacksmith in the old country who was engaged
in the practice of his vocation for many years,
thus earning a comfortable living for his family At the age of eighty-one years he bravely
broke all the old ties that bound him to the
land of his birth and sailed for America to
join his children who had established themselves in the great Land of Promise " After
seven years passed with the members of his
family he passed away here at the age of
eighty-eight years and four months

Gottfried Kamann was reared and educated
in his native province in Germany received
excellent educational advantages in the public schools of Germany which were at that
time conducted by the state and thus laid
the foundation for an excellent practical
education which has proved of great value to
him since coming to the United States and engaging in business independently He was a
far-sighted youth saw that with the land in his
native country largely owned and controlled by
the Junker class, there was little chance or opportunity for him to acquire land of his own
and he did not care to spend his life in the
laboring class For many years Mr Kamann
had heard of the many advantages to be had
in the United States and after attaining his
majority and his period of military service
over, he decided that his future would be
brighter in America and on March 6 1885 set
sail for this country For two years he was
engaged in varied occupations while learning
the customs of the country and the English
language and thus had an excellent opportunity
to hear of the different sections of the country
and decide which section would be the most
desirable for his home Mr Kamann chose
Nebraska locating on a homestead on the
Dutch Flats in 1887 He at once began im-

provements on the place, such as farm buildings, a house, and as water was the paramount need of every settler, drilled the first well in this locality, from which all the neighbors hauled their water for a long time. The country was sparsely settled at this period, farm houses being far apart with great stretches of prairie separating the primitive homes of these sturdy pioneers of civilization and Mr. Kamann says that he saw antelope, deer and other wild game running across his land and could go hunting from his dooryard for supplies of meat. What changes this man has seen in the brief span which has elapsed since he first drove into the country, for today the wide prairies which smiled with wild flowers in the sun have become a prosperous countryside of fine farms, dotted with thrifty and thriving communities that are as barometers of the country itself. Upon first establishing himself here Mr. Kamann had to drive to Sidney for his supplies but was glad that he had the money to get them as many of the first men to locate here had a very hard time, became discouraged and returned to their old homes farther east, but this German was determined that he would not be daunted by a few hard years, and his faith in the section has been justified, and today he is himself the possessor of a comfortable fortune won on these prairies of the west. There were two children in his family, himself and his sister Elizabeth, who is the wife of Valentine Thomas, who resides in Sioux county.

Money was a very scarce commodity in the west during the early eighties and as Mr. Kamann was a strong, healthy man, he found employment with the construction men when the railroad was built from Broken Bow to Alliance, and with this money was enabled to place many improvements on his land that other settlers had to do without or wait to establish at a later date, after they had managed to sell some of their farm produce at some distant market. From first locating in the valley, Mr. Kamann took active part and interest in all movements for the development of this section, being one of the first men to have the vision of what this land would become with water and as a consequence was one of the pioneers in irrigation, working on the construction of the first ditch which was to bring water to the thirsty earth and prosperity to the Morrill section. This was known as the "Farmers Canal," which has been such an important factor in the development of what is now one of the richest farming sections of the whole country as well as the most prosperous, for the river valley soil, with plenty of water

and the never-failing sunshine of the high prairies, has caused the valley lands of the Panhandle to become a veritable garden spot where the greatest returns are obtained from the labor placed upon the land.

In 1891, Mr. Kamann married Miss Wilhemina Bremer, and to this union five children have been born. Those living are: Henry, a farmer of Scottsbluff county, who responded to his country's call when the United States declared war against Germany and served with the rank of sergeant in the army, but has been discharged and is again at home; Arthur W., also a farmer in Scottsbluff county; Clara A., a school teacher in her home district; and Katharine, who is at home. Mr. Kamann is affiliated with the Masonic fraternity and has attained the thirty-second degree. The pioneers of the eighties and nineties know Mr. Kamann's early activities and hold him and his family in high esteem and today he is regarded as a prominent and leading spirit in the community.

JOSEPH C. WILLIAMS. — These lines concern one of the younger generation of business men—one just a decade beyond his majority; one who comes of sturdy, fine, old colonial stock, of a family that located on the Atlantic seaboard states during the period of settlement in the tide water region and their indomitable courage and characteristics that insure a high degree of success have been handed down to the man whose name heads this review. Mr. Williams is the owner and manager of the largest drug house in Henry, which he established himself and today it is one of the leading business houses in the valley of Scottsbluff county.

Mr. Williams is a southerner, as he was born in Allendale, Barnwell county, south Carolina, December 12, 1888, being the son of Joseph J. and Virginia (Wooten) Williams, the former also a native of South Carolina, while the mother was born, reared and received her early education in Florida. Three children grew up in the Williams family: Edgar L., lives at Greeley, Colorado; Joseph, and Lelia, who married James T. Pomeroy of Chicago. Joseph J. Williams was a physician who came west and located in Colorado when Joseph was a small boy. For many years Dr. Williams was engaged in the practice of his profession in Hotchkiss, Colorado, where he built up a lucratice practice which he enjoyed until he retired from active life, dying November 19, 1919, at Hotchkiss, Colorado. Mrs. Williams died in 1915. Dr. Williams was a member of the Baptist church, was for many years a Democrat in his political faith and fraternally was allied with the Masonic order

Joseph Williams received his early educational training in the excellent public schools of Hotchkiss, Colorado. After finishing the elementary grades he entered the high school, graduating after a four year course. Following this he entered the pharmacy department of the Colorado State Agriculture college, where he remained a student until he received his degree of Ph G. He was at once registered as a graduate in pharmacy under the state pharmacy board and admitted to practice his profession. Within a short time Mr Williams came to Scottsbluff, where he worked for the Great Western Sugar Company until 1915, when he believed he saw an excellent opening in Henry and located here. He opened a modern up-to-date store on the main street, equipped with every convenience to handle his trade and now enjoys a fine business. A good drug house is one of the necessities of a community and a pharmacist must use care to give safe and satisfactory results, as his business is regulated by strict laws of the state and nation and he is no less responsible for the health and life of his patrons than the physician whose prescriptions he is called on to fill. Henry has been fortunate in having Mr Williams in whom full reliance can be placed. He carries a fully equipped stock of the various medicines, patent medicines and all lines allied to the drug business which the public has learned to expect and demand. His store is very attractive and is one of the most prosperous representative business centers of the town

On June 25, 1912, Mr Williams married Miss Florence Wallace, a native daughter of Scottsbluff county, who was reared here on the high prairies, received her educational advantages in the public schools and here also met her future husband. Two children have been born to Mr and Mrs Williams, Joseph Wallace and Virginia Lee, attractive youngsters for whom a bright future is in store. Mr Williams is an independent in politics, and though he takes no active part in political affairs is a worthy and representative citizen who lives up to his own high standards of Americanism as the worthy scion of an old southern family should. Fraternally his affiliations are with the Masonic order as he has taken his 32d degree in that order

HERMAN G STEWART, who is one of the capable and progressive representatives of the farming and stock-raising industries of the Mitchell valley and Scottsbluff county, is not one of the earliest settlers in this section as he located in Sioux county when he first came to Nebraska, but since coming into the valley of the Platte has kept pace with the steady advancement that has marked this favored section of the state. Mr Stewart was born in Fond du Lac county, Wisconsin, September 29, 1854, being a child of Henry and Ruth (Grant) Stewart, the former a native of New York and the latter of Ohio. Henry Stewart was descended from a long line of colonial ancestors who had played an important part in shaping the growth and development of our great country when it was in its infancy and he himself as soon as he attained manhood's estate took an active and interested part in the councils of the Republican party, as he was one of the fifty men who organized it, when the new party began to take shape. He had a good, practical education in his youth and upon this excellent foundation he continued to build by wide reading along both political and business lines, until he was regarded as one of the best informed men of his day. Mr Stewart was one of the men who believed that a great future lay in store for the great Mississippi valley and early determined that he should have his part in the opening up and development of the country. As railroads were few and the price of transportation high he came west by way of the great lakes, making the trip from New York by boat. After arriving in Wisconsin he located on a farm in Fond du Lac county, where he immediately engaged in general farming, and stock-raising when that state had hardly been reclaimed from the wilderness, for Wisconsin was heavily timbered, especially along the water course, and Fond du Lac county is in the lake country. There were seven children in the Stewart family. Martha, the wife of W A Thornton died at Crawford, Nebraska, Martela, the wife of John Stewart, a distant relative is deceased, Celia L, married Thomas Jefferson Cummings, and now lives at Riverside, California, Henry, lives near Crawford on a farm, Heman G, A E, a farmer in Scottsbluff county, and John F, a farmer near Crawford

Mr Stewart was reared on his father's farm in Wisconsin and there received his educational advantages in the public schools. While still young he left home to start independently in life and removed to Iowa with his parents in 1867. They located on land in the western part of the state and later moved again to Mills county, where the father died. Mr. Stewart thought he saw an opportunity to get some good land in Kansas and took up a homestead in that state on which he proved up and after making improvements on the farm was able to dispose of it at what he considered a good price for those days. He then came to Nebraska and bought land in the Loup river valley, in Nance county, but learning that Indian land could be bought for eight dollars an acre in Sioux county, disposed of his holding near the river to locate in the extreme northwestern part of the state on the high prairies. He filed on a homestead, on the pre-emption plan and later homesteaded, so that he had considerable landed estate. After living on this new land for some time Mr. Stewart made fine improvements of a permanent character, had raised the land to an excellent state of fertility and was well and favorably known as a farmer who believed in modern methods in conducting his farm enterprise and was one of the first to place his land under irrigation, for his study of agricultural subjects had led him to know that on this western land any profitable crop could be raised provided there was assured rainfall or water and today he has five hundred and forty acres under ditch which is equal to more than three times that number of acres as far as making money is concerned if he was not able to get water, the right amount, and at just the right time necessary for the growing grains. Mr. Stewart, like his father has taken an active part in politics and the esteem in which he is held by his friends and acquaintances is shown by the fact that he has twice represented his district in the state legislature. Independent in his manner of life, early led to think and do for himself. Mr. Stewart is an independent in politics, and draws no close party lines when the question of the best man for office comes up. In 1876, Mr. Stewart married Miss Marie Clites, a native of Illinois and to this union five children have been born: Thadius a cattle rancher in Sioux county; George F., who lives at Wind Spring on a ranch which he owns; Herman C., a farmer of Sioux county; Mary R., the wife of Clyde Cross who owns a farm near Mitchell; and Mabel E., who married Fred Newell a Sioux county rancher. The Stewart family are Christian Scientists.

WILLIAM E. ALVIS. The proprietor and editor of a newspaper is a man of great potential power for good or evil in a community as he occupies a vantage ground from which he may make or mar a reputation, or build or tear down a cause worthy of public support. Not only the city of Morrill but Scottsbluff county and the western panhandle has reason for congratulation that the Morrill Mail is in such capable sagacious hands that are so thoroughly clean as those of the present owner. It is considered one of the best general news sheets published in the county, as well as an outspoken, fair play exponent of the best element of political elements in this district; in fact, it is in all respects well worthy of the thought and sound judgments displayed in its news items and editorial columns and reflects credit on its joint editor and publisher, William Alvis, one of the younger men of the newspaper fraternity, who are playing an important and able part in shaping the policies and destinies of Western Nebraska.

Mr. Alvis is descended from a long line of colonial ancestors on his father's side as the family located in the Old Dominion during the early days of its settlements and various members of the family have taken a prominent and active part in the life and politics of their community from that early date to the present time and it is but seeming that a descendent of such an illustrous family should now be carrying on the torch of progress, as civilization is westward making its way. William Alvis was born in Clark county, Iowa, April 26, 1891, the son of John W. and Ida M. (Thompson) Alvis, the former born in Virginia, where he was reared and received his early educational advantages and after attaining his manhood being a youth of ambition and action he came west, locating in Iowa, where he engaged in business as a farmer. The mother was born and reared in the west, and spent the greater part of her life in her native state of Iowa, where her three children William, John W., and Hazel were born. In 1906, the family removed to Scottsbluff county, where John Alvis took up a homestead. As he was already a practical and successful farmer he soon had the wild land under cultivation, had erected the necessary buildings on the farm and a comfortable home for his wife and children. Later he was able to dispose of the place at a handsome profit and returned to Iowa, where he again engaged in farming, an occupation

PETER VONBURG

in which he is still engaged being one of the well known, prosperous and substantial men of his community, standing high in the esteem of his friends, acquaintances and business associates by reason of his integrity, high moral level and the fact that his word is as good as his bond Mr Alvis is an adherent of the principles of the Democratic party while his fraternal associations are with the Modern Woodmen of America an organization in which he takes a prominent part

William Alvis received excellent educational advantages in the public schools near his father's farm in Iowa, thus laying the foundation for the higher studies which he has since pursued both in educational institutions and by himself as he is a wide reader of the best English literature, and the many periodicals of the country as well as the special lines connected with his editorial work Reared in an agricultural environment the boy early learned the practical side of farm life and was a youthful but expert farmer while still in school, as he early assumed many of the duties on the home place that his strength and age permitted William was only fifteen years old when the family came to Scottsbluff county, here he continued to reside at home and ably assisted in establishing a new farm on the land his father had taken from the government, but as he studied and considered the future, the boy did not see himself as a farmer, his tastes were literary not agricultural, and wisely following that profession toward which his mentality led, he entered the realm of journalism and after thoroughly acquainting himself with both the editorial and business departments of a newspaper, Mr Alvis purchased the Morrill Mail from George Mark At the time of the transfer the paper had a subscription list of only about two hundred subscribers, but the young manager at once set out to remedy this defect He introduced modern methods, replacing the old Washington press with a new power driven one, at once changed the style of the publication, inaugurating the latest manner of make up and soon the number of subscribers began to climb until today he has a circulation extending throughout Scottsbluff county and is producing a well printed, well edited sheet, with clean, live, authentic news, timely editorials and interesting locals His efforts to give the people a good, readable newspaper have evidently been appreciated, and he is well supported in an advertising way by the merchants and professional men of Morrill In connec-

tion with his newspaper, Mr Alvis has a well equipped job department, and turns out all manner of high class job printing

Mr Alvis was married to Miss Goldie Shofstall of Jefferson Nebraska, July 31, 1912, and to them three children have been born Melba L , Elden R , and Kathleen, all of whom have a bright future in store for them as their parents are determined to give them every advantage in a social and educational way that the state of Nebraska has to offer

Independent in his life work and business it is but natural that Mr Alvis should think independently along political lines and is an avowed Independent

PETER VONBURG, the subject of this record, has the distinction of being the third man to file on a homestead in the Morrill district for he came to Cheyenne county in 1887 and on the 3rd of April recorded his land entry for a claim on land where the town of Morrill now stands Mr Vonburg is descended from a long line of sturdy thrifty Scandinavian ancestors, as his father was a Swede It was these hardy Norsemen of the sea countries who first discovered America and it is to these countries that the United States is indebted for such a large element of her best immigrant population as they have been pioneer settlers in many of the best agricultural sections of the country and it is through their industry, hard work and foresight that so many broad acres of this land have been made to yield a bountiful crop where once were the rolling unproductive prairies, the "American Desert and Stake Plains" of the historian of an earlier day Peter Vonburg is a native of Illinois, born in Knox county, June 19, 1865, being the son of John and Sarah Vonburg, both natives of Sweden, where they were reared, educated and married The father was a stone mason, a trade he learned in his native land and followed there for some years before emigrating to America He had heard of the many opportunities to be obtained in the United States from some of his returned countrymen and when he perceived that there was little ahead of him in his native land but hard work for a bare living, he left the land of his birth to sail for the new world to there carve out a career and fortune for himself and his family After landing on our shores he came to Chicago, Illinois where he at once obtained employment at his trade, subsequently removing to Knoxville, Illinois, to follow the same vocation There were eight children in the family, five of whom are living Eber, who resides in Illinois, John also lives in that state,

Eli, whose home is in Knoxville, Tilda, the wife of Charles Bjlgren, who died at Wilmer, Minnesota, and Peter. The father and mother were both members of the Lutheran church, while John Vonburg was affiliated with the Republican party

The children of the family were sent to the excellent public schools which this generous land affords and in Knoxville. Peter received the academic training and laid the foundation for the good practical education that has been of great value to him in his business life. He remained in Illinois for one year after attaining his majority, but like all the youth of the land, in the early eighties heard of the many advantages to be had, as well as adventures, in the lands lying west of the Missouri, and as the country was well settled up in Illinois, and land high in price for that time, he came west in 1887, locating in Nebraska on the townsite of the present thrifty town of Morrill and little did he realize that his homestead was to become the location of so progressive a community within a few decades. Mr Vonburg proved up on his claim, put upon it good and permanent improvements, and engaged in general farming and stock-raising. He passed through the hard and discouraging years of drought, insect pests and the winter blizzards that killed his stock, but "stuck it out" and has had his reward for today he is one of the largest landed proprietors in this section so well known for its progressive and prosperous agriculturists, who have so nobly responded within the recent years to the demand for increased production. Mr Vonburg has made a deep study of farming and its allied industries and was one of the first men of the valley to advocate irrigation and put into practice intensive farming methods that have brought him such gratifying returns for his labor and the study which he has devoted to his chosen vocation. Today he owns 680 acres of land and leases 600 more, nearly all of which is under ditch, and as he believes that high-grade stock brings the greatest returns, has nothing else on his farm. He has fine substantial buildings for the various necessities of his business and a good convenient modern home. While Mr Vonburg has never had the time or inclination to take an active part in politics, he gives his support to the Republican party, cooperates with his fellow citizens in the furtherance of measures advanced for the general welfare of the community, and is loyal to all civic duties and responsibilities. His fraternal affiliations are with the Masonic order, as he is a thirty-second degree Mason. Mr Vonburg stands head and shoulders above the average farmer

of the west, for he is the representative of progress and thus is an example for the community in which he lives for by his very manner of life and the conduct of his business he has had great influence in introducing modern methods and equipment into the most favored agricultural section of Nebraska

ALFRED J STEWART, M D, a man of distinguished intellectual and professional attainments, high ability and ideals, came to Nebraska nearly a quarter of a century ago and it has been given him to wield large and benignant influence, not only as one of the early surgeons and physicans of this state but also as a man of affairs and a citizen whose civic loyalty and exceptional talents have made him a most influential factor in public affairs since locating in the Panhandle and especially in Mitchell and the county adjacent. Dr Stewart was born in Maquoketa, Iowa, March 22, 1868

While he was still a small boy the family moved to a farm near Marion, that state, so that Alfred was reared in a fine healthy environment, early learning the value of thrift and when his strength and age permitted began to assume many of the duties about the home place. He was sent to the public school nearest his home for his elementary education but as he early decided upon a professional career entered Cornell college, at Mount Vernon, Iowa, to complete the required studies for entrance at the medical college. The fall after finishing his college course at Cornell, Dr Stewart matriculated at the Hahnemann Medical college. Chicago, Illinois, took a three year course in that institution, receiving his degree in 1896. Within a sort time he had chosen a location at David City, Nebraska, where he opened an office and began the active practice of his profession. Dr Stewart had some of the early hard years that every physician does, but soon won the confidence of the people, had a sympathetic and courteous manner that won him patients and friends so that his practice grew rapidly and he was soon regarded as one of the leaders of the medical fraternity in David City and the surrounding territory. Desiring to keep up with the progress made each year in the medical profession, which could not be done by reading alone, Dr Stewart entered the graduate department of the medical school of the University of Illinois in 1905, where he took a year's course in the special branches in which he desired to further perfect himself, being graduated from that medical department of the university in the early summer of 1906 Almost immediately after returning to Nebraska, he came to Mitchell, opened an office and has been in continuous practice here since that time Within a short time he had built up an extensive practice ranging over a large radius of the surrounding country He devoted himself earnestly and unselfishly to the alleviation of suffering under conditions that in the early day involved arduous work He gained the affectionate regard of the citizens of Mitchell and the community which he served and today has one of the largest practices in the county

In 1895 Dr Stewart married Miss Natie Woodward, who died in 1901. leaving one daughter, Carol, who now is the wife of Paul Pattorf, of South Dakota In 1906 the Doctor was married again to Miss Harriet Platte. a charming and gracious woman who is the chatelaine of the hospitable home maintained by the doctor, where the latch string ever hangs out for their many warm friends

Since coming to Mitchell the doctor has taken an active and interested part in all public affairs though he has never aspired or had time to hold public office. as all his time and energy is demanded by the duties of his profession, but he is one of the fine citizens who is ever ready to help in promoting any movement for the development or benefit of the county and the city of Mitchell, giving freely of time and money in all good and laudable movements In political belief he is a liberal Republican, believing that the best man should be elected to local office the best fitted to serve the people well Prominent in Masonic circles, Dr Stewart has taken all the degrees up to the Shrine and is one of the prominent Shriners of the northwestern part of Nebraska He is a member of the Scottsbluff County Medical Association, the Nebraska State Medical Association and the American Medical Association

GEORGE E MARK, one of the self made men of the Panhandle who today is the owner of one of the journals of the northwestern section of Nebraska, which has had a large part in the moulding of public opinion in this section is the owner and editor of the Mitchell *Index*, one of the cleanest, most fearless and progressive newspapers of the state

Mr Mark was born in Chautauqua county, New York, March 14, 1866, the son of David and Delilah H (Durfee) Mark both natives of the Empire state so that Mr Mark is descended from a long line of colonial ancestors who took an important part in the development of the New England states His ancestors on both his father's and his mother's sides were colonial residents and the male members of the families were all soldiers of the Revolution With his parents he came to Nebraska in 1872, his father taking up a homestead in Thayer county, on which he proved up, developed it, and which he sold a short time before moving to Gering, in Scottsbluff county, in 1899 David Mark spent a useful and constructive life which closed in November, 1900 He is survived by his widow who now makes her home with her son George was reared on his father's farm early beginning to help around the home place in the summer vacations while during the winter terms he attended the district school nearest his home Subsequently he entered the Hebron high school, from which he graduated While still a youth of only eighteen years he began his independent financial career as a teacher in the country schools He saved some money from his salary and as he had already learned that the best equipment a man can have for his life work is a good education determined to take a higher course, and with this in view matriculated at Fairfield college where he took special courses along the lines which most interested him In the fall of 1893 still following his chosen profession of teaching, he moved to Gering in Scottsbluff county, having accepted the position of principal of the city schools At the same time he determined to take advantage of the fine opportunity to secure a good farm, so filed a homestead near Gering on which he proved up In 1896 Mr Mark purchased the *Nebraska Homestead* which had been published for some time at Gering, as he had a natural aptitude for journalism and the work was congenial to him, more so than the teaching profession and at the same time it gave him opportunity to bring to the attention of the people many things to their advantage Mr Mark acted as publisher and editor, which under his able management developed into one of the strong and influential journals In April. 1901 Mr Mark moved his printing plant to Mitchell and started the *Index*, as he saw there was a good opening here for a live, up-to-date paper and his wisdom in this move has been justified for he now has a large subscription list which extends all over Scottsbluff county In addition he has built up a large and profitable job printing business He conducts one of the cleanest, most independent and out-spoken sheets in the Panhandle, with well written editorials upon timely and interesting subjects, many good locals and the latest telegraph news During the war he was able to give the subscribers exceptionally good service with reference to the movements in Europe which they appreciated, especially after the United States entered the conflict

On Spetember 1, 1902, Mr Mark was married at Bayard, Nebraska, to Miss Maggie L Wells, a native of Missouri, where she was reared and educated   Three children have been born to them   Eldridge D, Margaret and George E, Jr In politics Mr Mark is an adherent of the Democratic principles, but is not bound in local affairs by strict party lines, as he advocates the best man for office when it comes to serving the people He is a member of the Independent Order of Odd Fellows and with his wife is a member of the Christian church

WILLIAM MARLIN, who may be termed one of the pioneers of Scottsbluff, as he built one of the first houses in the place, has been a farmer almost all his life, and a resident of Nebraska for forty-three years He has witnessed great changes in that time and has done his part in agricultural development

William Marlin was born in Franklin county, Indiana, February 19, 1852 His parents were Charles and Mary (Ralf) Marlin, the former of whom was born in New Jersey and the latter in New York The father was a farmer in Indiana and died there at the age of seventy-two years The mother came to Nebraska and died in Frontier county at the age of seventy years Their family consisted of three daughters and five sons, William being the second born

In the country schools near his father's farm in Indiana, William Marlin obtained his schooling He grew up on the home farm and remained in Indiana until he was twenty-four years of age, then came to Nebraska, settling first near Red Cloud, in Webster county Two years later, however, he moved to a safer section, in Frontier county and homesteaded there For twenty-two years Mr Marlin resided on his homestead, developing and improving it, and then moved to Scottsbluff, induced to some extent to come into town in order to give his children better educational advantages As one of the early settlers here, he has been concerned in many ways with the

town's progress He owns his own comfortable residence here and also has one hundred and sixty acres of irrigated land

Mr Marlin was married December 24, 1875, to Miss Amanda Ray, who was born in Decatur county, Indiana, and the following children have been born to them   John, a farmer living near Scottsbluff, Dore R, lives on a farm near Scottsbluff, Jesse H, also a farmer near Scottsbluff, Clifford, lives at Scottsbluff, Otis, who has seen six years service in the United States Marine corps, is a captain in rank and is now stationed in the Danish West Indies, Benjamin H, lives in Scottsbluff county, Cora, resides at home, Mrs Lenora Ashbough, lives in the county, William, a farmer; and Viola, lives with her parents Mrs Marlin is a member of the Presbyterian church In politics Mr Marlin is a Republican but has never sought public office

I W NEWSUM is a native of North Carolina, born January 31, 1852, the son of Gillin and Amanda (Spease) Newsum Both his parents were natives of North Carolina, and to them thirteen children were born, eight of whom are now living

After finishing his schooling in his native state, Mr Newsum engaged in farming Coming to western Nebraska in 1886, he took a homestead south of the present town of McGrew and lived through the experiences that were common to all the early settlers of those days He farmed and raised stock, and after a successful term of years in that enterprise he has taken advantage of a well earned rest and disposed of his ranch, now owning only ten acres of land on which is located a comfortable home

October 6, 1889, he was married to Mrs Mary Minces, whose maiden name was Mary Lee, and who had formerly been married to Isaac L Minces and had two children, Leonard, who now resides at Bayard, and Harry, who was killed by lightning in 1906

Mr Newsum is a Republican in politics and a member of the United Brethren church He recalls many experiences of the pioneer days before the Black Hills line of the Burlington Railroad was built, when he had to haul his grain to Sidney or Alliance In those times people were few and money was scarce, but he, like the others who stayed with it, has lived to see the time when population and wealth have multiplied many times over and the community that was once a sparsely settled range country is now one of the richest

JAMES R. RUSSELL AND FAMILY

agricultural sections of the United States with prosperous and growing cities and towns every few miles.

Mr Newsum stands high in his circle of aquaintances as a man who has been upright and enterprising. He has made a success of his life and has retired to enjoy the fruits of his labors, enjoying the friendship and respect of all who know him.

JAMES R. RUSSELL, a pioneer of Scottsbluff county and one of the energetic and progressive citizens of the Mitchell valley, is a representative of the spirit that in recent years has proved such an important factor in the advancement of the Panhandle. He is the owner of a valuable and productive farm located in section 35, township 23-57, and he has also been identified with the business interests of the valley from first locating here and his career has been marked by a versatility that has done much to make him one of the substantial and influential men of this locality, well known for its able agriculturists and progressive, successful business men.

James R. Russell is a native of the Badger state, born in Vernon county, Wisconsin, in 1868, the son of Calvin Russell, who was born and reared in Ohio. Mr. Russell grew to manhood on the farm owned by his parents in Wisconsin and acquired his education in the public schools. In 1888, a mere boy in years, he broke all the home ties and with high heart and the determination to success started for the west to seek and make his fortune. Coming to Nebraska in 1888 he soon looked the different localities over, became an embryonic farmer and pioneer of the Panhandle. He took a homestead of 160 acres. Twelve years later the young man married and from then on for several years he and his devoted wife encountered many hardships and weathered many storms, but they did not falter in courage, made the best of the circumstances and privations, without complaints, and manifested the faith that has been graciously rewarded in the later years. Industrious by nature, Mr. Russell in the early days obtained work wherever he could get it to tide over the hard years when crops were destroyed by grasshoppers or burned up by the droughts, and by means of such employment provided his family with the necessities of life and was able to retain his land and gradually carry forward the added improvements which he deemed necessary to become a successful farmer. He went 200 miles away to find work and assisted in building the railroad west of Alliance. This land, located in township 23-57, section 35, has been his home continuously during the long intervening period. He has added

to the original tract and is the owner of 400 acres, all now in a high state of cultivation, is well equipped for intensive farming and extensive stock-raising, with substantial buildings that have taken the place of the first placed there. Mr. Russell has been a deep student of agricultural methods and naturally was one of the first men of the valley to realize and advocate the value of irrigation. He has one hundred and twenty acres of his land under water and it is mostly a question of time before many more acres will be under ditch. Mr. Russell raises a good grade of stock on his farm and finds that branch of farming very profitable. He tells of the makeshifts the early settlers were forced to employ when they could not obtain necessary farm machinery and family supplies and laughs as he describes how the first postoffice of Mitchell, a frame structure eight by twelve feet square, was put up over night in the stress of necessity and that he became the first postmaster, cancelling thirty dollars worth of stamps the first month. Branching out into a pioneer merchant, Mr. Russell became owner of the second store in Mitchell where he handled everything needed by the farmers of the valley.

While Mr. Russell is an advocate of the principles of the Republican party, he is bound by no strict party lines when it comes to casting his vote in local elections and gives his influence to the man he deems best qualified to serve the community or county. He is a member of the Independent Order of Odd Fellows, the Fraternal Union, and the M. W. A., while the family are members of the Congregational church. In 1900 was celebrated the marriage of Mr. Russell and Miss Lena Ewing, a native of Pennsylvania, who accompanied her parents to Nebraska when her father settled in this state in 1887. Thomas Ewing is now deceased, after having been a potent factor in the development of the region, being a representative pioneer settler who shouldered his part in opening up the middle west for settlement and development. There are seven children in the Russell family: Eva, Lester, Thomas, James, John, Clem, and Amy, all of whom are at home and to whom their father and mother have given all the educational advantages that their children cared to avail themselves of.

From first settling in the Panhandle, Mr. Russell has been progressive in spirit and is the advocate for all movements that tend to the betterment of the county and community in which he lives. Both he and his wife take an active part in their church affairs and they are numbered among the sterling and honored pioneer citizens of Scottsbluff county.

PEARL M STONE, educator and agriculturist, furnishes in his career another exemplification of self-made manhood   He is one of the most prominent and prosperous exponents of farm enterprise in the Mitchell section of Scottsbluff county, is a liberal and progressive citizen who well merits recognition in this publication   Mr Stone claims the great Sunflower state to the south, as the place of his nativity and is a scion of one of the sterling pioneer families of that great commonwealth   He was born in Smith county, September 28, 1876, the son of W E and Madord (Duffie) Stone, the former a native of Illinois and the latter a descendant of a long line of New England ancestors, having been born in Vermont   The father was one of the successful farmers who settled in Kansas during territorial days when that was regarded as a "Land of Promise," and such it proved to be for him   There were three children in the family, two of whom are living Pearl and Edna, the wife of Thomas Maycock, who resides in Gilette, Wyoming

His father being a man of comfortable means Pearl was given all the educational advantages that this section of the country afforded as his parents removed from Kansas to Scottsbluff county in 1890, the father being instrumental in the great enterprise of securing the irrigation ditch which has made this county "bloom like the rose"   The boy attended the public schools at Gering and Lincoln, graduating from the Western Normal school in the latter city   The following seven years Mr Stone devoted to his profession as teacher and the success he gained in this field may be understood when we learn that he was then elected county superintendent, an office most creditably filled by him for two years   A highly educated man, he kept abreast of all the questions of the day and being of far vision saw that the free, independent man of today is the one who owns land, this man being his own master   The wide world must be fed and the farmers of this great country are carrying on the greatest agricultural business ever witnessed in history   Mr Stone had opportunity to observe the more than satisfactory results achieved by the farmers on irrigated land and having been reared on a farm in childhood was well qualified to take up this pursuit for life   During his scholastic years he had accumulated considerable capital and with this was able to purchase a large tract of irrigated land, consisting of three hundred and twenty acres in Mitchell township   Mr Stone has been remarkably successful in his farming which is diversified, though he carries on a large stock-raising and feeding business   Mr Stone keeps cognizant of all questions of the day and improved methods of farming and thus has come to be recognized as one of the leading exponents of this industry in his section of the country   He has demonstrated that a cultivated mind and fine instincts reach their highest development often-times amid rural surrounding, diffusing around them that refinement and peace which are the hall marks of the cultured   Mr Stone has for years been a supporter of the Republican party, is an advocate of every movement for the improvement of his community, he and his wife are members of the Christian church while his fraternal affiliations are with the Independent Order of Odd Fellows   In 1910 he married Miss Minnie Whittaker, native of Kansas, who became the mother of five children   Ellen, Maxine, Perl Hazen, Dorothy and Bernice, all of whom are at home attending school and all assured an excellent education because of their father's superior mental attainments

LAWRENCE A FRICKE — The successful men in Western Nebraska today, are by no means all of the older generation   Starting out to carve a career for himself, a young man undoubtedly is helped if his educational training has been thorough, but not education alone explains personal popularity, political prominence and keen business foresight   Possessing these qualities, Lawrence A Fricke has become a leading representative citizen of Bayard while still almost at the beginning of his career as a dealer in real estate

Lawrence A Fricke was born at Madison, Nebraska, January 8, 1889, and is a son of Herman and Johanna (Ruegge) Fricke, both of whom were born in Germany   They came to the United States as young people and were married in Illinois   In 1865 they came to Richardson, Nebraska, where he bought land and traded a horse for additional land   He followed farming in that section for some time, then moved to Omaha and went into the agricultural improved implement business and still resides in that city, being now retired Ten of his eleven children survive, Lawrence A being the youngest of the family   In politics the father is a Republican and both he and the mother are members of the German Lutheran church

After completing the high school course at Omaha, Lawrence A Fricke spent one year in the Nebraska State University and then entered on railroad work in the engineering department of the Burlington system In 1914 he embarked in the real estate business at Bayard, in partnership with his brother-in-law, Peter O'Shea, of Scottsbluff The firm handles both farm and city property and through choice locations and honest business representations, has built up a prosperous business

Mr Fricke was married in February, 1917, to Miss Eleanor Parks who was born at Greeley, Nebraska, and they have two children, namely Robert L, who was born in January, 1918, and Johanna Ruth, who was born January 16, 1919 Mr Fricke was baptized in the German Lutheran church He is a leading Republican of Morrill county, has served Bayard in the office of mayor with the greatest efficiency and is a city councilman at the present time He is a Consistory Mason of advanced degree Personally, with genial manner that shows sincerity, Mr Fricke impresses one favorably and he has a wide circle of friends

IRA BIGELOW, who has been a resident of Nebraska almost his entire life, owns and operates a fine farm in Morrill county, upon which he has placed substantial improvements Mr Bigelow has been prominent in the Tri-State Ditch project, and served three years as treasurer of this enterprise He was born in Wisconsin, May 27, 1868, and is a son of Reuben and Saphronia Bigelow

When Ira Bigelow was three years old, his parents left Wisconsin, moved to Iowa, settled there on a farm and remained until 1879 Another change was made and Mr Bigelow remembers the journey from Iowa to the new home in Holt county, Nebraska, where his father homesteaded He attended school there and later went to Omaha and came to Morrill county in 1910 Here he purchased eighty acres of wild land and immediately set about its development and improvement He now has a valuable farm and is in a position to feel well satisfied with conditions of all kinds as they are in Nebraska

In 1895, near Kearney, Nebraska, Mr Bigelow was married to Miss Esta Ford, who was born and reared in Iowa Her parents Samuel W and Angelina Ford came from Iowa in 1887 to Kearney, Nebraska Mr and Mrs Bigelow have had children as follows Mrs

Zana Warren, lives at Redington, Nebraska, Mrs Pearl Harms, lives near Bayard, Vera, lives at home, Ray died at the age of three years, and Hazel, lives with her parents Mr Bigelow is not identified with any particular political party but is a wide awake citizen and casts an idependent vote for the candidates of whom his own good judgment approves He takes a deep interest in the public schools and has served in the school board for fifteen years

HENRY MILLER, whose valuable irrigated farm in Morrill county, Nebraska, lies on section fifteen, town of Bayard, has been a resident of Nebraska for thirty-eight years and during that long period has witnessed many wonderful changes He has been a farmer all his life and has developed a fine property on which he lives

Henry Miller was born in Alsace Lorraine, then a province of Germany, March 10, 1860 His parents were farming people named Peter and Elizabeth (Schmidt) Miller, who emigrated from Germany to Canada, in 1866 The change of climate and manner of living did not agree with them and both died shortly after reaching their new home Henry was young at the time He remained in Canada, where he obtained a fair amount of schooling and learned to be a farmer, until 1881, when he came to Nebraska and settled in the eastern part of the state He followed farming there until 1909 and then came to Morrill county and bought two hundred and forty acres of wild land With accustomed industry he began the development of his land and soon had a crop started but a drouth ruined it and in the season of the following year, a hailstorm caused great damage to his growing crop Since that time, however, Mr Miller has been continuously successful, and with his large farm all irrigated may well be considered one of the county's substantial agriculturists He has excellent improvements, keeps standard stock, owns modern machinery and his entire place gives the pleasing impression of a profitable, well regulated farm

Mr Miller was married to Miss Alvina Going who was born in Germany, March 23, 1861 Her parents came to the United States from Germany in 1867 and settled in the eastern part of Nebraska, where the father carried on farming until his death The mother of Mrs Miller still lives on the old home farm and is now in her eightieth year Mr and Mrs Miller have had eight children

Henry and Willie, twins, both of whom live in eastern Nebraska; Louis, lives in Wyoming; Martha, the wife of L. J. Tilden, of Morrill county; Alvena, the wife of E. J. Tilden, of Wyoming; and Walter, Esther and Paul, all of whom are at home. The family is of the Lutheran faith. Mr. Miller is independent in his political views. His neighbors know him to be an honest, dependable man.

MARTIN J. KING. — In these days, to the ordinary individual, the ownership of vast tracts of land and thousands of cattle represents wealth almost inconceivable, yet there are men in Morrill county who go quietly about the ordinary affairs of life without ostentation, who can claim such possessions. A sale of 2,300 acres of land recently recorded by Martin J. King, one of the county's well known cattlemen, brings this to mind, although it is but an incident that may be repeated, for Western Nebraska men are apt to think and act in large figures. Mr. King has spent almost all his life in Nebraska but his birth took place September 12, 1878, at Creston, Iowa.

The parents of Mr. King were Valentine and Barbara (Hutchinson) King, natives of Ireland and truly worthy people there and later in the United States. They located first in Maryland but after the ways of the new country had become familiar, removed to Iowa and lived there as farmers until 1887, when they came to Cheyenne county, Nebraska. The father homesteaded and turned his attention to growing cattle, in the course of years becoming one of the big cattlemen of this section, at one time having 6,000 head. He was a good business man, attended closely to his own affairs, voted the Democratic ticket and brought up a large family in the Roman Catholic church. He died at Alliance, Nebraska, the mother of Mr. King passing away in the city of Omaha. Of their children Martin J. was the fifth in order of birth, the others being: William, lives at Alliance, is in the stock business; Patrick, a farmer near Blackfoot, Iowa; John, a farmer in Morrill county; Annie, resides at Alliance; Maggie, the wife of L. Jacobs, a farmer near Angora; Nellie, the wife of James Murphy, a ranchman near Alliance; and Thomas, lives on the old King homestead, and with one of his brothers owns no less that 14,000 acres of land in Morrill county and runs about 1,000 head of cattle. John King keeps 300 head of cattle, included in these being 150 pure-bred Herefords.

Martin J. King was nine years old when the family located near Alliance, Nebraska, and he remembers going to school in a little tent, for schooling was one of the first privileges the most of the early settlers endeavored to secure for their children, second only to religious instruction.   Later he had public school advantages. After his school days and until 1915 when he moved into Alliance, Mr. King engaged in ranching and became one of the county's well known cattlemen. He has a fine farm of 320 acres but has retired entirely from active farm and ranch life. For several years he carried on an automobile business at Alliance and then came to Bayard and bought the Bayard Hotel, and now occupies his time in managing this place of business.

In 1907 Mr. King was united in marriage to Miss Elizabeth L. Shetler, who was born near Kearney, Nebraska, and is a daughter of Lesley L. Shelter, who came to Cheyenne county in 1887 and now lives retired at Denver. Mr. and Mrs. King have three children: Lavern L., Catherine Barbara and Martin Carroll, the youngest being at the engaging age of three years while the older children are doing well at school. Mr. King and his family belong to the Catholic church. Mr. King follows his father's example in political membership but has never been willing to accept a public office. He belongs to the lodge of Elks at Alliance.

CHRISTIAN NUSZ, who owns a well improved farm situated on section 12 town of Bayard, Morril county, Nebraska, has not lived in the United States so very many years and still fewer in Nebraska, but he has demonstrated what a man of energy and enterprise can accomplish when given free opportunity. Mr. Nusz was born in Russia, in 1869, a son of Christian and Mary (Hass) Nusz. Both parents were of Russian birth. The father died on his farm in Russia and the son hopes that his beloved mother still lives there. The unsettled condition of his native land has made it impossible for Mr. Nusz to communicate with the old neighbors and eight years have passed since he had reliable news.

In 1908 Mr. Nusz came to the United States and made his way to Kansas. There he worked as a laborer until 1914, when he came to Morrill county and invested his savings in one hundred and sixty acres of land. He has not spared himself in developing this land and has improved it very well. Almost all' of

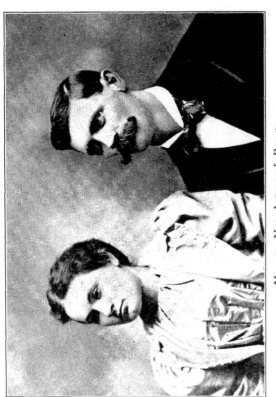

MR. AND MRS. ARTHUR J. BAILEY

his farm is now irrigated and his crops are abundant. Knowing the hard conditions of life for men with little chance to get ahead as they were when he left Russia, he feels that he has been fortunate in coming to Nebraska. He has found friends here, has acquired a beautiful home and assured comfort for old age and has been able to give his children the educational opportunities he has desired.

Mr. Nusz was married to Miss Latie Deins, who was born in Russia. Her parents were Jacob and Ella (Fogal) Deins, who never came to the United States. Mr. and Mrs. Nusz have had children as follows: Christian, a farmer in Colorado, Alexander, works on a farm in Colorado, and Jacob, David, Lydia Victor, Mary and Carl, all of whom are at home. The family belongs to the Russian church. Mr. Nusz in an American citizen but has never identified himself with any particular political party. In his neighborhood he is known to be a man of his word and is highly respected.

ARTHUR J. BAILEY was one of the pioneer cattlemen of western Nebraska who played an important part in the early development of this section during that period when the great cattle barons ranged their cattle from the Pecos on the south to the Yellowstone on the north, and was well and most favorably known throughout the Panhandle.

Arthur Bailey was born in Iowa, July 3, 1869, the son of J. P. and Julia (Birdsall) Bailey, who were farming people and who came to Colorado where he became the owner of land upon which the city of Fort Collins has since been built. The son grew up on his father's farm in Iowa and attended the public schools near his home. During the heigh-day of the cattle business on the high plains it had a lure for the young man of the period and many of them joined the great cow outfits that drifted from Texas to Wyoming with the changing seasons, as the pasture became used and burned up in the south the herds slowly drifted northward and were finally sold on the northern market at the close of the season. Mr. Bailey joined such a camp and by practical experience learned the live-stock industry as conducted at the time. After serving his apprenticeship as a cowboy his ability soon became recognized and he was offered the position of foreman of the Standard Cattle Company at Ames, Nebraska, where he soon demonstrated his ability. He proved so efficient that subsequently he was given charge of the vast business of the concern at North Platte

and later at Scottsbluff and thus learned at first hand the country of the western Panhandle and its future possibilities. His reputation as a manager became well known throughout the cattle country and the Paxton people of the Hershey Ranch made him such an advantageous offer that he accepted a position with them. Mr. Bailey kept abreast of the movement of the times, studied the markets and watched the increased settlement of the western part of the states bordering the great 'cattle trail,' and was one of the first to recognize the signs that pointed to the fact that the day of the open range was over and the future of the meat industry was to change from the great companies to the farmer who would raise and feed a high bred beef stock. As he had been raised on a farm he decided to avail himself of the fine government land still to be obtained in the rich Platte valley and in 1906 purchased 240 acres in township 23-57, section 35, Scottsbluff county, where he at once established himself as a farmer raising diversified crops and engaged in stock-raising. Water had been the paramount question of the cattlemen for years and having given considerable study to obtaining it while on the range when he bought his land he chose that which lay near the river and became one of the first advocates of irrigation. Three hundred acres of his estate were under water rights and much of the rest was rich pasture, a combination that worked out well for the various lines of business which he carried on. Mr. Bailey was a strong man and from first settling in the Mitchell district by reason of his force of character, was enabled to inaugurate many improvements and thus become a potent factor in the affairs of the locality and the lives of its citizens. He stood for progress and reform, served for many years as a school director, and stood behind all movements for the benefit and development of his district. In politics Mr. Bailey was a supporter of the principles of the Republican party and he and his wife were members of the Presbyterian church, of which they were liberal supporters. Fraternally his associations were with the Masonic order, the Elks and the Modern Woodmen of America. His death occurred at the farm home, May 12, 1916.

On April 18, 1898, Mr. Bailey married Miss Elizabeth Harvey at Webster, Nebraska. She was the daughter of Andrew and Margaret (Richie) Harvey, both natives of Scotland, who came to America many years ago and are now well known residents of Dodge county, where Mr. Harvey has been a successful farmer. Eight children became members of the

Bailey family: Idell, the wife of Lemuel Smith, who resides on the old home place; Lillian, who holds a business position in Fremont, Nebraska, being associated with the Hammond Printing Co.; Ruth and Julia, who are seniors in high school; Edna, Arthur J., Beryl, and Grace, who also are all taking courses in the public school. Mrs. Bailey resided on the farm till 1919 when she moved to Mitchell to give her children better school advantages.

GILBERT ROSS, who has been a resident of Nebraska since 1909 and who is the owner of an excellent ranch property in Morrill county, is essentially to be noted as one of the world's productive workers, for his advancement has been gained entirely through his own ability and well ordered efforts. In the thriving village of Bayard, Morrill county, he conducts a substantial teaming business, the while his family resides on the homestead which is eleven miles northwest of the town.

Mr. Ross was born in Westmoreland county, Pennsylvania, on the 26th of April, 1864, and is a son of Thomas and Elizabeth (Knox) Ross, both of whom passed their entire lives in the old Keystone state, the former having been a school teacher within the period of his early manhood. Gilbert Ross was still a child at the time of his father's death and thereafter he lived in the home of his paternal grandfather until the death of the latter. At this juncture in his career Mr. Ross, who was at the time a lad of but eight years, was taken into the care of strangers and he continued to work for his board and clothes until he had attained the age of fourteen years. He then received eight dollars a month for his services in hauling lumber with a four-horse team, and he continued to be identified with work of this order until he was twenty-five years old. He then obtained employment as locomotive fireman on the Pennsylvania Railroad, and within the four years of his service in this capacity he served as fireman on both freight and passenger trains. It can well be understood that his early educational training was limited to a somewhat irregular attendance in the public schools of his native state, but he made good use of the advantages afforded in the stern school of practical labor and experience.

After retiring from railroad work Mr. Ross was engaged in teaming in Pennsylvania until 1909, when he came with his family to Nebraska and located on a homestead eleven miles northwest of Bayard, to the general improvement and supervision of which he has

since given his attention, though he devotes the major part of his time to his prosperous teaming business at Bayard. His ranch comprises six hundred acres and is utilized principally for the raising of cattle and horses, the land being excellent for grazing and his average herd of cattle comprising about one hundred head. Mr. Ross is vigorous and ambitious and is the type of citizen that is most fully valued in this progressive section of Nebraska. He is a Republican in politics, is affiliated with the Fraternal Order of Eagles and he and his wife hold membership in the Brethren church.

Mrs. Blanche (Shaffer) Rose, wife of him whose name initiates this review, is likewise a native of Pennsylvania, as were her parents, Frank and Mary (Carus) Shaffer. Mr. and Mrs. Ross have three children: William E. and A. C., both of whom are engaged in farming in Morrill county; and Blanche Ione, who remains at the parental home.

REV. THOMAS C. OSBORNE has been a resident of Western Nebraska since his boyhood days, is a representative of one of the honored pioneer families of this section of the state and individually he has done well his part in the furtherance of civic and material progress. He has given most effective service in the ministry of the Presbyterian church and is known as a man of distinctive culture and broad and well fortified convictions. Since his retirement from active ministerial labors he has given his attenton principally to the supervision of his valuable landed interests in Morrill county, (where he is also proprietor of the *Farmers Exchange*, a progressive weekly paper, of which he is editor and publisher, at Bayard.)

Mr. Osborne was born in McLean county, Illinois, on the 9th of September, 1876, and is a son of Samuel H. and Emily (Benson) Osborne. Samuel Osborne was born in Steuben county, Ohio, and was a child at the time of the family removal to Indiana, where he was reared on the home farm and received his education in the common schools of the period. When the dark cloud of the Civil War cast its pall over the national horizon he loyally went forth in defense of the Union. He enlisted as a private in the Eighty-eighth Regiment of Indiana Volunteer Infantry, which was assigned to the Army of the Cumberland and with which he served until the close of the war. He lived up to the full tension of the great conflict and his military record is virtually coincident with that of the gallant regiment of which he was a member and with which he participated in many important engagements, including the battles of Chickamauga and Stone's River. After the war this valiant young veteran passed some time in Iowa and at Kearney, Nebraska, and about 1870 he returned to Illinois and engaged in farm enterprise. Later he conducted a general merchandise store at Colfax, that state, where also he served as postmaster. In 1887 Samuel Osborne came with his family to what is now Morrill county, Nebraska, where he entered claim to a pioneer homestead two and one-half miles northeast of Bayard. He duly perfected his title to this claim and then in 1890, entered a pre-emption claim three miles southeast of Bayard. He developed and improved this property, upon which he continued to maintain his home until his death, his name being held in gracious memory as that of one of the sterling pioneers of the county. He was a man of much prevision and progressiveness, was a loyal and liberal citizen and did much to forward the advancement of this part of Nebraska. In earlier years he was a Republican in politics, but he was actively aligned with the populist party during the period of its maximum influence in national and state affairs. He served for a long period in the office of justice of the peace and was otherwise accorded marks of popular confidence and esteem. Both he and his wife whose death occurred in 1917, were members of the Methodist Episcopal church. Mrs. Osborne was born in McLean county Illinois, and was one of the revered pioneer women in Morrill county, Nebraska. Of the four children Thomas C. of this review, was the second in order of birth. Dale B, the eldest, now resides upon the old home place of his parents, Eva June died at the age of twenty-seven years, and Dean H, who was for eighteen months in the government aviation service in connection with the world war, has been residing at Bayard since his discharge, after the close of the war.

Thomas C Osborne acquired his preliminary education in the public schools of his native state and was a lad of eleven years at the time of the family removal to Nebraska. After completing the curriculum of the high school at Crawford, Dawes county he entered the Nebraska Presbyterian college at Hastings, in which institution he was graduated in 1901, with the degree of Bachelor of Arts.

Thereafter he completed a three years' course in the Presbyterian Theological Seminary at Omaha from which he was graduated on the 1st of May 1904. During the autumn of that year he had charge of the church at Wayne, judicial center of the county of the same name, and there his ordination occurred. He retained this pastoral charge until the spring of 1910, and from March 1st of that year until March 1, 1918, he was pastor of the Presbyterian church in the city of Scottsbluff. Through his able and earnest labors the church was greatly advanced in spiritual and material well being and he gained recognition as one of the leading clergymen of his denomination in this part of the state. His retirement was forced through a throat disorder and generally impaired health, and under these conditions he established his residence on the homestead which he had obtained in 1902, the same being situated four miles northeast of Bayard. He has made excellent improvements on this place, which comprises one hundred and sixty acres, irrigated from the farmers' ditch and two miles east of Bayard he has a tract of eighty acres, with similar irrigation facilities. His land is effectively given over to the propagation of grain, alfalfa and sugar beets and his health has been recuperated through his outdoor life in the supervision of his farms. Mr. Osborne takes lively interest in all things pertaining to the moral, social and industrial advancement of his home community and state. He was a member of the Nebraska Constitutional convention of 1919-20. He and the members of the family are zealous workers in the Presbyterian church, the while he maintains an independent attitude in politics.

In the year, 1903, was solemnized the marrage of Mr. Osborne to Miss Julia M Jones, a college classmate of his at Hastings, and of this union have been born five children Emily L, Charles C, Clifford W and Howard B remain at the parental home and Roger C died in infancy.

LEON A MOOMAW, Cotner Uni A B A M — Although entitled to place after his name letters indicating hard won college degrees, it may be possible that Professor Leon A Moomaw of Morrill county, takes equal pride in the success that has attended his agricultural undertakings. Born and reared on a farm the memory of Nature's ever recurring miracle of seasons and plenteousness rewarding honest toil may have accompanied him through university life and subsequent intellectual effort in the educational field, for the time came when college honors were laid aside and the learned teacher became an enthusiastic farmer. Remembering that agricultural production is the basis of all production, the transfer of scientific knowledge from the professions to the fields must, with such earnest men as Mr. Moomaw result beneficially.

Leon A. Moomaw is a native of Nebraska, born in Scottsbluff county, December 27, 1887, and is a son of Austin and Agnes (Spriggs) Moomaw. The father was born in Illinois, fifty-seven years ago, and the mother was born in Missouri. In 1886 they came to Nebraska and homesteaded in Scottsbluff county, where the father has ever since been a general farmer. Later he secured a tree claim in Morrill county. Until he was twenty-two years old, Leon A. Moomaw resided on his father's farm, but in the meanwhile his education was attended to and from the local schools he entered Cotner at Lincoln, from which he was graduated with the degree of A. B. Later he entered the State University at Lincoln, from which he bore off the degree of A. M. He then entered the educational field, in no subordinate position, however, but as a member of the faculty of Cotner University, in which institution he was professor of history for three years.

In 1914 he was united in marriage to Miss Minnie E. Young, who was born in South Dakota, July 25, 1886. Her parents, Hiram and Sarah (Adams) Young, lived in Iowa. Since 1909 they have lived retired at Lincoln, Nebraska. Mr. and Mrs. Moomaw have two children: Evelyn and Robert. In 1912 both Mr. and Mrs. Moomaw homesteaded in Banner county and they are still holding their 1,180 acres of dry land there, the subsequent development of which may be stupendous. In 1913 they came to Morrill county and he took charge of his father's old tree claim, has 200 acres, and has devoted his best efforts to the development of this land ever since. All the land is now irrigated and under Mr. Moomaw's intelligent management is a wonderfully productive property. He has placed fine improvements here and has one of the spaciouse modern homes of this section. Both he and wife are members of the Christian church, and they have a wide social circle. Although not active politically, Mr. Moomaw is not an indifferent citizen, but on the other hand, every movement that promises to be of substantial and permanent benefit to the county, finds in him an earnest advocate.

JAMES A. CADWELL, who will long be remembered as one of the fine men of Morrill county, passed away at his home on the beautiful farm he had worked hard to develop and improve, on January 8, 1918. His birth took place in Saunders county, Nebraska, March 27, 1878. He was a son of John T. and Sarah E. (Gilbert) Cadwell, both of whom were born in Ohio. They were early settlers in eastern Nebraska and homesteaders, and they passed away on their farm in Saunders county.

James A. Cadwell grew to manhood in his native county and was educated in the public schools. With self-respecting independence and wise provision, as soon as his schooldays were over, he learned the trade of a carpenter and followed the same to some extent even after he became interested in farming. He was a man of high principles, and when his country became embroiled in war with Spain, he enlisted as a soldier and served all through the Spanish-American war.

In 1901 James Asa Cadwell was united in marriage to Miss Lulu Parks, who was born in Lancaster county, Nebraska, a daughter of Theodore and Florence (Spencer) Parks, the former of whom was born at Plattesmouth, Nebraska, and the latter of Massachusetts. They still reside in Nebraska and Mr. Parks continues his agricultural industries. The following children were born to Mr. and Mrs. Cadwell: Everett, Jessie, Clarence, Glenn, Dorothy, Florence, Vera, Eliza and Maxine. Mr. Cadwell was ever anxious concerning his children's welfare and gave them every advantage in his power.

In 1905 Mr. Cadwell come to Morrill county and homesteaded and his family joined him in the following year. He left them a well improved farm of one hundred and ten acres, eighty-seven acres of which are irrigated. He was never an active politician in the sense of desiring to hold public office, and was always a Republican, although he entertained a high personal opinion of William Jennings Bryan, who had been the colonel of his regiment in the Spanish-American war. With his family he belonged to the Baptist church at Ashland, Nebraska.

JOHN ROBERTSON. — Coming to the United States from his native Scotland, where he was born September 27, 1862, when but seventeen years of age, Mr. Robertson has spent almost forty years in Nebraska and a goodly portion of them in Morrill county, where he is widely known and much respected. He accompanied his parents to Quebec, Canada, but they never came into this country Mr Robertson spent three years at Schuyler, in Colfax county, Nebraska, and then made his way to Scottsbluff county, finding the present busy, prosperous, little city of Gering. a settlement of two log houses and one sod house Later he homesteaded in Morrill county and pre-empted an entire section of land, all of which he has sold

Mr Robertson was married at Columbus, Nebraska, to Miss Myrtle May Folnsbee who was born in Missouri, and they have had children as follows Mrs Mary Hays lives at Mitchell, Nebraska, Harry, lives at Whealand, Wyoming, Mrs Alta Detrick. lives in Missouri, Robert, a farmer, John, safely returned home from military service in France, Clifford who went to France with the American Expeditionary Force as did his older brother, spent fifteen months in France and saw hard service, later being attached to the army of occupation in Germany, and Myrtle, who resides at home Mr Robertson is an independent voter

LLYN O McHENRY, one of the leading merchants of Morrill, is distinctly a Nebraska product as he is a native son of Scottsbluff county Here he was born, here he was reared and educated, here he married, and here he has practically lived his life to the present time He belongs to a family which is well and favorably known in the county and also the Panhandle and which is highly respected for its contribution to the civic and material welfare and progress of this section of the state

Llyn McHenry was born in Scottsbluff county, May 26, 1891, being the son of Oliver O and Mary J (Hall) McHenry, of whom complete mention and portraits appear on other pages of this volume, to whom were born five children Elizabeth, who married John A Burton and now lives in California, Matthew H a resident of Gering, Lucy, who married John M Springer and now lives in California, Harry H, who resides at Springer, Wyo and Llyn O

Llyn grew up here in his native county attended the excellent public schools, and thus laid the foundation for his subsequent business career After his school days were over he accepted a position in the county court house as deputy clerk of the district court a position which he so ably filled that he remained in office for seven years The young man, however, had decided that he would enter business independently and with an idea of learning the intricacies of finance first hand entered the Gering National Bank of Gering, where he was

able to gain practical and theoretical knowledge of banking Two years later, while holding a lucrative and responsible position with this institution Mr McHenry responded to his country's call for men to enter the army and aid the United States and the Allies to make the world safe for democracy and rid it of the horror of the Hun After entering the army he was stationed at Fort Logan for a year and a half and after receiving his honorable discharge at the close of hostilities returned to Scottsbluff county Soon after returning Mr McHenry formed a partnership with R B E Quick and the two men established the Quick Drug Company at Morrill, Neb They have a fine store building, excellent and attractive equipment and are able to handle a constantly growing trade Mr McHenry's varied business experiences as well as those in the army supplement the natural ability and qualifications which mark him as an able executive in any line of business thus he and his partner are conducting successfully an establishment that has varied demands and requires far sight as well as work to keep abreast of the constantly changing demands and wider field That they are fully able to do this is demonstrated by their gratifying returns financially as well as the ever-increasing clientele which they enjoy

Mr McHenry is a Republican in politics and though he is the supporter and advocate of every movement for the improvement of Morrill and the surrounding district and in every way lives up to his own high standard of American citizenship he is now far too busy to take an active part in politics, but throws his influence to the man best qualified to serve the county and city He is a wide reader of the best literature of the day as well as a student of subjects allied with his business and thus keeps abreast of the times and for the firm we predict a prosperous and successful future

On May 29, 1912, Mr McHenry married Miss Delight Byers, who is also a native of our great commonwealth, born in Washington county, where she was reared and was given the benefit of an excellent education Mrs McHenry is a gracious woman of charming personality who has made many friends in the city of Morrill where the McHenry home is regarded as one of the most hospitable Mr and Mrs McHenry have one child Ina Corinne, who is at home The entire McHenry family are splendid people and well merit the high esteem of their friends and their business associates Scottsbluff county is the richer by the mere fact that it has such citizens who will hand down to posterity their traditions and high ideals of what true Americans should be

ANDREW J. DUNHAM, who is one of Morrill county's substantial farmers and stockmen, is also one who has built up his fortune through individual effort. Left an orphan when five years old, his memories of childhood and early youth have no home setting, and the opportunities that came to better his condition, were those he found for himself.

Andrew J. Dunham was born in Mansfield, Connecticut, November 5, 1860, a son of Ephraim and Mary (Little) Dunham. His young mother died when he was born, and his father when the boy was five years old. He was cared for in the city of Windsor, Virginia, but lived in Newhampshire for the next four years and before starting out on his own account, had some educational training, and lived in Massachusetts and Connecticutt until at the age of twenty-one years he began working in Minnesota remaining there until 1888, when he came to Box Butte county, Nebraska, and homesteaded near Hemingford and proved up on his one hundred and sixty acres. He lived there for about twenty years and then moved to Morrill county. Here, in 1907, he bought a relinquishment claim of one hundred and sixty acres and subsequently an entire half section of land. He now has four hundred and eighty acres of fine grazing land and feeds fifty head of cattle and one hunred head of hogs annually, and carries on general farming on his eighty irrigated acres. Mr. Dunham is quite modest over all he has accomplished, but undoubtedly it shows strong character and high principles and Mr. Dunham deserves great credit.

Mr. Dunham married Miss Lena Anderson, who was born in Norway, where her parents spent their entire lives. Eight children have been born to Mr. and Mrs. Dunham, namely: Mrs. Eva Fleheaty, of Bayard, Nebraska; Mrs. Myrtle Ross, of Bayard; Melvin, a farmer in Morrill county; and Verne, Opal, Hattie, Hazel and Anna, all at home. The children have attended school regularly and have been taught to prize an education. Mr. Dunham has been a member of the town school board for ten years. He is independent in politics.

EMMONS C. VIVIAN, who is a representative of one of the old and substantial pioneer families of Morrill county, Nebraska, has spent the greater part of his life here, being a youth of sixteen years when he accompanied his parents to this section. He was born in Cass county, Nebraska, August 24, 1872.

The father of Mr. Vivian, Richard Vivian, was born in England, in 1830. In 1844 accompanied by a brother, he took passage in a sailing vessel, which was on the sea for three months before reaching the harbor of New York. During the voyage the brother of Mr. Vivian disappeared and supposedly was accidentally drowned. Richard Vivian was a fine man and it would be interesting to know how the young English boy spent his time before he came to Nebraska, which was prior to 1872. In the meanwhile he was married to Miss Elizabeth Frazier, who was a native of New York. She died in Nebraska, March 10, 1896. They had four children, of whom Emmons Clarkson was the last born. In 1888 Richard Vivian removed from Cass to old Cheyenne, now Morrill county, and took a homestead of one hundred and sixty acres and a tree claim of the same extent. The entire purchase at that time was nothing but wild prairie, but Mr. Vivian lived to see great changes wrought through his industry. At the present time all that land is irrigated and wonderfully productive. Mr. Vivian died in 1911, having been very successful as a farmer and ranchman.

Emmons Clarkson Vivian remained with his father and grew to manhood well acquainted with farm and ranch life. In 1897 he homesteaded one hundred and sixty acres for himself, under better conditions perhaps, than had attended his father, but under no such favorable opportunities as at present are presented, when the homeseeker, if he has sufficient capital, may possibly secure an irrigated farm that will produce more abundantly than in any other state in the Union. Mr. Vivian's homestead is such a farm, all irrigated and finely improved. Adjoining his farm is the forty-nine acre farm of his wife, also improved and irrigated, and it is upon this tract that the comfortable farm-house stands.

In Morrill county, in 1900, Mr. Vivian was united in marriage to Miss Blanche Snider, who was born April 4, 1879, at Kirksville, Missouri. She is a daughter of Albertus and Armilda (Legan) Snider, the former of whom was born in Ohio and the latter in Indiana. They came to Nebraska some thirty years ago and homesteaded near Camp Clark. They now live retired in Idaho. Mr. and Mrs. Vivian have one child, Carol. Mrs. Vivian is a lady of intellectual requirements and has interested herself greatly in the matter of public education. Her work in this direction has been recognized by election to the school board, on which she has served faithfully and efficiently for six years Mr and Mrs Vivian are widely known and universally esteemed

ZIBA VALETTE CLEVELAND, who is a very highly esteemed resident of Bayard, came early to Nebraska and for many years was a substantial farmer He was born in the state of New York, April 24, 1844 His parents were S A and Ruth (Ferris) Cleveland, who spent their lives in New York Mr Cleveland's oldest brother died in the Civil War One brother, I A Cleveland, is a retired druggist living in Chicago, and a sister, Mrs Ida J Weed, lives in New York

From his native state Mr Cleveland went to Iowa and bought land on which he followed farming for twenty years In 1886 he moved to Banner county, Nebraska, homesteaded, pre-empted and secured a tree claim and subsequently proved up on all, when there were but three houses between his homestead and Kimball He sold out there and in Banner county and moved to near McGrew and from there retired to Bayard He built the first frame house in Hull precinct and the first school in the precinct was held in his kitchen for three months Later Mrs Cleveland taught three terms of school in a sod house

In Iowa, in 1875, Mr Cleveland was married to Miss Mary Warrington, a daughter of David and Sarah Jane Warrington, the former of whom was born in Indiana and the latter in Pennsylvania Mr and Mrs Cleveland had four sons born to them, namely Bert, who is a farmer in Scottsbluff county, Ralph, who lives at Spokane, Washington, Lee R, who lives at Bayard, and Roy, who died when aged twenty-five years Mr and Mrs Cleveland have a very comfortable home at Bayard He has always been interested in the public schools and has served on school boards for thirty years He has also been a justice of the peace In politics he has always been a Republican and has voted for fifty years He is independent as regards religious beliefs

VERT B CARGILL, the owner and managing editor of the *Western Nebraska Observer*, published at Kimball, is one of the most prominent members of the newspaper fraternity of the Panhandle, where he has been located nearly a decade A journalist in this twentieth century occupies a vantage ground from which great influence extends, he may build up a cause worthy of public support, may lead men to action in state and civic affairs and plays an important part in the development of the section of the country where his paper circulates Not only the city of Kimball, but Kimball county as a whole has large reason for congratulation that the *Observer* is in such skilled, safe, sagacious and thoroughly clean hands It is one of the best general newspapers published in the Panhandle, as well as an outspoken, fair play exponent of the best elements of the Republican party, it is in all respects well worth the care and sound judgment displayed in its columns and reflects credit on the owner-editor, Vert B Cargill

Mr Cargill was born in Iowa, July 14, 1884, the son of Ezra C and Stella E Cargill, was reared in his native state and received his education in the public schools of Shannon City Soon after graduating from the high school, the young man entered the employ of the *Shannon City Sun*, to learn the practical end of the newspaper business He worked in the printing department from 1900 until the following year, leaving to become associated with the Gravity (Iowa) *Independent*, where he finished his apprenticeship as printer Having mastered the trade, Mr Cargill rose rapidly in the printing business for so young a man and became the foreman of the Corning (Iowa) *Free Press* in July, 1905 This business connection continued for nearly five years during which time he learned all the varied intricacies of the publishing and newspaper business, became well and favorably known among the publishers of Iowa and in June, 1910, was offered and accepted the position of managing editor of the Afton (Iowa) *Star-Enterprise*. In July he took charge of that sheet, soon becoming a joint owner of it in partnership with Senator Charles Thomas, of Kent, Iowa Three years later Mr Cargill sold his interest in the *Star-Enterprise* to Mr O T Meyers After looking over the Nebraska territory, he came to the belief that there was a great future for men of the Panhandle and cast in his lot with this section when he bought the *Western Nebraska Observer* published at Kimball Taking over the management of the paper in August of that year, Mr Cargill has enlarged the original plant, has a good and lucrative job printing business which is run in connection with the paper which today is one of the live, up-to-date, progressive publications of the western half of the state, yielding a strong and wide spread influence in Kimball and adjoining counties, where it plays an important part in the moulding of public opinion

Mr. Cargill is one of the progressive men of the middle west who advocates personally as well as editorially all movements that tend to the development of the city and county. He is alert to present to the people the latest and best discoveries in agriculture, irrigation and education thus being a great force for progress.

He is a staunch supporter of the tenets of the Republican party and served as county chairman of the Republican County Committee and at all times takes an active part in local politics and affairs.

Mr. Cargill is a charter member of the Masonic Lodge No. 294 of Kimball and was Master of the organization from 1917 to 1919.

July 31, 1911 Mr. Cargill was united in marriage with Miss Belle M. McElroy, the daughter of Samuel and Mary McElroy, at Corning, Iowa. The McElroy family are of Irish extraction and Mrs. Cargill's parents were born in the Emerald Isle, coming to America many years ago. Two children have been born to Mr. and Mrs. Cargill: Mary Elzene and Wayne McElroy, seven and five years of age respectively. Since coming to the southwestern part of Nebraska, the Cargill family have made many warm friends in Kimball, where they are progressive and leading residents of a growing and populous city.

OSCAR E. FORSLING, sheriff of Kimball county, Nebraska, most efficiently fills an office of danger and importance. As long as unruly elements insist on breaking the law in a community, public officials must be elected to curb them in order to protect the innocent. These officials, in the nature of things, must be men of great personal courage as well as of close discernment and sound judgment. Such an official is the sheriff of Kimball county.

Oscar E. Forsling was born July 19, 1873, in Sweden. His parents were John and Inga Forsling. Their nine children all reached maturity, the following members beside Oscar E. being well known in this locality: Anna, who married B. A. Norberg, and their son, Ensign Thor Norberg, was an officer at Great Lakes training station, Chicago, during the great war; Alfred, who occupies his ranch situated eight miles west of Kimball; Clarence A., who is a large landowner in Kamball county, served two terms as county sheriff; Augusta, who married Rev. A. M. Breener, chaplain at Camp Taylor during the World War, they being the parents of three sons in the service, one of whom, Paul, died at Des

Moines, a victim of influenza; Frank, who lives at Kimball; and Emma, who is the wife of E. A. Hagstrom, a prominent farmer living six miles from Kimball. In 1883 the parents of Sheriff Forsling came to the United States and for one year afterward the father worked in the Pullman shops at Pullman, Illinois, then came to Nebraska and in 1889 filed on a homestead in Kimball county.

Oscar E. Forsling was twelve years old when he accompanied his parents to Kimball county, and grew up on his father's prairie farm. Some fifteen years of his life were spent riding range as a cowboy in Colorado, Wyoming and southern Montana, and thus his thorough knowledge of this western country can scarcely be overestimated, not only having knowledge of the configuration of the country, but of the people, among whom he has hosts of friends.

On November 25, 1900, Oscar E. Forsling was united in marriage to Miss Ethel Whitman, who is a daughter of Fred M. and Mary (Francis) Whitman. Sheriff and Mrs. Forsling are members of the Presbyterian church. In politics he has always believed in the principles of the Republican party and has taken a somewhat active part in its councils in Kimball county. After serving six years as deputy county sheriff, he was elected sheriff in the fall of 1907 and is still serving. For a number of years he has been prominent in the fraternal order of Knights of Pythias, in which he has passed all the local chairs and after serving one term as deputy grand chancellor, in 1918 was again elected, and on several occasion has attended the meetings of the Grand Lodge in an official capacity. He belongs also to the order of Modern Woodmen of America.

WILLIAM D. ATKINS, one of the prominent and representative men of Kimball county, has spent many useful years in this section, to which he came with his father, March 22, 1889. He was born in Davis county, Iowa, September 2, 1869, son of Peter L. and Delilah Atkins. He has one brother, Dallas K., who lives in Kimball county.

William D. Atkins grew up on his father's farm in Davis county and obtained his schooling there. When his father decided to move to Nebraska and secure a homestead, William D. determined on the same course and both father and son proved up on their land in Kimball county. They at first went into the

YORICK NICHOLS

sheep business and later raised cattle also The father died December 12, 1908, but the mother survives and lives on the old homestead adjoining that of William D His early years in Kimball county were mainly spent in herding cattle and working on the farm In later years Mr Atkins has extended his interests and is now one of the county's substantial farmers

In 1890, Mr Atkins was united in marriage to Miss Elizabeth Pywell, daughter of John and Mary Ann Pywell, both of whom are deceased Their family consisted of three sons and three daughters the survivors all living in Nebraska To Mr and Mrs Atkins three daughters and two sons were born Mabel, the wife of George Ketch, of Kimball, Arthur E, who served overseas in the great war, in the One hundred and ninth Engineers, was promoted to top sergeant, returned to America safely and was honorably discharged at Fort Dodge, July 1, 1919, Grace, the wife of Harley Neely, Ira, who assists his father, Mary, deceased, and Heloise, at home

For many years Mr Atkins has been active in the councils of the Democratic party Fully twenty-five years ago he was elected chairman of the town board and one many occasions since has filled important offices, in 1914, being elected a county commissioner, the only one of his party candidates elected, and in 1919 was re-elected for a second term of four years He has also served as highway commissioner He signed the franchise for the first telephone company in the county He is prominent in the order of Knights of Pythias, being past chancellor and also in the Modern Woodmen of America Mr Atkins and his family belong to the Methodist Episcopal church

YORICK NICHOLS was born in Tioga county, Pennsylvania, in 1863 the son of William A and Nancy (Mitchell) Nichols both natives of Pennsylvania The father was an attorney who practiced law at Wellsboro, Pennsylvania, until 1869, at which time he removed to Kansas, where he continued his law practice and at the same time took a homestead and proved up on it He did the first hardwood building in that county, having floated lumber across the Neosho river with an ox team He also dug the first cellar in Neosho county, started the town of Tioga, Kansas and practiced his profession there until he died in 1873, a successful man and one of the pioneers who helped to start the development of the great West following the Civil War

Yorick Nichols was the oldest boy of five children in the family The others were May, the wife of Henry Block, now deceased, Carroll who died in Morrill a few years ago and who together with his brother Yorick built the first substantial building in that town Willis, who lives at Sweetgrass, Montana, and is the owner of the townsite at that place, and Blanche, now Mrs Henry Russell at Mitchell, Nebraska

Mr Nichols still keeps as one of his most prized relics his father's commission as a captain in Hancock's corps of the Union army, signed by President Abraham Lincoln

In 1882 Yorick Nichols came to Wyoming and worked as a cowboy He took a preemption claim and timber claim in that part of old Cheyenne county, Nebraska, which is now Scottsbluff county, and also a homestead of 320 acres in Wyoming with 100 acres under irrigation He has followed stockraising for his main occupation all his life, feeding in the winter time, and raising a good grade of stock He ran cattle on 5 000 acres adjoining the present town of Henry Nebraska, which town was founded by him His place is now known as "Little Moon Ranch "

He was married first to Alice D Dyer, a native of England and a woman of literary talent who did quite a little writing She is now deceased They had an adopted son, Henry B Dyer who met an accidental death by drowning a few years ago It was in his honor the town of Henry was named

Mr Nichols' present wife was Maude Lawrence, a native of Nebraska She is a member of the Christian church

Mr Nichols is an independent voter He is one of the best known of the old-timers of western Nebraska and eastern Wyoming, and claims the distinction of being the first bona fide settler in this part of the North Platte valley

CALVIN NEELY, one of Kimball county's highly esteemed citizens, for many years was engaged in the stock business but now lives practically retired in his comfortable home at Kimball Mr Neely was born in Grant county, Wisconsin, December 27, 1861 His parents were Samuel and Anna Neely, who had eleven children and eight of these reached maturity Samuel Neely enlisted in the Forty-second Volunteer Infantry in the Civil War at Lancaster, Wisconsin, and was honorably discharged at Cairo, Illinois

Calvin Neely grew up in Wisconsin and obtained his schooling there As one of a large family he early had to assist in his own support by working on farms and herding cattle and remember how far he felt from home and how big the world looked to him, when his father sent the eleven-year old boy to look after the herding a dozen miles away. He accompanied his parents to Nebraska in 1886, when his father homesteaded in Cheyenne county. The family lived there about eight years, then moved to Cheyenne, Wyoming, where the father died, the mother returning then to Nebraska, where she lived with a daughter until her own death.

In 1887 Calvin Neely was united in marriage to Miss Ella M. Bliss, daughter of Ambrose K. Bliss. Mr. and Mrs. Bliss came to Nebraska in 1886 and located in Cheyenne county, where he homesteaded one hundred and sixty acres and proved up on the claim. In 1898 they moved to Eaton, Colorado. The mother died in Cheyenne, Wyoming in 1901, and the father died in Denver, Colorado in 1908. Mr. Bliss was a corporal in Company C of the Twentieth Wisconsin Infantry, and served for three years, being mustered out at Galveston, Texas, July 14. 1865. Mrs. Neely was one of a family of eleven children, like her husband, and all reached maturity except one who died when fourteen months old. The Bliss family lived in Wisconsin, but Mr. and Mrs. Neely were married in Cheyenne county, Nebraska. They became the parents of three sons and one daughter, namely: Charles Vere, Chester C., Harlan L. and Doris G. All three sons were soldiers in the great war that has left its black trail of sorrow in so many homes. The eldest son of Mr. and Mrs. Neely, Charles Vere Neely, was well and affectionately known all through Kimball county, for he had qualities that won him friends wherever he went. For about twelve years before entering the National army, he had lived at Golden, Fruitdale and Maple Grove, Colorado. He was sent from Golden in the draft contingent leaving April 27, 1918, to Camp Funston, where he was assigned to the Three hundred and fifty-fourth infantry and was sent overseas with the Eighty-ninth division. Although, through bravery exposed on hundreds of occasions to a soldier's hazard, he escaped injury until the practical ending of the war, receiving his death wound just fifty minutes before the signing of the armistice that ended the fighting. In a beautiful, touching letter received subsequently by the bereaved family, his closest comrade during their sojourn in France, says: "a better buddy in every way, a more fearless soldier, a quicker or more dependable runner, and a surer

guide, never lived." Neely Post No. 22, at Kimball was named in his honor.

Chester C. Neely, the second son, is an overseas soldier who is now at home, having been honorably discharged from military service, May 17, 1919, at Camp Lee, Virginia. He had twelve months of training at Camp Funston and Camp Cody, New Mexico, then went to France attached to company A, One hundred and ninth engineers, Thirty-fourth division, and served there eight months. Harlan L., the third son, was in training for some months at Lincoln, Nebraska, and Camp Sherman, Ohio, and was honorably discharged. The one daughter of the family, Doris G., a high school graduate, is employed in a Kimball business house as a bookkeeper and resides with her parents. Mr. Neely and his family are members of the Presbyterian church.

SAMUEL B. HANNA, who has the distinction of being the second oldest real estate dealer, in point of time, in Kimball county, came here in 1906 and has built up an extensive business connection in land and insurance. Mr. Hanna was born in Fayette county, Ohio, April 7, 1870. His parents were James and Tabitha Hanna. His mother died May 28, 1870, leaving a daughter, since deceased, and Samuel B., an infant. In 1904 his father came to Nebraska and bought land in the Wood River valley, on which he lived until 1907, when he moved to Oklahoma and his death occurred in 1915, at Hennessey, in Kingfisher county. Both parents were members of the Presbyterian church, solid, respectable people of good old Scotch-Irish stock.

Samuel B. Hanna had adequate school opportunities in his youth but no special advantages. In 1906 he came to Kimball county and on June 1, of that year, embarked in the real estate and insurance business, having secured an agency from W. F. Shelton, of Omaha, in the sale of Union Pacific Railroad lands. In fulfilling this contract Mr. Hanna has handled many thousand acres of land, disposing of the last tracts in this section in 1911. Perhaps no one in the business is better qualified concerning land of every description and value all through Nebraska, and many eastern firms consult him concerning investments. He also represents old line insurance companies. In every phase of his business Mr. Hanna has been found reliable and upright.

In 1892 Mr. Hanna was united in marriage to Miss Effie M. Briggs, who was born at

Greenfield, Ohio, a daughter of Jesse and Delilah Briggs, farming people To Mr and Mrs Briggs nine children have been born Charles Wesley, a minister in the Methodist Episcopal church, in Ohio, Elijah, deceased, was a farmer, Jesse, died in 1918, Effie and Elmer, twins, the latter of whom died in infancy, Clara, the wife of Edward Preston, Martha, resides at Greenfield, Ohio, Rebecca, the wife of William Roseboom, a retired farmer of Summitville, Indiana, and Emma, who was the wife of William Fisher The family home is a handsome modern residence on the corner of Fourth and Chesnut streets, Kimball Both Mr and Mrs Hanna are members of the Methodist Episcopal church He is prominent in the order of Knights of Pythias, being a past chancellor and deputy grand chancellor and master of finance in the local body, and belongs also to the Modern Woodmen

WILLIAM J CRONN, who has been a prominent citizen of Kimball for many years, active in business and foremost in civic affairs, was born at Millbrook, Ulster county, New Jersey, July 7, 1860 He was reared and educated in New Jersey and Pennsylvania, had public school advantages, and was twenty years old when he went to Wisconsin, his parents following about eight months later

The family lived in the above named state for five years and then the father moved to Nebraska, in 1885, locating in Colfax county In 1890 he took up a homestead in Banner county but at a later date sold it and moved to California, where they are still living near Los Angeles, being aged about eighty-six years Of their thirteen children nine are living They are members of the Methodist Episcopal church

William J Cronn was thirty-eight years old when he came to Kimball county and started into business in the village as a painter and paper-hanger Business prospects at that time were not very bright for the village and Mr Cronn remembers seeing three of the rather limited number of houses moved by their owners out on their ranches He found plenty to do however, as he was the only man in his line in the neighborhood and the most of the painting and paper-hanging jobs between Sidney and Cheyenne came to him He now has a paint and paper store at Kimball and is a contractor in this line of work He owns considerable property at Kimball included in which is his fine modern residence

In 1888 Mr Cronn was united in marriage to Miss Hattie Longworth, a daughter of William Longworth and wife, who reside in Schuyler, Colfax county, Nebraska The latter have three daughters and one son Ethel, the wife of Mr McGregor, has five children, Alice, who married Mr Wilson and resides at Kimball, Chester, a painter by trade, and Irene, her father's assistant in the store From the first Mr Cronn has been enterprising and progressive as a citizen He has been mayor of the city and is now serving in his third term as city alderman In speaking of him his fellow citizens say, "he is a fine man" Mr Cronn and his family belong to the Methodist Episcopal church

GEORGE W HARVEY, who is a highly respected retired resident of Kimball, came to Nebraska many years ago and in one way or another, has been identified with the substantial development of several sections of the state He belongs to that sturdy group of pioneers who blazed the way for those who later more comfortably followed the trail

George W Harvey was born in Hardin county, Ohio, March 19, 1849 His parents were Brice and Caroline Harvey, who were married on February 18, 1847 They had two children, George W and Mary D, the latter of whom died in infancy The father died on the old homestead in Ohio, June 8, 1856 The mother remained a widow six years, then married John Merritt, a fine man, a farmer and stockman of Jones county, Iowa To the second marriage of Mr Harvey's mother seven children were born and the mother died February 16, 1904 on the homestead situated three miles west of Olin, Iowa

After his father died and until his mother married again George W Harvey lived with his grandparents and an uncle, then went to Iowa with his mother and step-father, the latter treating his stepson very kindly and he remained at home until he was twenty-one years old In August, 1871, Mr Harvey was united in marriage to Miss Cora A Williams, a daughter of Harris and Louise (Young) Williams Mrs Harvey had one brother who died in infancy Her mother died in Illinois, while her father was a farmer and stockraiser near Joliet but he later moved to Iowa and died in Jones county To Mr and Mrs Harvey the following children were born Celesta born August 8 1872 died in infancy in Iowa Lillian born October 8, 1873, is the wife of John McKinnon, who owns a fruit farm in California, Charles born May 27, 1875, is a rancher in Montana, Ella born

March 8, 1877, is the wife of Frank O. Baker, who is a banker; Arthur, born June 18, 1878, is in business at Joliet, Wyoming, is married and has three living children; Earl, born March 18, 1881, conducts a stock ranch and farm in Banner county, is married and has seven children; and Nina E., born April 30, 1882, is the wife of William Deakin, of Omaha.

· In 1882 Mr. and Mrs. Harvey came to Nebraska and located in Burt county, the nearest market town being Decatur. They remained in Burt county for some years, Mr. Harvey buying three hundred and twenty acres, which he sold in 1888 and then they came to what is now Banner county, just prior to the contest over the county seat, details of which are found in the county annals. After taking up a pre-emption in Banner county he proved up, remained twelve years and from time to time bought land until he now owns seventeen hundred and sixty acres of fine land there. Mr. and Mrs. Harvey's first home was in a tent that served them for six months, when Mrs. Harvey's uncle, Ebenezer Williams, built them a stone house, Mrs. Harvey assisted in mixing the mortar, and a warm, comfortable residence was the result. When Mr. Harvey had his house ready to move into, he had just seventy-five cents in his pocket, and when his household goods arrived at Kimball he could not find any place to store them, so hauled them out to his homestead, covered them with boards and left them undisturbed until he had managed to put in a crop on forty acres.

Mr. Harvey at that time had to drive sixty miles to Cheyenne to find a market for his crop, and received thirty-five cents a bushel for wheat that took the first prize at the state fair and also the sweepstake prize for the best wheat grown in Nebraska. Like other settlers he faced hard times on many occasions, often worked for seventy-five cents a day at anything that offered, and during one winter Mrs. Harvey went to Cheyenne and did nursing in order to add to the family exchequer, and at other times she remained to look after the crops while Mr. Harvey and son Charles worked for John Gordon on Horse Creek. It was not all work and no play in early times, however, and Mr. Harvey had an enjoyable occasion, when settlers from all over the county met in Bull Canyon and got acquainted with each other, that being the first "get together" meeting they ever had, but not the last. At one time Mr. Harvey had an open range in Banner county of fifty-five sections and it took him three days to ride

around it. The first school house was built by the settlers, a log structure 16x24 feet in dimensions, which was used for some years as the taxes in the district did not prove sufficient to build a new one. In 1909 Mr. Harvey came to Kimball and bought fifty acres and a comfortable residence. He and wife are members of the Presbyterian church.

FRANCIS O. BAKER, who is a man of business prominence in Kimball county, where he has large land interests as well as in Banner county, is president of the Bushnell State Bank at Bushnell. Mr. Baker was born in 1863 in De Kalb county, Illinois. His parents, William and Mary (Newport) Baker, were born near Dover, England, and after coming to the United States lived for two years near Syracuse, New York, then moved to De Kalb county, Illinois. In 1877 they went to Nebraska, settling in Saline county, where the father yet lives, but the mother died in June, 1917. They had children as follows: an infant that died at birth; Mattie, who married Emmett Buckingham of Beaver Crossing, Nebraska, and they have five children; Charles, who is a retired farmer and stockman, lives at Lincoln; Francis O, who was fourteen years old when he came to Nebraska; and Addie, who is deceased.

Francis O. Baker had public school advantages in Illinois, accompanied his parents to Saline county, Nebraska in 1877, and remained at home assisting his father until he was twenty-four years old. At that time he went to Banner county, where he homesteaded one hundred and sixty acres, on which he proved up and to which he subsequently added. Mr. Baker now owns four sections of land in Banner county and one section in Kimball county, having lesser interests in other counties. In May, 1910, Mr. Baker came to Bushnell and embarked in the mercantile business, shortly afterward organizing the Bushnell State Bank. He served this institution at first as cashier, but in 1917 was elected president and has ably directed its affairs ever since, making it one of the sound, stable banks of the county, Mr. Baker having the full confidence of the public in his business sagacity and personal integrity.

Mr. Baker was married in 1893, to Miss Ella Harvey, a daughter of George W. Harvey, a prominent retired citizen of Kimball. Mr. and Mrs. Baker have three children: Charles, born in 1898, has finished his high school course and is assisting his father; Robert born March 8, 1904; and Alice, born

September 1, 1906 Mr Baker and family attend the Presbyterian church He is identified with the fraternal order of Modern Woodmen of America

PERRY BRAZIEL is one of the real pioneers of the West, the son of one of the pioneers of the middle states He was born in Madison, Wisconsin, in 1854, the son of Robert Braziel, who was born in Tennessee and died in the 70's His mother's maiden name was Steele She was a native of North Carolina, and died at the age of fifty years The father moved to Illinois in 1812, when that country was young He was a farmer and an Indian fighter in the Black Hawk wars a Democrat in politics, and a Methodist There were five children in the family, the subject of this sketch being the only one now living

Mr Braziel came to Kansas with his father in 1857 and settled in the Osage Nation, where they farmed and ran cattle In 1868 he went to Texas and worked on the trail and the Texas range as a cowboy In 1880 he came to western Nebraska — what is now a half-dozen counties being then all Cheyenne county — and in 1884 he took a homestead east of where the town of Haig now stands He proved up on his homestead and bought several different tracts of land until twenty-two years ago when he came to his present farm where he now owns 420 acres of irrigated land, well improved He has followed farming and stockraising and has been very successful

In 1888 he was married to Ida Rayburn, a native of Illinois, daughter of Thomas Rayburn who homesteaded in 1886 in what is now Castle Rock precinct of Scottsbluff county They have three children, namely

Robert, now in South Omaha, having lately returned from service with the American army in France, being with the Ninth Veterinary Corps overseas for eleven and one-half months

Thomas A, who is now at home after a service of twenty-two months with the colors He served on the front line in France fourteen months with the 148th Field Artillery and 116th Ammunition Train

George, the youngest son is at home

Mr Braziel has been a leader in the life of the county since it was organized He has held the office of county commissioner, and is a member of the board of directors of the irrigation district in which his land is situated He is an independent voter and belongs to the Masonic order Mrs Braziel is a member of the Methodist church

No man in the North Platte valley stands higher than Perry Braziel as an independent and progressive citizen and a man of integrity and honor He has prospered in this world's goods and in the opinion of his fellow man He has raised a family that is a credit to him and has a right to be well satisfied with the record that he has made

HORACE C AMOS, who is one of Kimball's representative business men and for a number of years identified with the Citizens State Bank, is not a native of Nebraska, but has spent the greater part of his life in this state He was born in 1877 at Racine, Wisconsin His parents were Arthur and Julia (McCumber) Amos, both of whom died at Kimball, his father having been a banker and stockman

Horace C Amos was nine years old when his parents moved to Kearney, Nebraska, where they remained two years and then came to Kimball Here Mr Amos attended the public schools and completed the tenth grade studies, then entered the Kearney Military academy and remained one year He then became interested in the stock business on his own account and continued in that line for twenty-one years, when he was elected county clerk After serving in that office with the utmost efficiency for three years, Mr Amos resigned in order to give his attention to the affairs of the Citizens State Bank of Kimball In the meanwhile he proved up on a homestead in Kimball county which he sold at a later date

In 1905 Mr Amos was united in marriage to Miss Ema Tracy, who was born at Pine Bluff, Laramie county, Wyoming, and they have two daughters Marjorie and Marian, aged respectively twelve and seven years, both of whom are attending school at Kimball Mr Amos and family belong to the Episcopal church He belongs to the fraternal order of Knights of Pythias and in that connection as in every other, is held in the highest esteem

ORLEY D PICKETT, who is an enterprising business man of Bushnell, Nebraska, where, in partnership with his brother Roy, he conducts a cream and produce station, was born February 25, 1886, in Nemaha county, Nebraska His parents are Frederick and Effie (Dickerson) Pickett who came to Kimball county March 3 1907 and now reside on a farm north of Kimball Of their six children Orley D is the eldest, the others being as follows Bertha, who is the wife of Henry Wright, a farmer and ranchman living northwest of Kimball, Clinton, who resides on his ranch north of Kimball; Ernest, in Bushnell; and Francis, who is a farmer south of Bushnell; and Roy, who is associated with his eldest brother at Bushnell.

Orley D. Pickett spent his boyhood days on the home farm and attended the public schools until the opportunity came to engage in farming for himself. In the fall of 1913 he came to Bushnell and accepted a position as clerk with Mr. Baker, a leading business man, and remained with this employer until September 4, 1915, when he purchased the business, and his prospects seemed so bright that he put in a stock representing the investment of $16,000. War clouds quickly gathered as time went on, America became involved in the great struggle and when the government found it necessary to issue calls for soldiers, Mr. Pickett and his brother Roy found themselves among those selected. With only four days' notice, Mr. Pickett sold out his business and prepared to answer the call, his brother Roy having made arrangements to leave Kimball for camp at El Paso, Texas, on the day the armistice was signed.

In 1911 Mr. Pickett was married to Miss Bertha Bower, who is a daughter of Franklin and Helen (Gross) Bower. Mrs. Pickett has one sister, Iva Maud, who is the wife of F. E. Miller, a stockman and farmer near Delaware, Ohio. The father of Mrs. Pickett still lives in Ohio, but her mother passed away August 24, 1883. Mr. and Mrs. Pickett attend the Presbyterian church. He is prominent in public affairs at Bushnell and has been a member of the city council since incorporation.

PHILIP NELSON, who is a well known and respected resident of Kimball county, where he has successfully carried on several business enterprises, was born on Chellon Island, between Denmark and Norway, and in 1886 accompanied his parents to the United States. Their names were Julius and Sophia Nelson, honest, hard-working people, who were very highly thought of in the neighborhood of Dix, where they first located. They homesteaded one hundred and sixty acres and also took a tree claim, proved up on their land, and the father died there in 1913. The mother then went to Blair and died in the home of a daughter, in 1917. They had nine children, five sons and four daughters as follows: Johanna, who died in 1903; Peter, who lives on his fruit farm in California; Christina, who was the wife of Melvin Tracy of Butte, Montana, died April, 1920; Hans, who lives in

California; Bina, who is the wife of John Hanson, of Council Bluffs, Iowa; Jack, who lives on a ranch near Dix, Nebraska; Philip, who is the youngest son; and Margaret, who is the wife of Ole Anderson, of Blair, Nebraska.

Philip Nelson's boyhood days were spent in going to school, and herding cattle for his father and other ranchmen. He had ambition, however, to be a business man in another line, and with this end in view went to Potter and engaged as a clerk in a general store. In 1910, in partnership with his brother Peter, he bought the lumber yard at Potter and operated it for two years, when he purchased the general store of C. E. Birt, thereby acquiring a stock of merchandise valued at $2,100. He continued in the mercantile business until 1919, selling his store at that time after a very satisfactory business season, his stock being valued at $15,000. In the meanwhile he had bought a half section of land in Kimball county which he rents out, the products of his farm and ranch being mainly grain and horses. Mr. Nelson is looking out for another investment, as he is too active to think yet of retiring from the business field. In 1915 he married Christina Peterson and they have two children, Louis and Ruth. Mr. Nelson is a fine example of what determination will do, for through it he has overcome many difficulties, has made an honorable business name for himself and has personal friends everywhere. He belongs to the Lutheran church.

ERNEST EUGENE GODING. — Although aspiring to no position of leadership in advancing the welfare of development of the growing little city of Dix, the successful activities of E. E. Goding have brought him to public attention. Since locating here he has been active, interested and useful in many ways.

E. E. Goding was born in Pawnee county, Nebraska, April 14, 1879. His parents were Rufus H. and Jessie F. Goding, who came to Nebraska in 1877, the year of their marriage, and Rufus H. Goding bought a quarter section of land near Pawnee, on which the family lived for five years, Mr. Goding engaging in farming and raising stock. He then removed with his family to Lincoln county, South Dakota, remaining there until 1907 and then moving to Morrill county, Nebraska. In 1917 the parents of Ernest E. Goding sold their Morrill county interests and went to California and now reside retired at Harper in that state. Of their family of six children, Ernest E. was the first born, the others being as follows William M, who is a prominent citizen of Cheyenne county, where he is a member of the board of county commissioners, Bertha E, who is the wife of Mark Myers, Edith, who died at the age of five years, Clara R, who is the wife of Dr Dayton Turney, of Los Angeles, California, and Flora, who is the wife of James Davison, who did own an extensive cattle ranch near Dalton, Nebraska, but now resides in Colorado The parents are members of the Baptist church

E E Goding attended school in Lincoln and Turner counties, South Dakota, then entered the South Dakota State University, where he spent four years, then entered the United States army, serving through a first enlistment in Company A, First South Dakota volunteer infantry He re-enlisted in Company I, Thirty-seventh United States volunteer infantry and with this unit went to the Philippine Islands, where he served almost three years, being promoted first sergeant of Company I He was honorably discharged at San Francisco, in 1901 Mr Goding has a fine record as a soldier in the Spanish-American War With his contingent he reached the Philippines on August 24, 1898, just eleven days after the battle of Manila After his safe return to the United States and his discharge from the service he had honored, he took up a homestead in Charles Mix county, South Dakota, on which he proved up After selling his homestead at a profit, he taught school in Charles Mix county for the next seven years, also being interested in ranching In 1909 he came to Kimball county, Nebraska, taking up four hundred and eighty acres of land under the Kinkaid act, located northwest of Dix He resided on that land until March, 1919, when he came to Dix and embarked in the real estate business made practical investments in the way of substantial business enterprises here, and through newspaper connection has been a valuable exploiter of the interests of this place He is interested in the Dix Mercantile Company, in the erection of a number of substantial business structures and in the laying of fine cement pavements Mr Goding still owns two sections of land in Kimball county but has them under rental

In 1909 Mr Goding was united in marriage to Miss Eva Parker, a daughter of Charles and Ella Parker The father of Mrs Goding died in Kimball county in 1918, but the mother survives and lives two miles west of Dix The Parkers came to Kimball county from South Dakota Mrs Goding is one of a family of fourteen children, seven boys and seven girls Mr and Mrs Goding have two daughters Mildred, V and Olive J, aged respectively eight and three years Mr Goding and his family attend the Presbyterian church He belongs to the order of Odd Fellows having united with this organization at Lake Andes, in Charles Mix county, South Dakota

L FRANK PRICE, who has been so prominently concerned in the development of the incorporated town of Dix, Nebraska, that it is difficult to mention any of its important enterprises without reference to him was a man of business prominence in other sections before coming to Kimball county

L Frank Price was born in Shelby county, Illinois, October 20 1877, and was young when the family moved to Decatur, where his boyhood and youth were spent, his public school advantages extending through the high school course He also completed a course in Brown's Business college at Decatur, following which he went into railroad work and continued in train service for five years He then became identified with the insurance business, in the Peoria Life Insurance Company, with which concern he remained for some years as superintendent of the Decatur district A change of climate being deemed best for some members of his family, Mr Price moved to Ogden Utah, and for two years was in the employ of the Short Line Railroad in the Ogden yards Removing then to Denver, he shortly afterward homesteaded in Weld county, Colorado, and proved up before moving to Cheyenne Wyoming, where he re-entered railroad service for two years afterward being with the Northern Pacific system

Mr Price, however, had not entirely retired from the insurance line and finding prospects encouraging in Wyoming became connected with the state agency of the Central State Life Insurance Company for the state of Wyoming, and continued as state agent until the spring of 1917, when he took over Western Nebraska for the same company, establishing his headquarters at Kimball His business experience had given him excellent training along many lines and before he had lived long in Kimball county his interest was aroused in the little village of Dix, at that time an insignificant country hamlet, with a population of not over 26 individuals all told Mr Price, however, was an experienced railroad man and many times had he witnessed a section of country developed almost over night by the coming of the railroad Hence he was liberal in his investments in land at Dix, although, at that time, he found little encouragement among the old settlers here, even such well informed men as the Gundersons, Phillip Nelson and E. J. Horrum. Mr. Price was not discouraged however and soon, through his vitalizing energy had wonderful development take place, culminating in the incorporation of the town of Dix on September 4, 1918. He has continued active in every business and public-spirited enterprise, in many of these being associated with Mr. Goding, the firm of Goding & Price, carrying on a large real estate business, being founders of the Dix *Tribune*, which issued the first newspaper here on May 12, 1919, foremost in other matters of business. During the time Mr. Price was chairman of the town board about twelve thousand feet of cement walks were laid. He is secretary of the school board of Dix and it is no secret that largely through his efforts the township high school was accorded Dix instead of Kimball. The laying of the corner stone of that handsome modern structure was a memorable event in the town, the exercises being under the auspices of the Masonic fraternity, the Grand Master of the state laying the stone.

Mr. Price was united in marriage to Miss Alice J. Griffin, who was born in De Witt county, Illinois, a daughter of B. C. and Judy (O'Brien) Griffin, natives of Ireland, who came when young to the United States with their parents. Mrs. Price has four brothers and two sisters. Her father died in 1916, but her mother survives and lives in Illinois. Mr. Price's mother, Mrs. Lodema Price, makes her home with her son, they two being the only remaining members of that family. Mr. and Mrs. Price have one daughter.

HANS GUNDERSON, who has the distinction of having been the original purchaser of the town site of Dix, has always been a man of discretion and foresight, and the present flourishing town owes much to his energy and practical enterprise. He is one of a group of earnest, public-spirited men, whose united efforts have brought about a wonderful degree of progress in a comparatively short space of time.

Hans Gunderson was born in Norway, August 12, 1866, came to the United States in 1873 with his parents and three brothers. The family lived in Omaha for fifteen years, during which time two more sons were added to the family. Of them all, Hans was the second born. He was twenty-two years old when he came to Kimball county and took a homestead, later selling it and taking a Kinkaid claim. Aside from his large holdings at Dix, Mr. Gunderson owns two hundred acres forty acres under water, of land situated three miles north, one and three-quarter section four miles south of Dix, and rented to good tenants. He has a $30,000 investment at Dix. When Mr. Gunderson came here first he associated himself with other enterprising men and his interest has been continuous. The first building he erected was a blacksmith shop, then a town hall, a restaurant building, two small houses, a carpenter shop and his own comfortable residence. His activities at present include the erection of two brick buildings on Maple street, in which the post office wil be located, also the Central telephone and city offices.

Mr. Gunderson was married to Miss Belle Snyder, November 16, 1891, at Harrisburg, Banner county, Nebraska, who was born in Iowa. They have four children: Aye, Effie, Mervin and Claria.

Mr. Gunderson has always been progressive in his ideas. He owned the first threshing machine in Kimball county and for years did all the threshing in Kimball and Banner counties, even as far as Bridgeport. This machine was a J. I. Case horse-power rig, 12 horses being used for power. He belongs to the Knights of Pythias lodge at Kimball, and assisted in the organization of the camp of Modern Woodmen at Dix, which now has a membership of seventy-five individuals. Few men in this and adjoining counties are better known than Hans Gunderson.

FRANK E. CAMPBELL, who is one of the enterprising business men of Dix, Nebraska, is a native of Illinois but has been a resident of Nebraska during the greater part of his life. He was born March 2, 1872, and is a son of most worthy parents, John and Catherine Campbell.

The parents of Mr. Campbell were born in Ireland and both came to the United States when young, and were married in the city of New York. They located afterward in Illinois and Frank E. was born while his father was a farmer there, one of a family of five daughters and four sons, all of whom lived to maturity. In 1884 the Campbell family came to Nebraska, first settling four miles east of Fairfield, but later moving to that part of Cheyenne county that is now included in Kimball. The father homesteaded in the northeast corner of Kimball county, in 1886, securing one hundred and sixty acres and later a timber claim and proved up on his land. He died there December 5, 1893, after which the

MR. AND MRS. WILLIAM L. WALLACE

mother moved to Cheyenne, Wyoming, where her death occurred in 1908

Frank E Campbell had country school advantages in Kimball county, but much of his time in boyhood was given to herding cattle Later he homesteaded in Lawrence Forks Valley, in Banner county, lived on his land for five years and proved up, residing for about eight years in that county In 1894 he sold his ranch to his brother-in-law, J E Bevington After leaving the ranch Mr Campbell went to Potter, where he started a billiard room and soft drink establishment, conducting this business there until March, 1918, when he came to Dix Here he went into the business on a larger scale and now has well equipped billiard parlors, which are well patronized by lovers of this form of pleasant exercise, and he also conducts a confectionery and soft drink business

Mr Campbell was reared in the Roman Catholic faith and has always been sincere in his church relations He has never married Not particularly active in a political way, nevertheless Mr Campbell is a good citizen and as a "booster" has done much of Dix, being ever ready to co-operate with other enterprising citizens in movements for the general welfare Mr Campbell is a member of the order of Modern Woodmen at Dix

WILLIAM L WALLACE, one of the early settlers of the North Platte valley, was born in Marshall county, Indiana April 8, 1867, the son of M F and Nellie (Ada) Wallace There were eleven children in the family, of whom six are living, namely Frank, of Scottsbluff, Nebraska, Etta, of Hastings Edward, living in California, John, of Alliance Nebraska, Julius, of Hastings, Nebraska and the subject of this sketch The family moved to Hastings, Nebraska, in the spring of 1873, and there the father took a homestead and followed general farming Both the father and mother are still living on the old home place near Hastings

Mr Wallace received his education in Hastings, and after completing his schooling he farmed for a year at that place then came west in 1886 and homesteaded on Snake creek in 1888 He followed the stock business for a number of years, moving his family to Scottsbluff after it was started In the cattle business he met with excellent success and a few years ago he bought the Henry State Bank at Henry, Nebraska, and now devotes his attention to the banking business With a capital of $10,000 00, this institution has deposits of $125,000 00 and a surplus of $2 000 00, and

Mr Wallace is on the fair road in the banking business to repeat his success in the stock business

He was married in October, 1889, to Nellie Gaddis, a native of Indiana, and their union has been blessed with eight children, namely Florence, now Mrs J C Williams of Henry, Nebraska where Mr Williams is in the drug business, Willo, now Mrs Arthur Selzer, of Scottsbluff Bessie who is employed in her father's bank Dorothy, Wilbur, Shirley, Neal, and Helen all at home

Mr and Mrs Wallace are attendants of the M E church Mr Wallace is an independent Democrat in politics and a member of the A O U W He has always been recognized as an enterprising and honorable man, taking an active part in public affairs and wielding the influence that goes with prominence and high standing in the community

CHARLES L BOGLE, who owns and conducts the leading general mercantile business at Bushnell, was born in Gosper county, Nebraska, May 4, 1889, a son of J W Bogle, for many years one of Kimball county's most respected citizens

Charles L Bogle attended the country schools near his father's ranch in boyhood, later taking a business course in a commercial college at Grand Island In 1908 he came to Bushnell and went into the mercantile business with his father, that association lasting until 1913, when, in partnership with his brother-in-law, he bought the elder Bogle interest In 1914 he bought his partner s interest and since then has been sole proprietor Mr Bogle carries a stock worth $20,000, consisting of general merchandise, additional features being a meat market and confectionery store Being energetic and a good business man Mr Bogle has continued the expansion of his enterprise and is doing a large and profitable business

In 1912 Mr Bogle was united in marriage to Miss Flora Snyder, and they have three children Charles L, Dora Mildred and John V

Mrs Bogle is a daughter of James M and Elizabeth (Shanks) Snyder, who now live comfortably retired in Furnas county, Nebraska The father of Mr Bogle was born near Columbus, Ohio, in 1842 and served as a soldier in the Civil War as a member of the Forty-fourth Indiana volunteer infantry His venerable mother still survives, residing at Columbia Indiana, in her ninety-ninth year The mother of Mrs Bogle was born in 1850 in Whitley county, Indiana, and her mother was a second cousin of Abraham Lincoln, through the Shanks connection Mr and Mrs

Snyder had a family of four daughters and two sons, namely Rosa, who is the wife of Frank Bogle, of Bushnell, Nebraska, Inez, who is the wife of Archie Deen, of Bushnell; Dora, who is the wife of Walter Rogers, a homesteader in Wyoming, Flora, who is Mrs Charles L Bogle, of Bushnell, Roy, who is a farmer, and Clarence, a returned overseas soldier of the great war He received his military training at Camp Cody before sailing for France, where he served bravely for sixty-two days on the front line Mr and Mrs Snyder are members of the Christian church at Edison, Nebraska Mr Bogle has never found time to be very active in politics but he is a good citizen and highly esteemed

EMORY C HOWE, who is a prosperous business man of Bushnell, owes his success in life to natural ability and also to his faculty of making and keeping friends He was born August 2, 1883, in Nemaha county, Nebraska, a son of Seymour and Ellen Howe The Howe family has been so highly regarded in that section, that one of the flourishing towns of Nemaha county bears the name of Howe

The father of Emory C Howe was born in the state of New York and the mother in Illinois Both came to Nebraska when young and were married in Nemaha county The following children were born to them Adelia, who is the wife of C R Russell, a farmer and ranchman in Nemaha county, Charles, who follows the carpenter trade in that county, Eugene, who is in the real estate business at Weatherford, Oklahoma, Ambrose, who is a traveling salesman with home at Council Bluffs, Iowa, and Emory C, who was educated and lived in Nemaha county until December, 1914, at which time he came to Bushnell Here Mr Howe went into the automobile business handling Buick cars, which has become a growing concern He is now erecting a first class modern garage of brick construction, two stories high, with many square feet of floor space and fine display room for the Buick cars, of which he is sole agent In addition to his other responsibilities, Mr Howe is deputy sheriff

On March 12, 1907, Mr Howe was united in marriage to Miss Geneva West, a daughter of Jacob and Alice West, who came from Missouri to Nebraska in 1900 and now reside at Salem

WOODFORD G JONES, who now lives in comfortable retirement at Bushnell, came to Nebraska thirty-five years ago During early years in the state, Mr and Mrs Jones saw much hardship and they had some losses which necessitated hard work and economy, but early conditions passed away and they have lived to enjoy the fruits of their industry

Woodford G Jones was born at Centerville, Iowa, April 28, 1857, a son of Woodford and Louisa Jones Attending school near the home farm in boyhood, Woodford G Jones took charge when his father died and operated the farm for his mother and sister as long as he remained in Iowa In 1884 he came to Nebraska and homesteaded eight miles south of Dix, in Kimball county There was but a poor shelter on the place and when a blizzard that lasted three days set in following their arrival, they had to take advantage of every expedient to keep warm, all remaining in bed until the snow was so deep there that Mr Jones had to shovel it off Happily they had enough beans and corn dodgers to keep them from being hungry and in that way were more fortunate than many of their pioneer neighbors Mr Jones had previously had a serious experience in one of the sudden blizzards that sometimes unexpectedly swept over the country, during which he stumbled and was lost in the snow through a whole night between Dix and Kimball, finally being rescued and cared for by Henry Warner During that winter Mr Jones could not find work and in the emergency he remained at home and took care of the children while Mrs Jones went to Kimball, fifteen miles distant, where she worked in the hotel for $4 a week, high wages for that day, and walked the distance home when she made a visit Water had to be hauled eight miles and when one of the team of horses died, Mr Jones carried half of the neck yoke and as much as possible eased the work of the remaining horse

At length Mr Jones accepted an offer and sold the homestead for $500, a property that would bring $8,000 today He moved then to Custer county and bought land for $10 an acre but through the lapse of a mortgage, they lost that farm but later bought another in Custer county, on which they lived for twenty-five years Mr Jones then sold that property and they came back to Kimball county near Bushnell, but one year later Mr Jones sold that farm to his son-in-law, Ralph Taylor, and came to Bushnell, where they are people held in high esteem

In 1875 Mr Jones was united in marriage to Miss Henrietta Rucker, who was born in Iowa, and they became the parents of the following children Eva, who is the wife of

Ralph Taylor, a rancher in Kimball county, Woodford Robert, who married Charity Hammond of Custer and now lives nine miles north of Bushnell and has one and one-half sections of land, Maud, Mrs Hammond, who is in charge of the Central telephone office at Bushnell, and May, Fay and Ray, triplets, the last named being deceased Both May and Fay are married, the former being Mrs Coons, and the latter, Mrs Stuckert Mr and Mrs Jones are members of the Methodist Episcopal church.

JACOB PEDRETT, who is one of Kimball's most highly esteemed citizens, is widely known in the state and is one of Kimball county's heavy land owners The story of his life since coming to America is exceedingly interesting

Jacob Pedrett was born in Kanton Graeubunden, Switzerland, November 7, 1856, a son of Ulrich and Fieda Pedrett The father was a farmer, cattle grower and dairyman, his milk business being important and profitable He had a contract with a local hotel for eighty gallons of milk a day and sometimes sold one hundred and twenty gallons, probably when the tourist trade was at its height The father died in 1887, having been an invalid for some eighteen years previously The mother of Mr Pedrett communicated the fact to her son in America but before the latter received the letter, she also had passed away, having survived the father only fourteen days Their family consisted of but one son, Jacob, and the following daughters Elizabeth, who died at the age of twenty-two years, Fieda who is married, lives in the old home in Switzerland and has five children Cristena, on the old homestead, Magdelenia is the only one who came to America, with four children, in June, 1920, and is living on the Pedrett farm, and Marie, who lived to the age of eighteen years All the children are well educated and before Jacob Pedrett came to America, he was proficient in the Italian and German languages

In his own land Mr Pedrett gave military service according to the law, between the age of twenty and thirty years, and he had reached the latter age when he came to the United States, on the ship Normandie, which landed him safely in the harbor of New York He came across the country to Hastings, Nebraska, where he found work as a cheesemaker, having brought his diploma as to his efficiency in this industry For one summer he worked in Webster county for his board in the meanwhile using every effort to learn the English language, but in the fall he returned to Hastings There he rented a dairy farm and went into the business of making cheese He operated with thirty-six cows and his bargain included one-half of the proceeds from his factory, together with the stock increases Mr Pedrett remained on that farm from 1887 to 1890, coming then to Kimball county, bringing along ten cows and two horses He homesteaded in the same district in which he served as a school director later on for twenty-nine continuous years

Here Mr Pedrett resumed the making of cheese, in 1891 he and his wife milking forty-one cows, some of them being rented, the rental being paid in cheese He found this arrangement profitable He has always grown some wheat but has given the most attention to thoroughbred Hereford cattle, at times having owned two hundred and seventy-five head of registered stock and also has fed unregistered, doing business under the firm name of Pedrett & Clarke At the present time he owns two full sections of land and other tracts, aggregating about sixteen thousand acres, general farming being carried on an an extensive scale

On March 31, 1887, Mr Pedrett was united in marriage to Miss Marie Louisa Grothaus, a daughter of William and Katherine Grothaus, who came to the United States and to Hastings, Nebraska, in 1885, from Westphalia, Germany To Mr and Mrs Pedrett the following children were born Ulrich, who was born December 28, 1887, who was in military training at Camp Funston, during the great war, was honorably discharged as quartermaster sergeant, Fieda, who is the wife of Clyde Taylor, is a graduate of the Kimball county high school, lives on the farm in Kimball county and has three children Harry, Ruth and James, Louisa, who is a student in the Nebraska State University, has been graduated in the department of typewriting and shorthand and taught in the high school at Superior, Nebraska, Willis who died at the age of six years, and Harry who is a graduate, like his eldest brother, of the high school and the Agricultural college at Lincoln Mr Pedrett and his family are members of the Presbyterian church He has always been active in public affairs and has served in many public capacities, holding such offices as road overseer and county commissioner He has assisted in the building of three school structures, the latest erected in his school district being a modern two-room building, two teachers being employed and sixty-five children attending

Mr. Pedrett is president of the State Potato Growers Improvement Association, and a director of Nebraska State Farm Bureau association, and is treasurer of the local farm bureau, and also president of the Beet Growers association of Kimball county.

EMORY HORRUM, whose interests cover farming, stockraising, banking and other lines of business, is one of Kimball county's most prominent young men of affairs. He is a native of Nebraska, born June 25, 1886, at Dunbar, where his parents now live retired. He is a son of Lyman T. and Claudia Horrum, and he has one sister, Della May, who is the wife of Montgomery Lowery, a substantial farmer near Dunbar.

After completing the high school course at Dunbar, Mr. Horrum went to Lincoln and completed a commercial course in a business college there. It was in February, 1915, that he came to Kimball county, where he has made heavy investments in land, aggregating over eleven thousand acres. This land is cultivated in a modern way, farm tractors being made use of together with improved machinery of all kinds. Mr. Horrum is much interested in raising thoroughbred Hereford cattle. He has been largely concerned in the development and improvement of Dix, owning one hundred acres in town lots, the Horrum addition to Dix, and has built and sold some handsome residences in this part of the rapidly growing town.

Mr. Horrum was one of the organizers of the Farmers State Bank of Dix, of which he is vice president. He is associated in this financial enterprise with Gus Linn, George Vogler and Philip Nelson. The original capital was $10,000, which has been increased to $25,000. Mr. Horrum was president of the Dix Mercantile Company, which occupies a handsome brick building of modern construction, and plans are under way for the carrying of one of the largest stocks of general merchandise in this section of the state.

On October 8, 1910, Mr. Horrum was united in marriage to Miss Esther Tell, who is a daughter of Francis and Catherine Tell, well known retired residents of Omaha. Mr. and Mrs. Horrum have had two children, both of whom passed away in their infancy. The Presbyterian church holds their membership. Mr. Horrum is a Thirty-second degree Mason.

JULIUS J. JOHNSON, who is one of the large farmers and stockraisers of Kimball county and one of the representative, solid citizens, was born in the province of Halland, Sweden, a son of Jons Larson and Johanna Johnson, who died on their farm in Sweden, the former in 1902 and the latter in 1912.

Julius J. Johnson was born May 21, 1859, grew up on the home farm in Sweden, in the meantime attending school as opportunity offered. A thoughtful, sensible young man, by the time he was twenty-two years old, he had made up his mind to emigrate to America, in which country, as he learned from others, there were many chances for a young man without any capital but his industry, to acquire financial independence. When he landed in the United States he had $12 in his pocket, which paid his way from New York City to southern Illinois, where he found work with a railroad company, going from there to Sheridan, Michigan, fifteen miles south of Big Rapids. Afterward for three years he worked in the railroad shops in Chicago, at the end of that time coming to Nebraska. He homesteaded in Kimball county on a part of the same ranch that he now owns, proved up, kept on adding one tract of land after another until he now owns over seventeen hundred acres. He runs about one hundred head of cattle, all good grade Herefords, and has two hundred acres under the plow. Mr. Johnson has every reason to feel satisfied with his determination made so many years ago, to become a resident and citizen of the United States.

In 1889 Mr. Johnson was married to Miss Ida C. Strandberg, a daughter of Jonas and Christina Strandberg, who came from Sweden to the United States in 1885 and homesteaded in Nebraska. The father of Mrs. Johnson died in 1900 and the mother in 1901. Five children have been born to Mr. and Mrs. Johnson, namely; Alma E., who is a student in the high school at Dix; Hilda M., who is also a high school student; Carl A., who has reached the high school also; and Leonard J. and Verner O., who are in the grade schools. Mr. Johnson and family are members of the First Lutheran church at Potter. He has many times been honored by election to public office by his fellow citizens. At one time, while working at Dix, he served as postmaster. For two terms he served as township assessor, for several terms was road overseer, and at present is treasurer of the school board. In every office he has proved efficient and trustworthy.

WHITCOMB BROTHERS. — There are not many people in Kimball county who have not heard of the Whitcomb Brothers, extensive wheat farmers, who operate so successfully their extensive property entirely by

JOHN E. FRENCH AND FAMILY

means of tractors and other modern machinery. The firm is composed of two brothers, Edwin and James Whitcomb, born in Illinois and sons of Edwin and Mary (Champlin) Whitcomb.

The father of the Whitcomb brothers was born in Virginia and the mother at Chatham, New York. After their marriage they lived in Illinois but later moved to Rochester, Minnesota, which town he assisted in building, as he made all the brick used in general construction. During his last illness he was attended by Dr. Mayo, the father of the celebrated surgeons at Rochester. After Mr. Whitcomb's death, his widow returned to Illinois, with her three sons, James, Herbert and Edwin, having lost a little daughter at the age of three years. The mother looked carefully after the rearing and educating of her sons, continuing her residence in Illinois until 1909, in which year her death occurred.

In 1910 James and Edwin Whitcomb came to Nebraska, locating at Columbus, after a stay there entering into a business agreement whereby they traded their Illinois property for a section of land in Kimball county, assuming a mortgage of $4,000. Nothing much was done until in 1914, when Edwin Whitcomb came to the acquired property, an unbroken tract of miles of prairie as far as the eye could reach. Mr. Whitcomb soon proved how practical he was in business affairs. At Denver he had bought a tent house and in that he and his brother lived until, later on, they had a bungalow erected, equipped with electric lights and a hot water system, the first residence of its kind in the county. Near the bungalow soon appeared other structures, including a garage and a work shop.

In the first year the Whitcombs put out two hundred acres in wheat, reaping 7,200 bushels, and every succeeding year they have increased their wheat acreage, and, carrying crop insurance, hail storms and early crop damages have not materially affected them. In the third year of their experiment they put four hundred acres in wheat, and sold their 10,000 resulting bushels for from $1.90 to $1.95 a bushel, in the market at Dix. It is their custom to summer fallow all their wheat land for a time, merely dragging it to keep it clear of weeds. On the whole estate they have no mules or work horses, all the work being done by tractors and Duplex trucks, the latter carrying the wheat to market. When Edwin Whitcomb came here he formulated plans that have been carefully carried out and successfully expanded. He invested $3,000 in modern machinery and equipments, these including the tractors and trucks, drills, disc drags and three steel grain houses, each one having a 1,000-bushel capacity. Since coming here the brothers have sold several tracts of land but none of the original purchase. They have clearly demonstrated what can be expected from Nebraska soil in Kimball county when intelligently cultivated.

JOHN E. FRENCH — Practical industry, wisely and vigorously applied, seldom fails of attaining success, and the career of John French, now one of the leading farmers of the Henry district, is but another proof of this statement as he is a worthy representative of the younger generation of agriculturists who have played such a constructive part in the development of the valley and demonstrated past all discussion that irrigation of the rich alluvium of the valley brings golden returns to the men who are devoting their energies and time to intensive farm industries.

Mr. French was born in Clay county, Illinois, in 1876 being the son of William and Hettie (Etchison) French, both born in Indiana. William French was a farmer, residing in Illinois until 1881, when he came to Nebraska to take advantage of the public lands which were to be had for the taking in the western section of the state. He first located in Dodge county, but five years later took up a homestead in Cheyenne county early in the fall of 1886. He proved up on the 160 acre tract and after he had broken the land, erected suitable and permanent farm buildings, as well as a good house, became one of the substantial and dependable men of the Panhandle; later he disposed of his farm at an attractive figure. Mr. French was a Republican in politics, and though he never accepted public office, was one of the most progressive men of the section and took an active part in every movement for the development of the county and the uplift of his community. He was one of the first men to realize what inestimable benefit water would be to the valley and helped in building the first irrigation ditch in his locality, now known as the Mitchell ditch. He lifted the first spade of dirt on its construction work. Later he promoted and built over a quarter of the well-known Steamboat ditch that opened up a rich district for intensive farming. The French family were members of the Baptist church in which they were active workers. He is now dead. The mother lives at Minatare.

John French accompanied his parents from Illinois when they came west and received his educational advantages in the public schools of this state, early learning to rely upon himself as all boys who were reared in the Panhandle during the pioneer days did With his family he suffered the hardships and privations incident upon settlement of a new region, and early learned the practical side of farm industry as carried on in this section and while a boy in years, was able to conduct much of his father's business, as he was the oldest of the family, the other children being Lorenzo, a ranchman of Big Trail, Wyoming, Jessie, the wife of Charles F White, deceased, and she now lives in Minatare, Edna, the wife of R M Woode, a farmer of Wyoming, and two children who died

As soon as he was old enough John French took up a claim in Wyoming, consisting of a quarter section of land, where he engaged in general farming and stock-raising made permanent improvements on his land, and by his industry, executive ability and hard work was soon enjoying a good income With increased capital he decided to branch out as a landed proprietor and invested his money in more land from time to time until he now owns nearly a thousand acres of fine, arable, valley property, most of which is under ditch Mr French has not devoted all his energies to one line but has carried on varied farm industries along with stock-raising. having good grades of animals He thoroughly believes, as did his father, in intensified farming on irrigated land, as the best proposition in farming and has ably demonstrated his theories on section 16 in township 23-58 Early in his life he became associated with his father in business, first on the farm and then in the contracting business when William French began construction work on some of the most important irrigation projects in the upper valley Mr French found that he could easily carry on both branches of his business, and while he has become one of the largest landholders near Henry and a representative farmer of the section he stands high among the business men and is rated one of the solid, reputable men of the financial circles of Scottsbluff county Mr French still owns the first land he homesteaded here over twenty years ago, to which he has so materially added with the passing years He recalls vividly the trials and early struggles which his parents and the other pioneers here encountered in contending for victory over the untried forces of a new land, and, notwithstanding the anxiety and arduous toil imposed, he looks back to those days as the happiest of his life

In 1898 Mr French married Miss Lowa Dickenson, the daughter of S S Dickenson, of whom a record appears elsewhere in this history Six children have been born to Mr and Mrs French, three of whom survive Doris, Warren, and Dorothy, all of whom are still at home.

Mr French is independent in his political views, voting for the man he believes best qualified for office, his fraternal associations are with the Modern Woodmen and the Woodmen of the World.

GEORGE H TURNBULL, a representative citizen of Kimball county, where his well improved stock and grain farm is located, was born in Page county, Iowa, January 9, 1880 He is one of a family of eight sons and eight daughters born to Robert A and Rebecca Turnbull, the latter of whom died in 1907. They were natives of Illinois, coming to Iowa, following their marriage, which took place after the father's honorable discharge from the Federal army He served through the Civil War for three and a half years, in the Nineteenth Illinois volunteer infantry, participating in such important battles as Cickamaugua, Lookout Mountain and Stone Ridge When he and wife went to Iowa they lived at first near Coin, in Page county, Clarinda is the county seat, much of the county being little settled The father engaged in farming in Page county during his active years and is now deceased, dying February 8, 1920

George H Turnbull was reared on his father's farm and attended the public schools in Page county In 1908 he came to Kimball county, Nebraska, and homesteaded where he now lives, adding to his original purchase until he had one and a quarter sections, later selling three-quarters of a section to great advantage He has placed excellent improvements on his land, pays close attention to his business, thrift and good management being in evidence on every hand

At Pawnee City, Nebraska, Mr Turnbull was married to Miss Frances Lillian Correll, whose parents were Ohio people Mr and Mrs Turnbull have an adopted son, William Gale, and a daughter, Erthel He is a stockholder in the Farmers Elevator at Dix and the Farmers Union store at the same place He has never been particularly active in politics, has never desired public office, but is one of the reliable, upstanding men of his community, whose good citizenship has never been questioned He belongs to the order of Odd Fellows at Sidney and both he and wife attend the Presbyterian church The mother of Mrs Turnbull was born in York state and the father in New Jersey They were the parents of eleven children, six boys and five girls

JOHN W ROBINSON, who for a number of years was a resident of Kimball county, was considered an able business man and good farmer and was highly esteemed for his sterling personal character Mr Robinson was born at Granville, in Putnam county, Illinois, in 1862, and died on his large estate in Kimball county, July 26, 1919

Mr Robinson had educational advantages in Illinois From there, in early manhood he went to Iowa, spent one year there as a farmer, then went to Gates county, Nebraska, where he rented land, moving from there to Chappel, in Deuel county, where he lived four and a half years In 1913 he bought a quarter section of land there, for $38 an acre, which he sold for $62 an acre and then came to Kimball county, where he purchased two sections, which land is still in the possession of his family At the time of his death Mr Robinson had one hundred and fifty acres under the plow and sixty-five head of standard cattle His death was occasioned by an apoplectic stroke

In 1901 at Blue Springs, Gates county, Nebraska, Mr Robinson was united in marriage to Nannie Murgatroyd, a daughter of John and Elizabeth Murgatroyd, natives of England The father of Mrs Robinson came to the United States at the age of fourteen years, and the mother was two years old when her parents brought her across the Atlantic ocean Both families settled in Racine county Wisconsin In the spring of 1867 the parents of Mrs Robinson drove in a covered wagon from Wisconsin to Gates county, Nebraska, with their one son and five daughters, and one son and three daughters born to the father's first marriage. The father died in Gates county, April 19, 1891, and the mother, June 9, 1903 They were members of the Christian Science church Two sons were born to Mr and Mrs Robinson, namely Robartus S, who was born June 15, 1902, and Edward Lee, who was born February 13, 1904 They are fine young men and are very successfully carrying on the farm industries their father started so well

JOHN R MANNING — To Nebraska's invigorating climate one of Kimball county's enterprising and successful young farmers is indebted for restoration to health He is Arthur Manning, owner of a half section of excellent land, a son of the late John R Manning, who for many years was connected with large business houses in St Louis and Chicago

John R Manning was born and educated in New Jersey He came as far west as St Louis, Missouri, and in that city was united in marriage to Miss Mary Ebling, in 1890, in which city she was born, reared and educated Two sons were born to Mr and Mrs Manning, namely Arthur, born June 20, 1892, and John R, born July 18, 1897 For a number of years John R Manning was manager of the Famous Clothing Company, St Louis afterward becoming a salesman for the Spicer National Shirt Company of Chicago, and subsequently was agent for this large business house in both Chicago and St Louis He was widely known to the trade and was held in high esteem, was a member of the order of Knights of Pythias and belonged also to the Royal Arcanum Mr Manning's death occurred in 1906

In 1904 Mr Manning had consented on account of his son Arthur's delicate health, that the youth should accompany R R Barnes to Nebraska to prove what the climate might do for him His improvement was so marked that in 1909 his mother and brother joined him and the family has lived in Kimball county ever since After Arthur Manning had bought the half section that is the homestead, the former owner supplied lumber and Arthur and John erected the farm buildings They have made improvements since then and now have everything comfortable around them General farming and stockraising are the industries carried on and the young men have proved equal to all the responsibilities they have undertaken

Arthur Manning was united in marriage to Miss Lulu Leverne Straub, daughter of Daniel and Phoebe Jane Straub, who came early to Nebraska, settling first near Fairfield but later moving to Kimball county The father of Mrs Manning is living but her mother died some twenty years ago Mr and Mrs Manning have two sons Glen Winfield Manning, who was born April 27, 1918, and Wayne Daniel, born October 25, 1919 Mr Manning belongs to the Modern Woodmen and the Farmers Union

JOHN F BOGLE, who is a prosperous farmer and stockman of Kimball county, is also a keen and successful business man and is closely identified with many of the important interests at Bushnell Mr Bogle was born March 3 1878, in Worth county, Missouri, a son of James W Bogle, extended mention of whom will be found in this work

John F Bogle was reared on his father's farm and in boyhood alternated herding cattle with attending school He was well trained in every agricultural industry and encouraged in every manly endeavor In 1907 he came to Kimball county and homesteaded under the Kinkaid law six hundred and forty acres, six miles north of Bushnell, proved up and then sold advantageously Mr Bogle further displayed business judgment in buying a quarter section north of Bushnell, a half section one mile east of Lodgepole creek, and ten acres adjoining the town of Bushnell, which, in the course of time will no doubt become a part of this thriving town Mr Bogle is engaged in general farming but gives a large part of his attention to his fine Holstein cattle and thoroughbred Poland China hogs, and additionally is doing a profitable land business

In 1905 Mr Bogle was married to Miss Rosa May Snider, a member of one of the prominent old pioneer families of the state, and they have three children, namely James F, who is employed in the Farmers Union store at Bushnell, Merlyn Alva, a student in the Bushnell schools, who is preparing to enter a commercial college at Grand Island, and Ada May, who resides at home Mr Bogle is a man of high standing in his community and while not unduly active in politics, has opinions on public matters that he is not backward in making known when occasion calls for such action

GEORGE A ERNST, owner and proprietor of a fine estate in Kimball county known as the Pleasant View farm, has been a resident of Nebraska for thirty-five years He was born near Hamilton, in Butler county, Ohio, December 10, 1862, a son of Jacob and Elizabeth Ernst They were both born in Bavaria, Germany and from there came to the United States in 1848, and after their marriage in Butler county remained there for many years They were members of the Lutheran church

George A Ernst remained in Ohio until he was twenty years old, attending school in Miami county, south of Dayton, then went to Illinois, and from there, five years later, came to Nebraska in company with his brother John on January 30, 1886, settling near Aurora in Hamilton county In the spring of 1910 Mr Ernst accepted a contract to break one thousand acres of land in Kimball county, for H A Clark of Columbus, and came with his tractor to accomplish what was a rather big undertaking He was a pioneer in the sod-

breaking business here and continued in that line for about three years In the meanwhile he had bought his present estate, a railroad section, and to its cultivation and improvement he has devoted much time and profitable effort He has about three hundred and sixty acres under cultivation, keeps some stock and takes pride and pleasure in his fine orchard He set out seven hundred and fifty tree, some of which he lost during a severe hail storm, but his plum and cherry trees weathered it well In addition to having an abundance of fruit for home use, he has had cherries to sell His experiment has proved that fruit will do well in Kimball county if proper precautions are taken Mr Ernst has erected a fine modern residence and his barns, out-buildings and fences are all substantial, the result being that Pleasant View farm justifies its name

On December 22, 1887 Mr Ernst was married to Miss Anna M Donner, a daughter of Jacob and Veronica Doner, who came from Illinois to Hamilton county, Nebraska, in the spring of 1883 Mr and Mrs Ernst have had four children, namely Ezra J, who was born April 25, 1893, is assisting his father and is a very reliable young man, Esther V, Mary E and Ruth E, all of whom have been afforded educational advantages, Ezra J being a graduate of the Aurora high school in the class of 1911 and the others from the Kimball high school

Mr Ernst and his family belong to the Christian Science church He has never been particularly active in politics, but in the interest of law and order is careful when he casts his vote, believing that the privilege of citizenship carries with it a large amount of responsibility Mr Ernst has been quite prominent in movements for advancing the welfare of the farming community, is a member of the Farmers Union, and is president of the Farmers Co-operative Company at Kimball At a meeting of the board of directors Mr Ernst was put in as manager on February 12, 1920, with the assistance of his daughter, Mary E, who had taken a course in state university commercial work and they soon restored the business in the confidence of the public, and from the time Mr Ernst has taken charge, the affairs of the company have been much improved

WOODFORD R JONES, who is a large land owner and prosperous grain farmer in Kimball county, is a worthy representative of an old American family of many genera-

Mr. and Mrs. Robert M. De La Matter

tions back, and he is justly proud to bear a Christian name that has been honored by father, grandfather and great-grandfather

Woodford R Jones was born in Iowa, January 10, 1882, a son of Woodford and Etta Jones, natives of Iowa, who came to Cheyenne county in 1885, homesteaded, then sold and bought a place near Dix, again selling seven years later They now live retired Extended mention of the family will be found in this work

After a happy boyhood on the home farm and a sufficient amount of school attendance for practical purposes, Mr Jones invested in a section of land in Kimball county and is paying much attention to developing a grain farm, he has one and one-half sections With three hundred and fifty acres in wheat and corn and with oats yielding ninety bushels to the acre, it probably is only a matter of time before he is one of the leading producers of "the golden food of the world," the bread that not only our own, but other lands are in such dire need of He is a careful, intelligent, well informed farmer and good business man

In 1903, Mr Jones was united in marriage to Miss Charity Hammond, at Mason City, in Custer county, Nebraska The parents of Mrs Jones came from Harrison county, Indiana, to Custer county, Nebraska, thirty-five years ago Mrs Jones is one of a family of twelve children Four children have been born to Mr and Mrs Jones, namely May, Fredy, Grace and Woodford, the son being the fifth in direct line of descent to bear the family name Mr Jones has no political aspirations but he is influential in business circles as a member of the Farmers Union He owns an interest in the Farmers Union store at Bushnell and also in the Farmers Elevator Company of the same place The family attend church services and have pleasant social connections at Bushnell

ROBERT M DE LA MATTER — Nearly forty years have passed since Robert De La Matter drove into Scottsbluff county in true pioneer style and settled on a homestead in township 22-57, section 36 This section of Nebraska at that time was mostly open prairie covered with the curly buffalo grass and prairie wild flowers, habitations were few and far between and civilization was still in its primitive form, so that today he belongs to that rapidly thinning coterie of men who blazed the way for the present great development of this favored section, and a worthy pioneer he has proved to be

Mr De La Matter was born in Illinois, July 29, 1852, being the son of Cyrus and Mary Ann (Rowe) De La Matter, a history of whose lives will be found in the biography of Judge De La Matter, of Gering

Robert was reared in his native state by attending the common schools, living the life common to most farmers' sons, as he assumed many tasks around the home place, and thus became a practical farmer When his school days were over he entered farming as an occupation compatible with his tastes and a vocation with which he already had an excellent working knowledge His business life progressed, but land in Illinois was high and as he was foot free, he decided to come west and on the high prairies take up enough land to make agriculture a paying business After considering various states west of the Mississippi where homesteads were yet to be had from the government, Mr De La Matter came to the Panhandle and he must have been endowed with a far vision of what the future held for when he located in Scottsbluff county in 1888, he filed on a claim which has since come within the irrigated district of the valley Soon the prairie sod was broken, crops planted, a primitive bachelor establishment in running order and buildings erected for the stock and horses Mr De La Matter was a good manager, he was young, not afraid of hard work, and in the early days was willing to turn his hand to any honest occupation that brought in a dollar and thus he was able to weather the hard years of the early nineties, when drought burned up his crops, blizzards killed some of his stock, and the grasshoppers took what was left, but he was not discouraged as were so many of the pioneers and did not, like them return east, but stuck it out, and his faith in the Panhandle has been fully justified as is evidenced by the comfortable fortune which the family today enjoy With increased free capital from the sale of farm products Mr De La Matter invested in other tracts of land adjacent to the homestead and today is the proprietor of a landed estate of 400 acres, all well improved with a part under ditch which makes a fine combination for the general farm industries and stock-raising which he conducts From first locating in the county he has devoted much time to a good grade of cattle and horses and specializes to a considerable extent in breeding them With the passing years new and better farm buildings have been built on the place and a fine, convenient, modern home erected which is one of the most hospitable in the Morrill valley, where the De La Matter family is regarded with great esteem by the most recent settlers, who look to this old-timer as an example of what industry, plus pluck and the modern methods he advocates and practices can accomplish in this favored farming community The home farm is one of the old school sections and is one of the well known places for miles around It need not be stated that Mr De La Matter has been a successful man and now that life's shadows are beginning to lengthen from the crimson west he can look back across the years and feel his to have been a life of achievement and it is this type of man to whom posterity owes a great debt as he helped in opening up what is today the very garden spot of the country

On March 1, 1890 occurred the marriage of Mr De La Matter to Miss Sophie Adair of Illinois, she died March 22 1899, and he married a second time in 1899, Mrs Mary E Blackburn Mr De La Matter is a Republican in politics, advocates all movements for civic and communal advancement, and is a citizen who stands high in the Morrill valley

Mr and Mrs De La Matter are members of the M E church

By former marriage he had two children Jesse, on a claim in Wyoming, William, has a claim in Wyoming He has just returned from France where he saw service in the United States army during the World War He enlisted September 22, 1917, and served until June 21, 1919

THOMAS L BOGLE, who owns and operates one of the big grain and stock farms of Kimball county, has developed this property from its original state, and in the process passed through many hardships in earlier days These are not forgotten but they have been overcome, and Mr Bogle is now one of the county's substantial men

Thomas L Bogle was born in Gosper county, Nebraska, October 21, 1883, a son of James Bogle, extended mention of whom will be found in this work Mr Bogle obtained his education in the country schools, and made his first money by herding cattle He was thoroughly trained in farm work and has never desired to enter into any other line of business In 1907 he came to his present homestead, under the Kinkaid law being able to take up all of section thirty-two, the greatest hardship of living on the land at that time being the necessity of hauling all water used a distance of six miles To this original homestead he kept adding land until he owned twelve hundred acres, of which he later consented to sell two hundred and forty acres, when the transaction was very advantageous At at the present time he has five hundred

acres in wheat, oats and corn, his 1918 harvest aggregating about two thousand bushels, the 1919 crop being considerably heavier Mr Bogle has over three thousand bushels of wheat this year and one hundred and seventy-five acres of fine oats, that are the best in this part of the county He keeps quite a few good cows for cream and home use While Mr Bogle has not entirely eliminated horses for farm work, a large part is now done with farm tractors

On February 15, 1905, Mr Bogle was united in marriage to Miss Edith Hanes, who taught school for five years before her marriage, and is a daughter of Harvey and Sarah Hanes Mrs Bogle has one brother, Ellsworth Her father was a wagonmaker by trade but the family lived on a farm and the father died near Des Moines, Iowa, in December, 1885 The mother of Mrs Bogle, who was born in 1858, still survives She has followed the profession of teaching since girlhood and is yet easily and satisfactorily going on with her educational duties, and resides near Stockville She is a woman of culture, education and refinement Mr and Mrs Bogle have had the following children Howard, Harold and Avis, and three who died in infancy Mr and Mrs Bogle attend the Methodist Episcopal church He is a member of the order of Modern Woodmen of Bushnell, and belongs to the Farmers Union at Bushnell, in which organization he is interested as a stockholder No family is more highly esteemed in this part of the county

JOSEPH H PHILLIPS, who is a prominent and representative citizen of Kimball county, has spent the greater part of a busy and useful life in Nebraska He was nine years old when his people came to this state and there is much unwritten history that Mr. Phillips knows through experience

Joseph H Phillips was born in Wabash county, Indiana, March 18, 1878 His parents were Henry and Eliza Phillips, farming people who came from Indiana to Kimball county, Nebraska, in 1887, removing then to Kimball, in Cheyenne county Of their eight children, there are four living, Joseph H Phillips having one brother O C who is a sheep man in Kimball county, and two sisters, Orpha, who is the wife of Louis Wayhouse, and Ruth, who is the wife of George Fast

Mr Phillips grew to manhood on the home farm and obtained a country school education, more practical than decorative, just what was needed for a young man starting out to find fortune as a farmer and stockraiser. Under the Kinkaid act he homesteaded and afterward added two hundred and forty acres this giving him eight hundred acres of fine land. He immediately began raising stock and has done remarkably well with cattle and horses, and at the same time has three hundred acres under fine cultivation, devoting it to general crop raising. Mr. Phillips is credited with being an excellent farmer and a good judge of stock, but public affairs claim a part of his time. For six years he has been a county commissioner of Kimball county and has made a record of which he may be proud.

In 1900 Mr. Phillips was married to Miss Jennie Green, a daughter of Abel and Martha Green, who came to Nebraska from England, settling in the neighborhood of York. Mr. Green is deceased, but Mrs. Green survives and makes her home with her children. Mr. and Mrs. Phillips have had four children, namely Mable, who is teaching school near her father's ranch, Bessie, who died October 18, 1918, at the age of fifteen years, a victim of influenza, Raymond, who is attending school, and Pearl, who is the youngest. Mr. Phillips and his family are members of the Presbyterian church, which they attend at Bushnell, where they have a wide acquaintance. At one time Mr. Phillips was active in the order of Modern Woodmen.

DAVID H. SONDAY, who is numbered with the substantial and representative men of Kimball county, an extensive land owner and for some years a business man of Bushnell, is a native of Nebraska born in Seward county, March 20, 1881, on his father's pioneer homestead.

David H. Sonday was one of a family of seven children, four sons and three daughters, born to Edward and Elizabeth Sonday, who were born, reared and married in the state of Illinois. From there they came as early homesteaders to Seward county, Nebraska, where the family lived about thirteen years. The father died at Oberlin, Kansas. After his death the family moved to Brewster, Kansas and the aged mother still resides there. Mr. Sonday has the following brothers and sisters. Nettie, who is the wife of Samuel Ayers, a retired citizen of Chappell, Amiel, who follows the blacksmith trade at Brewster, Kansas, Joseph, who is a farmer at Brewster, Kate, who is the wife of Ernest Calkins who is a farmer, Louis, who is a farmer in Kansas, and Lucy, who is the wife of Walter Stair, a merchant at Brewster.

Until he was twelve years old David H. Sonday lived in Seward county, but afterward until 1900, at Brewster, Kansas, where he attended school. He had his own way to make in the world and after coming to Lodgepole, in the above year, was variously employed until 1903 when he went to Cheyenne and entered the railroad shops, working there as a machinist. In 1907 he was sent to Philadelphia in the capacity of engine inspector for the Harriman system, and upon his return to Cheyenne, became shop foreman and continued there until August, 1908. He came then to Kimball county and homesteaded under the Kinkaid law, and now owns two sections of land. During 1916 and 1917, he engaged in the hardware trade at Bushnell, then returned to his ranch and since then has given close attention to his farm and stock. He has two hundred and fifty acres under cultivation, a large acreage being in grain, and raises some of the finest stock that reaches the great markets from Kimball county. His improvements on the ranch include a comfortable ranch home and he also has an attractive residence at Bushnell. He has some important business interests here also, and is a stockholder in the Bushnell State Bank.

In 1902 Mr. Sonday was united in marriage at Chappell, Nebraska, to Miss Etta Peters, who is a daughter of George J. and Catherine Peters, who were early settlers in Cheyenne county. Mr. and Mrs. Sonday have one daughter, Lucy, who is attending school at Bushnell. Mr. Sonday belongs to the order of Modern Woodmen and Mrs. Sonday to the auxiliary organization. He takes a somewhat active interest in politics and has served as a member of the board of county commissioners of Kimball county.

FRED MORBY who is one of the enterprising and prosperous young wheat farmers of Kimball county was born at Axtell, Nebraska, June 1, 1898. He is a son of Andrew and Caroline Morby who had other children as follows. Axel, Christina, George, Sadie, Lydia, David, Lena, Robert Lillian Leland, Harry Inez and Goldie. All are living except Axel and Lillian who died in infancy.

The parents of Fred Morby were born in Sweden and they were reared there and were married in that country. After they came to the United States the father settled in Phelps county, Nebraska later moving to Kearney county and followed the blacksmith trade until 1910, when he came to Kimball county and took up a section of land under the Kinkaid act, and at the time of death, in May, 1919, he owned seven hundred and twenty acres of land He was an honest, sturdy, hard working man all his life and was greatly respected wherever known The mother of Mr Morby survives and has recently moved to Ogden, Utah, in order to give her youngest daughter high school advantages in that city

As early as 1912 Fred Morby started out to take care of himself as an independent farmer and stockraiser, renting land and now farming eight hundred and fifty acres, and preparing to put seven hundred acres in wheat as his three hundred and seventy-five acres in wheat in 1918 gave a yield that was encouragingly profitable He keeps about fifteen cows but is not much interested in stock at the present time He carries on his farming according to modern methods, using both horses and tractor

On January 1, 1913, Mr Morby was united in marriage to Miss Grace Leeper, who is a daughter of Rev David A and Ella Leeper The father of Mrs Morby is a minister of the Methodist Episcopal church, now stationed at Hoisington, Barton county, Kansas His other children are as follows Adrian, Mable, Zoe, Paul, Murlin and John Mr and Mrs Morby have two fine, sturdy little sons named Charles F and John L Morby Mrs Morby is a member of the Methodist Episcopal church Mr Morby is not active in politics but he takes much interest in organizations intended to protect farmers and belongs to the Farmers Union and has stock in the Farmers Elevator Company

ANDREW F AHLSTROM, who is one of Kimball county's most highly respected residents, came to the county thirty-one years ago and has been identified with its material development in no small degree He was born in Sweden in 1848, a son of Gustavus Ahlstrom who was the father of seven children, five of whom were sons, two only coming to America, Andrew F and Otto

Early in 1888 Andrew F Ahlstrom and his brother set out from Sweden for the United States, the first stopping place on the way to Nebraska, being Newton county, Iowa, and also a short time in Minnesota and Indiana In the same year Mr Ahlstrom came on to Kimball county and homesteaded a quarter section as a beginning, later homesteaded a three-quarter section and now owns five quarter sections all of which is very desirable property He carries on general farming and is a large raiser of fine stock Mr Ahlstrom has been

honest and industrious all his life and has met with a large degree of success

In 1886 Mr Ahlstrom was married to Miss Josephine Swanson, a daughter of Pearson and Frelott Swanson Her father died in Sweden and her mother came then to the United States and lived in Lucas county, Iowa, until her death Of her nine children, six sons and three daughters there are but two living Mrs Ahlstrom and Mrs Matilda Hall, the latter of whom lives near Little Falls, Minnesota Mr and Mrs Ahlstrom have had three children David, a fine young man, who died in 1913, aged twenty-six years, Joseph, who is his father's right hand on the farm, and an infant that died unnamed. Mr Ahlstrom and his family enjoy a comfortable residence on their ranch He has numerous business interests at Bushnell, Kimball county, these including membership in the Farmers Union, stock in the Farmers Elevator Company and stock in the Bushnell State Bank

ERNEST JURBERG — When such thorough-going farmers as Ernest Jurberg, who is well and favorably known over Kimball county, apply themselves to agricultural pursuits, a high standard of excellence is set and maintained He has had almost a lifetime of experience and today is numbered with the leading, well informed farmers and stockraisers of Kimball county

Ernest Jurberg was born in Sweden, October 7, 1872, a son of Theodore and Anna Jurberg They came to the United States when their son was young When the latter was eighteen years old he began working on a cattle ranch in Kimball county and continued about eight years When the Kinkaid act became a law he took advantage of its provisions and homesteaded and proved up After holding the land for nearly fifteen years he was offered $30 an acre for it, which he accepted, afterward receiving the old home section from his grandmother Here he carries on general farming and stockraising, doing very well in both industries, usually keeping one hundred and twenty-five head of stock, of which long experience has made him an excellent judge

In 1912 Mr Jurberg was married to Miss Anna M Elmquist, a daughter of Carl J and Augusta W Elmquist The parents of Mrs Jurberg still own their farm situated four miles west of Axtell, Nebraska, but they now live retired in that town Mrs Jurberg had two brothers, namely Frank, who is farming for his father, and Albert, who died October

Edward P. Cromer

21, 1914 Mr and Mrs Elmquist are members of the Swedish Lutheran church at Axtell Mr and Mrs Jurberg have an attractive little four year old daughter named Evelyn Mr Jurberg is a good citizen but not very active in political matters except in relation to the guarding the interests of the farmers He is a member of the Farmers Union at Bushnell, has stock in the Farmers Elevator Company, and also is a stockholder in the Farmers State Bank at Bushnell

EDWARD P CROMER — Nearly thirty-five years have passed since Edward P Cromer drove up the valley in true pioneer style and settled on a homestead in what was then old Cheyenne county, and is now Scottsbluff, where habitations were few and far apart and civilization existed in a most primitive form Since that time he has lived and labored in varied vocations in this section, slowly and arduously improving his land in the early days and at the same time taking an important part in civic and scholastic developments of this section of the country

Mr Cromer was born in Indiana, January 18, 1860, being the son of the Reverend John B and Mary (Hedrick) Cromer, the former a native of the Keystone state, while the mother was a daughter of the Old Dominion, having all the gracious hospitality and charm which Virginia gives her children as an inheritance Both are now deceased The father was a preacher of the English Lutheran church, holding charges in various places in the middle west, where he labored as a shepherd of God's Kingdom all his days There were ten children in the Cromer family, six of whom are living Jas M , a Lutheran preacher, who for several years had charge of Grace church of Kansas City, before being called to Casper, Wyoming, John B , who for twenty years before his death was a train dispatcher at Ossawatomie, Kansas, Richard W , a farmer in Iowa, now residing in Des Moines, Emma J , the wife of Judge Scott M Ladd, a member of the Supreme Court of Iowa for twenty-eight years, now residing in Des Moines, Rosa H , married Samuel Wiley, deceased, and she now lives in Irving Illinois, Mary, deceased , Clara J , deceased, George C , has charge of a Lutheran orphanage at Louisville, Kentucky, and Effie, the wife of a Mr Nelson, editor of the Prohibitionist, of Turtle Lake, North Dakota Mr Cromer received his elementary training in the public schools of Illinois and after these courses were completed entered Carthage College, Carthage Illinois, where he pursued higher studies He at once

engaged in the teaching profession and became one of the well known and successful men of the pedagogic fraternity, but he was ambitious to become independent and knew that a man who owned land and was not dependent upon a salary was so He studied farming in his spare time and in 1886 came west As the railroad was not built up the valley at that time he drove overland from Sidney, landing in what is now Scottsbluff county the 22nd of February, 1886 Mr Cromer at once filed on a homestead and tree claim of 320 acres in section 30, township 21, range 54 As he had been a teacher before coming to the Panhandle and as men of his profession were scarce in this section in the early eighties, Mr Cromer, after he had made some improvements on his land, was induced to teach here and opened school in a "soddy" south of the present site of Gering, the first school in the valley Summers he devoted to working his land, putting up the necessary farm buildings and in time erected a home for his family, where they would be comfortable As these were hard and trying years on the settlers the money Mr Cromer made by professional work tided the family over a time when many of the residents of the Panhandle grew discouraged and returned east, but he and his wife had faith in the country and happily both have lived to see it justified Mr Cromer taught in Gering four years, in Minatare two years, and then in Harrisburg two years From there he went to Kimball two years, then to Mitchell to assume charge of the schools four years and returned to Gering for a period of two years The first class graduated from Gering consisted of L L Raymond and Mary Sayer, now Mrs E S Wood, while the last class to graduate under Mr Cromer consisted of seven boys Earl Neeley, Harry Barton, Earnest Moore, Lesley Moore, Amon Downar, Roy Leavitt, and Robert McFarland Since he resigned his post as teacher, Mr Cromer has devoted his entire time and energies to farming He paid twenty dollars an acre for his present land and home of 120 acres, December 1, adjoining the city of Gering Today he has a well-improved farm is engaged in raising beets general crops, forage and feeds cattle to a large extent He has been a breeder of pure-bred Percheron horses for some years and in this line has won an enviable reputation in the valley, as he took eleven prizes out of twelve entries at the county fair in 1918 Mr Cromer now owns horses that won prizes at the International fair at Chicago in 1919 as well as at the state fairs The head of his herd is an International winner He has seen great changes come to the valley which today is one of the richest farming districts in the world, and he has shared in the great wealth that has come with irrigation as 120 acres of his land is under ditch, a fertile and productive tract. He has seen this land advance from $50 a quarter section to $500 an acre. Sometimes Mr. Cromer can be induced to recount experiences of the early days, and they are not all hardships as he tells them. He remembers when their oldest child, Rowena, was but six months old, he and Mrs. Cromer drove to Sidney with a team and wagon, to have a picture of the baby taken, a trip that required two days; then he recalls the time when Robert Osborn came tramping up the valley and by the time he reached the Cromer home he had been so long on the open prairie that he had eaten nothing for two days, having run out of supplies on the way.

Within recent years Mr. Cromer has become convinced that the upper valley is adapted to fruit and now has a fine young orchard of four acres which is now bearing bountifully. He is one of the pioneers in this line.

In politics Mr. Cromer is an independent voter, while he and his wife were charter members of the first Methodist church in the valley.

In 1883 Mr. Cromer married Miss Ida J. Kerr, of Hillsboro, Illinois. Four children have been born to them: Rowena C., wife of Reverend E. M. Kendall, of Bayard; she is an accomplished musician, having taught in Wesleyan University after completing her musical education in Boston; George C., of whom personal mention is found on other pages of this volume, a farmer, who graduated from the agricultural course at the State University; Ida Gladine, the wife of George C. Coughran, who was a teacher in the Gering schools; and Miriam, who graduated from the normal course at Wesleyan University, is now employed in the Gering schools.

FRED D. RUTLEDGE, whose large ranch interest and success in the stock industry, mark him as one of the important men of Kimball county, was born January 14, 1886, in Wyoming, a member of a well known family, the representatives of which are responsible men and women leading useful lives in their communities. The parents of Mr. Rutledge were Thomas and Minerva Rutledge. the former of whom was born in Canada and the latter in Missouri.

The parents of Mr. Rutledge were early settlers in Laramie county, Wyoming, where the father was an extensive ranchman for many years. His death occurred in 1915, since which time the mother has alternated residing with her children. Of these Fred D. was the first born, the others being as follows: Frank, who is a farmer and ranchman near Pine Dale, Wyoming; Thomas and Richard, both of whom live at Pine Bluff, Wyoming; Harry, who is in business at Denver; Mary Elizabeth, who is the wife of Edward W. Peterson, living on the old Rutledge ranch west of Pine Bluff.

Fred D. Rutledge passed his boyhood on his father's ranch and attended the public schools, later entering the Wyoming State University at Laramie, where he continued his studies for two years. After his return he assumed the larger part of his father's labors on the ranch and in this way his training for his own important industries was thorough and practical. He came to his Nebraska home on June 12, 1917, purchasing his ranch of eleven hundred and twenty acres, a small part of which he is devoting to general farming, giving his main attention to ranching. He is an example of the sound sense and and good judgment that belong to an encouraging number of the well educated young men of the state, for in no field of endeavor could he have found a more useful or needed exercise of intelligent or generally remunerative effort.

In 1911, Mr. Rutledge was united in marriage to Miss Eliza E. Cook, a daughter of Charles and Anna Cook, who reside on their sheep ranch near Hayward, California. Mr. and Mrs. Rutledge have two children, James and Agnes, aged respectively eight and five years. The family home is at Pine Bluff, Wyoming.

THOMAS E. BOWERS, a widely known and highly respected citizen of Kimball county, now in the United States mail service, was born in Missouri, December 25, 1873. His parents were Charles and Matilda (Harris) Bowers. The mother was reared in Missouri, to which state the father came from Virginia, soon after the close of the Civil War, having been a soldier in the Confederate army. On the maternal side, two uncles of Mr. Bowers entered the Union army but both met a soldier's death before they were far from home.

Thomas E. Bowers lived in Missouri until he was fourteen years of age, when he accompanied his parents to Nebraska. They stopped first in the eastern part of the state but later came to Kimball county and settled four miles southeast of Dix, this then being included in Cheyenne county. The father died there in 1899, after which the mother lived with her children until her death in September, 1912 Mr Bowers has two sisters, both of whom live at Loveland, Colorado

During early manhood Mr Bowers worked for the Union Pacific Railroad, after which he homesteaded eighty acres, later acquiring four hundred and eighty acres under the Kinkaid law, recently selling the entire five hundred and sixty acres for $36 an acre and has made plans for the investment of his capital In the meanwhile he is taking care of a United States mail route

On July 26, 1894 Mr Bowers was united in marriage to Miss Olive Robinson, a daughter of William and Mary Robinson, natives of Indiana and early settlers in Adams county The father of Mrs Bowers is deceased but the mother survives Mrs Bowers has three sisters and one brother To Mr and Mrs Bowers the following children were born Alice Fay, who is teaching school in Wyoming, Eva, who is also a teacher in Wyoming, Estelle, who resides in Kimball, Leta who died aged eight months, Eathan, who is attending the high school at Kimball, and Charles William, who is at home Mr Bowers and family belong to the Methodist Episcopal church He belongs to the order of Modern Woodmen, and politically has always been affiliated with the Democratic party, although never blindly following any leader, being a thoughtful man capable of entertaining independent views

EDWARD E. LESTER — The passage of the Kinkaid law brought to Kimball county many men of ambition and enterprise who now are some of the county s most substantial citizens One of this class is Edward E Lester, who is a limited farmer here but an extensive stockraiser

Edward E Lester was born June 10, 1868, in the great state of Illinois his father's farm lying in Henry county His parents were James B and Barbara Sarah (Kemerling) Lester, the latter of whom was born in October, 1832, and died October 28, 1878 They had the following children Lucretia, who died in 1852 Cyrus Jerome, who was born July 10 1853 lives at Lamont, Iowa Frank Delos, who was born November 6 1857 is a farmer in Nebraska, George W, who was born in January 1859 lives at Omaha, Grant, who was born June 8, 1864, Edward Elbert, who is of Kimball county, Minnie, who was born July 29, 1871, is the wife of Alexander Carbaugh, of Iowa, Emma B, who was born in June 1873 and Jefferson Ella and James all of whom died in infancy

During Mr Lester's boyhood the family lived in Illinois, Iowa and Missouri He had fair educational opportunities and remained at home assisting his father until he was twenty-four years old and afterward was variously employed until January, 1899, when he came to Nebraska, which state has been his chosen home ever since He remained at Omaha until 1904, when he homesteaded in Kimball county and still lives on his original homestead to which he has added other tracts In association with his wife he now owns four entire sections in Kimball county, and they also own a quarter section in South Dakota Mr Lester has two hundred acres under the plow but is not doing a great deal of farming, bending his efforts more to the raising of fine stock, aiming to turn off a good number of head annually

On July 17, 1911, Mr Lester was united in marriage to Miss Joannah B Hulsebus, a daughter of Bernard Hulsebus, a substantial farmer of Shelby county Iowa Mrs Lester's father is still living in Defiance, Iowa but the wife and mother died October 1, 1918 Of their children Mrs Lester is the first born, the others being Julia, who was the wife of Frank McGuire died in Defiance, Iowa, February 10, 1912, Albert, died June 2, 1907, Gerhard, living at Eddieville, is a minister in the Evangelical church, Tillie, who is the wife of Benjamin Ahrenholtz a farmer near Defiance, Iowa, Marie, who is the wife of W W Jenkins, a merchant in Defiance, Iowa, Bennie, who died March 6 1896 Mrs Lester is a member of the Methodist Episcopal church at Kimball Mr Lester belongs to the Farmers Union In his political views he is independent, casting his vote for the candidate that meets the approval of his own excellent but unprejudiced judgment

JOHN CLAUSEN, Jr, who is successfully operating his large farm and ranch in Kimball county, is well and favorably known in his neighborhood for he was born on the site of the present thriving town of Dix, August 2, 1890 He was reared and educated here and his main interests have always been centered in this part of Kimball county

The parents of Mr Clausen John and Catherine Clausen now live retired at Tecumseh They came to Kimball county in 1883, the father being section foreman on the railroad He homesteaded a quarter section just south of Dix which he later sold for $200, the same land being now held around $70 an acre He then bought five sections for $1 25 an acre, which he later sold for $20 an acre, then purchased seven sections as pasture land Mr Clausen not only proved to be an able and enterprising business man during his most active years, but became influential in Democratic political circles and for several years was a county commissioner in Kimball county   He belongs to the order of United Workmen and Woodmen of the World, and both he and wife are members of the Presbyterian church They have had the following children   Minnie, who is a school teacher at Fairmount, Nebraska, Mary, who is the wife of Gustav Wendt, of Kimball county; Emma, who is the wife of Clarence Anderson, of Tecumseh, Nebraska, Annie, who is the wife of Glenn L Byers, of St Joseph, Missouri, John, who is of Kimball county, Hilda, who is the wife of Richard Rowe, of Tecumseh, Otto R, who is a railroad man at Buford, Wyoming, and Herman, who attends school and lives with his parents at Tecumseh, and is taking a course in agriculture in Lincoln

Before Mr Clausen started out as a business man for himself he had the opportunity of acquiring practical agricultural knowledge, which he has put to good account, as is evidenced by the success attending his present undertakings He owns a half section and has the other half section under lease, has one hundred and sixty acres under the plow and keeps about one hundred head of stock His place is highly improved with a really fine residence, a barn with dimensions of 42x90 feet, an abundance of well kept out-buildings, and a well two hundred and eleven feet in depth Progressive, intelligent and scientific, Mr Clausen as an agriculturist, is removed as far as possible from the old-time farmer who expected from his land more than he put into it   ffl

Mr Clausen was married June 1, 1916, to Miss Lillian M But, a daughter of Clarence E and Catherine (McRory) Birt, whose sketch appears in this volume Mr and Mrs Clausen have one son, Bruce John, who made his welcome appearance April 28, 1917 Mr and Mrs Clausen are members of the Presbyterian church, attending St John chapel He belongs to the order of the Woodmen of the World at Potter, Nebraska Mr Clausen lives up to the requirements of a good and useful citizen, but is identified with no political party, and has never been a seeker for public office

CLARENCE E BIRT, who is one of Kimball county's representative men, is widely known and is identified with many important

interests here Dependable and reliable, true to every trust reposed in him in boyhood, Mr Birt grew from youth to manhood with the sound, steady character that has ever since gained him recognition among those who cherish high ideals of the true value of life

Clarence E Birt was born in County Kent, England, January 12, 1868, a son of Alfred Nelson Birt, who came to America in 1888 The mother of Mr Birt never left England, her death occurring in the city of London Of the family of five sons and three daughters, Clarence E was the third in order of birth, the others being as follows   Alfred, who died in infancy, Henry, who resides in the city of London, Claude, who is a resident of Fargo, North Dakota, Herbert, who served as a soldier in a Canadian regiment during the great war, Maude, who is the wife of John B Kenyon, of Carlton, Oregon, and Agnes and Florence, both of whom died when infants   The father died in Humoldt county, Iowa, in 1903

When fourteen years old Clarence E Birt went to work as a clerk in the office of James Carr & Sons' Flouring Mill, at Waltham, England, and Mr Birt prizes highly a testimonial as to his character and efficiency that is signed by this great English firm, and, in fact, has similar testimonials from every business house with which he was connected while remaining in his native land For three and a half years he was employed by the great firm of J Jackson, clothiers, London, and still later traveled as a jewelry salesman, and in this capacity while in Ireland, during industrial troubles there witnessed the eviction of tenant farmers He also had mercantile experience in a Capital and Labor store, in London, where goods were sold for cash on a five per cent basis, the daily sales sometimes amounting to $4,000

In 1893, soon after his marriage, Mr Birt and wife left England, crossed the Atlantic ocean in the steamship City of Paris, landed in the harbor of New York and immediately joined his father in Humboldt county, Iowa Prior to this Mr Birt had not had agricultural experience, but, with his father's encouragement and substantial backing, he embarked in the business of farming, and finding the venture both congenial and profitable, continued to rent farms and operate them in Iowa for the next fifteen years In 1907 he came to Kimball county, Nebraska, and bought a relinquishment claim of four hundred and eighty acres, under the Kinkaid act, and proved up, in the meanwhile engaging in other business enterprises, especially merchandising, the details of which were familiar to him Railroad building

Farm Residence of George Ehrman

George Ehrman

was causing an increase in the value of property and a little settlement at Dix was expanding into a village. Mr. Birt took advantage of this opportunity and started a store, and his appointment as postmaster was brought about. He remained at Dix for four years and in addition to his general store, conducted a lumber yard, sold coal and handled grain. When he sold his interests there his yearly sales amounted to more than $12,000. He traded his store to Philip Nelson for a section of land that adjoined his homestead. He now owns sixteen hundred acres of land, carries on general farming and keeps about one hundred head of high grade White Face cattle.

In 1893 Mr. Birt was united in marriage to Miss Catherine McRory, a daughter of Richard and Elizabeth McRory. They were residents of London England, where the father followed the trade of harnessmaking. Mrs. Birt is one of a family of fourteen children, as follows: Charles Jones, who died in 1906, Richard Jones, who has been chief decorator of Windsor Castle for almost fifty years, Ellen, who is the wife of Charles Lonergan, Henry G. who follows his father's trade in London, Edward G., who is conducting a market business in London, Mary, who was the wife of Bert Chenney, a city policeman in London, Francis M., who lives in the Malay Straits settlements, the home of the Royal family, and Arthur, who is manager of a rubber estate in that part of the world, Catherine, who is Mrs. Birt, and five who died in infancy. Mrs. Birt has reason to be proud of the record made by her family in the great war. She had seventeen nephews who served on the battle fields of France, all of whom lived to return after being honorably discharged, although one was badly gassed by the inhuman enemy and another had an arm shattered, and still another returned with the D. C. M. shining on his breast.

Mr. and Mrs. Birt have children as follows: Lillian Maud, who is the wife of John Clausen, Jr., Arthur, who, at the time the United States entered the great war, offered his services to the sheriff of Kimball county who sent him to a training camp at Lincoln from which he subsequently was honorably discharged and is now operating his father's ranch, Alfred G., who is also on the home ranch, Hazel D. who is attending the high school at Kimball, and Audrey E. and Joyce O., who are at home. Mr. Birt and family are members of the St. John Presbyterian church.

Although Mr. Birt has by no means retired

from active life, he has shifted his ranch responsibilities to the capable shoulders of his eldest son, his time being largely occupied with other business affairs. He is a director of the Farmers Elevator Company at Dix, and is secretary and treasurer of the Farmers Union Co-operative store at that place. He is prominent in the councils of the Republican party, is precinct assessor and chairman of the Republican Central committee. He was chairman also of the board of regents of the Kimball county high school. For more than twenty years Mr. Birt has belonged to the order of Modern Woodmen, and is an Odd Fellow, his local conection being with the lodge at Sidney.

GEORGE EHRMAN — One of the younger generation of agriculturists carrying on operations in Scottsbluff county, whose progressiveness and industry are rapidly bringing him into a favorable position, is the man whose name heads this review.

Mr. Ehrman was born in Germany, June 9, 1884, the son of George and Katherine Ehrman, an account of whose lives will be found elsewhere in this volume under the name of Frederick Ehrman. George accompanied his parents to America when they emigrated from their native land and received his educational advantages in the public schools of Colorado where the family located after reaching the United States. While living at Brush, in the mountain state, he devoted himself to his studies and thus laid the foundation for a good practical education. After his school-days were over Mr. Ehrman began to farm with his father so that while still a youth he had a good working knowledge of agricultural business and farm methods. He remained in Colorado until 1910 when he decided to establish himself independently in business operations and that year in partnership with his brother Frederick came to Scottsbluff county and bought 160 acres. At that time no one believed the land was worth much and could not see where the brothers were to become successful but irrigation solved that problem and today most of the property is under ditch and that which is not makes fine grazing pasture. Eighty acres more land was added to the original holdings in 1913, at a hundred dollars an acre the final payments being completed in 1917. This joins the town of Gering and is worth $500 per acre. This now belongs to George personally and is one of the show places of the county, a rather fine thing for two young farmers to do when you consider that all the equipment they had ten years ago was their ability to work and a determination to succeed From the first the Ehrman brothers placed good and permanent improvements on their farm, these have been added to with the passing years and a fine comfortable home erected where the mother now lives The soil has been raised to a high state of fertility, they are engaged in general farm enterprises and specialize in thoroughbred stock, having pure-blooded Percheron horses, Short Horn cattle and Duroc Jersey hogs, shipping a large quantity to the eastern markets each year Mr Ehrman believes in modern methods on the farm and has inaugurated many that he believes are efficient in his business He is a shrewd buyer and good seller, due to his study of market conditions and today is one of the best and most representative members of the younger generation of the farming element of the valley who are making history for the Panhandle as one of the most productive sections of a rich state Mr Ehrman is an independent in politics and a member of the Lutheran church He advocates and supports all movements for the development of the county and his community and lives up to a high standard of citizenship In 1919 he erected a beautiful modern home and a large barn in 1917 On his place is one of the finest homes in the county His place adjoins the town of Gering

HANS C L LARSON — When the good people of Kimball county refer to their best and most useful citizens, they are considering such men as Hans Christian Lund Larson, a successful and enterprising farmer and stockraiser, who lives up to every requirement of law and order, sets an example of thrift and industry, and co-operates officially and otherwise with his township neighbors in work for the general welfare

Mr Larson was born in Polk county, Wisconsin, June 20, 1880, one of a family of two sons and six daughters born to Peter and Sophia Larson, the other members of the family being as follows Mary, who is the wife of Peter Nelson, a prominent resident of Kimball county, Annie, who was accidentally killed on the railroad in 1906, was the wife of Peter Nelson, Emma, who died at the age of eight years, Rose who is the wife of Guy M Fleming, of Kimball county, Emma, who is the wife of Jesse Rockwell, and Lillian and Clarence V, both of whom reside at Kimball The father of the above family was born in Denmark, May 11 1852, and died in Nebraska, September 21, 1910 He came to the United States when about twenty-one years old shortly afterward locating at Taylor Falls, Wisconsin, where he lived as lumberman and

farmer for eighteen years His marriage to Sophia Hanson took place in Wisconsin, and they lived on his farm in Polk county until they came to Nebraska, living at Potter at first, then homesteading a quarter section and securing also a quarter section tree claim in Kimball county, situated ten miles south and one mile east of Dix The father proved up and spent the rest of his life on this land, the mother, after his death, retiring to Kimball

Hans C L Larson was ten years old when his parents moved to Kimball county He worked on the home farm and had school advantages in both country and town When the Kinkaid law went into effect, he determined to take advantage of its provisions, with excellent business judgment securing his present farm, filing on section 2-12-54, proved up and built a comfortable farm cottage and a commodious barn, the dimensions of the latter being 48x48 feet He has made numerous other improvements that greatly enhance the value of his property which is kept in the best possible condition. He has 300 acres of his land under cultivation and keeps one hundred and fifty head of cattle and horses Mr Larson has a section of school land under lease as pasturage

On March 30, 1909, Mr Larson was married to Miss Minnie T Benson, who is a daughter of Gunder and Louise Benson, whose other children were as follows Helen, John, Ida, deceased in Canada, Elizabeth, Emma, Clara, deceased on the homestead, Alice and a son who died in infancy The father of Mr Larson was born in Norway and was only six years old when he accompanied his parents to the United States They settled in Iowa and Mr Benson grew up there and remained until 1907, when he came to Kimball county and located four miles south and east of Dix, where he died four months later The mother of Mrs Larson has a fine home in Dix. To Mr and Mrs Larson three children have been born, namely Glennie, who was born December 18, 1909, Mable, who was born October 8, 1911, and Marguerite, who was born April 3, 1914 Mr and Mrs Larson moved to Dix and built a fine residence in order to give their children advantages of schooling there, Mr Larson being a firm believer of education He has served six years as a faithful member of his township's school board and for three years has been school treasurer

HANS P NELSON — In times of great trouble and industrial unrest in a country, it is a relief to turn attention to such sturdy, self-reliant men as Hans Peter Nelson, who is one of the substantial and representative men of Kimball county. It probably would be a difficult matter to convince such a man that there is anything ignoble in the work of hand and brain, hard, continuous, honest work, through which he has been able to build up an ample fortune in one of the finest states in the American Union.

Hans Peter Nelson was born in Denmark, March 28, 1853, and is the younger of two sons born to Nels and Bertie Nelson. His older brother bore the name of Rasmus. Through a second marriage the father had two daughters, namely Sina and Bertie. The father had a small farm of four acres and spent his life in Denmark.

When Hans P. Nelson was a boy he helped his father till the little home farm, and later worked for neighbors who had somewhat larger tracts of land, during this time possibly hoping for a future that would enable him to cross the great ocean to a country where fertile land was easy to acquire. It is not probable, however, that in those days he ever dreamed of his present possession of hundreds of acres of richly productive land, of the fine stock in his pastures and investments in reputable business concerns. It was not until he had been married six years that the opportunity came for Mr. Nelson to come to the United States. After landing in the harbor of New York, he and wife soon were on their way to Linn county, Missouri, where he rented farm land and remained for ten years. It was on October, 18, 1882 that he and wife reached America, and it was in the spring of 1893 that they came to Nebraska. Mr. Nelson homesteaded one hundred and sixty acres in Kimball county, and after the Kinkaid law became a fact, took an additional three-quarter section to this large body of land adding gradually until he now owns two entire sections of some of the finest land in Kimball county, all of which he acquired through his own industry. Mr. Nelson carries on extensive general farming and stock-raising and has four hundred and eighty acres under cultivation. He is yet active in looking after his farm industries and is ably assisted by his adopted son, John W. H. Nelson, who is a very enterprising and capable young man. The latter married Miss Edith Whittaker, a daughter of H. A. Whittaker, and they have two little adopted daughters. Ruth who was born in 1917, and Mary, who was born in June 1919.

Mr. Nelson's marriage took place in Denmark, in 1876, to Kirsten Hanson, who accompanied him to America and bore her part in his struggles to make headway after reaching this country. Mrs. Nelson came with him to the homestead in Kimball county Nebraska, but remained with him only two years longer, her death occurring May 3, 1895. She had taken much pride and interest in the new home, and when she passed away Mr. Nelson laid her to rest within fifty feet of the front door. It was not until 1919 that her remains were removed to the beautiful city cemetery of Kimball. Mr. Nelson is not only interested in his land but has other investments, including stock in the Farmers Elevator Company at Dix.

JOHN N. RASMUSSEN — Not nearly all the interesting stories have yet been told of the pioneering days and people of Nebraska. There may be a similarity in many of these but nevertheless there is always a personal touch that arouses interest. To Howard county in its early days, came many strong and sturdy people who brought with them the habits of thrift and industry in which they had been reared in the native Denmark, and of these was the Rasmussen family which has continued true to type.

John Nels Rasmussen, who is one of Kimball county's upright men and substantial farmers and stockraisers, was born in Howard county, Nebraska September 7, 1878, one of the two children born to Hans and Maria (Nelson) Rasmussen. Mr. Rasmussen had one sister, Julia, who married James Miller, who lived near Greeley, Colorado, at that time, moved later to Idaho, where she died on her husband's pioneer ranch. Both parents of Mr. Rasmussen were born in Denmark. The father came to the United States when a young man, in 1863 and located near Green Bay, Wisconsin. The mother came in 1865 and for a time was in New York and a short time in Chicago, finally Green Bay, and there, in 1867 Hans Rasmussen and Maria Nelson were married. Until 1871 they remained in Wisconsin, but in that year decided to move to Nebraska in order to secure government land. They traveled by railroad to Omaha in which city they secured a prairie schooner and a team of horses with which they started across the prairie to Howard county, Nebraska, where the father homesteaded eighty acres. The wagon served as a home until other arrangements could be made and the family lived on the homestead for twenty-one years, during that time often facing hardships of all kinds.

The venerable mother of Mr Rassmussen, with her unimpaired memory, can tell of those days, of their trials and pleasures, in a very interesting way In the spring of 1892 they left Howard county and came to Kimball county, settling on section 10-13-53, and resided there until the father's death

John N Rasmussen remained at home and assisted his father both before and after coming to Kimball county In 1913 he was married to Miss Helen Benson, a daughter of Gunder and Louise Benson The father of Mrs Rasmussen died in 1907 but the mother survives and has many friends and acquaintances in Kimball county The entire connection belongs to the Lutheran church

Mr Rasmussen was one of the first to be prepared to file on land under the Kinkaid act when the opportunity came In 1904 he homesteaded under this law, on section 22-13-53, and now has three hundred and twenty-five acres of his extensive tract under careful cultivation, and gives much attention to stock, keeping one hundred and fifty head of cattle and horses He devotes the most of his time to looking after his farm industries, but has some other investments, included in these being stock in the Farmers Union Elevator at Potter Mr Rasmussen has always been deemed a good citizen, is widely known and belongs to that class of men of whom it is often said, "his simple word is as good as his bond " Mr and Mrs Rasmussen have one child, Elmer J, born December 9, 1919

ANDREW ANDERSON — There are few men better known or more highly respected in the neighborhood of Potter, Nebraska, than Andrew Anderson, who, for many years has been a large landowner here, an extensive farmer and stockman, and financially interested in a number of successful business enterprises at Potter and Dix

Andrew Anderson was born in Denmark, September 24, 1862 His parents were Peter and Maria Anderson, natives of Denmark, where the father died in 1911, at the age of eighty years and the mother in 1915, at the age og eighty-two years Andrew Anderson had one sister and four brothers, as follows Elsie, who died at Plano, Illinois, was the wife of Michael Johnson, an infant that died at birth, one who died aged three years, Jens, who is a farmer in Denmark, and Edward who lives at Brush, Colorado

In 1873, accompanied by his only sister, Andrew Anderson came to the United States After landing in the harbor of New York, they made their way to Plano, Illinois, where the sister remained the rest of her life, Mr Anderson, however, working in that vicinity for three years only He then came to Potter, Cheyenne county, finding employment in the village for a year, after which he began to accumulate land which now aggregates many hundred acres He took up a half section, homesteading one hundred and sixty acres, with tree claim of one hundred and sixty acres and has remained here ever since, and at the present time has eight hundred acres of deeded land, and a twenty-five-year lease on a half section of school land Of this he has two hundred and fifty acres devoted to general farming and is a heavy raiser of cattle and horses, formerly turning out as many as two hundred head a year Mr Anderson's improvements have kept pace with his financial progress He has an abundance of water which he utilizes according to modern methods, has erected one of the handsome farm residences of this section and barns and other farm structures equal to the best in the county

In 1892 Mr Anderson was married to Miss Elsie Johnson, a daughter of Jens Johnson Mrs Anderson was born in Denmark and grew up in the same neighborhood as Mr Anderson When he had a home prepared he sent for her and she came alone to America and joined him Mr and Mrs. Anderson have had but one child, a daughter, who is now the wife of Jacob Nelson ,and they live in Kimball county A son of Mr Anderson's brother, Christian Anderson, now a young man of twenty years, has lived with Mr and Mrs Anderson since he was left motherless at the age of two years He is now Mr Anderson's right hand man

Mr and Mrs Anderson are members of the Danish Lutheran church Mr Anderson has never been very active in local politics, but he has been an important factor in founding and carrying on business concerns of considerable magnitude in this section and has investments in the Dix Mercantile Company, at Dix, the Farmers Elevator at Potter, he also has stock in the Western Mortgage Company of Denver He owns a residence at Potter

JACOB M NELSON, who is a prominent citizen of Potter, Nebraska, and interested in business enterprises here, was born in Denmark a son of Julius and Sophia Nelson, who came to the United States in 1886

It was considerable of an undertaking for

MR. AND MRS. JOHN W. MORRIS

the parents of Mr Nelson to move from Denmark to America, for they had a family of nine children, the youngest being six months old They accomplished it however and safely reached Racine, Wisconsin, where a relative was comfortably established and remained there about four months The father, in the meanwhile, started out to look up a home, finally homesteading in Nebraska His death occurred at Racine, Wisconsin, and the mother died at Blair, Nebraska Their children were as follows Peter, who is a fruit-grower in California, Christine, who was the wife of Melbourne Tracy, of Montana, and died March, 1920, John, who died at the age of thirty years, Hans, who lives in California, Bina, who is the wife of John Hanson, of Council Bluffs, Iowa, Jacob M, our subject, who belongs to Potter, Nebraska, Margaret, who died at the age of two years, Phillip, who is in business at Dix, Margaret who is the wife of Ove Anderson, county clerk for twelve years and now in the real estate business at Blair, Nebraska The parents were members of the Lutheran church

On November 22, 1915, Mr Nelson was united in marriage to Miss Matilda Anderson, a member of a prominent family of this name in the county, and they have one son, Leonard, who was named in honor of Gen Leonard Wood, who, at present is an outstanding figure in political as he has been in military circles for many years

Mr Nelson owns two thousand acres of fine land, in addition to having an interest in the old homestead He has seventy-five acres under a fine state of cultivation and raises a few horses, but devotes his main attention to cattle, running annually about two hundred and fifty head He is interested in the Farmers Elevator Company at Potter He and wife are members of the Lutheran church at Potter

JOHN W MORRIS, pioneer, frontiersman, and early settler, is probably one of the oldest men now living within the confines of Scottsbluff county, having passed his seventy-sixth year He has the honor of having filed on the first claim in the Gering valley, then called Cedar valley His career has been one in which he has had varied and interesting experiences, from hunting buffalo on the western prairies of Nebraska to the civilized existence of these modern days, and few men, twenty years younger, bear so few of the scars of life Mr Morris and his faithful wife ran the full gamut of pioneer experiences and their reminiscences of the early days are most graphic and

interesting They made the overland journey to and through Nebraska with a team of oxen and a wagon and girded themselves with the indomitable valor and undaunted purpose that are ever the prerequisites of success under the conditions that must obtain in the opening of a new country to civilization and progress Mr Morris has been in the most significant sense the architect of his own fortunes and few men have played a larger or more important part in connection with the development and upbuilding of Scottsbluff county along both civic and industrial lines Of this no further assurance is needed than the statement that he has amassed a comfortable fortune, and has so ordered his manner of life as to merit and receive at all stages the unqualified respect and confidence of his fellowmen It is most gratifying to be able to present in this publication a tribute to Mr Morris, as a pioneer of pioneers and to enter brief review of a career that has been marked by earnest endeavor, and no history of this county would be complete without the name of the first white settler in the Gering valley Mr Morris now lives in gracious retirement in the city of Gering, and though venerable in age, the years rest lightly upon him, while he finds a full measure of satisfaction in reverting to the attractive social and material conditions and environment which he has aided in creating in Scottsbluff county John W Morris is one of the gallant sons of the nation who went forth in defense of the Union when the Civil War was precipitated on the country In response to President Lincoln's call for volunteers he enlisted in the First Delaware Cavalry and when his regiment was dissolved entered the infantry and with this gallant command served out the entire course of the war in the Army of the East, in the Petersburg campaign In later years Mr Morris has found pleasure in vitalizing the associations of his military career by affiliation with the Grand Army of the Republic, Gering Post No 169

John W Morris was born in Caroline county, Maryland, June 4 1843, the son of Vincent and Elizabeth Morris the father being born and reared in this state along the bay, and the mother in Delaware Mr Morris received his educational advantages in the public schools and while still a boy assumed many of the duties and much of the work on his father's farm After his schooling was over he established himself independently in farm industry, as that was the business with which he was most familiar and of which he had an excellent working knowledge but he was an ambitious man, and this old settled county offered few opportunities to a man of vigor who determined to branch out, "put fortune to the hazard," and seek out what the "Golden West" might have in store. Mr. Morris had read widely along lines connected with his business and knew of the offers made by the government of fertile lands on the high prairies of the middle west and in 1885 he and his wife severed all the old home associations and ties that bound them to the east and started for Nebraska, then considered a part of the "Great American Desert." Mr. and Mrs. Morris drove into the state in true pioneer style; they had a team of oxen, the best animals for breaking the sod, hitched to their wagon in which were carried their household goods. They drove their hogs and cattle along with them as settlements were few and far between in that early day. It was a long, tedious journey up the river route across the great commonwealth that today is one of the richest in the Union, but they were high-hearted and their faith in this new country kept up their courage. At last they reached Scottsbluff county and took up the first claim in the Cedar valley, later changed to Gering valley. At that time all this great plains country was the range of the great cattle barons, who owned vast herds that ranged from Texas in the winter to Wyoming in the summer and Mr. Morris tells that it was impossible for men or women to go out on foot for fear of cattle running them down, so were forced to go everywhere on horseback. He remembers very well the first day in the valley, when he was running out the line of his claim, that a man came along driving several horses through; they talked and it proved that he was H. M. Springer, who was one of the early residents of Mitchell, a friendship that has continued through the years. Mr. Morris says that he had to drive to Sidney for his supplies, a trip that took four days, and when he decided to replace his first sod house with a frame building he had to drive to Laramie Peak, Wyoming, and freight the lumber into the Gering valley. After getting settled and erecting a log house for shelter of the family and such primitive farm structures that were absolutely necessary, Mr. Morris began the laborious work of breaking the prairie sod with his team of oxen. Soon after arriving in the Panhandle, Mr. Morris put his previous farming experience to good use by buying cattle to stock his land and soon developed a paying business of it. He planted diversified grain crops, but the early years were hard ones in western Nebraska, due to drought, blizzards, crop failures, and the insect pests that destroyed the growing grain. However, the Morrises were not dis-

couraged and they have lived to see their faith in this section proved true, where were only unbroken rolling prairies when they first came is now a smiling countryside, green with the growing crops in the summer, dotted with prosperous, flourishing towns and villages, and with irrigation Scottsbluff county has become the garden spot of Nebraska. Mr. Morris improved his homestead, and when his capital permitted bought other land adjoining the original claim, until he was one of the heavy and substantial landed men of the section and for many years was actively engaged in the various branches of farm enterprise from which he reaped a well deserved return and today has given up active life, disposed of all his holdings but five acres where his beautiful home is located. Now in the sunset years of life he can look back and feel that life has been worth while for he can visualize the changes that have taken place in the thirty-five years since he drove up the valley. In politics Mr. Morris is an adherent of the Republican party but draws no tight party lines when it comes to local elections, believing that the man best fitted to serve the people should be elected.

October 17, 1872, Mr. Morris married Miss Elizabeth Haskell, born in Scott county, Illinois, February 7, 1847, and they became the parents of three children: Bertram, who lives in Tacoma, Washington; Bertha, who married Sam Lawyer, who died, and she now lives in Gering; and Benjamin, who is the deputy sheriff of Scottsbluff county.

The foregoing record, implying much to him who can read between the lines as well as appreciate the data of the context itself, will be read with great pleasure by the many friends of Mr. and Mrs. Morris in Scottsbluff county and will prove a definite and worthy contribution to the generic history of this favored section of Nebraska, as their names merit an enduring place of honor and distinction on the pages of the history of Scottsbluff county.

JOHN G. BAUR, who is a highly prosperous farmer and stockman in Kimball county, has lived here for eighteen years, and during that time has been a witness not only of great agricultural development in this section, but of the actual building of such busy and important towns as Bushnell and Dix. He has done his part in forwarding many of the enterprises that have contributed to this rapid expansion.

John G. Baur was born in Germany, November 4, 1862, one of a family of fourteen children. Six of the sons and five daughters came to America. The parents died in Germany, the mother in 1886 and the father in 1890 They were honest, virtuous people respected by all in their community and members of the Lutheran church

When John G Baur landed in the port of New York he was twenty-seven years old In his native land he had learned the shoemaking trade, but his aim in coming to America was to become the owner of a western homestead with material comforts for himself and family In 1901 Mr Baur came to Kimball county and settled near what is Bushnell at the present time, but then was represented by a little shed on the site of the flourishing town He lived there one year during which he was in partnership with his brother-in-law, Charles Snyder, in the cattle business He then moved to what is the present site of Dix and started into the cattle business for himself, in which he continued for three years and then homesteaded a three-quarter section on the main road three miles from Dix Thus Mr Baur succeeded in his desire that had brought him to America, in a comparatively short time He has placed substantial improvements here, has an attractive and comfortable farm house, commodious barns and other buildings and an air of thrift is everywhere to be observed Of his homestead he now has three hundred and fifty acres under the plow Since his first purchase, he has added the other quarter section and additionally has bought a three-quarter section east of the homestead

In 1892 Mr Baur was married to Miss Catherine Funk, who was born in Germany and accompanied her people to the United States They were very early settlers in Madison county, Nebraska and her father built the first blacksmith shop In the early days there the Funk family endured many hardships They lived a distance of fifty miles from a market and on many occasions the father or brothers of Mrs Baur would carry a dressed hog to town and exchange it for a bag of flour The crops were eaten up by the grasshoppers, the only fortunate son of the family being the blacksmith, for the insects could not eat the anvil Eight children were born to Mr and Mrs Baur and all survive except the eldest son, who died at the age of two and a half years The others are as follows Walter, who is engaged in farming, was honorably discharged from military service in the great war after training in camp at Fremont in New Jersey, and at Fort Lee, Virginia, Gertrude, who lives with her parents, Henry, who is manager of a cattle ranch in Wyoming, Otto, who is associated with his father, Frank, who is also a farmer, and John and Eugene, both of whom are attending school Mr Baur and family are members of the Lutheran church Aside from his land and stock, Mr Baur has other investments, one being in the Elevator Company at Dix Mr Baur is an honorable, upright citizen, a competent farmer and business man and a friendly, helpful neighbor

CHARLES G NELSON, who is prominent in business circles at Kimball and well and favorably known in other sections, was born at Stanton in Montgomery county, Iowa, February 22, 1872 His parents were Lars Peter and Louise Nelson, both of whom were born in Sweden Their marriage took place in Henry county, Illinois, in 1866

Charles Gustav Nelson remained on the farm with his father until he was twenty-five years old His father died at Stanton, Iowa, February 14, 1901 and his mother at Boone, Iowa, March 21, 1919 In 1897 Mr Nelson embarked in the real estate business at Stanton, three years later accepting a railway mail route and two years afterward was appointed assistant postmaster at Stanton On July 17, 1906, he came to Genoa, Nebraska and became identified with the insurance department of the Modern Woodmen of America and continued in that work for eighteen months He was then called to Omaha as state manager for the Monarch Land & Loan Company of Kansas City, Missouri He remained in that position for one year, then returned to Genoa and became associated with C W Kaley of Omaha, and became state manager for all of South Dakota and the northern half of Nebraska for two years for the Woodman Accident Association, after which he was with the Woodmen of the World for two years Mr Nelson then went into business of handling flour, feed and produce, which enterprise he turned over to his son in July, 1916, and then established the Monarch Land Company of Genoa with William E Martin On March 1, 1919, a third interest in the business was bought by Carl O Heart On April 1, 1919, Mr Nelson came to Kimball and established the real estate business in partnership with his son Wayne I which is operated as the Monarch Land Company A large land business is now being done in the western part of the county by this firm

On June 14, 1895, Mr Nelson was united in marriage to Miss Julia I Peterson a daughter of Gustav and Louise Peterson who had children as follows George who died in infancy, Amanda, who died in infancy, Lydia, who is the widow of Herman Anderson, Emily, who lives at Genoa, Nebraska, George, who is deceased, Julia J, who is Mrs Nelson, Annie, a twin sister, who died aged two and a half years   Gerhard, who is in the greenhouse business at Denver, Albert, who is a farmer near Genoa, Helga, who lives in Sweden, John, who is a merchant at Hult, Sweden, and Edith, who died when nine years old   The parents of Mrs Nelson died at Hult, Sweden

To Mr. and Mrs Nelson were born three sons and two daughters, namely Frances, who is the wife of Reuben Dawson, a farmer north of Bushnell, and they have a little daughter, Dorothy, Hazel, who died when fifteen years old, Wayne I, who is associated in business with his father, Morris, who is a farmer north of Bushnell, and Leland, who is attending school   Mr Nelson and his family belong to the Lutheran church   He belongs to the Odd Fellows, the Woodmen of the World, the Woodmen of America and the Royal Neighbors   Mr Nelson is one of the county's far-sighted, trustworthy business men

EDWARD L ROLPH, M D, physician and surgeon at Kimball, a man of wide professional experience, was born at Chautauqua Lake, New York, in 1859, a son of Lyman D and Willoughby (Crandall) Rolph, the latter of whom is deceased, but the father of Dr Rolph survives and resides at Pender, in Thurston county, Nebraska

Edward L Rolph comes of old American stock, the family name, properly Rolfe, belonging to early Virginia history tracing back to the marriage of the young Englishman Rolfe to Pocahontas   Like many names, the change of spelling came about for reasons now lost to the family, and for generations back the name has been Rolph   Dr Rolph enjoyed superior educational advantages in his native state and secured his medical training at Louisville, Kentucky   He engaged first in practice in South Dakota, in 1894 locating at Pender, in Thurston county, Nebraska, and it was during his years of professional work in eastern Nebraska that he so endeared himself to the Winnebago Indians, that they conferred on him the greatest mark of confidence and esteem making him a member of their tribe   In 1909 Dr Rolph went to Old Mexico, and in 1916 came to Kimball

In 1894 Dr Rolph was married to Miss Edith E Stebbins, of Pender, Nebraska   Although Dr and Mrs Rolph have had no chil-

dren of their own, that has not prevented their having young life about them, for out of the goodness of their hearts they have given shelter and parental affection to several orphan children   An adopted daughter is no longer living, but an adopted son has grown to fine young manhood and during the great war was in military training at Camp Dodge   Dr and Mrs Rolph are members of the Methodist church   He belongs to the Masonic fraternity and Mrs Rolph is a member of the order of the Eastern Star

CHARLES E JACOBY, proprietor of the only photographic studio at Kimball, has been in this line of business ever since he left school. Mr Jacoby was born at Wilton Junction, Iowa, in 1870, where he was reared   His parents died in Iowa

Charles E Jacoby was educated in Muscatine county and is a graduate of the public schools   From boyhood he manifested certain artistic tastes, and when nineteen years old, left to his own choice of profession, he decided to learn photography   He established his first studio at Sioux Rapids, Iowa, where he continued twelve years in the business   In 1910 he came to Kimball county, Nebraska, homesteaded and lived on his land until 1914, when he came to Kimball, erected a building suitable for studio purposes and occupies a large part of it for photographic development. He has kept fully abreast of the time in the photographic field, and his rooms are equipped with all necessary instruments and high priced lenses, together with draperies and settings that may be found in establishments of this kind in metropolitan cities

In 1894 Mr Jacoby was united in marriage to Miss Pearl Noll, of Wilton Junction, Iowa, where she was born in 1872   The father of Mrs Jacoby is deceased but her mother survives and lives at Walnut Grove, Minnesota   Mr and Mrs Jacoby have had four children, namely   Esther, Maurine, Phyllis and Charles E   Mr and Mrs Jacoby are members of the Methodist Episcopal church   While living in Iowa Mr Jacoby was active in the Odd Fellow and Rebekah lodges   He owns property at Kimball which includes his studio buildings and a handsome modern residence

CHARLES J OLDAKER, who is a widely known representative and worthy citizen of Kimball county, has been a resident of Nebraska for many years, and owns a large body of richly cultivated land in Kimball county   He was born August 10, 1860, in Johnson county, Iowa, where his parents were farming people His father died in September, 1896, and his mother in September, 1916

Charles J Oldaker obtained his education in the country schools and remained in Iowa until he was twenty-one years old, when he went to Bozeman, Montana, with the Northern Pacific Railroad He tried farming in the vicinity of Bozeman for a year, then gave it up and went back to Iowa Later on he again left Iowa and came on a visit to Frontier county, Nebraska, went then into Colorado and took up a pre-emption claim and proved up In 1887 he came to Kimball county and, pleased with the aspect of the country and the fine people he met among the earlier settlers, decided to remain, and in the following season homesteaded a half section located nine miles north of Kimball To his first purchase he added and now owns an entire section Mr Oldaker remained on his farm until 1917, when he came to Kimball to live having a comfortable residence here and an unlimited number of friends

Mr Oldaker was married at Bicknell, Nebraska, to Miss Clara C Kennedy, and they have the following children Roy C born April 19, 1889, Elmo born April 18, 1891, John G, born May 31, 1892, Fay, born July 2, 1894, Lola, born July 30, 1896, Clara, born June 15, 1898, Earl, born June 14, 1902, and Lynn, born July 15, 1907 John Gilbert Oldaker of the above family, is one of the returned heroes of the great war He enlisted in the United States navy on December 11, 1917, was sent to France and served seven months on the Flanders front, and was honorably discharged July 10, 1919 Both Mr and Mrs Oldaker belong to the order of Royal Highlanders and Mrs Oldaker is also chief matron in the Degree of Honor lodge, and both set a good example of thrift and foresight by carrying life insurance Mrs Oldaker was reared in the Christian church but she attends services in the Presbyterian church with her husband, of which religious body he is a member He belongs also to the Knights of Pythias

ALBERT HUBBARD was born in Randolph, Indiana, December 5, 1862, the son of Francis and Elizabeth (Meriwether) Hubbard His father was a native of Indiana, and his mother of Delaware The subject of the sketch was the second of five children born in this family, the eldest being a daughter, Lavina E, now living in Indiana, the wife of Riley Hinshaw Of the others, Ira is a resident of Scottsbluff county, Nebraska, and Elza and Riley live in Indiana The father was a farmer and was killed crossing a railroad track

August 19, 1913, at eighty years of age, the mother died May 25, 1890, at about fifty years of age

Albert was educated in the public schools of Indiana After completing his schooling he took up farming in his native state but heard the call of the great undeveloped West and came to Nebraska in 1886 In October of this year he took up a preemption claim and proved up on same He then took up a homestead of 160 acres in Scottsbluff county, developed and improved it through the years of pioneering, clerking in a store in Gering several years, and now owns 160 acres of well improved, irrigated land, of the kind that is fast coming to be known as the most valuable because the most productive land in the entire United States

On February 15, 1894 Mr Hubbard was united in marriage with Gertrude England, and to their married life has come the blessing of four children, all of whom are living at home They are Emery O, Edna V, Ralph, Waldo, and Laura E

Mr Hubbard is a member of the Christian church, and is a Republican in politics He stands high in the estimation of his fellow citizens, and is bound by the close ties of sympathy and common experience with the early settlers of this community who experienced along with him the struggles and trials of living in a new country during the period of drouth and hard times before the magical power of irrigation was invoked to turn the desert into a garden

JAMES W BOGLE, for many years one of Kimball county's enterprising and progressive business men now lives comfortably retired at Bushnell in which city he owns a large amount of valuable realty He is a native of Indiana, born and reared, in Washington county, July 13 1849 Both parents have long since passed out of life

James W Bogle attended the country schools in boyhood and grew up on a farm In 1871 he left Indiana and went to Jasper county Illinois working there as a farmer for two years and then found better opportunities in Clay county, where he remained six years Having a natural desire to see more of the great country in which it had been his good fortune to be born he kept making his way westward going from Clay county, Illinois, to Ringgold county, Iowa and six months afterward reached Missouri Mr Bogle engaged in farming in Missouri for a year and a half but in 1879 came to Nebraska, located in Gosper county, took a homestead and tree claim of a quarter section of land, settled down to its development and improvement, during the first two years living in a sod house, which afforded a great contrast to his modern residence in Bushnell. When he left his farm and came to Bushnell he went into the mercantile business, being a pioneer here in this line and continued for some years and then sold his stock but still owns his fine brick building, in which are located a number of other business firms. Mr. Bogle has shown great business foresight in his investments in land since he came to Nebraska and owns extensive tracts, including four hundred acres in Gosper county and tracts in Kimball county aggregating fourteen hundred and eighty acres.

In 1874 Mr. Bogle was united in marriage to Miss Mary C. Barnett, who was born in Washington county, Indiana, a daughter of Martin M. and Martha Elizabeth Barnett. The mother of Mrs. Bogle died in Indiana in 1860 but the father survived until April,1891, moving to Missouri in 1889 and engaging in farming there. To Mr. and Mrs. Bogle the following children were born: George D., who resides with his parents; John F., who is a farmer near Bushnell; Lauretta, who lives near Nampa, Idaho; Mrs. Eva May Meyerhoeffer, who lives in Gosper county; Thomas Leander, who is a rancher near Bushnell; Freddy, who was born September 12, 1886, died February 17, 1887; Laura Alice, who resides at Bushnell; and Charles L., who conducts a general store at Bushnell. While Mr. Bogle has never been unduly active in politics, he has always been an upright, forward-going citizen and has not neglected any of the responsibilities of good citizenship. Both he and wife are members of the Christian church, and benevolent movements of every kind find them interested and helpful when possible.

DAVID R. READ. — One of the substantial and prominent men of Bushnell, whose life story is filled with interest, because it tells of worthy effort bountifully rewarded, is David R. Read, now living retired in this beautiful little city. He may well be classed a first citizen, as he was one of the pioneer settlers of the hamlet of Orkney, which was the original site of Bushnell.

Mr. Read was born in Henry county, Missouri, in 1860. His parents were Joseph T. and Mary Anna (Gilbert) Read, the former of whom was born in Tennessee and the latter in Pennsylvania. The mother died in 1893 and the father came to Scottsbluff county, Nebraska, and died there in 1917. Mr. Read obtained his schooling in Henry county, Missouri. The Civil War undoubtedly had its changing effect on the fortunes of the family and he was not very old when he made his way to Kansas. He remained in that state for five years, and afterward lived in Nebraska and Missouri until 1906. He had met with financial misfortune before this and when he reached Kimball county in that year, his capital amounted to $22.65, which he had obtained by selling a cow. He took a homestead of four hundred and eighty acres in the northeastern part of what is now the Bushnell settlement, then borrowed money and sent for his family. Times were hard during the next two years but through the helpful assistance of a most estimable wife he made headway. While he worked in the town, Mrs. Read took care of the children and the affairs on the homestead, thus holding down the claim. From their present position of affluence, it may seem almost impossible to believe the difficult things they accomplished in those early days, when they had to carry all water used a distance of two miles, and when coal gave out, gathered buffalo chips on the prairie to use as fuel. After Mr. Read had proved up on his homestead he sold it to advantage and invested it in town property, and now owns one of the finest cement block buildings in the city, the first floor of which is used as a bank.

Mr. Read married Miss Willie Felts, and they had two sons: Ernest, who lives in Nevada; and Wm. T., who lives at Stanberry, Missouri. Twenty-three years ago he married Zora Van Gundy, daughter of George and Ruth (Minnick) Van Gundy, who were natives of Indiana, but Mrs. Read was born in Iowa. They have one son, Arthur T., who lives at Bushnell. Mr. Read has never been inclined toward great activity in politics, but he has always been a good citizen, and one proof of this may be cited in the fact that he not only invested in property but when the government called on loyal citizens to help, he bought $600 worth of Liberty bonds. Both he and wife are members of the Christian church. While living at Cameron, Missouri, he was an active member of Star Hope Lodge No. 182, Odd Fellows, and belonged also to the order of Patriarchs at the same place.

FRANK G. TANNER, who is a representative citizen of Scottsbluff county and a successful general farmer, has been a resident of Nebraska for thirty-two years. He is a native of Illinois and was born in Kankakee county,

July 24, 1864 His parents were E. M and Helen (Haskell) Tanner The father was born in the state of New York, while the mother's people were of New England and she was born in Connecticut Of the three sons and one daughter in the family, two sons are living. F G and H C, the latter being a resident of Wyoming

Mr Tanner obtained his education in the public schools of Guthrie county, Iowa, and remained at home until 1887 when he came to Nebraska and homesteaded in what is now Scottsbluff county, his eighty acres, on which he still resides, being located three miles east of Scottsbluff Mr Tonner has done well since he came to Nebraska and now owns two hundred and forty acres of irrigated land His home place is well improved, his stock is high grade and all his farm industries are carried on according to modern methods

In 1899 Mr Tanner was united in marriage to Miss Mary Ferguson who is a daughter of James Ferguson, one of the early settlers Mr and Mrs Tanner have the following children Cassius, Lawrence, Grace, Myron and Willow Mr and Mrs Tanner belong to the Methodist Episcopal church In politics he is a Republican

CORIE J HAIN — The young man to whom an easy life appeals should not locate in an arid section of country as was a part of Scottsbluff county when C J Hain came here, but neither should such a young man adopt farming as a vocation The men who have been successful in Nebraska have been workers with a large natural endowment of common sense In coming here such men have expected pioneer hardship and have taken pride in overcoming the most discouraging conditions Mr Hain came to Scottsbluff county with limited capital but is now the owner of one of the finest farms in this section

C J Hain was born at Lake City, in Calhoun county, Iowa, June 11, 1868, and is a son of Elias and Laura E Hain, the former of whom settled in early life in Iowa and died there at the age of seventy years The latter died when aged thirty-five years Mr Hain has seven brothers and one sister He obtained a good public school education and has been engaged in agricultural pursuits all his life When he came to Scottsbluff county in 1906 he homesteaded and his farm of one hundred and forty acres lies on section 24 town 23-54 thirteen miles distant from the city of Scottsbluff One of the greatest drawbacks to comfortable living when Mr Hain settled here, was a lack of water, and for the first six months he was obliged to haul all the water used a distance of a half mile By that time he had a well dug and with irrigation project well under way, there is little danger of this beautiful and naturally fertile section of country ever again suffering seriously from drouth Mr Hain has added to his possessions and he now has full three hundred acres, on which he carries on general farming His improvements are adequate and substantial and all his surroundings indicate thrift and plenty

In Pottawatomie county, Kansas, Mr Hain was married to Miss Flora Ettlinger, who was born in Kansas They have had the following children Raymond whose homestead farm is located near that of his father in Scottsbluff county, Cecil, who lives on his homestead in Wyoming, Mary, who has taken a homestead in Wyoming, Alvin, who lives at home, and Bessie Edna Orrie and Archie, all of whom are in school, and Effie Leona, who died when aged eighteen months Mr and Mrs Hain are members of the Baptist church Mr Hain has never desired public office and is an independent voter

G F HAAS — The financial interests of Minatare. Nebraska are well taken care of by stable and honorable business men here, the Minatare Bank being a trustworthy institution conducted along conservative lines by men of known probity and high commercial standing

Mr Haas was born in Shelby county, Iowa, February 5, 1876, a son of Frank and Elizabeth Haas The father was born in Switzerland and was brought by his parents to the United States when three years old He grew up in Wisconsin and was married there, and in 1872 started with his wife for Western Nebraska but never reached their proposed location When some distance on the way their party was overtaken by a band of savage Indians who were so threatening that the travelers turned back and Mr and Mrs Haas settled in Iowa and the father still resides on his farm in Shelby county G F Haas was reared there, attended the public schools and after graduating from the Shelby high school, went to Omaha and completed a business course in a commercial college In March 1909 he came to Minatare Nebraska, and in association with S K Warrick purchased the bank in which they have been interested ever since Their success has been phenomenal They have increased the original capital of $5 000 to $25 000, and the $40,000 to $350 000 deposits

Mr. Haas is also a stockholder in the First National Bank of Scottsbluff.

On June 25, 1906, Mr. Haas was united in marriage to Miss Evelyn E. Witter, who was born at Woodbine, Iowa, April 15, 1878, who is a daughter of Parry and Hattie Witter, who are natives of Iowa. Mr. and Mrs. Haas have one daughter, Frances E., an attractive little maiden of nine years. The family belongs to the Methodist Episcopal church. Mr. Haas is an independent voter in national affairs, but not from lack of interest but because he often feels that he can rely on his own trained preceptions in regard to men and their probable reactions in times of national emergency. There are few men who make a closer study of human nature than bankers. Mr. Haas has been identified with both the Masons and Odd Fellows for many years.

ADAM WALKER, who has proved a good citizen and has done exceptionally well in business since coming to the United States in 1906, still naturally takes a great deal of interest in his old home in Russia, but his interests are now mainly centered in Nebraska. He was born in Central Russia, August 17, 1879. His parents were Adam and Catrina Walker, both of whom died in Russia, the father when seventy-three years old and the mother when aged fifty-five years.

When Adam Walker was twenty-seven years old he came to the United States with the intention of becoming a farmer. His father had been a small shopkeeper but Adam desired greater opportunity and believing he could find it in America, left Russia and the other members of his family behind. After landing on the soil of the great United States, he came directly to Nebraska and for seven years worked at Lincoln. In 1913 he came to Scottsbluff county, by 1917 was able to buy one hundred and sixty acres of good land, and to the development and improvement of this property he has ever since devoted himself. He has a fine place here now, being not far from Scottsbluff where he has market facilities, and few farms show more careful tillage. He carries on general farming and raises some stock.

Mr. Walker was married to Christina Hurst, who was born in Russia and was brought by her father, Peter Hurst, to the United States when two years old. He people at present are in Russia, having twice visited the United States but make their home in the old country. Mr. and Mrs. Walker have six children, namely: Marie, Carl, Reinhold, Florence, Esther and Rudolph, all of whom will receive the best of educational advantages. Mr. and Mrs. Walker belong to the Lutheran church.

MAX SCHROEDER, who owns some of the most valuable farm property in Scottsbluff county, has been a farmer and stockman ever since he reached man's estate, and has made such practical application of his knowledge that he is one of the county's most substantial agriculturists, although not yet in middle life. Mr. Schroeder was born in Saunders county, Nebraska, March 1, 1883.

The parents of Mr. Schroeder, Henry and Tillie Schroeder, were born in Germany. The mother died at the age of fifty-five years but the father survives and notwithstanding a life of hard work bears well his seventy-seven years. He came to the United States in 1865 and prior to coming to Nebraska spent some time in New York, Pennsylvania and Illinois. After reaching this state he homesteaded in Saunders county.

Max Schroeder grew up in Saunders county and attended the public schools. He assisted his father on the pioneer homestead when many hardships beset the early settlers, and afterward became a farmer on his own account. In 1916 he came to Scottsbluff county and bought a quarter section of land, in 1917 bought a second quarter section, and in 1918 bought his last tract of eighty acres. He carries on general farming and makes feeding cattle a feature. His land is situated on section 26 town 22-53, and there are few farms in the county that have been better improved. Mr. Schroeder has common sense ideas in his work, providing the latest improved machinery and appliances for carrying on the same, and in consequence is able to take from his land much more than the less progressive farmer can.

In Saunders county, on June 9, 1904, Mr. Schroeder was united in marriage to Miss Mintie Clouse, whose father, Calvin Clouse, was born in Tennessee. Mr. and Mrs. Schroeder have three children, namely: Cecil, Evelyn and Harris. The family belongs to the Methodist Episcopal church. In politics Mr. Schroeder is a staunch Republican. He takes interest in the county's development, lends his influence to the betterment of the public schools and favors measures looking to improvements in regard to public roads, but has never consented to hold office. He belongs to the order of Modern Woodmen of America.

ALONZO THURMAN — Practical industry, wisely and vigorously applied, seldom fails of attaining success, and the career of Alonzo Thurman, now one of the leading farmers of Scottsbluff county, is but another proof of this statement. When he started out in life he had but few advantages but a practical education to assist him along the road to success, but his diligence and judicious management have brought him ample success as a reward for his labors.

Alonzo Thurman was born in Knox county, Illinois, May 29, 1864, the son of Wesley and Martha (Denison) Thurman, they both were natives of the Buckeye state. They had a family of seven children, Sylvester, deceased, Arizona, who married Joseph Casper, lives in Kansas, Alonzo, the subject of this review, Sheridan, a farmer living seven miles north of Scottsbluff, Dorothy, deceased, Lilly, the wife of George Wason, resides in Illinois, and Arthur who lives in Idaho. Wesley Thurman carried on a general farming business in Illinois where his family were reared. Alonzo grew up in his large family used to the give and take that exists where boys are growing up together and thus early became self-reliant and able to hold his own in the youthful affairs of life. He attended the school near his home and thus gained a good practical foundation in an educational way, which became of great value to him in his business relations in later life. After his school days were over, the young man decided upon agricultural pursuits as a life's vocation, as it was a business he had learned practically on the farm, with his father's supervision, until Wesley Thurman died when the boy was seventeen years old, then he began to work out the daily problems for himself, and that he was able to do so remarkably well is attested by the fact that by his own unaided efforts he has accumulated a comfortable competency. Mr. Thurman remained in Illinois until he was twenty-four years of age, but that state was well settled up, land was high in value, and so he determined to avail himself of the homestead plan of acquiring land and with this end in view came to Nebraska in 1888, locating on a claim in Cheyenne county, which at that time embraced the territory now known as Banner county, as the latter was carved from Cheyenne and erected into one of the divisions of the state later. He proved up on his homestead and at once began excellent and permanent improvements that greatly enhanced the value of the land, erected a comfortable home and good farm buildings as soon as his capital permitted and was soon established as one of the progressive and prosperous agriculturists of the district. In 1903 he sold his homestead and went to northern Iowa for five years, in 1908 returning and buying his present place. Mr. Thurman made a good choice in picking out the location of his claim as all his land today is under irrigation, and on his 160 acres under ditch he is able to raise many times the amount of crops that he could on unirrigated soil. In 1914 he bought eighty acres adjoining. His land is now worth $400 an acre. From first locating in this section he has advocated improved methods and used the latest and most modern agricultural implements in his farm work, making it much easier to conduct operations than it was years ago. Today he recalls vividly the trials and struggles which the early settlers of this county encountered in contending for victory over the untried forces of a new land, and notwithstanding the anxiety and toil imposed, looks back to those days as among the happiest of his life — a view that is fully shared by the other members of the family. He makes comparison between the high prices paid farm labor today, with what he as a young man first earned when he went to work just after leaving school and in the realm of retrospection, Mr. Thurman is duly impressed with the fact that "the world moves," and waits upon no man.

In 1896, Mr. Thurman married Miss Eva Palmer, a native of Iowa, who came to Nebraska when very young with her parents. Six children have come to brighten the Thurman home, Wesley, Stella, Iona, Lola and Viola, all at home, and Vera, who died at the age of sixteen. The Thurman home is one of the most delightful in the Gering valley and they keep open house to their host of warm friends. Mr. Thurman is a Republican in his political views while his fraternal relations are with the Modern Woodmen of America.

ROY SCHAFFER, who is one of Scottsbluff's enterprising young men and successful farmers, was born October 12, 1892, near Johnson, in Nemaha county, Nebraska. His father, Henry Schaffer, was born in Illinois and as a young man came to Nemaha county, Nebraska, and at Johnson was married to Miss Flora Able. The family home continued at Johnson, where Mr. Schaffer engaged in farming until 1909, when he came to Scottsbluff county and bought eighty acres on which he yet resides.

Roy Schaffer obtained a public school education in Nemaha county. He accompanied his father to Scottsbluff county in 1909 and is engaged in farming. He carries on general farming, making beet growing his main crop, and rents eighty acres, being situated within four and three-quarters miles of Scottsbluff The home farm is well improved, the father of Mr Schaffer attending to that as soon as the place came into his possession

At Scottsbluff on December 10, 1915, Roy Schafter was united in marriage to Miss Tessa Harrison, who was born at Fort Collins, Colorado, who is a daughter of Bert and Minnie (Yocum) Harrison, who were born in Missouri Mr and Mrs Schaffer have one daughter, Jean, who has passed her second birthday The family belongs to the Presbyterian church Mr Schaffer is not active in politics but takes an intelligent interest in local matters, particularly those pertaining to agricultural affairs, and voices approval of the high officials of the state who are seeking protective legislation along this line

MANUEL G WILSON, who is a successful general farmer and respected citizen of Scottsbluff county, Nebraska, was born in Indiana, February 13, 1871 His parents were Henry and Julia Wilson, natives of northern Indiana The father was a farmer all his life, first in Indiana and later in Nebraska There were four children in his family

M G Wilson was reared in Indiana and attended the public schools there His entire life has been devoted to agricultural pursuits and since 1914 he has been a farmer in Scottsbluff county, Nebraska He owns eighty acres of irrigated land on section 25 town 22-54, which is well improved Mr Wilson has no particular specialty, the natural soil, with irrigation, producing abundant crops of all kinds

In 1900 Mr Wilson was united in marriage to Miss Fay Galford who was born in Iowa, February 28, 1880 Her parents now reside at Burwell, Nebraska, where they are highly respected Mr and Mrs Wilson have one son, Norman, who lives with his parents Mr Wilson has never desired political office and takes only a moderate interest in public affairs except as they affect the farmers He votes independently

THEODORE CARLSON who is one of the representative citizens and successful farmers of Scottsbluff county, has spent the greater part of his life in the United States and many years of it in Nebraska He was born in Sweden, November 9 1866

The parents of Mr Carlson were Carl E and Christina Louisa Carlson The father was born in Sweden in 1830 and was a farmer all

his life The mother was born in Sweden in 1843 and still lives there

After his schooldays were over and when seventeen years of age, Theodore Carlson left his native land for America It required some courage to thus start out for himself to make his way in a strange land, but he soon found good friends in Iowa, where he lived from 1883 until 1888, when he came to Nebraska He located first in Banner county, homesteaded there and kept his quarter section of land until o good business opportunity came to sell at a profit Realizing that only irrigation was needed to make land in Scottsbluff county wonderfully productive, he had the good judgment to invest here, securing one hundred and seventy-two acres, the entire body now being irrigated and worth many times the price he paid for it In early days he faced the hardships that met all settlers here, but Mr Carlson is not the type of man to be easily discouraged and his persistency has been well rewarded He has everything very comfortable about him in the way of substantial buildings, and if he so desired, might take more ease that he does, but he has always been a hard worker and so continues

Mr Carlson married Miss Mary Peterson, who was born in Sweden in 1872 Her parents were also natives of Sweden Her father came to the United States in 1886, was a farmer in Nebraska and he died here Mr and Mrs. Carlson have had five children, four of whom died in infancy, the one survivor, Carl, living at home and assisting his father The family belongs to the Methodist Episcopal church Mr Carlson is a Republican in politics and takes considerable interest in public affairs as an intelligent citizen must, but has never been willing to accept any public office He is widely known and the entire family is held in high esteem

GUY C McPROUD, who is a general farmer in Scottsbluff county, owning one hundred and forty acres of irrigated land, came here in 1906 and homesteaded one hundred and sixty acres He was born at Farmland, in Randolph county, Indiana, December 7, 1858 His parents, Joseph and Sarah (Taylor) McProud, moved to Kansas in 1856 when he was young, and they spent the rest of their lives on a farm there

Mr McProud attended school through boyhood but after his schooldays were over, went to work on a farm and has followed agricultural pursuits ever since and has been very successful In early days he went through the usual pioneer hardships and remembers well when all these productive fields suffered every season from lack of water He carries on general farming and pays some attention to stock

On December 27, 1883, in Kansas Mr Mc-Proud was married to Miss Eva Baker, who was born in Kansas, August 12, 1864 and is a daughter of Ephraim and Ellen (Sweeney) Baker, who were farming people and now deceased Mr and Mrs McProud have had children as follows Nellie, who lives in Missouri, Ross, who operates the homestead, Garnett, who lives in Kansas, Hazel, who lives at home, Alta, who lives in the state of Washington, and two who are deceased Mr Mc-Proud has given his children every advantage within his power and they are all well educated and the most of them are married He and wife belong to the Presbyterian church He is an independent voter and has never accepted any public office except membership on the school board Mr McProud is known among his neighbors as an honest upright man, one whose word is as good as his bond He organized the first school district in this neighborhood and school was held in his house He had to go as far as the Platte River to get children in sufficient numbers to organize the district

OSCAR A CARLSON, whose well improved, irrigated farm is situated on Section 18 town 12, Scottsbluff county, has been a resident of the United States for thirty-five years and has prospered through industry and good management He was born November 31, 1860, in Sweden His parents were Carl and Johanna (Anderson) Carlson, both of whom are deceased Of their three children Oscar A was the youngest

Mr Carlson remained in his native land until he was twenty-five years old There he attended school and assisted his father who was a general farmer In 1884 he came to the United States, and in 1892 his parents came also For eight years Mr Carlson worked on farms in Kansas, helping through many bountiful harvests in that productive state He watched his opportunity, however, to secure a farm of his own and after coming to Scottsbluff county, Nebraska, in 1897 homesteaded forty-six acres He has remained here, continually improving his place and now has a valuable property and a comfortable and attractive home

Mr Carlson was married to Miss Ellen Anderson, who was born in Sweden, June 24,

1867, the ceremony taking place in Nebraska, in 1896 The parents of Mrs Carlson never came to the United States and still live on their farm in Sweden Mr and Mrs Carlson have two children, namely Edwin and Anton, both of whom reside at home The family belongs to the Methodist Episcopal church They have a wide acquaintance and are very highly esteemed in their neighborhood

ENOCH BOWMAN who is one of the representative men of Scottsbluff county and a prominent farmer, came to the eastern part of Nebraska and homesteaded as early as 1883 He was born in Boone county Iowa, March 4, 1855 His parents were Jeremiah and Elizabeth (Brown) Bowman, the former of whom was born in Pennsylvania August 18, 1829, and the latter in Kentucky

Enoch Bowman attended the district schools in early years and afterward assisted on the home farm A natural desire to own land of his own, led him to come early to Nebraska, and in spite of many hardships that faced all the pioneers he has never wished to leave the state In 1906 he came to Scottsbluff county and has a fine, irrigated farm of one hundred and sixty acres, on which he has placed substantial improvements He carries on general farming and deals in live stock

In Nebraska in 1889, Mr Bowman was united in marriage to Miss Ida Harter, who was born in Indiana Her parents were David and Mary (Weeks) Harter, the former of whom was born in Pennsylvania and the latter in Connecticut For many years Mr Harter was a general farmer in Indiana and both he and wife died there To Mr and Mrs Bowman ten children have been born, namely Ethel, who is the wife of L L Hewitt, lives at Scottsbluff, William E, who is his father's right hand man on the farm, and Edward, Kittie, Mary, Bertha Ella, Herbert, Bernice and Gordon, all of whom are at home Mr Bowman and his family belong to the Methodist Episcopal church In politics he is a Republican He has never cared for public office but takes a close interest in both outside and local affairs and is particularly concerned in the matter of public schools For many years he has belonged to the Odd Fellows and is a member also of the Order of Modern Woodmen

DANIEL BOWMAN, whose large well improved farm is situated on section 17 town 12 Scottsbluff county is considered one of the successful agriculturists of this section, and he is also one of the town's representative men. He has served on the school board for a number of years and his opinion is very often consulted in regard to public matters.

Daniel Bowman was born in Boone county, Iowa, July 16, 1870. His parents were Jeremiah and Elizabeth (Brown) Bowman, the former of whom was born in Pennsylvania, August 18, 1829, and the latter in Kentucky. Mr Bowman received his education in Iowa. In 1907 he came to Scottsbluff county, Nebraska, homesteaded, and now has an irrigated farm of one hundred and five acres that would command a high price should he place it on the market. Mr Bowman has improved his land with substantial buildings of every kind and his surroundings show thrift and good management. He carries on general farming and raises stock for his own use.

In Nebraska, in 1894, Mr Bowman was united in marriage to Miss Elizabeth Linton, who was born in Otoe county, Nebraska, February 19, 1876. Her parents were James and Elizabeth (McNiel) Linton, now deceased, who were farming people in Otoe county. Mr and Mrs Bowman have had the following named children: Cecil who is a farmer in Scottsbluff county, Maurita, who is married and lives in Colorado, Harold, who assists his father on the farm, and Merle, Allen Stanley and Delbert, all of whom are at home. Mr Bowman belongs to the order of Modern Woodmen.

MONROE J REED — When a man chooses any vocation in life it is satisfactory indeed to find that his judgment has not been at fault but that success has rewarded his earnest efforts. While there is no business so important as farming, not every young man can make it profitable when he leaves the home farm where everything is familiar, and starts out for himself. When Monroe J Reed, however, came to Scottsbluff county, from his father's farm in Sarpy county, he found he had been well trained and the effects of this training are seen in the excellent condition of his homestead which is situated on section 17 town 22-53, no great distance from a fine market at Scottsbluff.

Monroe J Reed was born in Clearfield county, Pennsylvania October 21 1873. He is a son of James Mitchell and Mary E (Read) Reed, both of whom were born in Pennsylvania. In the spring of 1880 they moved to Sarpy county, Nebraska and engaged in farming for the rest of their lives. Monroe J Reed obtained his education in the pub-

lic schools. He was reared on his father's farm and from there came to Scottsbluff county in 1910 and homesteaded. Almost all the substantial improvements which mark the place as the property of a careful owner, were put here by Mr Reed. His land is all irrigated and abundant crops result from his efficient methods of farming.

In 1919 Mr Reed was married to Mrs Minnie (Van Meter) Meyers, who was born in Illinois, December 4, 1885, and was carefully educated in Nebraska. She is a daughter of Henry and Mary (Hulbut) Van Meter, her father being a substantial farmer in Hitchcock county, Nebraska, where both parents have spent their lives. Mr. and Mrs Reed are members of the Methodist Episcopal church. In politics he has always been identified with the Republican party.

WILLIAM BOSTON MEEK, who is one of Scottsbluff county's successful self-made men, finds himself quite ready to believe stories often told the traveler through this section, of the pioneer hardships of seventeen years ago. Like many other young men, he came here with more courage and ambition than capital but all these were necessary during the early years. He came before the great irrigation projects were under way, and deserves credit as do others, because of the determination and persistency with which he earned the right to his land. His fine irrigated farm is enough reward.

William Boston Meek was born in Wayne county, Indiana, June 30, 1888 and is a son of John William and Wilda (Porter) Meek, the former of whom was born in Indiana, February 2, 1851, and the latter in the same state, November 6, 1853. They now live on their large ranch in Boxbutte county, Nebraska. When young Mr Meek's parents moved to Morrill county and he was reared there and worked on his father's farm until 1892 when he came to Scottsbluff county and homestead on one hundred and sixty acres. He then had this large tract of wild, unimproved land and set about developing it as rapidly as possible, in the meanwhile reducing his living expenses to the lowest ebb. It was a long season of hard work before he had completed his contract with the government and at times found it necessary to leave his own farm and work for others in order to get money to hold his claim. At first the land was not properly productive, his crops being injured on many occasions by protracted drouths, but since it is irrigated an entirely different story may

Will N. Randall and Family

be told He devotes himself to general farming and his neighbors call him very successful Mr Meek is not married but he is widely
known and his fellow citizens are all friends
He belongs to the order of Odd Fellows

GOTLEIB GRASSMICK, who is an enterprising young farmer of Scottsbluff county,
was born April 28, 1887, in Russia, where
he went to school for several years before accompanying his parents to the United States
His father, Carl Grassmick, was born in Russia and lived there until 1899, when he brought
his wife and children to America The mother, Mary (Helzer) Grassmick was born in
Russia and now lives in Colorado, aged forty-
five years The parents landed safely in the
United States and went west to North Dakota, where the father homesteaded Later
he moved to Colorado and died there when
aged sixty-four years

Gotleib Grassmick worked on his father's
farm in North Dakota until 1914, when he
came to Scottsbluff county, Nebraska Here
he is operating a farm of forty acres that belongs to his father-in-law, and is doing exceptionally well In 1915 he was married to
Katie Hohnstein, who was born in Russia,
November 11, 1894 She is a daughter of
John and Katie (Tellman) Hohnstein, all born
in Russia Mr and Mrs Grassmick have two
children Esther and Helen

WILL N RANDALL is one of the pioneer
settlers who has assisted in demonstrating the
splendid advantages of Scottsbluff county in
the development of the agricultural and live-
stock industry, and his well improved farm
property is situated in section 28, township 23-
55, about seven miles from the city of Scottsbluff Mr Randall was born at Des Moines
Iowa, on the 18th of October, 1863, and was
reared and educated in the Hawkeye state,
where he duly availed himself of the advantages of the public schools of the period He is
a son of Milo M and Adelia (Roberts) Randall, the former a native of New York state
and the latter of Vermont The father, who
was a wagonmaker and cabinetmaker by trade
became one of the early settlers of Iowa, where
he developed a fine farm property in Polk
county, the same being still in the possession of
the family and the place of abode of the venerable widow, who celebrated in 1919 the
eighty-fourth anniversary of her birth Mr
Randall having died at the patriarchal age of
ninety-one years Milo M Randall was a pioneer in the work of his trades in Iowa but
there he gained his major success through association with farm enterprise He purchased

the old home farm from a man named Randall,
but no relation, who had obtained the land
from the government, and thus it has changed
ownership only once In politics he was first
a Whig and later a Republican, and he held
membership in the Methodist Episcopal church,
as does also his widow He was long affiliated
with the Masonic fraternity and was a man of
sterling character and strong mentality Of
the six children the eldest, Emma, is deceased;
Will N, of this review, was the next in order
of birth, Sophrona M is the wife of Martin
Troup, a prosperous real estate dealer at Maxwell, Iowa, Charles is a prosperous farmer
near Lander, Wyoming, Rose is deceased, and
Cora resides at Maxwell, Iowa

Will N Randall early gained practical experience in connection with the work of the
home farm and he continued to be identified
with agricultural enterprise in Iowa until 1886,
when he came to Scottsbluff county, Nebraska,
which was then a part of Cheyenne county
Here he filed on a homestead of 160 acres and
also on a tree claim, both of which properties
he still owns He also owns land on the North
Platte river, in the same section of the county,
and the aggregate land holdings of the family
comprise 640 acres, one-half of which is supplied with excellent irrigation facilities Diversified agriculture and stock-raising have continuously engaged the attention of Mr Randall, and through these basic mediums he has
achieved substantial success, with incidental
status as one of the popular and representative
citizens of his community He has made good
improvements on his land and the family home
is one of attractive order

In politics Mr Randall holds aside from partisan lines and gives his support to men and
measures meeting the approval of his judgment He is affiliated with the Masonic fraternity, including the Order of the Eastern
Star, in which his wife likewise holds membership, as does she also in the Church of the
Brethren

The year 1891 recorded the marriage of Mr
Randall to Miss Margaret Pfoutz, who was
born in Pennsylvania, and they have four children Milo M who married Christina Hass,
and is one of the progressive exponents of
farm industry in Scottsbluff county as is also
Ora P, who married Catherine Bear, Ira N,
who entered the military service in connection
with the late World War and who was stationed at Camp Funston Kansas, is now conducting a well equipped farming and cattle
ranch near Du Bois, Wyoming, having married
Mabel Wilson and Charles Glenmore, who is
farming with his brother at Du Bois Wyoming

JOHN HOHNSTEIN, who is one of the highly respected residents and substantial farmers of Scottsbluff county, has been a resident of the United States since 1904 and of Scottsbluff county since 1910 He was born in Russia, January 16 1873 His parents were Henry and Lizzie (Schreere) Hohnstein, natives of Russia and good people The father was a farmer all his life and died in his native land when seventy-two years old, and the mother at the age of sixty-five years

John Hohnstein grew up on his father's farm in Russia and had some school training From early manhood he had determined at some time to come to the United States, but the chance did not present itself until he was thirty-one years old He landed in this great country in 1904 and found plenty of work and comfortable living on a Kansas farm, where he remained until 1910, when he came to Scottsbluff county He went to work again as a farmer and four years later had a farm of his own, buying eighty acres of land situated on section 20 town 22-53 This land is now irrigated and richly productive He now owns one hundred and eighty acres Mr Hohnstein has put many substantial improvements on his place and has comfort, plenty, and is well content

In Russia, in 1894, Mr Hohnstein was married to Miss Katie Tellman, who was born in Russia March 15, 1875 Her aprents were John and Katie (Loos) Tellman, natives of Russia Both parents died on their farm The following children have been born to Mr Mrs Hohnstein Katie, who is the wife of Gotlieb Grassnick, who is a farmer in Scottsbluff county, Henry, who assists his father on the home place, Maggie, who is the wife of George Snell, who is a farmer in Scottsbluff county, Anna, who assists her mother at home, John, who works for a neighboring farmer, and George, Christina and Asrey Mr Hohnstein is giving his children every advantage in his power, and all are doing well The family belongs to the Russian church and Mr Hohnstein helped to build the first church of that faith in the county

WILLIAM M LACKEY — The well improved farms for which Scottsbluff county is justly noted, indicate the type of people who live here, a solid, self-respecting class who desire comfortable and attractive surroundings as a part of the enjoyment of life Such a farm is the beautiful eighty acres of irrigated land that is the homestead of William Mitchell Lackey, and it lies on section 21 town 22-53, within easy communication with Scottsbluff for market, church or social purposes

William M Lackey was born March 17, 1873, in Ontario, Canada His parents were Andrew and Eliza (Campbell) Lackey, the latter of whom survives and lives at Gering, Nebraska The father of Mr Lackey was born in Ireland and was brought to the Dominion of Canada when two years old He grew up on a farm and followed agricultural pursuits all his life He was married in Canada and some years afterward moved to the United States and settled in eastern Nebraska where he bought three hundred and twenty acres of land He was a man of sterling character and was respected wherever known

William M Lackey was reared in Ontario province, Canada, and had the advantage of good schooling He chose farming as his vocation and has followed it in a thorough, practical way that has brought profitable results He came to Nebraska with his parents when seven years old and in the spring of 1892 came to Scottsbluff county and homesteaded one hundred and sixty acres At the present time he has his irrigated farm in Scottsbluff county and in addition owns four hundred acres in Banner county, range land Mr Lackey devotes himself pretty closely to the management of his large holdings, has all his industries well in hand, and undoubtedly is one of the county's level-headed, competent farmers

In 1889 Mr Lackey was united in marriage to Miss Bertie Jones, a daughter of W S and Mary (Smith) Jones, natives of Illinois The mother of Mrs Lackey is deceased, but the father is living at Scottsbluff To Mr and Mrs Lackey the following children were born Ethel, who is the wife of Evan Jones, lives in Colorado, Leo, who is deceased, Winnifred, who lives with her parents, Floyd, who is in the employ of the government, and Charles, Eva, Artist, Fred, Eugene and Lillian The family belongs to the Methodist Episcopal church He is a member of the district school board but is not unduly active in politics

JAMES A BAXTER, who owns two hundred acres of fine land in Scottsbluff county, was reared on a farm but for a number of years was a railroad man before coming here in 1916 Mr Baxter was born in Delaware county, Iowa, October 15, 1882 Extended mention of the Baxter family will be found elsewhere in this work

James A Baxter remained on his father's

farm in Iowa until he was twenty-one years old He then secured work in Nebraska and Colorado where the government had inaugurated a great irrigation project He worked at ditching in three state, and worked on the railroad all through Idaho railroading and ditching from Miles City to Spokane and from Spokane back to Idaho In 1916 he located permanently in Scottsbluff county, where he bought two hundred acres of land, eighty acres being irrigated He breeds White Face cattle and raises eighty head of livestock yearly He has had an oil lease on his home place for five years and with a one-eighth royalty, may realize a fortune

Mr Baxter was married to Miss Dora Bosh, a daughter of Henry and Tena Bosh residents of Utah Mr and Mrs Baxter have three children, namely Donald M, Murray C, and Marjorie, the eldest being a sturdy lad of six years, and the youngest yet an infant Mr Baxter is a stockholder in the Globe Life Insurance Company

WILLIAM T TENNIS — It is an interesting story indeed that can be told of early Nebraska by William T Tennis, who is one of Scottsbluff county's worthy and substantial citizens, for his experience covers the development of at least four counties of the state, and he was an active participant in many events of historical interest Mr Tennis has been a resident of Nebraska for almost forty years

William T Tennis was born in Marion county, Iowa, May 4, 1855, a son of John and Mary (Dawson) Tennis, who were born in the city of Richmond, Virginia They moved to Iowa in 1844 and both died there, the father when aged forty-seven and the mother when aged fifty-six years Mr Tennis has three brothers and three sisters, but he is the only member of the family living in western Nebraska

Mr Tennis had country school opportunities in boyhood and remained on his parent's farm until manhood, and in the state of Iowa until twenty-five years of age In 1880 he came to Nebraska and located near Newman Grove in Madison county There were but few settlers in that locality and the Indian menace was not yet over so that within six months he sought a more satisfactory location near Oakdale, in Antelope county He lived there four years and then moved into Sheridan county where he engaged in farming and cattle raising for fourteen years and then returned for a like interval to Oakdale Each county possessed

advantages over the other along some lines and Mr Tennis made many friends in every section in which he settled In 1914 he came to Scottsbluff county, where irrigation had increased both the value and price of land He bought eighty acres of ditched land and has a valuable farm property that he devotes to general farming and the feeding of cattle and hogs

Mr Tennis was married to Miss Lora Mullen, who was born in Iowa, and they have had the following children Dalbert who assists his father on the home place, Arthur, who lives in Oregon, Eliza, whose home is in the state of Washington, Maude, who lives in California, Percy, who lives near Minatare, Nebraska, and Araminta, who died at the age of twenty-three years Mr Tennis and his family belong to the United Brethren church He is an independent voter

CHARLES BRACKMAN, who is one of the successful farmers and respected citizens of Scottsbluff county, was born in Germany and obtained a practical education in the schools of his district He came to the United States in 1883, and to Scottsbluff county, Nebraska in 1887 He homesteaded one hundred and sixty acres which he still owns, and through other purchases has increased his holdings until he now has two hundred and forty acres of valuable land He carries on a general farming line and is an extensive raiser of alfalfa

Mr Brackman married Alice Holthusen, who was born in Colorado, and they have children as follows Mrs Clara Gilbert, who lives in Scottsbluff county, Edward, who assists his father on the homestead, Anna who is a school teacher at Morrill, and Martha, George, Karl and Alice all of whom live at home Mr Brackman has never desired political office but has been interested in maintaining good schools in district No 2, and for nine years has served as a school director

JAMES CHRISTIAN HANSON, whose well improved property lies in Scottsbluff county Nebraska where he is widely known and highly respected was born in Denmark, October 8 1856 His parents were Lars and Ann Hanson the latter of whom lives in Idaho in her eighty-third years The parents came to the United States and the father homesteaded in South Dakota His death occurred at the age of forty-six years

J C Hanson came to Scottsbluff county in 1887 and homesteaded one hundred

and sixty acres   He has always been engaged in general farming   In early days he
often found it difficult to provide for his family as he desired to do, as was a fact with almost every other settler in this section at
that time   To "make both ends meet" in
those days required hard work and good business management   Mr Hanson remembers
when, after taking two days to cut, two days
to split and consuming three days to and from
Alliance with his load of sixty-five posts, he
had to accept seven cents apiece for the posts
Notwithstanding the many hardships of those
early days, Mr Hanson says that social enjoyment was not absent and recalls the many
occasions on which the settlers would drive
miles in their lumber wagons to attend some
dance or other gathering, when everyone was
on an equality and a genuine feeling of friendship and good fellowship prevailed   Mr Hanson now owns two hundred and twenty acres,
practically all of it being ditched and is numbered with the prosperous citizens of the county

Mr Hanson married Ada L Roberts, who
was born in Iowa, the ceremony taking place
in Cheyenne county, Nebraska   They have
four children, namely   Mrs Ethel Fosberg,
who lives in Morrill county, and Guy J, Clarice and Ivy, all of whom reside at home   Mr
Hanson has always been a great friend of the
public schools and has served as a school director   He has never united with any particular political party but is a good, fair-mind-
ed, intelligent citizen and votes according to
his own judgment

ROBERT I FRANKLIN, who is a successful general farmer in Scottsbluff county, has
spent the greater part of his life here and is
well and favorably known over the county
Mr Franklin was born in Putnam county,
Missouri, September 30, 1882   Extended
mention of the Franklin family will be found
in this work

Robert I Franklin accompanied his parents
to Nebraska in March 1887, and grew up on
his father's homestead in Scottsbluff county,
obtaining his education in the public schools
At Sidney, Nebraska, in 1907, he was united
in marriage to Miss Beulah Rashaw, who was
born in Nebraska, and they have four children,
namely   Paul, Cal, Roberta and Betty   Mrs
Franklin is a member of the Roman Catholic
church

Mr Franklin has been a farmer all his life
and is considered a very competent one by
his neighbors   He owns one hundred and

eighty-one acres, all under ditch, and the improvements he has placed here are obvious
and substantial   Mr Franklin entertains very
decided opinions on many public questions
and votes according to his own judgment

HERBERT L CLEVELAND, who exten-
sive operations in farming and stockraising,
make him an important factor in the agricultural life of Scottsbluff county, is well experienced in this line, having devoted his entire life to such pursuits   He has proved himself a good business man and his neighbors
hold him a worthy citizen in every respect

Herbert L Cleveland was born in Story
county, Iowa, July 24, 1877, a son of Z V
Cleveland, a sketch of whom appears in this
work

Mr Cleveland came to Scottsbluff county in
1909 and homesteaded   He now owns eleven
hundred and eighty acres of farm and ranch
land, two hundred and twenty acres being
ditched and exceedingly productive   He averages one hundred head of cattle yearly

At La Grange, Wyoming, Mr Cleveland
was married to Miss Edna Miskimmins, who
was born in Iowa and brought to Nebraska in
1887   Their five children are as follows
Robert, Gladys, Harold, Nina and Grace
Mrs Cleveland is a member of the Presbyterian church and is active in its various benevolent missions   Mr Cleveland is intelligently
interested in the country's public affairs as
well as matters near home, but has never identified himself with a particular political group
and has never accepted public office   He belongs to the order of Modern Woodmen of
America

JOHN F MARSHALL, who is an experienced and successful farmer of Scottsbluff
county, came to this section of Nebraska
thirty-one years ago and has lived here ever
since   With other early settlers he experienced some hardships, but in the main has been
well satisfied with all the investments he has
ever made here   Mr Marshall was born in
Fulton county, Illinois, August 29, 1860   His
parents were Samuel S and Eliza Clannon
Marshall, both of whom were born in Fulton
county, Illinois   The mother died in Iowa
when thirty-five years old, but the father survived to the age of eighty-one years   He had
been a general farmer in Illinois and in Iowa,
later retired and removed to the state of
Washington and his death occurred there

John F Marshall obtained his education in
the public schools and remained on his father's

ELMER SCHOOLEY

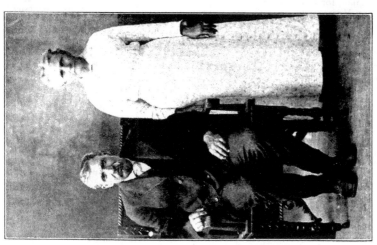

WM. H. SCHOOLEY AND WIFE

farm in Iowa until he reach his majority He came then to Nebraska and on April 11. 1889, took a tree claim in Scottsbluff county, remaining so well satisfied with his surroundings and his neighbors, that in 1908 he homesteaded one hundred and sixty acres in the same county Without attempting to acquire an extensive acreage, Mr Marshall has most sensibly devoted his efforts to the development and improvement of his homestead and has found both pleasure and profit in so doing

Mr Marshall married Mary Orin, and six children were born to them Edward, Charles, John, Bessie and Charlotte, and Earl who is deceased Mr Marshall has never been active in a political sense but is not an indifferent citizen where the welfare of the county is concerned and casts his vote according to his own well considered ideas on public matters

ELMER SCHOOLEY was born in Banner county, October 25, 1888, son of William H and Mary Jane (Wildman) Schooley, a record of whose lives will be found elsewhere in this publication He was reared in Nebraska and educated in the public schools of the state He joined, on July 25, 1918 the Thirtieth Balloon Company for service in the World War He was at Camp Dodge, Iowa, and went later to Ft Omaha and then to Newport News On the evening of sailing for France he became ill and died at Camp Morrison October 12, 1918 He was a young man of fine character, and a patriot

WILLIAM H SCHOOLEY, who has spent thirty-three years of his life in Nebraska, is well known both in Banner and Scottsbluff counties Coming to the state in 1886, he experienced many pioneer hardships, and his reminiscences of those early days are very interesting

William H Schooley was born July 2, 1851, in Martin county, Indiana, a son of Obed and Rachel (Morley) Schooley His father was born in Ohio and his mother in Indiana They never came to Nebraska The father died in Missouri, the mother in Indiana William H Schooley attended the district schools in Indiana and grew to manhood on his father's farm In 1881 they moved to Kansas, living there four years In 1886 he came to what was then old Cheyenne county, Nebraska, now Banner county, homesteaded 160 acres proved up and remained on that place until 1895, when he sold it and came to Scottsbluff county He engaged in general farming and raised cattle

On March 29, 1877, William H Schooley was united in marriage, in Indiana, to Miss Mary Jane Wildman, and they have had five children Nettie who lives at home Harvey, who is a farmer on the old homestead, James, who resides in Sioux county, Elmer, who died of pneumonia at Camp Morrison, Virginia, while in military service during the World War, and Levi, who went to France with the American Expeditionary Forces and returned in July, 1919 In 1900 Miss Nettie Schooley homesteaded in Scottsbluff county and her farm of 160 acres situated on section 23, township 23-54, is a very valuable property Mr Schooley has looked after his daughter's farming interests for some time but is now practically retired The family is very highly respected in this section Mr and Mrs Schooley are members of the Christian church In politics he is independent in local affairs, but in national elections is a Democrat

JOHN J BROWN, who is well known in Scottsbluff county as an enterprising farmer and worthy citizen was born at Westbury, England May 26 1878, and was reared and educated there He is a son of Henry W and Mary (Jackson) Brown, both of whom still live at Westbury where the father is a cloth manufacturer

John J Brown remained in his native land until twenty-eight years old and then came to the United States In 1906 he reached Scottsbluff county, Nebraska, and homesteaded one hundred and sixty acres He has placed substantial improvements here and as soon as the ditching project that is under way is completed, will have an exceedingly valuable estate He carries on general farming and has met with more than a moderate degree of success

Since becoming a citizen of the United States, Mr Brown has made one visit to England, where, on January 18 1917, he was married to Miss Ethel Grist, who is a daughter of Edward and Emma (Wheeler) Grist, who reside at Westbury England, where Mr Grist is a cloth maker Mrs Brown is a member of the Baptist church Mr Brown has never served in any public office although well qualified as to character and sound judgment Politically he is affiliated with the Democratic party

S S FOLMSBEE, is a pioneer not only of western Nebraska but of the entire west He was born in Hamilton county, Ohio May 30, 1833 He missed being born on Decoration Day by some forty years, but, of course, when he chose May 30 for his birthday he had no way of knowing that Congress would pick the same day, many years later, for Memorial Day.

The father of Mr. Folmsbee was Isaac Folmsbee, a native of Pennsylvania, who served his country as a major in the War of 1812. His mother, Debora (Swift) Folmsbee, was a native of Maine.

The subject of this sketch freighted through Nebraska over the old Oregon Trail to California in 1852, and encountered the hardships and adventures that were common to that dangerous journey, meeting many Indians, and stopping long enough to carve his name on the famous chimney Rock. Arriving safe in California, he spent five years in mining in the newly discovered gold field, but failed to make a big strike and ended his adventure by enlisting in the United States navy and cruising in Pacific waters for three years. In 1860 he returned to his home, and in 1862 was married in Indiana to Mary Quick, who was born June 27, 1843, in Franklin county, Indiana, the daughter of George and Susan (Lyons) Quick, both natives of that state. They lived together fifty-seven years.

Mr. Folmsbee moved to Nebraska in 1886 and located five miles south of where Melbeta village now stands. Here he made his home until about fifteen years before his death. Eight years ago he moved to Melbeta where his death occured March 20, 1919, at the age of eighty-six years. To him and his wife were born eleven children, namely: Leona, Jennie, Myrtle, Cora, Emmet, Harry, Clifford. Stella, Maude, Ethel and George; eight of whom are living.

He was a successful man and enjoyed the respect and esteem of those who knew him. His homestead in Scottsbluff county was improved by his own labor and remains as a monument to his industry and progressiveness.

JOHN E. CLURE, who is an enterprising and progressive young farmer of Scottsbluff county, owns a valuable irrigated farm and operates it carefully, intelligently and successfully. Mr. Clure was born in Dawes county, Nebraska, November 10, 1888.

John Clure, the father, now resides on a farm near Bayard, in Morrill county. He was born at Aurora, Illinois, July 2, 1849, son of Joseph and Mary (Burlaugh) Clure, natives of Canada. When he was ten years of age the family removed to Benton county, Iowa, later going to Iowa county. John Clure engaged in farming. He lived in Cass county, Iowa, until removal in 1881 to Dawes county,

Nebraska. Later he lived in Scottsbluff county, and then moved to Morrill county, where he has lived since. He married while in Iowa Sarah M. Parker, who was born in Lee county, Illinois, the daughter of Humphrey and Nancy J. (Cole) Parker, natives of Indiana. Mr. and Mrs. Clure endured in Nebraska, all the hardships and privations of pioneer life. They were the parents of eleven children, ten of whom grew to maturity.

John E. Clure lived in Dawes county until he was about eight years old, when he accompanied his parents to Morrill county, where he attended school and assisted his father on the home farm until 1908. In that year he began farming for himself and has demonstrated his competency. He now owns eighty acres of finely improved, irrigated land in Scottsbluff county and devotes it to general crop raising.

Mr. Clure was married to Miss Elsie Wood, who was born in Scottsbluff county, January 19, 1895, a daughter of J. P. Wood, a sketch of whom will be found in this work.

GEORGE B. DENTON, who is one of the substantial farmers and livestock men of Scottsbluff county, has been engaged in farm pursuits all his life, and for the past fifteen years has been operating for himself.

George B. Denton was born February 1, 1876, in Pennsylvania, and is a son of J. B. and Eliza (Bateman) Denton, both of whom were born in England. J. B. Denton was brought to the United States when eleven years old and his wife came here when nineteen years old. They were married in Pennsylvania and in 1886 came to Nebraska and settled in Box Butte county. The father is a retired farmer and both parents live at Alliance. George B. Denton accompanied his parents to Box Butte county and assisted his father until 1904, when he began farming on his own responsibility. In 1918 he came to Scottsbluff county and purchased two hundred and forty acres of well improved, irrigated land. Additionally he owns three sections of cattle land in Sioux county.

Mr. Denton married Miss Anna Lore, who was born in Kansas, November 10, 1880, a daughter of J. A. and Luella (Dunlap) Lore. The mother of Mrs. Denton is deceased but the father is yet an active farmer in Box Butte county. Mr. and Mrs. Denton have three children, namely: Arthur, Richard and Everett, aged respectively sixteen, thirteen and seven years. Mrs. Denton is a member of the Methodist Episcopal church. Mr. Denton is somewhat interested in politics and votes the Republican ticket He is widely known in this section and very well thought of

AUSTIN MOOMAW, has been a resident of the great state of Nebraska for over a half century He was born in Pike county, Illinois, July 29, 1860 His parents were Joel and Susan (Pence) Moomaw The father was born in Ross county, Ohio, engaged in farming all his life and died in Missouri when aged seventy-three years The mother was born in Pennsylvania and lived to be eighty-four years old In 1871 the family went to Missouri, but it was not until the spring of 1887 that Austin Moomaw filed on a claim and moved on the homestead in Scottsbluff county, on which he has lived ever since The early days here were full of trial and discouragement to the hardworking settlers and almost all of them lost crops and cattle because of unseasonable storms and unusual dry weather At that time there was not a house in sight and he lived for fifteen years in a sod house Farseeing men may have visioned a time when the arid land might be transformed into productive farms, but if so, their ideas came to naught for many years In the meanwhile sturdy, hopeful men like Austin Moomaw held on to their land and the time has arrived when the wildest fancies of those who believed in the country's great future have been more than realized Mr Moomaw owns three hundred and twenty acres of well improved irrigated land and is successfully engaged in general farming and crop raising

Mr Moomaw was married to Miss Agnes Spriggs, who was born in northern Missouri, March 15, 1860, a daughter of Thomas R and Luvenia (Carlin) Spriggs, natives of Westmoreland county, Virginia, the former of whom died on his farm when aged sixty-six years, and the latter when seventy-three years old Mrs Moomaw served two terms, four years, as county superintendent of schools in the early days Mr and Mrs Moomaw have two children, namely Leon and Vera, the latter of whom is the wife of Roy Walford, who is an attorney at Lincoln Nebraska, the son is married and lives in Morrill county Both the children were given college educations The family belongs to the Christian church Mr Moomaw has always voted the Democratic ticket but has never been willing to serve in public office

MELVILLE NEIGHBORS, who is an enterprising and successful young farmer of Scottsbluff county, operating on section 10, was born April 7, 1894, in Missouri His parents are Joseph G and Carrie A (Franklin) Neighbors who are mentioned elsewhere in this work

Melville Neighbors obtained his education in Nebraska He remained at home assisting his father until 1915, when he started out for himself and now operates eighty acres of irrigated land very profitably, devoting it to general farming He follows modern methods in carrying on his farm industries and uses improved machinery

Mr Neighbors was married to Miss Marie Peterson in 1915, who was born in Morrill county, November 3, 1898 Her parents are Arthur and Elizabeth (Phillips) Peterson, both of whom were born in Canada The father is still engaged in farming in Morrill county Mr and Mrs Neighbors are members of the Methodist Episcopal church

MISS MYRTLE HILL — There are fine, productive farms in Scottsbluff county and the histories of these read much alike because they all have been developed out of a wilderness through the industry of men and women who went through hardship and deprivation to make them what they are The owners of these farms are not the only pioneers who came here and made attempts at settlement, but they are, in almost every case where injustice was not done, those who were on the field early and labored hard to acquire what they have One of these fine properties is owned by Myrtle Hill, a well known resident of this county

Miss Hill was born in Sullivan county, New York, December 21, 1856 Her parents were Albert and Sarah L (Palmer Hill, both natives of Sullivan county, the father born August 12, 1825 and the mother, February 2, 1828 The father of Miss Hill was a farmer and also operated a sawmill In 1885 Myrtle Hill came to Scottsbluff county and took a homestead and a tree claim but as this claim was contested, she lost that property She still has the homestead, to which she subsequently added and now owns three hundred and sixty acres of the finest land in the county, all irrigated and well improved Miss Hill values her land at $300 an acre She has seen hard times in this section but never lost faith in the real fertility of the soil and has lived to see her ideas on irrigation carried out She carries on general farming and also raises some stock Miss Hill is one of the county's substantial women

ARTHUR A JEFFORDS, who is one of McGrew's most highly respected retired citizens, came to Scottsbluff county at an early date and has been prominently identified with its developing enterprises Mr Jeffords has been particularly interested in the great irrigation projects that have changed this once arid country into a section of agricultural profusion and has made it one of the richest counties in the state of Nebraska

Arthur A Jeffords was born in Muskingum county, Ohio, June 25, 1850 His parents were John and Nancy Jeffords, both of whom were born in Ohio The father was a farmer there until 1886, when he moved to Nebraska, settling near Broken Bow in Custer county. The mother died there when aged sixty-six years, but the father survived until in his seventy-ninth year

Mr Jeffords in 1886 drove from Iowa to Custer county, Nebraska, with a team of horses He traded the horses for oxen and after one of the oxen died, worked his land with a cow and the other ox He landed in what was then called Cheyenne county with not much more in worldly wealth than a sack of beans and $70 in cash He homesteaded one and a half miles south of McGrew, but at that time there was nothing to be seen but bare prairie which was the range for the Bay State cattle company He homesteaded one hundred and sixty acres and also secured a timber claim of one hundred and sixty acres, and for a number of years carried on agricultural operations, then sold and retired to McGrew, where he has since resided He was a director of the first school established in school district eighteen and continued to be interested in the schools as long as he lived in that district He served for four years as assessor Mr Jeffords was one of the enterprising men instrumental in getting the Castle Rock ditch project started, in 1889, and has been a member of the managing board ever since

Mr Jeffords was married to Miss Mary E Kating, who was born at Lexington, Kentucky, a daughter of Edward and Katherine Kating, the former of whom was born in Ireland The mother of Mrs Jeffords survives, but the father has never been heard from since he started for Pike's Peak in search of gold Mr and Mrs Jeffords have three children Ira a carpenter at Ogallala, Mrs Abbie Vandevere, of Ogallala, and Glenn, a ranchman in Wyoming Mr Jeffords has always been identified with the Democratic party

HENRY C BLOOD, who owns a valuable farm of one hundred and sixty acres in Scottsbluff county, at one time was quite active in its management, but is now practically retired from agricultural pursuits He is well and favorably known over the county, especially at Minatare, where he was in the hay, grain and coal business for ten years

Mr Blood was born in Portage county, Ohio, April 17, 1868, a son of Adorno and Hannah F. Blood, the former of whom died in his forty-fourth year and the latter when aged seventy-five years They came to Nebraska in 1887 and the father homesteaded in Sioux county. Mr Blood had two sisters, Mrs Ettie Yoey and Mrs Mary Hood, the former of whom is deceased and the latter resides at Melbeta, Nebraska

In 1887 Mr Blood homesteaded in Sioux county, Nebraska, and spent ten years on his homestead of one hundred and sixty acres there, then came to Scottsbluff county in 1897 and worked for others and rented land for several years In 1901 he bought property at Minatare He has put excellent improvements on his farm of one hundred and sixty acres and has seventy-five acres ditched He raises hay and grain exclusively He has always voted the Democratic ticket

JOHN BRADY, who is a representative citizen of Scottsbluff county, an extensive farmer, large landowner and successful cattle raiser, was born in Columbia county, Wisconsin, January 14, 1851 His parents were John and Rose Brady, both of whom were born in Ireland They came to the United States in 1842 and settled in Wisconsin The father served in the Mexican war He and wife died on his Wisconsin farm at advanced age

John Brady was reared on a farm but had excellent educational advantages and for nine years before coming to Nebraska was superintendent of schools of Fillmore county, Minnesota He came to Scottsbluff county in 1912 and homesteaded one hundred and two acres, and at the same time his two sisters and his mother-in-law also homesteaded He now owns three hundred and twenty acres of land irrigated by the Highland government ditch When the family came first to this valley there were few neighbors and no organized road system Mr Brady has very substantially improved his property, has a comfortable and attractive rural home place and all buildings needed for the carrying on of farm industries in a modern way He en-

MR. AND MRS. BECK

gages in general farming and raises registered Shorthorn cattle

Mr Brady was united in marriage to Harriet Elizabeth Janes   Her father was born in Illinois and her mother in Canada   They came early to Kearney, Nebraska, and Mrs Brady enjoyed excellent educational training and for a number of years prior to her marriage, was an instructor in the Kearney high school   Mr and Mrs Brady have one son, John H , who is in business at Seattle, Washington   Mr Brady has always given his political support to the Republican party   With his family he belongs to the Presbyterian church   For many years he has been active in Masonic circles and assisted in establishing the first Masonic lodge at Minatare

WILLIAM H BECK, who was one of Banner county's most respected citizens for many years, was born in Wayne county, Ohio in 1847 and died in Gering, December 1, 1904   He was an example of the honest, industrious, intelligent and conscientious pioneer settler, to which class Banner and other counties of this great state owe so much in the way of substantial development

The parents of Mr Beck were William and Mary Ann (Hartman) Beck, who were married February 29, 1844   At the age of thirty-five years Wilham Beck became a minister in the Methodist Episcopal church   He was born at Middletown, Dauphin county, Pennsylvania, June 23, 1817

William H Beck grew to manhood in Ohio and when the Civil War came on proved his loyalty to the Union by enlisting as a soldier in Company F, 186th Ohio Infantry, in which he served faithfully and took part in many battles   It was after the war was over that he went to Indiana, where on June 3, 1869, he was united in marriage to Miss Harriet Brown, who was born in Fairfield county, Ohio, December 22, 1842   Her parents were Travis H and Matilda (Banister) Brown, who were natives of Pennsylvania, and Mrs Beck was the fourth born of their seven children, six of whom were daughters   For four years after their marriage Mr and Mrs Beck lived in Indiana   In 1884 they came to Nebraska and settled in what was then Cheyenne county, later changed to Banner county, where Mr Beck preempted land and also secured a timber claim   He was a general farmer and at the time of retirement, about 1900, came to Gering and owned 320 acres of well developed land

The following children were born to Mr and Mrs Beck Mrs Nora McCoy, whose husband is a merchant in Oregon, Worthy, who died in

infancy, Mrs Grace Nelly Forman, who lives near Mitchell, Mrs Sadie Bell McCampsey, who lives in Oregon, Mrs Mary E Adcock, who is deceased, and Walter T, who resides near Gering Nebraska   Mr Beck was a Republican in politics   He was a faithful member of the Methodist Episcopal church, to which religious body Mrs Beck also belongs, in which she is quite active at Gering   She has a small residence in this city and has a wide acquaintance and many friends

FRED L BURNS, who has passed almost his entire life in the state of Nebraska, is a representative and prominent citizen of Scottsbluff county and is the owner of a fine farm but resides at Gering   He was born in Illinois, October 27, 1869 and accompanied his parents to Nebraska in 1871   Both parents, A S. and Elizabeth Burns, were born in Canada   The father homesteaded in Fillmore county, but now resides with his son at Gering   The mother died when aged fifty-five years

Fred Burns obtained his education in the public schools   In 1901 he came to Scottsbluff county and in 1906 homesteaded one hundred and sixty acres, eighty-eight of which are under irrigation   He married Miss Nan Fulton, who was born in Missouri   Her parents were J R and Mary Fulton, who were born in Ohio, came to Fillmore county, Nebraska, and homesteaded and died in that county aged respectively seventy-six and eighty-four years   Mr and Mrs Burns have an adopted son, Edward Burns, who is a bright and obedient youth now attending school at Gering   For some years before her marriage, Mrs Burns taught school in Fillmore county and also in Wyoming and both she and Mr Burns have been greatly interested in school development in school district number thirty-three   The first sessions were held in a dugout, in 1905, but in 1908 a schoolhouse, 18x24 feet in dimensions was provided and in 1910 an addition was built to the structure   Largely through Mr Burns's influence a commodious, modern school structure took its place in 1915   The first teacher was a Miss Elquist, who had six pupils, while now there are forty or more and when all are present two teachers are required   In politics Mr Burns is an old-line Republican   He has served in different local offices and was deputy county treasurer from 1907 until 1912, and county treasurer from 1912 to 1916, and Mrs Burns was deputy under him   Mrs Burns owns one hundred and sixty acres of fine land southwest of Gering

HARVEY HARWARD, who is a well known, enterprising and highly respected citizen of Scottsbluff county, a successful farmer and public official for a number of years, was born in Iowa, January 9, 1864 His parents were Charles and Nancy Harward, natives of Ohio The father was a farmer all his life and died in Missouri at the age of sixty-five

Harvey Harward lost his mother when he was eight years old He has one older brother He came to Nebraska and on April 13, 1886, homesteaded one hundred and forty-four and a half acres and secured a timber claim, and since that time has improved four farms and has in prospect another He now has eighty acres ditched and is making extensive improvements

In Scottsbluff county Mr Harward was united in marriage to Miss Ina Williams Her parents were T J and Lovina (Michel) Williams, the latter of whom died at the age of fifty years The father of Mrs Harward was born in Henry county Iowa, February 9, 1849 His parents were Henry and Leah (Stanbrough) Williams, natives of Ohio Mr Williams spent twenty-nine years in Iowa, then moved to Dakota and in 1885 to Cheyenne county and homesteaded near Bayard, one hundred and sixty acres of dry land all of which is now ditched He no longer is active on the farm but Mr Harward, with whom he lives, looks after his interests Mr Williams was county assessor from 1892 until 1896, held school offices many years and was a leading citizen in many ways His children were as follows Mrs Rose Williamson of Iowa , Mrs. Harvey Harward, of Scottsbluff county, Mrs Gatch, Mrs Davis and A O Williams, all of Scottsbluff county , R C , of Melbeta, and Guy, deceased The last named left two children, Lovina and Thomas James, and Mr Harward is rearing them as his own

In politics Mr Harward has always been more or less an independent voter For a number of years he has held school offices and offices in connection with irrigation projects and has also been assessor He has been very active in forwarding educational and church movements and on January 22, 1889, helped to organize the First Baptist church in Highland precinct, a charter member with J M Adams and L A Christian Both he and wife belong to the Methodist Episcopal church and he is superintendent of the Sunday school

FRANKLIN A REDFIELD was born in Livingston county, New York, November 25, 1834, and died August 26, 1904, in Johnson county, Nebraska His wife, Mary E Aldrich, is a native of Ray county, Missouri, where she was born February 4, 1839 She came to Scottsbluff county in 1916 from Johnson county, where she had resided from the time of her husband's death, and now at the age of 80 years she lives by herself in the town of Melbeta and successfully looks after all her affairs

Mr Redfield, after his marriage on February 4, 1858, lived in Illinois as a farmer until the outbreak of the Civil War When the call of the country came for volunteers he enlisted and served three years in the war He came to Nebraska April 1, 1870 and settled in Johnson county He was first a farmer and then a merchant there for ten years, was a very successful man, and widely known Mr Redfield was a member of the Grand Army of the Republic of Crabb Orchard Post G A R

To Mr and Mrs Redfield were born two children The elder of these, Lucien H. Redfield, was born in Illinois June 10, 1859 , came to Scottsbluff county in 1911 and purchased land which he has improved and upon which he has been successful in general farming To him and his wife, Alice (Worley) Redfield, a native of Iowa, eight children have been born, six of whom are living, namely Clark, now employed in Melbeta, was across the ocean nine times, being in the U S navy during the late war on the transport Wilemina, Clara, a nurse in the Midwest hospital in Scottsbluff; Lucy, the wife of Edgar Decker, a merchant in Melbeta , and Mary, John, and Arthur, at home

The other son of the subject of this sketch is William C Redfield, a banker at Haig, in Scottsbluff county He was born in Illinois on December 30, 1860 He was married to Mary E Barrett, a native of New York, and two children have been born to them, namely Franklin, who has recently been discharged from the United States navy, and Martha, deceased

Mr and Mrs Redfield were members of the Congregational church in Illinois, but after coming to Nebraska they joined the M E church in Johnson county, bringing a letter from the Crabb Orchard M E church to the Melbeta church

JOSEPH P WOOD. — The subject of this sketch is a native of Iowa, and was born December 11, 1857 His father was George Wood, who was born in Madison, Indiana, and followed the occupation of a blacksmith in Indiana, later moving to Kansas where he worked at his trade until his death at the ripe age of seventy-nine years. His mother, Artemisia (Austin) Wood, died at the age of fifty. She was a native of Kentucky.

Mr. Wood was united in marriage to Ella Johnson, who was born in Illinois November 24, 1859. Her parents were Horace Johnson, a native of New Hampshire, and Helen (Smith) Johnson, a native of Connecticut. Both Mr. and Mrs. Johnson are now deceased. They followed the calling of general farming in Illinois and never came to Nebraska.

Two children were born to Mr. and Mrs. Wood: Delmar, who is married and lives in California; and Elsie, now Mrs. J. E. Clure, living on a farm in Scottsbluff county.

Mr. Wood came to Nebraska in 1886 and settled on a homestead of one hundred and sixty acres. After improving his place, he found, like many other western Nebraskans in the early days, that money was not so plentiful in the short grass region, so he secured employment as manager of a ranch in Wyoming and held the position for twelve years. At the end of that time the prosperous growth of Scottsbluff county was beginning, and he returned to his place here and has successfully followed farming and stockrasing from that time. He is a Democrat in politics and is widely known among the early settlers as an industrious man, a good friend and neighbor, an upright American citizen. He has seen the country grow and has grown with it. He is one of the many who have proved that the main difference between success and failure is the ability to stick to it.

WILLIAM JOHNS is a native of Nebraska, born in Johnson county August 22, 1868, and has spent his life in the state. His father, Ferdinand Johns, was a native of Germany, but came to America in the spring of 1868, where he took a homestead in Johnson county and followed farming there until his death at the age of seventy-six years. His mother, Caroline (Bolt) Johns, also born in Germany, died in Johnson county, aged sixty-eight.

Mr. Johns was married to Louise Zinsmaster, a native of Ohio. Her father, Jacob Zinsmaster was born in Germany, coming to America when a young man. He married Maria Sutvarn, a native of Ohio. They were farming people and died in Johnson county, Nebraska.

To Mr. and Mrs. Johns eleven children have been born, ten of whom are living. They are: Earnest, Myrtle (now Mrs. Warren Dickinson, living on a farm in Scottsbluff county), Harry, Elnora, Wilber (deceased), Roy, Earl, Nellie, Bernice, Lorine, and Grace. All are living at home except the married daughter.

Mr. Johns has farmed in both ends of Nebraska, twenty years in Johnson county and ten years in Scottsbluff. When he came here he purchased four hundred acres of land and has been engaged in general farming and stock raising. He now owns seven hundred and ten acres. He is not inclined to be boastful, but modestly says that he considers himself successful. When a man at middle age has a family of ten children, a prosperous farm and ranch business, and seven hundred and ten acres of Nebraska land that is increasing in value every year — if such a man is not entitled to call himself successful, we must get a new definition of success.

C. H. BURK. — The subject of this notice was born January 29, 1856, in Fountain county, Indiana, the son of John and Mary Burk, both now deceased. John Burk was a native of Kentucky. He dealt in horses, and during the Civil War bought horses for the United States government. He never came to Nebraska, but died at the age of seventy. The mother lived to the advanced age of eighty years.

C. H. Burk came to Nebraska in 1883 and settled at Tamora, in Seward county, where he engaged in the lumber business for about two years. He was married at Phillips, Nebraska, on June 30, 1885, to Pet W. Wood, who was a daughter of James W. and Margaret (Showalter) Wood. She was born in Benton county, Iowa, where the father was an early settler and a prominent attorney, having come from London, England, directly to Iowa. Her mother was a native of Ohio. Both the parents are now deceased.

One child, Harmon J. Burk, was born to this union, and died at the early age of four years.

From 1885 to 1893 Mr. Burk conducted a hardware and lumber business at Phillips, Nebraska, and then entered the banking business at the same place. Since that time he has been a banker and has had the unusual experience of founding and successfully developing some half a dozen banks in western Nebraska, all of which are now prosperous and growing with the fast growing country. His first location in this vicinity was at Bayard, where in company with J. W. Wehn he opened the Bank of Bayard in January, 1900, with a capitalization of $5,000. Later the same men founded the Deuel County Bank, at Oshkosh, with $10,000 capital, and Mr Burk made his residence in Oshkosh for two years directing the affairs of this bank   Next he spent two years with the Bank of Lewellen, in the same county, another $10,000 institution   In 1909 the Broadwater Bank, in Morrill county, was opened with $10,000 capital, and in 1911 Mr. Burk went to McGrew, in Scottsbluff county, where he opened a bank with $15,000 capital. He sold this in February, 1914, and since that time has been retired from active business pursuits

Mr Burk is a Republican in politics, and belongs to the Woodmen and Highlanders. Mrs Burk is a member of the Presbyterian church

G R CONKLIN — Born in Polk county, Iowa, September 18, 1872, Mr Conklin came to Nebraska in 1886 with his parents   His father, Gilbert Conklin, was a native of New York and a farmer   His mother, Lucy M. Conklin, was likewise born in New York but was reared in Iowa   The parents took a homestead in Nebraska, and the father dying before final proof was made, the mother completed the proof   She is since deceased at the age of sixty-one years

The subject of this sketch was married at Gering to Florence Alberts, a native of Iowa Of the five children born to them, one son, Clifford, died at the age of eighteen months The others, all living at home, are Walter, Clayton, Charles and Leslie

Mr Conklin proved up on a homestead near McGrew, Nebraska, which he sold   He also bought and sold several other places before locating on his present farm where he is engaged in general farming and preparing to make extensive improvements   He has, like practically all early Nebraska settlers, raised cattle   Furthermore, like all the other early settlers, he took a hand at every side line that offered a chance to make an honest dollar in the days when dollars were as scarce as steamboats on these western prairies   He hauled the first load of freight into Gering from Sidney   He drove a prairie schooner through from Missouri, and saw all the hardships of the pioneer days   Some of those who took part in those early struggles gave up and left, others stuck it out, and they are now the successful and prosperous members of the community in its present days of great development and growing riches   Mr Conklin was one of those who stuck   He is road overseer of district number six and was a

member of the school board of district number eighteen   He belongs to the W O W , while his wife is a member of the Woodmen Circle and of the Presbyterian church

GEORGE KEIPER WHITAKER was born June 14, 1862, in Morgan county, Indiana   His father, Bland Whitaker, was a native of Kentucky, a farmer by occupation, and lived to the age of seventy years   His mother, Fanny Whitaker, died at the age of fifty-four

Mr. Whitaker was married at Kearney, Nebraska, to Miranda Carpenter, whose parents were early settlers in Buffalo county   Her father, E W Carpenter, came to that locality in 1872   Both he and his wife, Emily, are now deceased

Eight children have come to bless the home life of Mr and Mrs Whitaker, and all of them are living at home   Their names are, Nigel, Dell, Clarabell, Ruth, Harry, Emma, Jackson, and George K , Jr

The subject of this sketch came to Nebraska in 1880 with his mother, and both of them took up homesteads in Buffalo county, near Kearney   After living there twenty-eight years, engaged in extensive farming and stock-raising operations, Mr Whitaker came to Scottsbluff county in 1909 and bought his present home   This is a well improved place of three hundred and twenty acres, of which eighty acres is now under irrigation, and all the balance will be irrigated by the new Government ditch which is now being constructed Coming to this section of the state with the first railroad, Mr Whitaker has seen it develop from a sparsely settled range country to one of the wealthiest sections of the United States, all in a few years, and he has had his share of the prosperity   He now confines himself to a general farming business, although up until 1918 he had raised cattle   In that year he closed out his cattle and retired from stock raising   In politics he is a Republican, and is a member of the Methodist church   He is well and favorably known to all of the pioneers of the county, and there is between them the bond of friendship that comes from pioneering together in the days of hard times.

CHARLES E NEELEY was born in Schuyler county, Missouri, September 3, 1866 His father was Robert S Neeley, a native of Lancaster, Ohio, and his mother, Sarah M Neeley, a native of Kentucky. Both lived to an advanced age, the mother dying at eighty-five and the father at eighty-seven   The father, after living in Missouri for a number of years, moved to Colorado and engaged in the ranching business In a runaway accident he was thrown from a buggy into the icy waters of an irrigation ditch, and owing to his age he was unable to withstand the shock and his death resulted

The subject of this sketch came to Cheyenne county, Nebraska, in 1885 as a single man In 1906 he was united in marriage with Elizabeth J Baumer at Lancaster, Missouri, his wife being a native of that place Three children have blessed this union, namely Alpha, Joseph R, and Perry E, all of whom are living at home

Mr Neeley homesteaded and pre-empted three hundred and twenty acres of land in Mitchell Valley, in Scottsbluff county, which is now one of the richest sections of the great irrigated territory of western Nebraska Later he sold his holdings there, and moved to the old home in Missouri, but after an absence of five years he became convinced of the great truth that it is a crime to leave the North Platte Valley irrigated country, and an unpardonable crime to stay away, so he returned and purchased a farm of one hundred and sixty acres of irrigated and eighty acres of non-irrigated land near Gering, which he has improved himself It is now up to the standard of Scottsbluff county farm homes, which is one of the highest standard in the world

In politics Mr Neeley has been an independent voter He belongs to the I O O F, Masons, and M W A, and for a number of years has been a member of the school board of his district He is well known and stands high in the estimation of his neighbors

J J KIPP — The subject of this sketch was born in Germany on September 5. 1858, and came to the United States in 1862, and settled at Quincy, Illinois He is the son of Joseph and Elizabeth Kipp, both natives of Germany, and both now deceased When he was four years of age his mother died his father being at that time in America, and three children of the family made the long journey across the ocean to join their father The brother, Frederick, and the sister, Elizabeth, who accompanied the little one are now not living

Mr Kipp's first wife was Mary E Crane, a native of Iowa, and to them were born two children, Earl and Elizabeth, both living The son is a farmer at Reddington, Nebraska, and the daughter, whose name is now Elizabeth Davis, resides at Harrison, South Dakota

After the death of his first wife, Mr Kipp was married to Frace E Myers, who was born in Illinois, the daughter of Andrew and Ella Myers, of Independence, Missouri Four children have been born to this union, Joseph H, Ella, Dorothy, Mabel, and Victor

In March, 1888, Mr Kipp settled in Sioux county, Nebraska, and followed farming and cattle raising there until 1901, when he disposed of his homestead and purchased land in Scottsbluff county and has since made his home upon it He has eighty acres of fine irrigated land, which he has improved from its former condition of raw prairie into a modern and up-to-date farm, and in addition he owns a half interest in eighty acres near his home place In common with the other pioneers of western Nebraska, Mr Kipp endured the privations and hardships of early Nebraska homesteading and often found it hard to make both ends meet He is now prosperous and one of the substantial members of his community

JOHN HARVEY PFEIFER — One of the prosperous exponents of the agricultural and stock-raising interests of Scottsbluff county is the man whose name heads this review, who has been a resident of this section for a decade

Mr Pfeifer was born in Crawford county, Ohio, November 2. 1872 the son of Godfrey F and Emiline (Snyder) Pfeifer, both of whom were natives of the Buckeye state Seven children were born to them Chris, now a ranchman in Banner county, Nebraska, Laura, the wife of Frank Sears, a lumber merchant in Montana, Katy, who married Francis Whitman of Russell county, Kansas, and J H are living, the others are dead

The father was a farmer school teacher in Ohio, and his was also the distinction of having been a gallant soldier of the Union during the Civil War After the cessation of hostilities he became a farmer in the Buckeye state but being a man of excellent education and high attainments devoted a part of his time to communal affairs for the benefit of the rising generation, teaching in the public schools After some residence there he sold out and came to Kansas, taking up a homestead From there he came to Banner county, where he finished proving up on a homestead left by a son who died Subsequently he came still farther west to Scottsbluff county where he passed the remainder of his days The mother lived in this county until September, 1913 when she too sought her last rest In politics the father was a staunch supporter of the principles of the Democratic party but cast his vote independently when it came to county and municipal affairs, throwing his influence toward the man best fitted for each office He was a member of the Grand Army of the Republic, having been a member of a battery of light artillery from Ohio during that memorable conflict He was a Christian man of high standing in every community where he resided and had many friends who held him in high esteem

Instances are numerous in Scottsbluff county where men have arrived in this section with few acquaintances or friends and have worked their way to affluence and independence, Mr Pfeifer is one of the number; before he came here he had been unable to accumulate any large sum of money but this in no way discouraged him, for he was a man of energy, had faith in the future of western Nebraska and set out to become possessed of a share of the prosperity he believed was coming to this county and today his faith has been justified by the comfortable fortune he and his family enjoy

April 12, 1906, marks the day Mr Pfeifer became a resident of this great commonwealth, for it was then that he located in Banner county as a ranchman, where for three years he was engaged in developing and operating a farm He had already learned the best methods of planting and harvesting so that he was well equipped with practical experience to enable him to carry on agricultural pursuits in the new country which he had decided to make his future home Three years later he came to Scottsbluff to work for a cattleman where he gained valuable experience in handling stock on a large scale, feeding, buying and marketing that has proved of value to him in recent years Within a short period he bought his present farm of one hundred and forty-two acres, which at that time had few improvements but which he has brought to a high state of cultivation He is engaged in general farming and stock-raising, being highly successful in feeding and fattening cattle on alfalfa and beets, shipping to the great meat centers farther east

On November 9, 1910 Mr Pfeifer married Annie Hiersche, a native of Germany, and to them have been born four children Leonard, Emma H , Dean and Clyde Mr Pfeifer was educated in the excellent schools of this state and Kansas and is a firm believer in a good education for everyone and special training for any special vocation in life He says nothing shall stand in the way of his children securing the best educational advantages afforded by district, town and state, and as a result of his convictions he is a supporter of every movement for higher education, county farm bureaus, civic improvements in both local and state wide affairs He is today one of the progressive representatives of modern rural life

EARL W COLLINS, has identified himself most fully with the civic and material interests of Scottsbluff county, for he is not only a representative agriculturist of this section, but is also the owner of a well improved farm estate in section 30, township 23-55. He is a native son of the west and has exemplified its progressive spirit in the varied activities that have brought him a generous share of temporal prosperity

Mr Collins was born in Valley county, Nebraska, in 1878, the son of Warren and Amanda (Thurston) Collins The former a native of Allegheny county, New York, while on his mother's side he inherits traits from sturdy old New England ancestors, as she was born in the state of Maine There were eight children in the family· Oscar, a farmer in Valley county, Carrie, the wife of W J Seeley, a farmer of Milford, Helen, a trained nurse at Ord, Nebraska, Earl W , Ralph and Lynn, both farmers in Valley county, Rex, now engaged in farming in Washington, and Floyd, a student in a medical college who spends his vacations at home The parents came to Nebraska in 1872, when the only buildings known in the central part of the state and westward were composed of sod, and it was in such a home on a prairie homestead that Earl Collins spent his boyhood days, attending the district school during the winter and doing such work as was suitable to a boy on the farm He has watched with the eye of a proprietor, the various changes that have been brought by the passage of the years and the sturdy and progressive work of the big hearted pioneers, and has himself borne a full share of the labor of development He is now one of the landholders and successful agriculturists of the Mitchell community, of Scottsbluff county where his accomplishments entitle him to the respect and esteem in which he is uniformly held by his fellow citizens

Mr Collins located here in 1905, taking up a homestead of one hundred and sixty acres, on which he at once placed excellent improvements He now has all his land under cultivation, has a fine home and substantial buildings and has established himself as a progressive and skilled farmer who thoroughly knows his business and can make his labor pay him proportionately

In 1907 Mr Collins was united in marriage with Frances Hewett and they have two charming girls, Doris and Helen, both at home

Mrs Collins was a native of Plymouth county, Iowa, but since coming to western Nebraska, has learned to love the great wide open spaces of this section, where the skies are nearly ever sunny, and the country a wonderful picture with its great expanses of growing crops in the spring and of yellow ripened grain in the fall She like her husband, is progressive in ideas and is a worthy helpmate for such a man Mr Collins is an up-to-date business man, keeps abreast of all questions of the day, whether national, state or communal and favors every progressive movement in this section He is an independent voter, exercising his privilege of the franchise as his wisdom and conscience dictate, while his fraternal affiliations are with the Modern Woodmen of America

MICHAEL L KIESEL —Scottsbluff county has few finer citizens of finer fiber or more sterling worth than Michael Kiesel, whose field of operations is in the Mitchell district where he is located on a fine farm with well developed land, beautiful home, excellent and practical farm buildings and where he expects to pass many happy prosperous years Mr Kiesel is a Hoosier, born in Gibson county, Indiana, August 30, 1881, the son of Matthew and Lena (Whitman) Kiesel The mother was born in France and though a devoted wife and mother lived to spend but a few short years with her family as she died when quite young Matthew Kiesel Sr, was born in Indiana where he was reared and educated Upon reaching manhood's estate he engaged in farming, owning a fine, well developed tract of one hundred and twenty acres of land where he successfully conducted general farming operations for many years Today he is a sturdy, vigorous old man of seventy years He is a staunch supporter of the principles of the Democratic party and liberal supporter and member of the Catholic church, a faith in which he was reared from childhood

Michael Keisel, Jr, availed himself of the public school advantages afforded in his native state, by which he qualified himself for use-

ful citizenship, and such public service as he is called upon to perform His life occupation of his own choosing was farming, in which he has made a striking success His first practical work of this nature was as a boy on the old home place in Indiana, where he helped as much as his years and strength permitted, thus gaining a practical education along with theoretical studies in school Indiana was, however, well settled and there was little land available for the younger generation As the young man was a wide reader he learned of the opportunities afforded on the great rolling prairies of the middle west and yielded to the call of the open country, coming to Nebraska in 1907 After looking the country over he decided to locate in the panhandle and took up seventy-six acres of relinquishment land in the Mitchell district, Scottsbluff county, on which he has erected excellent buildings, a good farm home, placed the land in an excellent state of cultivation and everything around the farm indicates that the owner is one of the prosperous farmers of the county Mr Kiesel is modern in his methods he believes that the day of the open range is over and that the future meat producers will be the small farmer who specializes in thoroughbred stock, with this idea in mind he has made a specialty of raising nothing but pure breds His choice has been Holstein Friesian cattle Shropshire sheep, Poland-China hogs and barred Plymouth Rock chickens, all of which have a wide reputation for their uniform standard of excellence and have been widely distributed over western Nebraska and the surrounding states, as Mr Kiesel holds a public sale almost every year to which buyers come from all over the northwestern section of the country He takes great pride in the many blue ribbons won by his fine stock and chickens at the various county fairs where he has become a well known exhibitor

Independent in his ideas and methods as well as a far sighted business man it is but natural that he should follow along these lines in other matters and is an Independent politically, knowing no party lines when a good man is running for office, as he wants the best man to serve the people of the community He is a member of the Catholic church, the faith in which his ancestors were reared

June 17, 1913 is a day marked in his life, for on it was solemnized his marriage with Miss Maud L Kesler, and to them have been born three happy children Sylvester, Agatha and Ruth

JESSE FRANKLIN ENLOW, who is one of the representative citizens of Scottsbluff county, living in the Michell district, where he is a landowner of well known prominence, holds a unique position in the annals of this great commonwealth, as he was the first man to conduct a dairy in the capitol city of Nebraska. Today he is a worthy representative of the agricultural interest of the county, having been engaged in farming pursuits in this locality for nearly a quarter of a century, and so may be regarded as one of the pioneers who has played his part in the vital drama that has turned this section of Nebraska into a paradise for the homeseeker, developing the unbroken prairie into rich farms dotted with thriving communities. Mr. Enlow is a Hoosier, born in Indiana October 7, 1869, the son of James H. and Minerva (Hardsaw) Enlow, the former a native of the far famed Blue Grass state and the latter of Indiana. Both of them died in the prime of life at fifty years of age. James Enlow was a farm boy by vocation but a stone mason by trade, who divided his time between his land and business, in which he was markedly successful. A man of high standing in the community, well educated and read, his advice was sought on many matters of importance by friends and acquaintances. He was a hearty supporter of the Republican party and a member of the Methodist Episcopal church of which his family were also communicants.

Jesse Enlow was left an orphan at the age of twelve years and made his home with a family in that locality. He received his educational advantages in the public schools of his native state and graduated from the high school at Valley City, Indiana. He became a farmer but the business openings afforded in Indiana did not satsfy him, and with the idea of gaining a broader field he came to Nebraska in 1890. The same year he entered the dairying field in Lincoln. The capitol of the state was not then the thriving city of today; the population was small and when Mr. Enlow opened a modern up-to-date dairy it was an event of moment to the inhabitants who heretofore had never enjoyed such cosmopolitan a service. Subsequently there were other men engaged in the same business but he was the pioneer.

The life in Lincoln proved unsatisfactory, as it was so confining to a man used to the open, and he decided to take up farming in the western part of the state where homesteads were still obtainable and came out to Scotts-

bluff county, homesteading in the Tub Spring district, but later sold his relinquishment and purchased one hundred and sixty acres of land on section 16, and the remainder of this school section he operates under a lease. Part of his land is already under irrigation and it is but a question of time until more will be under ditch.

Here he has established his home, erected a comfortable house, fine farm buildings and now is the prossessor of fine improved farm land. From the first Mr. Enlow seemed to see into the future; that the profits were to go to the man who handled thorough-bred stock, so has specialized in blooded Hereford cattle, Duroc Jersey hogs, and also makes a business of feeding range cattle for market. His farm is out of the general run, being characterized by his own individuality, which makes it one of the most prosperous and interesting in the whole panhandle region. He planned all his improvements and that they are exceptionally fine, attest to his ability. He has set out a fine large orchard with many trees already in high state of production. It goes without saying that such a man has a complete line of farm and orchard equipment, with latest designed machinery. Mr. Enlow stands for the epitome of progressiveness in this section and is a worthy example that younger men in agricultural pursuits would do well to follow if they would journey rapidly along the highway to success. He is an independent voter and a keen student of political affairs from those of his immediate community up to ones of national scope. He is a sturdy supporter of the public school for he says that all his life he has been using the knowledge that he learned in both elementary and high school and that every boy and girl ought to count the diploma from the high school as a white milestone in life. Mr. Enlow married Miss Alice Lonsdale of Lnicoln, whose parents are deceased. She is a gracious and charming woman with a great big heart, as she gives a home to two of her nephews, George and Sterling, who are sturdy youths in whom she takes great pride. They have one daughter, Dorothy Elizabeth, aged five years, who at the age of twenty-three months won second prize, scoring ninety-seven and a half, at Better Babies contest at the State Fair in Lincoln, in 1916. Mr. and Mrs. Enlow are splendid neighbors on whom no one calls in vain at time of stress and trouble, and they enjoy the love and confidence of a large circle of friends.

WILLIAM T EVANS — Although variously identified with affairs in Scottsbluff county since his arrival here more than a decade ago, it is probably as a county agricultural agent, he will be longest and gratefully remembered More and more is it demonstrated that a cultivated mind and fine instincts reach their highest development oftentimes amidst agricultural surroundings, diffusing around them that refinement and peace that are the hallmarks of the born student To such a class belongs Mr Evans, who is now one of the land owners of Scottsbluff county, but who for years has been one of the most prominent men in civic affairs

He was born in Adams county, Iowa, February 4, 1883, the son of F E and Eva L (Roberts) Evans The father was a native of Wisconsin and the mother of Illinois, both came from a long line of sturdy eastern stock, as the earlier members of the respective families were pioneers in the middle west Both the parents are still living on their fine Iowa farm, the father being sixty-three years of age and the mother sixty All the family are hale and hearty in old age, living out the full Biblical span of "three score years and ten," for as a boy Mr Evans had eight living grand and great-grand parents while two of his children had seven grandparents and the others have five

The country school of Iowa furnished William Evans his early educational training, supplemented with the practical work a boy learns on the home farm He spent his youth and early manhood in Iowa, but soon after attaining his majority he determined to establish himself independently in business Looking the country over Mr Evans decided that there was the greatest future in the irrigated lands of the middle west where crops are always assumed with the plentiful supply of water and never failing sunshine of the rolling prairie lands With this in view he located in section 30 township 23-55, not far from Mitchell, in 1906, a young man of twenty-three with all the future before him, filled with optimism and confident that with hard work, study of climatic and crop conditions dame fortune would smile upon his efforts and he was not mistaken, for in thirteen years he has won a fine farm, good business and is considered one of the most representative and progressive agriculturists of this up-to-date farming community The improvements on his land are of the latest, he has a good home, adequate farm buildings with modern equipment, all of which are indications that he is a capable farmer, good citizen and progressive in business Mr Evans carries on general farming and stockraising, specializing in Duroc Jersey hogs That he stands high in the community is attested by the positions of honor and trust that have been confided to him by his friends and acquaintances, for he is director of the school board, secretary of the District Farmers Union, secretary of the Farmers' Union Local, and some three years ago was commissioned county agent of Scottsbluff by Governor Neville All the offices which take so much of his time have been filled to the great satisfaction of the community which he serves

Well educated, and a thinker who keeps abreast of all present day movements whether commercial or political, Mr Evans is an independent voter on both local and national questions, voting as his conscience dictates and for the best man for office He is a member of the Methodist church, to which he is a liberal donator

In Furnas county, Nebraska, on the 25th of December, 1910, Mr Evans married Pearl Converse, also a native of Iowa, and they have four children Ilda, Dale, Lura and Ronald, all residing with their parents

WILLIAM OTTE, whose well improved farm is situated on section 22, township 22-54, is one of the substantial men of Scottsbluff county He was born at New Bremen, Ohio, November 27, 1868, a son of William and Lizzie (Sollman) Otte The father was born in Ohio and spent his life there, dying at the age of seventy years The mother was born in Indiana but moved to Ohio with her parents when twelve years old

William Otte grew up on his father's farm in Ohio and attended the country schools In 1888 he came to Nebraska, locating at first near Talmadge and for some time worked for farmers in Otoe county He remembers an early experience in freighting, when he hauled posts to Alliance for ten cents each In 1891 Mr Otte came to Scottsbluff county and homesteaded and this property he still owns and has added to until now he has an entire section, which, through hard work has been well developed His buildings are substantial, his fields are cultivated with modern machinery, and his stock is standard He has been industrious and saving and has something to show for his years of labor

Mr Otte was married to Miss Mattie Schuyler whose parents came from Pennsylvania to Nebraska and located at Burwell in Garfield county Both are now deceased Mr and

Mrs Otte have three children Belle Clifton and Wilma Mr Otte has never been very active in politics and has never been a candidate for any office He is a good citizen, however, takes interest in the public schools, good roads and other general subjects, contributes his share to public enterprises, and may well be called a representative citizen of his county where he stands well with his neighbors

FRANK LINCOLN LOGAN — Though he is yet many years from the psalmist's "three score years and ten," and still possessed of his full amount of physical and mental vigor, Mr Logan has the enviable distinction of being one of the first permanent residents of Scottsbluff county, arriving here in 1890, and thus his memory compasses the entire gamut that has been run in the development of this section of Nebraska from a prairie wilderness to a populous and opulent district of this great commonwealth, and it is gratifying to him to have been able to play a part in the civic and industrial progress and upbuilding of the county

Frank Logan is descended from staunch Pennsylvania stock, as his ancestors located in the Keystone state at an early day He was born in Lee county, Iowa, September 4, 1865, the son of H R and Catherine C (McFarlane) Logan, both of whom were natives of Pennsylvania, the mother lived to the age of forty-five years, passing away in Iowa in 1885, while the father survived to be sixty-four years of age, passing away in 1906 He was a farmer in Washington county, Pennsylvania who came to Iowa at an early day to take advantage of the fine farming land in that state, and there he passed the remainder of his days Both the parents were members and supporters of the Presbyterian church and the father was a stalwart member of the Republican party

The subject of this review spent his boyhood days on his father's farm in Iowa, acquiring his early education in the public schools of his district and at the same time learning the practical business of farming from experience on the land in Iowa, where he remained some years after attaining his majority, but the country was getting well settled up around the home farm and he decided to establish himself farther west where land could be obtained by homesteading Accordingly he came to Nebraska in 1887 locating at York Then in 1890 locating in Scottsbluff county, in section 22, township 23-56, where he took a homestead and timber claim

which has never passed out of his ownership Mr Logan had a little capital, composed of $1,000 earned and saved in three years while at York, when he came to this section, combined with a sturdy determination to succeed, a healthy body and mental ability of a high order and these have prove enough for him to make a fortune, for from time to time he has purchased more land until today he is one of the largest holders of real estate in Scottsbluff county, as he now owns 2,850 acres, four hundred and seventy of which are under irrigation He has been a resident since the time when the only houses in this section were sod, half dug outs, and has watched with the eye of a proprietor the various changes that have been wrought by the passage of years and the sturdy and progressive work of the settlers, and has himself borne a full share in the labor and development, for all the improvements on his large holdings are the result of his own brain and muscle, all the trees on his property he set himself and he has literally made the "desert bloom like a rose" So that the virgin earth has become a fruitful mother to him and his Mr Logan was a man of foresight, thrift and diligence and with such qualifications it is but natural that his accomplishments have been of an unusual order They have won for him fortune, and the esteem and respect which are accorded him by his friends and associates

In addition to his own property Mr Logan leases three quarter sections of school land, he has fine buildings on his property, a beautiful home, the latest farm equipment and uses modern methods, having long been established as a progressive and skilled farmer who thoroughly knows his business He carries on general farming and stock-raising and his success attests to the soundness of his management and methods

Mr Logan was first married to Miss Bertha Akers, a native of Colorado, in 1895, she was the daughter of William Akers of Alliance, who was engaged in the land office of that town She died in 1900, and he married a second time in 1908, Mrs Ruth Etchison, the daughter of James Roberts There were ten children in the Logan family Emma, who died in infacy, Addie, the wife of J P Braden, of Arcadia, Nebraska, John, a farmer of Morrill, Frank, the subject of this review, Harry, now living in Iowa, Samuel, a farmer in Kansas, James, a resident of Wisconsin, where he is a school teacher, Alex, an Iowa farmer, Emmet, in the hardware business in Morrill, and Cora the wife of Harry Morris,

a lumber dealer of Morrill In politics Mr
Logan is a Republican

ALBERT ERNEST CURTIS — The man
whose life history these lines relate lives in
the Mitchell district, where the soil is productive and where the inhabitants are among
the finest and best people in the country From
the time of his arrival in Scottsbluff county
in 1906, to the present, Albert Curtis has been
demonstrating the possession of qualities of
perseverence, industry, and good citizenship,
which have combined to win him personal
success as an agriculturist and the esteem and
friendship of those among whom he has lived
and with whom he has been associated This
enterprising and energetic farmer and stock-
raiser of section 31, township 23-55, is a Wolverine by birth, having been born in Branch
county, Michigan, November 15, 1867, of fine
colonial stock, as his parents were Henry R
and Anna (Hepler) Curtis, the former a native
of New York state, born there in 1834, while
the mother was born in Pennsylvania in 1843
The father was a farmer who also engaged
in a meat and butcher business in Nebraska
He heard of the wonderful opportunities afforded men willing to win farms from the
prairies, and desiring greater advantages for
his growing family decided to locate in the new
country being opened up west of the Missouri
He removed to Nebraska, locating in Polk
county, where he purchased a relinquishment,
proved up on it, and made excellent improvements on his land for that day He soon was
actively engaged in general farming, which,
due to his thrift, hard work and perseverence
proved a most satisfactory investment and
there reaping the harvest of early endeavor,
he passed the remainder of his life

There were ten children in the Curtis family, four of whom are living Matilda, the
wife of William Root lives in Missouri, Nettie, who married Henry S Gerard, Viva, the
wife of Ernest Rogers, and Albert, the subject of this review The children grew up on
the prairie homestead, attending the district
schools such as were afforded in their day and
at home developing into fine men and women
by the assistance they rendered their parents
The father was a staunch supporter of the
Republican party and at one time was constable of Butler county, Nebraska, but he cared
little for the turmoil of political life though
he took active and interested part in all civic
movements for the uplift and improvement
of his community, state and nation He was

one of the leading members of the United
Brethren church and was known as a man
whose word was as good as his bond, while
his deeds were worthy of emulation by the
younger men who could have the benefits
of his precepts

Albert Curtis came to Scottsbluff county in
1906, bought a relinquishment of one hundred and sixty acres of land and today is the
owner of one hundred and twenty acres of
choice farming property This land at that
time was not in good condition, but he set
about remedying this defect, and he now has
one of the fertile valuable tracts that go to
make this section of the state one of the garden spots of Nebraska He is engaged in general farming, being equally at home in all
branches, and has made what may be considered a great success in his chosen vocation,
for he keeps only high grade stock and today
enjoys the benefit of the most modern equipment procurable on the market

In 1902, Mr Curtis married Miss Epsy
Harper, a native of Ohio, and they have two
children Jessie Blanch, the wife of Harley
Abbott, a farmer in Minnesota, and Minnie
Hazel, who is a trained nurse in the Edmonson
hospital at Council Bluffs Mr Curtis chose
for his second wife, Emerett T Banning, a
native of Wisconsin, a daughter of Newell
Bartlett and Cora E (McKeen) Banning He
born in Connecticut and she in New York On
the maternal side, one Sarah Miller was a
passenger on the Mayflower She married
Thomas McKeen who was a signer of the
Declaration of Independence

Newell Bartlett Banning was a descendent of Lady Cheswick, who was lady in waiting to the queen of England She married
Captain Ranson She died at the age of
twenty-seven years, leaving one daughter who
became the wife of Theopolis Banning, parents of Newell Bartlett Banning

Being a well educated man it but follows
that Mr Curtis is deeply interested in all civic movements and educational matters pertaining to the welfare of his community and
for three years served as director of school
district number 42 He is independent in his
politics, casting his vote as his judgment dictates He and his wife are members of the
Congregational church Mr Curtis enjoys the
respect of a large circle of friends and is
really a progressive son of the great west who
has exhibited the sterling qualities and characteristics of the hardy race from which he
springs

JOHN B DOUGLAS, is a well known citizen of Scottsbluff county who has given to farming the careful management that insures great returns for his labors Mr Douglas was born in Poweshiek county, Iowa, November 3, 1878, the son of Andrew and Julia (Timmins) Douglas, he born in Scotland, she born in Pennsylvania The father came to America when a lad of thirteen locating first in Vermont, then in Illinois, where he obtained excellent educational advantages to supplemen the schooling he had received in his native land After reaching manhood he removed to Iowa where he purchased land, conducting general farming operations for a number of years, at the same time engaging to a large extent in stock-raising Subsequently he went to Colorado where he became engaged in general truck farming, but the mountain country did not appeal to him as did the rolling prairies of his youth and disposing of his business he came to Nebraska to settle in the Mitchell district where he lived but five years before passing away at the age of seventy-three His wife died at the age of sixty-nine Mr Douglas had responded to the president's call for troops during the Civil War and served in the army during the last two years of that memorable conflict on the side of the Union, as an agent of the secret service He was a member of the Grand Army of the Republic, belonged to the Methodist church and his political affiliations were with the Republican party There are five children in the Douglas family Maud, the wife of Joseph Harden, is deceased, George, a farmer in Iowa, John B , Jessie, who died in childhood, and Myrtle, who married Harry Ashbrook In Iowa the youthful years of John Douglas were passed During the summer he was employed on the farm while the winter months he attended the country school, where he received a fundamental education that has served him well in his subsequent business life His life was not that of the twentieth century boy, for his chores were not done before sundown and he did not have the use of the family automobile after supper to go to town or visit his young friends In fact the farms of his day were operated on the eight hours plan eight hours in the forenoon and another eight hours in the afternoon, so while still a youth he had by experience fitted himself for farming in an independent way He remained in Iowa until his twenty-eighth year when he decided to try his fortune in western Nebraska locating in Scottsbluff county in 1909, where he purchased a relinquishment, proved

up on it and was the owner of a fine farm He immediately began practical and excellent improvements He was wise in taking up farming as all his training had been along that line and success crowned his efforts from the first Since his arrival he has greatly enlarged his holdings and is now the owner of seven hundred and twenty acres, all in a high state of cultivation, seventy-four being under irrigation and it is but a question of time until the whole tract will be under ditch, while there are fine substantial farm buildings, a good farm home and modern equipment Mr Douglas is fortunately the possessor of just those qualities which are essential to success in the business of farming, and having had considerable experience in this field of endeavor, he is accounted one of the able and progressive men of his community. He has not been active in public affairs save as a good citizen, devoting his time and energy to the exacting cares of farming and stock-raising, and that the time has been well spent needs no mention when we learn of his extensive lands and comfortable fortune won from the soil in such a few years In politics Mr. Douglas is a supporter of the Republican party while his fraternal relations are with the Yeomen

In 1897 was solemnized his marriage with Miss Bertha Kimbley, a native of Iowa, and they have five children Claude, Harold, Warren, Hazel and Leo, all of whom are at home Both Mr and Mrs Douglas enjoy a large circle of friends who delight in the good fortune that has come to them in their western home, and Scottsbluff county can well be proud of such a citizen

WILLIAM T SMITH — If you are familiar with the Mitchell valley, you have often heard the name of Smith There the subject of this sketch, a prominent farmer, operates one of the most up-to-date farms He exercises an energy and skill which put him well in the front among the food-raisers and producers of the county Self acquired prosperity, liberal ideas, ideals expressed in promoting agriculture, education and simplicity of living as well as unquestioned public and private integrity, constitute the fundamentals upon which rests the structure of his life and a firm foundation they have proved

William Smith is a son of the Old Dominion, and reflects the high ideas that have flourished in that state from its first settlements He was born in Virginia February 19, 1874, the son of W H and Callie (Boone) Smith, the former born there in 1844 and the mother at a somewhat later date By trade the father was a shoemaker while in the eastern states though he was also the owner of a small tract of land which he cultivated He was a man of vision and with a growing family on his hands decided that the only way to give them the proper start in life was to go west where there were greater opportunities both for himself and his family The family broke all the old home ties and started across the country for Colorado where the first new home was established, subsequently the Smiths came to Scottsbluff county, taking up a homestead in section 6, township 23-55 The farm homestead consisted of eighty acres of land on which the father still resides with his son William, who was the seventh child in a family of eight children The others were Mary, the wife of J D Whitworth now deceased, Minnie, the wife of William T W Smith, a farmer in Colorado, Effie, who married J V Striker, a farmer of Sioux county, Laura M, who resides in California, Callie M, married B Kirks and also resides in California, Pearl, married Clyde Elliott, a farmer near Mitchell, and E W, a farmer of Sioux county

William received an excellent elementary education in the public schools of Virginia, which has proved of great benefit to him in business since he became independent He has lived in this section for a decade and during that time has won an enviable place in the esteem and hearts of his many friends and acquaintances He became a resident of the Mitchell district in 1907, when he purchased a relinquishment of seventy-three acres He at once began the development of his land, installed modern improvements of an attractive and useful character, erected substantial buildings, raised his land to a high state of cultivation, and as a result soon accumulated surplus capital which he invested in more land, this plan he has continued until today he owns one hundred and fifty acres of the finest farming land in the district known far and wide for its thriving agricultural properties Nearly all the land is under irrigation and all will be within a short time Mr Smith has engaged in intensive farming, which has proved markedly successful for he is insured of fine crops with irrigation and wonderful climatic conditions of Scottsbluff county where sunshine is abundant throughout the year He is a student of his vocation and as such has attained results far in advance of men who

have not believed or adopted modern agricultural methods

1906 marks a happy years in the Smith home for it was then that Mr Smith was united in marriage with Miss Fannie Dillon, a daughter of the Old Dominion, who has brought to her western home many of the beautiful customs and traditions of Virginia Seven children have come to this happy family Ralph, and Ethel V, are at home, Irene, is deceased, Howard, and Helen, twins, Myrtle and Zelpha also are still members of the family circle

Mr Smith is a member of the Democratic party and is proud of the record it made during the war, he is not an office seeker but has always displayed a lively and intelligent interest in local, state and national affairs His fraternal affiliations are with the Modern Woodmen of America while he and Mrs Smith are helpful members of the Methodist Episcopal church

CYRUS H GODBEY —Among the prominent and progressive farmers and stock-raisers of Scottsbluff county, there are found many who make a specialty of certain departments of agricultural work, believing that in this way they reap the greatest amount of success from their labors, in that they are able to centralize their energies and attention upon one definite thing In this class is found Cyrus Godbey, of the Mitchell valley, who, while he follows general farming to a certain extent, has for some years devoted his time to specialization in well bred stock He is accounted one of the energetic and progressive men of his community and is among the early settlers in this locality he has won the confidence and respect of his business associates and friends

Cyrus Godbey was born in Mahaska county Iowa, December 10, 1870, the son of W M and Engeby (Ryan) Godbey The father was a Hoosier born in Indiana, who lived out the psalmist's "three score years and ten," as he passed away in 1913 aged eighty-four years, while the mother like her son was a native of Iowa, who was sixty-four when she died in 1909 For the last twelve years of their lives the parents made their home with their son C H Godbey

During the early part of his young manhood the father was a Whig, but later in life became an independent voter, casting his ballot regardless of party lines and using his influence for the man he believed best fitted to hold office regardless of whether he ran on a Democratic or Republican ticket.

The public schools of Buffalo county, Nebraska furnished Cyrus Godbey with his early educational training, and during the summer months he helped his father and brothers in the fields, while being trained in all the arts and methods of agriculture as practiced in that locality. His choice of an occupation when he reached mature years rested upon farming, and this he has followed in a methodical, careful and practical way, adopting modern methods whenever they have been proved to produce better results. Mr. Godbey remained in business in Buffalo county until 1900 when he determined to avail himself of the opportunity to secure land suitable for irrigation farther west and located in Scottsbluff county in the Mitchell valley in section 31, township 23-56, where he took up a homestead of one hundred and sixty acres, proved up on it and established his permanent home. When he first came to this locality the land was as yet virgin prairie and he turned the original sod to make it fertile and available to put in the seed for his first crop. Today Mr. Godbey conducts a general stock-raising industry in connection with his agricultural pursuits. From these operations he has been able to equip his farm with splendid improvements and contribute to all the war time days through which the country has so lately passed. He and his wife are widely known and highly rated in the community. Their religious affiliations are with the Methodist Episcopal church of which Mrs. Godbey is a member. Mr. Godbey is a member of the Farmers' Union and a stockholder in the Farmers' Union store at Mitchell, while his fraternal affiliations are with the Independent Order of Odd Fellows, the Modern Woodmen, the Rebeccas and he is of high standing in the Masonic fraternity, being a Thirty-second degree Mason. Mr. Godbey married, January 26, 1898, Miss Annie L. Hicks, a native of Virginia, a daughter of J. D. and Cornelia A. (Gibson) Hicks, both natives of Virginia. The father is living but the mother is deceased. There are three children in the family: James S., who was born in 1899; Elta, fourteen years of age, and Nettie in her eighth year, all of whom are at home, and their father says that every one of them is to be given advantage of every educational facility afforded by the district, county and state, as he is determined that they all shall have every avantage possible for success in life. Mr. Godbey is well informed on all social questions and current events, thinks for himself and exercises his franchise as his conscience dictates in local affairs.

WILLIAM JOSEPH SCHUMACHER.—

The Schumacher farm in the central part of Scottsbluff county, may be said to constitute one of the landmarks on which may be found evidence of almost every phase that has marked the progress of agricultural industry in this section during the past quarter of a century. This fine landed estate now comprises two thousand, one hundred and twenty acres — considerably over three sections — and includes the old homestead and family home on which the present owner located when he came to Nebraska. In addition to fulfilling its mission as a medium of financial profit, this farm property has been developed to a high state in which it compares most favorably with any other Scottsbluff county landed estate likewise accumulated through pioneer courage and determination. Some of the land marks the new era of agricultural activity in western Nebraska as two hundred acres are under irrigation, and it is the irrigated land that can be relied upon, year in and year out, to produce a big crop, for with the never failing sunshine of the middle west, assured water, properly cultivated soil, the seeds planted bring bountiful returns.

William Schmacher is a Canadian by birth, born in Grey county, Canada, March 3, 1876, the son of Martin and Mary C. (Wakeford) Schumacher, both Canadians, the former lived to be seventy-three years of age while the mother was seventy years old June 11, 1919, so that Mr. Schumacher may be said to come from a sturdy, long lived family. The father was a small landholder in his native country, who was engaged in general farming, which he conducted with considerable success. There were eight children in the family: Alexander, a cabinet maker in Canada; Mary Ann, the wife of Michael Schiestel; W. J., the subject of this review; George T., a farmer near Blackfoot, Idaho; Margaret, the wife of Philo Gallup; Angeline, who married William Preston, a ranchman of Montana; Walter, living on a farm in Banner county, and Sarah, the wife of John McCumpsey of Scottsbluff county.

W. J. Schumacher received in his youth the advantages of the schools of his native land, and he early learned the lessons of practical toil. As he grew to maturity he selected the vocation of farming as a life work. Determined to avail himself of the greater advantages that the irrigated land in the United States offered to the young men of ambition, in 1899, accompanied by his wife, he immigrated to this country and soon was located in Scottsbluff county. Like many another pioneer they settled on a homestead of one hundred and sixty acres in section 6, 22-56 township, after spending a period at Kimball, Nebraska, earlier in the year. A young man of twenty-three, Mr. Schumacher and his wife were not daunted by the great task before them. Locating amidst primitive surroundings, he and his devoted wife bravely fortified themselves for the trials and hardships that still were to be overcome in the Mitchell valley and though not as unfortunately placed as the settler of the seventies and eighties, they had to overcome many obstacles during the first years, but did not falter in courage, persistence or self reliance, with the result that they gradually made their way forward to the position of success and definite prosperity. As returns from his vigorous activities as an agriculturist and stock-raiser justified such action, Mr. Schumacher began adding to his original holdings until he has accumulated a section of fertile land, but that was not yet the goal of his desires and as the years passed he continued to add to his landed estate until today the Schumacher property is one of the largest tracts in the Panhandle, which is one of the most favored sections of Nebraska, commonly known as "The Garden Spot" of the west. From the first Mr. Schumacher erected good buildings on his farm and made other excellent improvements of a permanent order, and through his individual achievement, as well as his civil loyalty and liberality, he contributed his full share to the development and progress of his community and the county in general. Aside from the management of his farm properties Mr. Schumacher takes his share in the civic improvements of the county which are so closely related to its commercial success, for he is the superintendent of the Mitchell Irrigation District which supplies water to the farms of the Mitchell valley, the business and policies of the district have been well directed in his competent hands.

In politics, Mr. Schumacher is found aligned with the republican party, and while he has no desire for the honors of public office, he gives efficient service to the public as a patriotic citizen should. Fraternally he is connected with the Independent Order of Odd Fellows and is a Mason of high standing, having taken his Thirty-second degree.

In 1898 Mr. Schumacher married, in his native land, Miss Sarah Yeo, a Canadian. She died in Scottsbluff county, leaving a daughter, Mary Belle, the wife of Purl Campbell, a farmer of Scottsbluff and for his second wife he married Mrs. Fannie Springer, a native of Nebraska. Mrs. Schumacher has a daughter by her former marriage, Retta, the wife of Glen Foreman.

HUGO PIEPER has identified himself most fully with the civic and material interests of Scottsbluff county, being one of the well known farmers and representative agriculturists living near Mitchell. He is not only a native son of the west but also of the Sunflower State, and has exemplified its progressive spirit in the varied activities that have brought to him a generous share of temporal prosperity. His parents, J. T. Pieper and Bertha (Yellick) Pieper, claim Germany as the land of their nativity, they are respectively fifty-nine and fifty-seven years of age. The father received his early educational advantages in his native country but being ambitious and with a great desire to gain more of the advantages obtainable in the New World, he immigrated to America when a young man of twenty-one years. Soon after reaching the shores of this land of freedom and promise he located at Hanover, Kansas, to become engaged in agricultural pursuits, to which he had been trained in the old country. He became the owner of land, which by industry and perseverance he brought to a high state of cultivation and from which he gained a comfortable fortune and now is spending the sunset years of his life in well earned enjoyment and ease. Mr. Pieper has ever taken a keen interest in civic affairs that concerned the welfare of the community in which he has chosen to make his home for more than a quarter of a century, and though he has never aspired to public office, having devoted his energies to the cares of his business, he did consent to accept the duties of coroner of his county, well filling this position for eight years. Mr. Pieper thinks for himself but is a supporter of the Democratic party in nation wide questions. Brought up in the faith of the Roman Catholic church he is a devout member of the congregation.

There were five children in the Pieper family, of which Hugo is the second oldest. The others were T. H. Pieper, now thirty-one years of age, Hugo, Emil twenty-six, on a farm, Hedwig, who is working at home in the bank and Harry, twenty years of age, now employed on a farm in Scottsbluff county.

Hugo, as stated before, the second oldest member of the family, passed the period of his childhood and early youth under the sturdy and invigorating discipline of the home farm The public schools of the Sunflower state afforded to Hugo his early educational advantages following which he was associated with his father in agricultural business until 1912, when he determined to become the master of his own fortunes, and came to Scottsbluff county, purchasing eighty acres of fine arable land with the water rights of the same, or as expressed in this locality "under ditch " He he began vigorously the agricultural and live stock enterprise that has brought to him ever increasing success with the passing years, and as his finances were augumented he began to study on the subject of modern intensive farming with the idea of gaining the greatest returns possible from the soil He is a resourceful and progressive executive, being one of the first men of the community to use tractors for motive power in place of horses, and the increased production has justified this radical change in farm management

While devoting his greatest energy to the direction of his business Mr Pieper has not been unmindful of civic responsibilities He is a member of the Roman Catholic church in which he was reared, being generous in its support In a basic way he gives support to the Democratic party but in purely local affairs believes in electing the man best fitted for the office , his fraternal connections are with the Central States Yeomen, the C M B A and the Union Accident Society In 1915, Mr Pieper married Miss Clara Feckley, who was born in Knox county, Iowa, the daughter of Doris Feckley of Missouri They have three children Irene, Gladys and Lawrence, all of whom are at home

T H PIEPER — As one of the exponents of most modern and scientific policies as applied to farm industry, Mr Pieper stands forth prominently as one of the distinctly representative and influential agriculturists and stock growers of Scottsbluff county He is a member of a sterling transplanted German family that came to America to take advantage of the great opportunities that would be afforded their children in a new country and by his own energy and well directed policy he has made his way to the goal of success and prosperity Mr Pieper was born in Hanover, Washington county, Kansas, in 1887, the son of J T and Bertha (Yellick) Pieper both of whom were born in Germany and came to the United States thirty-eight years ago In

1881 Kansas was not thickly populated and the sturdy vigorous Germans took up land, nothing daunted by the fact that they must break the virgin prairie to raise their first crops and live in a sod house until better living conditions could be afforded Here the subject of this memoir was reared to manhood under the conditions and influences which marked the initiation of civic and industrial development in the Sunflower State, and thus he was more strongly fortified in mature years to carry on the important work which has made Scottsbluff county one of the opulent and attractive sections of western Nebraska He received his education in the public schools of the community in which his youthful years were passed, at the same time assisting to reclaim a frontier farm, remaining to assist his father until reaching his legal majority Being independent by nature the young man determined to engage in agricultural pursuits for himself, believing that this vocation was one best adapted to his tastes and in which he saw a great future He believed in the great basic industries of agriculture and stock-raising, which have yielded substantial returns to this man of enterprise and good judgment When twenty-seven years of age Mr Pieper came to Nebraska, locating on land owned by his father in section 3, township 22-56, in 1914, where he has been a large feeder of hogs and sheep, and the prosperity that has attended his efforts, mark him as one of the vigorous and resourceful farmers of the precinct Mr Pieper's father bought this land about seven years ago at $125 an acre, he at first assisted his son in its development and today the men would refuse $300 an acre for it, showing that they were foresighted in their investment and well deserve the material prosperity which today crowns their achievment In 1910, Mr Pieper married Miss Elizabeth Schneiderjans, a native of Kansas, the daughter of a farmer who owns a fine quarter section of land in that state Two children have been born to this union Lucille, aged seven and Dorothy, a charming child of three

Mr Pieper is a member of the Democratic party and though he is a man who enjoys the respect and good will of his friends and associates he has manifested no ambition for public office devoting his time strictly to business while always interested in any question pertaining to the uplifting of the community He is a member of the Roman Catholic church and a liberal contributor to its support, while his fraternal affiliations are with the Yeomen and the C M B A

R. O. CHAMBERS

ROBERT O CHAMBERS — Without its men of business enterprise no community could make much progress, and one of the leading factors in development is a good, live newspaper, particularly devoted to local interests Minatare, Nebraska, one of Scottsbluff county's beautiful and prosperous little cities, may owe much to its leading journal the Minatare *Free Press*, which is owned and ably edited by Robert O Chambers, long well known in the educational field

R O Chambers was born at Sidney, Nebraska, May 8, 1889, a son of Chas P and Susan (Sanderson) Chambers, natives of Indiana, who came to Nebraska and settled in old Cheyenne county in 1885 Mr Chambers was reared in that county, attended the public schools and after being graduated from the Sidney high school took a course in the Chadron Normal School From 1906 until 1916 he taught school very acceptably in different sections, but in the latter year bought the *Free Press*, which he has made one of the leading organs of the county It is Democratic in political policy but is mainly devoted to city and county affairs, Mr Chambers being a writer of ability and discretion

In 1912 Mr Chambers was united in marriage to Miss Helen Schroeder, who was born in Colorado, and is a daughter of Frederick W and Minnie (Brockmann) Schroeder and they have four children, namely Robert Frederick, Dorothy, and Glen Mr and Mrs Chambers are members of the Episcopal church He has additional business interests as he is in partnership with Smith Chambers in the real estate line, and at present is serving as clerk of the town board He belongs to the order of Knights of Pythias

H C BRASHEAR — Under the modern system of agricultural and live-stock industry the application of energy and good business policy insures success, and this has been significantly demonstrated by the man whose name heads this brief review H C Brashear is a native of the Keystone state born in Venango county, Pennsylvania, August 3, 1859, the son of R A and Sarah A (Seaton) Brashaer, the former a native of Brownsville, Pennsylvania while the mother was born in Butler county of the same state, and is still living in Harrisburg, Pennsylvania, at the mature age of eighty-three years The father was a noted civil engineer, gained great renown in this profession which he followed the greater part of his life To him was intrusted the first survey of the ground for laying out the Hoosac tunnel in Massachusetts and for

many years he held the important position of chief engineer of the Lake Shore and Michigan Southern Railroad He was a man of exceptionally high standing among men of the engineering profession, was a Mason of high degree, and a stalwart Republican in politics, was loyal and progressive as a citizen and his ability and popularity gave him marked influence in community affairs serving as city councilman of Franklin, Pennsylvania for a number of years There were six children in the family W G, who lives in New York, H C, the subject of this review, F L, who travels for a manufacturing house of Pennsylvania, Lillian, the wife of F B Wolf, lives in Norfolk, Nebraska, Eugenia, the wife of F L Wright of Harrisburg, Pennsylvania, and R A, on a homestead in Montana Belonging to a family of ample means and high education it was but natural that H C Brashear should follow in the footsteps of his father in this matter He received his elementary education in the public schools of his native town, his father seeing that the foundation was well laid This was greatly supplemented by his father's still more valuable information, technical knowledge and experience, which he imparted to his son The young man spent his youth and early manhood in Pennsylvania, but he wished wider fields and knowing of the many opportunities afforded in the west determined to establish himself independently and came to Nebraska in 1886 for that purpose He saw a great future in agricultural industries and stock raising, results today justify his far sighted vision Soon after reaching Nebraska Mr Brashear located in Cheyenne county, homestead one hundred and sixty acres pre-empted one hundred and sixty and purchased one hundred and sixty, an immense amount for one man to handle at that time While he did not share the hardships and vicissitudes of the pioneers of territorial days he had his full share of blizzards, droughts and insect pests to contend with On his fine estate he made the best improvements and soon was extensively engaged in general farm industry, including general agriculture raising of the grains suitable to this climate and altitude stock-raising and feeding He has always held that thorough-bred cattle paid the best and has adhered to this principle in stocking his land and before long all his holdings will be under the government ditch, so that every acre may be irrigated, thus insuring a crop each year This is a great contrast to the condition of the rolling prairies when he first came here as it was then necessary to break the virgin sod in order to put in a crop, today waving fields of grain are seen for miles where in 1886 there was nothing but wild grasses and prairie flowers. Mr Brashear has no communion with apathy or idleness, has been a productive worker and has been found busy at all stages of his career. Essentially a business man, he has had neither time nor desire to enter the turbulence of practical politics or to seek public office, though he is liberal and public spirited in his civic attitude and gives staunch support to the principles of the Republican party and is able to give sound logical reasons for his adherence to its tenets. Widely known throughout this section of the state he has by his earnest endeavors entrenched himself firmly in popular confidence and esteem, and this has contributed to his success as a cattle man and farmer.

Mr Brashear was married in 1884 to Miss Lenora Golden, a native of Pennsylvania, the daughter of John Golden, and two children have been born to this union. R A who is married and lives on one of his father's farms, and J W who still lives at home associated with his father in business.

A C DAVIS — Practical industry wisely and vigorously applied seldom fail of attaining success, and the career of the man whose name heads this review, now one of the progressive farmers of Scottsbluff county, is but added truth of the statement. When he started out in life he had but few advantages to assist him along the road to success but his diligence and judicious management have brought him ample reward for his labors.

Mr Davis was born in Tennessee, June 16, 1876, the son of J A and Margaret (Arrowood) Davis who had eight children John, who lives in Colorado, Laura, also a resident of that state, the subject of this sketch, Florence the wife of A P Jones, lives in Utah, S J, of Morrill, Nebraska, Elizabeth, the wife of Frank B Kelly lives in Colorado, and M F, also a resident of Morrill. When Mr Davis was but a young child his parents removed to Colorado in 1879 where the boy spent his youth and early manhood, attended the excellent public schools of his district and while a little lad began to assume many duties around the home farm. As his age and strength increased he assisted more and more in the labor incident to the operation of a farm and under the guidance of his father learned the best methods of planting and harvesting, so that when he reached manhood he was well equipped with the practical experience to en-

able him to became a farmer on his own account. He started out for himself at an early age, operating land in Colorado for some years before coming to Nebraska. It is nine years since he purchased his present fine, one hundred and fifty acre tract in Scottsbluff county, which since 1910 he has brought to a high state of cultivation and by the erection of suitable buildings he, today, has a very valuable property all of which is under irrigation. General farming and stock-raising form the basis of the enterprise carried on by the owner who makes the proud boast that "There is nearly everything on the farm but weeds and a mortgage." In politics he votes as an independent, while his fraternal associations are with the Independent Order of Odd Fellows and the Woodmen of the World.

In 1907, Mr Davis married Alice Brown, a southern woman born in Alabama, and to them have been born five children Ruth, Arthur, Margaret, Ellen and Mart, all of whom are at home under the careful training and guidance of their parents.

SHERIDAN GUMMERE — The pioneer families of Scottsbluff county who played their parts in the vital drama that has turned this section of Nebraska into a paradise for the homeseeker, developing the unbroken prairie into a garden spot of the earth, where thriving communities have grown up, have reason to hold themselves responsible for much of the present day progressiveness. While many of those who actually experienced the actual hardships of the early days have passed away, there still remain many who, through sheer force of will and energy, brought out of primeval conditions what have become twentieth century actualities. Among these is found in Sheridan Gummere, who was a homesteader of the year 1897. A member of such a family is Sheridan Gummere who came to Scottsbluff county nearly a quarter of a century ago and took up a homestead on the virgin prairie.

He was born in Champaign county, Illinois, June 3, 1870, the son of Jack and Elizabeth Gummere to whom were born ten children William, living in Oklahoma, Sheridan, Lottie, deceased, Anna, the wife of Arthur Draper, lives in Idaho, Otto, who for some years has been in Alaska, Daisy, the wife of Oskar Departee, also lives in Idaho, Ira, a resident of Montana, Nellie, who married Frank Frazer, Leonard, who now lives in the state of Washington, and Ida, who also is married. The father of this sturdy family was a farmer in Illinois who learned of the fine public domain in the middle west and decided to avail himself of a farm with the idea of giving his boys and girls greater opportunities than could be afforded in a more settled country east of the Mississippi river, and with this end in view came to Nebraska in 1897 pre-empted a claim in Scottsbluff county, established a home and engaged in general farming That he met with success goes without saying when we know that today he has retired from active participation in work to enjoy his sunset years in well earned and well deserved ease and comfort

Sheridan Gummere received his education in the public schools, attending during the winter when the exigencies of farm work permitted and while but a lad of thirteen became a practical farmer himself at an age when most youths are thinking more of sports than making a living, for he has supported himself by his own unaided efforts since that period, a thing which few men of this day can boast When the Gummere family came west, Sheridan had already decided to establish himself independently as soon as possible As soon as his age permitted he took up a homestead and began operations as a farmer and stockraiser, and through making the most of his opportunities, working industriously, managing his affars carefully, and applying all his knowledge to his daily labor, he has succeeded in accumulating a great agricultural estate of a thousand and eighty-five acres, so that today he is surrounded by the comforts and conveniences that serve in some measure to compensate him for the numerous hardships which he experienced during his early days in this section He has fine improvements on all his property, excellent farm buildings Mr Gummere is accounted one of the energetic and progressive men of his community and belongs to one of the families which is well known in the county His long experience in agricultural pursuits has made him more or less of an authority in western Nebraska and he is frequently called upon by his associates for counsel and advice Politically he is a Republican but politics and public affairs have had small share in his career but his actions have always shown him to be a public spirited citizen, ready to support good measures and a man who owns such a vast property with five hundred acres under irrigation has ample means to give liberally

In June, 1899, Mr Gummere married Miss Della Pense, and to this union one child was born Mildred, who lived a happy joyous childhood until her thirteenth year when she was taken away, leaving a saddened home and sorrowing parents

For years Mr Gummere has been a leading factor in every important, public-spirited movement for the promulgation of high standards in business circles, intensive and modern methods in farming and his influence is a valuable and valued one

HENRY M SPRINGER — The business career of Henry M Springer has been significantly characterized by courage, progressiveness, as well as by dynamic initiative and executive ability that brings normally in its train a full measure of success His resolute purpose and integrity have begotten the popular confidence and esteem that are so essential in the furtherance of success in the important lines of enterprise along which he has directed his attention and energies, and through the medium of which he has gained secure status as one of the representative figures in the financial circles of western Nebraska and stockmen of the Northwest During practically his entire business career Mr Springer has been closely associated with the live-stock industry and there is needed no further voucher for the precedence he has gained than the statement that today he is vice-president of the First National Bank of Mitchell the oldest banking institution of the city It has a capital stock of $25,000, surplus of $25,000 and deposits of about $727,000 Mr Springer has shown special constructive talent and through his effective policies and efforts he has furthered the success of every financial enterprise with which he has been associated As one of the representative business and stock men and progressive and public spirited citizens of Scottsbluff county he merits specific recognition in this publication

Henry M Springer is a Missourian, born in Sullivan county March 3 1860 the son of E F and Hollie A (Jones) Springer both natives of Illinois and to this union five children were born but two of whom are living The family consisted of Henry, John M, who lives in Goshen county Wyoming William A, deceased Flora the wife of Edgar M Sanders, a resident of North Powder Oregon is deceased The father of the family was a successful farmer in Illinois who removed to Missouri at an early day locating in that state in 1852 where he again followed agricultural pursuits until he felt the call of the great west and in 1879 fitted out a prairie schooner and with the members of his family either in the wagon or on horseback made the long overland trip to Idaho, where he engaged in stock-raising and general farming for five years He then returned to the eastern part of Nebraska, traveling through the Panhandle on the trip but the spirit of the west had entered into his blood and not contented in the more thickly settled sections he again turned his face toward the setting sun and came to Scottsbluff county Here he entered actively into the civic and business life of the community, soon became a figure in public and municipal affairs, gained the confidence of his associates which was demonstrated when they elected him to represent this district in the state legislature Mr Springer was active in the local councils of the Republican party, his fraternal relations were with the Masonic order while he and his wife were members and supporters of the Methodist church In 1910 he passed from life having been a successful business man, loving and devoted father and husband, leaving a sorrowing wife and family Mrs Springer still resides on the old home farm south of Mitchell

Henry Springer attended school in Missouri during his youth, then accompanied the family on their long and interesting wagon trip across the plains to Idaho and today he can recount many of the thrilling and interesting events which happened during the journey While in the west he was engaged to some extent in agricultural pursuits but had a natural love for horses and devoted more of his time to that branch of industry as he was a youthful cowboy, loving the free, open life of the range, even its loneliness and hardships could not dampen his enthusiasm In 1884 he, his father and brothers drove a herd of four hundred head of horses clear from Idaho to Bridgeport, old Camp Clark at that time, where he disposed of them and then determined to locate here permanently, took up a tree claim in 1885, making his home in Wyoming, set out the required trees to secure the land and in 1889 took up his residence He at once engaged actively in agricultural pursuits and stock-raising which has been given his greatest attention and time Soon after coming here he began improving his property, put his previous experience to good use by buying horses to stock his farm and soon developed a prosperous business Mr Springer has for a long term of years been recognized as the most progressive and substantial farmer, stock-raiser, feeder, and shipper in this section of the state When the First National Bank was organized it was but natural that such a man should be chosen to guide the financial policies

of the institution Under his careful and progressive regime the First National Bank has made a wonderful advancement in the volume of its business, it is regarded as the soundest bank in the west, doing a volume of business almost incredible for a banking house for its years While taking an active part in the Republican party Mr Springer has served as county commissioner when the county was young and is again serving in the same office at the present time He is an active member of the Masonic fraternity and belongs to the Eastern Star and Modern Woodmen

In 1882 Mr Springer married Miss Alice Boltenberg, who was born and reared until her marriage in Illinois Their family numbered five, three of whom are living Odessa, married to Edward B Deering, lives in Torrington, Wyoming, Ruby, is the wife of Lon D Merchant of Goshen county, Wyoming, and Vera, who is at home Two children died in infancy

THOMAS H YOUNG, rancher and business man is a Pennsylvanian who transferred his activities to the middle west when this section was still called "The Great American Desert," where he has made a record of which any man might be proud For forty-three busy years Thomas Young has been known in the Panhandle and today he is one of the best known farmers and stock-raisers of this section He has seen wonderful changes of all kinds since first coming here and has done his fair share in the development of the county's agricultural resources and in establishing such necessities of civilization as good roads, schools and churches When a community can claim a majority of such stable and dependable men as Mr Young its permanence and progress are assured

Mr Young was born in the Keystone state July 2, 1860, the son of William and Melissa Jane (Logan) Young, both natives of Pennsylvania To them were born five children but two of whom are living Margaret, deceased, Emma, deceased, Jerry, also deceased, James, who lives at Butler, Pennsylvania, and Thomas, the youngest The father was a carpenter and builder, being regarded as a master workman of his trade in the locality where he lived He was accidentally drowned in 1864 He was a member of the Presbyterian church and in politics a staunch Republican It was a hard struggle for the mother left alone with a growing family on her hands and in 1870 the battle proved too much for this fragile but loving mother and she passed from life leaving her little son to the care of his elder sister and brother-in-law. He attended the public schools in his native state gaining a good rudimental education, but at an early age was forced to begin earning a living, first as a newsboy so that he may be accounted among the self made men of his locality as he started out in life with only the equipment of ambition and ability as capital, and what he has achieved has been by his own unaided efforts. He has reared a family of children all of whom are splendid citizens, prominent in present day affairs, reflecting great credit upon their parents and the home from which they come, also constituting a valued asset to the community of which they are a part.

As a youth in the east Mr. Young read of the life on the western prairies and determined that he would go west to seek and make his fortune and when only seventeen years of age started across the country reaching Fremont in 1876, but a few years after Nebraska Territory had been admitted to statehood. Those were the days of the great cattle barons who owned vast herds of cattle that ranged free over the prairies, and after working on a farm for two years the young man joined a cattle outfit and became a cowboy, in Custer county for the Olive operators and was there when the famous desperados, Mitchell and Ketchum, were hung. For eight years he followed the life of the range, gaining invaluable knowledge of cattle, feeding, buying, and marketing, which induced him to decide upon a ranch for himself, and with this end in view he came to Scottsbluff county in 1886, took up a homestead and today is the owner and manager of his one hundred and fifty acre tract of highly improved land, where he has been engaged in farming and horse-rasing. Mr. Young has not confined his activities entirely to agriculture, but has branched out into other commercial lines, being the first man in the United States to erect a concrete ice house for storage purposes. This initial adventure received country wide notice for its unqualified success, being written up in *Popular Mechanics* magazine and other technical and scientific publications. The United States government sent to him information concerning plans, specifications and materials used in the construction of the building. Mr. Young also gained a small fortune by his keen business qualities, as what he plans, he executes and the success that has come to him is the result of good judgment and years of persistent labor.

In 1884, Mr. Young was married in Grand Island, Hall county, to Miss Ella M. Turpin,

who was a native of Minnesota, who had been reared out on the Pacific coast. Her father was surveyor general of Minnesota, himself and wife both natives of Pennsylvania. There are three children in the family, Rex, who has chosen agriculture as a pursuit and is on one of the home farms, James, who is in business in Denver, Colorado, and Daniel, who also is at home, in the law business. Mr. Young takes an interest in public affairs, keeps himself well informed and generally votes the Democratic ticket, while fraternal affiliations are with the Modern Woodmen.

AMOS ELQUIST, one of America's adopted sons whose energy and well directed efforts have within recent years in Mitchell resulted in the building up and development of a prosperous implement house and the placing of its founder in a position of financial independence. Mr. Elquist was one of the pioneer settlers of Cheyenne county where he was engaged in agricultural pursuits for many years before coming to Scottsbluff county and later locating in the city of Mitchell.

The subject of this review is a native of Sweden, born February 19, 1853, the son of Christian and Anna Elquist, who had a family of four children. Amos, John, Andrew, deceased, and Joahanna, who lives in Sweden. The father was a farmer in the old country, a vocation he followed all his life, passing away in 1909, the mother surviving him but two years. Amos received an excellent education in his native country where the schools are supervised by the government. He attained his majority before severing home ties and embarking for the new world to seek and make his fortune, landing in the United State in 1875. He came west with no definite idea of the country and first located in Roseville, Illinois, obtaining employment in a mine, but he had not come to this golden land of promise to spend his life away from the light and sunshine and determined to come still farther west and obtain land by government grant. With this in mind he came to Nebraska, settling in Cheyenne county on a homestead. He had a cheerful outlook on life, an inherent faith in his own ability to accomplish anything, a faith that no discouragement could dim and it was as well for this carried him through the years of hardship caused by the drought. This self confidence has been more than justified, for today Mr. Elquist is one of the leading business men of Mitchell, the owner of a hardware house and the possessor of a valuable farm which pays substantial dividends.

Mr Elquist is a tireless worker and from the day he became the proprietor of land began its improvement, at first in a primitive way as did all the pioneers, but he never let up, each year saw some new ground broken some new cattle bought or some new building erected so that in less than a decade, from 1887 to 1893, he sold his homestead for $200 an acre, having brought it to a high state of cultivation

The latter year he came to Scottsbluff county, bought land which he still owns on which he for a time carried on farming and stockraising before he opened his present mercantile establishment in Mitchell, where he conducts a general hardware and implement business, in which line he has shown himself equally as good a manager as on the farm He built up a good trade, in the development of which his pleasant personality has played a large part The family belongs to the Federated church Mr Elquist is a Republican in politics, and while he takes an active part in communal affairs has never desired to hold office but is interested in all civic improvements and gives liberally to their support

In 1875 occurred the marriage of Amos Elquist and Miss Anna Andersen, and eight children have joined the family circle Charles, living at Torrington, Wyoming, George, also a resident of that state, Theodore, of Torrington, Alvin, all in the hardware business in that city, Reuben, a soldier who served with the American Expeditionary Force in France, Fred, associated with his father in the implement business in Mitchell, Anna, married Lester Morgan of Livermore, California, and Aneti, married, living in Lingo, Wyoming

JOHN E KEEBAUGH has been a resident of Nebraska since his early childhood and his father was not only a pioneer of this state but had also previously gained pioneer experience in the state of Minnesota, where he was identified with large lumbering operations for several years He whose name initiates this paragraph is the owner and active manager of the Rexall drug store at Mitchell, Scottsbluff county and this has the distinction of being the largest and best equipped establishment of the kind in any town of not more than a thousand population in the entire state In addition to being one of the representative business men and influential citizens of Mitchell Mr Keebaugh is also the owner of a well improved farm property near this village

John E Keebaugh was born in Blue Earth county Minnesota, on the 15th of May, 1868,

and is a son of George W and Sarah (Ward) Keebaugh, the former of whom was born in Ohio and the latter in Illinois, from which latter state they removed to Minnesota in the pioneer period of the history of that commonwealth Of their six children four are living William, resides near Portland, Oregon, Ida, lives in Oklahoma, Mrs Mary Wood, resides in Oregon, and the subject of this sketch was the third in order of birth of these surviving children George W Keebaugh continued to be identified with the lumber industry in Minnesota until 1871, when he came with his family to Nebraska and took up a homestead of one hundred and sixty acres, in Butler county, He reclaimed this land from the virgin prairie and developed it into one of the productive farms of that county He had given valiant service as a soldier of the Union in the Civil war, as a member of Company C Forty-second Minnesota Volunteer Infantry, a regiment comprised of pioneers in the Gopher state In later years Mr Keebaugh vitalized the associations of his military career by maintaining affiliation with the Grand Army of the Republic His political allegiance was given to the Democratic party and he and his wife, both of whom died in the year 1908, were earnest members of the Christian church Their names merit enduring place on the roll of the honored pioneers of Nebraska

John E Keebaugh was about three years old when the family home was established on the pioneer farm in Butler county, Nebraska, and there he was reared to adult age, was afforded the advantages of the public schools of the locality and period and supplemented this educational training by a course in the school of pharmacy of Northwestern University, in the city of Chicago After completing this technical course he returned to Nebraska, and for about three years thereafter he was employed as a drug clerk in Butler county He then opened a drug store of his own, at Shelby, Polk county, and there he remained five years He conducted a drug store at Surprise, Butler county, during the ensuing five years, and he then went to David City, the judicial center of that county, where he was manager of the office and business of a local telephone company in which he had become a stockholder There he remained until 1906 when he removed to Scottsbluff county and took up a homestead, besides which he purchased the Rexall drug store at Mitchell, which well ordered establishment he has since conducted most successfully, as shown by the large and representative patronage which it commands He has been the owner of this drug store since 1911, and in the meanwhile he has made the best improvements on his homestead, which is now one of the valuable farms of the county and which is situated four and one-half miles distant from Mitchell

Loyal and public-spirited as a citizen and a staunch Democrat in politics, Mr Keebaugh has had neither time nor ambition for political office He is affiliated with the Masonic fraternity Mitchell Lodge No 263. Consistory, Scottish Rite and Shrine of Omaha Nebraska

In 1889 was solemnized the marriage of Mr Keebaugh to Miss Myrtle Hamm who was born in the state of Illinois, and of the three children only one is living — Clyde A, who is a field boss for the Great Western Sugar Company

EDWARD H REID is associated in the ownership of about twelve thousand acres of land in western Nebraska and is prominently identified with ranching enterprise upon a most extensive scale. his place of residence being the attractive village of Mitchell, Scottsbluff county, where he is known and valued as a loyal and progressive citizen of marked public spirit

Mr Reid is a scion of one of the sterling pioneer families of the Hawkeye state, and thus his earliest experiences were those gained in the vital and progressive western portion of our vast domain He was born in Page county Iowa, December 20, 1859. and is a son of Joseph A and Margaret (Long) Reid both of whom were born and reared in Greene county, Ohio Joseph A Reid became one of the successful agriculturists and stock-growers of Iowa, where he established his residence in the early pioneer days and where he won substantial prosperity through his well directed endeavors He was one of the honored pioneer citizens of Iowa at the time of his death, in 1906 and his widow there passed away in 1908, both having been active members of the United Presbyterian church They became the parents of seven children James Harvey, is deceased, John Franklin is a resident of Torrington, Wyoming, Anna Laura resides at Colorado Springs, Colorado, Edward H, of this review, was the next in order of birth, Julia Elizabeth, maintains her home at Casper, Wyoming, William Lincoln is a resident of Mileston, western Canada, and Jessie Wilson, is deceased

After completing the curriculum of the public schools of his native state, Edward H Reid completed a higher academic course at Amity college, located at College Spring Iowa After leaving college Mr Reid identified himself with ranch enterprise, and for some time he was associated with the celebrated Rankin Ranch in the state of Missouri Mr Rankin having been one of the most extensive cattle men of the west and having fed more cattle on his own land than any other man in the business as may be inferred when it is stated that at times he fed as many as ten thousand head of cattle on his own grazing land and from corn raised on his own land After severing his connection with this great enterprise Mr Reid went to Denver Colorado, where he became associated with the Continental Trust Company, in the position of inspector of cattle and farm lands From Colorado he went to Wyoming and about the year 1910 he established his residence at Mitchell, Nebraska, where he has since been a prominent and influential factor in the exploiting of ranch industry upon an extensive scale, his operations having contributed much to the advancement of agricultural and live-stock industry in western Nebraska and his broad experience making him an authority in all things pertaining to the development of new land He is a broad-minded and liberal citizen, is a Republican in his political allegiance and both he and his wife hold membership in the Federated church at Mitchell

The year 1882 recorded the marriage of Mr Reid to Miss Mary Elizabeth Maiden a native of the state of Missouri, and concerning their children, the following brief data is available Earl H, is a resident of Torrington Wyoming, Margaret is the wife of Lester Collins, of Mitchell, Nebraska, Jessie is the wife of Edward P Grant, who is in government service at Washington D C his home being at Kensington, Maryland, and Dorothy, is the wife of Lee Ashbrook, Jr, who is associated with his father and lives on the Ashbrook Ranch in Sioux county, Nebraska

WILLIAM C REDFIELD figures as one of the representative exponents of financial enterprise in Scottsbluff county where he is cashier of the State Bank of Haig of which institution he became the founder in June, 1916 and to the active management of which he has since given his attention This bank has developed a substantial business and affords facilities to a goodly number of appreciative patrons who are prominent in the various fields of industrial and business enterprise in the northern part of Scottsbluff county

The bank in incorporated with a capital stock of $10,000, its deposits are now in excess of $50,000, and it maintains a surplus of $1,800 It constitutes one of the most valuable adjuncts to the business life of the community in which it is established and its success is proving notably cumulative, a fact which attests its able management and also the confidence reposed in its able and progressive founder

Mr Redfield was born in Peoria county, Illinois, on the 30th of December, 1860, and is a son of F A and Mary E (Aldrich) Redfield, of whom a record appears elsewhere

William C Redfield acquired his preliminary education in Illinois and supplemented this discipline by attending the public schools of Johnson county after the family removal to Nebraska As a young man he became actively concerned with independent farm enterprise in Johnson county, and there he gained inviolable vantage place in popular esteem, as shown by the fact that in 1905 he was elected county clerk, an office of which he continued the incumbent four years, after which he served five years as county treasurer He continued his residence in that county until August, 1915, and moved to Lincoln and then in June, 1916, he removed to Scottsbluff county and founded the vital banking institution of which he has since been the executive head Mr Redfield has become a zealous advocate of the principles of the Republican party and has been influential in its local campaign activities Both he and his wife are members of the Methodist Episcopal church, as were also his parents

In 1885 was solemnized the marriage of Mr Redfield to Miss Mary E Barrett, who was born at Brockport, New York and who was a girl at the time of the family removal to Nebraska Mr and Mrs Redfield became the parents of two children Mattie, who became, in 1915, the wife of William W Lockwood, died in December, 1917 and her only child, Marguerite, is being reared in the home of her maternal grandparents Mr and Mrs Redfield, Franklin, only son of the subject of this review, entered the United States navy in connection with the nation's participation in the late World War his enlistment having taken place in June, 1917, and he having been assigned to service in the electrical department of the submarine work, at San Pedro, California where he continued to be stationed until the close of the war, he is now employed by the telephone company at Scottsbluff

HARVEY BEEBE, whose well improved ranch property is situated in the north central part of Scottsbluff county, where Mitchell is his postoffice address, is to be definitely credited with pioneer honors in this section of the state, where he took up his abode in 1887, when Scottsbluff county was still an integral part of old Cheyenne county His individual success and advancement have kept pace with the splendid development of this section of the state and he is one of the leading men of his community — a substantial agriculturist and stockgrower and a citizen who takes loyal interest in all things that conserve civic and industrial growth

Mr Beebe was born in Monroe county, Iowa, September 22, 1865, and he is a posthumous son of Jeremiah Beebe, who was one of the pioneer farmers in the Hawkeye state, where he died fourteen days prior to the birth of the subject of this review He was a native of Illinois and his wife, whose maiden name was Vaneta Chidester, was born in Virginia, she having passed the closing years of her life in Iowa and having been a devoted member of the Methodist Episcopal church, while her husband was a staunch Republican in his political proclivities Of their six children, the statement above implies, Harvey, of this sketch, is the youngest, Eliza, is the wife of Lorenzo Warner, and they were residing in Cloud county, Arkansas, at the time when Mr Beebe last heard concerning them, Eli continued to reside in Monroe county, Iowa, where he is a successful market-gardener, John is a prosperous farmer near Albia, that state, Richard is deceased and was a resident of the state of Washington at the time of his death, and Isaiah died at Gering, Scottsbluff county, Nebraska, in which vicinity he owned land and had instituted the development of a farm, his widow now being a resident of Wyoming

Reared under the conditions that marked the pioneer period in the history of the Hawkeye state, Harvey Beebe early gained fellowship with honest toil and endeavor, as is evident when it is stated that when he was but thirteen years old he began to provide for himself and assist his widowed mother by hauling coal with team and wagon He made good use of the advantages afforded in the public schools of his home county, but naturally his attendance in school was somewhat irregular, owing to the heavy responsibilities that were early placed upon him He continued his residence in Iowa until 1887, when,

Mr. and Mrs. Jesse Pickering

as an ambitious young man of about twenty years, he came to old Cheyenne county Nebraska, where he purchased land in what is now the north central part of Scottsbluff county. Here he has applied himself with all of vigor and determination, and the passing years have brought to him well merited success. His landed estate now comprises three hundred and sixteen acres, and of the tract one hundred and thirty-four acres receive irrigation from the Mitchell ditch, while one hundred acres are similarly provided for by a government irrigation project. He gives his attention to diversified farming and the raising of good grades of horses and cattle, while he has erected good buildings and made other modern improvements on his place.

Mr Beebe has been loyal in the support of all things tending to advance the general welfare of his home community, county and state, and is one of the popular pioneer citizens of Scottsbluff county. His political views are expressed in staunch support given to the cause of the Republican party, and while he has not sought office his civic loyalty has caused him to give efficient service in the position of school director and also that of road overseer. In a fraternal way he is affiliated with the Independent Order of Odd Fellows.

Mr Beebe wedded Miss Ida May Billings, who likewise was born and reared in Iowa, where her father, L. W. Billings still resides. In conclusion is given brief record concerning the children of Mr and Mrs Beebe. Ralph is engaged in farm enterprise in Wyoming, V. C. is a prosperous farmer in Scottsbluff, Nebraska, Walter and Clarence remain at the parental home and are associated with their father in the activities of the farm, Mabel is the wife of Carl Smith, of Mitchell, this county, and Bessie, Goldie, Leo and Alda are the younger members of the ideal home circle.

JESSE PICKERING, who for many years was one of Scottsbluff county's most worthy citizens widely known and highly respected for his sterling traits of character, was born in Fulton county, Illinois, March 1, 1847 and died on his Nebraska homestead September 26 1915. For thirty-three years he had been a resident of this county. He was a son of Curtis and Mary (Stroade) Pickering, who were natives and life-long residents of Ohio, until 1848 when they moved to Fulton county Illinois, settling in a neighborhood where there were already many Quakers, they being members of the Society of Friends.

In 1879 Jesse Pickering went to Cheyenne county and in 1886 came to Scottsbluff county, Nebraska, where he secured a homestead of 160 acres and a timber claim of equal extent. In 1915 his wife homesteaded and still owns her 120-acre homestead, together with other land, and pays taxes on 280 acres, all well improved and the greater part of which is irrigated. Mr Pickering was a judicious farmer and for many years he was also prominent in the public affairs of the county. He was a veteran of the Civil War, after which he came to the West and a measure of military dignity always lingered with him, together with personal courage and an irreproachable character. He was a Republican in politics.

Mr Pickering was married in Illinois to Miss Lucy Reese, who was born in Ohio, February 3, 1850. Her parents were Alexander and Rachel (Tingler) Reese, natives of Pennsylvania. They came to Nebraska at a later date and died on their farm here. Mr and Mrs Pickering had four children. William S., who manages his mother's farms. Sylvester, who married and then moved to Colorado, and Irena, who is now Mrs Scofield and lives at Gering, Nebraska. Jacob, who died at the age of twenty-eight years in Missouri, grew to manhood in Nebraska. The family belongs to the Society of Friends.

ALONZO L. MOON is to be credited with being one of the energetic and broad-guaged exponents of farm enterprise in Scottsbluff county, where he is a representative of the younger generation of those who are here contributing much to the industrial supremacy of the county along agricultural and live-stock lines. His farm is in the Mitchell neighborhood and is situated in section 6 township 23-55, where he is the owner of three hundred acres, of which about one hundred and twenty acres are provided with effective irrigation, so that his agricultural activities have the proper basis for assured success. Mr Moon established his residence here in 1908, when he acquired his present farm by purchasing a relinquishment to the claim. He has made many substantial improvements on the place, including the erection of good buildings, and here he diversifies his activities by raising the crops best suited to the soil and climate, and by breeding and feeding good types of live stock. He gives special prominence to the propagation of alfalfa and finds this a very profitable feature of his enterprise. He is one of the popular and influential men of his community, is a Democrat in his political allegiance, and has held for eleven years the office of school director and has otherwise given staunch support to those things that conserve the general wellbeing of the community. He is affiliated with the Masonic fraternity, and both he and his wife hold membership in the Congregational church

Mr Moon was born in Page county, Iowa, on the 14th of May, 1882, and is a son of M C and Frances (Anderson) Moon, both natives of the Buckeye state, where they were reared and educated and where their marriage was solemnized. The father was born in Ross county, Ohio, in 1850, and his death occurred in Valley county, Nebraska, in 1907, his widow, who was born in Greenfield, Ohio, in 1852, being still a resident of Valley county. Of the eight children, the subject of this review is the eldest of the three surviving. Dora, is the wife of Milo Russell, of Edinburg, Texas, and Stella is the wife of Charles Luedtke, a farmer near Arcadia, Valley county, Nebraska. In politics the father was not a strict partisan and his religious faith was that of the Congregational church, of which his widow has long been a devoted member. Mr Moon served as justice of the peace and also as school moderator. From Ohio he removed to Iowa, in which state he continued his activities as a farmer until 1883, when he came with his family to Nebraska and entered homestead and timber claims near Arcadia, Valley county, where he developed a valuable farm, where his death occurred. His widow is still living in that county. He was a man of sterling character and achieved worthy success through his association with farm enterprise in Valley county.

Alonzo L Moon was a child at the time of the family home was established in Valley county, where he was reared to adult age. After completing the curriculum of the public schools including that of the high school at Arcadia, he continued his studies at Crete Academy, where he finally entered Doane College, in which he continued his studies three years the following year having found him enrolled as a student in the University of Nebraska from which he received the degree of Bachelor of Arts, in 1906. In 1908, as previously noted, Mr Moon established his residence on his present farm, and he has since continued as one of the vigorous and successful devotees of farm enterprise in Scottsbluff county, where he and his wife are held in the highest esteem

The year 1906 recorded the marriage of Mr Moon to Miss Rhoda Whitman, who was born in the state of Nebraska, and they have four children — Frederick, Josephine, Millard C and David. In the home of Mr Moon was reared Miss Lucille Whitworth, who remained eight years and who is now married, and her sister Mary has been for six years in the Moon home circle

HENRY E RUSSELL, who is one of the live real estate men of Mitchell and Henry associated with his brother, H G Russell, in the vigorous little city of Mitchell, Scottsbluff county, is not to be denied pioneer distinction in this county, where he established his residence more than thirty years ago and where he became a successful exponent of farm development. He here took up homestead and tree claims, of one hundred and sixty acres each, in the year 1887, and with confidence and prevision he applied himself earnestly to the reclaiming of the land into productiveness, an enterprise in which he succeeded admirably. He gave his attention to diversified agriculture and stock-raising and he still owns one hundred and twenty acres of land in the Mitchell vicinity, this being one of the well improved farms of this part of the county and having excellent irrigation facilities from the Mitchell ditch. After having improved his land, a portion of which he eventually sold at a distinct profit Mr Russell finally retired from the farm and from 1905 until the spring of 1919 he was manager for the Carr & Neff Lumber Company at Mitchell, save that he passed a portion of the year 1912 in the state of Washington, to which he went principally for the recuperation of his health, which had become somewhat impaired, but while in Washington he was employed for a time by a lumber company. He is one of the well known and highly esteemed pioneer citizens of Scottsbluff county and well merits consideration in this publication

Mr Russell was born in Grant county, Wisconsin, on the 14th of October, 1865, and he is the son of Calvin W and Lydia (Spargo) Russell, the former of whom was born in the state of New York, in 1832, and the latter of whom was born in Ohio, in 1835. Mrs Russell passed to the life eternal in 1896, and her husband survived her by more than twenty years, his death having occurred in February, 1917, after he had attained the venerable age of nearly eighty-five years. Calvin W Russell became a pioneer farmer in Grant county, Wisconsin, where he continued his residence until 1888, when he came to Nebraska and established his residence in that part of Cheyenne county that is now comprised in Scottsbluff county. Here he filed entry on a homestead and a tree claim, to which he perfected his title in due course, and he developed a considerable portion of the three hundred and twenty acres into productive farm land the while he was honored and valued as one of the sterling pioneer citizens of the county, where both he and his wife passed the closing years of their lives, both having been consistent members of the Christian church and his political faith having been that of the Republican party. Of the ten children two died in infancy another died at the age of one year and Esther at the age of twelve years. Mary married Charles F. Peckham, of Gothenburg Dawson county, Henry E. is the immediate subject of this sketch. James R. is a successful farmer near Mitchell, Rose is the wife of Joel Jackson, of Mesa Arizona. Lawrence likewise resides at Mesa, Clarence is engaged in the real-estate business at Gilbert, Arizona, and Herbert G. is similarly engaged at Henry, Scottsbluff county, Nebraska.

Henry E. Russell was reared to manhood in the old Badger state, where he was given the advantages of the public schools and where he continued his association with farm activities until 1887, when he came to what is now Scottsbluff county Nebraska and took up government land as has been noted in a preceding paragraph of this review. He has been a true disciple of civic and industrial progress in this favored section of Nebraska, where he has achieved independence and prosperity through his own well directed endeavors and where he commands secure place in popular confidence and good will. His political allegiance is given to the Republican party and he has served as a member of the township board of supervisors, as precinct assessor and as school director. He is affiliated with the Independent Order of Odd Fellow, the Masonic fraternity, including the Order of the Eastern Star the Modern Woodmen of America and the Brotherhood of American Yeoman, both he and his wife being members of the Christian church at Mitchell.

The year 1896 made record of the marriage of Mr. Russell to Miss Blanche Nichols who is a native of the state of Kansas and they have three children all of whom remain members of the family home circle — Warren S., Ruth and Doris. The only son is giving his attention to the operation of a farm tractor and is one of the popular young men of Phœnix Arizona.

FRANK ALLISON has been a resident of Scottsbluff county since 1910 and is one of the enterprising and vital younger representatives of farm industry in the north central part of the county, where his well improved farm comprises eighty acres, all of which tract is to be covered by irrigation facilities provided through the government irrigation project in this locality. Further interest attaches to the career of Mr. Allison by reason of the fact that he is a native of Nebraska and has inherited the full measure of energy and progressiveness for which this commonwealth almost invariably makes provision.

Mr. Allison was born in York county, Nebraska on the 28th of January, 1888, and is a son of Frank and Rose (Jacobs) Allison, both of whom are now deceased, the father having passed away when fifty-six years of age. Of the four children the subject of this sketch is the youngest, Ray is a resident of Stratton, Lloyd is identified with real estate enterprises in the state of Washington, and Harry is a prosperous farmer near Mitchell, Scottsbluff county. The father became a successful farmer in York county and also gave attention to the teaching of music in which domain he had much talent. He was successful as an agriculturist and also in the raising of fine grades of cattle and horses the while he was a citizen who ever commanded the confidence and esteem of all with whom he came in contact in the various relations of life. He was a Democrat in politics and his wife was a devoted member of the Christian church she having assisted in the erection of the church of this denomination at York and also in the establishing of the college there maintained under the auspices of this denomination.

He whose name introduces this review gained his youthful education in the public schools of York county, where he early gained experience in connection with the varied activities of the home farm. He continued his residence in that county until 1910, when he purchased a farm in Scottsbluff county where he lived until 1919 when he bought his present farm where he is notably successful in his progressive activities as an agriculturist and stockgrower his being one of the model smaller farms of this section of the county.

In politics Mr. Allison is not constrained by specific partisan dictates, but votes in accordance with his judgment. He is affiliated with the Independent Order of Odd Fellows and the Royal Highlanders, his religious faith being that of the Methodist Episcopal church and his wife being a member of the Presbyterian church.

In April 1913 was solemnized the marriage of Mr. Allison to Miss Dessa Morrison, a daughter of Amos C Morrison, of whom specific mention is made on other pages of this work Mr and Mrs Allison have two children — Margaret and Frank, Jr

BARTON E LANE is one of the native sones of Nebraska whose fealty to the state has never faltered and whose good judgment has led him into successful enterprise as one of the representative farmers of the younger generation in Scottsbluff county, where he is associated with his brother Kent in the active management of a well improved ranch of one hundred and sixty acres in section 1, township 23-57, about four and one-half miles distant from Morrill, which is their postoffice address

Mr Lane was born at Crete, Saline county, this state, on the 11th of October, 1888, and is a son of William H and Mary J (Deems) Lane, the former a native of Indiana and the latter of Pennsylvania The father still resides at La Feria, Texas and the mother passed away in 1909, at the age of fifty years Of the children the eldest is Noble, who is now a resident of Sheridan, Wyoming, where he is engaged in the hardware business, Gertrude is the wife of F E Dewey, who is identified with coal mining enterprise in the state of Colorado, Neva M is the wife of S E Adams, a farmer in Morrill county, Nebraska, Kent, who is associated with the subject of this sketch in the control of the farm, has been in the military service of the government in connection with the World War, he having been assigned to instruction and duty at Camp Mills, New York, as one of the representatives of Scottsbluff county in the great war ordeal, and the two younger children died in infancy The father is a Republican in politics, has served in various local offices of public trust and is affiliated with the Independent Order of Odd Fellows He became a successful contractor in the construction of irrigation and drainage ditches and on his removal to Scottsbluff county he here became identified with the reclamation service, besides which he took up a homestead of eighty acres, which he has reclaimed and improved and upon which he erected in 1919, a modern house of seven rooms

B E Lane was reared and educated in Saline county, and in 1907 he came with his parents to Scottsbluff county, where he has gained a position of security as one of the substantial young farmers and stock growers of this section of the state, the greater part of the farm being provided with irrigation

facilities of excellent order. A loyal and progressive citizen, Mr Lane is an independent voter in political matters and is actively identified with the Nonpartisan League He is affiliated with the Masonic fraternity, in which he has received the thirty-second degree of the Ancient Accepted Scottish Rite, and also with the Modern Woodmen of America

May 13, 1913, recorded the marriage of Mr Lane to Miss Nellie J Bohl, who was born in Colorado, and their attractive home is brightened by the presence of their three children — Goldie Marie, and Barton and Francis, who are twins

AMOS C MORRISON is one of the distinctly substantial and representative figures in connection with agricultural and live-stock industry in Scottsbluff county, where he established his residence on the 15th of October, 1898, and where he is now the owner of a splendid ranch property of seven hundred and twenty acres There are many points of special interest in the record of his career He has the distinction of being a native of the historic Old Dominion state, was a pioneer cattle man in Wyoming, prior to which he had been a resident of Iowa, and since coming to Nebraska he has achieved large and worthy success in connection with farm industry, besides which he is vice-president of the American State Bank in the city of Scottsbluff

Mr Morrison was born in Frederick county, near Winchester, Virginia, on the 23d of March, 1860, and in the same fine old commonwealth were born his parents, Amos and Elizabeth Catherine (Miller) Morrison, who were residents of that state at the time of their death Amos Morrison was a prosperous agriculturist in Virginia at the inception of the Civil War, which great struggle brought financial and industrial distaster to him, as it did to nearly all other citizens on the southern stage of the war activities Within the progress of the war he served as a teamster in the Virginia militia and eventually he again placed himself in a position of definite independence and prosperity He attended and supported the United Brethren church, of which his wife was a member, and after the close of the Civil War he became an active supporter of the Republican party Of the children only two are now living James, the eldest of the number, was a resident of Virginia at the time of his death, Elizabeth is deceased, Anna is deceased, Belle is deceased, the subject of this review was the next in order of birth, and Milroy and John are deceased

A C Morrison was reared to maturity in Virginia, where he received his early education in the schools of the period, he having been born the year prior to the outbreak of the Civil War and thus having known as a boy the hardships through which his native state passed during the so-called reconstruction period After leaving school Mr Morrison continued to be actively identified with farm enterprise in Virginia until 1886, when, as an ambitious and sturdy young man of twenty-six years, he made his way west and established his residence in Iowa After remaining there about two years he went to Wyoming, where he engaged in the cattle business, in connection with pioneer agricultural operations The section in which he established his home was sparsely settled, and he relates that in the early days there he frequently traversed a distance of thirty miles without seeing a single house

Mr Morrison continued his industrial activities in Wyoming until October, 1898, when, as previously noted, he came to Nebraska and located in Scottsbluff county Here he bought land, and since that time he has added to the same until he now has a valuable landed estate of seven hundred and twenty acres, in the north central part of the county and not many miles distant from Mitchell, his attractive and admirably improved home being situated in section 30, township 22-55 He has brought a large acreage under effective cultivation and is one of the representative agriculturists and stock-growers of the county, besides being a loyal and progressive citizen who commands unequivocal esteem His political allegiance is given to the Republican party, but he has always been a worker and has had neither time nor desire for political activity It is especially interesting to record that Mr Morrison and John and William M Newell left Iowa on the same day and removed to the same locality in Wyoming, besides which all three came to Scottsbluff county about the same time, all having been successful here in their well ordered undertakings and their friendship being as close as that of brothers

In the state of Wyoming on the 10th of February, 1891, Mr Morrison wedded Miss Amanda M Warner, who was born in Nebraska and who passed the closing years of her life at the pleasant home in Scottsbluff county, this state, where her death occurred January 11, 1904 She is survived by six children Melvie is the wife of Robert Fuller, of Meadow Sarpy county, Dessie is the wife of Frank Allison, of whom personal mention is made on other pages of this volume, Clinton is a farmer

in Mitchell valley, Scottsbluff county, and in this enterprise he is associated with his next younger brother, Ralph, and Milroy and Mildred remain at the parental home

On the 31st of March, 1906, Mr Morrison contracted a second marriage when Sallie R Gross became his wife, she having been born in the state of New Jersey She is an active member of the Methodist Episcopal church Mr and Mrs Morrison have four children — Amos, Esther F, Harold and Paul

WILLIAM W NEWELL is a member of a specially close triumvirate of friends who have become prominent and successful in the field of agricultural and live-stock industry in Scottsbluff county, another of the number being his brother John, of whom specific mention is made on other pages, and the third being Amos C Morrison, of whose career a review likewise appears in this work The three went in company from Iowa to Wyoming more than thirty years ago, and from the latter state they came to Scottsbluff county about the same time, their success having been on a parity with their enduring friendship, which is marked by mutual appreciation and utmost harmony

William W Newell, the owner of seventeen sections of land in Sioux county, is here one of the most prominent and successful representatives of the cattle industry in the famous Panhandle of Nebraska, and he maintains his home on his well improved place in the vicinity of the thriving town of Mitchell He was born in Louisa county, Iowa, on the 7th of June, 1858, and is a son of Robert F and Christina (Newell) Newell, who were sterling pioneers of that section of the Hawkeye state Of the nine children the following brief record may consistently be incorporated at this juncture Elizabeth, Thomas, Caroline and Hugh are deceased, Mary is the wife of Edward Curtiss and they reside in the state of Iowa, John is individually mentioned on other pages as previously stated William W, of this review was the next in order of birth, Robert still resides in Iowa, and Harriet is the wife of Frank Sidnam, a resident of Kansas The father became a successful farmer in Iowa, where he reclaimed and improved a good farm and he was one of the honored pioneer citizens of that state at the time of his death He was a Democrat in politics and was affiliated with the Masonic fraternity

He whose name initiates this article was reared under the conditions and influences that marked the pioneer era in the history of Iowa, where he early began to aid in the work of the home farm, his educational advantages in his youth having been those afforded in the public schools of the locality and period In his native state also did he initiate his independent activities as a farmer, but after continuing operations two years he accompanied his brother John and his friend A C Morrison to Wyoming, all locating in the same vicinity and all engaging in the ranch business, their postoffice address having been the village of Spring Hill There Mr Newell conducted successful activities in the cattle industry for a period of eleven years, at the expiration of which he came to Nebraska, in 1898, and established his home in Scottsbluff county, where he has achieved large and worthy success as an exponent of agricultural and live-stock enterprise He is one of the extensive land-owners of this part of the state, as previously intimated in this review, and virtually all of his land is devoted to the cattle industry, though he raises the various fodder crops that are an essential adjunct His civic liberality has been in harmony with the success that has attended his earnest and well ordered endeavors and he is always ready to do his part in the promotion and support of enterprises tending to advance the best interests of the community He is allied with the Democratic party and Mrs Newell and family hold membership in the Presbyterian church

The year 1883 recorded the marriage of Mr Newell to Miss Sarah Parkin, and she has proved a devoted companion and helpmeet during the long intervening years To them have been born eight children Frederick is engaged in farming and stock-raising in Sioux county, Vinnie is county superintendent of schools in that county, Caroline remains at the parental home, Louis is a resident of Wyoming, Edith, William and Nellie are the younger members of the parental home circle, and Maggie died in infancy

CLAUDE R WATSON M D — Within the pages of this history will be found specific mention of a goodly proportion of those who are admirably upholding the prestige of the medical profession within the boundaries of the great Panhandle of Nebraska, and in noting the representative physicians and surgeons of Scottsbluff county is special consistency in according recognition to Dr Watson, who is engaged in the successful general practice of his profession in the north central part

of the county, with residence and headquarters at Mitchell He is known as a man of high professional attainments and is valued as a loyal and progressive citizen

Dr Watson was born in the city of Louisville, Kentucky, on the 15th of September, 1884, and is one of the seven children of Bufford and Margaret (Miller) Watson, both likewise natives of the fine old Bluegrass state, the father being still a resident of Louisville To the schools of his native state and city Dr Watson is indebted for his early educational discipline, which included his pursuing higher academic studies in the University of Kentucky In preparation for his chosen profession he was matriculated in the medical department of the University of Indiana, and after receiving from this institution his degree of Doctor of Medicine he served for some time as interne in the city hospital of Louisville, Kentucky, where he thus gained valuable clinical experience and further fortified himself for the successful practice of his profession In 1908 he established himself in practice in Buffalo county, Nebraska, where he developed a substantial professional business and where he continued his activities until 1914 when he was favored in securing the splendid experience incidental to active association with the work of the celebrated hospital conducted by the Mayo brothers at Rochester, Minnesota, the brothers of this name having gained national reputation in the domains of medicine and surgery and their Minnesota institution being one of the most noted of the kind in America Dr Watson was associated with the Doctors Mayo for fifteen months, at the expiration of which he returned to Nebraska and established himself in practice at Mitchell He he controls a large and representative practice, and he is specially well known for his skill in the surgical department of his profession He is affiliated with the Nebraska State Medical Society and the American Medical Association, has received Scottish Rite degrees in the Masonic fraternity, besides being identified with its adjunct organization, the Ancient Arabic Order of the Nobles of the Mystic Shrine, and also with the Independent Order of Odd Fellows and the Knights of Pythias Both he and his wife hold membership in the Congregational church and in politics he maintains an independent stand, by giving his support to men and measures meeting the approval of his judgment rather than being constrained by strict partisan lines

The year 1908 recorded the marriage of

Dr Watson to Miss Pfrimmer Zollman, who likewise was born in Louisville, Kentucky, and who presides most graciously as chatelaine of their pleasant home at Mitchell

JACOB GOMPERT has the sterling character and the industrious and persevering habits that tend to conserve personal advancement, so that it is not strange that in Scottsbluff county he has became a substantial representative of farm industry, his home being in section 9, township 23-56, about four miles distant from Mitchell which is his postoffice address The well improved ranch which he has here developed is given over largely to the raising of cattle and other live stock, though he is equally successful in the agricultural department of his farm enterprise

Mr Gompert was born and reared in Rhine province, Germany, the date of his nativity having been November 27, 1867, and his early education having been gained in the excellent schools of his native land He is a son of Johan and Marie (Hulser) Gompert, who became the parents of six sons and one daughter, four of the sons having eventually become residents of the United States, where each has stood exponent of loyal and appreciative citizenship Herman is a resident of Sioux county, Nebraska, Jacob of this sketch, is the next younger, Gerhard, likewise a substantial farmer of Scottsbluff county, is individually mentioned on other pages of this work, and John is a resident of Waterloo, Iowa The father, who was a farmer in Germany, come to the United States in 1899, and the following year marked his arrival in Scottsbluff county, Nebraska, where his death occurred in April, 1906, his wife having passed away a number of years previously

Jacob Gompert gained in his fatherland a valuable preliminary experience in connection with farm enterprise, and in November 1887, shortly before his twenty-fifth birthday anniversary, he made his appearance in the state of Texas, where he remained until 1889, in which year he came to Nebraska In 1899 he took up a homestead in Scottsbluff county and he forthwith began improving the place, to which he has since added, by purchase until he now has a valuable landed estate of three hundred and twenty acres, of which one hundred acres are under effective irrigation With characteristic energy and good judgment Mr Gompert has pressed steadily forward toward the goal of success and independence, and he now merits classification among the substantial agriculturists and stock-raisers of Scotts-

bluff county Here the consensus of popular opinion is that his success has been worthily won and well merited He is aligned as a supporter of the principles of the Republican party and he and his wife hold membership in the Federated church

In 1892 Mr Gompert wedded Miss Elizabeth Holbrook, who was born and reared in Germany and who came to America in company with the father of her future husband Mr and Mrs Gompert have five children — John, Gerhard, Gertrude, Mary and Fred None of the children have yet gone forth from the parental home circle with the exception of the eldest son, whose loyalty was shown by effective service in connection with the nation's participation in the late World War, he was assigned to service with a hospital corps and was sent to France, where he remained for some time after the signing of the historic armistice, his return to America having occurred in August, 1919

GERHARD GOMPERT came from his native Germany to America in the year 1889, and that year had not yet come to its close when he established his residence in the present Scottsbluff county, where he thus gained a measure of pioneer distinction The county has gained through his loyalty as a progressive citizen and his well directed activities in the furtherance of agricultural and live-stock industry Within a short time after his arrival in the county Mr Gompert took up a homestead in the north central part thereof, his success being evidenced by the fact that he has since evolved from this nucleus a fine farm estate of three hundred and forty acres that are under effective irrigation and two and one-half sections of dry land He is giving his attention primarily to the cattle business, in which line of enterprise he conducts operations upon a large scale In initiating this prosperous enterprise he began in the most modest way, — in fact, with only one cow, — but his energy, careful methods and mature judgment have enabled him to make consecutive advancement and gain precedence as one of the substantial men of the county of his adoption His home place is situated in section 6 township 23-56, and is about six miles distant from Morrill, which is his postoffice address and six miles from Mitchell He takes loyal interest in all matter pertaining to the general welfare of the community and is deeply appreciative of the opportunities that have been given to him in Nebraska He is independent in politics and both he and his wife hold membership in the Lutheran church.

Mr. Gompert was born in Rhine province, Germany, on the 29th of December, 1864, and is a son of John and Marie Gompert, of whose six sons four came to the United States; Herman is a resident of Sioux county, Nebraska; Jacob is a substantial citizen of Scottsbluff county and is individually mentioned on other pages; John is a resident of Waterloo, Iowa; and Gerhard of this review, is the other one of the four. The father, who was a farmer in Germany, came to the United States in 1888 and returned to the old country after one year. In 1900 he came again with his family and took a homestead. His death occurred in 1906.

Gerhard Gompert gained his early education in the schools of his native land and was twenty-four years of age when he came to America, where he felt assured of better opportunities for achieving independence and prosperity through his own efforts. The passing years have fully justified his hopes and ambition, and he has reason to be proud of the success which he has achieved in connection with farm enterprise in Scottsbluff county.

In 1889, shortly after his immigration to this country, Mr. Gompert was united in marriage to Miss Elizabeth Langenfurth, who accompanied him on the voyage across the Atlantic and who has been his loyal coadjutor in the labors and responsibilities that have marked the passing years. Of their seven children the eldest is William, who is associated in the work and management of the home ranch, he was in service of the United States during the late war; Henry is actively identified with farm enterprise in Scottsbluff county; Herman went forth as one of the loyal young men who served with the American Expeditionary Forces in France during the progress of the great World War, he having been assigned to duty in the veterinary corps and his return to his home state having occurred in the summer of 1919; Jacob, who was in the service of the United States during the war; John, Anna and Carl are the younger members of the parental home circle.

GEORGE IRELAND, who conducts a well equipped mercantile establishment at Mitchell, one of the vigorous and progressive little cities of Scottsbluff county, is properly assigned prestige as one of the representative business men of the county, and further interest attaches to his career by reason of the fact that he is a native of Nebraska and a representative of a sterling pioneer family of this commonwealth.

Mr. Ireland was born in Sarpy county, Nebraska, on the 4th of March, 1879, and is a son of George M. and Mary E. (Sexton) Ireland, the former of whom was born in West Virginia and the latter of whom was born in Iowa, where her parents were pioneer settlers and where her marriage to Mr. Ireland was solemnized. George M. Ireland and his wife now maintain their home at Arapahoe, Nebraska. They became the parents of eight children, all of whom are living.

He whose name initiates this review is indebted to the excellent public schools of Nebraska for his early educational discipline, and after leaving school he continued his association with farm enterprise until 1914, when he removed from the farm upon which he had been conducting operations, in Scottsbluff county, to Mitchell, where he has since been successfuly engaged in the mercantile business, with well selected stocks of groceries, shoes, furnishing goods, etc. He became a resident of the county in 1907, in which year he here took up a homestead and instituted the reclamation and improvement of the same. He developed a good farm and on the same conducted successful operation until 1914, when he sold the property and established his present business enterprise. He is a Republican in his political proclivities, is affiliatd with the Independent Order of Odd Fellows, and both he and his wife hold membership in the Methodist Episcopal church. They have an attractive home at Mitchell and the same is known for its generous hospitality — a place where their many friends are always assured of cordial welcome.

The year 1907 recorded the marriage of Mr. Ireland to Miss Mabel E. Henderson, who was born in the state of Iowa, and they have three children — Helen, Wilma and George, Jr.

JOHN W. PLUMMER. — In the northwestern part of Scottsbluff county to mention the name of John W. Plummer in even a casual way, is sure to bring forth words of commendation and good will, for he is a man whose characteristics are such as to make and retain to him the confidence and esteem of those with whom he comes in contact, besides which he has thoroughly proved his value as a loyal citizen and a man of productive enterprise. He is one of the representative agriculturists and stock-growers up in the Morrill neighborhood, his farm being situated in section 1, township 23-57, a few miles distant from the village mentioned. He has had his full share of the varied experiences that

CHARLES H. FLOWER AND FAMILY

marked the earlier period in the history of Nebraska and other western states, and he reverts with special satisfaction to the service which he gave as a cowboy on the open ranges He also was engaged in mining and prospecting, and is at the present time financially interested in mining and oil properties

Mr Plummer was born at Des Moines, Iowa, on the 17th of March, 1861, and is a son of Ezra and Lena (Garrett) Plummer, both natives of Ohio, where they were reared and educated They were very early pioneer settlers in Iowa, where they established their home in 1842 and where the father developed a productive farm, both he and his wife having been residents of Fort Collins, Colorado at the time of their death and both having been members of the Methodist Episcopal church In politics he was a stalwart Republican Of the children the eldest was Emma who died in childhood, John W, of this sketch, was the next in order of birth, Vernon is a resident of Fort Collins, Colorado, Luella is the wife of Harden Puckett, of Orange Junction, Wyoming, Denver, the fifth child, is a resident of Scottsbluff county, Nebraska, Effie is the wife of Frank Moore and they are residents of the state of Montana, and Rose Glenn resides at Adel, Iowa

John W Plummer was about seventeen years of age at the time of the family removal to Colorado, where he was reared to adult years and where he acquired his youthful education by attending the public schools whenever opportunity was presented His preliminary education was acquired in the schools of his native state As a youth Mr Plummer lived up to the full tension of the free and sturdy life of the cowboy on the range, his service in this capacity having been in Colorado In personality he stands as a fine type of the western pioneer citizen, strong and vigorous, keen of mind and honest and upright in thought and action

In the year 1905 Mr Plummer filed entry on a homestead in Scottsbluff county, where he now owns and occupies a finely improved farm of one hundred and twenty acres, the property having greatly increased in value under his admirable management, as evidenced by the fact that he was recently offered $37,000 for his farm, the same being provided with excellent irrigation facilities and otherwise having modern improvements that were installed by him He has given his attention to diversified agriculture and stock-raising, and has been one of the specially successful sheep feeders in the northern part of Scottsbluff

county, He is liberal and loyal as a citizen and in politics classes himself as an independent Republican, implying that in local politics he supports men and measures rather than holding close to partisan lines He and his wife hold membership in the Methodist Episcopal church at Morrill

On the 15th day of May, 1887, was solemnized the marriage of Mr Plummer to Miss Carrie B Seydell, who, like himself, claims Iowa as the place of nativity Of their four children the first born was Ethel, who died at the age of seven years, Eugene and Nellie remain at the parental home, Edith died in infancy, and Lavinia and Howard are the younger members of the cheerful home circle

CHARLES HENRY FLOWER, who has been a resident of Nebraska for thirty-two years, now owns a large body of land, giving his main attention to the cattle business He has been identified with many things of importance in the development of Scottsbluff county and in earlier years served as county commissioner

Mr Flower was born in New Hampshire, June 24, 1860, and accompanied his parents first to Iowa and later to Madison county, Nebraska They were Charles E and Mary A Flower, natives of New Hampshire, the former of whom lived to the age of eighty-six and the latter to eighty years The father was a veteran of the Civil War Mr Flower has two sisters and one brother In 1887 he secured a preemption and a tree claim in what is now Morrill county just north of Bayard, and proved up on both He attempted farming but drouth and storms destroyed his crops and as no irrigating had been done at that time, he decided to go into the cattle business and has continued to be interested ever since He did some freighting in early days and remembers when he carried goods from Ft Sidney to Minatare for twenty-five cents a hundred weight Times have changed vastly since then For the last eighteen years Mr Flower has been a member of the Ditch board and a member of the Drainage board since organization and has had much to do with irrigation projects in this section He came to Scottsbluff county seventeen years ago and owns 691 acres here and 2480 acres in Banner county He has three sets of improvements on his land all made by himself Mr Flower's present farm was school land and wholly unimproved

On March 25, 1883, Mr Flower was married to Miss Mary E Clark, who was born in Illinois, a daughter of A C and Mary Clark, and they have two sons Lorenzo, who conducts a store at Bayard, and Louis C, who is a farmer near the home place

EDWARD W TROUT — From a homestead upon which he filed entry in 1908 and to which he has added until he is now the owner of a valuable property of two hundred and twenty-four acres, Mr Trout has developed one of the fine farms of Sioux county, the same being situated in the southern part of the county, about five miles distant from Morrill and in section 31, township 24-56   As may well be understood the intrinsic and industrial value of the property is greatly enhanced by reason of its having excellent irrigation facilities — and it would be difficult to find anywhere more productive soil than these irrigated tracts in Platte Valley   Mr Trout has erected good buildings and made other modern improvements on his ranch and is here gaining substantial returns from his progressive activities as an agriculturist and stock-raiser   He is animated by loyal communal spirit, gives support to the various undertakings advanced for the general good of his home county and in politics he is not constrained by strict partisan lines   Both he and his wife hold membership in the Presbyterian church and are popular and valued citizens of the community

Edward W Trout was born in Atchison county, Missouri, on the 12th of January, 1875, and is a son of Abraham and Mary Trout   He was a boy at the time of his parents' removal to Valley county, Nebraska, where he was reared to adult age and where he profited duly by the advantages afforded in the public schools   He has been continuously identified with farm enterprise during his entire active career, and through the medium of the same he has achieved definite success

The year 1898 entered record of the marriage of Mr Trout to Miss Nettie Darrow, and the home circle is brightened by the presence of their four children — Merville, Virgil, Clifford and Wallace

PETER JANSSEN — It is pleasing to note the success which has attended the well directed activities of this sterling citizen since he identified himself with the great basic industries of agriculture and stock-growing in Scottsbluff county   He is the owner of a well improved irrigated farm of one hundred and sixty acres in the northwestern part of the county the place being in section 27, township 24-57, and about four and one-half miles distant from Morrill   Mr Janssen came to

Sioux county in 1907 and here filed entry on the homestead which has since been the stage of his progressive and successful activities as a farmer   He has not only brought his land under effective cultivation but has also erected good buildings and made other substantial improvements on the domain, the value of which has been greatly enhanced under his able management   He is always ready to lend his cooperation in the furtherance of measures projected for the general good of the community, is independent in politics, is affiliated with the Modern Woodmen of the World and is a citizen who enjoys unqualified popular confidence and esteem

Mr Janssen was born in Hanover, Germany, on the 6th of December, 1877, and is a son of Harm and Frauke (Peters) Janssen, of whose seven children he was the second in order of birth, the eldest, John, being now a resident of Idaho, Renske is the wife of John Heirl and they reside in Connecticut, Wilhelmina is a resident of Dubuque, Minnesota, and August, Fred and Henry live in northwestern Canada, as do also the parents   Harm Janssen, who was a farmer in Germany, immigrated with his family to the United States in 1890, and settlement was first made in Illinois, whence he later removed to Minnesota   Still later he became a resident of Colorado and finally he established his home in one of the provinces of western Canada, where he is successfully engaged in farming

Peter Janssen gained his rudimentary education in the schools of his native land and was a lad of thirteen years at the time of the family immigration to America   He was with his parents in Illinois and Minnesota and in the meanwhile he gained intimate experience in farm industry   He came from Colorado to Sioux county, Nebraska, in 1907, and here has worked zealously and indefatigably, regulating his activities with excellent judgment, with the result that he has achieved independence and prosperity as one of the representative farmers of the county   He is independent in politics and in a fraternal way is affiliated with the Woodmen of the World

The year 1903 recorded the marriage of Mr Janssen to Miss Clara H Schmutzler, who was born in Saxony, Germany, and who was a girl at the time of her parents' immigration to America   Mr and Mrs Janssen have three children — Emil, John and Minnie, — and they are being given the educational and home advantages that will fit them for honorable and useful citizenship

Otto A Smutzler, a brother of Mrs Janssen is making his home with this family He was born in Kansas, October 4, 1892 He enlisted at Wausau, Wisconsin July 15, 1917, and went overseas March 4, 1918, with the Thirty-second division He lost an arm and two fingers at the battle of Soissons August 3, 1918 He returned to the United States on October 16, 1918, and was discharged from a hospital February 11, 1919, at which time he came home He has surely proved his loyalty to the country of his adoption

JOSEPH G WOODMAN is a native of Illinois and his paternal and maternal ancestors were numbered among the early settlers of New England, that stern but gracious cradle of much of our national history He has been a resident of Scottsbluff county, Nebraska, since April 8, 1907, and has here reclaimed and developed one of the fine farms in the vicinity of the thriving village of Morrill, his homestead of eighty acres being eligibly situated in section 3 township 23, and being supplied with excellent irrigation facilities On his arrival in the county Mr Woodman took up his homestead which was entirely without improvements, and in the intervening years his vigorous efforts and good judgment have affected its development into a model farm that is devoted to diversified agriculture and the raising of high-grade live stock He gives his undivided attention to the management of the place and is making a specialty of raising fine Jersey cattle for breeding purposes He is one of the alert and progressive men of the county and is honored and valued as a loyal and public-spirited citizen

Joseph G Woodman was born in Kane county, Illinois, on the 3d of June, 1854, and is a son of Joseph and Eleanor (Barnard) Woodman, both natives of New Hampshire, where the respective families were founded many generations ago Of the six children the subject of this sketch was the sixth in order of birth, Freeman is a resident of Santa Anna, California, Juliet is the wife of Joseph Gray and they maintain their home at Portland Oregon, Iva was a resident of Ramona, California, at the time of his death, John, who met his death while serving as a youthful soldier of the Union in the Civil War, was a member of Company A Fifty-second Illinois Volunteer Infantry, and Elizabeth is the wife of Simon Chaffee, of Santa Anna, California The father, Joseph Woodman, became a successful farmer in Illinois, and died at the age of eighty-six years in

Orange, California, in 1896, his widow having likewise been venerable in years at the time of her demise, in 1901, both having been devout members of the Methodist Episcopal church and Mr Woodman having been an uncompromising advocate of the principles of the Republican party

In addition to receiving the advantages of the public schools of his native state Joseph G Woodman completed a higher academic course of study, in Northwestern University, at Evanston, that state Thereafter he resumed his active association with agricultural industry, but eventually he assumed the position of state grain inspector in Illinois, an office of which he continued the incumbent for seventeen years In 1907 he came from his native state to identify himself fully with the new and progressive county of Scottsbluff, Nebraska, and well has he justified his presence on this stage of activity, for he has achieved gratifying success as a farmer and has also been influential in those community undertakings that have conserved civic and material advancement He is a staunch supporter of the cause of the Republican party is affiliated with the Masonic fraternity and the Modern Woodmen of America, and both he and his wife are zealous members of the Presbyterian church — active in the various departments of its work Mr Woodman has been specially influential in Sunday-school work and was president of the Panhandle Sunday School state convention held at Scottsbluff in 1920

The year 1878 recorded the marriage of Mr Woodman to Miss Alice J Davis who was born in Milton Wisconsin, her grandfather (Joseph Goodrich) being the founder of Milton and Milton college, their marriage having been solemnized at Davis Junction, Illinois, a town named in honor of the father of Mrs Woodman Mr and Mrs Woodman have four children George is a successful grain broker in the city of Toledo, Ohio Harold J, a skilled civil engineer, entered the government service in connection with the activities of the American forces and was assigned to professional service in France honorably discharged at Camp Dodge Iowa, June, 1919, Genevieve J and Chester G are at the parental home both are students in Hastings College, at Hastings this state

WILBERN ROBERTS was born in Lee county, Iowa, January 10, 1857, and died at his home south of Bayard Nebraska April 9, 1916

He was the son of John and Mary (Gilchrist) Roberts, his father being a farmer in Iowa. There were eleven children in the family. He was educated in Iowa and took up farming in that state after completing his schooling, and stayed in his native state until he arrived at the age of 30 years, when he came to western Nebraska. This was in 1887. He took a government homestead of land that lies just south of the present town of Bayard. He followed farming and stock-raising and was very successful, being the owner, at the time of his death, of six hundred acres of land, having sold forty acres of the homestead to the town site of Bayard.

Mr. Roberts was married September 12, 1879, to Nancy A. Duncan, a native of Wayne county, Kentucky, who had moved with her parents to Iowa when she was seventeen years old. Then children were born to this union, namely: Elizabeth, deceased; Mary Etta, now Mrs. John R. Duncan, living in Wyoming; Edith Anna, who married Chester Morgan, and lives in Wyoming; Viola Ethel, now Mrs. Orville Smith, of Morrill county, Nebraska; Delia Amber, the wife of Charles Wilcox, of Morrill County, Nebraska; Raymond Rasson, a rancher, residing in Wyoming; Myrtle, deceased; Minnie U., now Mrs. Alvin Einsel, living at Bayard; Marie Ida, at home; Lila J., at home.

Mr. Roberts was a member of the Modern Woodmen and of the church of the Latter Day Saints. He was a Democrat in politics and took and intelligent and active interest in public affairs, although not engaged in politics himself as a candidate for office. He was a man who enjoyed a high reputation among all who knew him, as an industrious, honorable and progressive citizen, who prospered in his business and reared a large family who are a credit to his name and memory. His widow also survives him. Mr. Roberts lived to see the country that he found in the state of undeveloped prairie grown to a condition of wealth and prosperity, in which he received and was justly entitled to a considerable part. He leaves behind him a name that is honored and respected and his life work had proved a success when he laid it down.

STEPHEN SMITH, was born March 28, 1851, in Whippoorwill county, Kentucky, and died July 27, 1917. He was the son of Mark Anderson and Elizabeth (Taylor) Smith, who were both natives of Kentucky but moved to Indiana in 1860.

Stephen was educated in Indiana and came to western Nebraska in 1887 and took a homestead. He followed farming and stock-raising until failing health compelled him to retire from active business. He was married in 1872 to Amelia Wiley, who was born in Colfax, Illinois and who survives him. To them were born eleven children, namely: Pearl, now Mrs. William Pullen, living in Morrill county, Nebraska; Homer William, who lives southeast of Bridgeport; Orville, living in Morrill county, Nebraska; Clarence, deceased; John, residing at Liberty, Missouri; Mark, living in Morrill county, Nebraska; Elizabeth, now Mrs. Clarence E. Roberts; Nancy, deceased; Stephen, living in Morrill county, Nebraska; William, who lives at home; George, and Nannie, deceased.

Mr. Smith was a Republican in politics and a member of the Methodist church. He raised a large family and was respected as an upright and honorable man. He was one of the early settlers in his community and saw the country grow from a raw state to its present position of prosperity and wealth. He leaves an honored name and a family that is a credit to his memory.

GEORGE L. WHITMAN was born in Iowa on August 1, 1864, the son of George and Marie (Davis) Whitman, both the parents being natives of New York. There were ten children in the family six of whom are living. Their names are: Ella, now the wife of B. M. Odell, residing at Normal, Illinois; Martin, who lives in Fresno county, California; Charles, living in Morrill county, Nebraska; George L., the subject of this sketch; Ieora, deceased; Burton, living at Normal, Illinois; Ralph, living at Berwick, Illinois; the other three children died in infancy.

The family moved to Nebraska in 1871 and settled in Fillmore county, where the father engaged in farming. He was a Republican in politics and belonged to the Methodist church. He died in 1909, his wife having died in 1897.

The subject of this sketch, George L. Whitman, was educated in Nebraska, and after the completion of his schooling he engaged in farming and was also in the feed business. In the year 1900 he came to Morrill county and took a homestead, on which he followed farming and stock raising until 1915, at which time he and his persent partner started at Bayard the feed business which is conducted in the name of Walford & Whitman.

Mr. Whitman and his brother-in-law, Mr. Fulton, former a partnership in the farming business thirty years ago, being renters until they came to the Platte Valley, when they each took a homestead They now own three thousand four hundred acres and are still operating as partners

Mr Whitman was married in January, 1889, to Jennie E Fulton, who was a native of Ohio To them have been born five children, namely Leslie, who is a former stock raiser near Bayard, Nellie, now Mrs E R Lincoln, living at Bayard, J Benjamin, deceased, Mary and Edith both living at home

Mrs Whitman is a member of the Methodist church Mr Whitman is a Republican in politics and has always taken an interest in public affairs and is a leading member of his community He has been prosperous in his business and stands high in reputation with all who know him

JAMES WEBSTER was born in Ohio on May 27, 1859, the son of Marcus P and Anna (Taft) Webster, both natives of Ohio There were three children in the family

He was educated in the schools of Ohio and Minnesota, his parents having moved to the latter named state in his youth After finishing his schooling he engaged in farming for a few years, and in 1886 he came to western Nebraska and took up a homestead He followed the line of all the early settlers in the new country, farming and raising livestock and doing whatever else come to hand to do Among other things he freighted from Sidney to Alliance in the years before the railroad line was built After a successful period of ranching he sold his land and is now living in Bayard

Mr Webster was married in 1912 to Mrs Mary Boyer, whose maiden name was Mary White She was formerly the wife of J H Boyer and is the mother of four children, all of whom are living They are John E Boyer, who now lives in Marion, Iowa, James S Boyer, living at Angora, Nebraska, Melissa Jane, the wife of Thomas McCann, of Bayard, Charles H Boyer, living at Bayard

Mr Webster has no children He is independent in politics and is a member of the Odd Fellows He enjoys the friendship of a large circle of acquaintances and stands high in reputation in his community He takes an active interest in the growth of the country and of his home city, which he has seen develop from a little prairie village to one of the most progressive and prosperous young cities of the new West

JOHN L MUELLER was born in Council Bluffs, Iowa, the son of Gustav and Margaret Mueller His father was a native of Saxony, and his mother of Switzerland They came to America in 1854 and settled in Iowa, and there the father followed the occupation of a cigar maker Three children were born to them, one of whom, Henry, died in infancy The oldest child, Maggie, is the wife of William Bernauer and lives near North Platte, Nebraska The youngest is the subject of this sketch The father was a member of the Evangelical church He died in 1870, and the mother was married again to Frank Schram and they came to Morrill county in 1893 and settled on a homestead

Mr Mueller came west in 1888 and homesteaded He now is the owner of one hundred and sixty acres of fine irrigated land on which he was successfully engaged in general farming and stock-raising until a year ago when he moved to town and is connected with the Farmers Union

In 1892 he was married to Lena Arnold, and to this union six children have been born One child died in infancy The others are Edward, who lives on the home place

Mabel, married Ray Hunter, living near North Platte, Margaret, now Mrs Melvin Gund, residing at Aurora, Nebraska, George and Bessie, both living at home

Mr Mueller is a member of the Methodist church and of the Modern Woodmen He is independent in politics He has been successful in business and enjoys the respect and good opinion of a large circle of friends, being known as a man of upright character and industry He has lived through the period of early hardship and hard labor that goes with the development of a new country and now is in position to reap the fruits of his efforts in the prosperous condition of the community where his interests are centered

JONNIE S MAINARD is one of the pushing, energetic and enterprising business men of Mitchell, where he has built up a prosperous trade in the automobile and accessory business Born in Yankton, South Dakota, June 6, 1876, he is a son of Joseph and Mary M (Blade) Mainard The father was a Missourian by birth, born at St Joseph, while Mary Blade was born in Decatur county, Iowa, where she was reared and received her education She died in 1894 while still a young woman, being but thirty-eight years old when she passed away Joseph Mainard and his wife had six children John, Dora, the wife of Archie

Foster of Twin Falls, Idaho, Harry, in the livery business in Idaho, Slyvesta Anna, the wife of Edward Sparks of Twin Falls, Idaho; Verna, married Rube Ashby, a farmer of Cozad, Nebraska, and Forrest who runs a laundry in Twin Falls, Idaho Joseph Mainard, after attaining his manhood and completing his education engaged in independent business for himself as a farmer in Iowa, where he met and won his wife Subsequently he determined to avail himself of the free lands to be had in Nebraska and in 1886 came to Banner county where he took up a homestead and also preempted one hundred and sixty acres of land; he proved up on his holdings and on a tree claim so that he was one of the large land holders of the Panhandle After the first hard years he made good and permanent improvements on all his land, raised the soil to an excellent state of fertility and for many years was one of the progressive and successful farmers of this section He carried on general farming and stock-raising, gave all his family many of the advantages that he, as a youth, had been unable to obtain and in his later years was able to retire from the active management of his business and thus is spending the sunset years of life enjoying the comfortable fortune which he won from the soil by his own unaided efforts Today Mr Mainard is living quietly in Twin Falls, Idaho, where he takes an active interest in all progressive movements of the community and no one looking at him would believe that he had passed his sixty-fifth birthday, as his active out door life has kept him a man young in ideas and strength Both he and his wife are members of the Methodist Episcopal church, of which they are staunch supporters, and in politics he is lined up with the Democratic party, having cast his vote with it since first given the privilege of the franchise

John Mainard was reared and received his early educational advantages in the state of Iowa, where his father was a farmer for a number of years before the family came to Nebraska At the time the Mainards settled here Johnnie was a sturdy youth in his eighth year and after assisting his parents in breaking the farm land on the homestead in Banner county determined that he too, would start out on a career as an agriculturist but his young blood called for more activity and he did not locate near the family home but went to Wyoming where he filed on a homestead of one hundred and sixty acres, upon which he placed the improvements required by the government,

proved up and became a landed proprietor. During this time he was a foreman and rancher in Wyoming for eight years and spent fifteen years on the range, meeting with the well earned and deserved reward for the thought and labor required in the work on the farm, Mr Mainard decided that he would widen his field of endeavor He had always been interested in machinery, having a natural ability in a mechanical line After looking the country over for a desirable location he came to Mitchell in 1909, and engaged in a general livery business as an introduction to the business which he later planned to establish and after two years opened a general garage, carrying a complete line of accessories He is the district agent of the Hudson and Essex make of automobiles and today is one of the most substantial business men of the Platte valley His natural bent toward mechanics and the fact that he realized and seized the great opportunity offered in the wide field of motor transportation has led to his firm establishment as a progressive leader of the automobile field in the western Panhandle His company does a general electric repair work, a special feature is the vulcanizing of old tires that prolongs their life for thousands of miles of use He runs a battery, a service station, and now has an excellent patronage, not only among the citizens of Mitchell but all up and down the valley and from other parts of the state, the accurate and expeditious workmanship of the garage and the unfailing courtesy and consideration of Mr Mainard have given him a wide patronage and have combined to bring customers and make friends

In 1906, Mr Mainard married Miss Clara Green, a native of the Buckeye state, whose parents live in La Grange, Wyoming, where her father was for many years a farmer and rancher Three children have been born in the Mainard family Maude, Joella and Helen, all of whom are at home Mrs Mainard is a member of the Christian church in which she is a devoted worker Mr Mainard has carved his own career and it is but natural that he is an independent thinker on all subjects as he is a wide reader of the best books on professional and current subjects as well as the best of periodical literature Following out this line he is an independent voter, drawing no close line when the question of the best man to fill office is put before the voters, giving his influence to the one who can best serve the people In August, 1919 Mr Mainard sold his garage and at once started his present large building 100x140, brick, which will be the largest in Mitchell, which he expected to have completed by July 1, 1920

ORLA F COOK, a well known farmer and stock-raiser near Morrill, has been a resident of Scottsbluff county for fifteen years, and during that time has devoted himself to agricultural pursuits, his industry, thorough methods and good judgment bringing about very substantial results Mr Cook is descended from a long line of distinguished English ancestors as his father was a nephew of the famous English actor, George Frederick Cook Orla F Cook is a Hoosier, born in Orland, Indiana, February 3, 1875, being the son of George F and Lodaska (Rogers) Cook, the former born in 1843, died in Gresham, Nebraska, while the mother, a native of Indiana, born there in 1848, is still living at Gresham George Cook was one of the gallant sons of the nation who went forth in defense of the Union, when the Civil War was precipitated on this country In response to President Lincoln's call for volunteers he enlisted in the Union army and served three years, taking part in some of the hardest engagements of the war He was a member of the army which served under General Sherman and took part in the famous march through Georgia to the sea, as a member of Company B, One hundredth Indiana Volunteer Infantry He thus upheld the gallant traditions of his family, as there has been no war in which the United States has been engaged but that a Cook has served under the stars and stripes, and if necessary made the greatest sacrifice, of life, in defending his country At the close of hostilities, George Cook laid down the sword and took up the ploughshare to carry on the pursuits of peace and again engaged in farming in Indiana He was an ambitious man, and like many of the returned soldiers, wanted to get ahead in life, not only for himself but for his family and decided to avail himself of the cheap land to be had west of the Missouri In 1880, he came to Nebraska, purchased land near Ulysses upon which he at once began excellent and permanent improvements He was not daunted by the many privations and hardships incident to settlement in this, then, new country, but bravely set out to break the virgin soil, plant and till his crops, and erect a good home for his family So well did he carry out his work of improvement that in nine years he was enabled to dispose of the first holdings at a very satisfactory figure and with this money purchased more raw land, at once began to improve it, make improvements similar, but better to those on the first farm and as time passed carried on extensive general farming and stock-raising operations, which placed him in the front rank of the agriculturists of this section Mr Cook had great faith in the future of the Panhandle and remained through the years of drought, the invasion by insect pests and weathered the hard winter blizzards, when so many of the settlers gave up and returned to their old homes in the east and his faith was justified for with the passing years his land was raised to a high state of fertility and the three hundred and twenty acre tract in time returned him a comfortable and substantial fortune, due to his hard work and natural ability He was one of the brave men of high spirit and courage who played such an important part in the development of western Nebraska and thus paved the way for the present great agricultural production carried on in this section at the present time Early in life, Mr Cook was a Republican but after locating in Nebraska and studying the trend of events he became one of the leaders of the new Independent party and before his death swung over to the Democrats and the platforms advocated by that party He was a member of the Grand Army of the Republic, of the Masonic fraternity and the Independent Order of Odd Fellows, attending the Christian church of which his wife was a member There were two children in the family, James D, a farmer of Scottsbluff county and Arla who spent the first five years of his life on his father's farm in Indiana and when the family came to Nebraska was just a little lad He received his educational advantages in the public schools afforded on the frontier during his youth and thus laid the foundation for a good practical education to which he has been adding all his life as he has been a student of agriculture and the great questions of the day whether county state or national While on the home farm the boy early shouldered many of the small duties that are ever to be found around a country home and as his age and strength permitted he began to work on the land So while yet young he had learned by experience the practical side of farm business which has been of inestimable value to him since he began an independent business career In 1905 Mr Cook came to Scottsbluff county, as he had already realized that the ever successful agriculturist was the man who had water for his growing crops at just the right time and was not dependent upon the uncertain rainfall For some years he has been studying up on the question of irrigation and was of such keen vision that he realized that the rich soil of the Platte valley would produce any kind of crop and in great bounty with moisture. After looking this section over he filed on one hundred and sixty acres of land but subsequently released eighty acres of this tract, but later purchased another eighty that all might be under water. Today this land is one of the finest and most highly productive farms in the county as Mr. Cook conducts general farming and stock raising. Being a student of his profession he has decided that the greatest returns are made from a pure blooded strain of animals and has specialized in fine high breed Duroc Jersey hogs. From first locating in the valley he has introduced and practiced intensive modern farming, adopting the methods and machinery advocated by the state and government farm experts wherever he sees that such are of use to him and as a result Mr. Cook today is regarded as one of the leading prosperous and progressive exponents of the agriculturists in this section, noted for its able and productive farmers.

On May 28, 1912, Mr. Cook married Miss Augusta Rulla, a native of Nebraska, born in Johnson county and to them have been born two children: Opal and Clyde Coy, both of whom are at home and for whom a bright future may be forecast, as their parents have determined that both shall have every advantage in a social and educational way that the county and state affords, to equip them for life. Mrs. Cook has taken a keen interest in her husband's business and aided and encouraged him in all his business affairs. She is a member of the Lutheran church of which the family are helpful supporters. In principle, Mr. Cook believes in the platforms of the Democratic party but is bound by no strict lines when it comes to casting his vote, as he believes the man who qualifies best for every office should be elected to fill it, whether in county, state or nation. Fraternally he is associated with the Royal Highlanders, the Woodmen of the Word, Ben Hur and the Yeomen. He stands for progress and the uplift of the community in which he lives, advocating good roads, good schools and progressive business.

CURTIS M. CLUCK. — A resident of Nebraska and the Panhandle from the time when the only buildings known in this section of the state were composed of sod, Curtis Cluck has watched with the eye of a proprietor the various changes that have been wrought by the passage of years and the sturdy progressive work done by the first settlers, and has himself borne a full share of the labor in opening and developing the country to its present high state of production and prosperity. He is now one of the large landholders and successful agriculturists of the Morrill community of Sioux county where his accomplishments entitle him to the respect and esteem in which he is uniformly held by his business associates and many friends. Mr. Cluck is descended from a long line of worthy ancestors who settled in Pennsylvania at an early date and played an important part in winning that state from the wilderness and settling it up. He was born in Perry county, Pennsylvania, in 1862, being the son of William and Barbara (Kirck) Cluck, both born and reared in the Keystone state, the former living to be sixty-nine years of age, while the mother died in her forty-seventh year. William Cluck was reared and educated in Pennsylvania and after his school days were over learned the trade of blacksmith which vocation he followed in his native state for some years, but he saw little opportunity to give his family all the advantages he wanted for them in the east and being a wide-awake man realized that the greatest chances for himself and his family were to be obtained in the newer country west of the Mississippi valley and emigrating from the old home with his family, he took up a homestead in Banner county, filing on a hundred and sixty acre tract, upon which he proved up. Mr. Cluck broke the virgin soil of the prairie, planted crops to tide his family over the first winter on the prairies as settlers were few and far apart when the family came to the Panhandle and each family must supply itself the best it could in order to save the long wagon trips to the nearest market for supplies. Soon improvements were made on the land, permanent shelters for the stock and a comfortable house for the family which consisted of five children: C. M.; Katherine, the wife of Reverend Allen Chamberlain of Madison, Nebraska; Alice, deceased, was the wife of Thomas G. Granshaw, of Council Bluffs, Iowa; Anna, who marrier Lee England of Orient, Iowa; and Millard, who is a ranchman of Banner county. The father was a member of the Republican party, as he believed in its principles and with his wife was a member of the Evangelical Lutheran church, a faith in which the children were reared.

Curtis Cluck was reared on his father's farm

AARON P. FISHER AND FIVE GENERATIONS

and received the educational advantages afforded by the public schools of his neighborhood and then continued in the greatest of all schools, that of experience, which though a severe teacher, teaches lessons that are never forgotten, and he thus laid the foundation for the life work which has been fraught with struggle and success In 1895, Mr Cluck began his present career as an agriculturist by taking up a homestead in Sioux county, his first tract being one of eighty acres, where he at once began to make improvements in the way of excellent buildings for his stock and other farm operations and erected a comfortable home Here he carried on general farming operations, making a specialty of the dairy business There were few ranches in this section at the time Mr Cluck located and he was able to use the open range for his cattle for some years, and being a keen buyer and man of foresight was able to turn his cattle over often to good profit and thus laid the foundation for the comfortable fortune, which has come to repay him for the time, energy and toil that has been expended by him on the various branches of his profitable business When he first came to the Panhandle he saw little to encourage him, the country being practically undeveloped, while the few settlers, living far apart, were doing without conveniences and living in small houses if not sod dug-outs, but he had faith in the future of the section, both for farming and cattle and with his wife determined to remain and win fortune against all odds Today all his land is under cultivation, he has a fine modern home and excelletn outbuildings, and has established himself as a progressive and skilled farmer who thoroughly knows his business and can make his labor pay him proportionately, being known and highly respected throughout the Morrill section He still carries on general farming and raises all kinds of live-stock, the while the success which has attended his efforts is evinced by the consistency of his methods, as he is modern in thought and deed, employing the latest methods in his business as well as buying and operating the latest types of machinery to lighten the labors of farm work The family are members of the Methodist Episcopal church of which Mrs Cluck is an earnest worker, while Mr Cluck is a staunch supporter of the principles of the Republican party and while he has never aspired to hold public office he advocates all improvements for the community which tend to social and civic development

In 1883, Mr Cluck married Miss Isa Mc-Guffin in Iowa he was born in Pennsylvania, but accompanied her parents to Iowa when they came west to locate on a farm in the newer country There are two children in the Cluck family Elmer who lives in Morrill, where he is connected with the street department, and Blanche the wife of C J Goakey, a ranchman of Fulton, Wyoming

AARON P FISHER, an experienced and successful farmer in Scottsbluff county, is a native of Indiana, born in that fine old state, March 27, 1853 His parents were Samuel and Margaret (Huffman) Fisher, the former of whom was born in Virginia and the latter in Ohio They had eight children, Aaron P being the youngest The mother died aged seventy-six years and the father in his eighty-second year

Mr Fisher attended the public schools in Iowa in boyhood and remained with his parents and accompanied them to Nebraska in 1885 He settled first in Furnas county but conditions were hard there at that time and he moved into Kansas, but finally, after a trial of two years, decided that Nebraska offered better opportunities for the poor man, came back to the state and homesteaded in Cheyenne, now Scottsbluff county two miles west of McGrew Mr Fisher brought his family and household goods across the country from Kansas in a covered wagon with team and this wagon served as the first home Like other settlers of that time, Mr Fisher was forced to bear hardships that entailed loss of crops and stock, but his courage held out and now he is the fortunate owner of 309 acres of excellent land, 189 of which is under irrigation He confines his attention to general farming at present, but at one time he was in the cattle business

Mr Fisher was married December 28, 1884, to Miss Abby Foster, who was born in Kansas, a daughter of John and Sarah (Wilson) Foster, the former of whom was born in Ohio and the latter at Eddyville, Iowa, where they reside The maternal grandmother of Mrs Fisher was born in 1828 and still lives Mr and Mrs Fisher have three children Marion, married Sarah Howard and has one child Marie, Lilly, married Elmer Baquet and has two children, Berl and Myrl and Bert, who has always been at home Mrs Fisher belongs to the Baptist church Mr Fisher was independent in his political views He died December 28, 1919 The widow continues on the old place

GEORGE F TAPLIN, who is a well known resident of Sioux county, where he is engaged in farming, has been a resident of this state for fourteen years and during all this time has been interested in the agricultural development of the Platte valley

George Taplin was born in Hardin county, Iowa, in 1874, being the son of C R and Malvina (Harrington) Taplin, the former a man of sturdy vigor now in his seventy-first year, while the mother died a young woman at the age of thirty-five years in 1889 The father was a Canadian by birth, he received his educational advantages in his native country and there upon attaining his majority entered into independent business for himself as a farmer, but he looked to the south and when his family was fairly well grown decided to emigrate Coming to the United States the family located in Iowa, where C R Taplin bought a farm Here he at once put into use the practices of his earlier business life, conducting general farming operations and engaged in stock-raising After some years in that state he went to Colorado but did not find all conditions to his taste and decided to locate in Nebraska and took up a homestead in Sioux county He soon put many excellent permanent improvements on his eighty acre tract, erected good farm buildings, a comfortable home, and within a short time was regarded as one of the most reliable and substantial men of the district Mr Taplin was a well educated man, took a keen interest in the affairs of the community and as a reward for the public spirit he displayed his county elected his commissioner, believing that the affairs of this section would be well and honestly conducted in his capable hands Thus for seven years he not only conducted his own business but that of the county and he has the satisfaction of knowing that he holds the respect and confidence of its residents The Taplin family are members of the Methodist Episcopal church to which they give hearty support, while in politics Mr Taplin was allied with the Republican party There were three children in the Taplin family: Franklin, a farmer in Canada, Mary, the wife of F T Biddle, a foreman in an ore mill in Colorado, and George F, the subject of this review, who was educated in the excellent public schools of Canada before the family came to the United States Like all boys reared on a farm, he at an early age assumed many of the duties around the home place, learned from his father the practical side of farm business For a number of years Mr Taplin was engaged in farming in Iowa and Colorado, but he was ambitious to get ahead in the world on his own account and believed that with

his knowledge and experience in business he could do well on land of his own and determined to take advantage of the free land offered settlers by the government in Nebraska and in 1906 located on an eighty acre homestead in Sioux county He has raised the soil to a high state of fertility, believing in the best principles of intensive farming, has erected good and substantial farm buildings and a fine modern home for the family As the first land soon began to bring most gratifying results for the thought and labor expended upon it, Mr Taplin obtained capital with which he bought a second eighty, giving him a farm of a quarter section, all of which is under irrigation. He does general farming and stock raising.

On July 25, 1906, Mr Taplin married Miss Ellen Arbuckle, a native daughter of Nebraska, the daughter of J F Arbuckle, of Fort Morgan, Colorado, and to this union four children have been born Marion, Anna, Francis and Arthur, all of whom are at home The children will be given every advantage to equip them for life as both their parents believe in education as the best start in the world

The Taplin family are members of the Methodist Episcopal church of which they are liberal supporters Mr Taplin is allied with the Republican party and his fraternal affiliations are with the Masonic order Mr. Taplin is progressive in ideas and methods, stands for all communal and civic improvements and is one of the community's public-spirited men and one who has won the confidence and esteem of his business associates through his high ideals and integrity

GEORGE H GARRARD is one of the substantial farmers of Scottsbluff county He came to the Panhadle while this section was still new and here has lived to view the marvelous development of this favored section of Nebraska

George Garrard was born in England in 1870, being the son of James and Mary (Hill) Garrard, both natives of that country, where they were reared, educated and married The mother lived until her forty-sixth year, being survived by her husband who passed away at the age of sixty-two They had a family of ten children George was given all the educational advantages his family could afford in his mother country and after his school days were over became an apprentice to a carpenter and thus learned a profitable trade that was of inestimable value to him in his later life The young man was ambitious, he soon realized that there were few opportunities for him in his native country and while still young courageously severed all the dear ties binding him to his home and set sail for the newer land with its many openings on the west of the Atlantic. Soon after landing in this country Mr Garrard came west as her heard of the vast wide stretches of land that the government would give to settlers free and with the idea of becoming a landholder in 1891 he located in Kimball county. At this time he had just attained his majority but not daunted by the many privations he had to endure began at once to work at his trade as a carpenter and many of the early dwellings and business houses of that day were erected by his skillful hands. Subsequently Mr Garrard removed to the frontier village of Gering, where he engaged in business as a carpenter, gradually branching out into a modest contracting business, and he played no small part in the building and development of the present seat of justice of Scottsbluff county.

He desired land of his own and about 1905 bought land in section 22, township 23-57, where he at once began business as a farmer. Mr Garrard had devoted considerable study to agriculture problems before he purchased his farm so he at once began to inaugurate the latest methods of farming. He erected necessary farm buildings and a comfortable home, and has successfully engaged in general farming and stock raising. He and his wife are favorably known and highly esteemed in the community.

In 1906, Mr Garrard married Miss Margaret Schultz who has been a loving and devoted wife as well as brave helpmate in building the comfortable fortune to which they may look with pride as the result of their own unaided labors. In politics Mr Garrard is a supporter of the principles of the Republican party though he is bound by no strict party lines when it comes to local elections. Fraternally his affiliations are with the Ancient Order of United Workmen.

JOHN N HUDSON — As a follower of the oldest vocation of the human race, John Hudson has achieved that success which comes to a man who finds his work congenial and who invests it with determination, enthusiasm and natural ability. The agriculturist has ever before him the opportunity of making himself an enormously useful factor in the production of a community, and a realization of this fact has come to Mr Hudson in Scottsbluff county and the Panhandle, as he is one of the pioneer settlers of this section who have played such an important part in opening up and developing the western part of the state of Nebraska which for so many years was regarded as not only non-productive but a veritable wilderness by residents of the middle and eastern states. That his vision was far and keen is testified by the fact that today the valley of the Platte is one of the garden spots of the earth and during the recent World War, when all farmers were called upon to aid in feeding the starving hords of Europe, this section not long since regarded as worthless for producing anything but jack rabbits and grease wood, responded with food products in such quantities that it seemed incredible that land could produce such bountiful crops. But it was not the land alone it was the men who owned and operated it with their modern intensive farming methods, their hard work and determination to show the world what the American farmer could do and today the men who are engaged as agriculturists in the United States lead the world in every branch of this great enterprise. Mr Hudson is one of the best representatives of farming business in the middle west. He is descended from a long line of fine old Virginia stock that settled in the Old Dominion at an early day and its members have taken a dignified and active part in the history of the activities of that fine old state.

John Hudson was born in Marion county, Iowa, in 1866, just after the close of the Civil War, being the son of William J and Jane (Moreland) Hudson, the former born in Virginia and now living, hale and hearty at the advanced age of eighty years, while the mother was a native of the Hoosier state, born in Indiana in 1844, and now a woman of advanced years, having passed the psalmist's span of three score years and ten, but is still vigorous in mind and body as though she were twenty years younger. William J Hudson came west at an early day to take advantage of the many opportunities to obtain land on the then, frontier of Iowa and after locating in that state began his business career as a farmer planting the diversified crops which brought the greatest returns in that section and became known as one of the substantial and prosperous men of his community. Subsequently he became interested in a mining industry in Iowa, to which he devoted considerable time for some years. The following children were members of the Hudson family. Alice, the wife of William Sratton, who lives near Lake Alice; Clyde a farmer near

Mitchell and John   The family were members of the Baptist church to which they have ever been liberal contributors while Mr Hudson was a Democrat in politics

John N Hudson was reared on his father's farm in Iowa, attended the excellent public schools where he laid the foundation of his education which has been of priceless value to him during the many years of his business life   He grew up sturdy, self reliant and strong, early learning the practical side of farm industries and at an early age assumed many of the duties to be found around a country home   While still a youth he was a good capable farmer, so that when he decided to establish himself in an independent business of his own he wisely chose that vocation which he knew best and to which his tastes inclined and became an agriculturist   As land in Iowa was already high in price, the young man hazarded his fortunes farther west, where cheaper land was to be obtained from the government   Coming to Nebraska, Mr Hudson first located in Platte county, where he opened up and operated a frontier farm for some time before he again responded to the lure "of the farther on" and coming up the river settled in Buffalo county   He proved up on his land and became one of the well-to-do men of his section   Living near the river, Mr Hudson began to study the question of water for the land, for he had long since realized that with the fertile lands of the river bottoms, the never failing sunshine of western Nebraska that all the farmer needed to reap golden harvests was water, in sufficient quantity and at just the right time for the growing crops, for the only drawback to handcap the agriculturist of the high plains was his dependence upon the uncertain rains of this section   Becoming interested in irrigation and watching the success of men who were owners of land under ditch, after the canals had been dug in the Panhandle, Mr Hudson became convinced that the greatest future for him laid in such a locality and disposing of his holdings in the central part of the state came to Scottsbluff county in 1907   He filed on a homestead of one hundred and ten acres in the Morrill district, on section eighteen, township five   Being a man of experience, he at once began permanent and good improvements on the place, erected the necessary farm building for carrying on his business, built a comfortable home and soon had his land well worked   As his returns from his labor permitted he invested in more land, adjacent to his first homestead, on which he believed money could

be made, so that today he is the owner of a landed estate of two hundred and ten acres of rich farming land, ninety-six of which are under ditch, producing abundant and assured crops each year   Mr Hudson has not confined his energies to one line but is engaged in diversified farming and stock-raising, having a good grade of horses and cattle, he is a shrewd buyer and studying his business, is usually a good seller, shipping so that he obtains the benefit of the long side of the market in Omaha or Kansas City

Mr Hudson is a public spirited man who lives up to his standard of what an American citizen should be as he takes an active part in all local affairs, and being a man of education is well qualified to hold the office which he has been willing to accept as he is school director of his district, advocating the best and most modern methods and studies for the rural schools, as he fully unerstands that the education of the youthful farmer means a successful man in later life   In politics he is a staunch supporter of the principles of the Republican party while his fraternal affiliations are with the Independent Order of Odd Fellows and the Ancient Order of United Workmen

On April 14, 1887, Mr Hudson married Miss Minnie Beye, a native of Illinois, being the daughter of George Henry Beyer who came to Nebraska in the early eighties and located on a homestead in Platte county   Six children were born to Mr and Mrs Hudson Earnest G , a farmer in the state of Wyoming, William C , who served as a military police-man in the 338 field artillery 88th Division, A E F , in the United States army during the war with Germany , Olive E , the wife of J N Shaver, a resident of Scottsbluff county , and John W , who is at home   Bessie Alta died in infancy

In recording the lives of men who have been instrumental in the development of the Panhandle to what it is today, the historian would be remiss in his duties if he failed to give a prominent part to Mr Hudson who has been a pioneer across this great state as he helped its settlement at three distinct places along the great valley of the Platte

CHARLES N WEST, is one of the well known farmers of Scottsbluff county whose industry, energy and good management has placed his in comfortable circumstances and gained for him a reputable standing in this progressive section of the Platte valley

Charles West was born on a farm in Douglas county, Illinois, in 1873, being the son of A J and Sidney (Campbell) West The father was one of the gallant sons of this great country who responded to President Lincoln's call for volunteers to preserve the integrity of the Union when the United States was threatened with disintegration He was born, reared and educated in Indiana, but before the outbreak of hostilities had established himself as a farmer in Illinois and it was as a member of the Twenty-fifth Illinois Volunteer Infantry that he entered the Union army serving for three and a half years and thus participating in many of the hardest battles of that memorable conflict Today Mr West is still living a hale, hardy, well preserved man of seventy-seven and from the sunset years can look back along the decades and feel that he has played a worthy and important part in the history of his country Sidney (Campbell) West was born in Ohio in 1844, is living at Osceola, Iowa

After the close of the war A J West moved to Iowa where he was able to get land for a reasonable price and there established himself as a general farmer and stock man There are nine children in the West family John, a farmer in Iowa, James, a locomotive engineer who went to Japan when the great Trans-Siberian railway was built and has since been associated with that company Charles, Edward, the sheriff of Clark county, Iowa, L S, who resides in Iowa, Nellie who is living at home with her parents, Minnie, the wife of True Wood, an Iowa farmer, George, a carpenter who was in the service of the government during the war with Germany, serving twenty months in France with the engineering corps, and Walter, who is in the government mail service at Osceola, Iowa

Charles West gained his early education in the common schools of Illinois and while going to school also assumed many duties on the home farm, thus early becoming a potential business man while a youth in years After finishing school he chose agriculture as the vocation toward which his inclinations turned also as the business with which he was most familiar, but he was ambitious to get ahead in the world and as Iowa was so well settled up that land was high in price, decided to hazard his fortune in Nebraska and in 1905 came to Sioux county where he took up a homestead of eighty acres on which he proved up Later he purchased the Royce place in township 23-58, section 10, which gave him a landed estate of four hundred and fifty-three acres of which two hundred acres are under ditch Mr West

has made this one of the valuable farm properties of his neighborhood and engages in general farming and stock-raising

In 1912, was solemnized the marriage of Charles West and Miss Polly Johnson, the daughter of David Johnson, who was a resident of Iowa when his daughter was born Both Mr and Mrs Johnson are now deceased Mrs West is a member of the Baptist church, one of the substantial progressive women who are making their mark in this twentieth century and demonstrating that it is the woman at the wheel who helps make the world roll on to a greater and better development Mr West is a Republican in politics and his fraternal affiliations are with the Masonic order, being a Thirty-second degree Mason and also a member of the Odd Fellows

JOHN BOATSMAN — Definite efficiency has characterized the service of the Boatsmann in the responsible office of president of the Farmers and Merchants State Bank of Morrill, and his administration has done much to conserve the success that has marked the history of this important and representative financial institution of Scottsbluff county, the while his personality and civic loyalty have gained for him an enviable place in the popular confidence and esteem of the residents of the valley Mr Boatsman was born in Iowa, September 5, 1875, being the son of Deark M and Margaret (Menken) Boatsman, both of whom were natives of Iowa, where they were reared educated and later met and married There were five children in the Boatsman family four of whom are still living Mangel, Fannie, the wife of Menhard Ehmen, who resides at Sterling, Nebraska, John, the subject of this review Minnie, who married Henry Eilers of Sterling, Nebraska and Carrie, the wife of Edward Johnson Mr Johnson himself is now associated with the management of the Farmers and Merchants Bank of that city of which D M Boatsman is president The parents were members of the Evangelical Lutheran church, a faith in which the children were reared The father was and adherent of the principles of the Democratic party with which he usually cast his vote Margaret Menken Boatsman died November 12 1899, leaving a sorrowing family and darkened home, as she had been a devoted wife and mother and worthy helpmate for her husband during the years of their married life

John Boatsman was reared in his native state and there in the common schools laid the foundation for the excellent practical education that has proved of inestimable value to him during the years of his commercial life. At an early age the boy determined upon a career in the realms of finance and as soon as his school days were over entered a bank to there learn the practical side of the business and soon became a business man of marked circumspection and progressiveness, so that success has come to him as a natural perogative. That this choice of a vocation was a wise one needs no telling to the friends and business associates of this man who while yet young in years bears the responsibilities that so long were regarded as only to be held by the grey beards, but new blood was needed in financial circles and it has been such youthful bankers who have written history in the Panhandle and Nebraska. In 1909 Mr. Boatsman decided that a great future was in store for the irrigated sections of the Platte valley not only for the agriculturist but for allied interests and one of the most important of them was the banks, which play such an important part in opening up and developing any section of the country. With this idea in mind Mr. Boatsman located in Morrill where he became the prime mover in the organization of the Farmers and Merchants Bank and its heaviest stockholder assuming from the first the guiding hand in its policies as he was elected first president of the institution. The bank, founded the same year Mr. Boatsman came to the city, based its early operations upon a capital stock of $25,000, has surplus of $20,000 and deposits of $25,-000, an aggregate that is fairly large for a young institution and well demonstrates the able management that has established a policy which has given the bank the confidence and hearty support of the resident of the entire Morrill valley.

As a broad minded and progressive citizen Mr. Boatsman manifests lively interest in all things touching the communal welfare and the development of the many and varied industries of this rich, teeming agricultural district that has so well demonstrated what the American farmer and business man can accomplish when they set out to pace the world in production and modern business. Mr. Boatsman is too wide guaged a man to be tied down by strict party lines in casting his vote in local elections though nationally he advocates the principles of the Democratic party with which he usually voted. His fraternal affiliations are with the Masonic order of which he is a member of high degree, being a Thirty-second degree member of that order and is also associated with the local lodge of the Benevolent and Protective Order of Elks of Scottsbluff.

On September 5, 1899, Mr. Boatsman married Miss Emilie Ehmen, who was born in Illinois, and to this union one child has been born: Joy Sterling, who, after completing the high school course of four years entered the State University of Nebraska, where he expects to spend another four years and graduate with his Bachelor's degree. Mrs. Boatsman, a woman of refinement and culture is a leader in the social activities of her city, being the popular chatelaine of one of the most attractive and hospitable homes in Morrill, a position in which she is ably supported by her husband who has made a host of friends in the Panhandle and is held in high esteem by the many men of his wide and varied business acquaintance as well as the warm friends of Scottsbluff county, where he is winning an enviable position as one of the younger and progressive members of the banking fraternity.

HENRY G. KARPF, cashier of the First National Bank of Morrill, is one of Scottsbluff county's young and progressive citizens who have created a favorable impression in banking circles and established themselves in positions formerly held by men many years their senior. Mr. Karpf is not a product of the Panhandle though he is a middle western man and thus has the push and energy that is the most striking characteristics of men born and reared west of the Allegheny mountains. He is a native of the Buckeye state, born in Conneant, Ohio, July 22, 1891, being the son of Charles and Hattie (Daniels) Karpf, the former born and reared in Colorado while the mother like her son first saw the light of day in the Ohio valley and is a worthy representative of one of the hardy pioneer families that located in Ohio when that state was close to the American frontier. Three children formed the younger members of the Karpf family: Louise, who is married to O. K. Collerick, a resident of Ohio; Henry, the subject of this review; and Marion, who is still in Ohio. Charles Karpf was for many years a merchant in Ohio and still continues the business to which he has devoted his time and energies to a result that he is now one of the well known and substantial business men of his district. He takes an active interest in the civic and communal affairs of his locality and in politics is an adherent to the princples of the Republican party.

Henry Karpf was reared in his native state

and as his parents lived on a farm was thus enabled to secure the best educational advantages during his youth in the community's public institutions and soon after completing his schooling started in to attain his business training in local banking institutions, where his early career reflected credit upon the schools in the excellent foundation that he had been enabled to establish and upon which his later financial career was erected From first entering business life the influence of his early training has been of inestimable advantage to Mr Karpf and his career thus far has in turn reflected credit on those who instructed him and also upon his own ability and character

Desiring a broader field for his endeavors while still a youth of sixteen, he determined to enter the realms of finance and believed that there were many more opportunities in the newer country west of the Mississippi river, came to Nebraska in 1907, to accept a position in the First National Bank of Mitchell where he became acquainted with this section of the country and the banking business at first hand and within a sort time had become established as one of the rising younger men in banking circles of the Panhandle Devoting not only much time but deep study to the many problems that arose in his business Mr Karpf became an authority on many of the intricate problems that arose in the bank He was conservative in policy and at the same time was far sighted enough and also progressive in his ideas and methods that few opportunities for increasing the business and establishing the prestige of his institution escaped him so that in 1913 when he came to Morrill to accept the position of cashier of the First National Bank of this city Mr Karpf had an enviable reputation and his high standing among the men of importance in the financial circles not only of the Panhandle but throughout western Nebraska and the neighboring states to the west and north The First National Bank has a capital of $25 000, surplus of $5,000 and deposits of $400 000, which establishes it in the front rank of banks in the middle west Mr Karpf is a man of marked business ability, strict integrity, personal probity and sound citizenship, and he well merits the success and respect which he has won among the citizens of the Morrill valley and Scottsbluff county In his political views Mr Karpf is independent, not bound by close party lines when he votes in local elections, choosing the man best qualified to serve the community rather than a party candidate His fraternal associations are with the Masonic order in

which he has taken his Thirty-second degree

On August 21, 1916, Mr Karpf married Miss Lodicea Babcock, a native of Colorado, where she was reared and educated Mr Karpf is one of the coterie of young business men who are making history in the Panhandle today and it is to them that we must look to maintain the high standard set during the great war and which for the development of our country must be maintained in the future and this torch of success which they are so ably carrying will then be handed down the years to posterity

EDWIN A BEARD, M D, is one of the favored mortals whom nature launches into the world with the heritage of sturdy old colonial ancestry, a splendid physique, a masterful mind and energy enough for several men Added to these attributes are excellent intellectual and professional attainments and the useful lessons of a wide and varied experience which he has stored away and which are within call whenever an emergency arises He is the true type of the natural physician and gentleman, using the word in the older and best accepted sense, and today is a worthy representative of the best in professional and communal life, dignified and yet possessing an affability and abiding human sympathy that have won him warm friends among all classes and conditions of men in the Morrill valley and all over Scottsbluff county

Mr Beard was born in the state of Wisconsin, February 3, 1871, being the son of Abraham and Sarah F (Hays) Beard, the former being a worthy scion of one of the old colonial families of Virginia, where he was reared and received excellent educational advantages during his younth and upon attaining manhood's estate decided to seek his fortune farther west, he crossed the Allegheny mountains then passed along the southern shores of the Great Lakes, but did not find that which he was seeking until the great pine forests of Wisconsin were entered and there, in 1841, he became a pioneer of the Badger state Nothing daunted by the work required to claim a home and farming land from the wilderness Abraham Beard gradually cleared his land, planted his crops, erected a good though primitive home for his family and with the passing years became a man of prominence in his section He was a hard worker, had excellent executive ability and with the passing years accumulated a comfortable fortune and in his later life could look down the passing decades and feel that his was a life well spent, that he had played no unimportant part in opening up and developing the west for settlement and the culture and civilization that have since become so firmly established in that great commonwealth bounded by two of the greatest bodies of fresh water in the world It was such men as Mr Beard who blazed the way for Wisconsin to become a state in which some of the best and greatest reforms of our great country have had their inception and later been adopted by other states and the nation itself He was a staunch adherent to the tenets of the Republican party and a loyal member of the Presbyterian church, to which both he and his wife belonged Passing away in 1880, Mr Beard had lived long enough to realize that the work, privations and hardships of the pioneers had not been in vain and handed on to posterity, through his children a heritage of inestimable value Mrs Beard was born in Indiana where she was educated and passed her girlhood, she proved a woman of courage and resource, as she accompanied her husband to the pioneer home in the wilderness of Wisconsin and there became the helpmate every man needs who is combating nature and the primitive and for all the years of his life was the one to encourage in dark hours and enjoy in the happy ones She survived her husband, living beyond the psalmist's span of three score years and ten, as she was not called to her last long rest until 1903, a woman whose good deeds were as the number of her days

Edwin Beard was reared on his father's farm and attended the public schools where he laid the excellent foundation for his higher studies that proved of inestimable value when he entered college While attending high school he had already determined upon a professional career Having grown up in a frontier community he was mentally advanced far beyond his years due to the many decisions and occasions when he was required to think for himself and act quickly without the opportunity of consulting an older or more experienced person After graduating from the local high school Dr Beard entered the College of Physicians and Surgeons, of Keokuk, Iowa where he received his degree of M D in 1898, and the following summer located for the practice of his profession at Cobb Wisconsin, where he remained three years, built up a good practice, of which he disposed to advantage when he sought a wider field and removed to the city of Milwaukee For three years the doctor devoted his time, talents and energies to special branches of his profession in the city but came to realize that a young man had greater oppor-

tunities for varied practice in a newer settled country than Milwaukee county and decided to locate farther west In 1904 he came to Nebraska, looked many towns over before settling in Stanton, where he opened an office and was soon enjoying a most gratifying practice both in the town and the surrounding country Being a student of affairs as well as medicine and human nature, Dr Beard sized up the conditions in this state very accurately He realized that the independent man is the one who does not have to rely solely upon his fellow men for a living, that a land owner is freer than the city dweller, so after six years he determined to join the ranks of the landed class and being far sighted and a good business man availed himself of the practically free land yet to be obtained in the Panhandle and in 1910 came to Scottsbluff county to take up a claim in the Henry neighborhood He lived on the land, made the required improvements, broke the soil, planted crops for the necessary supplies for the family and his animals and at the same time engaged in the practice of his profession throughout the valley, supplying a great need for medical and surgical attention that the residents of the section had been forced to do without or else go a long distance to obtain During the period on the farm Dr Beard established himself highly in the esteem of his patients, his friends and business associates and within a short period was one of the most popular physicians in the Morrill district, enjoying a lucrative practice which many a city man would envy When the town of Morrill was started he became its pioneer physician and as the town has grown so has his business until today he is one of the most prominent men of the medical profession in Scottsbluff county and the Panhandle, being called long distances for consultation on difficult and complicated cases In his profession and as a man he is ever one to remember and aid "those who are forgotten," and he bears optimistic cheer and encouragement as well as professional administration to those in suffering or distress so it may well be understood that he is loved in this community in which he has lived and labored for a decade

September 8, 1902, Dr Beard married Miss Alice A Pooley, the daughter of Robert and Esther (Rapson) Pooley, both natives of England who came to the United States when small with their parents in about 1840 Robert Pooley was a farmer of considerable success and went overland to California in the early fifties Miss Pooley was born at Scales Mound, reared and educated in Illinois and to them

HARVEY WALLACE CRUME AND FAMILY

two children have been born Esther Alice and Robert A, who are growing up at home and for whom a bright future is in store as their parents will give them every advantage as to equipment for their life work whatever they may choose Mrs Beard is a woman of high attainments, and since her marriage has been the able and valuable coadjutor of her husband in his business and professional life Her gracious womanhood, gentle sympathy and deep interest in the movements for civic and communal uplift have given her a distinct place in the life of Morrill where she is the chatelaine of the generous hospitality dispensed by herself and the doctor, who have a host of warm friends throughout the valley The Beard family are active members and supporters of their Presbyterian church while the doctor's fraternal affiliations are with the Masonic order and politically he is a Republican, though not bound by party lines in local elections

HARVEY WALLACE CRUME — There are few residents of Scottsbluff county perhaps who are better informed than Harvey Wallace Crume as to the advantages, in a business way and otherwise, that belong to those who can claim Nebraska for a home Mr Crume has spent a half century in this state and has been identified with the development of different sections

Harvey Crume was born in Vernon county, Wisconsin March 25, 1847 His parents were John William and Rhoda (Griffith) Crume, both of whom were born in Illinois and lived into old age, the father dying in Nebraska at the age of eighty-seven years and the mother when aged eighty-eight years They had eight children, four sons and four daughters Harvey being the fourth born in the family The father followed the trades of blacksmith and wagonmaker

In 1869, when twenty-two years old, Harvey Crume came to Nebraska, locating first at Lincoln and afterward teaching school for some years at Dunbar, in Otoe county Afterward he lived in Nemaha county and for a number of years was employed at Peru Mr Crume has witnessed many changes and has met people from almost every part of the country and his reminiscences are very interesting In 1910 he came to Scottsbluff county and homesteaded 120 acres, placed substantial improvements on his property and has one of the best irrigated farms in this section

In early manhood Mr Crume was married to Sadie Laughlin who was born in Illinois, as were her parents John and Olive Laughlin Her father was a blacksmith by trade To Mr

and Mrs Crume the following children were born John W, who is a resident of St Louis, Missouri Gertrude, who is deceased, Pearl, who married Walter Carlson and is living in their home in Scottsbluff county, Roy, who attends to the operation of the homestead, Leona and Ernest Frame who live in Mitchell, and Ray, who is deceased Mr Crume and his family are members of the Methodist Episcopal church He has always voted the Republican ticket

ZINA PHINNEY — The experiences of Zina Phinney, now a member of the retired colony of Morrill, have had a wide range, as he is one of the pioneers of the Platte valley and has the distinction of being one of the first residents of the Morrill district For thirty-two years he has resided within the borders of Scottsbluff county, during which time he has seen the country grow from the wild untamed prairies into a smiling country side where the growing crops in the spring seem to be an endless emerald sea, dotted with prosperous villages and the mainstay of Americanism, the district school Mr Phinney has nobly assisted in all this development, his experiences having included the various conditions, incidents and rugged happenings of the frontier days to the refinements and conveniences of our modern civilized existence His career has been a long, useful and eminently successful one, and in his declining years, he may look back with a measure of pardonable pride over his accomplishments For many years he occupied a leading and prominent place among the progressive agriculturists and ranchmen of Morrill valley and at one time the large landholders of Scottsbluff county

Zina Phinney was born in the Empire state, descended from a long line of worthy ancestors who played an important part in the development of that state from its early settlement He was born in Green county, New York, October 22 1843, being the son of Henry and Mary (Carter) Phinney, both born and reared in the same great commonwealth, where they received the advantages of the excellent public schools and after attaining man and womanhood met and were married To them eleven children were born of whom five are living Henry Phinney after finishing school established himself as a New York farmer, being thrifty, hard working and an able manager he gathered a considerable competency and became one of the prosperous agriculturists of his section He lived until 1900, being survived by his wife one year

The early educational discipline of Zina Phinney was secured in the district schools of Green county, New York, where he was brought up under the training of his practical father, who instructed him fully in all the departments of agricultural work, so that when he reached manhood he began farming for himself, in partnership with his father, and this business was conducted by them jointly until Mr Phinney was in his twenty-fourth year, when the younger man decided to come west and obtain land of his own He first located in Iroquois county, Illinois, rented a farm The family moved to Iowa, where the father bought a quarter section of land upon which they lived twelve years But ever westward the tide of civilization takes its way and as Mr Phinney was a student of the times as well as his own occupation, agriculture, he knew that though Iowa is a rich farming state that a man who could grow good crops for a few years on new rich land could make more money than to remain in the older settled localities And willing to prove his faith, he disposed of his holdings in Iowa and after looking the country over located on a homestead in what is now Scottsbluff county in 1887 He also took up a tree claim and with the passing years proved up on both, which was the beginning of his later large landed estate The first year in the Panhandle he was alone as he did not wish to subject his family to the hardships until he had made some preparation for their reception The first year he broke what land he could, put in the first sod crops, erected some of his farm buildings, those absolutely required and lived in his own "soddy" warm and comfortable though the winter blizzards were severe In 1888 the family followed and there were comfortable quarters for them to live in, though nothing like what they had been accustomed to in their old home, but both Mr and Mrs Phinney were courageous souls, determined that nothing should daunt their courage, and truly nothing did, for they stuck throughout all the hard years of drought, insect pests and the severe winters when much of their stock died They firmly believed that the Platte valley was to become a rich agricultural section in time and that their faith was justified needs not be reiterated All the family worked in the early days the older members assuming the heavy, hard tasks but the little people did their share tending cattle, milking and driving plows and cultivators as soon as they were big enough to handle horses and as a reward the family prospered beyond other

settlers With the passing years as he amassed sufficient capital Mr Phinney purchased more land until at one time he owned sixteen hundred acres, mostly grazing land that he disposed of at $30 per acre From first establishing himself here Mr Phinney engaged in general farming operations, principally the stock-raising business, and as he came here at the time when the cattle business of the great rich companies was in its heigh-day he naturally became interested in that industry and as soon as his means permitted became an extensive stockraiser and feeder at a later date All his varied interests proved profitable due to his able management, his good judgment in buying and selling at the right time and the many years of experience he had before coming west Mr Phinney was a man who was willing to carry out any project that tended to the benefit of his community and in 1894 accepted the Star route from Gering, which ran up the Mitchell valley and for four years served the residents of that section with marked ability so that they were sorry to see another man take his place when he was obliged to resign in order to devote his entire time to the growing demands of his own farm and its allied business interests The Phinney family are active members and generous supporters of the Christian church, and Mr Phinney has been a supporter of the principles of the Republican party since he cast his first vote, and though he has never desired or had the time to hold public office he is a public spirited citizen who helps and advocates all civic and communal movements for the benefit of his community and is a worthy representative of the class who live their ideas of Americanism and by their precepts hand down to their children and posterity an example that it will be well for the rising generation to emulate

On March 8, 1865, Mr Phinney married Miss Laura Lake, a native of New York, born May 29, 1845, and to them five children were born of whom only two survive Delbert, living southwest of Morrill, and Daniel, also a resident of the same district, where they are considered two of the most substantial and responsible men of their community, and thus are carrying on the reputation of the family established in the valley so many years ago by their father

CHARLES E SWANSON, one of the early and prominent settlers of western Nebraska, is a native of Sweden, born December 3, 1860 He is the son of August and Matilda (Anderson) Swanson, both parents being natives of

Sweden who came to America and settled on a homestead in Cheyenne county, Nebraska, northwest of Sidney, in 1886 The father engaged in farming and stock raising He had been a colonel in the Swedish army, was a member of the Lutheran church, and after coming to America was a Republican in politics He died in Des Moines, Iowa The children who survive, in addition to the subject of this sketch, are John, living in Moline, Illinois, Minnie Anderson, in Des Moines Iowa, Matilda, in Oskaloosa, Iowa, and Ida, in Fremont, Iowa

Charles E Swanson came to Iowa in 1879 and worked on a farm and in the mines He came with his father to Cheyenne county in 1886 and took a homestead on which he later proved up, and four years later came to Scottsbluff county and bought a quarter section of land in Mitchell valley He now owns four hundred acres of the finest irrigated land in the world and devotes himself to general farming and stock-raising

Mr. Swanson was married September 17, 1883, to Freda Anderson, who was born in Gothenburg, Sweden, and to this union seven children were born They are Clem a farmer in Wyoming, Helen, now Mrs Theodore Elquist, of Torrington, Wyoming, Alex, who is proprietor of a meat market at Morrill Ruth, living in Denver, McKinley who is now at home after serving in the United States army in France, with the 15th Machine Gun Battalion of the Fifth Division, having been in the front line at the Argonne Forest and Meuse River battles, wounded and gassed and discharged in February, 1919, and Paul who saw thirty-eight months of service in France in the front line of battle with the Canadian army, having enlisted at the beginning of the war in Calgary Harry M is still at home

Mr Swanson has good reason to be proud of his family as well as of his own success as a developer of this western country He is well and favorably known for his industry and honesty and is justly regarded as one of the most reliable and substantial members of his community By enterprise and good judgment he has made himself comfortable for the future and is still in the prime of his life with many years ahead of him He is a Republican in politics, a director of the school district in which he lives, and has served as assessor for a number of years He belongs to the Modern Woodmen and is a Methodist

HARRY E BAIRD, was born in Osceola Iowa, in 1879, the son of Samuel Baird who was a native of Wisconsin, and Johanna (Carpenter) Baird, also a native of Wisconsin, now deceased The father was a general farmer in Iowa and owned land there devoting himself to farming and stock-raising He is now the owner of land in Scottsbluff county There were four children in the family, Harry being second youngest The other are Clarice, living in Iowa, Nellie, now Mrs Louise E West, of Scottsbluff county, and Frank, who is engaged jointly with Harry in farming operations here The father is a Republican in politics The mother in her life time was a member of the Methodist church

Harry came to Scottsbluff county in 1905 and took a homestead He now is the owner of 160 acres of Scottsbluff county land, all under irrigation, and is engaged in general farming in conection with his brother Frank The place is well improved and promises to keep on advancing in value as the western Nebraska territory is further developed, and Mr Baird is in position to carry on his farming and stock-raising enterprises profitably while growing with the country During his residence in Iowa he was a dealer in pure bred Shorthorn cattle and Percheron horses He has not yet branched out much in the pedigreed stock line in his present location, since it has been a man's size job to take the undeveloped land as he found it and bring it to its present high state of productiveness, but he may take up the old line in the future

Mr Baird is a member of the Woodmen of the World, is a Methodist in religion, and is an independent voter though Republican in principle He stands high in the opinion of those who know him and is well known as a straightforward and enterprising member of the community in which he resides

ROBERT CURRY — In recounting the life history of the men who have come to Nebraska from other states and acquired substantial standing in the Panhandle, attention may be called to Robert Curry, who is a heavy landholder in Scottsbluff county and one of the progressive and responsible citizens of the valley In business operations covering a decade and a half, he has built up a reputation for astuteness as well as integrity, and today typifies the American farmer at his best as he stands in the forefront of the progressive movements that have placed the Platte valley at the head of the productive sections of our broad land

Mr Curry was born at Cassville Missouri, in 1876, being the son of Frank and Martha (Horner) Curry, the former a native of Tennessee while the mother was a Missourian

Frank Curry was reared and educated in his native state and being ambitious in his youth looked for wider fields and as he had chosen farming for a life's vocation located on a farm in the rich Missouri bottom lands where he soon became established as a man of worth and ability He and his wife passed their lives in the state of their adoption They had five children W N , now a farmer in the state of Washington, Charles, in Missouri, A L, following farming in Nebraska, as he owns a landed estate near Morrill, Robert, and Mary Ann, the wife of George Skelton, living in Missouri Mrs Curry was a member of the Baptist church during her life while Mr Curry was an ardent supporter of the tenets of the Democratic party

Robert was reared on his father's farm in Missouri, attended the public schools near his home and after his elementary education was completed remained on the home place assisting his father in agricultural work for several years When the time came for him to become established in independent business, Mr Curry chose farming, as the vocation to which his tastes tended and also as the one in which he had already gained practical experience For a period he remained in his native state but Missouri had become well settled up and as he desired to become a land owner he decided to avail himself of the cheaper land farther west where he could take up a claim After looking over the various localities yet open to settlement he believed that the greatest future for the agriculturist was along the Platte where irrigation insured crops and farming was not quite such a gamble as many men found it who depended upon the uncertain rainfall In 1907 he came to Scottsbluff county and took a claim in section 4, township 23-57, where he at once began good and permanent improvements He erected the necessary farm buildings, broke the sod for the first crops and soon built a comfortable home for the use of the family Having had years of experience, Mr Curry soon had his farm in fine shape, the soil was cultivated and raised to a high state of fertility and as he gave time and study to his business, soon was putting in the crops that brought the greatest returns The first homestead consisted of eighty acres, this was proved up and as soon as his capital permitted Mr Curry purchased other tracts adjoining so that today he has a landed estate of two hundred acres of land, all under ditch and this irrigated land produces as much as a farm four times its size without water rights From first locating here Mr Curry

realized that a farmer did best by having several lines so that he has carried on diversified agriculture and stock-raising in all of which he has been markedly successful, due to his work, foresight in planting and selling, and his executive ability and today is rated one of the most substantial agriculturists of the Morrill district He is a man of public spirit, who has kept fully abreast of the times, and believes that the world moves and its citizens must do the same or fall behind and as a result has advocated all the progressive movements that have played such an important part in the development of this country He has served his district as school director for a number of years to the great satisfaction of the men who elected him, is a member of the Presbyterian church to which his wife also belongs, believes in the principles of the Democratic party though not tied by strict party lines in local elections He is a Mason of high standing, having taken the Thirty-second degree

In 1902, Mr Curry married Miss Dora B Skelton, the daughter of J N Skelton, of Missouri, where she was born The Curry family consists of two children Frank J and Julia Alice, both at home

The Curry family has become well known in the Morrill district, where they have a host of friends and as both Mr Curry and his wife are ever ready to do their full duty for the home locality and the nation, are the type of true Americans to whom the future of our country may well be instrusted

VINCENT A GARRETT, a son of the Golden State, who typifies the best in American manhood and today stands in the foremost ranks of that great body of men who are so prominently in the eyes of the world, the "Great American Farmer," has already scored a gratifying success for so young a man

Vincent Garrett was born in Nappa county, California, in 1879, the son of Samuel and Esther (Bodwell) Garrett, the former a Buckeye by birth, as was his wife They were reared and educated in their native state and later went to the Pacific coast where some of their children were born After some years in the Golden State Samuel Garrett removed with his family to Red Willow county, Nebraska and later to Colorado in 1890, where he located on land near Fort Collins and for sixteen years was engaged extensively in feeding sheep along with his general farming operations In 1906 he came to Sioux county, purchased land which he farmed and also engaged in stock-raising After his farm was in satisfactory condition Mr Garrett opened an office in Morrill, in partnership with Nichols and Carpenter, the firm handling real estate He continued to carry on these varied business lines until his death, being survived by his wife who still resides in Morrill Mrs Garrett is a member of the Methodist Episcopal church, while during his life her husband attended with her Mr Garrett was a Republican in politics There were five children in the family H R, on a farm near Morrill, Vincent, the subject of this review, John A, a Sioux county farmer, E E also farming in Scottsbluff county, and Roy J, engaged in farming with E E

Vincent lived with his parents on the farm in Colorado, attended the public schools of his district where he laid an excellent foundation for his higher education Both he and his father were ambitious for his success in life and after the boy graduated from the public school he matriculated in the State Agricultural college of Colorado as he had decided on farming as his life vocation and realized that the best equipment for it would insure a success not to be otherwise obtained Mr Garrett finished the long agricultural course and then, after his studies were completed, came to Sioux county in 1906, took up a homestead of eighty acres and there at once began to put into execution both the practical knowledge he had gained as a boy on the farm and the theoretical ideas of the college experts The land he chose is located in section 28, township 24-57 He placed the land under cultivation, established good improvements in the way of buildings and then erected a fine home as he was married the year he located on the farm All Mr Garrett's land is under ditch as he was too far sighted and too well up on the subject of his business to think of farming without assured success From first locating in the valley he has put into practice the modern methods advocated by the professors of the college and the state and government experts and as a result has achieved success far beyond that of most young farmers of the section which does not lack for capable and prosperous members of the agricultural profession Mr Garrett has specialized in raising potatoes and as his farm began to produce abundantly entered into the produce business in partnership with his brothers in Morrill, where he has erected a warehouse with a capacity of sixty thousand bushels and ship in car load lots He has the distinction of having shipped the first car of potatoes out of Morrill Now they are associated with the Albert Miller Produce Company of Chicago and ship from two hundred and fifty to three hundred carload lots of potatoes out of the valley yearly Mr Garrett well typifies the benefit to the farmer of a special course in his business and today is one of the pushing, energetic members of the younger generation of business men who are making history in the Panhandle and demonstrating that irrigation is to be the salvation of the farmer on the semi-arid reaches of the high plains of this and neighboring states

Having such an excellent academic training Mr Garrett is not a man to let others do his thinking for him and on state and national questions takes an independent stand in politics and is one of the most prominent members of the Independent faction He votes for the man he believes will honestly give the voters the best service His wife is a member of the Presbyterian church to which they are liberal subscribers

In 1906, Mr Garrett married Miss Grace J Kernohan, a native daughter of Nebraska, born at Grand Island Her father was J P Kernohan, now a resident of Delta Colorado Six children have been born to this union Esther, Ross, Samuel, Clarence and Clyde, twins, and Hubert all of whom are still at home and for whom a bright future is in store as their parents are well to do and will give them every advantage in a social and educational way that the state of Nebraska affords to establish them in life

CHARLES B FOSTER — Few of the pioneers of Nebraska have passed through more character building experiences, faced greater hardships, overcome more obstacles and in the end gained greater results than Charles Foster His life is typical of the courageous persevering spirit which brought about the settlement and development of the west, and he has won his way unaided from a man whose only capital consisted of his hand, a high courage and the determination to wrest a competence from the soil, directly or indirectly as the case might be In the early days when the Panhandle was yet the frontier he was a cowboy buffalo hunter, ranchman, and later became the owner of considerable landed estate which today places him among the prosperous and substantial residents of the Morrill valley

Mr Foster was born in Orange county, Vermont, January 26 1856 the son of Gardner N and Olive (Chapin) Foster both natives of

the same state where they were reared, educated and married The family lived in Union
Village, where the children were born, and
where the father plied his vocation of carpenter as long as he took an active part in business life The Foster family were members
of the Methodist Episcopal church in which
they were active workers Frederick, the
youngest child is deceased while Charles has
become a resident of Nebraska The latter was
reared in Union Village, and attended the public schools When the rumor of the discovery
of gold swept over the country many of the
youths of the eastern states came west to gain
a fortune as they hoped in the Black Hills
Mr Foster was one of these who determined
to hazard his fortune in the Dakota gold field
and joined a party of fifty men who were going there to dig but he stopped in Cheyenne on
the way, liked the town and remained there
for a year He began to enjoy the free open
life of the country and as this period was the
time when the great, rich cattle barons were
gaining fortunes from the vast herds that
ranged from Texas to Montana he became
enamoured of the cowboy's life and joined one
of the big outfits, the CY company with headquarters on Horse Creek, Wyoming Mr
Foster adapted himself to the life of the cattle
camps, road herd and spent many seasons drift-
ing cattle from the early spring range north
throughout the summer months to the markets
in the north in the fall He displayed great
ability in handling men and cattle and for
seventeen years was associated with the same
company He realized that the future of the
cattle industry was to be in the hands of the
farmer, that the day of the open range was
doomed, as settlement was ever encroaching on the pastures that were leased occasionally from the government and availed himself of the opportunity to get in on the ground
floor, so to speak and in 1892 filed on a claim
of one hundred and sixty acres of Scottsbluff
county land, located in township 22-58, section
3, a tract that had the advantage of being easily irrigated Mr Foster at once set about
breaking his land made good and substantial
improvements on the farm He engaged in
diversified farming, planting the crops best
adapted to the climate of this section and which
thrived with good water Having been in the
cattle business for so many years he also
established himself as a stock-raiser As he
prospered and his capital enabled him to do so,
Mr Foster bought more land adjoining
the original farm until today he has a fine,
well cultivated estate of four hundred acres,

all under ditch, which gives him greater returns
than two thousand acres would without water
rights Mr Foster has taken a commendable
interest in community affairs and has served
as school moderator of his district and has
served as treasurer of district number one for
many years Independent in politics, he casts
his vote for the man he thinks best qualified for
public office Mrs Foster is a member of the
Christian church Mr Foster was superintendent of the Mitchell ditch for fourteen
years

April 10, 1889, registered the marriage of
Mr Foster and Miss Eunice Ray, the daughter
of John and Caroline Ray She was born in
Indiana but accompanied her parents to this
state when aged eight years, settling in an eastern county In 1888, the family came to Scottsbluff county as Mr Ray was appointed postmaster of Caldwell Both parents of Mrs
Foster are now deceased Six children came
to brighten the Foster home circle Edward,
now at home, who was in government service during the World War, being attached to
an ambulance corps in Italy ; May, the wife
of Bernard Andrews, of Bridgeport , Earl, who
died in infancy , Harry, deceased, and Neal and
Emmett also at home Mr Foster and his
family are held in the highest esteem by a large
circle of friends and are people of real genuine worth

WALTER E BAKER — Among the men
who have contributed materially to the growth
and development of the Panhandle since the
pioneer settlement of this part of the state,
one who has been an eye witness of, and participant in this great growth and progress is
Walter E Baker, who lives retired in Mitchell
Mr Baker was one of the pioneers of the
late seventies and a homesteader of the early
eighties, as he accompanied his father to this
section as a youth and the greater part of
his business career has been passed as a farmer and stockman In whatever capacity he
found himself, he always carried on his transactions and conducted himself personally in
a manner that won and has held for him the
respect and esteem of his fellows, so that his
life history has been unmarked by stain or
blemish, and today he stands as an example of
what this wide western country can produce in
manhood

Walter E Baker was born in Crawford county, Pennsylvania, March 25, 1862
His parents were Cornelius and Jane (Bel-
knap) Baker, the former of whom was born in
Ashtabula county, Ohio, March 27, 1827, who

died October 23, 1880 The latter was born in this same county October 23. 1833 and passed away May 4 1868 The father was a farmer in Pennsylvania, emigrated from there to Indiana when that state was considered near the frontier and then to Illinois, engaging in agricultural business in both states, where he was known as a substantial reliable business man Mr Baker was a natural pioneer, one of the men who have blazed the way for civilization and settlement He was not afraid of the hardships and privations incident to a new unsettled country and played an important part in developing the various sections in which he located After spending some years in Illinois, Mr Baker disposed of his farm there to good advantage and knowing of the broad stretches of government land to be had for the taking in Nebraska, where not only he, but his sons might take up land, came to Gosper county, in 1878, locating on a homestead south of the Platte, where he lived but two years before his death occurred There were five children in the Baker family Medella C Bunce the wife of J P Russell. of Birmingham. Alabama, Walter, William C , Arthur C Bunce, a physician of Omaha. Nebraska, and Benny J, of Mitchell, who is a veterinary surgeon After the death of his first wife Mr Baker married her sister in 1869, and it was she who assumed charge of the family when the father died and ably she shouldered the burdens and proved up on the homestead

Walter came west with his parents in 1878, a boy in years, and yet a man in his ability to carry on farm work and shoulder responsibilities as he was the oldest boy, and as his age and strength permitted took more and more of the work on the farm He was ambitious to get ahead in the world and coming to Cheyenne county in 1885 saw the many opportunities open to him to secure good farm land so in 1889 he filed on a claim of one hundred and sixty acres, and took up a tree claim He placed the required improvements on the land in the first years, then made such others as he desired for himself, erected good and substantial farm buildings as were required for his stock and a good comfortable frontier home for himself and wife Mr Baker engaged in general farming and cattle-raising and met with success in both branches of his business The Baker family passed through all the hard years on the plains, those of drought blizzard and insect pests but both husband and wife were courageous, determined that they would win and "stuck it out," and that their faith in the country has been justified is attested by the comfortable fortune which they now possess With the initiation of irrigation in the Platte valley, Mr Baker was one of the first residents of the section to realize and advocate it as the salvation of this semi-arid section and today all his land is under ditch From first locating in this section of the Panhandle Mr Baker has taken an active and interested part in the development of the country and has advocated all civic and communal movements for the benefit of the community In politics he is a staunch adherent of the principles of the Republican party to which he has belonged since casting his first vote, and served as county commissioner for three years Fraternally he is affiliated with the Independent Order of Odd Fellows, while during the World War he was associated with all the movements inaugurated in the county to aid the government in the prosecution of the war

January 22 1891, marks the date of the marriage of Mr Baker and Miss Alva Ray, the daughter of John Ray who came west in 1888 and located in Scottsbluff county Mrs Baker was born in Indiana and died in Scottsbluff county September 19. 1915 having all the years of her married life been the able helper and loving wife and mother, who in her passing left a sorrowing family and many friends There were nine children in the Baker family Abbie E, the wife of William Gehrt Bessie J, who married Lester Fox, of Scottsbluff county, Ruby H married George Yocum, also of Scottsbluff, Ada M , the wife of Perry Wright who is at home keeping house for her father, Dora M , Alice F , Mary Bertha and Walter E , Jr , all of whom are still at home under the family roof tree

JOHN A LARSON — The ultimate and consistent reward that should prove the crown of years of earnest endeavor and effective toil is the prosperity that may be had by men and women who have arrived at the stage on life's journey where the shadows begin to lengthen from the crimson west, where the sunset gates are open wide Such reward has been granted to the sterling man whose name heads this review

Mr Larson is of Scandinavian descent born in Sweden in 1859 one of the members of a race that has contributed so largely to the better class of settlers in our country and to whom we owe a great debt for the examples of thrift and industry, characteristics which they brought with them when they crossed the stormy Atlantic to this land of promise, where they entered every walk of life and in so many cases made good. John Larson was the son of P. A. and Ann (Peterson) Larson, both natives of Sweden, where they were reared and received the excellent educational advantages provided by the government. They met and married in their native land and there some of their children were born before they determined to seek greater opportunities for themselves and their family across the sea, in the new country where land was to be obtained for the taking, an almost incredible idea to these people of a European country where land was high in price and held from generation to generation in one family, so that the younger sons must emigrate if they desired land for themselves. Mr. and Mrs. Larson were ambitious and severing all the sacred ties that bound them to family and native land, came to the United States in 1873, locating in Colorado on a homestead of a quarter section of land, on which they proved up. Here the father engaged in general farming and stock-raising for the remainder of his life. Mr. and Mrs. Larson were members of the Lutheran church, while Mr. Larson was an adherent of the Republican party.

Eight children completed the circle of the Larson home: John, the subject of this review; P. A., an implement dealer of Grover, Colorado; Emma, the wife of Peter Johnson, of Boulder, Colorado; Charles, a carpenter in Denver; Ida, is the widow of Edward Brubaker; Otto, a miner near Boulder; Jennie, the wife of Nels Nordquist, of Victor, Colorado; and Albert, the proprietor of a restaurant at Greeley.

John Larson attended the elementary schools in Sweden before his parents came to America, and after the family had settled here was given the educational advantages obtainable in the public schools. After finishing his schooling he began work as a miner, as did most of the young men of vocational age, as mining was the chief business of a greater share of the population for many years, when the gold boom was on. Mr. Larson was employed in several different places where mining operations were being carried on for about twenty years, but as he was a man who gave thought and study to many questions of the day he began to realize from his readings that the proprietor of land was the most independent man in the world, especially if his holdings consisted of farm land. He studied up on the question of agriculture though he had a good practical knowledge from his life on his father's farm, but methods had changed and he was far

sighted enough to realize that times do change. After looking the country where homsteads could still be obtained over, he decided to engaged in farming in Colorado. Twenty years passed while he carried on both farming and at times mining until in 1906 he came to Nebraska, locating in Sioux county where he took up a homestead of one hundred and sixty acres in township 24-57, section 22, where he began improvements as soon as a home could be built for his family and they became settled. Having had long and varied experience by this time in agriculture, Mr. Larson had chosen his land well as a hundred acres of his holdings are under ditch. He has made excellent improvements, purchased the most modern farm equipment to lighten labor and also assist in greater production from the land, and today is rated as one of the most well-to-do and prominent men of the valley. He carries on general farming industries and as his irrigated land yields generous forage crops has branched out in an allied business, sheep and hog feeding. in all allied business, sheep and hog feeding. He buys lambs and sheep off the range, feeds heavily for a period varying from sixty to ninety days and then ships to the eastern markets. Mr. Larson is a shrewd buyer and long seller so that his returns from this branch of his enterprises has brought in most gratifying results, especially since the beginning of the war, as meat has been so high.

In 1895, Mr. Larson married Miss Christine Peterson, a native of Sweden, the daughter of Peter Peterson. Mrs. Larson accompanied her father when he immigrated to this country and lived in Denver, before her marriage. Four children have been born to this union: Stella, the wife of Charles Hutchison, a farmer of Sioux county; Rose, who married Harold Gilbert, of Sioux county; Harry, and Edna, both of whom are at home with their parents. Mr. Larson is a Democrat in his political affiliations. Mr. Larson has never been willing to accept public office, but lives his citizenship every day in the manner in which he orders his life, and today is rated one of the reliable men of his section who advocates every movement for the benefit of the community.

WALTER D. HUFFMAN. — Here is a younger member of the agricultural profession, a self-made man who though young in years has already scored his initial success and whose life record deserves a place among the representative and progressive men of the Panhandle who are making history in this section of Nebraska.

Mr Huffman was born in Iowa, March 11, 1881, being the son of Lorenda and Sarah P (Kirk) Huffman, the father born in Illinois, and the mother in Iowa, where they were reared, educated and later married. To them nine children were born Marcus, living at Fort Collins, Colorado, Frederick, a resident of Julesburg, Herbert, of Scottsbluff county, employed in a sugar factory, Charles, living in Sedgwick, Colorado, Millie, deceased, Lennie, the wife of George A Monroe, of Sioux county, Ernest, also living in Morrill, and Ella, the wife of Ralph Bookout, of Fort Laramie. The father of this family was one of the gallant sons of the Union who responded to the president's call for volunteers during the Civil War, serving in the One hundred and forty-second Illinois infantry. After the cessation of hostilities he located in Iowa, where he was a well to do farmer, a member of the Grand Army of the Republic and of the Methodist church. He died in 1915, being still survived by his wife who lives with her children.

Walter was educated in Colorado, as the family moved there while he was a small boy. After his school days were over he decided to become a farmer and establish himself in independent business and followed this vocation in the Mountain state until 1909, when he decided to avail himself of the fine chance of farm land in the Panhandle where irrigation was well introduced and where at the time, there was still government land open for claims under the homestead act. Coming to Nebraska Mr Huffman filed on eighty acres in township 23-57, section 8, Scottsbluff county. Having had years of practical experience in agricultural pursuits, he soon had his farm in fine shape, erected good permanent farm buildings and a comfortable home which is one of the social centers of the Morrill valley, as he and his wife have made a host of friends and Mr Huffman is regarded as one of the enterprising and progressive men of the vicinity.

In 1905 occurred the marriage of Mr Huffman and Miss Louetta McCullough, who was born and reared in Iowa. To this union three children have been born Clarence, Edith and Ethel, all at home. The family are members of the Presbyterian church, in which they take an active part. Mr Huffman is a Republican, but has found no time to enter politics as a candidate for public office though he is public spirited and has marked civic pride, and his work in behalf of beneficial movements in his community is always of a constructive character. Fraternally he is affiliated with the Yeomen.

JOSEPH G NEIGHBORS, who owns one of the best improved farms of Scottsbluff county, has lived in Nebraska since 1887 and is a representative and respected citizen of this section. He was born in northern Missouri, September 21 1863, and is a son of Joseph and Nancy (Carter) Neighbors. The father of Mr Neighbors was born in Virginia and the mother in Ohio and both are now deceased. The father was a farmer before he became a soldier in the Civil War, and died while in service.

Joseph G Neighbors was an infant when his father died. He was reared in Missouri and from that state came to Nebraska some thirty-two years ago and homesteaded in Scottsbluff county. Like other settlers at that time, he was called on to endure many hardships, but he was industrious and resourceful and in the course of time made encouraging progress and now has a valuable property, his land being all irrigated. He carries on general farming and stockraising.

Mr Neighbors was married to Miss Carrie A Franklin who was born in Missouri September 11, 1869 a daughter of Thomas and Hannah (Minear) Franklin. Thomas Franklin was born in Kentucky and the mother in Virginia. They married in Indiana and moved to Missouri. Mrs Neighbors was seven months old when her mother died. The father of Mrs Neighbors was a physician. He came to Nebraska in 1886, locating at Gering, where he practiced his profession until his death at the age of sixty-three years. He and Frank A Garlock built the first hotel and the first store in the town. He owned the first drug store. He was a soldier in the Civil War, serving the full term and came out with the rank of captain. Mr and Mrs Neighbors have three children Grace who is the wife of Samuel Shove, lives in Wyoming, Thomas who is an attorney, resides at Bridgeport, and Melville who is a farmer in Scottsbluff county. They are members of the Baptist church. In politics Mr Neighbors is a Democrat.

FRANK E POWELL. — One of Sioux county's successful and well-to-do citizens whose present prosperous condition is due to his own industry and good judgment, is the gentleman whose name introduces this paragraph.

Mr Powell is a native of Ohio, born in Morrow county April 17 1862. His parents were Evan and Elizabeth (Everett) Powell, natives of Virginia, who were the parents of thirteen children, eight of whom are living, but only two came to Nebraska Walter and

Frank The latter was reared in Ohio, obtaining excellent educational advantages in the public schools of that state His father was a farmer and at an early age the boy learned the practical side of agricultural business as well as stock-raising as conducted in his native state The father died in 1904 but the mother survived until 1913

After his school days were over, Frank left home before he was twenty-one years old and went to North Dakota and took a homestead He had heard and read of the west and the lure of the wide spaces of the prairies called him He came first to North Dakota, where he established himself as a farmer on a homestead, subsequently he went to Kansas and was there fifteen years Mr Powell had read of the great strides agriculture was taking in the valley of the Platte with the introduction of irrigation so he came here in 1902, filed on a claim in Sioux county not far from Morrill and proved up on it, making good and permanent improvements on the place and erecting a comfortable home When his capital permitted he purchased land adjoining the home place and today owns one hundred and sixty acres of land, nearly all under ditch, where he is conducting a general farming business He is a thorough advocate of intensive farming and irrigation, gives study to the agricultural problems that arise, has the latest modern equipment for his land and is accounted a man of weight and means in the valley, where he is known for his integrity, high ideals and the care with which he carries out his business obligations Mr Powell is allied with no political party, voting as his conscience dictates, for the men best fitted to fill office, whether local, state or national in character

In 1889, occurred Mr Powell's marriage to Miss Dema Smith, who was born in the state of Michigan, and to this union two children have been born Frances and Winifred

While he has made a success of his undertakings Mr Powell has not been unmindful of his duties of citizenship and is held in unqualified esteem by all who know him

JOHN H KELLUMS — The subject of this record is one of the honored early settlers of Scottsbluff county and through his own efforts, marked by diligence and good management, he has achieved substantial success

Mr Kellums was born in Clay county, Illinois, October 29, 1859, being the son of John W and Margaret (Henry) Kellum, the former was a Hoosier by birth, the latter was born in Ohio They were reared in their native states,

received the educational advantages there in the public schools and later met and married To them eight children were born, of whom three survive John, Ferdinand, living in Illinois, and Elizabeth, the wife of John Frazier, a resident of Crosby, North Dakota The mother, a Baptist in faith, died in 1878 Mr. Kellums was a general farmer in Illinois, and also engaged in stock-raising, an occupation he followed all his active business life He was a Republican in his political views while his fraternal affiliations were with the Independent Order of Odd Fellows He passed from life at a hale old age in 1915, in Illinois

John Kellums grew up on his father's farm, received his educational training in the public schools of his district, thus laying the foundation for his subsequent business career, as he learned farming from his father at the same time he was under academic discipline Following in the footsteps of his sire he engaged in farming independently upon attaining his majority as this was a business with which he was familiar and one toward which he was inclined by temperament Hearing of the many opportunities a young man had to obtain good farms in the newer states west of the Missouri river he learned all he could about different sections of the country and then in 1887 decided to locate in Nebraska Coming to the Panhandle he filed on a claim in township 22-R 58, section 12, Scottsbluff county, also a homestead in section 2, land which has since come within the irrigated district of this section of the valley Mr Kellums was fortunate in the selection of his homestead as practically all his land is now under water rights and he is insured crops every season Being young and full of vigor and inured to hard work, the young man soon had his land under cultivation, made good and permanent improvements on the place and before his marriage had built a good, comfortable home As money came in from the sale of his produce and capital permitted, Mr Kellums bought other tracts near the home place, so that he now owns a full section of land

On Christmas day, 1889, occurred the marriage of Mr Kellums and Miss Mattie E Parish, a native of Iowa, and to them were born eleven children Felix Lockwell and Roy W, both at home, Hazel, married, Roy Shultz, living in California, Clarence, at home Blanche, the wife of James Robertson, living near home; Floyd, Ena Frances, Maude, John H, Arthur Temple, Hugh, all at home with their parents Mr Kellum is an independent voter, being bound by no close party lines, casting his influence in the scale for the man best qualified for office With his wife he attends the Seventh Day Adventist church while his fraternal affiliations are with the Independent Order of Odd Fellows Mr Kellums can truly be called a self-made man who has made good use of his oppotunities and today is rated as one of the substantial and public spirited men of Scottsbluff county

CHARLES H GATLIFF — As a living example of what resolute work, earnest endeavor, and perseverance will accomplish, Charles Gatliff stands prominent among the worthy citizens of Scottsbluff county, coming here in 1887, with little capital save that represented by his personal qualities and characteristics, he has worked his way uninterruptedly to a position of independence, and his status today is that of a substantial citizen and prosperous agriculturist, though now the sunset years are casting shadows from the crimson west he has disposed of most of his holdings and is living in semi-retirement enjoying the fruits of his long and arduous labors

Mr Gatliff was born in Missouri, September 21, 1856, being the son of Joseph and Rebeccah (Wakefield) Gatliff, the former a native of the Blue Grass state, while the mother was born and reared in Illinois Four children were born to them Charles, William, living in Custer county, Nebraska, George, who went to Brazil, South America some years ago, and Flossie, deceased The father was a farmer in Missouri where he was engaged in agricultural pursuits all his life, passing away in 1869, leaving the mother to shoulder the responsibilities of the family and rear her children She was a worthy woman who died in 1906, whose good deeds were as the number of her days The family were members of the United Brethren church while the father was a Republican in his political views

Charles grew up a sturdy lad on his father's farm, received his elementary educational training in the common schools of that state and after the family removed to Iowa finished his schooling there When old enough he chose farming as his vocation and engaged in that business in Iowa until 1887, but he was ambitious to get ahead in the world and desired land of his own and to obtain what he wanted decided to come west and take up government land where there was an opportunity of doing better for himself and his children Coming to Scottsbluff county he filed on a claim in township 23-58, section 21, where the family was established as soon as a house could be constructed and the necessary shelter provided for the stock Those were primitive days in the valley settlers were few and far apart, distances to trading centers many miles away, but these people of true pioneer stock were not daunted by the hardships and privations that they were forced to endure for a few years and with high courage wintered the blizzards and withstood the droughts and insect pests of the early eighties and nineties, and fortune finally smiled upon their efforts Irrigation, the great salvation of this semi-arid climate was established in the Morrill valley, crops were assured, railroads were built up the Platte, money came easier and prosperity was assured to the Gatliffs, who well deserved whatever they had accumulated in a material way, and today this excellent family is highly respected by their neighbors and friends, who regard them as examples of true American citizens

In 1878, Mr Gatliff married Miss Elizabeth Phillips, a native of Ohio, who has been a worthy helpmate to her husband during the many years they have been taking a prominent part in the development of this rich valley country Mr and Mrs Gatliff reared an adopted son named Carl Sears, who is now a stockman in Wyoming He married Estella Gatliff Mr Gatliff is a staunch adherent of the Republican party Mrs Gatliff is a member of the Christian church

GEORGE A MUNROE is a resident of Sioux county where he is well known as a representative citizen and prosperous agriculturist who is developing large interests

Mr Munroe was born in the Province of Quebec, Canada in 1887 being the son of John and Ann (Nixon) Munroe, the former born in Scotland, while the mother was a native of Ireland To them were born eight children, five of whom survive Edward, of Fort Collins, Colorado John, living in the Province of Quebec George, Clarence, also residing at Fort Collins and Hubert a farmer of Sioux county The father of the family was a farmer in Canada all his life he carried on general farming enterprises and also conducted a dairy business The mother died in 1904 being survived by her husband until 1911 when he was called to the last long rest They were members of the Presbyterian church while the father was a liberal in his political views

George was reared on the farm in Canada, attended the common schools near his home, and when old enough began farming on his own account He heard golden tales of the opportunities in the United States and as he felt the lure of the west as well as a spirit of adventure that called loudly, came to Colorado in 1905, but remained there only three years as he learned of the fine land to be obtained under the homestead act in the Panhandle and came to Sioux county, filing on a claim in township 22, 24-57 The first quarter he soon had under cultivation, made good and permanent improvements that have stood the test of time, though the buildings have been added to since and before his marriage Mr Munroe built a fine comfortable home He bought other land near the homestead and today is the owner of a landed estate of two hundred and forty acres of dry land and a quarter section ;all under irrigation This makes an excellent combination for general farming and the raising of cattle, a line in which Mr Munroe has specialized, as he raises nothing but high bred white faces, the best beef stock in the opinion of experts The high land makes excellent pasture, while that under ditch raises grains and the necessary forage crops Having an abundance of feed, Mr. Munroe has branched out in another line of agricultural industry and buying lambs and sheep from the western growers, feeds them in the winter for from sixty to ninety days, then ships them to eastern markets, making a quick turnover of his money which he has found exceedingly profitable, especially since the war when meat prices advanced to such a high figure He is not only an advocate of intensive modern farming but is rated as one of the most successful stock-men of the Morrill district which has already become well known for its able business men

Mr Munroe is a Republican in politics and though he takes no active part in political life, is a man who stands behind every movement that tends to the development of the county and has a high reputation for his integrity and the fulfillment of business obligations

On December 23, 1913, Mr Munroe married Miss Lena Huffman, and to them three children have been born Everett, Grace and George Edward, all at home

ROBERT G WALSH, whose standing as a prominent business man of Morrill and leading citizen of Scottsbluff county is high and who has been intimately identified with the material growth and industrial and financial development of the county for many years is now one of the partners of the most popular and largest automobile houses in the valley.

Mr Walsh was born in Kankakee county, Illinois, April 20, 1886, the only child of Robert and Harriet E (Richardson) Walsh, the former born in Ireland, while the mother was a native of Illinois Robert Walsh, Sr was a farmer in Illinois, also engaging in business as a railroad contractor In 1879 he came west to Colorado and later removed to Lingle, Wyomwhere he became established as a railroad contractor, building a part of the Chicago and Northwestern Railroad from Chashon to Harrison in 1886, he was a shrewd man, studied his business opportunities and became one of the rarely successful men of his section at that time As that was the period when great herds of cattle ranged over the prairies Mr Walsh became interested in the live-stock business as a side line and achieved a wide reputation as a man able to handle cattle industries as carried on along what was then the frontier Now that the sunset years have come and the shadows begin to lengthen from the west, he has retired from active participation in commercial activities and is now spending the later days in retirement, quietly enjoying at Fort Collins, Colorado, the fruits of his earlier endeavors Mr Walsh was one of Illinois' gallant adopted sons who responded to President Lincoln's call at the outbreak of the Civil War, serving through some of the hardest campaigns of that memorable conflict as a member of Company G, Twenty-fifth Illinois infantry and after the war was over returned to the pursuits of peace literally exchanging the sword for the plough-share He is a member of the Republican party in his political views, belongs to the Grand Army of the Republic while his fraternal affiliations are with the Masonic order and with the Benevolent and Protective Order of Elks With his wife he attends the Methodist church of which she is a member

Robert attended the public schools of Illinois and after the days of educational discipline were over became a farmer, as that was a vocation which appealed to him In 1886, when a youth of twenty he also came west, to Fort Laramie, as he had become ambitious to be a land owner and learn the cattle business first hand With this end in view he joined one of the cattle camps of the great baronial cattle companies and rode range for several years, becoming well acquainted with the livestock business, seeing much of the country and broadening his outlook on life Like many another man of vision, when settlement began to creep up the fertile valleys of the Platte and other great rivers of the prairie states, he read the doom of the great cattle firms and realized that the future of this business was to lie in the hands of the farmer with his smaller holdings who would produce a better grade of beef animal Reared to farming he now determined to engage in the cattle business on an extensive scale and the first step toward this end was taking up a homestead in the Morrill valley in 1891, and this original grant has never passed from his ownership Farming was a side line, as he specialized in horses and cattle exclusively, raising forage mostly on the arable land Mr Walsh became known in the upper valley as one of the phenomenally successful men of his profession, he made money and today is one of the most substantial men of the district With the pasing of the horse as a means of transportation Mr Walsh was too progressive to hold to the old ways and old days, and early realized that a great future was before the men who early entered the automobile business and while he still holds considerable land and raises stock he devotes much of his time to this line as he has formed a partnership with a Mr Williams and they own the largest garage in Morrill, carrying a fine line of accessories and maintaining a fine service, not only for the town but all the surrounding country Unfailing courtesy, prompt fulfillment of business obligations and integrity have won for the garage and its owners a most gratifying clientelle, so that it is a money making proposition The family are members of the Methodist church while Mr Walsh votes the Republican ticket and is a member of the Masons

In January, 1892. Mr Walsh married Miss Cora M Akers, of Iowa, and to them seven children have been born Irene G the wife of Everett Barclay of Seattle, John G, of Morrill, is an aviator, Margaret, a teacher at home, Mildred, at home, William, a student at the State University, Esther and Ruth, at home

ELTON GARRETT, is distincly a Nebraska product as he was born in Red Willow county in 1883, and today is representative of the best element of the younger generation of progressive farmers who are today making history in the Panhandle and demonstrating that intensive agricultural pursuits under irrigation together with modern methods and equipment is a paying business in this favored section

Mr Garrett is the son of Samuel and Esther (Bodwell) Garrett, a sketch of whom will be found elsewhere in this volume devoted to the life histories of prominent settlers of the Panhandle Elton received his early educational advantages in the public schools of this state, he grew up sturdy and self-reliant as most farm boys do and early learned the lessons of industry and thrift as demonstrated on the home place When his father went to Fort Collins, Colorado, the youth accompanied him remaining there until 1906 when he came to Sioux county and filed on a quarter section of land in section 20, township 24-57, as he had decided to take up farming permanently as a life vocation Mr Garrett made improvements on his land. erected necessary buildings and within a short period had his land under cultivation Subsequently he found it necessary to relinquish eighty acres but this gave him time to devote more attention to the remainder which was all under irrigation and upon which he has most successfully tried out and proved that intensive farming, as advocated by the farm experts, pays Mr Garrett entered into partnership with his brothers to specialize in raising potatoes, under the firm name of Garrett Bros, and they have become known widely for their success in this enterprise Though young in years the brothers are old in experience and their rise in the world as producers of the second great food product of our land has been due to their devotion to business keen foresight and executive ability During the one season of 1917-1918 they shipped more than three hundred carload lots of the tuber out of the valley, easily giving them first place as potato men, and today their products vie with the famous Wausau county potatoes of Wisconsin that have hardly a rival in the field With the passing years better buildings have been erected on Mr Garrett's farm, a comfortable, modern home is enjoyed by the family and in addition to his general crops and potatoes he raises pure bred Percheron draft horses and has a good grade of other stock on the place including Duroc Jersey hogs Mr Garrett's mother makes her home with her son, being a woman well advanced in years as she has passed her sixty-fifth birthday but is still keen mentally and no one would believe she was not many years younger due to her body vigor, which is that of a much younger woman

Mr Garrett is a member of a well known and highly respected family of Sioux county which has contributed liberally to civic and material progress and prosperity and is what may be called the true type of American farmer. a class that leads the world in production as demonstrated when America was called upon to feed the hungry world during the World War. Independent in his views of life it is but natural that Mr. Garrett should be independent as a voter and he is bound by no party lines when he casts his vote but gives his influence for the man best fitted to serve the people, whether county, state or nation.

LAWSON E. MEREDITH, is a representative Sioux county farmer who came of Hoosier stock. He lives in the Morrill district, where since 1905 he has been conducting farming operations on a more or less extensive scale and where he has made an enviable reputation as a substantial citizen and successful farmer.

Mr. Meredith was born in Indiana in 1865, and is a son of William and Haney (Fansler) Meredith, both natives of the Hoosier state, where they were reared, educated, later met and were married. They are both living today in Atlantic, Iowa, at the advanced age of seventy-eight years, having passed that allotted span of the psalmist, three score years and ten, but are mentally as vigorous as people years their junior. William Meredith owned land in Indiana, where he engaged in general agricultural pursuits and stock-raising, being successful along both lines and there was regarded as a successful man. During the Indian troubles on the frontier he enlisted in the service of the government, serving a part of this time in North Dakota, at the time of the uprising there on the reservation.

Lawson E. Meredith was reared on his father's farm, attended the public school near his home and thus gained a good practical education for his later life. After the school days were over he began independent business life as a farmer in Iowa. In February, 1905 he came to Nebraska and located on a homestead in section 31, township 23-57, Scottsbluff county, where the family resided until he moved to Mitchell and disposed of his farm. He is now operating a farm in Sioux county, which he rents and is meeting with good success.

In 1891 Mr. Meredith married Miss Mattie Chizum, of Iowa and to them one child was born, Frank, in the government shops at Mitchell. Mr. Meredith is a Republican in his political views, attends the Methodist church of which his wife is a member and his fraternal affiliations are with the Independent Order of Odd Fellows.

SAMUEL BARTON, one of the substantial and progressive farmers of the Gering district is a native of England, the tight little island

from which the first settlers came and which has furnished the greatest proportion of the best elements of our population, and while he has been a resident of Scottsbluff county more than fifteen years and of the United States since 1872, he retains all the excellent qualities of the English which has made them the colonizing race of the world.

Mr. Barton was born in 1864, being the son of Frederick and Mary (Tomlinson) Barton, both of whom were born, reared, educated, met and married in the Island of Great Britain. Frederick Barton was an ambitious man, and having a large family he saw no future for them in the mother country; he read and also heard of the fine opportunities to secure land for the taking in America, determined that both for himself and his children he would emigrate. Breaking all the dear associations that bound them to the land of their birth Mr. and Mrs. Barton accompanied by the children sailed for the United States, then the land of promise to so many people of European birth who desired land of their own. After reaching our shores the Barton family came west, locating in Iowa, where the father bought land and established himself in general farm industries. In addition to raising diversified crops he engaged in stock-raising as conducted at that period and became a man of substance and weight in his community, passing the remainder of his days there, as did the mother, both are now deceased. There were eighteen children in the Barton family, seven of whom are living today, so that young Samuel grew up on his father's farm sturdy and self reliant, used to the give and take of a large family and at an early age was well qualified to hold his own against anyone not his senior or stronger. He attended the public schools of his district, thus gaining a good practical education of inestimable value to him in later life. Having worked on the home place he acquired a practical knowledge of agricultural business methods and when he was old enough to establish himself independently in business chose farming as a vocation with which he was acquainted and also one compatible with his tastes. Mr. Barton became a well known and substantial representative of the farming element of Iowa where he remained until 1903. In the meantime, he had kept abreast of the progress being made in agriculture all over the country and when a satisfactory offer was made him for his home place disposed of it and came to the Panhandle, as he had become convinced that irrigation was to establish farming upon a stable basis, not possible when rain was depended upon for water Mr Barton purchased one hundred and sixty-four acres of land in section one, township five, Scottsbluff county, where the family were soon established and he himself engaged in general farming operations and the rearing of high bred stock, as he believed the greatest returns were obtained from thoroughbreds He has specialized in Holstein cattle, owning the grandson of Ragappe, the bull famous among registered stock and also the daughter of King of Pontiac, who sold a short time ago for $100 000 As a side line he raises hogs of Duroc Jersey breed, so that all his animals are either of pure strain or else very high grades Mr Barton has been markedly successful since coming to the Panhandle and today is one of the largest raisers and shippers of this section His farm is modern in every way as he uses modern methods and the last and most improved machinery for lightening labor and increasing production Mr Barton is a Democrat in politics while his fraternal affiliations are with the Ancient Order of United Workmen

On June 17, 1888, Mr Barton married Miss Mary Heft, and to them nine children have been born Harry Clay, with his father, worked for a time in the First National Bank, Mary, the wife of L R Wright of Scottsbluff county, Eunice, married R G Neely, register of deeds of Scottsbluff county, Daisy, the wife of Ray Irley, Chester, on the home place, Ruth in the county treasurer's office, Samuel, Robert and Helen, all at home

CYRUS D COOPER has proved himself the possessor of a large amount of that excellent manhood and that self-reliance, which united with perseverence and industry, have enabled him to become one of the valued men of Scottsbluff county

Mr Cooper was born in Union county, Iowa, in 1856, the some of Amos C and Ruth Amanda (Thurlow) Cooper, the former a Buckeye by birth, being reared and educated in the excellent public schools of Ohio, where he engaged in farming after his academic career closed Subsequently Mr Cooper removed to Missouri and later still to Iowa, where he bought land and became one of the prosperous and progressive farmers of his locality, being engaged in raising general farm products and a good grade of cattle There was a good stream of water on his land and he soon realized that this could be turned to profit in water power, with this end in view he constructed a dam across the stream, erected a mill and after the initial outlay had a good paying business

as a miller He and his wife are now deceased There were nine children in the Cooper family David, who enlisted at the outbreak of the Civil War in the Twenty-ninth Iowa Volunteer infantry, died at Little Rock, Arkansas John D deceased William F, a merchant in Des Moines Iowa, Amos C, owns a livery business at Thayer, Iowa Cyrus, James H, a Kansas farmer, Edward M, deceased, Ruth, the wife of Thomas Weeter of Union county, Iowa and Robert M, an osteopathic physician of Garden City, Kansas The father was a Democrat and Prohibitionist in his views while he and his wife belonged to the Christian church

Cyrus grew up on his father's farm, sturdy and self reliant as a boy in a large family on the frontier must be, to survive the hard knocks incident to life in a new country He received all the educational advantages afforded in the public school near his home and early learned the practical business of farming from his father When he grew to manhood he chose agriculture as a profession and followed this in Iowa until 1904 In the meantime he had studied up on intensive farming under irrigation and when he was offered an attractive price for his place in Iowa, disposed of it and came to the Panhandle to obtain cheaper land and more of it than before Mr Cooper purchased eighty acres in section 12, township twenty-five, all of which is under ditch since locating in the valley he has engaged in general farming to some extent but devotes most of his time and energies to general truck farming, which under his capable management has proved most profitable and he is one of the prosperous and responsible men of this line in the Gering locality, which does not lack for capable men

In 1879 occurred the marriage of Mr Cooper and Miss Elizabeth Poe, a native of Iowa, now deceased She became the mother of two children Edna the wife of James W Tillman of Stanford Montana and Lizzie, who married J W Reynolds, a ranchman of Wyoming For his second wife Mr Cooper married Mary E Miller a Canadian by birth, the daughter of William and Jerusha (Townsend) Miller both now deceased Five children were born to the second union, William M a farmer of Scottsbluff county, Alice M, the wife of J O Rose, a Banner county farmer Oran C, in Wyoming, farming in Campbell county, Mark T, who works for his brothers, and Ward M, also working for him

Mr Cooper is an Independent in politics, voting for the man best qualified for office, while he and his wife are members of the Methodist church. While the family resided in Iowa, Mr. Cooper served as assessor for his district and while he has never held office since coming west, is a public spirited citizen who advocates and supports all movements for the development of his community and the county.

GEORGE C. CROMER, who belongs to the progressive younger element of the agricultural fraternity of Scottsbluff county is a native son of Nebraska as well as this county, and already out of his labors he has worked the start of a successful career where the outlook of his future is very bright. His present property on one hnudred and thirty acres not far from Gering is under a high state of cultivation; he is engaged in general farming, is in partnership with his father in breeding and rearing thoroughbred Percheron horses, and since establishing himself in business has begun to raise irrigated fruit and is considered one of the rising horticulturists of this section, and as fruit growing has but recently been inaugurated in the upper valley as an indutry he may be regarded as one of the pioneers in this line.

Mr. Cromer was born in Scottsbluff county, March 3, 1890, being the son of E. P. Cromer, a history of whom will be found elsewhere in this volume. George was reared on his father's farm, attended the public schools where he laid the foundation for the higher educational advantages which he has enjoyed. He early began to assume many of the duties about the home place and thus while a boy had a good practical working knowledge of agricultural industries. He attended the high school in Gering, and after graduation having chosen farming as a vocation, entered the agricultural department of the State University at Lincoln, where he devoted considerable time to the study of plant life in addition to his general course of applied farming and its methods. After receiving his degree from the college Mr. Cromer returned to the upper valley to form a partnership with his father in the live-stock business, specializing in breeding percheron draft horses for farm work. He owns, independently, one hundred and thirty acres of land, more than half of which is under ditch, upon which he raises varied farm crops and forage as well as some special products which pay remarkably well when irrigated. He has already begun in a small way, as to acres, having only two planted to orchard yet, to devote time and work to horticulture and in 1917 sold $175 worth of plums from this one tract, as this line has proved so gratifying in bringing in returns Mr. Cromer contemplates increasing his orchard each year until it has attained considerable size. Plums are not the only trees as he has apples, cherries, pears, and black walnuts, all of which will bring an assured income. In the near future he is planning to erect a big barn which will also be a sorting and fruit house for use in the harvest season. Mr. Cromer has a pleasant and commodious home and other up-to-date buildings and through his progressive and energetic work he is attaining credible and gratifying success for so young a man.

November 28, 1918, Mr. Cromer married Miss Freda Henatsch, of Scribner, Nebraska, the daughter of G. H. and Anna R. Henatsch. Mrs. Cromer is a member of the Congregational church in which she is an active worker. Mr. Cromer belongs to the Methodist Episcopal church, being curator of the church at Gering; in politics he is independent, being bound by no party ties in casting his vote, but gving his influence to the man best qualified for office. He supports all movements for the development of the county and his community and is one of the type known as true American citizen, who lives up to the high ideals he sets.

CARL THOMAS is one of the native born inhabitants of the North Platte valley. He is the son of Valentine Thomas, a sketch of whom appears elsewhere in this work, and was born in Sioux county on June 15, 1894. After a preliminary schooling in the public schools he graduated from the Morrill high school and attended the University of Nebraska at Lincoln, then returned and engaged in the business of general farming and sheep feeding.

He was married May 17, 1919, to Mary Horn, who is likewise a native Nebraskan, having been born in Lincoln, the daughter of W. H. Horn who follows the trade of carpenter in that city.

Mr. Thomas was a member of Sigma Chi fraternity in his college days, is a Thirty-second degree Mason, and a member of the Elks. He is Republican in politics. Mrs. Thomas is a member of the M. E. church.

Though a young man and just well started in life, Mr. Thomas has the equipment of education and lifelong friend with the country in which his lot is cast, and faces a bright prospect for the future. His father is one of the best known residents of this community, and the son has a high standing among his friends and acquaintances as a progressive

Mr. and Mrs. William T. Walters

and enterprising young man who has the foundation laid for a great success in life and has the ability and energy to build on that foundation

WILLIAM T WALTERS, who came to Nebraska almost a half century ago, has witnessed many wonderful changes along every line of development During his long period of residence here he has done his part in this development and is well and favorably known not only in Scottsbluff county but in other sections Mr Walters was born in the southern part of Kentucky, August 21, 1855, a son of Isaac and Susan Jane Walters, natives of Kentucky The father was a farmer prior to the Civil War, in which he served until his death The mother died in Iowa

William T Walters was still young when he lost his father and in the main he is self-made and self-educated When he came first to Nebraska it was not as a homesteader but as one interested in lumber While many portions of Nebraska have never been timbered regions, there are sections in which many varieties of trees are found, and in earlier days the lumber business in Adams county was one of large importance For twenty-nine years Mr Walters was engaged in that business at Hastings In 1911 he came to Scottsbluff county and homesteaded and now owns a valuable irrigated farm of 120 acres, situated on section 3, township 22-53, and well improved

Mr Walters was married to Julia Lechleiter, who was born at Chicago, Illinois, February 3, 1872 Her father, a stonemason by trade, came to the United States from Germany in 1865 Both he and his wife survive and live at Jacksonville, in the state of Illinois

SYDNEY J DAVIS was born in Longmont, Colorado, June 6, 1881 He is a son of John A Davis

After completing his schooling in Colorado he entered the farming business and has followed the calling of general farming ever since He lived in Colorado until 1908, when he joined a good many other Colorado people in investigating the possibilities of the new irrigated country that was growing up in the valley of the North Platte river in western Nebraska and eastern Wyoming He became convinced of the superior opportunities for a young and energetic man in the new field and ended by coming to Sioux county and taking a homestead under the Reclamation Service North Platte project in 1908 He has so well taken advantage of the opportunities offered that today he is the owner of two hundred and forty acres of fine irrigated land in a good state of improvement He still sticks to general farming and is one of the most successful

Mr Davis was married in December, 1915, to Blanche Camp, a native of Iowa They have one child, James Keith, aged two years

As a public-spirited man, Mr Davis finds time for other pursuits outside of his agricultural duties He is a Scottish Rite Mason, and his wife is a member of the Methodist church In politics he is independent Although a comparatively new member of his community, he has won a place of respect and good will among his fellow citizens

FRED L YOUNG is a native born Nebraskan His birthplace was in Burt county, where he first saw the light September 19, 1880 He is the son of Andrew and Clementine (Lilly) Young, and is one of nine children in that family, all of whom are living, he being the oldest The other members of the family are Edvinna, now Mrs Louis Larson, living at Oakland, Nebraska, Lee who resides at Carroll, Nebraska, Dora, who is married to Waldo Christensen, Ben, living at Orchard, Nebraska, George at Craig, Nebraska, William, at Craig, Nebraska, who recently returned from service in the United States army with the famous Eighty-ninth Division overseas, Ethel, and Julius, who both are still at the family home in Craig, Nebraska The father is a general farmer and stock-raiser

After finishing the common schools course Mr Young engaged in farming, and realizing the advantage that the trained man has in farming as well as other pursuits, he entered the State University at Lincoln and was graduated from the agricultural course in 1904 He moved to western Nebraska in 1908 and took a homestead in Sioux county, and now owns a half section of well improved land, part of it under irrigation on which he does a general farming business and raises Hereford cattle

He was married February 14, 1912 to Pearl Kent, who is also a native of Nebraska Their family consists of two children William, aged six years and Andrew, aged two

Mr Young is well known as an honorable and progressive man, and is well started on a career of prosperity He has a deservedly high reputation among those who know him Mr Young is a Democrat and a member of the Presbyterian church

JOHN W YOUNGHEIM was born in Illinois on May 2, 1867, the son of Julius and Mary Youngheim He was one of eight children He and his sister, now Anna Predmore, were the only members of the family that came to Nebraska His father was a farmer in Illinois, and died in 1890 He volunteered for service in the Civil War, and served three years and nine months in Company F, regiment unknown, was wounded twice in service His mother died in his early youth, about the year 1870

Mr Youngheim, after finishing his schooling in Illinois, came to Nebraska in 1888, locating in Perkins county as a farmer In 1899, he removed to Fort Morgan, Colorado, the center of the great irrigated district in that state, and from there came to the North Platte valley in 1905 and took a homestead He is now the owner of two hundred and eighty acres of well improved land under irrigation, and has all he can attend to in tilling his land as a general farmer and stock-raiser

He was married in 1896 to Ethel Osler, a native of Iowa, and they have eight children Urith Fern, Zalla, Hazel, Saila, Winford, John and Lurine

Mr Youngheim has a good basis on which to found a judgment of dry and irrigated farming in the west, having had a number of years experience at both kinds, and having also tried the famous Colorado irrigated country, with all this knowledge to guide him, he has chosen the irrigated valley of the North Platte as the best of them all and has definitely cast his lot here In addition to being a successful farmer and stockman Mr Youngheim enjoys the reputation of being an upright and honorable man in his dealings and has made many friends in the years he has been in his present location His word is good and his judgment has been proved to be good by the success that has attended his efforts He is justly regarded as one of the substantial and reliable men of his community, and has yet many years ahead of him in which to enjoy his accumulations and add to them

CECIL FAY HUTCHINSON was born near Red Cloud, in the state of Kansas, January 10, 1883 His father was Valley Tan Hutchinson, and his mother Cora Belle (Potter) Hutchinson, both of whom were natives of Iowa The subject of this sketch was the oldest of six children, all of whom have located in Nebraska His sister Grace, now Mrs E T Purinton, and his brother Leonard reside at Wilcox, Nebraska, another brother, Charles,

has for the past three years made his home here with Mr Hutchinson, but is now married and lives on Leonard's farm, a sister, Beulah, the wife of Hubert Munroe, and the youngest brother, Earl, lives in Sioux county

Mr Hutchinson was educated in Colorado, and after completing his schooling began farming at Wilcox, Nebraska He came to Scottsbluff county and took up a homestead in 1910, and now owns a valuable farm of eighty-eight acres, well improved, and all under irrigation He does a general farming business

July 12, 1911, he was united in marriage with Edna Weston, who is a native of Franklin county, Nebraska Two children were born to them, one of whom was taken by death in infancy The other is named Merville Weston

In politics Mr Hutchinson classes himself as a Prohibitionist, and is a member of the Congregational church

Like every man who is the owner of an irrigated farm in the North Platte valley in a good state of development, Mr Hutchinson is to be called a successful man He enjoys the respect and good will of his neighbors and stands well in his community, and being still a young man he will no doubt achieve a still further measure of success, for he is well started in a business and in a location where the only requirements for success are industry and honesty of purpose, and these he has proved that he possesses

GEORGE W LAWYER is one of the well known men of Scottsbluff county He was born in Iowa, November 25, 1862, the son of William and Caroline (Jackson) Lawyer, being one of eight children in that family His father was a general farmer in Iowa, but came to Nebraska in 1884 and took a homestead in Custer county He died in 1916 The mother died in 1915

Mr Lawyer was educated in Iowa, and upon the completion of his schooling he took up farming and in 1886 came to Nebraska and homesteaded government land From that beginning he has continued to grow with the country until he now is the owner of two fine irrigated farms and a quarter section of other land His efforts have been devoted to farming and stock-raising and have met with an excellent success

He was married in 1885 to Stella Basin, who was born in Illinois but lived in her youth in Iowa Five children have been born to them four of whom are living The names of the children are Verne, who is married and lives in California, Rowland D, living in Des

Moines, Iowa, Chester C, who died at the age of 18, Manley M, now living at home after serving with the United States army in France, and Elgin, who has spent two years and a half in the United States navy

Mr Lawyer is a Republican in politics and is a member of the Methodist church He has always been identified with the spirit of progress and development and has been a part of the growth that has made such a remarkable change in the western part of Nebraska in the last thirty years He stands high in the opinion of those who know him While this section of the country has not lacked for good men no community can have too many of them and there can be no dispute that its future progress will be helped by having all it can get of such men as George Lawyer

HARRY B PATTISON is one of the younger men who have helped to develop the new country that is putting western Nebraska on the map He is one of the sons of Alfred M and Martha J (Goodman) Pattison, of whom mention is made on other pages of this volume, and was born in Indiana, on June 4, 1880, coming to Scottsbluff county, Nebraska, with his parents in 1893

He was educated in the common schools of Hamilton county, Nebraska, and attended the Gering high school after the family's removal to this state After completing his schooling he took advantage of the opportunity that was then offered to lay the foundation of future success by taking up a government homestead in Scottsbluff county, which he relinquished to the government In addition to this he now rents land from his father and is engaged in general farming

In 1903, Mr Pattison was united in marriage to Sylvia Whitis, who is a native of Furnas county Nebraska and, their household has been made happy by the birth of seven children, all of whom are living at home Their names are Belva, Laura, Thelma, Audrey, Hildred, Venita, and Beulah Mrs Pattison is a member of the Seventh Day Adventist church

Harry has proved himself a worthy member of the family name which he bears, and is known as an energetic and industrious farmer, one of those who have not stood idly by while the country moved ahead and left them behind, but went ahead with the country, helped to develop the productive lands that were waiting for development, and took an active part in public affairs while attending in an an honorable way to private enterprises His

family is an ornament to the home and to the community, and he is deservedly counted as one of the substantial and upright men of the county He still has a future to look forward to, and figures that he has not lived half of his life yet That the future for him will be in good old Scottsbluff county goes without saying He is independent in politics

JOE N PATTISON was born in Indiana on December 30, 1876 His father is Alfred M Pattison one of the old and respected settlers in Scottsbluff county, of whom mention is made on other pages of this work, and his mother Martha J (Goodman) Pattison, a native of Tennessee

After attending school in Hamilton county, Nebraska, Mr Pattison came to Scottsbluff county and took up farming and stock-raising, renting land from his father for that purpose He has a good home on a well improved place and is sharing in the general prosperity of western Nebraska, where it is only necessary for a man to be industrious and stick to it in order to be successful

On July 21, 1901, the subject of the sketch was married to Carrie M Snook, a native of Nebraska, and one child, Edward P, has been born to them Mrs Pattison is a member of the Christian church

Mr Pattison is independent in politics, but comes from a Republican family, his father having been active in that party in Civil War times, and his grandfather having been identified with it when the party was first started

Among his friends and Neighbors Joe Pattison is known as a man who stands for progress and industry In connection with his father and brothers he has achieved substantial success in the field that he has chosen, and being yet a young man he has a right to feel that the best part of his life is yet before him He enjoys the respect and esteem of those who know him A fitting compliment to him is to say that he is a worthy member of the family that is well known throughout the county as the Pattisons

ALLISON E STEWART — If the ability to do hard work cannot be designated as a talent, then it is one of the best possible substitutes for that desirable possession Things do not turn up in this work-a-day world unless someone turns them and industry and perseverence lead to the goal, success as in the case of Allison Stewart, a resident of Nebraska since 1885, so that he takes true rank among the hardy pioneers of the Panhandle

who have played an important part in opening up and developing this now favored section

Mr Stewart was born in Fond du Lac county, Wisconsin, being the son of Henry A and
Ruth (Grant) Stewart, the former a native
of the Empirestate, where he was reared and
educated, while the mother was a Buckeye
by birth   The Stewart family are of old
colonial ancestry, the first members having
come to America before the War of the Revolution and they played an important part in
the political and social life of their respective
communities while the tide water region was
being settled and developed   Henry Stewart
came west from New York State by way
of the Great Lakes on a boat at the period
when Michigan and Wisconsin were on the
frontier   He located on a farm in Fond du Lac
county, Wisconsin where he engaged in farming   He was a man of varied talents, being
a prominent member of the Republican party
and active in the social and political life in
eastern Wisconsin all the time he resided in the
Badger State   In 1867 he removed to Iowa,
ever the lure of the "farther west" urged him
toward the new open country, locating first on
the western border, he later removed to Mills
county, where the sands of his life ran out

Allison Stewart was reared on his father's
farm, early learning the lessons of reliance
on self and thus has ever been able to cope
with emergencies when they arose, as he was
quick of judgment and discernment   He received the educational advantages obtainable
in the frontier schools as his boyhood and early
youth were lived on the ever changing frontier   When a boy he displayed great ability
in handling stock on the home farm and after
coming to Nebraska in 1885, was employed as
a cowboy at different times by some of the
great cattle companies who ranged their vast
herds over the prairies at that period   Mr
Stewart filed on a claim in Dawes county, the
year he came to this state and after he gave
up life on the range, established himself as a
farmer and stock-raiser there   He put good
and permanent improvements on the place
but he knew that there was fine land to be
had in the upper Platte valley and having an
opportunity to dispose of his land at a good
figure in 1901 sold out and came to Scottsbluff county, buying two hundred and forty
acres in the Morrill district on section twelve,
township twenty-three-fifty-seven, where he
has since resided   As most of the farm is under ditch Mr Stewart has inaugurated intensive farming using modern machinery to light-

en the work on the place and also increase
production   He has erected good, permanent
farm buildings and a fine modern home for his
family where they keep open house to the
many friends they have made since coming to
the Panhandle   In addition to his own holdings Mr Stewart rents a quarter section of
land to enable him to conduct the varied industries which he finds are so satisfactory from
a financial point of view   He is an independent voter, being guided as his conscience dictates and giving his influence to the best man
for office, he is public spirited though interested in civic affairs only as a good citizen,
not as an office seeker   Mrs Stewart is a
member of the Baptist church, which the family attends while Mr Stewart's fraternal affiliations are with the Modern Woodmen

In 1882, Mr Stewart married Miss Elizabeth Owens, who was born in Wales and accompanied her family to America when they
came here to live, being married in Iowa
Twelve children have been born to this union
Matilda, the wife of T B Allcorn, Laura M ,
deceased , Paul, deceased , Edward M , a stockman of Wyoming , John F , connected with
the Chicago & Northwestern Railroad at
Chadron, A E , Jr , garage man of Riverton,
Wyoming , Frank, on a farm in Wyoming , and
Robert, Charles, Eleanor, Lester, and Henry,
all at home with their parents

RUNEY C CAMPBELL has been a resident of what is now Scottsbluff county for
nearly forty years — a period in which hos
been compassed virtually the entire development and upbuilding of this now favored section of the state, probably the first actual permanent settler in what is now Scottsbluff
county   He has availed himself fully of the
advantages offered in connection with agricultural and live-stock enterprise, and is one of
the representative exponents of these industries in the county, his large and well improved
ranch property being situated three miles east
of Gering and being devoted to diversified
agriculture and the raising of excellent types
of live stock, the while the place has good irrigation facilities   As one of the county's pioneer and honored and valued citizens Mr
Campbell is properly given recognition is this
history

Runey C Campbell is a native of the Hawkeye state and has the distinction of being a
scion of one of its very early pioneer families
He was born at Des Moines, Iowa, November
17, 1858, and is a son of Runey and Euphemia
(Fagan) Campbell, the former a native of

Ohio and the latter of Indiana Runey Campbell and his brothers were numbered among the very first settlers in the vicinity of the present fair city of Des Moines, Iowa where they built the first store, their first camp having been on the site of the present state capitol building They conducted the first mercantile establishment at Des Moines and Runey Campbell long continued as one of the leading citizens of that section of the Hawkeye state He was a man of fine character and excellent mentality — one well fitted to meet the demands placed upon the pioneer He died at the age of seventy-six years, and his wife passed away in 1894, he having been a Democrat in politics and she having been a zealous member of the Methodist Episcopal church The names of both have a place on the roll of the sterling pioneers of Iowa At this juncture is consistently entered brief record concerning the children Clarence is a prosperous farmer in Iowa, Runey C, of this review, was the next in order of birth, Joseph is deceased, Lillie F became the wife of Hubert Smith and is now deceased, William E is a farmer near Dallas Center, Iowa, Charles is similarly engaged near Camp Dodge, that state. Frank is a farmer in Polk county, Iowa and Clara is the wife of Delbert Blake, of Iowa

Runey C Campbell was reared under the conditions that marked the pioneer period in the history of Iowa, and there he received the advantages of the excellent schools for which the commonwealth early became celebrated As a young man of twenty-three years May 8, 1883, Mr Campbell left his native state and came to Nebraska, where he gained experience in connection with cattle raising on the frontier He arrived in the Panhandle district of the state in May of the year noted and here entered the employ of the Bay State Cattle Company, for which he herded cattle on the range for the ensuing three years He then turned his attention to the improvement and development of the homestead which he obtained east of Gering, and during the long intervening years he has continued his active and successful association with agricultural and live-stock industry in Scottsbluff county, where he is now the owner of a well improved and valuable landed estate of four hundred and seventy-one acres, his home place being in section four, township twenty-one range fifty-four, and three miles east of Gering, the county seat

Mr Campbell is a Democrat in politics and as a liberal and progressive citizen he has been influential in the activities that have forwarded the civic and industrial development of his home county He has never been a seeker of official preferment, but he served four years as sheriff of the county and gave a most satisfactory administration He is affiliated with the Gering lodge of the Knights of Pythias and is one of the well known and popular pioneer citizens of Scottsbluff county

April 21, 1886, recorded the marriage of Mr Campbell to Miss Etta A Thornburg, of Perry, Iowa, in which state she was born and reared To Mr and Mrs Campbell have been born seven children Carroll I is at the parental home and is associated with his father in the work and management of the ranch, Ada F is the wife of Roy Shaffer, of this county, and they have one daughter, Ralph E, who is now at home, was for nearly two years in the military service of the nation during the World War, having been a member of the One Hundred and Ninth Engineers, with which he served first at Camp Funston, Kansas and later at Deming, New Mexico, his honorable discharge having been granted after the signing of the historic armistice, having been overseas about a year. Agnes is the widow of Frederick Franklin and now resides at home, Allie O served as a member of the One Hundred and Ninth Engineers, at Camp Funston and Deming, accompanied his command to France and received his discharge after the close of the war, he being now at the parental home, Constance is at home and is attending the public schools at Gering, as does also Lorena the youngest of the children

There wasn't a white woman living in what is now Scottsbluff county when he came to this county, and they are still living on the homestead that they filed on in May, 1886 Mr Stewart also owns a pre-emption adjoining this property that he filed on in 1884 and proved up on same in 1886, then filed on his homestead and timber claim and proved up on both of them

CHARLES V GINGRICH is one of the sterling pioneers of western Nebraska for whom the state of Indiana is to be credited and well has he done his part in the furtherance of civic and industrial development, as a sturdy exemplar of farm enterprise His well improved farm property is situated one-half mile distant from Gering, the judicial center of Scottsbluff county and is located in section thirty-five township twenty-two range fifty-five

Mr Gingrich was born in Jefferson county, Indiana on the twelfth of December 1859 and is a son of John and Anna (Manning) Gingrich, his father having been a carpenter by trade and having been a successful contractor and builder in Indiana, as was he also in southern states, where he passed many winters. Of the four children George and Mary are deceased; Ada is the widow of Thomas Boland and resides at Kearney, Nebraska; and the subject of this review is the next youngest of the number. The father was independent in politics, was prominently affiliated with the Independent Order of Odd Fellows and his wife held membership in the Presbyterian church.

Charles Gingrich duly profited by the advantages of the public schools of his native state, and he was a determined and ambitious man when he decided to seek the better opportunities that were afforded in the west. He wisely chose progressive Nebraska as the stage of his future activities and first located in Dawson county, where he took a homestead of one hundred and sixty acres, which he improved and to which he perfected his title. After selling this property he came to old Cheyenne county, in 1886, and here took up a tree claim in what is now Banner county. He has here developed a model farm property and has been successful in his earnest and well ordered activities as an agriculturist and stock-grower. In 1905 he came to Scottsbluff county and bought one hundred and eighty acres and farmed and later sold and bought his present home. In politics he holds aside from strict partisanship and votes for the candidates that meet the approval of his judgment. His wife holds membership in the Methodist Episcopal church.

In 1891 was solemnized the marriage of Mr. Gingrich to Miss Francis Fitzsimmons, who was born in Ontario, Canada, and they became the parents of eight children, six of whom are living: Addie, at home; Mabel, married Louis Torgenson; Ray is the wife of Walter Leonard, of Scottsbluff; and Vinton, Robert and Grant are still members of the home circle.

J. S. RICE. — In section 15, township 22-55 about two miles distant from the city of Scottsbluff, is to be found the well improved farm which is the stage of the successful activities of Mr. Rice, whose parents were pioneers of Nebraska, the family prestige being well upheld in the civic and industrial status of him whose name introduces this paragraph.

Mr. Rice was born in Farnam, Dawson county, Nebraska, on the 17th of December, 1897, and is a son of Samuel D. and Jennie (Boyle) Rice, both natives of Illinois, where they were reared and educated, their marriage having been solemnized in Nebraska. Samuel D. Rice came to this state in 1884, and took a homestead near Farnam, Dawson county, where he developed and improved an excellent farm and where he continued his residence, as an agriculturist and stock-grower, until 1901, when he sold his property in that county and established himself in a similar line of enterprise in Scottsbluff county. He stocked his farm with excellent Polled Angus cattle and became specially successful as one of the enterprising agriculturists and stock men of the county. His wife died in 1905 and he passed away January 3, 1918 — a sterling citizen whose circle of friends was coincident with that of his acquaintances. Samuel D. Rice was independent in politics and both he and his wife held membership in the Presbyterian church. Of the three children, Lena holds a clerical position in the State Bank of Scottsbluff; J. S., of this review, has active management of the home farm, which comprises one hundred and forty acres; and Harry is a resident of Scottsbluff county.

June 12, 1917, recorded the marriage of J. S. Rice to Miss Maude Ansen, who was born and reared in this state, a daughter of Frederick and Theresa Ansen, the former of whom is deceased and the latter of whom resides at Scottsbluff.

JESSE H. MARLIN, who resides upon and gives his active supervision to his well improved farm of one hundred and sixty acres, in section 28, township 23-55, Scottsbluff county, entered claim to this land in the year 1900, and here the conditions of the present day demonstrate his ability as an agriculturist and stock-grower and his progressiveness and enterprise, for he developed the land from the raw prairie and has made it a farm notable for the excellence of its improvements. His knowledge of what to do and how to do it has been fortified by practical experience which he gained in Nebraska, for he is a native of this state and a scion of a sterling pioneer family of Frontier county. On the old home farm in that county he was born on the 2d of March, 1882, and he is a son of William and Amanda (Ray) Marlin, who were born and reared in Indiana and who became pioneer settlers in Frontier county, Nebraska, the father having there developed a valuable farm property and having become a very successful agriculturist and stock-grower, he having done much to improve the grades of live-stock in his county, by raising Polled Angus cattle, Poland-China swine and pure blood Shire horses, including a number of stallions of specially fine type. He finally disposed of his farm in Frontier county and is now living retired at Scottsbluff, both he and his wife being honored as worthy pioneers of this great commonwealth. In politics Mr Marlin is known as a stalwart advocate and supporter of the principles f the Republican party.

On the old home farm in Frontier county Jesse H Marlin was reared to adult age, and in the meanwhile he made good use of the advantages afforded in the public schools of his native county. There he continued his association with farm industry until 1900 when, as before stated, he came to Scottsbluff county and took up the homestead which has since continued the stage of his successful activities. In politics he is an independent voter and he takes lively interest in all things touching the communal welfare.

In the year 1902, Mr Marlin wedded Miss Verl Smith, who was born and reared in Knuckles county and who is a daughter of H W Smith. Mr and Mrs Marlin have two children — Evelyn and Elwin, who add joy and brightness to the attractive home.

CHARLES H UGLOW gives the measure of his ambition and ability in no uncertain terms when recognition is taken of the success which attends his progressive operations on his fine estate of two hundred and eighty acres, in township 23-54, five miles distant from Scottsbluff, where his residence, a model farm home, is situated in section twenty-seven. Irrigation facilities are provided for two hundred acres of this tract at the present time and the productivity of the soil is thus brought to the maximum, while Mr Uglow's activities as an agriculturist are scarcely less remunerative than are those which he brings to bear in the raising of excellent grades of live-stock. As one of the constructive workers and representative farmers of Scottsbluff county he is properly given consideration in this history.

Mr Uglow was born in Ringgold county Iowa, on the 22d of June 1875 and is a son of Nicholas and Mary (Cort) Uglow, the former a native of Pennsylvania and the latter of England. The father was reared and educated in the old Keystone state, which, as a young man, he represented as a valiant soldier of the Union in the Civil War. He enlisted in Company H, Two Hundred and Eighth Pennsylvania volunteer Infantry, which was assigned to the Army of the Cumberland and with which he

participated in many important engagements. Within a short time after the close of the war he removed to Iowa where he purchased land and became a pioneer farmer in Ringgold county. Later he removed to Kansas, and he passed the closing years of his life in the soldier's home in that state, an institution which he entered after the death of his devoted wife. Both were members of the Methodist Episcopal church, he was unwavering in his allegiance to the cause of the Republican party and was affiliated with the Grand Army of the Republic and the Independent Order of Odd Fellows. Of the family of eleven children seven survive the honored parents. Fannie is the wife of John P Benson, of Iowa, Mary is the wife of William Feeney, of Kansas City, Missouri, John likewise attained to maturity but is now deceased, Ella, the wife of Henry Miller and they reside in northwest Canada, Carrie is the wife of Peter C Fisher, of Oklahoma, Ethel is the wife of Jesse Johnson, of Keyapaha county, Nebraska, of the present whereabouts of Fred other members of the family have no definite knowledge, and Charles H.

The public schools of Iowa furnished to Charles H Uglow his early educational discipline, and he not only gained valuable knowledge in connection with the work of the home farm. In 1900 he came from Iowa to Nebraska and first located in Rock county, where he remained until 1907, when he removed to Scottsbluff county and took up a homestead of eighty acres, the same constituting an integral part of his now spacious and admirably improved landed estate, which is devoted to diversified agriculture and the raising of high-grade live-stock. Not only as pertaining to his individual affairs but also as a citizen is Mr Uglow liberal and progressive and he takes loyal interest in community affairs. In politics he is independent of strict partisan lines and his wife's religious faith is that of the Adventist church.

At Scottsbluff, on the 16th of October, 1907, occurred the marriage of Mr Uglow to Miss Eva C Johnson, who was born in Missouri and whose parents James and Catherine Johnson became early settlers in Rock county Nebraska, where her father died and where her mother still resides. Mr and Mrs Uglow have two children — Wilma and Homer.

HENRY J OTTE — Gauged only by evidences that are definitely observable in connection with his well improved farm estate in Scottsbluff county, it is certain that Mr Otte is to be consistently designated as one of the progressive and wide awake agriculturists and stock-growers of the county, while his loyalty to civic responsibilities, his sterling characteristics and his genial personality are the fortifying elements in his unqualified popularity.

Mr. Otte is a citizen who can claim the distinction of being a native of the old Buckeye state. He was born in Auglaize county, Ohio, on the 10th of April, 1875, and is a son of William J. and Elizabeth (Sollman) Otte, the former a native of Ohio and the latter of Indiana, the father having become one of the prosperous farmers of his native state, he is now deceased but his wife is still living.

Henry J. Otte passed the period of his childhood and early youth in his native state, where he gained practical experience in connection with the work of the home farm and where also he made good use of the advantages afforded in the public schools. There he continued to be identified with farm enterprise until 1907 when he came to Nebraska and established his residence in Scottsbluff county. In May of that year he filed entry on his present homestead farm, which comprises eighty-two acres and upon which he has made good improvements, including modern buildings and the supplying of the entire tract with excellent irrigation facilities. Here he is successfully carrying on diversified agriculture and raising good grades of live-stock. He is one of the live men of the community and always ready to aid worthy enterprises projected for the general good. He contributed liberally to the various agencies that upheld the government during the period of the World War and in local affairs he is independent of specific political partisanship. He is affiliated with the Masonic fraternity, as a member of the Blue Lodge at Mjamesberry, Ohio.

In 1896, Mr. Otte wedded Miss Mary Springman, a native of Pennsylvania, and they became the parents of five children — Ruth, Earl, Frederick, Ray and Ralph. All of the children are living except Ralph, who died in early childhood.

CHARLES HILLS, whose model farmstead is situated in section 28, township 23-55, Scottsbluff county, is one of the younger and representative exponents of agriculture and live-stock industry in the county and is a popular citizen who properly finds recognition in this publication. His home is about 5 miles distant from Scottsbluff, which is his post-office address.

Mr. Hill was born in Brown county, Illinois,

on the 2d of July, 1881, and in the same state were born his parents, George and Paulina (Green) Hill, of whom a record will be found elsewhere in this volume.

Charles Hill is indebted to the public schools of his native state for early educational training and he was about twenty-four years of age when he accompanied his father to Nebraska. In Scottsbluff county he filed entry on a claim of eighty acres, but this does not constitute his present farm, which has been brought under effective cultivation and is devoted to diversified agriculture and the raising of specially fine live stock. He sold his original claim and since that time he has accumulated a valuable property of five hundred acres, of which eighty acres are thus far supplied with irrigation facilities. In the cattle department of his farm enterprise Mr. Hill breeds from a registered Shorthorn bull and cow, besides which he has a specially fine Percheron stallion and a splendid jack stallion which is much in demand for breeding purposes. Mr. Hill is independent in politics and is essentially liberal and progressive in his civic attitude. His wife holds membership in the Presbyterian church.

The year 1904, recorded the marriage of Mr. Hill to Miss Gertrude Ford, who was born in Butler county, this state, and they have four children — Kenneth, Rena, Gertrude and Ahlean.

GEORGE FORD, whose death occurred at his home, in Scottsbluff county, November 24, 1914, was a man who had deep appreciation of the natural advantages of western Nebraska and he used his mature judgment when he made investment in land here, even as he did in the improvng and developing of the property. He was a citizen whose life was guided and governed by utmost integrity and honor and thus he commanded the high regard of those with whom he was brought into contact within the period of a career marked by earnest and fruitful endeavor. His widow still resides upon the farm property which he accumulated in Scottsbluff county, where her home is situated in section 32, township 23-54, about seven miles distant from the city of Scottsbluff.

Mr. Ford had the distinction of being a native of Granada, West Indies, where he was born in the year 1858, and he was a scion of fine English stock on both paternal and maternal sides. His parents, Charles and Harriet (Fish) Ford, were born and reared in England and he became a foundryman by vocation.

MR. AND MRS. WILLIAM DEBILY

He was sent by the English government to the West Indies, and land which he there purchased is presumed to yet be an asset of his estate He died in the West Indies and his widow subsequently contracted a second marriage, she too being now deceased The subject of this memoir acquired his early education principally in the schools of England and as a young man he came to Nebraska with Lord Jones, who owned a large tract of land in the vicinity of Crete, Saline county There Mr Ford gained in this connection wide and practical experience in the herding of cattle under the conditions of the pioneer days, and after coming to the state he had occasion to make five visitations to England, at varying intervals His stepfather came to Nebraska about two years after he himself had here established residence, and the former took up a homestead in Butler county, he and his wife having passed the remainder of their lives in this state Mr Ford became actively identified with farm enterprise in Butler county and there he continued to reside until 1908, when he moved to Scottsbluff county and took up a homestead of one hundred and sixty acres in section 20, township 23-54 Later he added greatly to his holdings, and at the time of his death he was the owner of a tract of two hundred and eighty acres, which is still in the possession of his widow and children and which is one of the well improved farm properties of this section of the state, Here are conducted successful operational along the lines of diversified agriculture and stock-raising, and the two sons of Mr Ford are here well upholding the honors of the family name In a general way Mr Ford was a Democrat in politics, but he was a zealous advocate and supporter of the cause of prohibition and did all in his power to bring about the obliteration of the liquor traffic He was affiliated with the Modern Woodmen of America and was an earnest member of the Presbyterian church, as is also his widow

In Butler county, this state was solemnized the marriage of Mr Ford to Miss Kate French, who was born in Illinois and who is a daughter of George and Mary (Wilson) French, who established their residence in Butler county in 1882, after having made the overland trip from Illinois by medium of team and wagon Mr French, who was a veteran of the Civil War, in which he served three years in defense of the Union, purchased land and improved a good farm in Butler county He passed to eternal rest in 1904 and his venerable widow still resides in Butler county In

conclusion is given brief record concerning the children of Mr and Mrs Ford Gertrude is the wife of Charles Hills, a farmer in Scottsbluff county, Grace is the wife of Harry Drawbaugh, of David City, this state, and Charles I and George I remain upon the old home farm of which they have the active management Charles is married, the maiden name of his wife having been Edna Stratton, George married Jesta Andrews

WILLIAM DeBELY, who owns one of the productive irrigated farms of Scottsbluff county, was born in Switzerland, August 9, 1860 His parents were Frederick and Sophia (Peret) DeBely The father was born in France but went to Switzerland when young and learned the watchmaking trade, which he followed until his death at the age of sixty-two years The mother was born in France, came to the United States in 1900 and is now deceased

William DeBely was reared and educated in Switzerland In 1888 he came to the United States and located in Scottsbluff county, Nebraska, where he was variously employed until 1892, when he homesteaded The hardships of early times in this section affected him to some extent but he never became discouraged and now feels well repaid for his hard work in the development of his property He carries on general farming and raises stock

In 1887 Mr DeBely was married to Elizabeth Lowe, who was born August 4, 1859, and died May 16, 1901 Her parents never came to the United States They have five children Frederick Pearl, Carl, Blanche, and Fannie Mr DeBely and his family belong to the Lutheran church

JOHN MATHSON — Admirably has this sterling citizen demonstrated the forceful energy and mature judgment that have made those of Scandinavian birth or lineage so potent a factor in connection with the industrial development of many of our western states, and in Scottsbluff county he has reclaimed and improved a fine farm property in the Mitchell vicinity his home place being in section 5, township 23-55 As one of the substantial and highly respected men of the county he is entitled to recognition in this history

Mr Mathson is a native of Norway, where he was born in the year 1879, and he is a son of Mathiason and Emborg Mathson, the former of whom still resides in Norway and the latter of whom is deceased

John Mathson was reared to maturity in his tages of the common schools, and in 1900 he native land, where he was given the advantages of the frontier in western Nebraska and became a pioneer of pioneers in that part of the Cheyenne, immigrated to America and established his residence in Wisconsin, where he found employment at farm work, besides which he attended school at intervals during the years of his stay in the Badger state. Thence he went to Denver, Colorado, where he found employment in connection with the manufacturing of brick, and from that state he continued his way westward and visited California and Nevada. In 1906 he came to Nebraska and took up a homestead of one hundred and sixty acres in Scottsbluff county, and this place has since continued the stage of his energetic and well directed enterprise as a successful agriculturist and stock-grower. He has made excellent improvements on the farm, which has good buildings that were erected by him personally, as he is a skilled workman at the carpenter trade. For eight years he was identified with the government reclamation service. In politics Mr Mathson is an independent voter and he has served efficiently as school director of his district. He had supervision of the construction of the consolidated school building in his locality, and he and his wife attend and support the Union church organization. He is affiliated with the Woodmen of the World. Mr Mathson is progressive and wide awake as an exemplar of farm enterprise and his valuable farm property has been provided with excellent irrigation facilities.

In 1905 occurred the marriage of Mr Mathson to Miss Hannah Mundel, who likewise is a native of Norway, and their pleasant home is brightened by the presence of their five children Erling, Judith, Helen, Lloyd and Ruth.

WILLIAM B SWINDELL is one of those valiant souls who braved the hardships of the frontier in western Nebraska and became a pioneer of pioneers in that part of Cheyenne county that is now included in Scottsbluff county. He has reclaimed one of the large and valuable farm properties in this section of the state and his well improved estate is situated four and one-half miles north of the village of Minatare. In a general way as well as an individual way he has been an artificer of development and progress, and he is known and valued as a sterling citizen to whom the highest pioneer honors are due.

Mr Swindell was born at Silver Creek, Delaware county, Iowa, on the 26th of October, 1857, and is a representative of one of the honored pioneer families of that section

of the Hawkeye state, where his parents, William and Isabel Swindell, settled in the year 1851 and where they passed the remainder of their lives, secure in the high regard of all who knew them. He whose name introduces this review acquired his youthful education in the district schools of his native state, where he was reared under the influences of the pioneer days, and at the age of twenty years he became a clerk in a general store at Manchester, Iowa. In 1885 he came to what is now Scottsbluff county, Nebraska, and entered a pre-emption claim four and one-half miles north of Minatare, which now a thriving village then had no semblance of communal dignity. He perfected his title to his original claim, as did he also to adjoining homestead and tree claims, and under his vigorous and well ordered direction this property has been developed into one of the fine landed estates of this locality, the improvements being of the most approved modern type. Mr Swindell has been somewhat of a leader in community affairs and has held three commissions as postmaster at Minatare — two under President Taft and one under President Cleveland. He is a Republican in his political adherence and has long maintained affiliation with the Ancient Order of United Workmen, in which he served twelve years as recorder, besides which he has passed the official chairs not only in this organization but also in that of the Modern Woodmen of America, in which latter he held for six years the office of clerk of his camp.

August 27, 1882, recorded the marriage of Mr Swindell to Miss Ida Johnston, of Manchester, Delaware county, Iowa, and they have two fine sons. Earl J, who was born December 27, 1887, married Miss Clara Smith and they reside at Hot Springs, South Dakota; Donald W, who was born January 18, 1889, was afforded the advantages of the University of Nebraska, and when the nation became involved in the great world war he entered the military service, in the officers' training camp at Fort Sheridan, Chicago. His command was not called into active service over sea and he gained therein the rank of sergeant; he is now residing at Minatare.

EDWARD F VANDERBERG has shown marked energy, enterprise and good judgment in his various business operations during the period of his residence in Scottsbluff county, and he is now the owner of valuable real estate in the city of Scottsbluff, where he is successfully established in the barber business, besides which he is the owner of property in the state of Wyoming His advancement has come as the result of his own energy and ability and he is one of the well known and popular citizens of his home city and county

Mr Vanderberg was born at Rock Island, Illinois, on the 25th of August 1867, and is a son of Leo and Mary Ann (Marshall) Vanderberg both of whom were residents of Nebraska at the time of their death Leo Vanderberg came from Belgium, in 1848, and was a young man when he came to America He resided in various states of the Union prior to coming to Nebraska, and the major part of his active life was devoted to agricultural pursuits Mrs Vanderberg was born in Kentucky, where her father was an agriculturist and slave-owner of no little importance prior to the Civil War, and she developed her talents in such a way as to become a woman of superior education and high intellectuality

Edward F Vanderberg was about seventeen years old at the time of the family removal to Nebraska, his early education having been gained principally in the public schools of Illinois, where also he acquired his initial experience in connection with farm enterprise At the age of seventeen years he was found independently engaged in farming and stockgrowing in Frontier county, Nebraska, where he thus continued operation eleven years He then removed to Maywood, that county, where he learned the trade of barber, which he there followed three years On the 15th of August, 1901, he established himself in the barber business at Scottsbluff, and he began operations with but one chair He eventually developed a large and representative trade, with a well equipped shop, and in 1909 he further manifested his progressive spirit by establishing and equipping the first thoroughly modern laundry in the city He sold this latter business after successfully conducting the same about two years and thereafter he again gave his attention largely to his barbering business In 1906, Mr Vanderberg took up homestead and preemption claims in Funston precinct, ten miles northeast of Scottsbluff, and the same year he sold his rights to this property for $400 It is worthy of special mention that on August 15, 1919, the west eighty acres of this tract sold for $18,000 The four hundred dollars which he received for his claims Mr Vanderberg invested in his business operations and since that time he has made substantial advancement, as shown in his ownership of his business property, his attractive residence in Scottsbluff and a farm at Goshen Hole, Wyoming June 13, 1903, his barber shop was

destroyed by fire, but he promptly erected a new and modern building on the site

In politics Mr Vanderberg maintains an independent attitude, and his popularity is indicated alike by his affiliation, as a charter member, with the local organizations of the Independent Order of Odd Fellows, the Modern Woodmen of America, the Benevolent and Protective Order of Elks, and the Knights of Pythias He has passed all of the official chairs in his Odd Fellows lodge and is also a valued member of the Scottsbluff Country Club

At Maywood, Frontier county, on the 27th of May, 1900 Mr Vanderberg wedded Miss Emma J Schnase, a daughter of Gustave and Rose (Meyers) Schnase, both natives of Germany Mr Schnase came to the United States in 1865, and he and his wife eventually became pioneers in Frontier county, Nebraska, where he became a prosperous farmer near Maywood Mr and Mrs Vanderberg have four children, all of whom remain at the parental home, — Doris S, Elsie May, George Edward and Rosemary Ann The eldest daughter was graduated in the Scottsbluff high school as a member of the class of 1919

PARVIN E GILBERT, who resides in the city of Scottsbluff and who is the owner of a well improved and irrigated ranch estate in the county, has here been prominently identified with mercantile and industrial interests for nearly a score of years, and he is now serving as salesman for a leading wholesale grocery house He has been active in progressive movements that have inured to the advancement of Scottsbluff county and is a citizen who is well entitled to recognition in this history Parvin Edson Gilbert was born in Van Buren county, Iowa, July 18, 1879, and has been a resident of Nebraska since his boyhood He is a son of David L and Minnie J (Stout) Gilbert whose marriage was solemnized November 22, 1877, and who were residents of this state, at the time of the father's death, on the 6th of May, 1887 On the 1st of November 1895, Mrs Minnie J Gilbert contracted a second marriage, by becoming the wife of William M Garrison, and she passed to the life eternal on the 17th of January, 1904

Parvin E Gilbert was afforded the advantages of the excellent public schools of Ogallala Keith county Nebraska, where he was graduated in the high school and he early gained business experience of practical and valuable order In 1901 he engaged in the general merchandise business in Scottsbluff, as successor of Mr. Kirkpatrick, who opened the first store in the new town. Mr. Gilbert was the first to initiate a free-delivery system in connection wth retail mercantile enterprise in Scottsbluff and in order to meet the demands placed upon his establishment by a constantly expanding trade he removed from the original store to larger quarters, on the present site of the First National Bank. In August, 1905 he sold his substantial business to the firm of William Rice & Company and turned his attention to the reclamation and improvement of his landed estate in Scottsbluff county. On the 14th of June, 1904, under the provisions of the reclamation act, Mr. Gilbert filed entry on a homestead, and to this place he removed with his family in January, 1906. He assisted actively in the building of the government irrigation canal and laterals and continued his service until the work was completed to Lake Alice. Through this medium he gained for his land excellent irrigation facilities, and he made excellent improvements on the tract, the most of which is seeded to alfalfa. He erected good buildings and brought the place to a status that marked it as one of the valuable properties of the county. In the autumn of 1910, Mr. Gilbert returned with his family to the city of Scottsbluff, and since 1911 he has given specially effective service as traveling salesman for a wholesale grocery house, the while he continued to give a general supervision to his ranch property.

The Democratic party finds Mr. Gilbert aligned as one of its loyal supporters, he is an appreciative and valued member of the Scottsbluff Commercial Club, and he and his wife are members of the Methodist Episcopal church.

July 14, 1904, recorded the marriage of Mr. Gilbert to Miss Golda Agnes Westervelt, daughter of James H. and Luranie A. Westervelt, well known citizens of Scottsbluff county. Mrs. Gilbert completed a course in the Scottsbluff high school in 1903, and she is a popular factor in the representative social activities of her home city. Mr. and Mrs. Gilbert have four children, and their names and respective dates of birth are here noted: Adelaide Bernice, November 5, 1905; Lucile Marie, February 29, 1908; Howard James, March 4, 1910; and Charles L., January 16, 1914.

THOMAS E. CHAMBERS. — More than thirty years ago, when Scottsbluff county was still an integral part of Cheyenne county, Thomas E. Chambers and his wife became pioneer settlers near Minatare, though that now thriving village was at that time not to be found "on the map." They lived up 'to the full tension that marked the pioneer period in the annals of Scottsbluff county, and in addition to reclaiming one of the excellent farms of the county Mr. Chambers long held precedence as one of the leading merchants at Minatare, where he is now living retired, in the enjoyment of the rewards of former years of earnest and successful endeavor. As one of the representative citizens and pioneers of the county he is consistently accorded recognition in this history.

Thomas E. Chambers was born in County Kent, England, on the 18th of December, 1858, and his early education was obtained in the schools of his native land. On the 18th of December, 1871, — his thirteenth birthday anniversary — Mr. Chambers landed in New York City. He proceeded to Wapello county, Iowa, and for thirteen years he was employed in connection wth coal-mining industry in the Hawkeye state. In 1884 he removed to Hamilton county, Nebraska, where for three years he was a coal merchant, and he then moved and numbered himself among the pioneer settlers of that part of Cheyenne county that is now included in Scottsbluff county. He arrived on the 23d of March, 1887, and settled on a homestead claim two and one-half miles northeast of Minatare. He made improvements on the land and in due time perfected his title to the property, which he eventually sold to advantage. It is but fitting to state that when Mr. Chambers came to Nebraska his resources were very limited, his equipment when he arrived at Sidney having comprised one hundred pounds of flour, one hundred pounds of corn meal (which was later stolen from him), one dozen chickens, two cows, and a few household effects. Upon arriving at the Camp Clark bridge he had no money to pay the toll, and he borrowed one dollar from L. C. Marquis to meet this emergency. He thus depended entirely upon his own exertions in making his way to prosperity in the new country, and it is needless to say that he and his loyal wife endured their full share of the hardships and trials that marked the pioneer epoch in the history of Scottsbluff county, while they pressed forward to the goal of definite success which should ever attend honest and earnest endeavor.

Mr. Chambers has always stood ready to do all in his power to further the advancement and prosperity of his chosen county and state and he has been influential in community af-

fairs. In 1890 he was elected assessor of
Tabor precinct, an office of which he was the
incumbent two years. In politics he has ever
been unfaltering in his allegiance to the Democratic party, and he is well fortified in his opinions concerning public affairs. Mr Chambers
helped organize school district No 2 and was
a director for eight consecutive years

In 1905, Mr Chambers established a general store at Minatare and he built up a large
and prosperous business, to the conducting
of which he continued to give his attention until impaired health led him, in 1915 to sell
the business and stock to his son-in-law, E
H Johnson. He still owns the building in
which this store is located and he also owns
and occupies one of the attractive residences in
the village of Minatare.

June 22, 1880, recorded the marriage of
Mr Chambers to Miss Jennie Wicks, of Des
Moines, Iowa, where she was reared and educated. In conclusion is given brief record concerning the children of this sterling pioneer
couple. Alfred B, who was born April 15,
1883, and who is a successful farmer in Montana, married Miss Olga Dalquist, of that
state, Frederick R, born May 18, 1888, married Miss Selma Dalquist, of Montana, and
they reside in Valley county, that state, Laura
Belle, born December 4, 1890 is the wife of
E H Johnson, a leading merchant at Minatare, and Leo L, born September 14, 1893,
married Miss Estella Duncan and they now reside at San Diego, California

ROBERT J HARSHMAN is a popular
representative of one of the sterling pioneer
families of Scottsbluff county, and the name
which he bears has been prominently and
worthily linked with the development of the
Minatare vicinity, where he himself has done
much to foster civic and industrial progress

Robert James Harshman was born in Tama
county, Iowa, March 22, 1865, this date indicating beyond peradventure that his parents
were numbered among the pioneer settlers of
that section of the Hawkeye state. He is a son
of Theodore and Rebecca (Thompson) Harshman, both natives of Fayette, county Pennsylvania, where the former was born in 1841
and the latter in 1844, their marriage having
been solemnized in 1861. About the year
1864 Theodore Harshman and his wife established their home on a pioneer farm in Tama
county, Iowa, and this property he effectively
reclaimed and developed. There he continued to reside until 1885, when he sold the
farm and came with his family to what is

now Scottsbluff county, Nebraska. One-half
mile north of the present village of Minatare
he took up homestead and tree claims, to both
of which he perfected his title in 1892. In
the meanwhile he had made excellent improvements on the property and in addition to his
activities as a pioneer agriculturist he had
also found much demand for his services as
a blacksmith. He established the first blacksmith shop in the valley, also in the village of
Minatare and in the building was maintained
also the local postoffice, he himself having
served as the postmaster. He and his wife
were honored pioneers of this locality, where
they ever commanded the fullest measure of
popular esteem and where they continued to
reside until their death

To the public schools of Iowa, Robert J
Harshman is indebted for his youthful education and he was there graduated in the Collins high school as a member of the class of
1886. He then came with his parents to Nebraska, where he likewise gained pioneer honors in the settlement of the portion of Cheyenne county that is now included in Scottsbluff county. In 1886 he took up a homestead
claim four and one-half miles northwest of
Minatare, and upon this he established his
residence. Later he entered a timber claim
adjoining, and he proved up on his claims in
1892. On his homestead he continued to reside fourteen years, successfully engaged in
agricultural and stock-growing enterprise, and
he then sold the property, after which he removed to Nine-mile Canyon, where he was
engaged in raising cattle during the ensuing
nine years. He then located in the village of
Minatare, where he was engaged in the hardware and implement business for five years.
After disposing of this business he was for five
years a traveling representative for the International Harvester Company, and for some
time after severing this relation he sold threshing machines for the great factory of M
Rumely Company. At the present time he is
giving his attention principally to the restaurant and soft drink business and is one of the
representative citizens of Minatare.

At intervals Mr Harshman has been actively concerned in the construction of irrigation
ditches in his home county. He and his father and brothers built about fifty per cent of
the Minatare ditch and he was identified also
with the construction of the Winter Creek
ditch. He has been loyal and vigorous in the
support of public enterprises including the
building of churches and school-houses such
undertakings in the early days having depended

entirely upon popular subscriptions for their carrying out   Mr Harshman and his brothers, with the co-operation of a few neighbors, built the first schoolhouse at Minatare   He was for five years treasurer of the Minatare irrigation ditch and for several years was secretary of the Winter Creek ditch   In politics he is a Democrat, and fraternally he is actively affiliated with the local organizations of the Modern Woodmen of America and the Knights of Pythias in each of which he has passed the various official chairs

At Wellsville, Cheyenne county, in 1892, Mr Harshman wedded Miss Mary Rosenbrook, and concerning their children brief record is given in conclusion of this review Roy T conducts a men's furnishing store at Bridgeport, Estelle, bookkeeper in a hardware store in this village, Fred W, who is now at the parental home, served eleven months in the national army during the period of the World War, and at Jacksonville, Florida, he was promoted to the rank of sergeant of his company, Alice and Leone, remain at the parental home, the former having been graduated in the Minatare high school as a member of the class of 1919

ARMENAG SIMONIAN is one of the comparatively few Armenians who can claim pioneer distinction in Scottsbluff county, and here he achieved a large measure of success in connection with the development and operation of one of the pioneer ranch properties of the county He is now living virtually retired at Gering and is a citizen eminently entitled to recognition in this history

Mr Simonian was born at Bitlis, Armenia, on the 19th of March, 1868, and he acquired his rudimentary education in the far distant country of his nativity   At the age of twelve years, in company with a younger brother, Hiram, he came to America and joined an older brother, Isaac, who was then living at Ludlowville, New York   There the two younger brothers were enabled to attend school at intervals, and there Armenag initiated his practical service by working on a farm, for eight dollars a month   In 1885 he came to Nebraska and after remaining one year with his brother Isaac in Lincoln county he made the overland trip to Cheyenne county, with horses and wagon   He and his older brother, Isaac, settled in the vicinity of Scottsbluff, where they engaged in farming and stock-growing Finally he made entry on a pre-emption claim, besides purchasing a relinquishment on a tree claim   To make the requisite payments he borrowed money from the bank at Gering and paid interest at the rate of four per cent a month   With unremitting energy he applied himself to the development and improvement of his land, and with the passing years abundant success attended his efforts as an agriculturist and stock-raiser, so that definite prosperity came as his just reward   He perfected his title to both claims and retained the property in his possession for many years, special attention having been given to the raising of horses and cattle   He still owns a well improved tract of seventy acres, northwest of Gering, and is now living practically retired His brother Isaac, who died in 1897, was one of the well known pioneers of what is now Scottsbluff county and commanded unqualified popular confidence and esteem   Mr Simonian is a Republican in politics, is affiliated with the Modern Brotherhood of America and is an active member of the Presbyterian church

GEORGE SOWERWINE is most consistently to be accorded consideration in this history, by reason of his being one of the sterling pioneer agriculturists and stock-growers who have aided greatly in the civic and industrial development of Scottsbluff county. He owns one of the well improved and valuable landed estates of the county and is a venerable citizen to whom is accorded the fullest measure of popular confidence and esteem in the county that has long been his home   He not only gained wide and varied pioneer experience in the west, but also rendered valiant service as a soldier of the Union during the stormy epoch of the Civil War   Mr Sowerwine is now living virtually retired, in a pleasant home at Gering, the judicial center of the county

George Sowerwine was born in Delaware county, Indiana, on the 19th of June, 1843, and is a son of Christian Sowerwine, who was born in Virginia, who became a farmer in Indiana and who later was a pioneer agriculturist in Iowa, where he died when about seventy-seven years of age   The subject of this sketch was about eight years old at the time of his mother's death and was a child at the time of the family removal to Iowa, where he was reared and educated under the conditions and influences of the pioneer days   In 1859 he equipped himself for the long and perilous journey across the plains to California   He was at the time a lad of sixteen years and from Council Bluffs he started forth with an ox team for the New Eldorado   In due time the plodding train of ox teams reached California, and there Mr Sowerwine continued in the quest of gold until a higher duty confronted him, when the Civil War was precipitated upon a divided nation At the age of eighteen years he enlisted in the United State Cavalry and with his command, under two enlistments, he served in the Indian campaigns in the west during the progress of the war between the north and the south He continued in military service four years, two months and seven days, and had many trying and hazardous experiences in campaign work By reason of this valiant service during a stirring period in the history of the nation he is eligible for and affiliated with the Grand Army of the Republic, in the affairs of which he has maintained deep interest as its ranks have been rapidly thinned by the one invincible antagonist, death

After the termination of his military career Mr Sowerwine returned to Iowa where he engaged in farming and where also he was identified with coal mining for some time In 1886 he came to that part of Cheyenne county Nebraska, that is now comprised in Scottsbluff county, and as a pioneer he filed entry on both homestead and pre-emption claims, to which he eventually perfected his title He here engaged in general farming and stockgrowing under the adverse conditions that marked the early period in the history of the county, but the passing years brought to him cumulative success and he developed one of the fine farm properties of this section of the state He has been specially active and influential in the development of the irrigation facilities of the county, assisted in the construction of the Winter Creek ditch and was president of the company which built the same He also aided in the construction of the Central ditch on the south side, and this supplies irrigation for his fine ranch of 422 acres, which he now rents He is also interested in an oil prospect in this part of the county and though he is living retired at Gering he still takes lively interest in all things that touch the welfare and progress of his home county and state He is Republican in politics At Gering he is affiliated with Post No 265, Grand Army of the Republic

December 27, 1868 recorded the marriage of Mr Sowerwine to Miss Elizabeth C Marquis, who was born in Michigan and who was reared and educated in Iowa where she was a popular school teacher prior to her marriage She is a daughter of John W and Margaret (Scott) Marquis, who were born and reared in Ohio, where their marriage was solemnized and when they removed to Michigan

where the father became a farmer, as he did later in Iowa, whence he came as a pioneer to what is now Scottsbluff county, Nebraska In 1888 Mr Marquis here took up a homestead, which he developed and improved and which he finally sold to Nellie M Richardson When well advanced in years he removed to Grand Island, and there he died when about eighty-four years of age, his wife having died at Palisade Hitchcock county when she was sixty-seven years of age Of the seven children of Mr and Mrs Sowerwine four are living Clarence and his wife live in Oklahoma and have one child, Morris is a resident of California and is the father of four children, Eugene, who resides at Gering, Scottsbluff county, has two children; Mabel is the wife of Elmer Sherman, of Gering, and they have five children Mr and Mrs Sowerwine lived up to the full tension of the pioneer days in Scottsbluff county, and here their circle of friends is limited only by that of their acquaintances

MELVIN MILLER was an infant at the time when his parents became residents of Nebraska, and as this removal occurred somewhat more than forty years ago it must be conceded that pioneer honors are to be ascribed to the parents Mr Miller has been actively identified with farm enterprise in various counties of the state since he arrived at maturity, he became successful as the owner of one of the best equipped barber shops in Scottsbluff county, and in the city of Scottsbluff he served two years as chief of the police department, an office in which he gave a most satisfactory administration and from which he retired several years ago

Mr Miller was born in Henry county, Iowa, June 24, 1877 and in the following year his parents came to Nebraska and located near Fairmon Filmore county, where the father was engaged in farming for a year thereafter Melvin Miller is a son of William J and Calfrenia (Welch) Miller, both natives of Scott county, Illinois, where the former was born April 7, 1843, and the latter on the 19th of January, 1848 — dates that show that the respective parents were numbered among the pioneers of that state William J Miller was reared and educated in Illinois and there became the owner of a farm, as he did later in Henry county, Iowa, where he remained until 1878 when he came with his family to Nebraska as noted above From Filmore county he removed to York county, where he purchased land and was successfully engaged in farming for fourteen years The following five years he passed in Gosper county and he then sold his farm in that county and removed to Phelps county, where he passed the residue of his life, he having been seventy-three years of age at the time of his death and his widow being now a resident of Scottsbluff, where she is a revered member of the family circle of her son Melvin, whose name introduces this sketch

The public schools of Nebraska gave to Melvin Miller his early educational advantages and he was reared to the sturdy discipline of the home farm At the age of eighteen years he initiated his independent career as a farmer in Gosper county, and in 1901 he came to Scottsbluff, where he engaged in the barber business, to which he gave his attention for four years For the ensuing four years he was again a devotee of the basic industry of agriculture, and at the expiration of this period he sold his farm, in Phelps county, and returned to Scottsbluff, where he was engaged in teaming for two years He then resumed his active alliance with the barber trade, and his ability and popularity enabled him to build up a very prosperous business, with a well equipped and essentially modern shop of five chairs After conducting this shop two years he was elected chief of police of the city, and of this office he continued the incumbent two years In politics he is a staunch Democrat and has taken an active part in the local campaigns of his party Fraternally, he is found affiliated with the Modern Woodmen of America

At Lexington, Dawson county, on the 26th of July 1902, Mr Miller wedded Miss Ethel J Godsey, a daughter of Samuel Godsey, now a resident of Scottsbluff Mr and Mrs Miller have two children — Donnell, born in July, 1903, and Louise Imogene, born November 19, 1904

LABANNAH A MONTZ, who is one of the progressive representatives of agricultural and live-stock enterprise in Scottsbluff county and who is a citizen of marked intellectuality and civic loyalty, is a scion of a family whose name has been worthily linked with Nebraska history for nearly two score years, his parents having come to this state in the year of his birth and he being a twin brother of Martie R Montz

Labannah Alvesta Montz was born in Clinton county, Missouri, on the 7th of February, 1884, and a few months later his parents established their residence in Hall county, Nebraska His father, Martin Montz, was born in

the state of New York, July 22, 1858, and was nine years old at the time of his parents' removal to Missouri, where he was reared to manhood and received the advantages of the common schools of the locality and period At the age of sixteen years he found employment as a farm hand, and in Missouri he continued his association with farm industry until his removal to Nebraska In 1881 was solemnized his marriage to Miss Gertrude Ossman, who was born and reared in Missouri, and of their children, three sons and three daughters are living In the summer of 1884 Martin Montz came with his family to Nebraska and engaged in farming in Hall county, but in 1886 he removed thence to Banner county, where he filed entry on a homestead and a tree claim, which he developed into a productive and valuable farm There he continued his activities as an agriculturist during a period of about fifteen years, at the expiration of which he removed to Scottsbluff county, in 1901, and established his residence upon the farm which was the home of himself and his wife for several years when they moved to Scottsbluff, where they now reside

Mr Montz was reared in Banner county, where he profited fully by the advantages of the public schools, and after the family removal to Scottsbluff county he was for one year a student in the high school in the city of Scottsbluff For more than four years thereafter he rendered effective service with a government corps engaged in making geographical survey in western Nebraska, and in 1905 he filed entry on the homestead which is the stage of his present successful activities as an agriculturist and stock-grower He has made good improvements on his farm, which comprises eighty acres and which is situated in section ten, township twenty-three, range fifty-five, about eight miles distant from Scottsbluff, which is his postoffice address

Mr Montz is a man of well fortified opinions concerning public affairs, is actively affiliated with the Nonpartisan League and is an influential member of the Farmers' Union in his locality, he having been one of the organizers of the same On his farm he has an attractive home that is known for its generous hospitality, with his wife as its gracious and popular chatelaine

On Christmas day of the year 1918, was solemnized the marriage of Mr Montz to Mrs Julia A Wallace, of Scottsbluff, she being a daughter of George Dunham, a well known citizen of Scottsbluff

Mr. and Mrs. James N. Howard

JAMES N HOWARD, who has been a resident of Nebraska for many years, an early settler and freighter, now owns valuable, well improved property in Scottsbluff county and still takes part in operating it although the heavy responsibilities are borne by his sons James N Howard was born in Owen county, Kentucky, May 17, 1851 His parents, John and Martha (Carter) Howard, were born in Kentucky Both are deceased but Mr Howard has three brothers and one sister

Mr Howard's early years were passed in Kentucky Afterward he went to Missouri, where he remained until 1879, when he removed to Sidney, Nebraska It was while living at Sidney that he engaged in the freighting business and took part in many adventures that were not unusual at that early day in that section He homesteaded in Cheyenne county, but later sold his first 160 acres and afterward bought another 160 in Scottsbluff county During his eight years at Sidney Mr Howard went through many of the hardships that are a part of the history of almost every pioneer, but he philosophically accepted them and in comparing those times with many of the conditions of present-day life, is disposed to think they were not so bad after all Like many other early settlers he remembers the mutual good feeling that existed, when every one was kind and neighborly Mr Howard has raised as many as 100 head of cattle yearly on the range

Mr Howard married Elizabeth Minshall, who was born in Wisconsin and died at the age of fifty-six years Their six children survive Mrs Lizzie Smith, who lives in Morrill county, John, who lives in Northport, Morrill county, George and Albert, both of whom live in Scottsbluff county, and Sarah and William twins The former married Marian Fisher and lives in Scottsbluff county, the latter at home Mr Howard and his family belong to the Presbyterian church He served four years on the school board and at times has otherwise been useful in public matters in his neighborhood

JAMES S ROSENFELT is the owner of an irrigated and well improved farm of 220 acres in Scottsbluff county where he has maintained his residence nearly twenty years and where he has so improved his land and so directed his productive energies as to gain a secure place as one of the representative agriculturists and stock-growers of the county his home being situated five miles northeast of the city of Scottsbluff

The native sons are accounting well for themselves in connection with industrial and civic affairs in the famous Panhandle of the state, and one of the number is James Samuel Rosenfelt, who was born in Lancaster county, this state on the 23d of February 1874 His father, Henry Rosenfelt, was born in Hamburg Germany December 10, 1824 and was a sturdy and ambitious youth of nineteen years when he immigrated to America He located near the city of Albany New York, where he engaged in the work of his trade, that of broommaker, in which connection he raised his own broomcorn He continued his residence in the Empire state and later in Illinois until 1871, when he came to Nebraska and took up a homestead in Lancaster county There he improved a productive farm and there he continued his successful activities as an agriculturist and valued citizen until his death, in October, 1900 In the state of New York, in 1850, he wedded Miss Sophia Newman, who is now deceased, and of their fine family of sixteen children all are living except three, there having been eleven sons and four daughters and the subject of this review having been the thirteenth child

James S Rosenfelt gained his youthful education in the schools of Lancaster county, where he remained until he was twenty-two years of age, when he removed to the western part of the state and established his residence in Frontier county There he followed farm operations until September 20, 1900, when he established his residence in Scottsbluff county In the following month he located on the homestead which is still his place of abode, and to the original tract he has added until now he has a well improved and valuable farm of two hundred and twenty acres devoted to diversified effective irrigation Mr Rosenfelt assisted in the construction of the government irrigation system in this county and has otherwise been progressive and public spirited in his civic attitude In politics he is a Republican, and he has been for twelve years a member of the school board of his district He is affiliated with the Independent Order of Odd Fellows and the Modern Woodmen of America, his wife holding membership in the Daughters of Rebekah and the Royal Neighbors and being also president of the Domestic Improvement Club of her neighborhood Both are active members of the Methodist Episcopal church Mr Rosenfelt is a stockholder in the Farmers' Union Exchange and is secretary of the Farmers' Union local organization No 963

August 30 1899 recorded the marriage of Mr Rosenfelt to Miss Dora E Marlin a daughter of William M Marlin now a resident of Scottsbluff Mrs Rosenfelt was born in Indiana, September 24, 1876, and was about one year old at the time of her parents' removal to Nebraska She continued her studies in the public schools until her graduation in the high school at Cambridge, Furnas county, and was a resident of Frontier county at the time of her marriage Mr and Mrs Rosenfelt have six children, who names and respective dates are here noted Laura A , April 1, 1901, Marion T November 18, 1902 , Cecil Carl, December 28, 1907, Alice Z . July 2, 1909, Mabel L , May 26, 1913 , and James L , June 1, 1919 All of the children remain at the parental home, which is a center of generous hospitality and good cheer

JOHN H HALL, a popular pioneer citizen of the city of Scottsbluff, where he is now living virtually retired has had a plethora of experience in connection with pioneer activities in western Nebraska and has been a resident of this state since childhood, his parents having come to Nebraska prior to the admission of the territory as one of the sovereign states of the Union As a cowboy on the great cattle ranges that marked the early period of the history of the Nebraska Panhandle, and as a ranchman and farmer, Mr Hall has contributed his quota to the march of progress in this section of the state, and both he and his gracious wife are specially entitled to recognition in this publication, Mrs Hall being a stockholder in the Western Publishing & Engraving Company of Lincoln, by which corporation this edition is issued

John H Hall was born in Clarke county, Iowa, on the 8th of March, 1861, a date that denotes that his parents were pioneer settlers in the Hawkeye state where his father, James Hall, served as sheriff of Clarke county from 1858 until 1862, besides which his was the honor of having been a member of a regiment of Iowa infantry that gave valiant service in defense of the Union during the climacteric period of the Civil War James Hall was born and reared in Indiana and was a young man when he numbered himself among the pioneers of Iowa, where he continued his residence until 1866 when he came with his family to Nebraska Territory and settled near Plattsmouth, Cass county, where he secured a tract of land and became a pioneer farmer He became one of the influential citizens of his county commanded unqualified popular esteem, was active in local politics and served two terms as a member of the state legislature He passed the closing years of his life at Elmwood, Cass county where his death occurred

in July, 1906, his wife having passed away in 1893, at Plattsmouth, she was a native of Kentucky and her maiden name was Elizabeth Castle

John H Hall was a boy of five years at the time of the family removal to Nebraska, and was reared under the conditions and influences that marked the pioneer period in the history of Cass county, his educational advantages having been those of the public schools of the locality and period He continued his residence in Cass county until October, 1885, when he came to old Cheyenne county and settled in that part of the county that now constitutes Scottsbluff county In this wild and thinly settled section of the state he filed entry on homestead, pre-emption and tree claims, to which in due course of time he perfected his title While making preliminary improvements and development work on his land, he added to his financial resources by riding on the cattle ranges of this locality, in which connection he made an excellent record for effective service as a cowboy In the spring of 1886 he drove through from Cass county to the present Scottsbluff county with team and wagon, and incidental to this trip he brought from Gothenburg a load of corn Eventually he made advantageous sale of his original claims, after which he purchased a quarter-section of land where the Scottsbluff beet-sugar factory is now situated, and there he continued to reside until 1903, successfully engaged in agricultural enterprise and the raising of cattle, horses and sheep In the year last mentioned Mr Hall removed to Scottsbluff where he became associated with George W King in the hardware and furniture business About six years later he sold his interest in the business to the firm of Ebert & Gamble, and for a few years thereafter he was actively engaged in the cattle business in Banner county, While residing on his farm in Scottsbluff county he assisted in the building of the Minatare and Winter Creek irrigation ditches

In politics Mr Hall gives his allegiance to the Republican party, and he is affiliated with the lodge, Encampment and Canton bodies of the Independent Order of Odd Fellows Mrs Hall is affiliated with the Order of the Eastern Star and the Ladies of the Maccabees, in which latter she is a past commander of the Women's Benefit Association Mrs Hall is an earnest and active member of the Christian Science church

March 3, 1895, recorded the marriage of Mr Hall to Mrs Ella (Fasha) Stone, widow of William E Stone Mrs Hall was born in

Ogle county, near Freeport, Illinois, and her early education was obtained in the schools of Illinois, Kansas and Iowa She is a daughter of William and Salina (Hanky) Fasha the former of whom was born on shipboard while his parents were crossing the Atlantic ocean to America, and the latter having been born and reared in Maryland, where she received her education at Frederick Mr Fasha was reared to manhood in the state of Illinois, and his higher education included preparation for the ministry of the United Brethren church as a clergyman of which he gave effective and devoted service in both Kansas and Iowa, he having been a resident of Story county, Iowa at the time of his death, when forty-six years of age, and his widow having passed the remainder of her life, she having been summoned to eternal rest at the age of forty-nine years They became the parents of two sons and eight daughters, of whom Mrs Hall was the second in order of birth On the 13th of March, 1886, Miss Ella Fasha became the wife of William E Stone, and they established their residence in Wyoming, where Mr Stone took up a homestead and engaged in the raising of cattle and horses His death occurred there in March, 1891, and he is survived by one son, Frederick W, of whom individual mention is made on other pages of this work After the death of Mr Stone his widow sold the Wyoming property and came to Scottsbluff county, Nebraska, where she established her home at Gering, of which place she was a resident at the time of her marriage to Mr Hall Mr and Mrs Hall have one son. Donovan G A, who is in the employ of the Great Western Sugar Company at Scottsbluff

JOHN KONKLE — Along manifold lines has this honored pioneer exerted good influence during more than a half century of residence in the west and he is now living virtually retired, his attractive home being in the city of Scottsbluff He is a man of broad intellectual, keen, high ideals, and gracious personality — a citizen who commands the fullest measure of popular confidence and esteem More than sixty years have passed since John Konkle and his parents drove into Page county, Iowa, in true pioneer style and settled on a homestead amidst a veritable wilderness where inhabitants were few and where civilization was still in its primitive form and thus he stands out as one of the prominent figures among the men who have played such an important part in the development and advancement of the middle west

John Konkle was born in Knox county, Illinois, September 12 1846, the son of Michael and Ann (Buller) Konkle, the former a native of the Buckeye state where he was reared and educated The mother was also a native of Ohio where she grew to womanhood, was educated and there met and married Michael Konkle, but she lived but a short time to enjoy her family as she passed away when John was a small boy of three years After attaining manhood's estate Michael Konkle engaged in farming in Illinois and in 1856 became one of the pioneer agriculturists of Iowa, as he located on a homestead in Page county when the country was thinly populated and the nearest town was St Joseph Missouri, ninety miles away, and it was to this river settlement that the early settlers of the locality had to take their produce for sale and there buy the necessary farm implements and supplies for the households and families After about eight years a railroad was built through the country nearer the farm and then life was a little less strenuous and conditions more favorable Michael Konkle was invigorated by his outdoor occupations and the struggle he had to make to meet the hardships and privations of frontier life and lived out the psalmist's span of three score years and ten, as he passed away in Page county at the age of seventy-five years, a man whose good deeds were innumerable

John Konkle was reared on his father's homestead in southwestern Iowa, where he grew up a sturdy, robust lad under the steady discipline of the farm learning at an early age to assume many of the tasks for a boy on a farm and as the early years passed worked to aid his parents as his strength permitted and thus while still a youth was an experienced, capable farmer He received excellent educational advantages in the public schools of Page county and this practical training has been of great value to him in later life, as it has been the foundation for still advanced studies, carried on by himself, along the special lines to which he has devoted his time and energies After completing school, Mr Konkle established himself in the business with which he was familiar and became a well known farmer and stock-raiser of Page county, where he remained until 1889 when he determined to take advantage of the offers of the government to secure land on the homestead plan and came to Nebraska locating in Frontier county, where he filed on a homestead and a tree claim He proved up on both pieces of land, made good and permanent improvements, erected good practical buildings for his farm industries, a comfortable home and raised the land to an excellent condition of fertility His main business was diversified agriculture and stock-raising, in which he was markedly successful and won most gratifying returns for his labor Mr Konkle was wide awake to every advantage, kept abreast of the times and general agricultural improvements and progress and as time passed he watched the progress irrigation was making in the western part of the state and concluded that with assured water, the abundant sunshine and clear atmosphere, the Panhandle was the place for the aggressive and progressive farmer of to-day and his faith in this section was so great that he sold the old homestead, severed all the old ties and associations to locate in this "Garden Spot" of the high prairies In 1905, having disposed of his holdings in Frontier county he came to Scottsbluff county and bought a quarter section of land four miles northwest of the city of Scottsbluff, where he at once engaged in intensive farming, meeting with the just and well deserved reward that a man does, who gives time and thought to the industry in which he is engaged As his capital increased with the prosperous crops he marketed, Mr Konkle again justified his faith in the future of Scottsbluff county by investing it in more land until he became the owner of a landed estate of nearly a thousand acres about two hundred acres of which are under irrigation and it is but a question of time until a still larger tract will be under ditch After living on his western land for twelve years, during which time he developed the soil to a very high state of fertility, made many improvements in the way of farm buildings for stock, erected a good home and introduced modern machinery and methods on the place, Mr Konkle retired from the active management of his property and now lives in the city of Scottsbluff, where he is enjoying the sunset years of his life in the well deserved and well earned rest and enjoyment that is the result of a long and worthy life

During the twelve years he farmed here Mr Konkle raised great quantities of feed and one of his paying enterprises was sheep feeding, as he secured sheep from the western ranges, fed them for about ninety days to fatten, and then shipped to the packing centers of Nebraska and Kansas, where the great cattle markets are located In this way he made a quick turnover of his money, which was soon reinvested to good profit Mr Konkle from first settling here has taken an active

part in the progressive movements of this section, having been superintendent of Winter Creek irrigation ditch for eight years and also was in charge of the east end of the Scottsbluff ditch His nine hundred acre tract of land lies eighteen miles northeast of Scottsbluff so that his interests are not centered in one particular section, but in a manner cover nearly the entire county, while he owns three residence properties in the city Mr Konkle is progressive in his ideas, putting them into practice, and he advocates every reform and improvement that tends to the betterment of the county and the community in which he lives Today he is regarded as one of the most substantial and wealthy citizens of this flourishing district In politics he is an adherent of the Democratic party, to which he gives allegiance, but has been too busy all his life to accept public office other than those concerning communal affairs

On December 5,   , Mr Konkle married Miss Sarah Ham, the daughter of Christian and Elizabeth (Hidlebaugh) Ham, the former a native of Germany, who came to America to engage in farming He and his wife were pioneer settlers of Iowa, where their daughter was married in Page county The mother, Elizabeth Ham, was a Pennsylvanian by birth and played her share in the development and growth of their home in the west She still lives in Iowa Mr and Mrs Konkle have had seven children Mrs Stella Kilbourn of Cambridge has three children, Mrs Jennie Kilbourn is deceased, Mrs Nettie Price of Torrington, Wyoming, has one child, Mrs Ada Marlin of Scottsbluff, has two children, Mrs Anna Lane of Scottsbluff, has three children, Mrs Grace Chapin, also of Scottsbluff, has one child, and her twin, Roy, who lives in Scottsbluff has four children

JOSEPH E KELLER, the efficient superintendent of the Central irrigating canal in the Gering district of Scottsbluff county, has been closely identified with the development of this and other irrigating ditches in the county and has been the incumbent of his present executive position since 1910 He has been known also for his activities as an agriculturist and has won no little prominence in the raising and exploiting of standard-bred horses

Mr Keller was born in Stephenson county, Illinois, on the 8th of February, 1878, and is a son of Eli and Mary (Harding) Keller, who were born and reared in the state of Pennsylvania Eli Keller was nineteen years of age when he left the old Keystone state and established his residence in Illinois, where he became a prosperous farmer and where he died at the age of fifty-two years. His widow, who was born in the city of Philadelphia, now resides in the home of her son Joseph E., at Gering, and as he is a bachelor he is favored in having his loved and gracious mother as the supervisor of the domestic economies of their pleasant home.

Joseph E. Keller is indebted to the public schools of Illinois for his early educational discipline and after the death of his father he continued to attend school until he was sixteen years of age, having been assigned a place in the Evangelical Orphans' Home at Flat Rock, Ohio. In initiating his independent career he was identified with farming for one year and thereafter, in 1893, he came to Scottsbluff county, Nebraska, where he has since maintained his home and where pioneer distinction is due him. Here he has contributed his quota to the advancement of farm enterprise, has given attention to the breaking and training of horses and has given efficient aid in connection with the construction of the various irrigating ditches in the vicinity of Gering. He is an enthusiast in regard to this section of the state and is a citizen who is well known and highly esteemed. In politics he gives his support to men and measures meeting the approval of his judgment and is independent of strict partisan lines. He is affiliated with the Modern Woodmen of America, as a member of the camp in his home city.

PETER THOMPSON — A varied and interesting career has been that of this vital and progressive citizen of Scottsbluff county, where he is prominently identified with the basic industry of raising and feeding cattle and where he is the owner of a valuable landed estate of large area. His fine cattle ranch is situated at the head of Nine Mile canyon and comprises twelve hundred and eighty acres, besides which he is the owner of a quarter-section of land four miles northeast of Minatare, in which village he maintains his residence.

Mr. Thompson was born in Denmark, April 11, 1872, and is a son of Andrew Thompson, who immigrated to the United States in 1879 and located in Kansas City, Missouri, where he died within a short time thereafter, his wife having passed away in Denmark, shortly prior to his coming to America. Peter Thompson was a lad of about six years when

he came with his father to the United States and after the father's death he was taken into the home of his aunt, in Kansas City, where he remained until he was thirteen years of age and here he was afforded the advantages of the public schools. At the age noted, he showed his juvenile independence and self-confidence by "beating his way" to Omaha, from which city he continued his journeying to the capital of Nebraska. At Lincoln he found employment in the railroad yards, and he was thus engaged about one year. He then made his way to Ogallala, Keith county, and for the ensuing period of about eighteen months he was employed on farms and ranches in that vicinity. Thereafter he again yielded to the wanderlust, and continued his travels into Montana and Alberta, Canada. Finally he enlisted in the United States Army, in which he served from 1891 to 1894, having been stationed at Angel Island, California. After the expiration of his enlistment he went to Cheyenne, Wyoming and in 1895 he returned to Nebraska and entered claim to a homestead in Deuel county. He eventually perfected his title to this claim and there he continued to be actively engaged in the cattle business until 1902, when he purchased his present extensive and well improved cattle ranch in Scottsbluff county. His energy and progressive policies have here brought him success and prominence as an exponent of the cattle industry and he is one of the substantial and valued citizens of the county. He owns his attractive residence property at Minatare, where he is one of the associate owners of the farmers' co-operative store, besides which he has the distinction of being mayor of the village at the time of this writing, in the winter of 1919. He is one of the loyal and representative men of of the town and has been a member of the village council for a period of eight consecutive years. He has done all in his power to further the civic and material upbuilding of Minatare and was especially active in the work that resulted in the erection of the present fine school building in the village. In politics Mr. Thompson designates himself an independent Republican and he has been affiliated with the Knights of Pythias for fully a score of years. In addition to his local interests he is a stockholder in the Higgins Packing Company of South Omaha, and it is specially to be noted that his success has come entirely as the results of his own ability and well directed efforts.

February 28, 1901, recorded the marriage of Mr. Thompson to Miss Caroline A. Douglass, of Alliance, Box Butte county, this state. Mrs. Thompson was born at Genoa, Nance county, Nebraska, and is a daughter of George and Ida (Merrill) Douglass, well known pioneers of western Nebraska. Mrs. Douglass passed to the life eternal in 1910, when about fourty-four years of age, and Mr. Douglass now resides in the state of Michigan. Mr. and Mrs. Thompson have six children, all of whom are at the parental home at the time this review is prepared, their names being entered here in respective order of birth: Noel G., Leon L., Peter, Jr., Cora, Corinne, and Lulu. The family is one of prominence and popularity in the home community.

ANTON HIERSCHE became a resident of Scottsbluff county in 1886, is thus entitled to pioneer honors and has done well his part in connection with the social and industrial development and upbuilding of this now favored section of Nebraska. He was a pioneer in the sheep industry in western Nebraska and through his well ordered activities he acquired substantial success. He is now living virtually retired from active business and maintains his home in the city of Scottsbluff. His loyalty to and appreciation of this section of Nebraska are of unequivocal type, for while he has traveled extensively both in this country and Europe and has been a close observer of conditions, he states impressively that these travels have but heightened and confirmed his opinion that for attractiveness as a place of residence and as a field offering great opportunities to the honest and earnest worker, western Nebraska stands pre-eminent. As a citizen who is well known and who commands unqualified popular esteem, Mr. Hiersche well merits representation in this history.

In the old ancestral home, at Hirschberg, Bohemia, a place that had been in the possession of the family for six generations, Anton Hiersche was born on the 14th of May, 1863. He is a son of Menzel and Anna (Engel) Hiersche, both of whom passed their entire lives in that section of Bohemia, where the father died at the age of sixty years and the mother at the age of seventy years. In the family were six sons and three daughters, two of the children having died in infancy. The daughters remained in their native land and four sons came to the United States and became American citizens.

In the public schools of his native town Anton Hiersche gained his early educational training under effective conditions, and at the age of fourteen years he there entered upon a practical apprenticeship to the butcher's trade. With this line of work he continued to be identified until he had attained the age of eighteen years, when he severed the gracious home ties and set forth to seek his fortunes in the United States. He landed in the port of New York city on the 2d of October, 1881, and incidental to his venturesome voyage he had assumed an indebtedness of seventy dollars. He was fortified, however, with ambition, courage, self- reliance and determined purpose, so that he bravely faced the problems that confronted him in the land of his adoption. After working as a farm hand in Clinton county, Iowa, for one year, at a wage of twelve dollars a month, he had so conserved his earnings as to be able to pay the debt above mentioned and also had a sufficient additional sum to enable him to remove to Sac county, Iowa, where he continued for a time as a farm workman, after which he rented land and engaged in farming in an independent way. His two years of experience in this connection brought him a negative success, and he accordingly determined to come to Nebraska, where he felt assured of better opportunities. In December, 1885, he arrived at Sidney, Cheyenne county, this state, with only thirty-seven cents as a starting capital. Of this amount he paid thirty-five cents at a stage station fifteen miles north of Sidney, for a breakfast consisting of a cup of coffee, a biscuit and a piece of bacon. From Sidney he made his way on foot to Cider Flats, south of Gering, Scottsbluff county, and he covered the distance — seventy-five miles — in two days. He passed through a wild range country, and when the cattle made charges at him he would frighten them away by waving his hat and coat.

In February, 1886, Mr. Hiersche filed a pre-emption claim to the northeast quarter of section twenty-one, township twenty-one, range fifty-five, and after completing proof of this property he went to the western part of Wyoming, where he found employment as a ranch hand, in the meanwhile having previously taken a tree claim near his pre-emption, this tree claim having been the north half of the southeast quarter of the south half of the northeast quarter of section eight, township twenty-one, range fifty-five. From Wyoming Mr. Hiersche made his way into Montana, where he worked on ranches and gained his initial experiences in connection with the sheep industry as conducted on an extensive scale. In 1894 he returned to Scottsbluff county, and for an aggregate of $725 he sold sixty-one head of cattle, jointly owned by himself and his brother Wenzel. They borrowed sufficient additional money to enable them to buy eight hundred and fifty lambs, which they purchased of Senator Warren, at Cheyenne, and for which they paid about $800. The brothers first handled these sheep on a ranch south of Gering, and in 1896 they removed to section twelve, township twenty-three, north of Scottsbluff, where they continued to be most successfully associated in the sheep business until 1910. Incidental to their somewhat extensive operations it became necessary for them to borrow money from time to time and on such loans they paid an average of two per cent a month, while at time they found it necessary to pay as high as four per cent a month. Wool was sold at five and one-half cents a pound, lambs at five and one-half cents a pound and ewes at four cents a pound, in the Omaha market.

In 1910 Anton Hiersche disposed of his sheep and landed interests and since that time he has not been engaged in active business, his well earned prosperity justifying him in the gracious retirement which he enjoys. Since his retirement Mr. Hiersche has traveled somewhat extensively, both in this country and Europe, but he has constantly looked upon Scottsbluff county as his home and has never wavered in his allegiance to and appreciation of the section in which his maximum success was gained as one of the world's productive workers. He takes loyal interest in public affairs, especially all things touching the welfare of his home county and state, and in politics he maintains an independent attitude.

WILLIAM I. KENDALL has been a resident of Nebraska from the time of his young manhood and here he has so effectively availed himself of opportunities that he has achieved substantial prosperity. He has concerned himself successfully with agricultural and livestock enterprise in Scottsbluff county and now resides in the city of Scottsbluff. — a citizen well worthy of recognition in this publication.

Mr. Kendall was born in Freeborn county, Minnesota, on the 18th of July, 1864, a date that indicates that his parents were numbered among the pioneers of that state. He is a son of Isaac and Christiana (Clark) Kendall, the former a native of New Hampshire and the latter of Ohio. Isaac Kendall settled in Minnesota in the early pioneer period of its history and he represented that commonwealth as a gallant soldier of the Union in the Civil

War. He continued to be engaged in farming and stock-raising in Minnesota until 1881, when he came with his family to Nebraska and settled at Chadron, Dawes county. In 1884 he took up homestead and tree claims in that county and he not only perfected his title to these properties but also improved the land and became a successful agriculturist and stock-raiser besides which he owned and operated a saw mill. He eventually removed to Oregon, and at Burns, that state, he died when seventy-eight years of age, his venerable widow being still a resident of that village.

William I. Kendall was reared under the conditions and influences that marked the pioneer epoch in the history of Minnesota where he made good use of the advantages afforded in the public schools. He was a youth of seventeen years when, in 1881, he accompanied his parents to Nebraska. In 1883 the family moved to Fort Worth, Texas, and in 1884 they returned to Nebraska, settling in Dawes county. On his land Mr. Kendall raised and fed cattle and horses in a successful way and he was associated with his father in the operation of the saw mill. In 1887 he removed to Harrison, Sioux county, where he remained about one year and in 1889 he located at Custer City, where he engaged in the lumber business in connection with which he operated a saw mill. There he remained until 1898, when he removed to Chadron and in 1901 he came to Scottsbluff county, where for the ensuing three years he conducted farming activities on the old W. H. Wright place. For the next two years he occupied the Gable farm and thereafter he farmed one year on the place owned by Michael Powers. Upon leaving this farm he removed to the city of Scottsbluff where he purchased a good residence property and where he has since maintained his home. Mr. Kendall has opened up and developed much farm land in this county and has been known as a vigorous and productive worker — one who has contributed his share to civic and material development and progress. For four years he had active charge of the feed yards at the beet-sugar factory at Scottsbluff. He has identified himself loyally with community interests, is a Republican in his political proclivities and is affiliated with the Modern Woodmen of America.

At Hay Springs, Sheridan county, Nebraska, on the 12th of November, 1885, was solemnized the marriage of Mr. Kendall to Miss Jennie Eltingsteb, theirs having been the second marriage performed in that place. Mrs.

Kendall was born July 3, 1869, at Brooklyn, New York, where she was reared and educated She was but four years old at the time of her mother's death and was reared by the paternal grandparents of Mr Kendall, having accompanied them to Wisner, Nebraska, when she was a girl In conclusion is entered brief record concerning the children of Mr and Mrs Kendall Charles died January 9, 1917, born August 29, 1887, at Scottsbluff, and was about twenty-eight years old at the time of his demise, Lucy passed away at Custer City, when but eight years of age, Ralph J is married and resides at Scottsbluff, Harry J likewise resides in this city, and he was a soldier in the national army in connection with the World War, having received his training at Camp Dodge, Iowa, and Augusta is the wife of Marshall Barnhart, of Bayard, Morrill county They have a baby boy, born October 29, 1919, named Berle

ORIN DELBERT CHAPEN — Though from the inception of its settlement and development the western part of Nebraska has offered manifold attractions and advantages, yet those who have proved successful here have been the men whose energy and judgment enabled them to take full advantage of the opportunities presented, for supine ease and listlessness never prove efficient, no matter how great the advantages offered Mr Chapen, who is now living retired in the city of Scottsbluff, has been a resident of Nebraska for approximately thirty-five years and here he has achieved success worthy of the name, — as a man of initiative and constructive ability and as a citizen whose loyalty has been shown in labors that have made for development and advancement

Orin Delbert Chapen was born at Sandusky, Ohio, December 25, 1851, and is a representative of one of the sterling pioneer families of the old Buckeye state, within whose gracious borders were born his parents, John and Elizabeth (Smithers) Chapen The father was a valiant soldier of the Union throughout the Civil War, in which he served as a member of an Ohio regiment of volunteers, and the major part of his active career was marked by close identification with the basic industry of agriculture He was a resident of Ohio at the time of his death, which occurred when he was about seventy years of age Mrs Elizabeth (Smithers) Chapen, mother of the subject of this review, was thirty-two years of age at the time of her death, her parents George William and Mary (Arnold,) Smith-

ers, having been born in Pennsylvania and having been pioneer settlers in Ohio, where the former died at the age of eighty years and the latter at the age of seventy-nine years Orin D Chapen was but seven years old at the time of his mother's death and he was taken into the home of his maternal grandparent, by whom he was reared and educated his scholastic training having been acquired in the public schools of his native state Thereafter he continued his association with Ohio farm enterprise until 1883, when he removed to Edgar county, Illinois, where he was engaged in farming until 1885, when he came to Nebraska and established his residence in Hamilton county There he put his former experience to good use by turning his attention to agricultural and live-stock industry, and there he continued to reside until 1890, when he removed to Deuel county and took a homestead, the same being not far distant from Garden, in Garden county. In Deuel county Mr Chapen developed a valuable ranch, which he devoted principally to the raising of cattle and horses, and thereafter continuing in the stock business about four years he removed to Alliance, Box Butte county, where he engaged in the livery business in 1905 he established himself in the same line of enterprise at Minatare, where he remained for a time, as did he also at Bayard, Morrill county Within these passing years he was successful in his business activities, and since 1907 he has maintained his home at Scottsbluff, in which city he is living retired

In politics Mr Chapin is a stalwart Republican, and while he has never sought public office he served very efficiently as assesor of Alkali precinct, Deuel county, in 1901-02 He is a member of the Protestant Episcopal church, while Mrs Chapen belongs to the Christian church

In Delaware county, Ohio, on the 20th of October 1875, Mr Chapen wedded Miss Lottie J Linnabary, and she passed to eternal rest when thirty-five years of age Of this union were born four children Mrs Maude L Van Deusen, of Stockham Hamilton county, Mrs Cora B Rutter, of Minatare, Scottsbluff county, Mrs Rose E Fulton, who is deceased; and Clarence L, who died when about twenty-eight years of age At Scottsbluff, on the 9th of October, 1902, was solemnized the marriage of Mr Chapen to Mrs Louise J (Hibbs) Baker, who was born in West Virginia, as were also her parents, Dr Stephen and Malinda (Yost) Hibbs, the former of whom, a successful physician and surgeon, at-

Mr. and Mrs. Seymore S. Dickinson

tained to the age of seventy-four years, and the latter of whom died at the age of thirty-one years. Mr and Mrs Chapen have an attractive home in Scottsbluff, and they delight to extend its hospitality to their wide circle of friends in this community

SEYMOUR S DICKINSON, who came

in early times to Nebraska, settled on his present homestead almost forty years ago and has lived here ever since. Like other pioneers in this section of the state, he was called upon to face many hardships, but these he had more or less expected and through courage and persistent industry, overcame them all. Mr Dickinson is one of the older residents of the county and is widely known and universally esteemed

Mr Dickinson was born in Saratoga county, New York, October 25, 1838. His parents were Nathan and Jane Dickinson, natives of New York who never came to Nebraska. The father died at the age of sixty-three years but the mother survived him many years and at time of death had passed her ninety-second birthday.

Seymour S Dickinson was reared on his father's farm, attended the country schools and afterward engaged in farming in Saratoga county. In 1879 he paid his first visit to Nebraska. In 1880 he returned and then homesteaded the 160 acres on which he has lived ever since. During his active years he improved his property and under his management the farm was profitable even before the irrigation projects were introduced, and now sixty acres of the land are benefitted by the Gering ditch

Mr Dickinson was married first to Miss Eunice Duncan, who was born in New York, and she is survived by one daughter, Mrs Fannie Curtis, a resident of Gering. His second union was with Miss Helen Safford, who was born in New York, a daughter of John L and Sarah Safford, both of whom are deceased. To this marriage the following children were born: Mrs Lowa French, who lives near Henry, Nebraska, Mrs Hattie McCue who lives at Gering, Emma, who died at the age of twelve years, Ray, who is a farmer near the homestead, Floyd, who is a farmer in Canada, Ernest, who farms at home, Warren, who was a soldier in training at Camp Funston for service in the great war, is now at home, and Mrs Helen J Girvin, who lives on a farm near the homestead. Mrs Dickinson is a member of the Methodist Episcopal church. At one time Mr. Dickinson served as assessor of township 21, but has never been very active in politics

EDWARD A WHIPPLE, one of the stur-

dy pioneers of Banner county, substantial farmer and ranchman, has spent many years in this section of Nebraska, and has borne an active and useful part in its development. He is a native of Indiana, born in Franklin county, September 28 1856, one of the two of his parents' nine children to locate in this state, the other being a brother, Frank

The father of Mr Whipple, Arnold J Whipple, was born in Rhode Island, in 1814, and was four years old when his parents crossed the country and over the Allegheny mountains in a covered wagon, to the new home, then in the far west, in Indiana. The greater part of his life was spent at Metamora, in that state, where he worked at the blacksmith trade from the age of sixteen years. He was a man of sound judgment, of sterling New England character, an honest worker and a conscientious member and liberal supporter of the Methodist church. He was also a Mason and lived up to the obligations of Masonic membership. He married Elizabeth Kennedy a native of Indiana, who was a faithful and devoted mother

Edward A Whipple attended the village school in boyhood but as soon as old enough began to be self supporting by working on the farm for four dollars a month, later doing much better but not well enough, in his opinion to keep him in his native state. In the spring of 1882 Mr Whipple came to Nebraska and settled in Otoe county, remaining there until the spring of 1887. In previous fall he had homesteaded in Banner county and in the following spring came by wagon to his new home, his household possessions at that time consisting of a little bedding, while the cash capital in his pocket was fifty cents. As many be imagined his early residence was not a palace, in fact it consisted of a one-room structure, of which he took possession April 6 1887, and on May 8, following, after he had started to break ground, he found himself almost snowed in because of a furious storm. This was a serious matter as he had to go a half mile for his meals and had made no provision for the same in his little bachelor home. For some years agricultural conditions were very discouraging. Mr Whipple remembers when he hauled wheat a distance of sixty-five miles, from Big Horn Basin to Sidney and sold it for thirty cents a bushel. It took three days to make the round trip. Mr Whipple says there were three things considered indispensable on these trips, namely cooking utensils, bedding and lariat ropes

The ropes were used to tether their horses at night and on many occasions he had been roused in the middle of the night to drive the coyotes away, these animals finding pleasure is using their sharp teeth on the ropes.

Mr. Whipple remained on his banner county homestead until 1907. In 1892 he mortgaged his place for two hundred dollars, with the prospect of doing better in the stock business, and bought heifers and steers, at six and five dollars respectively, and pastured them on the range, and in that way secured his real start. That spring he bought two cows of Hope Brown for eleven dollars and twenty-five cents each, but yearling heifers brought on seven dollars. In 1907 Mr. Whipple secured the Kinkaid tract on which he now lives and at present owns fourteen hundred and forty acres, about one-fifth of which is under cultivation. Mr. Whipple now has everything very comfortable around him and his present residence is a great contrast to his first home in Banner county.

On December 26, 1892, Mr. Whipple was united in marriage to Mrs. Mary (Carter) Hide, widow of Harry Hide, and they have had three sons, namely; Ernest, who died February 13, 1917; Earl who assists his father at home; and Edward, who died July 12, 1915. Mrs. Whipple had one daughter by her first marriage, Myrtle, who is the wife of Joseph Gregory and they live at Morrill, Nebraska.

Mr. Whipple was reared by a Democratic father and continued to support the candidates of that party until recent years. He now depends upon his own good judgment and votes accordingly. He has always been a firm friend of the public schools and has served in school offices in district number eight, and for eight years was a member of the board of supervisors, but beyond that has never consented to accept a political appointment. In his neighborhood Mr. Whipple is spoken of as a man whose word is as good as gold.

CLYDE O. WYATT, who is one of Banner county's most progressive agriculturists, having extensive farms and ranch interests, belongs to one of the fine old county families that was established here thirty years ago. Mr. Wyatt was born in Wayne county, Iowa, February 16, 1879.

The parents of Mr. Wyatt were William and Susan (Duncan) Wyatt, the latter of whom was born in Iowa and now resides at Harrisburg, Nebraska. The father of Mr. Wyatt died in 1896. Of their eight children,

Clyde O. was the third in order of birth. In 1889 the family came to Banner county, and the father homesteaded and secured a tree claim on secion 7-12-53, on which he lived until 1904, then sold and bought six hundred and forty acres northeast of Harrisburg. He continued in the cattle and stock business during the rest of his life. In politics he was a Democrat and served six years as a county commissioner. He belonged to the Knights of Pythias and the Modern Woodmen and ever lived up to his fraternal obligations.

Clyde O. Wyatt was ten years old when the family came to Banner county and he completed his public school education here. He assisted his father until ready to start out for himself when, with his brother Harvey, he leased the homestead, borrowing money in order to carry out this transaction, and together the brothers farmed and ranched for three years. Mr. Wyatt has been very successful as a business man. He began for himself in a small way, with no farm machinery and insufficient number of horses. He remedied the latter condition by taking horses from neighboring farms and paying for their use by breaking them. He now owns twenty-eight hundred acres of land and operates two thousand additional acres. Formerly he raised five hundred head of cattle, but in 1916 he sold his stock with the intention of taking things easier for a time, but as a loyal and patriotic citizen, when meat shortage faced the country on account of the great war conditions, he reconsidered and after a year of rest returned to his ranch. This is one of the best improved ranches in Banner county and Mr. Wyatt carries on his farming operations with modern farm machinery that includes a four-plow tractor. Although his farm buildings are all modern and first class there is one exception, the log barn, still sound and usable, that is probably one of the oldest structures in this neighborhood, has never given way to Mr. Wyatts progressive ideas and energetic improvements.

Mr. Wyatt was married November 9, 1904, to Miss Grace A. Waitman, who is the daughter of Price and Minnie (Kelty) Waitman, residents of Iowa. Mr. and Mrs. Wyatt have three children, namely: Perley, Earl and Inez. A well informed and earnest citizen, but not a politician, Mr. Wyatt votes with the Democratic party but has never consented to accept public office. He is known all over the county and has the respect of his fellow citizens generally.

RALPH DARNALL —While almost every state in the Union has contributed to the citizenship of Nebraska, it casts no reflection on any other state to assert that a particularly worthy class came from Illinois Many came westward from their fathers' Illinois farms, seeking the advantages that wider agricultural opportunities would give them, and ready to earn their reward through honest industry One of these sturdy farmers bore the same name of Darnall, a name that has been a respected one in the state for forty years

Ralph Darnall, who is a successful farmer and stock-raiser in Banner county, was born in Clay county, Nebraska, March 30 1883, and is a son of Walter Scott and Rosie (Tucker) Darnall, natives of Sangamon county, Illinois Prior to 1879 Walter Scott Darnall was a farmer in Illinois In that year he came to Clay county, Nebraska, where he leased and cultivated land until 1887, when he homesteaded For five years the family lived on that place and then the father moved near Kearney, not giving up his homestead however, in fact, he still owns it Discouraging circumstances that faced practically all the settlers at that time, led to Mr Darnall's return to Illinois, where he remained as a farmer for the next twelve years He had by no means forgotten his Nebraska land and in 1906 he returned and settled on his homestead in Banner county where he occupied himself profitably as a farmer until 1915, when he turned the most of his farm responsibilities over to his son Ralph and moved into his comfortable residence at Melbeta Although never particularly active in politics he has always voted the Republican ticket, and as an intelligent citizen has tried to promote the welfare of his county and state He is a member of the Farmer Union and the Modern Woodmen of America He and his wife have had seven children, namely Harry, who lives in Illinois, married Grace Billings, Arthur, who is a farmer in Banner county, married Lida Johnson, Ralph, who lives on section twenty, town one, Mable, who lives in Missouri, is the wife of David Johnson, Harvey, whose death occurred October 31, 1914, had been married just ten days previously, Goldie, who lives in Banner county, is the wife of Hugh Ridge, and Gladys, who lives at home

Ralph Darnall is company with his brothers and sisters attended the public schools and performed his share of duties on the home farm When twenty-one years old he went to Montana and worked through the summer on his uncle's ranch, then went to Illinois for the winter In 1905 he came back to Banner county Here he has since operated his own farm of one hundred and sixty acres, together with his father's twenty-four hundred acres, carrying on mixed farming and cattle raising The home farm was exceedingly well improved while his father lived on the place, and Mr Darnall is not one to let anything run down He makes use of modern farm machinery and his various industries are carried on systematically

Mr Darnall was married February 28, 1915, to Miss Gladys Kelly who is a daughter of Samuel Kelly, one of the solid, representative men of Banner county Mr Darnall votes the Republican ticket because he believes in the principles of that party He belongs to the Farmers Union Personally he stands well with neighbors and acquaintances because he is a young man of principle

ARTHUR DARNALL, who is a careful, competent and successful farmer and stockraiser in Banner county, is a native of Nebraska, born in Nuckolls county, March 30, 1881 His parents are Walter Scott and Rosie (Tucker) Darnall, who now live retired in the village of Melbeta They were born in Sangamon county, Illinois, and were farming people before coming to Nebraska in 1879 The family lived in Nuckolls and Clay counties prior to 1887, when the father homesteaded in Banner county Five years later he returned with his family to Illinois and twelve years passed before he came back to Nebraska He then took charge of his old homestead and greatly improved it He now owns twenty-four hundred acres and since 1915, when he pactically retired, two of his sons, Arthur and Ralph, have had the management of this property

Arthur Darnall obtained his education in the public schools and his business throughout life has been of an agricultural nature For some years he lived in Illinois and while there served in the office of county commissioner, being elected on the Republican ticket for a period of three years In 1905 he accompanied his father from Illinois, where the latter had been continuously farming, to Banner county, and now, as mentioned above, in association with his brother, has charge of the estate of twenty-four hundred acres, mostly range land and ranch The Darnalls are practical, experienced business men and they have been very successful in all their farm undertakings

On March 8 1905, at Barclay, Illinois, Arthur Darnell was united in marriage to Miss Lida Johnson, who is a daughter of William D and Louisa (Marshall) Johnston   The mother of Mrs Darnall resides at Riverton, Illinois, but the father died on July 27, 1915. Mr and Mrs Darnall have four children, namely· Nina, born November 3 1909; Frank, born September 13, 1912, Glenn, born October 12, 1913, Harvey, born October 28, 1915, this son was named in memory of Mr Darnall's youngest brother, who had died suddenly a year before, just ten days after his happy marriage, David Arthur born August 3, 1919

JOHN I. MUHR, who is one of Banner county's substantial farmers and ranchmen, has spent almost his entire life in this county, accompanying his parents here when three years old   He was born in Kansas, April 4, 1883, and is a son of John and Elizabeth (Milard) Muhr, natives of Illinois

The father of Mr Muhr still survives and he and the mother live retired at Fullerton, California   He is a veteran of the Civil War and at one time was a brave and valorous soldier, taking part in such battles as Chattanooga, Chickamauga and Stone River, being twice wounded, and marching to the sea in the victorious army of General Sherman   He came to Banner county, Nebraska, 1886, homesteaded and took a tree claim and later a Kinkaid claim, and followed farming and ranching until 1915, when he sold out and retired to California   Of his nine children, John L was the seventh in order of birth

John L Muhr attended the country schools and spent one year in school at Sidney   As he grew old enough he assisted his father, and knows something of the hardships that beset early settlers in the county   In contrast to the present day high prices, he remembers assisting his father to cut cord wood and haul it forty miles to Sidney and sell it for $3 a load, the round trip taking two days   Mr Muhr now lives on the old homestead and has improved the place with substantial farm buildings

Mr Muhr was married February 1, 1909, to Miss Grace Stroud, who is a daughter of Charles and Emma Stroud, now living at Bayard, Nebraska   Mr and Mrs Muhr are members of the Methodist Episcopal church Mr Muhr is an honorable, industrious man, in every way exemplary, and like his brother in Banner county, is held in much esteem   Mr and Mrs Muhr have three children Vivian, Garnold and Vernon, aged ten, eight and six years respectively

WALTER A MUHR, who is a practical and enterprising farmer and ranchman of Banner county, has a wide acquaintance and like other members of this family is held in esteem by all who know him   He was born on his father's homestead in Banner county, Nebraska, October 31, 1892, and is a son of John and Elizabeth (Millard) Muhr, who live retired in California

After a long and honorable service in the Civil War the father of Mr Muhr, a twice wounded soldier, came to the west, but did not locate in Banner county until 1886   He then homesteaded took a tree claim and later a Kinkaid tract and continued farming and ranching on his land until 1915 when he retired   Of his family of nine children, eight are living and Walter A is the youngest member

Walter A Muhr had excellent educational advantages at Sidney during the ten years the family lived there   Later he accompanied his parents in a year of travel, undertaken to benefit the mother, who was then in failing health   He has always been interested in agricultural pursuits   In July, 1914 he filed on a Kinkaid claim of four hundred acres and now owns it, in all having five hundred and sixty acres, having four hundred under cultivation and operating twelve hundred acres He breeds Aberdeen Angus Cattle and raises about fifty head a year

On July 25, 1914, Walter A Muhr was united in marriage to Miss Edna L Baldwin, who is a daughter of Frank A and Carrie (Kennedy) Baldwin, who are residents of South Omaha, Nebraska   Mr and Mrs Muhr have two children, interesting little sons of five and three years, Richard L, who was born December 6, 1914, and Winfred F, who was born October 26, 1916   Mr Muhr is independent in his political views and has always been fearless in defense of what he has believed to be right   He is a member of the Farmers Union

WILLIAM C MUHR — An old and representative family of Banner county bears the name of Muhr, and a substantial representative of the same is found in William C Muhr, who has been a continuous resident since he was seventeen years old and well remembers many seasons of pioneer hardships   He was born in Pike county, Indiana, September 18, 1869

The parents of Mr Muhr were John and Elizabeth (Millard) Muhr, natives of Illinois, who now live retired at Fullerton, California The father is a veteran of the Civil War, in which he served three years and eight months and during this time was wounded At that time work that is now quickly accomplished by a trained engineering corps, had to be done by the private soldiers and it was while assisting in laying pontoon bridges that he contracted rheumatism from which he still suffers He took part in such battles as Chattanooga, Chickamauga, Missionary Ridge and Stone River, and was with Sherman in the memorable march to the sea Before locating in Banner county, Nebraska, the family had lived for a time in Pike county, Indiana, and also in Kansas The father homesteaded in Banner county in 1886, took a tree claim and later a Kinkaid claim and at the time of retirement in 1915, owned nine hundred and sixty acres, which he sold before moving to California He has always been independent in politics Both he and wife are members of the Methodist Episcopal church They have had children as follows. William C, whose home farm lies on section elevent, town twelve; Emma, who is the wife of J W Mosier and they live in Oregon, George, who died in infancy, Allen, who lives in Banner county, married Angeline Broughton; Maggie, who lives at Orleans Nebraska, is the wife of Burr McConaughty, Effie, who is the wife of Homer Sickles of Banner county, John L, who lives on the home ranch, married Grace Stroud, Verdia, who is the wife of Clifford Houston and they live in California, and Walter, who is a merchant at Redington, Morrill county married Edna Baldwin

William C Muhr remained with his parents until he was twenty-nine years old He assisted his father from the first In recalling early conditions he says that for two years after the family came the crops were exceptionally abundant, but a long period of drought followed that brought discouragement and disaster to many settlers When his parents settled here they had one cow and four horses and just three dollars in money left after their long journey Within three weeks the mother was taken so ill that a doctor had to be brought from Sidney to attend her, and when he left he took one of the horses as his fee, seemingly a rapacious one even if the distance was fifty-one miles Later Mr Muhr and his father hauled wheat to Sidney and sold it for thirty-seven cents a bushel The Muhrs at first lived in a dugout and during one of the fierce blizzards of those days, at one time were imprisoned by the snow

In 1891 William C Muhr homesteaded and took a Kinkaid claim as soon as that law was passed He lives on this claim and has retained his homestead so that altogether he has ten hundred and forty acres, mostly ranch land He raises about sixty head of cattle yearly, formerly raising one hundred and fifty head and sometimes two hundred head He breeds Duroc-Jersey hogs and in earlier years raised Percheron horses but this industry is less profitable than formerly

On September 18, 1898, William C Muhr was united in marriage to Miss Alice L Broughton, who died May 6, 1916 She was a daughter of Orr and Ella (Miller) Broughton, old residents of Cheyenne county who now live retired in California The following children were born to Mr and Mrs Muhr Pearl, who is the wife of Bert Tinsley, of Morrill county, Opal, who is the wife of Lloyd Sample, and John, Cecyl, Alvin and Allen, twins, Theodore, Beryl, Gilbert and Ruby Mr Muhr's second marriage took place April 22, 1918, to Mrs Etta Sample, widow of John Sample and daughter of Stephen and Marian (Banister) Shaw, residents of McPherson, Kansas They have a beautiful home, for Mr Muhr has taken much pride in improving his property His residence is situated under the east side of the bluffs between Pumpkin Seed and Lawrence Fork creeks They are members of the Methodist Episcopal church He is an independent voter and has never been a seeker for office although he has served in minor township positions when he has felt it to be his duty He is a man who is universally respected

WILLIAM M WISNER, who is a well known citizen of Banner county, operates a large body of land here He accompanied his parents to Banner county when about twelve years old, has acquired land of his own but resides, at present, on the land his father homesteaded in March, 1888 He was born in Poweshiek county, Iowa, February 4, 1876

The parents of Mr Wisner are Seneca R and Delia J (Wells) Wisner, the former of whom was born in Rock county, Wisconsin, August 10 1851, and the latter in Mahaska county, Iowa, February 21, 1852 The father grew up from boyhood in Iowa, has always followed agricultural pursuits and was very active until he retired in 1913 On February 28, 1888, he homesteaded in Banner county and still owns six hundred and forty acres here He started on very limited capital and gained an ample fortune mainly through stock-raising He owns forty-five acres of land in Florida and some city lots in Palm Dale, near

Okechobee, where he and the mother of Mr Wisner have resided since the fall of 1913 While a resident of Banner county he was active in the councils of the Democratic party, and for years served in local offices He belongs to the order of Modern Woodmen of America and both parents of Mr Wisner are members of the Congregational church

William M Wisner, the only child of his parents, remained with them until February, 1907, when he took a Kinkaid claim and lived on it until he accompanied his parents to Florida in the fall of 1913 He remained in the South two and one-half years, returned then to his Kinkaid claim for eighteen months, and for a like period had been operating his own land and his father's also, devoting considerable attention to calves for market He recalls some early days of hardship on the homestead, caused mainly by the stringency of the money market, when he assisted his father to cut and haul wood as far as Sidney, when fine cedar posts brought only eleven or twelve cents apiece

In February, 1907, Mr Wisner was united in marriage to Miss Maude M Lease, who is a daughter of Asher and Lucy (Siemiller) Lease They have two children namely, Ethel, who was born August 5, 1907, and Fannie, who was born March 11, 1910 Mr and Mrs Wisner are members of the Congregational church In politics he is independent

FRANK W ABBOTT, general farmer and rancher who came first to Nebraska when sixteen years old, still owns the timber claim he secured in February, 1885, in Banner county Before locating permanently, Mr Abbott took in much of the adventurous life of this western section, and was acquainted with many men of prominence whose business or pleasure called them also to this part of the United States

Frank W Abbott was born at Jackson Michigan, February 7, 1857 His parents were Henry C and Eleanor (Harpham) Abbott, both of whom were born in England, the mother on November 9, 1821, and the father on June 16, 1822 They were married in England on June 16, 1846, and came to the United States in 1847 Henry C Abbott's father was a miller, carpenter, and contractor, and while in England, Mr Abbott was a miller but after coming to America was a carpenter and builder in Michigan until 1861, when he moved to El Paso, Illinois, to become land agent for the Illinois Central Railroad

For twelve years the family lived at El Paso, Illinois, while he was engaged with the Illinois Central After leaving the railroad he returned to England to settle up an estate left the family by Rear Admiral Fox Henry C Abbott was a man of education and fine presence and later in looking after real estate interests, visited Buenos Aires, South America, Havana, Cuba, and Louisiana, and died at New Orleans in 1907 He was influential in the Republican party and was a delegate in the convention that nominated Abraham Lincoln for the presidency Of his seven children, four died in infancy, and Frank W is the youngest of the two survivors, the others being Samuel H, who died December 18, 1890, and Jennie S, who has lived at Exeter, Nebraska, since 1873 She was married first to Schouler Roper, who died in 1885, and second, to Merritt Rogers in 1905, is living there yet

Frank W Abbott attended school at El Paso, Illinois, and began life on a farm In 1873 his mother and children, came to Exeter, Nebraska, bought railroad land and the mother died at Exeter in 1879 In 1881 Frank W went to Oregon and followed farming there until 1882, when he and his brother drove cattle and horses across the trail to Cheyenne, Wyoming, a journey that consumed five months After a winter spent at Exeter, Mr Abbott went west again and rode range for twelve years, driving over Idaho and Montana during that time In the spring of 1884 he came to Banner county and worked for C C Nelson who operated the Tusler ranch on Greenwood creek, and afterward for Dicky Brown At that time there were only three families living on Pumpkin Seed creek In February,1885, both he and brother took pre-emption and tree claims, and his sister, then a widow, also took a pre-emption and tree claim He lived on his pre-emption for ten years, then sold, but still has his tree claim and the heirs of his brother still retain their timber claims In 1894 Mr Abbott went to Wheatland, Wyoming, where he worked for two years for an irrigation company, from there going to Thermopolis, Wyoming, where he bought a hotel and bath-house, and operated it for twelve years Then Mr Abbot returned to his property in Banner county, where he owns one hundred and sixty acres and leases about one hundred and eighty acres. In addition, Mr Abbott has a comfortable residence and some lots at Woreland, Wyoming, and Mrs Abbott owns a homestead near

Thermopolis, Wyoming Mr Abbott has under consideration removal to a point on the Morrill and Banner county line

Mr Abbott was married July 12, 1912, to Mrs Ollie Quebbemam, who had one daughter, Carmel, who lives at home In politics Mr Abbott is a staunch Republican, and as did his father, belongs to the order of Knights of Pythias

In speaking of early days in this section, Mr Abbott recalls an occurrence that came under his own observation It was in 1870 that a band of government surveyors were making a sectional and sub-divison survey between Wright and Hubbard gaps and suddenly found themselves surrounded by a band of outlaw Indians, of the Sioux tribe, who would have killed them without mercy had not another Indian band suddenly appeared and drove the marauders away, the latter being under orders from the government agent The surveyors had to be men of courage and expedient, and among those whom Mr Abbott knew well was Frank Huber, who now lives at Custer, South Dakota Those old days of Indian danger have long since passed away in this vicinity

ARTHUR F BURNETT, who is a prosperous and enterprising farmer and ranchman in Banner county, was born in Jackson county, Wisconsin, June 8, 1881, and is a son of Sidney D and Kattie (O'Halloran) Burnett Both parents were born in New York, the father on March 16, 1843 The mother died in 1901 Of their seven children, two besides Arthur F are living, namely Mrs Belle Franklin, who lives at Bayard, Nebraska, and Archibald, who resides in Wisconsin

Arthur F Burnett was young when his parents brought him to Nebraska In 1886, his father moved to Box Butte county, Nebraska, homesteaded, and the family lived on the place until 1889, then moved to Bayard, where the father worked to some extent at his trade being a carpenter In 1902 he came to Banner county and has since resided with his son He is a Republican in politics and is a member of the Methodist Episcopal church

In the public schools at Bayard, and in Hastings College, Arthur F Burnett secured educational training that fitted him for almost any calling He chose farming and stock-raising, and after coming to Banner county in 1902 bought the land on which he lives, subsequently added to it until he now owns twenty hundred and ten acres It is mostly ranch land and Mr Burnett is very much interested in stock production He raises one hundred head of cattle, fifty head of horses, and forty head of Poland-China hogs annually His farm industries are carried on with the assistance of modern machinery and his methods have little in common with the old time ways of other days Mr Burnett has made may substantial improvements on his property and has recently completed a handsome up-to-date farm residence which is a credit in comfortable appointments, to this part of the county

On December 7, 1904, Mr Burnett was united in marriage to Miss Belle Skinner, who is a daughter of one of the best known and most highly esteemed pioneer citizens of Banner county Mr and Mrs Burnett have four children, namely Paul, Laura, Clifford and Nettie As a citizen, Mr Burnett has always been active and intelligently useful but has not sought political office, the proper management of his large ranch interests rather completely absorbing his time and strength He lives in great friendliness with all who know him

ARTHUR H HERMANN who is one of the solid reliable men of Banner county, has large farm and ranch interests He accompanied his parents to Nebraska in 1881 and has spent the greater part of his subsequent life in some part of the state He was born in Johnson county, Iowa, August 13, 1870, one of a family of ten children born to George M and Margaret (Wenkheimer) Hermann

The parents of Mr Hermann were born in Germany and Ohio respectively The father came to the United States when seventeen years old and followed farm life in Ohio until moving to Oskaloosa, Iowa, where he continued to farm until he came to Johnson county, Nebraska, in 1881 He bought land there and resided on it until the spring of 1890, when he removed to Morrill county in the hope that a change would restore the mother of Mr Hermann to health, but it did not avail and she died on March 15, 1891 The father, after her loss, went to live with his son Arthur H and continued there until his own death on December 13 1913 He was a Democrat in politics but never accepted any political office Of their seven surviving children, only two live in Nebraska, Arthur H and Lena The latter is the wife of William Schoemaker, of Morrill county

It was in Johnson county, Nebraska that Arthur H Hermann secured his schooling He started out for himself when twenty-one years old, working first as a ditcher in Morrill county, later in Wyoming, then worked on the railroad at Kimball, Nebraska In February, 1899, he left the railroad and then worked on the Amy Scott ranch, six miles north of Harrisburg for a year From there Mr Herman moved to a homestead he had previously filed on section 23-19-53, which land he still owns While not one of the earliest settlers, Mr Hermann experienced hardships and demonstrated his ability to overcome them Money was scarce at that time in this neighborhood and in order to get enough to pay necessary bills, the settlers had to watch for opportunities and work hard After cutting and then hauling a load of wood that took him two days to secure and deliver, with team, at Redington, he would get no more than three dollars, which was usually gladly accepted He also hauled freight to Sidney from Redington for fifteen cents a hundred weight When he worked at ditching his pay for himself and team was two dollars and fifty cents He bought three cows for twenty-seven dollars each and it took him three years to pay for them, paying interest on a note during this time

Mr Herman now owns and operates sixteen hundred and eighty acres, of which three hundred and fifty acres are devoted to farming and it is highly improved He breeds Polled Angus cattle, averaging seventy-five head a year, has a fair supply of Percheron horses and other stock in abundance He operates the farm with the help of his two older sons The scenery around the family residence is unusually beautiful, with hills in the background and the Pumpkin valley in front, with a clear view for many miles An artist would find here an irresistable subject for his brush

Mr Hermann was married at Kimball, Nebraska, to Miss Norah Skinner, a daughter of Richard Skinner, of whom extended mention will be found in this work Mr and Mrs Hermann were married on July 6, 1898, and they have six children and one grand-child, as follows Ervin, who was born April 20, 1899, Edward, who was born November 16, 1900, was married July 30, 1917, to Miss Ella Burkey, and their child was born February 15, 1918, Fred, who was born April 1, 1909, Carrie, who was born August 22, 1909, Lola, who was born December 18, 1911, and Harley, who was born February 11, 1916 In his political views Mr Hermann is a Republican He has never given much time to political office, with the exception of being road overseer for some ten years, but in the ordinary enterprises in which his neighbors engage for the general good, he is always ready to co-operate

JAMES JESSUP who owns two hundred acres of well improved farm land is situated on section thirty, town one, with postoffice at McGrew, did not come to Banner county with the earliest settlers but nevertheless had much to do with substantial development of this section of country It was Mr Jessup who built the first house at Melbeta, now a flourishing town, owned and operated the first lumber yard there and furthermore his son was the first infant born there

James Jessup was born in Orange county, New York, January 11, 1879, the fifth in a family of seven children born to James J and Delia A (Van Ostrand) Jessup The other children were as follows Elizabeth, who lives at New Castle, Wyoming, Lillian, who lives at Los Angeles, California; Alice, who lives at Palmyra, Nebraska, and Amzi A, Frank P and Stanley, all of whom live in Banner county The father, James J. Jessup, was born at Middletown, New York, October 31, 1845, and died August 23, 1912 The mother was born at Glenwood, New Jersey, in November, 1850; and resides at Minatare, Nebraska When he came to Nebraska, October 4, 1894, he bought land three miles east and a mile south of Minatare, on which place he lived for twenty-one years and died there He was a man of sterling integrity, intelligent and straightforward, and was locally prominent in the Democratic party, although he never accepted any public office except that of assessor The mother is a member of the Presbyterian church

James Jessup attended the public schools at Lincoln, in boyhood, then went to work in an upholster's shop and learned the trade His brother Amzi had the job of driving the stage from Camp Clark to Gering and Torrington, but later joined his brother James and both went to work on the Bayard ditch It was hard work to do for seventy-five cents a day and board themselves when not working, but times were hard, corn selling for ten cents a bushel, and therefore the Jessup brothers continued laboring for a time through frozen mud and water Of course on off days, Mr Jessup further relates, that during one fall and winter he made seven trips to Marsland, sixty miles distant often in frigid weather, where he exchanged wheat for flour, which he sold in his neighborhood for one dollar a sack of forty-eight pounds During one winter he herded two hundred head of cattle in Red Wil-

MR. AND MRS. D. E. WALLACE

low Canyon, northeast of Bayard He also hauled wood out of the hills, making sixteen trips one year, and sold at Bayard and Minatare Then corn was so cheap it sometimes was used as fuel

Mr Jessup homesteaded on section 30-20-53, on which land he now lives owning, as mentioned above, two hundred acres of fine farming land For three years Mr Jessup was a resident of Melbeta, then sold his lumber interest to Cox & Company and for three years was with the Proudfit Lumber Company at Minatare Since he has given his entire attention to his farm industries and has been very successful

On April 4, 1904, Mr Jessup was married to Miss Flossie M Learned, who is a daughter of Orlando and Lucy A (Davis) Learned who came here in 1888 The mother of Mrs Jessup died in November, 1916, but the father, who is a Civil War veteran, lives in the Soldiers' Home at Grand Island He was nineteen years old when he enlisted at Quincy, Illinois, in Company G, Sixteenth Illinois volunteer infantry, in which he served four years and three months, and was wounded March 19, 1865 He took part in many of the serious battles of the war In April, 1888, he homesteaded in Banner and Scottsbluff counties, and still owns the land Mr and Mrs Jessup have two children, namely Mildred O, who was born July 28, 1905, and James Loren, who was born May 6, 1912 Mr and Mrs Jessup are members of the Methodist Episcopal church He is an independent voter and has never accepted a political office He belongs to the Modern Woodmen of America

ELMER WALLAGE — The vigorous and progressive population of Scottsbluff county is made up largely of successful exponents of the agricultural and livestock industries In every part of the county farmers seem to thrive and as an able and honored representative of this oldest of industries, as well as being entitled to pioneer honors in Nebraska, Mr Wallage is specially entitled to consideration in this history He is descended from a long and honored line of ancestors who played an important part in the history of the Old Dominion, as his father was a native of Virginia who emigrated from the eastern part of the United States and became one of the hardy and rugged pioneers of the eastern part of Illinois, when the middle west was being settled up As a young man David Wallage located on a homestead just west of the Indiana line and there conducted general farming operations as his land was

cleared and he was able to put the rich soil under cultivation Not confining his operation to one line, Mr Wallage also raised stock and for many years was one of the representative agriculturists of the eastern section of Illinois where he engaged in business until his death at the advanced age of seventy-two years David F Wallage married Louisa Tweedy, born in Indiana, who was his faithful helper and devoted wife and survived him for some years passing away after a long and worthy career at the age of eighty-eight years Elmer Wallace was born on his parents' farm in Illinois, May 9 1875 He was reared in the healthy atmosphere of the country, attended the excellent public schools afforded by the state and as soon as his age and strength permitted began to assist his father in the farm work and thus at an early age was a good practical farmer, fully able while yet a youth, to conduct many of the enterprises carried on by the older members of the family He learned when to plant crops that were best suited to the climate and soil of the rich river bottom lands, what kinds of livestock brought the greatest returns and was an all-around business man For some years after attaining his majority Mr Wallage remained in Illinois but he was ambitious, and hearing of the free lands and the many advantages afforded farther west, determined that he, too, would share in the bounty of the government and in 1904 came to Nebraska, locating in Sioux county, where he at once began his promising career as a cattle man, for the great plains at that time were the Mecca of all the livestock men of the country Putting his knowledge of farming and the animal industry to excellent use, Mr Wallage soon was engaged in an extensive business that brought him well earned and well deserved success He soon became the proprietor of a well developed and cultivated landed estate on which he had an attractive and practical home, and through his association with the varied farm industries of the region was known as a vigorous and progressive farmer and stock raiser By hard work keen foresight in calculating the cattle market and the progressive methods he inaugurated on this western land, Mr Wallage achieved his own financial success and had the satisfaction of knowing that all his fortune was of his own making as he had no assistance of any kind From first coming to the west, he had studied on the great agricultural problems of the day and in his wide reading soon became interested in the question of irrigated land, and being a man of vision soon came to believe that the man who could have a plentiful supply of water on the high plains with the never-failing sunshine, would be the great factor in food production of the future. With this in mind, Mr. Wallage disposed of his large holdings in Sioux county, and in 1911, came to Scottsbluff county, purchasing eighty acres of land under ditch. This farm has been raised to a high state of cultivation under his careful management, is all irrigated and today is one of the finest farming estates in this section, noted for its progressive and prosperous business men. For this section is known not only in the middle west, but all over the country as the "Garden Spot" of Nebraska, which is one of the richest farming states in the Union. Since locating in this far western section with the never-failing water, Mr. Wallage has been engaged in diversified farming, which he believes pays the best, as large crops of all products find a ready market.

Mr. Wallage married Ella Wendle, a native of Indiana, who was born in Parke county, that state, on March 14, 1882. She was the daughter of a substantial Indiana farmer, who now lives retired in Parke county, her mother having died some years ago. Mrs. Wallage received her educational advantages in the public schools of her native county, grew to womanhood there and after her marriage accompanied her husband to the new home in the west, taking an energetic and active part in the upbuilding of the comfortable fortune which has been the reward of their joint labors and today is regarded as one of the broadest, and most worthy women of the community in which she and her family make their home. Three children have blessed this union: Oather W., Jarvis W., and Juanita, all of whom are still at home and are fortunate children, as their parents have determined that each one of them shall have every advantage in an educational way that county and state afford to equip them for whatever their life work may be.

The family are members of the Baptist church, of which they are generous supporters, while Mr. Wallage is a staunch supporter of the principles of the Democratic party.

JOHN McNETT, who is one of Banner county's most respected and best known pioneers, still lives on his ranch on section thirty, township twenty, range fifty-three, land which he pre-empted thirty-five years ago. Coming to this section originally in search of health, he not only long since attained his object, but at the present time is an example of vigorous robustness for his years, that reflects great credit on Nebraska's climate. He was born February 8, 1855, on the river Raisin, on the old Tecumach camping ground, Michigan.

The parents of Mr. McNett were William and Jane (Deming) McNett, the former of whom was born in Gattaraugus county, New York, in 1817, the latter of whom was born at Troy, New York, in 1820. As long as she lived she took pride in the fact that when a child of five years General La Fayette, shook her hand and that her grandfather was an aide on the staff of the great French commander during the Revolutionary War. Her father had charge of the Continental fleet and her mother was a niece of Benedict Arnold. The paternal grandfather was a soldier in the Revolutionary War and John McNett, an uncle, was made a guard at Buffalo, when he was but twelve years old, and Mr. McNett of Banner county was named for this uncle.

In 1883 the father of Mr. McNett located in Michigan. He was a cooper by trade. He grew up in hatred of human slavery and was a pronounced Abolitionist prior to the organization of the Republican party in 1857, at which time he united with it. When the Civil War came on he enlisted and has the unique record of serving one day. He was delegated a mechanic in the Fusileers, an organization that existed one day and was disbanded on the next. Thereafter, during the continuance of the war, he exerted his influence as a civilian. He survived until 1873, dying while on a visit to New York state, when aged fifty-three years. The mother of Mr. McNett lived into old age, passing away at Gothenburg, Dawson county, Nebraska, in 1911, having passed her ninety-first birthday. Of their seven children, three are living: a son and a daughter in California, and Mr. McNett, and his sister, Anna Miller, died in August, 1919, in Nebraska.

John McNett had educational opportunities in the common schools of both Michigan and Indiana. He remained at home and was the mainstay of the family, assuming responsibilities and over taxing his strength until twenty-eight years old. It was then he came to Nebraska, locating in Cheyenne county in 1886, now Banner county. He pre-empted land and now owns ten hundred and thirty three acres, much of it range land but two hundred acres in timber and farm land. He has raised as many as twenty head of calves a year, but is not as active in the stock business as at one time.

As noted above, Mr. McNett came to this section for his health and for some years hunting was engaged in both as a sport and for the wild game diet. He lived in a tent for a time. He tells of an early hunting expedition when he and his brother covered seventeen mountain sheep in a pocket, as they supposed, as the cliffs were so steep it did not seem possible the sheep could climb out, but since then he has learned more about the agility of mountain sheep, for when he returned in a few moments with his gun the sheep had vanished, having climbed an almost perpendicular bluff one hundred feet high Another hunting experience that Mr McNett tells of might have resulted fatally It was his early ambition to kill a mountain lion and one day after his skill as a marksman had been pretty well established, he discovered the tracks of a lion in the snow, followed them into a gulch in the Horse-shoe Horn and climbed the ridge He found his game so suddenly and unexpectedly that he had no opportunity to hide The lion emitted a yell that was very threatening and the situation was made worse by Mr McNett discovering that the snow had dampened the caps in his old-style rifle He had no time to replace, but with remarkable presence of mind raised the weapon in firing position and slowly retreated, in the meanwhile keeping a careful eye on the lion For some reason the animal did not take advantage of the encounter, but the occurrence was sufficiently alarming and Mr McNett was completely cured of desire to hunt mountain lions

During those early days in Banner county Mr McNett assisted in the digging of many wells, owning a regular outfit He had many experiences in this work, some of them being amusing and others partaking of tragedy He has been concerned in many movements of public nature and at one time was appointed a justice of the peace He qualified but afterward, finding that a large measure of his official duties consisted of performing marriage ceremonies, in which the happy bridegrooms took refuge in scarcity of money in the country to avoid paying a fee, and that most Judge McNett twenty-five cents to register, he resigned the office with its doubtful emoluments, and retired to private life

In 1886 or 1887, when the county seat fight came up, Mr McNett, as a resident of the north part of the county favored Ashford instead of Harrisburg Ashford was a village with a store and flour mill, named for William Ashford, who had established it, and the latter would have been pleased to have his village accepted and before the decision was not above diplomatically distributing a few deeds for land among those whom he considered would advance his ambitions When his own was not chosen he made no secret of the fact that he would like those deeds returned Finally a proper occasion arose and Mr McNett returned the deed he had received but never profited from

There is a grotto on Mr McNett's land that has a notoriety extending far beyond local circles It is not a cave although the sun never shines into it, but seemingly a natural configuration of the land For years visitors have come from far and near and among the many names and dates therein inscribed may be found notable signatures and dates as far back as 1861 Mr McNett has never married He lives a contended, independent life, looking after his ranch as suits his convenience A worthy visitor with proper credentials will meet a generous hospitality and find in McNett a jovial, genial host, whose natural friendliness has brought him a wide circle of well wishers In politics he has always been a Republican and his influence in county affairs is considerable

JOHNNIE T WYNNE, who is one of the enterprising and independent farmers of Banner county, is a native of Nebraska and was born in Adams county, March 22, 1885 He is one of a family of nine children born to John and Winifred (Marn) Wynne, who came to Banner county March 22, 1886 The father homesteaded and for many years lived on his land, afterward retiring to Pine Bluffs, Wyoming The family has been of recognized prominence for a long period

Johnnie F Wynne obtained his education in the public schools of Banner county On October 20 1910 he was united in marriage to Miss Ethel Gwartney, who is a daughter of Thomas and Mable (Ruc) Gwartney, and they have three engaging children, namely Elmer, Kenneth and Leon

Mr Wynne and his family reside on the Kinkaid claim on which he filed in 1906, and he now owns six hundred and forty-eight acres, mostly range land His industries are managed systematically and profitably He raises from forty to fifty head of cattle yearly and thirty head of pure Duroc-Jersey hogs He and wife are members of the Roman Catholic church In politics he is a staunch Democrat but has never entertained the thought of seeking public office Mr Wynne like all other members of his family is held in high regard in Banner county

CLARENCE WYNNE farmer and rancher who belongs to one of the fine old pioneer families of Banner county was born in this county May 11, 1888, and is one of nine children born to John and Winnefred (Marn) Wynne, extended mention of whom will be found in this work. The parents of Mr. Wynne now live retired at Pine Bluffs, Wyoming, but were residents of Banner county for many years, the father homesteading here in 1886.

Clarence Wynne was educated in the public schools of Banner county and remained with his parents until his own marriage which took place February 14, 1912, to Miss Alta Houser. The mother of Mrs. Wynne died when she was three months old but her father Nicholas Houser who was a homesteaded in Banner county, resides near Bushnell, Nebraska. Mr. and Mrs. Wynne have three children: Kenneth, Alta Loretta, and an infant.

Mr. Wynne owns six hundred and forty acres of land, mainly ranch land, and he puts it to good purpose. He breeds the best cattle, horses and hogs, Shorthorn, Percheron and Duroc-Jersey, and annually turns off from forty to fifty head of cattle, twenty-five head of horses and twenty head of hogs. Mr. Wynne takes a deep interest in his work and has been exceedingly successful. He and wife are members of the Roman Catholic church. In politics he is a Democrat and he belongs to the Farmers Union.

EDD S. CROSS, an enterprising, industrious, ambitious young farmer and ranchman of Banner county, is meeting with deserved success through his well directed agricultural efforts. He was born in Jasper county, Iowa, March 1, 1884, the only child of his parents, John W. and Carrie (Scoville) Cross, the latter of whom died when he was two years old. From that time until he was eight years old, Mr. Cross lived with his grandparents.

John W. Cross, father of Edd S., was born in Iowa, grew up in that state and followed farming there until 1887. He came then to Banner county, Nebraska, and homesteaded northwest of Harrisburg and remained on that property for five years and still owns it. He went back then to Iowa but in 1909 returned to Banner county and now owns and operates the Clay Springs ranch, a large estate situated seven miles west of Harrisburg. He is active to some extent in Democratic political circles but has never consented to hold office. His second marriage was to a Miss Hunt, who was born in eastern Nebraska, and they have had seven children, namely: Philip, Fred, John, Carrie, Lydia, Harvey and Benjamin.

Edd S. Cross obtained a country school education in Iowa. When eight years old he returned to his father and remained with him until his own marriage which took place October 26, 1909, to Miss Lillie M. Marshall. Her mother, like Mr. Cross's mother died when she was young, and her father, Stanley Marshall, is also now deceased. Mrs. Cross homesteaded where she and Mr. Cross now reside and they own two hundred and forty acres of fine land. Mr. Cross leases twelve hundred and eighty acres, and does an extensive business in livestock, raising fifty head of cattle yearly. In commenting on changed conditions that have effected every industry, Mr. Cross mentions that when he was seventeen years old he worked for farmers for fifteen dollars a month and now, even when offering seventy-five dollars a month, he is unable to secure sufficient help. He has always been a hard worker himself and obtained his first financial start by working on the government ditch at Torrington, Wyoming. Both he and Mrs. Cross have a wide acquaintance in the county and they are respected and esteemed by all who know them.

BERT WARNER, who was born in Henry county, Missouri, January 7, 1882, has spent the greater part of his life in Nebraska, and few native sons are more devoted to the best interests of this state. Mr. Warner and his people have grown prosperous here, while they, in turn, have been most worthy citizens.

The parents of Mr. Warner, Isaiah and Aletha (Smith) Warner, now live retired at Mitchell, Nebraska. The father was born in Michigan, in June, 1852, and the mother in Virginia, in 1851. From the time the father homesteaded in Banner county in the fall of 1887, he has been a man of prominence in this section. The family settled on the homestead in March, 1888, and that remained the family home until 1912, when the father and mother moved to Mitchell. The father still owns two thousand acres of land in Banner county and has property also in Michigan. Always a Republican in politics, while living in this county he frequently served in local offices and at all times was a man to be depended upon. He is a member of the Farmers Union and belongs to the fraternal order of Modern Woodmen of America. Both parents of Mr. Warner are active members of the Methodist Episcopal church. When Mr. and Mrs. Warner came to Banner county they were in better financial circumstances than many of the early settlers. They brought with them four milch cows and six head of horses, had wagons, farm implements and household goods, and had the forethought to bring along seventy-five bushels of shelled corn Later, however, Mr Warner rode on horseback to Wyoming in order to secure work in the hay fields as at that time there was nothing in this section with which an active man might connect himself and thereby earn even a small wage Industry, resourcefulness and good management, finally brought adequate reward

Of the ten children born to his parents, Bert Warner was the second in order, the others being as follows Vernie, who met death some twenty-five years ago from a stroke of lightning, Roland, who lives in Banner county, Bertha, who is the wife of Chester Cronn, of Kimball, Nebraska, Lawrence, who lives in Banner county, married Mabel Walker, William who lives in Banner county, married Minnie Palm, Jay, who lives in Baker City, Oregon, married Hazel Parker, Arthur, who lives in Banner county married Vina Mitchell, Mollie, who died in childhood, and Alice, who is the wife of Roy Hamilton

Bert Warner attended the public schools in Banner county until he was sixteen years old, then spent one year in Mitchell valley and two years at Gering After teaching one term of school at Gabe Rock, he worked on his father's farm near Gering for a year Then in association with a brother, he leased and operated the home farm for a year In 1904 he homesteaded the one hundred and sixty acre tract on which he lives and took a Kin-Kaid claim in 1906 Mr Warner now owns and operates sixteen hundred and twenty acres, all of which is good arable land suitable for farm purposes, but he only farms a small portion as he desires an extended range, being greatly interested in his fine stock He breeds Aberdeen Angus cattle, the only breeder of this variety in this section He finds them profitable and has about one hundred head the year round He keeps fifteen head of horses and has a fine flock of handsome Rhode Island Red chickens His industries are all attended to in a careful way, Mr Warner approving of regular discipline and routine on the farm just as in any other business

Mr Warner was married May 15, 1905, to Miss Lela Huffman, who is a daughter of Miles J and Julia (Armor) Human, the former of whom died at Gering, where he was a very prominent attorney The mother of Mrs Warner still resides there Mr and Mrs Warner have two children, namely Albert J, who was born June 16, 1906, and

Rowena, who was born September 9, 1913 They maintain a hospitable home and have a wide social circle Mr Warner has always been identified politically with the Republican party

WILLIAM E DUNN, who came as a homesteader to Banner county in the fall of 1888, did not live long enough afterward to fully realize how wise he had been in the choice of a permanent home for his wife and children His original homestead and tree claim have become valuable property Mr Dunn was born in Jasper county, Iowa, January 31, 1850, and died in Banner county, Nebraska, December 28, 1894

The parents of Mr Dunn were Philip G and Rachel (Culver) Dunn, the latter of whom died in February, 1899 The father died October 7, 1899 He was born in Ireland and in early life was a sailor He came to the United States and served in the Mexican War from beginning to end Afterward he went to Iowa and followed farming there, was married in that state and died there Three of his sons and one of his daughters are living, Sarah Plumer died in 1914

William E Dunn had good educational opportunities in Iowa, for his father was an intelligent well read man, and he remained on the home farm until his marriage, which took place December 6, 1873, to Miss Elizabeth Owens She is a daughter of Ryan and Elizabeth (Hutt) Owens, who were natives of Indiana Five children were born to Mr and Mrs Dunn as follows Othello M who resides in Banner county John H, who also lives in Banner county, Mary E, who is the wife of Emerson Faden, of Banner county, Philip, who is a resident of Banner County, Nellie G who is the wife of Frank Faden, of Banner county all of whom were born in Iowa except the two youngest

In the fall of 1888 Mr and Mrs Dunn came to Banner county and homesteaded on the place now owned by Mrs Dunn The original holdings consisted of a homestead and tree claim She now has several sections well stocked and well improved Her ranch is operated by her sons and she spends her summers here and her winters with her married daughters When Mr Dunn died she found herself at first in hard circumstances with little children to rear alone, but in her older sons she had strong, willing, industrious helpers Mr Dunn was a great believer in the value of churches in a community and he and Mrs Dunn assisted greatly in building up the Pleasant Hill Christian church, near Hull. He was interested also in the public schools and at all times used his influence to forward movements that would be of permanent benefit to the community. Mr. Dunn was a Democrat in politics. He was never a seeker for office but served as assessor and for many years was a school director. He was a man whose hearty good will toward everyone, brought him many friends and his death was considered a loss to his community.

Othello M. Dunn, the eldest son of the late William E. Dunn, was born March 28, 1876, in Iowa, and was educated in the schools of Burt and Banner counties. He accompanied his parents to Nebraska and homesteaded on the land he lives on, and in partnership with his brothers is operatings his mother's ranch as well as his own land. On November 25, 1908, he was united in marriage to Miss Mary Craton, who is a daughter of James and Almira (Sterns) Craton. Mr. and Mrs. Dunn have one daughter, Elva, who was born August 9, 1915. Mrs. Dunn is a member of the Methodist Episcopal church. In politics Mr. Dunn is a Republican. Like all other members of his family, he is a man of sterling character and has the respect and esteems of his neighbors.

John H. Dunn, the second son of the late William E. Dunn, was born in Mills county, Iowa, May 28, 1879. He was educated in the country schools in Nebraska and remained at home until fifteen years old, then adventured forth for himself, and during some years worked in Wyoming, Montana, and in North and South Dakota. In 1907 he homesteaded in Banner county but lives on the homestead adjoining, which belongs to his wife. He is interested with his two brothers in operating his mother's ranch of over thirty-six hundred acres. On April 3, 1912, he was married to Cecil Ogg, at Kimball, Nebraska, but her home was at Geneva. Her people still live there. Mr. and Mrs. Dunn have three children, namely: Willam, born November 15, 1914; Mildred, born July 8, 1916; Warren, born August 22, 1918. In politics John H. Dunn is a staunch Republican. He is considered a man of solid worth like his father.

Philip G. Dunn, the youngest son of the late William E. Dunn, was born in Burt county, Nebraska, August 23, 1882. He attended school in Banner county and in the city of Scottsbluff, and lived at home with his parents until his marriage, on April 28, 1910, to Miss Nora Adcock, who is a daughter of George and Ida (Evans) Adcock, very early

settlers in their section of Wyoming, where they still live. Mr. and Mrs. Dunn have one daughter Helen, who was born September 5, 1914. Mr. Dunn homesteaded where he now lives, in 1906, and in partnership with his brothers, operates about six sections of land, raising annually many cattle. He recalls the journey in the covered wagon from Burt to Banner county across the unfenced prairie. Although his admirable mother tried to be courageous, there was much in the surroundings to discourage her, although by that time the dangerous wild horses had been mainly driven off the range. She found the sand storms and the high winds hard to get accustomed to, but finally, in helping others, she apparently forgot her own annoyances. The sons revere the memory of their father, and they tenderly cherish one of whom they speak as "the best of mothers." Like his brothers, Mr. Dunn is an industrious, competent business man and good citizen.

WILLIAM C. SPAHR, whose personal recollections of Nebraska reach back thirty-three years, is one of Banner county's representative citizens and substantial farmers and ranchers. He is widely known for he has been active in public development in many ways and has been a useful factor in bringing about the acknowledged high standard of Banner county's citizenship. He is a native of Indiana, born at Portland, in Jay county, September 14, 1854.

The parents of Mr. Spahr were John and Exie (Hildreth) Spahr, both of whom were born near Xenia, Ohio. Of their nine children William C. and his sister Catherine are the only ones in Nebraska, the latter being the wife of William Louck of Loup City. Early in their married life the parents of Mrs. Spahr accompanied the Hildreth family when they removed to Indiana from Ohio because the latter state had become thickly settled and land had increased in price. They found cheap land in Jay county, Indiana, and seemingly none of the family would ever have need to go farther west to find either land or opportunity. The father became a farmer and stockman there, prospered and developed into a citizen of prominence. On the Republican ticket he was elected county commissioner, county assessor and county treasurer, and was held trustworthy in both public and private life. He and wife were earnest and willing workers in the Methodist Episcopal church. He, assisted by his son William C., cut and hewed the logs from which the first church

OLD FARM RESIDENCE OF MR. AND MRS. GREEN
Twenty-five miles south of Morrill, in Banner County, Nebraska

edifice was built in their neighborhood and the old structure stood until some twelve years ago when it came down to be replaced by a more pretentious building The mother of Mr Spahr died in 1869 but the father survived until 1889

William C Spahr attended the district schools and remained at home until eighteen years of age By that time Indiana like Ohio had become so closely settled that the young and ambitious man began to look westward, as had his father, for opportunity to expand It was this spirit that settled Nebraska Mr Spahr did not get father west than Illinois for the next six years, in the meanwhile working on farms in different sections of that state, then went to Nebraska and lived in Burt county until April, 1886, when he filed on a tree claim in Banner county, returning then to Fremont where he lived until the following spring, when with team and covered wagon he drove back to Banner county and here he has lived continuously ever since In spring of 1887 he took a pre-emption and lived there three years, then bought a relinquishment and on this land still makes his home

Mr Spahr did not find either comfortable or encouraging conditions when he first came to Banner county, but he was more resourceful and courageous than many of the early settlers and finally weathered the storm Lack of money as a circulating medium was a great hardship in the earlier days and Mr Spahr tells of working in the hayfield from sunup to sundown, for seventy cents a day When he looks at his own and his neighbor's trucks and tractors and other modern means of transportation, he remembered the long time it took him to cover sixty-five miles to Kimball, with his heavy load of wood that had to be sold for what it would bring in order to be exchanged for flour and needful groceries He frequently drove to Cheyenne in order to get long delayed mail There was no starvation in this section for deer and antelope were plentiful and every settler knew how to use a gun During the first winter Mr Spahr killed five deer and in the next spring when William Dunn, another early settler, visited with him while selecting a location, he gave Mr Dunn the hides The latter took them back with him to Iowa and had them tanned At the present price of hides, there are many the country over, who today would welcome such a valuable gift Mr Spahr now owns and farms almost five sections of land, and raises eighty calves and two carloads of White Face cattle annually With the help of his capable and industrious son and daughter, Mr Spahr continues to operate the ranch himself

On August 14, 1892, Mr Spahr was married to Miss Nellie Montgomery, who died in May, 1910, leaving two children Onno L and Vida In politics Mr Spahr is a staunch Republican He served one term as assessor and as long as he would consent to fill the office, was school director in his district Mr Spahr can relate many interesting incidents of early life here and has had business or social acquaintance with the most of the other early settlers

MILTON M GREEN, who was born in Pike county, Ohio, October 9, 1854, and died May 18, 1910, highly respected by all who knew him He was a son of Isaac and Martha (Waltz) Green, both of whom died when he was young He attended school in boyhood at Piketon, Ohio, and had a home with his two uncles until he was grown Afterward, until his marriage, he was employed as a farmer in different sections, but finding little opportunity to establish himself in Ohio, where land was dear, decided to go west and grow up with the country On October 1, 1878, Mr Green was united in marriage to Miss Louise A Arledge, of Circleville, Ohio where her mother yet resides Her father died November 11, 1911 Mr and Mrs Green left Ohio and moved to Missouri locating near Sedalia on a farm for which they paid a comparatively large cash rent Conditions were such that they could raise practically no crops and then it was that they determined to make another change that brought them to Banner county They then sold their best team of horses, as it was considered too good for traveling over such unbroken country as they knew their road lay They had five children by that time and and thus after reaching their new home they had an extension built on the covered wagon, and thus after reaching their new home they had some semblence of a dwelling and made use of it for some time With the friendliness that very generally prevailed among the early settlers, a neighbor ploughed a small tract and Mrs Green set about planting a garden In recalling that time Mrs Green says she was better able to see the bright side than was Mr Green He felt particularly discouraged on the day above mentioned as he had lost his best horse on the way, and seeing her actively preparing for the future by putting in garden seeds, asked her if she felt as if she really could stand living in so desolate a place as the homestead proved to be With a cheerfulness that might possibly have been assumed, she assured him she could and they "would make the best of it " He was permitted to live long enough to find "the best of it" very good indeed

Mr Green then bought an ox team and cut and hauled wood from the canyons The family reached Banner county May 1, 1889, and secured a pre-emption and tree claim in·Banner county and homesteaded just across the line in Wyoming Mrs Green, with thrifty eastern ways, raised chickens and turkeys during the first years and the eighty dollars she received for her poultry was a very welcome addition to the family purse About this time a Mr David, who owned the C P Ranch near the Greens, came to see if his new neighbors would be willing to go there and superintend the ranch as his foreman had left At that time the ranch had sixty-eight head of cattle Mr and Mrs Green accepted the offer and they remained with Mr David for thirteen years, operating the ranch so well and profitably that the owner was more than satisfied Great care had to be exercised at all times on account of cattle thieves, and Mrs Green suspects that they have not all been eliminated yet

Thirteen chidren were born to Mr and Mrs Green, and the following survive Clara, who lives at Mitchell, Nebraska, is the wife of John Manaird, Myrtle, who is the wife of Frank Sushatt, of near Chugwater, Wyoming, Anna, who is the wife of Ayre Chamberlain, near Kimberly, Idaho, Maud, who is the wife of James Shoemaker, of Kimberly, Idaho, Ernest, who lives near the home place, married Ida Rung, Hazel, who is the wife of Ruel Schindler, of near Gary, Nebraska, Orrin and Afton, both of whom live with their mother; Selina, who lives at home is the widow of Clarnece Stemler, and Waldo and Nellie, both of whom are at home This family has been carefully reared and well educated At first, the older children went to school when they had no accommodations, not even seats, in the schoolrooms and a great scarcity of books After Mr Green's death, Mrs Green filed a homestead claim and proved up on the same in 1917 She is now living on the Wyoming homestead and with the help of her sons farms and operates over two thousand acres, her shipment of cattle yearly averaging two cars Mr Green was a Democrat in politics He never accepted any public office except school director and served as such for many years There is no family in this section held in higher regard than the Greens

LEVI B SPEAR, who came early to Banner county and lived here peacefully and usefully for a quarter of a century, was well and favorably known He was an intelligent, industrious pioneer, a good neighbor and a man of sterling character He was born in Somerset county, Pennsylvania, January 11, 1845, and died on his homestead in Banner county April 17, 1912 Very little is known of his parents aside from the fact that they were James and Catherine (Bowlin) Spear The mother was born in Pennsylvania The father was born in Ireland and died when his son was seven years old, the latter being brought up by an uncle During his entire life he followed agricultural pursuits

On March 22, 1887, Mr Spear was united in marriage to Sarah M Spear, who was born in Chittendon county, Vermont, October 18, 1846 Her parents were Matthew and Gertrude (Forsythe) Spear. Immediately after their marriage, Mr and Mrs. Spear started for Banner county, Nebraska, in a covered wagon drawn by a team of mules, and arrived on the homestead, which Mrs Spear yet owns, on April 15, 1887 On the day following Mr Spear began plowing, and that year, in spite of blizzard and hail storms, they grew some corn and as fine a lot of potatoes as anyone could wish to see The notable May blizzard that brought suffering and distress to so many of the early settlers, found the Spears totally unprepared, with no well dug nor any barn or shelter put up As the storm grew more and more furious, Mr. Spear adopted the only possible way to preserve his stock, bringing the cows to the house and sheltering them with the wagon boxes and actually taking his own mules and those of Edward Campbell, into the house and there the animals and people had to stay during the twenty hours the storm raged It was not until August 8, 1889 that the Spears had an experience with a Nebraska hail storm, and their loss in growing crops was heavy, but they bore it courageously as they suffered no •more than their neighbors

Mr Spear continued to develop his homestead and then took a Kinkaid claim, on which Mrs Spear and their son Ralph still live, the latter operating all the land Ralph A Spear, an only child, was born May 23, 1889, was educated in Banner county and has always lived at home He now operates eight hundred and forty acres, raising from twenty to fifty head of cattle yearly, mostly Durhams, with a few good horses and some hogs Mr.

ERNEST ZEHNER

Spear, like his late father, is a Republican The latter took a great deal of interest in political matters and was always anxious for the success of his party because he believed in its principles, but would never accept any public office except in connection with the public schools He was active in the cause of religion and was a member of the Methodist Episcopal church at Hull, which church he had helped to organize and build

WILLIAM McCOMBER, who is one of the representative men of Banner county, has been interested here since the spring of 1887, a homesteader in that year He has been variously engaged in this section as farmer and rancher, and for years has been connected with development work in the irrigation areas

William A McComber was born in Lenawee county, Michigan, May 26, 1862 His parents were David and Rachel (Shippy) McComber, the former of whom was born in Rochester county, New York, and the latter in Vermont Both have been deceased for many years Of their eight children, William H is the only one in Nebraska He had common school training in Michigan and then helped on his father's farm

Mr McComber was married January 22, 1885, to Miss Ella Nelson, who was born in Michigan, where her people had settled early Her father died in 1904 and her mother lives with Mr and Mrs McComber The latter have had three children Bertha, who is the wife of Ray Holt, of Laramie City, Wyoming, Irene, who is the wife of Alvin Plaga, of Sibylee, Wyoming, and Averil, now deceased, who was the wife of George Phillips, of Laramie City

Mr and Mrs McComber came to Banner county in April, 1887, and still own their homestead, on which they lived over five years Their first home was a sod house, not to be compared to their present handsome, comfortable residence, but is served its purpose well as long as they needed it Mr and Mrs McComber were both musicians, and when they started for Nebraska she could not think of leaving her organ behind It had been a gift from her parents and she was an expert performer on organ and piano, while Mr McComber is a master of the violin He attended a dancing party given at the home of Mrs Lynch, at Kimball, a few nights after coming to the county, and after that, for two years, he and wife were engaged as musicians at every dance for forty miles around, Mrs McComber's organ being hauled from place to

place Their earnings from this source were not inconsiderable and wherever they went made friends

After leaving the homestead Mr McComber bought other land and they lived in the north part of the county for several years, when he accepted employment with the Swan Land and Cattle Company, on what was called the 2-Bar ranch, and he had charge of this ranch for five years Then he went to the Rock ranch and from there to Sebylee river, and worked there for over ten years Mr McComber then was engaged by the Wyoming Development Company on construction work, largely cement work, in the irrigation project areas and for five years rode the ditch at Wheatland, Wyoming He then went to the Laramie Water Company and was engaged in construction work for that organization for five years, also riding ditches In 1917 Mr McComber came back to Banner county and now owns and operate four hundred and forty acres on which he farms and raises hogs He has a well improved farm, well stocked as a first class farm should be and his buildings are adequate and substantial

Mr McComber was reared in the Methodist Episcopal faith Mrs McComber belongs to the Christian church She and daughter are members of the Eastern Star, of which she was local treasurer for three years For many years Mr McComber has been a Mason and in his home lodge has held all the offices except master In his younger days he was a Republican in politics but at present casts an independent vote He has never found time to serve in public offices even if he had had the inclination

ERNST ZEHNER is a broad-minded Scottsbluff county farmer of retiring disposition, who shrinks from undue publicity but whose story, nevertheless, should be recorded in a history of the county for the benefit of his children and also for the purpose of demonstrating to others the advantages that America has over European countries

Ernst Zehner was born in Germany, February 24 1860 being the son of William and Catharine (Maar) Zehner, both of whom were born, reared and married in the German Empire where the father was a general farmer all his life, passing away there some years ago They were sober, industrious and thrifty people who spent their lives in the quiet routine of agricultural life in their native land Ernest received his practical education in the excellent public schools of Germany which are maintained by the state and to which children must go for the required amount of study, and thus he laid the foundation for the broad and liberal education which he gained for himself through wide reading upon the subjects which interested him and also those connected with his business, for all his life Mr Zehner has continued to keep abreast of the questions of the day, both financially and politically, and today is regarded as one of the broad, liberal, wide-visioned men of the western section of Nebraska, where he has made his home for so many years As a youth he was ambitious and was far-sighted enough to see that class distinction and Junkerism in Germany would keep him from ever becoming a land owner or give his talents a fair opportunity While still a youth he began to read of the "Land of Promise" across the seas and before attaining his majority was brave enough to sever all the home ties that bound him to the old farm and the Fatherland and made the long ocean trip to America determined to carve a career and gain fortune for himself in the new world In 1880 Mr Zehner landed on our shores and though he was not well acquainted with the English language was nothing daunted by this great handicap, but started for the middle west where he knew that some of his countrymen had settled For a time he lived in Iowa, becoming versed in English and learning the customs of the country While there he engaged in general farming, the profession which he had learned on his father's farm in the old country and one which he was well adapted to carry on in the new Iowa was already well settled up at this time and the boy had a spirit of adventure which comes to every youth, so in 1886 he started for the west of which he had heard so much In the early eighties the great cattle barons were ranging their vast herds from the Pecos in Texas, to the Yellowstone in Montana, and Mr Zehner went to Wyoming to learn the cattle business by joining a cattle outfit Two years later he decided he wanted some land for himself and in 1886 took advantage of the offers made by the government to settlers and took up a homestead, also filed on a tree claim in Scottsbluff county, where he soon established a home, made good and permanent improvements on his land and was known as one of the younger and progressive men of the section He soon found that cattle was one of the paying industries and at times worked for the different companies who owned or rented range in this county and gained a wide and favorable reputation as one of the cowboys of the high prairies After the heigh-day of the cattle barons was past Mr Zehner engaged in general farming and livestock raising on his own account, and was accounted one of the most substantial operators in his county due to his practical experience on the range He was a close buyer, watched the market and usually disposed of his stock at a high price, he introduced modern methods in his agricultural industries and became one of the substantial men of the section For many years he was interested in the irrigation projects inaugurated by the government and private enterprise in the Platte valley and when he determined to retire from the active management of his business, bought a seventy-acre tract of irrigated land in Scottsbluff county where he now resides Mr Zehner is essentially the architect of his own fortunes and can look with pride over the wide acres which he has won by his own unaided efforts as he owns 762 acres of fine land in Banner county which is well adapted to stock raising, has 160 acres in Laramie county near Horse creek, and the homestead in Scottsbluff county where he lives All these varied tracts make him one of the largest landed proprietors in the Panhandle in addition to which he has considerable other financial resources

Mr Zehner married Cora Carter, a native of Scottsbluff county, born there May 9, 1879 Her parents were natives of Idaho, who are farmers of that state at the present time By former marriage Mrs Zehner has five children, Minnie, the wife of J Dickson, a farmer of the state of Oregon, Luty, deceased, Austin, Mottie and Twyla, all of whom have been given the best of advantages in schooling and who will have the advantages of practical experience in business Having worked out his own career and been independent since a young man it is but natural that Mr Zehner should be independent in his way of thinking, and in his political views he is an independent, being bound by no strict party lines when he casts his vote, but throws his influence toward the man best fitted to serve the people in whatever office is to be filled He is recognized as one of Scottsbluff county's public spirited citizens and in every relation of life measures up to the full standard of American citizenship, being a willing contributor and hearty supporter of every movement for the uplifting and improving of communal civic affairs

OWEN W LANGMAID, who is a representative of, perhaps, the oldest pioneer family of Banner county, has spent almost his entire life here and is one of the county's substantial and highly respected citizens He was born in Jasper county, Iowa, January 2, 1877

The parents of Mr Langmaid were Solomon

W and Eva (Badger) Langmaid, the former of whom was born in Caledonia county, Vermont, February 18, 1851, and the latter September 4, 1853 In his younger days the father was both carpenter and blacksmith, and after moving to Iowa, in 1876, he located on a farm on which he also had a blacksmith shop, and by working in the shop was able to add considerably to his income After ten years in Iowa, Mr Langmaid came first to Banner county, Nebraska, homesteading in the fall of 1884 and as far as recorded, he was the first homesteader in the county On three different occasions he visited his land and remained three months at a time, but in the spring of 1886 decided it was time to settle on it permanently Therefore a start was made from the Iowa home and after a tiresome journey by wagon, Kimball was reached on April 30, more than one remarking on the day of fine weather The family decided to rest over night at Kimball and start on the next long ride across the country refreshed by the night's sleep in comfortable surroundings Nature has a way of interfering with plans, sometimes, and on awaking the next morning the travelers found the ground covered thickly with snow and a blizzard raging, and conditions continued so bad that they were not able to leave Kimball for a week

At last, when the claim in Banner county was reached, the little family found a wide expanse of improved land and a dugout for a home It was not very inviting to the careful, tidy mother of the family, but she concealed her distaste as mothers often do, for a few days, but when she found a rattlesnake had claimed hospitality in the little home, she made her protests audible Fortunately they had brought a tent with them and this was set up in a convenient place and the family lived in it for almost a year, by which time the father had succeeded in erecting a log dwelling This first log house is still in use on the farm of J V Brodhead, a short distance southwest of Harrisburg At that time there was but one family living between them and Kimball, their house being located a mile east of the main road between Harrisburg and Kimball The parents of Owen Langmaid continued to live on this place for five years, then moved to Harrisburg, where the father operated the St James Hotel for two years and another hotel for the next six years In 1900 the parents sold out and returned to North Danville, Vermont, where the father resumed farming on the old homestead of his grandfather, which farm has been in the Langmaid family for

over one hundred years He continued on the old place until his death, which occurred December 15, 1918 In January, 1919 the mother returned to Banner county and now lives with her son Owen W, who is the eldest of her three children, the others being, Georgia, who lives in Canada, and Burl, who married and lives on the old homestead in Vermont

Owen W Langmaid went to school for a short time before leaving Iowa and later had some advantages at Harrisburg He remained with his parents until they returned to Vermont and then lived by himself on the homestead for a time On many occasions in early days he cut and hauled wood to town and exchanged it for flour Sometimes in boyhood he did not make a very profitable bargain, but money, in payment for work or service was gladly accepted instead of the universal trade He tells that on one occasion he was engaged to drive a team home from Kimball, a distance of twenty-five miles but after he accepted his employer decided that he would ride and the boy drive the cattle It took until nine o'clock at night with nothing to eat all day, and the youth was expected to be grateful for the wage of twenty-five cents he received for the job

On October 14 1903, Mr Langmaid was married to Miss Iva Bolen, a daughter of Dorsey and Nettie (Helmick) Bolen early pioneers in Banner county The father died in 1894 and the mother of Mrs Langmaid married second, Jerome S Rice, another old pioneer who died in March 1909 Mr Langmaid was county assessor for five years and is in his second term as precinct assessor He owns twelve hundred and eighty acres, one hundred and ten of which is farm and alfalfa land He lives on his well improved place of six hundred and forty acres which he homesteaded in 1905

GEORGE NOYES, who is one of Banner county's prosperous farmers and ranchmen, left his eastern home for what was then considered far West when twenty-four years old He was born in Wyoming county, New York, December 9, 1854, the only son of Leonard and Demis (Bailey) Noyes His mother died when he was six years old, a life long resident of Wyoming, New York

Leonard Noyes, father of George, was born in Vermont and was five years old when his parents settled on a pioneer farm twenty-five miles from Buffalo, New York Very often during his lifetime he told his children of early hardships, such as having to travel over an uncleaned country for sixteen miles to reach a corn mill, and working for fifteen years for a wage of eleven dollars a month   Father Noyes lived on one farm in southwestern New York, for seventy-five years   After the death of his first wife, in 1860, he married again and two daughters were born to him, one of these now living in New York, while the other died in 1915, in Oklahoma

The district schools of his native state provided Mr Noyes with a fair education   He was a well trained farmer when he came west to Kansas and followed agricultural pursuits in Brown county for the next twelve years.   In 1889, he took up a claim in the Cherokee Strip, in Oklahoma, having been in the Sac and Fox opening, but after looking the claim over decided some years later to give it up   For eleven years, however, Mr Noyes and his family lived in the strip, in Garfield county   The greater number of settlers had come there very poorly provided for farming, having no money and no farm utensils   Some of his neighbors carried on their farm work with a team made up of a bull and mule harnessed together   Mr Noyes worked with a yoke of oxen but became discouraged when his sowing of wheat only returned him five bushels of grain to the acre and this he had to sell for thirty-five cents a bushel   Every family felt the pinch of poverty and for weeks the regular diet would be Kaffir corn and sorghum molasses, with coffee (so called) brewed from the dried peelings of sweet potatoes

On May 22, 1909, Mr Noyes and his family came to Banner county and bought land on section twenty-five town six, on which they yet live   In partnership with his sons, Mr Noyes owns over four thousand acres of land here, ranching the most of it, breeding Hereford cattle   He has a fine well improved place and has prospered ever since coming to Banner county

Mr Noyes was married in September, 1879, to Miss Louise Shearer, who is a daughter of Joseph and Emma (Rickords) Shearer, residents of the state of New York   Mr and Mrs Noyes have had the following children James A , who lives in Mitchell, Scottsbluff county, married Louise Frane , Nellie, who lives at Flowerfield, Nebraska, is the wife of Lester Van Pelt , Ralph, who lives at home , Clifford J , who lives near the home place, married Janette Angel, and Lottie, whose home is at Pine Bluffs, Wyoming, is the wife of Charles Van Pelt   Mrs Noyes is a lady of much force

of character and mental capacity.   In 1904 she was appointed postmistress of the office at Hull, Nebraska, and continued in that office until she sent in her resignation in 1918   In politics Mr Noyes is a Republican and as an upright citizen exerts considerable influence in the county.

WILSON MITCHELL, who came to Banner county in October, 1887, has lived here ever since and now is one of the county's substantial men   He has identified himself with all movements promising to permanently benefit the county, particularly in relation to church and schools, and his reputation as an honorable, trustworthy man and good citizen is universal

Wilson Mitchell was born in Saline county, Illinois, December 26, 1866   His parents were Robert and Lydia A (Gaskins) Mitchell   They came to Illinois from Kentucky, and both died in the former state, the mother in 1871 and the father in 1883   On account of the family records not having been preserved, Wilson Mitchell knows little more of his ancestry   He is one of a family of six children, and he and a brother, Levi, were the only members to come to Nebraska   They reached Banner county in October, 1887, and homesteaded in 1888   Wilson Mitchell still owns his homestead but lives on the adjoining one that belonged to his brother, the latter going to Colorado in 1896

Mr Mitchell was married November 18, 1891, to Lila Michell, who is a daughter of Joseph and Julia (Covington) Mitchell, who were then residents of Illinois   The mother of Mrs Mitchell died May 16, 1900, but the father survives and lives in Oklahoma   Mr and Mrs Mitchell have the following children Lavina, who lives at Flowerfield, Banner county, is the wife of Arthur Warner, Sylvia, who is the wife of Roland Warner of Banner county, and Frank, Joseph, George and Rosamond, all of whom live at home

With the help of his son, Mr Mitchell now operates a large ranch   He has been a breeder of Hereford cattle and Poland China hogs from the first and believes these breed do best in Banner county   A busy ranchman finds little time to be very active in politics or to serve in public office and this is Mr Mitchell's attitude, but he is a sound Republican in his views and keeps well informed as to passing events   He has a comfortable, well ordered home and interesting family   All attend the Methodist Episcopal church, Mr Mitchell and wife both being interested in its many avenues of good influence and giving them financial support

LEMUEL M HOPKINS, farmer and ranchman in Banner county, is well and favorably known here For many years he has been a large landowner and responsible citizen, and in many ways has been identified with the substantial development of this section For a period of eight years, from 1906 to 1914, he carried the mail from Hull to Harrisburg, and during that time through faithful service made many friends Lemuel M Hopkins was born in Fulton county, Illinois, March 20, 1875, the youngest of three children born to Warren and Elizabeth (Barnes) Hopkins, the latter of whom was a native of Indiana The father spent the greater part of his life as a general farmer in Illinois and he died there in 1882 He was a Republican in his political faith but never accepted a public office Both he and wife were members of the Methodist Episcopal church She died in 1887 but the three sons survive Frank, who lives in Kansas City, Missouri, George, who lives at Denver, Colorado, and Lemuel, who belongs to Banner county

When Mr Hopkins losts his mother he came to his brother in Banner county and while here attended school, later accompanying his brother to Missouri and had further school privileges there He remained in Missouri until 1896 and was married there on September 3 of that year to Miss Myrtle Spear, a daughter of Clarence and Sadie (Spear) Spear, then resident of Missouri The father of Mrs Hopkins was born in Vermont and now lives retired at Mitchell The mother was born in Pennsylvania Her death occurred December 3 1916 Mr and Mrs Hopkins have had ten children and the following survive Lyndon, who lives in New Mexico, and Beth, Floy, Byile Kenneth and Hester, all of whom live at home a well educated, interesting family

Mr Hopkins homesteaded the place on which he lives on April 9, 1902, and now owns and operates eight hundred and eighty acres, all well improved, divided evenly between farm and ranch land He is profitably raising good cattle and general farm produce Mrs Hopkins is a member of the Methodist Episcopal church and the other members of the family attend For a number of years Mr Hopkins has been a member of the order of Knights of Pythias at Harrisburg He has never felt that he could afford the time to serve in a public office but he is a staunch Republican in his vote as a citizen in ranks

EARL HARVEY, who is one of Banner county's enterprising and successful farmers and ranchmen, has had much experience in this line of work, to which he has been accustomed since boyhood and engaged in since his school days ended Almost all his life had been spent in Nebraska, but his birth took place March 19, 1881, in Jones county, Iowa

The parents of Mr Harvey, George W and Cora A (Williams) Harvey now reside at Kimball Nebraska The father was a farmer and stockman in Iowa for many years In 1885 he moved to Nebraska and three years later came to Banner county, where he homesteaded and secured a tree claim and pre-emption He maintained the home there thirty years, then sold, bought property in Kimball and has lived retired ever since He votes the Republican ticket but has never accepted a political office Both parents of Earl Harvey are members of the Baptist church Of their seven children the following survive Lilian, who is the wife of John McKennon, of Merced, California, Charles, who lives in Montana, Ella, who is the wife of F O Baker, a banker at Bushnell, Nebraska, Arthur, who owns a garage at Gillette, Wyoming, married Mries Ricke, Earl, who belongs to Banner county, Nina who is the wife of W Deacon, who is employed in the post office at Omaha

Earl Harvey learned many lessons of thrift from his practical father in his youth During the first ten years after the family came here, the father worked out every summer and left Earl and his brother to care for the homestead They herded cattle on the home ranch and for the neighbors and in that way, all of them, for the times, earned quite a sum of money Mr Harvey encouraged the boys to save their money and buy calves, and with a small herd secured in this way, they gained their start In recalling those days Mr Harvey tells many interesting things He remembers when oxen were used in this section, and when a load of pine wood hauled many miles to town would bring two dollars, and a sack of flour could be bought for two dollars Frank Baker had the first horse cart in this neighborhood, and William Silvus was envied by everybody when his well was completed, operated by windlass and horse power, the well bucket holding a half barrel of water There have been many and wonderful developments since those days, and the present, when the public prints tell of Nebraska farmers being so progressive that on occasion, they do their threshing with electric motors not only in daytime but equally as well at night Speaking of the portable motors brings the fact to mind that it was the father of Mr Harvey's wife who brought the first portable sawmill into Banner county Afterward he went into partnership with Solomon Langmaid and it was in this first portable sawmill that the lumber was prepared for the building of the first bridge and the first buildings at Gering

On February 22, 1905, Mr Harvey was united in marriage to Miss Ethel Cross, a daughter of Benjamin and Cora A (Williams) Cross, early pioneers of the county Mr and Mrs Harvey have children as follows Inez, born September 13, 1906, Helen, born February 10, 1908, Elma, born August 25, 1909, Asa, born September 9, 1912, Darrel, born October 23, 1914, Vernon, born April 14, 1917, and Elsie, born February 6, 1919

With his homestead and Kinkaid claim, Mr Harvey now owns and operates seventeen hundred and forty acres of land, devoting one thousand acres to farm purposes and the rest to ranching He breeds thoroughbred Hereford cattle and ships a car load annually He has a good set of farm buildings and very comfortable home He is industrious and progressive in the management of his ranch and enjoys the respect and regard of his neighbors He belongs to the order of Knights of Pythias at Harrisburg, and in politics is a Republican

ROLLAND B BIGSBY. who is one of Banner county's substantial farmers, belongs to a pioneer family that came to this section in the spring of 1887 He was born at Omaha, Nebraska, July 14, 1879 His father, William Bigsby, was a veteran of the Civil War Further mention of the family will be found in the sketch of Everett Bigsby

Rolland B Bigsby obtained his education in Buffalo county, near Kearney His father homesteaded in Cheyenne, now Banner county, in 1887, thirty miles northwest of Kimball, preempted and secured a tree claim, the latter property now being owned by his daughter, Mrs John Heintz When the family came here in a covered wagon, they brought with them two cows, a calf and six chickens Mr Bigsby remembers that while on the way a windstorm came up that prevented their going forward for a whole day The father expended all the money he had while they stopped at Kimball, for something to eat, and Mr Bigsby remembers how lean the larder was for a long time after they were settled on the homestead Two meals a day was the rule and in summer time the choke cherry trees gave them their only fruit Money was scarce and the father was no longer a strong, able man, who could labor hard and continuously like many others Mr Bigsby tells of one occasion when he was returning from Sidney to which place he had gone to attend to some legal papers, and coming back on foot, was overtaken by exhaustion on the way, owing to his age and the long distance he had traveled In those days there were no settlers living near together, many miles intervening between the homesteads Had it not been that two women of the neighborhood happened to find him where he had fallen, as they passed picking berries, Father Bigsby might never have been revived With the water they gave him he was able to get within calling distance of his home and he was cared for by his family, but they were never willing for him to attempt such a journey again

In recalling those early days, fortunately so different in many ways from the present, Mr Bigsby tells of herding his father's cattle, fearlessly running barefooted through the cactus that grew thickly over the range, and his recreation was catching skunks, badgers and rattlesnakes, with the help of his dog His duties were only those of the majority of boys in that section at that time, but in recognition of the dangers and discomforts to which he was subjected, Mr Bigsby feels very thankful that his own children have not been through the same experience, that there has never been any such necessity

Mr Bigsby was married August 20, 1909, to Miss Clara Fuller, who is a daughter of William D and Elizabeth (Kimberly) Fuller, pioneers of the county who now reside at Bushnell Mr and Mrs Bigsby have six children as follows Clifford, Myrtle, Edyth, Ethel, Ruby, and an infant

In spite of many hindrances and discouragement, Mr Bigsby has prospered He has always been careful and industrious and possessed of good judgment He homesteaded under the Kinkaid law and now owns and operates four hundred and eighty acres Mr Bigsby is a well informed, practical man and good citizen He votes with the Republican party and performs every political duty incumbent upon him, but he has never desired public office of any kind

THOMAS W G COX, who is a representative citizen of Banner county, prominent both in public affairs and business enterprises, is satisfactorily proving today the fertility of Nebraska soil under proper cultivation and illustrating also the sure reward that follows persistent and well directed industry Coming to Banner county thirty-one years ago, a young man practically without capital, he now owns farm, ranches and stock, and is accounted one of the leading breeders of Holstein cattle and Shire horses in this part of the state

Mr Cox was born in Peoria county, Illinois, February 17, 1866, a son of Jacob and Herminia (Humphrey) Cox, the former of whom was born in Ross County, Ohio, February 18, 1830, and died in Nebraska, August 18, 1904, and the latter was born in Dallas county, Missouri, December 25, 1836, and died in Nebraska Of their eleven children, Thomas W G was the fifth in order of birth, the others being as follows Elizabeth A, who lives on the old home place in Nebraska, John, who died June 26, 1916, Rastus, who resides with his sister on the old homestead, Margaret A who is married, lives at Forest River, North Dakota, Charles H, who lives with his family lives in Banner county, Edwin, married, who lives in South Dakota, Jacob, married, who lives at Bushnell, Nebraska, and two who died in infancy The father of the above family worked as a carpenter and blacksmith in Ohio, Missouri, and later in Illinois From the latter state in March, 1880, he moved to Nebraska and for about ten years was engaged in farming in Cass county In the fall of 1889 he came to Banner county, pre-empted and homesteaded and resided here until his death He was a man of great enterprise and possessed the sound common sense that marked many of the sturdy early settlers He engaged in general farming and stock-raising, and surprised his neighbors by his farm success, as it was the general opinion at that time, that the soil here could not be made to produce crops on account of lack of moisture He did more than prove they were mistaken for he set out one of the first orchards and carefully cared for it, with the result that the trees are still bearing fine fruit In these directions he was really a public benefactor He was a staunch Republican and enjoyed taking part in political campaigns but declined to accept public office, but served in official positions in the Masonic order to which he belonged many years At the time of his death he owned eight hundred acres of land

Thomas W G Cox attended school in Illinois and later in Cass county, Nebraska, and remained at home until over twenty-two years of age On May 6, 1888, he came to Kimball, Banner county, and homesteaded in what was locally known as Bachelor's Bend, at the head of Bull canyon, so designated because all the homesteaders had been unmarried His nearest neighbor was three miles away After one year a brother joined him and secured a claim and they made a bargain by which the brother should take care of both claims and each should have half the crop, while Thomas W should accept work in Colorado and give his brother half of his wages With the money he earned in Colorado, Mr Cox bought three horses and two cows, and that was the nucleus of his present extensive horse and cattle industry At one time later he borrowed the sum of three hundred dollars and gave thirty horses as security

Mr Cox lived on his homestead and looked after his own domestic affairs for the next eight years during this time hauling all water a distance of eight miles, a task indeed when contrasted to present rapid methods of transportation Between 1891 and 1899 he raised wheat, about twenty bushels to the acre Again comparison is suggested because then Mr Cox had to pay ten cents a bushel for threshing, take his own time to assist the owner of the thresher to haul his machine back and forth, then wagon the wheat thirty-five miles to Kimball and sell it for twenty-five cents a bushel In 1897 he bought one hundred and sixty acres for one hundred dollars and his ability as a farmer has been shown by the production of twenty-eight bushels to the acre on that land, year after year When he was ready to sell that land he received thirty-five dollars an acre for it Mr Cox has proved his knowledge of land and has bought and sold numerous tracts advantageously He now owns nine hundred and sixty acres of land in Banner county, three hundred and twenty acres in Wyoming, leases one and one-half sections and breeds Holstein and Jersey cows, Shire horses and standard hogs Probably Mr Cox has the best improved property in Banner county He is making preparation to follow the example of his father and set out a fruit orchard and also a grove of shade trees, expecting to make a certain success of his venture, because he will use the same care and scientific knowledge with his trees that have proved so satisfactory in the growing of grain Progressive and enterprising men like Mr Cox are very valuable in a community

On December 25, 1898, Mr Cox was united

in marriage to Miss Eula V. Cronn, a daughter of Clarkson and Mary Runyon Cronn, pioneers of 1888 in Banner county, who homesteaded at Flowerfield, near Wild Horse corral. They retired in 1908 and resided at Kimball until 1918, removing then to California, and now residing in Oregon. They became the parents of thirteen children, of which family Mrs. Cox is the youngest. The others were as follows: Margaret, who lives near Harrisburg, is the wife of John V. Brodhead; Charlotte, who is deceased, was the wife of Jacob Kishpaugh; George, who lives in Washington, married Anna Campbell; William, who lives at Kimball, married Hattie Longworth; Sadie, who lives in Wisconsin, is the wife of Charles Park; Abraham, who died in Pennsylvania when eighteen years old; Scott and Wesley, both of whom died in infancy; Edwin, who lives in Wyoming; Carrie, who lives in California, is the wife of Edgar Morford; Florence, who lives in Oregon, is the wife of Harold Parker; and Chester, who lives at Kimball, married Bertha Warnor. Mrs. Cox says that when her father landed in Banner county with wife and five children, he brought with them a cow, a calf and five dollars in cash, and that when he retired he owned his homestead, one hundred head of cattle, thirty head of horses and other stock. Like others who have done well here, he lived a busy, frugal life for many years, but he brought up a large family in comfort and has the satisfaction of seeing them all well settled in life.

Mr. and Mrs. Cox have children as follows: Violet, born January 3, 1900; Florence, born January 21, 1901; Archie, born August 3, 1903; Doris, born August 28, 1904; Edna, born January 23, 1905; Ethel, born January 25, 1907; Una, born June 8, 1908; Josephine, born March 22, 1910; Edith, born August 16, 1913; Velma, born January 16, 1916; Gladys, born January 17, 1917; and Glenn, born June 10, 1919. This large and interesting family living in happy concord and unbroken ranks in their beautiful home, in the banner section of Banner county, under the wise and loving protection of father and mother, promise well for the next active generation that will push still further the car of progress in this section.

In his public attitude, Mr. Cox is an outspoken American and his influence in county and community matters is marked. He is a Republican in political affiliation and has frequently served on election boards and in other capacities and offices of responsibility, and for many years has been a useful member of the school board. He belongs to the Farmers

Union organization, and both he and wife are members of the order of American Yeomen.

ERASTUS W. COX, a prosperous general farmer and stockraiser in Banner county, belongs to an old pioneer family of Nebraska, and accompanied his parents to the state when five years old. He was born in Stark county, Illinois, January 10, 1863, and is a son of Jacob and Hermonie (Humphrey) Cox, both of whom died in Nebraska. The father was a native of Ohio and the mother of Missouri.

The parents of Mr. Cox came from Illinois to Nebraska in March, 1880, lived in Cass county until 1889, came then to Banner county and pre-empted and homesteaded. The preemption is now owned by Mr. Cox and his sister, Miss Elizabeth, and they reside on it. The father was a man of considerable consequence in the county and the entire family has always been held in high regard. Extended mention will be found in this work.

Erastus W. Cox attended the country schools in boyhood and lived at home until twenty-one years old, mainly occupied with farm work. In those days a farm of two acres was considered quite a grain field and when it was ready to thresh the neighbors all came to help, and as a harvester, Mr. Cox has travelled all over the country and at one time every face he saw was friendly and familiar. With the passing of years he often feels as if he only meets strangers now. At that time the farmers raised hogs only for their own use, often dressing as high as six hundred pounds, which, if sold, would not have brought more than six or seven cents a pound. Growers of hogs at the present day do not do business on any such basis. Mr. Cox had many head of Duroc-Jerseys on his one hundred and sixty acre farm and is making plans to breed more extensively in the future.

In the fall of 1892 Mr. Cox came to Banner county and homesteaded on section 20-18-58, Flowerfield precinct, now called Epworth precinct, and lived on that place for six years. He then became traveling representative of a threshing machine company of Oklahoma City, with which he continued six years, his territory reaching as far north as Winnepeg, Canada. He returned then and settled on his father's old homestead in Banner county and has resided here with his sister ever since. A year after his return, his old company made a flattering offer to him that would have taken him to Argentine Republic, South America, but he declined it, having tried commercial traveling and being well satisfied with his agricultural

MR. AND MRS. FRED EHRMAN AND RESIDENCE

prospects in Banner county Mr Cox is a Republican in his political views but has never accepted any office except membership on the school board He belongs to the order of Odd Fellows Miss Elizabeth A Cox is very well known all through this section For a number of years she was the accommodating and efficient postmistress at Epworth She belongs to the Methodist Episcopal church at Epworth, in which she is a devoted worker

FRED EHRMAN — Fred Ehrman's birthday is on the 12th of February, the same as that of Abraham Lincoln, greatest of American statesmen, and of Charles Darwin greatest of English scientists, so he should have gone into either politics or science But he wisely perceived that there is more money in farming Nebraska lands than in either of those two callings, so here he is

He was born in Germany in 1881, and came to the United States in 1890, settling with his father on a rented farm near Hastings, Nebraska In 1893 the family moved to Colorado, where the father, George Ehrman, died at the age of forty-eight years The mother Katherine Ehrman, lives near Gering, Nebraska

In 1897 Mr Ehrman went to Brush, Colorado, where he resided until 1910, when he came to Scottsbluff county and has since made his home here He was joined in marriage with Lola White, the daughter of W H White, a farmer in Idaho Two children were born to them, both of whom died in infancy, and a further bereavement came to Mr Ehrman when his wife was taken from him by death on May 27, 1918, at the age of thirty-five years

The home place is a fine irrigated farm of 160 acres, all of the improvements having been put on by the owner himself It was prairie when he took hold of it, so he is one of the builders of the country He has fed Shorthorn cattle quite extensively of late years, but one of his most important enterprises for the past fifteen years has been raising thoroughbred Percheron horses for the market This industry has not only been profitable to him, but has been an important factor in developing the country, for it is only a few years ago that large work horses were scarce in this section, a condition which means lack of efficiency and economy in farming

Mr Ehrman is independent in politics He has been a member of the irrigation district board for two terms and of the school board of his district for a number of years, and is a director of the Farmers' Union

GUST PEARSON, who is accounted one of Banner county's prosperous and successful farmers and stockraisers, was born in Sweden, May 21, 1869 His parents were John and Christina Pearson, who spent their entire lives in Sweden, where the father was a farmer, but never on so large a scale as his son and grandsons in Banner county

Mr Pearson came to the United States in 1900 and located in Banner county, Nebraska, to assist his wife's uncle, the late Carl E Hanson, who died in 1917 Mr Pearson homesteaded in Banner county, just north of where he now lives, and worked on the railroad until he had enough capital to stock his land His first house was half sod and half log, but it had a pine roof which many of the other houses in the neighborhood did not have at that time Mr Pearson's first residence did not cost him more than one dollar, a fortunate circumstance as his entire capital when he reached Banner county was twenty dollars In contrast, Mr Pearson has recently sold one of his farms for $47,000 He still owns three hundred and twenty-five acres, all fine productive land, after giving each of his four sons a farm Hard work and excellent judgment explain his success, which has been great

Mr Pearson was married in Sweden on June 17, 1894, to Miss Anna Hanson, a native of Sweden, whose parents still live in that country Mr and Mrs Pearson have six children, namely, Arland, David, Joseph, Mamie, Elsie and Carl All live at home in Banner county with the exception of Arland, who married Jennie Pearson and they live in Wyoming. Mr Pearson is not as active as formerly as his sons relieve him of many responsibilities, all being capable farmers and they not only operate their own land profitably but their father's also Mr and Mrs Pearson are members of the Swedish Lutheran church He is an American citizen but has never taken an active part in politics, but in voting an independent ticket, he gives support to candidates that meet with the approval of his own judgment Mr Pearson is well known throughout the country and is highly regarded

EMIL JOHNSON — One of the well known residents and greatly respected citizens of Banner county, a successful farmer and honorable business man, is Emil Johnson, who has lived in Nebraska since he was fourteen years old He was born in Sweden, February 19, 1879, a son of John F and Bettie (Larson) Johnson

The father of Mr Johnson preceded his family to the United States, reaching here April 5, 1884 and coming on to Wahoo, Nebraska, because a brother-in-law lived there He worked in that neighborhood until 1887 and then homesteaded until 1900, then came to live with his son Emil and his death occurred February 16, 1905 Of his six children five are living

Emil Johnson attended school in Sweden and before coming to the United States in 1893, had visited Norway, Scotland, and Denmark With his mother he took passage on the ship *Iceland,* sailing from Gothenburg, December 12, 1893 On New Year's night trouble developed in the ship's machinery and because of the high seas the vessel almost foundered, but good seamanship saved her and the passengers were safely landed in the harbor of New York on January 5, 1894 They joined the father in Nebraska and the mother survived until in November, 1906 Young Emil found hard times awaiting him in the new home He worked on his father's farm and for others and managed to save a part of his small wages His father's homestead was in an arid region and as there was no money to dig a well, which was a serious undertaking on account of the depth, water had to be hauled seven miles Mr Johnson remembers when game was plentiful and he has seen one hundred head of antelope at one time After proving up on their first claim of one hundred and sixty acres improved with a large barn, Mr. Johnson sold the place for seven hundred dollars, which was considered a good price, although the same property now would bring seven thousand dollars

From having nothing to start with, Mr Johnson has been remarkably successful in his business operations He owns eleven hundred acres of land suitable for farming purposes in the main, breeds Duroc-Jersey hogs and Hereford cattle, averaging about twenty-five head yearly He owns also a blacksmith shop and some building lots at Bushnell, Nebraska, where he formerly owned a livery stable For a number of years he has been land salesman for this part of Banner county and has succeeded well in this line He has the reputation of being a far-seeing business man but one whose word is as good as his bond

Mr Johnson was married January 14, 1906, to Miss Annie Olsen, who is a daughter of Lars Olsen, extended mention of whom will be found in this work Five sons and five daughters have been born to Mr. and Mrs Johnson, as follows Marie, Gertrude, Ivon, Lars Frederick, Ida, Helen, Alvin, Carl, Agnes and Leonard

In politics Mr. Johnson is a Republican He has served frequently in public office, for nineteen years being a school officer and every year on the election board, and at present is road supervisor He is a stockholder in the Farmers Union, and he belongs to the order of United Workmen During the World War he was a very active worker on behalf of the Y M C A, the Savings Stamp and the Red Cross drives

JAMES PATTON, who was the father of one of Banner county's large and most highly respected families, came here thirty-one years ago, and through hard work and good management, became one of the wealthy men of this locality He was born in Ireland, a son of John and Jane Patton, and although he never revisited his native land after coming to the United States in 1861, he never forgot it and throughout life always kept a warm place in his heart for any native of old county Down His death occurred December 7, 1912

At Newburgh, New York, in March, 1865, James Patton was united in marriage to Mary Carse She was of Scotch-Irish stock and was born at Ballymacrumble, county Down, Ireland In May, 1861, she came to the United States and entered domestic service and until her marriage worked in private families for a wage of six dollars a month To the above marriage eight children were born and the following survive Elizabeth, who lives at Dixon, Illinois, is the wife of James Bennett, Margaret, who also lives at Dixon, is the wife of Hugh Bennett, Isabella and John, both of whom live on the old family homestead, Mary, who is the wife of R N Biggs, lives in Colorado, and Letitia, who resides on the homestead, the brother and two sisters all having abundant means

After their marriage James Patton and his wife came as far west as Dixon, Illinois, where they lived during the next six years and then moved to Fayette county, Iowa, but a few years later went to Pottawattomie county, Iowa, where Mr Patton worked on farms and through his own industry and his wife's frugality, they prospered Desiring to secure a homestead still farther west, Mr Patton decided upon Banner county, Nebraska, and came here March 9, 1888, and homesteaded near where his children yet live He was accompanied by a brother and brought with him nine head of cattle, some hogs, six horses, six hundred dollars in money, which in those days in this section represented unusual wealth He brought also the household goods. The mother of the family, and uncle and the children, all came as far as Kimball on the railroad and from that time until the present has been prominent in many ways and at all times useful and worthy members of the community in which they have lived

During the first years after locating in Banner county the crops were poor but that did not mean to James Patton that his family should want for anything On the other hand, he hastened to Cheyenne and secured work in the railroad shops In the meanwhile his family carried on the home affairs well and wisely under the capable management of the mother and when Mr Patton returned and resumed farm work, he found conditions much improved and from that time on prospered in his agricultural industries As his children grew older they gave their parents dutiful assistance, and the mother for many years was as his right hand In early days both Mr and Mrs Patton would haul the produce of the farm to Cheyenne, sixty-five miles distant, where they would find ready sale Finally, however, the mother had a stroke of paralysis The family planned a trip for her to New York on a visit but before she could start she became so ill that the journey had to be abandoned She lived ten years longer, her death occurring June 21, 1909, But she had been an invalid add that time Both parents were members of the Presbyterian church when early meetings were held in the home of David McKee, who had been a neighbor in county Down, Ireland

Mr Patton was not only an industrious man but was able in business and when he died left a large estate, at that time owning fourteen hundred and twenty acres of land In early days almost all the settlers at one time or another put mortgages on their property and on one occasion Mr Patton mortgaged a horse and cow for twenty-five dollars in order to lay in seed corn, a temporary loan that was immediately paid back While the Pattons came up with such hardships as having to haul water a long distance, they never lacked sufficient food, nor did the children fail having educational opportunities Mr Patton was one of others in the neighborhood to put up a schoolhouse, a structure twelve by sixteen feet in dimensions The twenty-one children who attended that school had none of the helpful appliances of modern school children, few books, and had only boxes for seats The

first teacher was a resident of Kimball, who was a little dramatic as she dashed up to the little school house every morning, riding her pony without a saddle, but she evidently imparted knowledge to the children as their later progress showed Still later Miss Nettie McKinnon taught the school for two terms, driving back and forth a distance of nine miles every school day The Pattons have always been hospitable people and from the first have taken part in the neighborhood social life In politics Mr Patton was a Republican as is his son John, but neither desired political office

WILLIAM P MILLER — A truly interesting story is that of the Miller family of Banner county, which lost in the death of William P Miller, on September 23, 1908, not only its honored head, but one who, after fighting valorously in his country's battles, so faithfully turned his attention to the arts of peace, that no name in this community is mentioned years afterward, in terms of greater respect

William Palmer Miller was born September 18, 1831, at Alburgh, Grand Isle county, Vermont, a son of Duncan and Laura (Wiles) Miller, natives of Vermont The parents moved to Brookfield, Illinois, near Ottawa in the great Illinois corn belt, and there the father and his eight sons cultivated a great expanse of land and became wealthy and influential He died in 1870, and his eight sons have passed away also Their mother lived until 1877 William P Miller was educated at Alburgh and at Plattsburg, New York, and afterward he assisted his father until he enlisted for service in the Civil War, on August 13, 1862 He served in the One hundred and fourth Illinois volunteer infantry until July 12, 1865, following which was a long hospital illness from which he never really recovered

On October 12, 1869, Mr Miller was married at Duquoin, Illinois to Anna M Burbank, a daughter of George W and Mary J (Hatch) Burbank the former of whom was born in New Hampshire and the latter in New York To the above marriage the following children were born Charles P, who is in the furniture business at Gibbon, Nebraska, was the first barber in Kimball county and for two years taught school at Flowerfield, married Nellie Henline, Mary A, who is the wife of J W Hoke, lives in Colorado, Harry I, who lives in Banner county, married Grace Reynolds, Walter A, who is survived by his wife, Myrtle Bigsby, died near Harrisburg, Nebraska, when aged twenty-three years, George P, who died aged twelve years,

.one who died in infancy; Benjamin L., who lives near Stoneham, Colorado, married Mary Pennington; and Ward E., who lives with his mother.

Following their marriage William P. Miller and wife lived at Brookfield, where he taught school for a year, then moved to Ottawa, Illinois where he bought a cutlery factory and remained in business there for seven years, doing all the cutting himself. His health was poor, however, from the exposures endured in the army, and partly on that account, and partly for business reasons, he began to entertain thoughts of the great West. This resulted in his moving to Smith county, Kansas, in 1878, where he pre-empted land and during seven years of drouth, attempted to farm. In the meanwhile he taught school for one year, supposedly for a wage of fifteen dollars a month but had to discount all his checks. During his last year in Kansas he worked for the American Bible Society at Columbus, and he did well as his tastes were literary and his education solid. In 1887 he was offered and accepted a position as school teacher in Adams county, Nebraska, hastening home from Columbus in order to make preparations for removal. He found his wife sick but in order to fulfill his contract, he secured covered wagons and hastened on the way to his school district in the neighboring state, only to find, when he reached it, that the directors had become impatient and had hired another teacher. In after years Mr. Miller could smile over such a situation but at that time it was tragic.

In this emergency the family took shelter in a sod house that had been used for the storage of broom corn for a few weeks, then moved into a frame house which they secured rent free on condition that they would keep it comfortable for its owner and cook his meals. They had plenty to eat as the owner was a great hunter, and traded off game, quail and prairie chickens for other provisions. In the general discomfort, Mrs. Miller fell ill again and her recovery was slow as there were no physicians near, but in the following spring another move was made. Mrs. Miller owned an old family heirloom, a gold watch, and this she gave to her husband and he was able to trade it for a lease on eighty acres of school land in Webster county. They moved on that land and a year later it was sold for two hundred and fifty dollars and they came to Kimball county and homesteaded seven miles west of Kimball. On that place they lived seven years, then put their children in school in Franklin county. Mrs. Miller removed with

her parents who were getting old and needed her care and then settled twelve miles northwest of Harrisburg, where Mr. Miller died. For forty years of his life he had been a school teacher and faithful to his charges. There are many who remember him with feelings of gratitude because of his patience in instructing them.

After Mr. Miller died Mrs. Miller and her youngest son homesteaded in Nebraska near the Wyoming line, but when the great war called her son the homesteads were sold. She still owns four hundred and eighty acres in Banner county, is perfectly capable of looking after her own affairs and has always enjoyed social life wherever she lived. Highly educated both in books and in music, it was hard for this cultured lady to leave the comforts and associations of her eastern home and to bear with cheerful courage the hardships which later attended her. She has done so, however, and has, additionally, helped others with neighborly devices and loving sympathy, and is universally held in esteem. She has not entirely buried her talents, for she has been the highly appreciated correspondent of the Nebraska Farm Journal and Banner news for the past twenty years.

HARRY I. MILLER, who is numbered with the substantial men of Banner county, owns large bodies of valuable land and is a successful breeder of the famous White Face cattle. He is one of the progressive agriculturists of this section, operating with the latest improved farm machinery, and not only keeps in touch with scientific farm development for his own advantage, but as a contributor to the Nebraska Farm Journal of Omaha, has been the means for a number of years, of imparting vital information, clothed in interesting language, to those who need this knowledge.

Harry I. Miller was born at Ottawa, Illinois, April 28, 1875, and is a son of William P. and Anna M. (Burbank) Miller, extended mention of whom will be found in this work. He enjoyed excellent educational advantages, attending school at Franklin, Nebraska, the Methodist Episcopal College at Orleans, and Franklin Academy, while his home environment was always of a high intellectual standard. His father gave forty years of his life to the teaching profession, and his mother enjoyed, in her youth, both literary and musical opportunities.

In 1895 Mr. Miller started out for himself as a farmer in Franklin county. In 1897 he was appointed park superintendent of Greeley,

Colorado, where he remained until 1899, when he went to Denver and worked as a railroad fireman until January 1, 1904 He came then to Banner county, having borrowed fifty dollars to come and get established, and homesteaded on section thirty-four, town eight, on which he still lives To homestead was one thing but to pay for the land and provide for a family was another as Mr Miller and many other settlers here have reason to remember A lack of sufficient working capital embarrassed him for some years even though he did not have to contend with some of the disadvantages of earlier homesteaders His credit was good, however, and he was able to borrow enough to buy his first team, after which he worked for F. O Baker, afterward giving a note for the first well he had dug Food, especially meat, was not plentiful in those early days Corn bread risen by soda neutralized by vinegar was the daily fare Hunger has been found to overcome many prejudices, and Mr Miller tells of several proofs that came under his own observation On one occasion when a neighbor's steer was killed by lightning, there was no discussion as to the disposition of the carcass, the neighborhood enjoying strengthening meals to which many had long been unaccustomed Later, when Mr Miller found that coyotes had invaded his poultry yard and had killed twenty-four chickens by sucking their blood, the neighborhood had another and unexpected feast of meat These stories illustrate not a prevailing condition but a phase of pioneering that may well be brought forward in the way of contrast with the present Mr Miller has recently sold a tract of eleven hundred and twenty acres of land, still owns over twelve hundred acres and has no debts He breeds White Face cattle, shipping about one hundred and twenty-five head annually, and Duroc-Jersey hogs Additionally he owns stock in the Farmers Exchange at Omaha, is a stockholder in the Bushnell State Bank and owns considerable other property there His is a model farm and he operates with two farm tractors and other modern equipment

Mr Miller as married May 15, 1895, to Miss Grace D Reynolds, of Orleans, Nebraska, a daughter of Oliver F and Amelia (Reynolds) Reynolds, and they have had children as follows Lucile, who died April 19, 1917, was the wife of Andrew Miller, Hazel, who is the wife of Drew Davis, of Kimball, Nebraska, Dorothy A was married December 25, 1919 to George S Cerveny, a farmer of Banner county, and Paul, Carl, Helen, Grace and Walter, all of whom live at home, and Allen who died

in infancy Mr Miller and family belong to the Methodist Episcopal church and were instrumental in its organization and in the building of the parsonage In politics he is a Democrat and for a number of years has been a notary public and a justice of the peace He is one of the representative men of Banner county

ARTHUR G WARNER, who is ably upholding Banner county's reputation for enterprising, thorough going farmers and stockraisers, belongs to an old and representative family here, extended mention of which will be found in this work

Arthur G Warner was born in Banner county, Nebraska, May 20, 1891, and is a son of Isaiah Warner He attended school in Banner and Scottsbluff counties and later took a course in a Grand Island business college For seven years before marriage he lived by himself on a farm he operated, then bought a half section near his home for one dollar an acre Later he made an advantageous trade of this land for a half section near Flowerfield, on which he now resides He has a fine place here and understands the best methods of developing it He is particularly interested in breeding Aberdeen Angus cattle, averaging one hundred head yearly, and Duroc-Jersey hogs, raising sixty head annually

Mr Warner was married September 24, 1913, to Miss Lavina Mitchell, who is a daughter of Wilson and Lila Mitchell, both of whom were born in Illinois Mr Mitchell is a substantial citizen of Banner county, to which he came in October, 1887, and operates a large ranch Mr and Mrs Warner have one son, Martin She is a member of the Methodist Episcopal church Mr Warner belongs to the Farmers Union, of which he is a stockholder He is a Republican in politics but has never served in any public capacity except as school moderator, in which office he faithfully performed the duties incumbent for five years

ELMER E THOMAS, who is well known in Banner county where he has large land and stock interests, makes his home on his ranch and takes an active interest in civic affairs and general development

Elmer E Thomas was born at Ames Nebraska, July 10, 1887, and is a son of Henry and Belle (Zorn) Thomas They came to Banner county in 1887 and Elmer E was born shortly afterward He has an older brother, Roy Thomas, who married Cordelia McCullough, and they reside at Bushnell The parents homesteaded in the neighborhood of Gary and lived there until 1898, removing then to Harrisburg. Later they went to Chicago and still later to St. Louis, Missouri.

Elmer E. Thomas attended school at Harrisburg but has been interested in farming since sixteen years of age, when he rented land near Bushnell and operated it for three years. In 1907 he moved on his brother's homestead and managed the same until 1913, when he came to Bushnell and for a year the brothers were associated in a hardware business. Mr. Thomas returned then to the farm and remained until 1918, when he again took up his residence in Bushnell. He owns one section of land and leases another and has capable tenants. He breeds Shorthorn cattle and Red Jersey hogs, also a few horses, averaging very well.

On August 10, 1914, Mr. Thomas was united in marriage to Miss Mable Wilson, who is a daughter of J. M. Wilson, of Harrisburg, and they have one daughter, Theda. In politics a sound Republican, Mr. Thomas is much interested in political campaigns, but has never accepted any office for himself with the exception of one term as assessor of his precinct.

WILLIAM W. WARNER, who is an enterprising and progressive general farmer of Banner county, belongs to one of the fine old families of this section, extended mention of which will be found in this work. Mr. Warner was born at Brownington, Missouri, November 12, 1887, and is a son of Isaiah Warner, long one of the county's most substantial citizens.

William W. Warner enjoyed excellent educational advantages, first attending the country schools, later the public schools of Gering and still later completing a full business course in a commercial college at Grand Island. He remained on his father's homestead until 1908, in which year he homesteaded on section eighteen, town seventeen, on which he still resides. He has three hundred and twenty acres here, all fine farm land. Mr. Warner is following intelligent methods in his agricultural operations and no farmer in this section of the county makes use of better or more modern farm machinery. ,

Mr. Warner was married July 5, 1911, to Miss Minnie Palm, who is a daughter of John and Helga Palm. The parents of Mrs. Warner were born in Sweden and came to Banner county as pioneers. They now reside at Pine Bluff, Wyoming. Mr. and Mrs. Warner have three children, namely: Bernice, Raymond and Vernon.

In politics Mr. Warner is a Republican in national matters but votes independently on local affairs, his good citizenship leading him at all times to give support to movements that promise to be of permanent benefit to his own community. He has never been anxious for political preferment but has served as school treasurer. He belongs to the Farmers Union and is a stockholder in this organization, and he has been an Odd Fellow for many years. He has the respect and good will of the entire community.

WILLIAM VAN PELT, who is one of Banner county's veteran homesteaders, is well and widely known in this part of Nebraska, where he is a man of large estate. He was born in Highland county, Ohio, August 26, 1855, and is a son of Thomas C. and Nancy (Lucas) Van Pelt. His father was a soldier in the Civil War and died from exposure contracted in the service, in 1863. His mother, who was born in Highland county in 1825, resides in Banner county, venerable in years but not in fact, for she is alert both in body and mind, faithful in church attendance and entertaining in family and social circles. Mrs. Van Pelt homesteaded in Banner county and still owns her land, and is the oldest living homesteader in Banner county.

William Van Pelt was an infant when his parents moved to Iowa and he was educated and reared there, remaining with his mother until he was thirty-eight years of age. When a decision had been made in the family to move to Banner county, Mr. Van Pelt and his sister, Mrs. Johnson, came from Des Moines to Grand Island. There he loaded a car of stock and came with it to Banner county, March 25, 1887, having in the previous fall filed on a homestead at North Platte. He lived there several years, then sold and moved on the place on which he still lives. In 1904 he filed a Kinkaid claim of three hundred and twenty acres, mainly ranch land. He raises high grade White Face cattle, about thirty head yearly, and enough horses for his own use. Mr. Van Pelt has a valuable home property, commodious residence and well kept farm buildings.

On December 27, 1893, Mr. Van Pelt was united in marriage to Miss Blanche Snyder, who is a daughter of Mack and Adelaide Snyder, natives of New Cork. They came to Banner county in 1888 and homesteaded east of Harrisburg. The father of Mrs. Van Pelt died in 1905, but the mother survives and lives in Tennessee. Mr. and Mrs. Van Pelt have two children, namely Effie, who is the wife of Ernest W Pickett, who operates one of Mr Van Pelt's farms, and Roy E, who lives with his parents In politics Mr Van Pelt is a Republican, but the only public office he has ever accepted was that of school treasurer of district number sixteen, in which he served four years Like all members of his family, he is held in high regard in Banner county

F JOHN PALM, whose substantial success as a homesteader in Banner county, came about through good judgment and hard work, now lives comfortably retired at Pine Bluffs, Wyoming, but still retains possession of large estates and heavy stock interests in Banner county

Mr Palm was born in Sweden, July 21, 1857 His parents were Jonas and Anna (Andrewson) Johnson Palm The father was born in Sweden January 29, 1827, and died there February 9, 1883, and the mother, born March 4, 1834, died December 22, 1885 The father was a small farmer and stockraiser Both parents were members of the Swedish Lutheran church Of their children, F John was the eldest, the others being as follows Clara, who is the wife of Carl J Soderstrom, of Greeley, Colorado, Emma, who is the wife of Andrew P Malm, of Greeley, Gust, who lives in Banner county, Augusta S, who is the wife of Peter Lundberg, of Banner county, and Charlie E, who lives at Salt Lake, Utah

F John Palm attended the common schools in Sweden, afterward working as a railroad man for about eight years He was married November 27, 1885 at Gockhom, Sweden to Helga Anderson, whose people were lifelong residents of Sweden In April, 1886, Mr Palm and his wife came to the United States, locating at first in St Paul, Minnesota, where he worked until 1890, moving then to Lyons, Colorado, where he obtained work in a stone quarry In the meanwhile a brother of Mr Palm had become established in Banner county and it was his suggestion that F John and wife should visit him and see the country They set out in 1894 and drove from Lyons to the brother's home in this county The appearance of the country, although wild and a free range for cattle, pleased Mr Palm and he decided to file on a homestead and took a Kinkaid claim of three hundred and twenty acres He now owns eleven hundred and twenty acres of land, all valuable and improved with fences and substantial farm buildings Mr Palm has averaged one hundred head of cattle yearly and many head of Poland China hogs Mr Palm

carried on his farm operations with the assistance of his son until 1915, when he turned his industries over to Charles W, who lives on the homestead and is still further developing the property

The following children were born to Mr and Mrs Palm Minnie, who is the wife of W W Warner, of Banner county, Charles W, who married Hazel Mercer and they reside on the Palm homestead, Esther M, who is the wife of Victor M, of San Francisco, California, Ruth E, who is the wife of George Carlstrom, of Pine Bluffs, Wyoming, Anna S, who lives at Cheyenne, Wyoming, and Mabel M, who resides at home Mr Palm has large financial interests here, being a director and stockholder in the Pine Bluffs State Bank and also owning stock in the Farmers Union at Pine Bluffs He is a Republican voter Mr Palm is well known and is held in esteem for he has always been an honorable, upright business man and reputable citizen

PHILIP R BARKELL, who is a prosperous general farmer and a useful and representative citizen of Banner county, is a native of Wisconsin, born in Grant county, February 7, 1876, and accompanied his parents to Nebraska in the winter of 1888

Richard Barkell, father of Philip R Barkell brought his family to old Cheyenne, now Scottsbluff county, Nebraska, in the above year, having homesteaded at the head of Pumpkin creek in the preceding fall, and, under rather hard conditions the family lived on that place until 1893, when the father relinquished his Scottsbluff land and filed a claim in Banner county He died however, February 19, 1904, before he had proved up, but his son Thomas, extended mention of whom will be found in this work assumed responsibility and cleared the title The mother of the family passed away on April 5, 1912

Philip R Barkell attended school during boyhood and gave his father assistance on the homestead His own home place of one hundred and twenty acres is well improved and his entire three hundred and twenty acres is devoted to general farming

On June 24, 1902, Mr Barkell was united in marriage to Miss Harriet C Leftwich, who is a daughter of James E and Catherine (Nelson) Leftwich, residents of Missouri but natives of Kentucky Mr and Mrs Barkell have children as follows Mary G, Dewitt E, Edna M, Dorothy B, and Roseland In order to afford his children the best of educational advantages, it is Mr Barkell's intention to retire from the farm for a season, moving into one of the intellectual centers where school facilities may be best secured He is a Republican in his political views, but, aside from school offices, has never accepted preferment, serving many years on the school board in school district number nineteen He is one of the solid, dependable men of the county and during the World War proved his patriotism in many ways He was a member of the Council of Defense for the county and was vice chairman for his precinct

GEORGE W LEAFDALE — A family of considerable prominence in Banner county bears the name of Leafdale, and a representative member of the same is George W Leafdale, of section thirty-two, town twelve, an extensive farmer and enterprising young man He has many friends in this neighborhood and his appointment as mail carrier between Kirk and Harrisburg, on July 1, 1919, met with general approval

George W Leafdale was born at Central City, Nebraska, May 12, 1887, the sixth in a family of ten children born to Martin and Cecelia (Munson) Leafdale The father of Mr Leafdale was born in Sweden in 1849 and the mother in 1852 They came to the United States in 1886, and for some time resided at Central City, Nebraska In the fall of 1887 the father homesteaded in Cheyenne county, on the Banner county line, and the family lived on that land until 1909 He moved then into Banner county but still owns almost a section of land in Cheyenne county, and followed ranching for about seven years, in 1916 retiring to Potter, Nebraska Ever since coming to this state Mr Leafdale has been a man of extreme worth In his own country he had previously preached as an ordained minister of the Swedish Mission church, and after coming to America, as opportunity offered, he has traveled over the United States on mission work, and for ten years preached in Cheyenne and Banner counties He assisted in the organization of the flourishing Sunday school at Kirk While living on his homestead he served on the school board in his district, and also served in the office of justice of the peace In addition to George W , the following members of his family survive Anthony M , who lives in Cheyenne county, married Flora Johnson; Anna, who is the wife of Eric Anderson, of Cheyenne county, Emil, of Cheyenne county, who married Ruth Olsen, Selma,

who is the wife of Edward Swanson, of Cheyenne county, Alfred, who lives with his parents, Harry, who is a farmer on his father's land, Mable, who lives at home, and Esther, who is the wife of Carl Osberg, of Cheyenne county

George W Leafdale attended school in Cheyenne and Banner counties, and assisted his father until 1909, when he established himself as a farmer in Banner county. For eight years he rented farm land and carried on general crop production, then bought his present home place, where he carries on extensive farm industries He now owns eight hundred and eighty acres of fine farm land and additionally has a one-third interest in three sections of range land He is a member of the Farmers Union and is a stockholder in the Farmers Elevator Company at Potter

On December 18, 1918, Mr Leafdale was united in marriage to Miss Minnie Larson, who is a daughter of Swen Larson, a prominent settler of Banner county, extended mention of whom will be found in this work Mrs Leafdale, a lady of culture and intellectual attainments, is also a practical business woman   After completing her education, she taught the Lorraine school from 1913 to 1914, holding the sessions in a sod house, the only available structure in the neighborhood Later she taught one term at the Twin Tree school house In 1915 she assumed the duties of county superintendent of schools to which office she had been elected by a flattering vote, and served two terms, retiring from the office January 1, 1919 Ever since reaching womanhood she has been deeply interested in educational work and as her motives have been sincere and her ideas practical, she has accomplished a great deal for the public schools of the county She was instrumental in having the Junior Club work started in the common schools She also brought about that admirable system of education, the Home Demonstration movement Since girlhood she has been active in Sunday school work, and the love of reading good literature that now is a notable and encouraging feature in many homes, may, in part, be attributed to her influence As may be inferred, Mrs. Leafdale found, during the progress of the World War, ample causes to arouse her sympathy and patriotic interest, and she was particularly active and useful in Red Cross work, and in the war savings stamps and Liberty loan drives Personally she is a lady of attractive presence and engaging manner

HORACE W. GURNSEY

EARL F JONES, who is widely and favorably known all over Banner county, was born in this county, May 26, 1894, and is a son of John L and Dora M (Clayton) Jones, natives of Warren county, Illinois The father came from Iowa to Banner county, in 1888, homesteading near Hull He resided there as a farmer and ranchman until 1907, when he retired to Kimball, where he yet resides The mother died in the spring of 1901 Of their six children, Earl F was the fifth in order of birth

Earl F. Jones attended the country schools and lived at home until he was fifteen years of age, after which he resided with his brother Glenn for eight years During the World War Mr Jones served one year in the national army, and when released from military service in the fall of 1918, returned to Banner county He immediately leased the farm of his mother-in-law, Mrs Emily M Larson, and is carrying on farming there From boyhood Mr Jones has been fond of horses and he possesses a certain dominance over them, which combined with physical courage, has enabled him to be very successful in the work of breaking horses and he has the record of breaking sixty-five head of horses in one autumn for Mr Palmberg For some years he was in the employ of a horse buyer and in this line traveled all over the county and believes he knows every road and trail and the most of the people He left behind him many friends on these frequent trips

Mr Jones was married at Denver, Colorado, September 11, 1918, to Miss Josephine Larson, and they have a sturdy little son, Melvin D , who was born June 15, 1919 Mr and Mrs Jones are members of the Methodist Episcopal church In politics he votes independently, and fraternally is connected with the order of Knights of Pythias at Harrisburg On March 1, 1920, he moved to the Levinsky ranch, holding a two years lease

HORACE W GURNSEY, a well-known and popular citizen of Scottsbluff is a representative of the third generation of the Gurnsey family in Nebraska, within whose borders his paternal grandfather established a home in the very early pioneer period The name has been worthily linked with the civic and industrial development and progress of this favored commonwealth, and its honors have been well upheld by him whose name initiates this review

Horace William Gurnsey was born at Vesta Johnson county, Nebraska, October 7, 1870 and is a son of Phineas B and Susan Maria (Hartwell) Gurnsey the former of whom was born in Illinois, on the 25th of January, 1846, and the latter of whom was born in Ohio November 8, 1848, their marriage having been solemnized May 3, 1868, at Vesta, Nebraska — the year following that of the admission of the state to the Union Mrs Gurnsey passed to eternal rest March 6, 1896, and her husband still survives her, his home being at Scottsbluff The grandfather of the subject of this sketch came from Illinois and established his home in Nebraska in the territorial days He took up a tract of land near Sterling, Johnson county, and on this pioneer homestead he continued to reside until his death, at the age of fifty years Phineas B Gurney was reared and educated in Illinois, where he continued to be associated with his father in farm enterprise until the outbreak of the Civil War, when he tendered his aid in defense of the Union by enlisting in Company G, Seventeenth Illinois Volunteer Cavalry, with which he participated in many engagements and lived up to the full tension of the great conflict After the war he joined his parents in Nebraska and his name will ever be worthily associated with the pioneer history of the state

Horace W Gurnsey was afforded the advantages of the public schools of Lewiston and Tecumseh Nebraska, and as a mere boy he gained practical experience in connection with farm work In 1888 he became identified with railroad work and for three years — 1891-1894 — he was employed as a railroad bridge carpenter He was then transferred to the track department, and this connection continued about two years In 1898 he served as chief of police at Ulysses, Butler county, and on the 23d of May, 1899, he removed to Alliance, Box Butte county, out of which place he had charge of the west section of what is now the Chicago, Burlington & Quincy Railroad While he was thus engaged the county court house was removed on two flat cars from Marsland to Alliance September 10 1899 Mr Gurnsey was transferred to Whitman, where he had charge of the railroad section and coal sheds He there remained until 1900, when he was assigned supervision of a section out from Scottsbluff to which place he removed, and he retained this incumbency until the autumn of 1906 For the ensuing three years he had charge of the alfalfa meal mill at Scottsbluff, and the following three years found him here engaged in concrete construction work, as a contractor Mr Gurnsey became well known as a practical and successful apiarist, he having taken up bee culture in 1904, in connection with his other activities, and having had at one time 140 stands of hives He served three years — 1912-1914 — as chief of police of Scottsbluff, and his administration passed on record as one of the best in the history of this city At the present time Mr Gurnsey is giving his attention principally to the Great Western Sugar Co as engineer on generators In national affairs he gives his support to the Republican party, but in local politics he is not constrained by strict partisan lines He is affiliated with Scottsbluff lodge No 261, I O O F, of which he served three years as secretary

On the 31st of December, 1891, Mr Gurnsey wedded Miss Emeline Lockner, who was reared and educated at Bellwood, Butler county She was an earnest member of the Baptist church and was affiliated with the Daughters of Rebekah, the Royal Neighbors, and the Degree of Honor She passed to the life eternal on the 16th of April, 1914 Concerning the children of this union brief record is here offered Lloyd William, born October 30, 1892, was graduated in the Scottsbluff high school as a member of the class of 1913, and he now resides at Scottsbluff, being in the train service of the Chicago, Burlington & Quincy Railroad The maiden name of his wife was Edna Carpenter When the nation became involved in the great World War, Lloyd W Gurnsey enlisted, in April, 1917, as a member of Company G, Fourth Regiment of the Nebraska National Guard, and with his command he was sent from Fort Crook to Deming, New Mexico, where he entered the great training camp August 17, 1917, he was commissioned second lieutenant at Camp Taylor, Kentucky, and he received his honorable discharge December 10, 1918, about one month after the signing of the historic armistice Roxie Rachel, the second of the children of the subject of this review, was born August 1, 1894, and died July 28, 1900 George, born August 10, 1897, died December 19, 1913 Horace Harold, born July 12, 1898, was educated in the Scottsbluff schools and at the age of seventeen years he volunteered for service in the national army He was sent to the military training camp at Deming, New Mexico, and there he contracted pneumonia, after which he passed twelve months in the military hospital at Fort Bayard, his discharge having been received October 20, 1918, shortly after the close of the war Alice Olive, born February 26, 1901, is the wife of George Hays of Silverton, Colorado, their marriage having occurred June 17 1918 Jesse Ray, born March 6 1903 and Marian Gladys, born October 20, 1913, remain at the paternal home

The second marriage of Mr Gurnsey was solemnized March 28, 1919, when he wedded Miss Virginia L Dumphy, at Alliance, this state Mrs Gurnsey was born August 17, 1895, at Kearney, Nebraska, and completed her educational discipline in the Scottsbluff high school She is the popular chatelain of the attractive family home in Scottsbluff

LARS J HENDRIKSON — There were not so many permanent settlers in Banner county in 1887, when Lars J Hendrikson came here, as now, but nevertheless there were some fine people here and among these were natives of his own land Mr Hendrikson still resides on the homestead he secured thirty-two years ago, and long has been one of Banner county's foremost citizens

Lars J Hendrikson was born in Sweden, September 15, 1865, a son of Lars and Christina (Olson) Hendrikson, natives of Sweden The father was a small farmer in his native land In 1880 he came to the United States and remained two years, then went back to Sweden, but in 1884 returned to this country accompanied by his family, and settled in Kansas From there, in 1887, he came to Banner county and homesteaded in the eastern part of the county, and resided on the same until his death, September 4, 1911, the mother having passed away April 6, 1903 Considering that the father was a man advanced in years when he came here, he left an ample estate He identified himself with the Republican party but was never active in politics Both parents were members of the Baptist church Of the eleven children, the following are living Ida, who is the wife of Lewis Peterson, of Denver, Colorado, Lars J, who is of Banner county, August, who lives at Courtland, Kansas, Betty, who is the wife of Gust Anderson, of Pueblo, Colorado; Lottie, who is the wife of Frank Peterson, of Banner county, and Emma, who lives on the home place

Lars J Hendrikson obtained his schooling in his native land He came to the United States in 1884 and spent two years at Scandia, Kansas, but in 1887 he decided to locate permanently in Banner county, Nebraska, and secured a homestead adjoining that of his father Since then he has acquired much additional land and has become financially identified with a number of flourishing business enterprises His home ranch includes fourteen hundred and forty acres and he also owns a three-quarter section near Kirk He is a stockholder in the Farmers Elevator Company at Potter and

S. N. Larson and Family

also Melbeta, a stockholder in the Western Nebraska Telephone Company, and is also a stockholder in the Higgins Packing plant at Omaha He has always been a hard worker and a careful business man

On January 4, 1890, Mr Hendrikson was married to Miss Louise Rasmussen, a daughter of Lewis and Catherine Rasmussen The parents of Mrs Hendrikson were early homesteaders in Kansas, settling in Republic county in 1867, before the hostile Indians had been expelled to the reservations, or the rough element following the close of the Civil War, had been brought to a sense of law and order Seven children were born to Mr and Mrs Hendrikson, namely Henry, who resides at home, Ida, who is the wife of Alec Trostrum, of Denver, Colorado, Alfred, who resides at home; Agnes, who is the wife of Paul Hendrikson, living near Potter, Nebraska, and Isaac, Mabel and Emily, all of whom reside at home With the help of his sons, Mr Hendrikson carries on his large farming operations very profitably At one time he was interested in the Populist movement and served on that political ticket as county commissioner from 1892 until 1898, but since then has voted the Republican ticket He has filled numerous local offices and for many years has been road overseer and assessor of his precinct, and at all times displays such good judgment and marked public spirit, that his fellow citizens have confidence in his opinions

SWEN N LARSON — To mention the name of the late Swen N. Larson in Banner county, is to hear of one of the best known and most highly esteemed men who ever lived here He was a man of real worth in every relation of life Not only did he provide well through his industry for his own family, but took an interest in the welfare of others who had no claim on him except humanity, and was never so happy as when rendering assistance of some kind He was universally respected and trusted by his fellow citizens, and was greatly beloved by those who knew him best

Swen N Larson was born in Sweden, January 4, 1854 His father's name was Nicholas and both parents lived and died in Sweden, Swen N being the only one of the four children to come to the United States He had some schooling there, but his youth was one of hard work spent in assisting his father on the little home farm and in the charcoal pits, the burning of charcoal being the latter's main business In the meanwhile Swen learned the carpenter trade and made some money as a

salesman of timber By the time he was twenty-four years old he had saved enough to warrant his emigrating to the United States The Mattsons, old neighbors of the Larsons, in Sweden, had gone to America and were well established at Galesburg, Illinois, and this fact led to Mr Larson making his way also to Galesburg, where he remained for the next five years, finding employment on neighboring farms and in selling cattle

On September 4, 1883, Mr Larson was married to Miss Emily Mattson, who survives him Her parents, Benjamin and Pernelha (Anderson) Mattson, came from Sweden to Knox county, Illinois, at an early time of settlement there, making the voyage to the United States in a sailing vessel that was on the ocean for thirteen weeks They found no railroads completed to Galesburg, which was but a little settlement Some years afterward Mr Mattson removed to Knoxville, Illinois, where he operated a brick kiln and farmed Both parents of Mrs Larson died there, but one day apart, and their burial was in the same grave Of their nine children four are living At one time a brother of Mrs Larson filed on a homestead in Banner county but later relinquished it and returned to Illinois To Mr and Mrs Larson the following children were born Burt W, who was born June 5, 1884, married Edith L Neely, and they live at Kimball, Nebraska, Annette, who was born March 25, 1886, is the wife of George A Jones of Gering, Scottsbluff county, Ralph E, who was born September 29, 1888, married Margaret A Coleman, and they live at Kimball, Minnie, who was born September 20, 1892, is the wife of George W Leafdale, of Banner county, Josephine who was born June 29, 1895 is the wife of Earl F Jones, of Banner county, Josephine, who was born August 9, 1898 resides with his mother, and Pearl D, who was born June 7 1905, also resides at home

Following their marriage Mr and Mrs Larson moved to Kansas He engaged in farming there for five years, during which time the delicate health of their little son gave them uneasiness and when the family physician suggested moving into a different climate, Nebraska for instance they determined to take his advice In the fall of 1887 Mr Larson homesteaded in Banner county, Mrs Larson is still residing on that land, the southeast quarter of section six In early April, 1888, the family reached here, Mrs Larson and the children coming to Potter by railroad and stopping with the family of Alfred Olson until Mr Larson arrived with the livestock and household goods He was three weeks making the trip from Kansas with two covered wagons and teams, a saddle poney and a few cattle Mrs Larson says that in the drive from Potter to the Olson homestead, a distance of twenty-six miles, she saw but one house, that standing on the old Bracken place, now the property of Grant Brady

In speaking of the great blizzard that swept over this section shortly after the family came here, Mrs Larson gives many interesting details of how the pioneers had to meet such disasters, and in speaking of Mr Larson's heroic efforts, she mentions his going as far as twelve miles to drive home his bewildered cattle, and carried the young calves on his back, bringing them into the dugout to warm them by the fire In the following summer the family moved into a sod house, Mrs Larson losing all interest in the dugout after finding a mouse and a rattlesnake there, and later they had a comfortable log house constructed with a carpenter's skill by Mr Larson At first they hauled water either four miles from the south or the same distance from the north, but Mr Larson had about the first satisfactory well in this section He practically laid out all the roads here, for when he came there were no landmarks by which a driver could shape his course On his first trip to Kimball and other points he carried stakes with him and like other trailmakers before and since in a new country, thus "blazed" the path In early days he hauled his wheat to Sidney, a distance of forty-five miles and accepted twenty-two cents a bushel for it. He lived to see wonderful changes take place and to be the owner of over twenty-five hundred acres of fine land, all well improved This property is now known as the Little L ranch, and is managed and operated by Mrs Larson with the help of her son and her son-in-law, Earl F Jones It is a profitable enterprise both as to stock and grain In 1919 the five hundred acres devoted to wheat growing yielded from eighteen to twenty-five bushels an acre

In many ways Mr Larson was a prominent man in Banner county, not because of any claims he made himself, but on account of his upright, sterling character that marked all his dealings with his fellow men For twenty-eight years he served as postmaster at Heath, appointed under a Republican administration but never disturbed by political changes, and after his death, on December 20, 1917, was succeeded by his daughter Minnie, now Mrs Leafdale, who served until the office was discontinued a year later He operated a general store at Heath for many years also, and during the most of this time served as treasurer of the Heath school district While the family lived in Kansas, they were united with the Methodist Episcopal church, but Mr Larson was a man who needed no church creed to impel him to do deeds of justice and charity The only fraternal organization that he ever joined was the Modern Woodmen of America

RALPH E DUBBS, who is numbered with the substantial farmers and ranchmen of Banner county, lives on the place he homesteaded in April, 1888, to which he has from time to time added, until now he has one of the largest ranches in this part of the country. Mr Dubbs has made his own way in the world and is an example to which attention may be called, of the sure reward that follows persistent effort in a country where the work of one's hands, in itself, is honorable

Ralph E Dubbs was born in Columbiana county, Ohio, December 12, 1886 His parents were William and Mary (Coy) Dubbs, natives of Ohio Of their fourteen children, eleven are living, Ralph E. being the only one in Banner county The father was a farmer and stockraiser in Ohio until 1870, when he removed with his family to Hall county, Nebraska, where he died in 1909, having survived the mother for three years She was a faithful member of the Methodist Episcopal church In politics he was a Republican and after coming to Hall county served two terms as county supervisor

Only four years old when his parents brought him to Nebraska, Mr Dubbs has never had any divided state loyalty He was reared, educated, given business opportunities and married here, and it is in Nebraska that his interests are centered He remained at home assisting his father until twenty-one years old, then started out for himself and accepted work of all kinds that was not dishonorable, thereby providing well for his own necessities although because of small wages paid in those days, it was difficult to accumulate much capital He helped to build a number of log cabins after coming to Banner county and dug numerous wells, the deepest one of which he has recorded being one hundred and eighty-four feet in depth and twelve feet in circumference In April, 1888, he filed on his homestead in Banner county and then returned to Hall county to take care of a ninety acre field of corn, thirty-six hundred bushels of the yield being his share, which he sold for fourteen cents a bushel

Mr Dubbs returned then to Banner county, with a cash capital of fifty dollars, a team of mules and a buckskin mare In the first year on his homestead he broke sixty acres of ground and sowed wheat in the following spring, harvesting twelve hundred bushels, that brought sixty cents a bushel, and that crop was the nucleus of his present fortune In later years wheat fell to twenty-two cents, butter brought only ten and even six cents a pound, and he has hauled a load of wood many miles and exchanged it for one sack of flour During his first year here he planted sod corn and traded a part of his crop to the late Swen Larson for a hog In later years as he had the capital, Mr Dubbs added to his holdings and now has deeds for nineteen hundred and twenty acres, and leases three hundred and twenty acres of school land He breeds White Face and Polled Durham cattle, averaging fifty head yearly, fifty head of horses and many head of Duroc-Jersey hogs especially in recent years In addition to his well conducted farm and stock industries, Mr Dubbs has been a pioneer orchardist and small fruit grower He takes a justifiable pride in his beautiful fruit-bearing trees, cherries, plums and apples all doing well under his careful cultivation

On March 3, 1895, Mr Dubbs was united in marriage to Miss Mary Leftwich, who is a daughter of James E and Catherine (Nelson) Leftwich, and a sister of Mrs Philip R Barkell, of Banner county The parents of Mrs Dubbs were natives of Kentucky and later were residents of Missouri The mother died when she was young and when her father was obliged to leave home for a year to work at another point, he left little Mary with an elderly lady of the neighborhood This lady not only cared kindly for the child but provided her with clothes and school books while her father was away, and put here under such a debt of gratitude that Miss Leftwich felt she could never discharge it When the opportunity came some years later she did not hesitate to discharge this debt At that time she was employed by a firm near Lake Superior and was doing very well When she received a letter from the elderly lady mentioned, asking if she would not join the lady's lonely and homesick daughter, away out in Banner county, Miss Leftwich only waited long enough to make hurried preparation before she was on her way It was thus that she met Mr Dubbs and they were subsequent-

ly married They have the following children Harry E, Harriet C, Florence I, James W, Philip R, and Margaret E They reside at home, an intelligent, happy united family Mr Dubs is a Republican in politics but is no seeker for public office He belongs to the Modern Woodmen of America, attending lodge at Kimball

GEORGE N BENNETT, who is one of Banner county's solid, reliable citizens and good farmers, might also be named as the pioneer orchardist, for he was one of the first to take a practical interest in fruit growing and apply scientific methods for the preservation of his trees

George N Bennett was born in Hardin county, Ohio, March 5, 1861 His parents were John A and Mary (Roberts) Bennett, the former of whom was born in Logan county, Ohio, in 1845, and the latter in Hardin county in 1847 They had four children, namely George N, who grew up in Ohio, James E, who died when eight years old, John E, whose death occurred when traveling, on a railroad train near Alliance, Nebraska, and Jesse, who lives on the old home place in Ohio The father was a farmer in Ohio until 1884, when he came to Colfax, Nebraska, but not being satisfied with conditions there, in a few months returned to his old place in Ohio The mother died in 1893 and the father in 1912

Equipped with a country school education and excellent home training, when twenty-one years old George N Bennett started out for himself As a farmer he worked near Council Bluffs, Iowa, Omaha and Schyler, Nebraska, reaching the latter place on August 16, 1883 After working there for a time he went to Fremont, Nebraska, and ranged seven hundred cattle for their owners, then was employed by Frank Isabelle In the following year he started to farm for himself, but lost his crops through hailstorms, and somewhat discouraged sold his team for three hundred dollars — that is such was the price agreed upon, but after many years Mr Bennett is still waiting the payment of two hundred and fifty dollars After that Mr Bennett worked for others for four years In Colfax county at that time he could have bought land at from ten to twenty-five dollars an acre, but he did not take advantage of it and the price soon went higher In the meanwhile he married and came to Banner county Here he bought out John Trowbridge, which transaction placed him in debt and it was a number of years before Mr. Bennett could extricate himself and find the road to prosperity. Since then he has never had a complete crop failure but has had a few lossess of cattle from lightning and blizzards, but better than that, in his opinion, is the fact that in a residence of twenty-two years in Banner county, a physician has not been called in more than a dozen times to prescribe for the family.

In the fall of 1889, Mr. Bennett was married to Mrs. Ada E. Thinehardt, a widow and a daughter of William and Eveline (Stevens) Stevens, of Schyler, Nebraska. The father of Mrs. Bennett died in 1890 but the mother survives and resides with Mr. and Mrs. Bennett. The latter have three children, namely: Perlie V., who is the wife of John E. Johnson, of Banner county; Ethel, who is the wife of Alfred Sterner, of Madison, Nebraska; and Mary, who is the wife of Ralph Randall, lives on her father's farm. For three years after their marriage, Mr. and Mrs. Bennett resided on the Stevens homestead. Then he bought a hay baling outfit and worked with that for five years. On February 16, 1897, Mr. Bennett came to Banner county and bought one hundred and sixty acres for three hundred dollars. This property was well improved for the times, having a good well, comfortable· residence and sufficient fencing, and the family has resided on the place ever since. Ten years later Mr. Bennett bought another tract of one hundred and sixty acres, cornering on the first tract, for which he paid sixteen hundred dollars, and nine years afterward purchased three hundred and twenty acres, for which he paid twenty-eight hundred and eighty dollars. He carries on mixed farming, raises hogs for his own use and turns off quite a few head of cattle yearly.

Mr. Bennett is a lover of both shade and fruit trees. He has one of the best orchards of ten years standing in Banner county and succeeds where others fail, in growing fine cherries, apples, pears and plums, his cherry trees yielding fifty bushels of fruit last season, all being sold at the farm. His varieties of fruit are the best and his apples compare favorably with the great apple products of the Northwest. He protects his orchard with a surrounding grove of other trees. Mr. Bennett is intelligently interested in many other lines. He has a collection of Indian relics that would be creditable in a city museum, and scientists would be very apt to envy him the possession of the jaw bone and thigh of a mastadon, unearthed in Banner county. Mr. Bennett and his family belong to the United Brethren

church. In national affairs he votes with the Republican party, but in local issues uses his own excellent judgment in political action.

FRANK PETERSON, who is numbered with Banner county's substantial farmers and stockmen, was born in Sweden, August 1, 1864. He is a son of Nels and Bertha (Nelson) Peterson, who came to the United States in 1886. They lived two years in Iowa, then came to Banner county, Nebraska, in 1888, homesteading on section thirty-four, near Heath. The mother died July 11, 1891, when the father came to the home of his son Frank and resided with him until death, which occurred January 14, 1911. Both parents were members of the Lutheran church. Of their six children Frank was the fourth in order of birth, the others being as follows: Helena, who is the wife of Claus Peterson, of Banner county; Alma, who is the widow of John Carlson, lives in California; Claus, who lives at Greeley, Colorado, married Hilda Johnson; Victor, who lives in Banner county, married Emma Swanson; and Mary, who is the wife of Theodore Carlson, of Minatare, Nebraska.

Frank Peterson obtained his education in the common schools of his native land. He had some military training there also, spending one year in the Swedish army and afterward giving seventeen days of service each year. He worked on his father's farm in Sweden and in 1886 accompanied him to the United States. When he came to Banner county in 1888, he homesteaded and still owns this land, to which he has added much land, now owning sixteen hundred acres, two hundred and fifty acres being devoted to crop raising, the rest being fine pasture land. He breeds White Face cattle, about one hundred and forty head annually, raises also about twenty-five head of hogs and horses and poultry for home use. While Mr. Peterson is now financially independent, with well improved and well stocked farm, it was not always so, and his success may be attributed to his good judgment and continued industry. His first expense after locating on his homestead was the necessary purchase of a team of oxen and a breaking plow, the cost being seventy-five dollars. Considering the farm implements that he owned, his first crop of wheat was satisfactory and Mr. Peterson believes that just as bounteous a yield of grain could have been secured then if the pioneer farmers has such modern farm machinery as is used at present. Money was very scarce and prices for farm products were low, and after his father came to live with him, Mr Peterson left the farm in his care and went to Colorado and other points to work for wages Until he was able to have a well dug on his land, about three years after settlement, he had to haul water for all purposes a distance of four miles Like other homesteaders he passed through many hardships and has had losses of crops and cattle through the severity of storms In recalling the blizzards that swept over the county at different times, he mentions the unprecedent fury of the storm of April, 1912, in which his life was endangered With his family he had gone to Kimball one day with the intention of returning to the farm on the day following Although a blizzard was on the way attempted to make the trip with the large conveyance, but found the going was too heavy for that vehicle, returned to town and started again in a single-horse buggy He relates that if the horse had not known its way he would never have reached home on account of the cold and blinding snow

On March 5, 1902, Mr Peterson was united in marriage to Miss Lottie Hendrikson, a daughter of Lars J Hendrikson, extended mention of whom will be found in this work Mr and Mrs Peterson have four children, namely Harold, born January 27, 1903, Dorothy, born June 21, 1905, Wallace, born May 9, 1910, and Evelyn, born October 18, 1913 Although not church members, Mr Peterson and his family are regular church attendants, Mr Peterson arguing that no particular creed is necessary if the proper religious spirit is maintained He is a Republican in politics but has never sought any public office It is the pleasant custom of the family to spend the fall and winter at Kimball, where they have a comfortable residence, thereby giving the children better educational advantages than they could have in the country Additionally Mr Peterson owns three desirable building lots in this city, owns stock in the Higgins Packing plant at Omaha, and also is a stockholder in the western Nebraska Telephone Company Mr Peterson is highly respected by all who know him

CHARLES O JOHNSON, who came to Banner county and filed a pre-emption claim in the spring of 1887, has never regretted that step although for some years while acquiring additional land and developing it, he and family had to bear many hardships He was born in Sweden, April 12, 1850, one of eight children born to Johannes and Elizabeth (Swanson) Johnson They spent their lives in Sweden, where the father was a general farmer They were members of the Swedish Lutheran church

Charles O Johnson and a brother, Swen D Johnson, who lives near Kirk, Nebraska are the only surviving members of their parents' family Charles obtained a common school education, and according to the law of Sweden, served two years in a military training camp and during the following two years gave seventeen days of annual service He worked as a farmer before coming to the United States, in May, 1884, and in Kansas for three years afterward In May, 1887, he filed on a pre-emption in Banner county, and in 1890 secured the homestead on which he yet lives, situated on section twelve, range fifty-four, town seventeen Mr Johnson now owns eight hundred acres of fine range land and he raises about twenty head of White Face cattle annually for market and an abundance of other stock for his own use His ranch is situated near Sheep canyon and is well improved

When Mr Johnson entered Banner county, his covered wagon had been drawn over the rough country from Kansas by an old team of horses that accommodated their pace to the slow movements of the family cow, the latter helping to solve the food problem for Mr Johnson's entire cash capital was one dollar and fifty cents The first house was a dugout, with a sod roof, like those of the most of their neighbors They met with many discouragements and lacked many of the things that the present generation would consider essentials of life, but they were more fortunate than the majority of the settlers in the matter of water, for there was a spring on their land

Mr Johnson was married in Sweden, in March 1877 to Ida S Carlson, who was born in Sweden in 1854, and died in Nebraska, February 7, 1909 Her parents were Carl Samuelson and Mary E (Munson) Carson, who passed their lives in Sweden Mr and Mrs Johnson had six children born to them, as follows Charles, who lives on the homestead Oscar who is a highly educated man, is a professor in the college at Wahoo Nebraska, Frank who resides northeast of Minatare Nebraska, and Hannah, Joseph and Clara, all of whom reside with their father He is a member of the Lutheran church He votes with the Republican party and believes in the justice of its principles, but has never accepted a public office He has always been a good citizen of the county and has lent his influence to the improvement of the public highways and the establishment of churches and schools.

CYRUS W. RIDER, who is one of Banner county's most respected retired citizens, came very early to this section, in those early days bearing with his fellow pioneers as bravely as possible, the many vicissitudes that were often disheartening indeed, lending a helping hand whenever he could and never losing faith in the ultimate superiority of this beautiful part of Nebraska. Mr. Rider comes very near to going down in history as the earliest homesteader, having filed, with his brother Clinton O. Rider, in the first week in October, 1885, second only to the Cross brothers, who filed in September. There had been squatters on the land before this, but these were all permanent settlers.

Cyrus W. Rider was born in Geauga county, Ohio, September 20, 1843, a son of William S. and Martha J. (McElroy) Rider. The father was born in 1814, at Poultney, Vermont, and the mother in Geauga county, Ohio, in 1816. Her death occurred in 1905. The father was a farmer and from 1865 to 1894 lived in Iowa. He was active in the Republican party and for many years was a justice of the peace. Of the family of six children four are living, but Cyrus W. is the only one in Nebraska. In 1855 he accompanied his parents to Wisconsin, where he had winter school advantages, and assisted his father until he enlisted for service in the Civil War, entering Company D, Third Wisconsin cavalry, serving under General Steele in the Department of Missouri. He was a brave and cheerful soldier and escaped all serious mishaps of military life. From Iowa he came to Banner county to live and in 1887 homesteaded near Gabe Rock, west of Harrisburg, and lived on that property until 1905, then sold and bought land near Kirk Post Office. Recently he sold his farm and since then has resided with his youngest daughter, who is the wife of Clinton Trowbridge.

Mr. Rider was married April 6, 1877 to Miss Laura A. Clendening, a daughter of Hiram and Cynthia (Miller) Clendening, of Ohio. Mrs. Rider died November 12, 1886, survived by three children, namely: Alice, who is the wife of Wilbur E. Harmon, of Portland, Oregon; Charles E., of Freeport, Nebraska, who married Frances Turner; and Erma, who is the wife of Clinton Trowbridge, of Banner county. In politics Mr. Rider is a Republican.

MERVIN SNYDER, who has long been a representative citizen of Banner county, accompanied his people here when fourteen years old and, with the exception of seven months spent in Tennessee, has been a continuous resident of this county. He has been a foremost citizen in its development and has been identified with many of the worthy enterprises that have made Banner county what it is today.

Mervin Snyder was born in Scott county, Iowa, May 27, 1872. His parents were Melcoir and Adelaide (Labarr) Snyder, the former of whom was born in Pennsylvania in 1837, and the latter in New York state, March 22, 1841. The father spent sixteen years in Iowa, working at his trade of blacksmith, first in Benton and later in Hamilton county, and from the latter went to Banner county and filed on a homestead situated two and one-half miles southeast of Harrisburg, on July 6, 1886, settling on his land November 27, 1886. This land was still wild prairie and for many years after coming here, water had to be hauled several miles for all purposes. On account of Mr. Snyder's skill as a blacksmith and the urgent need of his services at that time C. A. Schooley deeded him a lot in Harrisburg, on which Mr. Snyder built a blacksmith shop, the first one in the place, which he operated for six years. After that he lived on the farm for two years, then moved to Tennessee and continued farming there until his death in November, 1907. He was a Democrat in politics and both he and wife were members of the Methodist Episcopal church. Of their fourteen children, Mervin was the seventh in order of birth, the others being as follows: Susan, who is the wife of William Pyle, of Kimball, Nebraska; Rosel, of Hamilton county, Iowa, who married Ida B. Stover; Helen, who is the wife of Martin Stover, of Mead county, South Dakota; Emma, who is the wife of William Scoville, of Banner county; Belle, who is the wife of Hans Gunderson, of Dix, Nebraska; Kate, who is the wife of Peter Clauson, now a resident of Odell, Nebraska, who once owned the land on which Harrisburg stands; Blanche, who is the wife of William Van Pelt; John, who is a farmer in Banner county, married Blanche Hammocks; William, who lives at Stanton, Haywood county, Tennessee, married Mollie Patted; and Amanda, who was the wife of Amos Patted, died in the spring of 1908; Bessie, who is the wife of Joseph Patted, of Stanton, Tennessee; Nettie, who is the wife of Robert Perry, of Stanton; and Ivy, who

Adolf Goos

is the wife of Myron Bants, of Odell, Nebraska

Mervin Snyder obtained his education in the common schools in Iowa He accompanied his parents to Banner county and when twenty-one years old filed on a homestead, but when his father moved to Tennessee he sold it to William Van Pelt and accompanied his parents to the new home After remaining in Tennessee for seven months he returned to Nebraska and shortly afterward bought his present home farm and has added to the same until he now owns eleven hundred and twenty acres of fine land He devotes six hundred acres to crop raising, and breeds Hereford cattle and Percheron horses When he came here Mr Snyder was faced with the necessity of immediately making improvements and from small beginnings he has developed one of the best improved farms in the county

On March 25, 1897, Mr Snyder was united in marriage to Miss Mary Koenig who is a daughter of Jacob and Anna (Fisher) Koenig, extended mention on other pages of this work Mr and Mrs Snyder have children as follows Alvin M, Pinkie A, James J, Glenn M, Arthur C, Bessie L, Ruby M, Herbert G and Susie A,

In politics Mr Snyder is a Democrat He holds no public office at present but has served efficiently in numerous positions at different times being road overseer for a long period and school moderator for twenty-two years He was influential in having the post office re-established at Harrisburg He owns stock in the Western Nebraska Telephone Company, of which he has been a director ten years and was president for three years In recalling the changes that have been brought about in this section of Banner county, Mr Snyder remembers when he hauled wood to Kimball and exchanged the same for flour and other provisions, and hauled his wheat to Gunderson's mill, an old water-power mill where grinding was done with no exchange of money He remembers the occasion of the first county fair It was held in the unfinished courthouse at Harrisburg and people brought interesting exhibits of all kinds his father showing an old heirloom in a seven foot clock, which was wound up by chains, this being a curiosity even then Mr Snyder participated in some of the sports, taking part in the mule race which was one of the enjoyable features A large crowd attended, visitors coming to Harrisburg from Gering, Kimball and towns still farther away

ADOLF F GOOS — More than thirty years ago when Scottsbluff county was still a part of Cheyenne county, Mr Goos here established his residence A young man of energy and determination, he was well equipped for meeting the labors that fell to the lot of the pioneer farmers in this section of the state, and the passing years brought to him a generous measure of prosperity, the while he gave his liberal support to measures and undertakings that promoted the social and material advancement of the community He developed a valuable landed estate and is today one of the substantial and popular citizens of his adopted country

Mr Goos was born in the province of Schleswig-Holstein Germany on the 27th of October 1861, and there he was reared and educated his youthful experience giving him close association with earnest toil and endeavor, for which he has retained the deepest respect In 1883 shortly after reaching his legal majority Mr Goos severed the home ties and set forth to seek his fortunes in America He arrived at Lincoln, Nebraska on the 14th of March of that year, and thence he made his way to Syracuse Otoe county in which part of the state he was employed at farm work for three years In the spring of 1886 he made his way westward to Cheyenne county the trip having been made overland with team and wagon and he became a resident of that part of the county that is now included in Scottsbluff county On the 18th of May of the same year he filed entry on a homestead — the southwest quarter of section 14, township 22 range 55 besides which he later took up a tree claim, to both of which properties he duly perfected his title With characteristic vigor he began the work of developing and improving his land and abundant success eventually rewarded his efforts He has retained possession of both of his claims and has developed the tract into one of the well improved and valuable farm properties of the county He was one of the early advocates and supporters of irrigation enterprise in the county assisted in the organization of Enterprise ditch in the construction of which waterway he was associated and he held sixteen shares of the stock besides being a director of the original organization For his share he received bonds at the time when the ditch was transferred to the district Mr Goos has shown loyal interest in community affairs, is a Democrat in politics and is a communicant of the Lutheran church

ROLLA W ALUMBAUGH who is one of the enterprising, progressive and successful farmers of Banner county operating a large body of land in association with Arthur J. Trowbridge, is a native of Kansas, born in Crawford county, October 27, 1883.

The parents of Mr. Alumbaugh were John W. and Rebecca (Baysinger) Alumbaugh, the former of whom was born at Bowling Green, Indiana, and died in California, November 24, 1905, and the latter in Illinois. Her death occurred at San Diego, California. The father engaged in farming and stock raising in Kansas and Missouri until 1887, then drove across the country and brought with him from Crawford county, Kansas, to Banner county, Nebraska, his household goods and thirty head of cattle. He homesteaded near Harrisburg and continued on his land there until 1905, when he went to California, where he died soon afterward. He was a Republican in politics and both he and wife were members of the Methodist Episcopal church. They had the following children: James, who lives in California, married Blanche Price; Elmer and John, both of whom live in California; Rolla A., who belongs to Banner county; Myrtle, who lives in California, is the wife of Elmo Carpenter; and Pearl, whose home is in Oregon, is the wife of Barton Irwin.

Rolla W. Alumbaugh attended the Twin Tree district school and remained at home assisting his father until he was twenty years old. In 1908 he homesteaded where he now lives, a Kinkaid claim, and in partnership with Mr. Trowbridge operates nine hundred acres, breeding White Face cattle, turning off about fifty head annually. Farm operations are carried on here according to modern methods and with modern farm machinery. Threshing is done on the farm with the use of wheat headers and farm tractors and the entire year round this is a busy place.

Mr. Alumbaugh was married August 3, 1918, to Miss May Timm, who was born in Iowa, a daughter of Claus and Mary (Hoeck) Timm, the latter of whom survives and lives in Banner county. Mr. Alumbaugh is serving as road overseer and also is treasurer of school district number one. Politically he has always been identified with the Republican party.

ARTHUR J. TROWBRIDGE, who is a prosperous farmer of Banner county, belongs to one of the old pioneer families that have been well and favorably known in other counties as well as Banner. He was born in Berrien county, Michigan, May 2, 1872 and is the only survivor of four children born to John and Emma

(De Long) Trowbridge. The father was born in Portage county, Ohio, and the mother in Berrien county, Michigan, is now Mrs. Cochrane living in Michigan. The father is living at Minatare, Nebraska.

In 1881 Arthur J. Trowbridge accompanied his parents from Michigan to Colfax county, Nebraska, and during the seven years they lived there attended the public schools. In 1889 his father homesteaded in Banner county on sections 6-17-54, near Heath, and resided there until 1897, then sold to George N. Bennett. During the next two years the family home was near Lone Pine Springs and then the parents of Mr. Trowbridge retired to Minatare. In 1888 Arthur J. Trowbridge filed on a Kinkaid claim of a full section of land and still owns this property. Since he has been associated with Rolla W. Alumbaugh in an extensive farming enterprise that is proving very profitable. Both partners are held in great esteem in Banner county, being good business men and trustworthy citizens. Mr. Trowbridge belongs to the order of Knights of Pythias.

STEPHEN W. TROWBRIDGE, who was born in Berien county, Michigan, November 1, 1857 has been a resident of Banner county since 1888, and now lives in his comfortable residence at Kimball. At times he has owned a number of tracts of land in the county but in recent years has sold the greater part of his property.

The parents of Mr. Trowbridge were Henry and Loretta (Hanchett) Trowbridge, the former of whom was born in Connecticut and the latter in Vermont. They were married in Ohio and from there moved to Michigan in 1850, and the father died there in 1881. The mother then came to Nebraska and resided with her son Stephen W. until her death, in the fall of 1896. Of the eight children in the family but two survive, John and Stephen W., the former being a resident of Minatare, Nebraska.

In boyhood Mr. Trowbridge attended school until he was fourteen years old, then began to take care of himself and during the rest of his active life engaged in agricultural pursuits. In 1877 he bought railroad land in Colfax county, Nebraska, but soon sold it and returned to Michigan. In 1879 he came back to Colfax county and from there, in March, 1888, to Banner county, homesteading and taking a tree claim in Lone Pine precinct, section four, range fifty-four, town seventeen. Although Mr. Trowbridge had some capital when he came to the county, he, like others of that time, went through some hardships His first home and all the buildings on his •farm were made of logs hauled from the nearby canyon

On June 17, 1877, Mr Trowbridge was married to Miss Emma Sherman, who died October 10, 1889 Her parents were Harry and Marion (Tubbs) Sherman, residents of Michigan Mr and Mrs Trowbridge became the parents of five children, the following surviving Lora, who is the wife of L W Cox, of Barclay, California, Stella, who is the wife of Frederick Gillman, of Reddington Nebraska, Blanche, who is the wife of William Thienhardt, of Banner county, and Grace, who is the wife of John Bybee, lives in Kimball Mr Trowbridge has been somewhat active in local politics and has served in precinct offices

G LEE BASYE, the popular and efficient county attorney of Box Butte county is a native son of Nebraska and almost one of the county as he came here when only six months old and belongs to the younger generation of professional men in the Panhandle who are carrying responsibilities and honors that used only to be accorded to the greybeards

Mr Basye was born near Minden Nebraska, December 1, 1886, the son of George S and Cora (Rhodes) Basye, the former a native of Ohio, while the mother was born in Illinois Lee was the elder of the two children in the family as he had one sister, Lenna George Basye was a ranchman who came to the Panhandle when his son was an infant, and took a pre-emption and homestead northwest of Alliance, where he became a well-to-do farmer and cattle-raiser On this farm the family lived for some ten years during which time the boy attended the country school two miles from his home, walking back and forth each day so that he grew up sturdy and healthy Mr Basye well remembers the first money he earned when only eleven years of age taking care of cattle for a neighbor a month in the summer, for which he received five dollars, a sum that looked very big to him then and it was, for money was not plentiful on the prairies The next winter a heavy snow fell throughout this section of the country and cattle perished by the thousands all over the plains and the boy went out when the storm had passed to skin the dead carcasses He sold the hides and thus secured enough money to buy himself a cowboy saddle and a pair of high heeled boots and when a kindly neighbor gave him a pair of spurs he felt himself to be a full fledged cowboy though only twelve years old, for he was an excellent rider and quite able to ride with the best of the boys on round-up and herd The first school he attended was the typical "soddy" of the plains, furnished with home made benches and desks, but there was a good teacher so the children really gained excellent knowledge In 1901 Mrs Basye moved into Alliance and Lee entered the city schools, finished the grammar grades then graduated from the high school He had early decided upon a professional career and though he would have to help himself to secure a higher education, was nothing daunted by this prospect He worked in the shops of the railroad company during the summer vacations and during the school year spent his afternoons in a shoe house, thus making a goodly sum of money In the fall of 1908, Mr Basye matriculated in the arts and science course at the State University, Lincoln. He specialized in public speaking and oratory, in addition to the general arts course and had the honor in his junior year of winning the Hastings prize in the State University Oratory Contest June 13, 1912, Mr Basye received his A B degree from the college of letters and science and the following fall entered the law school, pursued a two year course there and on June 11 1914, graduated with the degree of LL B and was admitted to practice During his senior year in the law school he was elected Ivy Day orator of the class of 1914 and took his part in the Class Day exercises and the planting of the class ivy Immediately after commencement he returned to Alliance and opened an office Prior to graduation Mr Basye had filed as candidate for county attorney on the Republican ticket and was elected by a substantial majority at the general elections in November 1914 He proved so efficient a county official that he was re-elected in 1916, and again for a third term in 1918, which proves not only his personal and political popularity, but testifies to his standing as a lawyer in the community During the first eighteen months Mr Basye was in office he tried more cases under the prohibition law than in any other county in the state excepting Douglas county in which Omaha is located When war was declared against Germany Mr Basye became a member of the Box Butte County council of Defense and chairman of the Box Butte Legal Council of Defense He was called in the first draft on Octobe, 22 1918 to entrain for Fort Kearney, San Diego California, but the call was cancelled on account of the Spanish influenza He was called again on November 12, but as the armistice was signed November 11, this call was also cancelled

June 16, 1915 Mr Basye was married at Lincoln, to Miss Alta M Kates, who was born near Hickman, Lancaster county, Nebraska, the daughter of Morris and Katharine (Stein) Kates, the former a native of New Jersey as was also the mother, On November 19, 1919, a son, Wendall Morris Basye was born to them, and who will possibly be a partner with his father in after years Mr Basye has great faith in the future of this section of, Nebraska which he has demonstrated by becoming a land owner, for he bought a hundred and sixty acre tract four miles north of Alliance, all of which is under cultivation He is forced to be a farmer by proxy as his duties keep him in the city, but he is progressive in his ideas and methods and keeps abreast of the development of agricultural industry and finds that his land is a paying proposition Mr and Mrs Basye are members of the Methodist Episcopal church In politics Mr Basye is a staunch adherent of the principles of the Republican party, while his fraternal affiliations are with the Masonic order and the Benevolent and Protective Order of Elks He owns a fine modern residence in Alliance at 614 Big Horn Avenue, where he and his wife and son dispense a cordial hospitality to their many friends

JAMES W MILLER, the present efficient and popular sheriff of Box Butte county came to Nebraska with his parents in the late seventies and thus must be accorded pioneer honors During his tenure of office Mr Miller has established and maintained a record for loyalty, fidelity to duty and courageous and diplomatic handling of the important work that has been assigned to him with the result that he has gained a secure place in the confidence of the citizens

Mr Miller was born in the heavily timbered country adjoining Toledo, Ohio November 23 1866, the son of William and Tabitha (Jeffers) Miller, the former a native of that land of hills and heather, who brought the rugged Scotch characteristics of his race to America when he immigrated to this country and in turn transmitted them to his son Mrs Miller was a native of the Keystone state, being reared and educated in Pennsylvania There were two children in the family, Anna, who married a man named Jeffers and James W William Miller was one of the gallant adopted sons of the Union who responded to President Lincoln's call for men to preserve the country at the opening of the Civil War, as he enlisted in Company I, Fourteenth Ohio Infantry in 1861 He participated in some of the most severe engagements of that memorable conflict and saw service under General Thomas His horse was shot from under him and he was wounded at the battle of Mission Ridge and he was again wounded, being shot through the neck at the Battle of Shiloh In the fall of 1865 there was an Indian uprising on the plains and Mr Miller with the members of his company started for the seat of disturbance but by the time they reached Omaha, the affair was over and being returned to Washington, D C, the company was mustered out of the service

James grew up on his father's farm early learned the practical side of farm business and during the winters attended the district school which was conducted in a log building two miles from his home While yet a young boy he began to earn money by riding a horse around on the barn floor for his uncle in order to thresh flax, later it was further threshed by men with wooden flails For this service he received a twenty-five cent "shin-plaster" note, the paper money of the war days, and spent it when his father took him to Toledo to buy a cap James remained on the farm with his parents, and when old enough assumed many of the responsibilities and duties of farming to aid his father The boy helped clear land for his father, hauled the rocks, stumps and brush away from the clearing when William Miller would break the sod, plant and sow the crops In 1879, the Miller family came to Nebraska, locating on a homestead near Phillips, in Hamilton county, where James continued to attend school during the winter time His father became an invalid soon after this and the whole burden of running the frontier farm and supporting the family fell on the slim shoulders of the son, but he was a stout hearted man and willingly put his strength to the task which would have daunted many an older person James remained on the farm taking care of his parents until his father died in 1895 He had been out to look the country in Box Butte county over four years previously, and when it became possible for him to move he sold the old place and with his mother came here permanently, locating in Alliance Almost at once he accepted a position in the railroad shops of the Burlington Railroad where he entered as an apprentice to the boiler maker's trade, served his time at it and became a master boiler maker For eighteen years he was connected with the company in various branches of construction work, then resigned to take up carpenter work, that he might be out doors more and not so confined as in the shops Mr Miller was a skillful mechanic and it was not long before he was a success in the new vocation from which he gained a very good income He bought land and erected a good comfortable home in Alliance and in 1918 was elected sheriff of Box Butte county on the Republican ticket His early training specially fitted him for such a position of responsibility which he has filled in a meritorious manner to the entire satisfaction of the citizens

On September 2, 1900, Mr Miller was married at Kearney, Nebraska, to Miss Esther Crowell, a native of that city, the daughter of Daniel and Sarah (Cassaday) Crowell, both natives of Pennsylvania One child has been born to this union, Herbert, a student in the Alliance high school Mr and Mrs Miller are members of the Methodist church while Mr Miller is a Thirty-second degree Mason and a Republican in politics

CALVIN I. HASHMAN, one of the popular and efficient commissioners of Box Butte county who is serving a third term in this responsible office was born near Mercer, Missouri, April 25, 1866 the son of William and Ruth (Mobley) Hashman, the former a native of Ohio Calvin was next to the youngest in a family of eleven children and grew up self reliant and strong William Hashman was a farmer, so his children were reared in the strict discipline demanded by farm life and each and every one of them began to assume duties around the home place when their age and strength permitted Calvin was sent to the district school near his home, and there laid the foundation of a good practical education that has been of value to him in his business life While yet a boy he was an experienced and practical farmer having learned the business from his father While yet a small lad he began to earn money, a habit which he has kept through life The first work consisted of riding a horse and hauling hay to the stack A rope was attached to the saddle and the other end was looped around a bundle of hay and then pulled to the desired location and for the work the boy received thirty cents a day Calvin remained at home until he was eighteen, then decided to establish himself independently as a farmer for he had chosen the vocation of husbandman for a life work, as it

was one to which his tastes turned and already a business with which he was well acquainted In 1884 he came to Boone county, Nebraska, one of the pioneers of this state and the next year located in Dawes county on a homestead S E one-quarter, section fifteen, township twenty-five, range forty-nine, ten miles north west of the present city of Alliance Later a part of Dawes was erected at Box Butte county and Mr Hashman found himself a resident of the new division He put his new land under cultivation built good and permanent buildings for his stock and erected a good frontier home for his family As fortune favored and he was able to dispose of his crops Mr Hashman demonstrated his belief in the future of this section of the Panhandle by buying more property until today he is the owner of a landed estate of two thousand acres of "good Nebraska land," all of which can be cultivated This magnificent fortune has been made by the unaided efforts of Mr Hashman himself who is the architect of his own prosperity and can look with well deserved pride upon the accomplishment He has been engaged in general farming and stockraising since first locating here In the early days when money was not so plentiful and his farm smaller, Mr Hashman employed his spare time as a freighter moving immigrants from Hay Springs, the end of the railroad, to their new homes on claims taken by them in the new country south and west When the railroad was being constructed he freighted for the contractors, bringing supplies to the construction camps established beyond the rail-head and at one time brought in all the necessities required when the tunnel was drilled through Pine Ridge He followed the building of the road as far west as New Castle, Wyoming

On November 29, 1883, Mr Hashman married Miss Cora Jay, in Mercer county, Missouri, and two children were born to them Bessie, who married Frank Vaughn, a farmer and stockraiser of Box Butte county, and Arthur C, who married Myrtle Heartley and has six children They live on a farm near Alliance where the father is engaged in general farming and stockraising Mrs Hashman died and in the early nineties Mr Hashman married Miss Ella M Lapham They raised a family of seven children Amy the wife of Floyd Tryne, a farmer near Hemingford and they have two children, J Leo, also married and now a farmer near Alliance, Ada, at home with her father, Frank C, Lester A, G Wesley and A Roy, all at home Mrs Hashman died February 16, 1916 and for his third wife Mr. Hashman choose Mrs. Jennie M. Condon nee Erickson, and one boy, Cecil B., has been born to the union.

Mr. Hashman was elected county commissioner in 1911, on the Republican ticket and at the present time is serving his third term in this office, which demonstrates his efficient service. He is a man of modern ideas and methods, is progressive in his ideas and takes and earnest and active part in all affairs of the county as well as civic and communal movements for the development of local enterprise. He supports with money and time every movement that will develop the county and Alliance. He was a member of the board when the Box Butte county court house was erected and it was due to his business sagacity and the fact that he was constantly supervising the work that the building was completed without a dollar's worth of graft, though he had many an upleasant encounter with the grafters to accomplish his end. On agricultural subjects the commissioner is an authority, especially alfalfa, and has contributed a valuable monograph on the raising of this valuable crop in Nebraska. He has one hundred and eighty acres himself in the county, runs an average of a hundred and fifty to two hundred head of cattle yearly and cuts about two hundred tons of alfalfa in addition to general farm produce, and two hundred tons of wild hay. Mr. Hashman is a true type of American farmer, the greatest producer on earth.

JAMES H. H. HEWETT, chief clerk in the United States Land Office at Alliance has filled this and many other public offices of the county with marked efficiency, honesty and loyalty that is unusual in this day of haste and hurry, when most men are chasing dollars rather than devoting themselves to the welfare of others.

Mr. Hewett was born in Brownville, Nebraska, July 23, 1862, the son of Obadiah B. and Mary W. (Turner) Hewett. The father was a native of Hope, Maine, and transmitted some of his sturdy New England qualities to his son. Mr. Hewett's one brother lives in Arkansas, where he has a thousand acre rice farm, while his sister, Mrs. Catherine L. Davis has for many years been a teacher in the schools of Los Angeles, California. Obediah Hewett was a lawyer and served as district attorney. When Nebraska was divided into three districts he was given charge of the southeastern district where he gained an enviable reputation as a jurist. He took an ac-

tive part in the affairs of his county and the community where he became a man of prominence. Mr. Hewett was a member of the state normal board for years, was first president of the Nebraska State Teachers' Association and was one of the prime movers in the organization and building of Hastings College. He was one of the country's gallant sons who served during the Civil War to help preserve the integrity of the Union as a member of Company M, Second Nebraska Cavalry. He was commissioned first lieutenant but as the captain was disabled Mr. Hewett filled his place as first officer of the company.

James Hewett finished the public schools of Brownsville, then matriculated at the Peru Normal School where he graduated in 1883. The next fall he entered Hastings College, receiving the degree of A.B. in 1885. That winter he found employment in the Government Land Office as clerk and served for a year before being transferred to McCook as clerk in the land office there. A year later he was again transferred to Bloomington, and on July 11, 1888, was married to Miss Maude L. Kelley of that city, who was born at Roanoke, Indiana, the daughter of James E. and Margaret J. (Lawrence) Kelley. Two children were born to this union: James K., a graduate of the Alliance high school in 1908, entered the State University in the fall and after pursuing an extended course in electrical engineering graduated with the degree of Bachelor of Science and Electrical Engineering in 1913. Before going to college the young man had learned the trade of printer. On his return home after commencement he joined the force of the *Alliance Times*, where he spent some years learning the editorial end of the newspaper business and in 1916, feeling that he was now in a position to handle a paper, he purchased the *Broken Bow Republican*, where he is now in business. He married Miss Anna M. Veith, the youngest daughter of Henry Veith of Lincoln, Nebraska, and they have one child, Helen. The second child of James H. H. Hewett is Helen Bernice, also a graduate of the Alliance high school, in 1915. She entered the State University that fall and graduated in 1919. While in the university Miss Hewett was a champion woman athlete and since leaving the university has been instructor in physical education in the Lincoln high school.

Mr. Hewett came to Box Butte county in the fall of 1888, locating at Hemingford for the purpose of practicing law, but when the United States Land Office opened at Alliance in 1890 he was appointed clerk and moved to that city to assume his governmental duties He had been chosen chief clerk of the office as he was the only man in the county who had had experience in a land office and had shown marked ability in conducting the affairs of the government while serving in the same capacity in other offices in the state Mr Hewett remained in charge here until the spring of 1894 when he resigned to engage in the practice of his profession in partnership with R C Noleman, a prominent attorney of this section who had made an enviable reputation for himself as a jurist Two years after opening his office Mr Hewett was elected county judge on the Republican ticket at the November elections of 1896 In addition to his judicial duties he acted as deputy county clerk, filling the two offices four years, then in January, 1900, he was appointed chief clerk of the land office, accepted, and has since remained there working for the government Mr Hewett is one of the prominent and progressive business men who have played an important part in the development of the Panhandle where he has seen the great changes that transformed the so called "American Desert" into a wealthy farming community that produces abundantly He is one of the best known and liked citizens of the county, having been a resident for nearly thirty years He has always been ready and willing to give his time and money to help with county or municipal affairs and as a result of this popularity had the honor in July, 1919, to be chosen by unanimous vote at a non-partisan mass meeting of the citizens of Alliance as a delegate to the Constitutional Convention of Nebraska, held in December of that year at Lincoln, to frame a new constitution for the state Mr Hewett is a prominent member of the Masonic order, having been the Worshipful Master of Alliance Lodge No 183 A F and A M, eight terms He is also a member of Sheba chapter No 54 R A M, is a Past Commander of Bunah Commandery No 26 K T and a member of Adoniram Lodge No 6 Scottish Rite and Nebraska Consistory No 1 A and A S R, Omaha For years Mr Hewett has been a member of Box Butte Camp No 733 of the Modern Woodmen of America Mr Hewett owns a fine modern home in Alliance where he and his wife dispense a cordial hospitality to their many friends

JESSE M MILLER, one of the prosperous and progressive business men of Alliance is the owner and manager of the Alliance Hotel and Cafe, enjoying a clientele of the best citizens of the town and the traveling public

Mr Miller was born in Peru, Indiana, January 13, 1879, the son of John A and Lucinda (Nell) Miller, the former born in Ohio and the mother a native of Peru Jesse was the second child of the four born to his parents As his father was a Dunkard minister the family moved from town to town as the minister assumed different charges and thus the children went to school in the institutions of the various localities where the family lived in Indiana, Illinois and Michigan, but they managed to lay the foundations for good, practical educations Jesse's first work was shoveling grain in an elevator After finishing the elementary schools Mr Miller attended a commercial college at night to further prepare himself for a business career, and his first position was with the firm of Marshal Field and Company of Chicago when he was twenty-one, where he remained eighteen months before going to New York to enter the service of the H B Claflin Company, wholesale dry goods house It was while working there that Mr Miller's whole career was changed, when Mr Claflin, president of the company said to him ' Jesse, a man of your temperament should be in business for himself, even if it was nothing but a peanut stand I would do that myself if I had nothing else" Mr Miller decided from that day never to work for another corporation, as a man on a salary never gets ahead in the world Soon after this he resigned, took a trip to England, Belgium and France, and while this was not a success from a financial point of view, as he arrived in New York broke, it was a great and broadening experience of value He worked his way back to his old home in Indiana, was employed on a farm for a month, then decided to head for the "land of opportunity,' the west and came to Nebraska locating in Crawford, July 28, 1905, with just a dollar in his pocket Almost at once he secured work on the ranch of Antone Mechem, where he remained a year and a half busy at all kinds of farm work He saw that the best thing was to secure land of his own and took up a homestead on Sand Creek in Sioux county, placed good and permanent improvements on his place and engaged in farming until 1908

October 3, 1908, Mr Miller was married at Chadron, to Miss Anna Lux born in Sioux county, the daughter of Carl Lux, a native of Germany, and Rena (Fellows) Lux, who was born in Illinois Mr Lux was one of the first settlers of Sioux county and held office as county treasurer four terms. One child has been born to Mr and Mrs Miller, Martha K, in school in Alliance. After his marriage Mr Miller went to Crawford, started the Owl Cafe, which he made a success but sold at the end of two years and then bought the Gate City Hotel there in 1910. After establishing it on a sound basis he disposed of it at a satisfactory figure in 1914 and came to Alliance. At first he rented the Alliance Cafe which he rebuilt in 1916 and bought the old Burlington Hotel that year. The following year he purchased the Alliance Hotel building and in 1918 rebuilt the Burlington which he uses as an annex to the cafe. Mr Miller has displayed his faith in the Panhandle by investing heavily in land as he is the owner of a quarter section of the finest land in Box Butte county, where he is considered one of the best all around farmers of the section, producing the butter, milk, and vegetables for his hotel and the cafe so that his guests get the benefit of fresh, home grown products, a very unusual thing now days. Mr Miller is an up-to-date business man in every respect, keeps abreast of all the latest movements and improvements in hotel business and adopts all that are of benefit to him. He is a supporter of all civic and municipal movements and very popular with the residents of his city. He is a Mason and Shriner and also belongs to the Elks.

ANTON UHRIG, county commissioner of Box Butte county, was born in the Province of Nassau, of the German Empire, February 2, 1847, and has through his life demonstrated the fine qualities of industry and thrift for which the German people have gained high standing. His parents were Franz and Anna M (Miller) Uhrig, both natives of Germany. The father of the family was a merchant who ran a grocery store and by this means supported his family of three boys and three girls, of whom Anton was the second oldest boy. In addition to his commercial affairs Mr Uhrig also owned a farm and thus was more able to care for the children than were many of his compatriots. Anton attended the public schools in his native land until he was fourteen years old when he was apprenticed to a harness maker, served three years at the trade and became a journeyman. It was then that he earned his first money by working for his employer for two dollars a week and board. The young man was ambitious, he had heard of the success of many of his countrymen who had emigrated from the native land and determined that he, too, would seek his fortune in the new world. When twenty years old, accompanied by his sister, Anna M, and a friend who had already lived in the United States for some years, George Gundlach, who had been sheriff in his adopted country and who afterwards was elected to the United States Senate from Carlisle, Clinton county, Illinois, Anton set sail for America. He had relatives in Clinton county, Illinois, among them an uncle who owned a soda factory. After his arrival the young man drove one of the wagons while he was learning the language and customs of the country. When the season for summer drinks was over Mr Uhrig worked at his trade of harnessmaker at various places and then secured steady employment at Council Bluffs, Iowa, for three years. During this time he attended night sessions in a commercial college to learn American methods which proved very valuable to him in later business life. Having laid aside some capital he opened a harness shop and factory of his own at Mondamin, Iowa, but disposed of it after three years to become a traveling salesman for a St Louis furniture factory and in 1885 came to Box Butte county. He was so favorably impressed with the western push and energy of the pioneer settlers that while in Valentine, Nebraska he decided to locate in this state, and filed on land near Hemingford in the fall of 1885. The towns of Hemingford and Alliance were not in existance then but Mr Uhrig, his brother Fred and brother-in-law shipped what supplies and goods they needed to Valentine and then with two teams drove to Box Butte across the country in true pioneer style, locating in what was nearly a wilderness. They proceeded to locations near the present town of Hemingford and were forced to use the cover of their "prairie schooners" for tents in which to live until they could haul logs and put up log and sod houses on their claims. Mr Uhrig took a pre-emption and timber claim and in the fall of 1885 returned to Mexico, Missouri, where he was married in January, 1886, to Augusta Basse. Returning almost immediately to Hemingford with his bride, Mr Uhrig brought with him a stock of goods bought while east, which consisted of hardware, saddlery, harness, farm implements and furniture and then opened the first store in Hemingford. This was really the start of the town of Hemingford, for from that time, the fall of 1885, people began to settle there and when the tide of immigration set in soon afterward, it became a flourishing village. Mr Uhrig owns about six hundred acres of fine, arable land

Asa B. Wood

near Hemingford of which thirty acres is laid out in town lots while he has a ranch of sixteen hundred acres eighteen miles southwest of the town, which is a valuable farm tract For about twenty-five years he ran a hardware and saddlery store in the town of Hemingford but seven years ago disposed of it and put the money in land which he began to see was a paying investment He owns a fine modern home just at the edge of the city limits where a delightful hospitality is dispensed to the many friends of the family, for they, being the oldest residents, are well and favorably known in the community which they have assisted so materially to develop Mr Uhrig is considered one of the most substantial men financially of Box Butte county and his high standing with the citizens has been demonstrated by the various positions of trust which they have placed in his capable hands He was elected county commissioner on the Democratic ticket in 1917 and has proved his efficiency in handling the vast business of the county in the past three years He is an enthusiastic believer in the future of this section of the Panhandle and gives time, energy and money for any project that tends to the development of county or community There are seven children in the Uhrig family Nettie, a graduate of the high school and the University of Nebraska, is now a teacher in the state normal school at Chadron, Frank is a farmer near Hemingford, Ida is at home keeping house for her father as the mother died in 1911, Otto A is a clerk in the Farmers State Bank of Hemingford, Wilfred is employed by a hardware firm of Hemingford and was one of the first boys to enlist from Box Butte county when was was declared against Germany, remaining in the service two years and eight months, most of that time being in France, he is also a graduate of the Omaha Business college, George, a farmer near Hemingford, and Margaret, who is attending the Chadron normal school Mr Uhrig has made a record for liberality and progressiveness since serving as commissioner and is highly thought of as he has ever the welfare of the county in mind when exercising his powers of executive

ASA B WOOD — In the compilation of a work of the province assigned to this publication there is both pleasure and consistency in giving more than cursory record of the career of Asa Butler Wood, the pioneer, the aggressive and successful newspaper man, the progressive and liberal citizen and the man who has wielded large influence in connection with the development and upbuilding of the western part of Nebraska At Gering, the judicial center of Scottsbluff county, he is editor and publisher of the Gering Courier which has the distinction of being the pioneer paper of the west half of the state and of which he has been continuously in active charge since April 1887, when as he himself has stated, "it was established virtually, on the raw prairie" Few devotees of journalism in Nebraska have been so long and continuously at the helm of newspaper "navigation" as has Mr Wood and he has a longer continuous business record here than any other man in the North Platte valley These years have been filled with large and worthy achievement on his part and he is to be honored for the service he has given in furtherance of the civic and material progress of this vigorous part of our great commonwealth

Asa Butler Wood was born in Wapello county Iowa August 26, 1865 and this date indicates emphatically that he is a scion of one of the pioneer families of the Hawkeye state His parents, Clay and Jane (Warren) Wood, were natives of Ohio The former passed the closing years of his life in the state of Iowa The mother died at Gering, Nebraska, where she had joined her sons after their location here She was a splendid woman who is remembered with appreciation by many earlier residents The father was a valiant soldier of the Union in the Civil War in which he served as a member of Company C Thirty-third Ohio Volunteer Infantry He was a man of sterling character and distinct individuality — one who was influential in community affairs and who commanded unqualified popular esteem He was a successful school teacher and in Iowa he served several terms as county superintendent of schools His political adherency was given to the Republican party and both he and his wife held membership in the Methodist church

Asa B Wood acquired his early education in the public schools of Iowa, and after completing the curriculum of the high school he eventually broadened his training by association with the various departments of newspaper enterprise — a discipline that has well been designated as the equivalent of a liberal education During his entire active business career he has never severed his allegiance to newspaper and allied work and in the same he has been definitely successful He has at the present time perhaps the best newspaper and job printing plant in the western half of the state and his prominence and influence in journalistic circles are evidenced by his having served as president of the Nebraska Press Association Mr Wood has made his paper not only an exponent of but also a leader in the work of development and upbuilding that has brought great prosperity to western Nebraska, and his loyalty has been on a parity with his enthusiasm and prevision as an apostle of progress along both social and industrial lines He has been unwavering in his allegiance to the Republican party and has given yeoman service in support of its cause He has served several times as chairman of the Scottsbluff county Republican central committee, has been a delegate to county, district and state conventions of the party and has twice served as delegate to the Republican national convention — once from his district and once as delegate at large He has not been ambitious for political preferment, however, and the only office he has held along this line was that of postmaster of Gering, a position of which he was the incumbent for sixteen years

With all things that have fostered communal interests Mr Wood has identified himself with characteristic loyalty and efficiency, and in this connection it may be noted that, as a pioneer, he erected the first frame building at Gering, this having been used as the office of his newly established newspaper The building is still standing, is used as a residence and is made the subject of an illustration in this history As a newspaper office Mr Wood has replaced this structure with a substantial and modern two-story brick building which is an ornament to Gering, the first floor being utilized for his printing establishment and the second floor being equipped and rented for general office purposes

Mr Wood has been president of the Gering Community Club and is secretary of the Gering Investment Company, which built the modern Gering Hotel He was specially active and earnest in the support of the various instrumentalities which furthered the work of preparation for the nation's participation in the World War and the upholding of the country's prestige in that great struggle He was secretary and a member of the executive committee of the Scottsbluff county Council of Defense during the war period, and also a member of the executive committee of the county's liberty loan committee He is affiliated with the Masonic fraternity, in which he served two terms as master of Scottsbluff lodge, No 201, Ancient Free and Accepted Masons, of which he is a charter member He is also a charter member of the local lodge of the Independent Order of Odd Fellows and has passed the various official chairs therein Mr Wood and his wife are active members of the Christian

church of Gering and at intervals during the past thirty years he has served as chairman of its official board

At Cozad, Dawson county, Nebraska, on October 11, 1888, was solemnized the marriage of Mr Wood to Miss Maggie Claypool, daughter of William and Sarah Claypool The five children of this union are Marie, Dorothy, Marjorie, Lynette, and Warren C The eldest daughter, Marie, is the wife of William B Sands, who served as sergeant with the American expeditionary forces in France and thereafter in Germany, subsequently to the signing of the armistice, and now operating farm lands owned by the subject of this sketch

Mr Woods, whose participation in public affairs, has made him one of the best known citizens of Scottsbluff county and western Nebraska, is a pioneer who has been a vital exponent of progress in this favored section of the state, and both in spirit and action he has commended himself to the confidence and good will of the community and county in which he has lived and labored to goodly ends What he has done speaks for itself, but even this brief review will offer a measure of assurance as to his achievement and to his high standing as a man and a citizen

JOHN C MORROW, receiver for the United States Land Office at Alliance, belongs to one of the fine old pioneer families that located in north central Nebraska in the late seventies He is one of the self made men of the Panhandle, whose experiences here have been diversified and interesting, ranging from the days of sod houses and Indians and the general frontier conditions of that period to the affluence and comforts of civilized modern life In all these changes and the development of the county Mr Morrow has taken an active and interested part

John Morrow was born in Lewis county, New York, February 25, 1867, the son of Thomas and Mary (McDonald) Morrow, the former a native of Ireland, who emigrated and located in the United States where he became a farmer John was the eldest child in a family of four boys and three girls and thus early learned responsibility When the boy was only twelve years of age the Morrow family left New York state for the west, as the father desired to give his children a good start in life and land was high in the east, while by coming west to the great prairie country he could obtain fine farming land for the taking The Morrows located on a homestead four miles west of Atkinson, Nebraska, in 1879 where the family lived for many years John attended school in the east but had not completed the elementary grades so when school opened in the "soddy" near his home he attended during the winter, helping his father in many ways during the summer, for on a frontier farm there was always plenty for agile feet and willing hands, herding cattle driving a plow or cultivator and thus the boy was a good practical farmer while a youth in years, and able to assume responsibility as the oldest child Mr Morrow recalls the first money he earned was for work on a farm, and he spent it for a pair of boots to wear to the district school that winter for he had to walk four and a half miles to Atkinson He says that sometimes he could catch a ride on one of the Black Hills Stage Coaches as their route lay the same as his to Atkinson, some of the drivers were kind and let the boys ride, others would not but Mr Morrow had exciting and interesting times as he rode with many of the famous Indian Chiefs going through to Washington to consult and confer with the President Among the well known Indians he met in this manner was Chief Red Cloud of the Sioux Mr Morrow was ambitious and after finishing the country school he entered the normal school at Fremont and subsequently the Western Normal school at Lincoln, thus gaining an exceptionally fine education, which has been of advantage to him in his business life After leaving school Mr Morrow was engaged in business in the eastern part of the state for many years, meeting with great success in his chosen vocation

On January 10, 1899, Mr Morrow married Miss Margaret Harrington, at O'Neill, Nebraska Mrs Morrow was the daughter of John B and Margaret (Carroll) Harrington, both descendants of a fine old Irish family whose members had taken active part in the public life in their country for many years Mrs Morrow's parents emigrated from their native land to locate in Canada, and she was born at Lindsay, in the Dominion, being the youngest in a family of six children, three boys and three girls, two of the latter are now dead Two of the brothers live in O'Neill, M F Harrington being one of the most successful criminal lawyers in the state, while J J Harrington was judge of the district court for many years and is a well known and prominent jurist of the north central section Mrs Morrow is a woman of high education and culture, having been trained during early years at the Loretta Convent at Lindsay, Canada, and also is a graduate of the O'Neill high

school There are two children in the Morrow family Theresa, who spent two years in the State University at Lincoln, graduating from the high school She was a teacher in the kindergarten school in Alliance for a year but is now at the university and expects to receive her degree in 1921, Edward, is at present a student in the Alliance high school

In 1906, Mr Morrow left Nebraska for the Pacific coast, locating in Seattle, Washington, where he engaged in the lumber business for five years gaining a wide business acquaintance up and down the coast Having been reared on a farm and always interested in agricultural pursuits he became a student of irrigated farming, being convinced that land with assured water was about the best investment a man could own He was practically a native son of Nebraska, having spent the greater part of his life in this commonwealth, so it was toward "home" that he turned when he decided to invest heavily in irrigated farming land In 1911 he came to the Panhandle, bought a relinquishment and took up a homestead seven miles northeast of Scottsbluff where the family were soon established Mr Morrow at once began the development of his property, erected good and permanent buildings and a comfortable home for the family Advocating modern methods, he applied them and at the same time invested in the best and latest machinery for the place The first farm proving such a good investment Mr Morrow later bought two hundred and forty acres of irrigated land nine miles northwest of Baird, which is only a mile and a half from the beet dump and he now has this land under process of development and anticipates that it will be in successful production within a short time

In 1915, Mr Morrow was appointed Receiver of the Government Land Office at Alliance and soon moved his family to this city, where they have since resided as he still holds the office At the present time Mr Morrow is constructing a beautiful new twelve thousand dollar home in Alliance, which will be ready for occupancy in 1920 The family have taken a prominent part in the civic and social life of the community since locating here and are well known among the residents for their hospitality Mr Morrow is a true American, living up to the high standards he sets for citizenship, he is liberal in contributions to all worthy and laudable enterprises which will benefit the community and a most enthusiastic booster for the Panhandle as he has great faith in the future of this garden spot of Nebraska

For many years Mr Morrow has given much time to public movements. being the chairman of the Box Butte Red Cross Association, is chairman of the League to Enforce Peace in the country, while his fraternal affiliations are with the Highlanders, the Modern Woodmen, the Elks and the Knights of Columbus In politics he is a staunch supporter of the principles of the Democratic party

JOHN W GUTHRIE — Prominent among the operators in insurance in Box Butte county is John Guthrie, a resident of this locality for more than fifteen years Long a business and insurance man, having given his time to these lines in Indiana until 1906, when he established himself in business in Alliance, and since that time he has so ably directed his activities and operations that today he is listed among the leading citizens of the county seat Mr Guthrie was born in Dubuque, Iowa, april 28, 1866, the son of Patrick and Emma (Mahar) Guthrie, the former a native of Ireland, who came to American to make his fortune, while the mother was descended from a fine old Pennsylvania family that located in the Keystone state at an early day She was born near the present city of Scranton Mr Guthrie is the oldest living child in a family of nine, except a half brother James B Gray, who was associated with Mr Guthrie in business from 1906 to 1913 Mr Gray died in Alliance in 1913 Patrick Guthrie, like many of his countrymen who came to the United States to get ahead in the world, engaged in railroad work soon after landing on our shores and later became a general contractor, erecting many buildings in the city of Dubuque, one of which is yet standing, the City Hall During the Civil War Mr Guthrie was postmaster One of the memorable pieces of his work was the construction of a section of a railroad known as the Dubuque and Western in 1856-57, in payment of which the company issued script, or as it was then known "Wild Cat" money For his work Mr Guthrie received about forty thousand dollars The company failed and as a result the script wasn't worth the paper on which it was printed John Guthrie still has some of this script which he keeps as a souvenir John Guthrie started his education in the public schools of Carroll, Iowa, following his course there he entered Notre Dame University, Notre Dame, Indiana, where he remained a student for eight years He took a special course in engineering, receiving his B S degree and then his degree of C E after higher study

as he had decided that civil engineering offered a broader field for his abilities than the other courses Soon after leaving college he accepted a position on the engineering staff of the Chicago and Northwestern Railroad, joining this organization in 1886 That year the road was built from Douglas, Wyoming, forty miles west of Casper, the same state, and the young engineer spent most of his time out along the right of way However, the next year he returned to Carroll Iowa, where he settled down to reading law in the office of Judge F M Powers, but within six months took over the management of a German newspaper, Der Carroll Demokrat, which he ran for three years This would seem to be rather a difficult task for an Irishman, but Mr Guthrie was quite equal to it and made the paper a financial success His bent was not in journalism and after disposing of the German publication, Mr Guthrie went to South Bend, Indiana, to become associated with the Birdsell Manufacturing Company, a relationship which continued ten years The Birdsell people manufactured wagons, clover hullers, and buggies In 1900 Mr Guthrie received an excellent offer from the New England Mutual Insurance Companies to become one of their Chicago representatives and he accepted, remaining in that city two years In 1906 he came west, locating in Box Butte county where he formed affiliations with various insurance companies in a similar line of work In April, 1906, he opened an offce in Alliance, began writing life insurance and has been the local representative of the Equitable Life Insurance since that date In addition he represents various lines of fire insurance under the firm name of Guthrie and Miller Today they are rated as one of the most substantial firms in their line in the Panhandle, writing insurance for many thousands of dollars annually

On June 11, 1890, Mr Guthrie was married at South Bend, Indiana, to Miss Flora C Sullivan, a native of Greencastle, that state She was the oldest of four children, having one brother and two sisters Five children constitute the younger members of the Guthrie family John M , a graduate of the South Bend, Parochial high school, was a reserve officer during the World War, being commissioned second lieutenant at Camp Taylor, Louiseville, Kentucky, August 15, 1918, and subsequently was transferred to Camp Jackson, South Carolina, being honorably mustered out of the service, December 23, 1918, Charles, also a garduate of the Parochial high school at South Bend, enlisted in the navy during the war, being stationed at Great Lakes, where he served as yeoman in the intelligence department until mustered out, February 21, 1919, Catherine W, graduated from the high school in South Bend and now is serving as claim auditor for the Northern Indiana Traction Company, Florence, is a student at St Mary's academy, Notre Dame, Indiana, and Mary a student at St Joseph's Academy, South Bend The Guthries maintain their home at South Bend for the present while the children are in school there but anticipate locating in Alliance when they have completed their educations

Since locating in Nebraska, Mr Guthrie has taken an active part in public affairs He took a prominent part in all works for the prosecution of the war in Box Butte county, serving faithfully in whatever capacity he was needed, being a member of the Red Cross, aided in the Liberty Bond drives and War Saving Stamp campaign, and for his energetic work in all lines was awarded a gold medal issued by the war service department of the Equitable Life Insurance Company, a considerable honor when it is known that only fifty of these medals were issued throughout the whole country, while Mr Guthrie's is the only one awarded in Nebraska Mr Guthrie is District Deputy of northwest Nebraska for the Knights of Columbus, is a member of the Elks and is past president of the Nebraska State Volunteer Firemen

MICHAEL F NOLAN, for many years connected with the Burlington and Quincy Railroad, and in years one of the oldest engineers in the service of the road in Nebraska is today retired and has become one of the well known and highly respected business men of Alliance where in association with his son he is conducting a coal, ice and feed house under the firm name of Nolan and Son which bids fair to become one of the most prosperous and flourishing concerns of the Panhandle

Michael Nolan is a son of the Emerald Isle born at Elphin, County Rosscommon, Ireland, April 18, 1864, the son of John and Nora Bridget (Tietman) Nolan, who had nine children, of whom Michael is the third in order of birth His father was a farmer and the boy grew up in the country attending the school of his neighborhood where he laid the foundation of a good practical education Mr Nolan recalls well the first money he earned in the old country, working in a grocery store for six cents a day and his board The boy

was ambitious and hearing of the good fortune that many of his countrymen had in the New World, broke all his home ties and dear association when only a lad of fifteen and came to the United States After landing on our shores he soon secured work driving a grocery wagon in Philadelphia for five dollars a week of which he paid out four for board and room, doing his own laundry work at night in the store Mr Nolan remained in the east eleven months, then struck out for the west where he heard good openings were to be had for a young, willing man who was not afraid of hard work In 1880 he settled in Albert Lee, Minnesota, working on a farm for fifteen dollars a month for a year and a half The Burlington Railroad was being constructed to the west at this time and he secured employment on the branch known as the Burlington and Cedar Rapids in the shops established at Albert Lee, but after a year left and came to Nebraska in 1884, and the same year, in the fall, located at Valentine, having been offered work in the round house of the F E and M V Railroad, a position he kept two years, when the road was extended to Rapid City Following this change Mr Nolan entered the railroad shops at Chadron, remaining until 1887 when he went to the Burlington, taking charge of the engines used in the construction work near Whitman, until the rails were laid into Alliance He then began to fire on one of the Burlington engines, being promoted to engineer in 1889 In 1897 he was given charge of the engines that pulled trains number forty-one and forty-two, known as the Portland and St Louis Limited and continued on this run until he left the service When Roosevelt was president Mr Nolan was recommended as engineer for the train on his division, hauling the president's special from Alliance to Ravianna In years of service Mr Nolan is one of the oldest engineers on the Alliance division, having been on this run for over twenty years and many are the thrilling tales he can tell of runs on stormy nights, when a broken rail or some obstruction of the right of way would have meant death and disaster for the heavy train and its passengers On May 2, 1917, because of disability due to injuries received while in the line of duty he retired Mr Nolan joined the Brotherhood of Locomotive Engineers in 1894 and still retains his membership in that organization and is also a Knight of Columbus

On August 31, 1888 Mr Nolan married Miss Mary E Tobin, a native of Iowa, the daughter of Martin and Margaret (Callan)

Tobin, both born and reared in childhood in that country There are four children in the Nolan family Ethel M, a graduate of the high school in Alliance and also of St Mary's Parochial school, now the wife of John O'Brien a merchant at Stewart, Iowa, Frances B, graduated from the Alliance schools and then entered the University of Nebraska where he received his degree and now teaches at Laramie, Wyoming, Martin J, finished the local school, then entered the Nebraska State School of Medicine, at Omaha where he graduated in February, 1918, following which he served as interne at Kings Hospital, Brooklyn, New York, two years, and Michael D, a graduate of the State University, at Lincoln, who was taking a course in the law school at the outbreak of the World War when he enlisted in the naval flying corps at St Louis, May 27, 1917 He took a four months special course at the Massachusetts Institute of Technocology, Boston, being sent from there to Key West for preliminary flying, then was transferred to Miami and to Pensacola, where he received his commission as ensign in the navy Michael became an instructor in teaching men to fly sea planes and remained in the service until honorably discharged when the war was over. On his return home he entered into partnership with his father in the coal, ice and feed business here in Alliance As both Mr Nolan and his son are good business men their prosperity is assured for they have a high standing in the community as men of integrity and care in filling business obligations, unfailing courtesy and modern methods, all qualities that presage a satisfactory return for the time, money and work they put into the business

EINAR BLAK, M D — Among the professional men of Box Butte county none is more worthy of being represented in its annals than Dr Einar Blak, a leading specialist in diseases of the eye, ear, nose and throat For six years he has lived within the borders of the state and for more than two years in this county, during which time he has been a prominent factor in its social, civic and professional affairs, having earned a well deserved reputation for high medical ability and straightforward dealings in all business affairs so that his name stands as a synonym for professional capacity and courtesy

Dr Blak is a native of Denmark, a country which has furnished such a high proportion of the best element of American settlers, as he was born in Copenhagen, March 17, 1886, being the son of James C, and Sine (Modeson) Blak, who had four children of whom the doctor stands third in order of birth James Blak was a lawyer by profession in Denmark and gave his children excellent educational advantages in addition to those provided by the state Einar attended the public schools in Copenhagen, the high school academy and then graduated from the University of Copenhagen after a scholastic course covering seventeen years In 1908 he received his degree of Doctor of Philosophy and the same year came to the United States In the fall of 1909, he matriculated in the medical school of Creighton University, Omaha, Nebraska, spending the following four years there in special medical studies and in 1913 received the degree of Ph D and M D from that institution The following years he spent in special higher courses of medical research in Denmark, devoting most of his time to eye, ear, nose and throat work America appealed to the Doctor and he decided to make this country his future home Returning to the United States in 1915, he came to Nebraska, locating at Gordon, where he opened an office, but desiring a wider field for his activities he removed to Alliance in 1919 and since coming here has made a specialty of lung diseases in addition to head work Dr. Blak has won a most satifactory practice since coming to the Panhandle and is regarded as one of the leading members of the medical fraternity in the northwest, being called into consultation in the most difficult cases and today enjoys a continually growing clientele among the citizens of Alliance and the surrounding countryside The doctor keeps abreast of all latest discoveries and advancements of his profession and being unusually well prepared for his work has attained a place in the very front ranks of his vocation

On December 27, 1909, Dr Blak married Miss Caroline Gemmell at Iowa Falls, Iowa She was born at Neligh, Nebraska, the daughter of John and Caroline (Torrance) Gemmell, both born and reared in the state of New York There are four children in the Blak family Kathleen, aged eight, Einar G, five, Paul S and Regina M, the baby For all of whom there is a bright future as their parents intend they shall have good educations and every social advantage obtainable for the equipment for whatever vocations they may choose

Dr Blak is affiliated with the Roman Catholic church, is a Knight of Columbus and also belongs to the Benevolent and Protective Order of Elks

EPHRAIM T KIBBLE — One of the most progressive and aggressive business men yet genial and dependable spirits of Alliance, who has made a success of both business and farming is Ephraim Kibble, who was born in Edgar county, Illinois, October 30, 1861, the son of James and Tempy (Bonser) Kibble, the former a native of the Old Dominion, where he was reared in true Virginia manner, while the mother was a Buckeye by birth, reared in her native state of Ohio. James Kibble was a true pioneer spirit as he immigrated to Illinois when that state was on the frontier and there cleared a pioneer farm, but when settlement began to crowd close, the lure of the west induced him to emigrate and the second time he came to Nebraska, locating near York in 1870, one of the hardy men with a brave and courageous wife who helped open this state to settlement and later development. When Mr Kibble first came to York county he was able to buy only eighty acres of land as that was the largest acreage allowed inside the railroad limit but later he bought another eighty from the railroad, giving him a full quarter section. Ephraim had attended school in Illinois as he was eleven years old when the family came west and after getting settled on the new farm he went to the district school near his home. After his education was finished he decided on farming as his vocation and remained at home working with his father until he married Miss Idella Hilton, in January, 1888. Mrs Kibble was born in Iowa, the daughter of John B and Anna (Bunsko) Hilton, being the oldest in a family of six children. To Mr and Mrs Kibble six children have been born. Effie, married Ivan Rogers, the manager of a mercantile establishment at Sheridan, Wyoming, Bessie, a graduate of the Alliance high school, spent one year at a business college in Lincoln and now is in her father's real estate office, Blanche, a high school graduate, also took a course in the Lincoln Business college, in 1908, and is secretary and treasurer of the Golden Fleece Mining and Milling Company of Denver, Colorado, Lloyd, is a farmer five miles east of Alliance, Marie, after finishing the Alliance school entered the University of Nebraska where she is specializing in music in order to teach, and Clarence, at home attending the high school. In 1907 Mr Kibble disposed of his holding in the eastern part of the state and came to the Panhandle as he believed

there was a great future in this section for a good business man and that his faith in the country was justified need not be doubted as he owns three thousand acres of land in the upper valley and has a business that brings in most gratifying returns for the time and energy which he devotes to it. Within a short time of opening his office in Alliance Mr Kibble was handling much of the business in this locality. He is a man who has won a well deserved reputation of always satisfying his customers. He is progressive in his ideas and puts these ideas into practice, which is the most important item in this hustling day. He was the man who originated the idea of building the potash plant at Antioch known as the Nebraska Potash Company, and it was due largely to his initiative that the plans were carried through to successful completion. The company established a large electric plant which furnishes lights for the town as well as the tenant houses of the employes of the concern. Mr Kibble is interested and takes an active part in every movement that is laudable for the benefit of the community, advocates good roads, good schools, and good and honest government. Since coming to the Panhandle he has invested heavily in land of this section and now is the owner of a large landed estate, holding sixteen hundred and forty acres five miles east of Alliance, of which one hundred and forty acres is planted to alfalfa, which is the best improved in the county, while he cuts a hundred tons of prairie hay a year. For some time he has dealt heavily in thoroughbred hogs. Recently Mr Kibble has purchased almost a quarter section of land adjoining the town site of Alliance which he will hold for the growth of the city in that direction before opening it up for building. The Kibble family has a fine modern home in the town where they are known for their open handed hospitality as Mrs Kibble is a charming hostess. Mr Kibble is so well fixed financially that when the shadows begin to lengthen from the golden west he can retire from active life and looking down the years, knows that his has been a life of accomplishment well worth while. In politics he is a supporter of the Republican party while his fraternal associations are with the Elks, the Workmen and the Eagles

ROBERT GRAHAM, the popular postmaster of Alliance and one of the progressive business and ranchmen of the Panhandle is the owner of large tracts of land, and has had large and varied experience in the west

Mr Graham is descended from a long line of Scotch forebears who located in the north of Ireland at an early day, and it is these Scotch-Irish families who have taken the greatest part in developing industries in the Emerald Isles and the descendents have contributed a large part to the best element of immigrants who have come to America to assist in opening up and developing this continent   Robert Graham is a worthy representative of such a family He was born near Belfast, Ireland, December 11, 1866, the son of William and Elizabeth (Davidson) Graham, the former born near Belfast while Elizabeth Davidson first saw the light of day on the Davidson farm, which has been held in the family since 1622, granted them by James I of England and VI of Scotland   While pride of ancestry is not a marked characteristic of the American citizens, it is nevertheless, not only natural but highly commendable that one should feel a just pride in the fact that he has descended from ancestors who were more than ordinarily distinguished in their day and generation With this thought in mind we have briefly sketched an account of Mr Graham's family, for it is the qualities which have descended from this long line that have enabled him to overcome many obstacles that would have been insurmountable to a man of less mental stamina   Robert Graham was the oldest child in a family of seven children consisting of five boys and two girls, and he thus grew up sturdy and early learned habits of self reliance and independence while he assumed many responsibilities while still young   The boy was sent to the public schools until he was seventeen and laid the foundation for a good practical education which he has continued by himself by wide reading and special attention to subjects connected with his business   As a youth Mr Graham heard much of life in the Australian bush and this fired his imagination with a desire to emigrate to that southern land, but he was dissuaded from this by his parents and induced to come to the United States to join two uncles who lived in Iowa Landing on our shores in the spring of 1883, he reached Iowa in June, and soon found employment on a farm, where he remained until 1886   By that time the young man had learned more about the country and determined to establish himself independently on land of his own Nebraska seemed a pretty good place to him so Mr Graham came out to Box Butte county to file on a homestead near the northern boundry   He made the necessary improvements, proved up and became a land owner in fact   In 1888, he removed to Cheyenne county, that part of which subsequently was erected as Morrill county, where he at once engaged in raising live-stock, as that was the heigh-day of the cattle business on the plains He made good from the start and from time to time, as his capital permitted, bought other tracts of land so that he is the sole owner of a landed estate of three thousand acres, a ranch from which he cuts over eight hundred tons of prairie hay each year. Thus for thirty-one years Mr Graham has been one of the leaders in this vast enterprise in the Panhandle In 1897 he formed a partnership with Eugene Hall, a business association which continues to the present time   Together they hold about thirty-eight thousand acres of ranch and grazing land in Morrill county, where they annually handle three thousand head of cattle and some horses, during the haying season the pay roll contains nearly half a hundred names of the employes engaged in carrying on the various farm industries Mr Graham has taken an active part in public life since first coming to Nebraska During the late war he was chairman of the Council of Defense for Box Butte county, chairman of the Liberty Loan Committee, chairman of the County Fuel Administration Board, chairman of the building Committee, and as he stand pre-eminently at the head of the live-stock producers of the state was appointed by Governor Neville, a member of the State Federal Live-Stock Committee, is also a member of the Live-Stock Sanitary Board, is president of the Nebraska Stock Growers Association, an office he has held for the past five years and to which he was re-elected for the ensuing year   Mr Graham is a broad minded man of far, keen vision and he early realized that the day of the open range was over so that he went in "on the ground floor," so to speak, in the cattle producing business, and his present large fortune is due to his ability of visualizing the future as well as his marked executive ability and willingness to work long and late when necessary, not only for his own business but in the interests of his county and the state, both of which he has served long and well

April 6, 1887, Mr Graham married Miss Lilla M Clark at Hay Springs, Nebraska Mrs Graham was born in Dewitt county, Illinois, the daughter of Reverend Samuel and Elizabeth (Brownlee) Clark, the former a minister of the United Presbyterian church

Mrs Graham is a great grand-daughter of William Brownlee, who was born at Strathavon, Lanark Shire, Scotland, who came to America several years before the Revolution, settling in western Pennsylvania He entered the Constitutional army and served until the close of the war, being promoted to quartermaster For his services he received a grant of several thousand acres of land near Beaver, Pennsylvania, on the Ohio River Nine children have come to join the family circle of the Graham family William F, a banker at Minatare, Nebraska, who graduated from the high school then took a course in the Lamphere and Mosher business college at Omaha, he is a Mason and an Elk, Elizabeth, after finishing the Alliance schools graduated from the Lincoln Business college, following which she served as secretary to the president of the Peru Normal school, but since the war has been in government service at Washington, D C in the office of the judge advocate, John R, a high school graduate, is foreman of the Hall and Graham ranch at Alliance, Margaret, after finishing school at home graduated from the Chadron Normal school and is now a teacher in Scottsbluff, and is a member of the Daughters of the American Revolution, Alice, graduated from the high school and then entered the training school for nurses of the University Hospital, at Omaha being in her third year, and will finish in 1921, Donald, after leaving school at Alliance took two years work at the State University and was matriculated in the medical college of the university at Omaha, in Omaha, in 1917, when he was called into government service, being stationed at the Auxiliary Remount station, Camp Funston, and after nineteen months in the army was mustered out in March, 1919, and has resumed his interrupted medical career, Katherine, after completing a four year high school course, graduated from the Chadron Normal school and now teaches in Box Butte county, Samuel B, is a freshman at the University of Nebraska, and Lilla, who is at home attending school

Mr Graham was appointed postmaster of Alliance in 1915, and proved such an efficient executive that he was re-appointed in 1919 He is a Scottish Rite Mason of Thirty-second degree, a Shriner, Knight Templar and also belongs to the Elks In politics he is a staunch adherent of the principles of the Democratic party, while his church affiliations are with the Presbyterian denomination, a faith in which he was reared

ANDREW BROADDUS BEARD, the senior member of the firm of Beard and Pickett, is one of the leading real estate men of Kimball county, who is well known as a farmer and merchant, as he has lived here many years and taken an active part in the development not only of his community but of Kimball county He was born September 12, 1867, the son of George W and Angie E Beard, whose biographies appear in this volume Mr Beard was reared and received his educational advantages in Bloomfield, Indiana He came west to Nebraska at an early day and as a pioneer newspaper man inaugurated his career in the Panhandle as editor of the *Western Nebraska Observer*, of Kimball, with which he was associated for eight years Later Mr Beard took up land, engaged in farming and then entered business life where he has attained gratifying success He is well known as a merchant and after engaging in the real estate business has broadened his financial field to cover all realty lines, being especially interested in the sale of farm ranch properties Entering the realms of finance, Mr Beard became the cashier of the Bank of Harrisburg where he ably filled such an exacting office He was elected county clerk in 1910, serving in that capacity two terms and in 1915, was elected county treasurer, holding office until January, 1917, which shows in what esteem he is held by the people of Kimball and the county who elected him to two offices of trust and responsibility Mr Beard is a Republican and has taken active part in local politics for some years He attends the Presbyterian church

September 1, 1892, Mr Beard married at Kimball, Miss Julia Wooldridge, the daughter of Samuel and Minerva Wooldridge, the father was a soldier of the Union army in the Civil War and after coming to Nebraska to settle was first county clerk of Kimball county, one of the pioneers who took an important part in the development of his locality The children of the Beard family are as follows Anna Bessie, married Leonard F Smith in 1912 and died in 1916, Daisy Irene, married D L Pickett in 1914, Marguerite, died in 1899, Lelia May, married Edward V Heffner in 1919, Marion Tyrone, married George W Van Aelstyne in 1917, Florence Jeanette, Vivian Delight, Rosalie Victoria, Mildred Verona, Charles Broaddus and Harold Eugene complete the family circle

TRUE MILLER, the owner of one of the best equipped business houses in the Panhandle, the Coursey and Miller Garage, of Alliance has been prominently identified with business interests of Box Butte county for thirty years. Here he has established a lasting reputation for ability in commercial affairs, integrity in transactions and engagements and probity in personal character. Mr. Miller has attained success through merit and not by chance or fortunate circumstances. He is a citizen of public spirit, progressive in his ideas and always willing to help with time or money every constructive movement for the general welfare.

Mr. Miller was born at Birmingham, Iowa, June 2. 1872, the son of Dr. William K., and Ellen S. (Elliott) Miller, the former a native of Illinois, while the mother was born in Ohio. Dr. Miller attended Rush Medical College, Chicago, after finishing his elementary education and from that institution received his M. D. degree in 1872. For a number of years he practiced in Iowa, but believing there was a wider field for his services in the newer country opening up to the west came to Nebraska in 1887, locating in Box Butte county, one of the first physicians in this locality. In connection with James Carothers the doctor engaged in mercantile business handling groceries as the fees a doctor received from the scattering settlers of that day would not provide a living for a family. This was one of the first stores established in Alliance, as the first goods were sold from a tent until a suitable building could be erected, for there was a crying need for merchandise in the district and they made service a point for their customers, rather than comfort and conveniences for themselves. Dr. Miller also engaged in the duties of his profession as needed, for physicians were few and far apart at that day on the prairies. For many years the doctor served the Alliance district faithfully and well, he built up a fine practice which he enjoyed until the shadows of life began to lengthen, when he retired from active life and now lives with his son, looking back along the years he can feel with honest pride and satisfaction that his life has been a constructive one, as he helped develop this part of the country and also alleviated the suffering of many.

True Miller attended the public schools in Winterset. Iowa, and again there in Alliance when the family came west, following which he matriculated at the Normal school in Chadron, and later, in the winter of 1890, and spring of 1891, took a special business course in the Lincoln Business college. Mr.

Miller recalls that his first business venture as a boy was a trade he made, exchanging a Waterbury watch and four dollars and a quarter for a colt, which was a most satisfactory investment, as he kept the colt and raised quite a herd of horses. After his father opened the store in Alliance the young man clerked there for some time, a year all together, then in 1893, opened a butcher shop of his own, but commercial life of this kind and its confinement did not appeal to him and in 1895 he purchased the Engelbreck and Ames ranch in Sioux county. There Mr. Miller entered actively into the live-stock business. He says that as long as he was in partnership with Uncle Sam and had free range for his many head of cattle his business was most satisfactory and life on the ranch was not only pleasant and exhilirating but exceedingly profitable; and were conditions the same today as they were for the twenty-three years he was a ranchman he would never have given it up, but with the free range gone business in cattle dealing was not so satisfactorily though Mr. Miller still owned three thousand, three hundred acres of land, which seems a rather large landed estate to most people, at the time he sold out, in May, 1918. A month later he bought an interest of James Keeler in the Keeler and Coursey Garage in Alliance, since which time the firm has been known as Coursey and Miller. They have a fine brick building, fifty by a hundred and thirty feet, three stories in height, where they handle Ford cars and all accessories, Ford tractors and Republic trucks, also doing a general repair business on all makes of cars. The company is distributors of Firestone tires for western Nebraska, southern Dakota and eastern Wyoming, a branch of the business which is constantly expanding and exceedingly profitable as a side line.

November 21, 1894, Mr. Miller was married at Alliance to Miss Lillie D. Stevens, born in Detroit, Michigan, the daughter of Frank P. and Jane (When) Stevens, the former a native of New Hampshire who emigrated from the Pine Tree state and located in Box Butte county in 1885, where he became the proprietor of the Grant House which he conducted for several years. Two children have been born to Mr. and Mrs. Miller, Harley, who received his elementary education in the Marshland schools and then entered the normal school at Chadron, but left his studies to enlist in Company G, Fourth Nebraska National Guard, from which he was transferred to the

One hundred and twenty-seventh Heavy Artillery, at Deming, New Mexico, and then was sent to Clintonville, Wisconsin, to attend a tank school, after six weeks being again transferred to Fort Sill, Oklahoma, and from there to Camp Upton, Long Island, soon to embark for overseas service He landed in Liverpool, England, first, then was sent to France, debarking at Brest, served during the great American offensive and upon his return to this country was mustered out of the service at Camp Dodge, Iowa, January 23, 1919, having been in the army nearly two and a half years Since returning to ways of peace Harley has been employed by the Alliance Potash Company in the plant at Antioch The second child, Mattie J, is at home a student in the Alliance high school

Mr Miller is a Mason of high degree and a Shriner He has won the confidence and respect of his business associates and is one of the commercial assets of Alliance which does not lack able men of affairs

CLARE A DOW, one of the younger generation of business men who has done much during a short business career to develop natural resources, both in Iowa and Nebraska, is now superintendent of the Public Utilities Light, Water, Power and Sewerage of Alliance and also the proprietor of one of the most modern and up-to-date electric contracting firms in the Panhandle

Mr Dow was born at Garner, Iowa, February 13, 1881, the son of Parker S and Oner S (Groom) Dow, the former born at Montpelier, Vermont, while the mother was a native of Polk county, Iowa Clare was the oldest of the two children born to his parents, as he had but one sister He attended the public schools at his home and after four years in the high school entered Iowa State College, at Ames, where he took a special course in electrical engineering Before this the boy had learned to be self reliant and recalls that the first money he earned was picking apples for a neighbor at fifty cents a day After leaving college Mr Dow went to St Louis where he held a position with the Union Electric Light and Power Company, which had the contract for wiring the World Exposition Grounds near that city When one considers the stupendous task of carrying power and light to the many buildings on the grounds and the complicated system that had to be employed that it is no wonder that a young man just out of college was rather overwhelmed

when placed in charge of part of the work However, the chief engineer had been young once himself and knew that all Mr Dow needed was a little practical experience and responsibility so told him that it was not necessary at first to comprehend the whole system but just to go ahead on each unit as it came up and soon all of it would work itself out for him This proved true for during the eight months he was at the grounds Mr Dow not only made good but won an excellent reputation with the company for his ability and management On leaving St Louis he came back to his home at Garner as foreman of the inside wiring department while the electric wiring was being installed by the new electric lighting company After the power was turned on his work was over so Mr Dow went to Des Moines, as stock keeper and assistant shop foreman in the meter testing department of the Des Moines Edison Electric Light and Power Company A year later in 1904, a fine offer was made him to go to Iowa Falls where he purchased the wiring and contracting department of the Iowa Falls Electric Light and Power Company On April 20, of that year, Mr Dow married Miss Vira F Walton, at Thompson, Iowa She was born in South Dakota, the daughter of John D and Juliet (Polhemus) Walton, the former a Pennsylvanian, being the youngest of their five children

Mr and Mrs Dow have two children Verna C, a student in the Alliance high school, and Vivian R, in the grades For three years after his marriage Mr Dow was one year wire chief and two years superintendent of the Iowa Falls electric plant, then was offered and accepted the position of superintedent of all public utilities, holding office until 1910, when he resigned to take up the same duties at Crawford, Nebraska While there he completed the contract for the electrification of Fort Robinson and in 1913 came to Alliance as superintendent of the public utilities here In 1916 he purchased the Pugh Electric Company, which he still owns and manages, being located in the Times block on Box Butte Avenue, where he also maintains a storage battery service plant Mr Dow is a general electric contractor, carrying on business all over the Panhandle, South Dakota, and eastern Wyoming, as he has won a well deserved reputation for excellent workmanship, fidelity to contracts and high business integrity He is a progressive citizen, living up to the standard which he believes every American should follow, is progressive in his business and ideas for civic welfare so supports all movements for the development of the county and his home city. The family are members of the Methodist Episcopal church while Mr. Dow's fraternal affiliations are with the Masonic order. The Dow family comes of a fighting race as both Mr. Dows' father and his wife's father fought on the side of the Union during the Civil War and the post of the Grand Army of the Republic at Garner, Nebraska, is named Parker S. Dows, in honor of his father.

JOSEPH H. VAUGHN, today one of the prosperous and progressive business men of Alliance, the man whose name head this brief review must also be given recognition as one of the early settlers and ranchers of the Panhandle for he has been a resident of this section more than thirty-five years, and has not only been an eye witness of the vast changes that have taken place but has also been an important factor in the development of this now favored section of the state.

Mr. Vaughn was born in Lafayette county, Missouri, January 6, 1860, during the disturbed period preceding the Civil War and it may be that some of the rugged determination and spirit of that day entered into his mental make-up for he has surmounted a handicap that would have discouraged a man of less moral fibre. He is the son of Benjamin and Anna (Williams) Vaughn, the former born at Lexington, Kentucky, while the mother was a Virginian, born at Norfolk. There were but two children in the family, Joseph and a sister. The father was a Missouri farmer so that the boy spent his life in the country attending the district schools until his thirteenth year, when it was found that his lungs were not as strong as could be wished and he was sent to Boulder, Colorado, and placed in charge of a Dr. Dodge, who had Mr. Vaughn drive him around on his calls in a one horse "chaise." This kept the boy out of doors and he soon became much stronger. While living at home he had already learned to work as he went in the wheat fields to gather the bundles of cut grain into chocks after the cradlers. The first money he earned he invested in a pig and thus early had become accustomed to frugality and industry. Reaching Colorado in 1873, Mr. Vaughn spent three years with the doctor, then accepted a position as stage driver from Boulder to Caribou, a distance of twenty-four miles, but as he says it was straight up or down the mountains the horses had to be

changed three times each way on the trip, four and six horses being used to draw the coach at one time. Passengers were carried as well as the government mails and express. For five years Mr. Vaughn was on this route, completely regaining his health so that when he was offered a position with the Marshall Coal Company, six miles southeast of Boulder, as buyer and barn boss, he accepted, holding this position three years. Realizing that a man who owns land is independent, Mr. Vaughn determined to take some up and with this end in view came to Nebraska in 1884, locating on a quarter section homestead in Cheyenne county and at the same time filing on a tree claim adjoining, where the Scottsbluff Sugar Beet Factory is now situated. On his homestead is the grave of Rebecca Wintersea, a Mormon woman who died on the way west and was buried, when the great migration of that section took place to Utah as they passed up the Platte Valley in 1859. Later Mr. Vaughn removed to the Sand Hills twenty-five miles north of Camp Clark land, through which the B. and B. railroad was built at a later date as the road now splits the Vaughn meadow, However, when the Vaughn family settled there railroads were not extended this far west and Mr. Vaughn had to drive to Sidney, eighty miles away, for supplies and provisions, fording the North Platte rather than to pay the tole of three dollars to cross the bridge at Camp Clark, for money was a very precious commodity on the plains at that early day as most of the trading was done by way of exchange. When Mr. Vaughn came to the Panhandle all his equipment consisted of a wagon, team of horses, two milch cows and calves, what household goods that could be loaded on the wagon while his responsibilities consisted of his wife and three year old son, and the determination to make a fortune as well as give his family comforts while doing it. That he has succeeded need not be questioned when one learns that he is the owner of a landed estate of four thousand acres. Soon after erecting his home and some buildings for the stock, Mr. Vaughn began to actively engage in the cattle business, as it was at the height of its prosperity at the time. He owned what was known as the "Old Hay Lakes," where the soldiers from Fort Robinson in early days cut wild hay, hauling it to the fort for provisions. With the passing years and as his capital permitted, Mr. Vaughn purchased land adjoining the original tract and carried on farm and ranching operations until 1911. His wife lived on the farm summers and in the winter the family came to town to educate the boy Mr Vaughn still has general charge of the ranch though he now has a foreman there so that he can alternate between the county and his business in Alliance where he is extensively engaged in running a jobbing house dealing in oil and gasoline and other allied products Mr Vaughn is a stockholder in the Reliance Refining Company of Eldorado, while his son, J Claude, is secretary and treasurer and salesmanager For some years Mr Vaughn and his son were engaged in a coal, feed, ice and oil business in Alliance but has disposed of this as their time is more profitably spent in carrying on larger concerns

September 12, 1881, Mr Vaughn married Luttia Clemmons at Boulder, Colorado She was the daughter of Jesse Clemmons of Kentucky, and Mrs Vaughn was born at Frankfort, that state One child was born to this union, J Claude, who was graduated from the Alliance high school then took a special commercial course in a business college at Kansas City, Missouri, and is now located at Eldorado in charge of his father's interests in oil production, as has been stated J C Vaughn was the youngest man ever to take the degree of Shriner in the Masonic order in Nebraska

The Vaughn family have a fine modern home in Alliance, where they are known for their generous hospitality to their friends, having made many warm and kindly associations since settling here at an early day and have been known throughout the valley for many years for their kind deeds and ready support for all worthy causes of the community Mr Vaughn owns a fine business block at 222 Box Butte Avenue, other residence properties in the city and has an annual income from six hundred tons of prairie hay he cuts each year that many might well envy, besides the general income from investments and the ranch so that his faith in the Panhandle has been fully justified

HERBERT A COPSEY, M D, president of the Alliance State Bank is well known among the progressive agriculturists of this section and for a number of years was a leading member of the medical fraternity of Box Butte county, thus he has been identified with numerous financial enterprises here and has established a high reputation for ability, judgment and general acumen His introduction to Alliance was as a physician in which

he served the community with great skill until he entered government service at the outbreak of the World War. Since leaving the army he has become connected with the large financial affairs of the Panhandle, his rise being rapid, sure and most substantial

Herbert Copsey was born in Crawford county, Wisconsin, January 21, 1880 the son of Alonzo H and Anna (Wallen) Copsey, who had a family of nine children of whom Herbert is the third in order of birth When the boy was only a year old his parents came to Nebraska, locating in Custer county near Westerville the country was so little settled at the time that supplies had to be obtained at Grand Island, eighty miles away The children were sent to the district school nearest their home, where they laid a good sound foundation for a practical education Young Herbert early learned to work on the home farm as well as all boys in the country do but well recalls the first money he actually earned by driving calves for a half day for a neighbor receiving fifteen cents for his work, but that number of cents in those early days looked as big to him as dollars did later After finishing the elementary courses in the local schools, Dr Copsey attended the high school at Ansley, followed by the teacher's course in the normal school at Broken Bow For three years he taught in the Custer county schools, but the life of a pedagogue did not appeal to him as a permanent vocation so he entered the Lincoln Medical college in 1902, graduating with the degree of M D in May, 1906 The following July he came to Alliance, opened an office and began his professional career Dr Copsey soon built up a good practice in Alliance as well as the surrounding country as he was a skillful physician courteous and sympathetic to those afflicted, and for thirteen years held a high position among medical men of the Panhandle When the president called for volunteers when war was declared against Germany, Dr Copsey volunteered was commissioned captain and placed in charge of the medical wards in the hospital at Camp Hancock, Sandy Hook, New Jersey serving from September, 1918 to January 4 1919, when he was mustered out at Camp Grant Illinois On his return to ways of peace Dr Copsey entered the financial field as business appealed to his tastes and temperament Soon after reaching Alliance he bought a large block of stock in the Alliance State Bank, becoming its president The bank had been organized in 1914 but its development was not decidedly marked until the present officers took charge of its affairs. The personnel of the bank at the present time is: H. A. Copsey, President; Charles E. Brittain, vice president and Jay O. Walker, cashier. Dr. Copsey's resolute-purpose, high integrity, won during his many years in Alliance, have begotten popular confidence and esteem that are so essential to the furtherance of success in financial circles and have materially aided in building up the clientele of the bank. As a banker Dr. Copsey is showing special constructive talent, and through his effective policies and efforts the Alliance State Bank is taking rank in the forefront of the financial institutions of western Nebraska, as it has a paid up capital stock of $35,000, surplus of $30,-000, with an authorized capital of $50,000, while the amount that it has grown may be gained from the fact that the deposits are well over $700,000, making it the second largest bank in the city. At the present time the board of directors consists of Herbert A Copsey, Charles E. Brittain, Jay O. Walker and M. C. Hubbell.

On January 20, 1909, Dr. Copsey married Miss Mabel C. O'Brien, at Broken Bow, and one child has been born to them, Mary Loretta. Dr. Copsey has ever been a believer in a great future for the Panhandle which he has demonstrated by investing his capital here for he is the owner of five thousand acres of land lying forty-five miles southeast of Alliance in Garden county, where he is actively engaged in agricultural business, as he annually runs about eight hundred head of cattle on his pastures and cuts a thousand tons of hay, a large and well paying business aside from all his other interests. In Alliance the doctor and his wife own a fine modern home where they dispense a cordial hospitality as they have a host of warm friends and acquaintances. Being progressive in his ideas for his own affairs the doctor also advocated progress in civic and municipal affairs and is a "booster" for every movement that will develop the county or city, giving liberally toward all worthy causes. In politics he is a Republican while his fraternal affiliations are with the Elks. He and his wife are members of the Roman Catholic church, while the doctor is a Knight of Columbus.

LYMAN A. BERRY, for more than a quarter of a century one of the leaders of the Box Butte county bar and for over thirteen years continuously on the bench, is the example of a life that has been worthily lived and as such bears its full measure of compensation. Now that he has passed life's meridian and the shadows begin to lengthen from the crimsoning west, he has stored up lessons of rich and varied experience, as one who has wrought wisely, justly and effectively. Each successive year to him must thereafter be radiant in personal contentment and gracious memories. Judge Berry, as he is best known in the Panhandle, is engaged in the practice of his profession in Alliance and his status as a citizen, a lawyer, and as a genial and popular man makes it specially pleasing to accord him recognition in this history.

The judge was born at Pompey, New York, May 15, 1854, the son of Matthias and Sylvia (Osborn) Berry, both natives of the Empire state, the former born at Morrisville, the latter at Fabius. Lyman was next to the youngest in a family of eight children and thus early learned to adjust himself to the welfare and comfort of others as is necessary among a number of growing children. Matthias Berry was a mill-wright by trade, a very skillful man in his vocation whose time was in great demand where a mill was to be constructed or repaired. During his boyhood the judge attended the public schools of his home town with his brothers and sisters, then entered the Pompey Academy and later matriculated at the Whitestown Seminery, Whitesborough, New York. Following his course in this institution he became a student at Hamilton College, Hamilton, New York, where he spent three years in college work. Soon after graduating he began to teach but after some time spent in the school room winters and working on a farm summers he began to read law in the office of Judge Parker at Marshalltown, Iowa, and on February 18, 1879, was admitted to practice in the district courts of Iowa. Two years later he located at Ida Grove, Iowa, and formed a partnership with Mat. M. Gray. The two youthful members of the legal profession opened an office under the firm name of Gray and Berry, an association that continued four years when Judge Berry formed a new partnership with Charles W. Rollins. For eight years the judge and Mr. Rollins enjoyed a fine practice but in 1893 the judge's health gave out and he returned to his old home in New York for a vacation of a year. After sufficiently recuperating he came west but this time located in the Panhandle in Box Butte county, establishing himself again in a law practice until he was elected county judge in 1903, on the Democratic ticket, he was re-elected several times serving thirteen and a half years on the bench before retiring to a private practice on January 1, 1917 During his long service as a county official Judge Berry established an enviable reputation as a jurist, few of his decisions being reversed by a higher court and since he has again opened an office for business in the Rumor Building at the corner of Third street and Box Butte Avenue, has had a continually expanding business that is most satisfactory from a financial view as well as showing the high place the judge holds in the esteem of the people of the community

On June 27, 1883, Judge Berry married Miss Minnie J Sparks, at Gilman, Iowa, the daughter of Lyman B Sparks, a native of Massachusetts Two children have come to make happy this union Leo M, who married Miss Florice Cook, is a ranchman near Lakeside, Nebraska, and they have one child, Grace, eight years old, and Llye S, who graduated from the high school in Alliance then entered college in Chicago, Illinois, where he took a special course in electricity, graduating with highest honors in just half the time that most men take for the course, in competition with twenty-five hundred students This was in the spring of 1908, and within a short time Lyle Berry accepted a position as electrician at the Boyson Dam Project, Wyoming, as superintendent of the light and power plant, where he remined two years before he was induced to go to Lusk, Wyoming as superintendent and manager of the light and power company of that city, where he was in charge until July 18, 1918 When war was declared with Germayn Lyle Berry entered the aviation division of the army, being sent to Elling Field, Texas, for general training He remained in the service until mustered out in March, 1919, then returned to assume his duties as manager of the electric company at Lusk He is a member of the Independent Order of Odd Fellows

Judge Berry came to the Panhandle more than a quarter of a century ago and before that was an old resident of Iowa, so that pioneer honors must be accorded him From first settling here he has taken an active and important part in the development of this section being ready with money, time and advice for every laudable enterprise that was of benefit to the town and community His faith in the future of this section has been unfailing and he has lived to see it justified He owns one of the finest and most hospitable homes in Alliance where, with his charming wife, open house is kept for their many friends He can recall many experiences of the early days, some of which will be found in the old settler's stories written by Judge Tash, on the historical pages of this book According to Judge Berry they were not all hardships, but he as usual shows us but the sunny side of frontier life on the high prairies Politically the judge is a staunch supporter of the tenets of the Democratic party while his fraternal affiliations are with the Masonic order, as he is a Scottish Rite Mason

HARRY P COURSEY the senior member of one of the largest automobile houses in the Panhandle, is a business man of marked ability and for the past ten years has become known as one of the specially successful auctioneers of northwestern Nebraska He is an adopted son of this great commonwealth but has the push energy and financial acumen usually attributed to one of our native sons

Harry Coursey was born in Grundy county, Iowa, December 22, 1877, the son of Septimus M and Sarah (Weatherby) Coursey, the former a native of Hagerstown, Maryland, while the mother was born in Indiana There were seven children in the Coursey family of whom Harry was the fourth boy His father was an contractor and builder by trade and followed this vocation in Iowa until 1879 when the family moved to Kansas, locating on a farm in Norton county, where they lived for some time Septimus M Coursey proved up on his homestead but during the years of the drought had such poor crops that he could not support the family from the returns of the farm and went to Topeka, to work as a builder and also undertook some contracting Harry had already attended school in the country and after moving to the state capital continued his studies in the schools there for three years, at the close of which time he accompanied his parents when they returned to the old home in Norton county They all remained on the farm until the spring of 1889, then removed to Lexington, where Mr Coursey engaged in work as a builder the boys worked at whatever they could out of school hours and during vacations in order to keep the family going for money was very scarce at that period in the plains country and many people became so discouraged they returned to their old home farther east but the Courseys were made of firmer fibre and determined that if they could "stick it out" would in the end be rewarded for their work, which has proved true. Harry Coursey well remembers the first money he earned while still a small boy, as he hired out to H. C. Stockey of Lexington to drive a cow to pasture for which he received a check of fifty cents, a goodly sum for a small boy at that time. In the fall of 1895, the Coursey family returned to Topeka and Harry went to work there, remaining in Kansas until 1907, when he came to the Panhandle, locating on a farm in Box Butte county in the spring of the year. His father and the rest of the family came at the same time and took a Kinkaid homestead of four hundred and eighty acres on which he lived and proved up, at the same time engaging in business as a contractor. In 1909 Harry Coursey bought the livery business in Alliance from C. C. Smith which he conducted two years then in 1912, purchased Forest Allen's livery business and merged the two, which was then known as the Keeler Livery, but receiving a good offer for this concern in 1913, sold at a very satisfactory figure. Just a month and nine days later on February 1, 1914, Mr. Coursey and J. R. Keeler formed a business partnership and took the agency for the Ford automobile for Alliance and the surrounding territory, established the Keeler-Coursey Company, which at once became a most prosperous concern with business covering a wide range of country. They were forced to move twice into larger buildings in order to have more room. June 7, 1918, True Miller bought Mr. Keeler's interest in the firm which became known as Coursey and Miller. They purchased the large brick building at the corner of Third and Laramie streets which has nineteen thousand square feet of floor space demanded by their rapidly expanding business, space to display their cars, tractors and trucks as well as their accessories and for tire space as they are the wholesale distributors of the Firestone Tires for northwestern Nebraska and southwestern South Dakota. Both members of the firm are men of initiative, executive ability and business foresight which is all contained in one word "hustlers" here in the Panhandle. They are aggressive in their business methods and development of Alliance and the local district and are a valuable adjunct to the citizenship of this progressive city. From the two years past we presage a great future for the firm which is so ably managed by its owners. For more than ten years Mr. Coursey has been an auctioneer in Western Nebraska, where he has built up a fine reputation for honest dealing, careful fulfillment of business engagements and personal integrity, while his courtesy and consideration are wining him a widening clientele in all his business ventures. Being a leading livestock auctioneer he is naturally thrown in association with the rural population where he has made many warm friends and business associates. Fraternally he is associated with the Elks, Modern Woodmen and Odd Fellows.

December 23, 1901, Mr. Coursey married Miss Laura M. Titus, at Holton, Kansas, the daughter of James W. and Novelle (McCormic) Titus, the former a native of Iowa. Mrs. Coursey was born and reared in Ottawa county, Kansas, where she still lived at the time of her marriage. Two children have been born to this union: Harvey P., in the high school at Alliance, and Novella, also in the high school.

M. S. HARGRAVES, is secretary of the Alliance Building and Loan Association, at 109 West Third street, Alliance Nebraska, and is engaged in a General real estate and insurance business.

CHARLES E. CLOUGH is one of the later settlers of Box Butte county who is now spending the sunset years of life removed from active business care in Alliance. His story which is about to be related proves, however, that the late-comers, if possessed of the same traits and business characteristics, make good practically as well as did the pioneers of the early days when the very best land in the county was open for homestead entry.

Mr. Clough is a son of the Empire state, born at Fabius, Onandaga county, April 30, 1844, the son of Ephraim and Emmaline (Fitch) Clough, the former, as his son, a native of New York state. Charles was next the youngest in a family of six children, one brother, Abel, being killed in action during the war of the Rebellion. Ephraim Clough was a farmer who kept a dairy herd, sold milk and also made cheese, an industry for which New York is famous the country over. Charles grew up in the country, attended the public schools and early began to earn money as his father paid him for milking. Later the family removed to Wyoming county, New York, and there the boy entered the Wyoming Academy, later he changed to an academy at Pike and after his elementary education was completed he matriculated at the University of Michigan, Ann Arbor, where he graduated

Enos S. De La Matter

with the class of 1864 After leaving college Mr Clough came west, locating at Ackley, Iowa, where he opened a law office being engaged in his professional duties there until 1873, when he removed to Boulder, Colorado, to engage in the live-stock business Somewhat later he moved his cattle to Laramie, Wyoming, establishing himself there in 1886 For a number of years Mr Clough continued to engage in handling and running cattle but he had looked the country over and decided he wanted to locate permanently somewhere in good grazing land and with this end in view came to Box Butte county in 1892, locating a ranch ten miles southwest of the town of Alliance, where he remained until 1918 Mr Clough prospered exceedingly here on the high prairies and as the shadows of life began to lengthen concluded that it was time for him to take life a little easier and enjoy his hard earned fortune He sold his cattle and rented his ten thousand acre ranch, moving into the city to live He built a fine modern home where he and Mrs Clough are prepared to live, travel and enjoy the remainder of their days in comfort and happiness

In 1869, Mr Clough married Miss Elizabeth Pardee, at Ackley, Iowa Mrs Clough was the daughter of Erastus and Sophie (Carter) Pardee, the former a native of Michigan She is next to the youngest in a family of six children born to her parents, received an excellent education in the public schools of her native district and since her marriage has become the mother of two children Charles E, Jr, who graduated from the Alliance high school and after engaging in business married Miss Abbigal Hill, and they now live at Minatare, Nebraska, where Mr Clough is in the hardware business, and Elsie Clough Estes, who also graduated from the high school in Omaha, and now lives in the country near Alliance.

From first locating in the Panhandle Mr Clough has had great faith in what is now becoming one of the most productive sections not only of Nebraska, but the entire country and he says that he is willing to go on record any day as saying that he believes no country can be found "that will beat the sand hills of Nebraska for stock-raising" He is a broad gauged, liberal minded man who keeps abreast of the progress of this country and firmly believes in progressive movements for Box Butte county and Alliance and in support of this is ever ready with his energy, time and money to "boost" an enterprise when convinced that it tends to the development and betterment of

civic and communal conditions In politics he is a staunch supporter of the Republican party while his fraternal affiliations are with the Elks and the Modern Woodmen

ENOS S DE LA MATTER, whose many years of service as county judge has but strengthened a reputation with the people of Scottsbluff county as a man of public spirit and sound judgment, has been a resident of Nebraska since 1884 Since then he has been one of her most loyal sons taking an interest in everything pertaining to her welfare, and identifying himself with movements that have helped in her development

Judge De La Matter was born in Grundy county, Illinois June 6, 1855 His parents were Cyrus and Mary Ann (Rowe) De La Matter the former of whom was born in Canada in 1820, and died in January 1890, and the latter in Indiana in 1825, and died in August, 1863 The paternal grandfather was Martin De La Matter, who spent his entire life in Canada where he was a farmer The maternal grandfather was Robert Rowe who was born in Edinburg Scotland, emigrated to the United States, and died on his farm in Illinois in 1878 The parents of Judge De La Matter were married in La Salle county, Illinois, and they became the parents of four children, the following three surviving Robert M who is a farmer in Scottsbluff county, Enos S, who resides at Gering and Sabina, who is the wife of Martin J Tubbs, who is a farmer near Missoula, Montana The mother of Judge De La Matter was a member of the Presbyterian church and his father of the Methodist Episcopal church The latter contracted a second marriage Rachel Barnes becoming his second wife and two of their four children are living namely Cyrus, who is a resident of Sterling Illinois and Mrs Mary Ann Terrill, who lives at Sheridan Illinois

Enos S De La Matter was reared on his father's farm and spent some years cultivating it He was afforded educational advantages, after attending the local schools becoming a student in the normal school at Valparaiso Indiana In 1884 he came to Nebraska and for two and a half years worked as a carpenter in Buffalo county In June, 1886, he homesteaded in Cheyenne county in what is now Scottsbluff county, and in November following he came to the county as a resident and has continued here ever since soon being recognized as a vigorous and dependable citizen, intelligently concerned in the material social and political development of this section He was soon called upon to accept public responsibility and at a time when the county needed strong, honest, constructive officials He served as justice of the peace from 1887 to 1895, when he resigned to become a member of the board of county commissioners, which position he held for six years In 1901 he was elected county judge and has been continuously re-elected since then, showing on the bench great legal ability, open-mindedness, and straightforward justice

In 1900 Judge De La Matter was united in marriage to Mrs Mettie A Lovell, who was born in Grundy county, Illinois Her parents came from Illinois to Buffalo county, Nebraska, where her father, Isaiah Casteel, was a farmer, but later moved to Scottsbluff county and died here Judge and Mrs De La Matter have two children, namely Mary and Ena Politically he is a Republican and an important party factor in the county, and fraternally is a Mason In all public matters during the past few years in which good citizenship was a feature, he worked in harmonious cooperation with his fellow citizens

JOHN C McCORKLE, manager of the Nebraska Land Company of Alliance, is a well known and highly respected citizen of Box Butte county, and has been one of its most substantial stockraisers and ranchmen, as he is one of the early settlers of the Panhandle, having passed more than thirty-four years in western Nebraska Wonderful changes have been wrought in that time and, in a way, they may be typified by the comparison between Mr McCorkle's first home, a tent in summer and a log house for winter, on a lonely wind swept prairie, with his present handsome residence in the city In all this change and development Mr McCorkle has taken an active part and today is one of the foremost factors in opening up the still available virgin farm lands of the county

Mr McCorkle was born in Columbia, Iowa, the son of John G and Susan (Rumsey) Mc-Corkle, the former a native of Indiana He was next to the youngest in a family of nine children born to his parents and remembers with a whimsical smile that the first money he earned was helping his brother herd sheep at a salary of ten cents a day John McCorkle was a farmer in Iowa so that the children were reared in the country, attended the public school near their home which required a walk of three miles back and forth each day The boy remained at home until his seventeenth year then worked for his brother driving oxen breaking hazel brush land in Iowa

On March 12 1878, Mr McCorkle married Miss Flora McMannis, born in Oskaloosa, Iowa, the daughter of Jawhue McMannis, and to this union four children were born Minet, died in infancy, Inez L (McCorkle) Dunning, a widow who has one child six years old and now resides in Alliance with her parents, was a graduate of the Peru Normal school and taught in the Alliance schools for eight years, Orville, deceased, and Norman A , a graduate of the high school at Alliance who later took a special course in the Boyle Business College of Omaha, then entered the superintendent's office of the Burlington Railroad, remaining three years before he enlisted in the navy when war was declared against Germany Mr McCorkle was assistant recruiting officer at Kansas City four months before being transferred to New York and shortly assigned to duty abroad the *Iowan*, a freighter He made one trip across the Atlantic on her when she was made into a transport, following which he made three trips to Europe safely, and was mustered out of the service in Kansas City in April, 1919, and at present is located on a two thousand acre ranch in which he holds a large interest

John McCorkle came to western Nebraska in July, 1886, on account of his wife's health They camped out at first in a tent on Pine Ridge, Dawes county, forty miles northwest of the present town of Alliance During that first summer he broke sod on the prairie for the amusement of it as he had become very expert in this method of plowing in his youthful days in the hazel brush of Iowa The country was thinly populated, settlers were few and far between and the Panhandle of that day deserved the name of "wild and woolly west" Early in the fall a heavy snow fell and Mr McCorkle and his wife were unable to make the trip back to the home at Superior, Nebraska, so Mr McCorkle built a comfortable log house, took a pre-emption claim, remained during the winter of 1886 and spring of 1887, proved up on it and was thus enabled to put a mortgage on the land and raised sixteen hundred dollars in cash for his living expenses and also bought a good team of oxen, the best animals for breaking prairie sod He then filed on a homestead, commuted this claim and raised another sixteen hundred dollars with which money he engaged in the live-stock business That was the heigh-day of the cattle business as there were vast stretches of open range over which the great baronial cattle companies ranged immense herds, while smaller dealers were also able to take advantage of the government privileges Mr McCorkle bought and sold live stock on a large scale from 1887 to 1890 living during that period on the homestead The latter year he moved into the town of Hemingford, there establishing a cattle and hog market of his own and at the same time began to deal in horses In 1893, he shipped two cars of the finest horse flesh ever gathered together in the northwest as they were from four to six years old and weighed from twelve to fifteen hundred pounds, selling at about seventy-seven dollars apiece in the St Louis market, a remarkable price for that time In 1895 Mr McCorkle sold his business to accept a position as foreman of the Carey and Arle Cattle Company at a salary of twenty-five dollars for himself and wife, who cooked for the outfit, but it needs not be said that the salary was very soon raised when the owners found what an excellent manager they had For nearly eleven years the McCorkles remained with this company who made it well worth their while to do so In the fall of 1905, they came to Alliance to live and Mr McCorkle went into the real estate firm of Watkins and Tagean, which then became Watkins, Tagean and McCorkle, though Mr McCorkle still had outside business interests as a live-stock man Two years later Mr McCorkle bought out his partners in the real estate branch and reorganized it as the Nebraska Land and Loan Company, becoming its chief executive and manager Under his management anyone can buy an interest in the company, as it will sell a customer a farm at an agreed price, the customer holding the deed and title in his own name and when the land is sold he receives one-half the advance on the land for his profit in the company, which incurs all the expense of handling and selling the land, thus putting the time and expense against the stockholder's investment, though he holds the security, which makes a most satisfactory and absolutely safe investment The office of the company is located on the first floor of the First National Bank, on Box Butte Avenue, where the books show more than a hundred satisfied customers At the present time Mr McCorkle and his son Norman are in charge of the business which has proved a success in every way and a most desirable investment Their farms and ranches offer many and varied opportunities to anyone who has money on which they desire a good return and at the same time an absolutely sound proposition

FREDERICK A BALD — In every community will be found quiet, industrious business men following different vocations, without whom the ordinary industries of modern civilized life could not go on, and very often it will be found that they are self made men, having unaided, built up their fortunes Such a type of the progressive and prosperous business men of the Panhandle is exemplified in Frederick Bald, the president of the Wyoming-Northeastern Oil Company, also engaged in the real estate business and in the practice of his profession

Mr Bald was born at Aurora, Nebraska, October 13, 1881, the son of Louis B and Matilda (Kemper) Bald, the former born and reared in the vicinity of Philadelphia, Pennsylvania He was an ambitious young man and after attaining his majority determined that he would win a fortune for himself and believing that the west offered more opportunities emigrated from his native state and settled in Grant county, Wisconsin, but did not prosper there as he wished and learning of the fine farms to be had in Nebraska came here in 1876, locating near Aurora on a farm he bought from the railroad company Mr Bald arrived in this commonwealth with very little of what we term "worldly goods," but was able to pay the required eight dollars an acre for his land, and that his vision of the future was a wise one may be gained from the fact that today this same farm would bring more than three hundred dollars an acre on the market were it for sale Frederick, better known to his friends as "Art," attended the public school near his home during the winters and worked on farms in the summer time and tells that at one time he worked an eight hour shift all right only it was twice a day, eight hours before noon and another eight hours after dinner, receiving for his labor fifty cents a day and well recalls that they always had salt fish for breakfast After completing the district school the boy took a four year course in the Aurora high school, graduating in 1900 The following two years he taught in the country schools then matriculated in the law school of the State University at Lincoln, obtained his degree of LL B and was admitted to practice in the courts of Nebraska, on June 1, 1904 He opened an office in Aurora in partnership with C P Craft a class mate from the university and they continued to be associated until 1909, when Mr Bald went to Watertown South Dakota, to engage in a real estate business In 1911, the year of the great drought he returned to Nebraska, locating at Central City, to engage in the practice of his profession seven years. In 1916, he was appointed to fill the unexpired term of Edward Patterson as attorney of Merrick county, and two years later came to Alliance. Since locating here Mr. Bald has entered into partnership with L. C. Thomas and they are engaged in business under the firm name of Thomas-Bald Investment Company. Within a short time they have been the moving spirits in the organization of the Wyoming-Northeastern Oil Company, which has located and leased some thirteen thousand acres of oil lands in the fields of eastern Wyoming. The company is incorporated for a million and is preparing to open up their business on an extensive scale. The officers are: F. A. Bald, president; C. M. Loomey, vice-president; L. C. Thomas, secretary; and A. M. Miller, president of the American State Bank of Hemingford, treasurer, while the board of directors consists of F. A. Bald, L. C. Thomas, C. M. Loomey, R, M, Baker of Alliance, A. M. Miller, P. Michael, of Hemingford, F. W. Melick of Hemingford, and F. T. Morrison and C. F. Gruenig of Omaha.

March 28, 1906, Mr. Bald married at Hampton, Nebraska, Miss Ella Kemper, who was born there, the daughter of Henry and Serepta (Smith) Kemper, both natives of Illinois. Mrs. Bald was an only child, she graduated from the Hampton high school and then took a special musical course at Nebraska Wesleyan College. Three children have been born to the Balds: Maurine, Warren and Helen L., all at home.

Mr. Bald owns a fine modern home in Alliance and also about five hundred acres of farm land which he gives general supervision and from which he says he gets a comfortable nest egg each year. In politics he is a Democrat, is a member of the Presbyterian church, while his fraternal affiliations are with the Elks and the Knights of Pythias.

JOHN O'KEEFE and WILLIAM L. O'KEEFE, the owners and managers of one of the progressive real estate firms of Box Butte county and the city of Alliance belong to a prominent and well known family that came here in pioneer days and helped make history in the upper Platte valley.

John O'Keefe was born in Fulton county, Illinois, April 28, 1867, the son of John and Sarah (Kelly) O'Keefe, the former a native of Ireland, who came to America and located in Illinois where his children were born. John was the second in a family of four children and spent his boyhood years on his father's farm, attending the public schools in the winter sessions and helping around the home place during his vacations, thus early becoming well acquainted with the business side of farm industries. He was an ambitious youth; desiring to establish himself independently he came to Nebraska where he could obtain government land in 1886, and took up a homestead five miles south of Hemingsford. Mr. O'Keefe put up the usual "soddy" for a house as lumber was almost unknown on the plains at that early day. He drove into Box Butte county in true pioneer style, freighting his goods from Hay Springs, sixty miles away, settled in a veritable wilderness where habitations were few and far apart and civilization was yet in its most primitive form. On the trips to and from Hay Springs he was forced to ford the Niobrara river, much of an undertaking when the water was at all high as bridges had not yet been built over the stream and a bitter cold experience for man and beast in the winter time. Feul and posts for the absolutely necessary fences on the farm were hauled from Pine Ridge nearly thirty-five miles away. Water was the great problem of the first settlers and to obtain some on his place Mr. O'Keefe dug a well a hundred and ten feet deep for his own use and to water his stock. It was one of the first in his section and neighbors came as far as five miles with tank wagons to secure water for themselves and their animals. In the fall of 1887, Mr. O'Keefe was elected treasurer of the newly created county of Box Butte, served two years and was re-elected. During the four years he held office John O'Keefe, Jr., acted as his deputy and in the fall of 1889, was elected to succeed his father and in the summer of 1892 the father was appointed postmaster of Hemingsford under President Cleveland and held office four years. Upon the close of his term of office John O'Keefe, Jr., came to Alliance in the spring of 1894 to enter the post office as deputy postmaster, serving in that capacity three and a half years, then joined his brother Dan in opening up and developing a ranch. They met with success and from time to time added to their original holdings until they owned ten thousand acres of land, on which they ran about a thousand head of cattle and horses, cutting annually two hundred and fifty tons of hay. As the country was new, for many years they had open range but as settlement continued this was restricted and after about fifteen years of successful ranching the brothers disposed of all but some three thousand acres of land lying ten miles southwest of Hemingford, but still own scattered and valuable tracts throughout the country At the present time they rent their ranch and Mr O'Keefe with his son William are engaged in handling real estate and insurance in Alliance, with an office in the Alliance National Bank Building Mr O'Keefe has invested heavily in city property and is active and energetic as most men many years his junior, he is progressive in his ideas and a "booster" for the Panhandle The O'Keefe family are members of the Catholic church while in politics he is a Democrat Mr O'Keefe's fraternal affiliations are with the Elks, he is a Knight of Columbus and a member of the Country Club

May 25, 1891, Mr O'Keefe was married at Nonpareil, Nebraska, to Miss Lucy Shipley, born near Bloomington, Illinois, the daughter of Robert S and Frances (Edwards) Shipley, the former a native of Kentucky Three children have been born to this union William L, Everett B, a graduate of the Alliance high school, took a course in dentistry at Creighton University, Omaha, graduating with the class of 1919, and is now established in practice at Alliance, an energetic purposeful young man for whom a bright future is in store, and Sarah F, a student at Loretta Heights Academy, Denver, Colorado

William L O'Keefe is a native son of Nebraska, born in the city of Alliance May 29, 1894, and inherits from his father all the excellent and versatile qualities of the Irish with a steadying influence from his mother's side of the family that have combined to make him a business man of far vision and full of resource yet conservative in his financial affairs, a combination which can not but bring success to the fortunate possessor He is the oldest of the three children in the family and in youth attended the public schools of Alliance After graduating from the city school in the city he matriculated at the Christian Brothers College at St Joseph, receiving his degree with the class of 1914 Having a brilliant scholastic record William O'Keefe was chosen as chief clerk to the secretary of state of Nebraska a year after leaving college, and served in that capacity until he responded to the president's call for volunteers when war was declared against Germany, when he enlisted in November, 1917, in the Signal Corps Balloon division and was a candidate for commission Later he was assigned to the Flying Cadet Corps, and stationed at Fort Omaha, serving in this branch of the service until after the signing of the armistice, when he was dis-

charged and returned to Alliance to form a partnership with his father in a real estate and insurance business Mr O'Keefe is one of the younger generation of business men who are infusing new methods and new blood into the financial affairs of the Panhandle and because of such are standing out pre-eminently as the leaders of progress in this section for which they are doing so much to develop

April 10 1918. Mr O'Keefe married Miss Pauline Golden a graduate of the Catholic schools of Milwaukee Wisconsin and Omaha, Nebraska, and also of St Mary's Notre Dame, Indiana She is a woman of gracious presence high culture and is the chatelaine of one of the most hospitable homes in Alliance Mr and Mrs O'Keefe have two children a daughter Alice Lucile, aged two years, and William John, an infant

Mr William O'Keefe and his father are both indefatigable workers, they, so to speak, are "on the job" all the time, both are well read men, who keep abreast of the movements of the day and seize every opportunity to support and promote movements for the civic and communal welfare of Box Butte county and Alliance A most brilliant future seems in store for the junior member of the firm, who is an Elk and a Knight of Columbus

ST AGNES ACADEMY of Alliance is the leading Catholic educational institution of Western Nebraska It is conducted under the direction of Mother Superior Henrietta of the Sisters of St Francis, whose mother house is located at Stella Niagara New York The Academy was erected in 1908, under the auspices of Reverend William McNamara, who for some years had cherished the idea of founding an institution of learning for the benefit of the children of the parish as well as to give exceptional advantages to non-resident students Although the building had not yet been completed, the school was opened to students in September 8, 1908, when eighty nine pupils enrolled The Academy met with the warm approval and hearty support of all the Catholics of Alliance and the surrounding country so that the number of pupils increased very rapidly and within a short time it was found necessary to erect another building The work on the new structure was begun in the fall of 1910, and the rooms were ready for occupancy the following spring Both buildings are of the highest type architecturally and are equipped with every modern appliance and improvement conducive to the health and convenience of the pupils They are heated with steam and lighted with electricity. The class rooms, study halls, gymnasium and dormitories are commodious, well lighted and ventilated and are suitably furnished. The laboratories are up-to-date in every way and are well equipped and offer abundant opportunities for class and lecture demonstrations and for individual work in physics, botany and agriculture. The extensive play grounds are well supplied with swings, teeters, slide, tennis-courts and general recreation space. During bad weather the gymnasium affords ample space for recreation for the students who are encouraged to take advantage of the excellent facilities to build up the bodies. As in all schools and colleges, basket ball is the favorite indoor sport during the cold season.

St Agnes Academy is accredited at the State University and enjoys all the powers and privileges granted by it to the leading educational institutions of the state. It is the aim of the faculty to maintain this high standard of scholarship, and we feel this is also the desire of all those who are interested in the welfare of the Academy. The Sisters endeavor to lay a solid foundation in the mind and to develop the character of the pupils upon which to build the super-structure of moral and religious life as well as purely scholastic attainments. Thus the students receive alike the purely scientific along with the higher training of religion. A close supervision is exercised over the students but in such a manner as to exclude all idea of harsh espionage. The rules and regulations of the Academy are enforced with mildness and consideration, but when there is question of the good of the students or the reputation of the Academy, great firmness is exercised.

The curriculum offered is as follows: primary, grammar, high school, commercial, music and art courses. Pupils who complete the grammar grades are awarded diplomas which admit them to any high school in the state. The courses in the high school are those of the usual college preparatory schools, normal training and commercial. The pre-college course qualifies the student to enter the State University or any standard college. The normal course is pursued during the junior and senior years and students who finish it receive a second grade certificate and after the completion of one year's successful teaching a first grade certificate will be issued to them without further examination. Commercial subjects may also be pursued during the last two years and while this course possesses all the advantages of a thorough business training,

it affords the student a more liberal education than that received in the ordinary business college. A special commercial training is offered for those who wish to prepare themselves for a higher place in the world of finance.

The course in music is under the able direction of competent teachers. Instruction in instrumental and vocal music is optional and is open to all students, provided they can carry this extra work without detriment to their regular school work. A complete course in china painting and decorating is also offered while oil painting, water color, and charcoal and pastel are also taught.

The enrollment for 1920, is one hundred and forty-seven day pupils and one hundred and four boarders representing the states of Nebraska, Wyoming, Colorado, Kansas and Washington. During the past two years it has been found necessary to limit the number of boarders because of limited accommodations. Owing to abnormal conditions resulting from the World War, it has been found impossible to begin the erection of the new building which has been planned but it is sincerely to be hoped that work on the contemplated new wing will soon be started.

Since June, 1920, the Academy is supervised by Mother Gerard who succeeded Mother Henrietta, when the latter was removed to Columbus, Ohio.

MRS NELLIE HARVEY — More than casual distinction attaches to the personality and record of this woman, for though she is a recent citizen of Alliance, she came to Nebraska as a child in pioneer days and had to endure her share of the hardships that marked the stages of development and progress in this now favored commonwealth. She is a woman of keen business capacity, broad minded and is typical of the women of post war conditions, as she like the thousands of other ambitious and educated women of our broad land, are stepping in and filling important business positions.

Mrs Harvey was born in England, the daughter of Edward and Penelope (Ellison) Weston, both natives of the tight little island that has given to the United States the greatest proportion of our best settlers. She was the second in a family of five children born to her parents. The father emigrated from England to the United States in 1878, with the idea of becoming a farmer on the broad western plains. With this end in view he came to Nebraska, located on land where Orleans is now situated but after remaining a short time left as there was no timber to shelter stock and no available water. He changed the filing of his homestead, going about five miles northwest of the original allotment, to a farm lying along the bank of the Republican river, where he soon established his family and lived until his death seven years later. The widow left alone with a family of young children soon sold out and moved into the town of Orleans which had been established on the old Weston homestead. Even then the settlers of the village had to haul their supplies and provisions from Kearney or Red Cloud, the nearest towns of any size where merchandise could be purchased. The mother remained a short time, then taking her little family returned to the old home at South Killworth, England, as the burden of being both father, mother and provider of the family was proving too much for her fragile strength. Later the Weston family, because of Mrs Weston's failing health returned to Hastings, Nebraska. Mrs Harvey attended the public schools in England and after she came to Hastings entered the high school, from which she graduated. In 1901 she came to Alliance, where she has since resided.

On August 5, 1918, Nettie Weston was married to Joseph C Harvey at Hot Springs, South Dakota. After their marriage Mr and Mrs Harvey managed the Silver Grill Cafe for about eight months but were made an attractive offer for the cafe business and sold out. June 18, 1920, they bought the fixtures for a cafe and rented the Vaughn building on Box Butte Avenue, and now are running a first class eating house. We predict for them a splendid business in the line as the cafe can seat over one hundred people.

GEORGE J HAND, M D — For a period of sixteen years Dr Hand has been established in the practice of his profession at Alliance and the unequivocal success which he has achieved in his exacting vocation fully attests to his high professional attainments and his faculty in the effective application of his technical knowledge. The doctor long controlled a substantial and representative practice and then took special courses in diseases of the eye, nose, throat and ear and now devotes his entire time to these specialties which gives him mostly office and hospital work, as he has many delicate and difficult operations. Dr Hand commands high place in the popular confidence and esteem of the residents of the city and is essentially one of the representative members

of the Box Butte County Medical Association.

The doctor was born in Vermillion, South Dakota, August 2, 1875, the son of Redmond and Mary (Keough) Hand, the former a genial son of the Emerald Isle, born in County Roscommon, who came to the United States when a lad of ten years, when his parents emigrated to the new world that they and their children might take advantage of the great opportunities afforded the new settlers in our broad land. Mary Keough was a native of the Bay State, born in Lowell, where she spent her early life and was educated. After landing on our shores Redmond Hand came west to locate at Dubuque Iowa, where he was married in 1861, and the same year went to Vermillion with his bride. They settled on a homestead where Mr Hand placed the required improvements and proved up. Subsequently they moved into the town and for some years Mr Hand ran a hotel and livery establishment then opened a meat market. All of his ventures were proving profitable. In 1881 the town of Vermillion was practically wiped out by the great flood, when an ice gorge in the Missouri river broke and the water which had been held back by the ice dam, swept down stream carrying all before it. Mr Hand lost practically everything he had as his hotel was landed on a sand bar three miles down stream from the town site. Following this disaster he engaged in contracting and for nearly eight years was constructing grades for the new railroads that were creeping westward across the great plains. As a consequence of this George attended school in various towns along the new road, for the family was continually moving to keep up with the advancing construction crews. After finishing the grade schools Dr Hand graduated from the high school at Hay Springs in 1897, then attended the Chadron Academy for a year. He had already decided upon a professional career but wished experience and also wanted to earn some money. When he was offered a school to teach he accepted and was a successful pedagogue until 1890, then entered the medical department of the Iowa State College. Three years later, in 1903 he won by competitive examination, the undergraduate internship in the Iowa Homeopathic Hospital at the University of Iowa and a year later graduated with high honors and the degree of M D. In October the doctor came to Alliance, opened an office and became established in medical practice, since which time he has continuously served the city and surrounding country. As a phy-

sician and surgeon he soon won a high reputation for skill and successful operations and
after enjoying a most gratifying general practice for several years he specialized in diseases
of the head, throat and ears and now has a
growing clientele along these lines At the
present time Dr Hand maintains his office
in the Imperial Theatre Building He is a
member of the Box Butte County Medical society, the Nebraska State Medical association
and the American Medical Association and is
United States Pension Examiner for his district He is also physician on the county board
of insanity During the World War the doctor was a member of the Medical Advisory
Board and at the present time is serving as
city physician of Alliance Dr Hand is a
widely read man not only along lines of his
profession but keeps abreast of all movements
of the day, he is progressive in his business
and thoroughly believes in supporting every
movement that tends to the development of
the county and improvements of civic and communal affairs In other words he is a "booster" for this section of the Panhandle The
doctor is a Mason and a member of the Country Club

HARRY A DUBUQUE, the owner and
manager of the Imperial theatre, the leading
moving picture house of Alliance, is a man
who during his business career has followed
various occupations in several parts of the
middle west as well as the Dominion of Canada Since starting out in life independently
he has been mill hand, foreman of a spinning
mill, foreman of a milling machinery contruc-
tion party, barber, ranchman, member of the
Canadian Northwest Mounted Police and now
is in the moving picture business and in all his
several fields of endeavor his versatility has
assisted him to well deserved prosperity He
is a native of the Dominion of Canada, born
September 21, 1880, at St John, the son of
Joseph and Elsie (Rainville) Dubuque, the
father being a native of the same country
Harry was next to the youngest boy in a
family of five children and when he was only
two years old his father died leaving the
burden of supporting her little family to the
mother She had little means to earn money,
removed to Pawtucket, Rhode Island There
Mrs Dubuque and the oldest girl found employment in a cotton mill and managed in
some way to keep the children with her Child
labor laws had not been passed at that time and
young children were not prohibited from working as they are today, so when Harry was

only six years old he too, went to work in the
J P Coats thread mill earning two and a half
dollars a week He worked nine months of
the year and attended school three, so his education was not entirely neglected He continued to work in the cotton mills and when only
fifteen was promoted to the positions of foreman over nearly two hundred operatives, the
youngest man to hold such a responsible position He had not, however, gained it easily
for he had worked hard, given study to the
machinery used and won the promotion on
pure merit Three years later Mr Dubuque
was made a most attractive offer by the
Whitesville Machine Company to go out on
the road setting up cotton milling machinery
as he had an exceptionally good understanding
of the machines and the manner of their operation Just three weeks after entering on the
new work, the boy, for he was nothing more
in years, was given charge of the men as foreman While still working in the mill the young
man began helping in a barber shop nights to
earn a little more toward the upkeep of the
home He learned the barber trade in this
manner and as the shop was mortgaged he
bought it, cleared up the mortgage in less
than a year and sold it at a profit of over sev-
en hundred dollars Soon after this he came
west, locating at Leads, South Dakota, and
opened a barber shop, but five months later
disposed of it to go on a ranch but three weeks
later left for Canada, where he joined the
Northwest Mounted Police at Regina He
remained with this organization four months,
but as the law provided that the personnel
of the Royal Police must be British citizens and as Mr Dubuque was not willing to give up his United States citizenship he bought his release for four hundred
dollars and returned to Belle Fourche, South
Dakota Near that town he took a position
as foreman of the Half Circle M ranch and
remained four years and a half

On October 10, 1909, Mr Dubuque was
married at Belle Fourche to Miss Myrtle Pengree, who was born at Mitchell, South Dakota,
where she was reared and later graduated from
the Methodist high school She was the daughter of Ira and Mary (Humphrey) Pengree,
both natives of the buckeye state, the youngest
in a family of seven children born to her parents After spending more years on the ranch
Mr and Mrs Dubuque came to Alliance in
May, 1911 Mr Dubuque looked the business
situation over, believed he saw a fine opening
in the moving picture theatre proposition so
bought the Majestic which he at once remodel-

WILLIAM MAUPIN AND FAMILY

ed and changed the name to "Empress." He was the pioneer movie man to give a change of program every day regardless of the initial expense and that he was far-sighted needs not be told when we learn of the phenominal success with which the business has met. The first theatre was soon playing to over crowded houses and then Mr Dubuque remodeled the old Charter Hotel, practically rebuilt it, and in 1919 opened the new fifty by one hundred and forty foot building as an up-to-date, fire proof structure with a seating capacity of nearly a thousand. He installed a ventilating system that delivers thirty-two thousand cubic feet of fresh air a minute so that it is absolutely sanitary. A fine seventeen thousand Robert Morton organ that plays twenty-three instruments is an added equipment for the pleasure of the patrons and they have responded generously in support of this most pleasurable project, as this theatre is the most modern in northwestern Nebraska. Mr Dubuque is a progressive man in ideas and his business is a great addition to the business circles of the city. He is an Elk, a Modern Woodman and Knight of Columbus. H A Dubuque is now interested in the Osage and Kansas oil field and sheep business at Belle Fourche, South Dakota, was also the first picture show man in the state of South Dakota.

WILL M MAUPIN, editor and proprietor of the *Midwest*, which he founded at Gering, Nebraska, in October, 1918, and which has proved a profitable enterprise, is not only a practical printer, but is a widely experienced newspaper man and somewhat prominent in labor circles. He was born in Callaway county, Missouri, August 31, 1863, the eldest of a family of eight children born to William Taylor and Sarah (Miller) Maupin.

Both grandfathers of Editor Maupin were born in Kentucky and from there moved to Missouri, the paternal grandfather, George Maupin, a farmer and slaveholder, settling in that state in 1804. William Taylor Maupin, father of Will M Maupin, was born in Callaway county, Missouri, and for sixty years was a minister in the Christian church. His death occurred in Hennesey, Oklahoma, in 1911. He was married in Audrain county, Missouri, to Sarah Miller, a native of Missouri, who died at North Bend, Nebraska, in 1894. Of their three surviving children, Will M is the eldest, the others being Kittie who is the wife of George L Burkhaltere an employe of the United States government at Galveston, Texas, and T W, who is in business with his older brother.

Will M Maupin attended the public schools of Holt county until sixteen years old when he entered a printing office at Oregon Missouri and, as old veterans of the case who understand the lure of the types would declare, "settled his fate." He has never since been able to escape from the atmosphere of the printing office, in which he has filled every position from the lowest to the highest. He learned his trade with old-time thoroughness and before he left Missouri owned his own newspaper at Craig. In 1886 he came to Nebraska for a number of years afterward being connected with some of the leading journals of the state, working first at Fall City. He then went to Omaha and for nine years was with the *World-Herald*, and for the next ten years worked on the *Commoner* at Lincoln. From the capital he went to York, and after three years of newspaper work there, came to Gering and in the same year founded the *Midwest*. It is a well conducted journal that has made its way into many homes and has become almost a necessity to the business men who give it hearty support. Mr Maupin has a circulation of 900 paying subscribers and the list is constantly increasing.

In October, 1895 Mr Maupin was united in marriage to Miss Lottie Armstead at North Bend, Nebraska who was born in Ohio. Their children are as follows: Louis who is in the banking business at Baggs, in Carbon county, Wyoming, Lorena who is the wife of L B Lewellen, of Lincoln, Dorothy K who resides at home, and Richard Metcalfe, Margaret B, Charlotte May, Jack Robbins, all of whom attend school, and Dan Whitmer, who is an infant. Mr and Mrs Maupin are members of the Christian church. He is identified with the Knights of Pythias, as was his father who was also an Odd Fellow and a Mason and with the Elks. In politics he has always been in accord with the Democratic party. In 1909 he was appointed state labor commissioner and served in that office for two years and during seventeen months of the World War he served as director of publicity, when he resigned.

CHARLES E HERSHMAN, M D, deceased, was one of the younger members of the Box Butte medical fraternity. Dr Hershman was a physician of wide experience as he had charge of the various branches of the medical service of the Burlington Railroad for a number of years and after coming to Alliance in 1911, succeeded in establishing himself firmly in a position of prominence in professional circles, as well as in the confidence of the public. He was a native of the Hoosier state born in Jasper county, Indiana, January 12, 1885, the son of Frank M and Mary A (Hofferlin) Hershman, the former a Buckeye by birth, while the mother, like her son, is a Hoosier, born at Evansville Charles was the oldest of the five children born to his parents He attended the public schools during the winter and worked on his father's farm in the summer time, growing up sturdy and strong and able when it came to farm business to do his share of the work and enjoyed going back to the old farm to visit his father and to recall the old happy days of childhood After graduating from the high school at Rensellar, the young man matriculated at Valpariso University, Valpariso, Indiana, spending three years in study, specializing in pharmacy After receiving his degree in his home state Dr Hershman decided that he preferred medicine to pharmacy for a life vocation and realized that his pharmaceutical training would help him in his profession He entered the Chicago College of Physicians and Surgeons, took a four year medical course and was granted his M D degree June 18, 1908 He passed a brilliant examination for the internship of St Joseph's Hospital, Joliet, Illinois, serving in that capacity for eighteen months In 1919 the doctor opened an office for the practice of his profession at London Mills, Illinois, but within ten months had been appointed medical examiner for the Burlington railroad with headquarters in Chicago A year later Dr Hershman was transferred to Alliance but resigned January 1, 1912, as medical examiner and immediately was reappointed to the position of surgeon for this district of the Burlington with headquarters at Alliance As soon as this change had been made the doctor opened an office for his private practice which grew with most gratifying rapidity He was a young man fully equipped in every way to take charge of medical and surgical work and his great success in Alliance gained for him a wide clientele here and throughout the surrounding district After locating in the Panhandle the doctor became a convert to the great possibilities of this section and took an active part in all movements that tended to the development of Alliance and the county He was a member of the Box Butte County Medical Society, the Nebraska State Medical Association, and the American Medical Association His office was located in the Guardian Trust Building on Box Butte Avenue In addition to his medical duties and its many calls, Dr Hershman be-

came interested in the business of Alliance and demonstrated his faith in the future of the city by investing heavily in bank stock and became president of the Guardian State Bank and Trust Company and it was due largely to his progressive and constructive policies that this institution holds the high place it does in the financial circles of Nebraska and the Panhandle, for while he was progressive in his ideas and methods he was conservative in all business dealings and by these qualities furthered the interests of the bank and won for it the confidence of the citizens of this district

On October 18, 1913, Dr Hershman was married at Alliance to Miss Dorothy Hoag, a native of Blue Springs, Nebraska She was the youngest of six children born to her parents Mrs Hershman received her education in the Council Bluffs, Iowa, high school and after graduation took a course in a commercial college; then accepted a position in the office of the general superintendent of the Burlington Railroad at Alliance where she remained for five years previous to her marriage Two sturdy, healthy boys became members of the Hershman family Robert, aged four and Paul F, past three Mrs Hershman is a member of the Episcopal church, while the doctor belonged to the Christian church In politics he was a Democrat, while his fraternal affiliations were with the Masonic order and he was a Shriner and a member of the Elks

December 20, 1920, Dr Hershman was almost instantly killed by an electric shock while treating a patient in his office with an X-ray machine Dr Hershman's death is a loss to the entire community as he was the type of man and citizen no town can afford to lose Though a young man his energy and enterprise had made itself felt in the commercial and professional life of Alliance He contributed both his talents and money to the upbuilding of the city, and gained a prominent place in public affairs As president of the Guardian State Bank, he stood at the door of commercial activity and as a physician and surgeon ranked high

LLOYD C THOMAS, is one of the younger generation of business men of the Panhandle who are making history in Nebraska Mr Thomas is a native son and thus far in his business career has displayed the push and energy for which the citizens of this state have won a high and enviable reputation He was born at Elwood, Gosper county, July 8, 1889, the son of John and Dora L Thomas Lloyd received his elementary education in the public schools of Maywood, Alma and Beaver City and a denominational school at Orleans In 1892, the parents with their six boys moved to Omaha, where Lloyd entered Boyles Business College for a special commercial course When only nine years of age the boy had started in to learn the printers' trade so that by the time he finished school he had served his apprenticeship as a printer and was competent to handle any kind of work in that line Upon leaving the commercial school he accepted a position with a piano house, remaining two years before going on the road as salesman for typewriter companies In November, 1907, at the age of eighteen Mr Thomas won third place in a sales contest conducted by the Oliver Typewriter Company among its three thousand salesmen, his territory covering western Nebraska and eastern Wyoming

In the spring of 1908, he accepted a position in Alliance with a real estate firm and in December of the same year purchased the Alliance *Herald*, of which he has been chief owner and editor most of the time He has found that his practical education as a printer of inestimable value in carrying on the paper On February 17, 1908, Mr Thomas married Miss Belle M Liveringhouse, of Wayne, Nebraska, the daughter of Mr and Mrs John Liveringhouse, who were pioneer settlers of O'Neill, Nebraska During a part of 1912 and and also the folowing year, Mr Thomas was the manager of an irrigation company at Lingle, Wyoming, managed a large irrigated ranch for the same concern and published a country newspaper, any one of these lines being considered a busy job for a man He still retained his newspaper interests and residence at Alliance, however, and afterwards returned here to make his permanent home In November, 1916, Mr Thomas, better known throughout the Panhandle and Box Butte county as "Lloyd," was elected representative to the Nebraska Legislature from the district comprising Box Butte and Sheridan counties He served in the regular session of 1917 and the special session of 1918, being one of the introducers and sponsors for sixteen bills which became laws, including the Nebraska prohibition enforcement law, the eighteen-milk-per-hour railroad stock transportation law, the twice-a-month pay roll law for railroad employes, the woman suffrage law, the mineral leasing law, and other enactments that are considered progressive and desirable legislation Of his work in the legislature the *Nebraska State Journal* said in part, "While many men are entitled to distinguished mention for their services in

securing prohibition for the state of Nebraska, there are three whose staunchness at a trying time gave them strong claims for the honor of making the prohibition bill the tower of strength that it is These men are representatives, Norton of Polk county, Flansburg of Lancaster county and Thomas of Box Butte county These three men composed the membership of the conference committee that fought it out with the three wet conferees sent by the senate and wrested from them by sheer bulldog tenacity all the vital things that the contest was over They were able to do this in part because they knew the House was back of them, but it required three men who knew what they wanted and who refused time and again to give an inch from their position to stand up under the tremendous pressure brought to bear to get near beer through for the brewers If near beer had been permitted, prohibition would have been impossible so far as beer selling is concerned Representative Thomas is one of the owners and editors of one of the liveliest papers in the state, the Alliance *Herald* He was elected by a big majority as a representative for Box Butte and Sheridan counties, in a Republican district He is a Democrat and was one of the leaders of the last session"

Mr Thomas has been given credit by many for securing more favorable advertising and publicity for western Nebraska than any other one individual through his work in the state legislature, by his editorials and newspaper articles and his speeches made telling of the resources of this section of the state His activities at one time led to his nickname, "Live wire Louis" At the present time Mr Thomas is devoting most of his time to his real estate and investment business in Alliance and to the development of western Nebraska potash fields and the oil fields of eastern Wyoming Although given a deferred classification in the draft during the World War, he volunteered and took the examination for an officers' training camp and had the war continued until December, 1918, he would have been in training at Camp Fremont, California, the last of the "six Thomas boys" to enter the army At the present time he is county chairman of War Savings for Box Butte county and takes an active part in all public affairs that tend to the development of this section Mr Thomas is secretary of the Potash Highway Association is publicity chairman of the Nebraska State Volunteer Fireman's Association and a member of the Alliance Volunteer Fire Department Travelers

Protective Association, Royal Highlanders, Eagles, Odd Fellows, and other organizations and fraternal associations. He is a member of the Methodist Episcopal church, takes an active part in the Commercial Club affairs and was at one time secretary of the Alliance Commercial Club.

JASON B. WADE, pioneer, frontiersman and early settler, is probably the only man now alive in Box Butte county to locate in Nebraska as early as 1872. His career has been one in which he has had varied and interesting experiences, from hunting buffalo on the western plains, and passing through two Indian wars without, as he expresses it, "seeing an Indian," to the civilized existence of these modern days, and few men many years his junior bear so few of the scars of life.

Mr. Wade was born in Michigan, March 15, 1848, the son of George W. and Lucy G. (Bass) Wade, the former a Buckeye by birth while the mother was a Pennsylvanian. He was the oldest in a family of ten children consisting of five boys and five girls. The family moved to Illinois by ox team in 1852, and in 1854, to Boone county, Iowa, in the same manner and Mr. Wade says that the first negro he ever saw ferried them across the Des Moines river. George Wade bought a farm of Captain Berry, of Civil War fame, located in Boone county and there Jasen earned his first money dropping corn for fifteen cents a day. He helped his father on the frontier farm in the summer and attended the district school winters. They had only two teams at the time they came to Iowa, one of horses and an ox team for breaking. They planted a hundred apple trees on the farm and people laughed at them for it, as they believed, they would not grow "in such a country" as they expressed it, but they were wrong. Two miles from the original farm George Wade bought swamp land for fifty cents an acre that today is worth two hundred and fifty dollars an acre. In 1855, the Sioux Indians rose and attacked Fort Dodge but the Wade family were not attacked though warned of the danger. The only Indians they saw were some peaceful ones of "Old Johnny Greene's tribe," Mr. Wade says. These hardy pioneers suffered untold hardships and privations in Iowa; one cold winter they lived practically on elk meat and made the shoes from the hides of the animals. During the Civil War the settlers were compelled to hunt for deserters in their district. In 1872, Mr. Wade and several companions came by ox team to the location where Orleans now stands

on a buffalo hunt, using a prairie schooner to live in on the trip. They killed wild turkeys along the Republican river and the next spring, Mr. Wade, accompanied by his wife and one child, a neighbor, his wife and child made the trip overland from Boone county to the homestead near Orleans on which Mr. Wade had filed on his first trip. They settled on the claims farmed a little, but supplies were so scarce and hard to get that the men wore shirts made from flour sacks. The drought came that summer and all the crops that were not burned up were destroyed by the grasshoppers and Mr. Wade says that "if it hadn't been for the buffalo, elk, deer, antelope, and jack rabbits and cotton tails, wild geese and cranes as well as the grouse and fish, together with the flour and money sent by friends in the east, we surely would have starved to death." In the fall of 1873, accompanied by three friends Mr. Wade went to McCook, Nebraska, which consisted of but three log houses with one store also serving as postoffice and the merchandise was a poor scant stock but they bought what was absolutely necessary, then camped up on the Republican river to hunt buffalo. One night Mr. Wade lost his companions and spent the night alone wrapped in the skin of a buffalo he had killed and skinned, sung to sleep by the coyotes, after a supper of buffalo sirloin cooked on a spit over a "chip" fire, that tasted as good as a meal worth a hundred dollars. They secured a number of buffalo, put the meat in barrels and sent the hides to Fort Wallace for sale, getting but a dollar and a half for each as the pelts were not yet in prime condition. On the return trip at Red Willow, Mr. Wade learned of the death of his child from a mail carrier. In the fall of 1873 there was an Indian uprising and Mr. Wade was appointed a corporal in the company organized to fight them but they did not come that far and this was the second Indian war with no Indians. April 1, the following spring the Wade family consisting of father, mother and the second child, left Orleans in a wagon drawn by horses for Boone county. On his return to the old home, Mr. Wade bought a quarter section of land where he engaged in farming for twenty years, but the lure of the western country had ever held him and disposing of his farm at a handsome figure he returned to Nebraska, locating on a Kinkaid homestead in 1908. The new claim was in section twenty-one township twenty-one, range forty-six, Garden county, which he still calls "home" but has his land rented. Since becoming a resident of Alliance Mr. Wade has taken

J. B. Wade and Family

the agency here for the Sherman Nursery Company of Charles City, which has a territory covering northwestern Nebraska He is a progressive and up-to-date business man and is making a gratifying success of this line which is financially most profitable Mr Wade is a member of the Masonic order

In April, 1872, Mr Wade married Miss Charity Lynch in Boone county, Iowa, a native of Delphi, Indiana, the daughter of Parker and Maria (Hughes) Lynch, both Hoosiers Six children have been born to this union Carrie, deceased, Jennie the wife of Edward Callahan a farmer near Woodward, Iowa, Arthur A, a coal miner, Ralph P, a merchant at LaJunta, Colorado, was head of the guard squad at Fort Logan, serving in the army eighteen months during the World War, Roy L a coal miner of Boone, Iowa, and also a talented musician, Mabel, the wife of Andrew Erickson, a station agent at Moingona, Iowa, and John S, on the ranch with his father, enlisted in the army in 1917, being located at Fort Logan before he was transferred to the Heavy Artillery and sent to France where he served three months After the war was over he returned home and went to farming

ROBERT O REDDISH, one of the younger members of the Box Butte county bar who is making an enviable reputation for himself in his profession and also in business circles of Alliance and the Panhandle, is a native son not only of Nebraska, but of this county, born April 26, 1889, the son of Frank E and Mary E (Fisher) Reddish, the former a Hoosier while the mother is a native of Nebraska, born in Nemaha county Robert was the oldest of the three children born to his parents Frank Reddish is a pioneer of the Panhandle as he came here in 1887, locating on a homestead fourteen miles west of the present city of Alliance on land that today is known by the name of Barrell Spring The boy grew up on his father's farm but was sent into Alliance to school After graduating from the high school in the spring of 1907, in the fall he matriculated at the State University While in college, Mr Reddish took a preliminary letters and science course and then entered the law school, from which he graduated in 1911 Having spent his boyhood in the country the young attorney realized the value of land of his own and soon after being admitted to practice started out to secure a farm He took a Kinkaid homestead adjoining the town site of Angora, in Morrill county and to prove up on it had to actually live there seven months of

the year for three years This he did, the other five months of each year he devoted to his real estate, loans and insurance business in Alliance, in addition to engaging in the practice of his profession

On October 21, 1914, Mr Reddish was married at Hastings Nebraska, to Miss Ruth Tibbits who was born in Angelica, New York, the daughter of Lemuel and Aurelia (Burr) Tibbits, the former a native of New York, as was the mother Mrs Reddish is a graduate of the Nebraska State University and after receiving her degree she taught for one year, filling the office of principal of schools at Pauline, Nebraska Mr Tibbits has taken an active part in public life in Nebraska and at the present time is clerk of the district court at Hastings Mr and Mrs Reddish have two children Mary Ruth, aged four and Robert O, Jr, just past two In 1914, Mr Reddish and Eugene Burton entered into a partnership for the practice of law opening an office over the First National Bank Both are full of energy and push in business that is the marked characteristic of the successful Nebraskan of the present day They have excellent executive ability, are students of human nature as well as the law and are handling many interesting as well as complicated cases which shows the high esteem in which they are held by the citizens Mr Reddish has made money on his farm and in his real estate business, he owns his fine modern home in Alliance and from the present outlook he and his partner have a bright future before them Mr Reddish is a Republican while his fraternal affiliations are with the Masonic order and he is a Thirty-second degree Scottish Rite Mason Night Templar and Shriner and is a member of the Elks

JAMES M KENNEDY, D D S, is one of the younger members of the medical fraternity of Box Butte county and Alliance where he has become well known to the people of the community as a skilled member of his profession since 1907, when he located in the Panhandle He was born at Caledonia Ontario Canada, November 2, 1881 the son of Donald and Ellen (Madigan) Kennedy, the former born and reared at Glengary, Canada James was the fifth in a family of seven children and spent his boyhood days in the locality where he was born He started his education in the public school near his home, then entered the Caledonia high school, graduating in 1897 Though young in years he had already determined upon a professional career, passed the entrance examination to Northwestern University, Evanston, Illinois and matriculated in the dental college where he spent four years in the special studies leading up to the degree of D.D.S., which was granted him by the university in June, 1901. For six years he was engaged in dental practice in Chicago, but desiring less confinement than he could secure in a large city and believing that there were good openings in the west Dr. Kennedy came to Nebraska in 1907, and after looking various locations over chose Alliance, where he has in less than a decade built up a practice that many an older dentist may well envy. He has a charming personality which counts so much in purely office practice. As in medicine and surgery, the science of dentistry is constantly developing new phases of usefulness, and in order to secure success today the dentist must keep fully abreast of the latest achievements of his profession. This Dr. Kennedy has done to a marked degree, and it is one of the reason he has attained such a large patronage in the thirteen years of his business life in the west and why he has retained uniform confidence in the minds of his patients and the citizens of the city. Dr. Kennedy has been an advocate of organization among the dental practitioners of the state and done general construction work in the association. He is a Democrat. In medical matters he has shown the greatest interest as he is a member of the National Dental Association, the Nebraska State Dental Association, while his fraternal affiliations are with the Elks and the Knights of Columbus

July 3, 1907, Dr. Kennedy married in Chicago, Miss Adelaide M. Forde, a native of that city, the daughter of James and Jane (Bodkin) Forde, both of whom were born and spent their childhood in Ireland and they have transmitted to their daughter that undefinable charm which is the happy heritage of the Irish woman, in the native land, or her descendents in America. Mrs. Kennedy is a graduate of the Sacred Heart Academy of Chicago· and for several years after leaving college was engaged in teaching until about the time of her marriage. There are six children in the Kennedy family: Donald, Mary, Jean, James, Elen and Virginia. Dr. Kennedy has taken an active and interested part in the civic and communal life of Alliance since first settling here. He owns his modern home where he and his wife dispense a charming hospitality to their many friends. He is a "booster" for the county and town and aids in every movement for the betterment of Box Butte county and his adopted city, whether it be good schools, good roads or social service.

FRANK E. REDDISH. — Closely identified with many important interests of Box Butte county, Frank E. Reddish, who has conducted a real estate business at Alliance for more than a quarter of a century, may be numbered with the early settlers and awarded pioneer honors for he came in 1887, settling on a claim about fourteen miles west of the present site of the city, when settlers were few, as the country was practically an unbroken wilderness. Since that time he has born his full share of responsibility in developing the natural resources of the county, has been truly public spirited, and has done much for movements favoring progress in every way.

Mr. Reddish was born on the banks of the Tippecanoe river, in White county, Indiana, July 13, 1860, the son of Noah and Almira (Bartholomew) Reddish, the former born at Baltimore, Maryland, while the mother is a Buckeye, born at Ashtabula, Ohio. Frank was the second boy in a family of five children and as his father was a farmer he grew up in the country, learning early to assume many of the duties about the home place and at an early age became a practical farmer. Noah Reddish was one of the gallant sons of the Union who responded to the call for volunteers at the outbreak of the Civil War and enlisted in Company E, Ninth Ohio Volunteer Infantry. He participated in some of the hardest fought battles of the war, but was never severely wounded and when peace was established returned to the ways of peace and his farm. The school which the Reddish children attended was a log structure put up in the early days of Ohio, about a mile from home and they walked back and forth each day during the term. After finishing the grades in the country Frank Reddish attended the Monticello high school and after graduating began to teach, a vocation he followed six years but he felt that the life of a pedagogue was not what he desired and as there were many opportunities for young, ambitious men in the west decided to migrate there. In the spring of 1886, he came to Nebraska, located at Hartwell where he taught school one term while he was looking the country over to find land in just the location he had in mind. He had been reared on a farm, was thoroughly acquainted with every branch of farm industry and coming to Box Butte county in 1887, selected a homestead fourteen miles west of the present city of Alliance, on the stream called Snake creek

On April 26, 1886, Mr Reddish married Miss Mary E Fisher in Nemaha county, a native of that county and the daughter of Owen and Mary (Tufts) Fisher, the former born in New York, who came west and was a pioneer settler of Nemaha Three children have been born to this union Robert O, an account of whose life will be found on another page of this history, is a lawyer of Alliance who married Mary R Tibbits, and they have two children, Howard E, attended the Alliance high school, the Lincoln academy and then entered the state University, but at the close of the first semester returned to Alliance and entered the real estate business handling loans and insurance as well, being associated with his father with offices in the Reddish Block on Box Butte Avenue, married Miss Gail Ericson at Lead, South Dakota in 1916 She was born at Atchinson, Kansas, graduated from the Lead high school and then entered the school of music at Northwestern University, Evanston, Illinois Her husband is a member of the Elks and also a Scottish Rite Mason, and they have one child The third child is Edith, who married Lyle Anderson, also in the real estate and loan business They have one child Mr Anderson graduated from the Alliance high school and then attended Milwakee-Downer College, Milwaukee, Wisconsin, for one year She is a member of the Eastern Star After locating on his claim Mr Reddish proceeded to build the regulation "soddy" for a home but it was warm and comfortable and with his wife settled down there and began the improvements on their farm There were no railroads here then and Mr Reddish had to freight in his supplies in true pioneer style from Hay Springs, sixty-five miles away Their neighbors were few and far apart as they were almost the first settlers of the section James Leith, settled a half-mile southwest of the Reddish homestead and James L Underwood lived a mile and a half to the east Lewis M Kennedy was next on a claim seven miles away, which shows how isolated the first settlers were, especially when one considers that there were no towns, no stores and no doctors in the entire district For seven years Mr Reddish remained on his homestead, made good and practical improvements and in 1894, moved into the town of Alliance, opened a real estate office where he practiced as a land attorney, helping many men locate on homesteads and was one of

the early surveyors of the section as men of this profession were hard to get and correct boundaries of claims were most important Mr Reddish bought and sold land, was agent for insurance companies and practiced before the United State Land Office for twenty years He had great faith in the future of the Panhandle which caused him to locate here and that faith was never allowed to falter throughout the years of drought, grasshoppers and winter blizzards he held on when other settlers became discouraged and returned to their home in the east, many never to return He bought more and more land until he accumulated a landed estate of three thousand, seven hundred and sixty acres of the finest farming land in Box Butte county, an achievement of which any man may well be proud, as this fortune has been accumulated by his own unaided efforts For many years Mr Reddish has been a heavy and successful speculator in western lands He bought the large brick building known as the Reddish block on Box Butte Avenue in 1907, which is a modern office building In addition he owns a modern home and numerous other investments here and other places in the middle west At the present time he is so well fixed financially that he can afford to spend the remainder of his life with no thought or care for financial affairs He can look back and see that his life has been constructive of benefit to his community and the many settlers who have come here since he and the other few pioneers blazed the pathway for the present progress, development and civilization of this favored section Part of his time is spent in Alliance and part at his home at Long Beach, California Mr Reddish is a sound Republican in his political faith, but of the most progressive type, while his fraternal association is with the Benevolent and Protective Order of Elks

Mary E Reddish is a descendent of Revolutionary ancestors on her mother's side of the family and a descendent of the well known Israel Putnam family Frank E Reddish also is a descendent of Revolutionary stock on his mother's side of the family Mrs Howard Reddish is a descendent of Revolutionary ancestors on her mother's side of the family and belongs to the Daughters of the American Revolution

W J MAHAFFY one of the gallant sons of our country who enlisted in the army to serve during the Spanish-American war and a man widely traveled over various portions of the globe, is a recent member of the dental fraternity of Box Butte county to become established in Alliance but within a short time he has been a citizen of the Panhandle, the doctor has won the confidence of the public and is enjoying a practice that many an older man might well envy

Dr Mahaffy was born in Red Oak, Iowa He attended the grade schools in his native town and then graduated from the high school, following which he matriculated in a college at Tarkio, Missouri, but his college course was cut short when he enlisted in the First Illinois Cavalry, in 1898, at the commencement of hostilities with Spain The same year he received his honorable discharge from that organization and in the spring of 1899, enlisted in the Sixteenth Regiment United States regulars and was sent to the Philippines during the "Philippine Insurrection." Within six days of arriving in the islands he was promoted to corporal-sergeant and later made sergeant, then first sergeant, all before he had attained his majority, a rather good record for a middle western boy just out of his first year in college, but it is such men who have made the records for the army of the United States and enabled the volunteers to win the chevrons and citations for the regiments of which they are members After being discharged from the army Dr Mahaffy visited China, Japan, Iberia, Corea, Australia, Judea, Persia, Egypt, Spain and England He made two trips around the Horn, and visited practically every South American country After this remarkable period of travel and adventure the young man returned to the United States and having had wide and varied experience had decided upon a professional career and entered the dental college of Northwestern University, in Chicago, in October, 1902 Three years later he received his degree from the college and entered into practice in Chicago, where he was continuously engaged in office work of his profession until he decided to come west In August, 1914. Dr Mahaffy came to Alliance, opened an office and became one of the well recognized members of the dental profession He rapidly gained a gratifying practice for so young a man As in medicine and surgery, the science of dentistry is constantly developing new phases and in order to secure success the dentist of today must keep fully abreast of the latest discoveries and achievements of his profession Dr Mahaffy has done this for in 1916, he took a post graduate course in one of the finest dental schools in the west at Kansas City, Missouri His office in Alliance is located over the Alliance National Bank That the doctor has "made good" and gained the confidence of the citizens needs not be told when we learn that he has earned two fine farms and owns a modern home in Alliance, and still has a continually increasing practice among the best class of the residents Since first locating in the Panhandle the doctor has taken an active part in civic and communal life for he is modern in his professional methods and advocates all modern improvements for the county and his home city He is a director of the Alliance Commercial Club, is a member of the Masonic order, the Knights of Pythias, the Modern Woodmen, the Macabees and the Benevolent and Protective Order of Elks

E B O'KEEFE, D D S, a native son of Nebraska, Box Butte county and Alliance is one of the younger generation of the dental fraternity to locate in this city where he has been known since childhood and already has built up a very substantial and satisfactory clientele for so young a professional man

Dr O'Keefe was born in Alliance, the son of John O'Keefe The doctor was reared here, attended the public schools and after graduating from the high school in June, 1916, matriculated in the dental college of Creighton University, at Omaha, as he had decided upon a professional career He spent three years of study in the dental department of the university, receiving his degree in 1919 and came back to his home town to practice dentistry He is a young man of pleasing personality, which is so important in an office practice and is well acquainted with the latest discoveries in dental surgery which have come about by the special dental work done since the World War opened a new field for the dentist, he is progressive in his ideas and methods Already he has inspired the residents of the city with the fact that he is one of the young energetic and painstaking men of his profession, able to cope with the many oral diseases as well as teeth While in college Dr O'Keefe was a member of the Xi Psi Phi National Dental fraternity and is one of its loyal alumni He is also a member of the Nebraska State Dental Society and fraternally belongs to the Benevolent and Protective Order of Elks, and Knights of Columbus Everyone predicts a bright future for this young professional man, whose parents are among the well known and prominent members of the pioneer colony of Box Butte county

Thomas L. O'Harra

THOMAS L O'HARRA, who may well be called one of the progressive men of western Nebraska because of his continuous efforts to advance her interests, has been a resident of this state since his eighteenth year, and of Gering since the spring of 1915, in which city his business is extensive and his political prestige important. He was first elected mayor of Gering in 1917

Thomas L O'Harra was born in Decatur county, Iowa, August 19, 1870, a son of Elijah and Jane (Peterson) O'Harra. His father was born at Columbus, Ohio in 1848 a son of Thomas O'Harra, who was born in Ireland and came to the United States from Dublin in young manhood. He was a farmer and stockraiser in Ohio until his death at the age of seventy-nine, in the early eighties. Elijah O'Harra married Jane Peterson, who was born at Galesburg, Illinois, in 1852 and died in Nebraska, February 27, 1909. Her father, Captain James Peterson, was born in Pennsylvania moved to Missouri later in life, served in the Civil War as an officer, and on a visit to his old home in 1863, died at Chillicothe before reaching his destination. Four children were born to Mr and Mrs O'Harra, two of whom survive Thomas L and Morris G. The latter was born August 22, 1883, and was graduated from the high school at Lexington, Nebraska. He holds the very important office of district manager of the western division of the United States Rubber Company, with headquarters at Minneapolis, Minnesota. In 1888 the father homesteaded in Gosper county, Nebraska, has been a farmer all his life and now lives retired at Grand Island

Thomas L O'Harra attended school in Iowa before the family came to this state, and in 1891 was graduated from the high school of Bertrand, Nebraska. For a few years afterward he followed farming, then spent several years in a general mercantile business, after which he went into the meat and stock business in which he is yet engaged. In March, 1915, Mr O'Harra came to Gering, opening up an extensive market, the first livestock market established at this place. For twenty years he has shipped to South Omaha, while his local trade is exceedingly heavy

In March, 1892, Mr O'Harra was united in marriage to Miss Lottie McDonald, who died in March, 1897, the mother of three children Earl L, who is associated with his father in the meat business, Milo P, also connected with his father in business and Fleta L, a kindergarten teacher. In 1902 Mr O'Harra married Miss Maud M Johnson a highly accomplished, educated lady, who had been a teacher in the public schools and for eight years had been county superintendent of Gosper county. Mr and Mrs O'Harra have two sons Morris E and Thomas both of whom are in school. The family belongs to the Christian church

Fraternally Mayor O'Harra is an Odd Fellow and a Mason of the Royal Arch degree. In politics he has long been active in Democratic circles and is in close accord with the best party elements. Since 1917 he has been vice president of the school board of Gering and in the above year was first elected mayor, subsequent elections following. His administration of this office has been very beneficial to Gering. A citizen of positive patriotism a true American he has helped to win success during the last few years of financial stress, for every war loan, assisting in putting the last one over the top by $10 000. He has given encouragement to many of Gering's worthy enterprises and is always ready to contribute and cooperate for the city's good, a very charitable man, portraits of whom, with one of his father and brother, appear on the opposite page

S B WRIGHT, cashier of the Guardian State Bank and Trust Company, is also a leading and prominent, loan, insurance and abstract man of Box Butte county. He has been identified with numerous financial enterprises here and established a high reputation for ability, judgment and general acumen. His introduction to Alliance was in the roll of district agent for a life insurance company but it was not long until he became associated with banking affairs and since that time his rise has been sure, rapid and satisfactory to himself and his many friends. Mr Wright is possessed of that push and dynamic energy that insures success for whatever undertaking he assumes and has caused him while yet a young man to become one of the prominent factors in the financial life of Alliance and the Panhandle

He was born on a farm near Springfield, Missouri, and was reared on the home place, attending the district school during the term and helping on the farm as soon as his age and strength permitted, thus at an early age he became acquainted with the business side of farm industry. Later the family left Missouri and moved to Columbus, Ohio, where the boy entered the Bliss Business College and also to a special commercial law course in the Young Men's Christian Association night school, which he finished in 1903. Mr Wright well remembers the first money he earned by pulling weeds in a neighbor's corn field at twenty-five cents a day and used the money to assist him through the business college Soon after finishing the commercial course he accepted a position with a life insurance company and within a short time was transferred to Colorado by the Missouri State Life Insurance Company There he assumed the duties of district agent and later was transferred to Nebraska in the same capacity but added the duties of loan inspector in 1914 with headquarters at Alliance A year later he went into business for himself, carrying general insurance, farm, ranch and city loans, meeting with marked success in this new venture and making a very satisfactory income from his varied interests Mr Wright displayed marked executive ability in all his business dealing here and in the county, which was quickly noted by other men engaged in financial affairs so that he was asked to become associated with the Guardian Trust Company in 1917 It had an authorized capital of $100,000, Mr Wright assuming office as secretary and treasurer of the concern In July, 1919, the Guardian State Bank was organized with Dr C E Hershman, president, Thomas Katen, vice-president and Mr Wright, cashier, practically the same personnel as to officers as the trust company has The new banking house opened with a capital stock of $50,000 and so far has been doing a most satisfactory business with rapidly increasing deposits, for the men who have charge of its affairs have long stood high in the esteem of the business men of the city, for they are conservative in their dealings, yet constructive in methods and let no opportunity slip that will benefit the trust company or bank

Mr Wright also is an active member of the firm of Wright and Wright, which does a general insurance and the returns from the various lines handled by them are bringing in satisfactory returns for the time, energy and thought put into the business

On November 10, 1906, Mr Wright was married at New Lexington, Ohio, to Miss Agnes Thompson, a graduate of St Alloysius Academy of that city The Wrights have a modern home in Alliance on one of the best residence streets, Big Horn Avenue, and in addition Mr Wright owns several desirable properties which he has built for sale to meet the rapidly growing demand for homes He has one of the rental agencies of Alliance which works in well with his loan and insurance business, his far vision in real estate and his sound banking methods, and is a citizen of whom Alliance may well congratulate itself, for it is the men of the younger generation with their untiring energy who are writing the history of the Panhandle in deeds that are making this one of the favored sections of the middle northwest

CHARLES W JEFFERS, now one of the leading real estate dealers of Alliance is a pioneer settler of Nebraska and has lived in the Panhandle for nearly three decades so must be accorded the honors due to those who located here when northwestern Nebraska was "the frontier" He has witnessed the many changes that have been brought about, while in the work of development he has aided in many ways

Mr Jeffers was born in Wood county, Ohio, June 15, 1864, being the son of Joseph and Elizabeth (Smith) Jeffers, the former a son of the Keystone state while the mother was a Buckeye Charles was the oldest of five children in his own family and the youngest of nine half brothers and sisters, as both his father and mother had been married before and had families of their own He was reared on his father's farm and attended the district school near his home to acquire educational advantages until 1873 That spring his mother died and his father passed away the same year, leaving the boy to his own resources and though just a little lad he was compelled to go to work He remembers well the first money he earned planting corn with a hand corn planter, for which he received fifty cents a day in 1872, and putting some more with the three and a half dollars bought a pair of boots with copper toes, which became quite the envy of the rest of the small boys After the death of his parents he worked for strangers, first shucking corn then went onto a farm where he worked vacations, Saturdays and late in the afternoons, going to school the remainder of the time For two years he was thus enabled to pursue his studies at school but after that all the education he could get was by studying at night but he was ambitious and even by this slow method managed to secure a good practical foundation upon which to build a business life Some of the money he earned had to go toward supporting his younger brothers and sisters Later Mr Jeffers came west to Iowa, where he worked for a man named Maine, who was an extensive dealer He was soon promoted to foreman of the farm and handled many thousand dollars for his employer, buying up cattle in the surrounding states He also handled many government contracts for Mr Maine and thus gained a wide and varied knowledge of business In 1883, he came to Central City, Nebraska, as foreman of a gang of men who were engaged in grading for a railroad contracting firm known as Owen Brothers, as they were then constructing the road bed of a railroad from Central City to Erickson with a branch running to Ord The next year Mr Jeffers took land in Hamilton county which he farmed until coming to Alliance on March 2, 1891, when he entered the service of the railroad in the shops Starting in the cinder pit he was advanced to engineer of a passenger engine on the main line but resigned in 1899 and engaged in carpentry, a trade he had learned some years previously and continued to work for the company repairing cars for a while In the fall of that year he took up a homestead in Garden county, twenty-eight miles southeast of Alliance and when the Kinkaid act was passed filed on a section He put good and permanent improvements on the place and remained engaged in agricultural industries for five years In the meantime, he had been appointed deputy sheriff under Ira Reed, serving two years before he was elected constable and also deputy sheriff, filling both these offices at the same time under Calvin Cox Four years later he was appointed chief of police of Alliance and was chief police officer of the city until May, 1917, having held the office of constable all these years During his terms in office Mr Jeffers had to handle the wild element brought into the town and county by the saloons which ran wide open and by his double office could pursue his man beyond the three mile limit and proved a most efficient official at a time when life and property in the Panhandle were held rather lightly, for Alliance was a junction town of the two railroads and rough, unruly men congregated there and he can tell many enlightening and interesting experiences of that day On November 17, 1892, Mr Jeffers was married at Phillips, Nebraska, to Miss Anna Miller, a native of Ohio, whose brother, James W Miller, was sheriff of Box Butte county in 1919 Mr and Mrs Jeffers have two children Gladys, a graduate of the high school is engaged as ticket seller at the Imperial theatre, and Vera, at home Mr Jeffers owns a modern home in Alliance while his office is in the Reddish Block where he conducts a real estate and insurance business which is very successful due to the many friends he made in public life and the reputation he has attained for square dealing and his ability to work He and his family are members of the Methodist church and fraternally he is an Odd Fellow November 20, 1920, Mr Jeffers was again appointed chief of police of

Alliance, which shows the confidence the citizens have in his ability to maintain law and order

FRANK ABEGG — More than eight year's connection with the banking interests of Box Butte county and Alliance, during which time he has risen from a clerkship to be cashier of the First National Bank, has made Frank Abegg one of the best known figures in financial circles of the Panhandle and western Nebraska He was born in Wapello county, Iowa, April 7, 1895 being the son of Walter and Catherine (Smith) Abegg and was the second child in the family, which consisted of five people When the boy was six years old his parents moved to Blakesburg, Iowa, where his father established the Blakesburg Savings Bank, with which he has been continuously associated to the present time Young Frank well recalls the first money he made for himself when but a small child for he was not much more than six years old, when he began to sell the *Chicago Tribune* in Blakesburg and was induced by his father to put the money he earned in the bank and thus start a small savings account, and this excellent habit formed in early youth has stood him in good stead he says throughout all his business career, and to this day he always puts all his spare cash to his account which grows without his really knowing it to most satisfactory proportions Mr Abegg attended the public schools of Blakesburg where he laid the foundation for a good practical education to which he has been so constantly adding all his life as is a wide reader of the best literature, and of course specializes in banking Walter Abegg believed that every child should learn to work and thus know the value of the money he earned, and while he was abundantly able to support his family not only in comfort but luxury he had the boys start in at some industry as soon as their age and strength permitted and so young Frank at the age of fourteen went to work carrying mortar for a stone mason at a dollar and seventy-five cents a day, working at this trade for two and a half years Frank was given his board at home but out of his earnings bought his clothes and was allowed fifty cents a week spending money He thought this rather hard but later learned that the surplus was always credited to his account in the bank and when he grew older he had a goodly sum to his credit so that his father was really a much better parent than one who would have let his children squander their money on useless and unhealthful pleasures, while they at the same time learned habits of thrift and frugality

When only sixteen years of age Mr Abegg held a position as time-keeper for the Milwaukee Railroad on some of their construction work and in 1912, came to Alliance to enter the First National Bank as bookkeeper   From that time to the present he has been identified with banking matters, always advancing, and today he is cashier of the First National one of the strongest and most substantial banking institutions in the Panhandle and throughout the state   Mr Abegg is considered one of the capable, careful and conservative bankers of Box Butte county, and is a man whose personal integrity and probity have done much to conserve the interests of the institution and to gain and hold the confidence of the public   Mr Abegg is a stockholder and director of the bank and has proved his worth by the enviable record he has made within little more than a decade   He is a man who can meet all kinds of people, is popular with all the business men of Alliance and the surrounding country and while yet young has attained a prominent place in the financial circles of the state so that his future looks most promising

On November 16, 1916, Mr Abegg was married to Miss Mary A Newberry, the daughter of Chenia A and Ellen (Brennan) Newberry, the father a native of Michigan   Three children have been born to this union   Walter E, Frank Jr, and Rita Ellen, for whom the parents intend good educations to fit them for life   Mr Abegg is a Republican in his political views, while his fraternal affiliations are with the Elks and the Knights of Columbus

S A MILLER, a pioneer of the state and also of Box Butte county who has materially assisted in the opening up and development of the commonwealth and the Panhandle, was following several vocations before he located in the city of Alliance as he was in turn, cowboy, ranch manager, farmer and now is a business man of prominence who has become one of the dependable and substantial citizens of this community, and in all his various lines of endeavor his versatility enabled him to make a success of whatever he has undertaken   Mr Miller was born May 27, 1864, in Knox county, Illinois, the son of Daniel and Elizabeth (Humphrey) Miller, the former a native of West Virginia   The boy was reared in Illinois and went to the country school, two and a quarter miles from his home until his tenth year when he was forced to go to work and

the education started in Knox county was finished in the good, practical but expert school of "hard knocks," which teaches its pupils well by a strict discipline but the lessons so learned are never forgotten   For six years the boy worked on a farm by the month but when sixteen years old determined to do better for himself and as he had learned of the fine opportunities in the plains country came west in 1880   He secured work in Dawson county on the Colton ranch under George Nelson   Mr Miller had only nine and a half dollars when he left Illinois and with his companions drove across country in true pioneer style using a covered wagon for the trip   He earned his first money on Nebraska soil making posts at Buzzard's Roost, near Eddyville and remained there about a year   When Mr Colton sold his ranch to Dayton and Company, Mr Miller remained with the new proprietors who owned another large tract of land on Dry Cheyenne river, where he and other members of the outfit spent the summers with a large herd of cattle   In the fall they returned with the herd to the home ranch to ship to the eastern markets   In 1887, Mr Miller came to Box Butte county and the following year, located a homestead twelve miles southwest of the present city of Alliance, where he built the regulation sod house of the pioneer, made suitable shelters for his stock and kept "bachelors hall," as he says   Money was very scarce on the plains at that early day and the first winter here Mr Miller made money to buy his supplies by gathering up buffalo bones of the animals that had died on the prairies and sold them to W W. Norton of Alliance, who shipped them east for fertilizer

On September 9, 1889, Mr Miller was married at Nonpariel to Miss Mary C Sipson, of Nemaha, county, a true daughter of the state, whose father Alfred, had been born in England and was an early settler of Nemaha   Mr Sipson spent most of his life in the United States as his parents brought him to America at the age of four   Mrs Miller is the oldest of the five children born to her parents. After coming to Alliance to live Mr Miller owned and managed a dray line for about sixteen years but disposed of it in 1907, and in May, 1911, bought the fixtures of a shoe store from John W McNamara and opened an up-to-date shoe house, located at 305 Box Butte Avenue, carrying a general stock of men's and women's shoes   In 1917, Mr Miller went to Chicago where he took a special course in practipedics on the science of giving foot comfort to people suffering from fallen arches and other foot troubles From the first he met with a cordial response in business from the residents of the city and now with his added line of orthapedic shoes and devices his trade has materially increased so that it is a most satisfactory concern from the financial point of view He has a man in his employ who is a graduate of the same school, so that Mr Miller can devote his time almost entirely to the management of the business in its larger aspects Mr Miller has a modern home and other real estate which he rents and today represents the true spirit of the Panhandle business man "progress." In politics he is a Democrat

FRANK J BRENNAN, is the owner and manager of one of the oldest drug houses in the Panhandle and has been established in professional life in Alliance for nearly twenty years He has built up a clientele and substantial business of which any man may well be congratulated, especially when his success is due to his own industry, high reputation for business ability and the courtesy displayed to his customers Mr Brennan was born in Bay City, Michigan, the son of Martin and Mary (Fitzpatrick) Brennan, both of whom were born in Ireland and came to the United States at an early day to locate in the northern central states To their son they have transmitted those rare qualities of the Irish which no other race on earth is fortunate enough to possess, humor, wit and a genial optimistic disposition which enables the possessor to overcome and surmount many a difficulty in life that would daunt and discourage others

Frank was next to the youngest in a family of twelve children and as his father died when he was only seven years old the burden of rearing and educating her children fell upon the mother, but she nobly put her shoulder to the task of being both father and mother to her little brood and that she was successful in this great undertaking need not be said when one looks at her sons and two daughters Mrs Brennan came to Alliance in 1888, one of the pioneer women of the section and town, so that her children were educated in the excellent public schools here Frank received his instruction in the old building which was later converted into a flouring mill, and then graduated from the high school He had early decided upon a professional career, and having chosen pharmacy matriculated in the pharmacy department of Northwestern University The college proper is located at Evanston, but the medical and pharmacy departments are situat-

ed on the south side of Chicago, where he received his degree and was admitted to practice in 1899 Knowing that there were more openings and chance for building up a good business in the west than in Illinois the young man returned to Alliance after commencement and bought the drug store owned by Fred H Smith He at once established it on a modern footing with the slogan on "service." The health of a community depends nearly as much upon the druggist who fills the prescriptions as upon the doctor who writes them and Alliance has indeed been fortunate in having such a capable and conscientious man for this work as Mr Brennan who has always personally handled or supervised the prescription branch of his business He is modern in his ideas and methods, keeps an up-to-date establishment in every particular, gives care to his displays and has added all the attractive side lines that people have learned to expect in the modern store His promptness, courtesy and consideration have gone a long way toward his success in building up a fine trade which is also a money making business The store is located advantageously on Box Butte Avenue where the better elements of the city congregate for purchases In addition, Mr Brennan owns a modern home in the town where he and his gracious wife dispense a cordial hospitality to their many old, warm friends

On September 1 1909 Mr Brennan was married at South Hampton, Canada, to Miss Madeline Carey, the daughter of John V Carey and two children have been born to them Helen, seven, and John F, a lad of two Mr Brennan is a member of the Benevolent and Protective Order of Elks and the Knights of Columbus

ALLEN D RODGERS — Of the men who have lent dignity of character, excellence of labor and largeness of general co-operation to affairs in Box Butte county and the Panhandle for more than three decades, none is held in greater esteem than Allen Rodgers, the owner and manager of the leading grocery house of Alliance Mr Rodgers is entitled to pioneer honors for not only was he one of the early settlers of the section but was a freighter frontiersman and broke timber claims for other men in order that they might prove up on them Thirty-two years have passed since Mr Rodgers drove into Cheyenne county in true pioneer style and settled on a homestead and timber claim For many years he lived and labored slowly and arduously developing a farm and establishing a home for his family, watching and assisting in the advancement and progress While yet living in the country Mr. Rodgers gained a reputation for industry and integrity which is being perpetuated in his mercantile establishment and which he has transmitted to his children as a wonderful heritage, never to be purchased by mere money

Allen Rodgers was born at Sigourney, Iowa, May 23, 1861, just at the opening of the Civil War His parents were Willis and Madelina (White) Rodgers, the former a son of the Blue Grass state Allen was the youngest of a family of twelve children and early learned to give and take where a number of boys and girls grow up together His boyhood was spent in the country on his father's farm, where he grew up sturdy, healthy and self reliant, able to cope with many an unexpected emergency He received his early educational training in the good public schools of Iowa, and while yet a small boy began a financial career in a small way, his first venture was trapping quail and selling them for five cents apiece When only seventeen the boy became impatient to become financially independent and established himself in business as a farmer On December 16, 1881, Mr Rodgers was married in Sullivan county, Missouri, to Miss Margaret A Perry, who was born at What Cheer, Iowa, the daughter of Jefferson Perry, a native of New York She was the youngest in a family of five children and became the mother of three children of her own Cora M, deceased, Minnie T, who married George Roach, a farmer near Alliance and they have five children, and Chester C, who married Grace C Watson and they are the parents of two children Mr Rodgers is associated with his father in the grocery business He is an Odd Fellow, Mason, Elk and Eagle

Not satisfied with their surroundings, and desiring a home with its many opportunities, on the newly opened frontier of Nebraska, Mr and Mrs Rodgers drove overland in 1888, locating in what was then Cheyenne county, so that the nearest post office and town was Sidney, thirty-two miles away Like all new settlers they were forced to meet and to overcome many obstacles and to endure numerous hardships Hardly were the family settled when Mrs Rodgers died This left Mr Rodgers with three small children on his hands, but he had already built a good, comfortable, warm "soddy" and at once began the task of improving his pre-emption and timber claim He says that people today with their automobiles do not realize what distance meant to the first settlers who were forced to travel long distances by horse, mule or ox teams and that the thirty-two mile trip to Sidney and back in the eighties would mean much more to them than a journey of several hundred miles does today on the good roads in a machine In recounting experiences Mr Rodgers tells of one trip he and some near neighbors made to Sidney for mail and supplies They took two teams to the wagon, and on the way had to shovel snow for six miles in the canyons to get the horses and wagon through an undertaking that took seven days to cover the sixty-four miles The Rodgers homestead was located on the south side of the Platte but settlers on the north side had to ford the river or make a detour of twenty miles to cross on the old Clark bridge built by the soldiers in the late sixties and many a team and man Mr Rodgers has pulled from the quick sand of the treacherous stream when they attempted to cross As money was scarce on the plains in the early days many of the settlers worked when they could find something to do at times when the farms did not demand all their time, and Mr Rodgers broke timber claims for cowboys who had taken them but were obliged to be with their outfits at just the time such breaking could best be done The cattle men were paid in money by their employers and they in turn paid gold and silver for the work on the land In this way Mr Rodgers was able to make enough money to buy his little family many of the necessities which other men had to do without Settlers lived far apart in those days and Mr Rodgers had been fortunate in having three other families take claims adjoining his in the valley at about the same time, as the women were kind to his motherless children and the men were able to help one another in many ways In 1892 Mr Rodgers married Miss Minnie White, the daughter of John E White and the next year he sold his farm in the valley to move twenty miles north of the Platte into the sand hills where he bought a thousand acres of land He built a comfortable home, placed good improvements on the place and was engaged in ranching and cattle raising for five years Desiring better educational advantages for his children, Mr Rodgers disposed of his ranch and came to Alliance that year, buying in October, the grocery stock of J J Lyons or "Daddy" Lyons as he was well and kindly known to the residents of the town Mr Rodgers was already well known to the residents of the city for his care in fulfilling business obligations and this gained him confidence both with his customers and the wholesale houses with which he dealt. He soon gained a reputation for handling the best lines of goods, and kept such attractive displays that his business grew very rapidly and soon became most satisfactory from a financial point of view. Mr Rodgers has the delightful cordiality of the southerner, inherited from his father. He manages to make every customer feel at home, no matter though the purchase of the moment be small, and by this faculty has gained many new customers and always holds the old ones. His trade increased so that he found it necessary to build the present fine two story brick building, with nearly two thousand square feet of floor space on each floor, where he has been located since 1901, at 122 Box Butte Avenue. From first locating in Alliance, Mr Rodgers has taken an active part in civic and communal affairs, advocating every progressive movement for the upbuilding and development of the city and the confidence and position he has gained may be understood when we know that in 1913 he was elected executive officer of the municipality, was re-elected mayor in 1914, and by his economic management of the city finances in the matter of the light and water plant saved the tax payers at least twenty-six thousand dollars. Mr Rodgers proved so efficient an official of Alliance that he was re-elected mayor in 1919 and is serving at the present time. He has shown such marked executive ability in his business and as municipal officer that he was unanimously elected as one of the executive officers of the Nebraska Retailers' Federation, another office which he has well filled. Mr Rodgers is an adherent of the principles of the Democratic party and while he is in sympathy with the party is too broad minded to be closely tied in local elections, believing that the man best fitted to serve the interests of the people should be placed in office. Mr Rodgers is a member of the Independent Order of Odd Fellows and the Benevolent and Protective Order of Elks, and has taken seventeen degrees in the Masonic lodge.

ROY BECKWITH, the leading men's furnisher and haberdasher of Alliance and one of its solid and reliable business men, is an early pioneer of Nebraska and one of that famous band of men, the "cowboys," who herded cattle in the Panhandle and along the great "cattle trail," of the seventies and eighties. He has had many and varied experiences in this great commonwealth since brought here as a small child by his parents and is representative of that spirit that has opened up this state to settlement and development. In all of these changes Mr Beckwith has taken an active part. He is a native of the Keystone state, born at Smethport, McKinn county, Pennsylvania, March 24, 1863 the son of Daniel E and Elizabeth (King) Beckwith, both natives of that state, where they were reared and educated, later married and lived there for several years. Roy was the second child in a family of seven children born to his parents. His father enlisted in the One hundred and Twelfth Pennsylvania Volunteer Infantry at the outbreak of the Civil War; he participated in many of the most severe engagements and battles of the conflict under the famous general, Roy Stone, and Roy Beckwith is named in memory of that officer. Mr Beckwith served at Gettysburg and in the Battle of the Wilderness, where he was under fire twenty-eight days out of thirty. When peace was declared he returned to peaceful pursuits, but like so many of the returned soldiers was not contented with conditions as they existed before the war, as he desired greater advantages for himself and family, and knowing that land was to be had for the taking on the newly opened frontier in Nebraska, came to this state in 1866, when the country was a veritable wilderness. He located in Saunders county and the first year on the plains was a sub-contractor in getting out ties for the Union Pacific Railroad, which was then pushing toward the west across the state. The next fall he took up a homestead on Pebble creek, twenty miles west of Fremont, near the present site of Scribner. Roy and his brothers and sister were reared on this frontier farm and attended the pioneer school nearest their home when a teacher could be secured for it. While still a small boy he began to be useful around the farm and his first money independently earned was herding cattle when only eight years old and that two dollars looked very large to him. In 1872 the family moved to Antelope county within about a half mile of the famous "cattle trail," that led north from the Pecos in Texas to the Yellowstone river, along which the great herds of cattle were drifted north with the summer season and in the fall sold at a northern market. Cattle for the reservations used this same trail when being driven up for the Indians' meat supply and sometime cows would hide their calves near the bedding grounds and they were left behind when the herd moved. Roy found one of these while herding for his father and brought it home on his pony. When it was

372 HISTORY OF WESTERN NEBRASKA

a year old Mr. Beckwith took it to market when he sold his grain at Columbus, sixty miles away, and Roy bought a suit of clothes with the money. The boy worked on the farm in summer while attending school in the winter until he was seventeen; but he heard such wonderful stories from the cowboys who were driving the herds of cattle from northwestern Nebraska to the settlements that his young blood was fired with the spirit of adventure and the picturesque cow men decked out in "chaps" and spurs, big hats and six shooter guns lured him on and as he expresses it, "I got the fever," and nothing would do but that he too must join a cattle outfit. He hired out on a ranch near the present site of Valentine, where he remained two years before changing to the "Boiling Springs" ranch owned by Major Mayberry, but the cowboys were a changing lot of men and the next year he joined the N Bar Cattle Company and while working there took up a homestead near Gordon.

In 1884, he joined the trail herd that the company was bringing from Texas to the Indian Nation and thus early traversed that famous highway that has gone down in history, unique in its inauguration and different from any trail in the whole world. They had about eight thousand cattle in the herd, branded on the Cimaron river near the Kansas boundary, then drove two hundred miles west of Miles City, Montana, to the Mussle Shell river. The country was a wild expanse with practically no settlers, filled with rough cattle rustlers, French-Canadian and Indian trappers who wintered where snow found them and set traps along the streams. For three years Mr. Beckwith trailed cattle then, in 1887, came to Sheridan county and tried farming with, as he expresses it, "a varied amount of success." In 1900, he disposed of his place and established himself in Gordon in the clothing business, meeting with gratifying success from the first. Four years later he came to Box Butte county, settling in Alliance, where he opened one of the finest men's furnishing houses in the northwestern part of the state. He has a most attractive store, with a stock that would be hard to beat in a much larger city, is a man who reads human nature and by his courtesy, tact and reputation for giving everyone a "square deal," has built up a fine trade, which is most satisfactory from a financial point of view.

January 9, 1891, Mr. Beckwith was married at Rushville, Nebraska, to Miss Emma Flexing a native of Pennsylvania and two children

have been born to this union: Ora Fay, deceased, and Blaine G., who went through the Alliance schools and then attended a military academy for three years. He is now associated with his father in business and married Helen Rice. For his second wife Mr. Beckwith married Miss Maude Howell, at Chicago, Illinois on September 23, 1911. Mr. Beckwith owns a hundred and sixty acre farm of good Box Butte county land, has a modern home in Alliance is an Elk, a Mason and a Shriner.

GRANT G. MELICK, one of the progressive business men of Hemingford, who, though a late comer to the town, is doing his part in the upbuilding of the town and county, is the owner and manager of the Miller Hotel and Cafe, one of the up-to-date hostelries of the Panhandle. Mr. Melick was born in Finney county, Kansas, November 16, 1892, which places him in the younger generation of business men of today who are making financial history in this section. His parents were Franklin and Christiana (Larson) Melick, the former a Jerseyman, born in Hunderon county, New Jersey, while the mother was born in Copenhagen, Denmark. Grant was the seventh in a family of nine children born to his parents. When the boy was two years old, the family moved to Nodaway county, Missouri, locating on a farm where he grew up in the healthy environment of the country, attending the district school in the winter time and helping on the home farm as soon as his age and strength permitted, so by experience he was a good practical farmer while yet a boy in years. All the boys worked at something when he was young and Grant says that the first money he earned was making the fires and doing janitor work at the school which he attended. For this he was paid a dollar and a quarter a month. Grant remained at home with his parents until he was fifteen, then began to work on farms near his home independently during the summer seasons and continued his education during the slack period of the winter months, and thus he laid the foundation for a good practical education that has been invaluable to him in later years. He was strong, and ambitious, not afraid of work and the year he was sixteen cribbed a hundred and twenty-five bushels of corn in a working day of nine hours, a feat that a full grown man would have been proud to accomplish. The next year he established himself independently as a farmer and while he gained quite a satisfactory remuneration financially the

Francis M. Troy

spirit of adventure inherent in every youth called out to him and he responded and for several years was as he termed it "a rambler." He was engaged in farming in several localities for a period of years and in the fall of 1910, came to Box Butte county to visit his brother Fred, but did not settle as the following year he went east to visit New York and New Jersey, thinking that perhaps he might locate there, but the lure of the west was in his blood, the east was too crowded with settlers for him and he returned to Nebraska in 1912, to take up land and begin agricultural pursuits in Box Butte county. Mr Melick soon became recognized in his section as one of the prosperous farmers, who was well qualified for his vocation and received gratifying returns from the soil for his labor. The family remained in the country until the spring of 1919, when they came to Hemingford, soon after which Mr Melick purchased the Miller Hotel. He had it thoroughly refitted, put in a cafe and now runs one of the best European hotels in the Panhandle where guests have every comfort and convenience. All his rooms are steam-heated with hot and cold water and electrically lighted throughout. The cafe caters to the residents of Hemingford, and the traveling public and all are to be congratulated on having such service in the hands of a capable and progressive man whose slogan since establishing himself here has been "service." From the volume of business already handled a bright and prosperous future is in store for Mr Melick.

In the fall of 1913, Mr Melick married Miss May Grimes at Hemingford. She was born in Lucas county, Iowa, the daughter of Sidney and Winifred (Patterson) Grimes, the former a native of Lucas county while the mother was born in Marion county, Iowa. Mrs Melick was the oldest in a family of six children, was reared and received her education in Iowa and later graduated from the Sheridan high school at Sheridan, Iowa. Mr and Mrs Melick have one daughter, Christiana.

FRANCIS M. TROY, a resident of Gering since 1898, and who has been a public official for many years, came first to Scottsbluff county in 1886, settling here permanently in the following year. Few of the settlers of that date brought a large amount of capital with them and Judge Troy was no exception, but in his case the lack of money was made up by the possession of rare business judgment which guided his early investments and in later life

has made him the choice of his fellow citizens for responsible offices. At present he is serving, as he has been for years past, as police magistrate and justice of the peace.

Judge Troy was born at Oskaloosa, Iowa, March 3, 1856, the second in a family of nine children born to Abraham and Miranda (Malona) Troy, both of whom came of Irish ancestry. The mother of Judge Troy was born in 1846 while her parents were voyaging to the United States, and died in 1909 in Iowa. The father was born in 1832, in Pennsylvania, a son of Benjamin Troy whose father died in Ireland. He died at Oskaloosa as did his wife, surviving her five years. Six of their children are living. Mrs Louisa Carter, of Hutchinson, Kansas, Francis M., of Gering, George M., a farmer in Iowa, Edward a farmer near Lacey, Iowa. Harry, a farmer in Iowa, and Abraham L., a farmer and stockman near Little River, Kansas.

Francis M. Troy attended school in Iowa and assisted his father on the farm until he came to Nebraska. Following an inspection visit to Scottsbluff county in 1886, he returned in 1887 and took up a homestead and tree claim and settled down determined to make a success of his undertaking. Evidence of the accomplishment of his purpose was shown ten years later, when his ranch fence extended seven miles in length and three miles in width, and beside other stock, he had seven hundred fine horses in his pastures. He says little about the hardships he encountered but on account of the conditions at that time they were numerous, but by 1898 he felt ready to give up so hard a life and in that year took advantage of an opportunity to dispose of his farm and stock interests. He came then to Gering and for a few years engaged in no particular business, although somewhat interested along a line which he has subsequently developed. Situated about five miles from Gering and east of the city, he maintains an apiary, with one hundred and twenty-five stands of bees, some of his colonies being pure Italian. It has been much more than a recreation with Judge Troy, although he enjoys caring for his bees, as last year he realized two tons of fine honey from forty-five stands. In the meanwhile however, he has led a busy and serious life in other directions. After serving for ten years as deputy sheriff, he was made police magistrate and justice of the peace.

In 1880 Mr Troy was united in marriage to Miss Lizzie Akins of New Sharon, Mahaska county, Iowa, and they have four children. C E operates a meat market at Minatare, Nebraska, Ida M, resides with her parents. Asa

S , has recently been welcomed home from military service, is cashier in the Union Pacific station at Gering, and Frances M , is the wife of R L Beeman, a foreman in the sugar plant at this place The family belongs to the Methodist church In his political views Judge Troy is a sound Republican

JOHN VOGEL, for many years one of Box Butte county's best known pioneer farmers, and one of the representative and substantial business men of the Panhandle, came to this section as a youth when settlers were few in the upper valley of the Platte He was a member of a family that drove into the county in true pioneer style in 1876, settling on a homestead He and his wife were residents of Nebraska from the time when the only buildings known in the central and western part of the state were composed of sod and they watched the various changes that have been wrought and the sturdy and progressive work of the settlers, and themselve bore a full share of the labor of development Mr Vogel was one of the large landholders and successful agriculturists of Box Butte county, and is entitled to the respect and esteem in which he was uniformly held by his fellow citizens during life To his sons Mr Vogel left the example of an honorable and useful life, to his family, the memory of his loving care as a husband and father will remain forever as a blessed inheritance And now in the beautiful city of the dead, he sleeps the sleep that knows no awakening, awaiting the Master's call

John Vogel was born in Dubuque, Iowa, November 16, 1862 He was the son of John and Mary Vogel, both born in the German Empire They were reared and received their early educational training in that country and later came to the United States John, Jr, was the second boy in a family of five boys born to his parents His boyhood and early youth were passed on his father's farm He early assumed the tasks on the home place that his age and strength permitted and under his father became a good, practical farmer while yet of tender years He attended the district school near his home until his fourteenth year, when the family came to Nebraska, locating in Stanton county Mr Vogel entered the frontier schools after coming west and helped his father on the farm out of school hours and in the summer vacations As the country was little settled at that period John, like most other lively boys, became a trapper of small game and the first money he

earned was trapping muskrats, which he sold and thus had spending money of his own Mr Vogel remained at home with his parents assisting his father after his schooling ended until his marriage in May, 1884, at West Point, Nebraska, to Miss Mary Jannuch, who also was born in Germany, being the daughter of Herman and Augusta (Newbower) Jannuch, both natives of that country, Mrs Vogel was the second child in the family which consisted of four boys and two girls

The father was a stone mason in his native county who died when Mary was ten years old She had already attended school and was very capable and of much help to her mother, who finding it difficult to gain a living for herself and her family in her native land emigrated, as she had heard of the many opportunities for ambitious you men and women in the new world After landing on our shores, Mrs Jannuch came west to Chicago, where she had old friends and knew a number of people, settling in that city in 1878 When the family arrived they had some money but the mother, with the well know and admirable thrift of the German people put this away as a nest egg and set to work to save all she could so as to have capital to establish the children in business when the right time came She secured a position herself and also had the oldest children work, so that they were self-supporting and helped her care for the younger ones who were still too small to do much and were of course sent to the good American schools The oldest boy, Otto, was established with a cigar maker to learn the trade, and he found it so congenial and financially satisfactory that he still follows that vocation in Chicago Mary, now Mrs Vogel, secured work in a tailoring establishment, learned the business and worked at it four years before making a visit to Stanton, Nebraska, to some of her relatives While there she met John Vogel, a fine sturdy young farmer at that time, who could not resist the charm of the attractive German girl and persuaded her that she would be happy out on the high plains with him and they became engaged On May 7, 1884, their marriage was solemnized at West Point, and the happy groom with his bride established themselves on his farm Mr. Vogel was an industrious man and after his marriage his beloved wife became his devoted companion and helpmate, sympathizing when the way was hard and long and not only encouraged her husband but worked with him to build up their fortune, being his mainstay for thirty years Mrs Vogel is a woman whose strength and good deeds are as the number of her days and who has had a remarkable share in pioneer experiences in the great west, for she has taken active part in the development of her husband's land Mrs Vogel found a great part of the Panhandle still known as the "Great American Desert," and has seen the marvelous transformation of what was considered prairie become valuable farm land Six children were born to Mr and Mrs Vogel Ida, deceased, William H, who married Rose Knapp and they now live on a fine farm which they own near Alliance, having three children, all girls, Walter, married Beulah Reeves and they live on his ranch not far from town, and have a son, Ervin E, John O, married Fern Johnson and they have twin girls They make their home on a farm and Mr Vogel also is a well known and prominent hunter and trapper not far from Alliance Edward is the next child, who is now a machinist and makes his home with his mother in Alliance and Herbert, a farmer who also resides with his mother Mr Vogel and wife lived on their farm for many years and there reared and educated their children in great prosperity They were a happy loving family and it was a great blow when he was hurt by being thrown from a hay stacker on his farm ten miles west of Alliance, August 30, 1914, and died from the effects of his injury on October 12, of that year leaving his widow the management of their twelve hundred and eighty acre ranch Help became very hard to secure after the outbreak of the war Mrs Vogel found that she and the boys could not possibly carry on such a large place alone and before long discovered that the strenuous work was undermining her health, so in 1918 she leased the ranch and came to Alliance, buying a fine home at 804 Big Horn Avenue She is resigned to taking life a little more easily after the many years of endeavor of which few women can show a similar record She is a charming, gracious woman who keeps well abreast of all movements of the day in both the social and business world For many years she has had a large circle of acquaintances in this section of the country and since coming to Alliance has made many more and in recounting the experiences of the early days she ever gives the humorous picture of the hardships and privations she endured rather than the true situations for the events have become mellowed by time and as she looks at her fine, upstanding boys, feels well rewarded for whatever may have happened in the early years

Mrs Vogel dispenses a cordial hospitality to all the old friends and is an addition to the residents of Alliance for which the city may well congratulate itself

MILLARD F DONOVAN, pioneer frontiersman buffalo hunter, early settler, ranchman and now one of the best known and wealthy real estate dealers of Alliance and Box Butte county is a man whose varied career has given him many and interesting experiences on the plains Few men today, twenty years his junior bear so few of the scars of life Mr Donovan is a Hoosier, born in Owen county, Indiana, November 15, 1857, the son of Harvey and Emaline (Berry) Donovan, the former a son of Ireland where he was reared and received his early education before coming to America It is from his father that Mr Donovan has inherited his ready wit, sense of humor and the ability to look matters in the face and then work out his problems of life Millard was the third in a family of five children His boyhood was spent on the farm and he was given but two terms in school, but his Irish genius led him to become a wide reader and he learned more from books, and newspapers than many a boy in years at a desk under a good teacher His mother died when Millard was twelve year old and from that time he had to shift for himself almost entirely His father moved into Indianapolis and the boy sold papers in that city and also worked in a furniture factory, but he loved the country and returned to it, working on a farm in Johnson county for three years at six dollars a month, but when he came to settle up with the farmer, the latter said Mr Donovan owed him money but gave him seventy-five cents When only seventeen years of age the young man joined the regular army at Indianapolis, was sent to Newport, Kentucky, and before long was transferred to Austin, Texas, and from there the new recruits marched a hundred and eighty-five miles to Fort McCavit with their heavy equipment and Springfield rifles Mr Donovan was assigned to Company I Tenth United States Light Infantry which was assigned to service at Fort McCauitt on the extreme frontier to hold back the Comanche and Kickapoo Indians, where he served until 1877 The spirit of the west and of adventure had entered into Mr Donovan's soul and wanting to see more of the country he joined a party of four other adventurous youths who went to the Texas Panhandle, near the south fork of the Sweetwater river They established a camp and built two tepees and began to hunt Arriving on the hunting grounds September, 1877, they engaged in hunting until March, of the following spring The buffalo were very plentiful on the open prairie and often overran their camp so that it was not necessary to go any great distance to secure them They sold the meat of the animals to the grangers from the settlements during the winter and when they broke camp had nine hundred hides for sale The grangers had willingly traded various provisions for the fresh meat so they remained at the one camp throughout the season When ready to leave they disposed of the hides to a traveling buyer of the Gurley and Company at Fort Worth, which made a business of sending men out across the prairies for this purpose When we consider what a buffalo robe brings today it seems hardly possible that they sold each hide for seventy-five cents Soon after leaving the camp Mr Donovan went back on the San Saba river near Minard, where he became established on a cattle ranch in the fall of 1879 After becoming successful in this business he later traded his property and livestock for a band of horses on San Saba river and in the spring of 1880 drove them, five hundred and fifty in all, over the western cattle trail to Ogallala, Nebraska, selling them to farmers who wanted Mexican ponies, for about fifty dollars each After disposing of all the horses Mr Donovan took a sub-contract to do grading on the right of way of the Union Pacific railroad which was being built from Julesburg to Evans, Colorado. On May 2, 1883, Mr Donovan married at St. Paul, Nebraska, Miss Zella Caven, a native of Marengo, Iowa, the daughter of Benjamin and Hannah (Strong) Caven, the former a Pennsylvanian, while the mother was a Hoosier Nine children have been born to this union. Arthur, a farmer, who married Addie Brown and they have three children, Jay O, an automobile mechanic in the Cousrey-Miller garage of Alliance, Floyd R, a harness maker, who married Nettie Nation and served eighteen months in the service during the World War as a member of the Thirty-third Battalion, receiving his training at Waco, Texas They are members of the Methodist church, while Mr Donovan is a Modern Woodman, Claude B, who owns a ranch in the sand hills, married Emma Bowers and they have two girls, Cecil M, a member of the Baptist church and also an Eastern Star, has charge of the parents' home at 423 Big Horn Avenue, and Edna, who married Charles Walters, manager of the John Deer Implement Company of Scottsbluff

He is a Mason and Mrs Walters is an Eastern Star. She graduated from the Alliance high school and later held several fine clerical positions in Alliance, was a popular member of the younger society set before her marriage and both she and her husband are members of the Methodist church Roy C is the seventh child, who when only thirteen years old went to visit his brother Jay on his ranch and while there was dragged to death by a pet pony The boy was a favorite in Alliance and the *Times* honored him by printing an extra edition announcing his sudden and sad death John H, the eighth in order of birth, is a senior in the high school, is a fine, athletic boy, a member of the football team, and when through high school expects to go to college, and Ruth I, also in school, is a member of the Methodist church

Mr Donovan came to Box Butte county in October, 1888, and located on a farm thirty miles west of the present site of Alliance He built a frame house, quite a departure from the usual "soddy" and in order to do so had to go to Pine Ridge, thirty miles away to cut out the logs and then hawl them two miles to a saw mill, paying six dollars a thousand feet for sawing, after which he freighted the lumber to his farm He broke his land, put on good improvements and three years later sold to buy a ranch twenty-five miles south of Alliance making a goodly sum of money by the deal The new home was located near Camp lake where he again put time and work on the place to bring it under cultivation and thus was enabled to dispose of it at his own figure After this sale he for a second time located nearer the county seat as the new tract of twenty-three hundred acres was only eighty miles west of Alliance, where for eight years he was engaged in farming and cattle raising, making a success of both lines of endeavor In 1907, he sold out at a gratifying and handsome profit and came to Alliance The same year he opened a real estate office which he has since conducted He buys and sells farm and ranch property and has built up such an excellent reputation of square dealing and honesty that his judgment on land and its value is taken by all the prominent men of the county Mr Donovan does not sell on commission as a usual thing, but actually purchases the property, holds it as his own and then transfers it to the purchaser For many years while living in the country, Mr Donovan took an active part in all public affairs of his community and since coming to Alliance has been equally identified with the movements for civic and municipal development and improvement He owns his own beautiful home and also two that he rents in the best residence section of the city He is a public spirited man who lives up to the high standard which he sets as an American citizen Mr Donovan is a Mason and his wife is one of the prominent members of the Eastern Star

CHARLES A BURLEW — There is no man more widely known or more closely concerned with public affairs and county development than Charles Burlew Early settler, pioneer newspaper man and politician, public official and merchant of Box Butte county, his numerous business interests and a high reputation for honorable dealing as a business man, have given him an enviable standing in this section of the Panhandle Throughout his life he has not only maintained the reputation noted, but has also sustained and strengthened it so that there are few who have been in public life in Box Butte county today who stand higher in public favor, esteem and confidence

Mr Burlew is descended from a long line of fine old Pennsylvania stock as his ancestors settled in the Keystone state at an early date and there took an active part in its development and political history He was born in Mifflin county, June 12, 1852, the son of Henry and Nancy (Davis) Burlew, both natives of the same state, where they were reared, educated and later met and married Charles was the seventh in a family of fourteen children consisting of seven boys and seven girls He attended school during the winter terms and worked on a farm summers, so that he early learned the value of money When twenty years of age Mr Burlew came west to Dane county, Wisconsin, locating at Mazomanie in 1872, and found work on a farm for fifty cents a day The next year he taught in a country school and as he did not feel satisfied with his educational advantages so far, entered the Platville normal where he remained until graduation After commencement he continued teaching as superintendent and principal of the city schools until 1886 On September 10, 1884 Mr Burlew was married at Eau Claire, Wisconsin, to Miss Margaret C Cogan, now deceased, who was born in Dodge county Wisconsin, the daughter of John Cogan, a native of Ireland Three of Mr Burlew's brothers had served in the Union army When he left Madison Wisconsin, in 1886, for the Black Hills of South Dakota, he was accompanied by Charles T David-

son, now a resident of Hemingford, but as they were not pleased with the country did not settle there but came to northwestern Nebraska and finally located in the Panhandle, February 28, 1886 near the site of Hemingford then in Dawes county At that early date the only structure was a sod house on the corner where Shindler's store now stands, which was occupied by Joseph Hare who was in the land business, locating settlers Mr Burlew and Mr Davidson located land, the former taking a pre-emption two miles west of the present site of Hemingford which at that time was platted around a public square but as the land was not proved up it was illegal but a little later a town site company was formed, Mr Burlew being one of the prime movers of the enterprise In 1887 he was sent to Broken Bow to procure a surveyor to properly plat the town and soon a fierce competition sprang up between the towns of Nonpareil and Hemingford for the location of the seat of justice as there was talk of dividing Dawes county and erecting a new state division in which these new villages were located Mr Burlew was one of the organizers and principal figures in having this new county formed and tried to have the division made in such a manner as to have Hemingford nearest its center He headed the Hemingford delegation in the fight for the county seat, while a man named Gene Heath was the leader of the Nonpareil party It was indeed a bitter feud and the partisans of either side hardly spoke to the men opposing them when they met at Chadron to vote on the permanent location but when Nonpareil was chosen, the two leaders rode home in a one horse cart together and have since been good warm friends Later the B and M railroad survey was made through Hemingford instead of Nonpareil the town was re-platted and finally the county seat was established there In 1886, Mr Burlew bought the pioneer newspaper of the section known as the *Box Butte County Rustler* and during the county seat war had to bear the brunt of the abuse from those opposed to Hemingford for the seat of justice, but it was due to this sheet that the fight was finally decided in favor of his faction when the railroad was built Mr Burlew owned and edited this paper until 1890 and he played an important part in shaping the policies and forming the views of the early settlers He was fearless in his denunciation of anything that was not fair and square in politics advocated and aided in the upbuilding of the towns of the county and agricultural development and became one of the leading public figures of the day In 1888, he was elected county clerk on the Democratic ticket and served two years During this time he maintained his residence in Hemingford, ran his paper by proxy though he took active supervision of it, for he had a capable man at its head in Thomas O'Keefe, whose official title was "Printer's Devil" but who was really manager Mr. Burlew often walked the six miles from Hemingford to Nonpariel while a county official and tells that during his term in office the primitive accommodations of the temporary court house were so poor that the safe in his office was too small for the legal papers so the chattel mortgages were "dumped in an empty salt barrel for safe keeping" From first locating in this section, Mr Burlew has been a prominent figure in political circles, having been called upon to serve on many important committees, and as chairman of the Democratic conventions He has been delegate to the Democratic State Conventions many times and in 1896, was elected as delegate to the Democratic National Convention at Chicago, when Bryan was first nominated for president The honor of organizing the first bank in the county is his for he established this financial institution in 1886, and continued to be an official until 1897, when he disposed of his stock to engage in the mercantile business in which he met with instant and gratifying success He owns his building, erected in 1910, and also a fine home and other good properties in the city, so that today he has a comfortable fortune of his own making, and is considered one of the most substantial men of the community Today Mr, Burlew does as much work in his establishment as any of his younger assistants and is active in many other ways He has been a consistent and constant "booster" for Box Butte county for thirty-five years and for Hemingford in particular He says he has passed through all the phases of development this country could offer, from drinking from a buffalo wallow and sleeping on the open prairies wrapped in a blanket to awaken in the morning with another blanket of snow, to the present development in all its many ramifications of twentieth century life He tells of the interesting relic he had, the first chair in Box Butte county, made of pine lumber cut on Pine Ridge by R H Hampton and says if he had it — unfortunately it was burned years ago — he would not part with it for five hundred dollars Mr Burlew is a Knight of Columbus and a consistent Democrat There are two children in the family Regina

C, a graduate of St Mary's Academy, Omaha and Fremont College, now associated with her father in the store, and Charles A , Jr , who graduated from the commercial course of Fremont College and was for some time a student at Creighton Univeristy, Omaha, now the manager of his father's business and who is displaying marked executive ability

IDA M ROSS — Nearly thirty-three years have passed since Mrs Ross and her husband drove to their farm in Box Butte county and settled on a homestead in a wilderness where houses were few and conditions primitive For years they lived and labored, slowly and arduously developing a farm and establishing a home for their family, watching and assisting in the advancement and progress which were making the country flourish and thrive Mr Ross, an honored pioneer has passed from earthly scenes, but the reputation for industry and integrity which he established is being perpetuated by his children and his wife, who are still the owners of the homestead and the large landed estate which she and her husband owned at the time he passed away Matured and invigorated through the labors and hardships of the pioneer days, Mrs Ross, though a pioneer of two states, Wisconsin and Nebraska, retains the mental, and till recently, the physical vigor of a woman many years her junior She was the devoted helpmate and companion of her husband for about twenty-nine years — a woman whose moral strength was as the number of her days and who had a remarkable share in pioneer experiences in the lumber regions along the Wisconsin river and on the Great Plains of the west, as will be attested by the statements yet to be made in this context

Ida Strobridge Ross was born in Marathon county, Wisconsin, in the little frontier town of Jenny, the daughter of George and Margaret (Pedrick) Strobridge, the former a descendent of fine old Pennsylvania stock and a native of the Keystone state They had three children, two boys and the one daughter, Ida Mr Strobridge was a lumber man by vocation and the first white man to locate in Marathon county He established his home in Jenny, later called Lincoln, and today known as the City of Merrill, Wisconsin on the banks of the Wisconsin river He engaged in lumbering industries, and built and operated the first saw-mill in the vicinity The trees were cut in the timber during the winter snows, hauled along logging roads by horse and ox teams to the banks of the river from one to two miles and stacked there on skidways until the ice broke up in the spring, when they were sent into the water with the first spring freshet and floated down stream to the mill to be cut into lumber Mrs Ross was reared in the family home in Jenny, received her education in the public schools of the town and there grew to womanhood resourceful, spirited and able to cope with any emergency due to her self-reliance and high courage developed by the life she led in this new and little developed section of the Badger state She has many interesting recollections of the great wooded country and pine lands of the northwest and occasionally can be persuaded to tell of how wild the country was during her girlhood, when the Indians would get into inter-tribal fights among themselves and at one time she recounts how her father hid an Indian in the cellar of the house for days to keep him from being killed by his Indian companions At one time Mr Strobridge found an old squaw tied to a tree out in the woods and left to die, as that was the Indian custom with members of the tribe who had passed their days of usefulness, but it seemed too cruel to this white man and he and his daughter Ida secretly took the old woman food, though they had to let her remain tied up as otherwise the Indians would have searched for her and found that she was being fed Mrs Ross grew well acquainted with the Indians near Jenny and often as a child would run away from home to some Indian encampment to play in the tepee of an Indian, as they grew very fond of the young white girl There her father or mother would find her with the papooses or other Indian children

After graduating from the schools of her home town Mrs Ross taught school in Wisconsin for two years before her marriage which took place at Wausau, when she was united in matrimony with Alexander C Ross, a native of New York He was next to the youngest child in a family of six boys After the marriage the young couple decided to seek their fortunes still farther west in the newer country beyond the Mississippi river and came to Nebraska, locating on a large farm near North Bend, where they lived for two and a half years Ida Ross, while reared on the frontier had never lived on a farm and the young wife had many things to learn on the plains and she tells today with a quiet smile, that all her first experiences were not so congenial as might have been and far from her taste on many an occasion but she was a brave, high hearted pioneer and was not daunt-

ed by the hard work, lack of comforts and refinement to which she had been accustomed, and energetically took up her share of the burden of establishing a home Mr and Mrs Ross returned to Wisconsin, where Mr Ross again engaged in his trade as the manager of a large saw-mill company with big operations, but the lure of the west had entered his blood and after a sojourn in the wooded country they again returned to the open plains, but this time they came farther west, being among the first white settlers of Box Butte county They reached the town of Alliance before the railroad had been built and located on a homestead about twenty-four miles northeast of that city As there were few trees on the plains and no mills to cut the logs in that day, Mr Ross shipped the lumber for a frame house from the east along with their household goods, two cows, two pigs, a span of horses, a flock of chickens and enough provisions to last them a year, until the prairie sod could be broken and crops planted and harvested Unloading their things at Hay Springs, Mr and Mrs Ross put their goods on their wagon, led the cows and freighted their worldly possessions twenty-five miles across the prairies to their new home and there settled down to pioneer life in earnest As water is such an important necessity and they could not rely on any stream for it, they were forced at once to put down a well and had to drill over a hundred feet to water Mrs Ross says that they had been more fortunate than many of the early settlers and had considerable capital when they started west and so had some money when they arrived in the new country but there were so many demands for each dollar that they looked at one a long time before spending it Mr and Mrs Ross passed through all the hard times incident to the new country, the grasshopper plague, the droughts and other troubles but they were determined that time would show their judgment of a location was good, and were not discouraged as so many settlers were who returned to their homes farther east and time has proved that fortune was to smile upon their joint efforts They were a thrifty and economical couple, made and saved money and as their capital permitted bought more land with the passing years While Mr Ross tilled the soil, cultivated the crops, harvested and raised cattle and hogs, Mrs Ross bravely shouldered her share of responsibilities and made butter sold it and eggs along with the chickens she raised during the summer season and made weekly trips to Alliance, a round trip of forty-eight miles, to market the produce. She drove her own team of spirited ponies. At first she exchanged the farm products for groceries and clothing while the corn, hogs, cattle and horses Mr. Ross grew and raised were driven to the nearest shipping point, all the money being turned into a common fund to purchase land adjoining the original homestead until they had a landed estate of two thousand acres. Mr. Ross worked very hard and in time his strenuous labors told on his health which was poor for some time before his death, which occurred in February, 1912. He left a widow and three children, a boy and two girls: Margaret, who married Joseph Wiseman, a gold miner of Republic, Washington, where he owns his own mines, and employs a large number of men. Mrs. Wiseman was a graduate of the Chadron Normal school and taught for some time prior to her marriage. Mr. Wiseman is a Mason and Shriner and the family consists of three boys and two girls. Chester A. Ross, the son, is a ranchman of Box Butte county, one of the young, aggressive and progressive men of his community, where he is well and favorably known, always ready to spend time and money for the civic improvements and uplift of the community where he has been reared and lives. He is a member of the Masonic order and with his mother is the owner of several thousand acres of land. In 1918, in response to the president's call for greater production to feed the world they raised ten thousand dollars worth of small grains on their ranches and in 1920, expect to farm five hundred acres, using large tractors. They are also engaged in an extensive live stock industry, owning large herds of cattle and horses which are shipped to the large markets in the east each year. Evangeline M., the third child of the Ross family married Chester H. Aldrich, the son of Governor Aldrich and lives on their farm near Ulysses, Nebraska. Mr. Aldrich is a graduate of the agricultural college of the University of Nebraska, while his father is a member of the supreme court of the state. Mrs. Aldrich attended Carrol College of Waukesha, Wisconsin, where she was graduated with honors.

During the war, while help of all kinds was very scarce, Mrs. Ross assumed too great a burden, worked beyond her strength and at present is contemplating taking a much needed rest and leaving the active management of her estate in the capable hands of her son. She has a beautiful country residence on the ranch with every modern convenience and is so enamoured of rural life that she prefers this home to any other and chooses to live on the farm rather than in Alliance, though she owns a valuable property on the corner of Fourth street and Laramie avenue where she expects to erect a modern apartment building in the near future at a cost of over one hundred thousand dollars. Due to her fine character, enterprise, many charitable acts and her interest in civic and communal affairs, Mrs. Ross is known throughout Box Butte county as a woman of high moral standing, Christian character, and is considered one of the most prosperous and substantial residents of the Panhandle. She is a member of the Presbyterian church and of the Eastern Star.

THOMAS A. GREENE. — The career of Thomas Greene, now one of the members of the retired colony of Hemingford, has ranged through varied conditions in the Panhandle as he drove into what is now Box Butte county in true pioneer style in 1885, settling on the prairie in what was then a wilderness, for far as the eye could reach spread the unbroken virgin sod covered with buffalo grass and wild flowers. He has lived to see what was known in the early eighties as "The Great American Desert," blossom like the rose and the Panhandle become one of the favored and most productive sections not only of Nebraska but of the whole nation.

Mr. Greene is a native of the Pine Tree state, born in Danby, Vermont, March 6, 1859, the son of Rowland R. and Harriet E. (Parmeter) Greene, the former a Rhode Islander, so that he is descended from a long line of New England ancestors who played an important part in the history of the eastern states. Thomas was the third in a family of five children, as he had three half brothers and a half sister. Rowland Greene was a carpenter employed in building operations in the town of Danby and the vicinity so that the boy spent his youthful years near that town, attended the public schools of the village during the winter and helped on the home farm, rented by his father, during the summer time. While yet a boy he began to earn money for himself driving oxen for a neighbor and remained at home until he was twenty-one years of age. Land was high in the east and not very productive at best and Mr. Greene read of the great stretches of fertile land to be had in the west for the taking and decided that there was the place for a young, ambitious man. In 1883, he started west, locating first near Cresent, Iowa, but two years later he came on up the valley of the Platte to Dawes county, which

HENRY M. THORNTON

later was subdivided and that portion in which Mr Greene settled became Box Butte county He took up land twelve miles southwest of the present site of Hemingford, his nearest neighbor being four miles distant Mr Greene freighted his first supplies from Camp Clark, fifty miles away, having to cross the river on one of the early toll bridges, and the goods he bought had already been freighted in from Sidney, at the railroad fifty miles further south Among the things Mr Greene purchased were the barrels to haul water for his stock and household use, which was obtained at a spring nine miles away Mr Greene built the well known sod house of the high plains for his first home and established himself as a bachelor Subsequently when money was easier and the railroad had been built through this section, he erected a fine modern home, but for twenty-eight years he remained unmarried, sometimes employing a man and his wife on the farm but much of the time he maintained his home alone Mr Greene was not discouraged by the years of drought, grasshoppers and crop failures and "stuck it out," so that today he is one of the oldest settlers of the section From time to time as he made money from his cattle and farm produce he invested the capital in more land until today he is one of the largest landed proprietors in Box Butte county, owning nearly twenty-five hundred acres of the best land in this vicinity, all of which is well improved and much is under intensive cultivation

On September 22, 1914, Mr Greene married Miss Mildred Best, who was born at Springfield, Missouri, the daughter of James N and Eva (Haseltine) Best, the former a native of South Bend, Indiana, while the mother was born in Richland Center, Wisconsin Mrs Greene was the oldest of four children She is a graduate of the Bertrand high school, the Holdrege Junior Normal school, and took two years work in vocal and piano music at Fremont Normal and became a teacher, a vocation she followed for twelve years before her marriage, the time being about evenly divided between county and village schools Since her marriage she has taken an active part in all communal affairs and during the World War was an active worker, being president of the Nonpariel Neighbors, a club doing Red Cross work Three children have come to the Greene home, Harriet Louisa aged five, Thomas Hascal, past two, and Baby Mildred Ruth, six months old In the spring of 1919, Mr Greene decided to give up the active management of his land and came into Hemingford

to make his home where he and his wife enjoy the social advantages of a town and later give the children the benefits of the educational advantages afforded there They have a beautiful home in the town, where they hold open house to their many old friends of the country and also for the people of Hemingford who know and like them well Mr Greene is so well fixed with worldly goods that he can afford to take life easy, enjoy the fruits of his labors and looking back across the years feel that he has earned a well deserved period for enjoyment and relaxation His wife is a member of the Congregational church of which they are liberal supporters

HENRY M THORNTON who is at the head of a large business enterprise at Gering, is one of the county's substantial men and representative citizens He came to Scottsbluff county in early manhood and has long been concerned in the development that has brought so marked a change in this section of the state, for, with principles of the sturdiest kind of honesty, he has invariably demonstrated local interest and a practical public spirit Mr Thornton is known all over the county and for four years served as county clerk

Henry M Thornton was born in Kane county, Illinois, February 29, 1864 His parents were Edward and Jane (Stewart) Thornton, the former of whom was born in New Hampshire and the latter in New Brunswick, both going to Illinois in youth, in which state they married Of the three survivors of their seven children Henry M is the eldest, the others being his two sisters Margaret the wife of O W Gardner, and Grace, a resident of Salt Lake City The mother of the above children was a member of the Methodist Episcopal church The father, the late Edward Thornton, came to Nebraska in 1886 and for a season the family lived in a tent Later he built a two and a half story log cabin, which for years was considered the finest house in the county Buying a team of horses and a few head of cattle and homesteading, he became a substantial resident of the county in the course of time He became prominent in Republican politics and was postmaster at Gering for a number of years He was also one of the early and actively interested Masons in this section

Henry M Thornton attended the public schools as a boy and later a business college at Dixon, Illinois In connection with his father, he conducted a creamery business in Illinois, prior to the family exodus to Nebraska After locating here he preempted a body of land borrowing $300 to prove up on it, and after paying $300 interest, sold the entire place for $100. Following this venture in business he went into partnership with his father and together they farmed and raised stock, but after about five years on the farm, he went into a bank, where he worked his way up to be cashier and finally president. After retiring from the banking business in 1914, Mr. Thornton founded his present large hardware and furniture concern, and has developed one of the leading business houses in this line in the county. To properly handle and display his immense stock, he erected a two-story brick building which has dimensions of fifty by one hundred feet, which is favorably situated in the best business district at Gering, and trade is attracted from all over the county. He is at the head of the business but is ably assisted by his two sons, Douglas and Kenneth E.

In 1897 Mr. Thornton was united in marriage to Miss Alice J. Johnson, a native daughter of Nebraska. Her father, D. D. Johnson, for a number of years was in the livery business at Salt Lake, Utah, and was a well-known horseman. He came to Scottsbluff county in pioneer days, took up a homestead, and has been somewhat prominent in politics. He now lives retired at Scottsbluff. Mr. and Mrs. Thornton have four children: Douglas A. and Kenneth E., who are graduates of the high school at Gering and former students of the state university at Lincoln, are twins, and both belong to Sigma Alpha Epsilon college fraternity; Janet, who is in her junior year at the state university, and Beth, who attends school at Gering. Mrs. Thornton and her daughters are members of the Methodist Episcopal church. Mr. Thornton is a Mason of advanced degree and a Shriner, belonging to Tangier Temple at Omaha. For several years he served as master of his own lodge. Both he and wife belong to the order of the Eastern Star.

Political questions have interested Mr. Thornton to some extent all his life and formerly he was quite active in Republican circles but did not often accept public office, although particularly well qualified for the same. For four years he was county clerk and his administration was satisfactory in the highest degree. Although he devotes the most of his time to the management of his business in Gering, he keeps a careful overseeing eye on his 1000 acres of irrigated land, which he has found a profitable investment.

SIDNEY A. D. GRIMES, one of the popular and successful auctioneers of Box Butte county is a partner of F. W. Melick in a produce business and feed yard in Hemingford.

Mr. Grimes is an Iowan, born near Russell, October 10, 1876, the son of Henry T. and Mary C. (Cain) Grimes, the former a native of West Virginia, while the mother was born in Iowa. Sidney was the second child in a family of six children, consisting of five boys and one girl. Three of his brothers are dead. Henry Grimes was a farmer in Iowa and the children were reared on the farm, attended the public school near their home where they laid foundations for good, practical educations. Sidney began to help at home when he was big enough and soon learned the practical side of farm business so that while yet a small boy he hired out to a neighbor to plant corn, using an old Keystone planter. He recalls with a whimsical smile that the field they were to plant was a round hill the top of which had never been broke, so he started at the outside of the field and went round and round till the sod was reached then turned back the same way unwinding, and only planted two rows in a day as they were so long. For this, his first independent business enterprise, he received twenty-five cents a day. Sidney was sent to the district school nearest his home in the country during the winter terms and after his preliminary education was completed graduated from the Missouri Auction school at Trenton, Missouri, in 1908. In the meantime he had been established as a practical farmer in Iowa, where he became one of the substantial representatives of the agriculturists of his section as he was wide awake to all modern methods and improvements, adopting those which he saw would be of value to him.

On January 29, 1893, Mr. Grimes was married at Weller, Iowa, to Miss Winifred May Patterson. She was born near Attica, Iowa, the daughter of William M. and Martha (Rogers) Patterson, the former a native of West Virginia. She was the third in a family of ten children born to her parents and since her marriage has become the mother of six children herself, they are: Verna A., who married G. G. Melick, of Hemingford, who owns and operates a hotel and they have one child, Christiana; Hattie P., the wife of Frederick Seirwein, the owner of a garage at St. Bardanoes, California; Desse M., at home with her parents; Gladys R. and twins, Wesley Franklin and Wilma Fern, in school.

After he had taken the course in the auction school, Mr. Grimes continued to engage in farming and also auctioneering in Iowa, but he knew that there were many and good opportunities for an energetic man to make money in the newer country of western Nebraska and determined to come here and learn what fortune had in store for him. In 1913, he came to Box Butte county and located in Hemingford, though he still retains the ownership of his fine farming land in Iowa. For two years after coming west he devoted his entire time and energies to auctioneering, but saw that there was a good opening here for a much needed produce business and feed yard where the stock of the surrounding fine farming district could be bought and sold. In partnership with F. W. Melick he bought the Spencer Lumber Company's ground which was well located for their business and there erected a fine feed yard where the company buys and sells live stock, hay, feed and grain. In 1919, Mr. Grimes wrote the largest check ever banked in Hemingford for a sale of hogs, it was for $5,582.20, payable to C. E. Wilsey for seventy-five head. Today Mr. Grimes owns a modern home in the town and has built up a reputation as one of the most competent and resourceful auctioneers in western Nebraska though he is called all over the state to conduct sales and says, "I will answer calls anywhere this side of the celestial realm." Mr. Grimes stands high in the Masonic order, having taken his Thirty-second degree and is Past Grand in the Independent Order of Odd Fellows

JAMES V. POTMESIL, president of the First National Bank of Hemingford, is one of the leading financiers of Box Butte county. Efficiency has characterized all of Mr. Potmesil's business and most especially his services in the responsible office of president of the bank, and his administration has done much to conserve the success that has marked the history of this important and representative financial institution, while his personality and civic loyalty have gained him a high place in popular confidence and esteem.

Mr. Potmesil is a native of Bohemia, born in that province of the Austrian Empire, December 26, 1872, the son of John and Rose (Sixta) Potmesil, both natives of the same country. James was the third in a family of six children born to his parents in their native country. He attended school there for about two years before the family came to America as their parents were desirous of giving their children more advantages than could be obtained for them in Bohemia. The father emigrated in 1880, and after reaching America worked his way to Cedar Rapids, Iowa, and then learning of the opportunities to secure free land from the government farther west

came to Nebraska in 1882 locating in Saunders county, where he secured work on a farm. The year before, having saved the money and learned about reaching the United States, Mr. Potmesil had sent for his wife to join him and bring the children. After they arrived he continued to work on a farm and had the three oldest children do the same. Young James, who was eight years old when he made the long voyage to the new world was duly initiated into the art of herding cattle as soon as his age and strength permitted. He did this for two years during the summer and today tells that "it was afoot, if you please, no cow ponies for me in those days." However, the boy seemed to thrive on this out door work during the summer and helped as he could at regular farm industries during the winter season receiving twenty-three dollars a year and his keep. During these hard and trying years on a frontier farm the boy had little opportunity for educational facilities but availed himself every chance of self improvement and when the family had some headway toward a competency he was sent to Chadron to school, then entered the Chadron Academy and after graduating entered the agricultural college of the State University at Lincoln, for this boy of foreign birth early realized that the best and most secure capital he could have was a good education of a high order. He had been reared on a farm and after coming to this great commonwealth had lived and worked on one so that he naturally decided on agriculture as a business. In the fall of 1884, the father had accumulated considerable capital for a man on the frontier and as he and some of his Bohemian friends learned of the fine land to be had in the Panhandle, a party of nine men purchased horses and made the long trip up the Platte valley to Valentine. They filed on land in what is now Box Butte county about eighteen miles northeast of the present site of Hemingford, and after locating the homesteads returned to Saunders county. The following spring the new settlers shipped what provisions they could buy to Valentine while the families drove across country in prairie schooners, locating on the new farms. Settlers were few and far apart and it was fortunate for these men that they located in the same locality so that they could help each other in erecting their first homes of sod and in general farm work where one men could do little alone. Mr. Potmesil made a number of trips to Valentine to freight in supplies and provisions for himself and his neighbors, a trip of one hundred and fifty miles which took sixteen days. During these trips the mother was forced to pack water from the river a mile and a half away to the house for family use, at a time when the country was all open range over which the wild, long horned cattle roamed at will so that her trips were made with great labor and danger. Thus Mrs. Potmesil was many times fearful of her life and consequently her little ones at home. In the fall the railroad was completed as far as Hay Springs and the following winter the father and boys earned most of their money for supplies gathering buffalo bones on the prairies to sell at this station. Another piece of hard work the father accomplished was taking down a fence around a pasture twelve miles wide by fourteen long. By these various endeavors the family accumulated money enough to buy the first cow in the country and they had milk and butter in abundance. Having a keen, far-sighted vision the father gathered together what money he could and began to buy cattle in a small way. That was the heigh-day of the cattle industry on the prairies and knowing from the settlement already taking place that the day of the open range was waning had cattle to tide him over the years of drought and grasshopper invasions as he managed to find pasture along the river bottoms. More fortunate than many of the settlers he was not driven out during those trying periods and by holding on, increasing his herds and buying land at low prices, the father and his sons prospered and felt well paid for the years of privation, hardship and labor. By 1913, the family owned and held some seven thousand acres of land in Box Butte county and ran from three hundred to six hundred head of cattle, having as it may be said, "got in on the ground floor," in the cattle business before most men realized that the future of the live stock industry was to be in the hands of the farmer. Both the father and mother died in 1913, leaving the large landed estate to the sons, the oldest of whom, John, still remains on the ranch as owner-manager, though the land is shared by John and James jointly. In 1912, the First National Bank of Hemingford was organized by C. J. Wilder, Mr. James Potmesil becoming one of the organizers with him and a heavy stockholder of the new institution, and eventually when the personnel of the officers was chosen was called to the head of the institution as president. The bank has a capital stock of $25,000, surplus of $20,000 and deposits of $414,031 in 1919, which shows its rapid growth in less than a decade. The management of the bank has from the first endeavored at all times to serve the public efficiently and to be of real benefit to its friends and patrons, all of which has been carried out under the guiding direction of the president, whose policy has been progressive yet conservative and since becoming a member of the Federal Reserve System, has had the financial backing of the county to which it may be likened as the backbone or financial support. Mr. Potmesil has well earned his high position in this county, for he is one of the first settlers, and by his thrift, executive ability, sound business principles built up a reputation of which he may justly be proud, for he has been most especially the architect of his own fortunes, and looking back down the years may feel that his life has been well lived for he has been a builder, and played an important part in the opening up and development of this favored section of a great commonwealth. On January 31, 1917, Mr. Potmesil was married at Beatrice, Nebraska, to Cecil Wilkenson, born in Pawnee county, the daughter of William N. and Emma (Seldon) Wilkenson, the former a Kentuckian, while the mother was a native of Iowa. Two children have been born to this union: James, and Jeane, a baby. Mrs. Potmesil is a graduate of the Beatrice high school and of the Peru Normal school and after her graduation from the last institution she taught for about ten years prior to her marriage. They have a lovely home in Hemingford, where they dispense a cordial hospitality to all their many friends for few people in the county do not know the Potmesil brothers, for they are the type of true American citizenship who are making history in the Panhandle, both financial and civic.

WILLIAM M. CORY, one of the substantial merchants of Box Butte county and Hemingford, the owner and manager of a men's furnishing and jewelry store, has been a resident of this section since 1890, and during more than a quarter of a century that has intervened he has been variously connected with agricultural and the rising industrial and commercial interests of the county, always to the benefit of himself and his community.

Mr. Cory is a native son of this great commonwealth, born in Brownsville, September 8, 1873, the son of Thomas and Janet (Slatterly) Cory, the latter a native of the state of New York. William was next the youngest in a family of five children. Thomas Cory was a farmer who moved to Washington county, Kansas, when the boy was about five years old and in the Sunflower state he spent his early youth living with his parents on a farm William was sent to the excellent public school near his home, thus laying the foundation for a good, practical education, which was continued in the Nebraska schools upon the return of the family to this state in 1886 when they became pioneer settlers of Custer county, locating near the town of Lomax The father improved his land, made an effort to establish his family in comfort and they weathered the privations and hardships incident to settlement in a new country as well as the years of drought, insect pests and blizzards William earned his first money pulling weeds when a small boy and also by hoeing corn so that he knew well its value When the Cory family came to the Panhandle in 1890, locating in Alliance, William found a position in a news stand conducted by the Miller brothers, where he began to learn business and really laid the foundation for his commercial career On December 22, 1902, Mr Cory was married at Old Canton, Nebraska, to Miss Mary E. Clayton, who is a native of Illinois, the daughter of George H and Emma (Forbes) Clayton, the former a Missourian by birth, while Mrs Clayton was born in Illinois Mr Clayton was a well known ranchman of this section and Mrs Cory after finishing the country schools came to Alliance and graduated from the high school Seven children have been born to Mr and Mrs Cory Leslie P, in the Hemingford high school, George T, and Merle, both in school, Jessie May, deceased and Wilma L, Glen E, and Ruthelm also deceased A year after his marriage Mr Cory took a homestead of six hundred and forty acres twenty-eight miles southwest of Hemingford, where he established his family erected a good home and made substantial improvements for farm work He soon had his land broke, under cultivation, and his boyhood experience as a farmer aided him in becoming one of the practical and prosperous men of his section, but he had been in business and desiring many more advantages for his family than could be had in the country, decided to again enter commercial life and after six years on the farm came to Hemingford in 1909, to establish one of the leading haberdashery shops and jewelry stores of the upper valley Naturally courteous and cordial Mr Cory soon built up a fine trade that was most gratifying from a financial point of view as he felt that he was being well repaid for the thought, study and work he put into the business He owns the fine brick building, twenty-five by a hundred

feet, which houses his stock, which is one of the representative business houses of the town, also owns a fine home where he and his wife dispense a cordial hospitality to their friends Mr Cory has become widely known for his thrift, foresight in financial affairs and for giving his customers a 'square deal" He and his wife are members of the Methodist church and his fraternal affiliations are with the Odd Fellows

FREDERICK W MELICK, one of the progressive business men of Box Butte county who is engaged in the commission business at Hemingford, where he handles flour grain and buys and sells potatoes and stock, is a native son of Nebraska, born at Bennett, Lancaster county October 21, 1878, the son of Frank and Christiana (Larson) Melick the former a native of Hunderton county, New Jersey Frederick was the second in a good old fashioned family of ten children so that at an early age he began to assume many responsibilities, such as looking after and protecting the younger children who were often placed in his charge Frank Melick was one of the pioneer settlers of southwest Kansas as he took up a homestead and the boy spent his childhood and early youth on the farm He grew up under the strict discipline of agricultural demands so that he was thrifty, industrious and well able to cope with the many emergencies that arise on a frontier homestead His father encouraged him in his early attempts to make money by making him a partner when they took cattle to graze, so that the boy was given charge of the herding which he usually had to do on foot and recounts with a smile that many a time when he was far from the house all he had for his lunch was a few cactus pears which he himself gathered from the prairie Mr Melick says that his early educational advantages were few and rather far apart, as he worked for about nine months of the year and was able to attend school during the three winter months when it was impossible to work on the land But he made the most of every opportunity and laid the foundation for a good practical education that has been of inestimable advantage to him in his business life When he attained his majority, Mr Melick determined to start out in life independently and went to work on the Santa Fe Railroad in Kansas, where he gained a knowledge of business which has proved of benefit On December 22, 1901, Mr Melick was married at Bennett, Nebraska to Miss Alice Canfield a native daughter of that town, whose parents were

Leman and Ida Mae (Barsock) Canfield, the former born in Hundertown county, New Jersey. She was the oldest in a family of twelve children. Mr and Mrs Melick have one daughter, Marguerite, a student in the Hemingford school. In 1901, Mr Melick came to Nebraska and located on a farm in Lancaster county, where he engaged in farming for six years, but he heard of the many good opportunities to secure land in Box Butte county so came here in 1907, buying a hundred and sixty acre tract one mile southwest of Hemingford. There he demonstrated what a good reliable man can accomplish who makes a study of his business and is determined to succeed. After about six years Mr Melick had accumulated considerable capital and gave up the active management of his farm and moved into town, where he invested in a good grain business, buying and selling flour, potatoes and live stock, and soon had a paying commission business established, which has brought in most gratifying returns. In 1917, Mr Melick purchased the Hemingford Rolling Mills, which he now manages in addition to his other business activities. The business has expanded rapidly and today Mr Melick is the largest wheat and potato shipper in the Panhandle and the northwest for he sells hundreds of car load lots each season. Mr Melick is popular in financial circles, where he has gained an excellent reputation as a wide awake, progressive man of affairs, while personally he has made many warm friends. He is a Mason of high standing, having taken his Thirty-second degree. No opportunity slips through his fingers that he sees in business while he is one of the true Americans who live their patriotism giving freely of time and money for the upbuilding of the community in which he lives and Box Butte county.

ROBERT C MILLER, one of the younger generation of business men of Hemingford, who are the makers of financial history in this section may truly be called self-made, as his present prosperity has come to him through his own efforts, and his life record exemplifies what may be accomplished by industry and perseverence. He is one of the gallant Nebraskans who responded to his country's call when war was declared against Germany, and enlisted in the aviation branch of the service.

Russell Miller is a native son of Nebraska and of Box Butte county, as he was born in Alliance, May 11, 1896, the son of Melvin L and Grace A (Shaffer) Miller, the former a native of Illinois. Russell is the oldest of the two children in the family as he had a younger sister, Irene. Melvin Miller was connected with railroad work which necessitated his moving to Martinton, Illinois, when Russell was only two years old. The boy was sent to the graded schools and after finishing the elementary courses spent four years at the Martinton high school, graduating near the head of the class in 1912. The following year the young man returned to Nebraska to accept a position with his grandfather who owned the Hemingford mill. For two years Mr Miller kept the books of the firm before accepting a very advantageous offer to become associated with the Farmers Lumber Yard here, of which Alexander Murhead was manager and remained with that company until he enlisted in the army on December 12, 1917, and was sent to Kelly Field, San Antonio, Texas, having been assigned to the Air Service, 662 Aero Squadron. He remained in this branch until after the signing of the Armistice, receiving his discharge January 30, 1919. During the terrible epidemic of Spanish Influenza that swept over the country the winter of 1918, Mr Miller was in the hospital from December 10, 1918, to January 10, 1919. After leaving the army he came back home and like so many of the discharged men desired to go into business for himself. While at Kelly Field he had studied the construction and operation of motors and as the automobile business is a flourishing one everywhere now days, Mr Miller saw a good opening in this line and on June 23, 1919 purchased the garage and stock of George Hedgecock. He at once added to the building and equipment, and now has a brick building fifty by one hundred and thirty feet, giving him a large floor space for storage, work shop and display rooms, as he is local agent for several of the best makes of cars, among them the Chandler, Cleveland and Ford. In connection with his sales force he conducts a fine up-to-date repair shop, sells gasoline and several lubricating oils and maintains one of the best and most prompt auto liveries in the upper valley. The Miller Garage handles several lines of the best tires on the market while his business of storing and caring for cars is rapidly growing. Mr Miller is a fine machinist and few troubles to which autos are subject can not be put right under his skillful direction. Due to his thorough training and ability his repair work has been on the increase from the first while his courtesy, consideration and reputation for prompt service have built up all branches of the business. From the fine start he has made in this chosen vocation nothing but a bright future can be in store for this young, energetic and far sighted man

On May 17, 1916 Mr Miller was married at Hemingford to Miss Edna Geiger, born in York, Nebraska, the daughter of Charles and Nora (Grass) Geiger, the former a native of Michigan Mrs Miller was the second of their three children and has one child of her own, Marjorie Aleene

PHILIP J MICHAEL, one of the younger generation of business men of Box Butte county is also a leading and prominent real estate and insurance man of this vicinity who has been identified with numerous financial enterprises in Hemingford where he has established a high reputation for ability, judgment and the "push" which characterizes the Nebraskan the country over He is a native son of the state and of Box Butte county and since he entered business his rise has been rapid, sure and consistent Mr Michael was born near Hemingford, December 9, 1888 the son of Philip and Etta (Strange) Michael the former a Hoosier by birth while the mother is a native of Illinois Philip was the sixth in their family of nine children Philip Michael, Sr, was a farmer by vocation who came to Nebraska in 1885, and became one of the pioneer settlers of Box Butte county, as he drove across country from Iowa settling southwest of Hemingford when this country was sparsely settled, and practically an unbroken wilderness covered with buffalo grass and prairie flowers The family settled on their pioneer farm near the frontier, built the regulation sod house for a home, made necessary shelters for their stock, broke their land as soon as possible and put in the crops that would provide them with some food Those first years here on the high prairies were hard one for the early settlers, money was scarce, drought killed much of the crops, and what was left in many cases the grasshoppers ate up and many cattle died during the winter blizzards, but Mr Michael was stout hearted, believed that there was to be a great future for this country and held on For the first years he made money to buy provisions by freighting from Nonpariel to Valentine, a hundred and fifty miles away, and in this manner managed to hold down his claim and prove up on his land when many of the settlers became discouraged or were forced to leave or starve Water was the great and important question of the pioneers of this region and Mr Michael had to haul his for many miles, both for his family use and to water his stock, until wells were sunk on his farm

Philip, Jr, was born on the old homestead, grew up in the new country resourceful and self-reliant as any boy on the frontier had to be, as many occasions arose when he had to take care of himself and also cope with unexpected conditions He was a hardy, healthy lad, who soon began to assume many duties about the farm, working as his years and strength permitted and so learned the practical side of farm business from his father while he attended the district school nearest his home in winter time and laid the foundation for a good practical education After finishing the elementary department in the country the boy entered the high school at Hemingford graduating from the four years course Early Philip had learned the value of money as he was but ten years of age when he began to work on a farm for ten dollars a month and after earning over seventy dollars invested it in a colt which he attempted to break but the horse became frightened and bolted through a four wire fence, broke its leg and had to be killed, but fortunately Philip came through without a scratch though he got a good shake up in the fall The colt had to be killed but the boy felt he was lucky to get off so well Philip remained at home with his parents, helping his father on the farm until he was twenty years old, when he determined to establish himself independently in business and accepted a position as yard man with the Forest Lumber Company, was soon promoted to manager and remained with this concern until 1911 He saw the many openings for building and real estate in this rapidly growing section and resigned With William M Pruden he formed a partnership in a real estate and insurance business, also handling life insurance as a side line These young men are both progressive in ideas and methods and have proved by their rapid rise in the financial circles of the county that they are able and competent men in business In 1919, they built a fine office building with over a thousand square feet of floor space where they now conduct their large business which can best be described as rushing," where the handling of realty is concerned The Michael family are members of the Methodist Episcopal church, while Mr Michael is a member of the Masonic order and Odd Fellows November 18 1909 Mr Michael was married at Hemingford to Miss Etta M Kinsley a native daughter of Nebraska, whose parents were Noah and Harriet (Kirkendall) Kinsley, both Hoosiers One child has been born to Mr and Mrs Michael Audrey now six years old

JOHN T GARVEY railroad contractor, miner, veteran of the Confederate army and early settler, is probably one of the oldest men living within the confines of Box Butte county who has witnessed the many changes that have taken place on the plains and especially in the Panhandle since the western part of the state of Nebraska was the frontier His career has been one in which he has had varied and interesting experiences from trailing marauding Indians who raided his camp to the development and civilization of modern days, and few men show so little the scars of such a hazardous life Mr Garvey was born in Ash county, North Carolina, May 20, 1845, the son of John and Polly (Doerty) Garvey, both natives or North Carolina, where they were reared, educated and later met and married John was the youngest of the two boys born to his parents His father was a farmer and the boy spent his childhood and youth in the healthy country environment, growing up strong and willing to work, for he helped his father in the summer time and attended the school near his home during the winter terms Like most farm boys he wished money of his own and to obtain it dug gensing, a root which is greatly prized by the Chinese as a medicine This he sold for fifty cents a pound, and some of his first money was spent for a spelling book which indicates that while a small boy he was ambitious John remained at home to assist his father with the work on the farm until 1861, when he enlisted in Company B, Sixth Confederate Cavalry, and took his part with the Confederacy in the Civil War Mr Garvey participated in many of the most bitterly fought battles of the war He was at the battle of Nebern, Fort Croxton and Kenston, then was transferred to Tennessee and Kentucky, took part in the actions at the King Salt Works Abington, Morristown and later in the engagements of Bluntsville, Rogersville, Pound Gap, Cumberland Gap, Strawberry Plain, Janesville, Thorn Hill, and the sieges of Knoxville, Chattanooga, and Nashville After peace was declared he surrendered with the other members of his company at Newbern, North Carolina, on February 25, 1865 Taking the oath of allegiance he was again a citizen of the United States and went to Louisville, Kentucky, but like so many of the returned soldiers he was restless and the vocation of pre-war times was not satisfactory, so he decided to seek what fortune had in store for him on the western frontier and came to Omaha making the trip on the boat that carried the first load of railroad iron for the

building of the Union Pacific Railroad Mr Garvey remained in Omaha until 1867, when he bought a large number of government horses and engaged as a contractor to do grade work on the Union Pacific on the right of way from Omaha to North Platte, under the supervision of Cane. Collins and Kennedy, constructing and contracting engineers For eight months he was engaged in this work when a band of Sioux and Cheyenne Indians rode into the herd of horses belonging to the graders at the camp on the present site of the town of North Platte, stampeded the two hundred and eighty head and drove them off in a northwestern direction toward the "sweet grass hills" Soldiers from Fort McPherson were sent out to regain the horses but failed as the Indians had planned the attack well, having drawn three days' rations at Braidy Island, the end of the railroad at that time This broke things up for Mr Garvey so he and his cousin, T C Garvey, bought a couple of pack ponies and decided to hazard their fortunes in the newly opened gold fields They joined an emigrant train which was crossing the plains and Mr Garvey says that he walked beside his pony all the way from North Platte to Helena, Montana, with the exception of three miles Arriving at the latter city on November 19, 1867, they went to work in the placer mines, staked out their claims, and were engaged in mining from 1867 to 1871 Mr Garvey then went farther south and worked in the "Ore Knob" copper mine for about six years, gaining valuable experience in mining He had returned to Nebraska at just about the time gold was discovered in the Black Hills and joined in the stampede to that locality A picture of his bull team taken just after Mr Garvey left Deadwood is to be found in this history The "diggings" did not prove as worthwhile as Mr Garvey anticipated and in disgust he left and returned to Deadwood where he was engaged in business for nine years, running a delivery and dray concern On June 11, 1872, Mr Garvey was married at his old home in Ash county, to Rachel A Johnson, the daughter of Aaron and Jane (Tomblin) Johnson, the former a native of Ash county, born in Blueridge Five children were born to this union James H, who is employed by the Standard Oil Company, is married and lives in Chicago, Laura who married James H Prophet, is dead, Walter T, Naomi E, and Grace E, are all deceased Mrs Garvey died December 28, 1879, and on December 28, 1900, Mr Garvey was married in Box Butte county, to Miss Johanna John-

MR. AND MRS. CHARLES B. HOTCHKISS

son, the daughter of Benjamin P and Mary (Curry) Johnson, both natives of North Carolina Mrs Garvey is the oldest in a family of eight children After trying mining in the Black Hills, Mr Garvey decided that a man who owned a good farm was about as well fixed as a miner could ever be and usually much more successful, so he came to the Panhandle and filed on a homestead in Sioux county where he at once began good and permanent improvements He became recognized as one of the prominent and responsible agriculturists of his section, made money from his land and felt that he was repaid for the labor he expended

After nearly seven years in the country Mr Garvey moved into Hemingford, bought a good home and has since been a resident of this city He has made a host of friends since he first located in the Panhandle and when the citizens of Hemingford desired a responsible man for the office of city marshall he was appointed, and as he was a man of excellent business ability was elected street commissioner and commissioner of the water works, all of which offices he has filled to the entire satisfaction of the citizens and to the benefit of the community He has not found that the city offices have taken all his time and has been conducting an extensive live stock business buying and selling, and shipping cattle to the eastern markets As he has been a well known and well-to-do farmer for years, this business has become a very lucrative one under his able management The Garvey family are members of the Congregational church

CHARLES B HOTCHKISS, who is interested in several business enterprises at Gering came to Nebraska thirty-five years ago and has been identified with much that has been valuable in the development of different sections In a large degree he is a self-made man, beginning early to make his own way in the world, learning the give and take of business when many youths of his years in easier circumstances, were yet in the schoolroom

Mr Hotchkiss was born at Bloomington Illinois, November 1, 1864 one of a family of four children, two of whom survive, born to Thomas and Hattie (Wright) Hotchkiss The parents of Mr Hotchkiss were also born at Bloomington, where his mother died in 1917 His father was a machinist and a contractor, did much building at Bloomington Later he bought a farm near Bloomington, but afterward went to Topeka, Kansas, and there his death occurred in 1913 He was a Republican in politics and a member of the order of Odd

Fellows in good standing Both he and wife belonged to the Methodist Episcopal church Mr Hotchkiss has one brother, Frank who is a merchant of Bloomington, Illinois

In early boyhood Charles B Hotchkiss attended school in Bloomington but was still young when he began to work on a farm on his own account In 1884 he came to Nebraska and bought railroad land in Dawson county, subsequently engaging in the lumber, coal and grain business at Gothenburg He was one of the enterprising business men of that prosperous town for fourteen years Circumstances then attracted Mr Hotchkiss to Grand Island in Hall county, where he bought the interests of the firm of Walker & Blaine, and for ten years afterward was associated with Thomas Bradstreet in the horse business It was in March, 1913, that he came to Gering and started his implement business, in 1917 erecting a commodious brick building He also owns a meat market here and also has a valuable ranch located within a short distance of the city His numerous interests have made him widely known and his business integrity has never been questioned

On November 1, 1904, Mr Hotchkiss was united in marriage to Miss Kate Cullison, who was born in Pennsylvania, and is a daughter of Judson and Caroline (Corcilious) Cullison, the former of whom was born in Virginia and the latter at Louisville, Kentucky They came to Nebraska in 1883 and settled at Central City Mr Hotchkiss has always been intelligently interested in politics and votes with the Republican party He interests himself in all civic matters at Gering as becomes a faithful citizen, and he has served on the town board He belongs to the order of Modern Woodmen

C RUSSELL MELICK is one of the younger business men of Box Butte county, who is taking a prominent place in the financial circles of Hemingford, is also a son of Mars as he enlisted in the army when war was declared against Germany and thus demonstrated his Americanism and patriotism Mr Melick was born in Hopkins, Missouri, February 6, 1895, the son of Frank and Christiana (Larson) Melick, the father was a native of New Jersey, while the mother was born in Denmark Their history will be found on another page of this work Young Russell was reared in the healthy environment of the country, where he grew up strong healthy and full of life and vigor He recalls with a smile that the first money he earned that was obtained by real work was earned when he went into the hay fields at harvest time for fifty cents a day This money he thriftily banked and thus started a commercial career which has eventually led to his becoming a banker If more American boys were only imbued with this idea of thrift and made proud to be the owner of a bank account there would be fewer failures and idle useless men in our towns and cities Russell attended school at Hopkins and after completing the high school course took a special course in the business college at Chillocothe, as he had already decided to be a business man Graduating from the college in 1915, he came to Box Butte county the same year to accept a position under his brother F W Melick, and was working for him when war was declared On May 1, 1918, he enlisted in the army, was sent to Fort Logan for his preliminary training, then transferred to Camp Fremont, California, and on May 7, was attached to Company D, Eighth Ammunition train On October 1, he was sent to Camp Mills, Long Island, New York, and then to Camp Lee, Virginia, for special training The Armistice was signed before he was sent over seas so he was mustered out of the service at Camp Dodge, Iowa, February 1, 1919, and returning to Hemingford he again entered his brother's office, working there until May 1, when the American State Bank was formed and as his brother was one of the heavy stockholders and became vice-president, Russell Melick was offered and accepted the responsible office of cashier of the bank Mr A M Miller is president while the board of directors consists of A M Miller, F F Melick and C R Melick The prominent stockholders are, G F Hedgcock and H H Rensvold Mr Melick was married at Hemingford on December 20, 1917 to Miss Elsie Green a native daughter of Box Butte county who is also a graduate of the high school of Hemingford, the daughter of Harris R and Margaret (Shindler) Green One child has been born to this union, Marjory Laveta Mr Melick has already made an enviable reputation as a banker, being conservative in his ideas yet at the same time progressive in method He has won the confidence of the citizens of the town and surrounding territory because of his integrity, courteous manner and business ability There is a most promising future in store for him and his family

ALEXANDER MUIRHEAD — For two years identified with the office of county treasurer of Box Butte county, then elected treasurer for the same period and nearly re-elected to a second term and now serving his second two

year term as the executive head of the city government of Hemingford, Alexander Muirhead has become well and favorably known to the people of this community as a hard working, efficient and conscientious public servant He has also won a high place in the financial circles of the county and the Panhandle as a business man for he has been identified with various industries which have tended to the upbuilding of this section and when a capable man was needed as manager of the Farmers Co-operative Association of Hemingford, he was unanimously chosen for the office

Mr Muirhead is a Canadian, born in the province of Ontario, January 11, 1872, the son of Gavin and Catharine (McPhail) Muirhead, the former born on the heather covered hills of "Bonny Scotland," while the mother is a native of the Island of Tyree The sons and descendents of Scotia have always been men of thrift and industry, wherever they have elected to make their homes, and practically without exception have been found an asset to any community Mr Muirhead is one of the men of Scotch rescent living in Box Butte county whom to a marked degree has lived up to the reputation of his worthy family and race and today is materially aiding in the development of large interests here Alexander was next to the youngest in a family of ten children, consisting of five boys and five girls As his father was an Ontario farmer, the boy spent his childhood days and early youth in the country attending the local school during the winter terms and assisting in farm work during the summer vacations and soon assumed many important duties which can be well performed by a small boy He early acquired a good business education along agricultural lines while his theoretical training was gained in the school house After finishing the elementary grades he was sent to Owen Sound Collegiate Institute, as his father was a very well to do man, and like most Scotchmen, believed that a good education was the best equipment a man could have for a start in life After graduating from the institute, Mr Muirhead taught school for one year He decided to emigrate and came to the Panhandle in 1894, locating in Hemingford As teaching was the profession that would bring in an assured and quick income, Mr Murhead accepted a position in the schools here while he looked the country over before choosing what business field he would enter During the five years he followed this profession he amassed a comfortable capital and in 1898, when offered the office of deputy county treasurer of Box Butte county, accepted, filling this office most efficiently for a two year term The subsequent two years were spent in association with the Mollering Brothers in their mercantile establishment Having made a creditable record as deputy Mr Muirhead was elected county treasurer in the fall of 1901, served two years, and in the fall of 1903 was again nominated for the same office his opponent in the race being Charles Brennan The contest was close, but the canvassing board declared that Mr Muirhead had a majority of two votes, Mr Brennan contested the election declaring the majority was his, the question was carried up to the courts and the decision given was a tie The two men were most friendly about it and rather than call another election drew straws for it and Mr Brennan won

On July 10, 1903, Mr Muirhead was married at Hemingford to Miss Georgia A Miller, who was born in Iriquois county, Illinois, the daughter of Alvin and Addie M (Pearson) Miller. the former an Illinoisan, while the mother was a native of the Empire state Mrs Muirhead was the youngest of the two children born to her parents as she had an older brother There are two children in the Muirhead family Ruth A, in the eleventh grade of the high school, and Fay E, in the ninth grade The year following his marriage Mr Muirhead established himself in business as a real estate dealer, in Alliance, where they went to live, but they were acquainted in Hemingford and liked the people here so much that when an opening with good prospects occurred in business circles here they returned and in 1905 Mr Muirhead became associated with his father-in-law in the flour and lumber business, a partnership that continued until 1912, when Mr Muirhead was appointed manager of the Farmers Co-operative Association, having built up a good reputation for executive ability, and honest dealing No more capable man could have been found and the association was fortunate in being able to secure him for this important office He has a wide circle of business acquaintances throughout the county from his many years service in county office, has warm friends and has won the confidence of the people In 1916, Mr Muirhead was elected mayor of Hemingford, and as in his other public offices, proved so efficient that in 1918 he was re-elected Mr Muirhead is one of the thrifty. farsighted men who keep fully abreast of the times, its changes and progress, and is one of the most capable and expect financiers in the Panhandle His personal popularity coupled with his able management

of affairs has won a high place for him in the financial circles of northwestern Nebraska He is a man who does much and says little of it Is progressive in his ideas and methods and since becoming mayor of the city has inaugurated many improvements He is public spirited and supports every worthy movement for civic and communal uplift and improvement both with his time and generous contributions of money Today Mr Muirhead stands as an excellent example of the American citizens upon whom this country must rely during the coming years of unrest and adjustment to take an important part in local and national affairs and so tide the country over a period when the so-called "melting pot" of the United States shall have melted up the pure metals, annealed and alloyed them into a compact whole and shipped the dross and dregs back to Russia, and the Balcans. whence they came The Muirhead family belongs to the Methodist Episcopal church They are a prominent family in the community and at their home dispense a true Scotch hospitality to friends and acquaintances

WILLIAM G WILSON, is one of the favored men who has so directed his affairs that he has been able to change from the strenuous activities of farm life and take up another line of endeavor while the best years of life are still ahead of him, a reward that comes to but a favored few During the time he applied himself to agricultural pursuits. he demonstrated the possession of marked abilities which have proved fully as successful since he engaged in commercial life, and the competency he accumulated has been available in his new field

William Wilson was born in Jefferson county. Iowa, April 19 1851. the son of Joshua and Frances (Templeton) Wilson, the former a Hoosier by birth William was next to the oldest in a good old fashioned family of twelve children and in consequence grew up self-reliant When the boy was three years of age the family moved to Lucas county, Iowa, where Thomas spent his boyhood He was sent to the public school near his home in the winter time and worked for his father in the summer After completing the local school he entered Auckworth Academy, spending two years in higher study at that institution which is located at Auckworth, Iowa then started out independently in life as a school teacher being engaged in that vocation four terms William remained at home with his parents until he attained his majority, but in 1873 started west to seek an independent fortune He first settled in Colorado, obtaining work as a ranchman near Boulder but a year later returned to Lucas county, where he again became a student at the academy, as he realized that the best equipment a man could have in the world was a good education He again taught school until 1882, leaving Iowa to go to the Black Hills, where he began a freighting business, this led eventually to his going to Boseman where his experience as a practical transportation man was of value Seeing a good opportunity to make some money by taking a contract for grading a section of the right of way for the Northern Pacific Railroad, which was building west, Mr Wilson became a contractor, going in the fall of 1883, to Yackima county, where he ran a grading outfit for the road until 1885 Having gained much valuable experience in this line of business he decided to continue in it and in the fall of 1885, went to Maysville, Missouri, where he had bid and obtained a contract for railroad construction. The next year he came to Nebraska, having a contract for grading on the Ashland Cut Off to Omaha Finishing this work he came to Sheridan county in 1887, to construct the road bed two miles each way from the town of Antioch, on the Burlington Railroad Coming to realize that the most independent man is the owner of land, Mr Wilson determined to avail himself of the opportunity to obtain some under the homestead act and in 1888 filed on a homestead one mile east of the present town of Antioch, which until 1912, was known as Reno, but in that year the railroad changed the name to Antioch, which had been the name of the postoffice since its establishment in 1889 Mr Wilson determined to become a good practical farmer and with this end in view and his experience at home on his father's place he began to study the best crops for the high prairies, placed the best improvements he could upon his land, and from the first he found that his success was assured He was optimistic concerning this section of Nebraska, and the years of drought, grasshoppers and hard winters could not discourage him He remained when many of the settlers gave up and returned east, but it paid him for today he is one of the large landholders and prosperous men of Sheridan county From time to time as he made money on his crops and cattle, Mr Wilson invested his capital in other tracts of land until today he is the owner of more than six thousand acres of fine arable and grazing land in a good location For many years, — thirty in all — he

had annually about five hundred head of stock on his ranch He was a short buyer, being far-sighted in his business, and a long seller so that he made money where many men who did not study business conditions lost and now feels that he was well rewarded

On October 11, 1904. Mr Wilson married at Granger, Iowa, Miss Mable Taylor, a native of Des Moines, Iowa and they had two children Mabel L , a student in the Antioch high school and Donald W , a student in Alliance who lives with Mr and Mrs J E Wilson while in school Mrs Wilson died on September 15, 1910, leaving a sorrowing husband and lonely home Since that time a niece of Mr Wilson, Miss Frances Wilson, has come to assume charge of his home and really made it a home for the father and daughter She is a most estimable woman, who has seen much of life and the world, having been a nurse in Denver, Colorado, ever since her graduation from the hospital where she received her training The family have been members of the Congregational church for many years At the present time Mr Wilson has just completed a modern home in Antioch, costing over $12,000 where the family will keep their usual hospitable open door for their many friends Having spent so many years in the country Mr Wilson decided to give up the active management of his land and in 1917 leased the whole tract as well as the stock to a nephew, J W Wilson, and came to Antioch to live He soon bought a half interest in the building and stock of the Antioch Mercantile Company where he has displayed the same business ability that has been characteristic of him since his days as a contractor Under his able management and guidance the business has largely increased and today is one of the prosperous and substantial business houses of the Panhadle Mr Wilson is one of the pioneers of the county where he has won an enviable reputation as a sound business man, estimable citizen and has been one of the best "boosters" this section has ever had, as he is ever ready to help with time and money any laudable enterprise for the development of the community and county He has great faith in the future of the potash industry of this section and has been active in the development of it here.

FRANK H SMITH may truly be called self-made, as his present prosperity has come through his own efforts and under a handicap that would have discouraged most men Mr Smith has had a varied career for he has in turn been farmer, school teacher, clerk in mercantile establishments merchant, was one of the pioneer men to believe and engage in the cement business, clerk of the county, postmaster, mayor of the city of Antioch, and is one of the representative ranchmen and merchants of Sheridan county In all these vocations he displayed the business ability, financial foresight and thrift which he applied to his industries, personal and public, and by these qualities won and kept the respect of his colleagues, while at the same time gaining a comfortable fortune for himself Mr Smith was born near Manlius, Illinois, July 18, 1854, the son of William H and Phlinda (Stickle) Smith the former a native of Wyandotte county, Ohio, while the mother was born in Muskingum county of that state Frank was the second child in a family of three born to his parents His father was a farmer so that the children grew up in the country In addition to his agricultural pursuits, the father became an inventor and made the first two wheeled corn cultivator on the market which was later manufactured at Galesburg, Illinois He, however, gained little for it as he sold the rights of the cultivator in 1859 or 1860 for thirty-five dollars, but the new machine revolutionized the corn industry, doing away with the old one horse, double shovel plow, but like so many men of genius, while he could invent the machine he did not realize its value and so let a fortune for himself and his children slip from his grasp Frank attended the district school near his home and thus laid the foundation for a good, practical education, which was fortunate as he lost his right hand below the elbow in his fifteenth year in a broom corn scraper This accident would have daunted a youth of less high heart and courage, but he determined not to let this ruin his life and has "carried on," as the soldiers say, and been on the job every minute of the time and perhaps made more of a success than if he had not been forced to use all his grit and determination to overcome a physical defect For fifty or more years this brave man has faced the world and by his determination "to do, and to conquer" has achieved a marked success and fortune all of his own making When Mr Smith was only eleven years old his father died and he was forced to go to work for farmers to help support the family He continued to study whenever he could get to school and in 1873 came west and accepted a position as teacher at Carleton, Thayer county, Nebraska, where he taught three years, but the life of a teacher did not appeal to him and he decided to enter another field where there seemed to be a future and resigned to go into a store in Fairbury, Nebraska, to remain five years, thoroughly learning the business Mr Smith was married at Wyanett, Illinois, to Miss Rosetta Aldrich born in Bureau county, that state Her parents were Fenner S and Martha (Mowery) Aldrich, both natives of the state of New York Seven children have been born to this union Ethel, at home with her parents Gerald A, engaged in the management of the pumping station of the Union Potash Company at Antioch, is also a partner with his brother, Fenner A in the ownership and management of a fine brick garage having a building covering a hundred by fifty feet, while his fraternal affiliations are with the Odd Fellows, Frederick V, a graduate of the Chadron Academy, married Miss Josephine Hoffand and now is engaged in the grocery and hardware business in Antioch being the father of two sons and is also a member of the Odd Fellows and a member of the Congregational church, Fenner A, a graduate of the Chadron Academy, with his brother is the owner of the Potash Garage and his wife was Miss Alice Jackson of Dawes county and they have two boys and are members of the Congregational church, Homer C, also a partner in the garage with his brother, married Miss Maude Coghill and they have one child, a son, Zilma V a graduate of the Chadron Academy, is a teacher in Antioch, and Frank J, who enlisted in the army when war was declared against Germany and spent a year at Deming

After his marriage Mr Smith returned to Wyanett, Illinois, and engaged in the mercantile business, but the call of the west was in his ears He longed for the freedom and opportunities afforded on the prairies and returned to Fairbury where he accepted a position in a clothing store remaining three years before he went to Emporia, Kansas, to engage in the cement business, but on account of his health returned to Nebraska in 1888 locating in Loup county and in 1892 availed himself of the homestead act and filed on a homestead From the first he took an active part in public affairs and was elected county clerk serving four years, and at the same time managing his farm Two years later he moved to Dawes county, farmed, and also engaged in the cement business, building many walks in Chadron In 1914, Mr Smith came to Antioch to engage in the mercantile business the same year he was appointed postmaster, serving in that capacity five years When Antioch was organized as a city he was elected the first mayor He resigned from office in 1918, because of the "flu," but is still in business, owning one of the finest men's furnishing houses in the Panhandle, housed in a building Mr. Smith erected for the purpose in 1917, having a floor space of forty by sixty feet. He also owns a fine, modern home in Antioch. Mr. Smith is one of the pioneer residents of this section as well as one of the oldest merchants and takes great pride in this town of potash fame, and is glad that he has been able to take part in the opening up and development of the northwestern section of the state and as he expresses it, "helped to put Antioch on the map." He is a liberal, broad minded American, ever ready to assist in the good works of county and town, giving liberally and is a typical western booster. The family belongs to the Congregational church while Mr. Smith is an Odd Fellow.

HENRY H. SMICE, one of the leading merchants of Antioch, is entitled to pioneer honors in Nebraska and for that reason should be accorded a record in this history of the Panhandle. He was born in Louisa county, Iowa, January 18, 1864, the son of William H. and Carlista (Day) Smice, the former a native of Germany. Henry was the only child born to his parents. At the outbreak of the Civil War his father enlisted in an Iowa regiment for eighteen months, but after his discharge from the service again enlisted at Memphis for the period of the war. He died and when Henry was eighteen months old his mother married a second time, a Mr. Daniel Wainwright, and Henry has thirteen half brothers and sisters from the second marriage. The family moved to Columbus, Nebraska, when Henry was a small boy so that his childhood was spent on a farm here. The trip was made by boat and rail as far as Columbus and the remainder in true pioneer style in a prairie schooner overland to the farm, situated south of Grand Island on the Wood river in the Alda district. While still a youth Henry well remembers that he shocked grain for fifty cents a day and obtained what educational advantages he could in the winter time. At the age of eighteen he started out independently to earn his living and from that day to this has done so. He began to farm for himself near Alda, but when the grasshoppers came and ate up two crops in succession he grew discouraged and sold out for a team of horses, land that today is worth two hundred dollars an acre. Mr. Wainwright built a mill on the Wood river and settlers came from great distances to have

their grain ground, sometimes more than a hundred miles as people lived far apart and mills were still farther in that early day. Mr. Smice remembers that hundreds of Indians came, some from nearly every tribe in the vicinity, making a stop at the mill on their yearly hunting trips and to barter furs and hides for meal and flour. One time when the father was away some Pawnee Indians came to trade and seeing no man just took possession of the place but the mother called to a woman to get a musket and when they saw it the Indians knew they were covered and vacated, contented with the flour and meal Mrs. Wainwright had given them. In 1882, Mr. Smice returned to Kearney county and accepted a position with the men who were building the branch of the Burlington Railroad from Hastings to McCook, remaining with this outfit until the cold weather stopped the work. He then went to Hastings and soon was offered the management of the old Hollard Ranch which he ran for two years, leaving it to establish a feed and livery business at Hartwell, Nebraska, being engaged thus for six years. In 1885, Mr. Smice married Miss Nellie Holland at Hastings, Nebraska, who died in 1887, and Mr. Smice was married a second time on December 31, 1887, to Miss Nellie Howard, at Bedford, Iowa. She was the daughter of Peter Howard. Mrs. Smice is a graduate of the Nodaway high school, Nodaway, Iowa, and was the oldest in a family of six children. Mr. and Mrs. Smice have two children: Bulah M., who married Joseph Rum, storekeeper for the Hoard Potash plant at Lakeside and they have two children; and Orville M., a little lad of five

In 1888, Mr. Smice became one of the pioneer settlers of Box Butte county, locating in Alliance where he held a position with the railroad for about eleven months before he located on a ranch about twelve miles south of Antioch where the family remained two years then went to Iowa, but five years later they were back in Alliance, where Mr. Smice was variously engaged until February, 1919, when he purchased a building fifty by thirty feet on Second street in Antioch and opened an up-to-date second hand store, for he saw that with the climbing prices there was a good opportunity to make money in handling used furniture. He also manages an excellent repair shop where he is equipped to handle all kinds of repair work. Within a short time he expects to add a line of fine new furniture and the prospects of a fine business are assured.

CLAUDE D RICE is one of the merchants of the Panhandle who, while a recent resident of this section, has made a fine record in commercial circles in Sheridan county and the town of Antioch Mr Rice was born near St Joe, Michigan, February 4, 1881, the son of Alonzo and Jennie (Aldrich) Rice, both natives of that state Claud was the fourth in a family of seven children The father enlisted at the outbreak of the Civil War in one of the Michigan Volunteer regiments and served under the stars and stripes four years, taking part in many of the hardest fought engagements and battles of that conflict He entered the army at sixteen years of age as a drummer boy but showed such ability and military strategy that he was rapidly advanced from one office to another, so held that of brigadier general when he was mustered out of the service in 1865 After the war Mr Rice learned the trade of millwright and followed this vocation in the great lumber mills of his native state for many years When Claude was three years old his father came west, locating on a homestead near Hay Springs and there the boyhood days of the child were spent As there were no district schools in that country at that early days Claud studied at home until he was nine years old The family lived at first in the typical "soddy" of the plains but later the father freighted logs from Pine Ridge, forty miles away, and erected a comfortable log house for his family About the time of Claud's ninth birthday the Rice family moved into the village of Hay Springs and he then entered the public school As he was an ambitious boy and wanted money of his own he caught fish in the Niobrara river, dug a hole in a sand bar and kept them there for sale When he earned his first money he felt quite a capitalist From the time the boy was twelve until his sixteenth year the family lived in Custer City as Mr Rice was engaged in running a saw mill there as timber was cut and sent to the mines Claud worked in the mill under his father as all-around-man and practically learned the milling business However, he was anxious to be financially independent and in 1897, went to Chicago, arriving in the spring he soon found a position in a piano factory where his knowledge of wood working stood him well After a year in the city he went to St Joe and worked in a paper factory there three years, learning the trade of paper making On leaving Michigan he returned to Chicago soon to work for the street car company, a position he held a year and a half In 1901, Mr Rice was married in Chi-

cago to Miss Nellie Allen, a native of Glasgow, Scotland Five children were born to this union, but in 1908 the family all came down with typhoid and when Mr Rice recovered all were dead but the youngest boy, John, who now lives with an aunt in Chicago In 1910 Mr Rice was married a second time in Alliance, Nebraska, to Miss Grace Culliver, of Box Butte county and one little girl was born to them Ten days after she was born the mother died and more than a year later, in 1911, Mr Rice was united in marriage with Miss Aurie Smith, at Alliance She was a native of Custer City South Dakota, and their family consists of one boy and two girls Jennie, a student in the kindergarten, Claud, Jr, and June a baby Mr Rice came to Antioch in 1917, to become a partner with his brother Alonzo, Jr, in a fine meat business They own and operate one of the finest markets in this section and have built up a most gratifying trade which brings in good returns Both brothers are progressive in ideas and methods and are fine citizens of the community

FENNER A SMITH — In the thriving community of Antioch, one of the live and enterprising business men is Fenner Smith, who with his brother Gerald, is actively engaged in the management of an automobile garage which they own Mr Smith also has the distinction of being the pioneer merchant of the town and though still a young man is deserving of pioneer honors and mention in this history of the Panhandle He is a native son of Nebraska, born in Loup county February 15, 1890, the son of Frank H and Rosetta (Aldrich) Smith, whose biographies appear upon another page in this history Fenner, named after his grandfather Aldrich, was the fourth in a family of seven children Frank Smith owned a ranch in Loup county and there the children spent many of their childhood days When Fenner was about seven his father left this place and moved to a new farm near Chadron The boy was sent to the district school nearest his home where he laid the foundation of a good practical education He early learned habits of thrift and under the tutelage of his father became a good, practical farmer When only eight years old Fenner began to earn money for himself by herding cattle, receiving the munificent sum of a dollar a week for the work After completing the course in the country school Mr Smith entered the Chadron Academy where he graduated in 1909 Soon after this he accepted a position with the Northwestern Railroad, as time-keeper at the stone quarry owned by that corporation at Hot Springs, South Dakota, where he remained a year then returned to Nebraska and took up a homestead of two hundred and forty acres in Dawes county, twenty-five miles southwest of Chadron. He was already a good farmer and studied agricultural subjects with the idea of becoming a still better one and succeeded to a marked degree, due to his diligence, executive ability and hard work. In addition to the home farm, Mr. Smith rented five quarter sections, raised cattle, hogs, and grains. He made a good living and for five years kept, as he expresses it, "bachelor hall." In 1916, having a good offer for his ranch Mr. Smith leased to advantage and came to Antioch to join his brother in the ownership and management of a general mercantile establishment, the first store in the town. Associated with him was his brother Frederick V. This partnership continued two years when Fenner sold his interest to his brother. About this time he concluded not to return to his ranch and sold his holdings near Chadron and formed a new partnership, this time with his brother Gerald. They built the Potash Garage, a brick and tile structure fifty by a hundred feet which is equipped with every modern convenience for service, repair work and sales. Three skilled mechanics are employed in the repair department, while the brothers themselves handle the sales department as they are agents for Ford cars and Republic trucks in the towns of Birsell, Hofland, Antioch, Lake Side and Ellsworth, also the contributing territory thereto. In addition to the above business they handle a fine line of accessories and tires, having a very valuable stock.

On October 20, 1917, Mr. Smith was married at Chadron, to Miss Alice Jackson, the daughter of Thomas W. and Anna L. (Hakenson) Jackson, the former a native of Iowa, while the mother was born in Sweden. Two children have been born to this union: Otis Cecil and Aldrich F. St. Clair. The family are members of the Congregational church. Mr. Smith is one of the second generation of pioneers in business in the Panhandle where he has surely "made good," for he has been successful in every line of endeavor in which he has been engaged. He still owns a four hundred acre ranch not far from Chadron which is a money making investment, while he is at the same time enjoying a most gratifying return for the money invested in his business in Antioch.

JESSE W. WILSON, one of the younger generation of successful farmers of the Panhandle has for many years been familiar with conditions here and has not, therefore, been called upon to change the manner and methods of his business. He has been successful and no further voucher is needed than the fact that he himself owns twenty-five hundred acres of fine land and in connection with his partner, who is his uncle, controls about twenty thousand acres of fine grazing and farm land in this vicinity. Mr. Wilson was, born in Madison county, Iowa, February 2, 1882, the son of James E. and Dora (Windon) Wilson, the former born in Lucas county, Iowa, while the mother was a native of West Virginia. Jesse was the second in a family of seven children; he attended the district school near his home in Iowa until his twelfth year when the family came to Nebraska, locating on a homestead north of the present site of Antioch, which is located on three of the forties owned by Mr. Wilson. Jesse early began to assist his father in the work about the farm, learned the practical business of agriculture under his father during the summer vacations, while going to school during term time. After completing the local schools he attended the Chadron Academy for two years and as he had decided upon a business career, entered the Broken Bow Business College, to take a special course for three years. Upon finishing his education he accepted a position under his uncle J. O. Wilson in Custer county, remaining there for a year. On November 25, 1905, Mr. Wilson was married at Alliance to Miss Ada E. Harrell, born in Madison county, Iowa, the daughter of Edgar and Etta (Hancock) Harrell. She was the third in a good old fashioned family of ten children. One child has been born to Mr. and Mrs. Wilson, Dorothy E., a student in the Antioch schools. After his marriage Mr. Wilson engaged in farming for a year before he and his brother Ray Wilson formed a partnership and opened a men's furnishing store in Merna, Nebraska. Receiving a good offer for the stock and business good-will sometime later, the brothers traded for a farm near Anselmo, then bought another farm on West Table making a fine ranch of the two tracts. Working the land for a year to put it under good cultivation they were offered and accepted a satisfactory price for the two farms and the next year were engaged in running a steam plow breaking land along the North Platte river. The following season Mr. Wilson decided to give both himself and his wife a vacation and spent a year in California in travel and pleasure. Mr Wilson's wife died October 26, 1918, and on September 6, 1919, he was married at Kansas City, Missouri to Miss Lavinda Cochran, born in New Canton, Illinois, the daughter of Ellis L and Ora I (Shipman) Cochran, both born at New Canton, Illinois. Mrs Wilson is the younger of two girls in the family. Mr and Mrs Wilson are Presbyterians. On his return to Nebraska, Mr Wilson operated his father's farm, situated west of Alliance, for about a year but was able to purchase about the kind of land he wanted a mile west of Antioch, and now makes his home there. Since locating in Sheridan county he has become one of the heavy stock raisers of this section, as he has been far sighted enough to see that the meat industry of the future is to be in the hands of the small operators now that the day of the open range is over. For some time Mr Wilson has been associated in this enterprise with his uncle, W G Wilson. As stated they own and control over twenty thousand acres of land in this part of the Panhandle, annually run close to a thousand head of stock and raise the necessary fodder and grain for maintaining this herd and the horses they raise and sell. Yearly they ship by carload lots to the eastern markets and are meeting with assured success. In addition to his ranch interests Mr Wilson holds stock in the Pioneer Potash Company of Antioch, the plant being located on his land and at the present prices and prospects the company will make much money. Mr Wilson is one of the influential men of his section, is progressive in his ideas and methods, and supports most liberally all movements for the upbuilding of the county and town. He is a member of the Elks club and takes an active part in communal and civic affairs.

EDSON GERING — There are few residents of Gering, Nebraska better known than Edson Gering, who has witnessed the growth of this city from earliest days to its present importance. It was founded by his father the late Martin Gering, who gave it his name. Edson Gering has lived here for over thirty years and has done his part as a good citizen. For sixteen years he was a trusted employe of the United States government as a mail carrier between Gering and Scottsbluff, with the remarkable record of never missing but one train in all that time. Mr Gering owns valuable real estate at Gering, where he conducts a private cab service.

Edson Gering was born September 20, 1861, in Susquehanna county, Pennsylvania, the only son of Martin and Sarah Jane (Slote) Gering, the former of whom was born in Germany and the latter in New Jersey, and they were married in Pennsylvania. In 1879 they moved to Omaha, Nebraska where Martin Gering first conducted bottling works and later a saloon. In 1882 he moved to Custer county and operated a restaurant at Westerville for a time and then opened a hardware and implement store. In 1888 he came to Scottsbluff county, invested in land, established the town that bears his name and afterward was a hardware dealer and general merchant here for years. He was a man of great business enterprise. He was a veteran of the Civil War, in which he served four years as a soldier and to the day of his death, which occurred in the city of Washington, he bore the marks of his wounds. In politics a Republican he was always influential in party councils. He belonged to the Masonic fraternity and both he and wife were members of the Methodist Episcopal church.

Edson Gering obtained his education in the schools of his native state, and afterward learned the ironmolder's trade. In 1887 he came to Scottsbluff county on a prospecting tour, which resulted in his returning as a permanent resident with his father, in 1888. During early years here he engaged in hauling and freighting and with reference to pioneer experiences in this section, Mr Gering is a repository of knowledge.

In the fall of 1888 Edson Gering was united in marriage to Miss Nellie M Winner, who was born near Rochester Minnesota. They have children as follows: Ralph W, a sailor on the United States battleship Yankton, has been in the naval service over two years and is now with his vessel in Cola Bay, Siberia; Clara the wife of Ora Martindale a farmer in Banner county, Nebraska; Harry a cattle feeder in Scottsbluff county; Charles who operates a draying business in Scottsbluff; Iva, the wife of Reuben Lobdell, a carpenter at Gering; Sidney is in the vulcanizing business at Lusk, Wyoming, and Gladys, who remains with her parents. The family attends the Methodist Episcopal church. In politics Mr Gering has followed his father's example. As one of Gering's first citizens he has always been mindful of her best interests and has served usefully as a member of the town board. Both he and wife belong to the Royal Neighbors and he is a member of the Modern Woodmen.

JOSEPH E WARREN, manager of the American and Western Potash Company and mayor of Antioch is one of the younger business men in the Panhandle who has had varied experiences in commercial life and today is well and favorably known in Sheridan county where his rise in financial circles has been rapid, but demonstrates that when the right man and the right position come together nothing can follow but brilliant success. Joseph E. Warren was born in Council Bluffs, Iowa, July 25, 1882, the son of Fritz H. and Mary A. (Cress) Warren, the former a native of Pawnee City, Nebraska, while the mother was born in the city of Milwaukee. Joseph was next to the youngest in a family of five children born to his parents. His father was manager of the New York Plumbing Company for eight years and also held office as county clerk of Pottawatomie county, Iowa, for two terms. Later he was appointed clerk of the United States District Court for the district including Iowa, serving in that capacity for a term, so that the children were reared in Council Bluffs, attended the public schools there, laying the foundation for good educations. Joseph graduated from the high school in 1897; he had already earned money for himself by working in the Nonpareil Printing Office helping feed the press, so had tasted the sweets of independence as well as learned what it was to work for his money. Soon after finishing school Mr. Warren came to Nebraska, locating with his grandfather, Colonel J. E. West, at Rushville, and together they engaged in running the Commercial Hotel until 1899, when Joseph went to Omaha to accept a position as time-keeper, then worked in the stock yards until 1909. That year he returned to Rushville to help his grandfather in the management of a general mercantile establishment of the town, which the colonel owned. The next year the young man purchased a confectionery store in Rushville which he operated for two years. Being a natural musician Mr. Warren realized the great lack of musical facilities in the town and was the prime mover in the organization of an orchestra which toured the western part of Nebraska, playing for concerts and dances until 1917, when they came to Antioch to play at a barbecue, which was given when the city was selling off lots in the newly platted town. This was in September, and Warren liked the town so well he decided to stay for a time. He began to work for the American Potash Company as a common laborer, but true ability will not stay down and at the expiration of a month he was put in charge of the evaporation department. He began to devote time and energy to studying the business and within a short time his salary jumped from twenty-five cents an hour to more than a hundred and fifty dollars a month. For eight months he had supervision of his department, then in June, 1918, entered the office of the company as time-keeper, but a month later was promoted to the office of bill and cost clerk. One month in this position he was advanced to chief clerk of the office and when, in the fall of 1919, the Western Potash Company consolidated with the American Company, Mr. Warren was offered and accepted the management of the new concern. This but shows what western energy, superimposed upon natural ability and the determination to get ahead and succeed may accomplish when there is the will to do so, and a man is not afraid to work. Big concerns are ever on the outlook for men of ability and Mr. Warren was chosen from a large number who could not quite measure up to the job.

On February 11, 1918, Mr. Warren was married in Antioch to Miss Reneta C. Bahler, a native daughter of Nebraska, born in Indianola. One child has been born to them, arriving early in the morning of Armistice day, November 11, 1918, and the parents are very proud of young Joseph Blumer, who is named in honor of F. L. Blumer of Lake Side, who is the general manager of the Hord Potash Company of that village. Mr. Warren has made a host of warm business associations and friends since he located in Sheridan county and it was a fine tribute to his citizenship when he was elected mayor of Antioch in the spring of 1919, with only four dissenting votes cast against him. He is one of the young, energetic, progressive business men who are making, not alone writing the history of finance and commerce in this new field in the Panhandle and the country at large. The Warrens are members of the Episcopal church, while Mr. Warren is also a member of the Elks.

JOHN L. JENNY, a native son of Nebraska who by his western "push," ability and hard work has built up a fine business along two lines in Antioch, is one of the younger generation who has had varied experience in business and financial circles since he entered upon an independent career. Mr. Jenny was born near Columbus, Nebraska, May 8, 1887, the son of John L. and Amelia (Becker) Jenny, both natives of that little mountain country Switzerland, known for its population of thrifty, industrious inhabitants. The father came from his native land to the United States in 1873, and Mrs. Jenny followed the next year. John senior was a farmer by vocation and after landing in America came west, locating on a homestead eighteen miles north of Columbus being one of the earliest settlers of that region. He made good and permanent improvements on his land, erected a comfortable home for himself and wife, and was one of the hardy, determined pioneers who had great faith in the future of Nebraska and was not discouraged and driven out of the state by the hard, dry years of the late eighties and early nineties but held on, and lived to see his faith in his land justified. John Jr., was the fourth in the family of six children. When he was old enough his parents sent him to the school nearest the farm and there he laid the foundation for his practical education. When he was yet a lad of fifteen John went to Dubuque, Iowa, to live with a relative and learn the machinist trade as well as novelty iron working. He was paid while learning this business, but not much, and of that sum had to pay four dollars a month for his board and laundry. For three years the boy remained in the city tending strictly to his business and had the satisfaction of knowing that he had thoroughly mastered the trade. Following this he went to Stillwater, Minnesota, to work for his uncle August Becker, being engaged in putting up saw mills in that locality for two years. During this time Mr. Jenny realized that a knowledge of electricity would be of value to him and to gain this went to Chicago to take a position where he worked on general electric equipment, learning the practical side of that line of business. Feeling that he had a working knowledge of this business he returned to the middle west, locating in Omaha, where he was employed in various garages as he could put to good use his knowledge of metal work and electricity at the same time when repairing cars, and also learn the automobile business from the ground up. On June 12, 1911, Mr. Jenny was married at Omaha to Miss Dorothy M. Garlich, born at Mineola, Iowa, the daughter of Fred and Odalite (Ollnwinkl) Garlich, both natives of Germany. Two children have been born of this union. Adaline Emma, eight years old and Myrtle May, just past three. In the spring of 1918, Mr. Jenny come to Antioch to assist in the construction of the Western Potash Company's factory and when it was opened for operation he was given the position of night foreman, which he held until the concern ceased to operate in July 1919. He then accepted a position at the Standard Potash Works, where he remained until November. In the meantime he had kept his eyes open for business opportunities and soon saw that there was a great

demand in the town for an European hotel, so with his wife's hearty approval and co-operation he invested in such an enterprise, opening up a first class rooming hotel in the Burr Block where he and his wife have from the first met with great success. On February 1, 1920, he leased a building and opened an up-to-date garage, where he is well equipped for repair work and is the agent for the Sells automobile, carries a fine line of supplies and accessories, and maintains a good service station. The Jenny family are members of the German Lutheran church and Mr. Jenny is one of the young men who is becoming recognized as a leading business man of Sheridan county.

SEWARD E CROSS, formerly county judge, has been a resident of Banner county, Nebraska, for more than thirty-two years and has long been one of the county's representative citizens and most substantial residents. He has been identified with much that has served to develop and build up this section of the state, and his name not only represents financial importance but personal worthiness as well. Judge Cross was born in Jasper county, Iowa, March 30, 1875, the son of Benjamin D. and Alice (Geist) Cross, the former born in Muskingum county, Ohio, May 29, 1954, and the latter in Jasper county, Iowa. Mrs. Cross died in 1883. The seven children of the family all survive, Seward E. being the eldest. The others are John, Pearl D., Ethel, Frank S., Mabel and Benjamin D., Jr. Ethel is the wife of Earl Harvey, of Hull, Nebraska, Mabel is the wife of Earl Callahan and they live on her father's ranch in Banner county. This ranch, known as the Bay State ranch, is one of the oldest in Banner county. Benjamin D. Cross located there in 1885, and carried on an extensive stock business until recent years when he retired. He has twenty-nine hundred and sixty acres in his ranch all fine hay land. When he came here Indians were numerous, but he knew how to keep them friendly and never had any particular trouble with them. He now resides near Harrisburg, Nebraska.

Seward E. Cross attended School at Newton, Iowa and from there came to Banner county, March 30, 1887. He assisted his father until ready to enter ranch life on his own account. He purchased a section of land for a dollar and eighty cents an acre and added to this land until he now owns nearly five thousand acres six hundred in alfalfa. After starting to operate his ranch he found himself with some spare time and was led to engage in a general mercantile business at the same time, in this enterprise he continued for eleven years. In March, 1919, he disposed of his store to Icheberger and Graves, also of Harrisburg, and then retired to his farm to resume its active management. In the meanwhile he had become a leading factor in county politics and in 1913 was elected county judge on the Republican ticket, an office he filled with dignity and efficiency. Since then he has served four years as county assessor. Since June 1, 1919, Judge Cross has been one of the stockholders of the Harrisburg State Bank.

On July 25, 1899, the judge was united in marriage with Miss Rachel M Grubbs, a daughter of August B and Jennie (Axford) Grubbs. Mrs Cross' father is deceased but her mother survives and lives at Flowerfield, Nebraska. Judge and Mrs Cross have four children John, Gladys, Alice and Dale. The family is socially prominent. Judge Cross has seen wonderful changes take place in this section since he first came here and perhaps might modestly deny that his constant good citizenship, his energy and enterprise and his hearty co-operation with others in sustaining progressive movements that had a substantial basis, had any large measure of influence in the county's evident advancement, but his many friends and admirers think otherwise.

ROY D WILSON — No medium of information is so complete as the newspaper, and in modern days no agency is more educating. Consistent readers of newspapers have right at hand whole libraries of travel and histories of scientific achievement, also have the ripened opinion of men learned in every profession and line of business, and in addition may keep in touch with local happenings that are generally interesting because of familiarity. The community in which live newspapers are supported, is a progressive one. Attention may thus be called to the *Banner County News*, an eight page, six column journal, published at Harrisburg, Nebraska, which is owned and published by Roy D Wilson, the present county clerk of Banner county, Mrs Roy D Wilson, and their son Roger Wilson, a recently returned overseas soldier of the World War.

He was born December 20, 1871, at Tolona, Champaign county, Illinois, a son of Samuel L and Mary M (Crane) Wilson. He has an older sister, Estella, who is the wife of Herbert W Royal of Rocky Ford, Colorado. The mother, a native of Pennsylvania, died in 1909. The father, who was born in Ohio, was a veteran officer of the Civil War, died in March 1904. He served five years in the

cavalry regiment during the war, and was a first lieutenant and until the close of his life maintained his membership in the Loyal Legion, an organization of officers of the Civil War. After retiring from military life he taught school, but in 1881 entered the newspaper field and organized, published and edited journals in different sections of the country until the close of his life. He was prominent in the Republican party and for many years was postmaster at Beattie, Kansas, but otherwise accepted no political honors, preferring the congenial, if not altogether remunerative work of the newspaper. He founded a newspaper at Manning, Iowa, was in the newspaper business at Cherokee, Iowa, at Wallace, Axtell and Beattie, Kansas, edited a mining journal at one time at Denver, and in association with his son Roy D, conducted a paper at Ponca, Oklahoma.

Roy D Wilson attended school in Iowa and later in Kansas, but a large part of his education was secured in his father's printing office, where he became a practical student when only nine years old. His beginning was made in the *Monitor* office, at Manning, Iowa. He learned the trade in the old time way, long before linotype machines were introduced, and he says he has set type on all the large daily newspapers from Chicago to the Pacific coast.

After his father's death, Mr Wilson was interested in a publishing business at Vermillion, Kansas, but in 1906 he bought the *Observer*, at Kimball, Nebraska, which he conducted for eight years. He then started the *Pine Bluffs Post*, at Pine Bluffs, Wyoming, which he continued for six months, then bought the *Review* at Potter, Nebraska, and operated it for a few months. In November, 1917, he bought the *Banner County News*, a paper with a circulation of three hundred and fifty copies at that time. The business has been so developed and expanded since then that it has become one of the leading papers of the state. Its main circulation is in Banner county, the interests of which are carefully looked after in its columns, its circulation also has been doubled, making it one of the best advertising mediums in the county. Since Mr Wilson was elected county clerk, the business has been managed by Mrs Wilson, who is a woman of great newspaper ability herself, and their son, since his return from war service. The office has been thoroughly renovated and a power cylinder press installed, new type bought, and the latest ideas in producing a first-class newspaper adopted. In its political policy the paper is Republican

CURTIS O. LYDA

In October 1883, Mr Wilson was united in marriage with Miss Ella B Brown, at Axtell, Kansas She is a daughter of Nathan L and Johanna (Brown), who reside at Kimball Nebraska Mr and Mrs Wilson have had four children Harry, who is with the Great Western Sugar Company, at Sterling, Colorado, Roger, who has returned from France, Frank, who is at home, and Robert, who died in February, 1918

Mr Wilson is prominent as a Mason, Knight of Pythias and Odd Fellow, has been through all local offices and helped to organize two lodges of Odd Fellows Mrs Wilson belongs to the Eastern Star, Degree of Honor, and the Rebekahs In November, 1918, Mr Wilson was elected county clerk and is one of the county's efficient and popular officials

CURTIS O LYDA — To establish one's self firmly in professional practice in a strange community is no easy task, and a young lawyer sometimes meets with many difficulties, especially when the field he has chosen already has many older and well known practitioners Fortunately, however the thoroughly trained collegian of modern schooling has courage as well as education and is apt to enter the race with confidence that is subsequently justified, for court records show that cases are not always won by experience, the vigor and enthusiasm of youth when combined with legal ability often carrying everything before it Attention may thus be directed to one of the youngest members of the bar at Gering, Curtis O Lyda who has but recently doffed khaki for civilian costume, in 1918 having put aside his professional prospects to enter military service at the call of his country

Curtis O Lyda was born at Atlanta, Macon county, Missouri, January 17, 1891 He is the only son of Thomas B and Sarah (Williams) Lyda His father was born in Virginia in 1846 and died in Missouri in 1911, and his mother, born in 1863, died in 1891 They were married in Missouri, having made the journey from Virginia by water The father of Mr Lyda followed farming all his life He was thrice married

After a thorough course in the public schools Curtis Owen Lyda entered the Normal school at Spearfish, South Dakota, from which he was graduated in 1912, when he entered the University of Nebraska, securing his A B degree in 1915, and was graduated from the law department in 1917 He opened an office for the practice of law at Gering in September, 1917, and had made satisfactory progress when a crisis arose in public affairs and loyal young

men all over the country hastened to put aside all personal ambitions in order to enter military training Mr Lyda entered service in July, 1918 and was located at Camp Dodge but was unassigned and was honorably discharged in the following December He has resumed practice at Gering and was appointed city attorney in 1919 He has made many professional and personal friends in this pleasant city

In August, 1917, Mr Lyda was united in marriage to Miss Iva Irene Eastman who was born in Iowa Her father, John Eastman who now resides in South Dakota, is a veteran of the Civil War, having served throughout its length in a New York regiment On several occasions he was wounded Mr and Mrs Lyda are affiliated members of the Methodist Episcopal Church at Gering, and they also take part in the quiet social life of the city, Mrs Lyda being also greatly interested in Red Cross and other benevolent activities Mr Lyda is a Democrat in politics He continues to be interested in his fraternity of his law school days, the Phi Alpha Delta

LARS OLSON — Seemingly thirty-four years is but a short period in which to climb from the old-time wage of fifty cents for a day's labor, to the ownership of thousands of acres of well stocked land, and the presidency of one of the important financial institutions of a sovereign state of the Union In Lars Olson, who is president of the Banner county Bank at Harrisburg, Nebraska, and whose interests cover many additional enterprises, is found one who has achieved such results and has done so with such honorable methods that he finds himself universally trusted and esteemed by his fellow citizens

Lars Olson was born in Denmark, March 14, 1857, a son of Ole Hansen and Anna (Larsen) Olson They were natives of Denmark who came to the United States in 1880, joining their son who had settled three years previously in Cloud county, Kansas The mother died in Cloud county and in 1898, the father came to Banner county, Nebraska, homesteaded, and lived on his land until the time of his death, which occurred January 10, 1904, before he had succeeded in proving up on his farm He had been a farmer all his life Of his twelve children all died in Denmark except the two survivors, Lars and Hans who live in Kansas

After attending the common schools, Lars Olson worked as a farmer in Denmark until the spring of 1877, when he came to the United States, in this move showing courage and enterprise for whatever may be the hoped reward, it is not easy for a youth of twenty years to break home ties and seek fortune alone in an alien land He located in Cloud county, Kansas, and lived there for eight years, mainly engaged in farm work He then came to Banner county and on October 10, 1885, homesteaded on section four, where he now lives He worked hard in order to get a start, during that time spending six months in Wyoming, and through his industry and frugality made rapid headway Mr Olson now owns seven thousand acres of fine land He breeds White Face cattle, is a large raiser of both cattle horses and mules, making a specialty of Shire horses His finely improved land has been developed from virgin prairie and Mr Olson devoted close attention to his farm for many years When he came here first the Bay State Cattle Company had over ninety thousand head ranging from Pumpkin creek to Kimball, as this section at that time was a great cow country

On April 2, 1877, in Denmark, Mr Olson was united in marriage with Miss Marie Hansen, who was born in Denmark and her parents, Christian and Marian Hansen, always lived there Mr and Mrs Olson have six children Christen, who married Netta Decker, Oloff, who remains with his father, Arthur, who married Myrtle Bixby, Albert, who married Doris Spize, Annie, who is the wife of Emil Johnson, and Sadie, who is the wife of John Nelson, all of whom live in Banner county except the youngest daughter, who resides at Omaha

For a number of years Mr Olson has been identified with banking interests in Banner county and since 1910, has been president of the Banner County Bank at Harrisburg, which undoubtedly has prospered through his careful, conservative policies He is a stockholder in the American State Bank at Kimball For three years he served on the board of county commissioners and at present is president of the board of regents of the county high school Mr Olson is also chairman of the Farmers Union

MILTON E SHAFTO, county judge of Banner county, Nebraska, a high office that he has filled continuously with the exception of one term, for twelve years is an honor to the bench and well deserves the high esteem in which he is held by his fellow citizens Judge Shafto came into the Wild Horseshoe valley in the summer of 1886, and few of the old pioneers he found there are yet living He was born in Clinton county, Iowa, December 23, 1859, the son of Thomas and Anna B (Forman) Shafto, both of whom were born in Monmouth county, New Jersey, the mother in December, 1822 The father died in 1904, but the venerable mother survives and is tenderly cared for by Judge Shafto and his family Of her two children he alone remains She may be the most aged lady in Banner county and still takes an active interest in home affairs and in the Congregational church, to which she has belonged since girlhood Judge Shafto's parents came to Banner county in 1892, and the father secured a homestead on section three and proved up In New Jersey he was a contractor and builder While never active in the political field, he was a staunch Republican

Milton E Shafto attended the public schools and also a private school in Iowa He started to provide for his future by learning the jeweler's trade and worked through a short apprenticeship but never engaged in the business In June, 1886, he came to what is now Banner county and pre-empted land and later homesteaded on section three, township nineteen, range fifty-five, and was living on his ranch at the time Harrisburg was established as the county seat and soon became interested in politics and public affairs and has been more or less identified with county development ever since In 1896, he was elected county clerk and served continuously until 1900, and in 1907, was elected to the county bench Judge Shafto in this position and in others of publicity and responsibility, has proved worthy of the confidence reposed in him He has been prominent in the journalistic field, from 1908 to 1917, being editor proprietor and publisher of the *Banner County News* In addition to performing his judicial duties, Judge Shafto includes abstracting

On April 16, 1896, Judge Shafto was united in marriage with Miss Nellie Dillon, at Cozad, Dawson county, Nebraska Mrs Shafto is a daughter of former well known residents of Dawson county, George D and Lucy Dillon Judge and Mrs Shafto have two sons Clarence and Paul, both of whom reside at home Mrs Shafto is a member of the Methodist Episcopal church For over thirty years the Judge has been on the school board and is regent of the county high school During the World War he was on the county war board and served as county food administrator from the time of appointment until September, 1918, when the pressure of other work made his resignation necessary He has likewise been interested and useful in the progressive movements that have been of substantial advantage to Banner county

ALFRED G DOWNER, who for thirty years was one of Banner county's well known and highly esteemed residents, was a native of Illinois born at Aurora, November 13, 1840, and died on his homestead near Harrisburg, Nebraska, August 5, 1919, aged seventy-eight years He was a worthy man in every relation of life and his death removed from Banner county not only a pioneer, but one whose efforts had always been given to advancing the best interests of the county, and whose quiet influence lent itself to the maintenance of law and order The honorable record of such a life is a precious legacy to his descendants

Alfred Galen Downer was reared on the old Downer homestead in Kane county, Illinois, and his schooling was obtained in the same county, where a brother, Abel Downer, yet lives He entered into business as a meat dealer, in the city of Aurora, and continued there until 1899 In that year he came with his family to Nebraska and bought at first a relinquishment claim, to which, as years passed, he added until at the time of death, he owned more than two thousand acres For his first quarter section of land, which had twenty-five acres broken and buildings standing, he paid three hundred dollars, another section without improvements, he secured for a hundred and fifty dollars, and for the remaining section and a half, without improvements, he paid fifteen hundred dollars He was careful about his investments as he was in relation to all his undertakings About 1890, he embarked in a general mercantile business at Harrisburg, but continued the operation of his farm adjacent to the city For several years he was alone and then admitted J M Wilson as a partner, and the firm of Downer & Wilson continued in the business field until 1907, when Mr Downer sold his interest to Mr Wilson and retired, during the rest of his life looking after his ranch He was a Republican in politics but never sought public office, serving, however, for several terms when elected, as a justice of the peace

On July 11, 1881, in Illinois, Mr Downer married Welthy Walker, who died June 29, 1918 Five children survive them, as follows Mamie, who is the wife of John A Brewer of Chicago, Illinois, Weltha L, who lives at home, Stella, who is the wife of J N Wyatt, has one child, Helen Marie, and they live at Elkhorn, Nebraska, Herbert A, who is a resident of East Harrisburg, married Mary Schaffer and they have three children, Marguerite E, Marine L, and Allerton G, and Winfred, who lives on the Downer homestead, married Florence Wynne, and they have two children, Robert W and Helen Eldora

A man of generous instincts and kind and helpful to all who appealed to his sympathy, Mr Downer left many to mourn his loss, but perhaps none will miss him more than his beloved grandchildren to whom he was particularly devoted He rejoiced to have his family gather round him in the old homestead and he was ever contented when he believed them happy For many years he had been a supporter of the Methodist Episcopal church at Harrisburg, and there his funeral was held, Rev C K Shackleford of Minatare conducting the services and he was laid to rest by the side of his wife in the beautiful cemetery at Harrisburg, to which many of his pioneer neighbors and friends had preceded him

ELMER S ZORN —Perhaps Banner county has few better known citizens than Elmer Scott Zorn, a leading business man of Harrisburg, for he has been identified with the development of this section almost continuously since he came here thirty-two years ago Mr Zorn was born in Logan county, Ohio, January 14, 1868, the son of Joseph and Amy J (Richards) Zorn, the former was born in Virginia, March 29, 1827, and the latter July 10, 1829 It was in the spring of 1870, that they left Ohio in a prairie schooner, on the long overland journey to Nebraska Fortunately they carried with them enough necessities to last them for five months, for it took them that length of time to cover the distance They traveled over land where the roads were mere wagon tracks at times and where unbridged rivers could only be crossed by ferry, but finally reached Fremont safely and the father invested in land in Dodge county He operated his farm and raised some stock but as soon as he considered Elmer old and capable enough, turned the farm industries over to the latter and went into the business of selling fruit trees for the Stephens Nursery Company of North Bend, Nebraska, in which he continued for five years He decided then to make another change, moving to Kimball on April 11, 1887, but lived there only until May 3, following, coming then to land in Banner county which he had pre-empted on sections 34-17-58. It was an unkind welcome that the elements gave the Zorn family, for one of the justly celebrated Nebraska blizzards set in that very day. Before they could reach their claim the storm compelled them to seek shelter, and for three days of its continuance, four horses besides the family of twelve persons, existed in a little shack eight by ten feet in dimensions. When the Zorns finally reached their claim they found no improvements had been made and they had to live temporarily in a tent. Money was scarce and conditions were very hard. Elmer S. Zorn remembers how he supplied the larder with meat during the first two years by hunting antelope. The family lived on the place until 1900, but the father never proved up, and in that year moved to Harrisburg where he bought a livery barn, which he conducted until 1912, when illness fell upon him and he disposed of it to Martindale and Lewis, and his death followed in May, 1913. He had been active to some extent in Republican politics and had served at times in local offices. The mother of Mr. Zorn survived until October, 1914. They had five children and besides Elmer Scott, the following survive: John, who lives at North Bend, Nebraska; William H., who is a resident of Harrisburg; and Belle, who is the wife of Henry Thomas, of Bushnell, Nebraska. The oldest child, Zora, now deceased, was the wife of H. L. Braucht. The parents were members of the Methodist Episcopal church.

In boyhood Elmer S. Zorn attended school at Fremont and gave his father assistance as far as he was able. He accompanied his parents to Banner county and well remembers some of the discouraging conditions that faced settlers at that time. The lack of water for stock and even home use was, at times so acute that the men of the family were forced to secure it even when they had to travel a distance of fourteen miles as was the case with the Zorns. On one such trip, when within three miles of home, with their brimming barrels, Mr. Zorn and his father were over taken by such a downfall of rain combined with hail, that the wagon sunk so deep in the loosened ground that the horses could not move it, but, as Mr. Zorn philosophically remarks, there was no necessity to haul water any farther.

In 1890, Mr. Zorn went to work in Cheyenne, Wyoming, for Governor Warren and remained there for two years and then came back to Banner county where he has resided ever since. For two years he conducted a retail furniture store and was also in the undertaking business and subsequently for several years was county coroner. He then embarked in a general mercantile business at Harrisburg, which he conducted until January 1, 1917, when he went into the garage business and since then has enlarged his business, at the present time operating a grocery and a very popular feature in this connection is a first class lunch counter. In these enterprises he has the hearty and competent assistance of his son and daughter.

On October 19, 1901, Mr. Zorn was united in marriage with Miss Ella Wynn, a daughter of John and Winnifred (Marn) Wynn, natives of Ireland, who were early settlers in Banner county, west of Harrisburg. In 1917, they removed from their homestead to Pine Bluffs. Mr. and Mrs. Zorn have two children: Raymond and Georgia. In 1901, Mr. Zorn served as deputy county clerk, and in many ways, officially and otherwise, has become known to the people of Banner county, by whom he is universally esteemed.

HARVEY L. WYATT, who is a progressive farmer and ranchman of Banner county, has lived here ever since he was five years old and resides on the place he filed on as a homestead when he started out for himself. He is a member of one of the substantial families of the county. He was born in Wayne county, Iowa, January 5, 1884, the son of William and Susan (Duncan) Wyatt, the latter of whom was born in Iowa and resides at Harrisburg, Nebraska. The father of Mr. Wyatt died in 1896. Of their eight children, Harvey L. was the fifth in order of birth, the others being as follows: John, who lives at Elkhorn, Nebraska; Clyde, who lives in the eastern part of Banner county; Essie, who is the wife of Charles Dick, living in the state of Washington; Edna, who is the wife of Harvey Harmon, living at Portland, Oregon; Calvin, who lives in Banner county; Jessie, who is the wife of Pearl Cross of Harrisburg; and Alice, who is the wife of Owen Brodhead, living near Harrisburg. Prior to coming to Banner county in 1889, Mr. Wyatt's father was a farmer in Iowa. After reaching here he pre-empted and secured a tree claim and lived on that land until 1904, then sold and bought six hundred and forty acres of and one mile northeast of Harrisburg and also leased a school section. He continued in the cattle and stock business until the close of his life. The Wyatts reached

EUGENE T. WESTERVELT

this section at a time when conditions were particularly hard, little money being in circulation because of crop failures from dry weather. William Wyatt was an industrious and resourceful man and came successfully through that time of hardship, but only by hard work and the exercise of great frugality. During this early time, for two years he accepted and completed hay contracts on the Hereford ranch, at Cheyenne, Wyoming, and each year put up about three hundred tons of hay. In politics he was a Democrat, and he served six years on the board of county commissioners. For many years he was connected with the orders of Knights of Pythias and Modern Woodmen of America.

Harvey L. Wyatt obtained his education in the public schools and gave his father assistance until he was eighteen years of age. He then started out for himself and has greatly prospered. At the present time he owns twenty-six hundred and eighty acres of land, four hundred of which he devotes to diversified farming, the rest being ranch land. He raises Hereford cattle, shipping from fifty to sixty head annually, and breeds Shire horses and Poland-China hogs. Mr Wyatt carries on his farm industries according to modern methods and his well improved farm reflects credit upon this section of Banner county.

On January 22, 1908, Mr Wyatt was united in marriage with Miss Bessie Wartman, the daughter of Price P. and Minnie (Kelty) Wartman, who are the parents of ten children, eight of whom live in Morrill county, and Mrs Wyatt and her sister Grace in Banner county. Mr and Mrs Wyatt have three children, Leo, Norma and Susan. While Mr Wyatt votes the Democratic ticket, he has never felt that good citizenship required his acceptance of tendered local office and he gives his support to measures that seem to him beneficial without any desire for political reward. Mr Wyatt is looked upon as a representative and trustworthy citizen of Banner county.

EUGENE T WESTERVELT, founder, editor, and proprietor of the *Republican* at Scottsbluff, has been a molder of public opinion and a vigorous, constructive citizen for many years. He has served effectively in numerous public capacities. In the close confidence of the Republican party, he has been an influential disseminator of its principles, but his loyalty and patriotism have not been bound by party ties at any time in his career. During the World War he stood bravely by his beliefs and parted with three stalwart sons to fight in another land for human liberty. Mr Westervelt is the pioneer builder of Scottsbluff in the sense that he erected the first permanent home here in April, 1900, at the corner of Sixteenth street and Second avenue where his present fine residence now stands.

Eugene T Westervelt was born at Greenfield, Franklin county, Massachusetts, January 16, 1865. His parents were James H and Loranna (Day) Westervelt, the former of whom was born January 6, 1840 at Patterson, New Jersey and the latter at Stamford Vermont May 4 1848. They were married in Vermont and five children were born to them. Of the four survivors Eugene T is the eldest, the others being James P a merchant at Gering, Claude H, a general blacksmith at Scottsbluff, and Mrs Parvin Gilbert who resides at Scottsbluff. The parents of the above family were members of the Baptist church. The mother died in 1912 and the father in 1908. He served three years as a member of the Sixty-ninth New York infantry in the Civil War. In 1868 James Westervelt moved with his family to Michigan, where he followed the trade of a general blacksmith, and in 1877 came on to Nebraska. He homesteaded in Custer county on the site of the present town of Westerville, and from there came to Scottsbluff county to take up a claim on which he proved up. He belonged to one of the first Masonic lodges established here.

Eugene T Westervelt attended school in Custer county and his first important work in the newspaper line was the operation for one and a half years of the *Western Echo*, at Westerville. After leaving the newspaper office he engaged in farming for eight years, retiring then to assume the duties of sheriff of Scottsbluff county, to which office he was elected in 1896 serving until 1900. Mr Westervelt then returned to the newspaper field on May 4, 1900, starting the publication of the *Republican* which was issued first as a weekly but later to satisfy the public demand, was made a semi-weekly. The paper has a wide distribution with an actual subscription list of 1800. Additionally Mr Westervelt operates a large job printing office. The political complexion of the *Republican* is as its name indicates. Mr Westervelt has long been acceptable in all party councils. He served many years as county central committeeman and at present is state central committeeman. During the administration of President Taft, he was postmaster of Scottsbluff.

In 1886 Mr Westervelt was united in marriage with Miss Laura B Amos who was born at Carlton, Ohio, and they have children as follows James William who is associated with A J Shumway in the abstract business, Murial, who resides at home, John McKinley, who assists his father in the newspaper business, received military training at Camp Dodge from June to August 1918, then, as a member of Battery A, Eighty-eighth Division, Three Hundred Thirty-eighth Field Artillery accompanied the American Expeditionary Forces to France, returned to his home in January, 1919; Lawrence Eugene, who is still in the service of the United States, enlisted with the Naval Reserves, spent four months at Great Lakes and three months at the Minneapolis naval training school, and was then assigned to duty on the transport Arizona, Mendel Eli, who enlisted in October, 1918, in the United States Marine Corps, spent three months in training at Mare Island, was discharged in February, 1919, and now is employed on government building works north of Mitchell, Nebraska, and Catherine, who attends school Mrs Westervelt is a member of the Methodist Episcopal church, but from boyhood Mr Westervelt has belonged to the Baptist church Fraternally he is a Mason, a Woodman, and a Knight of Pythias and was the first chancellor commander of Hannibal lodge No 40, Scottsbluff

ROY L HOWARD, whose extensive land transaction and many business interests have made him well known in several counties, now resides on and operates his ranch of a thousand acres, which lies on sections three and township three, Banner county Mr Howard was born in Buffalo county, New York, not far from Syracuse, January 11, 1877, the son of Julius A and Lillian E (Wescott) Howard In early life the father was a fisherman on the Great Lakes In 1880, he came to Nebraska and to Banner county in 1886, where he homesteaded in August of that year, continuing on his land until 1895, when he retired to Gering, continuing in business there, however, for a few years, as a member of the firm of J A Howard & Sons, meat dealers At different times he was elected to local offices by the Republican party He belongs to the order of Modern Woodmen of America and both he and wife are member of the Christian church Of their five children, Roy L is the eldest of the four survivors the others being Leon R, who lives at Omaha married Grace Northrup, Leola M, the wife of Walter Beck, resides near Gering, and Luella M, the wife of Claud M Brown, a farmer near Gering

Roy L Howard was educated in the public

schools of Banner county and afterward taught school for three years He went to Omaha, where he learned the butcher's trade, afterward operating a meat market at Scottsbluff for four years under the firm name of Howard & Troy, subsequently selling to the firm of Deulin & Son Mr Howard then came to his ranch and this has remained his permanent home This ranch covers the site of the old town of Ashford and when a part of Cheyenne county, the courthouse was located here Mr Howard also bought the original town site of Northport, at a later date reselling it to individuals He also owned extensive properties in Scottsbluff county, but has disposed of it and at one time had seventeen hundred and sixty acres in Banner county, which he has reduced to about a thousand acres Additionally he has a farm in Morrill county in the vicinity of Bayard Mr Howard does a large stock business He breeds Duroc-Jersey and mulefoot hogs, an advantageous feature of the latter breed being their abnormal growth to great weight Recently Mr Howard slaughtered one of this variety that weighed eight hundred and fifty pounds He raises about three hundred of this type annually

In 1900, Mr Howard was united in marriage to Miss Ida M Troy, and six children were born to this marriage, all of his children now residing with Mr Howard His second marriage took place on March 22, 1914, to Mrs Anna E Brown, a widow, a daughter of George M and Anna E (Marsh) Babbitt Her father was an early settler and homesteader on the North Platte His death occurred October 8, 1911, the mother of Mrs Howard passing away July 19, 1910 A Republican voter all his life, Mr Howard has been conscientious in his views on public questions, but has never desired political honors

VANCE J CROSS, one of Banner county's prominent and representative men, has been a resident for thirty-four years and during that time has been closely identified with the constructive measures and general development of this rich and beautiful part of Nebraska Whether in official or private life, Mr Cross has been enterprising and useful, and by the residents of Harrisburg in particular, he will long be remembered, for he, in association with three others, J B Forsman, Charles A Scholley and Samuel Fisher, laid out the town site, the proposed city being named Centropolis

Vance J Cross was born in Muskingum

county, Ohio, October 1, 1841 a son of Elias and Clemintine (Dickenson) Cross, the former was born in Baltimore county, Maryland, and the latter in Coschocton county, Ohio The father was a cooper by trade and followed that vocation all his active life, retiring in 1882, and moving to Iowa where his death occurred soon afterward He was a man of substantial character and was postmaster at Harrisonville, Ohio, for some years Both he and wife were members of the Methodist Episcopal church Of their six children, Vance J is the only survivor

In the common schools of Ohio near his father's little farm, Vance J Cross obtained a fair amount of schooling for the time, then helped his father and together they went to Iowa where he rented land at first and afterward bought He carried on general farming there until in October, 1885, when he came with his family to Banner county, Nebraska, where he took a pre-emption and tree claim two years later and homesteaded where he has since lived Mr and Mrs Cross had real pioneer experiences The nearest neighbor was four miles distant from their homestead and probably the wide expanse of country between them and friendly faces seemed much greater Their first shelter was a log structure without door or window, unplastered and unfloored When Mr Cross had an opportunity to go to North Platte to make hay on the river banks, it was too profitable a chance to decline, although he had to leave his wife and daughter with the responsibility of raising the small crop already in the ground They were protected by two dogs, not from human interference as might be the case today, but from range cattle, that not only would have damaged the growing crops but might have trampled the people No accident of this kind ever happened to the family although Mr Cross often had to be away from home in order to get mail and provisions from Kimball the nearest town thirty-three miles away That first log house still holds together, Mr Cross probably having a sentimental interest in it in comparing it with the handsome modern residence that long since has taken its place There are other interesting landmarks in the vicinity, Long Spring, often mentioned in local history, being on Mr Cross's land, and "Lover's Gap," around which romances have been written, is on land adjoining the Cross line fence Mr Cross and his son are heavy landowners, he having almost a thousand acres and his son an entire section, the greater part being grazing land Mr Cross raises about twenty-five head of cattle a year

On January 10, 1871, Mr Cross was united in marriage to Miss Francis S Cross, a daughter of Jesse and Eleanor (Ryan) Cross, natives of Maryland and Ohio respectively They located in Iowa in 1856 Mr and Mrs Cross have had seven children and the following survive Jennie, who is the widow of John N Fickes, lives at Kimball, Elias L and Henry W H , both of whom live at home, and Dolly A , who is the wife of Otte Yaege, lives at Kimball Mr Cross has always been a Republican and has taken part in many county campaigns He served as sheriff of Banner county for four terms and has been prominent in other capacities His reminiscences of early days in this section are exceedingly interesting

WILLIAM H ZORN — The mental power of constructing or creating something entirely new, needful and expedient, is a faculty by no means given to every one but it is through this gift that the world makes progress In William H Zorn, inventor, Harrisburg, Nebraska, is a man who has turned his talents to good account Mr Zorn was born in West Virginia, August 21, 1856, and is a son of Joseph and Amy J (Richards) Zorn, both natives of Virginia

The parents moved to Logan county, Ohio, and from there in 1871 started overland to Nebraska After many adventures they reached Dodge county, where the father invested in land He later became interested in a nursery business and for five years sold fruit trees for the Stephens Nursery Company of North Bend, Nebraska In May, 1887, he came to Banner county and pre-empted land on sections 34-17-38, and the family lived on the place until 1900 They endured many privations and very often the larder was supplied by the prowess of the sons who hunted antelope In the above year the father moved to Harrisburg and bought a livery stable conducting the same until 1912, and his death occurred the next year The mother survived until October, 1914 Of their five children, three sons and one daughter are living

William H Zorn had common school advantages in Ohio After the family came to Nebraska, he followed farming until 1888, at which time he homesteaded for himself in Banner county and lived on his land for twelve years, moving then to Harrisburg Since then he has devoted himself to perfecting his inventions, many of which have not yet been patented but several have been very generally accepted, among which are two automobile gates, an end gate rod and an equalizer for windmills. The value of all these devices have been proved by wide use and in the near future Mr. Zorn will have patents on others equally ingenious and useful. Mr. Zorn recalls pioneer experiences with pleasant memories. While there were hardships, it was a hardy, wholesome life for youths and he and his brother thrived on it.

Mr. Zorn remembers seeing as many as a hundred head of wild horses in a drove in what was called Wild Horse corral. They hunted antelope with trained horses and dogs. Mr. Zorn is a Republican in politics.

JOHN V. BRODHEAD, who is a progressive farmer and ranchman of Banner county, has become one of the substantial men of this section through persistant industry, careful saving and wise investing. He was born in Pike county, Pennsylvania, December 29, 1854, and is a son of David O. and Marie (Van Etten) Brodhead, both of whom were born in Pennsylvania, where the father died in January, 1912, and the mother in February, 1916. They were members of the Presbyterian church. Of their nine children five are living, John V. being the only member in Nebraska.

With a common school education and sound home training John V. Brodhead started out for himself when twenty years old. He worked as a farmer until 1886, in his native state then came to eastern Nebraska and in 1887, to Banner county and homesteaded near Flowerfield. He lived on that property for eight years, moving then to Kimball, but in 1904 he returned to Banner county and bought the land on which he lives on section sixteen, township nine, where he farms over a hundred acres and raises a hundred head of cattle yearly. He has owned at least four thousand acres and yet has twelve hundred acres of ranch land. Mr. Brodhead has prospered greatly since returning to Banner county, at which time his stock consisted of two cows and two horses. Mr. Brodhead is a stockholder in the Banner County Bank.

In Pennsylvania, in March, 1878, Mr. Brodhead was united in marriage to Miss Margaret I. Cronn, a daughter of Clarkson and Elizabeth (Runyan) Cronn, natives of Pennsylvania who homesteaded in Banner county, Nebraska, near Flowerfield, in 1887. They now reside in Oregon. Of the seven children born to Mr. and Mrs. Brodhead the following survive: Ada M., who is the wife of Ethan Tracy, living near Norwood, Colorado; David C., who lives in South Harrisburg, married Allie Wyatt; and John H., who has recently returned from overseas service as a soldier in the World War, married Helen Smith and they live at Norwood, Colorado. Mrs. Brodhead is a member of the Methodist Episcopal church. He has long been a member of the order of Knights of Pythias. In politics he has always been identified with the Republican party, believing thoroughly in its principles and never desiring public office. He was chosen a member of the schoolboard, however, and served with great usefulness for nine years.

In recalling early days in Banner county, Mr. Brodhead remarks that he and wife came to Nebraska in order to secure land and establish a permanent home. At that time land in Pennsylvania was held at a prohibitive price for the ordinary young man with a growing family. They met with many discouraging conditions after reaching here and went through some real hardships, but they have long since passed away. During those early years he worked hard, often riding the range for days without seeing a single other person, then entered the employ of Lambert C. Kinney and was his ranch foreman for nineteen years. Mr. Brodhead is widely known and is held in universal esteem and is numbered justly with the representative men of Banner county.

WILLIAM D. SHAUL, general farmer, stock raiser and bee man, belongs to that enterprising group, of pioneers who came to Banner county in its undeveloped days, homesteaded, endured hardships and in many ways paved the road for easier traveling by the younger generation. He was born in Jackson county, Kansas, November 13, 1864.

The parents of Mr. Shaul were Aaron and Matilda Shaul. The father was born in Madison county, Indiana, January 20, 1829, and died in Oklahoma, July 2, 1898. The mother was born in Missouri, January 18, 1838, and died in the home of a daughter at Minatare, Nebraska, November 5, 1918. Of their family of nine children, William D. is the third of the four survivors, the others being: Sylvester A., who lives at Minatare; Ida Belle the wife of Samuel Sprigs; and Lula E., who married Milton Riles of Gering. Aaron Shaul went from Indiana to Iowa when young and throughout life engaged in agricultural

ALLEN B. McCOSKEY

pursuits From Iowa he moved to Kansas, from which state he returned several times to Iowa before he established a home there but afterwards lived twenty-four years in Kansas In the fall of 1887, accompanied by his son William D, S E Sprigs and Calvin Marts, he came to Banner county, Nebraska and all of them secured homesteads in the same neighborhood, this being near the present town of Ashford, which village was started in the following winter Aaron Shaul remained in Banner county for eight years, then moved to Oklahoma and his death occurred there three years later It required courage and resourcefulness in those early days to secure means of subsistance William D Shaul remembers that on one occasion, when there was great lack of corn and flour in the little colony, that his father and George A Palmer started for Paxton with a load of pine posts, hoping thereby to get money with which to make possible the purchase of real food necessities They could find no buyers however, at Paxton and had to go seven miles further, near North Platte, where they sold their posts for twenty cents apiece With this money they stocked up with corn and flour and when they returned home had no trouble in disposing of it profitably This was in the winter of 1894-95

William D Shaul attended the district schools in boyhood but never had advantages such as he has been able to give his own children He lived with his people and helped them until his marriage, which took place October 14, 1892, when he was united with Miss Mary E Palmer, an early settler in Banner county The mother of Mrs Shaul is deceased, but the father survives and is a resident of Holly, Oregon Mr and Mrs Shaul have had children as follows Aaron A who assists his father, William E, who lives in Banner county, married Sadie Olsen, James G, who is in business at Gering, married Ethel Clarey, Etta M, who is the wife of Carl Philpot, of O'Neill Nebraska, Orrin, who lives at home, Lulu and Lola, twins, and Rose all of whom live at home, and one who is deceased

When his father went to Oklahoma William D Shaul accompanied him but returned to his homestead in 1897 One year later he moved near Gering, but three years afterwards came back to Banner county and has remained here ever since He still owns property near Gering, and during the past year has operated some six hundred acres of land as a general farmer and stockraiser He turns off from

twenty-five to thirty head of cattle a year and seventy head of horses He has seen many changes and witnessed remarkable fluctuations in price both of livestock and grain In 1892 he hauled wheat a distance of seventy miles to Sidney and sold it for fifteen cents a bushel In addition to his other industries, Mr Shaul keeps colonies of Italian and brown bees and secured over fifteen hundred pounds of honey in the past year He has always been a busy man, too busy to ever consent to hold political office, but he has never changed in his adherence to the principles of the Republican party

ALLEN B McCOSKEY county surveyor of Scottsbluff county Nebraska, to which section of the state he came thirty-three years ago, has been a continuous resident, with the exception of four years during which he performed public duties at the state capital Coming here at an early day, when settlers were hastening to secure homesteads and boundary lines were often suggested rather than proved he found instant demand for his knowledge of applied mathematics and probably has done more in the accurate settlement of land claims than any other surveyor in this part of the state

Allen B McCoskey was born August 6, 1851, in Washington county, Indiana, in the same house in which his father Robert McCoskey had been born The McCoskey family has been more or less a pioneering one The great-grandfather of Surveyor McCoskey came from north Ireland and settled in Virginia There his son James McCoskey was born and from there came to Indiana and as one of the earliest settlers built his log cabin in a belt of timber in Washington county and died there Robert McCoskey, one of a family of seven boys and four girls spent the greater part of his life in agricultural activities although he was a cooper by trade In early manhood he married Julia V Wilson who was born at New Washington Indiana and died in her native state in August 1865 leaving five children Allen Bruce was the eldest, the others being Elizabeth who lives at Niobrara Nebraska Moffet Alexander, a farmer and gardener near McMinnville Oregon Julia A, the wife of C M Hiestand a telegraph operator at Thermahto Butte county California, and Robert Henry, a clerk in a store at Scottsbluff Both parents were members of the Presbyterian church After the death of the mother, the father kept the children together until they were all grown up He came to Nebraska in 1891 and made his home with Allen about a year, then took a preemption near the present site of Scottsbluff and proved up on same His death occurred while on a visit to his daughter at Ogallala, Nebraska, November 26, 1893, at the age of seventy-two years

In the country schools Allen B McCoskey received his primary education but he had additional advantages for his parents sent him to Blue River Academy after finishing the district school course This was an excellent institution maintained by the Society of Friends, and with Quaker thrift and simplicity more attention was paid to such branches of study as higher mathematics, including the principles of surveying, than others of a less practical nature During these early years home environment made Mr McCoskey a farmer and perhaps choice, later on, a school teacher, and it is quite probable that he was successful in both lines However, after four years in the schoolroom in rural Indiana with no very satisfying prospect for a different future, he finally decided, like former pioneers of the family to go pioneering into the West Mr McCoskey was too practical to expect an easy life for himself in the somewhat unsettled regions of which he directed his course in 1882 When he reached South Dakota he found work as a carpenter ready for him and for two months he adapted himself to life there, doing his best but not feeling satisfied with conditions He then crossed the border into Nebraska and shortly afterward joined a dredging outfit, with which he continued two years During the following two years he was employed in an undertaking establishment In 1886 Mr McCoskey came into Scottsbluff county and had the business foresight to pitch his tent here He homesteaded and resided on his land for a time but sold it when the demands of his profession made removal to the town of Scottsbluff imperative, and this progressive little city has been his home ever since with the exception of the years spent in Lincoln as assistant state engineer from 1897 to 1901, which called him all over the state He has surveyed on all the ditches of this county He began surveying shortly after coming here and there is not much concerning the topography of the county that Mr McCoskey cannot make clear He was first elected county surveyor in 1892, and in 1897 was appointed assistant state engineer and spent four years in the engineer's office at Lincoln

In 1895 Mr McCoskey was united in marriage to Miss Ora Johnson a daughter of William Johnson, a homesteader in Nebraska Mr and Mrs McCoskey are members of the Presbyterian church He is a Scottish Rite Mason and past master of his lodge and belongs also to the Royal Highlanders In political circles

he has been somewhat active as a Democrat, but outside his profession has never accepted public office

UZELL T SNOOK, who is an enterprising and substantial farmer and stockraiser in Banner county, has spent the greater part of his life in Nebraska and is well and favorably known in different sections He was born in Jefferson county, Iowa, February 1, 1873

In the spring of 1887 the parents of Mr Snook came to Nebraska They were Gustavus and Arvilla (Miller) Snook, the former of whom was born at Fairfield Iowa, October 14, 1846, and the latter in Ohio April 13, 1850 Of their family of ten children the following are living Henry W, whose home is at Union Oregon, Uzell T, whose home farm lies on section thirty-two, township twenty, range fifty-five, Banner county, John V, of Redwood valley, California, Carrie M, who is the wife of J W Patterson, of Gering, and Archie, who lives in Scottsbluff county The mother of this family died May 5, 1908 She was a member of the Methodist Episcopal church, with which she united at Gering, Nebraska The father of Mr Snook was an experienced sawmill man when he came from Iowa to Nebraska Here he first operated a sawmill for its owners near Long Spring, and later a sawmill in Muddy canyon He sawed the lumber which went into the construction of the first bridge at Gering, which was the only bridge in Scottsbluff county for many years The logs came from Carter canyon While thus engaged Gustavus Snook homesteaded near Pierson's ranch and lived in Banner county until 1898, when he moved to Scottsbluff county and followed farming south of Gering until 1907, then took a Kinkaid homestead, on which he lived during the rest of his life, his death occurring there on July 9, 1913 He was a man of sterling character and firm convictions, a strong Democrat in politics and sound in his Presbyterianism

Uzell T Snook attended school after coming to Nebraska at Harrisburg and remained with his father until the age of twenty-one, when he started out for himself He homesteaded a Kinkaid tract in 1907, on which he now lives, having four hundred and eighty acres here and farms about one-third of his land He breeds Hereford cattle and has about fifty head a year He has witnessed many changes since he first came here and in commenting on these, says that one dollar a day was the general harvest wage and he has worked for less In 1887 all the wild animals had not been exterminated for on one occasion, with an ordinary shotgun, he killed a bob cat that measured four and one-half feet in length Mr Snook has everything very comfortable around him and has done a large amount of improving

On February 20, 1896, Mr Snook was united in marriage to Miss Sadie E Smith, who died June 25, 1917 She was a daughter of John E and Nancy (Sego) Smith, who came to Nebraska in 1888 The mother of Mrs Snook died January 1, 1917 The father lives at Gering The following children were born to Mr and Mrs Snook Don M who is at Syracuse, New York, was in military service for twenty-three months during the Great War, Avice, who is the wife of Edward N Tart, resided at home while her husband was in military service in France Harold, who is in business at Brush, Colorado, Ethel M, the wife of Albert Burkey, of Scottsbluff county, and Victor, Chester and Agnes, all of whom are at home In politics Mr Snook is a Democrat

TED KELLY, one of Banner county's substantial citizens, has spent his entire life here and is a representative of an old county family a descendant of one of the earliest settlers He was born July 9 1886, in his parent's pioneer home, a sod house, in Banner county, this primitive dwelling still being preserved on his farm

Mr Kelly's parents were William J and Martha (Felts) Kelly, who were born in Pennsylvania William J Kelly was brought in childhood to Republic county, Kansas and when twelve years old accompanied his employer to Texas and was brought up on a ranch He was twenty-two years old when he came from Texas to Nebraska with a cattle herd, and it was at Sidney that he met and later married Miss Martha Felts He then decided to stay in Nebraska and set about looking for a permanent location, stopping at first on the present site of Bridgeport, in Morrill county In 1881, he came to Banner county one of the first settlers, and homesteaded the land on which his son was born and on which he yet lives There were two children in his family, Bessie and Ted The former was the first white female child born in Banner county She grew to beautiful womanhood and was married to Everett Walter She died November 14, 1904, survived by three children, Clifton and Harry, both of whom live

with Mr Kelly, and Marvin, who lives at Indian Hill Wyoming The father of Mr Kelly lived on his homestead until his death, which occurred December 21, 1914 The mother died November 24, 1919, at Santa Ana California

Ted Kelly was sent to school as soon as old enough, his father believing in general education and being one of the men who brought about the organization of the public schools in Banner county By the time he had reached middle boyhood, conditions of living in this section had become easier, but he remembers, his parents telling of early hardships and lonely isolation Sidney was sixty-five miles distant and that was the nearest point where provisions could be bought or mail received Later on a post office was established one-half mile west of the homestead, known as Livingston, the first postmaster being Lee Livingston Mr Kelly has five hundred and twenty acres in his ranch He raises Duroc-Jersey hogs and White Face cattle having a hundred head annually

On July 25, 1910 Mr Kelly was united in marriage to Miss Tressie B Ridge, a daughter of Jesse H and Tillie (Owens) Ridge, early settlers in Morrill county The mother of Mrs Kelly is deceased but the father till resides at Bridgeport, Nebraska Mr and Mrs Kelly have two children Carl, who was born August 18, 1911, and Alice who was born July 30 1915 Mr Kelly has always been a Democrat in political faith, as was his father He has never accepted any public office except membership on the school board and is still serving He is interested in the aims of the organization known as the Farmers Union and owns stock in the same Mr Kelly is well known throughout the county as an enterprising business man He belongs to the Knights of Pythias at Harrisburg

WILLIAM A GRUBBS, a representative citizen and extensive farmer and stockraiser in Banner county, came here thirty-three years ago and few men have been more useful to his section His progressive spirit has been a leading factor in the founding of enterprises which have been of great and substantial benefit

William A Grubbs was born in Story county, Iowa, May 10, 1866, one of seven children born to Abraham and Margaret (Barnharde) Grubbs He has one sister, Mrs Josephine Carpenter, living in Nebraska The father was born in Pennsylvania and came west to

Iowa when young and lived there until his death, in October, 1865 He was a general farmer and was a man of upright character The mother of Mr Grubbs died in 1885

In the district schools in the neighborhood of his father's farm William A Grubbs obtained some educational training, and he remained with his mother until her death He was nineteen years old when he started out for himself, and in the spring of 1886, came to Banner county, Nebraska In the following year he homesteaded and still lives on the place which lies on section 2, town 18-57, in the vicinity of Harrisburg After a residence of twenty-five years he moved into Harrisburg and took the agency of the Ford automobiles, in 1912, and during the five years he held it sold eighty-seven cars He owned the first Ford car in Banner county and at the time of purchase many of his neighbors doubted the expediency of the investment He was confident, however, that he would be able to convince others of the value of the new means of transportation Denver was headquarters and all the cars he sold had to be driven from there Mr Grubbs relates that when he drove his first car from Denver, he was accompanied by his wife and Mr and Mrs A H Pierson, and they had several adventures, one when they missed the right road and ran into a herd of antelope and another being caught in a heavy storm, and running the risk of having to stay all night in the car The people did not yet look on automobiles in any other light than luxuries and farmers along the way when applied to for shelter, were not anxious to put themselves out for "wealthy eastern people," as they supposed, while there was no limit to their hospitality when they found the strangers were Banner county farmers After the second trip Mr Pierson entered into partnership with Mr Grubbs and the association continued for eighteen months, after which Mr Grubbs continued the agency for three years longer Notwithstanding the expressed fears of many of his former friends that the venture would never be profitable, Mr Grubbs made a distinct success and perhaps many of his farmer acquaintances who now own and drive cars, are ready to acknowledge his business foresight

After giving up the Ford agency in 1916, Mr Grubbs established an auto-truck line between Kimball and Scottsbuff, carrying both freight and passengers, and continued to operate it about two years Circumstances then seemed to require his return to the farm Three of his stalwart sons had entered military service in the World War and the fourth was making ready to depart, hence the responsibilities of conducting an estate of sixteen hundred acres fell upon their owner Since that time he has remained on his farm and carries on his extensive industries under his own supervision, setting the example of farming with tractors, which has been followed by many of his neighbors

Before returning to his farm, Mr Grubbs operated the telephone exchange at Harrisburg for seven months, the introduction of this necessary invention of modern life, being largely due, in this section, to his foresight and mechanical skill The subject had interested him for several years previously He put in the first telephone line in the county, beginning by borrowing two "series phones" as they were called, barb wire being used for connections Now there are few farm residences which do not have telephones and credit may well be given Mr Grubb's progressiveness for their introduction In many other ways than those mentioned he has been foremost in county matters Mr Grubbs was an early supporter of "The Herd Law," helped to organize school district number eleven and served as director many years, and assisted in establishing Gabe Rock cemetery, helping to dig the first grave therein

On January 23, 1888, Mr Grubbs was married to Miss Lucinda M Fuller, which was the first wedding in Banner county Mrs Grubbs is a daughter of William D and Elizabeth (Kimberly) Fuller, early residents of Banner county, now living retired at Bushnell To Mr and Mrs Grubbs the following children have been born Minnie, who is the wife of Leslie Barrett, of Harrisburg, Walter L, who lives on his homestead near Dubois, Wyoming, was a soldier in the World War, Vernon M, a farmer living thirteen miles west of Harrisburg, married Vivian Heintz, and Dallas B, Leon C, Nellie F, Violet, Arlo R, Alvin E, Barton J, Carrie M, Wilbur A, Junella M, and Franklin A, all of whom live at home a happy, wholesome family illustrative of sturdy stock Mr and Mrs Grubbs were active members of the United Brethren congregation when there was a church of that denomination here, but now attend other religious bodies, as for several years no minister of their own faith has been stationed in their neighborhood Mr Grubbs is a man of sound judgment and of wide experience but has not identified himself with any political party, for these reasons perhaps, voting independently For many years in early days here he served

H. Leslie Smith

in the office of constable and at that time his duties included those of sheriff Mr Grubbs is a mine of information in reference to much of the development of Banner county in the last thirty years, and a great many illustrations seen in volume two of this addition were taken by Mr Grubbs, a business he followed for ten years, being at that time a great source of the resources that were so needed He succeeded in obtaining a great many scenes of interest in the county

H LESLIE SMITH the acknowledged leading criminal lawyer of the Scottsbluff bar may easily claim inherited tendencies as well as choice in his selection of a profession On both sides of his ancestral line there have been men of unusual distinction at the bar He grew up in an intellectual atmosphere, and recalls the veneration he felt in boyhood for a father whose knowledge and practice of the law brought so wide a reputation and attracted to him so many men of high purpose and deep learning like himself Although Mr Smith came to Scottsbluff at a comparatively recent date, it was not altogether as a stranger, for his work in the lecture field for a number of years had made him a familiar figure to many His remarkable gift of oratory assists in the success which attends his efforts as a criminal lawyer, which branch of his profession he prefers before others Mr Smith was born at Aurora, Nebraska, December 28, 1879, the son of J H and Roseltha (Likes) Smith, both of whom were born and reared in Iowa, in which state they married The paternal grandfather was Thomas Smith, born in Pennsylvania, and was a cabinetmaker by trade and an early settler in Iowa The maternal grandfather, Philip Likes, also settled at an early day in Iowa and became one of the greatest criminal lawyers of that state The father of Mr Smith began the practice of law at Osceola, Iowa, from which place he moved to Nebraska and in 1878 located at Aurora, where he became a corporation lawyer, to which difficult branch of the law he subsequently mainly devoted himself He was district judge while in Aurora then moved to Lincoln and engaged in corporation law, having no practice except in district, supreme and United States courts Needing a wider field for his talents, in 1904 he moved to Lincoln and subsequently was made judge of the district court He practiced all over Nebraska and had many cases in the superior and United States courts While in practice at Lincoln, he was attorney for the Burlington Railroad, the Beatrice creamery, the Royal Highlanders, the Modern Woodmen, and many other corporations He was a Republican in his political views and was a member of the Christian church as is the mother of Mr Smith who yet resides at Lincoln The father died in 1912 Of their family of six sons, H Leslie is the second of the five survivors, the others being Herbert H of Lincoln, who is a well known musician and artist, Roscoe L, who also resides at Lincoln, is an X-ray specialist, Jerome H of Scottsbluff, is in the real estate business, and Philip P a veterinary surgeon, lives at Ogallala Nebraska The fourth son, Fred, succumbed to influenza, during the recent epidemic

H Leslie Smith received his elementary education in the public schools, followed by a course in Wentworth Military Academy After leaving the academy he entered the University of Nebraska for a literary course before taking up the special study of law in that department, from which he was graduated He received his A B degree in 1902, and that of LL B two years later He entered into practice at Estancia, New Mexico, where he remained one year, then accepted a flattering offer from a lecture bureau and for the next five years appeared on the lecture platform all over the country, filling engagements in Iowa Missouri, Florida, and Nebraska For several years afterward he devoted himself to commercial pursuits In December 1915, he came to Scottsbluff and has engaged in the practice of law here ever since He has always been quite active in politics and while in New Mexico was permanent chairman of the Republican territorial committee

In 1913 Mr Smith was united in marriage with Miss Beulah Garman, who was born in Iowa and is a member of the Christian church They have one son, H Leslie, who was born March 8 1917 Mr Smith is prominent in advanced Masonry and belongs also to the Royal Highlanders As a citizen he has been warmly welcomed and as a token of appreciation he was elected in January, 1919, president of the Chamber of Commerce of Scottsbluff

RICHARD SKINNER — A widely known and highly respected pioneer of Banner county, Richard Skinner, came here when most of the present flourishing towns were still prairie, and when the only pretension of Gering to being more than a hamlet, lay in the fact that it had a general store Mr Skinner has been a judicious farmer and at one time owned large bodies of land still retaining a hundred acres As one who has lived continuously in Banner county for thirty-three years, he has had many experiences, the recital of which would add value and interest to the county's official history

Richard Skinner was born in Perry County, Ohio, December 13, 1841, one of four children born to Eli and Emma (Allen) Skinner, three of whom are living Mr Skinner has a brother in Wyoming and a sister in North Dakota. The father was born in Ohio and the mother in Virginia The father died when Richard was three years old, but the mother, coming from a long lived race, lived to be eighty-six years old and her mother survived to the unusual age of a hundred and five years The father was a small farmer in Ohio all his life He was a Democrat in politics, and both parents were members of the Baptist church.

Richard Skinner had some schooling in Perry county when a boy, but work claimed him before he was very old, and after some years on a farm he went into the coal mines and spent fifteen years in that industry, in the meanwhile becoming bank and track mine boss Thus, he was already a man of business experience when he enlisted for service in the Civil War, entering the Sixty-second Ohio volunteer infantry, and served during the closing eighteen months of that war. He was never wounded or taken prisoner, although he was on the picket line between the two opposing armies at the siege of Richmond, when the cross fire for three continuous days so injured his sense of hearing that it still annoys him in the right ear, and he still has some rheumatic reminders at times of the long nights of army exposure. He relates a tragic incident of this time, and tragedy belongs to all war, that many of his comrades witnessed After a long period on duty the men were parched with thirst and when relieved hastened to the nearest well for water without carefully observing its location Mr Skinner's cousin was in the act of drinking when a Confederate sharpshooter shot him and he fell at Mr Skinner's side and later died

When the war was over, Mr Skinner returned to his old home in Ohio and from there moved to Missouri in 1873, farming there until 1886, when he came to Banner county, Nebraska, and homesteaded on the tract that adjoins his present home farm The first home of the family was a dugout They left Missouri with four horses and two covered wagons, and reached their new home with the two wagons, two horses and twenty dollars cash capital They had reached Sidney and remained over night there, but yet had to drive across the country a distance of forty-five miles, and on the way an electric storm came up Although there were nine people in the party none was injured to any extent but one of the horses was killed by lightning and the other so shocked that he never recovered Mr Skinner and family lived in the dugout for two years but it was never felt to be a safe place of residence because of the great number of wild cattle then on the range, that had to be continually driven away to keep them from trampling over the little home in the ground and breaking in on the inmates. At first they had a neighbor two miles away who afterward left the country, and then they had to go from eight to ten miles for a friendly little visit There was very little money in circulation in this section and Mr Skinner with others gathered bones on the prairie and wood in the hills and hauled to Potter and Sidney, selling bones for eight dollars a ton and getting eleven to twelve cents apiece for cedar posts The government at that time ran a stage line from Sidney to the Black Hills, crossing a bridge at Camp Clark, between Bridgeport and Bayard This bridge was owned by a man named Clark who charged toll of twenty-five cents for every wagon and an additional twenty-five cents for every person in the wagon He was said to have become wealthy through operating this toll bridge

After proving up on his homestead, Mr Skinner borrowed money on it in order to buy cattle for livestock had long been recognized as the foundation of a fortune in the west However, when sickness fell on the children Mr Skinner found his money had to be spent for medicine and doctors, Dr Lonquest visiting them from Bayard After two years in the dugout the family moved into a sod house and lived there for ten years, when Mr Skinner bought the land adjoining his homestead and built a still more comfortable sod house, in which they have since lived For a number of years Mr Skinner was an extensive farmer and at one time had three hundred and ninety acres in alfalfa alone, but with added years he gradually relieved himself of many responsibilities, keeping only a hundred acres, in the management of which he has his son Edward as assistant

On August 2, 1868, Mr Skinner was united in marriage with Miss Emma Powell, in Perry County, Ohio, the only one living in Nebraska of a family of eight children born to Moses and Elnora (Barnes) Powell, natives of Ohio and Pennsylvania, respectively, both of whom died in Ohio Mr and Mrs Skinner have nine children Darlington, who lives in Banner county, married Clara Sickels, and they have three children, Jennie, the wife of Samuel Kelly, has seven children, Nora, is the wife of Arthur Hermann, and they have six children, Laura Belle, is the wife of Arthur Burnett, and they have four children, Margaret, is the wife of Charles Hutchinson, and they have two children, Eunice, who is the wife of Bernard Hutchinson, and they have four children, Bessie, the wife of Omar Smith, and they have two children, Ona, is the wife of Harry Barthng, of Riverton, Wyoming, and they have one child, and Edward, who resides at home Mr and Mrs Skinner have the satisfaction of having their children all happily located and within an easy automobile ride When they celebrated their golden wedding anniversary, August 2, 1918, their guests included their nine children, twenty-nine grandchildren and one great-grandchild They are members of the Seventh Day Advent church Since the Civil War Mr Skinner has been a Republican He served many years as road overseer, as precinct assessor and as school director He belongs to the order of Odd Fellows

THOMAS U VAN PELT, has for years been identified with history-making events in Banner county, and it will be many more years before his name is forgotten by those who honored him in life or now benefit by the beneficent agencies he led in organizing and all his life labored to maintain He was the owner of D bar ranch in Banner county, later known as the Van ranch

Thomas U Van Pelt was born in Marion county, Iowa, November 12, 1860, and died in his home in Banner county, August 6, 1912 His parents were Thomas and Nancy (Lucas) Van Pelt, natives of Ohio The father was a soldier in the Civil War, enlisting at Iowa City, Iowa, under Captain Johnson, in the Fortieth Iowa infantry, and died while in the service His six children were Sarah J Johnston, who lives near Harrisburg, Mary, who is the widow of August Stanfield, lives at Greybill, Wyoming, Jonathan, now deceased, left a widow who lives at Omaha, Nebraska, William, who lives in Banner county, married Blanche Snyder, Cyrus, who lives south of Harrisburg, married Jennie McKee, and Thomas U, who came to Banner county in June, 1887

At Essex, Iowa, on February 22, 1880, Mr Van Pelt was united in marriage to Miss Lottie Brookheart, a daughter of Henry and Matilda (Middaugh) Brookheart, both now deceased

The father of Mrs Van Pelt died August 3, 1888, and the mother, April 14, 1910, having spent the last five years of her life with Mr and Mrs Van Pelt To the latter were born ten children, as follows Lester, who lives in Banner county, married Nellie Noyes, Charles, who lives at Pine Bluff, Wyoming, married Miss Lottie Noyes Schindler, who was accidentally killed August 10, 1907, Rachael, who was the wife of Lester Nighswonger, of Wheatland, Wyoming, Myrtle, who is the wife of A C Hottell, of Banner county, Lewis, who is a farmer in Banner county, married Hazel Grubbs, Alonzo, who lives on the old home place, married Frances Wilson, now deceased, Frank and Gertrude, both of whom died in infancy, Alice, who is the wife of Arthur Lundberg, of Banner county, and Brookheart, who lives at home on the Van ranch

In the spring of 1887, Mr and Mrs Van Pelt shipped their teams from Vincennes, Iowa, to Schuyler, Nebraska, to select a homestead in Banner county (then Cheyenne) Mr Van Pelt had two mule teams and two wagons and found not enough vegetation to picket the animals The had shipped their teams from Des Moines to Ashland, Nebraska, where they had spent the winter, during which time Mr Van Pelt and Mrs Van Pelt's brother Alexander, had worked on the B & M railroad The immediate necessity being water, the men borrowed barrels from the nearest neighbor after fixing up a shanty, started out to find water which they had to haul a long distance While they were gone Mrs Van Pelt saw men running wild horses and with a fear of Indians in her mind, got out a revolver The men came and asked for something to eat and gave their names as Frank Pearce and Charlie Hall After Mr Van Pelt had built a sod stable he went to Grand Island and bought cows, hogs and chickens, and also a year's stock of oats, corn and flour After he came back he built a dugout of two rooms Mr Van Pelt's filing on his homestead was at Sidney, July 2, 1887, and is the second filing recorded at that place

Mr Van Pelt hastened to get a well dug as water had to be hauled a distance of nine miles, and secured a good flow of water at a depth of two hundred and twenty-seven feet In the fall of 1889, he harvested a scant crop of wheat but the range cattle were so numerous that they had to be constantly driven away or they would have trampled every field Antelope came also and often provided meat for the larder Money was needed and as previously Mr. Van Pelt had been a locomotive engineer he secured work on the run between Cheyenne and Sidney, and Mrs. Van Pelt agreed to teach the school at Fowerfield, although that necessitated her driving a distance of seven miles morning and evening. The school house in district number twenty-two stood on the present site of Gary. At one time, when the Cherokee Indians were hostile, there was talk of building a fort here, but the necessity of it grew less and less and now, the beautiful place known as Flowerfield, would be about the last location where war or savagery could be imagined.

In fall of 1890 the state of Nebraska voted on the prohibition amendment for which Mr. Van Pelt had worked long, but at that time was lost, only to be victorious at a later date. All his life Mr. Van Pelt was a strong advocate of temperance. He helped to organize the United Brethren church and meetings were held at Gabe Rock school house. He was vice president of the Banner County Sunday School Association for a number of years. The early Sunday school history is interesting. At first each attendant brought along a chair or bench and the building was used for day school and other meeting purposes.

In 1894 a Baptist church edifice in Lorraine district was bought and moved by Mr. Van Pelt and his brother-in-law, and it is still in use and was known as the Long Ridge School, later the Van School and now the Flowerfield school. Mrs. Van Pelt was the first teacher in the new school house. Later she taught two years at Gary, one year at Clearfield, two years more at Flowerfield, two years at McKinnon and two years, 1915 and 1916, at Gabe Rock. During the World War she was county chairman of the Woman's Council of Defense; helped to organize the county W. C. T. U., of which she is president, has been secretary and treasurer of the Sunday School Association for six years and for a number of years previously was vice president. During the war she visited Camp Cody to investigate hospital conditions, on her own responsibility.

In 1895, Mr. Van Pelt built a sod addition to the log residence and in 1900, a very large, modern dwelling. All the farm buildings are complete and substantial, and the last government survey gave this homestead as the highest point of altitude, in Nebraska. Mrs. Van Pelt owns six hundred and forty acres, three hundred of which is farm land, and is now homesteading an additional one hundred and sixty acres of grazing land. Mr. Van Pelt belonged to the Modern Woodmen lodge, was

interested to some extent in the organization of the Farmers Alliance. He was a man of sterling integrity, after a life of useful effort, left an honorable name behind him.

FRANKLIN W. SCHUEMAKER, who is well known and highly esteemed in Banner county, was born at Osceola, Iowa, August 15, 1889. He is the seventh member of a family of eight children born to Martin and Mary C. (Wakeford) Schuemaker, the others being: Alexander, who lives in Canada, married Tena Heatthly; William J., who lives near Mitchell, Nebraska, married Sarah J. Yoe; Mary A., who is the wife of John Shiestel, lives in northeastern Canada; George, who lives at Blackfoot, Idaho, married Nancy Smith; Margaret, who is the wife of Philo Gallup, lives at Kansas City, Missouri; Angeline A., who is the wife of Ernest Preston, lives in Montana; and Sarah, who is the wife of John F. McComsy, lives northwest of Hull, Nebraska. The father moved from Iowa to Canada in 1890, and lived there seven years as a farmer. In 1898, he came to Kimball, Nebraska, then spent a year at Gering, before going to Dorrington, near Hull to live on his brother's place for several years, returning then to Gering, where his death occurred in 1900. The mother resides near Hull.

Franklin W. Schuemaker obtained his education at Gering. All his life he has been connected with farm and cattle, and for years, up to 1914, he rode range. After that he operated land for Mrs. Lottie Van Pelt and is a successful farmer and stockraiser on section four, town seventeen, with post office at Bushnell, Nebraska.

Mr. Schuemaker was married November 14, 1915, to Miss Luella Van Pelt, a daughter of Cyrus and Jennie (McKee) Van Pelt, who are well known pioneers of Banner county, and they have three children: Cyrus A., Juanita I. and a baby. Mr. Schuemaker is a Republican in politics, and is a member of the order of Knights of Pythias.

CARL A. WAGONER is one of the honored pioneers of Morrill county and one of the successful exponents of agricultural and business enterprises in this section of the state, an influential citizen who resides in the vicinity of Broadwater.

Mr. Wagoner was born in the Buckeye state December 14, 1857, the son of Thomas and Amanda (Miller) Wagoner who were residing in Coshocton county. Both parents were natives of Ohio and are still living at the ages

Sams and McCaffree.

of eighty-three and seventy-nine years respectively Thomas Wagoner was a wealthy landed proprietor in Ohio, who engaged in agricultural pursuits He was a staunch adherent of the Republican party of Ohio and during his prime took part in shaping party policy, he was an active participant in all movements that tended to develop the country, early adopted modern methods in his farming operations and was a man of civic reform and progress There were twelve children in the family, of whom Carl was the eldest The others who survive are Frederick, now a stockman in Colorado, Lawson and Nels, both farmers in Morrill county, Gaty, who married Ed Saunders, lives in Kansas, Clara McConnell of Overton, Nebraska, and Dora Stiles of Morrill county

While irrigation was yet a question of the future, Mr Wagoner came west and was farsighted in the selection of his location After looking the country over he selected land near the Platte river in Cheyenne county on a homestead of one hundred and sixty acres, he pre-empted another one hundred and sixty and filed on a tree claim of one hundred and sixty He at once began improvements which increased the value of his holdings until today his farm is one of the most fertile in the valley Mr Wagoner put his earlier knowledge of agriculture to good use on his prairie farm, became a careful business man, studied up on the best stock for this climate and decided that thoroughbreds did the best and brought the greatest returns on the market, and purchased pure bred Durham and Hereford cattle He engaged in diversified farming also and has been exceptionally successful in raising meat animals for the market With new methods introduced by the irrigation carried out along the river, Mr Wagoner has begun raising varied forage crops, as he was the first to promote beet culture and this season had about two thousand tons He has appreciated the advantages of this country even when suffering from the drought and insect pest of earlier days, and today is one of the wealthiest men of the Morrill section Mr Wagoner is an independent in politics With his family he is a member of the Episcopal church while his fraternal affiliations are with the Masonic order, as he is a member of the Blue Lodge, also the Ancient and Accepted Scottish Rite

In 1881, Mr Wagoner married Miss Mary A Durell, a native of Illinois, the marriage taking place in Nemaha county Close attention to business, thrift, prudence and able management of his affairs are the rungs of the ladder of fortune up which this able man has climbed to his present affluence Mr Wagoner is too broad gauged to confine his ambitions to one line of endeavor and as his fortune has grown he has taken part in the commercial enterprises of his community, as he is interested in extensive oil properties, is the owner of a large block of stock in the Union State Bank, the Globe Insurance Company, the elevator the electric light plant and the Wyoming Refining Company and may be considered one of the progressive capitalists of Broadwater

HARVEY L SAMS, prominent in the affairs of Scottsbluff has large private business interests that demand close attention, nevertheless he is one of the most active citizens in public effort His enterprise and progressiveness have been of vital importance in the development of Scottsbluff, and his judgment concerning civic measures is very generally consulted Mr Sams has been a resident of Nebraska since boyhood, and of this city for seventeen years He is president of the Sams-McCaffree Company, dealers in real estate and general insurance

Harvey L Sams was born at Anamosa, Iowa, October 21, 1869, the son of Stephen and Mary (Wagner) Sams, the former of whom was born in 1837 in Ohio, and the latter in 1842 in Pennsylvania, and both died in Nebraska, the father in January 1914 and the mother in 1910 Stephen Sams went to Iowa in 1844 and married in Jones county, and his children were as follows Mrs L R Porter of Bartlett, Nebraska, Albert E of Nowata, Oklahoma, Harvey L, of Scottsbluff, Milton A, of Blair, Nebraska, Mrs E E Games of Valley, Nebraska, Mrs Nellie Shoaf, of Randolph, Nebraska and one deceased Stephen Sams and his wife were members of the Methodist Episcopal church He came to Nebraska with his family and settled in Nemaha county, coming from there to Scottsbluff county in 1904 He was a farmer all his life

Harvey L Sams attended the public schools and the State Normal School at Peru afterward passed two years in the State University at Lincoln, and in 1900 secured his A B degree in the Wesleyan University In the meanwhile he had taught school, eight years in all, leaving the educational field in 1902 when superintendent of the schools of Red Cloud Nebraska It was then he came to Scottsbluff and for four years was identified with the First National Bank, first as assistant cashier and afterward as cashier Upon leaving the bank Mr Sams formed a business partnership, incorporated, with F. S. McCaffree, under the style of the Sams-McCaffree Company, in the real estate and insurance line, and they have been very prosperous. Mr. Sams owns a large amount of valuable property in the county, some 2000 acres, including a whole section south of Gering, where he feeds two hundred head of cattle for shipment to the eastern markets.

In 1905 Mr. Sams was united in marriage to Miss Cora V. Slates, who was born at Broken Bow, Nebraska, a daughter of John B. Slates, a pioneer of Custer county. They have five children: Eldon, Angeline, Lenore, Josephine, and an infant son, unnamed. Mr. and Mrs. Sams are members of the Methodist Episcopal church. He is a Knight Templar Mason, past master of the Blue Lodge, and past patron of the Eastern Star, to which organization Mrs. Sams also belongs. He has always been somewhat active in the Republican party, believing politics a necessary part of national life, but has never sought political preferment. The substantial development of the city has interested him from the time he came here and he has given encouragement to many worthy enterprises. He is president of the library board and is urging immediate action concerning the erection of a public library building that will be creditable to the city and fulfill the terms on which Andrew Carnegie made a very acceptable donation. It was Mr. Sams who was one of the leading factors in the organization of the Commercial Club here and was its first president, and was also instrumental in starting the Water Users association, of which he was first president. He is a man of sound, practical ideas, remarkably unselfish as to his personal interests, and it is evident that with its general welfare, the development into one of the state's great centers is his cherished hope for Scottsbluff, which from present indications looks very reasonable.

STEPHEN BURROUGHS SHUMWAY was born at Spring Hill, Bradford county, Pennsylvania, April 15, 1822. He came with his parents to Lee county, Illinois, at a very early date. With his father he farmed and hauled wheat to Chicago with ox teams, and once their loads were mired down in the mud of State street in that city.

In 1849, he quit Knox College, in his senior year, where he was working his way, shaving shingles, and went to California, passing through western Nebraska in 1850. He remembered a terrific rain the night they were encamped at Chimney Rock. They sat up in their tents for hours with their blankets over their shoulders and the water running under their feet.

After a year in California and accumulating about two thousand dollars worth of gold, He returned home by way of Panama. On the trip down the Pacific ocean, he was impressed with the fact that it was anything but a peaceful ocean. The stokers kept the pipes red hot and several times the vessel took fire. In the storms the ship would crawl up an advancing wave for several hundred feet, to the top, and ahead of them there would be a trough in the water that seemed like a bottomless pit. Down into this the ship would plunge and when it hit the bottom the water would sweep over the decks carrying away anything loose. Then around the vessel, the water would begin to boil and the ship would again begin its laborious climb up another wave.

Crossing the isthmus, he rode a Spanish mule in a path worn so deep that its banks were often as high as his shoulders, the verdure of the tropics w   tted overhead so dense that it was fairly   places.

On arriving in Illinois, .   Alson J. Streeter, afterward Union Labor   date for President of the United States, gai. red together one hundred head of cows, and in 1852, drove them to California. This trip was made through the North Platte valley, on the north side of the river to Fort Laramie. They had plenty of adventure, but no losses, and arrived at the golden mecca with more cattle than they started with, for some of the cows dropped calves, which were taken in the wagons and carried when they showed sign of fatigue.

On February 28, 1854, he was married to Lydia Jane Streeter, a sister of Alson J. Streeter. She was born at Rock Hill, New York, October 1, 1835, and removed to Lee county, Illinois, about the same time that the Shumways moved there from Pennsylvania.

Mr. Shumway went into mercantile business at Oxford, Henry county, Illinois and later retired to his farm adjoining the town. The old farm house where he lived, and which was burned recently, was the home for many years, and here the family of eleven children were born, three of whom died while quite young. Those who grew to maturity were Clara Grace, Roswell, George, Grant, Minnie Mae, Alson and Mabel.

For about ten years S. B. Shumway served as county supervisor in Henry county, being elected as an independent, he being an independent in politics since the greenback days. He was once independent candidate for Con-

Mr. and Mrs. S. B. Shumway

gress In 1886, he came to Banner county, then a part of Cheyenne county, and filed on a homestead After the organization of the new county of Banner, he was elected county judge and served two years

Mr and Mrs Shumway affiliated with the Congregational church in Illinois but there being none of that denomination here at that time, they attended Methodist services, and occasionally the Presbyterian He was a Thirty-second degree Mason before coming west and was for many years High Priest of the Chapter at Rio, Illinois

Clara A, the eldest daughter came to Banner county, and filed upon a homestead in 1887 She was married to George B Luft December 10, 1890 Mr Luft was a pioneer merchant, and continued in business until his death at Scottsbluff Clara Shumway was the first superintendent of public instruction in Banner county, being elected without opposition and served two terms She has always been active in educational, mercantile, lodge, club and civic endeavor

Grace F, who was inclined to literary pursuits, and prominent in the W C T U was married at Woodhull, Illinois to C L Burgess, and was killed by the accidental discharge of a revolver August 13, 1899

Stephen Roswell, the oldest son except Gano, who died in infancy, was married to Mary E Brown at Woodhull, Illinois, and has been engaged in mercantile pursuits for the greater part of his life He now resides at Oklahoma City He attended Monmouth college

George O, who attended Knox College, was married to Lena Hoadley at Galesburg, and took up the practice of law He was seven times mayor of Galesburg, and once candidate for lieutenant governor of Illinois He now resides in northern California

Grant Lee, the writer of the history of western Nebraska, and Alson J have sketches elsewhere in this volume

Minnie Mae, now Mae Shumway Enderly, resides at Los Angeles, where her husband, Fred W, has mercantile interests She is professionally an entertainer for chautauquas, singing, reading, and high class vaudeville

Mabel (Raymond) is at Los Angeles, and is a student and practitioner of therapeutics, and active in civic affairs

The ranch of nine hundred and sixty acres in Banner county was sold and Mr and Mrs Shumway moved to Gering, where he died August 5, 1897 Mrs Shumway lived at Gering, Scottsbluff and Los Angeles, until October 22, 1904, when she died at the home of her daughter, Mrs Luft at Scottsbluff

MARTIN L WEHN — One of the younger but prominent and truly representative men of Morrill county, who is president of the Broadwater Bank of Broadwater, and who has had other important interests in the county and city, is a native son of Nebraska, born in Aurora, October 5, 1885, the son of George H and Susan C (Hartzell) Wehn, who are descended from a long line of fine Pennsylvania stock, where both were born The father was a minister of the gospel and lived in different places in the middle west as he was called to various charges The Wehn family first came to this section of the country in 1865, when George Wehn assumed charge of a church in Beatrice and from that date to the present has been an influence for good living, progress and the advancement of religion and education wherever he has been located Having spent forty-five years in ministration to the spiritual life of mankind he lives a retired life, passing the declining years in the homes of his devoted children

There were seven children in the Wehn family Margaret, who married J W Beggs, now lives at Whiting, Iowa, Daisy, the wife of B C Brons also of that city, Susan, married Herbert Scott of Hampshire, Wyoming, Martin, the subject of this review, and three who died in infancy

Being the son of educated parents, Martin Wehn was given advantage of every educational facility afforded in the public schools of Iowa and Missouri where the family lived while he was a boy and youth After completing his practical education the young man entered the general mercantile business, opening the first store in Broadwater, but after building up an excellent patronage within the short space of some eighteen months his interest was directed to financial circles by his uncle J W Wehn Having made a marked success of a retail business and having great faith in the future of this section of Nebraska, Mr Wehn purchased a large block of stock of the Broadwater Bank from J W Wehn who had owned the controlling interest in that splendid financial institution, and became its president The bank has a capital of $25,000, surplus of $7,000 and deposits of over $250,000

Mr Wehn has become a leading financier of the county and as president of the bank enjoys the confidence of the public generally He has taken an active part in all local enterprises of recognized merit and his public spirit has many times been sufficiently exercised to warrant his reputation as an earnest and able citizen of this progressive community

In politics he is a Republican and while he

is interested in all political questions that pertain to the progress of Broadwater, Nebraska, and the nation has no time to enter political life as his entire energies are concentrated on the business of the bank and its allied financial activities   Fraternally Mr Wehn is allied with the Masonic order while he and his wife are liberal supporters of the Presbyterian church

On September 4, 1907, Mr Wehn married Miss Mary Evans, who is a charming, gracious woman and the family have a large circle of friends, while Mr Wehn stands high in banking circles due to his constructive yet conservative policy, his universal courtesy to depositors and borrowers and his interest in the civic improvements of Broadwater and the county

JOHN H ADAMS, who has been prominently known in business circles in Broadwater for a decade, is a broad minded citizen and progressive man of affairs who manifests lively interest in all things touching the communal welfare and the upbuilding of the city

John Adams is descended from fine old English stock and displays many of the excellent characteristics of that race   He was born in Kansas, November 30, 1884, the son of Benjamin and Elizabeth (Bluer) Adams, both of whom were born and reared in England   They were ambitious people who desired greater opportunities both for themselves and their children than could be obtained in the island of their nativity and immigrated to the United States   Soon after arriving in this country they came west, locating on a Kansas farm, where the father lived all his life, passing away in 1892, while the mother still lives residing in Wakefield, Kansas, a sturdy, worthy, old woman

The youthful years of John Adams were passed on his father's farm, where he assisted in the farm work during the summer and attended the public schools of that state, and to this preliminary education he has been constantly adding, by extensive and well directed reading   He was one of a family of twelve children and grew up inured to the happy strife of a big family until his majority was attained when he decided to start out on an independent career and learned bookkeeping, after completing this course he entered the employ of a hardware house at Bayard where, in 1905, he gained the practical knowledge of its administration   After residing in this state for two and a half years he felt the call of the west, and with an idea of looking for

business openings as well as seeing the country, Mr Adams left, going to Nevada, then on to California and the Pacific coast, returning by way of Idaho   No country looked as good as the Panhandle and as he had faith in the future of this section he settled in Broadwater on his return   The two years and a half spent in travel had broadened his vision and he saw that a great future was in store for this section of fertile soil, with its assured sunshine and the irrigation which is being developed in the valley of the Platte   A year after his return Mr Adams opened his present grocery and dry goods house   This business dates from 1910, and in the decade since its establishment the store has grown to most satisfactory and gratifying proportions   Today it is regarded as one of the leading commercial enterprises in the Panhandle, due to the excellent policy of its proprietor, who won and holds his trade by his unqualified personal popularity   He is ever courteous, is regarded as a progressive man in ideas, an enterprising citizen as well as a successful business man   Mr Adams is still young, the years of a broad commercial career are still before him and the past makes this more of a statement than a prophecy

In 1912, Mr Adams married Miss Bertha Rogers, a native of this state and they have one child, Harland   The family is connected with the Presbyterian church, to which they are liberal contributors   Politically Mr Adams is affiliated with the Democratic party but he is an independent Democrat, not drawing close party lines when voting in local elections as he picks the best man for the position to be filled   He takes pride in the record made by the party during the war

JESS R MINSHALL has identified himself with the civic, financial and material interests of Morrill county, for he has not only been a representative merchant of this section and a banker of prominence but has held a public position of trust in the county   He is a native son of the west and has exemplified its progressive spirit in the varied activities that that have brought him such a generous share of temporal prosperity

Mr Minshall was born in the Sunflower state, February 12, 1881, the son of James R and Mary (Hogan) Minshall, the former born in Wisconsin, while his mother was a native of Missouri   There were five children in the family   Ralph, who lives in Tacoma, Washington, Nola, who married H W Benett, Benjamin, who lives near McGrew, Nebraska, Jess, the subject of this review, and Dora, deceased

The family resided in Kansas for a number of years, then removed to North Platte, where the father was employed as a machinist, an occupation which gave him a comfortable income so that he saw his children had good, practical educations

Jess Minshall attended the public schools of North Platte and after graduating from the High school he entered the Fremont Normal school, spending four years there in advanced study, making a specialty of commercial subjects After completing his course in this institution he entered upon a financial career by becoming interested in a mercantile house, where for ten years he was engaged in varied commercial pursuits gaining invaluable knowledge in business methods He was ambitious and decided to establish himself independently and in 1906 came to the Panhandle, for he believed there was a great future for this part of the state, due to the new intensive farming being introduced with irrigation Mr Minshall initiated his career in this section by opening the Blue Creek Mercantile Company at Lewellen, on the north bank of the Platte His store flourished from its initiation but commercial life was not the goal of his ambitions and after studying the situation over he entered the bank at Lewellen, to learn in a practical manner, the administration and policies of the banking business Leaving this institution he came to Broadwater in 1909, and established a hardware store but gave up active participation in its management when he was elected clerk of the district court Mr Minshall performed his public duties as a worthy citizen who had the best interests of the community at heart, and at the expiration of his term of office returned to Broadwater, becoming one of the stockholders and organizers of the Union State Bank of this city From the initiation of this substantial and progressive banking house Mr Minshall has been cashier, with Mark Shanogle as president The institution is capitalized at $25,000, with deposits in excess of $100,000, which attests to the popularity of its officers as well as the high standing of the men who direct it and to the confidence the public has in their sound business methods Mr Minshall has been active and liberal as a citizen, is intensively patriotic and public spirited, contributing liberally to all progressive movements of the community, and commands high place in popular confidence and good will

October 1, 1910, Mr Minshall married Miss Mary R Clary and they have one child, Georgia They are members of the Episcopal church, while Mr Minshall is affiliated with the Masonic order and the Independent Order of Odd Fellows In politics he is a Democrat and takes an interested part to see that good men are elected to office in the community where he makes his home

FLOYD S McCAFFREE — The story of success that crowns determined effort in the face of discouragement, will never lose its interest in a free country like the United States with its sound, sane people, because it reveals those admirable qualities that are fundamentals of her strength Courage perseverance, industry, frugality and hopefulness, are the stepping stones that have led many a young man starting out in life handicapped by poverty to the heights of comfortable independence These reflections call to mind one of Scottsbluff's representative men and well known capitalists, Floyd S McCaffree, former mayor of this city, who has been extensively engaged in handling real estate for some years

Floyd S McCaffree was born at Spirit Lake, Iowa, November 15, 1882 the fourth in a family of seven children, born to Floyd J and Rachel E (Stratton) McCaffree These are old family names in Iowa, and the father of the family was the first white child born in Bremer county For twenty-five years he was in the active ministry of the Methodist Episcopal church, riding a circuit in early days but now lives retired at Scottsbluff, the mother also surviving Mr McCaffree had the following brothers and sisters Charles an assistant commissioner at Pierre South Dakota Grace, the wife of H H Smith, a farmer in Canada, Alice, who lives with her parents Mattie the wife of O C Smith who is a government engineer and reclamation agent at Grand Junction Colorado, Harry in the real estate business at Mitchell, prior to his death from influenza, in November 1918, and Rolie K, an automobile salesman at Mitchell, Nebraska

Educational advantages were not denied Mr McCaffree in his youth in fact he had a year of collegiate training at Morningside College Sioux City Iowa In 1905 he started for Nebraska making the journey in a covered wagon, accompanied by his bride formerly Miss Esther Scott, who was born at Marathon Iowa She died in February of the following year Mr McCaffree came to Scottsbluff county with the intention of securing a homestead which he accomplished in the same year and he still owns this property now exceedingly valuable, as it lies eight miles north of Scottsbluff As he had no capital as represented by money stocks or bonds he accepted the first work that offered itself, which was working on the government ditch and other odd jobs that brought a good cash remuneration for his labors. It was unaccustomed work but he kept at it and for a long time cheerfully worked side by side with laborers who had come from every section. When it came to breaking the sod on three hundred acres of land that he desired to seed in alfalfa, he was able to do a man's work with anyone. In 1907, two years after reaching the county, he began the business of locating homesteads. In 1910 he became associated in the real estate and insurance business with Harvey L. Sams, and they have developed one of the largest concerns of its kind in this part of the state and together own thousands of acres of land. Mr. McCaffree has one tract of four hundred acres of irrigated land, which he has operated by farmers who understand modern methods, with all kinds of improved machinery supplied, and here he feeds cattle extensively. In his partnership with Mr. Sams, the insurance line is also an important feature.

In 1910 Mr. McCaffree married Miss Gertrude D. McDowell, who was born at Omaha, Nebraska, and they have three children, Ruth, Robert, and Edwin, aged respectively, eight, six, and four years. Mr. and Mrs. McCaffree are members of the Methodist Episcopal church. He is a Republican in politics and since coming to this county has been active and public-spirited, showing an earnest desire to advance the interests of this section in every way. For four years he served as county assessor and when elected mayor of Scottsbluff, gave the city a fine administration. He has been a vitalizing force here in many ways. Like his father he is a Mason and Odd Fellow, and belongs also to the Yeomen and the Modern Woodmen orders.

GEORGE W. BEERLINE was born in Missouri, in 1866. He lived in that state with his parents until he was seven years of age, when they removed to Sarpy county, Nebraska, and in 1887, to Cheyenne county. He homesteaded one hundred and sixty acres, and now is the owner of six hundred acres of land in what is now Morrill county, about two hundred acres of it being under irrigation. He is a farmer and stock raiser. At one time he made a specialty of Duroc Jersey hogs, and at all times has made a specialty of livestock, handling nothing but high-class grades. He put the improvements on his place, which are up to date and of the kind that will be permanent and add permanent value to the land, including a good orchard.

He was married in 1902, to Margaret Chase, a native of Nebraska, a daughter of John and Mary (Miller) Chase. The father is now deceased, the mother living at Papillion, Nebraska. The father, John Chase, located at Bellwood, Nebraska, in 1856, and lived for forty-six years on his original homestead there, dying at the advanced age of ninety-one years. He was a close friend of Secretary J. Sterling Morton, and at the time of his death was the oldest Mason in Nebraska.

Mr. and Mrs. Beerline have two children, both living at home. Their names are John Chase and Helen Catharine.

Mr. Beerline was educated in the public school of Sarpy county, as was also his wife. He is an independent voter, belongs to the Modern Woodmen, and is a director of his school district.

Mr. Beerline is one of the leading men of Broadwater and vicinity, and stands high in reputation among his fellow citizens. He has always taken an intelligent interest in public affairs and is always found in favor of progress and up-to-date methods both in private business and public matters.

JULIUS GEBAUER — The agriculturist has ever before him the chance of making himself an enormously useful factor in a community, and a realization of this fact has come to Mr. Gebauer in Morrill county, where he has maintained his home for more that eighteen years. He is the architect of his own fortunes, and as such deserves the greater credit for the success-which he has achieved by his own efforts. A native of Germany, when he came to this country he brought with him many of the admirable traits of that sturdy race, and the fortune that has come to him has been won through legitimate business enterprise.

Julius Gebauer was born in the German Empire in 1852, the son of Trangott and Johanna Elizabeth Gebauer, both natives of that country. The father was a butcher, he and his wife spent their entire lives in their native land. Julius had two sisters, Augusta Ewalt, who lives in Grand Island, Nebraska and Pauline Crause, a resident of the Grand Island district. Julius spent his youth and early manhood in his native country where he took advantage of such educational facilities as were available to him, but he saw no future for a man without capital in the old country and determined to take advantage of the lands opened up for settlement in the new world that could be obtained with but little initial outlay. In 1879, he immigrated to the United States, and soon after landing came west as he knew of the many Germans who had located on the prairies beyond the Mississippi river. Mr Gebauer came to Nebraska and found employment with the railroad in the vicinity of Grand Island. He was frugal, careful in his habits and expenditures and in time had accumulated fifteen hundred dollars, which he intended to use in the purchase of desirable land, but the money was stolen from him in Grand Island. This did not daunt his spirits and he but worked the harder, for he had no equipment but his sturdy determination and the will to succeed. As soon as he became acquainted with the language and customs of the country he started his career as an agriculturist on a farm in Hall county. He first purchased two hundred and forty acres of railroad land, and made permanent improvements. This state was not then the smiling countryside of the present time and the early settlers lived in sod houses, broke the virgin sod of the prairie in order to put in the first seed for crops. After raising the land to a good state of fertility he was able to sell the first farm for nine dollars an acre, a good price for that day. Following this sale he purchased forty acres of land from the railroad and also school land so that he had a considerable holding. But as the returns from the land at this early day came slowly he decided he needed more capital and went to Grand Island where he put the knowledge he had learned from his father to practical use and opened a butcher shop, subsequently he went to Colorado, locating in Julesburg but after a short period returned to Grand Island, being employed in a sugar factory until he decided to return to the soil and began operating a truck farm where he raised fancy vegetables to supply the city markets. Meeting with success in this line Mr Gebauer accumulated a comfortable capital, and decided to engage in farming on a more extensive scale. He came to the Northport district in 1903, and located on a homestead north of the Platte. A comfortable home was erected, permanent and substantial farm buildings were soon established and he began to operate a diversified farming business. He believed that the most paying stock was pure bred cattle and hogs and he has specialized in these successfully and he now owns six hundred acres of the finest land in this section. From first coming here he determined to extend his business and as his farm produced good crops he sold at good figures and with the money so made bought more land until he is one of the largest landed proprietors in the valley. This finely culti-

vated and productive property shows what may be accomplished by a man of self determination, foresight and who is not afraid to work, who starts out in life equipped with nothing but a sturdy body and a determination to succeed.

In 1881, Mr Gebauer married Miss Amelia Krause, at Grand Island, and they have become the parents of eight children. Erna is married and lives in Kansas, Oscar is a locomotive engineer at Bridgeport, Paul, who spent nineteen months in the army in the remount service is not at home farming the home place, Martha, lives in Kearney, Olga, is a school teacher in Bridgeport, Ella, also teaches, Arnold teaches at Lisco, and Adelia, is only fourteen and attend school near home, and one boy who died in 1914. The Gebauer family are members of the Lutheran church, while Mr Gebauer is an independent in politics.

WILLIAM JOHNSTON was born in Iowa in 1852, and died November 6, 1907.

He was one of the early settlers in the vicinity of Bayard, having come to that locality in 1887, after farming in his native state of Iowa. He took a homestead that included part of the present site of Bayard, the residence house being now in the north part of the city. He followed farming and stock raising, and in addition to his other occupations he built the first hotel in that part of the country and conducted it himself for sixteen years. His business prospered, so that at the time of his death he owned two hundred and forty acres of valuable land which is still more valuable owing to its location at the edge of a growing city.

Mr Johnston was educated in Iowa, and before coming west he was married in 1883, to Anna Varrier, a native of Indiana, who survives him and makes her home at Bayard. Five children were born to them all of them living. They are Albert and Clarence, living in Bayard, Irma, who makes her home with her mother, Cora, now Mrs John Zook at Bayard and Otta who married Monte Fullerton and lives in California.

Mr Johnston, while he did not live to see the full development of the country in which he was a pioneer, still saw it on the way to wealth and fame. At the time of his death the great system of irrigation which is now in full operation was getting well started and the future of the North Platte valley was assured. He was a man who was widely acquainted and enjoyed an enviable reputation among the people of his community, being known as a public-spirited and enterprising citizen who took an intelligent interest in the progress of the country and was ever ready to support public movements and advance the interests of his fellow men

PATRICK ROWLAND is a native of Canada, the son of Michael and Ella (McDonald) Rowland The parents were natives of Ireland, now deceased They had eight children, four of whom are living Michael, a farmer in Canada, John a miner in Colorado, died August 14, 1918, a daughter, now Mrs Blake, lives in Minneapolis, Minnesota, and Patrick, the subject of this sketch The father owned land and did a general farming and stock-raising business in Canada Later he moved to Kansas and bought land there He was a Democrat in politics and a member of the Catholic church

Patrick came from Canada to Kansas when nine years old and worked for his father until he was seventeen, then went to Nevada and Wyoming and from there to Cheyenne county, Nebraska, twenty-five years ago He homesteaded a quarter section and now owns thirteen hundred and sixty acres of land, of which eleven hundred and fifty acres are under irrigation He follows general farming and stock-raising

He was married twenty-five years ago to Anna C Hagerty They have four children, all at home Their names are, Estella May, John F, Helen A, and William L

Mr Rowland is a Republican in politics, a director of his school district, also a director of the irrigation district in which his land is located, and a member of the Catholic church He is one of the leading successful farmers of his vicinity and stands high in the community as an energetic and progressive man He takes part in public affairs and keeps abreast of the times in public questions as well as in matters affecting his own business . In Broadwater district, his home, he is widely and favorably known among a large circle of friends and acquaintances

DANIEL W WARNER, the owner of a fine farm near Hull, is numbered with the substantial and representative men of this section From an experience of more than thirty years in Banner county and taking part in its wonderful development, Mr Warner feels proud of what he has accomplished in a comparatively short time

Mr Warner was born in Jasper county, Iowa, in 1857, the son of Joseph and Ingaba (Webb) Warner, the former a native of Ohio who lived to be fifty-four years of age, while the mother was also a Buckeye by birth lived out the psalmist's allotted span of three score years and ten, passing away in 1917, at the advanced age of eighty-four years The father was a farmer in Indiana and later removed to Iowa where he was extensively engaged in general farming and stock-raising There were seven children in the Warner family Henry, deceased, Thomas, a farmer of Iowa, Daniel, of this sketch, Dora E, the wife of James S Edger, a farmer in Colorado, Elmer, farming in Iowa, Josiah, in Iowa, and Frank, who lives at Crawford Nebraska Joseph Warner was a Republican while both he and his wife were members of the Methodist church

Daniel was educated in the common schools in his native state and early learned the practical side of farm industry from his father and after his school days were over engaged in farming with his father for a time, but he was ambitious to get ahead in the world, and having heard of the fine land to be obtained in western Nebraska on the homestead plan, came to the Panhandle in 1886, to establish himself independently in business He located in what was then Cheyenne county — known today as old Cheyenne — on a claim of one hundred and sixty acres of land, proved up and still owns the old home place After several years passed in farming here Mr Warner went to Colorado, where he took employment with the P. O outfit horse ranch, but after a time returned to his land, where he continued to improve the fertility of the soil, erected new and better farm buildings and a fine home for his family As he sold his crops and thus had capital available Mr Warner purchased more land adjoining his first holding and today is the owner of a full half section all under cultivation From first becoming established on the plains he has engaged in general farm business and also raises a good grade of live stock, which he has found to be a profitable line Mr Warner is an independent in politics

In 1891 Mr Warner married Miss Lillie C Ammerman, the daughter of Hiram and Martha Ammerman, who located near Hull in 1887 The father is deceased but is survived by his widow Nine children have become a part of the Warner family circle Mattie, the wife of Oscar Barkell, Frances, the wife of William Jones, of Scottsbluff county, Gladys, who married L W Hopkins, Ida, Eva, Anna,

Frederick J. Colbert, M. D.

Elsie, Jessie, Willard and one child that died in infancy. Mr Warner has ever proved himself a loyal citizen and is wide awake to every movement that will particularly benefit Scottsbluff county.

FREDRICK J COLBERT, M D, is a native of this state and a representative of several of its oldest pioneer families. He has been engaged in medical practice at Gering for several years with marked professional success and the high esteem in which he is held personally is evidenced in his election to the highest municipal office in the gift of his fellow citizens, on April 1, 1919. Mayor Colbert has entered upon the duties of his office with the hearty good wishes of every one, and with the determination to devote his splendid abilities to promote the best interests of Gering.

Frederick J Colbert was born in Cass county Nebraska, September 27, 1889, the eldest of three sons born to John W and Lucy (Frisbee) Colbert, the former of whom was born in Michigan and the latter in Iowa, and both came with their parents to Nebraska, in 1868. The maternal grandfather of Dr Colbert was a veteran of the Civil War, in which he had served in an Illinois regiment for four years and been seriously wounded, from the effects of which he died when aged forty-three years. He was a native of New York state. The paternal grandfather of Dr Colbert was James Colbert who was born in Huntingdonshire, England. He came to Nebraska in 1868 and homesteaded in Cass county. The parents of Dr Colbert were married in Nebraska and both live in comfortable estate at Weeping Water where the father has real estate interests and is a farmer and stockman. He is a Republican in politics and belongs to the Masonic fraternity. Dr Colbert has two brothers Harry E who has just completed post-graduate work in a dental college at Chicago, now located at Gering, and Horton R, who is a student in the high school of Weeping Water.

In the public schools of Cass county, Fredrick J Colbert received his early educational training. In 1913 he won his A B degree in the University of Nebraska, and in 1917 the degree of M D was conferred upon him at Rush Medical College Chicago. He came immediately to Gering and entered upon a general practice. It is almost impossible in modern days, for intelligent and progressive men to keep out of politics, and Dr Colbert was well schooled in the principles of the Republican party from boyhood.

On June 19, 1914, Dr Colbert was united in marriage to Miss Maude O Case, who was born in Cass county, Nebraska and they have one son Frederick Case Colbert, who was born December 22 1916. Dr and Mrs Colbert are members of the Methodist Episcopal church. He is somewhat prominent in local Masonic circles and both he and wife belong to the Eastern Star.

JAMES A CARD, a well known and respected member of the younger farmer element of Scottsbluff county, who stands high in the agricultural and stock-raising circles of this section, has passed the greater part of his life here and has seen the many changes that have transformed the Panhandle into a rich farming district.

Mr Card was born in New York State in 1881, being the son of John and Martha (Mary Haue) Card, both natives of the Empire state where they were reared received their educational advantages and after maturity met and were married there. Both are now dead. John Card was a farmer and as land was high in New York he determined to take advantage of the opportunity of land for himself and his children in the newer states west of the Mississippi river, and with this end in view came to Nebraska in 1885, took up a claim in township eighteen, section seven, Scottsbluff county, where he established the family. Mr Card broke his land, put up the necessarily primitive farm buildings of the pioneer days, and soon became a well known farmer and stock-raiser of the section. He was a Republican in politics and attended the Baptist church with his wife who was a member.

James A Card accompanied his parents to western Nebraska when they located on their frontier farm. He was but a small child at the time and as he grew up hardly realized that he was not a native son. Mr Card attended the public schools near his home and thus laid the foundation for a good education which has proved of great value to him in his subsequent business career. While still young he began to assume many of the small tasks on the home place and thus from childhood, began to acquire a practical knowledge of farm industry. There were six children in the family. Alice, married Charles White of Sioux City, Iowa, Lydia, the wife of Ira Nagel of Scottsbluff county, Fred who lives with James, Hattie, the wife of Frank Schumacher of Montana, and Merle, also of Sioux City.

Mr Card was not contented always to work for others, and when he was old enough took up a homestead of six hundred and forty acres, and in partnership with his brother now manages a ranch of four full sections, some under lease They carry on general farming operations but devote most of their time to the cattle business as a large part of their holdings is fine grazing land They are shrewd buyers and long sellers and today are recognized as two of the most prosperous and substantial men of the Hull district, where they have won enviable reputations as men of high standing and character

HENRY KASCHKA, has been successfully engaged in agricultural and dairying enterprise during the greater part of his residence in Garden county, where he now owns and gives his personal supervision to a well improved farm of five hundred and twenty acres, situated about five miles north of Oshkosh He came to America as a young man endowed with ambition and determination, but dependent entirely upon his own efforts in making his way to the goal of independence and prosperity He has succeeded well and has so ordered his course as to merit and receive the confidence and good will of those with whom he has come in contact

Mr Kaschka was born in Pormer, Germany, on January 27, 1851, a son of Frederick and Wihelmina Kaschka, who passed their entire lives in the native land the father, a weaver by trade, having died when about sixty years of age and the mother having attained to the venerable age of eighty years Henry Kaschka was educated in the schools of his native province and was twenty-six years old when he immigrated to America He passed the first five years in Illinois and then proceeded to Colorado, where he took up and perfected title to a homestead, which he developed and improved Eventually requisition for his land was made by the Yuma Ditch Company, which used it for a site of an irrigation reservoir, and Mr Kaschka was paid substantial indemnity when his farm was condemned for this purpose In 1910, he came to Nebraska and has since been continuously engaged in farming and the dairy business in Garden county He is a good citizen and a man of purposeful industry, his political support being given to the Republican party and both he and his wife being communicants of the Lutheran church

March 24, 1887, recorded the marriage of Mr Kaschka to Miss Wilhelmina Press They have four sons and four daughters Mrs Martha Blaissey, of Oshkosh, has two children, Carl was one of the gallant young Americans who entered the nation's military service in connection with the World War, and was trained at Fort Dodge, Iowa, and is now engaged in farming in Garden county; Mrs Emma Haxtell, who has three children, reside at Clair City, South Dakota, as does Ella, and John, Hannah, Henry and Ernest remain at home

ROY E SWANSON, is a member of one of the sterling pioneer families of this state He is one of the successful and popular young farmers of Garden county, where his well improved farm, of one hundred and sixty acres, is situated two and one-half miles northwest of Oshkosh

Roy Eugene Swanson was born in Saunders county, Nebraska, December 28, 1890, a son of August and Anna M (Rosenberg) Swanson, both of whom were born in Sweden but whose marriage took place in the city of Omaha, Nebraska August Swanson was a young man when he emigrated from Sweden to the United States, and upon coming to Nebraska he settled near Wahoo, Saunders county, where he took up a homestead and eventually developed a good farm He was one of the well known and highly honored pioneers of that county, where he won substantial prosperity and lived many years He and his wife now reside at Oshkosh, Garden county Both Mr and Mrs Swanson are members of the Lutheran church They had ten children Cyril, Gilbert, Roy, Lillian, Eva, Minna, Esther, Hazel, Wilma and Grace

Mr Swanson was a child at the time the family came from his native county to Platte county, where he was reared on the home farm and attended the public schools He continued to be associated with his father in farm enterprise until he was twenty-four years of age, when he formed a partnership with his brother Cyril E and they established a dray line at Oshkosh He also took up a homestead, two miles northeast of Oshkosh, a property on which he proved up and upon which he made good improvements During the year that he was associated with his brother in the dray business they entered claim to a quarter-section of land two and one-half miles northwest of the town, the present home of Roy E, who took possession after he had purchased his brother's interest and who demonstrated his splendid capacity for diversified agriculture, the raising of cattle, including the development of a prosperous dairy business, and the raising of hogs, of which he has an average of fifty head a year He is a pro-

Mr. and Mrs. A. M. Parmenter

gressive young business man and public-spirited citizen, is a Democrat in politics, is affiliated with the Woodmen of the World, and is also a member of the Odd Fellows. He and his wife are active communicants of the Lutheran church. Mr Swanson is an advocate of scientific methods and progressive policies in connection with farm enterprise and for a number of years past he has held the local agency for the *Nebraska Farmer*, one of the leading farm papers of the middle west.

March 4, 1918, recorded, in the city of Omaha, the marriage of Mr Swanson to Miss Frances May Twiford, who was born in this state, where she was reared and attended school in Frontier county, later attending the public schools of Ogallala, Keith county, and later was a student in the University of Nebraska. For a time she held a clerical position in a mercantile establishment in the city of Lincoln, and prior to her marriage had been an operator in the telephone exchange in the city of Omaha. She is a daughter of George and Aurora Twiford, who became early settlers in Garden county, the father having taken up a homestead eighteen miles northeast of Oshkosh. He became a successful agriculturist and stock-raiser and died when about fifty years of age, his widow, Mrs Aurora (Dollard) Twiford, is now a resident of Oshkosh. Mr and Mrs Swanson are most popular factors in the best social life of their home community.

ALBERT M PARMENTER has been a resident of Scottsbluff county for twenty-three years and of Nebraska for more than forty years, having crossed the Missouri river on New Year's Day, 1880. He may justly be called a pioneer as his home has always been on the frontier of the different parts of the state where he has resided.

Albert M Parmenter is a native of the Buckeye State and was born in Williamson county, Ohio, October 30, 1858. His parents were David and Lydia (Huling) Parmenter, both of whom were natives of Ohio. The father was a farmer and passed away in 1866, while the mother had died some time before. Young Parmenter was thus left an orphan at the age of eight years and was thrown upon his own resources at an age when most children are playing with their toys and being under the watchful care of parents. The boy found employment on farms in the neighborhood and had but limited opportunity to acquire an education. He was a young man of twenty-one when he decided to take the advice of Horace Greeley and "go west." He

crossed the Missouri river on the first of January, 1880 and that same month took a homestead and tree claim in Custer county, Nebraska, becoming one of the first settlers of that county and resided here until 1888, when he sold out and again went west, this time to the Pacific coast where he spent some time travelling and looking over the country. Returning to Nebraska he found employment in a saw mill in Sioux county and in 1897, came to Scottsbluff county where he took up farming. In 1901, he took a homestead where he began to make permanent improvements and which has been the stage of successful operations as a farmer ever since. When the Gering irrigation ditch was projected Mr Parmenter established the first construction camp on the work, moving there with his family in the month of February during a severe snow storm. But he had become accustomed to hardships and such an experience was not new to him.

In Custer county, Nebraska, April 12, 1885, occurred the marriage of Albert M Parmenter and Miss Mary Predmore. She was born in Winnishiek county, Iowa, August 6, 1859, a daughter of John and Jane (Peters) Predmore, natives of Ohio. They came to Nebraska in 1881, and were among the first settlers of Custer county where they both spent the remainder of their lives.

Two children have come to bless the home of Mr and Mrs Parmenter. Lyda Jane, born June 23, 1886 is the wife of Arthur Clure, residing at Minatare and they have seven children Florence, Verl, Marion, Virgil, Lawrence, Iola and Ila, the second child John Arthur, married Etta Bartow and resides at Gering. They have two children Marie and Mabel.

Mr Parmenter has a well improved farm of one hundred and sixty acres, seventy acres of it being under irrigation and the place is devoted to general farming. Though over sixty years of age Mr Parmenter would easily be taken for a man much younger and is still active in the operation of the place. He is one of the substantial men of the community and takes an active interest in all things pertaining to the welfare of the neighborhood. The family are members of the Methodist church and in politics he is a Republican.

ORVAL SMITH, who is an experienced and successful farmer near Bayard, Nebraska, came to this state with his parents when seven years old, was reared, educated and married here and probably entertains for Nebraska the feelings of a native son He was born, however, in central Illinois, in old McLean county, June 4, 1880

The parents of Mr Smith were Stephen and Amelia (Wiley) Smith, the latter of whom was born in McLean county sixty years ago, and now a widow, living at Bayard The father of Mr Smith was born in Kentucky and before coming to Nebraska, had been a farmer in Iowa and Illinois In 1887 he brought his family to this state and settled in old Cheyenne county, now Morrill, where he secured a homestead of one hundred and sixty acres and pre-empted one hundred and sixty more At that time it was all dry land but the entire extent is at present under irrigation He was a very industrious man and believed in substantial improvement, therefore this land is very valuable His death occurred in 1917

Orval Smith remained at home, attended the public schools, and assisted his father as a general farmer At present he owns a one hundred acre farm one and a half miles from Bayard, carrying on diversified farming, and owns four unimproved lots In Morrill county, in 1905, Mr Smith was united in marriage to Miss Viola Roberts, who was born in Iowa and is a daughter of Wilburn and Nancy (Duncan) Roberts Her parents were natives of Kentucky who moved to Iowa and from there, in the spring of 1887 came to Morrill county and homesteaded one hundred and sixty acres The father carried on general farming until his death The mother of Mrs Smith lives at Bayard Mr and Mrs Smith have had five children, namely Dora, Violet, Lyle, Louis and Orval, all of whom survive except Orval

GEORGE DE GRAW, one of the substantial farmers of Morrill county, has great reason to be satisfied with the land investment he made here in 1902, for he has a beautiful farm of one hundred and sixty acres, well improved and exceedingly productive Mr DeGraw developed his farm from wild prairie and the result is very creditable to his industry and good judgment

George DeGraw was born at St Paul, Minnesota, November 5 1863, a son of Frederick and Elizabeth (Todd) DeGraw His father, of French Canadian ancestry, was born in Canada, June 13, 1838, and was eighteen years old when he came to the United States He served as a soldier in the Civil War and later became a general farmer in Minnesota For some years he has lived comfortably retired in Wyoming Mr DeGraw's mother was born in Minnesoa and died when he was quite young He grew up on the home farm and attended school, remaining in his native state until 1887, when he came to Nebraska and secured a homestead in Cheyenne county, near Dalton, where he remained for six years, moving then to Sidney and buying a ranch in that vicinity In 1902, he came to Morrill county with the intention of buying land if he found a satisfactory tract, with the result that he became the owner of his present farm, situated on section 12-20-51

At Sidney, Nebraska, December 7, 1897, Mr DeGraw was united in marriage to Miss Stella Wymer, who was born in Minnesota, February 18, 1877 Her parent were Joseph and Anna (Havens) Wymer, the latter of whom lives at Gering, Nebraska Mrs DeGraw's father was of Pennsylvania Dutch stock and was born in Pennsylvania He was a general farmer and lived to the age of eighty-seven years The following children have been born to Mr and Mrs DeGraw: Fred, who is the home farmer, Hannah, the wife of Vern Dexter, of Gering, and Iris. May, Nellie, George, Eugene, Eunice, Lora, Pearl, Andrew and Alice, a sturdy family of which any parents or community may well be proud

COLE HUNT, whose thorough farm methods and general enterprise are making the old Hunt homestead one of the best farm properties in Morrill county, was born in Alliance, Nebraska, April 13, 1895, and has spent his life in his native state With the good judgment that marks many young men in modern days, he has chosen agruculture as his life work and is devoting his best energies to the further development of the excellent property left by his father

The parents of Mr Hunt were John and Lillie (Gilmore) Hunt, the former born in Ohio, June 9, 1848, and the latter in York county, Nebraska, June 3, 1864, the first white child born in Nebraska Territory, Class A Territorial Association In 1875, John Hunt came to eastern Nebraska where he bought land and followed farming for some years, then moved to Box Butte county and homesteaded and the family lived there for eight years In 1898, Mr Hunt saw what he considered better opportunities in Morrill county, came here and bought two hundred and forty acres of land which, at that time, were entirely unimproved He continued the practical development of his property until the close of his life He was widely known and highly respected To John Hunt and his wife the fol-

GEORGE L. WILCOX

lowing children were born Eva, the wife of Elmer Bennett, of Minneapolis, Minnesota, Lillie, the wife of Alexander Underwood, of Box Butte county, George, who resides at Spokane, Washington, Susie, the wife of Arthur Jones, of Grand Island, Nebraska, John E, who was a soldier in the American Expeditionary Force in Europe during the World War and was in the army of occupation in Germany until September 11, 1919 Omer, who is deceased, Cole, who operates the home farm, Nellie, the wife of Leslie Allen, a farmer in Morrill county, and Nettie, deceased The mother of the family still resides on the homestead She is a member of the Methodist Episcopal church

Cole Hunt took charge of the farm for his mother soon after completing his education He is a level-headed, serious-minded young man, a hard worker and close thinker To him farming is not merely an occupation but a business that is deserving of a man's best efforts He has been very successful in his work here, has all the land irrigated, makes use of the best modern machinery and makes every acre return a profit At present he is not particularly active in politics

GEORGE L WILCOX, who has many substantial interests in Scottsbluff county, came to Nebraska in 1889 and has ever since lived in the state and is favorably known in the several sections where he has resided Mr Wilcox is a self-made man and feels that much of his success may be attributed to the business opportunities he found so generously offered in Nebraska

George L Wilcox was born in Franklin county, Ohio, February 23, 1879, one of two children and the only son born to Charles and Samantha (Freeman) Wilcox Both parents were born in Ohio The father died in that state in 1882, at the age of thirty-two years, and the mother lives at Los Angeles, California Mr Wilcox has one sister, Mrs R B Farris, who is a resident of Chicago

Following his graduation in 1896, from the high school of Worthington, Ohio, in which city his father had been a merchant George L Wilcox spent some years there identified with the same line of business In 1889 he came to Nebraska and located at Springfield, where he was employed for some years as a clerk in a store, his previous experience in a mercantile house making his services valuable On leaving Springfield, he went to Gretna, where he embarked in the furniture business and conducted a furniture store until November, 1911, when he came to Scottsbluff Here he resumed dealing in furniture, adding a hardware department In 1916 he built a commodious two-story building for business purposes, moved in, enlarged the scope of his business with a complete stock of furniture, rugs, hardware and undertaking supplies and now has a first-class, up-to-date establishment that is a credit to himself and the city His time, however, is not entirely given to the affairs of Wilcox & Company, as the firm name stands for he also manages his large ranch and raises cattle He is a stockholder in the Platte Valley State Bank

In 1902 Mr Wilcox was united in marriage to Miss Sarah McCarley, of Illinois, and they have four children Helen, Gertrude, Georgia, and Charles The three daughters are attending school but the son is yet an infant, his birth having taken place March 26, 1919 Mr Wilcox's mother belongs to the Christian Science church, but he and family are members of the Methodist Episcopal church He has always taken due interest in politics as a part of intelligent citizenship, and, like his father before him, has always been a Republican The only fraternal organization with which Mr Wilcox is identified is the Knights of Pythias, in which he is quite prominent and has been chancellor and grand vice chancellor Mr Wilcox is a representative business man of Scottsbluff and more than that for he is an earnest, unselfish citizen one who is ever ready to cooperate in all movements for the general welfare, and as such enjoys universal respect

DALE B OSBORNE, whose extensive agricultural interests and activities make him well known in Morrill county, has spent almost all his busy life here, being only thirteen years of age when he accompanied his parents to this county

Dale B Osborne was born in McLean county, Illinois, November 2, 1874 His parents were Samuel H and Emily (Benson) Osborne, the former was born in Ohio and the latter in Illinois They had four children born to their marriage Dale B, who has always lived on the old homestead since coming to Nebraska, Thomas C, who is a farmer in Morrill county, north of Bayard, Dean H, who was in training in the aviation department as a machinist, during some months of the World War but never was called overseas, and June, who died at the age of twenty-seven years In 1887, the family came to Nebraska and the father homesteaded in Morrill county He became a man of wealth through farming and stock-raising, and also influential in the county's public affairs, serving for years as a justice of the peace, and in many ways assisted in the county's substantial development. He died at the age of seventy-six years and left behind him the record of an honorable and useful life. The death of the mother occurred in her sixty-fourth year, in 1917.

Dale B Osborne had educational advantages in the public schools. From youth he has been interested in farm pursuits and resides on the estate of four hundred and seventy acres left by his father, carrying on both general farming and cattle raising. He has two hundred and forty acres under cultivation. He is a farmer of modern type, uses scientific methods, and believes it is the best all round policy to use every labor saving device and the best of farm machinery possible to secure. He is numbered with the successful agriculturists of the country.

Mr Osborne was married at Minatare, Nebraska, Sptember 6, 1917, to Miss Zora Guvain, who was born at Prescott, Iowa. Her parents are August G and Mary (Witkowski) Guvain. Mr and Mrs Osborne have one son, Harold. In politics Mr Osborne is a sturdy Republican but has never sought any public office, belonging to that quiet, intelligent, thoughtful body of men who believe that general good citizenship brings greater rewards than political position.

JAMES N THOMPSON, who has met with a large degree of success as a farmer and stock-raiser in Morrill county, has been engaged in agricultural pursuits all his life. He was born at Shelbyville, Shelby county, Indiana, January 3, 1874, a son of George and Louisa (Heath) Thompson. The father was born in Indiana and the mother in Ohio. Her death occurred when she was fifty years old. The father engaged in general farming during his active years, but retired and lived in California until August, 1919, when he returned to Omaha, and in June, 1920 came to Morrill county, where he died November, 1920.

James N Thompson was reared in his parents' comfortable home in Shelby county and obtained his education in the public schools. Believing that land is the real source of wealth, he started out for himself with the determination to acquire it, locating first in Gentry county, Missouri, where he engaged in general farming until 1909, when he came to Nebraska and secured a homestead of two hundred acres in old Cheyenne, now Morrill county. There were many hardships to endure before Mr Thompson found himself in easy circumstances, but through cheerful courage and industry they were overcome. He im-

proved his first tract of land and has added until he now owns three hundred and sixty acres, ninety-six acres being irrigated and surprisingly productive and the rest fine grazing land. He does an extensive business in raising stock.

On September 24, 1902, Mr Thompson was married to Miss Emma Harmon, who was born in Missouri, October 2, 1867, a daughter of Peter M and Caroline (Coy) Harmon. Her father was born in Tennessee, where he was a general farmer before moving to a farm near St Joseph, Missouri, where his death occurred at the age of sixty-four years. Mrs Thompson's mother was born in Missouri and still lives in St Joseph. Mr and Mrs Thompson have two children, a son and a daughter. Harmon, who was born July 3, 1904, and Mildred, who was born October 19, 1909, both of whom are being given excellent educational advantages. Mr Thompson and family belong to the Presbyterian church. He has never felt inclined to take a very active part in politics but keeps well informed on all public questions. He is a man of honest motives and sterling character.

CLARENCE V McRAE — It is not always the amount of land owned that makes a farmer of Morrill county, financially independent, but rather its location, its intelligent tillage and improvement and its development in the way of irrigation. The eighty acre farm of C V McRae, situated on section fifteen, town twenty-one, is an example. Mr McRae has made the excellent improvements on the land himself and has fulfilled other conditions and now has one of the most productive farms of the county, where, as a citizen, he is held in high esteem.

Mr McRae was born in Doniphan county, Kansas, December 17, 1880. His parents were Alexander and May (Bender) McRae, the latter of whom was born in Missouri and now resides in Washington county, Kansas. Mr McRae's father was born in Toronto, Canada, came from there to the United state in 1860, and settled in Doniphan county Kansas. He was a general farmer and died in Washington county, Kansas, at the age of sixty-two years. Of his seven children, Clarence V is the youngest surviving.

Mr McRae had school advantages in his native state and was reared on his father's farm in Washington county. He continued a farmer in Kansas until 1907, when he came to Nebraska and for three years engaged in general farming in the eastern part of the state. In 1910, he came to Morrill county and homesteaded, and has prospered ever since through industry and good management. He has always been a hard worker and his steady, earnest efforts have brought deserved reward

In Washington county Kansas, Mr McRae was married to Miss Grace Grout, who was born there July 6, 1881. She is a daughter of Elmer and Mary (McAlister) Grout. The father was born in the state of New York and was a successful farmer in Washington county prior to his death. The mother was born in central Indiana and still lives in Kansas. Mr and Mrs McRae have the following children Pearl, Lester, Opal, Genevieve, Luella, Mary Jane and Amy, a fine, representative, Morrill county family. The Methodist church holds the family membership. An independent voter, Mr McRae gives his political support to candidates of whom his judgment approves, and always may be found heartily co-operating with other good citizens for the general welfare

ELWIN M SPENCER, whose well improved farm of seven hundred and twenty acres is favorably situated in Morrill county, is not one of the old settlers of this section but is a native of Nebraska, born at Wymore in Gage county. He is a son of William Isaac and Ida Bell (Henry) Spencer, who came from Fremont county, Iowa, to Nebraska in 1869. They settled first in Johnson county but later moved to Gage county

Elwin M Spencer was reared and educated in the neighborhood of Wymore, Nebraska. He was brought up on a farm and has always followed agricultural pursuits. He started for himself in Gage county and remained there for sixteen years, at the end of that time removing to Kansas, in which state he continued farming for ten years. In 1913, Mr Spencer came back to his native state and bought six hundred acres of land to which he has added until he now has seven hundred and twenty acres, all ditched. He has developed this land from the sod, has placed substantial improvements here and is prospering as a general farmer and extensive feeder of stock

On August 28, 1910, in Republic county, Kansas, Mr Spencer was married to Miss Ethel C Bonner, who was born in Kansas They have three children Lucile B, Otho D and Paul J, all of whom are at home, the older children attending school. Mr and Mrs Spencer are members of the English Lutheran church. He is one of the directors of the Farmers Irrigation Canal project. In politics he votes independently according to his own judgment, and fraternally is identified

with the Odd Fellows and Woodmen of the World. He is one of the county's representative men and good citizens

S W DANIELS, deceased, was a resident of Morrill county for more than twenty years. He stood well with all who knew him. He came here an experienced farmer, homesteaded on section twenty-six, greatly improved his property and was numbered with the substantial and dependable men of the county. He was born in Pennsylvania July 30 1854

Mr Daniels obtained his education in the public schools and grew up on a farm. When he left Pennsylvania he went to Iowa as a farmer later to Kansas and before coming to Nebraska, was a resident of Michigan. In February, 1900, he homesteaded where he lived and had one hundred and sixty acres of well developed ditched land. He carried on general farming and raised pure bred Shorthorn cattle. While in Iowa Mr Daniels was married to Miss Rhoda Eckley who was born in that state. At her death Mrs Daniels left three children Leo, who is a farmer in Morrill county, John, deceased, Mrs Sadie Eckhart, who lives at Edison, Nebraska, and Mrs Lyda Smith, who lives in Morrill county. In 1913, Mr Daniels was married a second time to Miss Ada Van De Venter. Mrs Daniels is a member of the Presbyterian church. In political life Mr Daniels was a Republican. Mr Daniels died January 30 1920 leaving a sorrowing family and friends. His funeral was held at the Presbyterian church, and he was laid to rest in the Bayard cemetery after a long and useful life

Leo Daniels, the eldest son of S W Daniels, was born in Menominee county, Michigan, August 7, 1884. He was educated in the public schools and remained with his father, accompanying him to Morrill county in 1900. In 1910 he homesteaded for himself and now has a very good property consisting of one hundred and sixty acres, all ditched. He carries on general farming and takes much interest in his Shorthorn cattle, all thoroughbred. He annually feeds many hogs and raises horses and poultry for home use. He is considered one of the very successful young farmers of this section of the county

Mr Daniels was married March 11, 1918, to Miss Josephine Hoag, who was born near Blue Hill, Nebraska. Mr Daniels has substantially improved his farm and they have very pleasant and comfortable surroundings. Like his father Mr Daniels is a Republican in politics

HARTSON A MARK has been progressive, both as a citizen and official in Garden county, and is most ably serving as county surveyor, with headquarters at Oshkosh

Special interest attaches to the career of Mr Mark by reason of his being a native of Nebraska and a member of one of the pioneer families He was born at Belvidere, Thayer county, on June 17, 1875, a son of David and Delilah H (Durfee) Mark, who were born and reared in the state of New York, where their marriage was solemnized and whence they came to the west about 1870 The father filed entry upon a homestead in Thayer county, Nebraska, and in due time proved up on the property and developed it He was one of the representative farmers and valued citizens of that county until he came to Scottsbluff county and established his residence at Gering, where he passed the remainder of his life and died at the age of seventy-three years, his widow now residing at Mitchell

Hartson A Mark was educated in the public schools of his native county, including the high school at Belvidere When about eighteen years of age he rented land and began to farm in Thayer county, where he continued until he became of age He then, in 1896, after having previously been employed in a blacksmith and general repair shop at Belvidere, came to the Panhandle country and settled at Gering, and became associated with his brother, G E Mark, who without any previous experience, had purchased the *Nebraska Homestead* at Gering and embarked in the printing business In the autumn of 1897, he, then having suffered from typhoid fever which left him too weak to continue the printing business, became apprenticed in a photographic studio at Gering, and finally purchased the establishment, which he ran until 1901 He then purchased a similar studio at Alliance, but in the following year he sold out, after which he was for a time manager of a photographic studio at Lead, South Dakota Upon his return to Nebraska Mr Mark located at Lincoln, where he was similarly engaged until the spring of 1905, in charge of the engraving department of the Cornell Photo Engraving Company During this time he devoted himself to the study of concrete or cement engineering, and by 1905, had become well informed in the practical and scientific details of this industry, and established himself in business, as a contractor in concrete construction work, at Mitchell, Scottsbluff county He erected the first concrete business block in the North Platte valley, at Mitchell Mr Mark

continued his contracting business until 1907, when he established himself on a reclamation homestead near Morrill, Scottsbluff county. In the meanwhile he did a considerable amount of concrete contracting work In 1907, he became one of the associate editors of the *Nebraska Farmer* In 1911 he perfected his title to his reclamation claim and in the same year he rehabilitated the Gering Canal and did important work in the surveying of the Alliance irrigation canal, near Bayard, Morrill county In the meanwhile his health had become somewhat impaired and in the spring of 1912, he found opportunity for gaining less strenuous open-air occupation, by going out on the land-office retracement survey in Thomas and Grant counties His health compelled him to abandon this work In 1912 he founded and became editor and publisher of the *Hammer*, "The Builders' Tool, not a Brick-bat," a semi-monthly paper which was issued at Morrill and later moved to Gering and which gave special attention to shedding light on the reclamation service In 1913, Mr Mark went to Indianapolis, Indiana, where he accepted the position of efficiency engineer in the office of *Up-To-Date Farming*, a farm paper and the official publication of the Farmers' Society of Equity Later he resigned to accept the position of engineer for the National Concrete Company, of Indianapolis He retained this place until physicians advised him to leave Indiana, on account of the condition of his health He then entered the service of the Illinois Bridge Company, as foreman of construction work He was assigned to work on an important contract at St Joseph, Michigan, and there, in August, 1913, he met with a serious accident, which disabled him and which led to his return to his home, at Gering, Nebraska

In October, 1913, Mr Mark established his residence at Oshkosh and after a few months in the office of the *Oshkosh Herald*, was empoyed to edit a magazine — *Production* — they were about to establish This new publication being delayed, he taught one term of district school in this locality He was then appointed deputy county surveyor, in June, 1914, and served until the following autumn, when he was elected county surveyor The best evidence of his efficiency in this office was shown in his re-election in 1916 and again in 1918, and holds the post to the present time In 1918, he became the owner of a tract of four acres adjoining Oshkosh, and has made good improvements there

In politics Mr Mark is a stalwart advocate

William H. Wright

of the principles of Democracy, and has served as secretary of its county central committee in Garden county During the progress of the World War Mr Mark gave every possible assistance in connection with the various national campaigns projected for the aiding of the government and its military and naval forces, his services as a public speaker having been enlisted as one of the "four-minute men" He was made chairman at Oshkosh and acting county chairman He was a volunteer reserve of Garden county during the war, being also chairman of the Smilage committee, secretary of the county council of defense, associate member of the legal advisory board, chairman of the civilian relief committee for the Oshkosh A R C, and vice-chairman Oshkosh A R C and public-service representative of the government's employment bureau The governor of Nebraska advised him to remain on the local field, where he could work to better advantage in supporting the nation's war policies than he could by enlisting for military service Mr Mark is descended from eight ancestors who served in the war of the Revolution The lineage of the Mark family traces back to staunch Scotch origin and the original American representative or representatives were numbered among the early settlers at Jamestown, Virginia, in 1666 The American founder of the Durfee family the maternal line of the subject of this review, was Thomas Durfee, who emigrated to this country from England, about the year 1665

October 12, 1905, the subject of this sketch was united in marriage to Miss Frances E Twiford, a lady of exceptional musical ability and assistant in the School of Expression of the Wesleyan University at University Place, Nebraska Possessed of an impulsive carefree disposition, so often found in musical prodigies looking forward with great enthusiasm to a life on one of Uncle Sam's reclamation homesteads, the stern realities of homestead life soon changed optimism to pessimism It was with difficulty that Mr Mark induced her to continue as a homestead resident until residence proof could be made Following the depressing experience of extreme illness and prolonged disability, in the spring of 1913, the convalescent husband went to a lower altitude, only to return that fall a disabled man on crutches This was too much for the sensitive wife She declared marriage was a failure and that she had no further use for a husband Finding the lady could not be changed from this attitude, Mr Mark disclaimed all interest in family property, both real and personal and

began life anew Following the statutory two-year limitation, Mrs Mark applied for and received an absolute decree of divorce They have two children, Harriet T and Arc Durfee At the age of thirteen Mr Mark made a profession of Christianity and at present is a member of St Mark's Lutheran church of Oshkosh

WILLIAM H WRIGHT (deceased), a pioneer of Scottsbluff city and county, was born at Whitehall, New York July 23, 1834 He spent his boyhood years at home upon the farm where he was born, and was educated in the public schools, and later at Oberlin college, Oberlin Ohio

On November 13 1857, he was married to Ellen J Clark, who was born at White Hall N Y March 31, 1839 Miss Clark was also educated at Oberlin college and was residing there at the time of her marriage Mr and Mrs Wright returned to Whitehall and for a number of years operated the old Wright farmstead In 1870 they came west and bought a farm in Mills county, Iowa, and this farm they still owned at the time Mr Wright was called home It was near old Pacific City For a time he maintained a real estate office in Glenwood, the county seat and Mrs Wright and the sons operated the farm

In 1890 they moved to Weeping Water, Nebraska, and here he left the family for a short time and came up to the Scottsbluff country The family followed the subsequent spring and they bought a small farm near where the city of Scottsbluff now stands Alliance fifty-five miles away, was then the nearest railroad town

To Mr and Mrs Wright there were born seven children Carlton C was legal adviser of the North Western system in Nebraska for several years He was elected city attorney of the city of Omaha which position he filled with exceptional ability and honor Then he became the general counsel for the North Western system and moved to Evanston, Illinois, with offices in Chicago This place he held until his death in February 1918 (2) Catherine the eldest daughter, was married to Rev F T McCollum a Congregational minister of Chicago She passed away at the age of forty-eight years (3) Fred A is an attorney and resides at Scottsbluff where in county and city affairs he has held many positions of honor, and has always been active in public affairs (4) Chas W was a son who died at the age of sixteen years (5) Flavel L spent many years of his life in this vicinity and was known over a wide range of territory as an auctioneer of exceptional ability He was active in public affairs and upon the right side on public questions. His ability attracted the attention of the North Western Life Insurance Company of Milwaukee, Wisconsin, and he now resides at Harrisburg, Pennsylvania, where he went to take charge of the company's business in the Keystone state. (6) Cullen N. is vice president of the Platte Valley State Bank, at Scottsbluff, and is president and resident manager of the Tristate Land Company, which company has large irrigation interests as well as other financial connections with a large part of the valley.

Mr. and Mrs. William H. Wright were Congregationalists, but there being no church here when the town was organized, they were among the eight charter members of the First Presbyterian church at this place, and they assisted in the building of the first stockade shack in which the first church services were held in Scottsbluffs city.

The vision of William H. Wright has been demonstrated. He came here and saw the magnificent stretches of land, and when the lean years came he sought to amalgamate the interests at home and finance affairs from the east. The panic of the early nineties checked for a time the development, and wrecked the first corporation, but with the revival of business, other and new men of affairs took hold of it. Mr. Wright took an active part therein, and every man who invested in the first enterprise was repaid with interest and profit. He knew that the new enterprise was to succeed, he saw into the future, a little of the wonderful city and empire about it, but he did not live to see the water actually running in the great canal, to which he had devoted so much time and energy. To him, more than any one person, is that canal a monument of ability. He died May 10, 1906, and rests in Fairview, on a beautiful plot irrigated from the Tristate canal.

Mrs. Wright is a woman of great energy and intelligence. She has been foremost in religious, educational and civic work. With a number of friends, she organized the first woman's club in the city, which was named for a very dear friend at Weeping Water, and still is active as the Laura M. Woodford club. The first lecture course put on in this city was through the instrumentality of this club. Although eighty years of age she retains her health, strength and mental vigor, and walks with the brisk step of a woman of half her years. Her contribution to the community will endure for many years to come.

GEORGE H. MORRIS, M. D., is a representative physician and surgeon upholding high standards and professional prestige in the Nebraska Panhandle.

Dr. Morris was born in Des Moines county, Iowa, September 4, 1866, a son of William H. E. and Mary A. (Yates) Morris, the former a native of Ohio and the latter of Kentucky. The parents became pioneer settlers in the Hawkeye state, where the father developed a farm and worked as a carpenter. He removed from Iowa to Carthage, Illinois, and in that state he passed the remainder of his life, his death having occurred in the city of Peoria, when he was about seventy years of age. He was one of the gallant representatives of Iowa in the Civil War, serving three years as a member of Company K, Second Iowa Cavalry. His widow was a resident of Hancock county, Illinois, at the time of her death, at the age of eighty years.

Dr. Morris was still an infant when his parents removed from Iowa to Hancock county, Illinois. He passed his boyhood years on the home farm and attended the public schools, including the high school at Carthage, Illinois. As a young man he was in the government service in connection with the U. S. lighthouse system of the Mississippi river. In this and other positions which he held, he carefully conserved his earnings, with a definite ambition to declare himself for the medicine profession. Thus through his own energies he provided the means which enabled him to complete the course in the College of Physicians and Surgeons at Keokuk, Iowa, where he graduated with the class of 1907. After receiving his degree of Doctor of Medicine he engaged in the practice at Okmulgee, Oklahoma, but within a short time came to Nebraska. He located near North Platte, Lincoln county, to engage in professional activities until 1911, when he moved to Oshkosh, Garden county, where he has since built up a large and substantial practice that clearly places him among the representative physicians and surgeons of this favored section of Nebraska. He has kept abreast of the advances made in his profession, and in 1908, completed an effective post-graduate course in the Chicago Polyclinic, and the preceding year took a graduate course in the medical department of the University of St. Louis, while in 1915 he graduated at the Chicago Polyclinic. Prior to his graduation in medicine the doctor had completed a course and been graduated at the Keokuk College of Pharmacy in 1906, and the next year served as a member of the faculty of that institution. He is an active member of the Lincoln County Medical Society, the Nebraska State Medical Society and the American Medical Association.

Dr Morris is a Republican and served about two years as chairman of the Republican central committee of Garden county, besides which he has given effective service as chairman of the Oshkosh Board of Health, and was examining physician on the local exemption board when the nation was recruiting for the army and navy in connection with the World War He is affiliated with the Masonic fraternity, the Odd Fellows and the Modern Woodmen of America, and he and his wife are members of the Methodist and Presbyerian churches respectively

DAVID F FICKES, who has given able service in the office of assessor of Garden county since 1915, became a resident of this section prior to the creation of Garden county, and he was one of the first men to be elected commissioner of the new county He has been closely identified with the industrial and civic growth and development of the county and his present official position gives him knowledge of values in the county, so that he is able to speak with emphasis and assurance of the wonderful advances that have been made within the period of his residence in this section of the state

Mr Fickes claims the Keystone state as the place of his nativity His parents, Isaac L and Margaret (Weyandt) Fickes passed their entire lives in that fine old commonwealth David Fickles was born in Bedford county, Pennsylvania, January 11, 1858, and after completing the public schools he pursued a higher course of study in a normal school in Pennsylvania He put his acquirements to practical application for a time as a teacher in the schools of his native state Thereafter he was identified with railroad construction work in the state of New York and later in eastern Canada, and later engaged in the operating of a steam shovel in the state of Tennessee, in connection with the work in the ore mines, where he worked for five years In the autumn of 1892, Mr Fickes came to Nebraska and became a pioneer settler in what is now Garden county Here he took up a homestead, which adjoins the corporate limits of Oshkosh, which he still owns He reclaimed and developed the land, made good improvements on the place, and devoted himself to diversified agricultural operations and the raising of live stock He now rents the farm to a reliable tenant Mr Fickes purchased an adjoining tract, so that his landed property in the county now comprises three hundred and twenty acres

Mr Fickles is a Republican and has been

influential in public affairs of a local order during the period of his resident in Nebraska He served one of the first county commissioners of Garden county, and in 1915, he was appointed county assessor to fill a vacancy The following year he was elected to this office, his mature judgment and accurate information making him a most satisfactory arbiter of real estate value He is one of the pioneer members of the Masonic fraternity in Garden county and is a citizen who commands unqualified popular confidence and esteem His wife is a member of the Lutheran church

In Ohio, in 1882, Mr Fickles married Miss Irene Fought who was born and reared in that state and the two children of this union are Howard and Orvin Howard was married April, 1910, to Daisy Pueschel of Columbus, Nebraska, and they have two children, David Vada, both at home He is successfully engaged in the hardware business at Oshkosh, and Orvin, is at Pueblo, Colorado He was one of the gallant young men who gave to the nation loyal service on the European battle fields during the World War He was a private in the Three Hundred and Fifty-fifth Regiment, Eighty-ninth Division of the American Expeditionary Forces in France, and while with this command he was wounded by a fragment of shell, in the battle of Argonne, November 3, 1918 He returned to his native land after the signing of the armistice

EDWARD S WOOD, who, in the winter of 1919, was treasurer of Garden county, needs no further voucher for his hold upon the confidence and good will of the people of his constituent county He is one of the progressive citizens who are proving alert and resourceful in furthering the general welfare of Garden county, where he is the owner of a valuable tract of land, besides having an attractive home at Oshkosh

Mr Wood was born at Sutton, Clay county, Nebraska on July 15, 1880, a son of Samuel and Gertrude R (Reeder) Wood The father was born and reared in Ireland and about the year 1859, came to the United States and found employment on a farm in the state of Illinois By his own efforts he provided the means for securing the ministry of the Methodist Episcopal church He was in the very flower of his strong and useful manhood when he came to Nebraska and here he rendered service in the ministry, serving as pastor in turn of the Methodist churches at Sutton, Central City and Tecumseh He retained his charge in the last mentioned places until his death when he was about forty years of age His widow, who now resides at York, was born in Ohio. where she received her earlier education. which was supplemented by further study after the removal of the family to Illinois. in which state her marriage was solemnized In the public schools of Nebraska Edward S Wood continued his studies until he completed a course in the high school at York, then was employed three years as clerk in a grocery store In 1900, he came to the Panhandle of Nebraska and established his residence in what is now Garden county During the ensuing winter he was employed on a ranch, and he then entered the service of John Orr, one of the large landholders and extensive cattle raisers of this locality Mr Wood continued his association with Mr Orr for five years, and thereafter he was in the employ of the Western Land & Cattle Company, and held the position of foreman the greater part of the time He gained broad and practical experience in connection with the cattle industry and is still manager of the company This concern has about five thousand acres of deeded land, used for the grazing of herds of cattle, besides which the company has four head of buffalo Mr Wood continued the management of the business of this corporation until he was made a Republican candidate for the office of county treasurer, a post to which he was elected by a majority that was gratifying, in view of the fact that Garden county normally gives substantial Democratic majorities

In addition to being a staunch and active supporter of the cause of the Republican party and an influential figure in its local councils, Mr Wood maintains affiliation with the Modern Woodmen of America, the Odd Fellows and Masonic orders In Garden county he is the owner of eleven hundred and fifty acres of excellent grazing land, which he rented upon assuming his present county office Mrs Wood is a member of the Methodist church

On June 9, 1909, was solemnized the marriage of Mr Wood to Miss Nora Beaver, of Des Moines, Iowa, and their son Orien is nine years old

REUBEN LISCO, is a sterling pioneer of the Panhandle of Nebraska, and has exerted influence in connection with civic, business and industrial affairs He came to the Panhandle country nearly forty years ago, when this section of the state was practically nothing but unbroken prairie land, with the raising and feeding of cattle as its sole productive industry He gained experience in herding cattle on the great open range, he was long and prominently identified with the cattle business when operations were conducted upon an extensive scale, he has witnessed and aided in the march of development and progress and he is today one of the leading citizens of Garden county, where he is president of the Lisco State Bank, in a town that was named in his honor He is direct and unassuming in his varied activities and is essentially one of the representative business men of the great country to which this history is dedicated

Mr Lisco claims the Hawkeye state as the place of his nativity and is a representative of one of its early pioneer families He was born in Worth county, Iowa, November 20, 1858, a son of William and Martha (Shields) Lisco, both of whom were born and reared in Ireland, where they were married and then came to America, imbued with a determination to find better opportunities for the winning of independence and prosperity They first settled in one of the eastern provinces of Canada before becoming pioneer of Iowa, where the father reclaimed a farm and became a successful agriculturist He lived to be eighty-four years old and his wife was about fifty-three years of age at the time of her death, in 1881 Of their eight children six are living The family came to Nebraska in 1873, and settled near Columbus, Platte county, where the parents spent the remainder of their lives, the father having secured land and developed a pioneer farm

Reuben Lisco attended the public schools of his native state and was a lad of about fourteen years at the time the family removed to Nebraska, his broader education having been here, gained principally in the school of practical work and experience He remained at home until of age and a year later, in 1881, he made his way to the Panhandle of Nebraska He found free and vital life in this great cattle district of western Nebraska, for he was employed as a cowboy by what is now the Rush Creek Land & Stock Company, of which corporation he is president, a statement which implies that he has been continuously associated with the cattle business here during the intervening period of nearly two score years When he began life as a cowboy the place on which he was employed was known as the Club Ranch, and his reliability and energy enabled him to make such advancement that when the operating company was reorganized he was given the position of manager In the meanwhile he accumulated considerable land

and owned a goodly number of cattle, both land and stock being taken over by the company of which he became a member of its reorganization, in 1897, and he has been the executive head of this corporation since 1910 Upon the organization of Deuel county in 1888, Mr Lisco was chosen the first sheriff of the new county, a position in which he served nine years In 1909, when the Union Pacific Railroad built through this section of the state, Mr Lisco laid out the town of Lisco, which was named for him and which now has a population of about two hundred inhabitants To foster the interests of the new town which he thus founded, Mr Lisco established a well equipped general store, giving it general supervision for some time, besides which he became the prime mover in the organization of the Lisco State Bank, of which he has been president from the time of its incorporation and which has become one of the substantial and popular financial institutions of the county Mr Lisco lives in the town named for him, but he is also the principal stockholder in the Oshkosh State Bank, which was organized in 1917, and he has been its president since that time, the general management of the institution devolving upon him, — a fact that adds to its popularity, as he is known as a substantial capitalist and able executive A substantial brick building, two hundred and forty-four by seventy feet in dimensions and two stories in height was erected an dis the property of the bank

Mr Lisco has had neither time nor inclination for political activities, but is a staunch Republican and a liberal and public-spirited citizen He is affiliated with the Masonic fraternity, the Rush Creek Land & Stock Company, of which he is president and manager, has a ranch property comprising about forty thousand acres, and an average run of ten thousand cattle Mr Lisco has won success through his own ability and well directed efforts He has kept pace with the general development and growth of the Panhandle country and has been big enough at all times to do big things and do them well

The year 1895, Mr Lisco married Miss Addie R Miller, who was born and reared in Michigan, near Kalamazoo, and she is the popular chatelaine of their pleasant home Mrs Lisco is a member of the Presbyterian church

FRED A WRIGHT, one of the most prominent and best known attorneys throughout western Nebraska and eastern Wyoming, is the dean of the Scottsbluff county bar, having been longest in continuous practice here of all the lawyers of the county He was born in Washington county, New York, February 11, 1869, the son of William H and Ellen J (Clark) Wright, of whom mention is found on other pages of this work Removing to Iowa in early life with his parents, he was reared on a farm near Pacific Junction, Mills county, and attended the public schools, high school at Glenwood, and college at Tabor, Iowa While in Glenwood he read law, and after leaving college he came to Scottsbluff county in 1892 and took a homestead three miles east of the present city of Scottsbluff, on which he later proved up

Being admitted to the bar in this county Mr Wright began practice in Gering in 1894 and has been continuously in the practice of law in Scottsbluff county since that time, having moved his office from Gering to Scottsbluff in 1906 He is the senior member of the legal firm of Wright, Mothersead & York, of Scottsbluff

Mr Wright has held a number of public offices He was postmaster at Gering during the second Cleveland administration, was elected county attorney in 1898, and was a member and president of the board of education in Scottsbluff for several years At the primaries for the nomination of delegates to the Constitutional Convention of Nebraska in 1910 he was nominated with the highest vote of all the candidates in the district comprising Scottsbluff and Morrill counties He has also been vicepresident of the Nebraska State Bar Association and president of the Western Nebraska Bar Association He is a Democrat in politics, is a member of Robert W Furnas chapter of the Masonic order a Knight Templar, and a member of the Elks

On October 29 1896, he was married to Miss Elizabeth Royer, who was a native of Nebraska and was educated at Weeping Water Elmwood, and Lincoln, Nebraska, and was a teacher in this state Six children have been born to them

1 Charles R, who enlisted in the service of his country as a volunteer in February 1918, while a student in the University of Nebraska, having completed two years of his course in college He was trained in the coast artillery in Florida and embarked for overseas service with the rank of sergeant in Unit A, S A R D, in September, 1918 He was taken ill with pneumonia while on the ocean and died at an American army hospital in Brest France, October 11 1918 There was no finer, manlier, or more promising young man among all the great host of American volunteer soldiers, and

he lived long enough to be a credit and an honor to his father and his mother and died that others might live in liberty.

2. Floyd E., the second son, also left the State University to enlist in the army. He was drill sergeant at Fort Rosecrans, California, and after the signing of the armistice with Germany was discharged and returned home to resume his college course.

3. Dorothy and (4) William H. are students in the University of Nebraska, and (5) Elizabeth and (6) Flavel A, are at home.

Mr. Wright has reached the enviable position of having so extensive a practice in his profession that it taxes his energies to take care of it, but he enjoys also the much rarer distinction of a universal reputation for absolute probity in his professional and personal life and devotion to the highest ideals of legal ethics. He is known as one of the half dozen most brilliant trial lawyers of the state, and is retained as attorney for practically all the important commercial interests of this section, among others the C., B. & Q. Railroad Co., Tri-State Land Co., Great Western Sugar Co., and Lincoln Land Co. He owns a beautiful home in Scottsbluff and a valuable irrigated farm adjoining the city. Mrs. Wright is a member of the Presbyterian church and is prominent in the social, educational and club life of the community.

JOHN ROBINSON, who is now living semi-retired in Oshkosh, has been a resident of Nebraska for more than half a century and he has gained his full share of distinction during the period of his residence in what is now Garden county. He was one of the representative pioneers and cattle men of this county, where he settled when it was still a part of Deuel county, and he has been a prominent and influential figure in connection with the industrial progress and civic advancement of this section of the state.

Mr. Robinson is descended from Scotch and English stock and was born in Jo Daviess county, Illinois, on November 25, 1950. His parents, John and Mary (Adkinson) Robinson, were natives respectively of Scotland and England, and in America the family home was established in Illinois, where the father eventually met his death while working in the coal mines.

John Robinson was educated in the public schools of Illinois and in 1868, as a sturdy and ambitious youth of eighteen years, he came to Nebraska. At Columbus, Platte county, he learned the butcher's trade, and later he conducted a meat market at that place. Still

later he was similarly engaged at St. Paul, Howard county, and in 1885, he came to the Panhandle. Mr. Robinson, in association with Alfred W. and Henry Gumaer and Albert Potter, came to Deuel county to engage in the cattle business. Each of the party took up homestead and pre-emption claims, on one of which is situated the present town of Oshkosh. This was all wild prairie at that time, and settlers were comparatively few, as the cattle business was virtually the sole industrial enterprise. Within two years after coming to the county these men laid out and founded the new town of Oshkosh, which became the county seat of Deuel county and later as that of the newly organized Garden county. The founders established a general store and otherwise gave impetus to the upbuilding of the village. Later they sold the store and the townsite of eighty acres to a townsite company, and Mr. Robinson and Henry Gumaer purchased the ranch and cattle interests of the other partners, their association continuing several years before the partnership was dissolved by an equal division of the property. Mr. Robinson thereafter continued operations in the cattle business until the rapidly increasing settlement of the country led him to dispose of his cattle, with the closing of the large range facilities. He is the owner of five well improved farms in Garden county, having an aggregate area of about eleven hundred acres, all valley land and nearly all under irrigation, which are rented. The accumulation of this valuable property, with its rise in value, has in itself placed Mr. Robinson in comfortable circumstances. He built the first house, a sod structure, in what is now Garden county, and the primitive building was on his pre-emption claim, a half mile north of Oshkosh. Mr. Robinson has given attention to the raising of fine types of horses and at the present time he is the owner of about one hundred head.

A man of strong individuality and alert mentality, Mr. Robinson was well equipped for leadership in thought and action in the pioneer community, and he has been influential in communal affairs during the entire period of his residence in western Nebraska. He was the first postmaster at Oshkosh, and served two terms as county commissioner of Garden county. He was one of the organizers and is past master of Oshkosh Lodge, No. 286, Ancient Free & Accepted Masons.

In the year 1882, Mr. Robinson wedded Miss Mary Doolittle, who was born in Wisconsin. Of their nine children Edward and

George are conducting large live-stock operations in Wyoming, and with them is associated their next younger brother, Floy S.; Carrie, Mabel, Lillian, Elizabeth, and Frank are all members of the family circle, while William married Mae Winters July 3, 1918, and conducts one of his father's farms north of town.

JACOB C. SCHLATER, is a member of the well konwn and reliable real estate firm of Day & Schlater, of Oshkosh, Garden county, a concern that represents the best ideals of this important line of business, its two principals being young men of marked energy and enterprise, so that their firm is constantly growing.

Mr. Schlater, junior member of the firm, was born March 3, 1893, in that section of Deuel county that now is comprised in Garden county. He is a son of William E. and Hattie M. (Roudebush) Schlater, who were early pioneer settlers in the famous Panhandle of Nebraska. Here the father took up a homestead in what is now Garden county, in 1888. Later he purchased and assumed active management of a ranch in the northern part of the county, and was accidentally killed there when a horse fell upon him causing injuries that resulted in his death, when he was about thirty-five years of age. His widow now lives at Oshkosh.

Jacob C. Schlater was but six years old at the time of the tragic death of his father, and went to live in the home of an uncle at Plattsmouth, where he was reared and received good educational advantages, in the public schools, where he studied until he completed a course in the high school. After graduation he spent one semester as a student in the University of Nebraska, and upon his return to Oshkosh he took the position of bookkeeper in the First State Bank, of which he later became assistant cashier. After holding this position about three years he was advanced to cashier, of which he continued the efficient and popular incumbent until the bank changed hands, in March, 1899. when he severed his connection with the institution and became associated with Robert A. Day in their present real estate enterprise, under the firm title of Day & Schlater. The firm has already built up a substantial business of importance.

Mr. Schlater is a Democrat; he and his wife are members of the Lutheran church and he is affiliated with Oshkosh Lodge, No. 286, Ancient Free and Accepted Masons, and with Plattsmouth Lodge, No. 686, Benevolent and Protective Order of Elks.

March 3, 1918. was solemnized the marriage of Mr. Schlater to Miss Marguerite Day, daughter of Robert A. Day, pioneer citizen and prominent banker at Oshkosh. Mr. and Mrs. Schlater are popular figures in the social activities of their home community, where their circle of friends is large.

ROBERT QUELLE, is one of the representative business men and valued citizens of Oshkosh, where his well equipped hardware and furniture establishment has a department devoted to the undertaking business, with facilities of the most modern and approved order. He is one of the progressive men of the younger generation in Garden county, and in the management of his flourishing business he has an able helper, Howard C. Fickes, the enterprise being conducted under the firm name of The Oshkosh Hardware Company.

Mr. Quelle was born in Schleswig, Germany, May 25, 1873, and he was educated in his native land until he was fifteen years of age when his parents came to America and established their home in Nebraska. He is a son of Franz and Minnie (Werber) Quelle, who immigrated to the United States in 1888. Upon coming to Nebraska the settled in that part of Deuell county that now comprises Garden county, and became pioneers of this now progressive part of the Nebraska Panhandle. Franz Quelle entered claim to a homestead, to the improvement of which he directed his attention with marked success. He developed his homestead into a productive farm, later added materially to its area, and there he continued his successful activities as a farmer until his age and prosperity justified his retirement. He removed from his farm to Oshkosh, where he died at the age of seventy-five years, and where his widow still resides.

The third in order of birth in a family of seven children, Robert Quelle, was fifteen years of age when the family home was established on the pioneer farm in Nebraska. Many hardship and vicissitudes were naturally to be endured by the pioneers, and as a youth Mr. Quelle not only assisted in the work of the home farm but also found employment elsewhere, at intervals, to aid in the support of the family and provide a good home. He became a clerk in the first mercantile establishment opened at Oshkosh, where he was employed four years. He then purchased a farm, near Lewellen, Garden county, but disposed of it within the same year, then became one of the organizers of the Oshkosh Lumber Company, which engaged in the lumber, grain and coal business and which erected, in 1909, the grain elevator at Oshkosh. In 1912, Mr. Quelle formed a partnership with his brother, Albert Quelle, they became the interested pricipals in organizing the Oshkosh Hardware Company and developing a substantial business. In 1916, Robert Quelle sold his entire interest in The Oshkosh Lumber Company and purchased his brother's interest in the hardware establishment which they had opened some time previously. The following year he sold a third interest in this business to Howard C. Fickes, with whom he has since been associated; they have materially expanded the scope of the store, in each department of which they receive a substantial and representative supporting patronage. One mile southeast of Oshkosh Mr. Quelle owns a well improved and irrigated ranch of three hundred and twenty acres, to which he gives supervision in a general way.

Mr. Quelle has been a worker and business man, and has had no desire to enter political life or to hold office. He is an independent Republican. He is affiliated with Oshkosh Lodge No. 286, Ancient Free and Accepted Masons, and also with the local camp of the Modern Woodmen.

September 14, 1900, Mr. Quelle married Miss Doris Rohlfing, who is a native of Germany and was seventeen years old at the time her parents immigrated to America, in 1894. To Mr. and Mrs. Quelle have been borne four children: Robert, Jr., died at the age of two years; and Alvena, Carl and Maude.

WALTER W. BOWER, who is now living in semi-retirement at Oshkosh, is one of the well known and influential pioneer citizens of this section of the Panhandle and is a man whose experiences in the west have been most varied and interesting. He was long and prominently identified with the cattle industry as a cowboy. He is the owner of valuable real estate in Garden county and at the present time conducts a substantial enterprise in the handling of Nebraska farm lands and other realty.

Mr. Bowers was born in the Lone Star state and is a representative of a pioneer family of that great commonwealth. He was born in eastern Texas, January 7, 1863, a son of Thomas and Elizabeth Bowers, natives of Missouri. Mr. Bower was but three years old at the time of his mother's death and when he

was a lad of six years his father passed away. The orphaned boy was taken into the home of and uncle and aunt, in central Texas, where he lived until eleven years old, when he ran away from this home and found employment while he attended school and supplemented the education which he had previously received. He continued to work and go to school until he was fifteen years old, when he became a cowboy and assisted in driving cattle along the trail from Texas to Julesburg, Colorado. Six months was required to make this trip, in autumn of 1878. After arriving in Colorado Mr. Bower found employment with the L. F. ranch, located in Colorado and Nebraska. He assisted in herding and driving cattle from Ogallala to Greeley, Colorado and also to Cheyenne, Wyoming, the ranch having shipped fully twenty-five thousand beef cattle annually. After four years Mr. Bower gained a new experience, as he spent about two years driving Texas ponies into Nebraska and Kansas, where he sold them. This venture prove financially profitable and for several years Mr. Bower gave his attention principally to the buying and selling of horses, besides which he became the owner of a ranch property in the Indian Territory. He continued in be extensively engaged in the buying and shipping of horses until 1890, when he came to the Panhandle of Nebraska and purchased about nine hundred acres of wild land, in what is now Garden county. He engaged in the handling of both horses and cattle, which he raised and fed on a somewhat extensive scale, until 1905, when he sold his live stock and rented his land, and engaged in a general merchandise business at Oshkosh, where he was in business until 1908, when he sold his general store and opened a drug business. He disposed of the drug store in 1916, and since that time he has lived retired, giving a portion of his attention to the buying and selling of real estate.

Mr. Bower is a Democrat, but he has invariably refused to become a candidate for office, with one notable exception, when he served one term as sheriff of Deuel county, before Garden county was formed. He is affiliated with the Modern Woodmen of America and attends and gives liberal support to the Methodist Episcopal church at Oshkosh, of which his wife was a founder. Mrs. Bower started the movement that resulted in the erection of the first church building at Oshkosh, and she has continuously been one of the most active, loyal and loved members of this church, on the list of whose charter members her is the first name recorded. She has served

Bert J. Seger

as a member of the board of trustees of the church from the time of its organization and is a gracious figure in the representative social life of her home community

On August 31, 1883, was solemnized the marriage of Mr Bower to Miss Sylvia Thomas, who was born in Pennsylvania, but was reared and educated in Kansas, where her parents settled in the pioneer days She is a daughter of Lorenzo D and Mary Thomas, the former a native of Massachusetts and the latter of Ohio, and both were residents of Kansas at the time of their death Mr and Mrs Bower have two children Zulah, is the wife of William F Gumaer, a merchant at Oshkosh, and Beulah is the wife of Charles L Tomppert, editor and publisher of the *Garden County News,* at Oshkosh

BERT J SEGER, who is officially identified with the great irrigating projects under government control in the Panhandle of Nebraska, came to Scottsbluff county in 1906, and took up his residence at Scottsbluff in 1910 He is secretary and treasurer of the North Platte Water Users Association, and secretary of the Farmers Irrigation District

Bert J Seger was born in Polk county, Nebraska, February 13, 1874, the eldest of the seven surviving children of Andrew and Julia (Palmer) Seger, both of whom were born in Illinois, and came from there to Nebraska in 1872 The father homesteaded near Osceola and lived there until the spring of 1910, when removal was made to southwestern Kansas In the fall of 1915 Andrew Seger and his wife established their home at Scottsbluff Besides Bert J, their children are the following Wilbur operates a garage at Kinnard, Nebraska, Elmer, a farmer northeast of Mitchell, Nebraska, Harry, a stenographer employed in a sugar factory at Scottsbluff, Charles, an engineer employed by the state in road building, Lester, a photographer at Lexington, and Vera, the wife of Howard Jennings The parents are members of the Methodist Episcopal church

Bert J Seger received his elementary education in the schools of Osceola, following which he took advance courses at the Stromsburg and Fremont normal schools, and for twelve years was a teacher in his native state He is a veteran of the Spanish-American War, entering military service in June, 1898 and accompanied his regiment to the Philippine Islands where he participated in numerous engagements He was discharged in August, 1899 In 1906 Mr Seger came to Scottsbluff county and took up a homestead In 1910, when appointed secretary of the North Platte Water Users Associa-

tion he came to the city and has lived here ever since At the present time this government project has 1300 water users He is also secretary of the Farmers Irrigation District, which includes about 65 000 acres These enterprises are of such vast importance that only men of proved ability are entrusted with their management

In 1900 Mr Seger was united in marriage with Miss Carrie C Camp, who was born at Ashland, Nebraska They have two sons Arthur H and Don, aged respectively seventeen and six years Mr Seger and family are members of the Church of Christ Fraternally he belongs to the Knights of Pythias and politically he is a Democrat of considerable local prominence Since 1915 he has been a member of the Democratic county central committee

ANGNST SUDMAN, an honored pioneer merchant of Oshkosh, to whose development and upbuilding he has contributed in large measure, has also assisted in the industrial advancement of the county which was still a part of Deuel county when he came here Mr Sudman is still extensively interested in ranch operations in this county and at Oshkosh is interested in a lumber business, besides being one of the chief stockholders of the First State Bank

Mr Sudman was born in the province of Hanover, Germany, on October 6, 1865, and was reared in his native land, where he received excellent educational advantages After he came to western Nebraska, as an ambitious and spirited young pioneer, he attended school at Lodgepole, Cheyenne county, in the winter of 1884-85, for the purpose of familiarizing himself with the English language At the age of seventeen years Mr Sudman set forth to seek his fortunes in America He landed at New York City on June 14, 1882, and made his way to western Nebraska where for the first year he was employed on a sheep ranch in the vicinity of Lodgepole, Cheyenne county Later he passed about a decade as an employe on a large stock ranch in that part of Cheyenne county now comprised in Deuel county, and during this period he assisted in driving two large bands of sheep over the trail from Utah to Julesburg, Colorado, six months having been required to make each trip, one time about eight thousand head of sheep were driven, and on the other sixteen thousand head

In 1888, Mr Sudman took a homestead of one hundred and sixty acres, six miles east of Chappell, and after proving up engaged in farming until 1897, when he sold the land, for twenty-two thousand dollars In 1893, he returned to his native land to visit his widowed mother and renew the associations of his boyhood days He remained only a few months in Germany and after his return to Nebraska went to live at Oshkosh, in May, 1894 He engaged in the mercantile business, as a member of the Sudman Company, his associates being his brother Fred and B E Fish A large general merchandise business was developed by this company, Mr Sudman being the manager of the enterprise for sixteen years He then disposed of his interest in the business The Sudman Company purchased land in the early days in which was included the townsite of Oshkosh, which was platted by the company The coroporation conducted extensive operations in the raising of cattle, horses and other live stock, and when the company sold the mercantile business to August Sudman purchased his partners' interests in the ranch enterprise The Sudman company established the first lumber yard and implement business at Oshkosh, and in the early days it was necessary to freight all goods overland from Chappell, about thirty miles away Mr Sudman was the second postmaster at Oshkosh and when he succeeded to the individual ownership of the land held previously by the Sudman Company he found himself in possession of about twenty-nine hundred acres, all within three miles from Oshkosh This included a thousand acres in the valley, the remainder being grazing land Mr Sudman has since disposed of a portion of his holdings but he purchased other land and has done much to further the industrial development of his home county Here he now has a half section of wheat land, a thousand acres of hay and corn land, and five hundred and sixty acres available for irrigation, his holdings in Garden county comprising two thousand, nine hundred and eighty acres Mr Sudman raises cattle, horses and hogs on a large scale besides dealing extensively in live stock, as a buyer and shipper He is a Republican and has served twelve years as a member of the school board of Oshkosh He and his family are members of the Lutheran church and he is affiliated with Oshkosh Lodge, No 286, Ancient Free and Accepted Masons, as well as with the local camp of Modern Woodmen of America He is a citizen who has done much for his home town and county, where he is valued and honored as a pioneer

On June 6 1897, was solemnized the marriage of Mr Sudman to Miss Anna Pearl Plummer, who was born in Missouri, and came to Nebraska with her widowed mother, who took a homestead five miles northeast of Oshkosh and whose son, W C Plummer, likewise made homestead entry in the same locality To Mr and Mrs Sudman have been born five children Carl August, died at the age of six months, Clyde H, is his father's valued assistant in the work and management of the ranch property, and the younger members of the home circle are Donald E, Glenn F, and Pearl Augusta

KENNETH W McDONALD, who has served Morrill county with rare fidelity as county attorney under three elections, is a leading member of the bar at Bridgeport, of which city he has been a resident since 1913 Mr McDonald has lived in Nebraska since boyhood but his birth took place in the Old Dominion, the home of his ancestors, January 18, 1874

The parent of Mr McDonald were James V and Emeline A (Gannaway) McDonald Both were born in Virginia, and both paternal and maternal grandfathers of Mr McDonald were born there also, the former, Solomon McDonald, on his father's plantation of seventeen hundred acres The McDonalds came originally from Glencoe, Scotland, located first in Deleware but became established in Virginia as early as 1685 On the maternal side the ancestry was English Long before the war between the states, the McDonalds and the Gannaways were large planters and slaveholders in Virginia

In 1881, Mr McDonald's parents came to Nebraska The father and his two brothers, Franklin and William McDonald, had been officers in the Confederate army, and after the close of the war, facing new conditions and necessities, the father learned the carpenter trade and worked at it after locating at Pierce, Nebraska, in 1881, until he was seventy-five years old He died eight years later He warmly supported the principles of the Democratic party and both he and his wife were faithful members of the Methodist Episcopal church Of their twelve children six are living including Kenneth W, the others being: Beauregard, who is postmaster at Pierce, Nebraska, Charles, who is a contractor at San Pedro, California, Estelle, who is the wife of Alonzo Glaze in the decorating business at Pierce, Grundy, who is a physician in practice at Long Beach California, and Solomon R, who is a United States mail clerk residing at Council Bluffs, Iowa

MILES J. MARYOTT

Kenneth W McDonald was educated in the public schools at Pierce and graduated from the high school in 1892, then read law under a local attorney and completed his course in the Boston Law School He entered upon the practice of his profession at Pierce, where he remained until 1913 a period of seven years, during which time he served Pierce county in the office of deputy county attorney for one year At first Mr McDonald was alone in his practice at Bridgeport but subsequently admitted his brother-in-law, George W Irwin, to a partnership The firm is considered a very strong one and handles a large percentage of the important court business In 1914, Mr McDonald was elected county attorney on the Democratic ticket, was re-elected in 1916 and again in 1918

At Creighton, Nebraska, in March, 1913, Mr McDonald was united in marriage with Miss Ethel G Irwin, who was born at Creighton Mr and Mrs McDonald have an adopted daughter, Margaret Elaine They are members of the Episcopal church Mr McDonald is an advanced Scottish Rite Mason He is a man of high personal character, has many times proved the sincerity of his citizenship in advocating worthy enterprises at Bridgeport, and during periods of great general concern and national stress, he has heartily cooperated with his fellow citizens in bearing the burdens

MILES J MARYOTT has achieved high reputation as an artist and taxidermist and as a painter his technical skill is remarkable because his talent as an artist has been developed without instruction in either coloring or designing Like Charles Russell and other celebrated western artists, he is entirely self taught, and his creative genius has found expression in many beautiful canvasses that have received the highest of critical commendation The career of this native son of Nebraska may well be said to have been far aside from the beaten path and he had added to his laurels a remarkable record as marksman and as a baseball player He and his widowed mother now reside in an attractive home at Oshkosh, Garden county, and it is gratifying to be able to give in this publication a brief review of his career Mr Maryott was born at Tekamah, Burt county, Nebraska, September 4, 1873, a son of Asahel K and Emily (Herrick) Maryott The former was born in Brookfield, New York and the latter in Chautauqque, New York in 1842 Their respective parents were early pioneer settlers at Hus-

tisford, Wisconsin Ashel K Maryott began to farm in his native state, and came to Nebraska in 1865, before the admission of the territory to the Union He was one of the pioneers of Burt county, where he took up a homestead, near the present village of Decatur, where he developed a productive farm He continued his activities as an agriculturist and stock grower until 1884, when he sold his farm and removed to the vicinity of Cozad, Dawson county, where he secured a tract of land and continued farming on a more extensive scale There he passed the residue of his life as his death having occurred in 1907 His political allegiance was given to the Republican party His marriage was solemnized in Wisconsin, and his widow now lives at Oshkosh at the age of seventy-eight They became the parents of five sons and four daughter, of whom the subject of this review was the seventh in order of birth

Miles J Maryott was educated in the public schools of Cozad, Dawson county He early developed marked skill in connection with the American "national game," and in baseball outside of Nebraska was made when he entered professional baseball activities, as a member of the team of Mankota, Minnesota He played with this team in the seasons of 1895 and 1896, was with the Galesburg, Illinois, team for the ensuing season and the Fort Collins, Colorado, team claimed him as a member for the season of 1898 For the three following years he played with the Kearney, Nebraska team, and for the first half of the season of 1902, was with the Keokuk team, of the Iowa state league, the remainder of that season he served with the Sioux City team, in the Western League He terminated his professional career in baseball with the Wichita team, (Kansas) of the Western Association

In the meantime Mr Maryott had not neglected his talent as an artist, and he has made a specialty of pictures of animals and birds, his work being principally in oils and many fine paintings stand evidences of his talent He is conceded to be one of the best artists in the state and that in spite of the fact that he never has taken a single lesson in painting The same condition holds good in connection with his skill as a taxidermist in which field of work he has been licensed both by the United States government and the state of Nebraska He has one of the finest collections of native birds in Nebraska, besides which he has assembled one of the largest and most interesting collections of Indian relics to be found in the state His prowess as a marksman led to his being retained in the service of the Peters Cartridge Company in 1907, in the capacity of expert marksman In june of that year, in a contest held in the city of Chicago, Mr Maryott tied for first place in the grand American handicap, and he has the reputation of being one of the foremost all-round marksmen in the world

Mr Maryott has made the passing years bear to him an interesting and varied tribute, and he has maintained his home at Oshkosh since 1909 Here he is the devoted companion of his loved and venerable mother, and here he finds ample demands upon his time and attention in connection with his art and taxidermic work He took four hundred and eighty acres of land under the provisions of the Kinkaid law, and has proved his title to this property, upon which he has made improvements that mark it as a valuable Garden county farm In politics he supports the Republican party

WILLIAM L LAW, who is now serving his second term as county commissioner of Garden county, established his residence here when the county was still a part of Deuel county, and has developed and improved one of the valuable farms of four hundred and eighty acres which is well situated twelve miles north of Oshkosh, the county seat Here Mr Law has secure vantage-ground as one of the enterprising agriculturists and stock-growers of the county and the confidence in which he is held by the community is demonstrated by the office to which he was elected As county commissioner he has advocated and supported measures that have furthered the civic and material welfare of the county, and he is known as a wide awake and progressive business man

William L Law was born in Van Buren county, Iowa, on December 21, 1876, the second in order of birth in a family of three sons, his elder brother, Charles E, and his younger brother, John M, both being residents of Seattle, Washington, the mother having maintained her home at Seattle and Sumas, that state, since 1899 Mr Law is a son of Lorenzo and Eliza (Meredith) Law, both were born in Iowa, where the respective families settled in the pioneer days The father prepared himself for the medical profession, and after his graduation from a medical college engaged in practice in Iowa until he removed with his family to Long Pine Brown county, Nebraska where he conducted a drug store until the time of his death, which occurred when he was fifty-one years of age Dr Law was a man of sterling character and high intellectual-

ity and he was successful in the exacting work of his profession, having continued in active practice after engaging in the drug business at Long Pine He was a Democrat in politics

William L Law was educated in the public schools of Iowa and Frontier county, Nebraska, and began his career by engaging in farm enterprise, to which he gave his attention for a period of about eight years For the ensuing two years he was engaged in the livery business at Lexington, Dawson county, and he then removed to Deuel county and took up a homestead in what is now Garden county On this homestead, to the area of which he has since added, he has continued his vigorous activities as an agriculturist and stock-grower and his able management has brought him success He has made the best of improvements on his farm property Mr Law is one of the influential men of his community, has served thirteen years as school director of his district and is serving at the present time — winter of 1919-20 — his second term as a member of the board of county commissioners Well fortified in his views concerning economic and governmental policies, he gives his political allegiance to the Republican party and is influential in its local councils in his county

In October, 1898, Mr Law wedded Miss Mamie A Sprague, who was born at Danville, Illinois, and who was a girl at the time of the family removal to Frontier county, Nebraska, where she was reared and educated, her father, John T Sprague, a native of Indiana, having been a pioneer in Frontier county, where he took up and improved a homestead and where he continued his farm enterprise until 1909, since which year he has lived retired at Oshkosh, Garden county Mr and Mrs Law have five sons and five daughters Connie, Arthur, Lola, Sydney, Vera, Clyde, Rex, Ralph, Virginia, Vivian

WILLIAM F GUMAER is a representative of one of the prominent and influential families of Garden county and the name which he bears has been most closely identified with the development and upbuilding of this section of Nebraska, as may be seen by reference to the sketches concerning his elder brothers, Judge Alfred W Gumaer and Henry G Gumaer, on other pages of this publication Mr Gumaer has been actively identified with industrial and business interests in Garden county and is now one of the principal stockholders and active executives of the Oshkosh Mercantile Company, which conducts an extensive and

AMBROSE E. SCOTT

well ordered merchandise business at the county seat

Mr Gumaer was born at Weyauwega, Wisconsin on January 11, 1873, and in the sketches of his elder brothers is given record concerning the family history He was young when he came with his brother Henry G to St Paul, Howard county, Nebraska, where he attended the public schools, and later took a course in the Bryant & Stratton Business College in the city of Chicago Thereafter he was for seven years running a grocery store at Ashland, Wisconsin, on the shores of Lake Superior, and the following year he engaged in the collection business in that city In 1897, he went to Alaska, but in the following year when the nation became involved in war with Spain, his patriotism and loyalty prompted him to return from the north and tender his services as a soldier He enlisted in Company L, Second Volunteer infantry, which did gallant service in Porto Rico during the Spanish-American War Prior to this he had been for five years a member of the National Guard After the close of the war and receiving his honorable discharge, Mr Gumaer became a solicitor for the Fox River Telephone Company, in Wisconsin, holding this position four years In 1902, he returned to Nebraska and became associated with his brother Henry in agricultural and live-stock operations in the present Garden county After three years he was made manager of the general store of W W Bowers, at Oshkosh, and held this position until 1909, when he removed to the new town of Lisco, this county, where he became secretary and manager of the Lisco Mercantile company, besides gaining the distinction of becoming the first resident of the town He lived in Lisco seven and one-half years, and then impaired health caused him to pass about a year in the city of Omaha, where he received medical treatment Upon his return to Garden county he became one of the interested principals in the Oshkosh Mercantile Company, to the affairs of which he has since given his attention and to the upbuilding of the substantial business of which he has contributed in large measure

Mr Gumaer is a Democrat, he is affiliated with the Masonic fraternity and his wife is an Eastern Star His religious faith is that of the Presbyterian church, Mrs Gumaer holding membership in the Methodist Episcopal church

December 11, 1906, was solemnized the marriage of Mr Gumaer to Miss Zulah Mae Bowers, who was born at Caldwell, Kansas, but was reared and educated in what is now Garden county, Nebraska She attended the schools of Alliance, Lodgpole, Chappell and Oshkosh Mr and Mrs Gumaer have two children Viola Ruth and Priscilla Mae

AMBROSE E SCOTT — In considering the wonderful advantages that have resulted from the application of irrigation to the arid lands in many sections it is no small honor to have been the originator of the idea, and such honor may be justly claimed by Ambrose E Scott, formerly of Banner county, Nebraska, but now one of the leading business men of Scottsbluff Furthermore Mr Scott was so thoroughly convinced of the value of his idea that he built the first reservoir ever constructed in Banner county

Ambrose E Scott was born at New Concord, Ohio in 1863 the sixth in a family of twelve children born to William S and Violet Jane (Scott) Scott the other survivors being as follows Charles, who lives in Texas Alexander, a farmer near Twin Falls, Idaho, and Alvah B a physician of the osteopathic school at Harrisburg, Pennsylvania The parents were natives of Allegheny county, Pennsylvania, from which state they moved to Ohio and died there The father was a superintendent of construction on the Baltimore & Ohio Railroad for many years He was a veteran of the Civil War having served four years in the Fifteenth Ohio volunteer infantry during which time he was wounded and was captured by the enemy and was incarcerated in Libby prison for three weeks Both he and wife were members of the United Presbyterian church The Scott family came to America from the north of Ireland and are descended from a long line of true Scotch-Irish ancestors, who played important parts in the history of Scotland and the Emerald Isle

Ambrose E Scott had some school advantages in Ohio In 1886 he came to Banner county, Nebraska and in 1887 took a homestead on which he afterward lived for thirty years There he figured prominently in the history of that county as will be mentioned in different subjects by the historian in the history of the county In 1917 he left his farm and came to Scottsbluff and later went into the farm implement business in partnership with Guy Carlson and they are doing a heavy business

In 1902 Mr Scott was united in marriage with Miss Ida Eckerson who was born in Iowa and they have one son, Charles H Mrs Scott also proved up a claim in Banner county While living there Mr Scott served two terms as county treasurer, elected on the Republican ticket He belongs to the order of Modern

Woodmen and to the Knights of Pythias, in which organization he has passed all the chairs. Mr. Scott is widely known and is deeply interested in the work of the different irrigation districts.

ELMER J. HARNESS has been a resident of Garden county from the time of his birth, is a representative of one of the pioneer families of this section of the state and has gained a secure status as a prosperous and enterprising agriculturists and stock-grower, his operations being conducted on the old homestead of his father, situated two miles west of Oshkosh.

Elmer John Harness was born at Oshkosh, this county, December 23, 1892, a son of John and Ella (Martin) Harness, the former of whom was born and reared in Illinois and the latter in the state of Connecticut; by a prior marriage Mrs. Harness has one son, William Brown, now a resident of Denver Colorado. In 1886, John Harness came to Cheyenne county, Nebraska, and became a pioneer settler in that part of the county that now constitutes Garden county. Here he took up, improved and perfected title to a homestead now owned and occupied by his son Elmer.. Mr. Harness continued his activities as an agriculturist and stock-raiser until his death, in 1904, at the age of fifty-two years. He was a sterling citizen who commanded the esteem of his friends and his name merits place among the pioneers of Garden county, where his widow still resides, her home being about three miles west of Oshkosh.

Elmer J. Harness attended the pioneer schools of Garden county and he was a lad of fourteen years at the time of his father's death. Thus as a youth he gladly assumed the burdens and responsibilities of making his own way and also providing for his widowed mother. For three years he was engaged in farming on rented land, six miles west of Oshkosh, then returned to the old homestead, where he has since held place as one of the representative agriculturists and stock-growers of the younger generation, specializing in raising hogs. Liberal and loyal as a citizen, he supports those agencies and enterprises that tend to advance the welfare of his home community and county, his political allegiance being given to the Republican party and his religious faith being that of the Lutheran church, of which his wife likewise is an active communicant.

At Alliance, Box Butte county, on October 15, 1913, was solemnized the marriage of Mr. Harness to Miss Gertrude Catron, of Bridge-

port, Morrill county, where she was born and reared, being a daughter of Isaac and Anna (Foote) Catron, who are pioneer citizens who reside on their farm, four and one-half miles northeast of Bridgeport. Mr. and Mrs. Harness have a winsome little daughter, Viola May, who was born April 27, 1919.

JACOB H. ROUDEBUSH is a pioneer citizen who merits recognition in this history not only by reason of the prominent and influential part he has played in connection with the civic and industrial development and upbuilding of western Nebraska, but also because of his service in the Union army during the latter part of the Civil War. He took part in the suppression of Indian outbreaks in the west, and gained an experience that but few survivors of his generation can claim. Mr. Roudebush is one of the extensive landholders of Garden county and well known farmers.

Mr. Roudebush was born in Licking county, Ohio, January 11, 1846, a son of Jacob and Salome (Kuhn) Roudebush, the former a native of Germany and the latter of the state of Pensylvania. Mrs. Roudebush died when forty-five years old. Jacob Roudebush, was eighteen years of age when he came from Germany to America and settled in Bedford county, Pennsylvania, where he engaged in agricultural pursuits until he removed with his family to Ohio. In the Buckeye state he continued to farm until 1848, when he became a pioneer of Iowa, where he reclaimed and developed a good farm and where he passed the remainder of his life. He was a resident of Salem, at the time of his death, which occurred when he was seventy-two years of age.

Jacob H. Roudebush was reared in Iowa, where he availed himself of the advantages of the common schools and where he early began to assist in the work on his father's pioneer farm. In 1864, at the age of eighteen years, Mr. Roudebush tendered his service in defense of the Union, enlisting in Company A, Seventh Iowa Cavalry, which was assigned to the command of General Sully, and took part in the campaign against the Sioux Indians on the Missouri river, in what is now South Dakota. Later Mr. Roudebush was sent with his regiment to Little Rock, Arkansas, then proceeded to Fort Kearney, Nebraska, being assigned to service at Fort Laramie, Wyoming. In Wyoming Mr. Roudebush did much scouting service and also participated in the fight with the Indians at Horse Creek, in which engagement Captain Fouch was killed. For a time the regiment was stationed at Fort Mc-

Pherson, then proceeded to New Mexico, where it was in active Indian service under Generals Curtis and Mitchell, as a part of a force of twelve hundred men From New Mexico, Mr Roudebush returned to Fort McPherson, and later he was stationed at Fort Leavenworth, Kansas, where he was mustered out on May 17, 1866, and received his honorable discharge

After the close of this service as a soldier, Mr Roudebush returned to Iowa, where he farmed about four years He then removed to Osborn county, Kansas, and took a preemption claim, but drouth caused his crops to fail, with the result that he left his claim and made his way to Missouri, where he was engaged in farming for two and one-half years In 1884, he became one of the pioneer settlers in that part of Cheyenne county, Nebraska, that now constitutes Garden county He perfected title to his original homestead, and with increasing prosperity he added gradually to his holdings, until he is now the owner of a ranch of more than thirty six hundred acres, as well as a section of school land and four acres at Oshkosh, the county seat of Garden county He has given special attention to the raising of live stock, his ranch at the present time showing an average run of two hundred head of cattle and about fifty head of horses

Mr Roudebush has been a leader in community thought and action during the long years of his residence in Nebraska, and prior to the creation of Garden county he served six years as a member of the board of county commissioners of Deuel county After the organization of Garden county he was chosen a commissioner, serving four years and did much to further the advancement and prosperity of the new county He is a Democrat, is affiliated with the post of the Grand Army of the Republic at Chappell and holds membership in Oshkosh Lodge, No 268 Ancient Free and Accepted Mason his wife being a member of the Lutheran church

The first marriage of Mr Roudebush was solemnized September 9, 1869 at New London, Henry county, Iowa, where Miss Anna E Brown became his wife, she having been born and reared in that county Mrs Roudebush died when but thirty-five years of age, and was survived by six children William E, who is a resident of California, has four children, Mrs Hattie M Sarser resides at Oshkosh, Garden county, and her only child, Jacob C Slater, was born of her first marriage, Frederick R, is married and resides at Osh-

kosh as does also Ira, Jacob C is a resident of Bayard, Morrill county and is the father of six children, and Mrs Laura B Vance, of Oshkosh, has four children

On September 21, 1891, Mr Roudebush married Miss Sarah Isabel Hunter, who was born and reared in Missouri and came with her parents to the present Garden County, Nebraska about 1889 Of this union were born nine children George and John are residents of Oshkosh, the former having one child and the latter being the father of two children, Frank remains at home, Daisy died in 1917, Harry Mary, Theodore R, Thomas E, and Emma

JOSEPH PEBLEY, is a well known farmer and stockman of Garden county, where he established his residence more than thirty years ago, when it was a part of Deuel county He has been the architect of his own fortunes, and began his career when he was a lad of sixteen years He is one of the popular pioneer citizens of Garden county and it is gratifying to accord him recognition in this history

Mr Pebley was born in Atchison county, Missouri, on May 30, 1866 a son of Jeremiah and Rhoda (Morgan) Pebley, the former a native of Missouri and the latter of Tennessee, where she was reared and educated Jeremiah Pebley a farmer by vocation, passed his entire life in Missouri and died at the age of Sixty-six years, at Craig His widow lived to be seventy-four years old, being a resident of Amorette, Missouri, at the time of her death

Joseph Pebley attended the schools of his native state and early learned the lessons of practical industry, and began to farm when sixteen years old, but two years later, in 1887, came to Nebraska and settled in Butler county, where he was engaged in agriculture for the ensuing nine years He then spent one year in Arkansas and returned to Nebraska to settle in that part of Deuel county that now comprises Garden county, where he has maintained his home since 1898 Here he proved up on a homestead claim He was engaged for the first few years in freighting from Chappell to Oshkosh, his earnings in this service enabling him to improve his land to which he has since added another quarter-section, so that he now has an excellent farm of three hundred and twenty acres, with good buildings and given over to diversified agriculture and the raising of cattle and hogs Mr Pebley is always ready to lend his co-operation in support of measures and enterprises tending to advance the communal welfare, and is a Democrat.

At Ulysses, Nebraska, January 29, 1888, was celebrated the marriage of Mr. Pebley to Miss Mae Horner, who was born in Richardson county, this state, a daughter of Ephraim and Mary Anna (Robbins) Horner, both natives of Ohio, though the latter was reared and educated in Missouri, where her marriage occurred and where she died at the age of forty-two years. Epraim Horner, a millwright by trade and vocation, was a young man when he established his residence in Missouri, whence he came to Nebraska and settled in Richardson county in 1887. In 1893, he returned to Missouri, but eventually he returned to Nebraska, and now resides at Syracuse, Otoe county, at the patriarchal age of Ninety-one year, in 1920. He was a soldier of the Union in the Civil War, throughout the entire course of which he served as a member of the Forty-seventh Missouri Cavalry. Mr. and Mrs. Pebley have five children: Earl J., who resides a Mumper, Garden county, married Miss Clara Johnson and they have four children; Rosa is the wife of Frank Shaw, of Oshkosh, this county; and Harry L., Blanche and Irene remain at home.

GEORGE F. ALLEN first came to what is now Deuel county in the early pioneer days, and thought at that time he took up a homestead, drouth and other unfavorable conditions made his farm venture a failure, with the result that he returned to the eastern part of the state. A number of years later he again came to Garden county and here ample success has now crowned his activities as an agriculturist and stock-raiser, his homestead place being situated ten miles northeast of Oshkosh, and it value being enhanced by a considerable acreage of timber. Mr. Allen has made good improvements on his farm, and each year records a definite advancement along this line, as well as in the returns from the energetic and well directed efforts which he is putting forth in his agricultural and live-stock industry.

Mr. Allen has the distinction of being a native of New York City, where he was born on the 6th of June, 1848. He was reared and educated in Ohio and Nebraska, as the adopted son of kindly fosterparents, Mr. and Mrs. James McLaughlin, and he was about ten years of age at the time of the removal to the latter state. Vital and self-reliant, he was not yet fifteen years old when he initiated independent activities in connection with farm enterprise, in which he thus continued about twenty years in Cass county, Nebraska. He then came to the western part of the state and became a youthful pioneer of that part of Cheyenne county that now constitutes Deuel county where he took up a homestead and attempted to develop a farm. Protracted drouth made his success of negative order and thus he was virtually compelled to abandon his claim and return to the eastern part of the state, where he continued his farm activities several years. In 1903, he came again to Garden county, where he has since resided on the homestead which he took at this time and where his success has caused him to forget the failure that attended his efforts in earlier years. He is a substantial and popular citizen of his community, is a Republican in political adherency.

In Cass county, this state, in 1870, was solemnized the marriage of Mr. Allen to Miss Rhoda Root, a daughter of Charles and Mary (Splittstone) Root, who were pioneer settlers in Cass county, as were they later in Garden county, Mr. Root having died when about sixty years of age and his wife having passed away when about fifty years of age, she having been a native of Ohio. Mr. and Mrs. Allen have three children: Charles Edward, of Oshkosh, is married and has two children; Willis, of Big Spring, Deuel county, is married and has two children; and Mrs. Clara V. Duval, of North Platte, Lincoln county, deceased, had three children by a former marriage, Delbert, Charles and William Emmerson.

AUGUST BUSKE, a representative pioneer and successful agriculturist and stockman of Garden county, has gained substantial success, is a large landholder, and stands high in popular confidence and esteem.

Mr. Buske was born in Stetten, Germany, May 14, 1852, a son of Charles Buske, who passed his entire life in Germany. August Buske was afforded the advantages of the schools of his native land, and there, at the age of twenty-six years, he began to farm and spent five years in this industry before becoming foreman on a large hay farm for one year. He then immigrated to America and established his residence at Ford River, Michigan, being employed in a saw mill for five years. In 1887, he came to Nebraska and for the ensuing five years he was employed in a stone quarry near Louisville, Cass county. He then, in 1892, became a pioneer of what is now

WENZEL HIERSCHE

Garden county, then a part of Deuel county, where he took up a homestead eight miles northeast of the present village of Lisco He proved title to his claim and continued his activities as an agriculturist and stock-raiser for five years He then removed to government land two miles north of his original claim and began operations in the cattle business, besides purchasing two hundred acres and developing the tract for agricultural uses Under the provision of the Kinkaid act he finally secured four hundred and eighty acres of land, which constitutes his present home place Drouth nullified his labors on this place the first two years, but later years have given to him a generous return, and he is now one of the substantial men of the county, where he owns twenty-five hundred acres of valuable land, two hundred of which are under cultivation He runs an average of a hundred head of cattle and thirty head of horses

Mr Buske has always been a loyal and public-spirited citizen, his political allegiance being given to the Republican party and both he and his wife belong to the Lutheran church He is a stockholder in the farmers' elevator at Lisco, and takes lively interest in all things pertaining to the welfare of his home county

In Germany, November 15, 1878, was recorded the marriage of Mr Buske to Miss Minnie Kruger, and she accompanied him to America, where she has been his companion and helpmate in the years that have marked his rise to a position of independence and substantial prosperity To Mr and Mrs Buske have been born six children August, Jr, of Lisco, is married and has one child, Mrs Emma Barnwell is deceased and is survived by two children, Henry met a tragic death when he was sixteen years of age, having been accidentally shot by a boy friend, Mrs Dora Morris and Mrs Rosa Onston reside at Lisco, and Harry married Gladys Lisco and now lives near Oshkosh

WENZEL HIERSCHE who has long been identified with important industries and worthy enterprises in Nebraska came to the United States in 1877 and to Scottsbluff county Nebraska, in 1885 He is now a resident of Scottsbluff and devotes a large portion of his time to the management of the Farmers Union Exchange

Wenzel Hiersche was born in Bohemia Austria September 2, 1859 His parents were Wenzel and Anna (Engle) Hiersche, who spent their lives in Austria They were the parents of nine children, seven of whom lived to mature years Besides Wenzel three others came to the United States, namely August, who homesteaded in Scottsbluff county in 1885 Frantz who is a farmer in Scottsbluff county and Anton of whom mention is found elsewhere in this volume In his native land Mr Hiersche learned the trade of glovemaking He was eighteen years old when he came to the United States where his brother August, had come the year before They located in Clinton county Iowa where they worked as farmers for their uncle until 1876 Subsequently they went to Texas where they engaged in farming until 1880 when they came to Nebraska Here they purchased some railroad land in 1882 and farmed until 1884, when they sold and returned to Texas In the spring of 1885 they preempted land and took timber claims five miles southwest of Gering in Scottsbluff county Mr Hiersche proved up on same and also on homestead He has sold all but the 320 acres he now owns On this farm 160 acres are irrigated and he also has an entire school section leased He has operated his land mainly as a ranch and at first raised cattle and horses, at one time having 180 head of the latter For some years before leaving the ranch he made a specialty of sheep, having 4000 head at one time, and still has 200 head of full-blooded Merinos on his land, which he has under rental

Mr Hiersche is an able, influential man in many directions Formerly he was active in the Populist party but later identified himself with the Republican forces and from 1893 until 1896 was a member of the board of county commissioners In late years he has become interested in such movements as the Non-partisan League and the Farmers Union, and on March 12, 1919, came to reside at Scottsbluff in order to take charge of the Farmers Union Exchange here, which deals in farm implements, feed and grain He has additional interests, being vice president of the Farmers Union Transfer & Storage Company, and vice president of the Water Users Association, of which he is a director He is prominent also in fraternal life being a charter member of the Scottsbluff lodge of Knights of Pythias of which he is past chancellor and was noble grand of the lodge of Odd Fellows at Gering, and a charter member and the first noble grand of the lodge at Scottsbluff, of which lodge he was secretary for one term Few men in Scottsbluff county are better known

GEORGE P RUMER who conducts a prosperous automobile garage in Lisco Garden county was about two years old at the time his parents came to Nebraska from Iowa He was born in Jefferson county, Iowa, September 1, 1878, a son of James P and Mary (Gilbert) Rumer, both natives of Iowa, where their parents settled in the early pioneer days James P Rumer was reared and educated in the Hawkeye state and there learned the painter's trade at which he became a skilled workman In 1880, he came with his family to Hastings, Nebraska, the county seat of Adams county, and began to work at his trade and still continues to make that city his home his wife being at the present time in Fresno, California

George P Rumer gained his early education in the public schools of Adams county, and at the age of eighteen years found employment at farm work, in which he continued until he had attained the age of twenty-five years Subsequently he engaged in independent farming in Custer county, and in 1905, filed entry on a homestead claim in Garden county He began the improvement of his land to which he eventually perfected his title, and actively engaged in general farming and stock-growing for a period of eight years He then removed to Comstock, Custer county, where he remained about four years, at the expiration of which he returned to Garden county and established his present business enterprise at Lisco, his garage being well equipped and controlling a substantial patronage In politics Mr Rumer is a Democrat, and in a fraternal way he is affiliated with the Knights of Pythias and belongs to the Masonic lodge at Oshkosh

November 9, 1904 Mr Rumer was united in marriage, at Comstock, Custer county, to Miss Lydia Allen, who was born in Iowa but reared and educated in Nebraska, her parents, Benjamin and Melissa (Cushman) Allen, having been pioneer settlers in Custer county, where the father developed a productive farm, both he and his wife being now residents of Comstock Mr and Mrs Rumer have three children Melissa Emma, Bruce E, and Alden William

WILLIAM BARNWELL, one of the influential citizens and representative agriculturists and stock-growers of the Lisco district of Garden county, was a mere boy when he came with his fosterparents to western Nebraska In his youth he was a cowboy on the range, besides which he was actively concerned in the construction of early irrigation ditches in this section of the state By his own energy and ability he has pressed forward to the goal of prosperity Mr Barnwell was born in eastern Pennsylvania, on January 1, 1876, a son of Patrick and Mary Barnwell, both natives of Ireland Mrs Barnwell died when her son William was but five years old, and at the age of seven years he was doubly orphaned, his father having been killed as the result of an accident at the smelter where he was employed, at Pueblo, Colorado

William Barnwell was young when he came with his fosterparents, Mr and Mrs James Broughton, to the wilds of western Nebraska, and here he was afforded the advantages of the pioneer schools Mr Broughton established a home in that part of Cheyenne county that now constitutes Morrill county, and there young William Barnwell finally found employment on the ranch of Mrs Elizabeth Smith, on Greenwood creek, twenty-five miles north of Sidney, Cheyenne county About six months later he began to work on the construction of the Belmont irrigation canal, remaining two months Later he was employed about a year in construction work on the Farmers' irrigation ditch, after which he put in one summer in farming on Pumpkin creek His next occupation was found on the horse ranch of Barnhart & Thompson, where he remained one year For twenty years thereafter he lived the free and vigourous life of a cowboy, in the employ of various outfits, his last employer, in 1904, having been Reuben Lisco, in whose honor the village of Lisco, Garden county, was named In 1904, Mr Barnwell took up a homestead claim about eight miles north of the present village of Lisco, and in this locality he has since risen to prominence and prosperity as a representative of agricultural and live-stock industry He is now the owner of eight hundred acres of land, of which two hundred are farmed, the remainder being pasture land In 1919-20, Mr Barnwell had on his ranch fifty head of cattle and twenty head of horses, and he takes pride in keeping his live stock up to high standard He is president of the Farmers Mercantile Company of Lisco, and a director of the Farmers Elevator Company He has taken a lively interest in community affairs had the distinction of being one of the first county commissioners of Garden county, and served thirteen years as school director of his district His political views are shown in his stalwart support of the cause of the Republican party, he is affiliated with Oshkosh Lodge No 286, Ancient Free and Accepted Masons, his wife is a member of the Presbyterian church

In June, 1905, Mr Barnwell wedded Miss Emma M Buske, a daughter of August Buske, of whom individual mention is made on other pages of this work Mrs Barnwell died in 1912, and is survived by two children Glenn, who remains with his father, and Erma, who lives in the home of her maternal grandparents February 16, 1916, was solemnized the marriage of Mr Barnwell to Miss Laura E Davis, who was born and reared in Buffalo county, this state, and who was a popular school teacher in her native county, as well as in Colorado, as was she also at Lisco, Garden county, at the time of her marriage Mr and Mrs Barnwell have a winsome little daughter, Ruby Jeanette

SEWELL E BENNETT, has been a resident of Garden county since 1908, and developed and improved one of the excellent farm properties of the county, consisting of three hundred and twenty acres situated about nine miles from Lisco Mr Bennett is one of the substantial citizens of Garden county and his career has been one of signal industry

Mr Bennett was born in Guernsey county, Ohio, May 12, 1849 His father, George W Bennett, was born at Little Washington Pennsylvania, in 1821, and was but six months old when his parents became pioneer settlers in Guernsey county, Ohio The father took up a pre-emption claim of one hundred and sixty acres and reclaimed the forest, making it into a productive farm On this old homestead his parents passed the remainder of their lives, his father having attained the patriarchal age of ninety-seven years George W Bennett was reared and educated in Guernsey county, and in 1852, when twenty-six years of age, he removed to Monroe county, Ohio where he engaged in agricultural pursuits and where he passed the remainder of his useful life, being seventy-one years of age at the time of his death His wife, whose maiden name was Olive Payne, was born in Vermont and was a child at the time of her parents' removal to Ohio, where she was reared and educated and where she was for two years engaged in teaching school at Little Point Pleasant Guernsey county, prior to her marriage Mrs Bennett died in November 1920, at her home

Sewell E Bennett was reared in Monroe county, Ohio, and after attending public school at Malaga, he was a student for one term in the normal school at Hopedale, Ohio After that he taught school one term, at Boston, Ohio, and the following summer he assisted his father on the home farm He then entered Hiram college, where General James A Garfield and many other distinguished men were graduated, and of which General Garfield was

for a time president many years prior to his election to the presidency of the United States In this institution Mr Bennett continued his studies one term, and for one year engaged in farming in Monroe county, Ohio, and four years in Guernsey county He then, in 1881, removed to Iowa and became a farmer in Adair county, where he remained for twenty-five years From Iowa Mr Bennet returned to Ohio but two years later he again responded to the call of the west by coming to Nebraska, in 1908 and taking up the homestead on which he has since resided and upon which he has developed a prosperous enterprise in diversified agriculture and the raising of hogs, cattle and horses He is a man of broad views, is a Republican in politics, and he and his wife are members of the Seventh Day Adventist church

On February 25, 1875, was solemnized the marriage of Mr Bennett and Miss Rachel Naylor, who was born and reared in Ohio, a daughter of Louis and Rachel (Bailey) Naylor, the former a native of Vermont and the latter of Ohio Louis Naylor was born in 1819, and was about four years old at the time of his parents' removal to Ohio and the greater part of his active life was devoted to work at the carpenter's trade, in connection with which he became a successful contractor and builder He died at the age of eighty-seven years, in Belmont county Ohio, and his widow died in January, 1919, at the venerable old age of ninety-seven years There were the following children in the Bennett family Roy L, who resides in Morrill county, Nebraska, is a widower and has two children, Albert Oscar and his wife reside in Morrill county, and they have seven children, Mrs Lena Patrick of Lisco Garden county, has two children and Mrs Belva L Carrigan of Lisco, has one child

CHARLES T SCHLOSSER has been a resident of western Nebraska since 1891, and is now one of the extensive landholders and representative agriculturists and stock-growers of Garden county

Charles Theodore Schlosser was born at Pittsfield, Illinois, August 5, 1876 and was about seven years old when his parents came to Nebraska and settled in Harlan county, where his father engaged in farming Mr Schlosser is a son of Earnest and Harriet (Lakin) Schlosser the former born in Germany and the latter born and reared in Illinois The father was a young man when he came to America and began to farm in Illinois whence he came with his family to Nebraska in the pioneer days, the remainder of his life being passed in this state, where he died at the age of seventy-seven years, his wife passing away in 1916, at the age of seventy-three.

Charles Schlosser gained his education in the Public school of Harlan county, and in the meanwhile began to assist in the work of the home farm. At the age of fifteen years he found employment at farm work in his home county, and after working one summer came, in 1891, to that part of Deuel county that is comprised in Garden county. For the ensuing five years he was employed by various ranch concerns in the north part of Garden county, and he then took up a homestead and began farming, giving special attention to the cattle business. In 1904, he obtained a Kinkaid claim, where he still resides, besides which he has purchased two thousand acres of ranch land and thus is the owner of two thousand six hundred and forty acres. He raises cattle on an extensive scale. He is a stockholder in the farmers' grain elevator and general merchandise establishment at Lisco, and also in the Higgins Packing Company at South Omaha. In politics he is aligned with the Republican party, and as a reliable and popular citizen he takes interest in all things pertaining to the welfare of his home county and state.

On October 9, 1918 Mr. Schlosser married Mrs. Anna (Vandford) Ames, who was born and reared in Missouri. They have one child, Virginia Evaline.

LOUIS M. MEYER, is one of the representative agriculturists and stock-growers of Garden county. He has proved alike his energy and his resourcefulness and has gained independence and prosperity.

Mr. Meyer was born at St. Charles, Missouri, on July 11, 1865, a son of Louis and Barbara J. (Hintzelman) Meyer, both natives of the province of Lorraine, France. Louis J. Meyer accompanied his parents on the immigration to America. They settled at St. Charles, Missouri, where the parents passed the remainder of their lives. Louis J. Meyer learned the carpenter's trade, to which he gave his attention in Missouri until 1868, when he came with his family to Nebraska and settled at Nebraska City, where he became a leading contractor and builder and where he continued to reside until his death, at the venerable age of eighty-six years. His wife was seven years old when she came with her parents to America, and was reared and educated at St. Charles, Missouri, where her marriage was solemnized. She died, at Nebraska City, when seventy-two years of age. This sterling pioneer couple became the parents of nine children, of whom four are now living.

Louis M. Meyer was about three years old at the time of the family removal to Nebraska City, where he was reared and received good educational advantages, including those of Albert Hall College, which was founded by J. Sterling Morton. At the age of twenty-one years, in 1886, Mr. Meyer initiated his pioneer experience in western Nebraska, which section of the state at that time was but sparsely settled and practically undeveloped. He came to Cheyenne county and took up homestead and tree claims in what is now Scottsbluff county, and in due time he perfected his title to this land, upon which he made the necessary development and improvements. He then entered the employ of the owners of the Cedar Creek ranch, and for six years he was foreman. He then purchased a ranch on the south side of the North Platte river, in what is now Morrill county, and resided there, engaged in the cattle business about six years. He then bought a tract of eighty-seven acres close to Lisco, where he has made good improvements and has since maintained his home, his attention being given to diversified agriculture and raising hogs, shipping about eight carloads annually. He is affiliated with and carries insurance in both the Modern Woodmen of America and the Woodmen of the World. He is a Republican. Mr. Meyer has been successful and among his investments is stock of the Lexington Milling Company, at Lexington, Dawson county, three hundred and twenty acres of land, and sixty acres in Texas. The religious faith of Mr. Meyers is that of the Catholic church, of which both he and his wife are earnest communicants.

At Weyerts, Cheyenne county, October 15, 1895, was solemnized the marriage of Mr. Meyers and Miss Alma C. Johnson, who was born in Sweden. Her family came to America and settled in Iowa, and in 1888, came to Cheyenne county, Nebraska, taking up land near Broadwater, in what is now Morrill county, where the father engaged in farming and stock-raising, he was seventy-five years of age at the time of his death, in 1915; his wife died in 1913, at the age of seventy-four years. Mr. and Mrs. Meyer have one daughter, Daisy D., who is at home. She graduated in St. Bernard Academy at Nebraska City, and later from the Nebraska State Normal School at Chadron, as a member of the class of 1919,

FRANK F. FISCHER

then taught two years in the city schools of Bayard, Nebraska

FRANK F FISCHER — At the present time there are few lines of trade that demand more keen and careful business ability than the buying and distribution of groceries At the close of a great war prices on all foodstuffs for a time are unstable and it requires good judgment and wise foresight to provide for present and future conditions In this regard Scottsbluff is fortunate in many ways and attention may be called to a prospering grocery enterprise that has but recently entered the commercial field here, the Fischer Grocery Company of which Frank F Fischer is president and general manager

Frank F Fischer was born at Humphrey Platte county, Nebraska, October 12, 1894 and is a son of Jacob and Antonette (Uphoff) Fischer The father of Mr Fischer was born in France and the mother in McHenry county, Illinois They have had six children Frank F being the youngest of the three survivors The others are Anthony who is a merchant at Humphrey, Nebraska, and Anna, who is the wife of Frank Hodgin of Sioux City, Iowa In 1873 the father of Mr Fischer came to Nebraska and homesteaded in the eastern part of the state, and the mother's people came about the same time and homesteaded in Harrison county Both parents now live at Humphrey, the father being a retired merchant, and they are faithful members of the Roman Catholic church The father is affiliated with the Democratic party

Frank F Fischer was graduated from the public schools of Humphrey in 1908 when but thirteen years old, and immediately sought employment, for a time actually earning his living by carrying brick He had a natural leaning toward merchandising and while clerking for three years in a store at Humphrey, devoted himself closely to the interests of his employers and learned business details In 1912 he came to Scottsbluff and was given charge of the Diers' Bros Grocery Company, with which he continued until he entered the National army He was sent to Camp Funston for training and served eight months, with rank of first sergeant in an infantry company After his honorable discharge in January, 1919, he returned to Scottsbluff and embarked in business for himself, organizing the Fischer Grocery Company This company is capitalized at $25 000 Mr Fischer is president and general manager and has experienced associates with him Excellent quarters have been secured, the stock is fresh and up-to-date, the management is courteous

and delivery prompt Mr Fischer is a member of the Roman Catholic church and he belongs to the Knights of Columbus He is intelligently interested in public affairs but has not been personally active in politics He is a member of the Elks lodge and was elected city treasurer in 1919

HIRAM W MAXWELL has proved himself a man of boundless energy and resourcefulness during the years of his residence in western Nebraska in which part of the state he is entitled to the fullest of pioneer honors, for here he has not only given forceful power to the promotion of agricultural and live-stock enterprises but has also been a successful contractor and builder, his activities along this line having greatly conserved material advancement in connection with the splendid development of the country He is still the owner of valuable farm property, was in earlier days influential in gaining many valuable settlers for this part of Nebraska, he is a stockholder in the farmers' grain elevator at Oshkosh, in which city he owns and occupies a modern residence, and he still continues to give no little time and attention to contracting and building He was the first man to bring full-blooded stock into the Garden county region of the Nebraska Panhandle and he has been prominent in community affairs of public order, as indicated by formers service in the office of justice of the peace and by fifteen years efficient service as deputy sheriff and constable In connection with his civic liberality and loyalty he has never swerved from close allegiance to the Democratic party, of whose principles he is a staunch advocate

Hiram Wise Maxwell may well take pride in claiming the old Buckeye state as the place of his nativity, and in being the scion of one of the old and honored families of that historic commonwealth, of which his father, Dr David Cyphord Maxwell, likewise a native, he having been reared and educated in Harrison county, that state, and having been a man of fine culture and professional attainments, as he was not only a physician and surgeon but also a dentist his degree of Doctor of Medicine having been received from one of the leading medical colleges of Ohio In addition to his successful professional activities Dr Maxwell was closely identified with agricultural enterprise, and he was a resident of Holmes county Ohio, at the time of his death, when seventy-three years of age His widow, whose maiden name was Christina Myers, was born in Pennsylvania but reared and educated in Ohio, she having attained the venerable age of eighty-five years. Of their children four sons are living at the time of this writing. A. D. is manager of the new employees of the Kelley & Springfield Rubber Company, at Cuyahoga Falls, Ohio, Robert M., a skilled mechanic by vocation is a resident of Holmes county, Ohio, Hiram W., of this review, is the next younger, and Allen, an engineer by vocation, is now a resident of Ocean view, California, he being a veteran of the Spanish-American War.

Hiram W. Maxwell was reared in Holmes county, Ohio, and there received the advantages of the public schools, including the high school. He was born in that county on the 28th of March, 1867, and there continued his residence until he had attained to the age of seventeen years, when he gained pioneer distinction in Kansas, where he continued to be engaged in agricultural pursuits until 1888, when he provided for himself more strenuous pioneer experiences by coming to the comparative wilds of western Nebraska and establishing himself in old Cheyenne county, which then included Deuel, Garden and other counties of the present day. In the following year he filed entry on homestead and pre-emption claims, in what is now Garden county, and he has made excellent improvents on this property, which he still owns. For eight years he was actively engaged in contracting and building in Dawson county, as a skilled workman at the carpenter's trade, and he then returned to Garden county and engaged actively in general farming and in the raising of horses, cattle and hogs, with which basic lines of industrial enterprise he has here continued to be identified since that time, the while he has continued also to give considerable attention to contracting and building, in which field he has gained high reputation.

In Cheyenne county, on the 4th of October, 1892, Mr. Maxwell wedded Miss Lottie Johnson, and she died at the age of twenty-four years, being survived by one child, Esther, who is now the wife of Walter Cooper, of Varney, Montana, and who has three children. At Lexington Dawson county, on the 4th of October, 1900, was solemnized the marriage of Mr. Maxwell to Miss Ida B Woody, who was born in Indiana but reared and educated in Nebraska, she having been a professional nurse prior to her marriage. Mr. and Mrs. Maxwell have still as members of the happy home circle their fine family of eleven children, and it may well be imagined that the home in one of superabundant vitality and happiness. The names of the children are here given in respective order of birth. Melvin B. Bessie L. Minnie M., Christena A., Myra E., Fern J., Isabelle, Ruth, John, better known as Jack. Margaret, and Lucille

WILLIAM I DYSON, who is serving his fourth term as sheriff of Morrill county, is well known all through this section of Nebraska for personal courage and his wise discrimination whereby many criminals have been brought to justice and the laws have been upheld. Aside from his high office, Sheriff Dyson commands the respect of his fellow citizens and enjoys their esteem. He homesteaded in Morrill county in 1906, and has been a resident of the state since 1883.

He was born in Ottawa county, Missouri, September 9, 1869, the son of Thomas Dyson, who was born in Ohio, a son of Thomas Dyson, who was a native of Indiana. The grandfather died at St. Louis, Missouri, during the great epidemic of cholera. Sheriff Dyson's mother was born and married in Iowa and both parents died on the home farm in Ottawa county, Missouri. The father was a man of importance there, a zealous Republican and for years a county commissioner. He was a veteran of the Civil War, having served over four years in that struggle, being a member of Company H, Thirty-sixth Iowa infantry. At Blundy's mill, Arkansas, he was captured by the enemy and for a year was confined in a military prison in Tyler, Texas. For over sixty years he was a member of one Masonic lodge. Both he and wife were members of the Christian church. Of their seven children five are living, William I being the eldest of the family.

William I Dyson obtained his education in the district school of Nodaway county, Missouri. His early years were spent on a farm and he both owned and rented land in Missouri before he came to Johnson county Nebraska, where his father had purchased a farm. Sheriff Dyson lived on that farm until he came to Morrill, old Cheyenne county, in 1906. He homesteaded in that year and lived on the land for five years, then traded for property in Alliance, and moved from his homestead to Bridgeport in 1912. For six years, and while living on his homested, he was in the employ of the Bridgeport Lumber Company. In 1911, he was first elected sheriff of the county and the high regard in which he is held is indicated by his subsequent elections to this responsible office.

On September 29, 1892, Mr Dyson was

united in marriage with Miss Anna M Stevens, who was born in Iowa Of their family of ten children the following survive Rena F, who is the wife of Eugene Hall a railroad man at Bridgeport, Amy who is a teacher, Alta, who is the wife of Glenn Brown, a farmer in Morrill county, Thomas Allen, who has returned home safely from two years of military overseas service in the World War, was a member of the Thirty-second Division, Mildred and Walter B, twins the former of whom is a student in the high school, and the latter of whom died at the age of three years, and Clifford, woodrow, Leota and Phyllis Sheriff Dyson and his family belong to the Baptist church He belongs to the Woodmen lodges and is also an Odd Fellow, having passed the chairs in the local lodge in the latter organization Ever since McKinley was elected he has been a supporter of the principles of the Democratic party

SAMUEL S GARVEY — The grain and lumber interests of Morrill county are of great importance and when ably handled by men of sound judgment and business acumen, bring large returns that add to the wealth and commercial standing of the county At the present time perhaps no one man at Bridgeport is more profitably identified in this way than Samuel S Garvey, who, in addition to having large individual interests is secretary and treasurer of the Bridgeport Lumber Company

Samuel S Garvey was born in Lake county, Indiana in December, 1861 the only son of Samuel and Julia (Halloway) Garvey The mother was born in Virginia but was married in Indiana, in which state the father was born and died, his death preceding the birth of his son Mr Garvey has one sister, Nancy Hecker, who is the wife of a railroad man at Porter, Indiana The father owned a farm in Lake county His father Duncan Garvey, was born in Ireland

Under his mother's tender care, Samuel S Garvey grew into sturdy boyhood attended the public schools and made himself useful on the farm In 1879 he came to York county Nebraska, where he was employed on farms for a number of years, then purchased land and carried on agricultural industries there until 1902, when he moved to McCook There he went into the elevator business in which he continued for fifteen years and did well In January, 1917, he came to Morrill county and soon bought an interest in the Bridgeport Lumber Company, becoming secretary and treasurer of the concern and also assistant

manager Additionally he owns an elevator and lumber yard at Dalton, Nebraska his trade territory being all through the valley

In 1886, Mr Garvey was united in marriage to Frances Kleinschmidt, who was born in Germany, in which country her parents died They have one daughter Cora, who is the wife of Hurley Dye a merchant at Wyola, Montana Mr and Mrs Garvey are members of the Methodist Episcopal church He is influential in county politics and in the fall of 1918, was elected a county commissioner He is a Knight Templar Mason and has served as high priest and eminent commander and belongs also to the Mystic Shrine Mr Garvey is one of Bridgeport's upstanding, representative citizens

OVA N THOSTESEN the fortunate owner of one of the best improved irrigated farms in Morrill county located just one-half mile from Bridgeport, is a well known engineer on the passenger run between Bridgeport and Morrill, on the Burlington Railroad Since he was nine years old Mr Thostesen has lived in Nebraska, but his birth took place in Illinois March 27, 1871

Mr Thostesen's parents were Zachariah and Annie (Miller) Thostesen, both born in Denmark They came to the United States in 1865, and for a period of fifteen years lived in Illinois, where the father was a farmer In 1880 they came to Nebraska and the father first bought a tract of land from the Union Pacific Railroad but in 1883 moved from Minden in Kearney county to Custer county where he homesteaded on what is known as West Table He lived there for nearly twenty years and then moved to Broken Bow where he and wife still reside He brought some capital with him from Illinois and moved into Custer county with three yoke of oxen and two teams of horses In politics he is a republican and he and wife are members of the Lutheran church Of their six children Ova N is the eldest the others being Miller who made a fortune as a miner in Alaska, lives at Seattle Washington, Barbara who is the wife of J A Myers, a farmer near Broken Bow, Marie who is the wife of Peter Hartvigson a miner in Washington Florence the assistant cashier of the First National Bank of Bridgeport is the Widow of Fred Crom formerly of Sargent Nebraska, who was in the cattle business, and John S, who is a railroad engineer working out of Bridgeport

When Mr Thostesen was a schoolboy in

Custer county he received instruction in a sod house, similar to the one in which the family lived, for this was the usual type of dwelling in that section in early days He worked on his father's homestead and in fact continued to farm until 1899, although in 1890, he had begun to work on the Burlington Railroad as a fireman Through faithful attention to duty he received promotion, in 1895 was made engineer, and since 1913, has been passenger engineer between Bridgeport and Morrill

In 1897, Mr Thostesen was united in marriage to Miss Katie Reeder, who was born in Iowa, a daughter of Henry Reeder, who came to Custer county in 1892, and now lives retired at Bridgeport Mr and Mrs Thostesen have the following children Zeta, who is in the employ of Armour & Co, at Denver, Ivol, who is attending college at Hastings, Nebraska, Elna and Uena, both of whom are students in the high school at Bridgeport, Dean and Byrl, both of whom are in school, and Gayl who has not yet outgrown babyhood The family belongs to the Presbyterian church Mr Thostesen belongs to the independent wing of the Republican party Since coming to Bridgeport, in 1907, he has taken an interest in public matters, especially in regard to the public schools and has served on the city school board He is a member in good standing of the Railway Brotherhood of Firemen and Engineers Mr Thostesen is an honorable and upright man and is very highly esteemed by everyone and is very popular with the patrons of the Burlington whom he carefully and safely carries back and forth daily

CLARENCE S CHAMBERS, one of the younger business men of the Panhandle who has made a name for himself as a ranchman and is now a manager of the Central Graineries Company of Lincoln, having charge of the branch in Sidney, is a native son born in this city December 25, 1885, the son of Judge Chambers, a sketch of whom appears in this volume Clarence Chambers was reared and educated here, attending the public schools until he graduated The life of the cowboy appealed to him and he engaged in the cattle business for several years until familiar with all its phases This led to his becoming a ranchman He began farming and carried on extensive operations upon a fifteen hundred acre ranch There Mr Chambers was successful, made considerable money and became a well recognized figure in farm and ranch circles He was a successful grain man and so was chosen to handle the extensive

grain business of the Central Graineries Company in Sidney when wheat became one of the great products of this section His able management has built up a fine business for he is naturally constructive, is a buyer of ability and today is one of the representative men in business circles of the Panhandle He has all the push, aggressive ability and progressiveness accredited to the native sons of Nebraska not only in business but in his efforts to build up Cheyenne county and Sidney

Mr Chambers was married November 27, 1909, at the bride's home near Sidney, Nebraska to Pauline Wolfe, who was born in Sidney She is the daughter of Frank and Caroline Wolfe, both of whom were born in Germany Mrs Chambers is the second of the three children in the Wolfe family, the other two being boys Mrs Chambers has four half sisters and one half brothers Her father died when she was two years old, and the mother married Otto Kurz and continued to live on the old home place, located southwest of Sidney Mr and Mrs Chambers have four children, two boys and two girls Mr Chambers is a member of the Odd Fellows and he and his wife are members of the Episcopal church They are building a beautiful, modern home in Sidney Mr and Mrs Chambers are well and favorably known and they are members of the younger social circles of Sidney and have a host of friends

THOMAS C. MINTLE has indubitably shown his potency for achievement in connection with the basic industries of agriculture and stock-growing, of which he was a well known and highly successful exponent in Scottsbluff county, where his well improved ranch property, comprising two hundred and forty acres was situated in section twenty-two, township twenty-three, range fifty-five, six and a half miles distant from the city of Scottsbluff

Thomas Clark Mintle was born in the exceptionally beautiful little city of Glenwood, Iowa, and the date of his nativity was September 28, 1867, which shows that his parents were numbered among the pioneer settlers of that favored section of the Hawkeye state He is a son of William H and Mary (Clark) Mintle, each of whom has passed the psalmist's span of three score years and ten, their home now being in Glenwood, Iowa, where the father and mother live in well earned retirement, after many years of earnest and fruitful endeavor William H Mintle was born in Trenton, New Jersey, and his wife

is a native of Ireland Mr Mintle settled in Mills county, Iowa, in 1867, and here he reclaimed and improved a pioneer farm, to the supervision of which he continued to give his attention for many years He is a staunch Republican in politics and both he and his wife are zealous members of the Holiness church

Thomas C Mintle was afforded the advantages of the excellent public schools of Glenwood Iowa, and in his native county he continued his active alliance with farm enterprise until 1909, when he came to Scottsbluff county, Nebraska, and purchased a relinquishment to the land that constituted his farm, one hundred and sixty one acres of it being supplied with excellent irrigation facilities and all improvements and accessories being well arranged to insure his success as a vigorous exponent of agricultural and live-stock industry in one of the best counties in the state of Nebraska In 1919, Mr Mintle moved to Zillah, Nebraska, and purchased a fruit farm, to which he is now giving his attention He is independent of political partisan lines but is always ready to support the various enterprises that tend to conserve the communal welfare and advancement He and his wife hold membership in the Holiness church in the faith of which he was reared

1891, the dual bonds of matrimony united the life destinies of Mr Mintle and Miss Stella Barbee, who likewise was born and reared in Iowa, and the three children of this union are Harry E, Everett and Floyd Everett Mintle was married to Miss Blanche Tanner March 1, 1919, and a little daughter was born to them, December 31, 1920

JOHN A ORR — To the stranger or interested visitor in Scottsbluff this busy, beautiful, growing city would seem many years older than it actually is, for it lacks none of the facilities for modern comfortable living and none of the opportunities for commercial enterprise Many might also be surprised to learn that the pioneer business man of the city is still in active business and still engages very successfully in large commercial affairs He is John A Orr who conducts one of the largest seed houses in this section and also sells more real estate than many of his competitors A man past eighty-four years of age, he is busy every day in his office, does his own writing and attends personally to the details of his various business affairs with perfect possession of all his faculties and enthusiasm

He was born on the Hudson river, in the state of New York September 9 1835 His parents were Benjamin J and Mary S (Folger) Orr, who spent their entire lives in New York Of their thirteen children but two are now living namely John A and Sarah E The latter's married name is Jackson and she lives at Glenns Falls, New York By trade the father was a shoemaker Until 1858 he was a Democrat in politics, but then became affiliated with the new Republican party and continued to approve of its principles as long as he lived Both parents of Mr Orr were sincere members of the Methodist Episcopal church

As one in a large family when money was not very plentiful John A Orr had but limited educational advantages in his youth, beginning to work on a farm when he was twelve years of age, and following that occupation until the fall of 1856 But he cherished an ambition for a wider field of effort, and when he was twenty-one years old he decided that he wanted to see the new western country in his native land He went to Chicago, then a small town, and continued on west to Galena, Illinois, the terminus of the railroad From there he took a Mississippi river boat to Minneapolis, Minnesota, which only had two or three houses at that time He continued his journey by stage to Sauk Rapids, Minnesota about eighty miles from Minneapolis There he worked a short time in a sawmill, and when that shut down he helped in the building of a church and the following spring he took a contract to build five houses His equipment for this task consisted mostly of good American nerve, for he had no capital and was not an expert carpenter but having put in a bid and had the contract awarded to him he knew no such word as give up Procuring the lumber and other material on credit he finished his contract and with this start he continued in the contracting business at that point for about two years, and then returned east to visit his mother

In 1860 Mr Orr went to West Pawlet Vermont, where he engaged in mercantile business for four years, then was appointed postmaster and served in that office and also as justice of the peace for eight years He was married there and made his home in Vermont for a number of years In 1888 he again came west and settled in Lincoln Nebraska where he was connected with an electrical supply house until he came to Scottsbluff county He came here in the interest of the Farmers Canal Company and conducted a store for the company, transporting goods from Lincoln This was a large irrigation enterprise which in later years has been carried to completion and has fully justified the ambitions of those who started it but in the early years of its progress it became involved in financial difficulties, and in 1896 Mr Orr turned his attention to farming in the new country where he had cast his lot He bought eighty acres of land and rented other lands Later he sold his land at a very substantial profit, and came to the place where now is the city of Scottsbluff He put up the first new dwelling house under contract that was built in Scottsbluff, operated the first lumber yard, and in connection with William H Wright conducted the first real estate and loan office He was one of the committee that raised the money to build the first Presbyterian church of Scottsbluff, that cost $5000, when there were only seven members He took the district warrants and built the first school building in the city, which is still standing In his loan business he has loaned money on land for outside investors, according to his own judgment, and although the amounts have been large and the loans many in number he has never lost a dollar for his clients He continues in the real estate business, and few men in the North Platte Valley are better informed along that line than he In addition he conducts his large seed business, the extent of which is proved by his books that show sales of seeds of all kinds in the present year amounting to over $12,000

Mr Orr was united in marriage with Miss Lucinda Whedon in 1862 She was a native of Pawlet, Vermont, and died at Scottsbluff February 1, 1907 She was reared as a Baptist, but on coming to Scottsbluff where there was no Baptist church she and her husband united with the Presbyterian church, of which she continued a devoted member during her life She was a loving wife and mother and an admirable woman in every relation of life To this union were born four children, namely Horace W , who is in the hardware and plumbing business in Boston, Massachusetts, Alice D , principal of one of the public schools of Omaha David A , who is in the plumbing business at Whitehall Montana and Andrew J , a prominent business man of Scottsbluff

Mr Orr is a member of the Masonic order and has always been a staunch Republican in politics He has always been a temperate man and a strong advocate of temperance and can point to himself as an example to prove his faith — hale and hearty and active in business in his eighty-fifth year

Andrew J Orr youngest son of John A and Lucinda Orr, was born at West Pawlet, Vermont, January 31 1877 He obtained his education in the public schools of Lincoln, Nebraska, after which he followed farm work in Scottsbluff county until 1910 He then embarked in the plumbing business at Scottsbluff, under the business style of the A J Orr Plumbing & Heating Company, which he has developed into a very important institution The company does all kinds of tin and repair work, furnace and steam pipe fitting, and plumbing, a general contracting business being done in this line

In 1904 Andrew J Orr was united in marriage to Miss Ethel Sawyer, who was born in Weeping Water, Nebraska, and they have four children, namely John Clifton, Vivian, Lucille, and Alice May, their ages ranging from twelve years to fifteen months Like his parents, Mr Orr is rearing his own family within the folds of the Presbyterian church He is a Republican in politics and belongs to the Knights of Pythias, Odd Fellows, and Modern Woodmen For seven years he served as a member of the city fire department

GEORGE H HILLS has been a resident of Scottsbluff county since 1906, and has reclaimed and improved one of the valuable farm properties, situated near Scottsbluff. In addition to giving his careful supervision to all department of his farm enterprise, in which he has proved very successful as an agriculturist and stock-grower, Mr Hills has served for the past five years as ditch rider and inspector in connection with the governmental irrigation system of the county, his own farm having excellent irrigation facilities He took up the homestead of one hundred and fifty acres in 1906, and he still continues the work of improvement on the place, which he intends to make one of the model farms of this section of the state He came here with practically no financial fortification and has achieved substantial and worthy success, while he holds secure place in popular confidence and esteem in his community He is a Democrat in politics and while still a resident of Illinois he served two terms as township assessor He was the promoter of the establishment of the first rural mail delivery route out of Scottsbluff and has been influential in connection with the affairs of the county water board, which has general control of the irrigation system in the county He holds membership in the Presbyterian church

Mr Hills was born in Brown county, Illinois, and is a son of Charles and Rebecca (Farrington) Hills, who passed their entire lives in that state, where the father was a successful farmer until his retirement, when he established his home at Mount Sterling, where he passed his declining years He was a staunch Democrat, served as county supervisor and assessor, was affiliated with the Independent Order of Odd Fellows and both he and his wife were zealous members of the Baptist church Of their eight children two are living George H Hills continued his residence in Illinois until his removal to Nebraska, and in the meanwhile he had availed himself of the advantages of the public schools, besides gaining practical experience in farm enterprise Upon coming to Butler county, Nebraska he engaged in farming, and there he remained until 1906, when he removed to Scottsbluff county, as has been previously noted in this context He is one of the substantial men of the county and his course has been such as to commend him to the good will of all with whom he has come in contact

In 1878, was solemnized the marriage of Mr Hills to Miss Pauline Green, who likewise was born and reared in Illinois and whose death occurred in 1915 and who is survived by three children Edna Meyer, of Scottsbluff, Myrtle, who is the wife of H O McKinnon, of Scottsbluff county, and Charles is a successful farmer in this county

GEORGE W MOORE, one of the prominent early day railroad men of Cheyenne county and the Panhandle, was born near Milwaukee, Wisconsin He was educated in the public schools and at Madison, Wisconsin While still a young man Mr Moore came west, arriving in Cheyenne county in 1870, where he became associated with the Union Pacific Railroad and was a railroad man until his death For a time the rush to the Black Hills was on when gold was discovered, Mr Moore engaged in the freighting business to the Black Hills He owned sixteen outfits for this business, all teams of oxen and made a success of the enterprise. He was so occupied for several years, then returned to the employ of the railroad, worked up and became one of the prominent men in this field of endeavor, being assigned to the locomotive department

In 1876, Mr Moore married Miss Jane Sweet, a native of Wisconsin They came west after their marriage to settle in Cheyenne county and became the parents of two children Arnold, deceased, and George W, who lives in Sidney Mr Moore died in 1880 He was a member of the Methodist church

Later Mrs Moore was married to Dr S W Boggs, who was born in Alleghany City Pennsylvania Dr Boggs was educated in the public schools of his native state and then entered Jefferson Medical College, where he graduated After completing his medical course the doctor was at Bellevue Hospital for special work to fit him for his profession He came west in 1874, and became the pioneer physician of Cheyenne county, locating in Sidney Dr Boggs became a prominent man in this community was one of the most popular physicians who ever located here and took an active part in the development of the county and town He devoted all his time to the practice of medicine but was also active in helping to build up Sidney

1882, occurred the marriage of Dr Boggs and Mrs Moore and they had one child, Samuel W, who is a railroad man living in Sidney being a conductor Dr Boggs was well and favorably known here until his death, which occurred after many years of service here He was a Mason and a member of the Presbyterian church

Mrs Boggs was later married to Arthur Ragon who was born in Indiana in 1852 He was reared and educated in that state and after finishing his schooling learned the carpenter trade which he followed Coming west in 1900 Mr Ragon engaged in business in Sidney and was active as a contractor and builder until the time of his death He built many of the fine buildings that are the pride of Sidney today and was one of the well known and well to do men of the community Mr Ragon was a member of the Ancient Order of United Workmen and of the Christian church He died in September 1907 Mrs Boggs is a member of the Lutheran church She is one of the pioneer women who has lived to see Sidney develop from a small town to the prominent city of the Panhandle that it is today and has been a part of this for many years

J H FERGUSON — In according recognition to the early settlers of Scottsbluff county, mention should be made of J H Ferguson for while he came here as a youth, he has been a resident of the county for nearly thirty-five years and has not only been an eye witness of the great changes that have taken place but with his father has been an important factor in the development of the industries of the valley

Mr Ferguson was born near Clarence, Cedar county Iowa, in 1875 being the son of James and Isabella G (Anthony) Ferguson, both natives of the Empire state where they were reared and received their education advantages before coming west James Ferguson learned the Photography business in his youth, a vocation he followed several years before he determined to avail himself of the cheap land to be had in the states of the west and emigrated from New York The family first located in Iowa where the father engaged in general farming operations, but in 1886, they came to the Panhandle, becoming pioneer settlers of Scottsbluff county, as James Ferguson located on a homestead of one hundred and sixty acres in old Cheyenne county He began to raise potatoes, being the pioneer man in this enterprise of the upper valley Later he disposed of the home farm and bought other land where he engaged in general farming and stock-raising Mr Ferguson and his wife were members of the Wesley Methodist church he was a Prohibitionist in political belief and a man who took active part in the affairs of the community, having been precinct assessor at one time There were seven children in the Ferguson family Charles D, a real estate dealer of Scottsbluff, William E, a railroad man located in San Francisco, Fanny, the wife of James McKinley, of Scottsbluff county, Mary D, married F G Fanner, a farmer near Scottsbluff, J H, of this review, C B, a contractor of Minatare, Nebraska, and Minnie the wife of C F Shawver, a farmer near Glendale, Arizona

Mr Ferguson was reared during early boyhood on his father's farm in Iowa, he attended the common schools of his district and accompanied the family to the Panhandle when he was eleven years old, then attended school in old Cheyenne county After finishing school and when he had acquired sufficient capital Mr Ferguson purchased three hundred acres of land near Gering in section twenty-four, township fifteen, most of which is under ditch He has devoted considerable study to intensified farming under irrigation and has adopted modern methods and is well equipped with the latest type of machinery for his business He now raises varied farm products and also is becoming a well known stock-raiser of his district Mr Ferguson is one of the progressive men of the Gering community who takes an active part in county affairs, he is a Republican in politics, served as county clerk from 1912 to 1916 and then assumed the office of register of deeds in 1917, serving one term and established a fine record as a public official

In 1897, Mr Ferguson married Miss Edna A Lovelace, a native of Wisconsin and to them one child was born, Ruth, deceased

JEFFERSON DAVIS FUGATE — The life of a professional or literary man seldom exhibits the striking or exciting incident that call public notice and fix attention upon a man His character is generally made up of the many qualities and qualifications which are necessary for the successful prosecution of his duties of his vocation, though such men are largely responsible for the formation of public opinion as they play such a large and important part in shaping it Jefferson Fugate, editor and owner of the Henry Mesesnger, may not deviate from this general rule but since attaining maturity his life has been a full one for he has had a varied career, has followed different occupations in several parts of Nebraska, for he has in turn been student, pharmacist, ranchman, promoter of a town, postmatser, and now it the popular editor of one of the best edited sheets in Western Nebraska In these several field his versatility has assisted him to well deserved prosperity

Mr Fugate was born in Missouri, March 31, 1863, during the closing years of the Civil War, and it may be that the spirit of that memorable conflict entered into his make up and given him the high courage, resourcefulness and determination that was a characteristic of the members of the armies, whether they fought under the Stars and Bars or the Stars and Stripes, for each fought for that high ideal which he believed was right Mr Fugate was the son of Elbert M and Nancy C (Hollcroft) Fugate, the former a native of the Old Dominion and a worthy representative of an old Virginia family that located in that state during its period of early settlement, while Mrs Fugate was a Hoosier and had all the gracious characteristics of the daughters of that state Seven children formed the Fugate family James T, who resides in Missouri, Charles W, living in Greentop, Missouri, Jefferson, of this review, Robert, now a resident of Iowa, Isabella, who married Jefferson D Fowler, of Greentop, Missouri, William E, who lives at Lovila, Iowa, and Drusilla, the wife of O E Campbell also of Greentop, Missouri Elbert Fugate was a well to do farmer of Missouri, where he was known as a man of means and weight in the community He lived to be an old man, passing away on December 21, 1918, having survived his wife who died in 1907 Both Mr and Mrs Fugate were members of the Primitive Baptist church

As his father was a prosperous man young Jefferson was given excellent educational advantages He first attended the public schools near his home in the country and being a farm boy grew up self reliant, resourceful and full of a determination to get ahead in the world At an early age he realized that a good educa-

Mr. and Mrs. Thomas M. Howard

tion was one of the best equipments for the furtherance of his life's plan, so after finishing the grades he entered the high school, pursued a full course and after graduating matriculated in the pharmacy department of Northwestern University in the city of Chicago Being a good and conscientious student he in due time received his degree and was registered as a graduate pharmacist and licensed to practice in 1895 Within a short time Mr Fugate became established in the practice of his profession in Iowa, where he engaged in business until 1906 During his years in business he realized that the man of property is more independent than one tied to such an exacting business as a pharmacist Being a student of affairs he came to realize that many men were making money in the west and at the same time becoming owners of valuable land He studied on the question and having been reared on a farm determined to follow in the footsteps of his father and took up a homestead in Scottsbluff county He at once began the improvement of his land, erected good and permanent farm buildings, built a comfortable home, proved up and for five years was engaged in agricultural pursuits, finding that the practical experience he had gained as a boy came in more than handy now that he was a landed proprietor Mr Fugate was alive to every opportunity to make money and at the same time to assist in the development of this western country, and realizing the necessity of a village where the residents of his section could obtain supplies and sell their produce he became one of the prime movers in the establishment of the town He moved into the village, one of the first families to take up their residence there where the family lived until 1912, when they removed to Henry, locating here July 13, Mr Fugate entered actively into the business interests of his new community, became interested in various enterprises and on September 14, 1914, was appointed postmaster, an office which he has efficiently filled up to the present time Since assuming office he has inaugurated many reforms which have tended to facilitate and expediate the handling of mail which has proved of great value to the residents Realizing of what benefit a well conducted newspaper would be in this thriving community in 1917, Mr Fugate established the *Henry Messenger*, which has grown rapidly and now has a wide circulation in the immediate vicinity of the town and also up and down the valley, showing what success it has attained under his able management It is bright, clean, enterprising and wholesome,

and has gained many fast friends among the reading public, being also well deserving of the support it receives as an advertising medium In connection with the paper Mr Fugate conducts an up-to-date job printing office

May 15, 1888, Mr Fugate married Miss Sunsan M Satterfield, of Pike county, and to them one child was born, Mabelle, who married William F Young and now lives in Brementown, Washington Mr Fugate with his wife is a member of the Methodist Episcopal church of which they are liberal supporters He is a member of the Democratic party in politics while his fraternal associations are with the Masonic order

THOMAS M HOWARD the founder of the business that is now conducted by his family in Scottsbluff under the name of the Howard Greenhouse and Flower Shop, was born at Perry Center New York, December 26, 1855 and died at Scottsbluff, Nebraska, December 4, 1918

His life was an illustration of the principle that while many people admire nature in the abstract, not so many feel the impulse that leads to the cultivation of flowers or the scientific study of the mysterious results of grafting and propagation As a business it may prove exceedingly profitable as has been abundantly proved by the Howards, but it is one that demands special talents hence every individual would not be successful at it The large florist business belonging to the Howards and ably managed by Mrs Howard and her sons includes plants at both Scottsbluff and Gering

Thomas M Howard was married at Batavia, New York, November 30, 1887, to Deborah Tompkins, a native of Ireland She was born July 2, 1860, and when eleven years of age came with her parents to the United States in 1871 Immediately after their marriage Mr and Mrs Howard came to Nebraska and located at Weeping Water, where Mr Howard engaged in the banking business and also handled real estate the latter enterprise he continued after coming to Scottsbluff in 1902 in partnership with John A Orr and the late William H Wright and after Mr Wright's death he continued the same business with Mr Orr for several years The firm name was Wright, Orr & Howard and it was the leading real estate firm in the North Platte Valley

Mr Howard was an active citizen in public affairs at both Weeping Water and Scottsbluff, serving on the town board in both cities He was a man of influence in the Republican party and both himself and wife were most

worthy members of the Presbyterian church
No man stood higher in the estimation of his
friends and neighbors than he, and he has left
behind him the memory of a man who was the
soul of honor, a loyal friend and a successful
business man A lover of flowers all his life,
he was endowed with the gift that enabled him
to handle them successfully He started the
florist business in a small way while engaged in
other enterprises, it being to a large extent a
recreation with him and Mrs Howard, and
they spent many of their happiest hours working together in the greenhouses, but as the community grew the business grew with it until it
became their major occupation

To Mr and Mrs Howard seven children
have been born, all of whom are living

Albert T Howard is the third in age of the
seven children He was born at Weeping
Water, Nebraska, April 6, 1893, was graduated from the Scottsbluff high school in 1910,
and afterwards taught school for one year He
then accepted the position of chemist with the
Great Western Sugar Company at their Scottsbluff factory resigning after two years to enter
upon the duties of assistant postmaster, in
which position he remained five years He then
resigned in order to enter the service of his
country during the World War, being a member of Company G, Fourth Nebraska Infantry,
Battery D, One Hundred and Twenty-seventh
Field Artillery, serving from September 18,
1917, to December 5, 1918 For ten months he
was at Camp Cody, New Mexico, having been
commissioned at Camp Zachary Taylor After
his return from the army he resumed his peaceful activities in the florist business and has ably
carried it on In politics he is a stalwart Republican, and he is a member of the Presbyterian church

S Morton Howard, who attends to the
Howard floral interests at Gering, was born at
Weeping Water, Nebraska, March 30, 1896
After completing his course in the Scottsbluff High School he took a course in stock
judging, and for one season was with Jesse
Harris at Fort Collins, Colorado From there
he entered the National army in Battery A,
Three Hundred and thirty-eighth Field Artillery, and was trained at Camp Dodge, Iowa
During his period of overseas service he was
stationed at Winchester England, and Bordeaux, France, between May 23, 1918, and
December 25, 1918 He was honorably discharged at Camp Dodge on January 26, 1919,
returned home immediately, and since then
has been located at Gering

Richard L Howard is pronounced by his
brothers to be the most capable member of the

family as a florist In 1918 his father had retired from active business and left the management of the flower culture and sale to the
three sons When the two older sons joined
the army Richard consented to remain at
home and look after the business, since he
was under the draft age, being only twenty
years old, although he was eager to go as a
volunteer and would have done so if his sense
of duty to his mother had not prevented
While his brothers were absent in the service
he shouldered the entire care and responsibility of managing the greenhouses and truck
gardens, and ably performed the duty An
additional burden was the death of his father,
which occurred during that time Richard
was born at Weeping Water, Nebraska, April
25, 1898, and was educated in Scottsbluff
Since boyhood he has been connected with
the floral business and expects to make that
his life work

The family now owns a five-acre tract of
land, with five greenhouses and an up-town
shop, in addition to the plant at Gering Mr
Howard left quite a valuable estate, the accumulation of a lifetime of concentrated effort
and honorable industry, and all his sons are
highly esteemed as good citizens and capable
business men Mrs Howard is also a good
business woman and expects to continue active in carrying on the business in connection
with her sons Her interest and pleasure in
life are her family, flowers, and home

The other children are four daughters R
Janet, who has been a teacher in Wyoming,
and now resides at home, Roxa L, the wife
of Mark M Patterson, superintendent of
schools at Hamilton, New York, Marianna,
now attending the University of Nebraska, at
Lincoln, and Lydia A, a student in the Scottsbluff high school

WILLIAM BARNES, who is now to be
designated as one of the progressive and successful representatives of agricultural and live
stock enterprises in Scottsbluff county, gained
in his youth experience as a cowboy on the
great open ranges in the Lone State state, and
the vigor and self-reliance which this service
involved have never left him in the later years
which he has given to productive enterprise
along other lines of activity He is one of the
substantial and popular citizens of the northern
part of Scottsbluff county, where his well improved farm is situated about five miles distant from Mitchell

Mr Barnes was born in St Clair county,
Missouri, March 8, 1857, and is a son of

Lindsey and Mary (Preston) Barnes, the former a native of Virginia and the latter of Tennessee. Of the family of nine children the first born was Martha, who is deceased, and the second is William, the immediate subject of this sketch; George Robert and Lindsey Francis are deceased; Ann Eliza is the wife of Charles Hilton, of Appleton City, Missouri; Mary Ellen is the wife of Edward Shrewsberry, of Osceola, that state; Ada and Elizabeth are deceased; and O. K., who has been prominent identified with government reclamation service in western Nebraska, maintains his residence at Mitchell, Scottsbluff county.

Lindsey Barnes, a scion of an old and honored Virginia family, became a prosperous farmer in Missouri, where he continued to be identified with agricultural pursuits until he had attained middle life. He was influential in political and public affairs in Clark county, that state, where he served some time as deputy county sheriff, the cause of the Democratic party having recived his unwavering support. He was a member of the Baptist church, as is also his venerable widow, who celebrated in 1919, the eighty-seventh anniversary of her birth and whose home is at Appleton City, Missouri, though she endeavors to pass a portion of her time in the homes of her several children, all of whom accord to her the deepest love and veneration.

The common schools of his native state provided William Barnes with his early educational advantages, and there also he gained his initial experience in connection with farm activities. As a young man of twenty years he made his way to Texas, where he found employment in herding cattle on the range, his service as a cowboy having continued for seventeen years and the free and vigorous life having given him a vigor that stands him well in these later years. He has been actively concerned in the construction of irrigation systems in Nebraska and other states, and in this connection he aided in the building of twelve miles of government irrigation ditch in Scottsbluff county. Here he took up a homestead in the year 1906, but was absent from the county thereafter until 1908, when he established his permanent residence on his land, to the development and improvement of which he has since given his close attention. He has a quarter section of land, upon which he has made good improvements, and practically the entire tract is under effective irrigation. He has erected good buildings on the place and has one of the attractive farm homes of the county. The land is most productive, and in addition to the propagation of the various crops best suited to this locality he gives attention also to the raising of good types of live stock. Mr. Barnes maintains an independent attitude in politics, and takes lively interest in community affairs, as a loyal and public-spirited citizen. He is affiliated with the time honored Masonic fraternity.

JAMES FINN, one of the large landholders and successful farmers of the Broadwater district, who is known for his business ability was born in Wisconsin, January 10, 1853, the son of John and Celia (McGuire) Finn, both born and reared in Canada. To them were born eight children, six of whom are living: James, of this review; Mary Jane, deceased; John, of Morrill county; Josephine, who lives in Mason City, Iowa; William W., of Wesley, Iowa; Jarvis, deceased; Frank, of Geneva, Iowa, and Bert of Dumont, Iowa. The father was a farmer in Wisconsin, who moved to Iowa, when James was sixteen years of age. The family lived there many years; the mother died in 1910 and the father in 1907.

James Finn spent his early youth in Wisconsin, attended the public schools for his education and when that was finished began to farm. After coming to Iowa he farmed some and then learned the carpenter's trade, at which he was employed for fifteen years before coming to western Nebraska, thirty-one years ago last March, settling in old Cheyenne county when it was little developed. During the early days of Sidney, Mr. Finn worked on many of the buildings of that growing town, spending three years as a builder there. He then moved two and a half miles west of Bridgeport to a homestead of eighty acres. Farming was well known to him, he made a success of it even at this early date, made money and from time to time bought more land until today he owns more than seven hundred acres, all well omproved land. For years he has been engaged in general farming and stock raising. He is a shrewd man of foresight and by short buying and long selling has made a comfortable fortune here in the west.

In 1888, Mr. Finn married Miss Louise Fuchs, and to them have been born two children: Frank, who died in infancy, and Albert, who is at home. Mr. Finn is a member of the Catholic church and is a Republican. He takes an active interest in local affairs and backs all movements to develop his section of the county.

HERBERT G RUSSELL — In every community will be found quiet, industrious, business men, following various vocation, without whom the many industries of civilized life could not go on, and very often it will be found that they are self made men, having, unaided, built up their own fortunes   One of the leading men of the generation who typifies this class is Herbert Russell, who not only has but still is, taking an important part in the development of this western section of Nebraska

Mr Russell was born in Vernon county, Wisconsin, February 28, 1877, being the son of Calvin W Russell, a pioneer settler of the Badger state, who became a prosperous farmer there   Herbert Russell received his elementary education in the public schools of Nebraska, after his family moved to this state   After completing the graded work he continued his studies in the high school, graduating from a four year course   Having been a good and rather brilliant scholar he was urged to take the teacher's examination, which he passed brilliantly and the following fall found him a full fledged pedagogue   Finding the scholastic profession to his liking and being so successful in this line, Mr Russell continued teaching for some years   He realized the full benefits to be gained from higher education and during the summer vacations attended the State University at Lincoln, thus gaining not only benefits for his year's work in school but at the same time broadening his view of life   While he had achieved an enviable reputation as a teacher, Mr Russell soon felt that there was not enough of a future in teaching and decided to enter commercial life where his constructive talents would have greater opportunities for use and development   Coming to Henry, Mr Russell opened a real estate office with an allied line, insurance, and to this business he has since devoted his time and attention   As correct surveys are of the utmost importance in land deals, Mr Russell became interested in surveying, studied that subject and has done a considerable amount of the surveying in the western part of the state in the vicinity of Henry   Of a pleasing personality, a man of high ideals both in his private and business life, it is but natural that Mr Russell gained warm friends in Scottsbluff county where he has a reputation for straight dealing, personal honesty and the careful execution of all business obligations   He has built up a lucrative business which brings in most gratifying returns and is well and favorably known throughout the Panhandle   Mr Russell is really a pioneer of this section as he first came here when the valley was thinly populated and it is most interesting to get him to recount his early experiences, which in many cases were privations and hardships but not so according to his telling   His brother, James Russell, was one of the earliest settlers of the Mitchell district, long before it was known by that or any name and he recalls how the postoffice of Mitchell, a frame structure of one room only, about ten feet square, was put up over night of necessity   James Russell became the first postmaster and Herbert was often in the office with him   Thus the youthful school master aided in shaping the future development of a section which he has seen change from raw, unbroken prairie to rich, arable farm land where bounteous crops reward the labors of the husbandman   Knowing the country as only an early settler could, has greatly aided Mr Russell in his business to the end that he is regarded as one of the best buyers and sellers of real estate in the valley   Mr Russell is an independent in politics, casting his vote for the man best qualified to hold office, whether in the community, state or nation   The family are members of the Christian church while Mr Russell has fraternal affiliations with the Independent Order of Odd Fellows and the Yeomen

In the fall of 1912, Mr Russell married Miss Nellie Robertson, a native daughter of Nebraska, and to them one child has been born, Ralph Keith

BERT R WEBER, who is numbered among the successful younger exponents of agricultural and live-stock industry in Scottsbluff county, has the distinction of being a native son of this country, his birth having taken place on the home farm of his father, near Gering, on November 14, 1890   Of the family genealogy adequate record is given on other pages, in the sketch of the career of his father, William Weber

Bert R Weber early gained his quota of experience in connection with the activities of the home farm and his youthful educational advantages included those of the public schools of Gering   He continued to be associated with his father in the operation of the latter's ranch until he had attained to his legal majority, when he initiated his independent career in connection with the same important line of industry   As a general agriculturist and stockgrower he is emphatically alert and progressive and he conducts operations on a well improved place of two hundred and seventeen acres, all of which will eventually have ample

Fred F. Everett

irrigation facilities and upon which he is making the best of improvements This ranch is situated in section one, township twenty-one, about a half mile distant from Gering which is the postoffice address of Mr Weber On his farm he has a fine French draft stallion, and he is making a specialty of breeding this type of horses

In 1914, was solemnized the marriage of Mr Weber to Miss Lucile Duff, who was born and reared in Nebraska, and they have two children, a winsome little daughter, Marvel, and a boy, Kenneth Ellsworth Mrs Weber was educated in the public schools of Gering and there she holds membership in the Methodist Episcopal church In politics Mr Weber gives allegiance to the Republican party in national and state campaigns, but in local affairs he maintains an independent political attitude

FRED F EVERETT — Nebraska owes a great debt to the hardy pioneers who came to this commonwealth when a large part of it was unbroken prairie It was such men who opened up the new country and blazed the way for the later development of a rich agricultural district To this honored class belongs Mr Everett, who has been a prominent citizen of Scottsbluff for the past ten years, and one of the extensive dairy farmers of Scottsbluff county He came to Cass county Nebraska in 1879, and has made this state his home ever since Mr Everett has been active in all irrigation projects and has been a director of the Winter Creek ditch since 1901 He is a man of much enterprise, and movements that give fair promise of being beneficial to city and county, always secure his approval Mr Everett was born in Woodford county, Illinois, March 10, 1854, the son of Willard and Frances (Dodge) Everett, the former born in Dedham, Massachusetts, in 1785, and died in 1873, and the latter at Littleton, New Hampshire, in 1810, and died in 1870 After their marriage in New Hampshire, in 1845 they moved to Illinois, where Mr Everett purchased land and engaged in farming until the end of his life By an earlier marriage he had six children, and eight were born to his second union, Fred F being the youngest of the survivors The others are as follows Samuel B, who resides at Santa Anna, California, is a Civil war veteran, having served two years and ten months in Company G, Fourth Illinois cavalry, Edward, a retired farmer, lives at College view Nebraska, and Roselle, the widow of Edwin Benedicite, lives

in Canada The parents were members of the Congregational church On the formation of the Republican party the father adopted its principles and voted the Republican ticket until his death

Fred F Everett attended school in Illinois and later at Tabor, Iowa, having gone to that state when fifteen years old He returned to the home farm after a visit in Iowa but when eighteen years old again went to Iowa where he began working by the month and attended school a greater part of the time for about ten years, when he moved to Weeping Water Nebraska For almost twenty years he farmed in Cass county but in the spring of 1901 he came to Scottsbluff county and bought land, subsequently adding other tracts, and now has three hundred and twenty acres in his dairy farm which his son Lee operates with from ninety to one hundred and twenty-five head of cows, being the largest dairy in western Nebraska Mr Everett is also one of the extensive beet growers of this section, there being about two hundred and fifty acres of that crop grown on his land this year In 1909 he moved into Scottsbluff, his eldest son taking charge of the dairy business Mr Everett is one of the directing board of the Farmer Canal, and is one of the directors of the Platte Valley State Bank

In 1880 Mr Everett married Miss Minnie Fitts, who was born at Gustavis, Ohio, and they have four children Lee A, who has charge of his father's agricultural interests, married and has had five children, Frederick, who is deceased, Richard, Donald and Margie and Betty, Clare E, who married and is deceased, left two children, Earl and Katheryn, Pauline, the wife of Joseph Spurgeon, a farmer in Scottsbluff county, has two children John and Robert, and Jean, the wife of Thomas Richardson, a farmer in this county Mr Everett and family belong to the Presbyterian church Politically he is a prominent factor in the Republican party and is a member of the city council in which position he has served several terms

WARREN BEATTY — Known and valued as one of the enterprising and representative agriculturists, stock-raisers and feeders of Scottsbluff county Mr Beatty is the owner of a well improved farm of twenty-seven and a half acres of land under the highest state of cultivation He was born in northwestern Iowa in 1870, the son of Martin and Lucy (Lampher) Beatty The mother was a native of Illinois and the father of the famous blue grass state, Kentucky The parents were married in Illinois where the father successfully conducted a farm for some years and they had a happy home, but the call of the west was loud in their ears and their rapidly increasing family gave them additional reasons for seeking more room and greater opportunities both for themselves and their children With this end in view they left Illinois, crossed the Mississippi river and located in Iowa, but within a few years they went to Missouri, but still the lure of the west was beyond and heeding its call they finally came to Nebraska and were satisfied, it was the goal of their ambitions The family settled in Banner county on a homestead which the father secured, here a home was established and the children reared After twelve years the father returned to Missouri where he died, but the mother rounded out her life in Banner county

There were twelve children in the Beatty family, four of whom are now living in Scottsbluff county, and of these Warren is the youngest The others are James and Charles, both retired farmers, and Wallace, a dealer in sand Warren, the subject of this review received such rudimentary education in the public schools as the child of pioneer parents with a large family to raise could afford While still a small boy he can recall that his duties consisted in herding cattle on the family homestead and such light work as children are always able to do on a farm Reared in the new country under pioneer conditions, the boy learned the lessons of industry and turned vigorously to help reclaim the family homestead from unbroken prairie and make it arable, productive land The call of the west descended to the son from the parents and after helping the family at home for some years the young man went to Wyoming where he gained valuable experience in stock-raising, feeding and buying and selling cattle on the great ranches there Mr Beatty was frugal in his habits, saved money and after a period decided to return to Nebraska as no other country looked so good to him or offered greater opportunities to a man inured to work, who in childhood and early youth had passed his life on a farm under the sturdy discipline of such an environment Mr Beatty purchased twenty-seven and a half acres of land in Scottsbluff township which he has brought to the highest state of cultivation, as he devotes his entire time and energy to his business He raises beets on a portion of his irrigated land, fodder crops on the remainder which he uses to feed cattle as he is a heavy stock buyer,

feeder and shipper Today he is enjoying the well deserved and well earned fruits that crown the success of a man who knows his business, who has by thrift, economy and industry amassed a comfortable fortune which all his family enjoy and with which he will be able to give his children many advantages not afforded him in childhood The Beatty home is a very happy one as there are three children in the family, Fern, Merle and Edward Mrs Beatty is affiliated with the Presbyterian church, to which she and her husband are liberal contributors They represent the best element in communal life, are people of high ideals and fine character well recognized by their friends and associates Mr Beatty is a leader in his community, his progressiveness extends also to his status as a citizen, and he takes deep interest in community affairs, although he has no ambition for public affairs believing that his energies are best expended in his business

THOMAS R EVERETT — Eminently successful in his farming operations, the life of Thomas Everett has been an expression of diversified activity and in its range has invaded the realms of ranch life and agriculture, in which he has successfully accumulated a large and comfortable property He is in every sense a self made man, in that he has built up his fortune through his own industry, strict attention to business, far sighted vision along agricultural lines and honest dealing Mr Everett is in a way a pioneer for he first came to this section in 1885, and while he suffered the vicissitudes, hardships and trials of frontier life they did not daunt his high spirit For many years he has contributed to the civic and industrial development of the community where he has made his home

Thomas Everett is a native of the Keystone state, born at Crawford, Pennsylvania, August 13, 1877, the son of William W and Sophia (Carter) Everett, the former a New Englander, born in Vermont, while the father was of southern origin, born in Mississippi To this union seven children were born Nellie, the wife of John Urban, living in Los Angeles, California, Robert, a resident of Scottsbluff county, Eunice, the wife of Washington Ferry, now lives in Cochanton, Pennsylvania, William H, also a resident of Scottsbluff county, Daniel, of Slayton, Montana, John of Baison City, and Thomas, the subject of this review In his native state, the boy gained his youthful education, attending the excellent schools of Harrisburg, until his parents moved to Nebraska, locating in Banner county, then he continued his studies in the frontier schools afforded at that period. The father made a study of crops and grains best suited to the new country and as a result of this was well rewarded by ample crops in the good years, though as other pioneer settlers did, he fought drought and pest, but at last rewarded with marked success. He died November 22, 1913, being survived by his wife until October 11, 1916.

While his father was not among the earliest settlers of our great commonwealth, he came while the country was still but thinly settled and Thomas recalls herding cattle on the prairies and sharing all the hardships, hopes and joys of a youth in a new country. His alert mentality, fine powers of observation and keen intellect make his reminiscenses of those days especially graphic and interesting. As soon as his age permitted the youth became a cowboy, thus gaining invaluable knowledge of cattle, feeding, buying and marketing. Mr Everett is a self made man as he had no help from anyone in establishing himself in business. Cowboy life did not appeal greatly to him after reaching maturity and he decided to locate permanently as an agriculturist. Nebraska was his choice, having known the uncertainty of crops in a semi-arid region he with great foresight determined not to be dependent on rainfall so located in Scottsbluff, purchasing land which would be under water rights. He had been frugal, made money and this was invested in a large tract of land one hundred and thirty acres of which he disposed of some time ago. Today he is the manager of some two hundred acres, well irrigated and has become a successful and progressive exponent of live-stock and agricultural enterprises in Scottsbluff county.

November 22, 1905, Mr Everett married Miss Anna Shire, a native daughter of the Sunflower state, born in Crawford county, Kansas, where she was reared and educated. Three children have come to make this a happy home Thelma, Ivan and Opal. Mr Everett is one of the valued member of the community in which he lives, is ever ready to take part in any movement for the development or improvement of its welfare but has never sought public office, as political life in no way appeals to him. In politics he is an Independent, while his fraternal affiliations are with the Independent Order of Odd Fellows.

EARNEST G ROUSE, one of the well known and progressive farmers of the Bridge-port district who has made a success of his business and is today recognized as a substantial agriculturist was born in St Clair county, Missouri, March 24, 1871, the son of George Rouse, who lives with his son Earnest and Nancy Elizabeth(Smith) Rouse, The father was born and reared in Canada while the mother was a native of Battle county Pennsylvania, born in 1844, and died in 1896. George Rouse was a farmer and horseman. he served four and a half years during the Civil War in Company A, First Iowa Cavalry. At one time Mr Rouse had his horse shot from under him which crippled him. He owned land in Missouri and later in Iowa but subsequently homesteaded near Litchfield where he proved up and farmed for a number of years. At one time he was the owner of a livery barn at Litchfield and was one of the well known business men of that town. There were four children in the Rouse family Eva, the wife of Charles D Logan, a farmer near Huntly, Montana, Earnest, of this review, Harry, deceased, and a child that died in infancy Mrs Rouse was a member of the Methodist church while Mr Rouse was a member of the Grand Army of the Republic and a Republican.

Earnest Rouse was reared in Missouri and Iowa, and received his education in the public schools. He came to Cheyenne county in June, 1900, and took up a quarter section homestead and began to farm. He placed good improvements on his place and later bought more land. Today, Mr Rouse has eight hundred acres in his ranch, where he carries on general farming and stock raising, breeding a good grade of animals, especially hogs. His duroc Jerseys have proved very lucrative in years past. Mr Rouse has set out a fine young orchard on his place which is doing well.

July 5, 1900, Mr Rouse married Miss Mary A Stevens, a native of Iowa, the daughter of the Reverend A Stevens and Demaris (Blum) Stevens. They were pioneer homesteaders of Litchfield, as the father came to central Nebraska as a Methodist missionary. Later he homesteaded near Court House Rock, both are deceased.

Nine children have been born to Mr and Mrs Rouse Georgia deceased, John and Joe, twins the latter deceased, Laura and Louis, twins, deceased, Mary at home, Eva and Nellie twins, at home and Phillip also at home. Mr Rouse is a Presbyterian and an Independent voter.

M A LONGAN, well known farmer, live-stock man and popular auctioneer of the Broadwater district, who has found irrigated land a profitable investment, was born in Kansas, April 5, 1886, the son of John and Sarah (Howe) Longan, to whom were born nine children M A of this review, Iva, the wife of Bert Fraser, of Kansas, Addie, the wife of Earl Dougles, of Garden county, Frank, living at Gilman, Kansas, Bessie, who married Oran McNerlan, Everett, living in Garden county, Bert, also of Garden county, Ernest, living at home at Galena, Kansas, and Charles, also at home

The father of the family was a successful farmer in Kansas, where he carried on general farming and did some stock-raising He belongs to the Baptist church

Mr Longan was reared on his father's farm and obtained his education in the public schools After finishing his education he began to farm and was so engaged until he came to the Panhandle in 1909, locating in Box Butte county he established himself as a farmer and stockman, meeting with success Already Mr Longon had gained an excellent reputation as an auctioneer, a vocation which he he follows along with his farming business Five years ago the Longan family moved down to the Broadwater section, and Mr Longan bought a quarter-section of irrigated land which he finds is profitable to farm He breeds a high grade of stock and has all his extra time filled by his professional work as an auctioneer He has gained the confidence of the people as a square man of business and is considered one of the most successful auctioneers in the Panhandle where he has conducted many large sales

In 1907, Mr Longan married Miss Maude Reed, of Kansas and one child was born to them, Marion, who lives with her grandparents in Missouri Mrs Longan died April 8, 1911, Mr Longan married Miss Lula Browning, a native of Iowa They have had four children Loraine. Gerald, Olive Judith and one that died in infancy Mr Longen takes an active part in civic and public matters of his locality and is a Democrat

OLMSTEAD BISHEY BROWN — No better illustration of the value of industry, perseverence and the intelligent management of one's resources can be found than in the career of Olmstead Brown, now one of the representative farmers of Sioux county Driving up the valley of the Platte with little capital but his team, wagon and native ability, he has made the most of his opportunities and has so directed his activities that today he is in a position of independence, being respected and esteemed as a prosperous and substantial citizen

Mr Brown was born in Lake county, Illinois, in 1854, he was reared in his native state, received excellent educational training in the public schools and after his school days were over studied photography, a business which he followed in Kenosha, Wisconsin, before removing to Chicago Later Mr Brown went south, locating in Atlanta, Georgia, and from there went to New Orleans, Louisiana During this period he had heard of the fine land to be taken up in the prairie states under the homestead acts and determined to own a farm of his own Coming to Sidney in 1886, he took a pre-emption near the present locality of the town of Scottsbluff, proved up on the land, placed the required improvements on it and lived there for some time, but the late eighties and early nineties were hard years on the pioneers and as money was scarce, the crops having been ruined by the droughts and insect pests, Mr Brown turned to his profession for a livelihood and opened a photograph gallery at Occola, Nebraska, and by this means tided over the period of financial stringency, but he still longed for property of his own and returned again to Scottsbluff county in 1893, driving up the valley to Gering in a wagon, true pioneer style This time he filed on a homestead of a quarter section in Sioux county, in the Morrill valley, where he has since resided Mr Brown soon had his land under cultivation, made good and permanent improvements in the way of farm buildings and a comfortable home and began again his business career of diversified farming and stock-raising Being a man of intelligence, Mr Brown gave considerable study to agricultural subjects, was willing to take the advice of farm experts, both state and national, and from the first has specialized in high bred stock, as he raises pure blooded Duroc Jersey hogs and Jersey cattle, finding that there is a greater return from these than from grades Every year he makes large shipments of stock to the packing centers in the eastern part of the state Mr Brown is a Republican in politics, has taken a prominent part in the affairs of his district, as he has been school director and also school commissioner, he is progressive in his ideas and has advocated the latest and most modern methods in the schools

Mr Brown was united in marriage with Miss Cartha G Marker, a native of Wiscon-

CHARLES II SIMMONS AND FAMILY

sin, where she was reared and educated To them ten children were born seven of whom survive Frank, deceased , Homer deceased , Harlie, at home, Hattie. a school teacher of foreign languages in the high school at Columbus , Chester, who has charge of the management of the home farm , Beatrice a trained nurse in Seattle, Washington . Rush D , a mechanician in a garage, Royal, at home and Florence M , who lives in Seattle

Mr Brown has lived through many experiences in the Panhandle, from sod houses and prairie fires to the present day prosperity and now that the sunset years have come and the shadows are lengthening from the west, he has given up active participation in business, leaving such affairs to the capable hands of his sons. to whom he has handed the torch or progress, which he with the other early settlers, carried forward for so many years, and today in looking back across more than three decades in Nebraska, can feel with pride that he played no unimportant part in the growth of this favored section

CHARLES H SIMMONS — To facilitate the important work of presenting to her citizens a reliable history of Scottsbluff county, it is helpful indeed to be favored with the reminiscences of such men as Charles H Simmons, an honored, retired resident of Scottsbluff An early settler in the county, and one of the first residents and business men of Scottsbluff. for years he was prominent in the development of the town, served a decade as postmaster, and gave encouragement and substantial aid to business, educational and religious enterprises The first Sunday school in Scottsbluff county was organized in Mr Simmons's home

Charles H Simmons was born March 28. 1858, at Hamilton, in Madison county, New York, being the son of G M L and Mary Ann Jane (Plumbly) Simmons, who were married at Hamilton, New York Mr Simmons' father was born in Madison County, May 22, 1825, and died in Nebraska. June 6, 1915 He was a son of Otis Simmons, born at Little Compton, Rhode Island, January 13, 1796 and served several months as a soldier during the war of 1812 The mother was born in England, in October, 1832, and resides at Scottsbluff, one of the city's most venerable residents Her father, Charles H Plumly, a shoemaker by trade, brought his family to the United States in 1832 and settled at Hamilton, New York There were three sons in the Simmons family William L , who died at Buffalo, New York, at the age of thirty-six

years, Otis Thomas, a train inspector for the Santa Fe railroad at Los Angeles, and Charles H , who was born March 28, 1858 The father of the above family was a carpenter by trade In 1881 he came to Nebraska and settled in Dodge county, purchasing a tract of land on which he lived four years, in 1886 bringing his family to Cheyenne county, where he proved up on a pre-emption claim in one year The parents returned then to Buffalo, New York, where they resided until 1896 and then came to Scottsbluff county, where they lived two years, removing then to Dodge county In 1907 they returned to Scottsbluff and here the father died In many ways he was a remarkable man When over eighty-five years of age he assisted in the building of three houses In early manhood he was a strong Whig and later became active in the Republican party He was a man of sterling integrity and a member of the Congregational church

Charles H Simmons attended school at Hamilton, New York. then went to work on a farm, afterward employed in an insurance office for two years, following which he was a clerk in a store for four years Mr Simmons came to Scottsbluff county in 1886 and located land and has been a permanent resident since 1887 Until 1898 he lived on his farm but in that year moved to Gering and conducted a grocery store there for two years In 1900 he came to Scottsbluff and went into the grocery business, moving two log houses from Gering across the river and setting them up in a cornfield on the present site of the town, Mrs Simmons being the first woman to have her home on this town site On March 28, 1900, Mr Simmons was appointed postmaster of Scottsbluff, an office he continuously filled until July 1, 1910 For some years Mr Simmons was also a justice of the peace

At Hamilton, New York, in 1880, Mr Simmons was united in marriage to Miss Alice M Sheldon. who was born in Madison county, and died at Scottsbluff, Nebraska, March 1, 1918 Her parents were William P and Mary A (Beebe) Sheldon. who died on their farm in Madison county, New York To this marriage the following children were born William Lafayette, a carpenter-contractor at Scottsbluff, Otis William, a contractor, with other interests, Charles Sheldon. a sign painter at Scottsbluff, Edith J , the wife of Elmo L Harrison, a carpenter at Scottsbluff, Robert G , whose individual sketch appears in this work, who was the first baby under a year old to live in this town, Adah May and Ida Alice, both of the unmarried daughters reside at home In politics Mr Simmons is a Republican Fraternally he is identified with the Modern Woodmen of America He has not been very active in business since retiring as postmaster, and mainly occupies himself looking after his Scottsbluff property and investments

Shortly after locating in Scottsbluff, the late Mrs Simmons began to concern herself in the building of a church She was of deeply religious nature and felt a great sense of responsibility in this matter With Mr Simmons and six others, a few rough boards were secured and a shack built that served for a Presbyterian church for a time, and as long as she lived she continued active in promoting church work In every relation of life she was an admirable woman and when she was called away the loss was not to her family alone In 1910 Mr and Mrs Simmons made a most enjoyable visit to New York and their friends in Madison county

EDWARD STEWART — One of the younger generation of farmers and stockmen carrying on operations in Sioux county, is today a representative of what new, young blood can accomplish He comes of an old, well known and honored family of this county, being the son of H G and Marie (Clites) Stewart Edward Stewart is a native of the Sunflower state, born in Pawnee county, Kansas in 1878 He accompanied his parents to Nebraska upon their removal from the south to this section of the Panhandle and received his educational advantages in the common schools of Sioux County, thus laying the foundation for his subsequent business career After Edward Stewart s school days were over he remained on his father's farm, working for him and at the same time learning the practical side of farm industries and assumed many of the responsibilities until he was twenty-four years old Mr Stewart began his life as a cattle man while living at home as he worked for Wallace Merchant, a man of extensive cattle interests in this section During the winter season the men Mr Merchant employed were engaged in feeding but with warm weather they were out on the range branding and attending the various round ups, where the cattle of the wide feeding grounds were gathered, for the sorting of the calves and their branding Mr Stewart was one of the men who early realized that the day of the open range was over and he determined to settle down and have a farm of his own With this end in view he filed on a homestead near

Mitchell in 1900, proved up on the eighty acre tract and was able to dispose of it to advantage, so sold out and then came to Sioux county where he filed on five hundred and sixty acres near Sheep creek On this tract Mr Stewart made the required improvements, built a comfortable house for his family and when the ground was broken engaged in general farming and stock-raising, handling only a good grade of animals As returns began to come in from the ranch he invested his extra capital in other tracts of land adjoining the homestead until today he is one of the largest landholders in the sourthern part of the county, holding five sections and an additional forty acres The range riding of his early days has given him a fine constitution and Mr Stewart is a man who can turn out an astounding amount of work in the busy season, and personally supervises all the work of the ranch also riding over his wide range Mr Stewart is progressive in methods and uses modern machinery to carry on all the farm work and today is accounted one of the best managers of farm property in the valley

In 1904, Mr Stewart married Miss Daisy Tvariegek, the daughter of James Tvariegek, of England, where Mrs Stewart was born Her parents were early settlers of the Panhandle both of them passing away in 1919 Mr and Mrs Stewart have two children Gordon and Agnes, both at home

Mr Stewart is an independent in his political views, voting for the man he believes will best serve the people, whether it be in county, state or national office His fraternal affiliations are with the Modern Woodmen

A J HAMPTON — As the final test of every man lies in performance rather than promise, there is full justification in the scriptural statement that, "by their works ye shall know them" In this publication the various sketches that are presented offer a gallant record of worthy performance — of achievement that means character and ability He whose name introduces this paragrph is not only a representative of one of the prominent pioneer families of what is now Scottsbluff county but he has also made for himself a secure place as a man who gained substantial success through his well ordered operations as an agriculturist and stock-grower He developed one of the excellent farms of the county and since disposing of the same he has lived practically retired, at Gering In the very prime of life, he is able to enjoy fully the rewards of his former years of earnest endeavor, and he is a citizen whose course has been so directed as to commend him to the fullest measure of popular esteem.

Mr. Hampton was born in Jasper county, Iowa, in 1872, and is a son of Wililam R. and Sarah M. (Deeter) Hampton. William R. Hampton was for some time engaged in the practice of law in Iowa, where also he became identified with the development of a coal mine. He encountered an appreciable financial loss in this connection, as water made it impossible to operate the mine successfully. Under these conditions he sought a new field of activity, and in 1886, he came with his family to the present Scottsbluff county, which was then a part of Cheyenne county. He became a pioneer member of the bar of this section of the state and wielded much influence in public affairs here. He built up a good law practice, served one time as county attorney of Banner county, and was prominently identified with the movement which brought about the division of Cheyenne county and the creation of Banner and Scottsbluff counties, besides which he filed entry and duly perfected his title to a homestead and a timber claim in Scottsbluff county. He was a supporter of the cause of the Greenback party during the period of its maximum potency, and after its decadence he was found aligned with the Democratic party. Both he and his wife held membership in the Methodist Episcopal church. Of their fine family of eleven children the subject of the review is the youngest, and concerning the other the following brief record may consistently be entered: Theodore is a resident of Hastings, this state; Cornelia lives in Banner county; Caroline and Dora are deceased; Ida is a resident of Iowa; Huldah, is deceased; Jennie L. maintains her home in Oklahoma; Commodore and William live in Scottsbluff county; and Russell is a resident of Iowa.

A. J. Hampton gained his early education in the public school of Iowa and was fourteen years old when the family home was established in old Cheyenne county, Nebraska, where he continued his studies in the public schools, besides which he had the advantages of a home of superior culture. In Scottsbluff county he finally took up a homestead of a hundred and sixty acre, and eventually he here became the owner of considerable property, upon which he made good improvements and which he made the stage of successful agricultural and live-stock enterprise. He sold the property some years ago, and since that time has lived retired at Gering. His political allegi-

ance is given to the Democratic party and while he is liberal and public-spirited he has had no desire for the honors or emoluments of political office.

February 22, 1901, recorded the marriage of Mr. Hampton to Miss Edna Smith, who was born in the state of Illinois, and the supreme loss and bereavement came when this devoted wife and mother died in March, 1918. She is survived by four children, all of whom remain with their father: Jefferson, Raymond D., Inda and Audrey.

L. P. WELLS is a venerable and honored citizen who had the will to dare and to do as a pioneer, his experiences have been wide and varied, as he has been a resident of Nebraska for forty years, and in 1900, came to Scottsbluff county, where he purchased half a section of land, as did also his son. There he has since given active attention to the improvement and general supervision of this valuable property, comprising section thirteen of township 16 and lying near Gering, the county seat, which is his postoffice address.

Mr. Wells was born in Madison county, New York, in 1848, and is a son of Harrison and Lavina (Stone) Wells, both likewise natives of the old Empire state and representatives of families established there many generations ago. Harrison Wells was a blacksmith by trade and was a specially skillful mechanic For six years he was chief agent for the celebrated Walter Wood farm machinery in Europe, and he was manager of the concern's exhibit which took first premium at the Paris exposition. He eventually came to the west and engaged in farm enterprise in Sac county, Iowa, whence he and his wife later came to Elm Creek, Buffalo county, Nebraska, where they passed the remainder of their lives. They became the parents of six children: George and Estes are deceased; L P., of this review, was the next in order of birth; D. H., resides at Mitchell, Nebraska; H. L. is deceased, and Florence E. became the wife of Munson Brown, both being now deceased.

L. P. Wells acquired his early education in the schools of his native state, where he continued to reside until 1879. He then came to Nebraska and established his residence at Kearney, where he taught school and also worked as his trade, that of carpenter. Finally he took a homestead claim north of Elm Creek, Buffalo county, and later he obtained land on Buffalo creek, that county, where he developed a good farm, upon which he made the best of improvements and engaged in raising cattle and hogs upon a somewhat extensive scale. On this place he erected what was at that time the largest barn in the state, and he continued to reside on the homestead until 1900, when he sold the property and came to Scottsbluff county, where he made investment in land, as noted in the opening paragraph of this article. His years rest lightly upon him, for he has lived a sane and useful life, and he takes great satisfaction in the fact that he is still hale and hearty, able to apply himself vigorously when need be, and well equipped for the management of his fine farm property, upon which he has erected building of superior grade, besides making many other improvements of permanent order. Mr. Wells is a man of fine intellectual keen and a citizen of broad views and progressive policies; he has contributed his full quota to civic and industrial advancement in the state of his adoption. He is well fortified in his views concerning governmental and economic affairs and his political convictions have led him to ally himself with the Socialist party. He has long held. membership in the Methodist Episcopal church, of which his wife likewise was a devoted adherent, her death having occurred in 1910, and her memory being revered by all who came within the sphere of her gentle and gracious influence.

On November 3, 1869, was solemnized the marriage of Mr. Wells to Miss Jennie M. Russell, who likewise was born and reared in New York, and their devoted companionship continued for more than forty years — until her death severed the gracious association. In conclusion is given brief record concerning their children: Floy D. is one of the representative agriculturists and stock-growers of Scottsbluff county; Jessie, Irma and Dallas are deceased. In 1891, Mr. and Mrs. Wells adpoted twins, Jesse and Tessa Gilespie, whom they reared to maturity and both of whom are in the employ of the government, in New York City, at the time of this writing, in 1919.

JOHN HEINZ, a broad minded Sioux county ranchman, who represents the best element in the farming industries of the valley, is a man whose record should become a part of this history of the Panhandle for the benefit of his children.

John Heinz was born in Germany in 1876, the son of Matthew and Catharine (Mick) Heinz, both of whom were born, reared and educated within the confines of the German Empire. The father in his youth learned the trade of blacksmith, a vocation which he followed all the years of his business life. There were seven children in the family, of these two live in the United State: Rose, now the wife of Louis Ganzer, lives in Sioux county, and John of this review, who was an ambitious youth and while he studied diligently in the fine public school of his native land, learned of the great new country across the sea, where land could be had for the mere taking, and like so many of the European people, believed that land of his own would be the height of his desire. He began to ask question about America, to learn how to go and what he could do so that when only fourteen years of age he severed all ties to sail for our shores, landing in 1890. Within a short time he came to Nebraska, where he knew many of his countrymen had settled, locating near Schuyler, Nebraska. He soon obtained work on the great ranch owned by Marshall Field. At first he could not understand much of the English language but he began to study and made such progress that better positions were found for him as he was not afraid of work, had a good head and used it to the advantage of himself and his employer, thus rising from one place of responsibility to another, and spent fifteen years on the ranch in Stanton county. But the longing for a farm of his own had never ceased and now that he had capital, Mr. Heinz looked around for the place which would be nearest to that he had dreamed about since the old days in Germany and in 1907, believed he had found it, for he took up a homestead of eighty acres in Sioux county. Having had wide and varied experience in American farming methods on the Field place he knew just what improvements to put on the farm, the kind of buildings for the farm industries as well as a good comfortable home. Coming here late the Heinz family did not have to suffer the hardships and privations of the early settlers and before long were well established. With his native German thrift as a foundation, Mr. Heinz was soon enjoying a good income from his land. When he had accumulated sufficient capital he bought more land adjoining the homestead. This plan he has carried out until today he owns some six thousand acres, a rather surprising record for a boy who had no equipment when he landed in the United States but his bare hands and a determination to succeer. Three hundred acres of the ranch are under water rights, much of the remainder is fine grazing land, a combination excellent for general farming and stockraising, two lines of endeavor which Mr. Heinz carries on.

TULLIUS C. HALLEY

Having been a cattle man for so many years with the Fields, he realized that blooded animals payed the greatest dividends in the long run and he has nothing but thoroughbred Herefords, a breed in which he specializes

All the arable land is highly cultivated, producing fine crops of all kinds and forage for the stock, for Mr Heinz is a firm believer in intensive farming under water rights He is modern in methods, studies the farm problems of the day, advocates and practices modern methods while there is not a ranch in the Panhandle where better and more modern equipment can be found, to increase production and lighten the work of the hands Mr Heinz is a high type of the American farmer, he is public spirited and lives up to his own standard of citizenship, taking an active and prominent part in public affairs, as he is a school director in his district During his residence here, he has been vice-president of the Water Users Association and promotes every movement for the benefit of Sioux county Mr Heinz does his own thinking and is an independent voter and with his family attends the Presbyterian church

In 1898, occurred Mr Heinz' marriage to Miss Cora Belle Clark, a native daughter of Nebraska, born at Stanton, where she was reared and received her education To this union six children have been born Carlisle R, Howard H, George M, John Everett E, and J C, deceased

TULLIUS C HALLEY — Perhaps there are men in the Platte valley who know more about the sheep industry than Tullius C Halley, an extensive breeder of this invaluable animal, in Scottsbluff county, but it is not probable, for he has devoted many years to this business under all conditions, and his reputation in this line of activity is widespread and substantial

Tullius C Halley was born in Callaway county, Missouri, January 22, 1876, the son of Thomas H and Mildred A (Craighead) Halley, both natives of Virginia but residents of Missouri from childhood, the members of both families being early settlers in Calloway county, extensive landowners and successful farmers and stock-raisers Mr Halley's father was a man of public consequence in Callaway county, a leading Democrat and served for eighteen years as county surveyor and road and bridge commissioner Both parents were members of the Baptist church Of their nine children Tullius C was the youngest in order of birth, the others being Fannie, the

wife of P H Smith, a farmer in Missouri, H C a farmer and stockman in Missouri, J J, who practiced medicine for twenty-seven years in Benton City, Missouri and Fort Collins, Colorado is now a farmer and stockman of Scottsbluff, Nebraska, C R, a physician and surgeon of Sheridan, Wyoming, Thomas H, a farmer and stockman near Fulton Missouri, is circuit court clerk of Callaway county, S C a physician of Fort Collins, Colorado, N G, a farmer and stockman near Fort Collins, and Georgia, the fifth in order of birth, is deceased

Tullius C Halley had excellent educational advantages, attending first a military academy at Mexico Missouri, and later at the normal school at Chillcothe, Missouri, after which, for two years he engaged in teaching school in his native state It was in 1899 that Mr Halley went to Colorado and first became interested in the sheep growing industry He started, with practically no capital, by maintaining a sheep farm and feeding for other people, at that time having much to learn about the business He was determined, however, to make a success of this venture, and continued his operations in Colorado and Wyoming until 1907, when he came to Scottsbluff county, bought land and began feeding and shipping sheep He has developed this business into a great enterprise and through hard work has not only built up a comfortable fortune for himself, but has given great encouragement to an industry that stands among the first in importance in supplying food to the world

In 1905, at Cheyenne, Wyoming, Mr Halley married Miss May Congdon, who was born in Michigan, a daughter of Patrick and Mary Congdon, both of whom are deceased Mr and Mrs Halley have the following children Mildred, Mary and Jean, all of whom are doing well in school, and Thomas A, who has reached his fourth birthday Mrs Halley is a member of the Roman Catholic Church Mr Halley owns four hundred acres of fine irrigated land and in the summer of 1917 completed a handsome modern residence that is one of the county's beautiful country homes In his political views he is a Democrat He has never accepted political preferment, the rapid growth of his personal interests pretty fairly occupying his time In a fraternal way he is identified with the order of Yeomen

LEWIS EMERSON UTTER, one of the pioneer settlers of western Nebraska and the Panhandle, who has lived here for more than forty years and seen the great development and changes that have taken place and who by his own unaided efforts has achieved success as a ranchman, is today one of the well to do and honored men of the Wheatland community where he stands high

Lewis Emerson Utter was born in Ulster county, New York, January 7, 1866, the son of Harmon and Mary M (Tubbs) Utter, the former of German and Dutch descent while the mother was of pure English stock They were both born and reared in Ulster county, New York, the father died in 1919 and the mother in 1913

Mr Utter was reared in New York state and attended the public schools of his community but a short time and says that what education he has he acquired himself, largely in the school of experience He remained at home until 1883, when he came west to Nebraska, locating in Custer county, but three years later came to Cheyenne county and took a homestead on the head of pumpkin creek which later became a part of the Pierson Ranch Mr Utter ranched some, taught school and farmed while on Pumpkin creek but in the early nineties sold his land to A H Pierson and moved to the Wheatland locality He experienced all the ups and downs of farming and ranching in the Panhandle of those early days but remained on his land until 1905, when he sold his property and moved onto a ranch in western Nebraska which is now known as the Last Camp Ranch, consisting of four thousand acres Mr Utter has made a success of his ranching and stock-raising and is one of the prominent men of his district, where he has attained comfort and fortune through his own business enterprises

May 9, 1902, Mr Utter married at Cheyenne, Wyoming, Miss Fannie E Sudduth, who died October 9, 1915 She was the daughter of Benjamin F and Mary A Sudduth, the former born in Kentucky, moved to Illinois in early life, while the mother was born at Lands End, England and came to the United States when a child One child was born to Mr and Mrs Utter, Aileen D

Mr Utter was raised a Democrat but says that he was forced into the ranks of the Republican party by the conviction that the fundimental priciples of this party are more American and safer for the country He is a member of the Masonic order belonging to the Wheatland lodge, is also and Odd Fellow and a member of the Woodmen of the World Mr Utter is a member of the Christian church

HAROLD S THOMAS — Prominent among the operators in real estate and insurance also doing a loan business, is a native son of this state, Harold S Thomas, who established himself in business in Alliance in 1909, and since that time has so ably directed his activities and operations that he is listed among the leading members of the younger generation of financial men of the county seat

Mr Thomas was born at Elwood, Gosper county, July 9, 1891 the son of John W Thomas of Lincoln, Nebraska The father was a minister of the Methodist Episcopal church who moved from one town to another as he was called to take charge of different congregations Some of the towns and villages were Wallace, Maywood, Wellfleet, North Platte, Beaver City, Alma and Orleans, and young Harold attended school in Beaver City, Alma and Orleans before the family moved to Omaha in 1902, where he enrolled in the Bancroft, Lincoln and Castellar schools for his elementary studies When only eight years old he had entered a printing office to learn the trade of printer and continued to work at this occupation during his vacations till he was a competent man of that exacting business In 1909, he came to Alliance in February and the following summer attended the Junior normal school course following the same plan the next summer, as he wished to take some special journalistic courses to fit himself for general newspaper work A year later he accepted a position in the office of the *Alliance Herald* and remained on the staff of that publication until 1914, when he resigned to accept a fine offer of the Reese Printing Company of Omaha While living there, Mr Thomas took a business course offered by the Young Men's Christian Association in the night school to perfect himself in business methods In 1916, he removed to Gordon, Nebraska, to take a position in the Fair Department store as stenographer and bookkeeper, remained two years and then in the spring of 1918, went to Lincoln to attend the radio course given by the University of Nebraska, and on June 20, of that year enlisted in the Signal Corps of the United States army for service during the war with Germany Mr Thomas was immediately sent to Plattsburg Barracks, New York, for training with the Three hundred and twenty-first Field Signal Battalion, with which organization he remained until the signing of the armistice On September 1, 1918, he sailed from New York on the Australian steamship *Katooba*, and landed at Liverpool, England, twelve days later From that city he went to

Southhampton by rail, crossed the English channel about September 20, landing in France at La Havre Proceeding to St Agnon by rail the battalion marched to Choussy, remained there ten day then on to Cormeray, on foot, where it remained and sent men to the front as replacements Here Mr Thomas was assigned to duty as Battalion Sergeant-Major Upon the signing of the armistice he was transferred to Headquarters of the Third army, and immediately started for Dun-sur-Meuse and managed to catch up with the Third army which was on the march toward Germany to become a part of the Army of Occupation, the day after it arrived at the city of Luxemburg After several days spent there, Mr Thomas went by truck to Mayen, where another pause in the march was made before the army continued on to Coblenz, where it arrived December 15, 1918 Here he became Chief Signal Corps supply clerk for the Third army, assigned to duty in the office of the chief signal officer On April 1, 1919, Mr Thomas was taken sick and rushed to the hospital to be operated upon for acute appendicitis, and having received orders for return to the United States, left the hospital May 1 On May 14, he left Coblenz for St Angon and sailed from St Nazaire June 2, on the United States Steamship *Suwanee*, formerly the German steamer *Mark* Landing at Charleston, South Carolina, June 16, he was discharged from the service three days later at Camp Jackson, and arrived home in Lincoln on the 22d Mr Thomas remained in the capital city visiting friends for several days and then came on home, reaching Alliance on August 4 From that time he has been connected with the Thomas-Bald Investment Company He is a member of the Yeoman Lodge at Alliance, the American Legion and the Methodist Episcopal church

On June 28, 1911, Mr Thomas married Miss Ivy Hale and two children have been born to them Paul Creighton, at Alliance September 13 1912, and Claudia Pearl, at Gordon, August 5, 1917

JOHN T WATSON, deceased, was one of the best known and most picturesque characters of the early life of northwestern Nebraska and the Panhandle He was one of the early residents of Box Butte county when this section was little settled, was a typical pioneer and gained fame as a wolf hunter when these animals threatened the stock of the ranchers

John Taylor Watson was born in Rome Henry county, Iowa, December 25, 1848, and when two years of age was taken by his parents to Madison county, Iowa, and two years later to Cass county, Iowa, where he was reared and received his education in the schools of that locality In 1886, he came to Box Butte county, Nebraska, where he homesteaded and lived on his ranch for fifteen years, later moving to a ranch in Sioux county His first home was near Box Butte postoffice, some sixteen miles northeast of where Alliance now is located and made many friends there as well as at his home in Sioux county In later years Mr Watson sold his ranch and moved to Mitchell but visited Alliance and Box Butte county frequently For more than twelve year he lived in Mitchell before his death Mr Watson was married three times but is survived by no children His last wife the widow of the late Alexander Noble, of Mitchell, died a few months before her husband

In the early days here Mr Watson had many thrilling adventures and experiences, among them that of freighting during the Indian war, when in 1890 and 1891, the Sioux began their ghost dances and the government had to take steps to quell the uprising as the Indians believed that they could only save their lands by a wholesale massacre of the whites The Box Butte county men who needed money took work freighting the supplies to the soldiers at the Pine Ridge agency and Mr Watson was one who made more trips than anyone When the wolves threatened the stock of the ranchers he was employed by the northwestern cattlemen's association to kill them and given a bounty of twenty-five dollars for each wolf killed He employed dogs in the capture of the wolves and was so successful that he followed the vocation for months at a time In later years Mr Watson was greatly in demand for public celebrations as he was able to handle large crowds and was selected for marshall of the day For more than twenty-five years, the commanding figure of Mr Watson on his horse, Jack was sought on state occasions at Crawford, Fort Robinson Alliance and Mitchell, as well as other places to lead processions also the county fair, stock growers conventions and public gatherings

Mr Watson died at his home in Mitchell March 1, 1921, and was buried at Hemingford He was a member of the Independent Order of Odd Fellows which had charge of his funeral He was a man who possessed a true heart who had many friends of long standing in the Panhandle where he had taken a prominent part in development and progress

RICHARD H ARNOLD, one of the hardy pioneers and early settlers of Dawes county who came here at an early day and took his part in the opening up and settlement of the country, has for years been one of the well known ranchers and live-stock men who is widely known as a prosperous business man As one of the first men to locate in this section he deserves a place in the history of Dawes county for he tells many interesting incidents of its early history which he helped to make

Mr Arnold was born near Washington, Washington county, Iowa, September 6, 1862, the son of William M and Rebeccah J (Merchant) Arnold, the former a native of Fayette county, Ohio, was the second of the nine children born to his parents His father was a farmer and the boy was reared in the country, attended the public schools in the winter and worked on the home place in the summer time He early learned to work and make money for himself and at the age of fourteen went to stay with an uncle, where he continued to attend school In the fall of 1883, Mr Arnold came west to Beatrice, Nebraska, secured work on a farm but remained only two months before coming to the Panhandle and Sioux county He located on a pre-emption claim close to Sheridan Gates, on Bevcar creek, April 22, 1884 To earn money he worked for Nick Messenger, who ran a saw mill on Bordeaux creek, and so secured the lumber with which to put up a house on his claim Later he secured employment with W M Hudspath, where he earned enough money to prove up on his claim The Hudspath land was afterward relinquished to William Braddock In the winter of 1885, Mr Arnold and two other men walked the hundred and fifty miles to Valentine to prove up on their claims There was a foot and a half of snow on the ground and it took eight days to make the trip They carried their food with them and each meal had to thaw it out Mr Arnold had a snow bound experience in the winter of 1884, when he was freighting corn from Valentine to the P T Nelson ranch, just east of the present site of Chadron He left Valentine December 21, with a large load, it had been threatening to storm and when he was eighty miles on his way, about the location of the Fline Buttes at the crossing of the Little White river, the storm broke Mr Arnold made his bed by the freight wagon, with tapaulin and a pile of blankets and in the morning found himself snowed in, he burrowed out of the drift and saw it was still storming, a real blizzard: his dog was frozen and the horses had drifted with the storm, as he was eighteen miles from the nearest habitation he crawled back into his drift where he had considerable room and could keep from freezing He remained under the snow that day and night without any food and on the second day, finding that the storm was over started for the Jack Carlow ranch, on Spring Creek The snow covered the ground and all he had to guide him on his way were the sunflower stalks planted along the trail He discarded everything that would weigh and hamper him and started on his trip with overshoes on his feet and blanket strips wrapped around his legs as protection against the snow and cold Mr Arnold walked all day and was exhausted, but kept up as he saw an animal following him and learning it was a wolf, managed to keep awake and staggered on He finally reached the Carlson home at eleven o'clock at night, where he was fed and remained until the blizzard was over In the spring of 1885, Mr Arnold took a sub contract to carry the mail from Pine Ridge Agency to Fort Robinson, sixty-five miles away, remained on this route three months then went to work for the Ox Yoke Ranch, north of Fort Robinson, and rode the range three years From 1889 to 1890, he was employed by a contractor who supplied beef to the Pine Ridge and Rosebud Indian Agencies

June 20, 1892, Mr Arnold married Miss Laura A Churchill who is a native of Iowa, the daughter of Matthew M and Nancy (Bosser) Churchill He father was a carpenter and cabinet maker by trade who moved from Iowa to Sheridan county, Nebraska in 1885, and took a homestead in Beaver creek valley, where he farmed and worked at his trade Mr and Mrs Arnold have five children Edwin M, who married Amanda Brandt is a mechanic for the Mid West Oil Company, of Casper, Wyoming, and has two children, Grace M, the wife of Sprague O Smith, of Chadron, who is the son of Captain Fred Smith, a captain of the militia at the time of the Indian war in 1891, has one child Mr Smith served in the navy during the World War The third child is Eunice R, who graduated from the Chadron Normal School and is employed in the Citizens State Bank, Norma A, is at home and Hazel, a student is also home

After his marriage Mr Arnold settled down to ranch life and the cattle business He bought more land, dealt heavily in cattle and became one of the best known stockmen in the western part of the state Being on the

W. W. White

ground early he had the pick of the land and when he sold his interests in 1918, had nine hundred and sixty acres of land which brought in a comfortable fortune. With his wife he now enjoys life in their comfortable home in Chadron. Recounting the early days, Mr Arnold tells that he kept a few acres on some of his land where the Indians could come and camp, he traded extensively with them and says that they were very honest and were friendly to him and his family. He with his family learned to talk the Indian language and thus understood them well.

Mr Arnold is a Chapter Mason and takes interest in all progress in Dawes county and Chadron where he is considered one of the substantial and reliable men of the community.

WILLIAM W WHITE, attorney, whose professional reputation extends all over Western Nebraska, has been a resident of Gering for a number of years, but has, at the same time, filled numerous offices of public responsibility in other sections. He is recognized as an able, forceful lawyer, years ago was the pioneer newspaper man of Banner county, Nebraska, and yet finds time to give attention to large and valuable farming interests.

William Wallace White was born at Adrian, Missouri, February 17, 1866. His parents were George M and Sallie (Hughes) White, the former of whom was born at Knoxville, Tennessee, in 1837, and the latter in the state of Kentucky in 1841. They are honored residents of Amsterdam, Missouri. Their marriage took place February 14, 1861, at Sedalia, Missouri, and seven children were born to them. In addition to William W, three others survive, namely, Warren, who is a farmer-banker, at Amsterdam, Missouri, Walter, who is in the automobile business at Carnegie, Caddo county, Oklahoma, and George, who is in the auto business at Anadarko, in the same county.

George M White is a son of Bloomer White, who was an early settler in Missouri an "old-timer," who accompanied a cattle train from Independence to Santa Fe, New Mexico, in the days when great herds of buffalo roamed over the western plains. He moved to Missouri from Tennessee, when his son George M White was six years old. The latter now owns a beautiful farm near Amsterdam and has been both farmer and merchant. At one time he owned eighty acres of land on the bluff where Kansas City, Missouri, now stands. He is a Democrat in politics and for many years was a justice of the peace. Both he and wife are members of the Baptist church in which denomination her father, William F Hughes, a native of Kentucky, was long a minister. When he went to Missouri he established the town of Crescent Hill, now Adrian, and he also owned farm land that is the site of the present city of Sedalia. He was one of the early Masons in the state and belonged to the Commandery at Kansas City, and when he died the Knights Templar brought his body to the cemetery on a flat car. George M White is a veteran of the Civil war, his services being given to the Union during the entire period. One brother was his comrade in the Federal army, while two other brothers fought on the side of the Confederacy.

William W White received his early educational training at Adrian and later was graduated from the high school at Eldorado Springs. He was also a student in Butler Academy, at Butler, Missouri and studied law in the office of well-known attorneys. In 1892 he was admitted to the bar at Cheyenne, Wyoming, and in October of the same year to the bar in Nebraska. His first two years of practice was in Cheyenne as a member of a law office force. He then went to Oklahoma for a time, then to Galesburg, Illinois, for three years, and in 1903 established his residence at Gering. In the meanwhile he found opportunity to make a record for himself as the pioneer newspaper man on Pumpkin Vine Creek, now in Banner county. For ten years he served as county attorney for that county, but resided at Gering, and from 1911 until 1918, (at which time he resigned), he was county attorney of Scottsbluff county. During his time of residence at Pawnee, Oklahoma, he served as chairman of the board of county commissioners. Since 1910 he has been local attorney for the Union Pacific Railroad, and he has looked carefully after the interests of this corporation. His practice has been of a general nature and has been steady, honorable and satisfactory financially. For some ten years he was associated with Judge O W Gardner who retired in order to attend his personal interests, and the firm style now is White & Heiss.

In 1894 Mr White was united in marriage to Miss Alice D Everett of Cheyenne, Wyoming, and they have four accomplished daughters namely Ruth who is an instructor in the high school at Gering, Grace who is a teacher in the public schools of Scottsbluff, Carol who is a graduate of the high school, class of 1919 and Elizabeth who is yet in school. Mrs White and her daughters are members of the Episcopal church   During the long period of war needs, Mr White and his family were active in the various movements, their efforts never ceasing as long as there was urgent demand   Mr White is a Mason and belongs to the Knights of Pythias and the Modern Woodmen   In politics he is a pronounced Republican

EUGENE A HALL was born in Freeport, Illinois, and was the third in a family of four children — three boys and one girl, there was also a half sister by a former marriage of his father   Their father was George H Hall, born in Hubble, New York, the mother, Mary A (Coltman) Hall, was born in Poughkeepsie, New York

George H Hall was a farmer, speculator and stockman, and in 1866, drove over the Texas trail eight hundred steers to Council Bluffs, Iowa, and sold them to freighters for oxen   The same fall he took a mile contract, grading on the Union Pacific Railroad at Broakville, Kansas   After finishing this work he took five miles more at Russell Springs, now Russell, Kansas, and on June 20, 1867, a band of thieving Indians swooped down on the little camp of about forty men and teams and stampeded and drove off all their stock, except a black stallion and Indian pony belonging to Mr Hall   This was a hard fought battle, in which one of Mr Hall's men was shot through the head and killed, and two more seriously wounded

The leader of these marauding savages was Charlie Bent, a half-breed Indian, as Mr Hall learned a few years ago from "Charlie-the-Crow," an Indian scout under General Custer and the only survivor of the Custer Massacre The half breed was riding here and there, urging his followers forward, when Mr Hall shot his horse from under him and shot him before he hit the ground   A couple of Indians rushed in and carried their leader out of rifle range, and it was not known at that time whether he was killed or wonded, but "Charlie-the-Crow" said that Mr Hall had made him a "good Indian"   Be that as it may, he was never known to steal any more horses in that country

Mr Hall then hitched the black stallion and Indian pony to a wagon and, taking his wife and four children and wounded men, started for Fort Harker, about forty miles east, the teamsters traveling along afoot, having to abandon all their wagons, tools, harness and supplies, which it was afterward learned the indians destroyed by fire the next day

On arriving at the fort, he left the wounded men and, with his family, continued to Salina, Kansas, which was the end of the railroad, about thirty-five miles further on, where he put them on the cars, they went to Junction City, Kansas, to an uncle, where they stayed until fall, while the father returned to Ellsworth, Kansas, to recruit another outfit to finish his contract, but the Asiatic Cholera was raging in the country, and in about two weeks Mr Hall contracted the terrible disease and died July 16, 1867

That fall Mrs Hall sold all of their worldly possessions, being a few pieces of furniture and a piano which she had stored when they went to the railroad camp, and with the small amount of money realized from these she went to Freeport, Illinois, where her parents still lived, but she again returned in a short time to Ellwsworth, Kansas   When Eugene was four years old he was placed in school at Junction City   To amuse himself the first day he was throwing paper wads at some other boys, and in punishment the teacher placed him on top of her desk, when young Eugene very gravely informed her that his mother did not send him to school to "that up here," which of course put him in the limelight stronger than before   The teacher wrote a note to his mother, saying he was too young to go to school, but the mother did not agree with her about this and brought him back and he went to school just the same

The first money he earned when a boy was driving a fine team for a farmer harrowing in wheat   It was a new job for the boy and the team was a high spirited one, and he turned them loose and, as he expressed it, "let 'em go"   The owner came out and seeing the foam-flecked horses said, "Don't you ever get tired?"   Upon receiving a negative answer he said, "Well, if you do not, the horses do, you rest awhile"   Geene said nothing but watched the old fellow out of the corner of his eye, and as soon as he was out of sight continued his interrupted work   He was making fifty cents a day, which was top wages at that time

He grew up and finished his education in the Ellsworth schools   The last two vacations he herded cattle for Texas "Cattlemen" who brought up beef herds for summer grazing, and he seemed to take to the business as naturally as a duck to water

In 1878, he went to Buffalo station and hired to a man who had a herd of horses he was selling to farmers for work purposes   He worked for him about a month, leaving him at Beloit, Kansas, then taking his saddle horse, he rode about one hundred and eight-five miles

E. A. Hall

to Dodge City, Kansas, and hired to help bring a trail herd of beef cattle through to Spotted Tail Agency, Dakota, (now the Rosebud Agency) Being on the "drag" end of the herd he had to eat the sand and alkali dust, all the way up the trail

From 1886 to 1888 Mr Hall was a foreman for the Ogallala Land and Cattle Company In February 1885, this company bought the Bosler Brothers & Company brands and became one of the largest outfits in the northwest their books calling for seventy-seven thousand head of cattle They had three foremen Dick Bean on the North Platte River east from Blue Creek, Gene Hall from Blue Creek west and north to Bronco Lake, near which Alliance has since spring up, and Mac Radcliffe, west from Blue Creek on the south side of the river

In the fall of 1888, Mr Hall drove a beef herd of three thousand and thirty head for this company to the Rosebud agency trailing the cattle about three hundred and fifty miles, and weighed them over the scales to the United State Government without a loss of a single critter This was one of the most remarkable trips in the record of the cattle industry, and was the talk of cowmen, and could not have been put across by a less competent cow man than Gene Hall

After finishing his contract in driving the herd of Texas long horns to Spotted Tail Agency, February 1 1879, Mr Hall hire a half-breed Indian with a team and wagon and tent, to take him from the agency to Yankton, South Dakota As it was in the dead of winter, it would have been as much as a man's life was worth to have tried to go on horse back After a hard and strenuous trip they arrived at Yankton, and he took the train for Ellsworth, Kansas, the home of his mother at that time, but when spring came he returned to Ogallala, Nebraska, being fascinated by the country he had passed through the fall before Mr Hall started on this trip horse back, but at Orleans Nebraska, about eighty miles from Ogallala, his horse died forcing him to make the balance of the trip on foot He arrived in this little frontier town "broke" Mr Hall says he came to northwestern Nebraska a stranger without a friend or a dollar, afoot and alone

He obtained credit at a hotel for a week's board and started in search of work, he found a man that wanted him to work on a farm and when he found the boy's mind was fixed on becoming a "cow puncher," attempted to dissuade him from the venture and said, "he looks like too good a boy to work with such a wild lawless set of men" Gene replied that "he had found them pretty good fellows," thanked him for his offer but declined to give up his intention of hiring to some cattle company Not finding work he decided to go to Sidney Nebraska Being without money, on a moonlight night the boy crawled into a car of lumber but found it a pretty cold berth Everything went smoothly however until he reached Julesburg, Colorado when the brakeman discovered his feet sticking out, and pulled him out Gene told him this was the first trip of this kind he had ever taken and hoped it would be his last but he was out of money and wanted to get to Sidney, and hoped to get work on a cattle ranch He was taken to Sidney in the way car unmolested On arriving at the Minor Hotel a prominent hotel at this early day frequented by ranchers and cattle men he again asked for credit that he might appease hunger Again he obtained a week's credit (We stop right here to say that later, Mr Hall paid both his obligations at Ogallala and Sidney) Another week passed without being able to obtain employment wandering around the town "blue" and disheartened, he saw a warehouse he had not noticed before He entered the office to ask for work, and looking around he noticed a man with small black eyes staring at him, without speaking This angered the boy and he turned and left

He squatted on his toes against the south side of the building The man came out and walked backward and forward before the boy, still staring at him He finally stopped in front of Gene and said, You look to me like a boy who is looking for work" These words were never forgotten by the penniless boy And it was fortunate for both Tom Lawrence, who was foreman of the Bosler Brothers Company, and Gene Hall that he spoke when he did as the boy was ready to explode with anger at the continued stare of the stranger but his words were as balm of Gilliad, and the boy and man formed a friendship never to be broken

Gene hired to Tom Lawrence and worked on this herd ten years and as Gene puts it he was transferred in the deal, when in 1885, the Bosler Cattle Company was sold to the Ogallala Land and Cattle Company From the time Gene Hall hired to Tom Lawrence, as a "Cow Hand" during his ten years on the Texas trail and cattle range his life was one long series of adventure and "dare devil" exploits, that would rival the stories of "Buffalo Bill" if all were written.

Mr. Hall is of a retiring and modest disposition almost beyond belief for a man of his wealth and position, as he is one of the best known and influential "cow men" in the west today (1921). He is adverse to giving his personal experiences for publication, although often urged to do so, but he likes to sit down with old time friends and review the past in a reminiscent way, and go over again, some of those stiring times, when real "red blooded" men roamed over these western plains and accumulated fortunes, and where they had no use for the "tender foot" (unless he could be broken into the "harness") and take his share of hard "knocks."

As the time for beginning the round-up drew near, the "cow hand" would be found busily engaged in washing his clothing and blankets, his saddle and bridle were cleaned and oiled, bits, spurs and six-shooters were polished, and saddle broncos curried and given extra attention. Among these men found the same diversity of character, temperament, energy and intelligence common to mankind everywhere. A reputation for courage was a necessary requisite to good standing in a cow camp. He who could display the greatest recklessness, or assume the role of the greatest dare-devil, stood foremost and was the leader of the flock.

This desire for notoriety often led the "Waddie" into serious difficulties and gave rise to the general opinion that he was without feeling or regard for the rights of others, and naturally cruel. This opinion was erroneous, as a rule the cow hand was true to his friends and it was a religious principle to never desert one in a "tight place." Trailing cattle from Texas to northern ranges was fraught with all kinds of experiences. Sometimes crossing swollen streams, after forcing the cattle into the water, an experienced cow hand, had to swim his horse alongside the lead cattle, let them drift down the stream splashing a little water in the faces of the leaders, gradually working them to the opposite bank. Anyone not acquainted with the cattle business might think this a very simple affair, but it took a level head and daring men to do this, as a false movement would have set the cattle "milling and piling up" in the deep water and hundreds would have been drowned.

When sanding guard at night and a heavy thunder storm came up and stampeded the herd the poor "cow hands" had to turn out, and perhaps ride all night through the driving rain so dark you couldn't see a hundred feet away except by the lurrid flashes of lightning making the heavens glow with a wierd ghostly light and thunder crashing like a bombardment, still further adding fear to the already frantic, fleeing herd.

Cattle generally follow a leader, stringing out in single file, and they follow the leader as long as he runs. It was the duty of the "cow hand" to out run this leader and head him off, and get the cattle to running in a circle and by singing to them quiet them down and stop their mad flight. At other times they were unable to overtake the leader and had to take their chance of riding their bronco into a badger hole or over an embankment in the darkness and endanger the life of the rider, Perhaps the men had to ride all night in the storm and next morning find they were twenty miles from camp and the mess wagon.

Along in the early eighties Mr. Hall had a thrilling adventure roping a huge buffalo bull, about forty miles below where Alliance now stands, on Alkali lake; it was at the finish of the August roundup, about five o'clock in the evening when Mr. Hall located two lone buffalo in a sand "blowout." He rode over to the "mess wagon" to get his winchester, but found he was out of cartridges. He and two other "cow hands" (one a Mexican) decided to lasso one of them, each one of the boys eager to get the first rope on the buffalo. Gene was riding a beautiful black horse, called "Diamond L" in whom he had great faith and he believed he could hold anything on the end of his rope, short of a steam engine. He had the advantage of the other two boys as he knew exactly where the two animals were located and beat them "to it." "Diamond L" dashed over a little hill and down on the buffalo before either of the buffalo were aware of what was happening. With a frightened "snort," the horse shied so violently to one side when he saw the kind of animal his master was going to rope, that he threw Gene off his balance and he missed his game. The buffalo was as much frightened as the horse and bounded swiftly away in another direction, but Gene was soon on him again and this time was more successful and fairly caught the old fellow around the neck. He soon found to his amazement that "Diamond L" had at last found more than his match and struggle as hard as he pleased, he was too light for Mr. Buffalo, who pulled him over the prairie wherever he pleased, and wasn't doing it in a very gentle manner. Just as Gene was figuring to cut his rope and let him go, the Mexi-

Mr. and Mrs. Anthony Kennedy

can came to his rescue and roped him by one foot, still they had more than their share of buffalo meat, and it was only after the third man had swung his lasso on another leg, that they were able to control the huge beast, and they consumed nearly two hours before they had him tied down. Their horses were reeking with sweat and nearly exhausted. By the time he reached camp it was dark. They went back the next morning and butchered the buffalo and had a rare treat of the meat.

At another time Mr Hall was out in the hills with a bunch of boys rounding up cattle when a huge bear was routed out of the sage brush, and took to his heels at a lively pace followed by the boys. Gene was leading the bunch and threw his rope over the old fellow and soon found he was in more trouble than he ever dreamed possible from the innocent sport of roping a bear. With an angry roar bruin turned on the cow man and when he found he was tangled in the rope till he couldn't get loose, he grabbed the rope, hand over hand, like a sailor, and began pulling horse and rider up to him in spite of all that could be done to get away, one of the boys shot the bear with his winchester, and thus relieved the mind of a very much scared 'cow puncher'"

In 1887, W A Paxton, then general manager of the Ogallala Land and Cattle Company purchased from James E Boyd, ex-governor of the State of Nebraska, the O-C brand herd of cattle in Wyoming, and the Charles Campbell brand, which was the H-O brand. They became the nucleus in establishing their brands in Wyoming and the moving of their cattle into the Ogallala herd on account of the encroachment of settlers on their range. Mr Hall was directed by Mr Paxton to begin moving this herd to the newly purchased lands in Wyoming and August 1, 1887, he left Camp Lake on the western part of their range with the lead trail herd. Mr Hall remained with this cattle company until October, 1888.

A peculiar thing about his experiences is that he ended his career as a cow hand, so far as early day history is concerned, at the very place where he had helped drive his first herd ten years before, and the same month in the year.

After severing his conection with the Ogallala Land and Cattle Company, he came to Alliance, Nebraska, and has lived here continuously since that time. In the fall of 1889 Mr Hall was elected sheriff of Box Butte county on the Democratic ticket and served six years.

Eugene A Hall was married at Ellsworth, Kansas, May 17, 1893, to Miss Minnie E Bak-

er who was born in Leavenworth, Kansas. She was the daughter of John F and Elizabeth (Powers) Baker. Mrs Hall was the second in a family of seven children, five girls and two boys. Mr and Mrs Hall had only one child, a son, Albon B Hall, who graduated from the Kearney Military Academy in 1914 and continued in that school one year after graduation. Albon was also in the Officers' Training Camp at Waco, Texas, when the armistice was signed. He is a Thirty-second degree Mason and was married January 1, 1921 to Miss Esther Bevington, born November 1, 1899. Esther Bevington Hall was the only daughter of William Bevington, born at Centerville, Iowa, and May (Brooks) Bevington, born at Leon, Iowa.

Mr Albon B Hall is a live wire and we predict a successful career for this estimable young couple. He is connected with his father in the ranching and live-stock brokerage and sales business, with offices on the corner of Box Butte Avenue and Fourth Street, south of the postoffice.

Mrs Minnie E (Baker) Hall died July 26, 1905, and Mr Eugene A Hall married Miss Sadie E Fickell at Alliance, Nebraska, August 4, 1909. She was born in Ohio. Her parents, Joseph Fickell and Hanna E (O'Hara) Fickell, were also born in Ohio. Mrs Hall is the third child in a family of eight children, seven girls and one boy.

After a strenuous life such as few men can boast, Mr Hall has settled in a beautiful modern home in Alliance, Nebraska, is on "easy street" and also owns a fine business building in the heart of the business district. Mr Hall and Mr Robert Graham, the latter the present postmaster at Alliance, have a finely equipped ranch near Alliance of forty thousand acres run three thousand head of cattle, and cut four thousand tons of hay annually. Mr Eugene A Hall is a Royal Arch Mason, Knight Templar, and a member of Tangier Shrine at Omaha, Nebraska, a booster for Box Butte county, Alliance and her people.

ANTHONY KENNEDY — It probably would be a very difficult matter to prove to Anthony Kennedy, one of Scottsbluff county's most substantial citizens, that a youth with industrious inclination and good habits needs any other capital to start with in the building up of his fortune. Mr Kennedy can look back over his own career as example. Instead of waiting for helpful opportunity to come to him when his schooldays were over, he sought and found it even when it entailed the cross- of an ocean, kept right on and saved his money Mr Kennedy at present is not only one of Scottsbluff's most respected citizens, but is the owner of a large tract of some of the finest irrigated land in this part of Nebraska

Anthony Kennedy was born in Ireland, January 5, 1842, the son of John and Ann (Right) Kennedy, natives of Scotland but long residents of the north of Ireland, where the were married and spent their lives They were members of the United Presbyterian church Anthony grew up on his father's little farm and attended the local school When sixteen years old he went to England in search of employment, and found work with contractors who were building a cut-off dam in the sea His work was hauling and, although according to the present high wage scale of laborers, he was paid but a pittance, that money was saved and when he returned home a year later had amounted to enough, as he hoped, to pay his passage to the United States When he reached Liverpool he found himself able to still further add to his savings and he remained in that city for six months, in the meanwhile making friends, with one of whom, in 1864, he embarked for the United States

After landing the youths made their way to Pittsburg, Pennsylvania, where Mr Kennedy soon secured work as a drayman, later as a warehouse man, and then entered the employ of the Kilpatrick Grocery Company, with which house he remained connected for twelve years In the meanwhile Mr Kennedy married and later decided to embark in farming He removed to Deleware county, Iowa, and rented land there for eight years, then came to Madison county, Nebraska For the next four years he rented land there and carried on farming and stockraising very profitably, coming in 1886 to Scottsbluff county and secured a homestead near Minatare He now owns one hundred and ninety-two acres of irrigated land, the agricultural possibilities of which are incalcuable In 1908 the family moved into Scottsbluff but later returned to the farm for three years when Mr Kennedy bought a lot and erected a comfortable residence at Scottsbluff, which is now the family home

At Pittsburg, Pensylvania, December 20, 1870, Mr Kennedy married Miss Martha Baxter, who was born August 20, 1848, in the north of Ireland where she was reared She was of Scotch ancestry, her forebears having settled in Protestant Ireland at an early day Mr and Mrs Kennedy were school mates in youth, and after the young man came to America and

could support a wife he sent for his boyhood sweetheart to join him in this country Martha Baxter crossed the ocean in 1869 and the marriage occurred the next year Three sons and three daughters have been born to them, as follows William, who resides in Texas, Alexander, a stockman in Montana, Maggie, lives with her parents, Mary the wife of John Jensen, who operates Mr Kennedy's farm six miles northeast of Minatare, John, in the stockraising business in Montana, and Sarah who resides at home, is employed in the Platte Valley State Bank

Mr Kennedy has been an important factor in public affairs since coming to Scottsbluff county For ten years he was a justice of the peace and his decisions were never reversed He was one of the first elected county commissioners and his general popularity was shown by the returns when it was found that his majority was greater than that of any other county candidate Mr Kennedy served in other official capacities and was county assessor for four years He has always been a staunch Republican, and is a Royal Arch Mason With his family Mr Kennedy belongs to the Presbyterian church Few men in Scottsbluff county are better known than 'Squire Kennedy

FRANK L. BLACK, one of the earliest settlers of Dawes county who has taken an important part in the development of this section, deserves a place in the Dawes County History, as a man who has done much for his locality, as he has lived to see the many changes and improvements that have taken place here since he came in 1884 Today, Mr Black is one of the heavy land owners here and is considered one of the most substantial and reliable men of the county and Chadron He was born in Henry county, Iowa, October 9, 1854, the son of Layfette and Anna (Johnson) Black, the former born in Ohio, while the mother was a native of Kentucky Frank Black was the oldest child in a family of nine children but only one of the girls is now living His father was a farmer who enlisted in the Union army and served for three years during the Civil War, under Captain Little and Colonel Hardesty. The Black children were raised on the farm and Frank began to earn money while still a boy trapping opposums He attended the public schools for three months for a while in the winter time, but most of his schooling consisted of lessons in the hard but excellent school of experience where he learned well, for he has succeeded in becoming one of the influential and prominent farmers and producers of his section of the plains country

Mr Black remained at home until he was about twenty-one years of age and then began to work as a farmer, his day being usually sixteen hours long After three years Mr Black began to split rails and says that he often split seven hundred in a day He was married in Gentry county, Missouri, January 13, 1876, to Miss Mary A Green, who was born at Wayne, Michigan, the daughter of Henry and Matilda (Freeman) Green being the fifth in a family of nine children Twelve children have been born to Mr and Mrs Black Sarah M, the wife of Henry Miller, a farmer near Chadron, has nine children, Thomas, a farmer near Clifton, Wyoming, married Amanda Pell, and they have two girls, Victoria, married M L Mitchell, a farmer near Chadron, Samuel, a farmer near Pine Ridge, South Dakota, married Erna Gorton, and they have four children, Anna, Married Dan Claflin, a teamster in Chadron and they have one boy, Laura, married Ralph Munkers, a farmer near Chadron and has one child, William, who served in France during the World War, married Fontine Johnson, Maud, is the wife of Mark Jensen, a farmer near Newell, South Dakota, Ida M, is a student in the Chadron high school, while the other are dead

After his marriage, Mr Black farmed about eight years in Missouri and then joined a colony that was coming to Dawes county in 1884 He came as far as Valentine with the colony and located on a pre-emption claim ten mile southeast of the present site of Chadron, near Bordeaux creek The first six months he lived in a tent then built a log cabin, using straw and slabs for the roof with a coating of clay above that Though the home was primitive, Mr Black says those were happier days than these The country where he settled was so wild that a man could go out and kill all the deer he wanted to for meat any time For some time Mr Black worked by the day and ploughed sod for his neighbors to obtain money for supplies, when not busy with his own farm work He cut and hauled stove wood to Chadron, taking one day to cut the wood and another to drive to town At first the crops were fairly good, until 1890, when the drought killed everything For eight years he raised vegetables and ran a huxter wagon and made money He says that he raised four hundred dollars worth of water melons on an acre, and often brought in a load of vegetables that sold for thirty-five dollars One year his cabbages beat all those

shown at the State Fair thirteen head weighing ninety-six pounds From time to time, as he made money, Mr Black invested in land and now owns fourteen hundred and forty acres, all well improved, which he has cultivated and so became one of the influential farmers of his locality For some time now, Mr Black has given up the active management of his land and lives retired at Chadron where he is enjoying the fruits of his many years of labor He can look back and review the years of trials and hardships and feels that he has done his share in the upbuilding of Dawes county He is a man honored and respected by all who know him, has many friends and the Black family is one of the old pioneer stock that has made the present prosperity of the county possible

WILLIAM TOBERT STOCKDALE dean of the Chadron Normal School Chadron Nebraska, is a man of high culture and education who is well and favorably known throughout the educational circles of the state and holds his position of responsibility because of his attainments and marked ability

Mr Stockdale was born in Springfield, Illinois, November 4, 1866, the son of Jonas and Rachael Stockdale, being the second of the five children born to his parents The father homesteaded in Saunders county, Nebraska, in 1868 The summer preceding that the family lived on an island between the Platte and Elkhorn rivers just above the point where the Elkhorn empties into the Platte During the harvest season the father swam the Elkhorn river each morning to his work, pushing a tub ahead of him which contained his clothes Late at night on his return, his wife would go down to the river and by calling, guide him across the stream on his return Mr and Mrs Stockdale lived for two years on the homestead without a team, carrying water from a spring a mile away Then an ox team was secured with which William Stockdale of this sketch did his first days plowing when a little over seven years of age William Stockdale received his elementary education in the rural schools and took higher work in the Lincoln Normal University, receiving his first degree in 1898 he took still more work at Fremont College from which he received the degree of Bachelor of Science in 1899, and in 1921, was granted the degree of Master of Arts by the University of Nebraska

Mr Stockdale lived at home until he was about twenty years old and then began to teach in the rural schools where he continued several years. One memorable occasion was the blizzard of 1888, when he had to stand against the door to keep the children from rushing out into the storm when it struck the building. After serving in the district schools he taught in the Arlington, Nebraska, schools for several years; then became city superintendent of Wisner, Nebraska, where he remained eight years, followed by two years at Madison. Mr. Stockdale was appointed to teach in the State Junior Normal School at Valentine during summers, filling this position five years, from 1906 to 1910 inclusive, acting as principal the last year. He had gained a high reputation as a teacher and superintendent and when the new normal school was established at Chadron he was appointed dean of the institution, June 5, 1911, and assigned to the head of the department of education and teacher training, a position which he still holds. During all these years Mr. Stockdale

had continued his studies in the various institutions from which he received his degrees and kept abreast of the latest educational movements. He became recognized as an able executive and it was his high scholarship coupled with this that led to his present important position where he is winning more laurels as one of the foremost educators of the state,

October 9, 1890, Mr. Stockdale married Miss Ida May Vorse, of Saunders county, Nebraska, the daughter of Amos and Sarah Vorse, pioneer settlers of eastern Nebraska, who came to this state in 1878. Mrs. Stockdale is one of their four children and before her marriage was a teacher of recognized merit. Mr. and Mrs. Stockdale have had two

children: Alva Percy, who was a graduate of the Wisner high school, the Peru State Normal School and the University of Nebraska, who was principal of the high school of Alliance, Nebraska, when he died, January 3, 1919, and Irma Lucile, who was graduated with advanced credit from the high school department of the Chadron Normal School in 1919, is now in the senior class of the normal school and in addition to receiving her regular state diploma will be granted the Bachelor of Arts Degree in 1921.

Mr. Stockdale belongs to the Rotarian Club, to the Congregational church and is a Past Master in the Odd Fellows Lodge. The Stockdale family is one of the well known and prominent one in Chadron, especially in educational circles.

WILLIAM A. POTTS, the owner of the Chadron Steam Laundry, and one of the substantial and progressive business men of the town is a native son of Nebraska, born in Hamilton county, April 20, 1882, of an old pioneer family. He is the son of Abraham W. and Harriet C. (White) Potts, the former a native of Pensylvania, while the latter was born near Carbondale, Illinois. William Potts is the eldest of the four children born to his parents. After leaving school he obtained employment in a sugar factory at Grand Island, then he went to Wyoming. With another boy he walked from Clearmont to Buffalo, some forty miles. They worked on Crazy Woman creek on a ranch for some time. From the ranch he went to Billings, and worked in a laundry but left for Townsend to work in the hay fields. Returning to Nebraska he obtained employment in a laundry at David City, remaining six months before moving to Hastings to canvas for pictures but did not like that and again tried the laundry business which he now learned from the bottom up, working over twenty years, becoming an efficient laundryman. He was made manager of the Aurora Steam Laundry in 1906, holding that position until the laundry was sold. He then went to Holdrege for the same company as foreman of a laundry in that town and on August 20, 1907 was married there to Miss Luella H. Tatum, the daughter of William and Margaret F. Tatum, being the oldest of the two children in the family. Mr. and Mrs. Potts have two children, Margaret and Roscoe A.

Mr. Potts came to Dawes county and located in Chadron April 1, 1913. He bought the old Chadron Steam Laundry which he remodeled and equipped, and in 1919, he built his new

GEORGE LAUCOMER

laundry of reinforced concrete with a floor space of forty-five by one hundred feet, and two stories high This is one of the modern and best equipped laundries in the state

GEORGE LAUCOMER, one of Scottsbluff's capitalists and a prominent citizen has been identified with many important undertakings since he came to Nebraska thirty-five years ago, having already behind him years of success, failure and again success in the Pennsylvania oil fields Bearing with him, also, Mr Laucomer had testimonials of which he might well be proud, proving how valiantly he had fought as a boyish soldier in defense of his country and won promotion because of his gallantry It is the duty of the biographer to recall these days from the past and give them a true setting

George Laucomer was born in Lancaster county, Pennsylvania, October 4, 1848, the son of Jacob and Elizabeth (Reed) Laucomer, who were born in Pensylvania of German parents They spent their lives in Lancaster and Mr Laucomer remembers that his father for many years was a man highly esteemed, acting as sexton of the German Reformed church, an office of great dignity Of his ten children but two are living, George and Caroline, the latter being a widow and living at Sheffield Pennsylvania

A schoolboy when the Civil War was precipitated, George Laucomer determined from the first that he would enter the Union army and defend the liberties endangered, and possibly when he succeeded in becoming a member of the One hundred ninety-fifth Pennsylvania volunteer infantry, in October, 1862, he was one of the youngest soldiers in the entire army organization Nevertheless he bore himself as a man, serving under gallant General Sheridan in the Shenandoah Valley and with such efficiency that when he was honorably discharged in June, 1865, he was corporal although the youngest member of his company

Mr Laucomer did not return to school to complete his interrupted education, but he went back to Lancaster and served an apprenticeship of three years to the blacksmith trade and, according to the trade rule of the time worked one year as a journeyman Those were the busy days of oil development in his state and he soon drifted to Oil City, where at first he worked in a refinery, but later ventured into the business on his own account and made a fortune that a further venture caused him to lose just as quickly He remained in the oil fields for fifteen years and during that time

again made a comfortable fortune, after which he came to Nebraska, buying a ranch in 1884, in Frontier county, and for many years he was engaged in the stock business there In 1907 he came to Scottsbluff county, bought a cattle ranch and additionally a large acreage, and continued in the stock business until 1909 He still owns a large amount of land that is worth $200 an acre Mr Laucomer is financially interested in the Independent Lumber Company and for some years was president of this important concern

In 1872 Mr Laucomer was united in marriage with Miss Catherine Snader, who died in 1905, the following of their five children surviving George, superintendent of an oil company in Montana, Charles, a prominent ranchman in Sioux county, Minnie, the wife of Henry Clingman, of Ogden, Utah, and Dora, the wife of Arthur Jack, a merchant at Tekamah Nebraska In 1907, Mr Laucomer married Miss Emma Frances Shroad, who was born at Lancaster, Pennsylvania a daughter of Fred E and Frances (Roth) Shroad natives of Pennsylvania Mrs Laucomer's father was born April 6, 1846 and died in 1904 He served in a cavalry regiment in the Civil War Of their nine children only two live in Nebraska, Mrs Laucomer and Samuel, who farms a portion of Mr Laucomer's land Mr and Mrs Laucomer have one son Franklin George who was born June 15 1914 In politics Mr Laucomer is a Republican and at times has served in local offices and for six years was a justice of the peace For some time he was president of the Farmers' Alliance of Dawson county He has always taken great interest in the Grand Army of the Republic and has served as quartermaster of his post Mr Laucomer is known in many sections of Nebraska but since 1907 has resided in Scottsbluff, where he is held in universal esteem He has always liberally supported local enterprises has done much in the way of charity and has been generous to the Lutheran church of which he is a member

WILLIAM ALLEN WHITE one of the owners of the Chadron Furniture Company of Chadron is a practical furniture man of many years experience as he has been associated with the manufacturing and sales department of the furniture industry nearly all his business life and though a new merchant in this section is making good in his special vocation Mr White was born at Salem Dent county Missouri, January 7 1885 the son of Ransom A and Martha (Buckner) White both natives of Missouri William was the second of the five children born to his parents His father was a veteran of the Civil War, having served in the army three years under General Marmaduke and participated in Price's Raid in Missouri and many battle and skirmishes He farmed for several years and the family then lived in the country Later the family moved to St Louis and then to Springfield, Missouri, where the children attended school After completing his course in the public schools Mr White went to work for the Springfield Furniture Company in the factory and thus learned the business from the bottom up He first swept sawdust up, advanced from that to more important work and then became foreman of the shipping department After holding that position three years he entered the main office of the company and in six months was made time-keeper and assistant bookkeeper Following this he went on the road as a salesman for the company and continued to travel for eleven years His territory covered Missouri, Oklahoma, Kansas, Arkansas, Iowa and Colorado In this line he was very successful and built up a fine trade which led him to engage in an independent business in 1917, traveling for different kinds of furniture supplies but selling on a commission basis He liked the independence of an enterprise of his own so well that in 1920, Mr White came to Chadron where he already was well acquainted with the commercial outlook and purchased a third interest in the Chadron Furniture Company From selling in this territory he knew that business was good and located here Since coming to Dawes county he has taken an active part in the management of the store and inaugurated many policies that have increased business and though one of the later residents of the town has a bright future from the present outlook

July 4, 1920, Mr White married Miss Louise Horine, who was born in Springfield, Missouri She is a graduate of the high school there and Loretto Academy, specialing in languages After finishing school she was assistant in the Carnegie Library in Springfield for a year

Mr White is a member of the Knights of Pythias having taken the highest degree of that organization He is a man of energy, knows the furniture business thoroughly, both manufacturing and salesmanship and is an addition to the commercial element that is building up and developing Chadron

MICHAEL CHRISTENSEN, pioneer settler of Dawes county, prosperous ranchman and a man who has served as a public official with benefit to the county and credit to himself, is today one of the large landholders of the Panhandle who is regarded as one of the substantial and progressive men of his day He has played his part in the opening up and development of Dawes county both in his private business and as county commissioner, and deserves place in the history of the county, where he has been well known for more than thirty years and is highly esteemed Mr Christensen was born in Denmark, October 27, 1859, the son of Chris and Christine (Mikkelsen) Christensen, both of whom were born, reared and spent their lives in their native land Michael was the fourth in a family of seven children born to his parents His father was a small farmer and brick maker The boy was reared on his father's land and early displayed business ability as he would secure some article and trade it for something better and keep this trading up, always gaining on each deal He attended the public school in the winter time and worked in the summers At the age of twelve years he began to work in the brick yard and learned the trade of brick maker He remained at this occupation, going to school at the same time, until he was fifteen then obtained his parents' consent to work for a large farmer for two years He did not draw his salary but let it accumulate but made some money in trading deals When seventeen years old, Mr Christensen's father called him home to take charge of the brick yard where he was engaged two years He then ran a small farm of his father's for a winter before attending college for a time He had heard of the many opportunities for a young man in America and in 1882, came to the United States Coming west, Mr Christensen reached Cleveland, Ohio and as he did not understand English found it hard to secure work He found a Dane with whom he could talk and who promised to help him, but his money was gone and he finally found a place in a stone quarry mill, where the stone was cut in blocks At first he worked for very little, but soon learned some English, showed his ability and within a short time he was advanced The second year he was made foreman of the mill

December 6, 1882, Mr Christensen married Miss Catharine M Albertson, at Cleveland, who was born in Germany, of Danish parents, Christian and Catharine (Smidt) Albertson and was the fourth of their six children Mr

and Mrs Christensen have had eight children
Catharine, deceased, George C, a farmer near
Chadron, Albert, a farmer near Chadron,
Christiana M, in New York learning the mil-
linery business, William B, married Alice
Manchester of Chadron, and spent over a
year in France during the World War, is now
running his father's ranch on White river,
Mike F, who farms with his brother, Frank-
lin R, married Vernice M Robinson and owns
a farm southwest of Chadron, and Marion
B, deceased

Remaining in Cleveland until 1886, Mr
Christensen came to Chadron, then the end
of the railroad and took a pre-emption on
White river on which he proved up, then took
a homestead adjoining He at once began improvements on his ranch where he lived until
1898 The Christensens passed through all
the hardships and privations of life in a new
country Wheat was introduced, he became
one of the fine farmers and also a stock man
With increased capital he began to buy more
land, as he saw there was a great future in
the Panhandle and now owns about five thousand acres, used for farming and grazing purposes After accumulating a comfortable fortune he retired from the active work of the
ranch, leaving that to his sons and now lives
in his comfortable home in Chadron, honored
and respected by all his business associates and
friends With his sons Mr Christensen is
interested in raising registered Hereford cattle
of the Anexiety Fourth strain They have
from three to four hundred head all choice
registered cattle on the ranch and will have
about the same amount the next year, as they
find them profitable and make money from
beef cattle In addition there is about four
hundred acres planted to alfalfa, and many
acres to other farm crops The Christiansen
ranch is one of the fine properties in Dawes
county, it is well watered, has good buildings
which have cost about fifteen thousand dollars, is fenced and cross-fenced and the hay
land cuts from five to six hundred tons a year

Mr Christensen has not confined all his
time to his personal business but has taken
an active part in public life, as he was elected
county commission in 1914, on the Democratic
ticket and served four years During his term
of office he was instrumental in putting in
the first road grading in the county and also
took part in other county matters of benefit to
the people, for he is progressive in his own
business and applied the same methods to
the affairs of the county, giving them the
benefit of his shrewd business judgment and

executive ability The various public works
he inaugurated in Dawes county stand as a
monument to him, for he is a booster for his
county and community and is every ready to
assist every good movement that will benefit
the people Mr Christensen is a man who
has made good in this section of the state and
proved that a man of energy and the determination to succeed can overcome all obstacles and has builded wisely and well

JOHN H GLENN, a retired farmer of
Chadron, who for many years was engaged in
the teaching profession in Ohio and later here
in Nebraska, is a man who has won a high
place in the esteem of the people whom he
served long and faithfully He was born in
Cleremont county, Ohio, November 11, 1861,
the son of John W and Martha (Creamer)
Glenn, the former a naive of Scotland, while
the mother was born in Cleremont county,
Ohio John Glenn was the youngest in a family of six children and is the only one living
His father was a United Brethren minister
who preached for fifty years John Glenn
spent his boyhood in his native county in Ohio
and attended the public schools there until he
was nine years old, when his parents moved
to Clinton county near the town of Blanchester, where they lived the remainder of their
lives When only seventeen years old John
Glenn had a certificate to teach, but as it was
against the law to allow anyone under twenty
to have charge of a school he attended the
State University, at Lebanon, Ohio, for two
years and began his first term of school when
twenty years old and has taught thirty-two
years altogether Mr Glenn progressed in
his chosen vocation and was in charge of
various schools in Ohio where he won a reputation as a fine teacher

December 24, 1882, occurred the marriage
of John Glenn and Miss Isabella Goodwin, at
Blanchester, Ohio Mrs Glenn was the
daughter of Levi and Hannah (Runyon)
Goodwin both American born but of Irish
parentage Mrs Glenn was next to the oldest in a family of ten children all of whom are
living and there were eighteen grand-children
but one small child died and Raymond Goodwin was killed in action in France, in 1918
Another brother, John Goodwin was wounded
by the shell that killed his brother Mr and
Mrs Glenn have one child, Bertha, and Eva
Hazel Bowen who was adopted by them
when five weeks old

The Glenn family moved from Ohio to
South Dakota in 1908 and located near Buffa-

lo Gap on a homestead They remained there about twenty months, proved up on the land and sold it to advantage Coming to Chadron Mr Glenn accepted a position in the railroad yards of the Northwestern railroad in the summer and continued his old profession of teacher in the winter Remaining in Chadron until 1914, the Glenns then took a homestead eight miles south of the town, which they still own Mrs Glenn says that she feels she earned her share of the farm as they had a well a hundred and seventy-six feet deep where she assisted in pumping the water for nine head of stock, as well as for household use Mr Glenn continued to hold his position with the railroad in Chadron walking eight miles back and forth from the farm each day for a time Then they bought a horse and buggy and later a car In 1918, the Glenn family moved into Chadron and rented the Old Foster building where Mrs Glenn and her daughter kept a rooming house They made a success of this business and in 1920, bought the building known as the McFadden Building on Eagan street, where Mrs Glenn and her daughter are running an up-to-date rooming house and now serve excellent meals in a cafe which is connected with it They are capable business women and have won a high place in the town since engaging in hotel business They have many friends in Dawes county and are substantial reliable citizens The family are members of the Methodist Episcopal church while Mr Glenn belongs to the Masonic order, is an Odd Fellow and a member of the Knights of Pythias as well as the Modern Woodmen, having filled all the different chairs in these lodges

WILLIAM H REYNOLDS, the mayor of Chadron and well known dealer in real estate here is one of the pioneer settlers of this section whose figure stands out prominently in political and municipal affairs of the county as he has taken an important part in the settlement and development from first locating here Few men are better known and liked and few have willingly assumed so many of the burdens of public life for the benefit of the citizens of Dawes county and the town of Chadron Mr Reynolds was born in Morgan county, Illinois July 13, 1849, the son of James M and Amelia (Hand) Reynolds, the former born in Virginia, while the mother was a native of Illinois To them twelve children were born, William being the tenth in order of birth He spent his younger days in Morgan county, worked on the farm in the summer time and attended school during the term His mother died when the boy was six years old and his father when he was eleven years old He worked as a farm hand during the summer months, and attended country school in the winter time. Later he attended Whipple Academy and Illinois College, paying his way by working in a store After leaving college at the age of twenty-three, he went to Missouri, locating on a farm in Harrison county, where he engaged in farming for himself Two years later, in February, 1875, Mr Reynolds married Miss Elizabeth Waltz, a native of Ohio, a daughter of John W and Susan (Swan) Waltz, the latter born in Ohio One child was born to this union, Eleanor B who married Reverend Fred Hall, a Presbyterian minister of New Jersey Mrs Hall is a graduate of the normal school at Chadron and Doane College, of Crete, Nebraska Mr Hall is also a graduate of Doane College, and took a graduate course at Yale University During the World War he spent eighteen months in government service with the Y M C A, being assigned to work in Ireland, England, and was one of the few Y M C A men to accompany the American forces to Arch Angel, Russia

Mr Reynolds came to Dawes county in 1884, while it was still unorganized territory, being secretary of the Missouri Colony which settled near the present site of Chadron He built a log house on Bordeaux creek and was fortunate in always having an abundance of water With his wife he suffered all the hardships and privations of frontier life and in 1884 and 1885 the settlers were forced to get supplies from Valentine a hundred and thirty-five miles distant During that winter there was a heavy fall of snow which laid long, the mercury was very low and the people ran short of provisions so that it was necessary to make the trip for supplies On the way back the teams were stuck and Mr Reynolds and H S McMillan took a four horse team and scoop shovel and helped rescue the party In the spring the settlers gathered enormous supplies of bones from the prairie which they exchanged for groceries

For seven years, Mr Reynolds lived on his homestead, then in 1890 became clerk in the Government Land Office at Chadron, in 1892, he was elected county clerk of Dawes county on the Republican ticket and served two terms When Dawes county was organized in 1885, Mr Reynolds was active and played an important part in county affairs and organization He was elected State Senator from dis-

Robert H. Willis

trict twenty-eight, which contained eight counties, in 1898. served in 1899, and was re-elected in 1910 and 1912, which shows the esteem in which he is held in Dawes county and the district He is a man of great executive ability and proved competent to fill every office to which he was elected or called In 1919. Mr Reynolds was elected mayor of Chadron, a city to which he had so largely contributed in its first settlement and organization He is always ready to help in any movement for the benefit or development of the county or city and is a well known and well loved man in his city. Mr Reynolds owns a quarter section of land near Belmont, Nebraska, and has a modern home in Chadron For some years he has been actively engaged in the real estate business here, also carrying on loans and investments He has a partner, Fred A Hood, they have offices in the Citizens State Bank Building Mr Reynold's long association with the affairs of the county places him in a position to give his customers the greatest advantages in all real estate and land deals and he has a clientele that is not only satisfied but ever increasing

During the World War Mr Reynolds again took a front place in assisting its prosecution, as he was Fuel Administrator for Dawes county and took an active part in the work of the Red Cross Mr and Mrs Reynolds are members of the Congregational church while he belongs to the Odd Fellows and is a Republican

ROBERT H WILLIS, whose identification with important public affairs in Morrill county has been long and continuous, is now chief of the Bureau of Irrigation Power and Drainage of the Department of Public works of the state of Nebraska Mr Willis was born at Cheyenne, Wyoming, March 22, 1869, the son of John G and Cecelia J (Beck) Willis The father was born at Saratoga, New York, a son of Robert and Mary (Toner) Willis, the former of whom was born in Ireland and the latter in Scotland, both dying in the state of New York The mother was born in Iowa Her father, William Beck was a native of Scotland, who came to the United States, worked in the west as a miner and freighter and died in Montana Robert Henry was the first born of his parents' seven children, the others being as follows Cecelia Mary, the wife of William B T Belt president of the Northwestern Group Bell Telephone Company, Omaha, William H is in the implement business at Bridgeport, Blanche

I the wife of F W Smith, a merchant at Minatare Nebraska Edith, L, is an Episcopal Church missionary in North Dakota, Beatrice who resides with her parents, and Margaret J who teaches in an Indian school in South Dakota

John G Willis enlisted from Illinois for service during the Civil war serving four years under two enlistments in company K Seventeenth Illinois cavalry After the war he came west located first at Omaha, then went to Cheyenne Wyoming was married there and started the first general store of the town He continued in the mercantile business at Cheyenne until 1884 and was a leading citizen and as a member of the first city council was sworn in by the adjutant-general of the United States army of the Territory of Wyoming He is active in Masonry belongs to the Consistory and is a Shriner

Robert H Willis attended school at Omaha and later received a technical education as civil engineer at Rensselaer Institute, New York, from which he was graduated in 1890 His first important work was done in connection with the city engineer's office of Omaha For two years he was associated with Douglas county engineer's office, then he spent a year with the Union Pacific railway at Pocatello, Idaho, after which private contracts engaged him until 1895, when he was appointed a water commissioner of Morrill and Scottsbluff counties, in which office he continued until 1910 He then became water superintendent of District No 1 and now has charge of all irrigation work in Nebraska He came to old Cheyenne county in September, 1894, locating at old Camp Clark, and since 1901 his home has been at Bridgeport He has been identified with many of the developing agencies in this section and is one of the widely known men of the Panhandle With other enterprises of merit with which he has been connected, Mr Willis owned and conducted the Bridgeport *Blade*, the first newspaper issued here, for five years

In December, 1891 Mr Willis was united in marriage to Miss Carrie Lee Melius a native of Rensselaer New York who died September 14 1911 leaving one daughter Cornelia E, who was born in September, 1893 She resides in New York and is connected with the State Industrial Commission On January 6, 1914 Mr Willis was married to Miss Anna E Hascall, who was born at Grand Island, Nebraska They are members of the Episcopal church Mr Willis was early instructed by his father in the principles of the Republican party and they have always seemed to him just, wise and adequate for the governing of the United States. He is a Mason and belongs to the Lodge of Perfection at Alliance, Nebraska.

JAMES H. BANKS, was born August 29, 1864, near DeWitt, Clinton county, Iowa, one of a family of nine children of David I. and Mary (Smith) Banks, the father being a native of York State and the mother a native of Pennsylvania.

Growing to manhood near State Center,

Marshall county, Iowa, he attained a good practical education at the district school which was attended during the winter terms, working on the home farm during the summer; James' first financial venture was to contract with a neighbor to hoe broomcorn at fifty cents a day when not occupied at home. After becoming of age he farmed for himself two years; and then started a furniture store in his home town, in which business he has been identified practically ever since. Coming to Nebraska in the early nineties Mr. Banks first located at York, becoming associated in the furniture business there and later at Fremont, Omaha and Lincoln, Nebraska.

Mr. Banks was married to Miss Pearl Mann August 29, 1901, at Pittsburg, Pennsylvania. Miss Mann was born in Cleveland, Ohio, being a daughter of Horace and Priscilla (Cook) Mann and related by marriage to President McKinley. Mr. and Mrs. Banks being at the Buffalo Exposition with him two days previous to the assassination. A son James E. was born to them, who at this time is being edu-

cated in Chicago, Mrs. Banks having died in 1903.

Coming to Chadron in the fall of 1912, Mr. Banks became manager and part owner of the Chadron Furniture Company. For a number of years he was secretary of the Nebraska Furniture Dealers Association and a member of the executive committee of the National Association and has for a number of years been a familiar patron of the leading furniture markets. With a thorough knowledge of the business and of home needs, under his efficient management the Chadron Furniture Company has grown from a small stock in one room to one that occupies six times the original floor space, being the largest home furnishing establishment in northwest Nebraska, and is considered one of the substantial business houses of Chadron. Mr. Banks has further verified his faith in the future development of the city by the purchase of the two story and basement brick building occupied by his company, located at the corner of Second and Morehead streets, together with other real estate, and he is identified with other commercial interests here and in Crawford, Nebraska, all of which afford employment to a number of heads of families which help materially in the making of a city. He has always responded loyally and liberally to every progressive movement where the aim has been for the betterment to the home town; is an honored citizen and a loyal M. W. A.

BENJAMIN A. BREWSTER was born in Omaha, Nebraska, June 1, 1870, the son of George W. Brewster, born in Cleveland, Ohio, and Elizabeth (Barton) Brewster, born in Mason, Illinois. Benjamin was the second child in a family of three children. His father was a publisher and printer, publishing the first agricultural paper in Nebraska, at Omaha in 1871, and Benjamin has the bound volumes of this paper published while his father was in charge, from 1871 to 1883.

Benjamin Brewster was reared in Omaha until about fourteen years of age. The first money he remembers earning was setting type for his father, for which he received fifty cents, and Ben says he has been sticking type ever since. He went through the common schools in Omaha and in Oakland, Nebraska, where his father moved in 1882, and established the Oakland Independent. The father then went to Blain county, Nebraska, to start a town in the sandhills, believing the B. & M. railroad intended to build into this country, and his intention was to form a county about twenty-four miles square and have the town of Brewster the county seat. He succeeded in establishing the county all right and the town at the present time, 1921, is still the county seat; but the railroad has not materialized.

Benjamin's father was a Civil War veteran, serving in the Eleventh Illinois Infantry during the war, and was wounded at the battle of Shiloh. When he went to Blain county in 1885, he took a tree claim and pre-emption and built the regulation sod house, and established the *Brewster News*, when this county was unorganized territory of Sioux County. At that time there was not a house within three miles. Mr. Brewster is credited with having done

more than any other man in putting Brewster and Blain county on the map.

Benjamin A. Brewster went through business college and together with his brother, Win F. Brewster, took over the *Oakland Independent*, running the same till 1890. He then went to Deadwood, South Dakota, and went to work on the *Daily Pioneer* as a printer; stayed there until 1891 and then went back to Omaha and worked at his trade until 1893. He then bought the *Ord Journal*, a Democratic paper which he run about a year, and then went to Craig, Nebraska, and established the *Craig Times*, which he sold in 1885. He then went back to Omaha and worked at his trade until 1897, when he went

to Deadwood, South Dakota, and engaged in the newspaper work about two years, then went to Lincoln and was with the Newspaper Union until 1901. In 1901, he went to Shoshone, Idaho, and edited the *Shoshone Journal*, in partnership with F. R. Gooding, present United States Senator from Idaho. He remained here two years, returning to Crawford, Nebraska, and establishing the *Crawford Courrier* in 1905, which he run for about a year, and then went to Amarillo, Texas, and ran the *Amarillo Daily Panhandle*, where he remained until he came to Chadron, Nebraska, and established the *Chadron Chronicle* in 1909, which he published and edited until October, 1920.

During his residence in Chadron he ran on the Democratic ticket for State Senator of Nebraska, from the twenty-eighth senatorial district, and was defeated by a small majority. He was appointed postmaster under Woodrow Wilson, March 23, 1916, and re-appointed June 4, 1920, and still holds the appointment at the present time.

Mr. Brewster was married October 11, 1908, in Deadwood, South Dakota, to Ida Austin, who was born in Texas. She was the daughter of William Austin, born in Mississippi, and Nannie (Lewis) Austin, born in Tennessee, of an old southern family. Her ancestors were among the large slave holders of the south. Mrs. Brewster is the eldest in a family of six children — three girls and three boys. She finished her education at the Mary Nash School at Sherman, Texas.

Mr. Brewster owns a modern home and business house combined, on Bordeaux and Second Streets, besides a lot of valuable property in Chadron, Nebraska. He is an Odd Fellow, Modern Woodman and an Elk, and one of the live wires of his little city.

JAMES R. SCHOOLEY, one of Banner county's progressive farmers and ranchmen, is also one of her worthy and dependable citizens. He belongs to a very prominent pioneer family of this region and has lived in the county since he was nineteen years of age. He was born in Neosho county, Kansas, October 30, 1870.

The parents of Mr. Schooley were Levi and Mary J. (Ellis) Schooley, natives of Illinois. For some years prior to coming to Nebraska, the family lived in Kansas and the father was a farmer in Neosho county. On July 25, 1886, he homesteaded in Banner county, seven miles southeast of Harrisburg, and resided on his place for eighteen years. He was then elected county judge and moved into Harrisburg He was honored with re-election and served three more terms on the bench, and afterward, for many years was a justice of the peace For several years before moving to Harrisburg, he had been road supervisor of Lone Spring precinct He was a man of such sterling integrity that he enjoyed the confidence of every citizen, although he was elected to different offices on the Republican ticket Failing health caused Judge Schooley to remove in 1902, to Palisade, Colorado, in the hope that a change of climate would restore him to health, but it failed to do so and his death occurred there January 24, 1908 His widow resides at Fruita, in Mesa county, Colorado Of their five children, James R is the second of the four survivors, namely Jennie, the wife of A L Smith, who lives at Palisade, Colorado, Sidonia, the wife of Frank Lane, of Fruita, Colorado, and Francis Nathan, who is in business at McGrew, Nebraska, married Emma E Campbell Judge Schooley was a member of the Christian church, to which religious body his widow also belongs

James R Schooley attended the public schools in Kansas, somewhat irregularly on account of delicate health, and he was by no means robust when he accompanied his parents, across country in a big covered wagon, to Banner county His experience may, perhaps, be compared with that of the late beloved, great statesman and patriot, Theodore Roosevelt, who, under like impairment of health, came to the great west and found healing here Mr Schooley soon began to improve, and considering his present heavy business responsibilities, and the able way in which he is handling his various undertakings, good health is one of his best assets He has never been entirely separated from Banner county, although he has had interests near McGrew and Minatare, but since March, 1919, he has operated his own farm with vigor and profit and is making plans for extension of his industries

Mr Schooley was married March 15, 1899, to Miss Sarah J Harshman Her parents were Theodore and Rebecca (Thompson) Harshman, for many years residents of Minatare The mother of Mrs Schooley died at Albany, Oregon, May 26, 1916, but her burial was at Minatare The father survives and lives in Idaho Mr and Mrs Schooley have two children Theodore, who lives at McGrew, and Levi, who lives at home, and a step-daughter, Ethel R, whose father, John R Lendrum, was accidentally drowned in the

Platte River, July 10, 1891 She is the wife of Cecil Prettyman, of Broken Bow, Nebraska Mr and Mrs Schooley attend the Christion church In politics, like his father, Mr Schooley has always been a staunch Republican

SHADRACH PETERMAN, who has been a farmer and ranchman ever since coming to Banner county, for many years before locating in Nebraska, was actively identified with the lumber business He was born at Nordmont, Sullivan county, Pennsylvania, February 3, 1876

The parents of Mr Peterman were James and Hannah J (Hunter) Peterman, the former of whom was a native of Columbia and the latter of Sullivan county, Pennsylvania The mother still resides at Nordmont, but the father died January 19, 1896 Of their ten children, Shadrach is the only one living in Nebraska The father was a farmer and lumber dealer and was influential in Democratic political circles in Sullivan county For many year he served in such offices as school director, road overseer and supervisor

Shadrack Peterman attended school at Nordmont until fourteen years old, then started out for himself, finding employment with lumber companies, with one of which he continued for seven years Lumbering is hard work and there are many accidents in the woods from one of which Mr Peterman suffered, which resulted in a broken hip That closed his career in the lumber industry as heavy labor was no longer possible Mr Peterman then came to Nebraska and remained at Kimball for a short time before coming to Banner county He worked for about four years for farmers and then bought land and went into the business of breeding Shire horses He has done well in his investments and is looked upon as one of the county's substantial and reliable men He recently sold two hundred and forty acres of land but has a tract of eighty four acres in Pennsylvania and his ranch in section twenty-two, town nine, Banner county In politics Mr Peterman is a Democrat, and in national matters is loyal to party candidates, but in local affairs votes according to his own judgment He belongs to the order of Knights of Pythias

THEODORE F GOLDEN, president of the Farmers State Bank of Crawford, and owner of a large body of fine, irrigated land in Dawes county, has been a resident of Nebras-

Eng by E G Williams & Bro NY

ka for thirty-three years Of sterling personal character, he has built up a business reputation that is impregnable, and any financial institution that carries his name at the masthead, will be sure of a large measure of public confidence

Theodore F Golden was born in Vermillion county, Illinois, March 23, 1856 His parents were William and Mary (Collison) Golden, both of whom were born in Virginia The father died in 1911 and the mother in 1914 Of their ten children, Theodore F was the only one to come to Nebraska They were members of the Presbyterian church although the father was reared a Quaker In 1858, he moved to Iowa with his family and engaged in farming there on a large scale, during the rest of his active life First he was a Whig and later a Republican

Reared on his father's farm and attending the common schools in Iowa, Mr Golden remained in that state until March, 1887, when he came to Sioux county, Nebraska, and homesteaded, later securing a pre-emption and tree claim He lived in a log cabin with a sod roof and went through with about the same experiences as fell to other early settlers In recalling those early days, Mr Golden speaks of the thousands of acres of land that fell to the loan companies on account of many of the homesteaders mortgaging their land and never becoming able to pay the loans In 1900, he came to Dawes county At that time there was not enough hay to feed the stock and it was a curiosity to see even a load of hogs shipped out, while now one of the county's exports is hay, and both hogs and dairy cattle are shipped and even cream is bought and shipped to Alliance Wonderful changes have been brought about within the short interval of twenty years Mr Golden now owns over a thousand acres of land He specializes on cattle and alfalfa, his land being irrigated, and he sees a great future in alfalfa

Mr Golden was married October 11 1876, to Miss Anna Lane, a daughter of George and Tillie (Walker) Lane, who are deceased To Mr and Mrs Golden children were born as follows Mary, who is the wife of Frank Wright, of Dawes county, Bertha, who is the wife of Edwin Raum, of Crawford George who is a farmer in Dawes county, Florence who is the wife of Howard Glaze, of Sioux county, Chester, who is on the home place, Ralph, who is in Sioux county Hugh who lives at Lusk, Wyoming, Elizabeth, who is the wife of Henry Raum, of Sioux county, and Victor and Ida, both of whom live at

home Mrs Golden is a member of the Presbyterian church

The Farmers State Bank of Crawford was established March 20 1919 by F M Stapleton with a capital of $25,000, when Mr Golden became president The first cashier was H P Gipson, the present cashier being F M Stapleton, William Sutherland being vice president The bank has prospered from the first and its statement on December 12, 1919, showed deposits of over $100,000 By March 1, 1920, the bank expects to be housed in its own building a handsome structure, well located from a business point of view and equipped with modern conveniences and bank safety devices, being almost completed Mr Golden has never accepted any public office but he has always been a sturdy Republican in national affairs from principle, in local matters quite often depending on his own excellent judgment

JUDGE GEORGE J HUNT, who has been exceedingly prominent in important affairs in Morrill county for a number of years and one of the earliest professional men of Bridgeport, is numbered with the leading members of the bar of that county The judge is a native of Maryland, born at Baltimore, September 18 1856 He made his first visit to Nebraska immediately after his graduation from college in 1876

His parents were Thomas H and Julia A (Dorsey) Hunt, both of whom spent their lives in Maryland Thomas H Hunt and his father once mayor of Baltimore, established the Eutaw Savings Bank in Baltimore, which still is one of the soundest financial institutions in Maryland The Hunts, as well as the Dorseys, came to Maryland before the Revolutionary war Judge Hunt is thus a direct descendant on maternal side of Captain Bachel Burgess who was a British officer on eastern shore of Maryland prior to the Revolution and was captain in the American army during that war Many of the Maryland eastern shore Dorseys were leaders in the Society of Friends their descendants being among the most highly esteemed residents of Philadelphia today George J Hunt has one brother, Herbert M who is associated with the great packing firm of Libby McNiel & Libby at Baltimore One brother Frank N whose death occurred from accident in 1885, was cashier of the Eutaw Savings Bank at that time The parents were members of the Episcopal church

George J Hunt had both social and educa-

494   HISTORY OF WESTERN NEBRASKA

tional advantages in his youth. He attended Washington College, a branch of the State University of Maryland, from which he was graduated in 1876 and immediately came to Omaha, where he spent one year as an employe of the wholesale firm of Morgan & Gallagher, and then returned to Baltimore. During the next two years he applied himself to the study of law and in 1878 was admitted to the bar at Belair, Maryland, following which he came back to Omaha and as a member of the law firm of Congdon, Clarkston & Hunt, became well know in that city. In the spring of 1893 he came to Morrill county, being interested in the completion of the Belmont Ditch, in which he had invested largely. He bought 16,000 acres of land for the company, bonded the ditch for $250,000 and floated the bonds in the east. The ditch was completed in 1894 but Mr. Hunt was obliged to handle many suits against the company before the business was entirely settled. He protected every claim and never lost a quarter section of the land and paid off all the bonds. In 1902 when the town of Bridgeport started, he opened his office and has continued in active practice ever since, at one time being urged for the Supreme Court bench. Mr. Hunt owns a large amount of property in city and county and built many of the houses on the land under the Beemont Canal.

In 1883 Mr. Hunt was united in marriage to Miss Margaret Bouldin, who was born at Belair, Maryland, and they have three children: Harriet, the wife of T. B. Still, who is assistant cashier of the Merchants Bank of Denver, Colorado; Julia D., who married Rev. Philip S. Smith, an Episcopal clergyman, and Frank N., in the real estate business at Bridgeport. Mrs. Hunt is a member of the Episcopal church. Mr. Hunt has always been affiliated politically with the Democratic party. During the period of the World War, he took a very active part in helpful movements, was chairman of the Council of Defense and gave liberally of his time, means and professional advice when all these seemed necessary.

AUGUSTUS L. MOYER, who has been a prominent business man and representative, dependable citizen of Crawford, Nebraska, for over a quarter of a century, was born in Union county, Pennsylvania, November 12, 1859. His parents were John W. and Violetta (Meixell) Moyer, natives of Pennsylvania, in which state both died. Of their thirteen children, five are living, Augustus L. being the only one in Nebraska. The father followed agricultural pursuits all his life, was a Republican in politics and served frequently in county offices.

Augustus L. Moyer completed his educational training at Bucknell University, Lewisburg, Pennsylvania. Gifted with musical talent, he started out as a teacher of music and conducted a retail music business until 1884, when he came to Omaha and engaged as commercial traveler for Max Meyer & Brothers, music dealers. After two years he came to Grand Island, Nebraska, and conducted his own music store for five years. After selling his Grand Island store he established a similar one at Norfolk, Nebraska. In 1893, he came to Crawford and bought a drug business here, for which he paid sixteen hundred dollars: It was a small store at the time but capable of expansion, and under Mr. Moyer's careful but progressive business direction, grew into the largest establishment of its kind in Western Nebraska. In November, 1919, Mr. Moyer sold the store for sixteen thousand dollars. In the meantime he has been interested in other enterprises of modern character. In association with his son, he built the Crawford Telephone system, which he sold in 1912, to the Wyoming and Nebraska Telephone Company, now the Nebraska Telephone Company. He is about completing a modern garage at Crawford, a fireproof one-story and basement structure, fifty by a hundred and fifty feet in dimensions, of modern construction and equipment. Mr. Moyer is agent for the Cole-Hudson, Nash, Overland-4 and Chevrolet cars, and by the first of March, 1920, a complete repair shop will be in operation.

In the state of New York, on November 26, 1883, Mr. Moyer was married to Miss Myra L. Walter, a daughter of Abraham and Mary (Shively) Walter. Mr. and Mrs. Moyer have one son, John Walter Moyer, who is married, lives at Crawford, and is in business with his father. The entire family belong to the Methodist Episcopal church, and both father and son are Republicans in political affiliation. Mr. Moyer has always been a loyal party man but has never been willing to accept any office, it being his contention that a man so engrossed as he has always been in his own business affairs, could not find time to properly and honestly discharge the duties of a public office. He is a thirty-second degree Mason and has been an official in the Blue Lodge. In addition to the property above described, Mr. Moyer owns a business block and several residences at Crawford, also the telephone headquarters, and some exceedingly valuable farm property, about forty acres, near the city limits, for which he has recently refused six thousand dollars

GEORGE M ADAMS, long a prominent and representative citizen of Dawes county, enterprising in the field of business and honored in public life, has belonged to this section of Nebraska for thirty-five years He came into the county on foot, a boy of thirteen years with his gun over his shoulder, prepared to meet with courage and self defense any foe he might find in the long, lonely, almost unsettled country over which he had trudged, all the way from Valentine That same spirit of fearlessness combined with prudence has been manifested by Senator Adams as he has met with the problems of life in a somewhat extended business and political career

George M Adams was born at Tekamah, Burt county, Nebraska, October 16, 1872 His parents were John J and Clara (Kerr) Adams, the former of whom was born at Washington, District of Columbia, and the latter in Fayette county, Illinois The father served through two enlistments in the Civil War, a member of Company H, Forty-fourth Iowa Volunteer Infantry For some years he filled a very important position with the Chicago and Northwestern Railroad as buyer of right of way and coal mines for that company between Boone, Iowa, and Blair, Nebraska He was married at Boone, Iowa, and from there, came as a prospector and homesteader in Dawes county, in 1884, returned then to Iowa, and in 1885 came with his wife and seven children and settled on the homestead Mr Adams continued on his homestead near the old Mayfield ranch, until 1894, when the family moved into Crawford, where he lived more or less retired until his death on May 23, 1912 The mother of Senator Adams still resides at Crawford Of the family of seven children, George M was the second in order of birth, the others being as follows Charles F, whose home is at Casper, Wyoming, is an engineer on the Burlington Railroad, Carrie, who is the wife of Frank Wheeler, of Dunbar, Nebraska, Jennie, who is the wife of Charles Triplett, a resident of California, John J, who is postmaster at Crawford, Nebraska, Emma, who is the wife of Riley D Richard, of Harrison, Nebraska, and Harry, who served five years as treasurer of Dawes county, is in the oil business at Casper, Wyoming

George M Adams obtained his education in the public schools When the family left Iowa

for the new home in Dawes county, in the spring of 1885, he accompanied the other on the railroad as far as Valentine, the terminus From there it was necessary to transport family, household goods and provisions by wagon over the many intervening miles to the site of the homestead It was then that the youth determined to relieve the overloaded wagon by undertaking to walk the distance, fortunately finding a fellow pedestrian in Charles Spearman He made the trip without any serious mishap and at first gave his father assistance in getting settled

Mr Adams sought employment, as his services were not actually needed on the homestead, and found it with the railroad men constructing the line from Chadron to Lusk, Wyoming, as water boy After returning from this job he sought another and found it at Crawford, where he was a clerk in the Graves' drug store until 1893 In that year Mr Adams embarked in a general merchandise business for himself, in which he continued at Crawford until 1900, in the meanwhile becoming interested in other important business enterprises that resulted in his establishing a lumber yard and the organization of the Crawford Fruit and Cold Storage Company

As a citizen, Mr Adams early made his usefulness apparent first in civic life as a member of the school board and the city council, and later in the state legislature, to which he was elected a member of the lower house in 1907, and to the upper house in 1917 He has proved as wise in statesmanship as able in business He was reared in the Democratic party and has always loyally maintained its principles

Senator Adams was married at Crawford, June 20 1900, to Miss Nellie Johnson, whose parents were Sylvester and Eliza A (Welborn) Johnson, natives of Kentucky, he former of whom died August 5 1896 and the latter April 19 1906 Senator and Mrs Adams have one son, John J Mrs Adams is a member of the Methodist Episcopal church, while he was brought up in the Episcopal church For nineteen years he has been a Mason and is a member of the Crawford Lodge No 181 A F & A M During the Wounded Knee Indian uprising Mr Adams was one of a company that set out from Crawford to give assistance but a furious snowstorm prevented their reaching the place in time

MRS FREDA ROMINE who is one of Chadron's well known residents, prominent in social life and active in business, has a wide acquaintance not only in Nebraska but over the

United States and Canada As Freda Hartzell, for some years she was a noted theatrical attraction while traveling with her parents, her father being a man of outdoor life and connected for a long time with the familiar and popular entertainments known the country over as the "Buffalo Bills" shows, under the management of the late Colonel Cody

Freda Hartzell was born at Omaha, Nebraska, December 20, 1891 Her parents were James and Mary C (Boruff) Hartzell, her paternal grandfather being Solomon Hartzell The latter came to Dawes county as a member of what was called "the Sweat Colony," from Missouri, and homesteaded about four miles east of Chadron He still survives, residing at Soldiers' National Home, Hot Springs, South Dakota

The father of Mrs Romine was born at Mankota, Minnesota, and the mother at Bowling Green, Kentucky Of their three children Mrs Romine is the only survivor, twin sons dying in infancy The mother survives and lives at Los Angeles, California, but the father died January 16, 1915 As a cowboy and as owner and producer of "Wild West" shows he was long in the public eyes before he engaged with Colonel Cody He was a man of fine physical present, was a crack pistol shot and expert horseman It was Mr Hartzell who organized the novel "Cowboy Race," in 1893, which aroused national interest in connection with the World's Fair He homesteaded in Dawes county about seven miles east of Chadron At that time he might have acquired many acres of surrounding land but much of it was considered so valueless that he considered the most of it not worth paying taxes on

Mrs Romine was only five years old when she began to travel with her parents and received her first training in marksmanship She attended school at Deadwood, South Dakota, for a time and had private teachers that looked carefully after her education while traveling She soon developed remarkable skill with the rifle even when her childish hands could only hold a weapon two and a half feet in length She has used guns of all kinds but her favorite was a 22 caliber rifle and two of these she yet preserves in memory of old days She had many experiences in her exhibition life, her first public appearance being at Crawford, some thirty miles from Chadron, and her skill with her rifle not only interested the audience but aroused the suspicion of at least one in the crowd, that some trickery must be on foot While all the others

applauded, this one demanded proof that she could bring about results with other fire arms She therefore accepted the loan of a rifle from a soldier present, although its caliber was 30 30, and as soon as she had found the range, hit her target just as easily as before She was also taught fancy and trick riding, her thoroughbred steeds being known in the profession as "High School" horses Mrs Romine continued to travel and exhibit with her father, visiting almost every large city and every state in the Union, until 1910, when they returned to Chadron, where her father afterward lived practically retired During these years of professional life, Mrs Romine found appreciative audiences everywhere, has many pleasing memories to recall, and met people of distinction and culture whom she can yet claim as friends

On June 26, 1911, Miss Hartzell was united in marriage to Guy Romine, who for thirteen years had been connected with the Chicago & Northwestern railroad as a freight conductor In June, 1918 they purchased the City Garage, a business founded by Frank Plummer Since then Mr and Mrs Romine have conducted this garage, enlarged space and increased its scope and probably do the largest garage business in all this section They have car storage space for sixty cars, handle the Oakland, the J I Case and other machines, do a general auto repair business and give constant employment to four machinists and in busy seasons increase the number Mr Romine acts as salesman and general manager, while Mrs Romine attends to the office work They have one daughter, Catherine Mr Romine belongs to the Elks and the Order of Railway Conductors Mrs Romine takes a deep interest in all matters of public concern and at Chadron is prominent in the work of the Woman's Club

MRS GERTRUDE H ROMINE, whose name is prominently associated with social and political life, with ever ascending effort, at Chadron, fairly represents the nobility and substantial character of intelligent, progressive womanhood as it is presented to the world today Mrs Romine came to Nebraska from a home of culture and refinement, one of old tradition, in the east, and no subsequent unexpected hardships ever interrupted her natural aspirations toward development or lessened her interest in those things which make for social betterment and diffusion of knowledge

Gertrude E Hatch was born at Strafford, Vermont, secured her early education in her native state and completed her intellectual training at York Seminary and the Nebraska State University at Lincoln. She was the youngest in a family of twelve children born to R and Mary (Bacon) Hatch, the former of whom was a native of Massachusetts where he was prominent in business and the law and served as judge of the Probate Court. He died in Massachusetts and was laid to rest near his Puritan ancestors. The mother of Mrs Romine came to York, Nebraska where a son was engaged in the coal business, moving later to the home of a brother, J D Bacon, who was a broker at Chadron for many years, and she died here in 1907. Mrs Romine's only surviving sister, Mrs Hebbard, resides in California. This sister was the first teacher at Plattsmouth, Nebraska. Her husband was the first land commissioner at Seward, Nebraska, and was president of the first Nebraska State Fair, which he was instrumental in organizing. Other members of the family have been very prominent in state affairs in many directions. A brother of Mrs Romine established the Citizens Bank at Leigh Nebraska, which was the family home for a time. A cousin was an early president of the B & M railroad and probably did more for the development of Lincoln than any other man, while an uncle, George Harris, was one of the first land commissioners in the state.

On June 1, 1887, Miss Hatch was united in marriage to James S Romine, a dentist by profession. They moved on Mrs Romine's homestead situated four miles from Gordon, in Sheridan county, their first house being superior to many others in the county, as it was of frame, but it had an earthen floor and is described as being so full of open spaces that a cat might have been thrown through at any time. Mrs Romine had some valuable pieces of old furniture and still owns them, prizing them highly as she has every reason so to do, but there was great lack of the ordinary conveniences of housekeeping and the homestead was many miles distant from any source of supply. Pioneer resourcefulness came to the front, however, and when the family cook stove could not be adjusted to the chimney because of a lack of stove pipe, the difficulty was overcome by setting the stove on the table. In early days they suffered great inconvenience from an insufficiency of water, which had to be carried a long distance or bought from those who made a business of selling it.

When Dr Romine located his timber claim he had to drive to Valentine and at that time there was not a single house between the homestead and Hunter's ranch. At first they kept on adding to their property until they had several thousand acres, but hard times fell on the settlers of Sheridan county from different causes, and they sold one tract of land after another at a great loss, some of which adjoined Gordon and at the present time commands a high price. In 1890, they came to Chadron and invested here and now own large holdings in this section and have property interests also at Danville Illinois. During the early years after marriage and for some time afterward, Dr Romine practiced his profession in neighboring towns. He belongs to the Odd Fellows and organized the first Modern Woodmen of America lodge in the state of Nebraska.

Mrs Romine has a wide acquaintance with people of prominence and has pleasant recollections of former Governor Dawes, for whom the county is named, as he at one time resided in the home of her aunt at Lincoln. She has taken a very decided stand in Republican politics and assisted in bringing about suffrage legislation in Nebraska. She established the first Literary Club at York and later the first at Chadron and has been an inspiring influence in both and in such other organizations as the Chadron Culture Club, the Woman's Club both of the city and the Sixth District and has been president of all these bodies. She is a member of the Republican Women's committee and a director of the Home Service work at Chadron and is past noble grand in the order of Rebekah's. Many of the citizens of Chadron who enjoy the grateful shade of some of its streets, may not know that a large number of there beautiful trees were set out by Mr and Mrs Romine, both of whom are lovers of Nature. Her first flower garden aroused interest and admiration as did the well kept lawn about her home. She is held in esteem and affection by those with whom and for whom she has labored so long.

WILLIAM H WILLIS one of the representative business men of Bridgeport and a large property owner takes an active and public spirited part in the civic and community life of Bridgeport. He was born at Sioux City Iowa in 1873, being a son of John G Willis, extended mention of whom will be found in this work.

William H Willis attended the public schools at Omaha, the military school at Faribault, Minnesota and the Omaha Commercial College, Omaha these laying an excellent foundation upon which his later commercial career has been built After finishing his education, Mr Willis engaged in the real estate business at Omaha, operating there for five years with the Willis Land Company In 1899 he came to Morrill county, bought a ranch and established himself in the live-stock business Later he took up a homestead, continued to buy land until he had a fine estate and lived on his ranch for eight years Mr Willis was then appointed purchasing agent for the government which necessitated his moving into Bridgeport to live After retiring from government service eighteen months later, he embarked in the farm implement and automobile business Shortly afterward he purchased an excellent business site, tore down the dilapidated building and erected another which gives him floor space of 6,000 square feet and has a business structure modern in every particular. He has ample room in which to display the Studebaker and Oakland cars, for which he is agent, and does a general implement and automobile business Mr Willis owns a large amount of city realty and also holds property at Bayard and other places in the county In large measure Mr Willis is a self made man, his ample fortune having been built up through his own business enterprise and he can congratulate himself with pardonable pride upon the comfortable fortune of which he has been both architect and builder

In 1900 Mr Willis was married to Miss Eva J Young, who was born at Springfield, Ohio, and they have one daughter, Laura C , who is making rapid progress at school Mr and Mrs Willis are members of the Episcopal church A sound Republican all his political life, he has been somewhat prominent in the county organization, and at Bridgeport has served on the town council and as city clerk He is a Scottish Rite Mason and also an Odd Fellow Utilizing his early military training, Mr Willis was the prime mover in organizing the Home Guards, a military company that reflects credit on Bridgeport, of which Mr Willis is captain During the World war he was generous with his time, labor, money and influence to aid the government to "carry on" and with our Allies make the world a safer place for the coming generations

FRED J HOUGHTON —In the material upbuilding of Chadron and the substantial development of this part of Dawes county, no present resident deserves more credit than Fred J Houghton, who continues to be a representative citizen Judge Houghton came to Chadron when it was a village of tents and

unsightly shacks, invested in land and erected the first comfortable dwelling house in the block in which he still lives With practical ideas, he entered wholeheartedly into the business of development of this section, and for thirty-five years has been a prominent and useful factor

Fred J Houghton was born at Woodhull, Henry county, Illinois, July 23, 1853, the eldest of the three survivors of a family of eight children born to Calvin C and Lucy E (Johnson) Houghton Judge Houghton has one brother, Hugh, who is a resident of Hot Springs, South Dakota, and one sister, Mrs. Winnifred L Oliver, who lives at Packwood, Jefferson county, Iowa

The parents of Judge Houghton were natives of Chester, Vermont, where his father was born in 1816, and his mother in 1824. The father survived until 1874, while the mother lived to the advanced age of ninety-four years In many ways the father was a remarkable man, possessing business ability of a high order and a spirit of enterprise that made him prominent and useful as a pioneer in Illinois, to which state he went in 1848 He drove the whole distance and sold goods along the way, and when he reached Henry county, Illinois, had capital with which to take up a large amount of government land A man of sturdy principles, throughout life he maintained his views in relation to slavery, assisted in the operation of the underground railroad to assist slaves escaping to Canada, and when the Civil war came on, although not able to serve in the ranks, was a liberal contributor to the cause The land acquired so long ago in Illinois, is still in the possession of the family. Following the close of the war, he engaged extensively in raising of cattle, horses and mules

Fred J Houghton attended the country schools and then entered Knox College, at Galesburg, but did not complete his college course because his assistance was required on the farm, and when only fifteen years old, Mr Houghton had the oversight of from fifteen to twenty men He remained in his native state until he became convinced that many business opportunities could be found in the great west, and being particularly interested in Nebraska, came to the little railroad hamlet of Chadron, September 13, 1885 He found here other men of enterprise and vision, and alone and in co-operation with them, soon put the aspiring little city on a sound business basis and has remained here ever since He opened a real estate office, subsequently adding a general line of insurance, and has handled many thousands of dollars and has opened the way to a large amount of the outside capital that has been helpful in building up many important business concerns here

It must not be supposed that the early settlers in frontier towns especially with the type that came to Chadron were so given over to sordid business that amusements did not appeal to them, and in reminiscent mood, Judge Houghton has been heard to declare that the Fourth of July celebration at Chadron, in 1886, was one of the most interesting he ever attended The Indians in this section of the country were numerous and in the main friendly, and it was to the interest of the white settlers that they should remain so Responding to an invitation to come to Chadron and have a good time, they came about fifteen hundred strong, and Mr Houghton was one who sat in the circle with them and smoked the pipe of peace The town gave them a whole beeve for food, and they assisted in the entertainment with their exhibition of dancing foot and horse racing, fancy roping, and even the Indian children showed their expertness with bow and arrow

At Fairfield, Iowa, on September 8, 1879, Mr Houghton was united in marriage to Miss Margaret R Benn, who died August 9, 1898 Her parents were Alexander P and Phebe (Couger) Benn, residents of Iowa but natives of Virginia and Pennsylvania, respectively Three children were born to this marriage, as follows Inez M, who died April 20, 1920, Hugh Manly, who was accidently killed in the railroad yards at Chadron, and a babe that died unnamed

After locating at Chadron Mr Houghton soon recognized the advisability of acquiring a knowledge of law, therefore applied himself to its study in the office of E S Ricker and in 1889 was admitted to the bar While he has found it helpful in his business he has never engaged in the practice of his profession, although his knowledge has contributed much to the soundness of his opinions in his many years of official life A Republican in politics, on that ticket he was elected in early days city attorney and served several terms, afterward was city clerk for seven successive years, following which he was police judge for a long period, and subsequently, as long as he consented to serve, was a justice of the peace He has always been actuated by patriotic motives and has never failed in his devotion to the welfare of Chadron He and daughter were members of the Episcopal Church

CHADRON STATE BANK — For the large volume of business transacted at Chadron, Nebraska, to be adequately taken care of, here has arisen an urgent demand for sound, safe, ably directed banking institutions and it was in answer to this demand that the Chadron State Bank was organized This business was incorporated October 1, 1915, and the bank opened for business on January 8, 1916

The first bank officials were the following H A Copley, of Alliance, president, Ray Tierney, vice president, E K Reikman, cashier, and C A Drews assistant cashier The present officers are Ray Tierney, president, R L Isham, vice president, C A Drews, cashier

The bank structure is one of the most impressive in the city, handsome and dignified in design It is situated on the northwest corner of Main and Second streets, is constructed of white brick with terra cotta and marble inside finish, and is equipped with handsome fittings and most modern protective bank devices A strictly banking business is carried on, present deposits are $500,000, careful, conservative methods are in force and all the officials are sound, reliable, representative business men

CHARLES U COOPER stock raiser and well known general farmer of Garden county who is essentially a self-made man as he began in the Panhandle with nothing and today is a prosperous man, is a native son of Nebraska born in Lancaster county October 9, 1864, the son of Ephraim and Theda (Hitchcock) Cooper, the former a native of Pennsylvania while the mother was born in Ohio The father was a carpenter by trade who came to this state and settled in Richardson county in 1889, from there he moved to Lancaster county and later to Keith county in 1884 The mother died in Lancaster county in 1882, and the father in Denver in 1916 They were the parents of six children of whom four survive but Charles is the only one in this locality Mr Cooper was a Republican, was judge of Keith county four years and a prominent man He belonged to the Congregational Church while his wife was a member of the Baptist denomination

Charles Cooper was educated in the public schools of Lancaster county and started in life independently when seventeen years of age as a farmer Wishing land of his own he came to Garden county in 1885 took up the homestead where he still lives and engaged in farming

November 4 1896 Mr Cooper married

Miss Libbie Bundy at Lewellen; she was the daughter of Charles and Mary (Stubbs) Bundy, the father being a resident of Keith county, the mother is deceased Eight children have been born to this union Theda, the wife of Ben Grasy, of Garden county, Ina, the wife of Ray Southard, of Deuel county, Clara, deceased, Ephriam, Myrtle, Carl, and Ida, all at home, and Mary, Deceased

When Mr Cooper came here he drove across the country from Lancaster county, had many experiences of frontier travel as he was stuck in an alkali hole, had to pay fifteen cents a pail for water for his horses and found few roads any good The first house was the usual "soddy" and as he did not know just how to build it the wind partly blew it down while he was away from home but neighbors helped repair it and he later rebuilt At first water had to be hauled eighteen miles but later Mr Cooper built a well and the nearby settlers came to him for water The first summer he went north into the hills to hunt buffalo bones for money to buy supplies Money was scarce and hard to get and Mr Cooper at one time was in debt but managed to get out and today is one of the prosperous men of his locality for he believed in the future of this section, remained and has been well rewarded for today he owns nearly five hundred acres of good land for which he would not take a hundred dollars an acre, having raised as high as fifty dollars worth of wheat an acre For some years he bred Jersey cattle but lately has given more attention to general farming which he finds is more profitable He has fine, well equipped farm buildings, uses modern machinery and is a progressive business man In politics Mr Cooper is a Democrat, he has served as road overseer two terms, is a member, stock holder and director in the Farmers Cooperative League at Lewellen, and a public spirited man who advocated progress in his district

WILLIAM G HATTERMAN, one of the prosperous farmers of the Big Springs district, is a native born son of York county, this state, born December 2, 1880, the son of Anton and Minnie Hatterman, who came to Deuel county when the boy was only six months old William lived at home, attended the public schools for his education and though the Oregon Trail had been abandoned when his family came here he recalls finding log chains, parts of oxen yokes, Indian beads, arrows and other relics of the early days The wind and weather had not then destroyed the deeply rutted tracks of the trail After leaving home Mr Hatterman worked on farms near Lexington four years and then went to Cherry county where he found employment on the ranches February 16, 1905, he married at Day Post Office, Miss Martha Sonnenberg, the daughter of Henry P and Caroline (Lewine) Sonnenberg, natives of Germany who came to Deuel county in the late 80's where they farmed but now live near Sterling, Colorado Three children have been born to this union, Joseph, Roy, and Vera, all at home

Mr and Mrs Hatterman live on the homestead on which Mrs Hatterman filed and proved up She made the thirteenth person to file on this same piece of land, all the others failed to prove up, grew discouraged and gave up Conditions were very discouraging in the early days for there had been droughts, there was no work to be obtained as many men were trying to get jobs, then a few years came with good crops followed by the poor years of the early 90's, but the Hattermans retained their land and today it is worth nearly a hundred dollars an acre The change in weather conditions, the general raise in land values and the improvements on the farm have placed them in easy circumstances Today Mr and Mrs Hatterman own and operate three hundred and twenty acres of land, for many years they raised cattle but of late have devoted more attention to scientific farming, using the latest machinery and modern methods Today Mr Hatterman has only well bred Hereford cattle and Red Jersey hogs Since his marriage he says that he and his wife had few of the hardships to contend with of the early days They then had to go for water from a mile and a half to six miles and this continued eighteen years as there was only one well in the vicinity, W W Waterman's and a spring at Ash Hollow Mrs Hatterman's family lived in Gage county but decided to come west and take up a homestead, locating in Lincoln county They made the trip in true pioneer style, driving across country in the spring of 1890, using two cows to draw their wagon with two yearling calves on leads The cows were hitched with the usual yoke for oxen On reaching the homestead near Maxwell, the family was practically out of money and breaking the sod was slow work, only a little being put under cultivation each year As a result they suffered privations as crops were poor some years Finally the father secured horses to work and then farming became easier When they drove through to Deuel county horses were used and the family lived on the land

CHARLES F. MANNING

here for which they had traded until they sold out and went to Colorado Mrs Hatterman and her older sister had to help break the land and do other work on the farm so that she has experienced all the privations and hardships of frontier life to the full Today Mr and Mrs Hatterman are well fixed with a good productive farm Mr Hatterman is independent in his political views, has been treasurer of his school district for the past fourteen years, is a member of the Methodist Church at Day, and also belongs to the Farmers Union at Big Springs, holding stock in the Farmers Elevator and Farmers Store He is a substantial and reliable business man of progressive methods and ideas

CHARLES F MANNING — While many business enterprises are desirable and even necessary in a community and may be carried on with more or less care as to safety and satisfactory results, the business of the druggist is regulated by stern laws and he is no less responsible for the health and life of his patrons than is the physician whose prescriptions he puts up Bridgeport is fortunate in having a pharmacist in whom full reliance can be placed, in Charles F Manning, who has been a druggist all his business life, and since the summer of 1914 has been established in this city Mr Manning was born in Mercer county, Missouri, December 9, 1864, the son of Marshall Green and Carolina Virginia (Myers) Manning, who were reared and married in Missouri They had four children, the two survivors being Charles F and his older brother, Oscar G, who is a retired merchant living at David City, Nebraska In 1870 the parents of Mr Manning came to Nebraska and homesteaded in Butler county For a number of years the family lived on the farm, then moved to David City where the father died and where the mother still lives Quite recently she sold her farm of eighty acres for $20 000 Both parents were members of the Christian church In politics the father was a Democrat During the Civil war he was a member of the Home Guards in Butler county, and at one time held an official position

Charles F Manning attended the public schools in Butler county and when sixteen years old became a clerk in a drug store and a student of pharmacy and later owned a drug store at David City, which he sold in 1902, but bought another store at Bayard In the meanwhile he homesteaded in Morrill county and resided on his farm for four years, keeping in touch, however, with his profession, and in

1908 he came to Bridgeport and entered Dr Anderson's drug store, where he remained five years, then purchased and operated a drug store at Lincoln for ten months After disposing of his Lincoln business he bought another at Nebraska City which he moved to Bridgeport on August 1, 1914 He carries a full assortment of the most reliable goods in his line, pure drugs and patent medicines, together with the articles which the public has learned to expect in a druggist's exhibit He has a very attractive store and it is one of the representative business centers of the city

In 1884 Charles F Manning was united in marriage to Miss Paulina E Hamline, who was born in Illinois Her parents, Obed Hamline and wife now reside in Los Angeles, California The father formerly was a successful farmer in Illinois Mr and Mrs Manning have had two children Thornton Benton, who was born September 10 1885, at David City and educated there, is now his father's assistant in the drug store, and Jesse F, who died at Bayard, Nebraska, at the age of thirteen years The family belongs to the Christian church In his political views a staunch Democrat, Mr Manning has always been a loyal party man While a resident of David City he served as city treasurer and at the present time is a member of the town council of Bridgeport Mrs Manning is a member of the Eastern Star, in which order she has been very prominent For fifty-one years her father has been a Mason and she is past matron of Bridgeport Chapter No 260, having held that office two years and also served as the third matron after the chapter was established here

FRANK E JOHNSON, well known stockman of Garden county, is a member of a pioneer family that ran the full gamut of privations and hardships incident to life on the frontier He was born in Sweden, January 17, 1871 the son of John G and Lottie Johnson, both natives of Sweden where the father was a farmer The family came to the United States in 1882, locating first in Iowa but two years later came to Nebraska and took up a homestead in old Cheyenne county that part which has since been erected as Garden county where they lived until 1915 when Mr Johnson retired and went to Denver to live He died in May, 1916, and his wife still resides in that city There were nine children in the family Emma, the wife of Harry Geier, of Denver, one deceased, Charles of Denver, Lennis deceased, Gust deceased, Teckla, the

wife of August Geier of Denver, one deceased
in infancy, Oscar, of Garden county, and
Frank of this review

When the family came here they began their
farm work with a borrowed pony team, there
were no crops the first year so the father went
to work as a section hand on the railroad,
while the mother took care of the family and
hauled water from three to sixteen miles and
managed things as best she could   Once she
was caught in a blizzard and spent the night
alone in the manger of a barn to keep from
freezing   They suffered much from droughts,
grass hoppers and hail storms, but remained
and became prosperous by raising cattle

Frank Johnson was educated in Sweden
until the family came to this country when he
was twelve years old and then attended the
frontier schools when he could   He took up a
homestead but soon sold and went to Denver,
worked as a farmer and dairyman there for
four years before becoming a motorman on
he street railway   Three years later, returning to Garden county, Mr Johnson bought a
farm and now is the proprietor of more than
eleven hundred acres, five hundred of which is
in pasture   He has gained a reputation for
breeding Hereford cattle and Belgian horses,
also Poland China hogs, and today is regarded
as one of the substantial business men of his
district   His farm is well improved and for
years he has used the latest machinery   In
politics he is a Republican and for fourteen
years has been treasurer of school district No
40   He is a member of the Farmers Union of
Garden county

November 15, 1899, Mr Johnson married
Miss Anna Anderson, of Minnesota, and seven
children have been born to this union Gladys,
Mildred, Ellen Violet, Evelyn, Loraine and
Loes, all at home

ANDREW J WALRATH, well known in
western Nebraska and eastern Colorado during the early days as rancher, range rider, Indian scout and sharp shooter, is one of the few
men left in this section who took part in the
stirring events before settlement and railroads
had civilized the great plains   He was born
at Lodi, Boone county, New York, in August,
1850, the son of Andrew J and Amanda
(Stulphen) Walrath   The father was a horse-
buyer, knew his business well and commanded
a large salary   He and his wife spent their
lives in New York   They were the parents of
two children but Mr Walrath is he only one
in his locality   His father and mother were
members of the Baptist Church and highly
respected in their community

Andrew Walrath was educated in the public
schools of Cherry Valley, New York, and began his independent career when only sixteen
years of age, going to Cheyenne, Wyoming,
and then to Laramie   He began to work for
the Union Pacific Road in 1867, at Evans,
Colorado, the end of the line, also checking
freight for a line that ran cars to Denver, the
motive power being mules and oxen   Later
the railroad was constructed to Denver and
the Evans office discontinued   In 1871, Mr
Walrath established his first ranch, the ' Walrath Double OO Ranch," six miles east of
Julesburg Mr Walrath took an active part in
the frontier life, became a well known figure
as ranchman, range rider, Indian scout and
cow boy, while his reputation as a sharp shooter was hardly excelled   His honesty was unquestioned and he became known as a man
whose "word was as good as his bond," and
holds that reputation to this day   Julesburg
was then a stage station but was abandoned
when the railroad was built and the town
moved to its present site   The Double OO
Ranch consisted of some three thousand acres,
Indians and buffalo were thick in those days
and Mr Walrath tells most interestingly of
hunting on the ground now covered by Julesburg   He learned the Indian's methods of
hunting and preserving meat and hides and as
the Sioux often came to his locality to hunt he
accepted an invitation to go on one of their big
hunts   On the trip they served him dog meat
and worst of it all it was his own dog, but he
could do nothing as there were hundreds of
Indians   This banquet was served at Walrath's Ranch   In 1873, Mr Walrath bought
another ranch which gave him two fine properties, one sixteen miles east of Julesburg
and the other near Weir, then known as Old
Hay Bottom, which was used as a head quarters for both properties   Here were the corrals, stables and house for the men who kept
the cattle on the six mile range   From this
ranch the supplies were taken to the other,
where a foreman had charge   The Indians
burned this place just before Mr Walrath
reached it on one trip with supplies   He
started for help and the Indians saw him and
a running fire was kept up while he rode for
his life   His foreman and helper had returned
to their ranch, concealed themselves in a hole
for the purpose and drove the Indians off so
that Mr Walrath was saved   Mr Walrath
recalls the sensational train hold up at Big
Springs in 1876, when he took part in tracking
the bandits who committed the crime, following the trail until he came up with soldiers

stationed in this locality for such purposes and turned the work over to them while he returned to the ranch. Collins was the leader of the gang and finally was captured with twenty of the sixty thousand dollars of the loot. Mr. Walrath has had varied experiences with thieves, one stole his horse and when he trailed and found him was forced to surrender and when Mr. Walrath brought him in the citizens hung him from a telegraph pole, for horse stealing was the worst crime on the calendar in the west.

In 1880, Mr. Walrath was married at Canterbury, Connecticutt, to Miss Ida T. Appley, the daughter of Lyman and Bethia (Pember) Appley, the mother being a cousin of President Fillmore. Four children have been born to this union: A. Judson, lives in Detroit, Michigan, being employed in the Ford Automobile Factory; Ida, a widow; Bessie, the wife of Robert Adams, of Morris, Minnesota, and Robert, at home. When Mr. Walrath brought his wife home to the ranch they left the train at Barton's Siding, the Walrath ranch station and while Mr. Walrath had been gone so many cattle had died of starvation the ground was covered with them. The winter had been cold, there was little feed and the cattle for which he had been offered fifty thousand dollars were nearly all dead. The Walrath home was fifty miles from Sidney, it was twenty-five miles to the nearest doctor and eighty-eight to North Platte, where it was necessary to go to get horses shod. Mrs. Walrath was the only woman for miles around but she had brought an organ with her from the eastern home and it helped while away many lonely hours and the cow boys came from great distances to hear her play, repaying her kindness by giving impromptu wild west shows. Later Mr. Walrath took up a homestead five miles east of Julesburg and now owns four hundred acres of hay land on the south side of the railroad. All other land he has sold with the exception of his city home. In 1911, Mr. Walrath retired from the active management of his ranches but as he had always been busy could not entirely settle down and has been carrying on a real estate business with success, though it is for diversion as he is a rich man from his accumulated property.

Mr. Walrath is a Democrat; he served as county commissioner sixteen years when his land was in Cheyenne county; was treasurer of the school board twelve years and helped organize the first school in his locality. Mrs. Walrath is a member of the Methodist Church and Mr. Walrath has belonged to the Masonic order many years, having held all offices but that of worshipful master. He has been prominent in western Nebraska and eastern Colorado for many years and is today regarded as one of the prominent, leading and substantial citizens of Julesburg.

CARL PIDGEON, one of Deuel county's most progressive and prosperous farmers and stockmen, is a native son, born here December 31, 1893, the son of Lincoln W., and Frances (Pindell) Pidgeon. The father was a school teacher who had studied for the ministry but was never ordained. He also took a law course and is now a well known lawyer in Sumner, Nebraska. The mother died in 1910. There were the following children in the Pidgeon family: Mable, wife of Maurice Johnson, of Deuel county; Guy, Paul, Carl, of this review, all of Deuel county, and Roy, killed by a horse in 1908. The Pidgeon family came to the Panhandle in 1890 and have lived in Nebraska ever since.

Carl Pidgeon received his educational advantages in the public schools of this county and when his schooling was finished he engaged in farming, a vocation which he still follows. Meeting with success in his business, Mr. Pidgeon has from time to time increased his original holdings until today he is the owner of nine hundred acres of fine arable land. Eight hundred he farmrs, using the other hundred as range for cattle. For many years he dealt extensively in cattle but for the past three years he has devoted more land to intensive farming and finds that it pays. With a brother, Mr. Pidgeon owns a threshing outfit which they run in Deuel county and the surrounding country. The Pidgeon farm has modern buildings and equipment, the latest machinery and a fine farm home.

On May 22, 1913, Mr. Pidgeon married Miss Elsie Ward at Chappell, the daughter of Jennie L. (Johnson) and Cyrus J. Ward, the former lives in Chappell while Mr. Ward resides at Elm Creek. Three children have been born to this union: Francis, Vera and Doris. Mr. Pidgeon is a Republican, while his wife is a member of the Methodist Church. He is also a stockholder in the Farmers Elevator of Julesburg, and a member of the Independent Order of Odd Fellows and with his wife belongs to the Rebecca Lodge.

JAMES D. PINDELL, a member of one of the old pioneer families of the Panhandle and today one of its progressive and prosperous farmers, was born in Bonaparte, Iowa, October 15, 1869, the son of Presley and Mary (Fox) Pindell, the former born in Brown county, Ohio, in 1834, died in 1916, the mother was born in Iowa in 1839 and died in 1915 The father was a wagon maker by trade, who came with his family to Nebraska in the fall of 1882, settled first in York county but came to Deuel county in the spring of 1885, and took up a homestead on section 22-14-43 Later he filed on a Kincaid claim where his son James now lives Mr Pindell was a Democrat but never held office Both he and his wife were members of the Baptist church

The Pindell family drove across country to their new home with a span of mules and a wagon, the second year all the horses were lost when James and a brother went to Ash Hollow for wood and on the way home the brother had a sun stroke, fell from the wagon and the mules ran away This was a serious loss as water had to be hauled more than three miles Finally a well was dug, started on Christmas Day and finished in April The family first lived in a tent but the wind was so strong it always was dangerous and a dug out was built, the roof of that was blown off and then a frame home took its place Denver Junction was the nearest town, fifteen miles away, and there was only one house in that distance Since then, when the town consisted of a few houses, a store and station, the name has been changed to Julesburg Six children made up the Pindell family of whom four are living Frances, deceased, Charles, deceased, Albert, of Cheyenne, Wyoming, Nellie, of North Loupe, Nebraska, James, and George of Big Springs

Mr Pindell was educated in the publics schools of Iowa and Nebraska, he accompanied his parents to Deuel county in 1885, and when old enough took up a homestead in section 28 where he lived until he sold and moved to Big Springs to engage in drilling wells, a vocation he followed ten years Buying his father's farm, Mr Pindell returned to agricultural pursuits and has become one of the well known farmers here

February 21, 1906, Mr Pindell married Miss Pearl Nelson, daughter of Nels and Rosetta (Van Aken) Nelson, residents of Big Springs and have two living children, Charles and Frances

Mr Pindell is an Independent Democrat in politics, he and his wife attend the Presbyterian Church of which she is a member For some years Mr Pindell has been a member of the Farmers Union and is one of the progres-

sive farmers and stockmen who is progressive in ideas and methods

ADAM H ZIMMERMAN, pioneer settler of the Big Springs district who came here in the early days, passed through all the hardships and privations of frontier life and now is one of the substantial and prosperous farmers of the Panhandle, all this won by his own untiring effort and courage in the face of seemingly insurmountable difficulties He was born in northwestern Ohio, June 10, 1859, the son of Adam and Catherine (Schott) Zimmerman, both natives of Germany, who came to the United States about 1855 Mr and Mrs Zimmerman located in Defiance county, Ohio, where the father engaged in farming He was a carpenter by trade but never worked at it except for himself The mother and father both died in Ohio, the former in 1872 and the father in 1886 They were the parents of five children but Adam is the only one here The father remarried after his first wife died and had two more children, one of whom still lives Mr Zimmerman helped organize several schools and churches in Ohio

Adam Zimmerman was educated in the public schools of Ohio, helped on the home place and at the age of twenty-one started in life for himself, as a farmer, a vocation he has followed successfully all his life October 8, 1882, he married Miss Caroline Kurtz, the daughter of Jacob and Christina Kurtz, both natives of Germany Ten children have been born to this union Samuel, of Keith county, Frederick, deceased, James, deceased, Ella, the wife of George Brown of Colorado, Frank, of Deuel county, Adam, Ida, Annie, George and Ralph all at home

Mr and Mrs Zimmerman came to Nebraska in 1885, took up a homestead on April 5th, of that year, filed at North Platte and proved up at Sidney When they came here they had little of this world's goods but the determination to succeed and that has been enough Mr Zimmerman and a neighbor bought a team, wagon and harness in partnership to work their land but for a time the crops were poor Wood had to be gathered in the canyons and distances were great The family suffered severely from blizzards and drought but stuck to the land and won out Mr Zimmerman says that several times he was completely out of money and had to work to secure even a small amount and work was almost impossible to get Settlers were few and far apart and many grew discouraged and left the country

At one time Mr Zimmerman went to Colorado to work, received twenty dollars for the month but when he had paid for his labor at home and railroad fare was only two and a half dollars ahead which tells us of the difficulties of that day Today, Mr Zimmerman has a well improved farm of several hundred acres, modern machinery, raises horses and cattle and for the past ten years has had an abundance of the world's goods He is a Republican, a member of the Farmers Union and has served as school director of his district since he came here while Mrs Zimmerman is treasurer Today he has retired from active life and is enjoying the fruits of his labors

HARVEY K BALD — During the years that Harvey K Bald has been prominent in the financial field at Bayard, he has proven his trustworthiness as a banker, his usefulness as a citizen, and his worth as a man. A leading factor in two large financial institutions, his commercial influence is wide and his public responsibility great Mr Bald was born in Hamilton county, Nebraska, May 15, 1888

The parents of Mr Bald, Louis and Matilda (Kemper) Bald, were of German parentage but both were born in Philadelphia, Pennsylvania, where their fathers, Henry Bald and George Kemper, settled probably early in the forties Both moved to Wisconsin and died there in advanced age The maternal grandmother of Mr Bald still survives and lives at Aurora, Nebraska, in her ninety-fifth year. The parents of Mr Bald were married in Wisconsin and lived in Grant county prior to moving to Hamilton county, Nebraska, in 1876, making the journey with team and covered wagon At first the father provided for his family by freighting between Lincoln and Aurora, at a time when the business was fraught with considerable danger, as the Indians were numerous and sometimes savage Later he bought a farm and engaged in its cultivation, but subsequently retired to Aurora and yet lives there He is a Democrat in politics but has never been unduly active, and belongs to the order of Modern Woodmen Both he and wife are members of the German Evangelical church Of their family of six children, Harvey K was the fifth in order of birth, the others being as follows Eda who is the wife of D H Oswald, a farmer in Hamilton county, Frank, who is a farmer in Hamilton county; Frederick A, who is an attorney at Alliance, Nebraska, Arno, who is a physician and surgeon at Platte Center, Nebraska, and Harold, who lives on the old home place

in Hamilton county There are eighteen grandchildren in the family

Following his graduation from the Aurora high school, in 1906, Harvey K Bald went to work in a drug store at Aurora where he remained through the summer On January 1, 1907, he entered a bank at Aurora as bookkeeper, and continued in the institution gaining banking experience until he came to Bayard Here, in April, 1917, he organized the Farmers State Bank, of which he is cashier, and the Farmers Loan and Investment Company, of which he is secretary and treasurer Both institutions are capitalized at $25,000, and both are in a very prosperous condition, careful and conservative methods being used in the management that assure safety under every condition of public finances

On July 3, 1916, Mr Bald was united in marriage to Miss Frances Savage, who was born at Aurora, Nebraska, and they have two little daughters, Loyola Eda, born July 28, 1917, and Frances Patricia, born August 24, 1919 Mrs Bald is a member of the Roman Catholic church, but Mr Bald early united with the Presbyterian church He is a Royal Arch Mason and has served as secretary, senior deacon and junior warden of his lodge Mr Bald has always been identified with the Democratic party but at no time has he felt inclined to put aside the claims of congenial business in order to find leisure to serve in a political office, for a man of his character would not feel justified in assuming such responsibility without devoting his time and attention to its duties

HENRY C HATTERMAN, is a native son of Deuel county and one of the successful farmers of this district who has grown up in the west and made good though he took part in many frontier privations and hardships Mr Hatterman was born May 20, 1888, on the farm where he now resides in section 6-14-41 He is the son of Anton W and Johanna H (Claner) Hatterman, both natives of Germany who came to the United States in 1877 They settled first in Iowa and Mrs Hatterman tells of crossing the Mississippi river in boats as there were no bridges near their home A year later the family came to Nebraska, locating in York county, near Waco and remained there ten years In March 1888 Mr Hatterman brought his family to the Panhandle and filed on the homestead where the son Henry now lives Here the Hattermans passed through all the hardships and privations incident to life on the frontier the father worked in order to supply his family with the necessities of life, many times receiving only twenty-five cents a day Conditions were bad, crop failures frequent, work was hard to get and the large family had to be fed The first team Mr Hatterman used was a yoke of oxen and they were also used to haul water for family use and stock Much of the time the trip was twelve miles, but later Mr Hatterman made a cistern in a draw where the rain collected and this helped out on the water supply Butter and eggs were exchanged for work Hard work undermined the father's health and he died in 1909 The mother still lives with her son. There were six children in the family, of whom five are living Edward and William of Deuel county, Sens, the wife of John Curley of Brule, Nebraska, Tona, the wife of James Fenwick, of Keith county, and Henry of this review The father was a Democrat and in his early life belonged to the Lutheran church but he and his wife later joined the Methodist denomination Henry Hatterman received all the educational advantages afforded in this locality when he was a boy, which was not much As soon as he was old enough he began to work on the farm While still a lad he was sent to the canyons to gather wood for the home and once after he had gather more than he could load, left the rest for another trip and on his return found that someone had stolen it While the father was working on the railroad to earn money for supplies the boys carried on the farm work as best they could As Mr Hatterman grew older he assumed more and more of the work of the farm and when his father died took entire charge of the place and has gained a high reputation as a progressive and prosperous farmer

February 25, 1914, Mr Hatterman married Miss Effie C Stewart, the daughter of August and Martha (Coates) Stewart, pioneer settlers of Deuel county, living near Lodgepole, and three children have been born to the union Floyd A , and Lloyd A , twins and Bertha I Mr Hatterman is a Democrat and for two years has served as treasurer of the school board of his district

AGNEW R RYBURN, one of the well known hotel owners and operators of Big Springs, where he has been in business for more than a decade, is a pioneer settler and ranchman of Deuel county who has taken a prominent part in the opening up and settlement of this section of the Panhandle He was born in Fayette county, Indiana, November 20, 1856, the son of John and Diantha

(Gray) Ryburn, the former a native of Virginia, the latter of Indiana They were the parents of five children but Agnew is the only one in western Nebraska The father was a farmer who located in Indiana when a boy, near Brookfield, later bought land near Bushville where he died in 1872 The mother is still living at the advanced age of eighty-nine years Mr Ryburn was a prominent man in Indiana, engaged in general farming and stock-raising, was a Republican and a member of the United Presbyterian church

Agnew Ryburn was educated in the public schools of Indiana, his father was injured when the boy was fifteen years old and he took charge of the home farm and has been in business for himself since that time In January, 1876, Mr. Ryburn married Miss Ella M Murray, at Oxford, Ohio, and they became the parents of five children the oldest is deceased, Cora, the wife of Charles Morrow, lives in Indiana, Carthrine, the wife of Earl Hinchman also lives in Indiana; Murray, of Wyoming, and Hinsey of Tacoma, Washington Mr Ryburn came to Deuel county in the fall of 1886, on a visit and remained The family followed him in 1889 At first he worked for the State Line Horse Ranch, as foreman, but his wife and oldest daughter died the first year and after a few years Mr Ryburn became associated with Frank Doran In January, 1900, he bought the hotel in Big Springs which he has since operated During this time he has made many warm friends and gained a high standing in the community Mr Ryburn was clerk of the election board when Deuel county voted for the county seat, the rivalry between Big Springs and Chapell The first election was fraudulent and everyone knew it as thousands of votes were cast when there were not more than five hundred voters The case was contested and Mr Ryburn called to the stand, an effort was made to place the blame on him but he cleverly evaded many questions Later another election was held and Chappell was chosen as the seat of justice Mr Ryburn says he will never forget the many interesting incidents created by this affair and now laughs about them He lived in the Panhandle at the time when the last of the Texas range cattle were driven through and since that time cattle have been raised on the ranches Big Springs at that time was known as Lone Tree, later the name was changed The great cattle trail crossed the river at this point splitting up on top of the table land north of town so that he has seen all the marvelous changes from the old trail days, and the old pioneers can hardly realize it is the same country Mr Ryburn has been a member of the Independent Order of Odd Fellows since he was twenty-one years of age and filled all chairs

GOTLEIB C MANSER, pioneer settler and well known business man of Big Springs, is one of the essentially self made men of the Panhandle who came here with little but the determination to succeed and two willing hands with which he worked, and he has made good Today, he is one of the most substantial men of the community and has the distinction of being the first blacksmith of Big Springs Mr Manser was born in Germany, December 24, 1860, the son of Jacob and Rosina (Meister) Manser, both natives of that land The father was a blacksmith in his native land who came with his family to the United States in 1886, settled in Amherst, Colorado, where he engaged in business at his trade The mother died there in 1911 and the father in 1913, leaving a family of nine children, but Gotheb is the only one living in the Panhandle He was educated in the public schools of Germany and also learned the blacksmith's trade from his father Mr Manser came to this country ahead of the rest of the family, sailing from Europe in 1882 Soon after reaching our shores he located in York county, Nebraska, where he engaged in business as a blacksmith two years before coming to Deuel county to file on a homestead northeast of Big Springs in Keith county, but lived their only two years as he came to Big Springs in 1886 and opened a blacksmith shop, the first in the town For eight years Mr Manser carried on this business then moved back to his farm to engage in agricultural industries for nearly a quarter of a century He was sober, industrious, not afraid of hard work and by these qualities became a successful rancher When he came Mr Manser says that all he had was his two bare hands and today he has two sections well improved He has made his way independently, is essentially self-made by steady work his native ability and perseverance and is still a young man in years At first he did not do much farming as the country was not yet adapted to that but raised cattle and fed some having at one time over a hundred head Since he retired from the land his sons have charge of the farm Mr Manser recounts that at first the settlers had no wood and burned buffalo chips, wagons were the only means of transportation and for three years he hauled water six miles for family and stock, paying

five cents a barrel for it Since returning to town Mr Manser has again turned to his trade as he feels he is too young to give up all active life He owns a fine home in Big Springs

February 2, 1888, Mr Manser was united in marriage with Miss Anna Miller, the daughter of John and Anna miller, pioneer settlers of Deuel county, and seven children have been born to the union Otto, at the home place, Charles, married, lives on a home farm, Tillie, Emma, Lillian, Bennie and Mattie all at home

Mr Manser is a Republican and a member of the Methodist church He is a progressive man in his business and ever ready to help any movement for the development of his community and county

RILEY FORD early settler, well known ranchman and today a member of the retired colony of Big Springs, is one of the few men today who knew and associated with the cowboys who had charge of the great herds of cattle that ranged over the anhandle in the early eighties He was born in Rockford, Illinois, June 15, 1862, the son of Cebra and Harriet (Gates) Ford, the former a native of Ireland while the mother was born in France Mr Ford came to the United States to engage in farming, settled first in Ohio, then Illinois and from that state moved to Iowa in 1865 Mrs Ford died there in 1872 and her husband in 1892 He was a general farmer, a Republican in politics but never held office There were four children in the family but Riley, of this review, is the only one living He received his education in the public schools of Iowa and when old enough determined to have a farm of his own Learning that there was plenty of cheap land in the Panhandle he came here in 1885, locating in Deuel county in June of that year The trip was made across the country in true pioneer style in a wagon drawn by horses living in the covered wagon on the way Locating on a homestead five miles south of Big Springs, Mr Ford at once erected the usual frontier home — a sod house — a sod stable and was ready for his family when they come five months later At that time there was only one habitation between Big Springs and Julesburg the land being unbroken prairie The family was discouraged many times due to the poor crops so many years from drought and at first they had to haul water over two miles for family use and the stock However they could not sell stuck it out and in the end won out with a comfortable fortune Settlers made the best of the situation in those days, held parties in the sod houses and Mr Ford says had a better time than people do today He has seen the many changes in Deuel county which today is rich farm land well settled, and has taken his part in this development During the early days Mr Ford made friends of the cow-boys who were with the great cattle outfits and they often gave the family meat when they butchered Accepting an invitation from one outfit to go to Julesburg with them he had an adventurous time as they shot up the town, but the foreman paid all damages and all had a good time They put on an impromptau ' Wild West Show" with a colored man doing the riding in the dark on a rainy night and all he spectators saw was the glow of the point of the cigar in his mouth as the horse bucked In the spring of 1919, Mr Ford sold his farm land and retired from active life and now makes his home in Big Springs He made a comfortable fortune from land that was thought worthless and is enjoying the sunset years of life

May 28, 1882, Mr Ford married Miss E Miller, the daughter of John and Anna Miller, at Decorah, Iowa, and they became the parents of four children Hattie, the wife of Harry Jones, of Lawrence, Wyoming , Charles, on the home place , Claude of South Port, North Carolina, just mustered out of government service and John, of Big Springs Mr Ford is a Republican and attends the Methodist church

GEORGE F. RICHARDSON, one of the early settlers of Big Springs district who came here when this country was unbroken prairie and has experienced all the vicissitudes incident to life on the frontier and made good as a rancher, is an Englishman, born in Lincolnshire, January 13, 1861, the son of George and Annie (Jackson) Richardson, both natives of England where the father was a farmer and veterinary The family came to the United States in 1873, settled near York, Nebraska, where the elder Richardson took a homestead and passed the remainder of his days on the farm He died in 1899 The mother passed away in 1875, leaving a family of three children Arthur, of York, Nebraska , George of this review, and Herbert of Holt, Nebraska

George Richardson was educated in the common schools of England until he was twelve years of age then accompanied his parents to their home in the United States He remained at home until his twenty-second year then filed on a homestead ten miles southeast of Big Springs in Keith county This county was later divided and the subdivision called Perkins county, where Mr Richardson lived until 1895, when he moved near the town of Big Springs In 1916, he retired from active participation in business and since then has made his home in Big Springs He sold out his original homestead, but still owns an irrigated farm near the city and a home in the town When Mr Richardson came to this section in 1884, he drove across the country with a yoke of oxen, following along the right of way of the Union Pacific Railroad As teams became stuck in the mud some other traveller would help pull them out, thus Mr Richardson helped others and they in turn assisted in pulling his wagon from some mud holes His equipment consisted of a cow, plow and a few supplies For some time he made money breaking land for other men and secured capital for a start The second year he had a good crop of corn but range cattle broke into his fields and ate it all With a neighbor he cut hay with a scythe — there were no mowers then — and had to watch the hay stacks to see that range cattle did not break the fences and eat it too There were some wild horses, a few buffaloes on Frenchman's creek and numbers of coyotes in this section then and Mr Richardson killed many of the latter He is a Democrat in politics, but took no active part in political life for many years , has been one of the best known ranchmen in the Big Springs section where he is considered one of the substantial and prosperous farmers who is progressive in his ideas and methods

JOHN C STEWARD, one of the early settlers of Deuel county and today one of the most successful and prosperous farmers of this section, was born in Henry county, Illinois, September 4, 1859, the son of Martin and Mary (Woodruff) Steward, natives of New York. The father was a farmer who came to Illinois and at the outbreak of the Civil War voluntered, serving in the One Hundred and Twelfth Illinois Infantry from 1861 to 1864 In 1884, Mr Steward came to Nebraska, locating in Dawson county, then homesteaded five miles northwest of Big Springs, where the family lived until his death Mrs Stewart later sold that place and took a homestead as a widow, proved up but later married a minister from Iowa, I M Flyng, and now lives in Chappell There were eight children in the family, of whom five are living, but John, of this sketch, is the only one in Deuel county Mrs Steward was a member of the Methodist church, while her husband was a Republican in politics

John Steward was educated in the public schools of Illinois and was married in Bureau

Gus. Linn

county, that state, on December 17, 1882, to Miss Sarah M Estabrook the daughter of David and Mary (Ferrell) Eestabrook, both New Englanders by birth Ten children were born to this union, of whom seven are living Alice, the wife of Frank Thomas of Lexington, Nebraska, Blanche, the wife of Loren D Root, of Sedgwick, Colorado, Jesse, deceased, Archie, who lives on the home farm, Gertrude, the wife of John Ford, of Big Springs, Pearl, deceased, Clarence L, at home, Ruba, Ruth and David M.

Mr Steward came to Deuel county county in 1884, and took a claim, the northwest quarter of section 34-15-43 which he sold in 1903, and bought the land where he now lives, as he foresaw the great future of irrigated land and purchased what could be watered Since that time he has been engaged in general farming and stock raising Mr Steward is a self-made man, having had little when he came into the Panhandle but the determination to succeed, for at that time it was necessary to haul wood from the canyons of the Platte He was near the famous California Trail and saw the early prairie schooners let down the steep hills with windlasses One year he worked his land with an ox team, then purchased a pony horse team but today has the most modern machinery obtainable Mr Steward has specialized of late years in the breeding of full blooded Belgian horses, in which line he has met with gratifying success He is a Republican, has served as assessor and road overseer of his district and is one of the substantial men of his community

GUS LINN. — If any proof were needed as to the success that is reasonably sure to follow wholesome living and persistent, well-directed industry, it may be found in the career of Gus Linn, who is one of Kimball's foremost business men and highly respected citizens and public officials He is president of the Bank of Kimball and for years has served in positions of trust and responsibility

Mr Linn was born in Sweden, December 6, 1862, third in a family of five children His mother is deceased and his honored father died in October, 1910 Mr Linn attended the public schools but had no further educational advantages, and when nineteen years old decided to come to America to seek better business opportunity He reached Pennsylvania in June, 1882, remained in that state two months and then went to Michigan, working in lumber regions there for two years In 1884 he came to Nebraska and worked for a lumber com-

pany at Omaha for two years, then came to Kimball to take charge of the Kimball Lumber Company, with which concern he continued until 1893, during these years building up a wide acquaintance and many friendships based on the integrity of his character

In 1893 Mr Linn went into business for himself, purchasing the Kimball Lumber Company interests and adding hardware and farm implements to the goods he handled In 1907 he sold the lumber branch of his business to the Foster Lumber Company but continued the other lines, and now owns the largest hardware, wagon and farm implement business at Kimball While Mr Linn is recognized as an able business man, his achievements in this direction have been equaled by the progress he has made in public life, for he has been one of Kimball's most useful and public-spirited citizens in numerous ways He has never been unmindful of his responsibilities since he became an American citizen and, intelligent and thoughtful, he soon became interested in political questions Public confidence was shown in his integrity when in 1892 he was elected county treasurer of Kimball county, in which office he served with fidelity for four years In 1897 he was further honored by being elected county judge, serving in that capacity for two years In 1905 he was elected county commissioner and served until 1907, when he became mayor of Kimball, and served two years He has given encouragement to many of the worthy enterprises that have proved so beneficial to Kimball, has made investments here and stands today as a most worthy citizen in every sense of the word

On June 18, 1888 Mr Linn was united in marriage to Miss Anna C Willing, at Sidney, Nebraska, who was born in Sweden, but has lived in the United States since girlhood They have the following children Oscar G, Vernon E, Herbert E, Ernest A, John T Frank W, and Kenneth Mrs Linn died January 22, 1918

Since 1907 Mr Linn has been president of the Bank of Kimball and built a fine bank building He also has the largest hardware store in Kimball county and also built a fine building in Dix, in which he conducts a hardware store He owns an implement business at the town of Bushnell and Potter At all of these places he has done approximately $200,000 worth of business He is also a thirty-second degree Mason, belonging to No 294, Kimball and also is a member of the Knights of Pythias and Woodmen

JOHN M ANDERSON, deceased, for many years one of the well known and highly respected farmers of Garden county, a pioneer who played his intelligent part in the opening up and development of his locality, was born in Sweden, April 11, 1857, and died on his farm near Chappell, December 11, 1913 He was reared and received his early education in his native land, where he later learned he capenter's trade Mr Anderson was the son of Andrew and Marie (Johnson) Hanson, both natives of Sweden, where they spent their lives In 1882, Mr Anderson came to the United State and soon after reaching our shores located at York, Nebraska, where he engaged in work as a carpenter On his way west he had spent a few months in Omaha, working while there on a bank building but wanted a farm of his own eventually and came to the Panhandle In 1885, Mr Anderson came to Deuel county and took up a homestead which his widow and children now own, the farm being operated by the son When he came into this section there was no railroads as today and Mr Anderson drove to his land with a team and wagon, he and Mr Odell being the only settlers within a radius of ten miles Soon the usual frontier home — a sod house — was erected and Mr Anderson spent a few months on the place before returning to York On his return to the claim the sod house had been torn down by the cattle but he managed to save the wooden roof out of which he made a cupboard which the family treasures today as a relic In 1886, Mr Anderson again returned to his land bringing a cow boy with him Between them they had a team, wagon, and a cow pony They lived in the wagon until the first crops were harvested and again built a sod house for warmth during the winter The cow boy tried to make trouble for Mr Anderson and take the team but the latter met his attempts with a shot gun and the horses remained his property Mr Anderson was a good farmer, worked hard and by his ability was able to make a good living during the hard years He had become most skillful in handling and repairing machinery, doing a great amount of work for his neighbors and had he been able to have a technical education in his youth would have engaged in the agricultural implement business At one time Mr Anderson made water tanks, the lumber being furnished by the purchaser

April 30, 1888, Mr Anderson married Miss Ida Carlson at Sidney, Nebraska She was the daughter of Carl and Christina Carlson, natives of Sweden and was born there She came to this country in 1885, lived in Omaha until 1888, then came to Deuel county Two children were born to this union Arthur, at home and Huldah, the wife of Cornelius Diehl, now homesteading near Douglas, Wyoming Mrs Anderson has many reminiscences of the early days in this section, the hardships and privations, but she and her husband became prosperous, they bought their first land for forty dollars a quarter and became the owners of two sections which today are valuable

Mr Anderson was a Republican and a member of the Swedish Lutheran church

WILLIAM G MELTON, one of the earliest settlers of the Lodgepole district and well known stockman has lived in the Panhandle for more than twenty years, passed through all the hardships and trials of frontier life and today with his brothers is one of the prominent and prosperous ranchmen of western Nebraska He was born in Harrison county, Indiana, October 20, 1869, the son of James H and Catharine (Snyder) Melton, both natives of this same county The father was a general farmer who died in Indiana in 1877 He was a prominent man in his community, belonged to the Methodist church and was a Republican There were eight children in the Melton family, of whom four are living Harvey B, Jonathan K, George L, of York, Nebraska, and William of this review, who received his early education in Indiana and later in the public schools of York, Nebraska, as he came west with his brother and sister-in-law in 1878, because of poor health Mr Melton lived in York until 1881, before returning to Indiana for three years In 1884, he came back, remained in York county a short time before taking up a homestead in Deuel county in December, 1890 This land was near Lodgepole and he still owns the original tract Later he purchased his present farm in section 14-14-46, where he lives with his brothers, Harry and Jonathan For many years the Melton brothers have raised cattle and dealt in live stock They first bought short horn cattle but saw that high grade animals paid the best and in later years have changed to Herefords, keeping as good a grade as is to be found in this section of the country Starting with only thirty head, Mr Melon had more than five hundred head when he sold out in 1907, with a hundred and thirty head of calves A nephew now has the active work of the farm in hand but Mr Melton supervises the management They specialize in general farming and have but thirty head of cattle The six hundred acres are cultivated, all the brothers helping in the work of looking after various agricultural branches The brothers together own about four sections near Lodgepole, using the latest methods in farming as well as modern equipment in machinery and buildings Mr Melton is a Republican He and his brothers are the oldest settlers in this locality and in the early days they tell of using all the land for a general range, fencing only that used for crops Land sold then for a dollar an acre and many times they felt it was not worth the taxes, could never be farmer and Mr Melton once remarked that he was paying "good money for bad land," and today it is very valuable

Mr Melton and his brothers are of high standing in the community, being regarded as some of the most substantial men in Deuel county Mr Melton is a member of the Masonic order, having held nearly all the offices in his lodge

AUGUST G NEWMAN, a member of the retired colony of Chappell, who has spent more than forty years in the Panhandle and made many warm friends in Deuel county, was born in Waukesha county, Wisconsin, November 29, 1854, being of German extraction, as both parents were born in the German Empire and subsequently came to the United States The boy was educated in the public schools and worked on his father's farm until he started in life for himself in 1878 That year Mr Newman came to Nebraska and bought a ranch near Lodgepole where he raised sheep for eight years, four years of that time he lived entirely alone, camping out with his herds both winter and summer In 1886, he bought twenty-six hundred acres of land southeast of Chappell and devoted his time to raising horses and cattle, in which enterprise he met with success Many excellent improvements were made on the place, while the ranch had a naturally beautiful location on Lodgepole creek Trees were planted and the Newman ranch was known throughout this section Operating the place until 1893, Mr Newman then moved to Chappell, becoming the owner and manager of the Johnson House, a hotel which he operated thirteen years, during which time he became very popular with the people of the town and the traveling public

In 1889, occurred the marriage of Mr Newman and Miss Mary Barrett, the daughter of Harry and Jane (Barchard) Barrett, the former a well known railroad man having charge of the yards at North Platte, where he located

on coming to Nebraska from Missouri in 1866 The Barretts later moved to Lodgepole and died there Mrs Newman died in 1917, leaving four children, Guy C, of Chappell, Grace A, the wife of John Burgstrum of Chappell, Harry C, of Omaha, and Mary B at home with her father

Mr Newman has never disposed of all his real estate holdings and is still the owner of a fine ranch of nineteen hundred and twenty acre He is a Republican in politics, has taken an active part in the community life, serving as commissioner of Cheyenne county from 1885 to 1887, and later filled the same office in Deuel county Being public spirited and progressive he assisted in establishing several schools in his locality and held important school offices, both in county and city Mr Newman came here when the country was practically virgin prairie and has not alone viewed the changes but partaken in the development of this section He assisted in the partition of old Cheyenne county into six other counties, for when he came to the Panhandle there were no towns between North Platte and Sidney, while Chappell had only one house and the section house, so that he has witnessed the opening up of one of the finest sections of the country, seen villages and towns of this county platted and settled up, which few in the future will ever again witness Mr Newman is a charter member of the Masonic lodge

WILLIAM N DIEHL.— The energy, resourcefulness and initiative required to succeed in any line of endeavor today has been an integral part of the mental equipment of William Diehl since he first entered upon an independent career in business as an agriculturist While still a school boy he began laying plans for his chosen vocation and his faith in himself has been fully vindicated, for today Mr Diehl is one of the successful farmers of Banner county, where he owns land in section 34-57-17

William Diehl was born in Las Vegas, New Mexico, November 9 1893 the son of John S and Mary E (Goldborough) Diehl, the former was born in Snyder county, Pennsylvania, in 1858, while the mother was a native of Illinois, born there in 1861 Seven children formed the Diehl family of whom the following five still live, Cornelius, born in 1888 married Haldah Anderson Chappell in 1919, William, of this review, Hazel K born in 1895 is the wife of Benjamin Kipinger Elmer E, who married Rosa Kuhnpufter, lives at Priest river, Idaho, Alma Iva, the wife of John Miner, of Banner county. The two children who died were Lee, aged twelve years and Charles, aged eleven months.

When William Diehl was only two years of age his parents left New Mexico and came to Nebraska, locating at Nickerson, where they lived until the boy was fifteen years of age. He was sent to the excellent public schools of that town and early learned the practical side of farm industry by working during the summer vacation. After his elemenary education was completed Mr Diehl attended college at Fremont, Nebraska, before starting in business for himself and thus laid a firm foundation for his career as a farmer.

When the United States declared war against Germany and the president called for volunteers, the young man enlisted in the army October 3, 1917, remained in the service of his country until honorably mustered out on May 3, 1919, and while he did not succeed in crossing the ocean to fight was more than anxious to go and his services rendered here were as valuable as though he had been on the battle front. After his discharge Mr Diehl located on his present farm near Bushnell where he has since been actively engaged in general farming and stock raising. He is one of the enthusiastic, capable and progressive exponents of agricultural industry in this section and is earning a just reward for the time, energy and study he devotes to his vocation.

In March, 1918, Mr Diehl married Miss Clara Garrison of Ohiona, Nebraska, who died the same year.

Mr Diehl attends the Methodist Episcopal church and is a member of the National Socialistic party and the farmers' Union. His father, John S , is a member of the Independent Order of Odd Fellows. The family is well and favorably known in Banner county as good business men and upstanding citizens who advocate and support all movements that tend to the development of their community and the county.

WILLIS B JORDAN, who is one of the progressive farmers of the younger generation in Scottsbluff county, is a scion of red-blooded American stock that has been conspicuous in the pioneer annals of the nation, his paternal grandparents having been early settlers in Illinois and later in Iowa, and his father gained a due measure of pioneer distinction in Nebraska. That the subject of this sketch has lived up to the family traditions of loyalty and spirit, needs no futher voucher than the state-

ment that he was one of the gallant young men who represented Nebraska on the European battlefields within the period of the late World War, a service that shall ever reflect honor upon his name.

Mr. Jordan was born on the old homestead farm of his parents near Alliance, Box Butte county, Nebraska, April 26, 1891. He is a son of Jacob David Jordan and Florence Levisa (Snow) Jordan, the former a native of Illinois and the latter of Iowa, their marriage having been solemnized in the latter state, when Mr Jordan was twenty-eight years of age. Jacob D Jordan was born in 1847, in Illinois, and this date shows that his parents were numbered among the pioneers of that state, where his father became one of the honored members of the clergy of the Methodist Episcopal church, in the service of which he later was a pioneer in Iowa, to which state he removed when his son Jacob David was a lad of seven years. In the Hawkeye state Jacob D Jordan was reared under the conditions and influences of the pioneer days, and he was signally favored in having the advantages of a cultured and refined home, as well as those of the public schools of the locality and period. After reaching adult age he continued to be actively identified with agricultural pursuits in Iowa until he came to Nebraska and took up a homestead claim in Box Butte county. He not only perfected his title to this land but also to a tree claim, the property being situated twenty-five miles southeast of Alliance. He there proved successful as an agriculturist and stock-grower and devoted considerable attention to buying and shipping cattle. After having been thus engaged five years he removed to Scottsbluff county, where he became the owner of a ranch of twelve hundred acres, upon which he made further improvements and where he developed a substantial business in the raising of cattle and horses. He also continued to buy and sell stock upon a rather extensive scale. After remaining on the ranch about five years he disposed of the property and purchased a farm one mile north of the city of Scottsbluff. There he gave attention to agricultural enterprise, though he continued to raise stock upon a minor scale and after the lapse of six years failing health prompted him to visit California. He did not, however, recuperate his physical energies, and died in that state. His widow removed to Hastings, Nebraska, to give her children the advantages of the public schools and college at that place, where she remained until her removal to her present home. She is a zealous member of the Methodist Epis-

MR. AND MRS. HENRY VOGLER

copal church, as was also her husband Concerning their children a brief record may consistently be entered at this juncture Clara B and Oral D were born in Iowa, Willis B, of this review was the first to be born after the removal to Nebraska, Amber and Amy were born in Box Butte county, while Hazel, the youngest of the surviving children, was born at Alliance Amber attended college at Hastings for three years and is now the wife of Albert Catron, of Gering, Scottsbluff county Amy was graduated from the Hastings high school, and then took a two years' course in domestic science at a college, in that city, and is now principal of a school at Fraser, Colorado Hazel was graduated from the high school at Grand Junction, Colorado, and is a student in domestic science at college

Willis D Jordan acquired his early education in the public schools of Nebraska and was about eighteen years of age at the time of his father's death As the only son, he was thus called upon to assume responsibility, and he continued to have charge of the home farm until he had attained the age of twenty-two years, in the meantime finding employment in the beet sugar factory at Scottsbluff during the winter seasons In order that the younger children might obtain better educational advantages, the farm was finally rented for one year During the ensuing years Mr Jordan again had charge of the place, where he maintained "bachelor's hall," and the next year found him identified with the cattle industry in Sioux county. When there came to him the call of higher duty, in connection with the nation's participation in the great World War, Mr Jordan sold his cattle in 1917, and enlisted as a volunteer, his assignment having been to the medical department of the service in the First Division, First Army Corps He arrived with his command in France in December, 1917, and thereafter was with his division in the front lines during the progress of three heavy battles He continued at the front, loyal and valorous, until the signing of the historic armistice brought the great conflict to a close He then returned to America and he received his honorable discharge on May 5, 1918 This gallant young soldier of the republic is now actively engaged in farm enterprise in Scottsbluff county, where he is conducting operation on his well improved farm, eleven miles northeast of the city of Scottsbluff He is earnest in his political allegiance and his religious views are in harmony with the tenets of the Methodist Episcopal church, in which faith he was reared

HENDY VOGLER, ranchman and prominent citizen of Kimball county, is well known in this section of Nebraska where, for years, he has been engaged in successful business enterprises, and has also served frequently and efficiently in public capacities He was born in Baden, Germany, October 8, 1864 His parents were Peter and Anna Vogler, who had a family of ten children, three of whom, Peter, George, and Henry, came to the United States

Henry Vogler grew to the age of eighteen years in his own country, where he received a public school education In 1871 his older brother, Peter Vogler, had emigrated to the United States and established himself in Cass county, Nebraska, and when Henry reached this country, in 1882, he made his way across the continent as far as Nebraska and joined his brother During the first winter in Cass county he attended school, but left his brother's farm in the spring of 1883, went to Lincoln and there secured employment in a grocery and bakery store, where he worked faithfully for the next three years In the meanwhile he had prudently saved his wages, in prospect of going into business for himself and this he was able to do after coming to Dix, the town then being in old Cheyenne county Mr Volger not only embarked in an enterprise of his own, opening a general merchandise store, but he became a moving force in the development of the place Through his efforts Dix became a postoffice, and he was the organizer of the first school there

Mr Vogler continued in the mercantile business at Dix until 1895, when he was elected county clerk of Kimball county, removing then to Kimball, subsequently serving three terms in that office At Kimball he again went into the mercantile business with a partner, B K Bushee, and this partnership continued until 1908, when Mr Vogler was elected county treasurer, which office he held two terms, this being marked evidence of the confidence with which he had inspired his fellow citizens In 1908 Mr Vogler, in association with Gus Linn and J J Kinney, bought the Bank of Kimball, and Mr Vogler continued in the banking business at Kimball until 1914, when he sold his interest and since that time has mainly devoted his attention to affairs pertaining to the development and improvement of his ranch in Kimball county, a valuable property of large extent

In 1887 Henry Vogler was united in marriage to Miss Clementine Neeley, who is a daughter of Samuel and Anna Neeley, and nine children have been born to them, one of whom died in infancy, the others being as follows Olive, who is the wife of W T

Young; George, who is cashier of the Bank of Kimball, Shirley, who owns and operates a moving picture theatre, Robert, who is also employed on the ranch, and Margaret, Bonita, Ruth, and Donald, all of whom reside with their parents   Mr Vogler and his family are members of the Methodist Episcopal church He has always been a busy man but is social by nature and is a valued member of the Modern Woodmen and the A O U W, and other fraternal organizations

B F ROBERTS, gained in earlier years a broad experience in connection with the cattle industry, in the activities of which he was engaged for more than a decade in Wyoming, and of the same important phase of industrial enterprise he is a prominent and influential exponent in Scottsbluff county at he present time   He is one of the substantial landholders of this county and has here done much to promote the raising of the best types of cattle and horses, besides which he is similarly progressive in the agricultural department of his ranch enterprise and is. liberal and public spirited as a citizen — one of the wholesome and well fortified "boosters" of whom the Nebraska Panhandle may well be proud

Mr Roberts was born in Woodford county, Illinois, on the 10th of April, 1870   Both of his parents were born and reared in England. where their marriage was solemnized The father served with marked valor in the Crimean war, in which he was wounded in action, and the English government awarded him several medals for gallantry on the field of battle   He came with his young wife to America in the year 1865, and thereafter they continued their residence in Illinois until 1887, when they came to Nebraska and numbered themselves among the pioneer settlers of what is now Scottsbluff county   Here the father took up a homestead, to which he perfected title in due time and upon which he made good improvements   He and his wife merit place on the roll of the sterling pioneers of Scottsbluff county

B F Roberts gained his rudimentary education in the public school of Illinois and was a lad of fourteen years at the time of the family removal to Nebraska   Settlement was first made near Sidney. Cheyenne county, and from that locality the family removed to the present Scottsbluff county when the subject of this sketch was still young   He attended school at irregular intervals after coming to Nebraska   At the age of eighteen years he went to Colorado, where he was employed two years on a horse ranch, after which he initiated his career as a cowboy in Wyoming, where he was employed for a period of eleven years, and with characteristic thrift he saved his earnings, with a view to establishing himself in an independent business enterprise   In 1890, he took up a homestead in Wyoming and stocked the ranch with cattle, besides making special effort to bring his stock up to the highest standard   Mr Roberts likewise followed the same plan in the raising of fine horses, and his father had supervision of the homestead while he himself continued his activities on the range, his services as a cattle herder having been expert and having commanded from him the maximum wages About 1890, he took charge of the Castlerock irrigating ditch, to the supervision of which he continued to give his attention about six years In the meanwhile he had accumulated on his ranch a fine herd of about five hundred whiteface Hereford cattle, including some of the best sires that had been introduced in Wyoming and western Nebraska, besides which he had developed some of the best prize-winning Percheron horses in this section   After leaving Wyoming Mr Roberts returned to Nebraska, and in Scottsbluff county he is now the owner of a valuable ranch property of over two thousand acres, a considerable part of the tract being supplied with excellent irrigation facilities and making it list with the most productive and valuable agricultural land in the Nebraska Panhandle   Mr Roberts still conducts large and successful operations in the breeding and raising of the best types of horses and cattle, and in every way he has shown his capacity for doing big things in a big way   His civic liberality and public spirit are of the most consistent order   He owns the controlling stock in the McGrew State Bank. of which he is president, the village of McGrew being some miles distant from his finely improved ranch, the equipment of which is of the most approved modern standard   Mr Roberts has never had time or inclination for the activities of the political arena, devoting his time to business

The year 1897, recorded the marriage of Mr Roberts to Miss Ida Davis, whose father was a pioneer settler in western Nebraska Mr and Mrs Roberts have a fine family of seven children   Edith G, James William, Alonzo, Walter W, Rollin R, and Benjamin F.   Miss Edith Roberts was graduated in the high school at Gering and is, at the time of this writing, in the winter of 1919-20, a student in the University of Nebraska

CHARLES J McCUE has lived in various localities and followed differing lines of occupation, and should be given pioneer distinction in what is now Scottsbluff county, for he came here in 1885, and took up a tract of land on the line of Cheyenne and Keith counties, as then constituted He proved up on his homestead, to the improvement and cultivation of which he continued to give his attention until 1893, when he abandoned farm activities and went to Pueblo, Colorado, where he entered the employ of the Denver & Rio Grande Railroad Company With this corporation he continued in railroad work ten years and for the ensuing seven years was engaged in coal mining, at the Oakdale mine in Colorado He then returned to Nebraska and engaged in farming and dairying enterprise in Furnas county, which continued to be the stage of his operations until October, 1913, when he established his residence in the city of Scottsbluff, where he has since been engaged successfully in well-building He has developed a substantial business in this important line and his services are in requisition in the construction of wells in the most diverse sections of the county Mr McCue is known as a reliable and progressive business man and loyal and public-spirited citizen He has never sought or desired political office of any kind but gives his allegiance to the Democratic party

Charles J McCue was born February 16, 1863, at Garden Prairie, Iowa, where his parents were pioneer settlers He is a son of John and Anna (Davidson) McCue, both natives of Ireland John McCue was young when he accompanied his parents on the immigration to America, and the family first settled in Rhode Island The father eventually became a farmer in Illinois, later was similarly engaged as a pioneer in Iowa, and in 1881 came with his family to Clay county, Nebraska, where he purchased land and developed a good farm He later removed to Furnas county, and died there in 1903, when about seventy-one years of age His venerable widow still resides in that county and at the time of this writing, in the winter of 1919, she is eighty-six years of age She was fourteen years old when she came with her parents to the United States, and she was reared in McHenry county, Illinois In her native land she was denied educational advantages and she did not learn to read and write until she was forty-five years old, though her alert mentality has largely enabled her to overcome this handicap along specific educational lines

Charles J McCue was reared and educated in Iowa, whence he accompanied his parents to Nebraska in 1881 In 1885, shortly after attaining his legal majority, he took up a frontier homestead in what is now Scottsbluff county as previously intimated in this sketch, and there he remained until his removal to Colorado, as already noted It is worthy of special mention that while he was out on a hunting trip in 1886, he killed deer at a point within ten miles of this present city of Scottsbluff

At Ogallala Nebraska, on the 22d of February, 1893, Mr McCue wedded Miss Katie Meyer, who was born in Iowa, where she was reared and educated in Washington county She came to Nebraska and entered a homestead claim in Chase county, where she eventually perfected her title to the property She died in 1910, at the age of forty-two years, and is survived by three children Viola Rowe of Lander Wyoming, has one daughter, Orson W, who is now with the American forces in the Army of Occupation in Germany (November, 1919), saw eighteen months of active service in France and felt the full tension of the great conflict he was severely gassed while with his command in the trenches, and as a result was confined to a military hospital from July 22, 1918, until the 1st of the following November, his military membership having been in Company D, Third Division of American Expeditionary Forces, Alva B, the younger son, remains with his father at Scottsbluff

RICHARD S KNAPP was a lad of about eight years when his parents came from Iowa to Nebraska, and eight years later the family became pioneer settlers in what is now Scottsbluff county Thus from his youth he has witnessed and participated in the development and progress of this county, which was then a part of Cheyenne county, and he is now numbered among the substantial and representative exponents of agricultural and livestock industry, as the owner of a large and well improved landed estate, six miles northwest of Bayard

Mr Knapp was born in Pottawattomie county Iowa, August 28, 1871, and is a son of Philo P and Hattie C (Otis) Knapp, the former was born in the state of New York in 1838, and the latter was born in Ohio in 1840, their marriage having been solemnized in Iowa, in 1860 In the Hawkeye state Philo P Knapp continued his activities as a farmer until 1879 when he came with his family to

Nebraska and settled in Cass county There he continued his agricultural operations until 1887, when he became one of the pioneer settlers of that part of Cheyenne county that is now included in Scottsbluff county Here he perfected his title to a homestead and a tree claim, and with the passing years he made good improvements on this property, and gained a place as one of the sterling pioneers and influential citizens of his community He died at Plattsmouth, this state, in 1904, and his widow passed away there in 1917, the names of both meriting enduring place on the roster of the honored pioneers of Scottsbluff county

Richard S Knapp acquired his early education in the public schools of Cass county, this state, and was a lad of about sixteen years at the time of the family removal to what is now Scottsbluff county Here he assisted his father in the reclamation and development of the home ranch, and in 1893, entered claim to a homestead of his own He proved up on this place and continued successful operations as a farmer for thirteen years He then sold the property and purchased his present valuable estate, which comprises two hundred and eighty acres well situated six miles northwest of the village of Bayard In the work and management of this finely improved farm he has the effective assistance of his sons, and the place is supplied with excellent irrigation facilities Mr Knapp was one of the builders and officials of the Bayard irrigation ditch, and he has been specially prominent in connection with the construction of the irrigation ditches which have added so greatly to the success of agricultural enterprise in the county Scarcely a ditch has been built here without his active assistance in construction work, and he is a recognized authority in local irrigation enterprise He has been loyal and liberal also in the promotion and support of schools and churches, and has served as a school officer in his district for fifteen years He helped to build school houses and churches when such improvements were compassed only by the donation of work on the part of the citizens, and no worthy community undertaking has failed to enlist his earnest support His political views are shown by his alignment in the ranks of the Democratic party, and in a fraternal way he is affiliated with the Modern Woodmen of America, both he and his wife are members of the church

In April, 1896, at Gering, was solemnized the marriage of Mr Knapp to Miss Emma M Jamison, whose parents were numbered among the first settlers in the Minatare district of this county Mr and Mrs Knapp have six children, the two older sons having both graduated from the high school at Bayard and now are their father's able assistants in the work and management of the home ranch The names of the children are here entered in respective order of birth Rufus R , Harold E , Mabel G , Philo D , Ambrey, and Frank The younger children are still in school at the time of this writing and the family is one of prominence in the community

W M. BARBOUR, is one of the representative agriculturists and stock-growers of Scottsbluff county, where he established his home in the early pioneer days and where his ability and energy have conspired to win to him a generous measure of prosperity, as well as secure status as an influential citizen of the county He is associated with his venerable mother in the ownership of a well improved and valuable landed estate of six hundred and forty acres, and on this splendid property he has demonstrated most fully the splendid natural advantages of this favored section of the state

Mr Barbour was born at Spava, Fulton county, Illinois January 25, 1860, and is a son of William and Cynthia (Carter) Barbour, the former was born and reared in Ohio and the latter was born in Rock Island county, Illinois, where her parents were pioneer settlers and where she was reared to adult age, her educational advantages having been those of the common schools of the locality and period William Barbour was a young man when he removed from Ohio to Illinois, in which latter state his marriage was solemnized During the major part of his active career he followed the basic industry of agriculture, and he was about seventy years of age at the time of his death He was a soldier of the Union in the Civil war, during which he served as a member of an Illinois volunteer regiment In politics he was a staunch Republican

The only one of his parents' children who grew to adult age, W W Barbour passed the period of his childhood and youth on the home farm in Illinois, where he gained valuable experience, and made good use of the advantages afforded in the public schools In 1887, after the death of his father, he came with his widowed mother to what is now Scottsbluff county, Nebraska, and both entered homestead claims, which were adjacent Here they have resided during the long intervening years that have brought marvelous advancement and unstinted prosperity in this section of the state, and they are associated in the ownership of a fine landed estate of a full section of land, Mr Barbour having added materially to the area of the original homesteads by wisely investing his surplus funds in the land whose value he had proved through personal experience His land is now very valuable, and this statement in itself bears great significance He has been the apostle of progress and has done much to support those activities that have conserved the advancement and prosperity of his county along both civic and industrial lines He was one of the original nine men who projected and brought to completion the Enterprise Irrigation Ditch, and was concerned also in the construction of the Winter Creek Ditch His devoted mother, who celebrated in 1919 the seventy-eighth anniversary of her birth, remains with him in their attractive home and she is revered as one of the gracious pioneer women of the county She has long been a devoted member of the Presbyterian church and is still notably active in its work

Mr Barbour has been active and influential in the local ranks of the Republican party and has served two terms as county commissioner —a part of the time as chairman of the board In this office he did much to further public improvements of important order, and he has been chairman of the board of the Enterprise Irrigation Ditch from the time of its organization to the present The high popular estimate placed upon his character, his ability and his civic loyalty, was shown in 1918, when he was elected to represent his county in the state legislature, his district including also Morrill county As a legislator he is fully justifying the wisdom of the popular choice and is an active working member who is alert in protecting and advancing the best interests of his constituent district He has served locally also as chairman of the Scottsbluff Drainage Ditch, and few citizens have been more zealous in the promotion and support of well ordered irrigation projects He was one of the organizers and is still a stockholder of the Independent Lumber Company of Scottsbluff In a fraternal way he is actively affiliated with Scottsbluff Lodge, No 26 Independent Order of Odd Fellows

Mr Barbour married Miss Nellie M Andrus, who was born and reared in Nebraska and died in 1888 She is survived by two children Neale married Miss Louise Owen and they have one child, Jane Aldreth Neale Barbour is one of the successful agriculturists and stock-growers of Scottsbluff county

Charles B, the younger son, likewise is upholding the honors of the family name as one of the representative young farmers and stock-raisers of the county

FRED REYNARD MORGAN, one of the prominent and progressive business men of Kimball and the Panhandle, is distinctively a Nebraska product He was born, reared and educated within the boundaries of the state and his business life has been bound up with the progress and development of the southwestern section He is a member of a fine old pioneer family of Hall county and the thirteen children have played important parts in the industrial and civic life of the state They inherited from their father the sturdy traits and high ambitions of the English as he was born in the Island of Great Britain Today Fred Morgan is numbered among the substantial business men of Kimball, has been practically the architect of his own fortune, and having based his life's structure on firm foundations has builded soundly and well When he entered upon his commercial career he was possessed of little save a good elementary education, inherent ability and a determination to succeed, and these have been sufficient, through their development to enable him to become the leading drug merchant in a flourishing community that does not lack for able men

Mr Morgan was born in Grand Island, Nebraska, February 15, 1879, the son of Daniel and Elizabeth A Morgan The former came to Nebraska during the pioneer days of the early 'seventies, as he located in Grand Island in 1870 and opened a grocery store in what was then a frontier town This business proved good and Daniel Morgan being a well educated man of affairs entered actively into the civic and communal life, in which he took an important part for many years Thirteen children made up the Morgan family of whom the following are living in Kimball county Fred R, of this review, Arthur an electric engineer, and Walter, a well known farmer of this section

The children grew up in Grand Island and Fred with the others was given advantage of the educational facilities in the town As soon as his education was finished the young man entered a leading drug store as apprentice, applied himself to the study of pharmacy and passed the state pharmacy examination in 1902, becoming a registered druggist and licensed to practice in Nebraska In November of that year Mr Morgan was offered and accepted the management of a drug store owned by D. Atchison, here in Kimball, holding this position until 1906 when he purchased the business outright and has since been sole owner. In 1917 he built a fine new store, carries one of the most complete stocks in the Panhandle and has gained a high reputation for his prescription business. A druggist has the life of the community as much in his care as the physician who writes the prescriptions, and Kimball has indeed been fortunate in having a man of such ability and skill to take part in the general welfare of the community. As a business man Mr. Morgan has shown keen foresight, executive ability and resourcefulness, has known just when to expand his growing interests, has inaugurated many new and attractive innovations in his store which today is one of the largest in the southwest and has a large and increasing trade, both in city and from the surrounding country.

From first locating in Kimball, Mr. Morgan has taken an active and interested part in the welfare of the community and its progress; he is modern in his own business methods and believes that civic affairs should be run on a good, sound business basis and since being elected to the city council in 1914, where he served four years, was the means of establishing its affairs on a substantial basis, which is proving of benefit to the citizens.

In 1907 Mr. Morgan married Miss Margaret Wilkinson, of Pine Bluffs, Wyoming, the daughter of John and Margaret Wilkinson. The daughter, born in England, came to the United States with her parents in 1884. The family first located at Ainsley, Nebraska, then moved to Wyoming, where John Wilkinson was a successful and well-to-do sheep and cattle raiser.

Mr. Morgan is a sturdy adherent of the tenets of the Republican party and is a Mason in high standing, having joined that organization in 1904, he passed from one lodge to another rapidly, becoming a member of the Consistory in 1916 and a Shriner in 1917. He is a Presbyterian and his wife an Episcopalian. Since coming to Kimball Mrs. Morgan has made many friends and they have a charming home where their friends enjoy their cordial hospitality.

WILLIAM A. HALE merits special consideration as one of the honored pioneer citizens of western Nebraska and he is now living virtually retired, in the city of Scottsbluff. He was one of the pioneer teachers in the public schools of this state and is a man of fine intellectuality, as well as the possessor of those personal attributes and characteristics that ever beget unqualified popular esteem.

William Albert Hale is a scion of a family that was founded in America prior to the war of the Revolution, but no relationship has ever been claimed in connection with the historic patriot, Captain Nathan Hale. The subject of this review was born in Sullivan county, Indiana, September 3, 1852, and is a son of Stephen Clark Hale and Ann (Howard) Hale, the former of whom was born in Greene county, Virginia, April 10, 1810, and the latter of whom was born in Kentucky, in 1824, she having been, in the maternal line, a descendant of the Boone family of which the renowned Daniel Boone, historic frontiersman, was a representative. Stephen Clark Hale did active scout duty in some of the numerous Indian wars that occurred nearly a century ago, and as a boy William A. Hale listened with avidity to the interesting tales which his sire related concerning his experiences in making his way on horseback from old Vincennes, Indiana, through Kentucky and onward to his old home in Virginia.

He whose name initiates this sketch acquired his early education in the common schools of his native county and that he made good use of his scolastic advantages is demonstrated by his having become as a youth a successful teacher in the schools of Indiana, where he thus rendered service four years, in district and graded schools. His youthful ambition was to prepare himself for the legal profession, and he according prosecuted his technical studies under the preceptorship of John T. Hayes, of Sullivan, Indiana. He was admitted to the bar in 1878, but the causes which led to his deviation from the line of his profession have thus been stated by him: "Ill health and natural modesty kept me from following the profession. I concluded that I would rather be a live roustabout than a dead lawyer, so I packed my war sack in February, 1880, and landed west of the Missouri river." Upon thus coming to Nebraska Mr. Hale found ample demand for his continued service in the pedagogic profession, and for ten years he applied himself with characteristic ability and fidelity to teaching in the public schools of our state, his record being marked by exceptional success in his work in both the common and higher branches. He became a resident of Scottsbluff years ago and has given every possible aid in the furtherance of communal advancement and prosperity. His political allegiance is given unreservedly to the Democratic party and he and his wife are earnest members of the Methodist Episcopal church In January, 1873, Mr Hale became affiliated with the Independent Order of Odd Fellows, and in 1880, he took out a withdrawal card from his lodge He never deposited this card with any other lodge—for the excellent reason that during the most of his experience in Nebraska he was, until of recent years, in advance of established organization of this fraternal order in the communities in which he lived

In December, 1894, at Harrisburg, Banner county, Nebraska, was solemnized the marriage of Mr Hale to Miss Mary Genet Crosslen, whose father, the late Asberry Crosslen, came to Nebraska in pioneer days Mr Crosslen was a gallant soldier of the Union in the Civil war, in which he served three years, as a member of a cavalry command Mr and Mrs Hale have four children—William Howard, Ralph Dewey, Charles Casper, and Alice William H Hale, the oldest son, was one of the valiant young men who gave service to the nation during the great world war, and he was for one year in active service over seas, as a member of the Four Hundred and Seventy-fifth Aero Squadron

HERBERT R FULLER, vice-president of the Liberty State Bank, of Sidney, Cheyenne county, is one of the county's young and progressive citizens who have created a favorable impression in banking circles and established themselves in positions formerly held by men many years their seniors He is a native son of Nebraska, as he was born in Jefferson county, November 6, 1879, the son of Walter M and Sarah R (Wakeman) Fuller The father was the first male white child born in Webster county, Iowa, as his parents were pioneers of that state The Fuller family lived there many years and there Walter Fuller grew to manhood, received such educational facilities as were afforded in his locality at that early date and grew to maturity on his father's farm Upon attaining his majority he selected farming as his life work and was engaged in that occupation for some years before coming to Nebraska, where he located in Jefferson county about 1877, but Iowa had been his home for many years and his heart yearned for the old associations and with his family he returned to Webster county, where he passed the remainder of his days, passing away at the age of fifty-seven years, really a man in the prime of life He is survived by his wife who now resides at Fresno, California

Herbert Fuller received his early educational training in the public schools of his community, graduated from the high school and then entered Sabin College, Fort Dodge, Iowa, where he graduated In 1901, at the age of twenty-one Mr Fuller came to Nebraska, locating at Sidney as his father had extensive interests in the St George Cattle Company with headquarters in this vicinity This concern was one of the immense cattle outfits that held a large tract of land, owned many thousand head of cattle and was one of the great commercial meat enterprises that became well known in this section of the state For several years Mr Fuller was in the employ of the company but he desired to become an independent operator and with this end in view he purchased four hundred and eighty acres of land, leased a section of school land and was soon actively engaged as a progressive agriculturist and stock-raiser He subsequently disposed of his first farm but with no intention of giving up his business as he immediately began to acquire more property He developed in to a man of excellent business ability, and sound citizenship, qualities which insured his rapid success and prosperity As his business increased and his capital permitted Mr Fuller purchased land on which he believed money could be made until today he is the owner of a three thousand, six hundred acre ranch A great part of it is in pasture as he saw at an early date that the live stock industry was to become one of the greatest interests of this state where irrigation was not feasible for the intensive farming methods carried on along the Platte Most of the Fuller property lies about ten miles from Sidney, where extensive feeding is carried on, fattening stock for the great packing centers farther to the east Both his farming and cattle enterprises have yielded Mr Fuller a substantial income, he has made judicious investments which turned out well and today is rated one of the wealthy men of Cheyenne county However he was too full of life and energy to confine his business entirely to one field of endeavor and after his ranch was established on a firm foundation he entered the realms of finance buying a large block of stock in the Liberty State Bank of which he was soon made an officer Mr Fuller is one of the younger generation of bankers, progressive in his ideas and actively interested in all measures tending to advance the general welfare of the community His policy in shaping the affairs of the bank has met with general approval by the directors and stockholders, while he is held in high esteem by the many depositors and has won many additional accounts for the bank by his interest, kindness and courtesy in business circles.

On September 18, 1906, Mr. Fuller married Miss Jessie L. Forbs, a native of Webster county, Iowa, with whom he had attended school. Since locating in Sidney the Fullers have made a wide circle of acquaintances and friends. Independent in politics, Mr. Fuller has taken an active interest in political affairs but has never aspired to public office as he believes his energies and time should be devoted to his various business interests, though he is a man who aids in all movements for the improvement of the community and is a liberal supporter of all civic progress. His fraternal relations are with the Masonic order, as he is a member of the Blue Lodge.

JAMES W. McDANIEL, the capable and popular sheriff of Cheyenne county, furnishes in his career another exemplification of self-made manhood. Thrown on his own resources at the tender age of fourteen years, he has won his way to business success and public influence, and at the present time he is extending the scope of his usefulness to every line of civic affairs in the community which he has served so faithfully and capably for so many years. He well illustrates an encouraging example of success gained through the proper use of every-day abilities and opportunities, which the rising generation might well emulate on its climb up the ladder of fortune.

Mr. McDaniel is a Hoosier, born in Adams county, Indiana, January 20, 1863, the son of William and Mary (Foreman) McDaniel, the former probably a native of Ireland while the mother was an Indianian. William McDaniel was a farmer in Indiana who responded to President Lincoln's call for volunteers during the Civil war, enlisting in the Twenty-eighth Indiana Volunteers. He saw service in many of the hardest fought battles of the war laying down his all, his life, for his country that the Union might be preserved. This gallant man was shot in seven places during a skirmish at Marietta, Georgia, in 1865, and died from the wounds received during this comparatively small fight.

James was a small child of two years when his father was killed, leaving his wife to face the battle of life alone with small children to support. She subsequently married again and James realized what it was to be a step-child. He says that he never knew what it was to have a real home for he was compelled to work for his board and clothes as soon as he was able to hold farm implements or herd cattle. He was a high spirited boy, did not

like his step-father and when but fourteen years of age left home, working on farms or at whatever employment he could get that would pay for his food and but barely sufficient clothing to cover him and never enough food to satisfy his hunger or keep him warm during the severe cold of winter. As all boys do he heard of the then great west which stretched on and on beyond the Mississippi river and started out to seek what fortune might have in store for him at the age of sixteen. Soon after crossing the Missouri river he made his way to Cottonwood Falls, Kansas, where he worked on a large ranch for about five years, then married and came to Lincoln, Nebraska. For a time he was employed on a farm near the capital city, then moved into Lincoln and secured a position with the Lincoln Ice Company, which he retained until offered a city office driving the patrol wagon. Following this he entered the city service as patrolman and later as plain clothes man or detective, remaining a public official for about eight years before resigning from the force to embark in business independently, becoming a partner in a large ranch of Cheyenne county. Here he was again in the country and enjoyed the life of the open. The cattle business in which he was engaged for twelve years thrived under his able management and Mr. McDaniel became a prominent figure in agricultural and stock-raising circles of western Nebraska. He held a position of responsibility and demonstrated his worthiness for it by his capacity in discharging the many and varied duties of manager. He had taken an active part in all communal affairs, was a man of fine character, high ideals and personal honesty hard to parallel in the whole county. It was these sterling qualities that induced the voters of Cheyenne county to elect Mr. McDaniel sheriff in 1895, as his personal fearlessness was well known throughout the entire western part of the state. So well did he execute the duties of his office that he was re-elected in 1897, 1899, 1902, 1905, 1909, 1911 and 1913; was defeated in 1916 but was again elected to office in 1918 and is now serving his sixth term as the duly accredited official of the county, a record rarely to be equalled in the western section of the country. During his period in office he spent eleven months in hunting down Ernest Duligan, a cattle thief, and this trailing led him through thirty-two states before he secured his prisoner, but he got him. Mr. McDaniel was a progressive ranchman for many years, is today a progressive representative of his community who keeps well abreast

MR. AND MRS. GEORGE W. BEARD

of public affairs whether local, state or national, he is still a good judge of farm land and live-stock and on these subjects his expert advice is often sought Today he is an upstanding official of trust and enterprise of whom the citizens of Cheyenne county may well be proud

On December 17, 1885, was solemnized the marriage of Mr McDaniel and Miss Lula Pierson, a daughter of the Sunflower state They became the parents of ten children, five of whom died in infancy, the others are Chester P, a conductor on a train running from Sidney to Cheyenne, James R, who owns a transfer business in Sidney, volunteered for service in the army of the United States during the World war, being assigned to the quartermasters department at Camp Funston, Helen, Frances and Robert, all of whom are still at home

Mr McDaniel is a sturdy adherent of the Democratic party and is especially proud of the record it made during the war He is a Mason and also a member of the Modern Woodmen He is giving his children the many advantages that he himself was unable to secure in his childhood and youth, and assures them that if they desire they are to have full advantage of the wonderful educational advantages afforded by city and state to equip them for the business of life

GEORGE W BEARD, pioneer frontiersman and early settler of Kimball county, who now resides near Greeley, Colorado, has had a career of varied and interesting experiences, for he has been a resident of Nebraska for nearly half a century and has seen what was then called "The Stake Plains" of the western half of the state develop into one of the garden spots of the country In this opening up of the panhandle Mr Beard has played an important part, for he has been the owner-manager of several newspapers in which for years he published items of help to the settlers and others which told of the many advantages of this section Mr Beard is one of the gallant sons of the Union who enlisted in the northern army at the outbreak of the Civil War and so did his part in preserving the integrity of the nation and today is one of the brave, gallant figures of that great organization, the Grand Army of the Republic, the ranks of which thin so rapidly each year

George W Beard was born in Harrison county, Indiana, near the Ohio river, December 28, 1836, the son of Jesse and Charlotte Beard, the former a native of Bolling Green, Kentucky, while the mother was born in Georgia, one of the gracious and charming southern women of her day Both parents were of Scotch-Irish extraction, handing down to their children many of the admirable traits of both races They became the parents of eight children Laura, Jeremiah, John P, Samuel B, Ferdinand W who studied medicine and later became a practicing physician in Lawrence county, Indiana, and died at Vincennes, George W, of this review, Lucy F and William D, all deceased Mr Beard was reared in his native state, was given the educational advantages afforded in the new country where he lived in his childhood and youth Living near the Ohio river, which was the highway of commerce in the nineteenth century, it was but natural that he became interested in the river traffic and became a boatman Large barges were built upon which the provisions and commodities for sale were loaded and the long trip to New Orleans commenced Oftentimes all the products of the upper Ohio valley were sold before the barge reached New Orleans but the men finished the trip sold or left the barge and many times returned home overland, thus learning much about the country

In 1854 Mr Beard gave up his career on the river to enter professional life, entering the office of the *Western Argus*, at Corydon, Indiana, to learn the printer's trade and become a newspaper man Corydon was one of the early capital towns of Indiana, as the seat of government was moved there from Vincennes in 1813 and its newspapers were the leading ones of the section For five years Mr Beard was connected with the *Argus*, became an efficient printer and business man as well as learning the profession of newspaper man He was near enough to the border to be fully conversant with all phases of the slavery question and at the outbreak of the war, in 1861 enlisted in the Third Indiana Cavalry, commanded by Colonel Baker Within a short time was transferred to serve under Lieutenant-Colonel Scott Carter and later General George H Chapman Mr Beard and his troop were sent to Washington, D C, for equipment and instruction When ready for active service they were assigned to duty on the east shore of Chesapeake Bay in Maryland trying to catch blockade runners Subsequently the Third Indiana Cavalry took part in the battles of South Mountain, Antietam, the siege of Fredericksburg, Chancellorsville and many minor skirmishes before being assigned to

General Beaufort's corps to take part in the battle of Gettysburg. Mr. Beard and his comrades participated in the memorable Battle of the Wilderness, then served under General Phillip Sheridan on the march toward Richmond, destroying the railroads of the southerners and laying waste the country. It was on this march that Mr. Beard was wounded by a grape shot in his hip, which has caused him trouble ever since. He was removed to a hospital at Point Look Out, on Chesapeake Bay, where he remained until his third period of enlistment expired, was then sent to Indianapolis, Indiana, and given his honorable discharge though still walking on crutches. Mr. Beard returned to Lawrence County, Indiana, and within a short time became engaged in merchandising but soon sold his store to buy the Bloomfield *News*, which he ran six years with great success. Like so many of the returned soldiers Mr. Beard came west and in 1885 located in Kimball county; took up a homestead, proved up on it in a year and a half having a soldier's warrant and then opened a hotel in the town of Kimball. Later he engaged in the grocery business but sold out when appointed postmaster under President Harrison, serving during the Harrison administration. Mr. Beard had the honor of being elected the first judge of Kimball county, proved a most efficient official and was re-elected twice, serving in the judicial capacity three terms. His tastes and inclinations were for professional life and upon leaving the bench Mr. Beard became the owner-manager of the Kimball *Observer* which he conducted several years before disposing of the paper and moving to Central City to assume charge of the *Nonpareil*, the leading publication of that locality. Five years later he disposed of his interests in Nebraska, to locate at Las Cruces, New Mexico, near the Rio Grande and for two years was the editor of the *Republican*. Nebraska was really home to Mr. Beard and he returned to Kimball for a period before locating at Greeley where he has since resided.

In February, 1864, George Beard married Miss Angie E. Broaddeus, the daughter of Andrew and Jeanette Broaddeus, the former a native of Virginia and the latter of New York City. Six children were born to this union: Lena, deceased; A. B. Beard, a sketch of whom appears in this volume; Jennie, the wife of L. W. Bickel, of Greeley, Colorado, an early banker of Kimball county; Stannard, an attorney of Billingham, Washington; George, deceased, and two infants that died.

Mr. and Mrs. Beard have been members of the Presbyterian church for many years and active in church work. Mr. Beard has been a man of active and energetic life, ever supporting all movements for the development of every community where he has lived and is notable as one of the pioneer journalists of Kimball county and the Panhandle. He is now living retired, enjoying the sunset time, and as the shadows lengthen from the crimsoning west can look back across the years and feel that his has been a worthwhile and worthy life for he offered the greatest gift a man has— his life — for the defense of his country and later devoted his talents to newspaper work and the owner of such publication has great influence in moulding public opinion and opening a vast field of information to readers. Mr. Beard lead wisely and well.

WALTER CLARK, is not only to be ascribed pioneer distinction in Garden county, where he established his residence prior to its creation, but he also was called upon to assume the office of first sheriff of the new county. He is the owner of a valuable tract of land in the county and to this well improved cattle ranch he gives his personal supervision, besides which he is engaged also in real estate operation, with residence and business headquarters at Oshkosh.

Mr. Clark was born in Floyd county, Iowa, March 1, 1871, on his father's farm, near Charles City. He is a son of Thomas and Hannah (Smith) Clark, who were born and reared in Ireland, where their marriage was solemnized and whence they came to America as young folk, in the early '60s. Soon after their arrival in this country they established their home in Iowa, where the father devoted his attention to farm enterprise until 1886, when he came with his family to Nebraska and took up a homestead claim of a hundred and sixty acres, in Keith county. He developed a productive farm, upon which he made good improvements, and continued his activities as an agriculturist and stock-grower until he was well advanced in years. He then retired and joined his children in the vicinity of Lewellen, in what is now Garden county, where he passed the remainder of his life. He was about sixty-eight years of age at the time of his death and his wife was sixty-three years old when she passed away, both having been earnest members of the Presbyterian church, his political support having been given to the Republican party. Of the eight children five are living.

Walter Clark acquired his early educational training in the public schools of his native state and was a lad of fifteen years at the time the family came to Nebraska, where he did his part in connection with the work of his father's farm and attended school as opportunity offered. He continued his association with the work and management of the old homestead farm until he was about twenty-six years of age, when he came to what is now Garden county and purchased a quarter section of land, near Lewellen, this land having been measurably improved and also provided with irrigation facilities. He paid at the rate of fifteen dollars an acre, and the land is now conservatively valued at two hundred dollars an acre. When he instituted active operations on this ranch Garden county was still a part of Deuel county, and of this county Mr Clark was elected sheriff in the year 1906, when he removed from his ranch to Oshkosh, the county seat. He held the office of sheriff of Deuel county some years, and when, in 1910, Garden county was segregated therefrom and duly organized, with Oshkosh as the county seat, Mr Clark was elected the first sheriff of the new county, a special election having been called for the selection of the officials for the new county. Giving a most efficient and acceptable administration, Mr Clark retained the office of sheriff four consecutive terms—until January, 1919. He had refused again to become a candidate for the office in the fall election of 1918, as his private business and property interests demanded his time and attention. Alive to the enduring value of Panhandle land, Mr Clark wisely made further investments in real estate during the period of his service as sheriff, and now he owns nineteen hundred acres of excellent grazing land, all in Garden county, all fenced and constituting one of the extensive and well equipped cattle ranches of this section of the state. He rents his original farm, but to the management of his cattle ranch and other business interests he gives his personal supervision, and also maintains his residence in Oshkosh, the county seat. As a dealer in farm land and other real estate in Garden and other counties of this section of the state Mr Clark has conducted numerous transactions of important order, thereby contributing to the further advancement of the civic and industrial prosperity of the Nebraska Panhandle. His political allegiance is given to the Republican party, he is affiliated with Oshkosh Lodge, No 286, Ancient Free & Accepted Masons, and his wife is a communicant of the Protestant Episcopal church.

April 24, 1899, recorded the marriage of Mr Clark to Miss Cora Paisley, who was reared and educated in Polk county, this state. They have no children.

ROBERT A BARLOW, the cashier of the Liberty State Bank of Sidney, Cheyenne county, from the time of its organization, is a resourceful and progressive executive and has yielded primary influence in shaping the policy of this substantial financial institution. During practically his entire business career Mr Barlow has been associated with banking enterprises, having shown a special constructive talent for finance which he has used for the upbuilding of each bank with which he has been associated.

Robert Barlow was born in Canton, Illinois, August 29, 1883, the son of Frank M and Jennie (Carter) Barlow, both natives of Illinois, who were reared, educated married and spent their younger days in that state. In 1900, the family came to Nebraska, locating in Webster county on a farm not far from Red Cloud, where Frank Barlow and his wife still reside. Since locating in this state the father has been actively engaged in agricultural pursuits, in which he is meeting with well deserved success. He is a Democrat in politics and while he takes no active part in political affairs is intensely interested in all civic movements for the benefit of his community, and while making a success of his own affairs and enterprises, has not been remiss in duties of citizenship.

Robert Barlow acquired his early education in the public schools of Illinois and after the family removed to Nebraska, when he was about sixteen years of age he continued his studies in the country school and subsequently took a course in the business college at Grand Island as he had early determined to enter business life. After graduating from the college in 1904, he entered the American Bank of Sidney in May of that year holding from the first the responsible position of cashier of the bank. Mr Barlow played a large part in the upbuilding of the substantial business of this institution and remained cashier until November, 1918, when he resigned to become one of the organizers and heavy stockholders of the new bank which was chartered as the Liberty State Bank thus commemorating by its name the great struggle for the liberty of the world in which our country took such an important part. The original capital of the institution was $25,000, and it opened its doors for business April 5, 1919. The business has grown by leaps and bounds and the story of its initial success reads more like a fairy story of old than a modern prosaic business statement, for on August 22, 1919, the report of the bank—then little over four months old—shows that there were $172,000 on deposit. It is needless to say that the bank has made phenominal progress even in this country of great and speedy success. Since becoming associated with these representative banking houses of Nebraska, Mr. Barlow has been a potent factor in furthering their advancement in financial circles and it has been under his careful and progressive regime that the Liberty State Bank has made such a wonderful advancement in the volume of business and has already become one of the leading financial institutions of western Nebraska.

Though a young man, Mr. Barlow is recognized as a natural born financier; he is a broad-gauged and liberal citizen and has always shown vital interest in public affairs whether of his community, state or the nation, and during the war was one of the hardest workers in all movements for the furtherance of its prosecution in his town and county. Like so many of the younger generation, who today are holding the positions of responsibility that formerly were held only by men of advanced years, Mr. Barlow is an independent thinker on all subjects as he keeps abreast of all questions of the day whether financial, political or domestic and as a result is an independent in politics, drawing no strict party line when he casts his vote as his influence is ever with the man best fitted to administer public office.

In 1907, Mr. Barlow married Miss Grace Hart of Sidney, a native of this city who is the mother of four children: Virginia, Gretchen, Marjorie and Robert A., Jr.

Mr. Barlow has entered actively into the business and social life of Sidney since he first came here and has a host of good friends. His fraternal affiliations are with the Modern Woodmen, the Ancient Order of United Workmen and the Woodmen of the World. He has won many depositors to his bank by his policy of courtesy, his sympathy and general good fellowship.

W. H. BUETTNER. — Prominent among the energetic and progressive agriculturists and business men of the younger generation whose activities have been centered in Cheyenne county, one deserving of more than passing mention is W. H. Buettner. During his comparatively short business career he has become identified with the best men of his community and has demonstrated the possession of sound ability and practical knowledge, qualities developed through experience and training. He is a Badger by nativity having been born on his father's farm near Chilton, Wisconsin, November 9, 1881, and is a son of Joseph and Lucy (Bolz) Buettner. The father also was a native of Wisconsin. He was reared and educated in that northern state by beautiful Lake Michigan and enlisted in the Union army during the Civil war. Following the close of hostilities he returned to his home and was actively engaged there in farming pursuits for some time then came west, locating on a farm in Platte county, Nebraska, where he spent the rest of his life. Joseph Buettner was a hale, hearty man whose outdoor activities inured him to hard work and lived to be old, passing away after attaining the psalmist's span of three score years and ten, as he was seventy-seven at the time of his death. There were six children in the Buettner family, three boys and three girls, all of whom grew up in Wisconsin and Nebraska on their father's farm. They were given all the advantages obtainable in an educational way in the excellent public schools of both states, as their parents were anxious that they have good start in life. Mr. Buettner passed his childhood and early youth on the home farm, early acquiring valuable knowledge of farm business, the best crops to put in, the best time and methods of harvesting, and while a young man was capable of carrying on farm industry independently, due to his practical experience at home. Almost a decade ago he established himself in the Panhandle, renting a three hundred and twenty acre tract in Cheyenne county and demonstrated that his success was assured in his chosen vocation as he at once became one of the prominent·men of the community, who by able management, modern methods and good hard work and a plenty of it, was soon recognized as a capable producer of all farm products and a stock-man who thoroughly understood his business.

Independent in ideas, he is also independent in politics and gives his influence to the man most worthy and best fitted to fill an office. With his family he is a member of the Roman Catholic Church.

Mr. Buettner married Miss Josephine Korth, the daughter of John Korth of Platte county, where Mrs. Buettner was born. The

John I Filer

father was a successful farmer of his locality where he is still residing engaged in the management of his farm

JOHN I FILER, who is one of Kimball county's representative citizens, substantial farmers and highly respected men, was brought to Nebraska in childhood and has spent his life in this beautiful western country He was born in Fulton county, Illinois, April 26, 1876, the first born in a family of six children His parents were Joshua O and Lusetta (Morgan) Filer, natives of Illinois Their other children were as follows Joseph L, Asa L and George, all of whom are living, Myrtle, who is the wife of Fred McBride; and Bessie, who married John Carston, both of whom are deceased

In 1880 the parents of John I Filer moved to Nebraska, settling at Gibbon, in Buffalo county, remaining there seven years They then moved to Brady Island, from there going to a place on Pumpkin Creek, then in Cheyenne county but now Kimball There the father homesteaded a quarter section timber claim and proved up on 320 acres as his homestead He then went to Colorado and died there in 1899 but the mother of Mr Filer survives, making her home with her youngest son, George

John I Filer attended school during boyhood and then worked for the Brown Iliff Cattle Company until 1896, when he went into the cattle business for himself, branding three calves the first year He continued in this business and in recent years has branded as many as two hundred and fifty calves At the present time he has seven hundred head of cattle, owns nine hundred acres of land and leases a large acreage, carrying on both dry and irrigated farming, also a cattle ranch of about seventeen hundred acres in Carbon county, Wyoming Mr Filer has additional business interests as he is a stockholder and a director in the Bank of Kimball, at Kimball, Nebraska, and is also a stockholder and a member of the board of directors of the Farmers State Bank at Dix, Nebraska

In March, 1895, Mr Filer was united in marriage to Miss Eleanor Lingelbach, and they have had five children, three daughters, Maria, Helen and Mildred, and two sons, the latter of whom died in infancy Mr Filer has never felt inclined toward political activity but has always been known as a sound, sensible, upright citizen, ever ready to do his full duty in a public way He and wife are members of the Methodist Episcopal church

CLAUDE E GRISHAM—The power of the press is recognized throughout the length and breadth of our fair land, and today we are fully aware of the power yielded by newspapers of the nation One of the strong resourceful young men who is doing his part in connection with the civic and industrial development of Cheyenne county is Claude E Grisham, the owner and editor of the Lodge Pole Express, one of the progressive publications of the Panhandle, which under his careful guidance plays a large part in shaping opinion in this section and in formulating the policy of the voters of the Lodge Pole District

Mr Grisham is a Missourian, born in Lowndes Wayne county, September 25 1892, the son of Thomas and Ellen Grisham, both of whom were natives of the state Thomas Grisham was accidently killed in 1920 and his widow, Ellen lives in the southeast part of the state Claude Grisham was reared in his native state and there received the advantages of the excellent public schools After reaching manhood's estate, he determined to establish himself in business and with the idea of an independent career he went to St Louis, believing that there were many opportunities for a young and ambitious man in a large city After looking around for some time he entered a newspaper office where he began to learn the practical side of journalism as he first became a printer, which today, with the highly technical work required on the presses which print a whole newspaper at one time, color cuts as well, required time, concentration and many long hours of hard work, but he persevered and within a short period became an efficient pressman As there seemed to be no opening in this city, at that time, Mr Grisham decided to start west but going no farther than the Panhandle as he stopped in Scottsbluff where he entered the office of the Star-Herald as foreman of the composing room, being an expert printer For three years he held this position and then resigned to accept a similar position on the Republican force in that city After four years in that city on the North Platte he left to accept a position as foreman of the York New Teller, but remained there but four months before he was offered a still better opening in Potter He leased the Potter Review for one year in 1918 He then came to Lodge Pole on March 1, 1919, and purchased the Lodge Pole Express a weekly paper, formerly owned by James C Wolfe who had published it for thirty-three years

Under Mr Grisham's efficient guidance and the new policies inaugurated by him the paper has increased in volume, the circulation is much greater than he believed it would become in so short a time, and today is one of the well known newspapers of the Panhandle.

In connection with the paper he has established a job printing business with which he is meeting with well earned and deserved success.

The southeastern part of Cheyenne county had long been in need of an up-to-date printing establishment and the citizens of this section owe many thanks to this progressive business man who is now supplying this long felt want.

Mr. Grisham is largely the architect of his own fortune, for while his parents gave him the educational advantages they could, he has made his way since graduating from the high school, and today is one of the promising younger generation of the "Knights of the Quill" who are spreading the gospel of improvement, modern methods and intensive farming throughout the middle west. By his paper he enables the busy agriculturist and stock-raisers to keep in touch with the many important questions of the day. That he has won success at such an early date with the greater part of his life still before him, augurs well for him and his family in both material and intellectual happiness.

June 21, 1914, was solemnized the marriage of Claude E. Grisham and Ida Mae Osborne, who is also a Missourian, being the daughter of Mr. and Mrs. John Osborne, prosperous farmers of Fredericktown, Madison county, Missouri. Mr. Grisham has never sought public office though he takes keen and active part in all civic public affairs pertaining to the welfare of Lodge Pole and his status in the community is that of a reliable, constructive and public spirited citizen. He is an independent in politics, allowing no strictly drawn party lines to prevent him casting his vote for the best man. While living in Potter, he served as City Clerk of that village, and ably discharged the duties of that office. He and his wife are members of the Presbyterian Church.

MARSHALL L. BIGLER, a well known farmer and stock-raiser of Morrill county and a most worthy citizen in all that statement implies, has been a resident of Nebraska since boyhood and obtained his education at Hastings in Adams county.

Marshall L. Bigler was born in La Salle county, Illinois, February 19, 1873, a son of Logan and Etta (Ferguson) Bigler. The father was born in Pennsylvania and the mother in Canada. She still resides in Illinois

but the father is deceased. During his active years he was a railroad man.

It so fell about that Marshall L. Bigler came to Nebraska in 1886, and this state has been his chosen home ever since. He grew up a reliable, industrious, intelligent youth and for a time worked in a bank at Hastings. In 1904, he acquired a farm of a hundred and sixty acres in Scottsbluff county, on which he carried on general farming but mainly stockraising, until 1911, when he came to Morrill county and acquired another farm of a quarter section, situated just north of Bayard. He now owns three hundred and twenty acres and makes a specialty of fine livestock.

On October 4, 1899, Mr. Bigler was united in marriage to Miss Bessie Nowlan, who was born in La Salle county, Illinois, October 25, 1874. She is a daughter of Patrick and Mary (Perry) Nowlan, natives of Ireland, who came to the United States in 1841, and settled first in Knox county, Illinois. The father of Mrs. Bigler was a merchant but is now deceased. The mother is living and resides at Hastings, Nebraska. Mr. and Mrs. Bigler have three children: Lawrence, Francis and John. As a citizen interested in the public welfare, Mr. Bigler favors many public-spirited movements and encourages worthy enterprises. He is secretary of the Bayard school board, and in politics is a Republican.

GEORGE MAX HANNA, whose enterprise along agricultural lines has resulted in profit to himself, has also brought credit on Morrill county, because increased agricultural production is one of the greatest needs of the country. A hard-working, intelligent farmer and stockraiser of today, giving close attention to his own business no matter what turmoils may agitate the country, is nobly performing a duty that deserves recognition and praise.

George Max Hanna is a native son of Morrill county, born August 3, 1888. His parents were John and Martha Jane (Bower) Hanna, the latter of whom lives in Clark county, Illinois. The father, John Hanna, was born in Illinois. He came to Nebraska in 1884, and homesteaded a quarter section and also took a tree claim and was one of the earliest settlers in Morrill county. He met with many hardships during his earlier years, but persevered and became independent. His death occurred October 13, 1905, at the age of fifty-two years.

In the public schools of Morrill county, George M. Hanna secured his education, and on his father's land his farm training. In 1912 he homesteaded for himself in Morrill county and now owns six hundred and forty acres of fine land He runs about a hundred head of cattle, and his well cultivated land make his home surroundings comfortable and convenient

Mr Hanna was married to Miss Gertrude L Imus, who was born in Greeley county, Nebraska, October 21, 1895, a daughter of Leslie and Mary (Horn) Imus, the latter of whom was born in Greeley county and still survives The father of Mrs Hanna was born in Iowa and when he came first to Nebraska settled in Greeley county, but later came to Morrill county and homesteaded north of Bayard, where he and wife still live Mr and Mrs Hanna are members of the Christian Church They have had three sons Max Elwood, born June 8, 1916, John Raymond, born April 17, 1918, was drowned August 23, 1919, and Elden Leslie, born April 10, 1920 Mr Hanna is a good citizen, interested in all that concerns the best interests of Morrill county, but he has never taken any very active part in politics He votes according to his own judgment.

PRICE P WAITMAN, deceased, was one of Morrill county's best known citizens For some time before his death he had practically retired but for many years was extremely active here as a raiser of cattle, horses and mules, his land affording fine pasturage He came with the settlers of 1888, and passed through the hardships that brought discouragement and ruin to many of the pioneers of that period, escaping many of these himself through good management and former experience

Price P Waitman was born in Henry county, Iowa, March 4, 1853 His parents were Washington and Susanna Waitman, who were born in Ohio The father was a farmer in Iowa at the time of his death, when but thirty-five years old The mother lived to the unusual age of ninety-one years, passing away in Iowa Neither parents ever came to Nebraska Mr Waitman was only six months old when his parents moved to Benton county, Iowa, where he attended the country schools in boyhood and remained on the farm In 1888, he homesteaded in Cheyenne county, now Morrill, shipping his goods to Sidney and freighting the forty miles from there to his homestead on section thirty-two As soon as he had proved up on his homestead, Mr Waitman proved up on a tree claim Later he owned two thousand acres, all in Redington precinct While he raised some corn for many years, he gave but slight attention to small grain, devoting himself mainly to livestock, in which line he met with much success His homestead was well improved and he said that for thirty-one years since he came here, there has never been twenty-four hours of that time that some one of his family has not been on the place

Mr Waitman was married in Iowa, in 1877, to Miss Mary Kelty, who was born in Ohio, and the following children were born to them William, who lives in Morrill county, Pearl, who is Mrs Limberg, lives at Bridgeport, Dale W, who lives at home, Vernon, who lives in Morrill county, Mrs Grace Wyatt, who lives in Banner county, Mrs Bessie Wyatt who lives in Banner county, Mrs Hazel Eaton, who lives in Wyoming, Bryan, who enlisted in the United States navy at Denver, Colorado, July 28, 1918, is now at home, Freda, who is Mrs Ray Morrison, lives at Denver, Colorado and John, at home Mr Waitman was a Democrat in politics but has never sought public office He belongs to the Modern Woodmen of America Mr Waitman died at the Methodist Hospital in Omaha, September 19, 1919, from the effects of an operation

GEORGE R SICKELS, who belongs to that group of early settlers who came to Morrill county in 1885, and through hardships and crop and stock losses, courageously persevered and is now able to enjoy the fruits of hope and toil Mr Sickels has long been a representative citizen of his section of the county and took a useful part in its development He was born in Sussex county, New Jersey, July 9, 1854, and is a son of Jacob and Elizabeth Sickels, neither of whom ever came to Nebraska

George R Sickels was five years old when his parents moved from New Jersey to Iowa, in which state he grew to manhood as a farmer In 1880, he was married in Taylor county, Iowa, to Eva I. Drake, who was born and reared in New Jersey, a daughter of Owen and Martha Drake, who later were farming people in Iowa Mr and Mrs Sickels have six children Homer W S, who lives in Banner county, Guy W, who lives near his father in Morrill county, Glenn G, who lives in Morrill county, Warren L, who lives in New Mexico, Mrs Clara M Skinner, who lives in Morrill county, and Roland O, who assists his father

When Mr Sickels and his family came to this section they traveled in a covered wagon and brought household goods with them

Their first home was a dugout with a sod roof The range was free then and the settlers with unfenced land had little chance to get ahead with crops while cattle and antelope ran wild over the country On many occasions Mr Sickels hunted antelope and deer and it was fortunate at that time as provisions were hard to secure and money was hard to earn Many times has Mr Sickels hauled a load of wood after cutting it, a distance of forty miles to Sidney and selling it for two and a half dollars, and has hauled post to Julesburg, Colorado Mr. Sickels still lives on his homestead and also has a Kinkaid claim, eight hundred acres in all, of which he devotes eighty acres to alfalfa He has mainly devoted himself to general farming and has done well He has substantial buildings and a productive orchard and a grove he set out makes an attractive background for the comfortable farmhouse In politics he is a Republican, and he is a member of the order of Modern Woodmen of America

HERMAN RINNE — Among the many good farmers highly respected citizens of Morrill county, none can show more evidence of prosperity than Herman Rinne, who owns a large body of land here and carries on general farming and stockraising He is a native of Nebraska and was born at Steinauer, in Pawnee county, January 7, 1875 His parents were Henry and Annie (Kamen) Rinne, who were born in Germany and died in Nebraska Of their four children Herman is the only survivor

Herman Rinne attended the country schools near his fathers farm and remained at home until prepared to undertake farming on his own account He remained in Pawnee county until 1916, and still own three hundred and sixty acres of fine land there When he came to Morrill county he bought more than fifteen hundred acres of ranch land and is an extensive breeder of white face cattle and Percheron horses He makes use of modern farm machinery and all his farm industries are carried on with the good judgment that many years of practical experience have given him There are few farms in the county better improved than Mr Rinne's, all the farm buildings being substantial and well kept, the family enjoying many modern comforts and conveniences in the farm home

Mr Rinne was married to Mary Albers, who was born in Germany and was brought to the United States when two years old Mr and Mrs Rinne have children as follows

Ernest, Otto, Walter, Ida and Wilber Mr Rinne and his family belong to the Evangelical church A hard working man all his life, Mr Rinne has never been anxious to serve in public office He keeps well informed on all public questions however, and cast and intelligent vote according to his own judgment

CHARLES J CADWELL, a representative of one of the early settler families of Saunders county, Nebraska, has been a resident of Morrill county since 1905, owns a large body of valuable land but rents out the greater part of it Mr Cadwell is giving a part of his attention to duties pertaining to several public offices

Charles J Cadwell was born at Ashland, Nebraska, March 8, 1880 His parents are John Thomas and Sarah L (Gilbert) Cadwell, the father being a native of Ohio and the mother of Illinois For many years they have been highly respected resident of Saunders county Charles J Cadwell grew up on his father's farm and attended the public schools In May, 1905, he came to Morrill county and homesteaded He now owns a hundred acres of land, ninety-one acres being ditched and very valuable While Mr Cadwell operated his land himself and grew alfalfa, beets, oats and corn with great success and continues to operate a small body of land, he is not interested as formerly

Mr Cadwell was married to Miss Grace M Wireman, who died at the age of twenty-seven years, leaving one son Francis Wireman Her parents were F E and Mary (Rarick) Wireman, of Toulon, Illinois, who never came to Nebraska Mrs Cadwell was a member of the Methodist Episcopal church and Mr Cadwell belongs to the same He has always been an active citizen, although an independent voter, and has been called on several times to accept public office At present he is moderator of school district number fifty-eight, and also is serving in the office of road overseer in town twenty-two, showing much efficiency in both capacities

FRANK DEAL, an enterprising and successful general farmer and stockraiser of Morrill county, was born in Grundy county, Iowa, March 28, 1883 His parents are Milton and Jane Deal, both of whom were born in Pennsylvania They came to Nebraska in 1887, purchased farm land and still live in Aurora, Nebraska They have a family of fourteen children, the fourth in order of birth being Frank It is not often that the biographer has

Mr. and Mrs. John Ewbank

the pleasant task of recording that in so large a family, not only the parents survive but all of the children, the eldest being aged forty-two years and the youngest twenty-one It speaks well for sturdy original stock and wholesome living

Frank Deal was four years old when his parents settled in Hamilton county, Nebraska He grew up on his father's farm and attended the public schools, afterward following the life of farmer and stockman in Hamilton county until 1914, when he came to Morrill county He secured a homestead of eighty acres situated near Bayard, which he afterward sold to advantage, then purchased three hundred acres, ninety-one of which are irrigated and the remainder is pasture He has done remarkably well raising and feeding cattle and hogs, turning off sixty head of cattle and a hundred head of hogs annually He also raises grain and potatoes and other standard vegetables, has some fruit and gives more or less attention to poultry, in fact all the industries that go to carrying on of a fine modern farm, may be found here

In 1906, at Aurora, Nebraska, Frank Deal was united in marriage to Miss Ocie Castle, a daughter of Frank and Margaret Castle, the former of whom was born in Iowa and the latter in North Carolina, both being deceased Mr and Mrs Deal have four children, two daughters and two sons Mildred, Howard, Morris and Fay Mr and Mrs Deal are members of the Baptist church A good citizen, in the sense of aiding all worthy enterprises in the county to the extent of his ability, Mr Deal has never found it necessary to identify himself with any particular political party and he votes independently according to his own judgment Both as a business man and as a neighbor Mr Deal is highly esteemed

JOHN E EWBANK, a man of large possessions and business and social importance in Kimball county, has been a resident for twenty years, during which time he has been worthy in his citizenship and prominent in many lines of useful activity He is known in other sections of Nebraska as well as in adjacent states, and wherever he has lived, enterprise and practical progressiveness have marked his way

John E Ewbank was born in Yorkshire, England, August 15, 1856, the fourth in a family of five children born to Thomas and Mary (Falshaw) Ewbank, the others being as follows Christopher, Robert, Richard and Margaret All three brothers are deceased The

sister survives and is the widow of John Metcalfe and the mother of four children Mr Ewbank's parents died in England He obtained a fair education and worked in various ways in his native land prior to 1881, when, in company with William Percival, Ernest Stamp and William Metcalfe, he came to the United States He located first at Bismark, in North Dakota, but found the climate unexpectedly cold causing him to remove to Wisconsin where he worked for eight months in a meat market, in Madison county, learning in the meanwhile of the advantages offered in Nebraska He came to this state in 1883 and homesteaded a tree claim near Ansley, in Custer county, subsequently selling the same and locating eight miles east, in Laramie county, Wyoming There he remained about five years, then went to Colorado, where he purchased a homstead, and bought ten thousand acres of ranch land and engaged extensively in the cattle and sheep business

In 1899 Mr Ewbank came to Kimball county and embarked in the same business together with engaging in general farming He now owns 2,240 acres of fine land, operates it judiciously and raises some of the best stock that is marketed from Kimball county His improvements have brought his property to a high standard of value, and the family home, with its modern comforts and conveniences, compares favorably with a city residence and has the added advantages of beautiful country surroundings

In 1889 Mr Ewbank was united in marriage to Miss Alice A Wilkerson, who is a daughter of John Wilkerson, of Pine Bluff, Wyoming, and they have the following children Mary, who is the wife of Guy Graves, a merchant at Kimball, and they have two children, Elizabeth Ann and Virginia Jane, Isabel, who is the wife of Herbert Linn, Margaret, who is the wife of Arthur E Atkins, a sketch of whose father appears in this work, John and Robert, who attend to their father's ranch, Avelyn and Helen, who reside at home, and Richard, who is the manager of the home farm Mr Ewbank and his family are members of the Episcopal church attending services at Kimball Since coming here twenty years ago, Mr Ewbank has witnessed many remarkable changes in the development of this section and with his progressive ideas, sensible and practical, he has done a useful part in the same and may take justifiable pride in the fact that he has reared sons who are capable and willing to carry on the business he founded He has additional interests aside from those mentioned, being a director of the Kimball Lumber and Supply Company, and a stockholder and one of the board of directors of the Citizens Bank at Kimball

R F DURNAL, who carries on extensive agricultural operations in Morrill county, is one of the solid, dependable citizens and successful business men of this section Mr Durnal was born in Ohio, December 1, 1863, and is as son of Samuel and Sarah (Johnson) Durnal

Samuel Durnal and his wife were born in Ohio They were farming people there and when they removed to Illinois in 1869, settled on farm land in Clark county, and in the course of time Mr Durnal became the owner of large farm properties His death occurred at the age of sixty-six years His widow still survives and resides in Clark county Of the family of four children, two sons and two daughters, R F of this review was the first born

R F Durnal atended the public schools in Clark county, Illinois, and grew up on the home farm When he determined to seek a permanent home for himself in the western country, his first intention was to locate in Kansas, and with wagon and team crossed the country until he reached his objective point in the Sunflower state He considered a number of locations, but finding none to altogether satisfy him, drove on into Cheyenne county, now Morrill, homesteaded and still lives on the place which he first pre-empted and secured a tree claim Mr Durnal now owns a large tract of land and devotes it to general farming and stockraising His home place is well improved and his progressive and systematic methods have made his industrial undertakings very profitable Mr Durnal is financially interested in the Bayard Bank

Mr Durnal was married to Miss Mary Semiller, and they have four children Harry, who resides in Morrill county , Fred, who also resides in Morrill county, both sons being married, Mrs Alta Ericson, who lives at Bayard, Nebraska, and Hazel, who lives with her parents Mr Durnal is a member of the order of Modern Woodmen As a good and intelligent citizen, he has always taken some interest in politics and casts his vote with the Republican party

EDWARD M QUINN, one of the younger generation of ranchers in Morrill county, has set a pace in his agricultural undertakings that few will exceed His enterprise and energy have placed him among the prosperous men of the county, and his sterling traits of character have made him universally respected. Mr Quinn was born June 2, 1892, in the state of Iowa but has spent almost his entire life in Nebraska, as his parents moved here when he was young Edward M Quinn grew up on his father's farm in Cheyenne county and had educational advantages there He remained at home assisting his father and, in 1916, came to Morrill county and in preparation for ranching on a large scale bought two thousand acres of cleared land He does a limited amount of general farming but devotes the larger part of his attention to raising cattle and horses, and at the present time has on his ranch seven hundred hear of cattle and fifty head of horses

In February, 1914, Mr Quinn was united in marriage to Miss Hazel Thornburg, who was born in Cheyenne county, Nebraska, November 7, 1897, and they have one child, Vivian The parents of Mrs Quinn are George A and Minnie (Hensen) Thornburg, the former of whom was born in Missouri and the latter in Nebraska They now live comfortably retired, at Potter, Nebraska, Mr. Thornburg having been a farmer and ranchman in that neighborhood for a number of years Mr and Mrs Quinn are members of the Roman Catholic church

W H RHOADES —For more than forty years W H Rhoades has been a resident of the state of Nebraska, coming here with his parents in boyhood, and few residents of Morrill county can recall early conditions in this section and vicinity in a more convincing or interesting way Life in 1919 on an improved, irrigated, farm, is very different from what it was in Nebraska forty years ago Science and progress combined with pioneer enterprise, have brought about remarkable changes all over the state, and Morrill county in particular has been fortunate in attracting agriculturists of experience and good judgment, and one of these is W H Rhoades, whose farm is situated on section 6 town 20

W H Rhoades was born in central Illinois, July 26, 1864, the youngest of six children born to Thomas J and Lena (O'Neal) Rhoades The mother was born in Illinois and died in Nebraska, January 10, 1891, aged fifty-nine years The father was born in Johnson county, Indiana In his earlier years he worked at coopering but later was a farmer in Illinois, and afterward in Nebraska He came here and homesteaded in Kearney county, May 15, 1878, and lived on his farm until his death, on February 5, 1903, being seventy-eight years old. His six children all survive. Mary Ellen, who is the wife of Nelson Basye, Cora, who is the wife of George Basye, Joseph, who lives at Alliance, William, who lives in Colorado, and Charles and W. H., both of whom are farmers in Morrill county. W. H. Rhoades accompanied his parents to Nebraska and assisted his father in Kearney county until 1887, when he came to Morrill county. He now owns a hundred and sixty acres of fine land, irrigated and substantially improved, and devotes himself to general farming.

Mr. Rhoades was married to Miss Lena Albershardt, who was born July 27, 1874, in Delaware. Her parents were born and died in Germany, where her father followed the trade of cigarmaker. Mr. and Mrs. Rhoades have had the following children. Grace, who is the wife of Fred Durnal, of Morrill county, Charles, who is a farmer in this county, Mattie, who is the wife of Samuel McCormick, Moses, Louie, Walter, Ida, Kermit and Hazel, all of whom live at home, and three who are deceased. Mr. Rhoades is a Republican in politics, and is a widely known and respected citizen.

JOSEPH C. CHRISTENSEN, whose business is general farming and ranching, is one of the substantial men of Morrill county. He was born February 6, 1879, in South Dakota. His parents were Nels and Hannah Hansen Christensen, who were born in Wisconsin. They had a family of four sons and one daughter.

Joseph C. Christensen grew up on his father's pioneer farm in South Dakota. His father died at the age of forty-six years and his mother came to Morrill county, Nebraska and homesteaded the place on which Mr. Christensen now lives. He took a Kinkaid homestead in 1911, six hundred and forty acres, and owns additionally several large tracts near the homestead. During the first three years that he lived in Morrill county, he worked at ditch contracting and put in nearly all the ditches in this section of the county.

Mr. Christensen married a Miss McKenna, who was born in Ireland and came to the United States when eight years old. They have no children.

LORENZO FLOWER, who is one of Scottsbluff county's representative men and an early settler, was born in New Hampshire, October 23, 1864, and is a son of A. C. and Mary Flower. Mr. Flower has one brother, Louis C.

When the parents of Mr. Flower came to Nebraska they lived in Madison county until 1887. Then removal was made to Cheyenne, now Scottsbluff county, but one year later Mr. Flower returned to Madison county and engaged in farming there for two years, removing then to Box Butte county, where he remained in the farm and stock business near Alliance, for nine years. Mr. Flower then returned to Scottsbluff county and pre-empted land and still owns his homestead right. He has made other investments here and at present owns three hundred and forty-four acres of irrigated land, a large part of it being as productive as any in this part of the state. Mr. Flower is a careful, methodical farmer and good business man, and is well and favorably known all through this section.

In Madison county, Nebraska, Mr. Flower was married to Miss Eva Cunningham, who was born in Indiana and was one year old when her parents brought her to Nebraska. Her father, Oliver Cunningham, homesteaded in Madison county and both parents still live there. Her mother, Mary (Mangle) Cunningham was born seventy-five years ago, in the state of New York. Mr. and Mrs. Flower have four children. Charles, who lives on the homestead, Mrs. Mary Willard, who lives in Montana, Mrs. Blanche Bradley who lives in Scottsbluff county, and William, who is a rancher in Wyoming. All the children have enjoyed educational advantages and are useful and respected members of the communities in which they live. Mr. and Mrs. Flower are members of the Methodist Episcopal church. For many years he has been a member of the order of Odd Fellows.

CLARENCE E. ROBERTS, who is one of Morrill county's successful farmers and dependable citizens, was born in the western part of Iowa September 15, 1872. He is a son of Thomas and Mary (Bateman) Roberts, both of whom were born in Canada. The father was nine years old when he was brought across the boundary line between the Dominion of Canada and the United States, and grew up in Iowa. He was a general farmer there until he retired from active life, and now resides in Morrill county, being in his eighty-fourth year. The mother died in Iowa at the age of sixty-nine years.

Clarence E. Roberts was educated in Iowa and was a farmer there until 1887 when he came to Nebraska and homesteaded in Morrill county Like other pioneer settlers of that date, he met with many hardships and some losses, but he kept up his courage, as is very likely to be the way with the descendents of Canadians, and in the course of time changes came about that made his one unproductive homestead as valuable land as can be found in the county Mr Roberts now owns an entire half section, all of which is irrigated The fine improvements he placed here himself General farming and some stockraising keep Mr Roberts a busy man

Mr Roberts was united in marriage to a daughter of Stephen Smith, and they have three children Ila, Elmer and Velma Mr and Mrs Roberts are members of the church of the Latter Day Saints In political sentiment he is a Democrat

CHARLES SCHNIEDER, who is a substantial and highly respected citizen of Morrill county, has devoted himself closely to agricultural pursuits since his school days ended Since coming to this county he has purchased and improved a large body of land and is proving that this section of the state is particularly well adapted to stockraising

Charles Schnieder was born in Grant county, Wisconsin, October 31, 1877, and is a son of John and Susan (Grass) Schnieder John Schnieder was born in Germany He came to the United States when twenty-one years old and became a worthy citizen After locating in Missouri he followed merchandising for a time and also dealt in livestock He was married to Susan Grass, who was born in Missouri, and they moved to Grant county, Wisconsin, where he engaged in general farming until his death His widow still resides in Grant county

Charles Schnieder remained on the home farm and had considerable farm experience in Wisconsin before he came to Morrill county in 1909 Here he purchased a quarter section of unimproved land and immediately began the work of development In this section as in every other pioneer territory the early settlers have had to endure some hardships but Mr Schnieder had expected them and never became discouraged He made the best of things accepted a few losses of crops and stock, but in the main has been exceedingly successful He carries on general farming and stock-raising, at the present time having about a hundred head of fine cattle fifty head of hogs and the same number of lambs Mr Schnieder has always been a steady worker and is making his work count for much in the way of production

In Minnesota, in 1909, Mr Schnieder was married to Miss Etta Lipski, who was born in Grant county, Wisconsin, November 6, 1886, a daughter of Henry and Mary (Whitesh) Lipski The father of Mrs Schnieder was and early homesteader in Morrill county He met an accidental death, being killed by a lightning stroke on his farm The mother lives in Montana Mr and Mrs Schnieder have two sons Henry and John The family belongs to the Methodist Episcopal church Mr Schnieder votes the Republican ticket but has never desired a public office, although well qualified for many

HENRY NIEHUS, one of Morrill county's hard-working, successful farmers and cattlemen, was well known all through this section of the country and was very well thought of He was born in Germany, came to the United States in 1871, and died at Redington, Nebraska, aged fifty-eight years His parents were Claus and Jerusha (Dulling) Niehus, both of whom died in Germany

Henry Niehus was born December 24, 1855, hence was only sixteen year old when he came to America He was strong and willing and after he reached Grand Island, Nebraska, found no difficulty in finding employment on farms in Hall county, and as he was careful with his money, before long had accumulated enough to buy some land for himself and did very well as a general farmer In 1881, he went to Wyoming and for the next nine years was connected with ranches there In 1890, he came to Morrill county and homesteaded a quarter section, which he improved, and later bought more land until he owned four hundred and eighty acres

In 1879, Mr Niehus was married at Grand Island, to Dora Foellmer, who was born in Germany July 25, 1860 Her parents were Werner and Drusilla (Deitrick) Foellmer natives of Germany, who came to the United States in 1870, and homesteaded near Grand Island, Nebraska The mother of Mrs Niehus still survives, being in her eighty-fifth year, but the father died when aged sixty-five years Five children were born to Mr and Mrs Niehus, as follows Anna, who is the wife of R H Willis, of Bridgeport, Hilda and Joseph, both of whom are married and live at Redington, Dessie, who is the wife of William Barton, of Bayard, Nebraska, and Clause, who is deceased

VERNON WAITMAN, who is a representative citizen of Morrill county and an extensive raiser of fine cattle, is well known over

Mr. and Mrs. Frank J. Bellows

the county in which he has spent the greater part of his life. He was born in Benton county, Iowa, October 4, 1882, and is a son of P. P. and Minnie (Kelty) Waitman. Mr. Waitman's father lived retired on his homestead in Morrill county before his death and the mother still resides there. The father was born in Iowa and the mother in Ohio, and they had the following children: William who is a farmer in Morrill county, Mrs. Pearl Limburg, who lives at Bridgeport, Nebraska, Dale W., who lives on the homestead, Vernon, whose well improved property is situated on section twenty-one, town nineteen, Mrs. Grace Wyatt and Mrs. Bessie Wyatt, both of whom live in Banner county, Mrs. Hazel Eaton, who lives in Wyoming, Byron, who served in the United States navy, is at home, Freda, now Mrs. Ray Morrison of Denver, and John at home.

Vernon Waitman accompanied his parents to Nebraska when they removed here from Iowa in 1888, and remained on the homestead in Morrill county assisting his father, until 1908 when he homesteaded for himself. Mr. Waitman owns six hundred and forty acres of land and devotes the greater part of his attention to stockraising, usually turning off two hundred head of cattle annually. He gives the most of his time to his business as politics have never interested him beyond the responsibilities of good citizenship, although his fellow citizens have given evidence of their appreciation of his sterling character by electing him treasurer of the school board.

Mr. Waitman was united in marriage to Miss Alice Ridge, who was born in Benton county, Iowa, March 6, 1886. Her people came early to morrill county and homesteaded and now live retired at Bridgeport. Mr. and Mrs. Waitman have two children, Timothy and Rex. Mr. Waitman casts an independent vote.

FRANK J. BELLOWS, county judge of Kimball county, Nebraska, has filled this important judicial office with dignity and efficiency since he took his seat on the bench in January, 1915. Judge Bellows is an old resident of the county, owning his homestead for thirty-three years, and during that time has won such respect and personal esteem, that his election to the judgeship twice came about with no opposing candidate, he being the universal choice.

Judge Bellows was born in Cass county, Michigan, January 4, 1854, and is a son of James C. and Mary E. (Osborn) Bellows, the former of whom was born in New York and the latter in Indiana. The grandparents were James and Hannah (Smith) Bellows, and Elijah and Sarah Osborn, all people of many sturdy virtues, now long since passed away. Of his parents' family of children, Frank Jefferson was the second in order of birth, the others being Louis, who died in infancy, Charles P., who is deceased, was a farmer, Elmer E., who is also deceased, Emma J., whose married name is Mead, resides at Kimball, and Carleton J., who is a farmer in Michigan.

Despite the fact that Judge Bellows is a man of learning and thorough knowledge of the law, his education was obtained in a country school near his father's farm in Michigan. He had few of the advantages now considered essential for one preparing for public position, but reading, a habit of observation, sound judgment and association with his fellow men in many a crisis have been great teachers, and his decisions lack nothing a college degree could have given them. He came west in 1886, remained in Kansas for six months, then came to Nebraska and homesteaded, and ever since has been one of the county's useful and respected citizens.

In Michigan, in December, 1880, Judge Bellows was united in marriage to Miss Katie A. Renninger, daughter of Charles and Barbara (Swinehart) Renninger, who were old settlers in Elkhart county, Indiana. Judge Bellows is a Thirty-second degree Mason and a Shriner, his local membership being in Kimball lodge No. 294 A. F. & A. M., Kimball Nebraska. He was first elected county judge in 1914 and is still serving.

ROBERT H. FAIRBAIRN, Jr. who is extensively and successfully engaged in farming and stockraising in Morrill county, has not been a continuous resident since he first came here, many years ago, but has never lost interest in this section, and though property belonging to his mother, felt somewhat bound here before he invested on his own account. He is well known and highly respected citizen of Morrill county.

Robert H. Fairbairn, Jr. was born in Green Lake county, Wisconsin, October 23, 1867. He bears his father's name, who was also a native of Wisconsin and a well known minister of the Congregational Church in that state. After moving to New Hampton Iowa he was editor of the New Hampton Courier for thirty-five years. The mother of Mr. Fairbairn, Mrs. Lucy (Beshee) Fairbairn, was

born in Wisconsin and now resides at Redington, in Morrill county, Nebraska

Mr Fairbairn was reared in Wisconsin and
Iowa and had excellent school advantages
He accompanied his mother to Nebraska in
1893, who homesteaded here, and he lived on
her homestead and did some farming and improving, then went to Chicago, where he was
employed until 1909, when he once more returned to Nebraska and again took charge of
the homestead He has devoted himself quite
closely since then to general farming and
stockraising, owning eleven hundred and sixty
acres and employing two hundred and twenty-five acres in dry farming

Mr Fairbairn was married to Augusta Becker, who was born in Buffalo county, Wisconsin, October 11, 1834 Her father was born
in Germany, came to the United States in boyhood, was a soldier in the Civil war and afterward a farmer in Wisconsin, where he died in
1893, the mother of Mrs Fairbairn also being
deceased They have three daughters Ruth,
who is the wife of Earl Perkins, of Bridgeport, Grace, who is the wife of Charles W
Newkirk, a farmer in Morrill county, and
Ora, who is the wife of George Newkirk
Mrs Fairbairn is a member of the Christian
Church While Mr Fairbairn has never accepted a public office, having no desire for
political honors, he is an active, interested and
useful citizen in private life He is a Republican

THOMAS B LANE, Jr, a representative
citizen and successful general farmer of Morrill county, owns well improved land situated
on section 22, town 22, to which he came in
1914 Mr Lane was born in South Dakota,
December 9, 1885, and is the son of Thomas
and Carrie (Foster) Lane Both parents were
born in Illinois and the father carried on farming there for a number of years before moving
to South Dakota Later he came to Valley
county, Nebraska, and both he and his wife
now live there

Thomas B Lane, Jr, obtained his education in the public schools and has been a farmer all his life In 1914, he came to Morrill
county, Nebraska, from South Dakota, and in
the same year homesteaded eighty acres of
unimproved land Considering how short a
time has elapsed since then, Mr Lane has
made remarkable progress in the way of developing and improving and he now has forty
acres irrigated He feels well repaid by its
added yield for all it has cost him and probably the other forty will soon be equally productive

In 1911, in South Dakota, Mr Lane was
united in marriage to Miss Annie Smith, who
was born in that state and is a daughter of
Conrad and Mary Smith, natives of Wisconsin The father of Mrs Lane was a farmer
and both parents are deceased Mr and Mrs
Lane have three children Vernon, Dorothy
and Orville Mr Lane is an independent
voter

CHARLES H SMITH, prosperous farmer
and respected citizen of Morrill county, has
been a resident of Nebraska for nineteen
years He has been a general farmer all his
life, first in Wisconsin and later in Banner and
Morrill counties, Nebraska, and his long experience has served to make him a pretty fair
judge of what it means to engage time, energy
and money in agricultural industries in order
to make them profitable He was born on his
father's farm in Green county, Wisconsin,
June 1, 1872

The parents of Mr Smith were James H
and Polly (Baily) Smith Both were born in
Wisconsin where their parents had been pioneer settlers For many years the father followed general farming in Green county, Wisconsin, where he was a man of more or less
local importance In 1900, he became impressed with the opportunities offered in Nebraska and with his family came to Banner
county and homesteaded He continued a
farmer during the rest of his active life and
lived near Gering in Scottsbluff county at the
time of his retirement and subsequent removal
to California He still resides there, being in
his seventy-second year, but the mother of
Charles H died in the California home in
1919, aged sixty-four years

Charles H Smith was reared in Wisconsin
and obtained a public school education He
accompanied his parents to Banner county,
Nebraska, locating first near Gothenburg in
Dawson county, from which place he came to
Morrill county and homesteaded At that
time his hundred and sixty acres was wild
prairie and for the first six years he found
little compensation for all the hard work he
put on the land, the lack of moisture being a
continual setback Then came irrigation and
with the lifegiving water the really fertile
soil was able to respond to cultivation and Mr
Smith now has one of the best farms in the
county He has a hundred and forty acres in
his home place and all of this tract has been
well improved His buildings are commodious
and substantial and on every side may be observed provision made for the carrying on of
large industries in the best possible way

Mr Smith was married to Miss Lizzie Howard, who was born in Kentucky, December 15, 1885 Her parents, James and Eliza (Minshall) Howard, were also born in Kentucky In early manhood Mr Howard worked on the railroad near Sidney, Nebraska, but later homesteaded in Banner county and still resides on his farm there The mother of Mrs Smith is deceased Mr and Mrs Smith have had children as follows Lelia, deceased, Mary, who is the wife of Ernest Foster, a farmer near McGrew, Nebraska, Henry, who assists his father on the home place, Arthur, deceased, and George, Roy, Luella and Leslie, all of whom reside at home Mrs Smith is a member of the Methodist Episcopal Church In politics Mr Smith is a sound Republican Although interested in all matters concerning the welfare of the county, he has never been willing to accept a public office, contenting himself with setting a good example of sensible, practical citizenship

CHARLES R CHURCHILL —From every state in the Union young men have come to Nebraska, and Kansas has many worthy representatives here who have bettered their fortunes through the opportunities she has offered There was a time when certain sections of the state were called too dry to prove profitable for farming purposes, but now that the great irrigation projects have succeeded, no land in the country could be more desirable This was the judgment of Charles R Churchill when he invested in Scottsbluff county land and he is well satisfied with the decision he made

Charles R Churchill was born in Phillips county, Kansas, September 21, 1876 He is the son of Henry and Sarah (Brown) Churchill, the former of whom was born eighty-four years ago, in Ohio, and the latter, seventy-nine years ago, in Illinois They still reside on their farm in Kansas Mr Churchill obtained his education in his native state and grew up on the home farm In 1906 he came to Scottsbluff county, Nebraska, and homesteaded one hundred and sixty acres and has lived here ever since, now owning two irrigated farms, both of which he has improved, aggregating three hundred and eighty-nine acres Mr Churchill has been very successful in his agricultural undertakings and is numbered with the substantial men of the county In 1892, Mr Churchill was united in marriage to Miss Olive L Dickson, who was born in Kansas, October 12, 1882, and died at the age of thirty-six years Her parents were Frank and Lucy (Bruner) Dickson, the latter

of whom still lives in Kansas, but the former died in 1918 To Mr and Mrs Churchill the following children were born Edith, Maxine, Clyde, Richard, Bernadine, Arthur and Alice Mr Churchill is an independent voter and has never accepted any political office He belongs to the order of Modern Woodmen

GUSTAVE WIKSTON, a well known and highly respected resident of Morrill county, resides on his valuable farm in the Bayard district, situated on section 5, town 21, to which he came in 1900 Mr Wikston was born on his father's farm in Sweden, February 1, 1858 His parents were Peter and Mary Wikston, neither of whom ever left their native land They were honest, hardworking people who brought their children up to be frugal and industrious

While Mr Wikston had no educational advantages in his boyhood to compare with those he has been able to give to his own children, he had some schooling and was a well informed youth of nineteen years when he came to the United States As he had had farm training in his own land, it was on a farm that he sought and found employment in Howard county, Nebraska, and continued to work there, in the vicinity of St Paul, from 1877 until 1888, when he moved into Box Butte county and homesteaded From there he came to Morrill county in 1900 and shortly afterward bought a quarter section of unimproved land It is an interesting story that Mr Wikston can tell of what hard work it was to get his land properly developed and substantially improved, and of the wonderful advantage that irrigation has been His land is favorably situated for crop growing and this property investment in Morrill county has made him financially independent He has practically retired from active labor, his eldest son having taken over the management of the farm

At St Paul Nebraska, in 1882 Mr Wikston was married to Miss Matilda Olson, who was born in Sweden October 12, 1859 Her parents were Olaf and Anna Olson, both of whom spent their lives in Sweden, where the father was a general farmer Mr and Mrs Wikston have had children as follows Elmer, who operates the home farm in Morrill county, Oscar, who is a farmer in Morrill county, Ira, Thomas and John, all of whom live at home, and Edna, who is deceased Mr Wikston and his family are members of the United Brethren Church at Bayard For many years he has been an American citizen and early identified himself with the Democratic party in politics Wherever he has lived in the great country to which he came in youth, he has found good friends, for he has been peaceful, helpful and neighborly, ever endeavoring to do his full duty

COLE HUNT, whose thorough farm methods and general enterprise are making the old Hunt homestead one of the best farm properties in Morrill county, was born in eastern Nebraska, April 13, 1895, and has spent his life in his native state. With the good judgment that marks many young men in modern days, he has chosen agriculture as his life work and is devoting his best energies to the further development of the excellent property left by his father.

The parents of Mr Hunt were John and Lillie (Gilmore) Hunt, the former of whom was born in Ohio, June 9, 1848, and the latter in York county, Nebraska, June 3, 1864. In 1875, John Hunt came to eastern Nebraska where he bought land and followed farming for some years, then moved to Box Butte county and homesteaded and the family lived there for eight years. In 1898, Mr Hunt saw what he considered better opportunities in Morrill county, came here and bought two hundred and forty acres of land which, at that time, was entirely unimproved. He continued the practical development of his farm until the close of his life. He was widely known and highly respected. To John Hunt and his wife the following children were born. Eva, who is the wife of Elmer Bennett, of Minneapolis, Minnesota; Lillie, who is the wife of Alexander Underwood, of Box Butte county; George, who resides at Spokane, Washington; Susie, who is the wife of Arthur Jones, of Grand Island Nebraska; John E. who was a soldier in the American Expeditionary Force in Europe during the World War and was in the army of occupation in Germany; Omar, who is deceased; Cole, who operates the home farm as mentioned above; Nellie, who is the wife of Leslie Allen a farmer in Morrill county; and Nettie, who is deceased. The mother of the above family still resides on the home farm. She is a member of the Methodist Episcopal Church.

Cole Hunt took charge of the farm for his mother soon after completing his education. He is a level-headed serious-minded young man, a hard worker and close thinker. To him farming is not merely an occupation but a business that is deserving of a man's best efforts. He has been very successful in his work here, has the land all irrigated makes every acre return a profit. At present he is not particularly active in politics.

HARRY G GREEN, who is a general farmer and stockraiser in Kimball county, is a highly respected, reliable citizen, and a good influence in his neighborhood. He was born in Maryland, February 14, 1870. His parents, Thomas and Rebecca Green, came early to Maryland and located in Harford county. The mother died in 1886, and the father in 1893. He followed farming and blacksmithing. Of his children, Harry Giles was the fourth born, the others being as follows: George and William, both of whom are deceased; Thomas, who is connected with the creamery at Kimball; Robert, who is a farmer, is also in the monument business at Stewartstown, York county, Pennsylvania; and Margaret, who lives in the city of Baltimore.

Mr Green obtained his education in Maryland, from which state, when fifteen years old, he came to Cass county, Nebraska. In the following spring he began farming with a cousin and continued a farmer in Cass and Kearney counties until 1913, when he came to Kimball county. Here he bought a half section of land, situated two and a half miles west and one and a half miles south of Bushnell, where he has been engaged in general farming and stockraising ever since, carrying about thirty head of stock a season. Mr Green owns also a tract of five acres in Columbia Heights, a choice residential suburb of Lincoln, which ultimately will be a part of the city.

In 1904, at Plattsmouth, Nebraska, Mr Green was married to Miss Levonia Bell Vanscoyor, who is a daughter of Owen Kinney and Rosa Ann (Lucas) Vanscoyor. The father of Mrs Green was born in an eastern state and the mother was reared in Kansas. Both parents died at Lewisville, Nebraska. Mrs Green has the following brothers: Charles H, who is a stonemason living at Lewisville; John F, who is in business at Lincoln; William H, who is a carpenter at Lewisville; LeRoy, who has farm interests in Colorado lives at Dix, Nebraska, and Darwin J, who is a railroad elevator builder. Mr and Mrs Green had one child born to them but it did not survive infancy. They are members of the Methodist Episcopal Church. Mr Green has never taken a prominent part in politics, his idea of good citizenship being the faithful carrying out of everyday duties and neighborly helpfulness and good will.

Mr. and Mrs. F. O. Wisner

RAY A WISNER, editor and proprietor of the Bayard Transcript, of Bayard, Nebraska, has a fine property which he built up from the bottom The paper came into his hands practically worthless a little over a decade ago, and now it is known and valued all over the county, is an influential political factor and recognized advertising medium along progressive lines, while its large subscription list is constantly growing By inheritance and training Mr Wisner is a newspaper man

Ray A Wisner was born at Kilbourn, Adam county, Wisconsin, May 26, 1883 He is a son of Francis O Wisner, who was a pioneer in the North Platte Valley and was one of the first and ablest newspaper men in Nebraska, and is well remembered for his journalistic enterprises and his determination to make a newspaper what he believed it ought to be His standard was high but he maintained it as long as he lived When he came to Bayard, the nearest railroads touched Alliance, forty miles away, and Sidney, sixty miles distant Journalism was his chosen work and he had already started the first newspaper in Dakota, before the division of the state was made He found at Bayard, which village was to be his home, a small newspaper which, for six months had been printed under difficulties, in a sod house This paper he bought in the hope of developing it into a great journal and conducted it as the Bayard Transcript, along the line of his ambition, almost until the end of his life He was a man of intellectual strength but he was ahead of his times in thought and action, and never lived to see his journalistic hopes realized, nor to know that in his son there would finally be a worthy successor

Ray A Wisner attended the public schools of Bayard, Gering and Hastings He was brought up in the printing office learned the trade and has been connected with the gathering and distributing of news all his business life When his father bought a newspaper at Oshkosh, Nebraska, he disposed of the Transcript, but the new owner failed to maintain the standard of journalism Mr Wisner had set, and in 1907, when Ray A Wisner took charge of the Transcript it was necessary to bring about a complete reorganization Mr Wisner proceeded with considerable vigor and now has a newspaper plant that is a credit to his enterprise and to the city After plans of his own he had a substantial brick building erected which houses his newspaper and job presses, together with the modern machinery that belongs to a first class plant and with this equipment he is doing a very large amount of business Mr Wisner has quite a reputation also as an editorial writer and does not hesitate to call attention in his columns to needed local improvements while, at the same time, he discusses calmly and intelligently the great problems in public affairs that concern everyone

In 1913, Mr Wisner was united in marriage to Miss Gertrude Clifton, who was born at Ewing, Nebraska, and they have one daughter, Gwendolyn, who was born June 24, 1914 Mrs Wisner is a daughter of Rev C W Clifton, pastor of the United Brethren Church at Elgin, Nebraska

In politics Mr Wisner is a staunch Republican He has been an active citizen in many ways and has served as city clerk In his busy life he has not found much time for recreation, but he enjoys his fraternal memberships in the Modern Woodmen, the Woodmen of the World and the Odd Fellows being past grand in the last named organization Mr Wisner is in the Wisner Investment Company at Bayard

ISAAC ROUSH — In times of general unrest, the thoughtful citizen is sometimes led to consider how the average substantial men of his acquaintance have attained their comfortable competencies and their positions of public confidence This friendly inspection generally leads back to hard industrial activity in youth, and in most cases to continuance of the same until financial independence has been secured Prominent as an example among Kimball county's substantial and honored citizens is Isaac Roush, county treasurer of Kimball county

Isaac Roush was born in Snyder county, Pennsylvania, in August, 1863 His parents were George and Caroline (Page) Roush, the latter of whom died when Isaac was ten years old She was a member of the Lutheran church Isaac had one brother, Frederick, who died on his farm in 1888 For his second wife the father married Elizabeth Bolich, and the three children live in Pennsylvania, where the father died in 1896

The son of a hard-working father, Isaac Roush early learned to be useful, but had country school opportunities until he was thirteen years of age His home was in a mining district and it was no unusual matter for boys, even at that early age, to go to work in the iron ore mines in which work he spent two years Afterward he worked as a farm hand for three years, but as wages were better in the mining district, he went back to mine work for a time, although he had made up his mind to seek employment in a western state, having settled in Indiana, of which he had heard much He had labored incessantly to the age of twenty years, yet, when he started westward had been able to save just enough to pay his railroad fare to Elkhart county, Indiana He immediately went to work on a farm near Bristol and remained there two years, in the meantime making plans to move still further westward, working hard to earn the means to carry them out In 1886 he left Bristol with the good wishes of many friends he had made there following him, and came to Kearney county, Nebraska He was not looking for an easy job and during the next two years worked hard on a stock ranch, and also in a lumber yard for a time

It was in the spring of 1890 that Mr Roush came to Kimball county, which has been his home ever since Here he worked in a lumber yard for three years and then began clerking in the general store of L A Schaeffer until 1898 As his acquaintance widened, his business ability was further recognized and ere long public confidence was evidenced by his being mentioned for public office and his appointment by President McKinley as postmaster of Kimball met with universal approval Mr Roush served in that office for seventeen years, retiring then for a season of rest after his many years of strenuous private and public effort Very soon, however, he was called back to public life, being elected county treasurer of Kimball county in 1916, on the Republican ticket As a resident of Kimball he has given encouragement to many worthy enterprises He is a member of the Methodist Episcopal church

CLYDE E MEGLEMRE, who has numerous and valuable interests in Morrill county, is one of Bridgeport's respected citizens, where he is well known in business and also in official life He has been a hard worker all his life and through his industry has gained well earned financial independence He was born in Harrison county, Missouri, November 15, 1870

The parents of Clyde E Meglemre were John E and Sarah A (Richardson) Meglemre, the former of whom was born in Indiana and the latter in Virginia Of their seven children Clyde E is the third of the five survivors The mother and two sisters of the family are deceased but the father survives and lives at Oxford Nebraska He served three years in the Union army during the

Civil war, being a member of Company D, Twenty-third Missouri Volunteer Infantry, and was captured by the enemy but was exchanged only to fall into the enemy's hands again at Atlanta, after which he was confined in the dreadful prison pen at Andersonville, Georgia, for nine months At the close of the war Mr Meglemre became a farmer, a vocation he followed during most of his active life

Clyde E Meglemre went to school in Harlan county, Nebraska, then went to work on a farm and in 1888, came to Cheyenne county, where his mother had homesteaded He continued a farmer for many years, in fact still overlooks his irrigated farm of a hundred and sixty acres, on which he raises cattle and horses After coming to Bridgeport, in 1907, he embarked in additional enterprises, in 1908 beginning to work for the Standard Oil Company and continuing to the present time, and also started a draying business Mr Meglemre is a practical man and was the first to arrange for the delivery of ice, a business in which he has the whole field at Bridgeport, having made ample provision for supplying this necessity of life

In 1896, Mr Meglemre was united in marriage to Mary Rew, who was born in Wisconsin, and they have had seven children as follows Cecil, who works for his father as his right hand man, Treva, who is the wife of Frank Richards, who is in the oil business at Bayard, Sadie, Dela, Clyde Jr, and Vera, who are at home, and one who is deceased The family belongs to the Church of the Latter Day Saints In politics Mr Meglemre is a Republican He has led too busy a life to have found time to give to public office to any extent, but he served as one of the town's most efficient marshals for a period of three years He may well be numbered with the representative men of Morrill county

FRED R LINDBERG, one of Bridgeport's most substantial citizens, for many years has been extensively interested in raising cattle and horses, and has also been identified with banking enterprises Mr Lindberg has won his way to fortune and prominence through his own efforts and his whole career from boyhood to the present, may advantageously be studied by other youths who find themselves forced to start out early with neither capital nor influential friends

Fred R Lindberg is a native of Sweden, born in 1866, and brought to the United States the following year by his parents, Abraham and Anna Louisa (Boxtrom) Lindberg They

Isaac Roush

settled in Missouri and his father was a laborer in a stone quarry and on railroads for some years but later bought forty acres of land in Missouri and died on that farm. Three of his four children still survive. Annie, the wife of Andrew Feruquist, Peter, of Republic county, Kansas, and Fred R., of Bridgeport, Nebraska. After the father's death the mother moved to Republic county, Kansas, where she subsequently was married to Mr Lindberg, and two children were born to that marriage. Ellen, who is the wife of Jonas Johnson, a Kansas farmer, and Joseph L. who lives on the old home farm in Kansas. Mr Lindberg's parents were members of the Lutheran Church. His mother died in Republic county, Kansas.

Mr Lindberg attended the country schools for awhile, first in Missouri and later in Kansas, but as soon as old enough to make his services valuable, went to work on farms. It was hard work, for at that time in that locality, farmers made use of little labor-saving machinery, but he kept sturdily on and remained with one employer for six years. In 1888, he came to Nebraska and decided to remain in this state, shortly afterward homesteading in what was then old Cheyenne but now Morrill county. He remained on that homestead for fifteen years, then went to work by the month, on the Lyon Brothers' ranch and for three years of the five he remained there, had sole charge. When the owner of the ranch died, Mr Lindberg bought both the ranch and the horses, keeping the property until 1909, when he sold to advantage. In the meantime, through good business judgment, Mr Lindberg had acquired other property and still owns a ranch of six thousand acres near Reading, where he feeds and grazes three hundred head of cattle and a hundred and fifty head of horses. After coming to Bridgeport he became interested in business enterprises here and assisted in the organization of the Bridgeport Bank in 1901, of which institution he was the first president.

In February, 1906, Mr Lindberg was united in marriage to Miss Lillian Pearl Waitman, who was born in Iowa, and they have one son, Garland Frederick, who was born January 30, 1909. The family attends the Presbyterian Church. Mr Lindberg has been an active citizen for many years and has been prominent in the Democratic party. He has served honestly and efficiently in different local offices, as assessor and school director, and for six years was a member of the board of county commissioners of old Cheyenne county and afterward served for seven years on the Morrill county board, this service covering a period of great public responsibility. For many years he has belonged to the order of Odd Fellows, also to the Modern Woodmen, and in the latter organization has passed through all the chairs of the local body.

HEYWARD G LEAVITT, who may, perhaps, be called a founder of the sugar beet industry in Nebraska, and also the inspiration and financial support of the earliest irrigation projects in Scottsbluff county, is pre-eminently a man of action and his life for two decades past has been devoted to such useful effort that it amounts to public beneficence.

Heyward C Leavitt was born in New York City, March 22, 1861. His parents were Henry S and Martha A (Young) Leavitt, both of whom were born at Brooklyn, New York. The father died in New York City in 1904, at the age of seventy-eight, while the mother still resides there. Of their seven children five are living, Heyward G being the only one making his home in Nebraska. A sister, Emma, is the wife of William Fellows Morgan, who is in the cold storage business in New York City. For fifteen years he was president of the Y M C A there and Mrs Morgan is president of the W C T U.

On both sides of the family Mr Leavitt came from solid financial ancestry. His paternal grandfather, David Leavitt, who was a native of Goshen, Connecticut, was president of the American Express National Bank for many years in New York City, and his maternal grandfather, Henry Young, was also a banker there and the financier that lent the money to complete the dredging of the Sanitary Canal, Chicago. He was the builder of the first gas plants in New York and Chicago. Henry S Leavitt, father of Hayward G, was a banker in New York City during the greater part of his life. He was a Democrat in his political views, and both he and wife belonged to the Episcopal Church.

Hayward G Leavitt was fortunate in his early environment and educational advantages of an excellent character were his while growing up. He prepared for college under tutors and in private schools, then entered Harvard and was graduated in 1882. Two years later he was graduated from the Columbia Law School and entered upon the practice of his profession in his native city, making a specialty of patent law. Some years later he turned his energies in another direction, becoming to some extent interested in his grandfather's gas plant business and after attending to the installing of such plants in many eastern cities, came west to Chicago, in the same business, and subesquently to Grand Island, Nebraska Many things contributed to Mr Leavitt's then becoming deeply interested in the beet sugar industry He began at the bottom, distributing beet seed to farmers in Hall county, Nebraska, where he bought a farm for experiment taught farmers how to make the tests and during the eight years he lived on this farm designed different implements for the extraction of sugar, and implement men from all over the country visited him to learn of their value

In 1900, Mr Leavitt organized and financed the Standard Beet Sugar Company and erected a factory in the village of Leavitt, where the earlier operations of the company were carried on before the plant was moved to Scottsbluff No less interested was Mr Leavitt in the great subject of irrigation In 1902, he came to this county and after a thorough inspection of the valley, assumed charge of the Farmers Irrigation project, then in the hands of a receiver As previously indicated, Mr Leavitt has never been an idle dreamer He has "the vision" and with it has the sound judgment that insure his dreams coming true His first practical move was the purchasing of thirty-six thousand acres of land Four years later he organized the Tri-State Land Company which he financed in the construction of the canal at Scottsbluff At that time Scottsbluff and Bridgeport had a hundred and fifty families and representatives of only eight of these are here now Mr Leavitt at one time owned the controlling interest in the Winters Creek Irrigation Company was concerned in developing the plants on the Republic river and the Pathfinder Dam, and in all progressive enterprises that have done so much for this section in a substantial way, Mr Leavitt has assisted by the expenditure of time, money and legal advice At present he is looking mainly after his extensive agricultural interests holding large land leases, although in earlier days he leased at one time as large a tract as thirteen thousand acres

In 1899 Mr Leavitt was united in marriage to Miss Alvina Weller who was born in Saxony, Germany, a daughter of Conrad Weiler, who was an early settler near Grand Island and an extensive farmer and stockman Mr and Mrs Leavitt have one daughter and three sons Martha, who has just completed a three years course at Radcliffe College, Cambridge, Massachusetts, Heyward Lathrop, who has just entered Harvard College, David Henry,

who is attending school at Omaha, and George Conrad, who is in school at Scottsbluff Mr Leavitt still preserves his Greek letter fraternity memberships and also belongs to the Harvard A D club, of which he was president while in college He and his family are members of the Episcopal Church at Omaha

FREDERICK ALEXANDER, to whose energy, adaptability and progressiveness Scottsbluff is largely indebted, has been identified with business enterprises and city development here since 1902, and perhaps no citizen could be named who cherishes a deeper sentiment of civic pride, or is actuated by more unselfish motives Mr Alexander was the first mayor of Scottsbluff and it was under his able administration of that office that the present admirable public utilities were installed At present he is secretary and general manager of The Platte Valley Telephone Company

Frederick Alexander was born at Norwalk, Connecticut, May 30, 1875, and is the only surviving son of Louis F and Helen Louise (Curtis) Alexander, both of whom were born near Hartford, Connecticut The mother of Mr Alexander died in New York City His father lives retired at Scottsbluff For a number of years he had been in the life insurance business in New York City prior to coming here quite recently The Alexanders are of Scotch-Irish descent, but the family is old in Connecticut the paternal grandfather, Louis Robert Alexander, having been born and passed his life in that state On the maternal side, the name of Curtis is equally well known in the "land of steady habits," Curtisville, a flourishing town perpetuating the name of Frederick Curtis, Mr Alexander's grandfather, who was a manufacturer of silver ware

When Frederick Alexander was a boy, the family frequently spent a part of the year on his father's estate in Florida There he went to school and remembers earning his first money by herding a flock of sheep On account of the prevailing malaria on the Florida plantation, it was thought best for young Frederick to leave there and he was sixteen years old when he went to New Mexico, where he remained until 1900 He had become interested in several small telephone companies in New Mexico, but in the above year began to desire a wider field for his business energies, and after disposing of his interests in New Mexico, came to Nebraska He resided first at Gering, but came to Scottsbluff to make his home, in 1902, and put in the first telephone in Scottsbluff county From that beginning an

PETRUS PETERSON AND FAMILY

immense business has been developed In April, 1902, the Platte Valley Telephone Company being incorporated with a capital of $50,000, E. H. Price, of Whittier, California, became president, and Mr. Alexander secretary and general manager. The assets of the company reach three hundred and twenty-five thousand dollars, and they operate a hundred and fifty miles through the valley, from Bridgeport, Nebraska, to Guernsey, Wyoming, and serve twenty-two centers in Wyoming and Nebraska, including Guernsey, Heitville, Sunrise, Single and Torrington, Wyoming, and Morrill, Mitchell, Gering, Beard, Minatare, Melbeta and Scottsbluff.

At Clayton, New Mexico, in 1900, Mr. Alexander was united in marriage to Miss Catherine W. Jost, who was born in Missouri, a daughter of John Jost, a farmer near Kansas City, Missouri. The parents of Mrs. Alexander were born in Germany, came young to the United States and were married here, and now reside in California. Mrs. Alexander is a lady of education and charm. They have one child, Louise Roberta, who was born in October, 1911, and is attending school.

Mr. Alexander is a Republican, trained in this party faith by his honored father, but in no sense has he ever been a politician, and has never asked for a vote or sought for an office. Nevertheless he impressed his fellow citizens so favorably that he was elected the first mayor and re-elected and subsequently served a third term, his thorough business administration of affairs resulting, as mentioned above, in the installation here of the water, sewerage and light plants. Mr. Alexander is somewhat prominent in Masonry. A man of travel and broad-minded citizenship, he is a very entertaining conversationalist.

PETRUS PETERSON, who is in the real estate and insurance business at Dix, Nebraska, is also a progressive farmer and substantial citizen of Kimball county, is widely known through his official association with agricultural organizations at Dix. Mr. Peterson came to Nebraska in 1915 and few men have been equally successful in a material way in so short a time, or have more entirely secured the confidence of their fellow citizens in their trustworthiness.

Petrus Peterson was born in Denmark, June 14, 1879, a son of Christian and Mary Peterson. Mr. Peterson has two brothers and one sister, namely: Henry, who is a farmer near Belgrade Nebraska, Chris, who is a farmer near Millarton, North Dakota, and Anna, the wife of Peter Olsen, a farmer near Boomer,

Iowa. The parents came to the United States and settled in western Iowa where they were farming people, both dying in 1914. Petrus Peterson attended the country schools and assisted on the home farm until he was twenty-seven years old, when he embarked in a general store business. In 1915 he sold his store and came to Nebraska, buying 160 acres of land in Kimball county, which he has increased to 800 acres. He raises some stock but gives his main attention to grain farming, his yield in 1918 being 6,000 bushels which he doubled in 1919. Much of the work of the farm is done by tractors.

In 1903 Mr. Peterson was united in marriage to Miss Annina Jensen, whose parents died in Denmark, after which she came to the United States and made her home with a sister in Iowa. Mr. and Mrs. Peterson have four children, namely: Elsie, Chris, Henry and Donald, all of whom are attending school.

Mr. Peterson, as a good citizen, has made his influence felt in his section of the county, for he has taken an active interest in all movements for public welfare. He is particularly interested in the public schools and is serving as treasurer of the high school district. He is a stockholder and secretary of the Farmers' Shipping Association. Fraternally Mr. Peterson is a Mason, belongs to Danish Brotherhood at Council Bluffs, Iowa, and to the Odd Fellows lodge at Honey Creek, Iowa.

FREDERICK H. ROBERTS, who has played an unusually active part in the development of some of the most important public utilities of Scottsbluff, is officially identified with large enterprises at other points, and throughout a considerable portion of the west, is known as a man of ample fortune. Mr. Roberts is yet scarcely in middle life and twenty-six years ago he was working as a factory boy at ten cents an hour. His business success, however, has not been achieved through any spectacular methods but by the old-fashioned path of patient, steady industry, helped in his case, by ambition and a quick understanding.

Frederick H. Roberts was born at Winterset, Iowa, August 13, 1877. His parents are Hugh M. and Cordelia M. (Bowers) Roberts, the former of whom was born at Racine, Wisconsin, and the latter in Pennsylvania. They were married at Marshalltown, Iowa, and now live retired and highly respected, at Norfolk, Nebraska. They have two sons, Samuel R. and Frederick H. The former lives at Hamilton, Montana, where he has charge of the agricultural department of the Great Western

Sugar Company Hugh M Roberts is a veteran of the Civil war, having served during the last six months, enlisting as soon as his age permitted his acceptance as a soldier He has been active in the G A R post at Norfolk, Nebraska, and he and wife belong to the Methodist Episcopal Church there. When he came first to Nebraska he homesteaded in Brown county, lived on his farm for eight years and for one year on another tract near Ainsworth In 1891, the family home was established at Norfolk

After his school period was over, Frederick H. Roberts went to work in a small grocery house at Norfolk, where his duties were those of a general clerk, then found employment out-of-doors, cultivating sugar beets, later secured a job herding cattle at six dollars a month, and then accepted a place as electrical helper in the sugar factory He continued in the factory from 1893 until 1907, making rapid progress and became superintendent of the plant at Sterling, Colorado, and later became interested in the factory financially His electrical training in the meantime had borne fruit and when he came to Scottsbluff in 1911, he established the C & R Electric Company that supplied Gering and Scottsbluff, and in 1913 he bought a partnership interest, which he retained until March, 1916, when he sold it to the Inter-Mountain Railway Light & Power Company Mr Roberts is president of an investment company at Scottsbluff, is president of the light plant at Riverton, Wyoming, and is financially interested in the electric plant at Loyal, Wyoming His business sagacity has also been shown in the purchase of rich farming areas, and he owns valuable land in Scottsbluff county

In 1905, Mr Roberts was united in marriage to Miss Clara Runge, who was born at Bridgeport, Connecticut, a daughter of Herman and Lena (Schriner) Runge, both of whom were born in Germany, came early to the United States and were married at Bridgeport The mother of Mrs Roberts is deceased but the father survives and, in association with his one son, W M H Runge, is engaged in the hardware business at Denver Mr and Mrs Roberts have six children, Morris, born in 1906, Esther, born in 1908, Hugh, born in 1911, Carl, born in 1912, Gretchen, born in 1913, and Carroll, born in 1914 Mr Roberts and his pleasant family have a beautiful home at Scottsbluff and they are people of social importance They belong to the Presbyterian Church, are foremost in all general charitable movements, and both Mr and Mrs Roberts were very active during the World War in Red Cross work, Mr Roberts being president of the local Red Cross board In politics he is a Republican, takes a hearty interest in everything pertaining to the city's welfare and at present is a member of the school board He is a Thirty-second degree Mason and a Shriner, belongs to the Knights of Pythias and the Elks and is past chancellor commander of the Knights of Pythias Personally Mr Roberts is a man of genial presence and he not only commands respect from business and casual acquaintances, but wins their friendship as well

HARRY T BOWEN, who has been very active in the business affairs of Scottsbluff for over a decade, and a leader also in civic matters, is the main factor in an enterprise of great importance carried on as the Bowen Investment Company Mr Bowen was born at Beacon, in Mahaska county, Iowa, January 8, 1873

Mr Bowen's paternal ancestors came from Wales and the maternal, from England His father, John W Bowen, was born in Wales, a son of John Bowen who was a mine worker in Wales before he came to the United States He died on a farm in Iowa The mother of Mr Bowen, Ellen (Burdess) Bowen, was born in England and died in January, 1902 Her father, John Burdess, brought his family to the United States and located in Mahaska county, Iowa, where he was a mine worker near Oskaloosa for a number of years John W Bowen was reared in Iowa and early in the Civil war enlisted as a soldier in Company E, Fifteenth Iowa Infantry, in which he served three years and was wounded at Atlanta In 1885, he homesteaded in Gage county, Nebraska, residing at Lincoln from 1887 to 1900, during which time he was in the oil business From 1892 until 1898, he was city clerk His present place of residence is Portland, Oregon Of his six children, Harry T is the second in order of birth, the others being Albert L, who is in the livestock business at Denver, William E, who is associated in business with his brother Albert L, Herbert J, who is a farmer and stock feeder near Gering, Minnie L, who is a widow, resides at Scottsbluff, and Lola, who is the wife of B J Jellison, of Scottsbluff The above family was reared in the Christian Church The father is prominent in G A R circles and is a Mason, a member of the order of Woodmen, and is also an Odd Fellow

Harry T Bowen obtained his education in the graded schools of Lincoln He entered the business world in that city as a bookkeeper, later worked on text books, and for five years was bookkeeper for the city treasurer, then went into the First National Bank of Lincoln and continued there until 1908, when he came to Scottsbluff. In this city he became assistant cashier in the First National Bank, and in 1909, became cashier, following which change he remained with that institution until 1916, when he went into the farm loan business and inaugurated the Bowen Investment Company which now controls a large acreage. Mr Bowen is a heavy feeder and extensive dealer in livestock, while the company has handled various properties and estates. It owns the Ford garage at Scottsbluff and formerly owned the First National Bank building, which it recently sold for seventy-five thousand dollars. In 1916, Mr Bowen purchased the First National Bank of Gering and sold the same in 1917. His business sense has always been acute and finance his favorite field of effort.

In June, 1898, Mr Bowen was united in marriage to Miss Anna Pike, who was born in Illinois but at the time of marriage was a resident of Lincoln. They have two sons, Ralph and Wayne, aged respectively eighteen and fifteen years. Mr Bowen and his family are members of the Presbyterian Church. He has always been intelligently active in politics, is a leader in Republican circles and for three years has served as a city councilman. For the past two years he has been president of the Scottsbluff Commercial Club, and is always to be found among those who are promoting the best interests of this city. Fraternally he is identified with the Masons and the Knights of Pythias. As a business man Mr Bowen has always inspired confidence, and socially and publicly is a man of sterling character and high ideals

JENS C PEDERSEN.—Building operations at Gering have been extensive during the past six years, and that much of the work has been notably satisfactory, both in design and substantial character, may, in all justice, be attributed to Jens C Pederson, architect, and a practical builder and contractor. Mr Pedersen has now reached a point in his successful career, when he can devote his entire time to his profession after performing the duties of a public official. Mr Pedersen is city engineer at Gering

Jens C Pedersen was born September 8 1883, in Denmark. His parents are Eric and Christian (Rasmussen) Pedersen both natives of Denmark, where they still live. Of their nine children six survive, but only two

have come to the United States, Carrie and Jens C. The former is the wife of Andrew Christiansen, a farmer near Ottumwa, Iowa. The parents are members of the Lutheran Church. The father followed the carpenter trade in early life and later became a builder and contractor, and when fourteen years old Jens C began to assist his father. He attended the public schools and when he displayed special talent his father afforded him a course of instruction in a technical school at Aalburg, Denmark, where he studied architecture

In the meantime, Mr Pedersen's sister Carrie had come to America and was comfortably settled in Iowa. After completing his studies in the Aalburg School of Design, he decided to join his sister in the United States and engage in the practice of his profession in this country. He reached these shores in 1901, and went to his sister's home. Although there did not seem to be any great demand in Wapello or adjoining counties for the beautiful architectural designs he had in mind, there were many farm houses and barns to be built and he found plenty of employment. Although Mr Pedersen made no fortune while building in the agricultural sections, he has the satisfaction of knowing that the structures he erected were substantial in character and as attractive in appearance as circumstances permitted. In 1913 he came to Gering and opened an office and since then has prospered in every way, his professional reputation firmly established by the designing and building of the head gate for the Castle Rock Irrigation canal. He continued his building operations both in Gering and Scottsbluff until 1915, his last building contract being the Christian Church at Scottsbluff, a structure that is greatly admired. Since then he has devoted himself to architectural designing exclusively and among the fine structures erected from his plans may be mentioned a number of beautiful modern residences at Bridgeport, the Ideal Laundry at Scottsbluff and numerous residences, a modern school house, store buildings and residences at Beard, and a seventy thousand dollar school building at Gering, also a hotel and many residences. He carries on his work with the help of two assistants

In the fall of 1906, Mr Pedersen was united in marriage to Miss Marie Larsen who was born in Denmark, and they have two children Myrtle and Meyrna. The family belongs to the Lutheran Church. In politics Mr Pedersen is identified with the Democratic party, and fraternally he is a Mason and an Odd Fellow

JOSEPH L REEL—Two occupations, ranching and banking, have attracted the interests and energies of Joseph Reel, and in both fields of endeavor he has won standing and prosperity, being at this time the owner of about fifteen hundred acres of valuable land in Cheyenne county and is president and manager of the Farmers State Bank of Dalton Also he has been a prominent figure in public life and in several positions of marked responsibility has demonstrated his worthiness for such honors and his capacity for discharging the duties of his position

Mr Reel is a native of the Buckeye state, born in Pickaway county, June 21, 1881 the son of Aaron and Symantha (Lewis) Reel His father was also a native of Ohio, was reared there and educated in the public schools After his schooling was over he became an apprentice in a meat house and learned the practical side of that business, and in due time when his term of service was over engaged in an independent business of his own as the owner and proprietor of a meat shop He was a man of excellent habits and was just in the prime of life when called by death in his forty-fifth year Mrs Reel, like her husband, was an Ohioan, where she grew to womanhood She met and married Aaron Reel and was left a widow when her son Joseph was a boy of eight years She survived her husband and now resides at Vermillion Illinois

Joseph Reel was reared in Ohio and given excellent educational advantages in his youth in the public schools He was an enterprising boy and early determined that he was going to help his mother and himself along in the world He felt the call of the great west but at first did not go far from his native state as he established himself as a farmer in Illinois where he remained for six years, becoming recognized as one of the well-to-do men of the locality and won general public confidence by his straight forward manner of handling business affairs Mr Reel came to Nebraska in 1915, locating in Cheyenne county The first year he rented a farm to give him time to look around and opportunity to select just the tract that would be his idea of a permanent home, and then purchased his land where he at once took up the active management of farm industries Mr Reel possesses the kind of energy, resource and initiative required of the young man who would succeed in any profession, and is a prominent factor in the development of new methods of farming and stock-raising in the Dalton locality However, he was too broad guaged and had so active a mind that

all his abilities could not find expression in the country so he branched out into financial circle as his capital had become considerable, and in 1918, just three years after becoming a resident of this great state, organized the Farmers State Bank of Dalton From its inception this sound, progressive institution has won the confidence of the people, due to the policy inaugurated by Mr Reel as president and manager He is keen of vision, a natural financier and keeps abreast of all banking business of state and nation and under his skillful guidance it is but natural that the bank should have a most phenominal growth Since locating in Dalton and becoming a member of the financial circles of the Panhandle, Mr Reel has invested heavily in other commercial and civic enterprises that are playing a great part in the development of this section of the state, as he is director and stock owner in the Farmers Elevator of Dalton and also of the Dalton Trading Company Elevator A man of excellent education, high character and marked ability, it is but natural that the people of the county had confidence in him, and when it became necessary to elect a county commissioner to manage the tremendous business of this growing district he was chosen and elected by an overwhelming majority, and today is filling that office Since becoming a resident of this section Mr Reel has established a reputation for carrying on all his operations, of whatever nature, in a progressive and capable manner, and as stockman, farmer, banker and county official stands high in the esteem of the people, his business associates and friends In politics he is a Republican

On September 4, 1907, Mr Reel married Miss Mary Ernest at Omaha, Nebraska, a native daughter of this state, who was reared and educated here and is a member of the Roman Catholic Church There are three girls in the Reel family Minnie L, Myra, and Isabelle, all of whom are at home

BERNARD F DAILEY has passed virtually his entire life thus far within the borders of western Nebraska, and has become thoroughly imbued with the progressive spirit which marks this section of the state He is one of the successful agriculturists and stockraisers of the younger generation in Garden county, and concerning the family history adequate mention is made on other pages, in the sketch dedicated to his father, Robert F Dailey

Bernard F Dailey was born in the 12th of February, 1883, and his early education was obtained in the schools of Deuel, Garden

MR. AND MRS. WILLIAM BARKHOFF AND FAMILY

and Cheyenne counties, all of which were originally included in Cheyenne county Mr Dailey initiated his active career by obtaining employment as a cowboy and cattle herder, and he was thus employed by Reuben Lisco about two years He then began farm operations on the homestead which he had secured, and to which he later added by taking a Kinkaid claim, with the result that he now has eight hundred acres of the valuable land of Garden county, four hundred being devoted to general agricultural enterprise and the remainder of the land being used principally for pasturage and general forage purposes Mr Dailey is a vigorous worker in all departments of farm industry and his advancement shows that he has made the best possible use of the opportunities that have been afforded him in connection with farm development in the Nebraska Panhandle He is a stockholder in the farmers' grain elevator at Lisco, which village is his postoffice address, his political allegiance is given to the Democratic party, he is affiliated with the Knights of Columbus, and both he and his wife are active communicants of the Catholic church

At Sidney, Cheyenne county, February 22, 1911, was solemnized the marriage of Mr Dailey to Miss Anna Vacik, who was born and reared in Nebraska, her parents, Mr and Mrs James Vacik, having been pioneer settlers in Cheyenne county, where the father is now living virtually retired, at Sidney, both he and his wife being active members of the Catholic church Mr and Mrs Dailey have two children — Mary Bernice, born August 24, 1912, and Robert James, born December 20, 1914

WILLIAM BARKHOFF — While industry undoubtedly is one of the fundamentals of material success, yet sound judgment is equally important, and in considering the rapid progress of some men on their way to financial independence, it will usually be found that their efforts have not been haphazard but directed by intelligent foresight and matured judgment An example may be presented in William Barkhoff, a prominent and respected citizen of Kimball With many other settlers he came to this county in 1912, an investor in land Seven years later this land has more than doubled in value and he has entered the ranks of large wheat growers and has made a record as a stockraiser in Kimball county

William Barkhoff was born in Harrison county, Iowa, March 10, 1877 His parents were Henry and Christena Barkhoff, who were born in Germany They were married in that country from which they came to the United States in 1873 settling in Harrison county, Iowa, where they spent the rest of their estimable lives After coming here to make this land their permanent home, they became Americans in every sense of the word and during many years thereafter Harrison county had no more worthy, useful or loyal citizen than Henry Barkhoff They were members of the German Lutheran church They had children as follows Augusta and Frederick, both of whom live in Iowa, William, who belongs to Kimball county, Emma, who married a Mr Davis, of Seattle, Washington, Bert and James, both of whom are farmers in Montana, Edward, who is a farmer in Iowa, and Mary, who married a Mr Wilkins, of Missouri Valley, Iowa Both parents are deceased

William Barkhoff grew up on his father's farm in Harrison county, Iowa, obtaining his education in the public schools It was his father's wish that his children should learn the English language thoroughly, never permitting the use of the German tongue even in the family circle On April 1, 1906, he was united in marriage to Miss Augusta Pehrs, of Harrison county, a daughter of Julius H and Mary Pehrs, the former was a farmer and blacksmith near Denison, Crawford county, Iowa, the latter of whom survives and lives in Idaho Mr and Mrs Barkhoff have two children Wallace who was born October 25, 1908, is doing well at school, and Alice, who was born April 29, 1919

In February, 1912, Mr Barkhoff came to Kimball, buying a lot and comfortbale residence After some weeks of prospecting he purchased a section of land situated four miles east and three miles north of Kimball, for which he paid $17 75 per acre, later adding a half section, for which he paid $1,800, and had scarcely completed the transaction when he was offered $5,600 for this tract He carries on general farming and stockraising and in the latter industry found exceeding profit during the current summer, having sold $1,195 worth His seventy acres of wheat has returned so great a yield that the acreage will be largely increased in the future Mr Barkhoff has done exceedingly well since coming to Kimball county, and in spite of a season of ill health, has devoted himself closely to his business affairs In 1918 he was prostrated by an acute attack of appendicitis, that called for operation, and on the very day of his mother's death on the farm in Iowa, he was under the surgeon's knife in the Methodist Episcopal hospital at Omaha Mr and Mrs Barkhoff are members of the Methodist Episcopal church at Kimball

ROY ALLEN BABCOCK, banker, financier and large land holder is one of the younger and rising generation which today holds the responsible positions that formerly were filled only by men of middle age and many years of experience The short life record of Roy Babcock is an illustration of what may be accomplished by a laudable ambition and a determination to succeed From the modest circumstances in which he found himself when he arrived in Cheyenne county, twelve years ago, Mr Babcock is today the owner of three thousand acres of fine land in the vicinity of Potter, in the Pole creek valley, has large interests in many of the prosperous commercial enterprises of this section and is a heavy stockholder in several banks and the vice-president of the Citizens State Bank of Potter Little more than a decade's connection with the banking interests of Potter, during which time he has risen from a clerkship to his present position in banking circles, has made Roy Babcock one of the best known figures in the financial circles of western Nebraska and eastern Colorado

Mr Babcock is a native son of the west, born in Fillmore county, Nebraska, December 23, 1893, the son of Elmer and Ann (Corbin) Babcock, the former native of Indiana, where he was reared and educated and after attaining manhood's estate engaged in business there as a farmer In 1880, he came to Nebraska, becoming one of the later pioneer settlers of our great state He was a comfortably successful farmer until he retired from active participation in business and now lives quietly in Atkinson, Nebraska, enjoying his sunset years in ease and comfort Mrs Babcock was born in Iowa, where she was reared, receiving a good practical education in the public schools and after her marriage became a sturdy helpmate and devoted mother

Roy Babcock was reared and educated in Sherman county, Nebraska, where his parents were living during his childhood and early youth He attended the common schools of his district and after graduating from the high school desired to widen his educational facilities and attended a business college in Grand Island for two years, devoting particular attention to commercial subjects His first position was in Potter, as he entered the Potter State Bank as bookkeeper February 28, 1912 At the start, Mr Babcock received the magnificent salary of twenty-five dollars a month

and paid out practically all of it for his mere living expenses, but he had faith in himself and felt that it would be only a question of time until the financial outlook would brighten He applied himself to the business, learning practical banking from actual experience and as time passed his income increased, until at the close of his third year he drew fifty dollars a month, but he had greater ambitions than to remain a bank clerk and after looking the financial field over decided that just then one of the best propositions on which a quick return could be made was land, so he bought a four hundred and eighty acre tract on time Mr Babcock believed that the best way to get a thing done was to do it himself and went out on the farm to see that the improvements he made were of the right character, and evidently they were for after nine months he disposed of the property, making a profit of four thousand dollars, rather good returns for less than a year's time and for so young a man His business ability had already become well known in Potter and as soon as he was foot loose was engaged to take charge of the Farmers Elevator, holding the position two years, during which time its business was handled in an able and efficient manner Mr Babcock had not, however, given up his idea of becoming a banker and having by 1917 accumulated considerable capital by his various business ventures he became the prime mover in the organization of the Citizens State Bank, was one of the original stockholders, and from its inauguration as a banking house has been the efficient cashier of the prosperous, sound and progressive institution The bank was opened in 1917 That the men who have shaped the policy and methods have been successful need not be said when we learn that today the Citizens Bank is the third largest in Cheyenne county and has deposits of over a quarter of a million dollars This rapid growth and progress has been largely due to the initiative and executive ability of the men who are devoting their time and energies to its management and Mr Babcock has played an important part in furthering the precedence which it has gained, not only in the county but in the Panhandle As one of the representative business men and public-spirited citizens of Potter, he merits special recognition in a history of the county

Mr Babcock has become a stockholder and vice-president of the Peets State Bank, at Peets, Colorado, and also a stockholder and director of the Gurley State Bank so that he is widely known not only throughout western

Nebraska but in Colorado for his able executive control of the varied banking institutions with which he is associated He is no narrow guage man as he is the owner of the elevator at Egbert, Wyoming, owns a half interest in the Bennet Grain Company of Potter and a half interest in the elevator at Dix, Nebraska Mr Babcock has ever had great faith in the future of this section of Nebraska and to demonstrate this has from time to time invested largely in land of Cheyenne county, and is today the owner of three thousand acres on which he has had excellent improvements in the way of a fine country home, numerous and substantial farm buildings, while the cultivated land has been brought to a high state of fertility For some time he has been a heavy investor in the Coulter Live Stock Company, owning a third interest in that concern which runs over a thousand head of sheep and five hundred head of cattle annually

Having carved an independent career for himself, it is but natural that Mr Babcock has developed into an independent thinker as he is a widely read man and directs his interest to the election of the best man who will serve the people honestly and well His fraternal relations are with the Masonic order, while with his wife he is a supporter of the church which they attend in Potter

November 3, 1913, was solemnized the marriage of Mr Babcock and Miss Olga Cords, at Omaha, Nebraska She was a native of Grand Island, reared and educated in that city, the daughter of Emil Cords, who now resides in Rockville, Nebraska Mr and Mrs Babcock have one little girl

CHARLES S ANDERSON — Cheyenne county has no resident more widely known in either private or public life than the prosperous and progressive farmer and public spirited citizen whose name initiates this paragraph He is a native son of Nebraska and this county, having the distinction of being the second white child born in this region He first saw the light of day down in the Lodge Pole valley near Bronson, on September 15, 1874, the son of John and Mary (Johnson) Anderson His father was a native of Denmark who came to America when a young man with his wife, and brought with him many of the admirable qualities and traits of the people of that sturdy little country, and success came to him through honorable participation in legitimate business enterprises Mrs Anderson also was a native of Denmark where she was reared, educated, met and married her husband They were courageous souls who desired to make head-

way in the world and to this end broke all the dear home ties and sailed away for the United States, the Land of Promise' to so many European emigrants Soon after landing on our shores Mr and Mrs Anderson came west, locating in Cheyenne county Nearly a half century has passed since these sturdy, confident young Danes came into this section and settled on a homestead in the Lodge Pole valley on a homestead in pioneer style amidst a veritable wilderness In addition to the homestead the father took up a tree claim and in due time proved up on all his land For fifteen years he labored slowly and arduously developing the farm and establishing a home for his family and watching and assisting in the advancement and progress which were making the countryside flourish and thrive As soon as his capital admitted Mr Anderson gave up his work on the railroad and devoted his entire energies to his farm business and stock-raising, becoming recognized as one of the leading and most prosperous exponents of agricultural industry He was ever active in all civic improvement and lived to see this country most wonderfully transformed from the prairie to a veritable farm paradise, as he lived to be an old man, passing away in November, 1917 Mrs Anderson survived her husband and now makes her home with her daughter, Mrs Sherwin, of Sterling Colorado

Charles Anderson was reared on his father's farm near Bronson and received his education in the public schools provided at that early day, growing up as most farmers' sons on the frontier When old enough he assisted in the development of the homestead and thus in a practical way learned farming and stock-raising as conducted in this section and at that period After completing his course in school he helped his father for a time and then embarked for himself as a farmer to raise cattle and horses He carried on some general farming for ten years, then went west to Utah where he remained five years, but the Panhandle seemed to him a far better country than that west of the mountains and he returned to the state and county of his birth He now manages the original homestead of his father where he specializes in high bred cattle and hogs, being one of the largest shippers of meat animals to the packing centers of Nebraska and Kansas, as he prepared at least two car load lots each year

While a good and progressive citizen and the supporter of all worthwhile movements he has been contented to remain a farmer, though he has filled his civic duty as a member of the community by filling the office of county assessor for eight years, has been a member of the school board for fifteen years and road overseer about that same period, and today can feel that he has done his best for the community which he has served faithfully and well In politics Mr Anderson is an adherent of the Democratic party and has been proud of the record it made during the recent war He and his wife are members of the Lutheran church Mr Anderson has not confined his entire energies to his land alone but has engaged in extensive financial affairs of the locality as a stockholder in the Liberty Oil Company of Potter, owns a large block of stock in the Farmers Elevator of Bronson and stock in the Liberty State Bank of Sidney

On June 10, 1903, Mr Anderson married Miss F E Yoakhm, at Kimball, Nebraska She was born in Johnson county, Iowa, but her parents came to Kimball county, Nebraska, where she grew to womanhood and was educated in the public schools After completing her studies she became a teacher, a profession she was engaged in for seven years before her marriage, and today she is a worthy descendent of her pioneer parents who played their part in the development and upbuilding of this great commonwealth There are six children in the Anderson family Susie Lucille, Russell R, Ella, Bernadine Max and Jack, all of, whom are still under the happy family roof-tree

MRS LYDIA WALLACE, one of the best known and most popular pioneer women of Garden county, has marked her residence in this section of Nebraska with large and noteworthy achievement in connection with agricultural and live-stock industry, and her gracious personality, her fine intellectual attainments and her distinctive business acumen have given her no slight degree of leadership in community affairs She was one of the early school teachers in what is now Garden county, and in many ways she has contributed to the civic and material development and progress of the county in which she is the owner of a large and valuable landed estate and to which her loyalty is marked by deep appreciation

Mrs Lydia (Morgan) Wallace was born in Shropshire, England, and was twelve years of age when she came with her widowed mother to America, her rudimentary education having thus been received in her native land She is a daughter of Thomas and Emma (Timms) Morgan, the former of whom passed his entire life in England, where he was a farmer by vocation, having been only thirty-seven years of age at the time of his death After the death of her young husband Mrs Morgan finally came with her eight children to the United States and settled in Rock Island, Illinois, where later she became the wife of Francis Bailey, one of the sterling pioneer citizens of that county There she passed her life, and was about seventy-two years of age when she died, secure in the affectionate regard of all who had come within the compass of her gentle and kindly influence In Illinois Miss Lydia Morgan was afforded not only the advantages of the public schools, the high school at Moline, but also those of Knox College, from which she graduated After leaving college she took a course in stenography and typewriting, in a leading business college in the city of Chicago A young woman of spirit and ambition Miss Morgan had abundant confidence and self-reliance when, in 1889, she came to western Nebraska and numbered herself among the pioneer settlers in that part of Deuel county that is now comprised in Garden county Here she filed entry on a homestead a pre-emption and a tree claim, to all of which she perfected title in due time In the meantime she was instrumental in the establishment of the Orlando postoffice and was made its first postmistress While developing her land and engaging in the raising of live-stock, Miss Morgan had the distinction of teaching the first term in the pioneer school of her district With increasing prosperity she purchased an additional tract of eight hundred acres, but this she later sold to Charles Avery In 1900, she purchased the Spohn Creek ranch, and here, for the ensuing five years, she ran an average of three hundred head of cattle She then decreased the stock to one hundred and fifty head and after disposing of the ranch she purchased her present well improved place, which comprises six hundred and forty acres, which is given over to successful agriculture and stock-raising Mrs Wallace has taken very deep interest in everything pertaining to the social, moral and industrial development of the community, and is essentially liberal and public-spirited Under the woman-suffrage law of Nebraska she is duly registered and takes advantage of the franchise, and is a Republican in her political allegiance, her religious faith being that of the Presbyterian Church She still maintains affiliation with the sorority with which she identified herself while a student in Knox college

In 1896, was solemnized the marriage of Miss Lydia Morgan to William B Wallace, a

ROBERT GARRARD

sterling pioneer of whom individual mention is made on other pages of this work No children have been born of this union The achievement and standing of Mrs Wallace mark her as one of the representative pioneer women of western Nebraska, and both consistency and justice are observed in according to her a tribute in this history

ROBERT P GARRARD, who is widely known and universally respected, has spent the greater part of his busy life in the state of Nebraska, where his industrious efforts have been amply rewarded Mr Garrard owns many acres of rich land and a large amount of stock, for years having been one of the most extensive producers of wool in the state

Robert P Garrard was born in Canada, January 7, 1872 His parents were Joseph and Sophia (Pipe) Garrard, the former of whom was born in England, January 13, 1837, and the latter in England, July 29, 1837 Their children were as follows Hatsell, who joined the United States army in 1881 and was in the Spanish-American War, returned from service in Cuba and died in a hospital in San Francisco, Laura M, who is the widow of Albert Foster, lives with a son on a ranch in Kimball county, John, who died in infancy, Emma, who was born in 1859, died in infancy, Lovinia, who died in infancy, Jonathan J, who was born November 17, 1862, died at the age of ten years, Adelaid, who was the wife of John W Alexander, Ellen Alice, who was born in Canada, November 23, 1869, Robert P, who lives in Kimball county, and Frederick George, who was born in Michigan, May 1, 1875, is a farmer in Banner county, Nebraska

From Canada the Garrard family moved to Michigan and lived there until 1882, then moved to Nebraska, settling first in Gage county, but moving in 1883 to Thayer county The father died there July 23, 1886, leaving the mother with a family to look after in a strange country She was not only a faithful mother but was a resourceful woman After being left a widow she went into Banner county and took up a claim, on which the family settled in the spring of 1888 She was spared to her children some years longer, her death occurring February 7, 1899

Robert P Garrard was ten years old when the family came to Nebraska In his boyhood he made himself useful by herding cattle and assisting on the homestead after moving to Banner county, it requiring the efforts of the whole family to make the payments and secure the land In 1895 he took a homestead

of a quarter section for himself, in Kimball county, and in 1903, under the Kinkaid law, secured the balance of the section, and owns five other sections, his possessions aggregating 3,200 acres, and also he has 240 acres adjoining the city of Kimball, about half of this tract being under water irrigation For seventeen years he has fed his sheep from the products of the ranch, finding this a most profitable method At the present time he has 22,000 head of sheep on his ranch, which will return him 6,000 pounds of wool, the extremely high price of wool in 1919 being something he can consider with satisfaction

Mr Garrard is erecting a fine residence on his land near Kimball, the estimated cost of which will amount to about $10,000 In this connection, the biographer cannot refrain from repeating a remark made by Mr Garrard in reminiscent mood displaying an emotion that will find an echo in many a heart, although not always translated into words Mr Garrand says there is only one thing that would make him the happiest man in Nebraska, and that would be the having of his mother in his beautiful new home where he could surround her with comforts all her life denied her

On March the first, 1921 Mr Garrard was married to Mrs Annabel Lewis, a sister of Fred C Overton of Kimball, a daughter of Albert C Overton of Council Bluffs Iowa Mr and Mrs Overton were the parents of three children, two boys and one girl Mrs Garrard being the eldest The father died when Mrs Garrard was only four years of age The mother, after the loss of her husband, moved to Grand Island, Nebraska, in the fall of 1874 Her mother was again married two years later to a Baptist minister and they came to Gibbon in 1882 where Mrs Garrard finished her education in the Gibbon Baptist Seminary She was a teacher for several years and has the distinction of being the only lady on the State Reception Committee in the Nebraska State Building from the Sixth Congressional District in the Trans-Mississippi Exposition held at Omaha, Nebraska

ARTHUR AND HAROLD LYNHOLM — That surety of vision and judgment that makes for definite success in connection with the practical affairs of life is being signally exemplified in the business and farm career of the Lynholm brothers, who are young men well entitled to classification among the efficient and progressive exponents of agricultural industry in Cheyenne county They are native sons of the west and of this state, as Arthur was born near Sidney, May 2, 1891, while Harold

was born on the same farm January 3, 1893
Their parents were Nels P and Anna (Chris-
tison) Lynholm, both natives of Denmark
Nels Lynholm was born in the little sea girt
country of Denmark, and was reared and educated in his native land where he grew up
hardy, thrifty and unafraid of work, traits of
the inhabitants of that land which have won
golden opinions of them as settlers in the
new world He was an ambitious youth who
desired to make the most of his capital, energy, and determination to succeed, in the
United States He crossed the ocean in 1878,
and soon after landing on our shores came
west as he knew that many of his countrymen
were making money out on the plains After
reaching Nebraska, Mr Lynholm obtained
work on the Union Pacific Railroad and later
in Grand Island He was careful in his expenditures, saved his money, and in 1881 returned to Denmark to marry his sweetheart,
Anna Christinson This important ceremony
was performed on March 14, 1882, and the
happy couple had a honey moon trip across
the ocean to the home in a new country They
determined that a home of their own was the
thing and as land was to be obtained by homesteading took measures so to secure a farm
near Sidney, where they lived until they had
proved up on the land, making many permanent improvements, establishing a home and
engaging actively in farm industries For a
short period after first returning to Nebraska
Mr Lynholm had run a livery stable in Grand
Island, but they had not liked it in the town
and that was what induced them to take up
the land near Bronson on April 26, 1883
Soon they also filed on a tree claim, set out
the required number of trees and it was then
incorporated with the original farm Mr
Lynholm became known throughout the valley
as one of the honest, thrifty and prosperous
men, as he was not daunted by the many and
severe trials of this new country and was
helped and assisted in his work and ambitions
by his fine wife who was ever at his side to
encourage when the days looked dark and the
prospect of agriculture in this region seemed
doomed by blizzard in winter and drought in
summer Mr Lynholm was not much beyond
his prime when he was called to his last rest on
December 13, 1904, at the age of fifty-six
years Arthur and Harold Lynholm were
reared on their father's farm near Bronson,
where they grew up in a healthy family atmosphere, learning at an early age the lessons
of self reliance, usefulness and the strict discipline of farm life, and were of great help to
their parents in the lighter labor about the

place that farm boys can assume without detriment to health or growth The boys were
sent to the excellent public schools of the district and that they made good use of the
advantages thus afforded needs not to be said
when we realize the progress they have made
as business men and the high standing they
command in the community After completing their educations the brothers assisted their
father on the home farm, but they were independent in ideas and desired to establish themselves in an independent endeavor and purchased land of their own and with the passing
years substantial success has crowned their
activities as agriculturists and stock-growers,
for today they are recognized throughout the
valley as representative, progressive and up-
to-date exponents of the oldest profession in
the world From time to time as their capital
increased from the sale of cattle and farm
produce the Lynholm brothers have added to
their original holdings until today they are the
owners and managers of over two thousand
acres of the best and most fertile land in
Cheyenne county, a remarkable showing for
such young men as they still have the greater
part of their lives before them Since their
father's death the sons have taken charge of
the homestead where their mother still resides
with them Though they have never been active in political matters, the brothers are independent voters, and have always supported
sturdy and honest candidates for office They
are progressive in the methods employed in
the development and business of the ranch, taking the admirable advice of the state and national farm experts in regard to the best crops
for this semi-arid climate, and as to what cattle are most profitable, they stand for good
roads, good schools, and any movement that
tends to the uplift of communal life The
entire family are members of the Lutheran
church Aside from their rural interests,
Messers Lynholm are stockholders in the
Farmers Elevator of Bronson There are
four children in the Lynholm family as the
boys had two sisters Elizabeth Dorothy, who
lives in Chicago and Marie, who is now Mrs
Jones, and lives in Wyoming

JESSE CAMPBELL, a prosperous young
agriculturist and stock-raiser of Garden county, was born and reared in this section of Nebraska and has been gratified to do his part in
furtherance of the splendid advancement that
has been made here since the pioneer days,
when he was a child and gained his initial
experience in semi-frontier life
In that part of Deuel county that now com-

prises Garden county, Jesse Campbell was born on April 19, 1889, and is a son of Thomas and Josephine Campbell, the former a native of Iowa and the latter of Kansas, in which state her marriage was solemnized Thomas Campbell became one of the pioneer settlers in that part of old Cheyenne county that is now included in Garden county Here he established his residence in 1887, took a homestead, and became one of the substantial agriculturists and stock men of this section of the state, his activities in connection with the cattle business having covered a period of many years He has lived virtually retired since the spring of 1919, his wife having passed away in 1907, and he passed the most of his time at Oshkosh

Jesse Campbell profited by the advantages afforded in the public schools of this section of Nebraska, and early became associated actively with his father's farming and live-stock enterprise, so that he was well prepared for independent operations in the cattle industry, when he was about twenty five years of age He confined his activities largely to the cattle business for a period of about five years, and then purchased his present farm, which comprises three hundred and twenty acres, which is well improved Mr Campbell gives special attention to the feeding of hogs during the winter seasons He is one of the loyal and liberal citizens of Garden county, and in politics is an independence Democrat

July 5, 1911, at Bridgeport, Morrill county was celebrated the marriage of Mr Campbell to Miss Blanche Rand, who was born and reared in Pensylvania, and was still a girl at the time of the death of her father, John Rand, who had come to Nebraska with his family and established himself as a pioneer in old Cheyenne county His widow, Mrs Sarah Rand, returned to Pennsylvania, where she still maintains her home Mr and Mrs Campbell have two children — Florence, born March 16, 1915, and Thomas Edwin, born March 10, 1917

ELI F NASH, who was born in Madison county, New York, March 12, 1862, and is a scion of old and honored families of the Empire state, was reared and educated in Missouri, and as a youth he was there employed three years in a harness shop His ambition and self-reliance then led him to seek better opportunities in the progressive state of Nebraska, and after his arrival within the borders of this commonwealth he was engaged in farm enterprise in Seward county for a period of about eight years Though he had been

successful in that section of the state he had prevision of the greater advantages that were to attend the development of western Nebraska, and accordingly, in March, 1890, he came to this locality, where for six years he found employment as a cowboy and proved himself resourceful and expert in the herding of cattle on the great ranches that then marked the Nebraska Panhandle With good judgment, he then filed entry on a homestead, on Blue creek, and while making improvements on this land, incidental to perfecting his title to the same, he added to his financial resources by conducting for three years a livery business at Oshkosh Since that time he has given his undivided attention to the management of his well improved farm, which he has brought under effective cultivation, besides causing it each season to provide properly for the select live stock which he raises, special attention being given to the feeding of hogs during the winter seasons Energy and good management have made Mr Nash one of substantial and representative farmers of Garden county, and his pleasant home is situated near the village of Lewellen, which is his postoffice address He has had no ambition to enter the arena of practical politics or to seek public office, but he has been liberal in the support of progressive communal enterprises and is found arrayed as a stalwart champion of the cause of the Republican party

At Oshkosh, on August 8 1892, Mr Nash wedded Miss Rhoda Hunter who was born and reared in Missouri, and whose father, William Hunter, became one of the pioneer settlers in the vicinity of Oshkosh, Garden county Nebraska Mr and Mrs Nash have given hostages to fortune by presenting to Nebraska their fine family of eleven children Mrs Mabel Copley, of Lewellen, has one son, Mrs Viola Orr of Lewellen, Mrs Martha McConkey, of Oshkosh, Mrs Mamie Campbell of Oshkosh, Hazel resides in the village of Lewellen and Orpha, Lucille, Reuben, Howard, Waldo and Herbert remain at home, which is a center of cordial hospitality and good cheer

In conclusion it may be stated that Mr Nash is a son of Eli F and Nancy Nash, both of who were born and reared in the state of New York and the latter of whom died when her son Eli F, of this review was but two years old In 1864, the father removed with his family to Missouri where he became a substantial farmer and where he passed the remainder of his life, having been about sixty years of age at the time of his death

WILLIAM B WALLACE — The labors and hardships that fell to the pioneers of western Nebraska gained the full compliment of fellowship on the part of this well known and representative agriculturist and stockman of Garden county, and that he had the instincts, the endurance and the determination that make for success, is demonstrated in the substantial prosperity which he enjoys at the present time

Mr Wallace was born at Beaver Dam, Wissonsin, April 25, 1850, and is a representative of a staunch pioneer family of that section of the Badger state His father, John Wallace, who was a native of Indiana, followed various lines of endeavor within the course of his active career, as he turned his attention to farm enterprise, to mercantile business and also to service as a commercial traveling salesman He was about fifty years of age at the time of his death His wife, whose maiden name was Amanda Bushnell, was born in Ohio and was a girl at the time of her parents' removal to Wisconsin, where she was reared and where her marriage was solemnized She passed the closing period of her life in Minnesota, where she died when about seventy years of age

William B Wallace acquired his early education in the schools of Wisconsin and Minnesota and was about two years old at the time of his father's death, his mother thereafter moving to Minnesota At the age of eighteen years he purchased forty acres of land in Dodge county, Minnesota, the tract was covered with underbrush and he cleared this away and prepared the soil for cultivation There he continued to farm about five years, and he then bought a farm of eighty acres, in the same county, where he lived twelve years He then sold the property and, in 1887, in company with the late Asa C Mills, who became a prominent citizen of Minatare, Scottsbluff county, came to Western Nebraska, the two driving through to this section of the state about two hundred head of cattle and twenty horses In that part of Cheyenne county that now constitutes Garden county Mr Wallace took up homestead, pre-emption and tree claims, and this land he made the center of his activities as a stock-raiser, having continued about twelve years in the cattle business, besides raising horses He then sold his land and settled about thirteen miles northwest of Oshkosh, where the ensuing period of about four years found him engaged principally in the raising of sheep He then established his residence at Oshkosh, where he was engaged in the real-estate business about five years, and did not a little to further the settlement and advancement of this part of the state. For the next six years Mr Wallace resided on a claim which he secured under the provisions of the Kinkaid act and which he then relinquished to his son Hazen B He then returned to his farm, where he was engaged principally in the raising of hogs, about four years, at the expiration of which he virtually retired, by establishing his residence on a tract of forty acres, thirteen miles northwest of Oshkosh, which is his present place of abode and which is one of the model small farms of Garden county Mr Wallace has been liberal and progressive as a citizen, is a Republican in politics but has manifested no desire for public office of any kind

In 1870, Mr Wallace wedded Miss Mary Whitaker, who was born in the province of Ontario, Canada, whose death occurred at Alliance, Box Butte county, Nebraska Of this union were born six children Frederick D and his wife are residents of Oshkosh, their children being two in number, Erie, who now resides at Alliance, Box Butte county, was one of the young men who represented Nebraska in the ranks of the national army during the late World War, his service including seven months with the American Expeditionary Forces in France, Mrs Cora E Morris, of Broadwater, Morrill county, has five children, George W, of Lisco, is the father of three children, Mrs Carrie Allington, of Sunol, Cheyenne county, has four children, and Hazen B, now a successful young farmer and stock-grower of Garden county, was in the nation's military service in the late war, he having received his training at Camp Funston, Kansas

The second marriage of Mr Wallace was solemnized in 1895, when Miss Lydia Morgan became his wife, and of this well known pioneer woman specific mention is made on other pages of this publication

JOHN FINK, who is now living retired in the city of Scottsbluff, is one of the venerable and honored pioneer citizens of western Nebraska and has been a resident of Scottsbluff county since the time when it was still a part of Cheyenne county He reclaimed and developed one of the valuable farm properties of the county and continued to give his active supervision to the same until 1913, when he retired and removed to Scottsbluff

Mr Fink was born in Germany, on June 25, 1846, and is a son of William and Henrietta (Saite) Fink, the latter of whom passed her entire life in Germany, and the former of whom finally came to the United States, the

MR. AND MRS. HOLLADAY AND SON

closing years of his life having been passed in Illinois

John Fink acquired his early education in the schools of his native land and was an ambitious youth of twenty years when, in 1866, he immigrated to America and settled in Rockland county, New York, where he remained five years. During the ensuing seven years he was actively identified with farm enterprise in the state of Illinois, and he then, in 1879, came to Nebraska and settled near Seward, Seward county, where he became a pioneer exponent of farm industry and where he continued his vigorous activities until 1887, when he numbered himself among the pioneers of what is now Scottsbluff county. Here he eventually perfected title to the homestead and tree claims which he had taken up, and he developed the land into a state of high productiveness, the while he erected good buildings and made other permanent improvement of excellent order. On this original farm he successfully continued his enterprise as an agriculturist and stock-grower until 1913, since which time he has lived retired, his attractive home in Scottsbluff being known for the gracious hospitality there extended to the many friends whom he and his wife have gained during the many years of their residence in this section of the state. In politics Mr. Fink is a Republican, and both he and his wife hold membership in the Presbyterian church.

In the state of New York, on December 1, 1867, Mr. Fink was united in marriage to Miss Frederica Weishoff who likewise is a native of Germany and who was a girl at the time of her parent's immigration to America. To Mr. and Mrs. Fink were born ten children. Mrs. Louise McClalahan resides in Salt Lake City, Utah; William is a resident of Seward, Nebraska; Charles maintains his home in Attica, Kansas; Henry lives at Park City, Montana; George is a resident of Rose, Wyoming; Mrs. Minnie Wright is deceased; Mrs. Amy Harrison is a resident of Scottsbluff; Mrs. Clara Lewis is a resident of Salt Lake City; and Edward and Elmer reside at Park City, Montana.

JOHN B. HOLLADAY — In the death of John Bernard Holladay, not only his family met with an irremediable loss, but a wide circle of friends, many acquaintances and many important business interests suffered through the withdrawal of a strong, invigorating, helpful presence.

John Bernard Holladay was born near Marshalltown, Iowa, February 2, 1884, and his death from influenza occurred December 3, 1918. He was a son of J. A. and Mary Holladay, and was fourteen years old when he accompanied his parents to Omaha, Nebraska, where he had school advantages. When he reached manhood he homesteaded in Kimball county, seven miles south of Bushnell, under the Kinkaid act; he then bought three one-quarter sections one mile south of his homestead, eighty acres west and one-quarter section ten miles southwest of his homestead, which gave him a total of 1300 acres. He also owned hotel property at Pine Bluff and a twenty-acre fruit farm in California. Not only was he enterprising and active in the development of his property, having 400 acres of his land under cultivation and keeping 150 head of cattle and horses, but he was able to give attention to other important enterprises. He was agent for the Dempster, the Fairbanks and the U. S. Supply Company, operated well-drilling outfits and dealt in all the equipments that went with this kind of business, and in addition was extensively interested in the buying and selling of land. He was a young man of acknowledged business capacity, and success met him in the most of his undertakings.

On February 3, 1908, Mr. Holladay was united in marriage to Miss Luella Carlson, a daughter of Oscar and Huldah Carlson, who came to Pine Bluff, Wyoming, in 1902, from Red Oak, Iowa. Mrs. Carlson died at Pine Bluff, in 1906. Mr. Holladay is survived by his widow and one son, Walter Theodore, who is attending school. Mr. Holladay served ten years on the school board and it was through his efforts that the schoolhouse was built in his school district. He was one of the organizers of the telephone line from Bushnell to southern parts of the county, and was treasurer of the telephone company until within a short time of his death. He belonged to the order of Royal Highlanders at Pine Bluff.

ROBERT F. DAILEY has been for more than thirty years a resident of what is now Garden county and is one of the genial and popular pioneer citizens and is now the owner of a large and well improved landed estate, which represents the splendid fruitage of past years of earnest and honest endeavor, as he experienced to the fullest extent the trials and burdens which fell to the lot of the sturdy pioneers of western Nebraska. His reward has been ample, he has unbounded confidence in the still greater progress of the Nebraska Panhandle, and takes just pride in his success and in the country with whose development and upbuilding he has been closely associated.

Mr. Dailey is a scion of staunch Irish lineage and is a representative of the second generation of the family in America, his parents, John and Mary (Casey) Dailey, having been born and reared in Ireland, where their marriage was solemnized, and were young people when they immigrated to the United States and became pioneer settlers in Stephenson county, Illinois, in the year 1820 There the father took up a pre-emption claim of forty acres of government land, and eventually he accumulated a valuable estate of about three hundred acres, which he reclaimed into a productive farm His first crop of wheat he hauled to Chicago, which was then an insignificant little city, and to reach this market he was compelled to traverse a distance of eighty miles He continued his farming enterprise in Stephenson county until his death, in 1849, and his widow passed away in 1851, when about forty-nine years of age, both having been devout communicants of the Catholic Church

Robert F Dailey was reared and educated in Stephenson county, Illinois At the age of twenty-one years he established himself in the livery business in Butler county, Iowa, and after conducting this enterprise six years came to what is now Garden county, Nebraska, the county having been at that time still a part of Cheyenne county and having later become a part of Deuel county, prior to the erection of Garden county He arrived August 27, 1888, and forthwith selected a homestead and a tree claim, upon both of which he proved up in due course of time, in the meantime he began practical operations in the raising of cattle and horses With the passing years Mr Dailey developed his original claims into a productive and well improved farm, and in 1902, he took three-fourths of a section under the provisions of the Kinkaid act, besides which he added an equal amount by purchase, so that he now has a total of twelve hundred and eighty acres, four hundred acres being under cultivation and three hundred and twenty acres receiving effective irrigation through the medium of the Lisco ditch Mr Dailey has proved himself a man of energy and resourcefulness, and is one of the loyal and substantial citizens of Garden county His political support is given to the Democratic party and both he and his wife are communicants of the Catholic Church

On April 17, 1884, was solemnized the marriage of Mr Dailey to Miss Nellie Noonan, who was born in the state of New York but reared and educated in Butler county, Iowa, where her parents, John and Winifred

(Hayes) Noonan, were pioneer settlers, both having been natives of Ireland, where their marriage was celebrated Mr Noonan was about seventy years of age at the time of his death, and his widow attained to the venerable age of eighty years Mr and Mrs Dailey became the parents of three sons and three daughters, and of the number four are living Bernard F, of Lisco, Garden county, is married and has two children, and John Edward, Margaret and Mary remain at home, which is a center of genial hospitality

WALLACE DWIGHT BEATTY, who during a long and successful career, has followed various occupations in several parts of Nebraska and Wyoming, is now a well known resident of Scottsbluff, although his operations are by no means confined to the borders of the city or county During his residence in this state he has been in turn farmer, cowboy, foreman of a cattle company, live-stockman, contractor, irrigation superintendent and then a combination of several of these enterprises at one and the same time, and in his several fields of endeavor his versatility has assisted him to well deserved prosperity He is a native of Iowa, born in Howard county, December 1, 1866, the son of Martin and Lucy (Tamplin) Beatty The father was a son of the Blue Grass state, who was one of the honored pioneer settlers of Cheyenne county, locating there in the very early days, when there were no railroads in this section of the state The family suffered all the hardships and privations incident to life on the frontier Lucy Tamplin Beatty was born in Ohio but came west to Iowa as a child and there was reared and educated in Clayton county She married Martin Beatty and together they bravely started out to establish a home in the new country They improved the government land on which they first settled and there Mr Beatty engaged in general farming and stock raising Mr Beatty was sturdy and for many years he retained the mental and physical vigor of a man many years his junior, while his wife, who had a remarkable share in the pioneer experiences, was remarkably vigorous In 1901, Mr Beatty removed to Rockport, Missouri, where he died at the advanced age of eighty-one years Mrs Beatty lived out the psalmist's span of three score years and ten as she passed away in 1914 in her eighty-ninth year

Wallace Beatty was reared and educated in Howard county Iowa where he attended the public schools and thus laid the foundation of a good practical education In 1876, he moved with his parents to Missouri and remained there for nine years He then located in eastern Nebraska but two years later came west He filed on a claim in Banner county on which he proved up and became one of the early settlers of the Panhandle At that early day he was farsighted enough to realize that this section was to have a wonderful development and future After proving up on his land, Mr Beatty accepted a position with the Ogallala Land and Cattle Company, one of the firms that had wide ranges and thousands of head of cattle on the plains He remained with this concern for ten years, a large part of the time being spent in Wyoming, as he was foreman for them and had charge of moving the great herds from Nebraska to the higher summer pastures farther west in 1886, 1887 and 1888 In 1893, he was manager in the field of the Ogallala company and the following year helped catch the last bunch of wild horses that roamed the valleys of Montana and western Nebraska When the Burlington and Missouri River Railroad and the Missouri Pacific Railroad were being built through Nebraska, Mr Beatty entered independent business as a contractor for railroad grading Meeting with gratifying success he branched out from his initial effort and when the Chicago, Milwaukee and St Paul Railroad was constructed was engaged in building it through Wyoming When the government started its irrigation projects in the middle west, Mr Beatty was one of the first men to bid on the contracts for the Tri-State ditches and put in the new intake for the Winter Creek ditch in Scottsbluff county His standing as a contractor and specialist in irrigation won him a fine reputation among the people of the Panhandle country and in 1913, he was unanimously chosen superintendent of Enterprise ditch For some time Mr Beatty has had the contract for grading the roads in the Winter Creek district and it has been due to his care that this precinct has one of the best systems in the county In 1913, Mr Beaty again returned to the soil and engaged actively in farm enterprise, as he had been extensively engaged in feeding for ten years, shipping large consignments of stock to the great packing centers of Kansas and Nebraska each year At the present time he owns over a hundred acres of land just south of Scottsbluff, some of which is within the city limits, and has fine residence property in Scottsbluff and Gering While in the contracting business Mr Beatty realized the necessity of a fine, high grade sand in all construction work, especially where cement is used and in 1917, he decided that there was

money in supplying just such a commodity for concrete and now has a prosperous business in this line in the Panhandle, shipping to all points in the western part of the state Opportunity and years are yet before Mr Beatty, and his friends prophecy splendid things for him in the coming decades that must pass before he reaches the years to which his parents attained Mr Beatty is a Republican in his political views, and though he has never had time to take an active part in more than local affairs is a loyal and public-spirited citizen whose support is never withheld from helpful enterprises and good civic movements

Mr Beatty's first wife was Miss Blanche Draper, who died in Gering, in 1913, leaving one child, Wallace Dell, who is living with his father in Scottsbluff, a young man of seventeen years who is making the most of the many advantages afforded him On February 25, 1915, Mr Beatty married Mrs Jane Peters of Alliance, Nebraska The family are members of the Congregational Church

FRED W STONE has had a specially eventful career through his loyal and efficient service in the United States Navy, which he again entered when the nation became involved in the great World War He was a child at the time when he came with his widowed mother to Nebraska, and thus the honors which are his in conection with the Navy are reflected upon the state which has represented his home and in which he now resides in the city of Scottsbluff He is one of the popular young men of Scottsbluff county and is specially entitled to recognition in this work His mother, whose maiden name was Ella S Fasha, is now the wife of James H Hall, of Scottsbluff, and concerning her further mention is made in the sketch of the career of Mr Hall, on other pages of this work

Fred W Stone was born at Hubert, Wyoming, on November 5, 1888, and is a son of William Edwin and Ella S (Fasha) Stone His father was born at Dansville, New York, and was a boy at the time of the family removal to Iowa, where he was reared and educated under the conditions that marked the pioneer era in that state In 1863, when but fifteen years of age, his youthful patriotism led him to run away from home, in order to enter the Union service in the Civil war He managed to enlist in an Iowa volunteer regiment that was assigned to the Army of the West, and before the close of the war he had gained an amplitude of experience in connection with the great conflict, including his participation in the campaigns of General Sherman, with whom he took part in the historic march from Atlanta to the sea After the close of the war he resumed his educational work and finally completed a course in civil engineering During one winter he was engaged in trapping in Minnesota, and thereafter he spent four years in Kansas and New Mexico He was still a comparatively young man at the time of his death, which occurred in the state of Wyoming, where he had acquired land and engaged in farming and stock-raising Of his first marriage were born two children—Pearl H of Mitchell, Nebraska, and Mrs Edna Gatels, of Juliette, Wyoming—the subject of this sketch being the only child of the second marriage

After the death of the father of Fred W Stone, the widowed mother came to Scottsbluff county and established her home at Gering, where was later solemnized her marriage to James H Hall Thus Mr. Stone gained his early education in the schools of this county, including the high school at Scottsbluff, besides which he was afforded the advantages of the Kearney Military Academy, at Kearney, this state At the age of seventeen years he joined the United States Navy, in which he received his training course at Goat Island, California, and on the United States ship Pensacola He was on military guard duty in San Francisco at the time of the great earthquake and fire which devastated that city in 1906, and in December of that year he was assigned to service on the steamship Milwaukee, at Mare Island, California On this vessel he was in service on the west coast from Seattle, Washington, to Callao, Peru In April, 1908, he was transferred to the United States steamship Wisconsin, at Bremerton, Washington, and on this vessel he served while it was the flagship of the Fourth Division of the North Atlantic Fleet in the celebrated trip made by this fleet around the world Thus it was his privilege to visit Honolulu, New Zealand, Australia, the Philippine Islands, Japan, China, Ceylon, Egypt, Malta, Algiers, Gibralter and many other important ports, in 1908-9 The fleet arrived at Hampton Roads, Virginia, February 22, 1909, and with his command Mr Stone thereafter participated in the inaugural parade attending the induction of President Taft into office, in the city of Washington He received his honorable discharge from the navy on November 4, 1909, at Portsmouth, New Hampshire, and thus was released from service the day prior to his twenty-first birthday anniversary He then returned to Nebraska, and became associated with his stepfather, James H Hall, in the operation of a ranch in Banner county, where they ran a large number of cattle, horses and mules In December, 1910, Mr Stone sold his interest in this enterprise to his partner, Mr Hall, and then went to Long Pine, Brown county, and entered the employ of the Chicago & Northwestern Railroad Company, with which he continued his service until March, 1912 He then extended his already exceptional experience, by going to Davidson, Saskatchewan, Canada, where for the first year he devoted his attention to farming He then engaged in the moving-picture business at Prince Albert, but about one year later returned to Davidson, where he was engaged in the same line of enterprise until July, 1915, when he sold out and returned to Scottsbluff, Nebraska Here he became associated with Clarence L Chapin in carpentering and contracting, in which he continued successfully until the United States became involved in the war with Germany, when he entered the service of his country On the 9th of July, 1917, he returned to the United States Navy, as a volunteer, his enlistment having taken place in the city of Omaha He was sent to San Francisco, and in the following August was assigned to duty on the United States Steamship Standard Arrow, which loaded with oil, at Point Richmond, that state, and proceeded, by way of the Panama canal, and the Gulf of Mexico, to Hampton Roads, Virginia The vessel then made its way to New York City and Sydney, Nova Scotia, from which latter point it crossed the Atlantic to Portsmouth, England Mr Stone served as chief carpenter's mate on this vessel during the remainder of the war, and within the period the boat made twelve trips across the Atlantic, carrying oil, ammunition and airplanes to different ports in England, Scotland and France, the service being especially hazardous, in view of the submarine activities of Germany After the signing of the armistice Mr Stone was transferred to the United States Steamship Maumee, on which he served from January 23, 1919, until he received his discharge, on the 23d of the following May, at Portsmouth, Virginia He then returned to Scottsbluff, Nebraska, where he is to make his permanent residence

In politics Mr Stone is a Republican, and he is affiliated with the Masonic fraternity and with the lodge, encampment and canton bodies of the Independent Order of Odd Fellows His wife holds membership in the Ladies of the Maccabees

At Bridgeport, Nebraska, on October 17,

MR. AND MRS. PETER NELSON

1910, was solemnized the marriage of Mr Stone to Miss Reta H Williams, of Scottsbluff Mrs Stone was born at Creston, Iowa, and came with her widowed mother and her brother to Scottsbluff in 1902, having been graduated in the local high school in May, 1910, a few months prior to her marriage She is a daughter of Frank J and Catherine (Cluck) Williams, the former a native of Illinois and the latter of Pennsylvania, their marriage having been solemnized in Iowa Mr Williams, who was a commercial traveler out from Omaha, died at the age of forty years, and his widow later removed with her children to Scottsbluff, and she is now the wife of Rev Allen Chamberlain, pastor of the Methodist Episcopal Church at North Platte, this state Mr and Mrs Stone have a winsome little daughter, Ellen Ruth, who was born in the year 1911

PETER NELSON — Among the representative citizens of Kimball, there are few better known perhaps than Peter Nelson, who has been a resident of Nebraska for thirty-six years and of Kimball county but a few years less He has accumulated an average fortune and has become a man whose business judgment is consulted in many matters of public importance Should Mr Nelson be questioned as to the way in which he has managed to be so successful, in all probability he would answer, as have the greatest of social economists, "work and economy form the basis of prosperity"

Peter Nelson was born in Denmark, February 10, 1862, one of nine children born to Nels and Anna (Larson) Nelson, and one of three to reach maturity His parents died in 1887 In 1883 he came to the United States, located in Kearney county, Nebraska, worked there as a farm hand for two years, then came to Kimball county and homesteaded five miles north of Dix He lived on his homestead and tree claim until he had proved up on 320 acres, then went to work on the railroad, three years later coming to Kimball Here, in 1894, he purchased a dray and went into the hauling business, working early and late, and by 1904 was prepared to invest $250 in a tract of sixty-seven acres that lay within the city limits The purchase of this tract not only demonstrated faith in the future of Kimball but was a mark of business foresight that is characteristic of Mr Nelson This land is now valued at $200 an acre, and promises to be one of the city's handsomest residential sections He has jut completed laying out four blocks in town lots, and for the choicest of these he will probably realize more than he paid for the entire original tract His own handsome residence stands here When Mr Nelson first took possession of this land he raised a few horses, but soon realized the beter profits in dairying and supplies dairy products to the most of Kimball Mr Nelson also owns 1120 acres that he is farming in a small way, holding the property mainly as an investment and ready to sell when a satisfactory offer is made

In 1897 Mr Nelson was married first to a daughter of Peter and Sofie Larson, who came to Kimball county from Denmark, and March 20, 1890, homesteaded and took a tree claim ten miles south and one mile east of Dix In 1910 Mr Larson was killed by a stroke of lightning while stacking hay in his barn, during an electric storm, this being the second tragic death in the family, as Mrs Nelson had been accidentally killed by a railroad train, May 8, 1905 Mrs Larson resides at Kimball To Mr Nelson's first marriage two sons and two daughters were born, namely Paul E, Harold E, Mable E, and Helen E Both sons were educated at Kimball and both entered military service during the great war Paul was sent to the training school at Camp Sherman, Chillicothe, Ohio, in the 47th Aero Squad, Signal Corps, then was transferred to the 163rd, was sent to France in August, 1918, and was honorably discharged July 3, 1919, and returned home Harold E was in training for seven weeks at Bellevue, then was honorably discharged because of the end of the war

In 1906 Mr Nelson was married to the sister of his first wife and they have had children as follows Myrtle, Minerva, Ralph, Stanley, and Iris and Ira, twins Mr and Mrs Nelson are members of the Methodist Episcopal church Mr Nelson has served on the town board of Kimball

MORSE P CLARY is another of the representative men of Garden county to whom are justly to be ascribed pioneer honors in the great Nebraska Panhandle, to which this history is dedicated He came to this county while it was still a part of Cheyenne county, his original land claims later were to be found in the new county of Deuel, and finally became a part of the still newer county of Garden He has been continuously identified with agricultural and live-stock industry on his original land claims since the year 1886, and has been a leader in community thought and action He is now president of the Farmers' State Bank of Lewellen but still gives a general supervision of his farm enterprise

Morse Powell Clary has ample reason to be proud of his American ancestral line, for he is a scion of a family that was founded in this country in the early colonial era His father, Dennis B Clary, was born in the city of Baltimore, Maryland, and was of the fifth generation in line of direct descent from the original American ancestor of the name, the latter having accompanied Lord Baltimore's first colony from England and having taken up a "tomahawk claim" of about three thousand acres of land, twenty miles distant from Baltimore This property has in part continued in the possession of the family to the present day, and four cousins of the subject of this sketch still reside on the ancestral place In 1866, was erected in this locality the historic Strawbridge Church, notable as being the first Methodist Church edifice built in America The original structure, of sturdy oak logs, had no doors and no windows at the start, but in this unfinished condition it was used for religious services The ancient building remained as a landmark in Maryland for many years

Morse P Clary was born at Quincy, Adams county, Iowa, October 20, 1858, a date that indicates conclusively that his parents were pioneer settlers in that section of the Hawkeye state He is a son of Dennis B and Rachel M (Cooper) Clary, the former of whom was born in Baltimore, Maryland, as previously noted, and the latter of whom was born in Indiana, her father having been a native of Scotland and her mother of Indiana

The subject of this review acquired his early education in the public schools of Indianola, Iowa, and during the first fourteen years of his independent career he was engaged in farm enterprise in Warren county, that state In 1886, he came to western Nebraska and became a pioneer settler in what is now Garden county—at that time a part of old Cheyenne county Here he took up homestead and tree claims, about five and a half miles from the present village of Lewellen, and here he not only reclaimed his farm land from the virgin prairie but also made good improvements in the matter of buildings and other accessories, and became one of the successful agriculturists and stock-growers of the county He still owns the old farm property, which comprises two thousand acres, but is now living semi-retired in Ash Hollow

Mr Clary has been influential in community affairs under the three different county governments, Cheyenne, Deuel and Garden—and that without changing his place of abode He served two terms as county commissioner of

Deuel county—1892-9—and was county commissioner of Garden county from 1910 to 1915 Under these conditions it is apparent that he had much to do with the progressive movements and official agencies that were potent in the civic and material advancement of the community in which he has lived and wrought to worthy ends In 1915, Mr Clary became one of the organizers and incorporators of the Farmers' State Bank of Lewellen, and he has served continuously as a director of this substantial institution, to the presidency of which he was elected in January, 1919

In politics Mr Clary is a staunch Republican, he is a charter member of Oshkosh Lodge No 286, Ancient Free & Accepted Masons, at the county seat, and his wife holds membership in the Methodist Episcopal church

At Indianola, Iowa, on January 17, 1881, was solemnized the marriage of Mr Clary to Miss Louisa C McNaught, a daughter of Ezekiel and Roxanna (Durand) McNaught, her father had been a blacksmith by trade but became a prosperous farmer in Iowa, he was a gallant soldier of the Union in the Civil war, as a member of a regiment of Illinois Volunteer Infantry In conclusion of this brief sketch is given record concerning the children of Mr and Mrs Clary March D, who is a prosperous farmer in the Lewellen neighborhood, supplemented his common-school education by a course in the Grand Island Commercial College He married Miss Nora West, of Omaha May R, who received the advantages of the University of Nebraska, is the wife of Jess Minchall, of Broadwater, Morrill county, Frank L, who is now living in the city of Omaha, served as a member of Company K of the Coast Artillery during the World war, he was educated at Fremont College Josie J, who completed a thorough course of study at Fremont College, is principal of the public schools of Bingham, Sheridan county, at the time of this writing, in the winter of 1919-20 Nona, who was afforded the advantages of the Nebraska State Normal School at North Platte, is now the wife of Ray Brown, of Lewellen Ray S, who is now home, saw nine months of service in France during the late World war, having been bandmaster with rank of sergeant, with the One Hundred and Ninth Engineers, his assignment having been to Company E of this command Oren V, who profited duly by the advantages of the public schools of Lewellen, is still a member of the home circle, and Cora M, the youngest of the children, is still attending school in her home village

JAMES C FOSTER, a pioneer ranchman and venerable and honored citizen of Garden county, where he is now living retired, in the city of Oshkosh, has been a resident of Nebraska for more than forty years, and his memory and experience compass much of the stirring activities that marked the pioneer history of this commonwealth

Mr Foster was born in Jefferson county, Pennsylvania, March 5, 1847, and though he has passed the psalmist's span of three score years and ten, he retains the vigor and optimistic outlook that come as the heritage of right living and right thinking, during the course of an earnest and useful career Mr Foster is a son of Daniel and Elmira Antoinette (Williams) Foster, both natives of the state of New York and representatives of families founded in America in the colonial era of our national history Daniel Foster was reared and educated in the old Empire state and as a young man he removed to Pennsylvania, where he secured a tract of heavily timbered land, in Jefferson county, and instituted the reclamation of a farm He was captain in a cavalry command in Pennsylvania, when the state militia was kept in training largely for the repelling of the Indians His father and brothers were killed and scalped by the Indians, who took his mother prisoner at the same time She was held in captivity fourteen years, and her release was effected by her son Daniel, father of the subject of this review This remarkable woman, who endured with fortitude her experience as a captive, died about eight months prior to the time of her one hundredth birthday anniversary James C Foster was reared to manhood in the old Keystone state, where he received the advantages of the common schools of the period, and whence he and his two brothers, Hiram T and William M, went forth as valiant young soldiers of the Union when the Civil war was precipitated on the nation Hiram T was killed in battle, about one week prior to the surrender of General Lee William M was captured by the enemy and was held for some time in the historic old Libby prison, whose name is odious in the record of the war between the north and the south He finally made his escape from the prison, largely through the aid given him by a negro

James C Foster was twenty-four years of age when the dark cloud of civil war cast its pall over the national horizon, and he forthwith manifested his patriotism by enlisting as a private in a company of Pennsylvania Volunteer Infantry, with which he participated in many engagements and lived up to the full tension of the great conflict In later years he has perpetuated the more gracious associations of his military career by retaining affiliation with the Grand Army of the Republic

After the war Mr Foster was for some time identified with the lumber business in Pennsylvania, but in August, 1875, he left his native state and set forth for the west On the 11th of that month he arrived at Columbus, Platte county, this state, and for eight years was engaged in farming on the Pawnee Indian reservation

After his marriage he continued to farm four years, and then came with his family to the western part of the state, in 1887, and took up a homestead thirty miles east of Alliance He developed and improved a productive farm and to the active management of the place he continued to apply himself for twenty-eight years He then sold the property, which had become valuable and to the area of which he had added, and it was at this juncture in his career that he retired and established his residence at Oshkosh, where he and his wife delight to extend the hospitality of their pleasant home to their wide circle of friends In politics Mr Foster is found arrayed as a staunch supporter of the principles of the Republican party

The year 1882, recorded the marriage of Mr Foster to Miss Carrie M Douglass, who was born and reared in Wisconsin, and came to Nebraska when she was eighteen years of age She shared with her husband in the pioneer experiences in western Nebraska, and their ideal companionship continues as shadows of their lives begin to lengthen from the golden west In conclusion is given brief record concerning their children Mrs Nettie Miller, died in 1917, Mrs Rosella Ross, resides in Wyoming, Mrs Mary Lindley is a resident of Lakeside, Sheridan county, Nebraska, as is also Mrs Maude Hyland, William M, who is now living at Oshkosh, was one of the gallant young men who served with the American Expeditionary Forces in France during the progress of the World war, having been with his command in France for a period of one year, and James C, Jr remains at home

RICHARD CLARK, who has prestige as one of the pioneer farmers and stockmen of Garden county, is the owner of a fine farm property of 320 acres, three and one-half miles northwest of the village of Lewellen, and is known and valued as one of the sterling citizens and successful agriculturists and stock-growers of Garden county He was born in Butler county, Iowa, December 12, 1873 and on other pages of this work, in the sketch of the career of his brother, Walter Clark, is given ample review of the family history Richard Clark was the fourth in order of birth in a family of five children, and concerning the others a brief record may consistently be entered William P is a resident of Lewellen, Garden county, Mrs Mary Cushman resides in the state of Michigan, Walter is a resident of Oshkosh, judicial center of Garden county, Nebraska, and Mrs Birdie Anderson resides at Ogallala, Keith county

Richard Clark acquired his early education in the schools of the Hawkeye state and was a lad of about eleven years when, in 1885, he came with his brother Walter to Nebraska They made the trip in a box car and left their palatial quarters upon arriving at Ogallala In Keith county Richard found employment on a farm, and in a service of nine months he received a compensation of seven dollars a month Finally he came to what is now Garden county and joined his parents, who had taken a farm homestead on the table southwest of Lewellen He remained at the parental home until 1899, and in the meanwhile availed himself of the advantages of the public schools of the locality, besides doing his share in connection with the development and other work of the home farm In 1899 he purchased the half section of land which constitutes his present well improved farm estate, and here he has not only been successful in his operations as an agriculturist and as a breeder and grower of cattle and hogs, but has also erected modern buildings and made other improvements that mark his property as one of the model farms of Garden county His farm is well irrigated and he is one of the substantial stockholders of the Blue Creek Irrigation Company His political allegiance is given to the Democratic party, his wife is an active communicant of the Grace Lutheran church at Lewellen and Mrs Clark is an appreciative and popular member of the Woman s Club of Lewellen The attractive home of Mr and Mrs Clark is known for its generous and gracious hospitality, and it is a favored resort for their wide circle of friends

On the 20th of March 1900, was solemnied the marriage of Mr Clark to Miss Emma Paisley, of Concordia, Kansas Mrs Clark was born in Nebraska and is a daughter of John H and Addie Paisley, both of whom were born and reared in Iowa, where their marriage was solemnied and whence they came to Nebraska, where the death of Mr Paisley occurred Mrs Paisley later became the wife of John W Wilson, who likewise was born in Iowa and who represented that state as a valiant soldier of the Union during the Civil War Mr and Mrs Wilson finally removed from Iowa to Kansas, and he became a successful farmer near Concordia, that state Mr and Mrs Clark have one child, Ivor Cecil, who was born in the year 1901

MARTIN BRISTOL, a well known real estate dealer of Mitchell who has been engaged in commercial life in the Panhandle for many years and is known here and at Gering as a successful business man, was born near Peoria, Illinois, November 21, 1860, the son of John E and Anna (Martin) Bristol The former was a native of New York The family originally came to America from England about 1860 The mother was born in Kentucky John Bristol settled in Illinois in 1827, and was an officer in the company commanded by President Lincoln during the Black Hawk War He received a land warrant for his services at that time

Martin Bristol was reared on his father's farm, and attended the public schools until his sixteenth year when he was apprenticed to a carpenter to learn the trade The apprenticeship lasted five years and when completed, he accepted a position with the Rock Island railroad as a carpenter During his services with that road, Mr Bristol contracted asthma In 1886, he came to North Platte, Nebraska for his health and within ten days was able to get around and soon began to work at his trade He took up a homestead in the Gering valley, four miles south of the present town of Gering and remained there until 1888 He built the first log house in Gering for Dr Franklin and another for Frack Garlock in April, 1887 That was the beginning of the town Mr Bristol was busily engaged as a carpenter until 1890, when he took a trip through Utah, Colorado, and Kansas, working in different towns as he went along In 1893, he returned to Gering, where he built many of the buildings of the growing town Coming to Mitchell in 1901, Mr Bristol took charge of a lumber yard for Car & Neff, from 1902 to 1903, he clerked in a hardware store here then spent two years as

Fred Chew and Family

deputy county clerk In 1906, he became interested in a hardware and furniture store in Mitchell, having charge of the business until 1911, but sold to manage a hardware store in Lingle, Wyoming, running it until 1912 In the fall of that year he went to Henry, Nebraska, then took a vacation until 1914 Returning to Mitchell, Mr Bristol opened a real estate office where he has since been in business This enterprise has grown most satisfactorily In 1916, he was elected county assessor and served until 1920 then resigned to take care of his real estate interests which were increasing continually and demanded his time Mr Bristol owns a hundred and twenty acres of land in the valley, eighty acres of which adjoin the town of Mitchell and is very valuable Besides that he has other property in Mitchell as well as his home

March 1, 1905, Mr Bristol married Miss Mary Bowman, of Schuyler county, Illinois He is a staunch Democrat and has taken an active part in politics For more than thirty-five years Mr Bristol has been a Mason, has taken his 32d degree, is a Knight Templar and a Shriner He is one of the constructive men of Mitchell who is ably helping in the development of the valley

FREDERICK C CHEW, who is one of the large wheat growers of Kimball county, has built up his ample fortune by his own efforts and is recognized as one of the substantial men of the county

Frederick C Chew was born in Warren county, Iowa, August 1, 1876 a son of William F and Mary (Purvis) Chew, the latter of whom was born in Pennsylvania and died in Iowa, in 1897 She was the mother of three children, of whom Frederick C was the eldest He has a sister, Cora, who is the wife of Frank Thorp, a farmer in Iowa, and a brother Haymond, who is in the street car service at Omaha William F Chew maried Patience Lawrence for his second wife, who died November 25, 1908, leaving one son, John, who is a traveling agent for a stock food concern William F Chew is in the land business and resides at Anita, Iowa For about fifteen years he served as a justice of the peace at Adair, Iowa, where he was prominent also in the Masonic fraternity

Frederick C Chew attended school in boyhood and found himself generally usefully employed and recalls the first money he earned was when he was paid fifteen cents a day for digging potatoes He remained in Iowa as a farmer until 1909, when he came to Kimball

county and homesteaded sixteen miles southwest of Kimball, under the Kinkaid law securing a section of land When the transaction of this business, including removal of his family and settlement in the county, was completed, he found his cash capital reduced to $30 With A B Beard he discovered that his credit was not impaired and through him he bought cows, rented a tract of land from Mr Beard and engaged in farming and cattle raising for several years, when he located for a time on his homestead, since which time he has carried on dry farming, finding that some land, under proper conditions, will yield amazing crops of wheat He sold his land for $17 50 an acre, then bought another farm for $12 an acre, selling one-half of that land for $45 an acre, and balance at $55 per acre, later buying 400 acres for $20 an acre and 640 acres at $33 per acre, and 80 acres at $26 per acre, and on this land he is growing, by dry farming, wheat, the yield from which brought him $8,000 last season At the present time Mr Chew has 900 acres in wheat, with prospects fair for a large yield

On January 4, 1899, Mr Chew was married to Miss Ruth Evans, one of a family of eleven children born to Morgan L and Jane (Lewis) Evans The parents of Mrs Chew were born in Wales and came to the United States in 1865, with their five older children, the six younger being born in America The parents reside at Adair, Iowa, aged respectively, eighty-three and eighty years Mr and Mrs Chew have had four children, two of whom died in infancy, the survivors being Merle, who is farming for his father with a traction engine, and Marie, who is attending school at Kimball Mr Chew has purchased a fine residence which is beautifully located on an elevation overlooking Kimball and the surrounding country Mr and Mrs Chew are members of the Methodist Episcopal church, with which they united while living at Adair, Iowa He belongs to Beulah lodge, A F & A M, at Adair While not active politically, Mr Chew is very much interested in everything pertaining to farm and community welfare and advancement and is a member of the Kimball County Fair Association

Mr Chew bought a fine residence just south of the city of Kimball on a fine elevation overlooking the city, which he later sold at an advance in price and has since bought of A B Beard a residence in the southeast part of Kimball He now owns eleven hundred acres of farm land, all of which has doubled in value since he purchased it

WILLIAM QUIVEY, one of the well known members of the Mitchell bar, is a lawyer of ability who has been in active practice in the Panhandle for more than a decade and during that time has taken a leading part in legal affairs of the Platte valley. He is also a business man who stands high in this section as he is naturally constructive and has assisted in the development of this section since making Mitchell his home

Mr Quivey was born at Charleston, Coles county, Illinois, July 19, 1842   When he was six years old the family moved to Wisconsin and, in 1850, to West Union, Iowa, where the boy was reared and educated He attended the public schools, then graduated from the high school about the time of the outbreak of the Civil War.   August 24, 1861, Mr Quivey was attending the university but gave up his college course and with more than eighty other young men enlisted in the Union army as a private, being a member of Company C, Twelfth Iowa Infantry   He participated in the battles of Fort Henry and Fort Donaldson, where he was wounded and sent to the hospital for four months   He then returned to his company for further service and in 1864 enlisted in Battery K, First Missouri Light Artillery to serve until the close of the war.   During that time he was in armies that took part in the battles of Donaldson, Shiloh, Corinth, Helena, Arkansas, and many other skirmishes and light engagements   He was mustered out of the service at St Louis, August 4, 1865, and returned home to take up his interrupted education   For two years he was a student in the academy, then began to teach, a vocation he followed for ten years   During that time he was principal of the school and at the same time began to read law, being admitted to practice at West Union, Iowa, in 1880   Having served as county superintendent of schools three terms, Mr Quivey was well acquainted with the people and had gained their confidence   He went to Humbolt, Iowa, and opened an office to practice law, being engaged in professional work there five years before moving to Pierce, Nebraska, and was soon elected prosecuting attorney of Pierce county, was re-elected two terms and remained in practice there until 1910 Mr Quivey then came to this county and took up a homestead two and a half miles north of Mitchell, where he lived a year to get his patent to the land   Later he sold some of the property   After this year on the farm he came into Mitchell, opened an office and has been practicing here ever since   Today Mr Quivey is recognized as one of the able and prominent men of the legal profession in the Panhandle, a place which he has won by his hard work, high standards and ability.

November 1, 1868, Mr Quivey married Miss Jennie C Babcock, a native of Ohio, who was reared in Iowa   She was a teacher before her marriage   Mrs Quivey died in 1913, leaving three children   L. A., professor of literature at the University of Utah, Grace, the widow of C F. Montrose, now makes her home in Scottsbluff, where she has taught the past year, having taught in Pierce previously, Zoe Marie, the wife of J. R Ummell, of Mitchell   There was one child that died in infancy.

Mr Quivey is a Mason, is a member of the Methodist church, and four years ago entered the Federated church here   He is a Republican   From first coming to Mitchell he has taken an active and interested part in civic affairs and supports the movements for the improvement of the county and town

ISAAC CATRON, one of the energetic farmers of the Panhandle, who came to western Nebraska more than thirty years ago, is today a substantial business man who has made good in Morrill county and stands high in his community

Mr Catron was born in Kentucky in 1866, the son of M and Susan (Roberts) Catron, the former born and reared in Kentucky, while the mother was a native of West Virginia   They are both living at the age of eighty-three years   Five children made up the Catron family   Henry L, a farmer in Missouri, Martha, the wife of Thomas Brown, in the mercantile business in Oklahoma, Isaac, of this review, J S, a farmer in Missouri, and Eliabeth, the widow of Smith Graham, lives in Jefferson City, Missouri   The father left Kentucky and settled in Missouri where he became a farmer, though a Baptist minister by education. He later bought a bank in Brumley, Missouri, which he ran for some time, but returned to general farming   Mr Catron is a Republican and while living in Kentucky during the Civil War, the Confederates burned a mill that he owned there   When the Confederate soldiers were in his vicinity he hid out as he did not wish to be captured.

Isaac Catron was reared in Kentucky and Missouri and received his education in the public schools of the latter state and learned farming under his father at home In 1888, he came to Nebraska and homesteaded a quarter section of land in old Cheyenne county, that part which is now Morrill county Later he Kincaided three quarter sections which gives him six hundred and forty acres He has placed good and permanent improvements on his farm, and to-day has a good grade of live stock Mr Catron has carried on general farming with success, is one of the early men here who finds that Panhandle land pays

In 1886, Mr Catron married Miss Anne Tolle, a native of Missouri, the daughter of Fidelia Tolle, and nine children have been born to their union Ethel, the wife of Fred Homan, of Springfield, Missouri, Bessie, the wife of Clint Park, who works at the sugar factory in Gering, Gertrude, the wife of Elmer Harness, of Oshkosh, Nebraska, Roy, of Sterling, Colorado, who is a railroad man, Fay, Frank, Florence, Grace, Charles and Margery, all at home

Mr Catron takes an active part in the civic affairs of his home community and has served as director of his school district during the past fifteen years He is a Republican in politics but does not care for political life for himself Mrs Catron is a member of the Baptist church

PATRICK J DUNN, a pioneer settler of Morrill county, who came here when this was old Cheyenne county, is today one of the successful farmers of the Bridgeport district, where he has by his own efforts made a comfortable fortune Mr Dunn is one of the men who has seen the great changes that have taken place in the Panhandle and has shared in the work of developing and opening up this locality to settlement He was born in Omaha, Nebraska, September 4, 1863, the son of Michael and Mary (McBride)) Dunn, both natives of Ireland They were the parents of eleven children Thomas, of Omaha, Samuel, lives in Morrill county, Patrick, of this review, John C, and Mary J, are on the home place near Omaha, Ignatius J lives in Omaha, Michael, deceased, Clement, deceased, Cletius, lives in Montana, and Ita is the wife of Edward Keating The parents came to the United States in 1850 Mr Dunn was a hard worker and soon found employment in Illinois on the Burlington railroad, which was being built at that time He recalled for his children the fact that men used wheel barrows and ox teams in the work of

filling and excavation, quite different from today Later he came west to Omaha, to work with a government surveying party in Douglas and nearby counties He was a member of the Catholic church and during his early life a Democrat, but later became a Republican He died in November, 1900, being survived by his wife who lives on the home place near Omaha, a woman eighty-five years old

Patrick Dunn was reared near Omaha and received his education in the public schools of that locality When he was old enough the young man started out in life as a farmer, coming to Alliance in April, 1888, when there were only a few tents to make up the town Mr Dunn believed there was a future here for a young, energetic man and within a short time took up land in Cheyenne, now Morrill county The old sod house he built on his claim was the third school in all the wide reaches of old Cheyenne county, which has since been cut into seven counties, each large Mr Dunn suffered all the vicissitudes and hardships of the early days, but kept his land and was not discouraged He began to make money when the country settled up a little by farming and stock raising and when irrigation was established his success as a farmer was assured Today he owns five hundred and four acres of land all under water rights, which produce large crops Mr Dunn is one of the settlers to whom pioneer honors are due, he stands high in his community where he has lived so long, as a business man and one of the old timers who has played an important part in developing the country

November 9, 1897, Mr Dunn married Miss Gertrude A Wood, a native of Illinois, and to them have been born eight children Mary Helen, is cashier in a store in Bridgeport, Daniel F, lives in Bridgeport, Patrick J, is at home, William C., Kathleen, Ignatius and Josephine are also at home and James who died in infancy

Mr Dunn is a member of the Catholic church and an independent Republican in politics

HENRY WALSH — The vigorous and progressive population of Cheyenne county is made up largely of successful exponents of the agricultural and live-stock industries In every part of the county, farmers seem to thrive, and an able and honorable representative farmer as well as one of the progressive and prosperous citizens, is the man whose name heads this review Mr Walsh is one of the native sons of Cheyenne county, being born on his father's farm down near Potter in the Lodge Pole valley, October 26, 1885, the son of Stephen and Francis (Bartel) Walsh. The father was born in England and came to the United States when a young man to take advantage of the greater opportunities afforded here than in the tight little island of his birth, where a man with little or no capital had no chance of getting ahead in the world. Stephen Walsh came west after reaching America and in 1870 took up a homestead and tree claim in Cheyenne county on which he proved up and soon engaged in agricultural business. He put many permanent improvements on his place and though he suffered many of the hardships and privations of this new country during the eighties and nineties, was determined to stick and believed that the future of this section would in the end make him a fortune and his judgment was correct. Mr Walsh raised cattle and horses in addition to his general farming, and when money was scarce worked at his trade of carpenter and as such skilled artisans were scarce on the plains at that early day found all that he could do and more. He had the honor of taking an efficient part in the erection of many of the buildings of Sidney and Potter, and they in a way may be regarded as a monument to his memory. He became a well-to-do farmer and stockman of this section, standing high in the esteem of his business associates and many friends. About twenty-eight years ago he was killed in a runaway, leaving a sorrowful family. Mrs Walsh was born in Bohemia but came to America when young with her parents who settled in the locality of Schyler, Nebraska, where she was reared and educated, passing away while a young woman at the age of thirty-five.

Henry Walsh remained on the farm in Cheyenne county, obtaining his education by attending school near his home, working on the farm as all country boys do and so was well acquainted with the practical side of farming while still a youth. He was naturally fond of horses and other animals and as a boy helped break the colts that were raised by his father. As soon as his school days were over he became ambitious to establish himself independently and naturally chose the line to which his inclination turned and also that with which he was familiar and began to work while still a young man as a horsebreaker and to this day is regarded as one of the most efficient

men in this business in all Cheyenne county. September 3, 1914, Mr Walsh rode Buster, a noted outlaw horse of Cheyenne county, and won the championship of western Nebraska, in a bucking contest. This he still holds, having defended it for six years against any Nebraska man who has entered the bucking contests. For a number of years he was a cowboy with various outfits in Cheyenne county and along the Platte river, as that was the heigh-day of the range cattle industry when the great baronial cattle companies owned their herds, numbering many thousands that ranged from Texas and New Mexico to the Yellowstone river in Montana. For one year Mr Walsh was engaged in farming in Wyoming, then returned to Sidney for three years, being employed by a cattle concern in this county. Following this he was made an excellent offer to go to Colorado and accepted, but returned to spend nearly a year in this locality as a cowboy before going to Scottsbluff for a year, being engaged in feeding cattle there. From his father Mr Walsh had learned the trade of carpenter during his boyhood, and after locating near Sidney permanently has been following this vocation a part of the time. Reared on a farm and having spent many years in the cattle business it was but natural that he should desire land for himself and when he decided to make this valley his home, bought a farm where he has been actively engaged, raising diversified crops, especially forage as he is a stock-raiser and heavy feeder. Mr Walsh, like many of the scions of the pioneer families, possesses those qualities which have made it possible for him to meet and overcome obstacles, and as the years have passed, success has crowned his efforts. Thus today he and his family are surrounded with the comforts and blessings which they justly deserve. From his first entrance into business life Mr Walsh has taken an active part in the civic and communal life.

In politics Mr Walsh draws no strict party lines in casting his vote, as his influence is exerted to place the most practical and best man in office that the people may be well served, and thus he is known as an independent. With his wife he is a member of the Lutheran church.

March 16, 1906, Mr Walsh married Miss Ella Peterson, at Sidney. She was a daughter of the Sunflower state, where she was reared and educated. There are three children in the family: Morton E, Ralph H, and Ruby Evelyn, all of whom are at home

Mr. and Mrs. John H. Campbell

with the parents   Mr and Mrs Walsh are
leaders in their community where they have
been the active promoters and supporters of
all helpful enterprises, as Mr Walsh has
ever advocated good roads, good schools
and the most advanced methods in farming

JOHN H CAMPBELL, who is one of
Kimball county's best known men, now living
retired at Kimball, was born in De Kalb county, Illinois, September 17, 1849   His parents
were William R and Sarah Ann (Helmer)
Campbell   Of their children, John H was the
youngest, the others being as follows   James
W, William R, Orlando, and three daughters,
all of whom are deceased, and Mrs Adele
Hulbert, of Pine Bluff, Wyoming, and Mrs
Anna Eckerson, of Aurora, Nebraska

John H Campbell grew up on his father's
farm in De Kalb county, occasionally attending school in the winter seasons but the whole
sum of educational opportunity was small
When nineteen years old he married Elizabeth Murry, daughter of a neighboring farmer
in De Kalb county, who died early, leaving
three children, namely   Alice, who is deceased, Iva, now Mrs Pheiffer, who lives in
Banner county, and William R, who is deceased   In 1880 Mr Campbell was united in
marriage to Miss Minnie Murry, and to this
marriage the following children were born.
Ora, who is deceased, Roy L., who is a resident of Kimball, Mrs Etta F Reiseberg, who
lives in Kimball county, and William R, who
is deceased

Mr Campbell came to Nebraska in 1871
and homesteaded in York county, coming
from there to Kimball county, April 17, 1886,
which county has been his home ever since
He homesteaded and still owns his ranch of
two sections, where he engaged in general
farming and raised cattle and sheep for a number of years   In 1912 Mr Campbell came to
Kimball in order to give his children better
advantages than those afforded by ranch life,
and for four years conducted a land business,
he with his son owing 4160 acres of fine land
He has recently retired, selling his modern
office building to Edward Larson   Although
never very active in politics, Mr Campbell as
one of the solid, dependable men of the county,
has often been called upon for public service,
and he was one of the first county commissioners of Kimball county   He takes pride in the
fact that two of his grandsons served with
honor in the great war   The family attend the
Presbyterian church   They all belong to the
fraternal order of Highlanders, and Mrs
Campbell has the degree of honor in this or-

ganization   The Campbell name is one held in
very high esteem in Kimball county

FRANK STEARNS, president of the
largest mercantile establishment of Morrill
and one of the best known business houses
in the Panhandle, is numbered among the
progressive and substantial men of western
Nebraska   He is a self-made man and has
won to his present position by his own efforts and hard work   Mr Stearns is an energetic man of great personal magnetism,
and these qualities combined with his great
capacity for handling people and an eye for
detail have aided him in building up a prosperous business house within a short period,
for today he is one of the substantial business men of the Platte valley

Mr Stearns is descended from old New
England and Pennsylvania stock   He was
born in Chippewa county, Wisconsin, January 11, 1871, the son of Eugene M and Eva
(Brown) Stearns, the former a native of
Vermont and the latter of Pennsylvania
There were three children in the Stearns
family, but Frank is the only one living
Eugene Stearns received a good education
in his youth, and as there were not many
good openings for a young man in the east,
he came west, locating in Wisconsin, where
he established a general mercantile house
which he conducted with marked success
for many years   Having accumulated a
comfortable fortune, Mr Stearns retired
from active business some years ago and is
now enjoying the mild climate of the Pacific
coast, as he lives in Portland, Oregon

Frank Stearns passed his youth in his
native state, attended the public schools until he was fourteen years old, when the family came to Nebraska, locating at Loup
City, Sherman county, where he continued
his education graduating from the high
school and later attended a commercial college at Omaha, taking special courses to
prepare for financial life   Mr Stearns had
already learned the mercantile business in
his father's store which helped him in the
choice of his studies   After returning to
Loup City for a time Mr Stearns went to
Alliance in 1888, and became one of the
prime movers in the establishment of the
town of Bayard, Morrill county   as it was
through his energy and the money he supplied that the Nine Mile irrigation ditch,
known as the Bayard canal, was constructed, which, with other canals has brought
prosperity to a large and rich agricultural
district

Mr Stearns had been bookkeeper in Loup City after leaving college In the bank he became familiar with finance and later took charge of his father's bank at Bayard He then engaged in a mercantile business in Scottsbluff but sold the store to come to Morrill in 1915, as he believer there was a good opportunity for an establishment here He soon opened the "Stearns Store," which is the best known mercantile house in the town and one of the representative ones of its kind in the county Mr Stearns carries a large and representative line of goods, he is up-to-date in his business methods as his policy is service, square dealing and he now enjoys a trade covering a wide territory in the valley and surrounding country, where his reputation for honesty and integrity have won him customers as well as a high place in the esteem of the people Mr Stearns typifies the progressive spirit of the west, while he is conservative in financial matters He is a Republican, takes an active part in politics and was three times state representative in the legislature from the seventy-fifth district in 1913, 1915 and 1917 Mr Stearns has headed every movement for civic and communal progress that will develop this section and Morrill, and it was through his efforts that the Irrigation School at Scottsbluff, the only one of its kind in the world, was secured for this section In January, 1920, Mr Stearns sold his store in Morrill to Harry M Stearns and Roy C Craig and he now devotes his time to the management of his six hundred and forty acre ranch in Morrill county Most of the land is under irrigation and is rented

Mr Stearns married Miss Ada Flanagan, of Illinois and they have three children Harry M, living in Morrill, Irene M, and Leslie, both at home

RALPH E. CAPPER, one of the native sons of Nebraska, who is a prosperous farmer and stock-raiser of Scottsbluff county, has been well and favorably known here the greater part of his life Mr Capper is an example of what may be accomplished by the application of intelligent energy in one direction, his success in his agricultural activities depending upon his natural leanings and the study which he has given to his business problems

Ralph Capper was born at West Union, Nebraska, in 1882, being the son of Howard and Susan (Predmore) Capper, the former a native of Illinois, while the mother was born in Iowa The father was a farmer who located in Iowa at an early day and later removed to Custer county, Nebraska, where he engaged in farming and stock-raising until 1900, when he with the family came to the Gering district to locate upon a homestead Here in the valley Mr Capper continued the general agricultural pursuits with which he had been engaged farther east and soon became one of the substantial and progressive men of this section There were five children in the family Grace, the wife of Charles Johnson, of Dalton, Montana, Ralph, John, in Gering, Amy, married Walter Beebe, a farmer of Wyoming, and Howard, on a farm in Scottsbluff county

Mr Capper was a member of the Modern Woodmen and with his wife was a member of the Christian church

Ralph received his early academic training in Custer county, where he attended the public school of his district After locating here he finished school and then engaged in farming, first with his father and later independently as he took up a homestead of a quarter section of land in section eighteen, township fifteen Mr Capper has placed fine improvements on his place, has a good comfortable home, and is demonstrating the success attained by the adoption of modern methods in general farming and also raises a good grade of live-stock He is progressive in his ideas and also advocates the same in all lines of life, he boosts for all movements that tend to the development of his county and community and has taken active part in the affairs of the valley for some years, as he is school director of district No 12 at the present time Politically he is a member of the Socialist party, is a member of the Modern Woodmen and participates in general social activities

In 1906, Mr Capper married Miss Anna Cook of North Platte, Nebraska, and to them seven children have been born Grace Dulce, Warren, Claudia and Carroll, twins, Percy and Vesper, all at home These children have a bright future as their parents intend to give them every opportunity in an educational way to fit them for life

FRANK M SANDS is a conspicuous example of the faith that conquers In common with most of those who chose western Nebraska for a place of abode in the early days, faith was about all that he brought with him But he had an abiding faith in the ultimate development of this section of the country, and had the tenacity of purpose that wins Mr Sands and the late William H Wright, of whom mention is found elsewhere in this volume, were the two men most prominently identified with the promotion of irrigation in the valley of the North Platte river Mr Sands has lived to see the realization of their early plans for making one of the most productive sections of the country in the place which was then only a semi-arid, sparsely settled prairie, and to share himself in the rich harvest that has come to those who put their faith and their efforts into the development of western lands

He was born in Stanfordville, Duchess county, New York, on December 25, 1850, the son of Isaac G and Sarah A (Badgley) Sands Both his parents were natives of New York state, and the father farmed in that state all his life He also ran a freight boat from Poughkeepsie to New York City Hannah Griffin, the grandmother of Frank M Sands, saw the first steamboat, Fulton's "Clermont," on one of its voyages Six children were born in the family, two of them, Charles B and Elma, being now deceased The others, besides the subject of this sketch, are all living in New York Their names are James H, who owns land in New York and also in eastern Nebraska, Nettie, the wife of M B Cole, and Henrietta, the wife of Charles F Bishop, an attorney at law in New York City The father was a Whig in politics, and on the formation of the Republican party became a member of that political faith and was active in politics The mother was a member of the Christian church

Frank M Sands took a course in civil engineering at Cornell University, Ithaca, New York, and set out to make that profession his life work In 1872, he was in Chicago when the great fire that destroyed that city broke out, and the hotel in which he was staying caught fire He came on west and worked as an engineer for the government and in railroad construction in Arizona, then started in the sheep business in New Mexico and Kansas, handling and feeding sheep for market In 1886, he was married to Miss Phena Rogers, a native of New York and the daughter of Silas and Ada (Chamberlain) Rogers

Following their marriage Mr and Mrs Sands came west and located at Nonpareil, in Box Butte county, Nebraska Here Mr Sands embarked in the banking business, opening the Bank of Nonpareil He still remembers driving to Nonpareil with a team of ponies, tying his team, and going into the postoffice to inquire for mail The postmaster was Gene Heath, and near the window the accommodating postmaster maintained a public plug of Horseshoe chewing tobacco, the idea being that patrons of the office should help themselves to this while waiting for their mail

Two years later, in 1888, Mr Sands removed his bank to Alliance, where he bought the first business and residence lots and built the first brick building He continued in business there for four years, then came to Gering in 1892, and bought about two thousand acres of land Part of the house in which he still lives is built of lumber hauled from Pine Ridge Here he turned his attention to irrigation construction work, and for the next ten years engaged in the building of irrigation ditches in the capacity of contractor and engineer He was superintendent and financial manager of the Gering canal, and also engaged in general farming and stockraising, which latter calling he has since followed He fed five hundred head of cattle last year He has developed his land to a high state of productiveness, and recently sold a quarter-section for three hundred and fifty dollars per acre, the price received for one acre being about what the entire quarter was worth when he came to this section of the country

To Mr and Mrs Sands five children have been born Charles B, who now manages the home farm, Henry H, a farmer in Scottsbluff county, Sally, who lives at home, Antoinette, a teacher, and William B, who has lately completed a term of fifteen months' service in the American army in France, being in the motor truck service with the front line of troops

Mr Sands in politics is an independent voter, Blue Lodge Mason He has held the office of county commissioner and every office connected with the Gering Irrigation District He is a man who is abreast of the times, well informed on public questions, and his judgment commands the respect of all his acquaintances He stands high in the community as a man of integrity and has the satisfaction of knowing that his life of honorable and industrious effort has been rewarded with success

DEAN E RANDALL, who has spent the greater part of his life in Nebraska, is a substantial farmer and highly respected citizen of Scottsbluff county He was born in Trempealau county, Wisconsin, September

9, 1871 His parents were James M and Lucy (Hassan) Randall His father was born in Michigan and in early manhood worked in the great logging camps of Michigan and Wisconsin He was a man of fine constitution and still survives, living at Gering, Nebraska, in his eighty-fourth year In 1885, he moved to Nebraska with his family, settling in Butler county, but, in 1886, he pre-empted and took a tree claim in Scottsbluff county The mother of Mr Randall was born in New York and died at the age of thirty-six years

Of his parents' family, Dean E Randall was the second in order of birth He obtained a public school education and grew up on a farm and has followed agricultural pursuits all his life In 1890, he homesteaded, and his wife did also, in this county, and now owns and operates five hundred and twenty acres This land has been well improved and is a fairly representative farm property of this county

Mr Randall was married to Miss Sadie Belden, who was born at Jamestown, Kansas, and is a daughter of Wallace and Amanda (Dugger) Belden The parents of Mrs. Randall came to Nebraska thirty-three years ago, homesteaded near Bridgeport, and there the mother yet lives, the father being deceased Mr and Mrs Randall were married December 24, 1904, at Chimney Rock, in Morrill county, and they have four children, namely Kenneth, Darrell, Glenn and Rena, all of whom have had educational advantages In politics Mr Randall is a Republican He has never cared for political honors but has been interested in having good schools and served on the school board for four years

WILLIAM F PULLEN, who is an enterprising young farmer of Morrill county, belongs to an old pioneer family of this section that has been known and respected here for many years His father, William Pullen, homesteaded here among the first settlers, developed his land and has a good property Later he moved into Bayard and has continued to live there

William F Pullen was born at Bayard, Nebraska, June 25, 1896, and grew up in Morrill county, receivingg his education in the public schools He assisted his father on the farm and ranch until 1915, since which time he has been working for himself, renting his mother's farm of eighty acres, and also a hundred and sixty acres of hay land He has the reputation of being industrious,

provident and successful, and wherever known is held in high esteem He owns modern machinery and carries on his various industries according to the latest methods methods

Mr Pullen was married to Miss Pauline Worick, a daughter of John Worick, who now lives on a farm northeast of Bayard They have two children Frances and Dorothy Mr Pullen carries life insurance in the Mid West Life Insurance Company In politics he is careful and cautious and votes independently, having more trust in his own judgment than in the often conflicting statements of party campaigners

CHARLES W McFERON, an industrious farmer and respected citizen of Morrill county, was born in Washington county, Illinois, May 12, 1877 His father, J L McFeron, was born in Perry county, Illinois, seventy-three years ago He came to Nebraska in 1898 and settled at Sidney, Cheyenne county He resides with his son, Charles W His wife is deceased.

Charles W McFeron came first to Nebraska in 1895 His first summer was spent on a farm but for twelve years afterward he worked on ranches in different sections, then went to Texas There he engaged in general farming until 1917, when he came to Morrill county with the intention of investing in land In the meantime he is operating a rented farm of a hundred and twenty acres of well improved land and when an advantageous opportunity presents itself, he will become a landowner here He will be a welcome addition to the county's permanent citizenship

In 1910, Mr McFeron was married to Miss Zella Van Degrift, who was born in Saline county, Arkansas, and is a daughter of William and Nettie (Kawk) Van Degrift, natives of Arkansas, who now reside at Texas City, Texas Mr and Mrs McFeron have three children Alma, Elbert, and Calista

FRED BREYER, who is one of the representative men of Morrill county, resides on his valuable irrigated farm of two hundred acres, which lies in section 7, township 20-51 Mr Breyer was born in Michigan, January 4, 1867 His parents were Edward and Lucy (Kostenbader) Breyer His father was born in Scotland and came to the United States in 1858 locating first at Baltimore, Maryland, but later went to Michigan and married there His wife was born

Dr. A. E. Moss

in Pennsylvania They were theatrical people and as a family troupe appeared in various eastern cities during the sixties

In 1889, Fred Breyer came to the west and lived in Colorado and later in Mexico for some years When he settled permanently he homesteaded in Morrill county, Nebraska, and his hundred and sixty acres have been highly improved and to his original farm he has added until he now has two hundred acres, all irrigated Mr Breyer is one of the county's substantial men

Mr Breyer was married to Miss Mary Lyman, who was born in Hudson county, Wisconsin, February 6, 1866, and they have three children Bessie, Aaron and Daniel In his political views Mr Breyer is a Socialist At the present time he is a member of the school board of District No 44, but otherwise has accepted no public office

ALBERT E MOSS, D O, who has met with remarkable success in the practice of his profession, has been established at Kimball for twelve years, and during this time has not only built up a substantial practice in the face of competition, but has also won the personal esteem of his fellow citizens

Albert Edward Moss was born in McDonough county, Illinois, April 21, 1876 His parents, Samuel and Sarah A Moss, are deceased He attended the public schools in his native county, the high school of Centerville, Iowa, and later the Wesleyan University at Champaign, Illinois, from which he was graduated after a theological course, in 1896 For eighteen months after leaving the university he filled ministerial appointments Circumstances then turned his attention to the field of medicine, in which his reading and investigation soon aroused deep interest in the line of osteopathy, resulting in his becoming a student at Kirksville, Missouri, from which institution he was graduated with his degree in 1898 Dr Moss began practice in Iowa, where he remained until in February, 1907, when he came to Kimball, practicing alone until March 26, 1919, when he admitted a partner the firm name now being Drs Moss & Bonnell, they being the only osteopathic practitioners in Kimball county During the somewhat recent epidemic of influenza, Dr Moss had 166 patients prostrated with this disease and with one exception was able, by his methods, to restore them to health

In 1902 Dr Moss was united in marriage to Miss Mollie Florence Wood, and they have had four children Lucile, William E, Florence, and Victor Harold, William E being deceased Dr and Mrs Moss are members of the Presbyterian church

FRANK HOLLOWAY, deceased, who resided on his valuable irrigated farm of a hundred and sixty acres, which lies in Morrill county, spent thirty-three busy years here and was one of the county's well known and highly respected citizens Mr Holloway was born in California, August 18, 1858

The parents of Mr Holloway were Marcus and Eliza (Neal) Holloway, the former of whom was born at Dayton, Ohio, and the latter of Indiana They lived for a while in Iowa and then concluded, in 1852, to cross the plains to California It was a tiresome and even dangerous journey at that time on account of the Indians, many of whom were savage and revengeful Before they started they provided themselves most fortunately with trinkets and other articles to be used as gifts in case they had need to seek friendly help on the way, and thus they had no misadventures but on the other hand lost fear of them because they reciprocated kindness It was not so, however, with all the white travelers One boastful man in the wagon train following that of Mr and Mrs Holloway, declared that he would kill the first Indian he met and followed out his threat In a very short time his train was surrounded by a violent savage band of Indians demanding the slayer be delivered to them, which had to be done and he was killed by the most inhuman torture the Indians could conceive The parents of Mr Holloway reached California safely, the mother riding a pony almost half the distance Of their five children Frank was the first born

Frank Holloway was reared in California In 1886, he came to Nebraska and homesteaded and owned his original farm till his death It was entirely unimproved and for a long time he lived in a dugout as did the most of his neighbors Like other settlers about that time, he saw much hardship in the way of unusual storms and loss of crops and stock from floods and drouths In those early days there were no life-giving ditches with flowing water, the arid plains seemed entirely given over to the high wild grass that extended for miles like a swaying sea No early settler can ever forget that sight It is wonderful to note the contrast between then and now Mr Holloway was a very hard worker all his life but he had in his beautiful farm something to show for his industry His place was well improved and he had a very attractive home

In October, 1916, Mr Holloway was married to Miss Minnie Eding, who was born in Hendricks county, Indiana Her parents were William and Lora (Porter) Eding, both of whom were born in Hendricks county While in Indiana Mr Eding followed the carpenter trade, later was a farmer in Iowa, and in 1888, homesteaded in Morrill county, Nebraska At the time of his death he was conducting a meat market at Bridgeport. The mother of Mrs Holloway resides in that place, being now in her seventy-seventh year Mr Holloway has always taken a good citizen's interest in public matters but has never desired a political office He has consistently voted the Republican ticket since reaching manhood, and takes pleasure in the realization that he has helped in the election of many men of sterling character in both national and local bodies

ISAAC WILES, whose years have covered some of the most remarkable events in the history of the United States, at times has borne an important part in this great period of development Pioneer, cattleman, miner, soldier, statesman and farmer, all his life he has shown the sturdy qualities that are truly American He is widely known in Scottsbluff and in other counties of Nebraska, and his name belongs on the list of men whose memory is worthy of perpetuation because of their usefulness in their day and generation Mr Wiles has been a resident of Scottsbluff county for thirty-three years, and there have been few substantial developments in this section in which he has not taken interest

Isaac Wiles was born October 25, 1830, in Henry county, Indiana His parents were Thomas and Elizabeth (Hobson) Wiles, both natives of North Carolina, in which state these family names are numerously found Thomas Wiles was born in 1804, and died May 12, 1873 Elizabeth (Hobson) Wiles was born in 1797, and died March 26, 1878 Thomas Wiles was a carpenter by trade In 1852, he removed with his family to Mills county, Iowa, took up land and engaged in farming until the close of his life Of his family of children Isaac displayed unusual mental ability and was given educational advantages in a college at Savannah, Missouri This was in the days when public excitement ran high on account of the discovery of gold in California, and young Isaac partook of the unrest and instead of remaining to complete his college course, decided to try his luck with others who were flocking to the western land of promise It was in 1852, after the family had become established in Iowa, that he started across the plains with the intention of driving cattle as far as possible, on the way to the coast He passed through Nebraska and was so well pleased with the appearance of the state that he determined to locate here

Mr Wiles was a resident of Nebraska during the Civil War period and from this state enlisted as a soldier, becoming first lieutenant of Company H, Second Nebraska Volunteer Infantry, and later was captain of Company B in the same regiment When he returned to civil life he settled in Cass county, Nebraska, where he soon became prominent in Republican politics and was elected a member of the state legislature at a later date, in which body he made an impression through the efforts he made to bring about valuable legislation It was in 1886, that Mr Wiles came to Scottsbluff county and secured his homestead of a quarter section Like his fellow settlers of that time, he passed through hardships innumerable but never became discouraged as to the final value of land through this section He was one of the first to accept the idea of irrigating the arid tracts, did all in his power as a citizen to advance the great government projects, and still owns his old homestead, which under irrigation is proving how bountiful Nebraska soil may be Although the weight of years has deprived him of the vigor of youth, he still takes an interest in his beautiful farm and feels well compensated for years of toil

Mr Wiles was married October 30, 1856, to Miss Nancy Elizabeth Lunville, who died October 10, 1918 Her parents were Henderson and Ursula (Day) Lunville, the former of whom was born in Tennessee and the latter in Kentucky They were farming people long ago in Mills county, Iowa To Mr and Mrs Wiles the following children were born Mary Jane, who is the wife of D B Dean, of Mills county, Iowa, Ursula, who is the wife of Davis Everett, of the state of Washington, Abraham Lincoln, who lives at Plattsmouth, Nebraska, Jessie, who is the wife of J H Hall, of Cass county, Nebraska, Edward M, who resides at Minatare, Nebraska, Grace, who is the wife of M A Hall, of Mills county, Iowa, William, who is deceased, Thomas F, who is an attorney at Omaha, Luke L, who resides at

Plattsmouth, Nebraska, and Isaac R, who lives at St Louis, Missouri Mr Wiles is a member of the Christian church

MARK R HOOKER, who was one of Scottsbluff county's substantial men and good farmers, lived in this county many years and assisted in its agricultural development Mr Hooker was born in England, February 14, 1858 His parents were Robert and Mary Hooker, both natives of England They came to the United States and to Nebraska in 1879, and the father took up a tree claim in old Cheyenne, now Scottsbluff county Later he sold his property here and moved to California where both he and his wife lived until they died

Mark Hooker came to the United States first in 1876, later went back to England but in 1879 returned to America with his parents At first he did not engage in farming as he was master of a good trade and found profitable work at it in the city of Omaha, where he was one of the first practical plumbers In 1883, he took up a tree claim in Scottsbluff county at about the same time that his father filed on more land Mark Hooker's farm has never passed from the possession of the family and is now owned by his widow Thomas Hooker took up a homestead about this same time which he later traded for one a mile north of the original farm and still owns this property Mark Hooker had a large part of his land under cultivation and made many excellent improvements on the place For some years he was also interested in the plumbing and heating business, but later failing health caused him to give up all active life and with his wife he went to California, where he lived until his death from cancer of the stomach He passed away in 1919, at the age of sixty one years and was buried at San Bernadino, December 22d, with impressive services by the Masonic order to which he belonged

At Lincoln, Nebraska, in 1881, Mr Hooker was united in marriage to Miss Nellie A Barnes, who is a daughter of Addison G Barnes, and they have three daughters, namely Mrs Grace Pierce, who lives in California Mrs Leonora Miller, who lives in Scottsbluff county, and Ethel, who resides with her parents The family belongs to the Episcopal church Mr Hooker is an Elk and a Thirty-second degree Mason

FRANK SYDNEY FADEN — Credit this farmer of Irish extraction to the Kimball district, but give credit to the man himself for his thrift and enterprise and let it be known that Kimball county, in common with most other sections of this great country, is deeply indebted to Irish-American blood It is said that the Irish-Americans always succeed, in whatever line of endeavor they elect for a life work and Mr Faden is no exception to this rule He is a Badger, born at Salem, Wisconsin February 19 1876, the son of James and Elizabeth Mary (Armstrong) Faden, both natives of Ireland, the former born there in 1820 and the mother May 18, 1835 As young people they left their native land to seek fortune in the new world and were married in New York in 1852 Ten of their children are still living Sarah, of Kimball, Nebraska, John, of Lawrence, Kansas, Lizzie, who resides at the old home, Salem, Wisconsin, Elmer and Charles, twins, also of Salem, Wisconsin, Henry, of Burlington, Wisconsin, Emerson, of Banner county, Nebraska, Elizabeth and Carrie of Silver Lake, Wisconsin, and Frank, of this review

Mr Faden was reared in the invigorating climate and environment of his native town, received an excellent practical education, and while still a young man determined to come west to take advantage of the fine land offered at low cost west of the Mississippi river Coming to Nebraska he located in Banner county, August 3 1898, taking up a homestead which he still owns Mr Faden endured many of the hardships of the late 90's but was determined to stick to his land, as he had faith in this section of the country and his belief has been justified He managed to live through the lean years, prospered in his agricultural business as he is one of the good practical men of the Panhandle who has not been afraid of hard work He adopted modern methods of farming as he saw that they increased production, and from time to time bought more land until today his holdings consist of two and a half sections

For the first ten years he was in the Panhandle, Mr Faden specialized in sheep raising, then took up the cattle business in which he has met with gratifying success and now runs a hundred and twenty head

March 25, 1908, Mr Faden married Miss Nellie Gertrude Dunn, a history of whose family will be found elsewhere in this volume One child has been born to this union, Frances Elizabeth, a little over a year old

Mr Faden is a self-made man, having won his present fortune through his own unaided efforts; he and his wife are good neighbors and enjoy the confidence of their many friends socially and in business as well

WILLIAM CHRISTIAN EBER — Still rated among the younger generation of agriculturists of Banner county, William Eber is one of the progressive business men and energetic tillers of the soil upon whom much depends for the future prosperity of the Panhandle His varied business career and intimate knowledge of conditions prevailing here — a knowledge gained through experience while running a threshing outfit — is knowledge which is assisting him materially in his own business and material advancement

William Eber was born in Macoupin county, Illinois, the son of Jacob and Eliza Eber The former was born in St Clair county, Illinois, where he engaged in farming when old enough There were four children in the Eber family Frederick, who died at the age of nine years William, of this sketch, Aneda and Elsie, both at home with the parents William Eber's birth occurred December 7, 1889, so he is still one of the young men engaged in agricultural industry in his district 'He spent his boyhood on the home farm in Illinois, attended the public school and worked on the place during the vacations, thus gaining a first hand, practical knowledge of farm business which has proved of value to him since locating in Banner county Soon after his twenty-first birthday Mr Eber went to the Dakotas where he remained two years before returning to Illinois and within a short time went to Detroit, Michigan, where he entered the leading automobile school of the country and spent four months in the institution learning the construction and technique of motors Returning to Illinois, he became the manager of a steam threshing crew which covered a large section of the wheat country threshing grain It was on May 5, 1917, that Mr Eber reached Potter, Nebraska, and within a short time put his training in the automobile school to practical use running a tractor for Lee J Peterson, breaking new land For a year or two he again managed a threshing crew in the Panhandle before actively engaging in farming on rented land Two years later Mr Eber, in partnership with Leo Young, bought eight hundred acres of land in Banner county, as they believed in the future of this rich section of the state They have been actively engaged in general farming and stock raising until

recently, meeting with well deserved success in their chosen line of endeavor, and today are rated as two of the progressive and prominent men in agricultural business who are important factors in the development of the section Mr Eber and Mr Young turned their land back and now Mr Eber rents a half section

On October 20, 1919, Mr Eber was united in marriage with Miss Bertha A Johnson, the daughter of Magnes and Carrie Johnson, both natives of Sweden They were early settlers of Kimball county where Mr Johnson died in June, 1916 His widow still lives on the home place Mr Eber is a member of the Modern Woodmen

CONRAD A JOHNSON, who is well known at Pine Bluff, Wyoming, and throughout Kimball county, Nebraska, has long been numbered with the substantial farmers and stock-raisers of this section of country He has been particularly successful in the stock business and since disposing of some of his land in Kimball county, is making preparations to enter into very extensive stock-raising when he secures a large enough range

Conrad A Johnson was born in western Nebraska, November 2, 1885 His parents were Peter and Sophia Johnson, both of whom were born in Sweden, the father on January 27, 1845, and the mother on December 10, 1843 They spent their childhood and youth near Linkoping, were married there about 1869, and shortly afterward came to America, landing in the harbor of New York City From there they came to Nebraska and the father rented land in the eastern part of the state until about 1886, when he decided to homestead in Wyoming Taking his oldest son, Charles, with him, he drove across the country in a prairie schooner and homesteaded about two miles south of Pine Bluff His family came by railroad and joined him at that place and there they all remained until 1900, when the father sold his homestead and retired to Pine Bluff for the rest of his life The mother died there December 5, 1901, and the father February 10, 1913 Of their seven children Conrad A was the seventh born, the others being as follows Anna, who died in infancy in Sweden, John and Betty, both of whom died in eastern Nebraska, Charles, who is in the hardware business at Pine Bluff, Elmer, who is a farmer and stockraiser in Kimball county, Erich, who

JOHN NEWELL AND FAMILY

is also a farmer in Kimball county, and with his brother, Conrad A, owns a large part of section 14-13-58 in Kimball county

Conrad A Johnson assisted his father and went to school until about fifteen years of age, since which time he has largely engineered his own affairs and has done well He began with a small bunch of stock and herded it on the free range as a beginning After the Kincaid act became a law, he homesteaded section 22-13-58, improved his property with fine buildings, expanded his stock business and now has a hundred and fifty head, mostly cattle, and about a hundred and seventy acres under farm cultivation

On May 10, 1911, Mr Johnson was married to Miss Esther Ekstrom, a daughter of Mathias and Josephine Ekstrom, residents of Pine Bluff Mr and Mrs Johnson have two children Maxine, who was born June 9, 1913, and Conrad W, who was born February 12, 1917 Mr Johnson belongs to Kimball lodge No 294, A F & A M, Kimball, Nebraska, and also to the order of Royal Highlanders at Pine Bluff

JOHN NEWELL — An indomitable energy that has triumphed over seemingly great obstacles, as well as varied fortunes, is that which has dominated John Newell during the varied stages of a remarkably earnest and productive business career in which he has rallied to his cause splendid initiative ability and has spurred fortune until she has smiled upon him Depending entirely upon his own resources he has pressed forward along the line of worthy ambition and that he has arrived at the goal of substantial success and influence in connection with business operations needs no further voucher than the fact that he is now the owner of many thousand dollars worth of fine arable land in Scottsbluff county Energy, progressiveness, and correct business policies have enabled Mr Newell to achieve unqualified success in the different lines of farm industry, which he early chose as life vocation John Newell was born in Louisa county, Iowa, April 8 1856, the son of Robert F and Christiana Newell, both natives of the Buckeye state Matured and invigorated by hard labor, Robert Newell had a fine physical constitution, was sturdy and lived his full "three score years and ten," passing away in his seventy-fifth year, while his beloved wife who had been his devoted companion and helpmate for nearly a half century — a woman whose strength has been as the number of her days and who had many pioneer experiences in the great west, lived to the mature age of ninety-three Robert F Newell was for a long term of years recognized as one of the most progressive and substantial farmers, stock-feeders, and stock-shippers in his state, owning four hundred acres of land which was highly improved, having an ample supply of the farm implements of his day which greatly facilitated his operations They lived in a day of great piety, large families and plain living were much more common than in the twentieth century day of hustle and progress, the olive branches around their table numbered eight Elizabeth, the wife of Henry Cushman, a physician in Tacoma, Washington, Thomas, who died in Iowa, Caroline, the wife of Lewis E Riley, who also died in Iowa, Mary, the wife of E S Curtis, living retired in Iowa, John, the subject of this review, William, living in Mitchell, Nebraska, Robert, a resident of Iowa, Hattie, the wife of Frank Sidman, has a happy home in Kansas

Robert F Newell received his elementary education in Ohio, his native state, but being an ambitious youth determined to broaden his education and opportunities and became an expert bookkeeper He stood high in Masonic circles and in politics was a recognized stalwart in the ranks of the Democratic party Because of integrity and superior attainments he stood high in the esteem of his fellow citizens, which was shown by them in selecting him to serve as county supervisor, an office he filled with marked capacity John Newell was reared on his father's farm in Iowa, where he early gained experience in herding cattle and recalls some of the pioneer experiences of that locality He was afforded the advantages of the public schools and for some years continued to be associated with his father in farm industry His father having been a cattleman, it was but natural that the son should find that ranch life appealed to him and believing that greater opportunities were offered in the west he went to Wyoming in 1885, where for eleven years he was engaged in stock raising on a large ranch before coming to Scottsbluff county in 1896 Upon reaching Nebraska Mr Newell purchased one hundred and sixty acres of fine land in section 4, township 9, where he soon established a home Ten dollars an acre was the purchase price and we may well appreciate the success that has crowned the work of this man who today is the exponent of agricultural and live stock enterprise in this section, when we learn that recently he refused three hundred dollars an acre for his holdings Not only is Mr Newell a farmer he is a man of the world, having found or taken time to study political and economic affairs that not only affect his own business but that of the whole country He has never had time to devote to public office but gives earnest and active support to the Democratic party Mr Newell has been a devoted father to his family of six children, to whom he has given every advantage They are Elizabeth, the wife of J Fadney, a farmer of Saginaw, Oregon, Caroline, the wife of William Howard lives on a ranch near Mitchell, Hattie May, the wife of Robert Newell, whose death in the officers' training camp at Fort Snelling, Minnesota, was a heavy blow to the entire family; who, however, have the satisfaction and comfort that he cheerfully and bravely gave his life for his country, Mary, the wife of Walter Nunn, on a ranch in Wyoming, Robert, now living on the home place south of Mitchell, and Maud, the wife of Novrel Laberten, a ranchman of Wyoming

Mr J Newell passed away on the 22nd of April, 1920, at Eugene, Oregon, but was buried in Scottsbluff county

JENS JENSEN, who has been a resident of the Panhandle for more than a decade and is well and favorably known in Kimball county as a farmer and in Dix as an extensive dealer in real estate, is one of the younger business men of southwestern Nebraska who has nobly taken part in the development of this section and who responded to the call of his adopted country when was was declared against Germany Mr Jensen is a native of Denmark, that country which has supplied the United States with so many of her sturdy and worthy citizens, who are known for their industry and ability

Jens Jensen was born in the Danish peninsula, July 27, 1886, the son of Chris and Annie Jensen, the father having devoted his life to agricultural pursuits in his native land There were eight children in the family, five boys and three girls, along with the other children Mr Jensen attended the excellent public schools maintained by the government in Denmark and thus laid the foundation for a good practical education which has been of great value to him in business He remained at home with his parents, working on the farm summers until his education was completed and then devoted all his time to farming until 1911, when he broke all the home ties to sail for America, as he had determined to come to this land of promise to seek his fortune, for

here, free land could be obtained at a low cost Landing in New York Mr Jensen soon started west, came to Nebraska City, Nebraska, where he remained four weeks looking over the country over and to learn of the different sections His brother, Ollie, had already located in Kimball county on land situated twelve miles south of Dix and there Mr Jensen began his career in the new country He first worked on the farm to learn the American way of doing things, but in 1914, rented a place of his own four miles east of Dix where he remained a year Realizing the advantages of a commercial training, Mr Jensen then entered a business college at Grand Island for a special course which he believed would assist him Some time later he returned to Kimball county and bought a quarter section of land south of Dix which he farmed until October, 1917, when he enlisted in the army at the President s call for volunteers Mr Jensen was sent to Fort Riley for his preliminary training, having been assigned to the Three Hundred and Eleventh Cavalry Six months later he was transferred to the Field Artillery He remained in that branch of the service until honorably discharged December 22, 1919, at Camp Stetson, Kentucky Returning to Kimball county, Mr Jensen again engaged in the active management of his land and began to carry on a successful real estate business, a vocation in which he is meeting with well earned and deserved success He is today one of the leading members of his community and well demonstrates that a young, ambitious man with no other equipment than his long head, two capable hands and the determination to succeed can accomplish in the Panhandle Mr Jensen has a brother, Christian, who is a successful farmer near Gridley, California, which shows that the training the young men received in their native land has been well applied in the land of their adoption Mr Jensen is a member of the Woodmen of the World and will join the Masonic fraternity at Potter in the near future He is progressive in his ideas, believes in adopting modern farming methods and finds they pay, he supports and contributes to all movements for the development of his town and Kimball county

GEORGE LESTER VOGLER, cashier of the Bank at Kimball, Nebraska, has been connected with banking institutions since he completed his education and has advanced to his present important position through industry and personal merit His career has been the expression of well directed and well applied principles, and he has thus succeeded in building for himself a reputation as an astute man of business and adherent of sound, conservative banking principles, which leads to confidence with the public

Mr Vogler was born at Dix, Nebraska, September 23, 1889, the son of Henry and Clementia Vogler, a complete sketch of whom will be found elsewhere in this volume As a youth Mr Vogler was sent to the excellent public schools for his elementary education, graduated from the Kimball high school in 1906, and then took a special commercial course in the Lincoln Business College, to prepare himself for a business career Soon after leaving college he accepted a position with the First National Bank of Lincoln, where he spent two years mastering the varied branches of banking and made such good use of his time that, in 1909, he was offered the position of assistant cashier of the Bank at Kimball, which he accepted From the first Mr Vogler showed marked ability among the banking fraternity of the Panhandle, he had sound business sense and retained this office six years He was elected cashier of the bank in 1915, and has continued in that capacity to the present time Mr Vogler in his official capacity has contributed materially to the growth and prosperity of the institution, at the same time advancing his own standing in banking circles, as a capable and thoroughly informed banker

From first locating in Kimball Mr Vogler has taken an active interest in community affairs and served as a member of the village board, where his foresight and progressive ideas have helped in the development of the town Mr Vogler is also a director of the bank in which he serves

July 23, 1913, occurred the marriage of Mr Vogler and Miss Harriet E Rollings, at Lincoln, Nebraska Mrs Vogler is the daughter of Eunice Rollings She is a graduate of the Lincoln high school, the State University, where she specialized in physical training and after receiving her degree taught in Evansville, Indiana two years before her marriage There are three children in the Vogler family Dorothy, aged six, John Rollins, aged four, and George Lester Jr, a small boy of about two years The Vogler family attend the Methodist Episcopal Church, Mr Vogler is a 32d degree Mason and a Shriner, he also belongs to the Modern Woodmen of America and the Knights of Pythias In politics he is a liberal Democrat, not being bound by strict party lines when a good man for office is considered Though young, Mr Vogler has gained the confidence of the people of the southwest Panhandle as a banker, devotes his entire time and attention to the duties of his office and is regarded as one of the rising men in commercial and banking circles in western Nebraska

GOTFRED JENSEN has been numbered among the vigorous representatives of farm industry in Garden county for more than a quarter of a century, and thus it becomes evident that he here encountered his quota of the hardships and responsibilities that marked the pioneer epoch in this section of Nebraska He has brought to bear the perseverance, energy and sterling honesty of purpose that so emphatically characterizes the race from which he sprung.

Mr Jensen was born in Denmark, on May 2, 1864, and received his early education in the excellent schools of his native land There he continued to reside until spurring ambition and self-reliance led him to come to America and seek the better opportunities that are offered here. He was twenty-three years of age when he arrived in this country, and made his way to Council Bluffs, Iowa, where he was identified with business activities for about five years He then came to that part of Deuel county, Nebraska, that is now comprised in Garden county, and in the same year, 1893, he filed entry upon a homestead of a hundred and sixty acres, ten miles northeast of Oshkosh He not only perfected his title to this claim and developed it into a productive farm, but he also added gradually to his holdings, until he is now the owner of a well improved and valuable farm estate of nine hundred and sixty acres, in the work and management of which he has been ably assisted by his sons, who are now his partners in the well ordered industrial enterprise, which involves general agricultural industry and the raising of excellent grades of cattle and hogs After many years of earnest and prolific industry, attended with success, Mr Jensen felt justified in laying aside the heavier labors and responsibilities that had long been his and after assigning the active management of his farm property to his sons he retired and removed to Oshkosh, in the spring of 1919. He has supported those measures that have tended to advance the general welfare of the community, is affiliated with the Modern Woodmen of America, and he and his wife are zealous communicants of the Lutheran church, his political allegiance being given to the Democratic party.

At Council Bluffs, Iowa, November 10, 1889, Mr. Jensen wedded Miss Petria Fredericksen, who likewise was born and reared in Denmark and who was eighteen years of age when she came to America and established her residence at Council Bluffs. Mr. and Mrs. Jensen have three sons and five daughters: Anna is the wife of Roy Bentz, of Oshkosh, and they have three children; Fred C., is again at home in Oshkosh, served in the Seventeenth Hospital Corps of the American Expeditionary Forces in France during the progress of the late war, and he was with his corps in France for a period of nine months. He married Daisy Riley and lives on the home farm, they have one boy, Harley; James T., who likewise resides at Oshkosh, is a widower with one child, Bernard; Clara is the wife of Carl Hendricks of Oshkosh, and they have one child, Floyd G.; Martha M. married Irvin McConkey and lives on a farm; Frank L., Olga and Eva are all at home. As previously stated, the sons are associated with their father in the operation of the latter's fine ranch, which is situated near the village of Hutchinson, where Mr. Jensen served for a number of years as a member of the school board.

CALEB W. GAULT.— Though he made four removals, from one locality to another, within a period of fourteen years, there has been in the career of Mr. Gault no deviation from the line of purposeful energy in connection with the basic industries of agriculture and stock raising, of which he is now a substantial and popular exponent in Garden county.

Mr. Gault is one of the valued citizens contributed to the Nebraska Panhandle by the old Hoosier state. He was born in Montgomery county, Indiana, December 24, 1853, and thus, with infantile nonchalance, passed his first Christmas day as a new arrival in the home of his parents. His father, John Gault, was a millwright by trade and vocation and was sixty-three years of age at the time of his death, in 1894 He was a native of Kentucky and a

scion of a pioneer family of that commonwealth.

Caleb W. Gault is indebted to the public schools of his native state for his early educational advantages, and he has been continuously identified with farm enterprise since he was twenty-three years of age. He came to what is now Garden county, Nebraska, years ago, and to the homestead which he entered he has not only perfected his title but has also added, under the provisions of the Kinkaid act, until he now has a well improved and valuable farm estate of six hundred and forty acres, all fenced and cross-fenced, equipped with good buildings and modern working facilities and elegantly situated not far from Oshkosh. On his farm Mr. Gault is carrying forward successful operations as an agriculturalist and stock grower, and he is one of the representative citizens of his community. His political allegiance is given to the Republican party and his civic loyalty has been shown in ten years of effective service as school director of his district.

November 9, 1879, recorded the marriage of Mr. Gault to Miss Rose L. Dowling, a native of Missouri, and she has proved a true helpmeet to him in his advancement to the goal of independence and prosperity. In conclusion is given brief record concerning their children: Nora is the wife of Edward Hamlin, of Bellefourche, South Dakota; Allen F. wedded Miss Lucy Embell, in March, 1910; Mamie is the wife of Ogden Fought, of Oshkosh, Nebraska; Ava wedded Miss Lavina Adking; Albert and Vera continue to infuse youthful joy in the parental home.

MRS. WILLIAM BRODRICK, one of the later settlers of Alliance who is winning a fine reputation as a business woman, is the owner and manager of the American Hotel which was opened to the public during the late summer of 1919, and already has a growing and most satisfactory clientele among the traveling public which patronize the many hotels of Nebraska.

Mrs. Brodrick was born in Griswold, Iowa, the daughter of John H. and Mary D. (Thompson) Carroll, the former a native of the Dominion of Canada, while the mother was born at Griswold, Stark county, Illinois. Mrs. Brodrick was the oldest of four children born to her parents and when she was only a year and a half old her father removed to Omaha, as he was a civil engineer and his business caused the

Mr. and Mrs. C. E. Peterson

family to take up their residence in Nebraska. The little girl was sent to the public schools when old enough and she continued her studies until a four year high school course was finished and she graduated with honors in 1899, and the following year she attended a commercial college, finishing the course in a year. In 1909 she was married at Boseman, Montana, to William Brodrick, who was born at Hubble, Nebraska, the son of William and Mary (Stark) Brodrick, the father being a native of Wisconsin. Mr. Brodrick was the fifth child in a family of eight children and as his mother died when he was a small boy eight years of age, he was from that time compelled practically to shift for himself. He managed in some way to earn his living while yet a little lad and finally by pure grit and perseverance worked up to a good salary and as best he could studied and secured a practical education. One child has been born to Mr. and Mrs. Brodrick, William H., a student in the public schools of Alliance who lives with his mother. Mrs. Brodrick purchased the American Hotel located over the Harper Store, on August 1, 1919. This location on Box Butte avenue is an ideal location for an European Hotel, and she has thirty-two fine rooms for the accommodation of guests, with every modern convenience for their comfort. The location is central to all parts of the city and of great convenience to the traveling public, and already its patronage is bringing in a fine return for the investment. From present indications Mrs. Brodrick has invested in a most satisfactory business which has a great future. She is a woman of great charm, has unusual executive ability and knows how to look after the many guests to whom she is a considerate hostess.

CLAUS E. PETERSON, who is one of Banner county's sturdy pioneers and still living on his original homestead on section 2 has made Banner county his home for thirty years, and is widely known and universally respected. He was born in Sweden, September 8, 1853, the second in a family of five children born to Peter and Clara (Anderson) Johnson Peterson, the others being John, who lives in Sweden, Fred and Ephraim, both of whom live at Minneapolis, Minnesota, and Amanda, who has never left Sweden. The father was a farmer and stockman in Sweden. Both parents died there, the mother in 1915 and the father in 1916. They were members of the Swedish Lutheran church.

Claus E. Peterson attended school in his native land in boyhood and afterward worked as a farmer until 1888, when he came to the United States. After one year in Iowa he came to Banner county and on January 1, 1889 filed on his homestead. He later took a preemption and secured eighty acres on a Kinkaid claim. Mr. Peterson now owns 720 acres, mostly range land. He raises about 50 head of White Face cattle annually and other stock. He has one of the best improved ranches in the county and it is a contrast indeed to compare his present fine residence with running water, electric lights and all other modern comforts, with the little dugout in which he sheltered himself in his earliest days here when he had so much to contend with that he gladly accepted a wage of fifty cents a day, paid in trade, in order to sustain himself. With the first money he was able to save, by working in the stone quarry at Fort Collins, Colorado, he bought a team of oxen, paying $75 for this team, which was a necessity in order to break up the tough prairie sod. His first crop was not very satisfactory, in great measure from lack of moisture in the soil. He was too poor then to have a well dug and had to haul all the water used on the place. He struggled on however in spite of discouragements and in the course of time had a fine flock of sheep, about 500 head. A sudden blizzard swept over the land and in its fury drove the sheep before it. When Mr. Peterson was able to search for them he found them all piled together, about two miles distant, all covered with the drifted snow and very few of them alive. This was a serious loss, although he was able to sell the pelts for $1 each. In seasons following Mr. Peterson, through hard work, brought his land to a profitable state of development and for a number of years has been financially independent, but he has not forgotten those old days, nor his old neighbors, and his reminiscences of actual facts are interesting sidelights on the county's history.

In Sweden, September 8, 1881, Mr. Peterson was married to Miss Hilma Peterson, a daughter of Nels and Bertha (Nelson) Peterson, who came to the United States in 1886 and homesteaded in Banner county in 1888. The following children have been born to Mr. and Mrs. Peterson. Helge S. who was born April 29, 1882, resides with her parents, Ernest L., who was born April 16 1884 lives in Banner county, Richard M., who was born November 10, 1885, assists his father, Gertrude A., who was born October 24, 1888, lives at Minneapolis, Minnesota, Esther N., who was born January 18, 1891, is the wife of John Cedarburg, of Minneapolis, Edith E, who was born May 24, 1895, is at home, and Minnie V, who was born September 17, 1898, lives at Minneapolis Mr Peterson and wife were members of the Swedish Baptist church Mrs Peterson died February 25, 1920 He has been very active in politics but has always voted with the Republican party

CHENIA A NEWBERRY — The business career of Chenia A Newberry has been significantly characterized by courage, self-reliance, progressiveness His unchangable purpose, high integrity have won the confidence and esteem so essential in the important mercantile enterprise to which he has devoted his attention and energies, and through which he has gained high standing in the financial circles of western Nebraska, eastern Colorado and South Dakota During practically all his business career Mr Newberry has been associated with the hardware business and no other vouchers are needed to attest his success than the substantial establishments of which he is the owner in Alliance As one of the representative business men and progressive public spirited citizens of Box Butte county he merits specific recognition in this publication of the great commonwealth of Nebraska and its counties

Chenia A Newberry is a Wolverine, born in New Baltimore, Michigan, April 9, 1869, the son of Norman and Fannie (Morris) Newberry, the former having been a farmer Chenia was the youngest of two children, as he had one sister The boy passed his childhood and early youth on farms, acquiring the rudiments of an education by attending the common schools of Michigan in the winter time, and after he accompanied his parents to their prairie farm in Buffalo county, Nebraska, was sent to school in a sod house, and though it contrasted to what he had known in Michigan the boy made the best of his advantages This was in 1881, when the conditions of life and education were far different from those of the present day On the homestead twelve miles north of Kearney, Chenia Newberry grew to early manhood, enured to the hardships of frontier life which held no tears for him nor daunted his spirit Until the age of seventeen he remained on the farm, then determined to seek a commercial career The youth found employment in a hardware store in Ravenna in the northern part of Buffalo coun-

ty where he laid the foundation for his subsequent commercial career, which reads almost like a fairy story, as few men achieve such great success within such a short time In 1888, Mr Newberry came to Alliance to found what is today one of the large, important, and interesting industrial enterprises of this city and Box Butte county, and this publication gives special recognition to this representative corporation, for in the upbuilding of the business has been exemplified the splendid energy and initiative ability of its founder, who has made of his individual success a medium of leverage for the uplifting of civic and material prosperity in his home city and county Of the inception and growth of the business founded by Mr Newberry in Alliance, a brief record will be given

The present enterprise had a humble beginning Mr Newberry first established himself with a hardware stock in connection with which he ran a tin shop, his business grew rapidly, by leaps and bounds, within six years after coming here he found it necessary to build a large structure known as the Glenn Miller building with floor space twenty-five by sixty-four feet, this soon was too small so the building was extended back at the rear for greater floor space Then followed a tin shop built in 1898, and two years later he erected a brick warehouse of practically the same size near the railroad, but the growing business demanded still larger quarters to house the various departments To accommodate these a large three-story and basement structure was erected on Fourth street between Box Butte and Laramie avenues, which is entirely given over to the wholesale business, which has expanded to such proportions that at present Mr Newberry has three salesmen on the road covering a large territory which includes western Nebraska, eastern Colorado and western South Dakota Five years ago the retail business also was established in a new home, this fine building covers a ground space of fifty by one hundred and fifty feet, is built of brick and fire proof From basement to roof it is filled with the finest goods of the hardware trade The basement is used for storage, the main floor for retail trade, the second floor is at present used as a sample room for display of goods, while an up-to-date harness factory with the latest designed machines for this work has been installed on the third floor Fifty men are on the payroll of the Newberry Company which amounts to over $8,000 a month, which gives some slight idea of the immense amount of business transacted within a year

Mr Newberry is progressive in policy which implies up-to-date service to all his customers, and his equipment has all the modern facilities for this purpose His various establishments are a source of pride to the city of his adoption and its many residents In other words, "He has put Alliance on the map" Today this man is one of the best known residents of the northwestern section of our great commonwealth, because of his own reputation as a keen, shrewd, farsighted man and a wholesaler he is full of life and energy which is well displayed in his business houses for he is the busiest one of the fifty or more men to be found on the premises every working day, putting in many more hours at his desk than any of his many employes

July 23, 1893, Mr Newberry married Miss Nellie Brennan, the daughter of Martin and May (Fitzgerald) Brennan the latter being a native of Michigan Mrs Newberry was the sixth oldest in a family of twelve children, consisting of ten boys and two girls There are five living children in the Newberry family May, the wife of Frank Abegg, the cashier of the First National Bank of Alliance, Norman M, who graduated from the high school and then engaged in business with his father, Agnes, a student at St Agnes Academy, of Alliance, Helen, also attending the same school, and Master Bill, a sturdy boy of eight While Mr Newberry is interested in all movements for the civic and public uplift of the community, he has been far too busy with the many responsibilities of his business to take public office, believing the man best fitted to serve should be elected, and votes with this idea in mind Fraternally he is connected with the Woodmen and Knights of Columbus

NATHAN A ROCKEY, the senior member of one of the progressive business houses in Hemingford which is doing an important part in the development of this section is one of the best known automobile agents and dealers in the Panhandle Nathan Rockey was born in Green county, Pennsylvania, in June, 1868, the son of Samuel and Rosanna (Ernst) Rockey The former was descended from a long line of early Pennsylvania settlers Nathan was

the youngest of the seven children born to his parents and his father died when he was but a year old, from diptheria and scarlet fever, which also took some of the children The father had been a farmer and the mother assumed her burden of raising the remainder of her family She remained on the farm until the oldest son reached maturity and married some seven years later After that she felt that he was provided for and that she would not have to work so hard, and was just beginning to take life comfortably when she married a man named Daniel Ackley Young Nathan continued to live with his mother attended the public schools during the winter terms, helped on the farm during vacations and Saturdays and while yet a small boy had learned the practical side of farm work He early learned the value of money by working for it, for he helped a neighbor hoe and tend his corn receiving half the crop, and when just a boy earned his first dollar, which proved to be eleven as his fair share of the corn crop Like all boys he desired a fine saddle and that is what the money was used to purchase When only seventeen years of age the boy started out in an independent business as a liveryman with one team of horses and a buggy He made good on this venture from the start and as soon as his older brother saw this he backed Nathan so he bought another team and from this start in West Alexander Pennsylvania His business grew rapidly and for eleven years he was engaged in this vocation At one time his stable burned and as it was not insured the fire caused him a total loss, but notwithstanding this calamity having only one team and buggy left, he began again and when he closed out his business brought a handsome profit of $5,000 and had made a good living all those years On October 31, 1889, Mr Rockey married at Washington Pennsylvania, Miss Minnie Lloyd a native of the Keystone state, the daughter of Joseph and Amanda (Long) Lloyd She was the oldest in a family of six children After disposing of his business interests in West Alexander, Mr Rockey moved to Claysville Pennsylvania purchased a grocery store and became a merchant a business in which he engaged for ten years but he had heard of the many fine openings for an energetic, resourceful man in the west and after looking the country over came to Nebraska in 1904, locating in Box Butte county where he returned to the calling of his childhood and engaged in farming for two years, then as the country looked good to him, he took a homestead of four hundred and eighty acres and bought a quarter section five miles north and two miles east of Hemingford For eight years he remained on the farm, made good and permanent improvements on his land and became favorably known as one of the prosperous men of his community Wishing to give his family the many advantages to be obtained in a town, Mr Rockey traded some of his land for a store property in Hemingford in 1912 This ground has a frontage of a hundred and twenty-five feet on Main street and is a hundred and thirty feet deep One store on the street has a frontage of fifty feet and is the main part of his garage while the remainder of the ground is covered with a store building Associated with him in his business Mr Rockey has his son Earl S, who married Gladys Danbom, a native daughter of Nebraska, and they have one child, a boy Earl Rockey and his wife are both graduates of the high school in Hemingford and Mrs Rockey also took a business course in the Lincoln Business College, Lincoln, Nebraska The Rockey Company is agent for the Chevrolet, Ford and Chandler automobiles, and in connection they run an up-to-date repair shop, carry a fine line of accessories and maintain an excellent service station The Rockey family belongs to the Methodist Episcopal Church

THOMAS J O'KEEFE, register in the United States Land Office at Alliance, was one of the pioneer newspaper men of Box Butte county and the Panhandle, and as an editor occupies a vantage ground from which to mould public opinion, and the community has reason to congratulate itself that for so many years one of the leading papers of this section was in such safe, sagacious and thoroughly clean hands Mr O'Keefe is a genial Irishman, versatile with his pen, to whom the people of Box Butte county owe much for his fearless discussions of public and domestic questions while he owned the Hemingford Herald

Mr O'Keefe was born at Fulton, Illinois, June 12, 1874, the son of John and Sarah (Killey) O'Keefe, the former born in Ireland, where he was reared and received his early educational training and then emigrated that he might take advantage of the many opportunities offered a young and ambitious man in the new world There

were four children in the O'Keefe family, three boys and a girl, all of whom were sent to the public schools Thomas attended with his brothers and sister until the family came to Nebraska in the spring of 1886, locating in Box Butte county, when he entered the employ of Jean Heath, the owner and editor of the Jean Heath Grip, a paper edited and printed at Nonpariel, Nebraska Mr Heath was one of the old school newspaper men and Mr. O'Keefe says that he was very eccentric in character but a writer of marked ability and his paper was very popular among the early settlers of the county Thomas remained in association with Mr Heath three years and there laid the excellent foundation for his newspaper career. After severing his connections with the Grip, Mr O'Keefe accepted a position with C A Burlew, the publisher of the Box Butte County Rustler, which was issued at Hemingford where he remained for several years gaining still broader experience in journalistic work Desiring further experience, Mr O'Keefe went to Omaha to take a special course in a commercial college, returning to Hemingford in June, 1893, to become deputy postmaster of the town, as his father had just been appointed postmaster under President Cleveland A year later Mr O'Keefe purchased the Box Butte County Democrat published at Hemingford, changing its name to that of the Hemingford Herald which he continued to edit and publish at Hemingford seven years, removing to Alliance in 1901 and at the same time bringing his printing establishment with him Here the paper was printed until 1909, when it was sold to Thomas Pierson Mr O'Keefe had established a fine job printing business in connection with the paper and received good support from the citizens of the community in both enterprises After disposing of his business Mr O'Keefe opened a real estate office which he conducted until 1915, when he was appointed Register of Deeds in the Government Land Office at Alliance, succeeding Judge W W. Wood in this office In 1894 Mr O'Keefe had been appointed Government Land Commissioner and transacted a great amount of business in regard to land and land laws from his office in Hemingford and Crawford and it was, no doubt, due to the excellent service he rendered at that time that he was appointed to the vacancy in the land office where he is again demonstrating his executive ability as a government official

RANCH OF ED. J. McKINNON

Mr. and Mrs. Ed. J. McKinnon

Mr O'Keefe has taken an active part in all affairs both communal and civic since locating in the Panhandle, he is well and favorably known throughout this section where he has a host of warm staunch friends He is a man of great public spirit and gives liberally of time and money to every movement for the betterment of the city and country. He is one of the most enthusiastic "boosters" of the upper valley Mr O'Keefe is a member of the Elks, the Modern Woodmen, the Royal Highlanders and Knights of Columbus

In 1911 Mr O'Keefe married Miss Edith Snodgrass, a woman of high education and marked talents She is a graduate of the State Normal School and at the time of her marriage was supervisor of music and drawing in the city schools of Alliance Four children have been born to Mr and Mrs O'Keefe Edith M, John R, Helen Lucille and Jean Francis

EDWARD J McKINNON, who is one of Banner county's foremost busines men, is president of the Farmers Elevator and Store Company at Bushnell, and is one of the extensive breeders of Percheron and Shire horses, Duroc-Jersey hogs and White Face cattle in the county He was fourteen years old when he accompanied his parents to Banner county and this has been his home ever since

Edward J McKinnon was born at Muskegon, Michigan, June 19, 1875 His parents were Hugh and Elizabeth (Meikle) McKinnon, natives of Scotland, his birth taking place in 1830 and her birth a little later They were married in the city of Glasgow, Scotland, January 1, 1867, and in the same year came to Canada and in 1872 to the United States He had served an apprenticeship of five years to the machinist's trade in Scotland and after coming to this country worked at the same in Michigan and later in Iowa As a family of sons began to grow up around him he felt it to be his duty to look out for their future welfare There was little to encourage him in Iowa as the land was rapidly being settled, with consequent advance in price, and this led to his moving to Banner county, Nebraska He homesteaded in 1889 near where his son Edward J McKinnon now lives, shipping his goods from Iowa and being in somewhat better financial condition than many of his neighbors He remained on his homestead until 1904 when he sold his farm to his sons Edward J and Norman N, retired and moved to Harrisburg His death occurred June 19, 1904

After becoming a widow, the mother filed on a Kinkaid claim near the other property and resided on it five years Her last days were spent in the homes of her children, and her death occurred May 3, 1918 Of the ten children the following survive Robert, of Armour, South Dakota, John, of Merced, California, Edward J, of Banner county, Harry C, of Scottsbluff, Norman N, of Scottsbluff, Hugh A, of Bucyrus, Ohio, Mrs Annie R Kelly, of Harrisburg, Nebraska, and Mrs Nettie Schumway, of Scottsbluff The parents were members of the Presbyterian church The father assisted to organize school district No 31, and served as school director for some years Like other early settlers Mr McKinnon met with discouragements and losses, mainly occasioned by drouths, and for long periods had to haul water a distance of from five to thirteen miles At one time he went to Cheyenne and began work in a machine shop there but illness caused his return home and after that he continued on the farm He was a man of sterling character and is yet mentioned in terms of respect

Edward J McKinnon attended the public schools in Iowa and in Banner county and later took a busiess course at Shenandoah, Iowa, from which he was graduated with a diploma testifying to his excellence in penmanship He remained at home until 1905 On November of that year he was united in marriage to Miss Mary Barfoot, who is a daughter of Enos Barfoot, a prominent early settler of the county, extended mention of whom will be found in this work Mr McKinnon owns 640 acres of fine land, his home farm being exceptionally well improved and his operations are carried on with improved farm machinery As a breeder of fine stock he has been very succesful and averages from 75 to 100 head of cattle yearly In addition to the business enterprises above mentioned, he is a stockholder in the Banner County Bank of Harrisburg and also a stockholder in the Farmers Union Politically he is a Republican, but aside from school offices, he has never accepted a public office He has always been interested in school district No 31 and as school director for some sixteen years has looked after its best interests Mr McKinnon is a man of principle, upright, honest and dependable under all circumstances

BERT R WEBER, who is numbered among the successful younger exponents of agricultural and live-stock industry in Scottsbluff county, has the distinction of being a native son of this county, his birth having taken place on the home farm of his father, near Gering, on November 14, 1890 Of the family genealogy adequate record is given on other pages, in the sketch of the career of his father, William Weber

Bert R Weber early gained agricultural experience in connection with the activities of the home farm, and his youthful educational advantages included those of the public schools of Gering He continued to be associated with his father in the operations of the latter's ranch until he had attained to his legal majority, when he began his independent career in connection with the same important line of industry As a general agriculturist and stock-grower he is alert and progressive, and now conducts operations on a well improved place of two hundred and seventeen acres, all of which will eventually have ample irrigation facilities and upon which he is making the best of improvements This ranch is situated in section one, township twenty-one, about a half mile distant from the county seat which is the postoffice address of Mr Weber On his farm he has a fine French draft stallion, and he is making a specialty of breeding this type of horses

In 1914 was solemnized the marriage of Mr Weber to Miss Lucile Duff, who was born and reared in Nebraska, and they have two children, a winsome little daughter, Marvel, and son, Kenneth Elsworth Mrs Weber was educated in the public schools of David City, Nebraska, and she holds membership in the Methodist Episcopal church In politics Mr Weber gives allegiance to the Nonpartisan party in national and state campaigns, but in local affairs he maintains an independent attitude

HUGH F MANTOR — In promoting general efficiency along all lines of human endeavor there has come in these later days a distinct recognition of the supreme value of effort and concentration This is especially true in the medical profession and such exponents find the maximum success and are able to give the most benignant service through devoting their attention to perfecting themselves in some one special branch of medicine In Cheyenne county, Dr Mantor has gained exceptional prestige by specialization, as he devotes his time and attention primarily to surgical work and roentgenology He maintains his residence and professional headquarters in the city of Sidney, and has become well known because of his estimable character and high

professional attainments, as one of the representative physicians and surgeons throughout the Panhandle

Dr Mantor was born at Sheldon, Illinois, December 2, 1874, the son of Lyman and Mary (Cole) Mantor and is descended from old eastern stock, as his ancestors located in the Atlantic states at an early day His father was a native of Ohio, while his mother claimed New Jersey as the place of her nativity The father was employed during his early manhood as contractor and in 1885 came to Nebraska, with the idea of establishing himself independently in business Soon after arriving here he purchased railroad land in Dawson county, being one of the pioneer settlers of that section, where he began at once to make improvements on his land, put up a comfortable home as soon as possible, erected permanent farm buildings and began to engage in general farming, stock-raising and feeding Being a man of education Mr Mantor gave deep study to agricultural questions, kept abreast of all improved methods, buying the latest machinery to lighten the work on the farm and was rewarded with unusual success as the result of his labors After gaining a comfortable competency he retired from active participation in farm affairs and lived the last years of his life in a leisurly manner, enjoying the modern books and periodicals He passed away at the age of seventy-seven years, being survived by his wife who now makes her home with her son in Sidney

Hugh Mantor was but ten years of age when his parents came to their new prairie home in Nebraska He had already attended school back in Illinois but from his tenth year he spent his boyhood days and early youth on the home farm, learning there the excellent qualities of self reliance and service Coming from a family of unusually well educated parents they saw that their son had every advantage obtainable, as they had sufficient means to send him to school at all times At first he attended the district school in his home locality, then the high school at Lexington, where he graduated At an early age the boy had decided upon a professional career and with this end in view he entered the medical department of the State University at Lincoln completing his course in 1902, receiving the degree of M D The same year he opened an office at Cozad, Dawson county, and at once engaged in practice

For seven years he served the town and county around with great ability but he desired high work as he already realized that the twentieth century is one of specialization and having pased the time of his general professional work, went to England where he entered the West London Hospital for a course of post graduate study Spending the winter of 1909 and 1910 abroad the doctor on his return to America came to Sidney where he has since been continuously engaged in practice While in England Dr Mantor had given most attention to all phases of surgical work and since his return to this country this has been his special branch of medicine, so that he is called to points not only outside of Sidney but throughout the Panhandle for this kind of work He has equipped his office with the most complete and modern X-Ray equipment and Electro-Therapeutic appliances which he uses in the diagnosis and treatment of disease The doctor controls a large and representative clientelle, keeps in close touch with the advance made in medical, surgical and roentgenological science, and is unremitting in the study of the best standard and periodical literature of his profession For a number of years he has been district surgeon for the Union Pacific Railroad, is a member of the American Medical Association, the Nebraska State Medical Society, the County Medical Association, while his fraternal affiliations are with the Masonic Order, the Knights of Pythias, the Independent Order of Odd Fellows and the Elks He is an independent Democrat in politics and lets no party lines dictate the casting of his vote While he is a progressive citizen and promoter of every movement for the benefit of Sidney he does not care for public office and cannot be persuaded to hold one, devoting all his time and energies to the close study and many demands of his chosen calling, medicine

In June, 1907, Dr Mantor married Miss Stella Batdorf a native of Omaha, where she was reared and educated

GEORGE R BUCKNER, banker, financier, ranchman and oil magnate, is a man whose business career has been significantly characterized by marked executive ability and creative initiative His unchangeable purpose and integrity have gained the popular confidence and esteem that are most essential in the successful execution of the important and varied enterprises to which he has devoted his attention and energies, and through the medium of which he has secured high standing as a representative figure in banking and financial circles in Nebraska From first entering business life Mr Buckner has been identified with banking enterprises and his administration has done much to conserve the success attained by the institutions with which he has been associated His aggressiveness, keen foresight, honesty and civic loyalty have gained him a high standing in the communities where he has been engaged in building up his business to its present extensive proportions

Mr Buckner was born in Macedonia, Iowa, September 2, 1882, the son of Perry and Dora B (Starts) Buckner, the father being a native of Missouri and the mother of Illinois They were reared and educated in their respective communities, but after their marriage removed to Iowa, being pioneers of that state, as the father took up a homestead in Pottawatomie county, about thirty-five miles east of Council Bluffs, but died at the age of thirty-two years, leaving his wife with four small boys to rear and support After her husband's death Mrs Buckner put her capable shoulder to the wheel and though but a woman alone in a new country, she was equal to the task, she capably managed the frontier farm, paid off the money due on the land as it came due, devoted such time as she could to her family and became a successful farmer within eight years after she was left alone in the world Being a woman of good education and desiring every advantage for her fatherless boys Mrs Buckner decided to leave the farm and locate in some place where the children could have the best educational advantages After looking over the various locations she chose University Place, near Lincoln, Nebraska, a college town where youths would grow up in an atmosphere of culture and refinement The four boys were Charles L of Red Oak, Iowa, Louis J, who lives at Elliott, Iowa, George, and Wilbur G who died at the age of six years

George Buckner was about ten years of age when the family left the farm in Iowa and located at University Place he had already attended the public school near his home and after coming to Nebraska entered the grammar school After finishing the grades he entered the academy of Nebraska Wesleyan University of University Place He was a sturdy youth and realized the struggle his mother had had to rear her little family and decided to start out for himself and earn some money For four years he worked on a farm, then on a railroad for a year but was too ambitious to be contented with what he could earn in this position and having determined upon a commercial career, entered the Nebraska Business College at Lincoln where he completed a course in business methods and bookkeeping In 1901, he entered the Farmer's and Mechanics Bank at Haverlock as bookkeeper For one season he played professional baseball with the Burlington Railway shop men, but while in the bank had decided that banking was to be his chosen vocation and having accumulated considerable capital, in 1903, he became one of the organizers of the bank at Davey, Nebraska, where he was given the position of cashier and manager Mr Buckner had a special constructive talent for banking and it was through his liberal policies that he furthered the success of the first bank with which he was associated His foresight, progressive, yet conservative methods won the confidence of the residents in the vicinity of Davey who became heavy depositors of the bank In 1907, Mr Buckner severed his relations with the Davey bank, disposed of his stock and helped organize the Lancaster County Bank of Waverly, Nebraska, and two years later was made president of the institution By this time he had become recognized as one of the leading bankers and financiers of eastern Nebraska, for within the short period of two years he has promoted from cashier of the bank to its executive head The bank had become a flourishing institution under his capable direction, changes came about and still greater expansion ensued, but Mr Buckner had not been contented to devote all his energies to one line and had become interested in valuable oil and gas properties, which grew with such astounding rapidity that in 1917, he sold his stock in the banking house to take active management of his other interests, as he had already organized the Independent Gas and Oil Company with headquarters at Sidney, having established his home there in 1918 The following year he bought a block of stock in the American Bank of Sidney and became its vice-president All depositors in this bank are protected by the depositors' guaranty fund of the state of Nebraska In 1919, Mr Buckner was the prime mover in the organization and incorporation of the Grain Belt Oil Company, with a paid up capital of $85,000 and from its initiation has been president of this progressive concern The company operates in eight towns of Nebraska, Sidney, Chappell, Ogalalla,

Grant, Gurley, Dalton, Scottsbluff and Mitchell Reared during his boyhood on the farm Mr Buckner naturally has ever taken an interest in land and as his business met with its phenominal success and his capital increased he has invested from time to time in country property, having become the owner of the celebrated St George Ranch of some nineteen hundred acres only three miles from Sidney, in the Pole creek valley While working on farms after leaving school he had learned practical farming by experience and also had taken great interest in the cattle business so that when he acquired a large landed estate of his own he began to apply his knowledge to its management His operations have been most successful as he believes in modern methods, uses the latest machinery for work on the ranch and specializes in forage crops, feeding and the raising of high bred cattle and hogs, so that today he is one of the heavy shippers from this district The land out at St George has been raised to a high state of fertility where cultivated and the pastures are usually in fine condition, so that Mr Buckner may be said to be a progressive and exceedingly prosperous farmer aside from his many other varied interests Such ranches as his are of inestimable value in a community as he can become a leader in all new farm movements and when he has tried out the methods advocated by the farm experts of state and nation they can be adopted by men who have not the capital for such investigation who live in the district where the St George is well known In a way Mr Buckner's farm is doing for the Panhandle what the famous Rockfeller ranch did for the stockmen of Kansas, down in Kiowa county From first locating in Sidney Mr Buckner has taken an active and prominent part in all the activities of the city and county During the war he was president of the city and county committees of the Red Cross and was one of the largest salesmen of Liberty Bonds In politics he is a staunch supporter of the Republican party and takes an important part in shaping its policy in county and states As a young man, soon after he became independently established in business he began his political career, for, in 1913, he was elected as representative of his district to the state legislature on the Republican ticket In 1915, he was nominated for county treasurer, but was defeated He cares nothing for political preferment but has felt that his duty as a citizen called him to such office as he could creditably fill, today he is far too fully occupied to even think of accepting the nomination for any office Fraternally Mr Buckner s rela-

C. T. Grewell and Family

tions are with the Masonic order, standing high in its councils as he has taken the Thirty-second degree, he is also a Shriner and a member of the Modern Woodmen He and his wife are members of the Methodist Episcopal church and Mr Buckner was a delegate to the general conference in New York

In 1905, Mr Buckner married Miss Lola M Danforth, a native of Jefferson county, Nebraska, the daughter of George Danforth, who was one of the gallant soldiers of the Civil War He was a well educated man of high culture and attainments Soon after the close of the war he laid aside the sword for the pursuits of peace, and while he did not take up the plowshare did his part even more worthily as he was a member of the surveying party which laid out the route of an early railroad across the state of Nebraska Later, when the pioneer construction of the road was completed he became one of the first merchants of Alexandria, where he was engaged in business for many years He lived until 1917 Mrs Buckner was educated in the excellent schools of her native town and since her school days were over has ever continued to broaden and cultivate her mind by wide and judicious reading, she is a gracious and charmwoman who has a host of friends and is the able and capable chatelain of the beautiful Buckner home, which is one of the most hospitable houses in the Panhandle There are three children in the family Frances A, twelve years of age in school, Wilbur G and Harold B

CHARLES T GREWELL — Included among the substantial pioneer farmers and stock-raisers of Scottsbluff county, Charles Grewell is also known as a progressive, useful, and energetic citizen whose public spirited services have contributed in no small way to the advancement and development of his community's interests For thirty-six years he has been a resident of the Morrill valley, and there has identified himself with many of the movements that have served to aid in progress both agricultural and along civic lines; for he has seen pass before his eyes the panorama of change that first showed the rolling prairies covered with the curly buffalo grass and wild flowers, then it was covered with the vast herds of cattle during the heigh-day of the cattle business, following that came the few farms far apart and scattered, then closer settlement, and finally to crown all, irrigation was introduced so that the wilderness countryside now blossoms like the rose, and today he sees the Panhandle the garden spot of our great

state, and in all and to all this marvelous progress he has liberally contributed

Charles F Grewell was born in Illinois, February 18, 1863, the son of Christopher and Mary J (Hewett) Grewell, to whom were born ten children but only two of them came west. Charles and his sister, Mary, who married Henry Rose, and now lives in Wyoming The father was a farmer and stock-raiser in Illinois, who came to Nebraska in 1892, one of the pioneer settlers of this section, 1896, as he located on section 34, Sheep Creek township, the place which Charles today owns The home place consisted of a quarter section on which Christopher Grewell carried on farming occupation until his death in 1907 Charles grew up on his father's farm in Illinois, attended the district schools near his home where he laid the foundation for an excellent practical education As soon as his age and strength permitted he began to assume many tasks that are ever to be found in the country and while still young had a good practical knowledge of farm business When the family came west he accompanied them, assisted in the establishment of a home in the new country, broke the prairie sod and assisted his father in every way to make the good and permanent improvements on the farm He was far-sighted, shrewd in his business dealings, and was one of the first men to realize that the day of the open range on the prairies was doomed and that that the future of the cattle business was to be in the hands of the farmer who would raise fewer but high grade animals for the market With this end in view he early induced his father to add stock-raising to his general farming so that with the gradual withdrawal of the range cattle they began to deal more and more in farm bred animals, and thus were some of the first men of the valley to begin shipping to eastern markets

Upon his father's death, Mr Grewell took the old home place He had already become a large landholder on his own account and today is the proprietor of an extensive landed estate of six thousand, four hundred and forty acres Today Mr Grewell is not only one of the richest farmers of the Platte valley, but he is also one of the largest stock-raisers of a district known for its wealthy, prosperous cattle men He is modern in his ideas of farm management and has inaugurated the theories and practices of the best farm experts of the state and nation and finds that they pay From the first he has held that high bred stock brought in the greatest returns and has specialized in White Faced cattle on his ranch, shipping carload lots out of the valley each year In February, 1913, Mr Grewell married Mrs Mertle Fuller, and to them one child has been born, Wesley C Mrs Newell had two children by her former marriage, Von and Mona Mr and Mrs Grewell are highly respected in the Morrill community, active and gracious in social circles, and it is not surprising that they have a host of friends Mr Grewell is a Democrat in politics while his fraternal affiliations are with the Modern Woodmen and the Independent Order of Odd Fellows

WILLIAM P MILES, prominent pioneer, leading member of the bar of the state of Nebraska and well known real estate man of Sidney has seen nearly of half century pass since he came to a veritable wilderness, where habitations were few and where civilization was still in its primitive form Mr Miles is one of the worthy and sturdy pioneers who came to Nebraska just a decade after the territory was admitted to statehood, the first period of his residence within the borders of this commonwealth having been passed in Lincoln county He has contributed to the civic and industrial development and progress of every community where he has resided, representing the best in communal life and spirit and has borne with fortitude and unwavering faith and confidence the hardships and trials of frontier life Mr Miles is one of the far-sighted men, filled with energy, who had the vision to make the most of the opportunities offered in connection with the development and growth of a new country and has achieved success through his professional and business interests Of Irish descent he was born stirring of decision, judgment and with pronounced self-independence During all of his life he has had a dislike for the affected or pretentious, and despised hypocracy, deceit and dishonesty

Within a period of some thirty years of his professional activity, in this state, Mr Miles has won, and still maintains, for himself a reputation of being one of the strongest and most resourceful lawyers in western Nebraska No member of the Cheyenne county bar has participated in more contested cases and with such great success His whole aim in the work has been not so much for the material gain as to obtain justice for his clients His judgment of men is recognized by all, and this attribute alone has never failed him in selecting a jury, and in questioning the witnesses His mental make is about as follows he is honest, he is keen with a bright mind stored with legal lore gentle in spirit and retiring he yet stands as one of the central fig-

ures, he has a liberal education of his own winning, and is an able advocate

Mr Miles is a native of the Bay state, born in Middlesex county, Massachusetts, May 23, 1856, the son of Thomas and Johanna (Tooney) Miles, the former a native of County Limmerick and the mother of County Cork, Ireland They were reared and given such education as could be afforded them in their native land and there married Both were young, ambitious and had heard from many of their returned countrymen of how they had been able to get ahead in America, so they too decided to hazard fortune in the New World Thomas Miles and his wife landed in the United States in 1846, and were soon located in Massachusetts where he worked in one of the many shoe factories of that state, but he was not to live to enjoy long the country of his adoption as he died June 11th, of that year This left the mother and oldest children to shoulder the burdens of the family of nine, as there were eight children, but where there is a will there is a way and all grew up to become fine citizens of our great country and well-to-do men and women

William P received an excellent practical education in the fine public schools of Massachusetts, where his boyhood was spent He realized the necessity of helping himself and worked at any honest employment when his years permitted and after leaving school started out independently in life He was nineteen years of age and at the zenith of his physical and mental powers when he came to Nebraska in 1877 to accept a government position as teamster on one of the government routes to the west For a time he remained in Lincoln county then came to Sidney, where he worked while reading law Mr Miles had opportunity to look the country over and he at this early date had a vision of what the Panhandle was to become This was in 1888, when central and western Nebraska was very different in appearance from today then the rolling prairies stretched on and on unbroken for miles by habitation or fence Soon after settling in Sidney Mr Miles entered the law office of Norvall and McIntosh, where he diligently applied himself to study, soon mastering enough of the intricacies of the law to go up for his bar examination which he passed with brilliancy in 1888 a record that is not often attained today Within a short time after being admitted to practice he was elected county attorney of Cheyenne county, capably serving in this office four years He rapidly achieved success as a lawyer and became affiliated with the rising Republican party, and to this day has remained a loyal member of that great national organization, honoring it and frequently being honored by it being eminent and influential in its councils for many years. Soon after locating in the Panhandle Mr Miles became interested in business life being one of the prime spirits in the organization of the Home Land Company of Sidney. The business grew with gratifying rapidity and soon assumed large proportions under the able guidance of Mr Miles who eventually bought the controlling interest in the concern and in 1917 reorganized it under the name of the William P Miles Land Company. From time to time as the capital permitted Mr Miles invested in land in Cheyenne county and later in the Panhandle until today these holdings are extensive and very valuable.

For over thirty years Mr Milles has been attorney for the Union Pacific Railroad and has been a potent factor in its affairs in this locality during the quarter of a century which has marked the great development of this section of the state. For many years from his coming to Sidney Mr Miles took an active part in politics of western Nebraska, he was a Taft man but after that campaign ceased to take an active part in political affairs. He was Republican delegate to the National Convention in 1904, was a member of the committee on rules. Fraternally he is associated with the Knights of Columbus the Benevolent and Protective Order of Elks and with his family is a devout member of the Roman Catholic church, a faith in which he was reared. The Miles home is not only one of the finest in Sidney but of the state and there the many friends of the family enjoy the generous hospitality dispensed by both Mr Miles and his charming wife. From year to year Mr Miles has been accumulating a fine library as he is an omnivorous reader on a wide range of subjects and as a result of this wide range of literature which he has pursued he is today one of the highly educated and cultured men of the legal profession. His library is one of the largest and most select collections in Nebraska.

In 1901, Mr Miles married Miss Eva Whitman, who was born in Galesburg, Illinois, but reared in Iowa and Dodge county, Nebraska. She was attending school when Mr Miles met and married her. They have two daughters. Eva, who is the wife of M J Flintzer, a prominent business man of Sidney, and Mildred, who is the wife of R W Bauer, of the Sun Drug Company, of Lincoln Nebraska.

Time may bring additional honors to Mr Miles, it may enlarge his field of activities and usefulness, it may broaden his acquaintance, but it cannot augment the esteem, confidence and affection with which he is regarded by those who really know him.

ESTHER M JOHNSON — Few states in the Union have manifested in connection with their education systems as full and merited appreciation of women in scholastic executive office as has Nebraska and numerous counties in the state have gained unequivocally through selection of women for such responsible and exacting academic and executive positions as that of county superintendent of schools. Thus Garden county has had much to gain and nothing to lose in the able administration which Miss Esther M Johnson is giving in the office of county superintendent of school. She is possessed of high intellectual attainments a definite prerequisite for the incumbency that is now hers and is showing also a really remarkable constructive and administrative powers in the systematizing and advancing of the important work in her jurisdiction.

Miss Johnson was born in Red Willow county, Nebraska, and is a daughter of Alfred and Hannah (Pierson) Johnson, who were born in Sweden and who were young folk when they came to the United States. After coming to Nebraska Mr and Mrs Johnson became pioneer settlers in Red Willow county where he filed entry on a homestead and where she took up a pre-emption claim. With undaunted courage and faith they essayed the task of reclaiming their land and developing a productive farm — a work in which they succeeded admirably. They continued to reside on their land, which they had equipped with good improvements until the death of Mr Johnson, on June 24 1904 and within a short time thereafter Mrs Johnson rented the farm and removed with her children to Lebanon that county. There she remained until 1918 when she removed to Garden county and has since made her home with her daughter Esther M whose name initiates this review.

Miss Johnson passed the period of her childhood and early youth on the old home farm, and in the meanwhile she fully availed herself of the advantages of the public schools of her native county, including the high school at Lebanon. In furtherance of her higher academic education she entered the Nebraska State Normal School at Kearney from which institution she was graduated as a member of the class of 1914. She passed the following year with her widowed mother, and thereafter was for three years a successful and popular teacher in the public schools at Lewellen, Garden county, where she was a teacher in the high school at the time when she was appointed county superintendent of schools, to fill an unexpired term This appointment was made on March 1, 1918, and in November of the same year she was elected, on a non-partisan ticket, to fill this office for a term of four years, her election having shown that in the intervening months her administration had met with unqualified popular approval Under election Miss Johnson initiated her administration in January, 1919, and the results of her work since that time have fully demonstrated the wisdom of the popular vote which placed her in office She has the earnest co-operation of the teachers of the county, as well as of the people who support the excellent schools, and she is zealous and indefatigable in her work Miss Johnson was reared in the faith of the Methodist church, of which she is a communicant

NICHOLAS F ZEHR, the owner-manager of a popular barber shop of Chappell, was born in Livingston county, Illinois, March 30, 1871, the son of Christian and Catharine (Roth) Zehr, the former a native of Alsace-Lorraine, (France) while the mother was born in Ohio The father was a farmer who came to the United States in 1856, locating first in Illinois but in 1880, he came to Nebraska, settled in Seward county where he lived until his death in 1907 Mrs Zehr returned to Illinois after the death of her husband, where she resided until she died in 1910 Mr Zehr was a Democrat and both he and his wife were members of the Mennonite church There were thirteen children in the family but Nicholas is the only one in this locality, a brother Joseph, lives in Arthur county

Mr Zehr was educated in the Public schools of Seward county and recalls the great blizzard of 1888 when he and the teacher helped many of the pupils to get home and even then some were forced to stay in the school house all night In 1893, he came to Deuel county, beginning to work on a farm in March, remained engaged in that work several years and then accepted a position with Wertz Brothers on the ranch Five years later Mr Zehr came to Chappell to work for them in a hardware store and was associated with this business until the Wertz Brothers sold out in 1908 Soon after this he bought a barber shop where he has been engaged in business to the present time Mr Zehr has made many friends in

Chappell due to his courtesy and kindness, has built up a good business and today is regarded as one of the reliable and substantial men of the town

November 9 1893, Mr Zehr married Miss Nancy Roth, at Chappell, the daughter of Jacob and Lydia (Stutzman) Roth, pioneer settlers of Deuel county Mr Roth now lives at Nampa, Idaho, his wife having died in 1908 Three children have been born to this union William, Edna and Nicholas, all at home Mrs Zehr and the children are members of the Methodist church while Mr Zehr belongs to the Modern Woodmen of America He is a Democrat, was assessor of Deuel county two years and served as precinct assessor several years He is progressive in his ideas and one of the substantial men of Deuel county and the Panhandle

SARAH ROSELLA STALNAKER, widow of the late Charles Stalnaker, came to Nebraska in the early days and suffered all the hardships and privations incident to life on the frontier and her reminiscences of that period are interesting She was born in Jasper county, Iowa, August 16, 1866, the daughter of James R and Rachael (Cline) Thomas, both natives of Illinois, who had a family of ten children Warren A, of Washington, Sarah of this review, Anna, deceased, James A, lives in Canada, Ira E, of Oregon; Alice, of Seattle, Washington, William Charles, deceased, Nora, the wife of George Givens, of Shaw, Oregon, and two deceased Mr Thomas was a farmer who owned and operated a threshing outfit for many years, in Hamilton county, Nebraska He served as a private in Company D, One Hundred and fifteenth Illinois Infantry during the Civil War and was very ill, after his discharge from the army in 1865, he moved to Iowa and in 1869, to Hamilton county where the family lived sixteen years Later Mr Thomas moved to Dundy county, Nebraska, to the state of Oregon and Ellensburg, Washington, where he and his wife died Mr Thomas was a Republican, a member of the Grand Army of the Republic and the Woodmen, while he and his wife were members of the Christian church

Sarah Thomas Stalnaker was educated in the public schools of Hamilton county, where she came with her parents when four years of age Her father took up a homestead near Marquette and there she experienced many frontier adventures She well remembers the trip overland from Iowa, as they drove through with a span of mules, eight cows and

JOHN H. ORR

lived in their covered wagon on the way. The first home was a dugout which soon fell in from rain, they then tried a tent and it blew over, then a cave was found and later a frame house was built. Indians were frequent visitors and the children were afraid at first but learned they were friendly. Water had to be hauled four miles, Central City, known then as Lone Tree, was nine miles north of Marquette and it was from that town that the lumber for the home had to be carried by wagon, fording the river, as there were no bridges. Crops were poor the first years and the family suffered from want of food, provisions and even clothes. Mrs Stalnaker was much with her father in those days. When the railroad was built through near them the mother boarded some of the men for money to keep the family. Mrs Stalnaker herded cattle where the present village of Marquette is and tells of the terrible prairie fire she saw there when the flames were sixty feet high.

When Sarah Rosella Thomas was eighteen years old she married Charles Stalnaker, on September 30, 1884, he was the son of Samuel and Elizabeth (Ryan) Stalnaker, residents of Hamilton county, now deceased.

Mrs Stalnaker lived in Hamilton county forty-seven years and has seen all the marvelous changes that have taken place in this state covering nearly half a century. She became the mother of seven children, two of whom died in infancy, Cleveland, deceased, Grace, the wife of Ira Williams of Deuel county, Sylvia, the wife of Grover Moist, of Crawford, Nebraska, Vancil and Wilma, both at home. Mr Stalnaker died November 23, 1913, and soon afterward Mrs Stalnaker came to Deuel county with the two youngest children. She bought land seven miles north of Chappell where they ran a farm until the son enlisted in the army during the World War. Mrs Stalnaker sold out to move to Chappell where she bought a home and building lots but recently traded the lots for a quarter section of land in Wyoming. She is a fine woman of great ability and resource has played her part in the development of Nebraska and is an ardent worker in the Methodist church while the children belong to the United Brethren church. For years she has been affiliated with the Royal Neighbors.

JOHN H ORR is another of the sterling pioneer citizens of Garden county, where his energy, ability and progressiveness have gained for him substantial status as one of the representative agriculturists, dairymen and stock-growers of this favored section of the state. He is the owner of an extensive and well improved landed estate in Garden county, and is a man whose individual success and advancement have been attended with loyal and liberal support of those measures and enterprise that have been for the general good of the community.

The honored pioneer whose name initiates this review was born in County Tyrone, Ireland, December 25, 1862, of staunch Scotch-Irish lineage. He is a son of John H and Mary (McCormick) Orr, the former of whom was born and reared in Scotland and the latter in Ireland, where their marriage was solemnized. John H Orr, Sr, was a young man when he removed from Scotland to Ireland, and on the fair Emerald Isle he continued to follow his trade, that of wheelwright, until 1867, when he came with his family to the United States and settled at Rahway, Union county, New Jersey, where he passed the remainder of his life and where he continued to work at his trade during the remainder of his active career. He passed away at the age of seventy-six years and his wife was sixty-seven years of age at the time of her death. They became the parents of nine children, and three of the sons, Calhoun, John H and James W became residents of Nebraska.

John Orr acquired his early education in the public schools of New Jersey, and after coming to Nebraska he supplemented this training by completing a course in a business college at Crete. His first independent service was rendered by taking charge of one of the wards of the New Jersey State Insane Asylum at Morris Plains, where he remained three years. He then, in 1882, came to Nebraska and established his residence near Crete, where for four years he was employed on the Pleasant View stock farm of W H Smith. In 1886, he removed to Keith county, where he took up homestead and tree claims. He perfected his title to this land and met with excellent success in his activities as a farmer and dairyman. He was associated with A D Remington in establishing the first milk-skimming station in western Nebraska. On his original claims he continued in the dairy business twelve years, and in the meantime purchased the Spring Canyon Ranch, more familiarly known as the old Brand Hoover horse ranch. This he stocked with cattle and horse, and to this ranch, situated in the part of Deuel county now included in Garden county, he removed in 1894. He continued to reside on this place until 1912, since which time he has lived virtually retired, in his pleasant home in the village of Lewellen His ranch comprises eighteen thousand acres where an average of one thousand head of cattle and one hundred and fifty head of horses are kept Mr Orr also has a well improved farm of three hundred and twenty acres that is given over to diversified agriculture, with irrigation from the Meeker ditch He also has twelve hundred acres on the South Table, which is under the active management of one of his sons

Mr Orr is one of the most substantial and influential pioneer citizens of Garden county, and his splendid success has been worthily achieved He is vice-president of the Citizens' Bank of Ogallala, is president of the Orr-Spindler Mercantile Company, at Lewellen, is treasurer of the Farmers' Life Insurance Company of Denver, Colorado, is one of the heavy stockholders of the Meeker Irrigation Ditch Company, and is a stockholder of the Blue Creek Light & Power Company, of Lewellen All these associations indicate not only his aggressive and vital energy but also his civic loyalty and genuine public spirit In politics Mr Orr may be designated as an independent Democrat, and while he has had no desire for public office he served as justice of the peace and notary public during virtually the entire period of his residence in Keith county, as well as a member of the school board He is affiliated with the Masonic fraternity, the Woodmen of the World and the Modern Woodmen of America His wife is an active member of the Methodist Episcopal church In the light of his broad experience in connection with industrial life in western Nebraska, Mr Orr gives advice to those who are developing homesteads or Kincaid claims at the present time to take up the dairy business to a sufficient extent to defray expenses, and thus to save the increase and become a success

March 4, 1885, recorded the marriage of Mr Orr to Miss Eleanor E Smith, of Crete, this state Mrs Orr was born at Flint, Michigan, and is a daughter of Dr William H and Mary (Gordis) Smith, the former a native of Michigan and the latter of Holly, New Jersey Dr Smith, as an able physician and surgeon, continued in the practice of his profession in the state of Michigan until the early seventies, when his impaired health lead him to come to Nebraska, the family home being established near Crete, Saline county where he engaged in the breeding and raising of thoroughbred stock, including Percheron horses Durham cattle and Poland-China swine He successfully followed this line of enterprise until his death, at the age of sixty-five years He was one of the influential and honored citizens of Saline county His widow, a woman of superior education and gracious personality, received collegiate training, and her memory is revered by all who came within the sphere of her influence, she was sixty-nine years of age at the time of her death Mrs Orr is the eldest and only surviving member of a family of three children, and was reared and educated in Nebraska, where she was graduated in Doane College, at Crete Mr and Mrs Orr have a fine family of twelve children — nine sons and three daughters — and all have received the best of educational advantages, the sons having attended the Grand Island Business College and all being in partnership with their father — in connection with his large and varied agricultural, live-stock and business operations Joseph C, the eldest of the children resides at Lewellen, is married and has two children, John Wesley, of Lewellen, is married but has no children, Warren H, and his wife have one child; and the other children are not married at the time of this writing, their names being here entered in respective order of birth James A, Burton F, Edith E, Andrew G, Mildred M, Nellie G, William T, Edward C In thus giving twelve children as "hostages to fortune" Mr Orr has demonstrated his repugnance to "race suicide" and he has reason to be proud of his children, as well as of the success which he has won during a life of signal activity and usefulness

PHILIP McCORMICK, the well known and popular owner of Sunnyside Ranch in Deuel county, and its present manager, was born in Loraine, Illinois, April 24, 1879, the son of William and Jane (Taylor) McCormick, the former a native of Illinois while the mother was born in Missouri The father died in 1881, but the mother still lives at Chugwater, Wyoming Mr McCormick was a farmer all his life After his death the boy, Philip, was reared by a Mrs Craga, and saw little of the other two children in the family He was educated in the public school of Loraine until he was fifteen years old, his education being cut short by the fact that he ran away then and went to Dakota to work on the farms of other for two years Mr McCormick came to the eastern part of Nebraska then and a year later to Julesburg, spending a few months there in 1897, before coming to Deuel county Here he herded cattle for ten dollars a month and sheep for twenty, feeding himself To add to this income he played a violin for dances and made as high as fifteen dollars a month that way Following this Mr Mc-Cormick accepted a position to ride the irrigation ditch south of Big Springs, being the second man employed in that capacity after the ditch was completed and remained with the irrigation company over two years

October 9, 1904, Mr McCormick married Miss Maude Morrison, at Big Springs, the daughter of John and Amanda (Tigard) Morrison pioneer residents of Deuel county, who homesteaded southwest of the town and now live retired in Big Springs Two children have been born to this union Hattie and Harvey, both at home

When Mr and Mrs McCormick came to the Sunnyside ranch in 1906, Mr McCormick says he had just eighty dollars and cleared eight hundred the first year He entered the employ of Peterson and Loveland, proprietors of the ranch, as foreman but a year later bought a third interest in it and in 1914 a half interest which he still owns Sunnyside ranch consists of over twenty-seven hundred acres, where a specialty is made of rearing and breeding Hereford cattle, Belgian horses and Poland China hogs The herd of cattle is considered one of the finest in this section of the country Mr McCormick has made a study of his business, is naturally able as a business man and it has been largely due to his policies that the ranch is one of the highly developed and profitable one of the Panhandle He is the manager and ably handles all the business of the place as well as carrying on the necessary farming industries

Mr McCormick is a Republican and his wife is a member of the Methodist church He is a progressive man in his ideas and methods uses the latest machinery and devotes his entire time to the many branches of his business

WILLIAM H WINTERBOTHAM the owner and manager of the largest and most popular grocery and feed business in Julesburg, Colorado, is a man of varied experiences as he crossed the plains in the early days to seek gold in California but did not go that far as the party stopped in Colorado Mr WinterBotham was born in Savannah, Missouri, December 26, 1845, the son of Samuel and Mary (Strouble) Winterbotham The father was a farmer in the early days, then a miner and prospector, the mother died in 1847 and the children were left with friends while the father went west Later he return-

ed and the family moved to Iowa where he was employed in the State Penitentiary When the son William was old enough he and his father became partners in a mercantile business in Iowa moved from there to Columbus, Nebraska where the father died in 1876 William Winterbotham continued in the hardware business in Nebraska until 1902 when he located in David City to open a store which he conducted until 1909, the year in which he established himself as a grocer in Julesburg, where he is regarded as one of the leading merchants of the city Mr Winterbotham has always been in the mercantile business since first forming a partnership with his father, except for a few years which he devoted to mining He is a man of great bodily and mental vigor and looks twenty years younger than he is He owns the largest and best equipped store in this section of the country which he runs with the assistance of his two sons-in-law and daughter Maude, who is the bookkeeper

In 1874, Mr Winterbotham was married at Columbus Nebraska, to Miss Lillie Hudson, the daughter of H J and Sarah (Shefford) Hudson, the former being judge of his county and clerk for many years Three children were born to this union Maude, the wife of Ray C Johnson of Julesburg, Blanche, deceased, and Hazel, the wife of C W Larabee, of Julesburg

Mr Winterbotham is a Republican and in years past took an active part in politics, attending the state and county conventions but never would accept public office, devoting his time to his growing business

In an interesting manner, Mr Winterbotham recounts how in the spring of 1859, he with his father and brother started from Fort Madison, Iowa for the gold fields of California to prospect and mine They joined a large party going west, each small section driving a wagon drawn by oxen and he says that he walked most of the way from Iowa to Pike's Peak He was only a boy of fifteen at the time and enjoyed the adventure The route lay over the famous California Trail the "Crossing" of which lies just a short distance from Big Springs, Nebraska, where the trail split the north branch led to Cheyenne and Laramie the south to Denver Ash Hollow was on the first and there the party camped being visited by friendly Sioux Indians The party continued by way of Cheyenne Pass and on to Golden Gate — Golden, Colorado — Manitou and South Park where they mined for two weeks but did not get much gold so that the gold seekers began to doubt if there was gold and sought the advise of Horace Greely who was in the locality The elder Winterbotham was elected spokesman and Greely gave him much encouragement and the party struck off for Poke's Peak, went to Denver, then a tent city, then on to Gregory and Gold Hill, camping at Boulder creek, where they built a dam and mined some gold but there was not enough to pay them and they started back east following the Arkansas river route They saw many buffalo, killed some and met with a few hostile Kiowa Indians but had no serious trouble Returning home the family became established in Iowa where the father and son later entered business

HOMER J SPILLMAN, one of the prosperous general farmers of Garden county is a native son of Nebraska, born in Lancaster county, October 23, 1873, the son of Henry and Hannah (Dolcater) Spillman, the former a native of Pennsylvania while the mother was of German birth He died in 1889 and she in 1918 They were the parents of five children Emma, deceased, Harry, of Julesburg, Homer, of this sketch, Curtis, deceased, and Irwin, of Garden county, now in partnership with his brother The father was a farmer and stone mason when a young man The family came to old Cheyenne county — now divided and the part where they settled is known as Garden — in April, 1886, they took up a homestead where they lived until Mr Spillman died The mother then sold the farm and moved to Julesburg Mr and Mrs Spillman started for their homestead from Big Springs, as there were no roads or trails Mr Spillman knew only the general direction in which to drive With them they had five cows, four horses, a span of mules and the Wagon A storm came up while they were on the way which clogged the wheels making progress slow, finally they reached the shack which was only sixteen by twenty feet and were snow bound for some days The first year in the new home their crops consisted of watermelons, potatoes and squash which grew so large that Dr Babcock took some to the State Fair at Lincoln where they took the prize

Mr Spillman was a Republican, with his wife was a member of the Methodist church and they helped organize the Froid church and school, which were the first in their part of Garden county Mr Spillman had practically no education and knowing the necessity of schools, it was through his efforts the first

school was established William Barbee was clerk and Mr Spillman director, serving until his death The first church services were held in his house Water had to be hauled five miles at that time but later the Spillmans had the first well in the locality At the time of his death Mr Spillman owned a homestead and tree claim, was just nicely started to raise cattle and make money when overtaken by death The first plow lathe ever sharpened in Julesburg was owned by Mr Spillman and he did his first threshing by horse power

Homer Spillman was educated in the public schools of Garden county, worked on the farm and when old enough took up his present homestead northeast of the Froid church He and his brother Irwin are now in partnership, owning eight hundred acres of arable land where they engage in general farming, have well built and convenient farm buildings, use modern methods, and are regarded as two of the progressive business men of this section of Garden county Mr Spillman is a Republican and a member of the Methodist church

March 24, 1891, Mr Spillman married Miss Grace Daniels, the daughter of John and Isabella (Kearney) Daniels The mother is deceased and the father now lives with Mr and Mrs Spillman Mrs Spillman was a school teacher in Garden county before her marriage, teaching part of the time in a dugout, part sod, known as the Simpson school She drove to the school from her home with a pony and cart and sometimes was lost but the pony would always find the way The Spillmans have one child, Hazel

OLA CHRISTENSON, one of the progressive farmers of the Panhandle whose business ability has placed him in the front rank of farmers in the Chappell district where he has made good as a stockman, was born in Sweden, February 23, 1863, the son of Christian and Uellreka (Landeen) Christenson both natives of the same country, where the father was a butcher They passed their entire lives in their native land Nine children made up the Christenson family but Ola, of this review, is the only one in this part of the United States He was educated in the public schools of Sweden and came to this country when twenty years of age, locating in Polk county, Nebraska, in 1883, where he lived until 1908 then came to Deuel county to buy land The first quarter cost fifteen dollars an acre but that purchased later was higher His property is well improved, fine buildings have been erected and the Christenson farm is one of the finest in the locality, for Mr

ERNEST C. HODDER

Christenson has accomplished more in the way of improvement than many of the late settlers From time to time more land has been purchased until today Mr Christenson is the owner of three hundred and twenty acres and leases two thousand acres, which he runs with the assistance of his sons When he came here he had a car load of cattle, horses and hay and one of household goods and machinery, borrowing he money for the freight and all his property has been accumulated since that time, due to his far sight in business native ability and hard work Today the Christenson farm is equipped with modern machinery, tractors, threshing machine, auto trucks and an automobile for family use Mr Christenson first planted spring wheat, but now has more of the winter variety and some for spring so that there never is an entire crop failure The first two years in the Panhandle he raised some two thousand cattle but sees a great future in farming and is engaging in it more extensively each year He does his own harvesting and threshing which pays

In 1887 Mr Christenson was married at Osceola, Nebraska, to Miss Laura Cole, the daughter of Albert and Mary C (Van Brunt) Cole and five children have been born to them Allie, the wife of Vena Christenson, Archie, Louella, the wife of Peter Nelson, Archie and Carl

Mr Christenson is one of the prominent men of the Chappell district, is known for his good judgment and business qualifications, who ever looks ahead for larger opportunities A man of native ability rarely met He built a modern home in 1920, equipped with all modern conveniences

ERNEST C HODDER, a representative member of the bar of the city of Omaha, has wielded large and benignant influence in connection with the industrial development of western Nebraska and especially Garden county, where he has large and important interests in connection with farm and ranch enterprise He was one of the foremost in the development of the irrigation facilities of this section of the state and his close association with the interests of the Nebraska Panhandle entitle him to specific recognition and tribute in this publication

A scion of staunch English stock on both the paternal and maternal sides, Mr Hodder has been a resident of Nebraska since his boyhood, but he claims Newfoundland, Canada, as the place of his nativity In that maritime Canadian province he was born March 15, 1873, a son of Richard and Jemima (But-ler) Hodder, the former was born in England, in 1825, and the latter in Newfoundland, Canada, in 1829, her parents having come to that province from England in the opening year of the nineteenth century Her father became manager of a large wholesale and retail mercantile corporation Richard Hodder was reared and educated in his native land, then he immigrated to Newfoundland in the year 1842 He established his residence at Burin, where he became a representative merchant and prominently identified with the codfish industry, for which Newfoundland has long been celebrated In 1881, he came with his family to Nebraska and established a home in Omaha, where he accepted a position in the motive-power department of the Union Pacific Railroad and continued to reside there until his death, in 1894, at the age of sixtynine years, his wife passing away at the age of eighty years

Ernest C Hodder was a lad of eight years at the time the family home was established in Omaha, and after completing the curriculum of the public schools he entered the legal department of Bellevue College In continuing preparation for the work of his chosen profession he matriculated in the Omaha School of Law, and was graduated as a member of the class of 1898 He received at this time the degree of Bachelor of Laws and later the degree of L L M He graduated from the University of Omaha and received his degree of L L B He was a lecturer of the Omaha School of Law and was president until it became a part of the University of Omaha He continued lecturer in the Law Department of the University of Omaha and was also lecturer of medical jurisprudence at the Creighton University until compelled to resign by reason of his legal and personal business which occupied all of his time and attention

Contemporaneously with his graduation Mr Hodder was admitted to the Nebraska bar, and engaged in the practice of his profession in Omaha, where his ability and close application soon enabled him to develop a substantial and representative law business As a skilled corporation lawyer he finally became attorney for the Western Land & Cattle Company, in the interests of which he came to Oshkosh, Garden county, in 1903 He became so deeply impressed with western Nebraska and the splendid development possibilities of this section, that upon his return to Omaha he became a stockholder in the Western Land & Cattle Company, and at the en-

suing election of officers of this important
corporation he was elected its president, in
1904 At that time the company owned about
twelve thousand acres of land in western Nebraska besides controlling about eighteen
thousand acres of leased land, lying on either
side of the North Platte river, between Lewellen and Oshkosh The company was conducting extensive operations in connection with
the cattle industry, with an average run of
about twelve hundred head, besides which its
ranches had a contingent of hogs ranging from
five hundred to a thousand head Mr Hodder thus became a prominent figure in connection with the live-stock industry in this
section of Nebraska, and he early became a
vigorous advocate and initiator of irrigation,
the great value of which he clearly perceived,
as a medium for the normal and maximum
development of the country His interest was
one of decisive action, as shown by the fact
that he took over and finished the construction
of numerous irrigation projects, including the
Paisley Canal, the West Side Canal out of the
Blue river, and the Overland and Signal Canals out of the North Platte river While he
had these important enterprises under way
very little alfalfa was raised in Garden county,
and, with characteristic vigor and enterprise,
he gave a distinct spur to the propagation of
this important forage crop On the north side
of the North Platte he put in somewhat more
than a thousand acres of alfalfa in 1907-1908,
and it was about this time that the company of
which he was president began to dispose of its
extensive land holdings in the Nebraska Panhandle With prevision as to the possibilities
for further advancement with the construction
of the railroad through this section, Mr Hodder individually purchased about two thousand acres of the fine bottom land on the north
side of the river, a portion of this tract being
situated between Oshkosh and Lewellen, and
the rest being to the east of Lewellen He
has continued the improvement of these valuable holdings and has contributed much to
the development and progress of agricultural
and live-stock industry in Garden county He
still continues to reside at Omaha, where he
has a large law practice of important order,
but he makes about ten trips to Garden county
each year, in the supervision of his interests
here, besides which he and his family custo-
marily pass about two months of each summer in Garden county None is more enthusiastic in exploiting the great advantages and
future of western Nebraska than Mr Hodder
and he exemplifies his faith in both word and

action — greatly to the benefit of this favored
and progressive section of a great commonwealth

In connection with his professional activities
Mr Hodder served eight years as city attorney of Benson, prior to that city's becoming a part of the Greater Omaha, and for four
years he was a member of the Insanity Board
of Douglas county He was a director of the
Farmers and Merchants Bank of Omaha from
1908 to 1917, is at the present time a director,
as well as attorney, of the State Savings and
Loan Association, of Omaha, a position which
he has held since 1909, he is a director of the
Alfalfa Irrigation District of Keith county,
the Paisley Irrigation District of Garden county, and is an active member of the Nebraska
State Bar Association In politics Mr Hodder is found arrayed as a stalwart and effective advocate of the principles of the Republican party, and both he and his wife hold membership in the Methodist Episcopal church, in
the affairs of which he has long been active,
especially in connection with the Epworth
League and the work of the Sunday-school
In the time-honored Masonic fraternity Mr
Hodder has received the highest degrees of the
York Rite, is a member of the Mount Calvary
Commandery, Knights Templars, of Omaha,
and he has received the Thirty-second degree
of the Ancient Accepted Scottish Rite, besides
being affiliated with the Ancient Arabic Order
of the Nobles of the Mystic Shrine, and with
the lodge, Encampment and Canton bodies of
the Independent Order of Odd Fellows He
holds membership in the Omaha Young Men's
Christian Association the Omaha Chamber of
Commerce, the Omaha Athletic Club, and
Knights of Ak Sar Ben   •

At Council Bluffs, Iowa, June 6, 1900, was
solemnized the marriage of Mr Hodder to
Miss Bessie Huntington, and of the seven
children of this union the eldest Sherman
Huntington, died at the age of four years, the
surviving children, all of whom remain at
home, being Ernest C , Jr , Florence A ,
Charles H Donald R , Esther and Bessie C
Mrs Hodder is a daughter of Ephraim and
Elizabeth (Lamb) Huntington, honored pioneer citizens of Council Bluffs, Iowa, where
they celebrated July 23 1919 the fifty-sixth anniversary of their marriage, Mr Huntington
being a native of England Mrs Huntington's
father was the first station agent at Julesburg
Colorado, where he was residing with his family in the late fifties, and when they fled to
escape from hostile Indians of whose purposed
visit they had been informed by a friendly

Indian Ephraim Huntington has long been one of the leading and influential citizens of Council Bluffs, and is a venerable pioneer whose circle of friends is limited only by that of his acquaintances

CHARLES F HAGERTY — Credit for this thrifty farmer of King precinct of the Bridgeport region, must be given to Iowa, but give this section and Morrill county credit for his enterprise and ability and let it be known that the Panhandle, in common with most other sections of this great country, is deeply indebted to Irish-American blood, for Ireland has contributed such a large share to the best element of the population of our broad land

Charles Hagerty is a native of Iowa, born in Decatur county, in 1878 the son of John and Anna Gallagher) Hagerty, both of whom were born in the Emerald Isle The parents spent their youth in their native country, received what educational facilities were afforded by that land and then being ambitious decided to follow in the footsteps of so many of the countrymen and women who had come to America and here won a comfortable living or fortune as the case might be After coming to America John Hagerty became a farmer, live-stock dealer and stock-raiser He had the distinction of being one of the pioneer settlers of Cheyenne county, removing from Iowa to the Panhandle in 1887 He acquired a large amount of land by taking up a homestead, pre-empting another one hundred and sixty acres of land and filing on a tree claim Having learned in the old country the value of good stock Mr Hagerty bought the best grades obtainable in this section at the time he opened up his farm and the great success which he achieved in this line proved the truth of Ideas In 1888 he landed in Alliance with two cars of stock, the first ever unloaded in that place He was one of the courageous men who was not daunted by pioneer hardships He had the trials to contend with that any settlers in a new country must overcome drought insect pests, lack of adequate farm machinery for many years and often difficulty in obtaining seeds, as the nearest railroad was forty mile away either Sidney or Alliance, but he never was discouraged and with the passing years his faith in western Nebraska was justified for he lived to see what was known as the "Great American Desert" become one of the fertile and productive sections of the country upon which not only America but Europe depended for food during the great World War Mr and Mrs Hagerty spent the last

years of their life in great comfort and happiness, won by them from the soil The mother passed away in 1910 while the father survived her until 1918

Charles was educated in the frontier schools of Nebraska attending classes during the winter while he worked on his father's farm during the summer, where he early learned the lessons of practical industry He was one of a family of ten children born to his parents, the others being Annie, the wife of Pat Roland who lives near Broadwater, Mike H, a resident of this county north of broadwater, John D, of Bridgeport Charles the subject of this review, Mamie, who lives in Bridgeport, Simon, a railroad conductor, Margaret, the wife of Jack Riordan, Norine, the wife of Clyde Fairman, deceased, and Katherine who lives in Los Angeles, California

John Hagerty gallantly responded to the call of his adopted country at the outbreak of the Civil War and entered the Union army, where he served until the cessation of hostilities when he returned to his home and took up the plowshare in place of the sword, turning his energies to peaceful pursuits He was a faithful member of the Roman Catholic church the faith in which he was reared from childhood In politics he was a sturdy adherent of the policies of the Republican party, was a worthy, progressive citizen who ever advocated every movement for the progress of his community

Charles was but ten years of age when he accompanied his parents to Alliance in 1888, he remained on the farm and early learned the care of cattle, the best crops to be sown in this climate and altitude, the time and manner of harvesting from his father When he attained manhood's estate he naturally turned to agriculture and stock-raising as a life vocation as his tastes were directed along such lines from childhood and it was the business he knew best, that the choice was a wise one needs not be acclaimed when it is learned that to-day, while yet a young man in years with the future largely before him he is the owner of seven hundred acres of the finest farm land in Morrill county, all in one body a part of which is under irrigation and it is but a question of time until more will be placed under ditch Mr Hagerty has kept abreast of the times in his farm methods for he is one of the first men to adopt the latest machinery and the new crops which are making of the Panhandle one of the garden spots of the world, and it was to this district that the world and country looked and on which it called for food when the shortage arose in Europe and today it is to western Nebraska that the United States is looking for sugar during the world famine of that product Well may the men who are farmers here be proud of the trust placed in them to raise the greatest crops possible at this crisis Mr Hagerty has an attractive residence and substantial farm buildings on his place, as well as being provided with modern implements and machinery with which to lighten farm labor and add to the success of his operations While general farming has been his principal business, he has been very successful As a boy Mr Hagerty was one of the sons his father would call on when he had any long ride to make as he would often say, "that lad can put a horse over more ground and not hurt him than any man in the state," and on this account he was mostly sent to look for stock that had strayed or had been driven away, as he was hard to beat at reading brands, it being the custom for every one who owned stock to also own a brand that did not conflict with another man's brand, but on account of some being dim or if some animal had long hair, brands at times were hard to distinguish but by turning the animal at different angles to the sun a brand may easily be traced, this Mr Hagerty learned to do while riding with old ranchers and punchers as they all took a great interest in him as he was so young and willing to learn They taught him the tricks of handling cattle and horses on the range and how to approach and when to withdraw as they were all experienced stockmen Mr Hagerty mingled with the good and bad but always played his part well, and when yet a youth was often called on to take an important part of cutting out of different brands and as he knew them all where they belonged for many miles within his boundary He always took an active part in branding, as he was hard to beat at foot roping calves and delivering to the wrestlers at the fire in a cool, deliberate manner He always held the good will of all the old ranchers as he did many a good turn for most of them in locating and sending word and even bringing back to them stock they may have never found For years he worked on many of the largest ranches in western Nebraska, in so doing he could tell many a history of a lone cow trail He saw the free range without a fence for hundred of miles, saw quarrels and disputes in divisions of ranges when fencing began, he also witnessed the burning of powder on different occasions Those were the good old days where a man carried the law in his own hands Then came the building of irrigation ditches and Mr Hagerty hitched himself up in the shaves of a ship behind a team of big, long eared mules in the construction of Brown Creek Canal in the interest of his father's place, which covers nearly four hundred acres being the first selections along the North Platte river, Cheyenne county In 1887, with range cattle everywhere you might look and find but one other house to be seen besides the Hagerty home The Belmont was another development of the south side of the river in which Conden Hunt and McShane of Omaha started to construct, then came the Burlington surveyors, then the building of the road By this time a number of other ditches had been constructed along the river followed by the government canal, then the Union Pacific came creeping up the river and put in a nice little station just a quarter of a mile from the Hagerty place It would now make Alliance ashamed of itself for at the time Mr Hagerty's father shipped to that point, in 1888, he had to build chutes to unload his stock from the cars The biggest change of all in this section is the changing of the little alkali lakes over around Antioch and Hoffman, which have been turned into mints Mr Hagerty would like to see the land owners get a fair deal in this great sugar beet industry and at least get the fifty-fifty plan, as undoubtedly there is as much value in the land that the beets are raised on and machinery and horses and other expenses as in the factory Then all the little towns will keep a thrifty growth and will no longer remind us of little Indian village of the old days

W W WATERMAN, one of the pioneer residents of Deuel county who came here at an early day, suffered all the hardships and trials incident to a new country and who today is one of the responsible and substantial men of his district Mr Waterman was born in Jefferson county, New York, March 2, 1851, the son of W T and Rachael (Remington) Waterman, both natives of New York state The father was a carpenter by trade who in later life owned and managed a box factory and saw mill in connection with which he ran a blacksmith shop Some years before his death Mr Waterman engaged in farming He was a Republican served as supervisor of his community and died in the early seventies Mrs Waterman was a member of the Baptist church and survived until 1882 There were five children in the family, of whom two are living

Mr and Mrs Chris McCormick

Edd W McCormick

James McCormick

Jack McCormick

Jennie McCormick

Robert A McCormick

Mr Waterman was educated in the public schools of New York, then spent several years in the oil fields of Pennsylvania In 1884, he came west, locating in Cheyenne county, Nebraska He first took a tree claim, returned to Pennsylvania for the first winter but came back in 1885, to take up a homestead nine miles north of Big Springs, at the Day post-office He still owns the homestead but sold the tree claim for five hundred dollars, which today is worth a hundred dollars an acre The winter of 1885, was very mild so that the Watermans did not suffer from cold They shipped their goods from Pennsylvania and though Mr Waterman had money to buy supplies they were not in the country to buy and they lived on what he terms "sow belly" and water gravy until the next summer passed and they harvested some crops

October 12, 1869, Mr Waterman married Miss Libby King, the daughter of Henry and Lydia (Powers) King, natives and residents of New York and one child was born to this union Henry, who married Sophia Grass and now lives in Big Springs

Mr and Mrs Waterman have witnessed many changes in the Panhandle since they come here more than thirty-five years ago, they have seen the country grow and develop and towns dot what was a wilderness Mr Waterman wishes he had been forced to buy and keep more land He experienced the vicissitudes, up and down of all early settlers but is glad that he stayed to win prosperity Mr Waterman still owns the old home farm but rents the land, he also owns a fine business block in Big Springs In politics he is a Democrat and also belongs to the Modern Woodmen of America

CHRISTOPHER McCORMICK — If his youthful ambition were to achieve worthy success, that ambition has been realized in generous measure, if he were determined so to order his life as to win and retain the high regard of his fellow men, that reward has been granted to him, — and thus this sterling citizen of Garden county is a man who is most consistently accorded tribute in this history, especially in view of the fact that through his well ordered activities he has contributed definitely to the social and industrial progress and prosperity of the famed Nebraska Panhandle, to which this publication is dedicated

Mr McCormick was born in County Tyrone Ireland, August 24, 1854, and is a son of John and Anna Jane (Graham) McCormick the former of whom was born in Ireland, in 1826, and the latter of whom like-wise was a native of the Emerald Isle, though her father, Robert Graham, was born in Scotland The parents passed their entire lives in the native land, where John McCormick was a farmer by occupation, he having attained to the venerable age of eighty-six years and his wife having passed away at the age of forty-five years Of their family, Christopher, of this sketch, is the eldest, and five of his brothers likewise became residents of Nebraska, Calhoun, who maintains his home in Garden county, Andrew G, and Thomas, who reside at Lewellen, Garden county, Robert, who lives at Bigspring, Beuel county, and James, who is deceased

The schools of Ireland afforded to Christopher McCormick his early educational advantages, and the discipline proved effective in the developing of his alert mental powers and fortifying him for the practical duties and responsibilities of life At the age of twenty-two years, he received from Queen Victoria a ticket which provided him transportation to New Zealand, and in that Island colony of England he became actively identified with railroad construction, though his major occupation was that of farming After living in New Zealand four years he went to Australia, where he remained about two years, during which time he gave his attention to farming He still further broadened his experience on the return voyage to Ireland, for he made the trip by way of Honolulu and San Francisco, and thus incidentally made his first visit to the shores of the United States He remained in Ireland about three years and then, in 1883, came with his family to America and settled at Crete, Saline county, Nebraska, near which place he was engaged in farming about three years Mr McCormick then became a pioneer settler in that part of old Cheyenne county that now constitutes Garden county, having taken up homestead and tree claims in 1884, though it was not until 1886, that he removed with his family to the pioneer farm In due course of time he perfected his title to his claims, and developed them into one of the productive, valuable and well improved farms of the county There he continued his activities as a successful agriculturist and stock-raiser until 1916, when he removed to the village of Lewellen Here he has since lived virtually retired, in the enjoyment of the rewards of former years of earnest toil and endeavor He has been liberal and public spirited in his civic attitude, is a Republican in political allegiance, served eight years as a director of the Lewellen board of education, and is one of the principal stockholders of the Farmers State Bank of Lewellen, of which institution he is vice president. He and his wife are earnest and active members of the Methodist Episcopal church in their home village, and they have the high esteem of the people of the county in which they have maintained their home for more than thirty years.

May 29, 1883, in County Tyrone, Ireland, was solemnized the marriage of Mr. McCormick to Miss Margaret Wood, a daughter of Samuel and Jane (Wilson) Wood, both of whom passed their entire lives in Ireland, where the former died in 1862, at the age of fifty-two years, his widow having attained to the age of seventy-six years. They became the parents of three sons and four daughters, all of whom came to the United States. Allan Wood, the eldest, served as a member of an Illinois regiment during the Civil War, and the youngest son, Samuel, became a clergyman of the Methodist Episcopal church, his ministerial services having been initiated in Illinois and thereafter continued in Nebraska. In conclusion is entered brief record concerning the children of Mr. and Mrs. McCormick: Jennie is the wife of Calhoun Orr, of Lewellen, and they have one child; Jack, married Lotta Davis and served in 1919-20 as treasurer of Deuel county, with residence at Chappell, has two children; Edward W., of Lewellen, married Birddie Eggers and has one child; Robert A., of Lewellen, served in the United States Signal Corps during the nation's participation in the late World War; and James C., of Lewellen, married Maude Beddoe, has one child.

S. AUGUST FORNANDER, early settler and successful farmer of Deuel county who came here when this country was unbroken prairie, was born in Sweden, July 26, 1864, the son of Carl and Greta (Ankerberg) Carlson, both natives of that country, where the father was a farmer, who died in 1869, being followed by his wife in 1899. The mother came to the United States in 1887 to Knox county, Illinois, remained there until 1893, when she located in Phelps county, Nebraska, but five years later, in 189 8, moved to Cheyenne county. She lived with one of her sons the rest of her days. There were six children in the family, all of whom live in this county.

August Fornander was educated in the public school of his native country and came to the United States in the Spring of 1881. He first located in Knox county, Illinois, but two

years later came farther west to Phelps county, Nebraska, then to Deuel county in 1885. He took up a homestead a mile south of his present farm as he traded the original place to his brother in order to have all his land in one section. Mr. Fornander's first team in this country was a pair of bulls which were broken to the yoke. Later he bought horses. As soon as he would get a crop in on his homestead Mr. Fornander would leave for Colorado to work and earn money for supplies and one time worked on a ranch there. While on the homestead the first years he "batched" it by himself. Water had to be hauled from the tanks often times six to twelve miles, costing fifteen cents a barrel at first, this price was lowered and finally water was free. Sometime when Mr. Fornander and his brother got up early for water they would find other had been ahead of them and the water gone from several tanks making a long tiresome journey. In the early days the people of this section raised many cattle as numerous homesteads had been abandoned and there was range, but within the past decade great changes have come and today Mr. Fornander carries on general farming due to his improved methods and modern machinery. He tills some hundred and fifty acres; is a stockholder in the Farmers Elevator at Chappell and has been treasurer of school district number sixty-three since it was established.

July 10, 1914, Mr. Fornander married Miss Anna Simmons, the daughter of August and Sophia (Johnson) Simmons, natives of Sweden and two children have been born to this union: Neal and Joel. Mr. Fornander. is. an independent Republican, has never aspired to office beyond those of local affairs as he helped organize school districts number thirty-two and sixty-three and acted as trustee. He and his wife ar members of the Lutheran church and assisted in organizing Batesta church and later Berea church, in which they are active workers.

JOHN W. GRANNELL, one of the retired colony of Big Springs who has made a success of varied lines of endeavor since he came here in the early days was born in Vego county, Indiana, January 1, 1859, the son of Noah and Elizabeth (Baulding) Grannell, the former a native of Ohio while the mother was born in Indiana. The father was a cooper by trade and followed that vocation most of his life, having a shop of his own. He was a Democrat and a member of the Masonic order and with his wife was a member of the Presbyterian church. There were seven children in the family of whom four are living but John W is the only one in the west He received his education in the public schools of Indiana Mrs Grannell died when the boy was fourteen years of age and he began to work among strangers by the month until he was married on December 7 1881, to Miss Margaret Clugston, also of Indiana After that Mr Grannell farmed in his native state until the family came to Hastings Nebraska in 1884 where they lived until 1900 That year Mr Garnnell took a homestead in Perkins county ten miles south of Big Springs Good improvements were made on the place and the farmer it for seventeen years having bought three hundred and twenty acres in 1899 He then sold the land and retired to Big Springs At first he worked as a carpenter so as to fill his time as he had always been busy, for when the Grannells came here it was necessary to haul water for their use and that for the stock in addition to the general farm work This first farm had no water but the second had a well During the first years when crops were poor Mr Grannell worked out for other men to secure money to buy supplies, then took care of cattle but later managed his own farm After coming to big Springs Mr Grannell bought a meat market, managed it two years and sold to engage in the restaurant business but disposed of that and purchased a pool hall and cigar store where he remained six years Following this he ran a meat market again, a cigar store and then retired from the mercantile business entirely working as a carpenter just a little When Mr Grannell started in business in Big Springs it had but a few buildings so that he has seen its great growth and development Mr Grannell is an Independent in politics, is a member of the Presbyterian church, served on the school board for thirteen years and was assessor of his district many years He is well known and liked in Big Springs where he has taken his part in the upbuilding and growth of the town

EDWIN A PHELPS, Sr, was an early railroad man and resident of Big Springs for more than thirty-five years, there are few better known and popular men in the Panhandle Mr Phelps was born in Milford, Oakland county, Michigan December 24, 1842 the son of Aaron and Mary (Armstrong) Phelps both natives of Batavia, New York The father was the owner of a saw mill and flour mill and at one time of a distillery in addition to managing a farm He was a prominent man in his community until he moved to Chicago in

1853, where he owned and operated a hotel The mother died in 1854, being survived by her husband until 1894 Mr Phelps was a Republican There were eight children in the family, of whom only two survive Eliza, the wife of Henry Crane of Chicago, and Edwin of this review, who received his educational advantages in the schools at Milford Michigan, and Chicago Illinois When only eighteen years of age the young man began his independent career as a railroad man For a few years he was brakeman, then became a machinist and in 1865, went to Tennessee to build railroad bridges for several years In 1870 he came west locating in Nebraska to engage in erecting bridges from Omaha to North Platte He came to old Cheyenne now Deuel county, early in 1883 as bridge foreman for the Union Pacific Railroad spent three years in that work and then filed on a homestead and timber claim This land he proved up and farmed for some time before returning to the employ of the railroad, retiring in 1912 Since then Mr Phelps has not been actively engaged in business though he did not sell his ranch until 1919

Mr Phelps was the first assessor of Deuel county, elected on the Republican ticket but now votes with the Democratic party He is a member of the Presbyterian Church though he was one of the organizers of the Methodist Church at Big Springs and the first Sunday School was held in his home in 1884

In October 1869, Mr Phelps married Miss Sarah E Grote and they became the parents of two children Edwin A Jr and Arthur L, both living in Big Springs When Mr and Mrs Phelps first came to Big Springs the town consisted of a section house, depot and their home built out of bridge timber was the third structure The first store was opened in 1884 and a drug store was built and operated two years later On Christmas eve 1885, Mr and Mrs Phelps gave a dance in their new hotel building which was formally opened the next day and this structure has been in constant use up to the present time While Mr Phelps worked on the railroad his wife managed the hotel and we learn the conditions of the country at the time from the fact that the railroad furnished all its hands with guns to protect themselves against the Indians and kept a small detachment of soldiers at every station Mr Phelps is a member of the Masonic lodge and he and his wife are highly respected and well known from their long residence in Big Springs They have seen the many towns grow up and the country develop from a wilderness to fine productive farm lands

JOHN JOHNSON, one of the pioneer settlers of Deuel county who located here when settlers were few and far apart, has the distinction of having lived in three counties and never moved off the original homestead, as he came before the new counties were erected and as they were formed became a resident of each in turn Homesteading in Cheyenne county it was split and Deuel county formed then again divided and Garden county was erected Mr Johnson was born in Sweden January 25, 1857, the son of Joseph and Johanna (Junison) Johnson, both natives of the same land where the father was a farmer all his life He also operated a flour and saw mill and ran a blacksmith shop on a large scale and at one time ground all the wheat from the surrounding seven counties Mr Johnson took a prominant part in political life, holding several public offices, he died in 1878 and the mother in 1909 There were ten children in the Johnson family, but Gust and John are the only ones living in the western part of Nebraska, as the former resides at Oshkosh The family were members of the Swedish Lutheran Church of which the father was a deacon

Mr Johnson received his education in the public schools of Sweden and then worked on the farm until he came to the United States in 1879 He first settled in Iowa where land was selling at three dollars an acre but in 1885, came to Deuel county to secure a homestead for a permanent home This he sold in two years and then bought other land, a section of which he still owns though he has retired from active life, and now lives in a fine home in Chappell, where the family located in 1917

It can honestly be said that Mr Johnson is a self-made man as all he had when he came to Deuel county was a cow, calf and a few chickens, but he was not afraid of work and soon became established as a well-to-do farmer At the time of his settlement wild horses and antelope were common on the table northeast of Chappell where he located, cow boys rounded up the horses and corralled them for market When Mr and Mrs Johnson stopped at Julesburg, Colorado, on their way to the new home they were forced to sleep in the telegraph office on the floor, as the hotel, section house and all buildings were filled with the cowboys asleep for Julesburg was a shipping point for cattle In September, 1883 Mr Johnson married Miss Christine Anderson, the daughter of John P and Stenguta (Samuelson) Anderson, who settled in Iowa at an early day One

child has been born to this union, Jennie, the wife of L R Nelson living northeast of Chappell

Mr Johnson is a Republican and has been a hard worker for his party, has been elected to several public offices and was delegate to the county conventions a number of times With his wife Mr Johnson is a member of the Swedish Lutheran Church He has taken active part in public affairs, helped organize the church and several public schools and served as school director of his district nine years When the railroad came through this section he was appointed appraiser of land for the branch running to Oshkosh

PAUL SWANSON, one of the substantial and prosperous farmers of the Chappell district, Deuel county, is a native of Sweden, that country which has given the United States such a large proportion of its best settlers He has demonstrated that a man who is determined to succeed in this country can do so Mr Swanson was born in Sweden May 16, 1851, the son of Swen Olson and Swembo Swanson, both born and reared in Sweden where they spent their entire lives The father was a general farmer and followed that vocation all his days There were six children in the family, but Paul, of this sketch, and his brother Ola who lives in Minnesota, are the only ones in the United States

Mr Swanson was educated in the excellent public schools of his native land, served in the army two months every year until he was twenty-one years old, as is required in Sweden, and then engaged in farming Learning of the many opportunities for a man to secure land in America he came to the United States in 1885, locating first at Galesburg, Illinois, but two years later moved to Iowa where he remained a year before taking up a homestead on the divide in Deuel county Coming here in 1888, he is one of the early settlers of this region and has seen the many changes that have taken place in the opening up and development of the Panhandle, also taking his part in the agricultural industry of Deuel county Mr Swanson made many improvements on his farm and lived there until 1915, when he moved to land south of Chappell, which his wife had homesteaded He now is engaged in general farming and has a large amount of pasture land on the two sections

In September, 1916, Mr Swanson married Mrs Bessie Jacobson, the widow of Andrew Jacobson, a pioneer of Deuel county Mrs Swanson came to western Nebraska about two years after her husband and took up a home-

DICK BEAN EMMET JAMES ASA REMSBERG

stead Mr Jacobson died in 1913, leaving a family of eight children Angeline, the wife of Herman Runquist, of Deuel county, Amanda, the wife of Ed Olson, of Deuel county, and the following children who are single, Homer, Archie, Edith, Floyd, Gladys and Bertha When Mr Swanson left Sweden he brought with him an eight year old boy whom he adopted, Nels, who now lives in California

Mr Swanson is a Republican and a member of the Swedish Lutheran Church He has been successful in his business and today is regarded as one of the good solid men of his district

RICHARD E BEAN commonly known as Dick Bean, was born in the state of Arkansas in 1853 He moved with his father to Heneretta, Texas, when he was two years old, and lived with his father on a farm until he was about fifteen years old He then ran away from home, and hired to a trail herd coming to Ellsworth, Kansas He returned to Texas and worked on trail herds between Texas, Kansas, and Nebraska until 1874, working on Texas ranches during the winter months and leaving for the north on trail herds about the 1st to the 15th of March, during this period The last trip he made up the trail he drove to Lowell, Nebraska

He then began working on the northern ranches, and from then on became identified with the northwest Nebraska country From 1874 to 1876 he worked on various ranches between Lowell and North Platte City After this date he accepted employment with Bosler Brothers & Company, located on the north side of the North Platte river Their range was the river front from Brown's creek to Blue creek and north to Bronco Lake where Alliance is now situated This company ranged about forty-five thousand head of cattle In this herd were about twenty different brands of cattle and several brothers interested in different propositions which made it difficult for the ordinary brand man at the shipping season It was in this line that Dick Bean rapidly developed, he and Gene Hall working together He developed this into a science which was phenomenal He developed a facility wherein he could remember cattle like some men today remember men Whenever a cow passed in front of his vision it was not uncommon for him to remember this cow for a period of a year and would not need to look at the brand to identify it These cattle were always classified in what was commonly called Lone Tree Bottom about twelve miles northwest of Ogal-

lalla, where they were rounded up and sorted as to brands

This job of classifying cattle was handled the same as cattle are counted on the range today, the counter or classifier took his place at the head of the string and as the cattle passed by him would call out the different brands to two tally men, (usually owners), this position was always occupied by Dick Bean in this outfit He was always backed up by one top hand, usually Gene Hall, a personal friend and partner of Dick Bean It was his job to keep the stringing cattle from breaking back of the classifier, hence the following little story by Gene Hall today a prominent cattleman residing at Alliance, Nebraska These cattle were wild as deer, while working them, and in spite of all he could do, there were times when as high as seven to ten head would break around him and it would be Hall's job to spot these cattle His correct position would be on the left side, and as all these cattle were branded on the left side it makes the following statement regarding Bean even more remarkable When a break of this kind occurred, whenever possible Bean would block the string and glance over the cattle from the right side and call the cattle such as "four B bar" "three half circle B," and *three O B O* " and holler out to Gene Hall "let 'em go " Mr Hall used to try to catch Dick Bean in a mistake on this work as he would always be on the left side, (and which was his job to do so), but makes the statement in all their work together he never knew him to make a mistake in this way, which shows Dick Bean's wonderful memory for individual cattle

He developed into the most valuable employee on the Bosler Brothers & Company ranch and was often tendered the foremanship, but declined every overture of this nature On account of his mother's death when he was three years old he was deprived of an education and did not feel he was capable of handling this outfit, but whenever anything of importance came up Dick Bean was consulted and his judgment and ideas were always followed out by the management In December, 1884 the Ogalalla Land & Cattle Company was organized the first herds purchased being the W A Paxton and Ware herds, and the Shiedly Brothers herd In August, 1884, Dick Bean left the employ of Bosler Brothers & Company and was immediately employed by W A Paxton and Ware and when the Ogalalla Land and Cattle Company was organized was placed in charge of their outfit on the north side of the Platte river A peculiar instance occurred here wherein Bean refused to accept this position unless they agreed to furnish him a bookkeeper to handle this end of the work, as all that Bean could write at this time was his own name. In February, 1885 the Ogalalla Land & Cattle Company bought out the Bosler Brothers & Company outfit, which threw Bean back in direct control of the old brands and a herd with which he won his reputation as a cowman. This outfit branded into one of the biggest cattle outfits in the northwest, their books calling for seventy-seven thousand head of cattle. This outfit was operated by three foremen. Dick Bean from Blue Creek east, Mac Radcliffe on the south side of the North Platte river, and Gene Hall from Blue creek west. Dick Bean continued in the employ of the Ogalalla Land & Cattle Company until the latter part of 1888 then took a homestead on White Tail creek and went into the ranch business for himself and became well to do. He intended to be married in the year 1894, and went to the town of Ogalalla, and on the way home with a load of lumber to build a new home for his prospective bride, his team ran away on what was known then as Seven Crook Hill. The lumber not being bound on the wagon came loose and a two by four caught in the wheel, whirled around and struck him on the side of the head, killing him instantly, which ended the life of one of the greatest cowmen that was ever in the northwest.

EMMETT JAMES was born in Goliad, Texas. His parents died when he was a small boy and he had his living to make alone. He rode race horses for four or five years and in 1877, came up the Texas trail with a herd of cattle and upon his arrival at Ogalalla, entered the employ of the Bosler Brothers Cattle Company, and remained with this concern until the fall of 1883. He was a great friend of Dick Bean. After leaving the Bosler Brothers, he engaged with the Heart Cattle Company which was owned by W. A. Paxton and W. H. Parker, as foreman and then was taken over by Paxton and Bosler. He remained foreman for this company until the summer of 1886, when this company was bought out by the Ogalalla Land & Cattle Company. This was Mr. James last experience as a range "Cow man," he located on a small ranch in the sand hills about fifteen miles east of Alliance, Nebraska. Having saved about thirty-six hundred dollars in nine years of range work he bought about a hundred head of cattle as a start in the live stock business. Mr. James

was of a peculiar temperament, and very quick to take offense, often without any provocation, and just as quick to recover his equanimity, and apologize for his hasty temper and was a very conscientious man, being honest to the penny, and above reproach. He, like Dick Bean, had only education enough to sign his name. Dick Bean was level headed, and never got excited without cause, but both men were well liked and courageous to the limit. Emmett James was married in June, 1889, and died the following November. He was buried in Alliance. His death resulted from an injury caused when he fell from a wagon.

ASA REMSBERG was reared and educated in the historic old state of Maryland, where he was born in Frederick county, September 24, 1853, but in following out the course of his youthful ambition he came to the west when about twenty years of age, his first definite action being to assume as soon as possible the prerogatives and dignities of a cowboy. He was thus employed about one year in connection with the herding of cattle under the conditions of the great open ranges in the Lone Star State, but after an experience of about one year he trailed a "bunch" of cattle through from Texas to Ogalalla, Nebraska, and shortly after his arrival he entered the employ of the Chadley Brothers Cattle Company, of Big Springs, now in Deuel county. He continued as a valued member of the outfit of this pioneer cattle company from 1878 to 1884, and thereafter he was in the employ of the Ogallala Company until 1887, — principally in the capacity of wagon boss. He then made a trip to Texas in the interest of the Rush Creek Cattle Company, and at this time he assembled and shipped up to Nebraska headquarters of the company about thirteen hundred head of cattle.

In 1887, Mr. Remsberg, who had gained full knowledge concerning this western country and its possibilities, decided to make his appreciation one of somewhat more definite order, and accordingly he took a homestead claim in that part of old Cheyenne county that is now comprised in Morrill county. He proved up on this original homestead, as did he also on a pre-emption claim and a tree claim in the same locality, and there he gave his attention principally to the raising of cattle and horses. Finally he sold his land on the south side of the North Platte river, and on the ranch which he purchased on the north side of the river he continued his successful cattle and horse industry until 1903, when he sold his stock to the Rush Creek Cattle company, in the employ of which representative corporation he has since continued, his position being that of foreman of the outfit and his broad and varied experience make him an authority in the multifarious details of the live-stock business Mr Remsberg now owns some land in Garden county, and on the Belle Fourche river in South Dakota he owns about seven hundred acres of valuable grazing land Mr Remsberg owns the Ford garage buildings and the pool hall building in the village of Lisco, as well as an attractive residence property in this thriving town He is well known throughout this section of Nebraska and is a pioneer citizen who can muster his friends in veritable battalions His political support is given to the Republican party and he is a popular and appreciative member of Alliance Lodge, No 963, Benevolent and Protective Order of Elks, of Alliance His name remains boldly emblazoned on the list of eligible bachelors in Garden county

In conclusion it may be stated that Mr Remsberg is a son of John and Adaline Remsberg, both of whom passed their lives in Maryland, where the father was a farmer by vocation, he having been about eighty years old when he passed from the stage of mortal life, and the mother of the subject of this review having died when fifty years of age

JOHN R WERTZ, president of the First National Bank of Chappell, well known pioneer business man and successful real estate dealer of the Panhandle, is one of the practical, self-made men of this section and Deuel county who have so materially helped in the development of western Nebraska He was born in Bedford county, Pennsylvania, January 1 1872 the son of John W and Jane (Oliver) Wertz, both natives of the same county, the former born there in 1828 and the mother in 1832 The father was a farmer in his native state, spent his life there and died in 1872 The mother survived until 1904 Mr Wertz voted the Democratic ticket but never held office He and his wife were members of the Lutheran Church There were seven children in the family of whom four are living but John R is the only one in this part of the country

In his youth Mr Wertz had but little opportunity to obtain a good education as he attended the public schools for only about two months each year At the age of eighteen he started out for himself, coming west in 1890, he located here in Deuel county on a home-

stead on which he never proved up After locating the claim he erected a sod house with board roof bought a load of furniture which was installed and had his home ready for occupancy, but during an absence of ten days some other settler stole the roof and furniture so he never lived there, came to town and never went back For about two years Mr Wertz owned and ran a livery stable but sold the barn, house, three lots, teams, buggies and horses for eight hundred dollars and later the same property was sold for eight thousand In 1892 Mr Wertz opened a hardware store in Chappell, carrying a line of furniture, farm implements and for sixteen years was known as one of the leading progressive merchants of the county He made many warm friends and was elected county treasurer on the Democratic ticket in a county that is known to be overwhelmingly Republican, but political lines were not thought of when his business ability and high standing in the community were of value to the citizens Soon after leaving this office, in 1909 Mr Wertz opened a real estate business in which he has since been engaged Since first locating in the county he has taken an active part in public affairs and for more than seven years has been the president of the First National Bank holding a large block of stock and it has been due largely to his able direction of the policies of the bank that it has gained the confidence of the people as a sound financial institution For several years Mr Wertz served on the town board and during his office advocated many improvements for the municipality

November 15 1909, Mr Wertz married Miss Blanche Francoeur, the daughter of Adolphus and Alma Francoeur, early residents of Deuel county, the former now deceased but the mother still makes her home here Two children have been born to this union John D and Paul R

Mr Wertz is a member of the Masonic order has held all offices in the organization and for the past two years has been past master of the lodge at Chappell He is one of the self-made men of this section who has made good in the Panhandle, for today he is recognized as one of the leading bankers and best known real estate men in western Nebraska where he conducts a large and profitable business

JACK G McCORMACK, the efficient county treasurer of Deuel county is a native son of Nebraska of Irish extraction and the saying that the Irish-American always succeeds" in whatever line of endeavor he may choose for life work has been ably exemplified in the career of this official, who while still a young man stands high in his community and holds a position of trust

Mr McCormack was born at Crete, Nebraska, November 15, 1885, the son of Christ and Maggie (Wood) McCormack, both natives of Ireland who came to America while still young people Soon after landing on our shores they came west and in the spring of 1886 took up a homestead near Big Springs, Deuel county The father had been a farmer in his native land, so followed the same vocation here, meeting with many hardships but still persevering After eleven years on the homestead the family moved to Lewellen where the parents still live, retired Two of the sons now manage the home place Mrs McCormack was a well educated woman, taught school in her home several years and later had charge of the district school a mile away, taking her younger children along with her and often times carrying one that distance Mr and Mrs McCormack are members of the Methodist Episcopal Church, they helped organize the earliest in this section and have been two of its main supporters Mr McCormack is a Republican in politics but has never cared for public office

There were six children in the family, of whom the following are living Jennie, the wife of Cal Orr, of Lewellen, Jack, of this sketch, Edward, who married Birdie Eggers, lives in Lewellen, Robert, on the old homestead, and James, who married Wanda Beddo, also on the old farm In the early days the family suffered all the trials of drought and had only one crop in eleven years, but kept the farm and made a living by keeping twenty milk cows and selling the cream

Jack McCormack was educated in the public schools at York and later took a business course at a business college at Grand Island When only twenty-two years of age he started out in life for himself and can be said to be self-made He taught school two years before being elected county clerk, a position he efficiently filled seven years, acting in the capacity of judge for nearly two years in addition to his other duties On January 12, 1919, Mr McCormack was elected county treasurer of Deuel county on the Republican ticket and is still in office For about eight years he has been abstractor of the county, keeping most careful records, and giving entire satisfaction

On August 8, 1913, Mr McCormack married Miss Charlotte Davis, the daughter of David and Emily (Chambers) Davis, pioneer settlers of this county Two children have been born to this union J Weldon and Byron Eugene Mr and Mrs McCormack are members of the Methodist Church and the latter of the Ladies' Aid Society and Home Craft Society, both church organizations Mr McCormack is a member of the Masonic Order and the Independent Order of Odd Fellows, having been through all the chairs of this lodge For some years he has been treasurer of the county high school and is regarded as one of the substantial and rising men of his community

HARVEY I BABCOCK —There has been naught of austerity, doubt or indirection in the progressive career of the able and popular cashier of the First National Bank of Chappell, Deuel county, for he has been content to employ effectively the means at hand and has by his own ability and efforts risen to a place of prominence as one of the representative figures in financial circles in western Nebraska, which has been his home since he was a lad of about sixteen years and in which he is a scion of a sterling pioneer family

Mr Babcock was born at Ridgeport Iowa, September 19, 1870, and is a son of Reverend William H and Luceba (DeWolf) Babcock, the former a native of the state of New York and the latter of Illinois The father was a man of high attainments, having not only been a skilled physician and surgeon but also having long given devoted service as a clergyman of the Methodist Episcopal Church He was engaged in the practice of his profession in Iowa until he came with his family to Nebraska and engaged in practice at Bradshaw, York county In 1886, he came from that county to that part of old Cheyenne county that now constitutes Deuel county, and became a pioneer physician at Chappell, where he also established and conducted a drug store A man of noble character and unbounded human sympathy and tolerance, he endeared himself to all with whom he came in contact in his various ministrations and other activities in the pioneer community, and a deep feeling of personal loss and bereavement was felt by the community when he died June 18, 1895 In the year of his arrival in the present Deuel county he organized the first church and Sunday School within its borders, and he served several years as pastor of this church, now the First Methodist Episcopal Church of Chappell He was a staunch Republican and by virtue of his broad views and sterling character was well equipped for leadership in community thought and action He served several years as coun-

EDWARD M. REYNOLDS AND FAMILY

ty coroner and was otherwise influential in public affairs of a local order His wife, a woman of gentle and gracious personality, is now residing in Chappell, and of their six children, two are living—Elizabeth, the wife of George W Gordon, of Haxtun, Colorado, and Harvey I, of this review

In the pioneer school at Chappell, Harvey I Babcock pursued his studies after the family home had been established here, and these he supplemented by a course in a business college of, Denver, Colorado He initiated his business career by taking a clerical position in the Deuel County State Bank of Chappell, in 1889, and his original stipend was only ten dollars a month He applied himself diligently and effectively and in due time his ability received fitting recognition In 1892, he was made assistant cashier, and in the following year he was advanced to the office of cashier, of which important executive position he continued to hold after the bank was reorganized as the First National Bank of Chappell It may well be understood that he has been an influential factor in the development of the large and substantial business of this institution, the deposits of which now aggregate almost one million dollars, its operations being based on a capital stock of $25,000 and its surplus fund being $45,000 In addition to being cashier of the bank Mr Babcock is one of the principal stockholders of the Chappell Telephone Company and is the owner of several valuable farm properties near Chappell In politics he gives allegiance to the Republican party, but he has been primarily and essentially a business man and has had no ambition for public office, though he has shown his civic loyalty by effective service as a member of the village council and school board He assisted materially in securing to Chappell its waterworks system, as well as in the development of the excellent public schools of the village He assisted in the organization of the Farmers Elevator Company of Chappell, and was the prime mover in selling the stock which insured the success of the important enterprise He is a member of the board of regents of the Deuel county high school, is affiliated with the blue lodge, Chapter and Commandery bodies of the Masonic fraternity, in which he has held various official chairs, and is identified also with the adjunct organization, the Ancient Arabic Order of the Nobles of the Mystic Shrine, as well as with the local camp of the Modern Woodmen of America

May 10, 1892, recorded the marriage of Mr Babcock to Miss Helen Johnson, daughter of Joseph C and Laura (Lewellen) Johnson, of Chappell, and Beryl, the only child of this union, died at the age of eight years

EDWARD M REYNOLDS, one of the prosperous and representative business men of Hemingford, who has been a resident of Box Butte county for more than a decade and during that time has taken his share of the work of opening up and developing this section, was born at Buffalo, New York, October 3, 1856, the son of Milton and Alxina (Jacobs) Reynolds, both natives of the Empire state, where they were reared, educated and later met and married Edward was next to the youngest child in a family of five, as he had four sisters Milton Reynolds was a contractor who, when Edward was small, had a large and successful business as a builder, and at that early day he had to get his mouldings and interior finishings from the mills in the city of New York, as such things were not manufactured so far west as Buffalo As soon as the boy was old enough he began to play in his father's shop and before long was carpentering things for himself, such as little sleds, and other children's toys, small wagons and the like, which he sold to earn spending money, when he was yet only a lad of twelve The family came west to Chicago at an early day and there Edward was sent to the public schools, laying the foundation of a good practical education His father remained in the contracting business which caused him to locate in Harrison county, Iowa, Edward learned the carpenters trade and continued to help his father until about 1875, when the other members of the family returned to Chicago, leaving Edward in Iowa He soon went to Carrol City and engaged in contracting for himself, specializing in carpenter work for about three years but gave it up to accept an advantageous offer from the Northwestern Railroad to build the new stations they were putting up along the line as it was built west In addition, he took on contracts for the erection of bridges for the road After this he gave up his contracting, moved to Wall Lake, Iowa, and engaged in the furniture business Within a short time he established a branch store for the same business in Sac City, turning over the management of this store to his father In February, 1881, Mr Reynolds was married at Sac City to Miss Cora L Hatfield, a native of Massachusetts, the daughter of Charles and Levina Hatfield, both natives of the Bay state Three children have been born to this union Charles M who married Leona White, is superintendent of the Imperial Sash and Door Company of Omaha has three boys, while he and his wife are members of the Baptist church, Cora A, the wife of Percy Daily a carpenter at Gering, has one small daughter, Helen, and Roy, who married Ethel Price of Box Butte county, is a farmer, who responded to his country's call when war was declared against Germany and enlisted in the remount division in October 1917, and received his training at Camp Funston He has been honorably discharged and is now following his former vocation, as a farmer near Hemingford In 1886, Mr Reynolds accepted a position in the general repair shops of the Union Pacific Railroad at Omaha where he was employed for ten years, but the confinement of indoor work did not appeal to him for a life vocation so he resigned and came to Box Butte county in 1908, remained in Hemingford for a year then took up a homestead in Sioux county, as he knew of no man so independent as a land-owner He proved up on his six hundred and forty acre farm, engaged in agricultural pursuits for six years, being interested mostly in stockraising He still retains possession of this land which has turned out a lucrative investment Coming to Hemingford Mr Reynolds bought property here, owning a fine store building twenty-four by sixty feet, two stories high, where he conducts a furniture and second hand business, carrying both new and used goods, all of a high class He now contemplates building another store about the same size as the one he has and then will put in a first class up-to-date repair shop for all kinds of furniture, upholstery and fine finishing In addition he anticipates taking contracts for all kinds of job work as he is not only a skilled carpenter but a fine cabinet maker as well, a thing much needed in Hemingford Mr Reynolds says that he may branch out into the lines allied with furniture when he has the new store and space for it He and his family are members of the Baptist church while his fraternal relations are with the Modern Woodmen

HENRY C PETERSON was about five years of age when his parents came to Nebraska and assumed the responsibilities and labors of pioneers in that part of Deuel county that now comprises Garden county Thus he is virtually "to the manner born" in all that implies familiarity and association with the progressive activities that have marked the development and upbuilding of western Nebras-

ka, and that he has kept pace with the march of progress is shown in the prominent part he has played in connection with the agricultural and live-stock industries, the influence he has wielded in public affairs of a local order, and the secure vantage-place which he now holds as one of the representative business men of Chappell, where he is president of the Chappell State Bank, one of the substantial and well ordered financial institutions of the Nebraska Panhandle

Mr Peterson was born in Pottawattomie coutny, Iowa, February 15, 1882, and is a son of Peter S and Lena (Hansen) Peterson, both natives of Denmark, where the former was born May 6, 1850, and the latter October 8, 1858, their marriage having been solemnized in the state of Wisconsin Peter S Peterson came to America in 1872, shortly after attaining his legal majority, and after remaining for a time in the city of Chicago, he went to Racine, Wisconsin where he met and married Miss Lena Hansen, who had come with her parents from Denmark to the United States in 1865, the family home having been established at Racine For several years Mr Peterson was employed in the Mitchell Wagon Factory at Racine, and thereafter he was employed for a period at Council Bluffs, Iowa He finally turned his attention to farm enterprise in Iowa, and in 1887, he came to what is now Garden county Nebraska, where he took up homestead, pre-emption and tree claims, not far distant from Oshkosh, and where he developed one of the large and valuable farm properties of this section of the state, his widow still remaining on the old home place, and the landed estate owned by the heirs of Peter S Peterson now comprising eighteen hundred acres, Mr Peterson having been one of the honored and influential pioneer citizens of Garden county at the time of his death, which occurred September 29, 1916 He was an earnest member of the Baptist Church, as is also his widow, and his political allegiance was given to the Republican party Of the four children the eldest George M, resides in Garden county, Henry C, of this review, was the next in order of birth, Clarence W, likewise remains in Garden county, Dora C is the wife of Rev Robert C Sharp and they reside near Oshkosh, that county, and an adopted son, Charles Peterson, remains with his foster-mother on the old Peterson homestead

Henry C Peterson passed the period of his boyhood under the conditions that marked the pioneer epoch in the history of the present Garden county which was then a part of

Deuel county, and in the little sod school house he pursued his studies during three months of each year until he had attained to the age of seventeen years. In the meantime he had learned farming on the home ranch, and at the age noted he found employment on a neighboring ranch. After two years of application under these conditions he devoted a similar period to independent farm enterprise in Garden county. There he took up a homestead about the year 1903, and later he sold the property to one of his brothers. In the preceding year he had married, and after this important event in his career he continued his activities as an agriculturist and stock-grower in Garden county for three years. In buying a ranch he assumed an indebtedness of twenty thousand dollars. After continuing individual operations one year he formed a partnership with his brother Clarence, with whom he was associated in the development and improvement of the property, which they cleared of debt by 1911, besides having added materially to the area of their landed estate. This property Mr. Peterson finally sold to his brother Clarence. In the autumn of 1909 he was elected county treasurer, which, as a matter of course, involved his removal to Chappell, the county seat. He continued in the office of county treasurer five years and his administration was characteristically efficient, thereby justifying fully the popular confidence reposed in him. In the meantime he had become a stockholder of the First National Bank, now the First State Bank, of Oshkosh and on November 1, 1912, he effected the organization of the Chappell State Bank. The bank was incorporated with a capital stock of $15,000 and he served as vice-president of the institution from the time of its re-organization until 1917, when he was elected its president, an incumbency which he has since retained. He has been a resourceful power in developing the business of this representative banking establishment the capital stock of which has been increased to $50,000 and the deposits of which now aggregate about $750,000. In 1919, Mr. Peterson extended his banking interests by becoming one of the organizers of the Lakeside State Bank, at Lakeside, Sheridan county and he has been its president from the time of its incorporation. An enthusiast in all that pertains to western Nebraska and its great future, Mr. Peterson is a staunch supporter of progressive movements and enterprises tending to advance the interests of this section of the state, and this is further shown in his owner-

ship of a valuable tract of three hundred and twenty acres of irrigated land lying west of Oshkosh, Garden county. On this ranch, which is rented out on shares, special attention is given to the raising of hogs on an extensive scale. Mr. Peterson is also a stockholder in the Farmers Lumber & Hardware Company of Chappell, and is essentially one of the vital and progressive business men and liberal citizens of the fine country to which this history is dedicated.

In politics Mr. Peterson is found aligned as a stalwart in the local ranks of the Republican party, and while he has held no distinctive political office save that of treasurer of Garden county, he is now serving as treasurer of school district No. 7 and as chairman of the board of regents of Deuel county, besides being treasurer of the village of Chappell. He is affiliated with Golden Fleece Lodge, No. 205, Ancient Free & Accepted Masons, at Chappell, in which he has passed the official chairs up to that of senior warden, and he is identified also with the local camp of the Modern Woodmen of America, his wife being a member of the Order of the Eastern Star and also of the Methodist Episcopal Church of Chappell.

On September 24, 1902, was solemnized the marriage of Mr. Peterson to Miss Alice M. Atkinson, daughter of George and May (Miller) Atkinson who still maintain their home in Garden county. Of the five children born to Mr. and Mrs. Peterson, four are living—Chester A., Walter W., Henrietta I. and Wayne M.

Mr. Peterson has been like so many other representative business men of western Nebraska, the architect of his own fortune, and a significant comparison may be made between the modern and beautiful brick bungalow which he has recently erected for a home for his family and his first dwelling, which was a sod house of one room, with dirt floor. His first team comprised a horse and a bull, and in the early days he went a distance of fifty-five miles for mail to Ogallala as no bridge had yet been constructed across the North Platte river to give access to Sidney, a point much nearer. Within a few years after the family came to this section of the state the father of Mr. Peterson became grievously crippled by rheumatism, which was a contributing cause of his death so that much of the responsibility and work of the pioneer farm fell to Mrs. Peterson and her children. The father retired about five years prior to his death, but continued to reside on the old home place until he passed away.

HUGH RIDGE, one of the later homesteaders of Banner county, is a progressive and enterprising farmer and stockman and a citizen who takes an intelligent interest in the welfare of his community. He has lived in Nebraska since childhood but his birth took place in Ringgold county, Iowa, August 28, 1884.

The parents of Mr. Ridge, Jesse H. and Matilda (Owens) Ridge, had seven children born to them, Hugh being the eldest and the others as follows: Alice, who is the wife of Vern Waitman of Morrill county; Tressie, who is the wife of Ted Kelly, of Banner county; Forest, who went to France as a soldier with the American Expeditionary Force, died there October 4, 1918; Lydia, who is the wife of Walter D. Stewart, of Banner county; Lloyd, who lives with his brother Hugh, and Violet, who resides with her sister, Mrs. Waitman. The mother died in October, 1904. The father had been a farmer and stockraiser before he came to Nebraska in 1887, when he pre-empted near Steamboat Rock, about three years later moving to Cheyenne county and homesteading. In those early days he met with many misfortunes through loss of crops, and at times found himself willing to work for fifty cents a day and walk a long distance to and from his home, in order to provide for his family. He assisted in the construction of the railroad at Alliance. He lived on his homestead in Cheyenne county until 1910, then sold out and spent two years in Kansas engaged in farming, and one year in Florida. Mr. Ridge then returned to Nebraska and now lives retired at Bridgeport. He is a Republican voter but has never been active in politics.

Hugh Ridge attended school in Cheyenne county and remained with his father until he was twenty-one years of age. In 1908, he came to Banner county and homesteaded and lives on the place, a well improved tract of a hundred and twenty acres, to which he has added, owning at present three hundred and fifty-five acres, both farm and ranch land. General farming and stockraising engages his attention and he is in very comfortable circumstances. On November 29, 1908, he was married to Miss Goldie Darnall, a daughter of Scott and Rosie (Tucker) Darnall, and they have three children: Louise, born July 7, 1909; Nelson, born January 21, 1911; and Viva, born April 5, 1913. Mr. Ridge is an independent voter.

CHRISTIAN JENSEN, who was one of Banner county's well known and highly respected citizens, lived quietly, peacefully and industriously on his homestead on section 12, town 15, for more than fifteen years. He was an intelligent, upright man and his neighbors trusted and esteemed him. He was born in Denmark, February 11, 1855. His parents were Jens and Marian (Poulsen) Christinsen, according to family nomenclature in Denmark. Both parents lived and died there, Christian being the only member of the family to come to America.

In May, 1888, Christian Jensen came to the United States and settled at Dunlap, in Harrison county, Iowa, where he worked for five years. In 1892, he filed on a homestead in Kimball county, Nebraska, and in the winter of 1893, settled on his land and improved it, in 1903 trading his Kimball county property for the homestead in Banner county, on which his widow lives. In 1908, he filed on a Kimball claim and Mrs. Jensen owns and operates at the present time four hundred and eighty acres, about equally divided into farm and ranch land. She has shown herself a very capable business woman and carried on her farm and stock industries with extreme profit. Until 1919, she continued to reside on the land in the original old log house, where Mr. Jensen passed away on April 22, 1918, moving then to a comfortable frame dwelling newly erected. Her property is well fenced and all the farm buildings are substantial.

Mr. Jensen was married July 11, 1902, to Miss Julia Netvick, a daughter of Andrew and Anna (Jorgensen) Netvick, who lived and died in Norway. Mrs. Jensen and her brother Hans, who lives in Minnesota, were the only members of the family of eight children to come to America. Mr. Jensen met with some hard times after he located in Kimball county, prices on produce being low and drouth destroying his crops. For a time he hauled water from fourteen miles away and later bought water for a few cents a barrel. He was a Republican in politics, and served two terms as township assessor of Julian precinct, Kimball county. He assisted in building the Danish Lutheran Church there, to which both he and wife belonged. Mrs. Jensen is kind and hospitable and has many friends in the neighborhood where she has lived so long.

CHARLES V. WEBBER, whose life has seen many changes, came to Nebraska in 1904, with practically no capital. Here, through his energy and enterprise, he has acquired large properties and has continually broadened his efforts as opportunity has arisen and has become a leading business man of Banner coun-

Mr. and Mrs. Oscar O'Bannon and Son

ty It is needless to say that Mr Webber has been the architect of his own fortune

Charles V Webber was born in Pennsylvania, August 10, 1874, the only child of W M and Jennie (Heffley) Webber His young mother died when he was an infant She was born at Wimta, Indian Territory, but of her people he knows nothing, nor anything of his father's people Practically an orphan boy, he made his home with friendly people until he could provide for himself He began to work as a newsboy and through early youth sold newspapers along the Ohio river, at Wheeling and at Parkersburg After spending one winter in a lumber camp in the Virginia mountains, he decided to try his fortune in the west He was only seventeen years old when he reached Chicago, Illinois, and later he went to work on a farm near Aurora, Illinois, where he met with kindness and appreciation He was married on February 15, 1898, to Miss Mary A Milton, who died in Banner county, July 4, 1919 She was a daughter of John and Helen (McConicle) Milton, former residents of Illinois, now deceased Mr and Mrs Webber had two children, Harry and Peryl, both of whom reside with the father

Mr Webber and family came to Nebraska from Illinois, in 1904, he filed on a claim near Hull and the family lived on that homestead for five years, during this time Mr Webber leased several sections and ranged cattle on the land, later filed on three quarter section at Eagle's Nest and still owns that valuable property, still later buying two more sections He then leased land in Garden county but on finding conditions uncomfortable for a home, bought a lease on the A H Pierson ranch in Banner county and still resides there He does an extensive business in buying and selling cattle and rents pasture land to others His business success has been remarkable and considering his early handicaps, quite unusual Mr Webber is widely known and much respected He has never desired any public office, his business interesting him to a greater extent He is a Republican of many years' standing

OSCAR O O'BANNON — The qualities of adaptability, common sense, persistence and good judgment have prevailed in the energetic life of Oscar O'Bannon, winning for him an enviable place in the business circles of Alliance, where since 1908, he has been a member of the firm of O'Bannon Brothers, operators in real estate and now two of the largest producers of potatoes in the upper Platte valley Their property in the Alliance district may well serve as an example of good management and practical results in farming, as they are among the most progressive agriculturists in Box Butte county The O'Bannon brothers are essentially self-made men and may well look upon their success with pride, as the present large fortunes are the work of their own brains and hands as they accumulated all their property by their own unaided efforts and have builded solidly and well Oscar O'Bannon was born near Mattoon, Illinois, June 5 1876 the son of Oscar F and Sarah (Colson) O'Bannon, the father being a congenial son of the Blue Grass state, who had all the delightful charm of the typical Kentuckian, while the mother was born in Ohio Oscar was the third in the family of six children and as his father was a farmer in Illinois, he spent his early childhood in that state, early learning self reliance and thrift and able to take and give and look out for himself as all children in a large family do When the boy was about seven years old the family came west to Nebraska, locating in Seward county, where Oscar's boyhood days were spent He was sent to the nearest district school where he gained a good practical education and as soon as his age and years permitted began to assist his father in the work on the home farm While still a small boy he earned his first money by trapping mink, as the country was little settled and game plentiful From that time to the present he has always been a great hunter and fisherman and when William Jennings Bryan was running for president the first time a banquet was given in his honor at the Lincoln Hotel, Lincoln, Mr O'Bannon furnished four hundred quail, which shows his prowess as a huntsman All his educational advantages were obtained in Seward county and as the father had poor health, Oscar remained at home, really having the active management of the place until his marriage which occurred on December 31, 1901, at which time he was united in wedlock at Atkinson, Nebraska, with Miss Emma Schrader, a native daughter of Seward county whose parents were August and Johanna Schrader, the father being a native of Germany who became one of the early pioneer settlers of Nebraska Mr and Mrs O'Bannon have one child, Charles Oscar, a fine boy of seven years After his marriage Mr O'Bannon was engaged in farming for about six years, then came to Alliance in 1908, for he had a natural talent for business, which had made him so successful in carrying on his farming Though his capital was small at the time he landed in Box Butte county, this did not deter him from at once launching out into the real estate business and it seems that fortune has smiled upon his efforts most graciously This, he claims, is due largely to the aid and support he received from his capable wife She is an excellent manager of the home and when difficulties arose Mr O'Bannon often talked them over with her, obtaining timely and wise council with regard to investments and sales From first locating in this county Mr O'Bannon has carred out the policy of buying all the land he handled, and rarely works on a commission basis, thus giving the greatest satisfaction to his clients From time to time he has bought land in association with his brother, which they have held for their own agricultural business until today they have six well improved farms within three miles of Alliance, consisting of about fifteen hundred acres of the finest arable land in this section It is especially well adapted to raising potatoes and that is the use to which they have placed it In 1907, the O'Bannon brothers responded to the president's call for increased production by raising them by the carload — and we might say trainload — lots That year they realized a profit of $333 an acre from the land they had planted to the tubers This shows what may be accomplished by good management when men with good heads, who are willing to study farming can obtain by buying cheap land in Box Butte county The two brothers are partners in their varied lines of endeavor, not only do they raise but they also buy potatoes in the surrounding district In 1914, they shipped four hundred and fifty car loads, having seventeen cars on one train The property they now hold in the Alliance district is worth at a most conservative estimate $150,-000 In addition to this they have a large business building, enormous warehouses for handling the potatoes and have built a series of caves for storing them which adds about $50,000 to the value of the estate In Alliance they run a general feed, flour and coal business which in itself would be about all that one firm could handle This year—1920—they intend to use two thousand bushels of potatoes for seed alone and the prospects are that they will have a bumper crop, which at the present prices of potatoes, will mean a small fortune in itself Oscar O'Bannon is just completing a modern home in Alliance at the corner of Emerson Avenue and Fourth Street, at a cost of $22,000 where he and his hospitable wife will soon be at home to their friends

Mrs O'Bannon is a woman of great personal charm and talent She is well known in this section for her work in social and charitable circles and is also an active worker in the Methodist church to which the family belongs Mrs O'Bannon was president of the Methodist Aid Society in 1920, which is one of the most interesting aid societies of Alliance, with a large membership The O'Bannons have made a host of warm friends since coming to Box Butte county while the high standing attained in business by Mr O'Bannon has given him a place of prominence in the financial circles of the Panhandle They are citizens of whom Box Butte county may well be proud, ever ready to give of time and money for the improvement and upbuilding of the community and town, and are representative of the best element of Americans on whom the very existence of the nation is to depend during the trying years of social and financial adjustment which are looming large on the horizon The O'Bannon brothers do not say much but they are writing large on the pages of the financial history of the northwestern part of Nebraska, and with other prominent citizens of Alliance have struck pay dirt in the old fields of Wyoming and it would be hard to estimate the value of their holdings which reach into the hundreds of thousands of dollars

GEORGE A JONES, a representative citizen of Banner county, who has attained to some eminence in the county's practical affairs, was brought to this section in childhood, and has never desired to change his environment or seek a home in another part of the country Mr Jones feels that he is almost a native son His birth, however, took place at Kent, Iowa, March 8, 1885

The parents of Mr Jones were John L and Dora M (Clayton) Jones, natives of Warren county, Illinois The father was a farmer and stockraiser in Illinois and in Iowa From the latter state he came to Banner county, Nebraska, in 1888, homesteading near Hull, and he resided on that place until 1907, when he retired to Kimball, Nebraska, where he is pleasantly situated In politics he is a Republican He belongs to the Methodist Episcopal church at Kimball, and also to the order of Modern Woodmen of America at the same place George A Jones' mother died in the spring of 1901 He is the second in his parents' family of six children, the others being Arnold, who died June 15, 1911, Glenn, who lives at Dubois, Wyoming, married Louise Barfoot; Grace, who is the wife of W R Grant, of Banner county, Earl, who lives at Heath, Nebraska, married Josephine Larson, and Fern who is the wife of J J Smith, of Ogden, Utah

George A Jones attended school at Hull of which his father was one of the organizers, and later at Gering and Kimball, and before assuming personal business responsibilities, completed a course in a commercial college at North Platte Mr Jones served one term as deputy county clerk under County Clerk C S Page For five years he was identified with the Banner County Bank at Harrisburg serving that sound financial institution for three years as assistant cashier and two years as cashier In 1906, he homesteaded where he now lives His home surroundings indicate thrift and plenty and his well regulated farm operations are profitable in the extreme Mr Jones owns twelve hundred and eighty acres of land and runs about a hundred head of cattle, fifty head of Percheron horses and from thirty to fifty head of Poland China hogs annually His extensive industries are carried on according to scientific methods, and the latest improved farm machinery will be found on his place

On June 15, 1916, Mr Jones was married to Miss Nettie Larson, whose people were pioneers in Banner county, and the have five children Eldon Lois, Morris Helen and Merna They attend the Methodist Episcopal church In politics Mr Jones has always been a Republican He is a member of the Farmers Union and belongs fraternally to the Masons and the Knights of Pythias He was patriotically active during the World war, furthered to the best of his ability the various local war movements that included the selling of Liberty bonds, was a member of the Legal Advisory board and county chairman of the Council of Defense

In August, 1920, Mr Jones bought a home at Gering and moved there on account of school advantages for his children

THOMAS C BARKELL, who has been a resident of Banner county for thirty-one years, courageously bore his part in times of early hardship, and as opportunity came later on, did his best in assisting to bring about present conditions that make this section of Nebraska one of the finest in the state He was born in Grant county, Wisconsin, November 4, 1870

The parents of Mr Barkell were Richard and Mary (Ralph) Barkell the former of whom was born in England and the latter in Wisconsin Richard Barkell was but one year old when his parents came to the United States and settled in Wisconsin in 1842, and there he

lived until 1887, when he came to Nebraska and homesteaded northwest of Harrisburg, in Scottsbluff county He lived on that place for five years, removing then to Big Horn Springs, north of Kimball in Banner county and remained there until he retired to Kimball in 1899 There both he and his wife died, his death occurring February 17, 1904 and her death April 5 1912 Of their four children, Thomas C was the third in order of birth, the others being John H, who was accidentally killed in 1894 by a fall over a fifty-foot cliff when cutting timber Richard J who lives at Ingleside, and Philip R, who lives near Kirk, Nebraska The father was a Republican in politics Both parents were members of the United Brethren Church, but after removing to Kimball they joined the Methodist Episcopal Church

Thomas C Barkell attended school in Wisconsin and one term in Scottsbluff county, Nebraska When twenty-one years old he started out for himself and made headway by working on ranches mostly in Wyoming The eight hundred and forty acres he now owns in Banner county is mainly ranch land He breeds Durham cattle, and also raises hogs In connection with ranch work he has always looked about for opportunity to be usefully employed in other directions, and, from 1912 to 1918, he operated a mail route between Harrisburg and Big Horn Otherwise he has never accepted any public position but in the capacity of a private citizen has done his full duty

On December 20 1899, Mr Barkell was united in marriage to Miss Cora C Jackson, a daughter of Loren G and Margaret (Sloan) Jackson, who reside in Wisconsin Mr and Mrs Barkell have one son, Roy G who was born September 14 1906 They are members of the United Brethren Church and take active part in its work, Mr Barkell assisting in the organization of Sunday Schools at Kirk, Heath and Big Horn In politics he has always been affiliated with the Republican party Mr Barkell is looked upon as one of the dependable and representative men of the county

ROBERT L OSBORNE, Jr, who is well and favorably known in Banner county, Nebraska, was born here April 14, 1888 and is a son of Robert Osborne, an early settler here He attended the district schools and took a course in the Grand Island Business college He has lived in this county all his life was born in a dugout started out for himself without capital and strove hard to earn the money with which to pay for his homestead Today

Mr Osborne owns six hundred and forty acres of well improved land, raises thirty head of cattle yearly and is one of the most successful farmers of his section

On June 7, 1916, Mr Osborne was united in marriage to Mrs Elsie M Christ, widow of George N Christ and a daughter of Christian and Elsie M (Witmer) Siebert, who died in Europe Her father was a farmer near Neufchatel, France Mrs Osborne is one in a family of twelve children and is the only one living in Nebraska After coming to the United States in 1904, she lived three years in New York City, then came to Denver, Colorado In that city she was married to George N Christ, who died December 20, 1915, leaving two children Caroline, who was born October 30, 1908, and George S, who was born July 7, 1907 Mr Christ was a reclamation surveyor for the United States government Mrs Osborne has been a resident of Banner county since 1907 She homesteaded four hundred and eighty acres on sections 11 and 14 in township 19 Mr and Mrs Osborne have no children They have everything very comfortable about them In early days wild game was very plentiful in this section but is now so seldom encountered that when three antelope were found on the farm recently interest and curiosity were aroused all over the neighborhood Mr Osborne is not active in politics and takes more interest in developing his farm industries than in holding any public office They maintain a hospitable home and both he and Mrs Osborne have many friends

EWING E BARRETT, who is a well known resident of Banner county where his life has been spent, bears a name that has been honorably connected with farm, ranch and official life here for many years Mr Barrett was born in Banner county, July 6, 1886

His father was Isaac N, and both parents were born at Belle Valley, Ohio Isaac Barrett came west and was a ranger for some time and worked on different ranches In early days in Banner county he freighted from Sidney to Hull a distance of eighty miles He was one of the first homesteaders in the vicinity of Hull and lived on his land for six years, then sold and moved to Kimball, afterward working for several years on the Lewbickel ranch Mr Barrett was then appointed mail driver between Kimball and Harrisburg, in which position he served for eight years, when he was appointed on the mail route to Colorado He was equally faithful and efficient, during the three years preceding his death, performing his duties so carefully and completely as to bring commendation from all He died at the post of duty, probably from heart disease, in June, 1918 Of his five children, Ewing E is the fourth in order of birth, the others being Charles, who resides near Hull, Nebraska, married Lila Olsen, Weldon and Leslie, twins, both of whom live in Banner county, the former unmarried and the latter married to Minnie Crubbs, and Guy, who lives at Kirk, Nebraska

Ewing E Barrett attended the Banner county schools with his brothers, and all were reared on the farm In 1915, he engaged in freighting for a time, hauling potatoes, onions and cabbages from Greeley, Colorado, to his own neighborhood, where these vegetables are not generally grown, and the venture proved profitable He gave this up to engage in farming, and now, in association with his brother Charles, Mr Barrett is operating four hundred and eighty acres of leased land near Hull They raise some stock but devote the most of the land to general farming and are doing very well Mr Barrett has always taken an intelligent interest in public matters and has been a staunch Republican ever since he reached manhood He has a wide acquaintance and many friends, but has never married

GEORGE W INGLES, one of the progressive and enterprising young farmers of Banner county, was born in Scottsbluff county, Nebraska, December 24, 1891 He belongs to an old and prominent family of this section, being a son of William H and Edith (Richards) Ingles, extended mention of whom will be found in this work

George W Ingles obtained his education in the public schools of Banner county He has always followed farm pursuits, farming and ranching, and is thoroughly experienced For five years he worked on the Airdale ranch, where he had the practical training that he has since found very useful For the past three years Mr Ingles has been operating his father's farm of seven hundred acres in Banner county and has met with the success that close and careful attention to business usually brings about

On September 6, 1916, Mr Ingles was united in marriage to Miss Nellie W Jones, who is a daughter of Edward J and Almeda (Bond) Jones The parents of Mrs Ingles homesteaded in Scottsbluff county in 1913, but later sold their property and moved to Brush, Colorado, where they yet reside Mr Ingles has never accepted any political office, but he is staunch Republican

Alfred Gumaer

ALFRED W GUMAER — In manifold directions has the forceful individuality of Judge Gumaer been exerted beneficently in connection with civic and material progress within his forty years of residence in Nebraska, to which state he came when a young man and in which he has been prominent and influential in public affairs and business enterprise He has served in various offices of distinctive public trust, is now judge of the county court of Garden county and was one of the founders of Oshkosh, the judicial center of the county, — a place named in honor of his native city in Wisconsin The record of achievement on the part of the judge during the years of his residence in Nebraska reflects credit not only upon him but also upon the state, and as one of the essentially influential and honored citizens of Garden county it is imperative that he be accorded a definite tribute in this history

Alfred W Gumaer was born at Oshosh, Wisconsin September 1, 1853 and his native place, now one of the beautiful and thriving cities of the Badger state, was a straggling lumbering town at the time he was there ushered onto the stage of life He is a son of William G and Priscilla (Weed) Gumaer, both natives of the state of New York On the paternal side Judge Gumaer is a scion of a sterling French-Hugenot family whose original location in America was in Ulster county, New York, and in all succeeding generations likewise has the name Gumaer been honorably linked with the annals of our nation The original American progenitors made settlement in Ulster county about the year 1685

Judge Gumaer was a child at the time of his parents' removal to Weyauwega, Wisconsin, where his father engaged in the operation of a flour mill and a saw mill There the subject of this review attended the public schools of the period, and later he continued his studies in the high school at Oshkosh, after which he pursued a course of study in the University of Wisconsin at Madison On leaving this institution he joined an engineering corps, with which he assisted in the surveying of the line of the Wisconsin Central Railroad He was thus engaged about two years and then became associated with his uncle, J H Weed, in the lumber business at Oshkosh, his native place The concern operated several saw mills in Wisconsin and conducted a large business Judge Gumaer continued his association with this enterprise for five years, and in 1879, he came to Nebraska and opened what was known as the Oshkosh

lumber yard at Grand Island, being associated with his brother, Henry G They also opened a yard at St Paul, this state, and conducted both enterprises about two years They then sold the Grand Island yard, and in Valley county opened a yard at Ord, the county seat, as well as a yard at North Loup, these being pioneer lumber yards in that county and having been successfully conducted by the founders for several years

Within the period of these activities in the eastern part of the state Judge Gumaer and his brother, in association with Herbert W Potter, John Robinson and George T Kendall, came to what is now Garden county and founded the town of Oshkosh, which was destined to become the county seat All took homestead and pre-emption claims in this vicinity after which they proceeded to fence the whole valley and to lay out the town of Oshkosh Here they engaged in the cattle business upon a large scale and the company erected the first frame store building in the new town the lumber having been transported across the Platte river by fording, as no bridge had yet been constructed in this locality In 1894, these pioneers sold to August Sudman & Company the town site of eighty acres, and each of the original founders then assumed individual control of his own land claims

During all these years of pioneer operations in the Panhandle country Judge Gumaer retained his interests at St Paul, where he continued the operation of the lumber yard In 1884, as a resident of St Paul, he was elected to represent Howard county in the state legislature, and he gave yeoman service in the furtherance of wise and timely legislation, as one of the active working members of the lower house He was specially vigorous in his support of the act commonly known as the Slocum law

Prior to coming to Nebraska Judge Gumaer had traveled up and down the Mississippi river for three years, in the interest of the Weed Lumber Company, a Wisconsin corporation In 1886, after the death of his father he became president of the Weed & Gumaer Manufacturing Company, which was engaged in the manufacturing of baskets at Weyauwega Wisconsin, where it also supplied the town with electric-lighting facilities As the eldest of the children Judge Gumaer had general supervision of the family interests until the estate of his father was settled

Reverting to the political or official activities of the subject of this review, it may be stated that in 1889, he was elected treasurer of Howard county, a position in which he continued the incumbent four years and in which he gave a most efficient administration of the fiscal affairs thus consigned to his charge, his election to this office marking his candidacy on the Democratic ticket, and the same party having previously elected him to the legislature, as noted above. In 1896, bringing his technical skill as a civil engineer once more into play, Judge Gumaer assumed the position of deputy United States surveyor, with Robert Harvey, and entered into a contract to complete the survey of the counties as yet unsurveyed in the northeastern part of Nebraska, said counties having been a part of the old Rosebud Indian reservation. This work received his careful and able attention until it was completed. At this juncture reference may be made to still another phase of the interesting and varied career of Judge Gumaer. In February, 1898, at the suggestion and request of George D. Meiklejohn, Assistant Secretary of War under President McKinley, Judge Gumaer took the reindeer expedition for the United States government from New York to the Alaskan Klondike, this expedition involving the transportation of five hundred and forty-three reindeer to the far north, these all being animals previously trained for work. The well broken animals were in direct charge of one hundred and thirty-four Laplanders, and when the outfit arrived at Seattle, Washington, the expedition was halted, owing to contingencies incidental to the Spanish-American war. Under these conditions the reindeer were finally sent forward to J. S. Jackson, superintendent of education at Nome, Alaska, this herd forming the nucleus from which the government has developed a large number of the deer, which are used in northern Alaska for both work and food purposes, the experiment having proved most successful and extremely valuable in results. After arriving in Alaska Judge Gumaer joined the military exploring expedition that set forth under the leadership of Captain Glenn, to make extensive explorations in Alaska, at the instance of the government. Judge Gumaer served as chief guide and special agent of the war department on this historic expedition, by which was made the first survey from Valdez across the Copper river and thence onward to Forty-mile river. This hazardous and trying trip was made with pack train, and its work proved of importance and enduring value. Judge Gumaer returned to "the states" on the last boat that left Alaska prior to the close of navigation in the fall of 1898. He returned to the national capital, where on January 1, 1899, he was appointed deputy collector of customs, under General Tasker H. Bliss, for the port of Havana, Cuba. He retained the post of military collector at this port for three and one-half years, — or until the government of the island was restored to the Cuban people, in 1903. After leaving this governmental post Judge Gumaer returned to Nebraska, but about two months later he was appointed inspector of immigration at Ellis Island, New York, where he remained three and one-half years, at the expiration of which he was transferred to the position of immigrant inspector for the port of New Orleans. After retaining this incumbency thirteen months he resigned the office and returned to Oshkosh, Nebraska, to which place the railway line had just been completed. Here he has since remained as a vital factor in the development and upbuilding of this section of the Nebraska Panhandle. He was one of the most active worker in the movement that led to the division of old Deuel county and the organization of Garden county, in 1909, and he had the distinction of being elected the first county judge of the new county. The high popular estimate placed upon him has been shown in his continuous retention of this important office, by successive re-elections, and he has had no opposing candidate save on the occasion of the first election.

Judge Gumaer retains his original homestead in Garden county, the one hundred and sixty acre tract being provided with excellent irrigation facilities and being under a high state of cultivation; he has erected good buildings and made other improvements upon the property. The judge rents the farm to Italian contractors, who utilize the land for the propagation of sugar beets, this farm being situated one mile east of Oshkosh. In addition to this valuable property the Judge owns six hundred and forty acres twenty-five miles north of Oshkosh, and this range is rented for grazing purposes. In 1888, Judge Gumaer was a member of the company which construct the Oshkosh irrigation ditch, one of the first projects of this kind in what is now Garden county.

As previously intimated, Judge Gumaer is a stalwart advocate and supporter of the principles of the Democratic party, and he has been influential in its ranks during the years of his residence in Nebraska. He is actively affiliated with the Masonic fraternity, including Mount Arrartt Commandery, No. 27, Knights Templars, at St. Paul, this state, and Jordan

Chapter of Royal Arch Masons, in the same town besides which he is a charter member of Oshkosh Lodge, No 286, Ancient Free and Accepted Masons, in his home village, and a life member of Tangier Temple, Ancient Arabic Order of the Nobles of the Mystic Shrine, in the city of Omaha

GEORGE E STREEKS who is a descendant of one of the earliest homesteaders in Banner county, still lives in the comfortable old log house on the homestead, in which he was born, May 18, 1886 and has the historical distinction of being the third white child born in Banner county

The parents of Mr Streeks were Christian C and Ellen V (Ashford) Streeks The mother was a native of Canada and of English ancestry The father was born at Washington, D C, came in early manhood to Nebraska and became widely known on the cattle ranges and for many of the old-time big ranches he served as foreman He was one of the first pioneers of Banner county to secure a homestead and afterward lived on his land until his death which occurred October 9, 1909 He was somewhat interested in politics, was a Democrat and loyal to his party but never accepted any public office except in relation to the public school He had two sons, George E, of this sketch, and Frank The latter maintains his home at Akron, Ohio, and travels in the interest of Goodyear Tire Company The mother of Mr Streeks passed away May 2, 1902, mourned by many who had been benefitted by her sympathy and kindness through life

George E Streeks attended the public schools in Banner county He has always lived on the old homestead, forty acres of which he inherited, and it is on this portion of the land that the old house stands Although the body of the house is constructed of logs a frame addition has been built, the lumber for which had to be hauled from Sidney, sixty-seven miles distant Mr Streeks now owns three hundred and twenty acres and devotes it mainly to ranch purposes He breeds Hereford cattle and has about thirty head for market each year, and about a hundred head of Duroc-Jersey hogs He carries on his farm operations carefully and systematically and has always succeeded in his undertakings

On April 14, 1909, Mr Streeks was united in marriage to Miss Hazel M Lalley, a daughter of Marin and Jennie (Ellis) Lalley, the former of whom is deceased Mrs Lalley lives in Scottsbluff county Mr and Mrs Streeks have children as follows Ellen, born Feb-

ruary 5, 1910, Edgar, born October 25, 1912, Fred, born January 3, 1914, Marie, born February 2, 1916 and Marjorie born July 16, 1918 Mr Streeks has never held political office and is an independent voter

CHRISTIAN PFEIFER —One of the representative citizens and substantial residents of Banner county is Christian Pfeifer, heavy landowner and extensive breeder of White Face cattle Mr Pfeifer's residence is magnificently located, directly at the foot of Wild Cat mountain, which is five thousand feet above sea level and rises eight hundred feet above the house He was born at Gahon, in Crawford county, Ohio, December 28, 1868, the son of Godfrey F and Emma (Snyder) Pfeifer, the former of whom was born in Germany, March 31, 1839, and died in Nebraska, January 8, 1916 The mother of Mr Pfeifer was born in Ohio, August 17, 1835, and died September 19, 1913 The father was an educated man and after coming to the United States taught school first in Ohio and later in Kansas, coming to the latter state when it was opened for settlement In 1875, he homesteaded in Kansas and remained on his land until 1901, when he retired and came to Nebraska, after which until his death he lived with his sons He was a Democrat in politics and both he and his wife were members of the Christian Church Of their seven children Christian is the eldest of the survivors, the others being as follows John, who lives in Scottsbluff county, Laura, who is the wife of Frank Sears, of Saint Regis, Montana, and Katie who is the wife of Francis Whitman, near Paradise, Kansas

Christian Pfeifer attended school until nine years old and occasionally through a winter term later on and was also, to some extent, instructed at home by his father, but since the age of fourteen years has practically looked after himself in every way He has had many experiences and has seen wonderful development in this section He remembers away back in boyhood his father going to work on the railroad in order to get enough money to buy provisions for the family after the grasshoppers had ruined the crops, and on one occasion carried a bag of flour on his shoulders a distance of six miles to his home He has seen many settlers gathering the dried bones of cattle and buffalo that had died from lack of food and the severe winter weather, and at one time saw these bones piled eight feet high at Kimball He witnessed the hauling of all the lumber for the building of Gering it being freighted from Kimball He had experience in catching wild horses. He would often chase them for three days before he could catch them. There would be from fifteen to twenty in one bunch. Stallions of the bunch, on noticing the approach of a rider, would whirl and come to meet the rider and when within a short distance suddenly snort and turn, race back to the herd and chase them away by biting them.

On one occasion when chasing horses Mr. Pfeifer glanced behind him and saw a horse covered with sweat rapidly approaching, with a grey wolf close behind him; he turned and gave chase to the wolf, but his horse being tired, he lost both wolf and horses. Another recollection of his boyhood is of the use of oxen in the old days for farm labor. Practically no farm machinery was in use and no one had ever dreamed of such an astonishing contrivance as a farm tractor of modern days.

Mr. Pfeifer worked and rode range for the L. F. Cattle Company, which owned sixty five thousand head of cattle, for nine years, saved his money and since then has ranched for himself. He came to Wyoming in May, 1884, helped to take out the first ditch on Rawhide creek, remained one season but in the fall returned to Kansas and remained in that state until 1886, when he came back to this part of Nebraska. For several years he worked all around Kimball and from Greeley, Colorado, to Cheyenne and Julesburg, during the first year runing cattle as far as Ogallala, but after settlers began to come in, the cattlemen had to hunt other ranges. There were yet a few buffalo left but Mr. Pfeifer never shot any but on many occasions brought down antelope. He has trailed catttle from Texas to Wyoming. the trail being from a quarter to a half mile wide. On one occasion, on the way from Texas, at Pine Bluff the herd started for water and a railroad train ran into the stampede and killed sixty head. Mr. Pfeifer was in Banner county at the time of the battle of Wounded Knee. On account of the Indians so successfully running off the horses of the settlers, the cowboys succeeded in training their horses to run into the corrals when the Indians appeared.

In recalling these and many other early hardships, it is no wonder that Mr. Peifer declares these days in Nebraska better than the old ones. He now owns twelve hundred and twenty acres of farm and grazing land, cultivating about five hundred acres and raising three hundred cattle annually in addition to horses and some hogs and is interested also in buying and selling hogs.

On January 18, 1894, Mr. Pfeifer was married at Kimball, to Miss Iva B. Campbell, who is a daughter of John H. and lizabeth (Murray) Campbell, early settlers in Kimball county. The mother of Mrs. Pfeifer died when she was a child but her father survives and is a very prominent business man of Kimball. Mr. and Mrs. Pfeifer have children as follows: Floyd C., who served in France during the Great war; Agnes Fay, was married to Aaron Shaul September 17, 1919, and has a child, Roena Pearl Shaul, the first grandchild, born September 9, 1920; Inez B., Elsie Fern, Gladys Irene, Edythe M., and Christian John. The family attend the Christian and Baptist churches. Mr. Pfeifer has never accepted a political office although he has been identified with much that has substantially developed this section. At present he is an independent voter. He remembers the winter of 1919-20 as a very hard winter, and that on April 17, 1920, a severe storm starting with a rain and killed more cattle than was ever before known.

EVERETT BIGSBY, who is a prosperous general farmer living on his homestead situated on section thirty-two, town nine, Banner county, is well known in the county to which he came when twelve years old. He has always been engaged in agricultural activities, on his own account and for others, and is numbered with the successful farmers of this section. He was born in Sanilac county, Michigan, December 11, 1876.

The parents of Mr. Bigsby were William and Isabel (McClellen) Bigsby, the latter of whom was born in Ireland and died in Banner county, Nebraska, in September, 1916. The father of Mr. Bigsby was born at London, Canada. He came to the United States prior to the Civil War, in which he served for eighteen months and afterwards settled on a soldier allotment in Michigan. From there he moved to Buffalo county, Nebraska, and homesteaded, and five years later, on May 10, 1887, came to Banner county, took a pre-emption and tree claim, both of which he subsequently sold, but he resided in Banner county until his death in 1910. Of his family of ten children Everett was the fifth in order of birth, the others being as follows: Georgianna, who is the wife of John B. Hentz, county treasurer of Banner county; Chester, who lives in Iowa; Stella, who is the wife of Leonard Ball, lives in Michigan; Fred, who lives in Idaho, married Edith Case, who is now deceased; Roland, who lives in Banner county, married Clara Fuller; Myrtle, who is the wife of Arthur Olsen, a farmer in Banner county;

and three who died young. The parents were members of the Baptist church. The father was a Democrat in politics but never accepted any political office

Everett Bigsby completed his education in Banner county and then went to work, being employed for a number of years on different ranches in the county. In May, 1905 he was married to Miss Leola Hoke. Her mother died many years ago, but her father resides at Palisade, Colorado, where he is engaged in a real estate and parcel post fruit business. Mr and Mrs Bigsby have four children Rupert, Voyne, Derward and Keith, all of whom enjoy all the advantages a careful father can give Mr Bigsby homesteaded in 1909 and has a well improved, well stocked farm. He is a Republican in politics but he is not an office-seeker. As a neighbor and citizen he is held in esteem

H G GUMAER — The community was indeed shocked and saddened Monday morning to learn that during the small hours of the morning one of its most beloved and highly respected citizens had passed to the great beyond after a brief illness, death being caused from heart trouble. Practically all the immediate family were present at the home of his brother, William F Gumaer when this grand man gasped his last breath at two o'clock. The deceased had been ailing off and on for the past month but his condition was not thought serious until about ten o'clock Sunday night, at which time the family began to realize that the end was not far off. A few seconds before two o'clock he moved over onto the lounge from his chair unassisted, sat down, made the remark to those present that he was "a mighty sick boy," closed his eyes and passed peacefully away, sitting upright, to that undiscovered country from whence borne no traveler returns

Henry G Gumaer was born in Waupaca county, Wisconsin, on October 31, 1855. Died at Oshkosh Nebraska, October 20, 1919, and would have been sixty-four years old his next birthday

He was the second child in a family of seven, three boys and four girls and both father and mother are now dead. The former died while the family still lived in Wisconsin, about 1887, and the latter's death occurred when the deceased was about nineteen years of age. Both parents originally came from New York state

Mr Gumaer left Wisconsin in 1879 and came to Grand Island, Nebraska, and in 1880, in company with his brother, Alfred W Gumaer, he moved to St Paul, Nebraska, where they engaged in the lumber business, their's being the first venture of its kind in that town. In March 1885, in company with John Robinson, he left St Paul with an ox team and trailed four hundred head of cattle. Upon their arrival here they formed the old Oshkosh Land and Cattle Company, the members being H B Potter, George Kendall, John Robinson, A W and H G Gumaer. They erected a set of ranch buildings on the old Mills quarter, which now adjoins Oshkosh on the southwest corner. They also platted the town and named it after Oshkosh, Wisconsin, and erected the first store in Oshkosh which is now known as the old Miller Hotel and stands on the corner of Fish street and Gumaer avenue. This company continued in operation until about 1894 when three of the members dropped out and left the affairs of the company in charge of John Robinson and our subject. This partnership was continued several years thereafter after which, by mutual agreement the stock and land was equally divided and each went his own way, both accumulating large estates by staying with the country through all the trials and tribulations of the early settler. Up until several years ago the deceased had the ranch well stocked with hogs, cattle and horses. He was one of the first to see the advantages of beet raising and since his move in the agricultural field this section of the county is being known as one of the best beet producing communities in the valley

Mr Gamuaer was unmarried. He was a staunch Democrat and was elected the first county commissioner from the north district of Deuel county (which is now part of Garden county) and again in 1903 was elected to the same office and served in that capacity for six years. He was prominent in all county and state affairs. He was also prominent in Masonic lodge circles being a charter member of Golden Fleece Lodge, of Chapell, and Oshkosh Lodge No 286, holding the office of treasurer since the latter's organization

Thus ends the life on earth of one of Nebraska's early settlers and one of the founders of Oshkosh, a man who was beloved by all who knew him and whose loving way and kind disposition will be sorely missed throughout the years to come. "Hank,' as he was commonly known, was a man highly respected and one who had not an enemy in his large circle of friends and acquaintances. He was honest and upright in his dealings with all mankind, and it can truthfully be said that he lived a man's life among men.

He leaves to mourn their loss two brothers, Alfred W. and William F. Gumaer, of Oshkosh, Nebraska; and four sisters, Mrs. H. B. Potter, of Oshkosh, Nebraska; Mrs. A. L. Covey, of Omaha; Mrs. H. L. Cook, of Lincoln; and Mrs. H. B. Van Decar, of Los Angeles, California, the latter not being able to make train connections in time for the funeral, besides a large circle of friends and acquaintances. It is needless to say that the bereaved ones have the sympathy of the entire community.

All business houses closed their doors and the wheels of industry were stopped for two hours during the funeral services, which were held at the Methodist church Wednesday afternoon at two o'clock, Rev. Benjamin Kuhler presiding, in the presence of a large concourse of people who gathered to pay all due respect to the man who had been a friend to practically all of them. Immediately following the church service the body was placed in charge of his Masonic brethren who accompanied their beloved brother to the Oshkosh cemetery where the beautiful and impressive Masonic burial service was performed over the remains of our most worthy brother. His body has finally returned to the earth and his spirit to God who gave it, but his memory will ever remain with us until time immemorial. "The will of God is accomplished, amen." So mote it be. Farewell brother, we bid thee a long farewell. Yours was a life on earth that makes for good citizenship; you have been a friend in need as well as in deed; your column is broken and your good counsel and timely advice will be with us no more. Farewell brother, farewell!

The pall bearers were: K. A. McCall, Floyd Jones, August Neuman, R. Lisco, Mr. Persinger and J. F. Crane.

SAMUEL KELLY, whose extensive farm and stock operations have made him prominent in Banner county for many years, was born in La Salle county, Illinois, October 20, 1862. His father, John B. Kelly, was born in Lorain county, Ohio. He was a farmer in Illinois but moved from there to Missouri about 1865, where he bought a hundred and twenty acres of land and resided there until his death on November 8, 1871. A Democrat in politics but never an office-seeker, he was a self-respecting citizen and good man. The mother of Samul Kelly was a member of the Free Will Baptist church. Their children were as follows: Eunice, who is the wife of J. C. Iker, lives in Page county, Iowa; Samuel, who lives in Banner county; Rebecca, who is the wife of Frank M. Stockton, of California; John, who lives in Banner county; Georgia P., who lives in Kansas City, is president of the company operating the largest salt mine in the world; and Emma, who married a Mr. Bessy, lives in Colorado. The mother of the above family by a second marriage became the wife of D. M. Davis and at the time of her death, January 21, 1907, left four children: Charles, who lives in Worth county, Missouri; Alonzo, who lives at La Grange, Wyoming; J. E., who lives also at La Grange; and Ruth who is the wife of George Jennings, of Redding, Iowa.

Samuel Kelly was about nine years old when he lost his father and started out for himself when sixteen years old. He attended school in Worth county, Missouri, and worked on farms there and later in Iowa. In the spring of 1887 homesteaded in Banner county Nebraska, locating near Minatare, and in the spring of 1887, homesteaded in Banner county and has lived here ever since. Like many other homesteaders he found times hard and took advantage of every opportunity to earn money. Sometimes he had to leave his wife and children alone on the homestead for weeks together. For some time he worked on a ranch south of Cheyenne, in Wyoming; also cut timber from the hills and sold it at Scottsbluff, and thus, in many way of honest contriving and persistent industry, kept his family well and comfortable. In 1890-91 he remembers hauling his wheat a distance of thirty miles, to Kimball, and selling it for twenty-six cents a bushel. To contrast those days with the present brings pleasurable emotion. Mr. Kelly now owns and operates, including his wife's homestead of a hundred and sixty acres, about three thousand acres. His sons are farming seven hundred acres and the rest is ranch land, Mr. Kelly has bred Hereford cattle and standard hogs for many years. Progressive methods are made use of on this farm and the various industries are carried on very profitably.

Mr. Kelly was married first to Miss Idella Bowen, who died September 23, 1889, leaving two children, the one survivor being Laura, who is the wife of Ralph Darnell, of Banner county. On February 15, 1893, Mr. Kelly was married to Miss Jennie Skinner, who is a daughter of Richard and Emma (Powell) Skinner, natives of Perry county, Ohio, early settlers and prominent people of this county. The father of Mrs. Kelly is a veteran of the

Civil War Mr and Mrs Kelly have had the following children Glenn, who lives on the home farm, Earl, who operates the Peter Hanson place, Roy Earnest, who homesteaded at Glenrock, Wyoming, and Vern, Lillian Lavera and Clifford, all of whom are at home Mr Kelly and his sons are staunch Republicans but none have ever desired public office Mr. Kelly belongs to the A O U W at Scottsbluff and W O W at Minatare

MARK H CROSBY, deceased Not every early settler who came poor into Banner county improved his prospects by remaining, but some of them did, and one of these is Mark H Crosby, who is one of the county's most substantial farmers and ranchmen of today An old team and wagon and thirty-five dollars in money was his fortune when he came here in the fall of 1887, before his death it is probable that Mr Crosby's name on a bit of bank paper would have been acceptable in any financial institution in the country He started out for himself when fifteen years old and has made his own way in the world

Mark H Crosby was born in Ashtabula county, Ohio, November 20, 1851, a son of Almon and Abigail Hall Crosby, the latter of whom died when he, one of the fourteen children, was but three years old The father married again and the stepmother died in January, 1911, the mother of six children Of his brothers and sisters, Mark H Crosby is the only one in Nebraska The father lived until 1875 He was a general farmer and owned a sawmill at the time of his death in Henry county, Ohio He was a Republican in politics but never aspired to public office

Mark H Crosby's education was largely gained in the school of experience When fifteen years old he went to work for farmers at eight dollars a month In 1875, he went to Jasper county Iowa, and made his home there until 1884, for two years before moving to Harrison county, Missouri, he worked land on shares He remained in Missouri for three years, then came to Banner county, as above stated, and in the fall of 1887 homesteaded where he now lives Mr Crosby has eight hundred acres and devotes the most of it to ranch purposes, averaging from thirty to forty head of cattle yearly In early days here and before that, he worked hard at anything that would gain him an honorable living and has had much to contend with He hauled freight for twelve and a half cents a hundred weight from Kimball, worked from daylight to dark for seventy-five cents, accepted, in fact, almost any work and any wage, in order to get enough to make payments on his land and the cows with which he was trying to start a herd When making trips to Kimball and other markets he slept on the ground under his wagon in order to save hotel expenses To sell his first wheat he hauled it thirty miles to Kimball and then received twenty-seven cents a bushel He has seen many hard times but has lived through them and set an example that deserves emulation His first home in Banner county was a small structure of which he was very proud because it had a board roof and Nebraska shingles (Butte Clay) In the fall of 1911, Mr Crosby visited his relatives and they could scarcely believe the experiences he had passed through and in thinking them over, Mr Crosby was led himself to greatly marvel

Mr Crosby was married on November 12, 1882, to Alice Campbell, who died without issue, on June 21, 1911 His second marriage was to Mrs Cornelia (Hampton) Kimberly, widow of Edward Kimberly, and daughter of William R and Sarah M (Deter) Hampton, who came to Banner county in March, 1887, and homesteaded near Hull Several years later they moved to Harrisburg and there Mr Hampton engaged in the practice of law until he retired A short time previous to the death of Mrs Hampton, on December 4, 1903 they had moved to Gering, where he died December 6 1904 They had celebrated their golden wedding anniversary on July 3 1903 They had eleven children, the survivors being Theodore, living at Hastings, Nebraska, Mrs Crosby, Ida, wife of Emory Signs of Iowa Jennie L., wife of Henry Highes of Mapleton, Kansas, Commodore, living at Gering, William, living at Gering, Russell R, of Baxter, Iowa, and Albert J, living at Gering All of the five sons homesteaded in Banner county and Mrs Crosby also homesteaded after the death of her first husband but later sold, retaining, however, valuable residence property at Gering

Mrs Crosby came to Banner county with her first husband, Edward Kimberly, and they homesteaded ten mile northwest of Harrisburg Mr Kimberly developed a serious catarrhal infection and they moved from the homestead to Cheyenne, Wyoming, in the hope of improvement, but his death followed there, on August 16, 1913 He was one of the earliest school teachers in Banner county where he taught for a number of years When Mrs Crosby first came here she frequently enjoyed horse-back riding, but even for so daring and skilled horsewoman as herself there was great danger from the range cattle She always rode bareback and on many occasions when herding her cattle, they would get mixed with the wild range cattle and it was only through her dexterity that she escaped injury from them Mrs Crosby, like her husband has had many pioneer experiences and their relation is more interesting that a book of romances

Mr and Mrs Crosby were members of the Baptist church, to which he belonged for thirty years He was always a Republican in his political views and remained identified with that great old American organization At different times, when time and opportunity justified it, Mr Crosby accepted public office and for many years served on the school board and for three terms was eleced school treasurer Mr and Mrs Crosby were very highly esteemed by all who knew them and the acquaintance was very wide   Mr. Crosby died May 27, 1920

GEORGE SCHINDLER, while by no means the first homesteader in his section of Banner county, has been one of the permanent residents since he came here thirty-one years ago, never leaving the county as did many others, to seek work at intervals, in neighboring states Mr Schindler's object in coming to Banner county was to secure for himself and family a home, for which he was willing to face necessary hardships and work to the extent of his ability He is now one of the county's most substantial citizens and respected trustworthy men

George Schindler was born in Allen county, Ohio, September 18 1850 His parents were Dr George and Susanna (Thompson) Schindler His mother was born in Pennsylvania and died in Ohio in 1898 His father was born in Germany, was educated there as a physician, took part in the revolution of 1848 and afterward came to the United States and settled in Ohio Dr Schindler practiced there until his death in 1861 He was a man of scientific learning, a chemist who compounded his own medicines and was a specialist in the treatment of milk leg fever and cancer In those days in his section of Allen county, many rivers and other streams were yet unbridged but that fact never deterred Dr Schindler from going to the relief of his patients, and on many occasions he came home with his clothing frozen about his body from having to ford these unbridged waterways He belonged to the Masonic fraternity and was a Democrat in politics Of his eight children there are four surviving, George being the only one of the family to come to Nebraska

George Schindler attended the common schools in Ohio and remained in his native state until he was thirty-five years old. Then, with wife and children, he came to Nebraska and lived in Cass county for three years but did not feel satisfied there Therefore, in the spring of 1888, Mr Schindler and his family boarded emigrant cars at Weeping Water and by that means reached Kimball, and from there came to his present homestead in section ten, town eighteen, at the same time securing a tree claim, and he still owns all this property It would be foolish to declare that Mr and Mrs Schindler had no hardships to face for these were the portion of every settler of that date and were of such a nature that no foresight or excellence of judgment, could have prevented them During the first ten years in Banner county the family lived in a dugout, and for some of these years had to haul all their water, for family use, for stock and crop irrigation, a distance of nine miles Mr Schindler provided himself with a span of strong mules as water carriers, but even then four barrels was the limit that the team could haul up a very steep cliff

Mr Schindler devoted himself to the development of his land and found that in spite of climatic conditions, he could make it very productive He has sown as little as two pecks of wheat to the acre and harvested forty bushels of grain In early days lack of farm machinery was a great handicap, but Mr. Schindler was resourceful and ingenious and did better in this way than many of his neighbors In early days he paid considerable attention to raising mules and had the reputation of having the best animals in the county Later he became more interested in Shorthorn cattle and Poland China hogs and during his active years raised many head annually He has practically retired and has turned the farm industries, to a large extent, over to his son, who is a capable and successful farmer much interested in every line of farm development

In Putnam county, Ohio, September 27, 1874, Schindler was united in marriage to Miss Mary E Chandler, a daughter of Truman and Ruth (Gillett) Chandler, the former of whom was born in New York and the latter in Vermont and both are now deceased Of their six children three survive, Mrs Schindler being the only one in Nebraska The following children were born to Mr and Mrs Schindler Charles F , who was accidentally killed in 1907, by a runaway team married Rachel Van Pelt, who lives at Wheatland, Wyoming, Truman, who lives in Colorado, married Alma Cox , Edward who lives in Banner county, married Pansy Humphrey , Grace, who lives in Wyoming, is the wife of John Adcock , Ruel, who manages his father's properties, married Hazel Green, and Myrtle, who is the wife of Maunse Sandberg, of Banner county, all well known and highly respected in their several communities

In addition to his large estate of seventeen hundred acres of finely developed land, well improved, Mr Schindler has additional assets, being a stockholder in the Harrisburg State Bank, a stockholder in the Farmers Union Store at Bushnell, a stockholder in the Farmers Elevator Company at Pine Bluffs, and own a block of stock in the proposed Farmers Union store soon to be established at Flowerfield Mr Schindler has always voted the Democratic ticket He is a member of the Farmers Union In his neighborhood no other citizen has done more to further the establishment of schools and churches nor to advance the best interests of the county generally In a visit back in Ohio, he was led to make comparisons that convinced him that in the matter of agricultural opportunity, this section far exceeds the older portions of the country He has always maintained a home of hospitality and sometimes is led to believe that the old days, in their general friendliness and somewhat crude methods of enjoyment, were more wholesome and satisfying than is social life in many communities at the present time

JOHN T WOOD, president of the First State Bank of Oshkosh, the oldest bank in Garden county, had for years been identified with banking interests in Custer county In 1919 he bought the controlling interest in the First State Bank and came to Oshkosh as president and manager of the institution

John T Wood, son of Thomas J and Betsey J (Deans) Wood, was born in Greenville, Montcalm county, Michigan, June 19, 1868, and was still a lad when he accompanied his father to Nebraska He completed his education in the public schools of Custer county and commenced his independent career on the home farm near Ansley, but subsequently went to the depot of the Chicago, Burlington & Quincy Railroad, where he learned telegraphy Later he was transferred to the station at Edgemont, South Dakota, where he worked for six months, and in all was connected with railroad offices for two years In 1891, he came to Mason City, where, with his father, he started a general store business, which was conducted until 1895, at which time he returned to farming Mr Wood was identified with various enterprises, one of which was the poultry business, in which he

was engaged during 1894 and 1895, during which time he made a trip to New York with a car of poultry In the fall of 1910, Mr Wood went to Mason City to become an organizer of the Farmers State Bank, which at the outset had a capital of $10,000, but which later increased to $20,000 A beautiful bank building has been erected, and the institution, having shown itself substantial, safe and conservative, has attracted a large patronage, and now has average deposits of $125,000 Mr Wood acted in the capacity of cashier and manager, and a member of the board of directors, and proved himself capable, energetic and courteous, so that he accomplished the dual purpose of gaining the friendship and confidence of the depositors, and attained a place for himself among the capable bankers of Custer county In 1918, Mr Wood disposed of his banking interests in Custer county and bought the controlling interest in the First State Bank of Oshkosh, Garden county, taking charge of the institution March 1, 1919, and today is recognized as a leading financier here Mr Wood is a Thirty-second degree Mason and a Shriner, and was secretary and treasurer of his lodge at Mason City, he is a member of the Modern Woodmen of America, and also holds membership in the Independent Order of Odd Fellows, in which he is past grand Like his father, he has long been interested and actively prominent in Republican politics For eight years, from April 1902 to 1910, he served acceptably as deputy register and register of deeds of Custer county, and was his party's candidate for the Nebraska Legislature His work as a citizen has always been of a progressive and constructive order With his family, he belongs to the Christian church

John T Wood was married April 6, 1892, to Belle Bryan, who was born at Taylorville, Christian county, Illinois, the daughter of Joseph Bryan who fought as a Union soldier during the Civil War and came to Lincoln county, Nebraska then to Custer county in 1879 Mr Bryan homesteaded nine miles north of Mason City, where he died in 1892, and where his widow still resides Five children were born to Mr and Mrs Wood, of whom three are living Iowa, who married William A Runyan, who engages in real estate and insurance business in Oshkosh, and they have one son, — Roger Wood Runyan, Marie, the wife of Leslie Airhood, manager for the Nebraska Telephone Company, at Farnum, Nebraska, died January 9, 1919, leaving one son, John Pershing Airhood, and George

Clark, a graduate of the Mason City high school, where he established a creditable record both in his studies and in athletics, assisting his school to win several competitive cups, and now assisting his father in the bank. He enlisted in October, 1918, in the Reserve Officers Training Corps at Lincoln, Nebraska.

John T. Wood took an active part in all war work after the United States declared war on Germany, being chairman of the Field Committee in all the Liberty Bond drives in both Custer and Garden counties. Today he stands high in the financial circles of the Panhandle, as a progressive and substantial banker, who has won and holds the confidence of the public.

MRS. LOUISA HEWITT, one of the brave and courageous pioneer women of the Panhandle who took up a homestead here in 1887, has run the full gamut of frontier experiences and her reminiscences of the early days are graphic and interesting. The final steps of her journey to this section were made on a train, though railroads were far apart in that early day. Girded with undauntable purpose and the valor which are necessary to succeed under the conditions existing in a new country just opening to civilization and progress she established a home on the frontier, proved up on her land and became one of the well known residents of the Lodgepole district.

Louisa Saddington Hewitt was born in Appleby-Magna, Leicestershire, England, April 7, 1846, the daughter of Edward and Eliza Saddington. She was reared and educated in her native village and lived there until her marriage. Mrs. Hewitt grew up self reliant and with marked executive ability. March 1, 1877, was solemnized her marriage with Walter Hewitt, the service being red in the cathedral at Manchester. Walter Hewitt was born at Edgefield, England, June 1, 1853, the son of William and Fanny Hewitt, and he died at Chicago, Ilinois, August 1, 1885. After their marriage Mr. and Mrs. Hewitt remained in England until 1884, when they came to the United States. Soon after landing on our shores they came west and the next year Mr. Hewitt died in Chicago, leaving his wife with the one child Henrietta.

Mrs. Hewitt learned of the good land to be obtained in Nebraska for a small sum and determined to have a farm of her own. In 1887, she came to this state and filed on a homestead in the Lodgepole district and today is the owner of a quarter section in 30-15-45. She was brave and had a great ambition which helped tide over the terribly hard early years when the first settlers in the Panhandle suffered from drought and the winter blizzards, but she had determined to succeed and did so. By sticking to her land in time crops were raised, the railroad came through this section and she is now the owner of a valuable farm. It was through her ability and well organized energies that she was enabled to withstand the hardships and privations incident to life on the frontier — a woman whose strength has been as the number of her days and who has had a demarkable share in pioneer experiences in the great west. For more than thirty years she has viewed the many changes and the rapid development of a country that was virgin prairie when she first viewed it and has been known far and wide for her kindness and good deeds.

HERMAN KUEHNE, pioneer ranchman and successful farmer of Deuel county, is one of the men who came into the Panhandle in the early days, believed in the future of this section and has lived here to see his faith justified and profited by the many changes and development of the Big Springs section. He was born in Germany, November 14, 1858, the son of Frederick and Wilhelmina (Opferman) Kuehne, both natives of that country, where the father was shipper and sailor on the river boats plying between Saxony and Hamburg. He died in 1864 and his wife in 1910; they never left their native land.

Herman Kuehne was educated in Germany and then learned the roofing business, made slate roofs, built chimneys and allied work. In 1883, he came to the United States and while in Cumberland, Maryland, contracted malaria fever. His doctor ordered him to a higher altitude and learning of the fine opportunities to secure cheap land in western Nebraska, Mr. Kuehne came to Deuel county in April, 1885. His first home was a sod house with sod roof. He put in crops and says that in all the hard early years they never had an entire failure; some product was always raised and when there was not enough of one commodity they ate others, such as oats or any grain that had done well. Mr. Kuehne says men did not farm then to make a fortune but a living. With the passing years he has adopted modern farm methods and machinery and has a finely equipped farm. He has been a Democrat many years; served as road overseer, is a member of the Independent Order of Odd Fellows of Big Springs and now fills the office of Noble Grand.

July 8, 1884, Mr. Kuehne married Miss Wilhelmina Hemken, the daughter of Frank and Wilhelmina (Loeneker) Hemken, natives and residents of Germany Eight children form the family Fred, at home, Wallie, of Deuel county, Hattie, the wife of John Leff, of Colorado, Frank, lives in Wyoming, Carl William also live there, Marie, the wife of Ernest Koberstein, of Nebraska, and Herman, at home

When Mr Kuehne first came to this section he hauled water from Big Springs and Ash Hollow for family use and the stock, a distance of twelve miles, which shows what persistance he had The nearest neighbor was five miles away, when visiting people cut across the open prairie Soon after he located here other settlers came, though wild antelope were common and deer were to be found in the hills Mr Kuehne recalls the fraudulent county seat election and tells of many other pioneer experiences He and his family have warm friends and are highly respected here

SYVER JOHNSON, pioneer settler of the Panhandle who has worked hard and overcome all the trial and privations of life on the frontier and today is one of the prosperous farmers of the Big Spring district, was born in Norway, May 12, 1848, the son of John Erickson and Carolina Olson, both natives of Norway The father came to this country and located in Hamilton county, but remained only five years and then returned to Norway where he spent the remainder of his life There were eight children in the family, six sons and two daughters, four of whom reside in this country

Miss Martha Kardesen, daughter of Kardesen Peterson and Gjoren Jensen was born in Norway, February 15, 1852 Her parents were both natives of Norway There were six children in the family, one daughter and five sons, all of whom reside in this country and Canada

On October 3, 1875, Mr Johnson and Miss Kardesen were united in marriage in the old country, where they had grown to manhood and womanhood, both having been educated in the excellent public schools of that country

During their residence in Norway Mr Johnson worked at the mason trade but desiring to own land of his own he migrated with his family to the United States in 1881, locating in Hamilton county, Nebraska, in May, remaining there three years before coming farther west to take up a homestead in Deuel county In 1884, Mr Johnson broke a little of his land but did not bring his family out until the following year When he came to the Panhandle, Mr Johnson drove across the coutry with a team and wagon behind which were hitched two cows

The first home was a sod dugout with a brush roof and when it rained out it also rained in The first two years were exceptionally rainy and everything thrived without but not so well within There was no water on the place and it was hauled for ten years for family and stock

Having but a small acreage broken the first year it was planted to melons, crooked neck squash and gourds, the seed having been brought from the east These grew in abundance and were of enormous size and were stored in cellars or caves, forming the chief forage crop for the stock the first winter

Prospects were soon blighted by the dry years that followed Crops were very poor and Mr Johnson went east to look for work in order that the family might have the necessities of life He worked at the mason trade in Hamilton county, also in Omaha and Lincoln He would often be gone several months at a time while Mrs Johnson and the children stayed on the homestead and fought against fate for sustenance Mrs Johnson also worked for her neighbors whenever she could She would walk two and one-half miles to town, carying one youngster and leading another, do a hard day's work, washing clothes or cleaning house and then walk back at night Often after her return she would have to go a mile and a half to the spring for water She had nothing to drive except a runaway team hitched to a wagon, and the way they traveled over the black-root knolls wasn't slow One night when she reached home there was but one spoke left in one wagon wheel

For years the forage crops had to be cut with a scythe and raked with a hand rake The brunt of this fell on Mrs Johnson and the children, as Mr Johnson was often away in search of work at the harvest time Many were the times when there was nothing left in the house to eat Mrs Johnson would take a muzzle-loading shot gun (which many men wouldn't understand the loading of these days) and go hunting rabbits Often she would have to walk for miles over the sand hills before she would be able to scare one up, but when bunny made a move he was a dead rabbit The meal for the bread was ground in a coffee mill which was a long and tedious task

The rattle snakes were quite a menace in the early days It was not an uncommon sight to see them disapear into the walls of the sod house or find them basking in the sunlight on the door step The older boys would go snake hunting and a trophy of two dozen rattles wasn't an unusual day's hunt A snake spear was used for killing them, which consisted of a spearhead on a long handle None of the family was ever bitten by a snake, but the oldest daughter was one time bitten by a tarantula The forethought and quick action of the father saved her life Centipedes were very numerous and would often drop from the ceiling to the floor

During the drought years of 1893, 1894 and 1898 the family suffered severely, the father became discouraged but the mother never did Provisions were sent from the east for the relief of suffering humanity, but through mismanagement very little of it ever reached the places where it was most needed Many of the old timers left during the drought years, but through the courage and industry of the mother the Johnson family stuck it out and today are in comfortable circumstances

For the past ten years Mr Johnson has raised cattle, but with the loss of the open range finds farming profitable

Mr and Mrs Johnson raised a large family, there being eleven children in all, eight of whom are living, Lena, the wife of James Brown, former Commissioner of Deuel county, Johann, deceased, Charles, a farmer of Deuel county, Fred, a farmer of Deuel county, Selma, the wife of Dan Trolan, ranchman of Oregon; John, a merchant of Big Spring, Joseph, a farmer, of Deuel county; Edward, deceased; Nora, postmistress, Big Spring, Morton, ex-service man, Oregon, and George, who gave his life on the battle fields of France in the recent war

Mr Johnson is a Republican, his wife is a Presbyterian, and they are both members of the Farmers' Union

JOHN ELMQUIST, old settler, well known farmer and successful cattleman, has well demonstrated that a man who came to the Panhandle in the early days could succeed if he had the grit and perseverance He was born in Sweden May 8, 1858, the son of Peter and Sarah Elmquist, natives of Sweden The father was a farmer who came to the United States with his family in 1883, settled in Polk county, Nebraska, lived there over three years and then came to Deuel county, both the father and son John taking up homesteads here When the mother died in 1906, the father returned to Sweden to pass the remainder of his days and died there in 1917 They were the parents of four children, but John is the only one here Mr and Mrs Elmquist helped to organize the South Swedish Church

nine miles northeast of Chappell and school district No 24 The father was a Republican and a member of the Swedish Lutheran Church

John Elmquist was educated in the public schools of his native land and has been a farmer all his life He came to the United States in 1882; lived in Polk county until the fall of 1886 and that year took up a homestead here which he still owns The first seven years he hauled water, lived in a sod house, pastured cattle and had a team of mules with which to farm In 1889, he returned to Polk county to work and make money, later went to Wyoming while his wife remained on the farm with the children, though the nearest neighbor was two miles away. They burned buffalo chips for fuel as that was all they could get, and sold calves for their provisions Conditions were bad and it was hard to get along but the Elmquists were not to be discouraged and today can hardly realize the changed conditions, for they now own three sections of land, have three tractors, a well improved farm and breed white faced cattle for the stock market Mr Elmquist also breeds high grade horses and enough hogs for his own use

June 6, 1885, Mr Elmquist married Miss Anna Anderson, the daughter of John and Johanna Anderson She was born in Sweden in 1886 and came to this country when a child of less than three years of age, and after her marriage became the mother of thirteen children, twelve of whom survive: Joseph, at home, Oscar, on the old homestead, August, at home, Emilie, the wife of Charles Carlson, of Deuel county, Harry, deceased, Frank, Fred, Selvia, Ruth, Arthur, Elvin, Annie and Arvid, all at home

Mr Elmquist is a Republican, for years has been director in school district No 24, helped organize and build the Celia Swedish Church and is now helping to organize and build the Abreva Swedish Church north of his home He is a stock holder in the Farmers Elevator at Chappell and one of the well known and highly respected men of the district

GODFREY M ZALMAN, one of the large landholders and substantial farmers of Deuel county who came to the Panhandle with little and today is a well to do and progressive business man, was born in Cook county, Illinois, April 8, 1860, the son of Henry and Eunice (Bower) Zalman, both natives of Bavaria, the former born in 1819, died in 1900, the latter born in 1828, died in 1918 The father was a farmer who came to the United States about 1841, settled in New York where he lived for some years then moved to Illinois in 1860, locating an a farm in Shelby county He had been in Cook county, over the site of the present city of Chicago, then far from a great city There were ten children in the Zalman family, eight of whom survive, but Godfrey, of this review, is the only one in Deuel county The parents were members of the Lutheran Church

Mr Zalman was educated in the public schools of Illinois, and learned farm business under his father When only twenty-two years of age he began his independent financial career as an agriculturist, a vocation he has followed all his life November 15, 1882, Mr Zalman married Miss Anna Barth in Shelby county, Illinois, she was the daughter of John and Margaret (Hauk) Barth residents of Illinois, now deceased Six children were born to this union, of whom five live Bertha, the wife of Jack Bottles, of Oshkosh, Nebraska, Louis, of Deuel county, Myrtle, the wife of Nicholas Kollsen, of Denver, Colorado, Maggie, the wife of Martin Wendt, of Garden county, and Edna, at home

Mr Zalman came to Deuel county in the fall of 1893 took up the homestead where he still lives near Big Springs, starting with little equipment There was a small shack on the place and a well, which was important He says that he started farm life with a team and old wagon, one horse soon died and he later bought a span of mules He admits that he is self-made, had practically nothing to start with and today has a well equipped farm with good buildings, is comfortably situated financially, and all is due to his own good work and determination to succeed The first years were hard, then came a few years of good crops, but he had to work during haying time to help buy supplies in the poor years Their fuel was buffalo chips and all had to work hard They milked cows and sold butter at ten cents a pound, which shows what the times were Today Mr Zalman operates a farm of sixteen hundred and eighty acres, mostly range land, but he has the latest machinery for that farmed He is a Democrat, has been moderator of his school district for many years, is a member of the Farmers Elevator Company in which he holds stock, at Big Springs, and with his wife is a member of the Lutheran Church

CHARLES P CHAMBERS — There is no man more widely known and no man more closely concerned with educational and public affairs and county developments than the man whose name is announced by this title line

In a wide circle of the county he is known to everyone as "Judge Chambers" No history of Cheyenne county or the state of Nebraska would be complete without his name For more than a third of a century this scholarly man has made his home in the vicinity of Sidney where he has engaged in his educational profession He has been connected, from first locating here, with the educational interests of the county, and by sheer force of character, learning and ability of a high order he has imbedded his name deeply and permanently in the school history of the county and the Panhandle

While pride of ancestry is not a marked characteristic of the American citizen, it is, nevertheless, not only natural but highly commendable that one should feel a just pride in the fact that he has descended from ancestors who were well known in their day and generation With this thought in mind it is hoped that a brief account of the parents of Charles Chambers will not be deemed inappropriate in the sketch of their son

His father, Jobe Chambers, was a native of New Jersey, descended from some of the first settlers who located in this eastern state at an early day, of excellent European blood, while his mother Anna Jones, was born in England, and represented some of the best traits of the inhabitants of that tight little island who have played such a great and important part in colonizing the entire world Mrs Chambers was but eleven years of age when brought to the United States by her parents, and though she had already attended school in England completed her education after landing in America Jobe Chambers was born in New Jersey and was but six years of age when his parents removed to the then wilds of Indiana, and there he passed his childhood and youth, receiving his educational advantages in the state of his adoption, grew to manhood and when his school days were over engaged in business He came of healthy sturdy stock and has already passed beyond the psalmist's span of "three score years and ten" as he is still living in Indiana at the advanced age of eighty-three years and takes an active interest and part in the life of his community

Charles P Chambers was born in Bowling Green, Indiana, November 12 1858 His boyhood was passed on his father's farm, where he attended the public schools of his locality in the winter and during the summers found occupation on the farm His father being an educated and cultivated man he gave his son every advantage in training to fit himself for the battle of life. After completing the course in the local institutions the young man matriculated at the Central Normal College of Danville, Indiana, and upon graduating from the college, animated by an ambition to give fullscope to his abilities, he entered a business college in Indianapolis, Indiana. At the close of this course of special study, Mr. Chambers entered the pedagogic profession by accepting the position of teacher in Indiana, where he was engaged in professional service for three years before removing farther west to the state of Iowa, where he again became the guide and instructor of the rising generation of that commonwealth. Having been reared on a farm and being experienced in the practical management of agricultural business Mr. Chambers after coming west decided to become a land holder himself, and, in 1885, chose Nebraska for his future home. After looking about for the best location he settled on a homestead eleven miles southeast of Sidney in a valley of a branch of Lodgepole creek. Thus he became one of the pioneers of this section. More than thirty-four years have passed since Mr. Chambers first came to Cheyenne county, which at that time, while not the veritable wilderness of the early seventies, was still sparcely settled, habitations were few and civilization was still in primitive form. He first establshed a home for his family and made such permanent improvements on his land as was possible so far from sources of supply. Money was a scarce commodity at this period on the plains and very soon he saw that it would be necessary to have more capital to carry on the improvements he had planned for his farm. Having been a successful teacher farther east, he turned to his profession in this time of stress and assumed the duties of preceptor in the district school and later in the town of Lodgepole. His marked ability could not pass unnoticed and within a short time Mr. Chambers was elected county superintendent for four years, where he displayed such admirable qualities that the voters of the county showed their appreciation of his ability by placing him in office as deputy clerk of Cheyenne county. Following close upon the discharge of his duties in this capacity he was again elected county superintendent of schools, serving two years before coming to Sidney to accept an excellent position in the city school, but after four years here was forced to give up professional life as his health had been materially assailed by the indoor life and close confinement incident to his duties to the community, and being ad-

vised by his physician to live out doors, Mr. Chambers went out onto his ranch. As his capital had increased the original Kincaid homestead had been augmented by other tracts which Mr. Chambers purchased in the vicinity, so that by this time his landed estate consisted of over twelve hundred acres. He at once assumed active management of the farm which already had excellent improvements in the way of a comfortable home, excellent farm buildings and late and improved farm machinery for the lightening of labor about the place. For years Mr. Chambers had given much attention to farm industry, had studied up on the best crops for this semi-arid climate and what kinds of stock thrived best and were the best money makers. All this store of knowledge he applied to his farm business upon locating in the country and within three years his marked success as a diversified farm producer and stock-raiser was recognized throughout this district which is well known for its progressive agriculturists and stockmen. After regaining his health Mr. Chambers returned to Sidney to establish his home as he wished to give his children the benefit of the excellent educational advantages afforded in town, though he still continued the active management of his ranch and thus gained the benefit of the outdoor life which this entailed. In 1915, however, he rented the ranch as the year before he had been elected county judge, and in order to devote his time and energies to the duties of this exacting office gave up the management of the place. Judge Chambers was elected on a non-partisan judicial ticket, which shows how high is his standing in the community which has called upon him to serve it, and filled the office with such credit that he was re-elected in the fall of 1916 and again in 1918, and still again in 1920, with no opposition.

The public school has been a favorite realm of the judge as we know and since coming to Sidney as a permanent resident he has served on the school board, the advocate of every advanced educational movement for the permanent betterment of system, school buildings, good salaries for good teachers and modern methods. No man in Cheyenne county has responded more liberally to the calls of the public than this pioneer of the Panhandle; no man has contributed more cheerfully to every public enterprise, or lent a greater service to the development of his county and community, for he has been a public figure since first coming to our great state. During the war his influence was given to the furtherance of its prosecution and to the assistance of the government in every manner as he was chairman of the Council of National Defense of the county, was a heavy buyer of Liberty Bonds and took an active part in every Liberty Bond drive in Sidney and Cheyenne county.

All his life the judge has been a supporter of the Democratic party and has taken an active part in shaping party policy in his community, and it was on the Democratic ticket that he was elected county superintendent; but his popularity was so great that his election to the bench was due to no party influence, but a spontaneous appreciation of the community as he ran on a non-partisan judicial ticket with no opposition. Fraternally the judge is affiliated with the Independent Order of Odd Fellows and the Modern Woodmen of America, while he and his wife are members and liberal supporters of the Episcopal church in which both are active workers.

On November 7, 1880, Judge Chambers married Miss Sarah Stevens, a native of Indiana, who died in 1884, leaving one son, William Ezra, a young but prosperous ranchman of Cheyenne county. February 3, 1885 the judge married Miss Susan Sanderson, also a Hoosier by birth, and they have nine living children, one having died in infancy: Clarence F., the manager of the Central Elevator Company of Sidney; Robert O., the editor of a newspaper at Minatare, Nebraska; Guy Cleveland, who volunteered for the service of his country when the United States entered the World War and saw service in France as a lieutenant, but has already returned, received his honorable discharge and taken up the pursuits of peace as assistant attorney for the Chicago, Rock Island and Pacific Railroad; Anna Rebecca, the wife of J. D. Emrock, of Alliance, Nebraska, where she holds the important position of superintendent of schools; Eunice Viola, who married Oscar A. Olson, a rancher of Cheyenne county; Charles Allen, who enlisted in the army at the age of sixteen and was sent to a training camp in Maryland and though so young was advanced to the position of corporal of his company, but peace was declared before he saw service in Europe; Ray C., Vera Ellen and Arthur Dale, all of whom are still under the happy family roof tree.

PETER SODERQUIST, well known ranchman of Deuel county, is one of the early settlers of this region who has suffered all the hardships and privations incident to living in a newly opened country in the way of drought,

blizzards and lack of water, as well as finding it hard to get money to buy supplies, but he believed in the future of this western country, stuck and today is the owner of seventeen hundred and twenty acres of land, over seven hundred of which are under irrigation as the land lies along the Lodgepole creek.

Mr. Soderquist was born near Malmo, Sweden, February 24, 1867, the son of Lars and Hanna (Monson) Soderquist, both natives of that country, where the father was a farmer. While serving in the army he was regimental saddle and boot maker, a trade which he learned from his father. After leaving the army he farmed in the summer time and followed his trade in the winter season, and died in 1911, aged eighty-two years. There were seven children in the family, of whom six survive: Andrew, of Montrose, Colorado; Lewis, of Vernal, Utah; Peter, of this sketch; Betty, the wife of Nels Martinson, of La Mar, Colorado; Ola and Anna, in Sweden, and one that died in infancy. The parents were members of the Lutheran Church. Peter Soderquist was educated in the public schools of Sweden and came to the United States in 1883, locating in Wahoo, Nebraska, but remained there only one winter before going to Julesburg, as one of his brothers had gone there shortly before. He and another brother started for the south divide, driving some cattle, when a storm came up and they spent the night in the snow just a few rods from the dugout prepared by the first brother, the storm preventing them knowing it. They erected sod buildings and then Peter and one brother went to Colorado seeking work. They had a hard time as so many men were out of work and few jobs were available; suffering from cold and lack of food on the way to different towns, they finally secured work with a farmer who watched them with a gun to see that they did not run off. Several months later Mr. Soderquist returned to Saunders county, Nebraska, as foreman of a ranch, remained two years before going west to make charcoal in the mountains but in 1892, took up land near Julesburg which he bought a year before. On November 4, 1892, Mr. Soderquist married Miss Hildah Anderson, the daughter of Jens and Anna (Nelson) Anderson, of Deuel county, and they became the parents of the following children: Louis, deceased; Anna A., Percival A., Addie E., Edith E., Edna N., Selma N., Grace L., and George J., all at home. Mrs. Soderquist was born a half mile from the home of her husband in Sweden but they did not meet until years later here in Deuel county　After living on the farm a year Mr and Mrs Soderquist became disgusted as they were so far from town and water, and in the spring of 1893, came to Deuel county to buy land, living here since　With prosperity Mr Soderquist bought more raw land, erected good buildings, sold and again bought, so that today he is one of the large landholders of the section　He put in his own irrigation system and is a progressive man, using the latest machinery

Mr Soderquist is a Republican, he ran for county commissioner being defeated by only two votes　He has been school trustee of three districts and was instrumental in securing the new court house for Chappell　He reads widely and is well posted on public affairs and can tell many reminiscences of being out in storms in the early days and the different experiences of those times

WILLIAM W FAUGHT, who is now living retired at Oshkosh, Garden county, is to be attributed with the distinction of pioneer honors in this county, where he established his residence more than thirty years ago, when it was still an integral part of Cheyenne county, and where he reclaimed and improved the valuable ranch property of which he is still the owner　A pioneer citizen of substantial worth and unqualified popularity, he well merits recognition in this history of the Nebraska Panhandle

William Wilson Faught was born in Sandusky county, Ohio, January 17, 1850, and in the old Buckeye state were also born his parents, who were representatives of pioneer families of that commonwealth　He whose name introduces this review is a son of Levi and Rosanna (Miller) Faught　The father was born and reared in Perry county, Ohio, and in his native state he became a prosperous farmer　In 1864, he removed with his family to St Joseph county, Michigan, and there continued his activities as an agriculturist about twenty years, as one of the honored pioneers of the county　Finally he removed to Goshen, Indiana, where he continued to reside until he was venerable in years, when he came to Nebraska to make his home with his son William W, of this sketch, and three months after coming to Oshkosh he died, at the age of eighty-six years　His wife passed the closing years of her life at Three Rivers, St Joseph county, Michigan, where she died at the age of eighty-eight years, both having been earnest members of the church, and his political faith having been that of the Republican party　Of

the six children two sons and two daughters are now living

William W Faught acquired his rudimentary education in the public schools of his native state and was about fourteen years old at the time of the family removal to Michigan, where he continued to attend school, at the same time he contributed his quota to the work of the home farm　At the age of nineteen years he rented his father's farm, in St Joseph county, Michigan, and there continued operations five years　He then removed to Fremont, Ohio, and for the ensuing two years was engaged in selling the celebrated J. I Case threshing machines throughout his assigned Ohio territory　Thereafter he developed a prosperous contracting business, in the construction of bridges and drainage ditches, with which line of enterprise he was identified in Ohio, until the spring of 1887, when he came to Nebraska and became a pioneer settlers in what is now Garden county　He shipped his household goods by railroad to Lodgepole, Cheyenne county, whence he transported them overland to the present Garden county, then a part of Cheyenne county. Here he filed entry on homestead, pre-emption and tree claims, to which he perfected his title in due course of time, and here girded himself most gallantly for the work of reclaiming and improving his land, to which he eventually added, by purchase, until he figured as the owner of a fine ranch property of eight hundred acres, improved with modern buildings and showing every other evidence of thrift and prosperity as the passing years contributed their largeness　He continued in the active supervision of this ranch until 1910, when he rented the property to his son and son-in-law and removed to Oshkosh, where he has since lived virtually retired, although he retains the ownership of the valuable farm which he developed as a pioneer agriculturist and stock-grower of the county

Influential and loyal in community affairs Mr Faught served one term—three years—as county commissioner of Deuel county, prior to the erection of Garden county, and he gave several years of effective service as a member of the school board of his district　His political allegiance is given to the Republican party, and he gives liberal support to the Methodist Episcopal Church, of which his wife is an active member

March 12, 1869, recorded the marriage of Mr Faught to Miss Eva Stults, who was born in Sandusky county, Ohio, a daughter of James and Matilda (Johnson) Stults, the

Oscar R. Lovelace

former a native of Ohio and the latter of Indiana, the father having died at the age of sixty-three years, and the mother at the age of seventy-four years Nellie, elder of the children of Mr and Mrs Faught, became the wife of Curtis Farris, who farms a portion of the Faught ranch, and she was called to the life eternal on December 16, 1916, being survived by two children The only son of the subject of this review, has active charge of the portion of the father's estate that is not operated by the son-in-law, Mr Farris, and he is well upholding the civic and industrial prestige of the family name

OSCAR R LOVELACE, vice president of the American Bank of Mitchell, Nebraska, is one of the younger bankers who are making financial history in the Panhandle as his management of the bank has placed him in the front rank in the commercial circles of the northwest His business career has been characterized by self reliance, initiative and executive ability of a high order, all qualities which bring normally in their train a great measure of success His integrity and conservatism have begotten public confidence for the bank and he is held in high esteem by the citizens of Mitchell, which is necessary in the furtherance of success in the line of enterprise which Mr Lovelace has chosen As a banker, Mr Lovelace has shown great constructive talent and it has been largely through his policies and efforts as manager, that the American Bank has made such rapid progress since its organization in 1919

Mr Lovelace is practically a son of Scottsbluff county, though he was born in Grant City, Missouri, February 4, 1880, the son of N R and Alice (Lore) Lovelace, the former a native of the Badger state, while the mother was born in Pennsylvania In 1883, N R Lovelace came to Dawson county, Nebraska, being one of the earliest settlers of that section He took a pre-emption on which he proved up but learning of the fine land to be had in the upper valley of the Platte came to Scottsbluff county in 1889 After arriving here he took a tree claim near Gering but gave it up in 1893, to go to California where he engaged in fruit raising near Fresno The family, however, did not like California as well as the high prairie country so returned to Nebraska, when Mr Lovelace filed on a homestead and bought an additional quarter section of land adjoining, six miles southwest of Gering, where the family still make their home This farm has been highly developed and to-

day is one of the most valuable and productive tracts of land in the county

Oscar was practically reared in Scottsbluff county as he was but a small child when the family first located here He received his educational advantages in the excellent public schools and then graduated from the high school at Gering As the cattle industry was still one of the largest in this section of the country at that time, the boy naturally craved to join in the free life, with the adventures afforded by the cow camps of the great cattle barons and became a cowboy, spending four years on the ranges in western Nebraska and Wyoming When he had exhausted all the various phases of out door life he began to give thought and attention to the question of a career By the life in the open Mr Lovelace had become robust and hearty and he soon saw that muscle alone did not bring in the greatest returns for the amount of work expended but when guided by a well trained mind, the combination won Realizing the value of a higher education he deserted the cattle business and went to Lincoln, Nebraska, to take a special course in the Lincoln Business College, having decided that a business career best suited his inclinations On the completion of his studies, Mr Lovelace returned to Scottsbluff county and Mitchell, accepting a position as bookkeeper in the Mitchell State Bank, when it was organized in 1907 Here he displayed such marked ability in finance, learning the banking business from the ground up, that when the American Bank was organized in 1919, he was made executive officer and vice-president of the institution, in fact, he was one of the prime movers in its inception and organization He became one of the heavy stockholders and so has their interests ever at heart From its start the bank has had a phenominal success, due to the men who have shaped its policies and especially Mr Lovelace, who has proved a most efficient manager Mr Lovelace has made many worth while business acquaintances and associations which have been valuable to the bank, he is popular personally and when the bank was organized told all the employes that the slogan of the institution was to be "Service" Patrons are treated courteously, their interests are considered as well as those of the bank and as a result the deposits are constantly increasing From first entering business Mr Lovelace has also entered actively into the civic and communal life of the county and town of Mitchell, he advocates and helps "boost" everything that tends to the development and up-building of this section of the valley, giving liberally of time and money in support of all the progressive movements that are placing the Panhandle "on the map," and is the originator of many of them. He firmly believes in living his Americanism and citizenship and personally attains the high standard that he sets for others. During the war he aided in every way in its prosecution, being one of the prominent figures in raising money during all the drives for Liberty Bonds and for the Red Cross.

In politics he is a member of the Republican party, while his church affiliations are with the Federated church. For some time Mr. Lovelace has been secretary of the school board, which has led to many modern equipments being introduced and he is now serving as a member of the city council.

On April 15, 1909, Mr. Lovelace married Miss Gertrude Ford, born in Colorado, but practically reared in Scottsbluff county, as she is the daughter of C. W. Ford, who was clerk of the county for more than six years, and is one of the prominent pioneer settlers. Three sons have been born to Mr. and Mrs. Lovelace, Ross C., Charles Ford, and Joseph Nelson.

CONRAD LINDERMAN, owner and editor of the Crawford Tribune, one of the well known journalists of Nebraska, is a man whose life has been eminently useful and is a fair example of the "average man" in our American citizenship. His education was procured through close application to reading the newspapers and books obtainable by or accessible to the studious young man in the American country printing office. His efforts and influence have always been devoted to bettering the conditions of his fellow men from his viewpoint, but always finally submitting to the will of the majority, which is necessary in a true democracy. From young manhood he has contended that the two main levers of our civilization, the medium by which we transport our productions and the medium by which we exchange them, should be under full control of the people. He also contends that we should provide some method for the exchanging of our products which would not require the production of an equal value of our products in order to create a medium for the exchanging of the same.

Conrad Lindeman was born at Hazelton, Pennsylvania, November 28, 1853, his parents having come to the United States the previous summer. They were George and Maria S. (Woelner) Lindeman, natives of Germany, and the father was a locksmith and blacksmith by trade. He died when Conrad, the youngest of his five children, was still young, but the mother lived to the age of eighty-five years. Of Mr. Lindeman's four brothers, only one is living, George, whose home is in the city of Vancouver, British Columbia. Henry, who conducted a hotel at Crawford, died here, William died in manhood, and John died in infancy.

Until he was about ten years old, Conrad Lindeman attended the common schools of Luzerne county. That is a great coal mining section and in those days it was the custom to send little boys into the coal pits to act as drivers on the mule cars, and it was in that way that Conrad assisted in providing for the family for three years. He was an ambitious boy, however, and had a distinct leaning toward the printing trade, and after his coal mining experiences, succeeded in getting into a newspaper office, where he served an apprenticeship in the old time way. Completing his apprenticeship at the age of sixteen, Mr. Lindeman went from Hazelton, Luzerne county, to Shenandoah, Schuylkill county, Pennsylvania, and accepted a position as foreman in the mechanical department of the Shenandoah Herald, the first paper published in that town, then a village of several hundred inhabitants.

In 1872, Mr. Lindeman took Horace Greeleys advice and went west, locating in Cass county, Iowa, where he tried his hand at farming, and where in April, 1878, he was married to Miss Mary Meister, who was born in Pennsylvania, a daughter of Martin and Anna M. (Kreitzberg) Meister. Eight children, five sons and three daughters, have been born to Mr. and Mrs. Lindeman: Martha, who is the wife of George Patton, of Billings, Montana; Otto, who lives at Hill Crest, Montana; John A. and Daniel W., who are ranchmen of Dawes county; Helena, who is the wife of Major C. J. Gaiser, of Camp Dodge, Iowa; Amelia, and Emma W., both of whom assist their father in the printing business, and Clarissa R., who is the wife of Gilbert F. Marrall, of Crawford.

After several years on the farm Mr. Lindeman moved to Atlantic, the county seat of Cass county, Iowa, and secured a position as assistant foreman on the Atlantic Daily Telegraph, published by Lafayette Young, where he spent a year, and then, in company with W. H. Saunders, published the Peoples Advocate, at Atlantic, Iowa, for a short time, being succeeded by J. R. Soverign, later master workman of the Knights of Labor. He then engaged in the news and stationary business and in 1886, came to northwestern Nebraska, locating on a homestead in Sioux county, three miles west of Montrose Mr Lindeman and his family suffered the hardships of early homesteaders and took a leading part in the struggles of the early settlers against opposition by the ranchmen

In 1889, Mr Lindeman was elected clerk of the court and county clerk of Sioux county, in which offices he served until 1894, when he came to Crawford to enter the employ of C E Wilson, a clothing merchant, and later in January, 1897, went to work at his old trade, establishing the Crawford Bulletin, which he published until 1904

Selling the Bulletin plant to W H Ketchem, then publisher of the Crawford Tribune, which was established by Mr Ketchem in 1887, Mr Lindeman spent several years in the offices of the Tribune and Crawford Courier then purchased the Tribune in 1910 Under Mr Lindeman's ownership and direction the paper has become the chief organ of the Progressive Democratic party throughout Dawes and Sioux counties, while his subscription list shows that it circulates all through the country as far as Canada on the north and Mexico on the south He does a general printing business and has a well equiped plant On numerous occasions his friends have urged Mr Lindeman to accept political honors, but he has never accepted any office except those mentioned His views on public questions have sometimes been modified through changing conditions, but no one can ever accuse him of ever misrepresenting facts or, for his own benefit, concealing public matters on which should be thrown the light of publicity through the press In addition to his printing plant, he is interested in a ranch of over a thousand acres, lying southwest of Crawford, in association with his two sons, John A and Daniel W Mr and Mrs Lindeman are members of the Lutheran Church, but all the children are members of the Congregational Church

WILLIAM SPOHN has been a resident of Garden county for nearly thirty-five years, and this statement in itself marks him as one of the pioneers of the county, which was still a part of Cheyenne county when he took up land and instituted the development of a productive farm During the long intervening period he has been a foremost actor in the drama of civic and industrial development and progress, has stood exemplar of loyal citizenship and held precedence as one of the representative agriculturists and stock-growers of

this favored section of the Nebraska Panhandle

Mr Spohn is a scion of sterling Swiss ancestry and of a family that was founded in America in the colonial era of our national history He is a direct descendant of Philip Spohn, who was born in Switzerland and came to America when he was a boy His loyalty was shown by his gallant service as a soldier of the Continental Line in the war of the Revolution, serving under the command of General Washington

The honored Nebraska pioneer to whom this review is dedicated was born in Sandusky county, Ohio, June 26, 1851, a son of Daniel and Catherine (Bachman) Spohn, the former of whom was born in Ohio, in 1826, while the latter was a native of Alsace, France where she was born in 1828 and was about eight years old at the time her family immigrated to America, in 1836, and settled in Sandusky county, Ohio, where her father, John Bachman, became a pioneer farmer, his wife, whose maiden name was Mary Mathena, died at the age of sixty-six years and he likewise having been a resident of Sandusky county at the time of his death Daniel Spohn was reared and educated in Ohio, which state he represented as a gallant soldier of the Union in the Civil war, his service having been as a member of Company K, One Hundred and Sixty-ninth Ohio Volunteer Infantry After the war he continued his active association with agricultural industry in the old Buckeye state during the remainder of his active career, and he was seventy-five years of age at the time of his death His father, Henry Spohn, was born in the state of Maryland, March 12, 1787, and upheld the military prestige of the family by serving as a soldier in the war of 1812 He thereafter became a pioneer settler in Ohio, where he farmed in Perry county until 1829, when he became an early settler in Sandusky county, where he reclaimed a farm from the sylvan wilds and there passed the remainder of his life, having been eighty-six years of age at the time of his death The mother of the subject of this sketch continued her residence in Ohio until her death, which occurred when she was about seventy-six years of age Both she and her husband were earnest workers of the United Brethren Church They became the parents of seven children, of whom two sons and one daughter are living at the time of this writing, in the winter of 1919-20

William Spohn was reared to the sturdy discipline of the home farm and acquired his early education in the public schools at Fremont, Ohio He thereafter devoted about one year to independent farm enterprise in his native state, and then he came to Nebraska, in 1875, and settled on a pioneer farm near the present town of Wood River, Hall county, where he continued his operations as an agriculturist until 1885, when he became a pioneer settler in that part of old Cheyenne county that is now comprised in Garden county Five miles northwest of the present county seat, Oshkosh, he took up homestead, pre-emption and tree claims, all wild prairie land, and here he has continued his activities as a farmer during the intervening years, his property gives evidence that he has kept in full touch with the march of development and growth in this section of the state He has been successful as an agriculturist and stock-raiser and also has the distinction of having been a pioneer in the now important field of sugar-beet propagation in Garden county His enterprise was shown by his construction of the independent irrigation ditch which affords excellent water facilities for his farm, which is known as the Spohn ditch its supply of water being derived from the North Platte river On his farm, which comprises eight hundred acres, he has erected good buildings and provided other accessories that mark the model farm of the present progressive day

During all the years of his residence in Garden county Mr Spohn has been liberal in his support of enterprises projected for the general good of the community, and while he has had no ambition for political preferment he is found aligned as a staunch supporter of the cause of the Republican party He is affiliated with the Independent Order of Odd Fellows and both he and his wife are zealous communicants of the Lutheran Church

February 19, 1877, recorded the marriage of Mr Spohn to Miss Florilla Thomas, the ceremony being performed at Grand Island, Hall county Mr Spohn was born in Dodge county, Wisconsin, March 22, 1847, and is a daughter of Joshua Thomas, who was born in the state of New York, but who passed a considerable portion of his life in eastern Canada, where he was actively identified with lumbering operations He was a resident of Iowa at the time of his death, at the age of fifty years, and his wife, whose maiden name was Matilda Borden, was born and reared at Brooksville, Ontario, Canada, where her marriage was solemnized, she having been about forty years of age at her death, which occurred in Wisconsin Joshua Thomas became active in the field of lumbering in Wisconsin and

after the death of his wife he removed, in 1855, to Iowa and became a pioneer settler in Winnebago county, where he passed the remainder of his life Mrs Spohn was a girl eight years of age when she accompanied her father to Iowa, in which state she received her educational training in the public schools of the pioneer era, and in 1870 she accompanied her sister Sophia to Hall county, Nebraska, where they settled in the Wood river district and where her marriage was solemnized a few years later In conclusion is entered brief record concerning the children of Mr and Mrs Spohn Loren Rutherford Spohn, who was born in Hall county, this state, July 22, 1879, and completed his youthful education by a course of study at Hastings College, and was one of the patriotic young men of Nebraska who served as a soldier in the Spanish-American War, having served as a soldier in Company K, Second Nebraska, Volunteer Infantry Mr Spohn was not yet twenty-six years of age at the time of his death, which occurred November 13, 1905 He married Miss May Bowman, and their one child Loren W, now resides in the home of his paternal grandfather, the subject of this review Mabel Edith Spohn, who was born in Hall county, in 1880, passed away at the age of three months, so that the second generation of the Spohn family in Nebraska now has no representative, while the third generation has but one member

ANSON B ALLEN — The good old term, diligence, has been strictly applicable to the career of this well known and highly esteemed citizen of Garden county, where he has wrought out a goodly measure of success through his activities as a stock-raiser and agriculturist, besides which he is entitled to pioneer honors incidental to the history of this section of the state

Mr Allen was born in Clarke county, Iowa, February 8, 1858, and is a representative of one of the early pioneer families of that commonwealth, as the date of his birth discloses His father, Saunders Allen, was born in Clarke county, Ohio, and was sixty-five years of age at the time of his death, which occurred in Taylor county, Iowa Upon removal to the Hawkeye state he took up land, which he developed and improved and became one of the substantial citizens of that state His wife, whose maiden name was Nancy Rich, was born in Warren county, Ohio, but was reared and educated in Indiana, where her marriage was solemnized, she was about fifty-five years of age at the time of her death

JAMES T. WHITEHEAD

On the home farm of his father in Taylor county, Iowa, Anson B Allen passed the period of his boyhood and early youth, and in the meantime he profited duly by the advantages afforded in the public schools of the locality Soon after attaining his legal majority he there began farming operation in an independent way, and thus continued for five years, at the expiration of which he came to Nebraska and engaged in farming in York county Three years later he came to the part of old Cheyenne county that now comprises Garden county, and filed entry on a tree claim, in 1887, and a homestead the following year He perfected title to this land, upon which he has made the best of improvements including the erection of present attractive residence, on the homestead claim Energy and good management have brought to him prosperity in his farm operations, which include the raising of the crops best suited to this section and to the breeding and raising of horses and hogs as well as a due contingent of cattle He is now the owner of a valuable landed estate of twelve hundred and eighty acres, of which six hundred and sixty acres are maintained under effective cultivation, the remainder being range land He is a stockholder in the farmers' grain elevator at Lisco and also in the Farmers Mercantile Company at Lisco, which is his postoffice address, his home being situated about nine miles northeast of this thriving village Though he has had no desire for official preferment, Mr Allen has been liberal and loyal in support of those measures that have conserved the general wellbeing of his home county, and in politics he gives his allegiance to the Republican party

JAMES THEAKER WHITEHEAD, president of the Mitchell State Bank, of Mitchell, Scottsbluff county, of which he was the most prominent organizer, has since coming to the Panhandle been identified with the important financial enterprises of the upper Platte valley, and has also taken a place in the front rank of the men who are building up and developing this section of the state, as he is one of the officers of the Water Users Association which is placing water on the land, so that what was once known as the "Great American Desert," has become one of the most productive spots of the whole country Mr Whitehead has established a high reputation for ability, judgment and general acumen and since his connection with banking affairs his rise has been rapid, sure and consistent

The sons of Britain have always been men of thrift and industry and it is to "The tight little Island," that America owes the greatest proportion of the best element of the population Wherever they have settled the English have been found to be an asset to the community, their sturdy traits of racial character contributing to the locality's development Mr Whitehead is descended from a long line of forebears who played an important part in the mother country and after reaching our shores they became prominent men of their localities and to James Theaker Whitehead has been given this great heritage of many generations and he has run true to type for today there is no man in the upper valley who is doing more for the community than this unassuming banker

Mr Whitehead was born at Wataga, Illinois, November 17, 1867, the son of Abraham and Dora (Brunt) Whitehead, the former a native of England, who came to America with his parents James and Hannah (Theaker) Whitehead about 1840 They had been born, reared, and educated in Great Britain and after reaching man and womanhood met and were married, but as they desired greater advantages for their children determined to emigrate, and set sail for the new World, the land of promise to so many Europeans James Whitehead came west to Illinois soon after reaching the United States, took up land near Wataga, Knox county, and there engaged in farming until his death which occurred when he was about sixty-five years old Abraham Whitehead was reared on the farm in Illinois, attended the public schools of that locality and at the outbreak of the Civil War responded to the president's call for voluteer's to preserve the Union and enlisted in the Eighty-third Illinois Volunteer Infantry in 1861 He served until the close of hostilities and was neither wounded nor captured, though he participated in many severe engagement and battles, as he was a non-commissioned officer at the time he took part in the siege of Fort Donaldson Abraham Whitehead was a staunch believer in the tenets of the United Presbyterian church of which he and his wife were members, while his fraternal affiliations were with the Independent Order of Odd Fellows and the Masonic order Dora Brunt Whitehead was born in Columbia, Missouri, in 1850, the daughter of William Brunt, a native of England who came to the United States as a child, and served his adopted country as a captain in the Union army during the Civil War Mrs Whitehead was reared in Blackhawk county, Iowa, where her parents moved when she was a small child After the close of the war Abraham Whitehead returned to farming in Illinois, but like so many of the men returned from the army he was restless This is seen today in the men returned from France, so he left Illinois and in 1871 took up a homestead in Cloud county, Kansas, on which he made many good and permanent improvements but lost his life in 1886 when fifty-one years old, trying to stop a runaway team Mrs Whitehead survived her husband until April 8 1918, when she too was called to her last rest There were four children in the Whitehead family James T, the subject of this review, Reginald H, a resident of Oakland California, Edith E, who married Ray G Slater, and Abraham L, of Denver Colorado

James spent his boyhood on his father's farm, early learning habits of thrift, energy and the practical side of the farm industry, as he was the oldest of the children many tasks fell to him as soon as his age and strength permitted him to work Assisting his father on the farm during the summer time, he was sent to the district school nearest his home during the winter terms and laid the foundation of a good practical education so that when the family moved into Concordia, Kansas, when he was fourteen he was able to enter the high school from which he graduated When James was twenty he accepted a position with the Burlington Railroad, as he had learned telegraphy after leaving school Being assigned to work at Omaha, he became familiar with railroad work and also had an opportunity to learn of the business advantages and openings in Nebraska After a few years he came to Alliance and entered the lumber business, for at that early day Mr Whitehead believed there was a great future for a business man in the newly developing country here in the Panhandle For six years he remained in Box Butte county, thoroughly learning the lumber business and such was his application that his ability was recognized in lumber circles He was offered and accepted a much better position as secretary and general manager of the Forrest Lumber Company of Kansas City, Missouri, remaining with that concern from 1900 to 1906, when he sold his interests and returned again to the prairie country Coming to Mitchell Mr Whitehead interested several other moneyed men in the proposition of organizing and establishing a new bank to serve the city and contributing territory In 1907, the Mitchell State Bank opened its doors, with Mr Whitehead as one of the heaviest stockholders and the executive head of the institution The surety of his vision and judgment have been demonstrated in connection with the practical side of the banking business, for while yet a young man he is well entitled to classification among the efficient and progressive men of the business and financial circles of Nebraska Under his able guidance and policies the bank has gained the confidence of the people of Scotts Bluff county and the surrounding country, has a fine class of depositors, and the deposits have steadily and rapidly increased which shows in what high esteem the personnel of the banking house is held From first locating in the valley Mr Whitehead has taken an active part in the development of this section and he is one of the best known and prominent figures in the irrigation enterprises in the northwest He has been president of the North Platte Valley Waters Users Association for the past ten years, during the period when this country has been opened up to new methods of agriculture, intensive farming and the introduction of the sugar beet industry along the Platte river, so that he is a recognized authority on irrigation questions and as such was called to Washington, D C to confer with seven governors of the states interested in irrigation, being the representative from Nebraska at the conference in 1919 Mr Whitehead is essentially a self-made man, his clear vision, ability to see and grasp the business opportunity of the minute and turn it to advantage as well as his visualization of the future of this section have made him one of the unusual figures in the financial world of the middle west, where his far sight in finance and his integrity have given him a well deserved, well earned and yet enviable position In addition to his duties as banker, and irrigation manager, Mr Whitehead is also treasurer of the electric light plant of Mitchell, which under his management has been an efficient public utility

In 1897, Mr Whitehead married Miss Amelia Shetler, a native of Johnson county, Iowa, the daughter of Christian and Elizabeth Shetler, both natives of Pennsylvania, of German extraction Mrs Whitehead was reared and educated in Harvard, Nebraska Today the Whitehead farms, consisting of close to four hundred acres not far from Mitchell, are considered some of the most productive properties in the valley as they are both under water rights and so produce bountiful crops Mr Whitehead is progressive in his ideas as to farming and has introduced the latest and most improved methods on the farms, those advocated by the state and government experts and his places look much like the beautiful and scientifically run farms of the agricultural stations, and really are demonstrating places so far as the efficiency of their management and results are concerned In this manner Mr Whitehead has encouraged the other men of the locality to adopt measures that assure increased production

In political views Mr Whitehead holds with the tenets of the Republican party but has never aspired to public office beyond taking the part he believes a duty in local affairs, having served on the city council and as a member of the school board He is not bound by close party lines in local politics, believing that the man best suited to serve the people should be elected to office He is one of the public spirited men who is living his patriotism, took an active part in assisting the government in the prosecution of the World War, assisting in the sale of Liberty Bonds and also in raising money for the Red Cross He sets a high standard for an American citizen and lives up to it In Masonic circles Mr Whitehead stands high, belongs to the Scottish Rite Chapter, is a Shriner and also a member of the Independent Order of Odd Fellows With his wife he is a member of the Congregational church There are five children in the Whitehead family Mildred, who married H Roscoe Anderson, and now lives at Mission, South Dakota, Dorothy Elizabeth, Helen, James Theaker, Jr, and William Shetler, all at home The Whiteheads have a charming modern home in Mitchell, where they keep open house to their many warm friends

GEORGE O CURFMAN has been for more than a quarter of a century a resident of what is now Garden county, and this fact in itself proclaims him eligible for pioneer distinction in this section of the state He has been a forceful power in connection with industrial development and progress, has stood as an exponent of loyal and liberal citizenship, and has become one of the large landholders and representative agriculturists and stockmen of the county, has an admirably improved homestead farm, being situated fourteen miles northwest of Oshkosh, which is his postoffice address

Mr Curfman was born at Huntington, Mifflin county, Pennsylvania, August 27, 1855, and is a son of Joseph D and Eliza (Van Zant) Curfman, both likewise natives of the old Keystone state There the father was born in 1820, and, in 1857 he removed with his family to Illinois, where he became a suc-

cessful farmer and where he died in 1905 at the venerable age of eighty-five years, his wife having been seventy-five years of age at the time of her death, in 1903 Of their children, two sons and three daughters are living

George O Curfman was about two years old at the time of the family removal to Illinois, where he was reared and educated in Pike county Upon attaining legal majority he engaged in independent farm enterprise in that county, and he thus continued operations nine years He then, in 1885, came to Nebraska, and for the following eight years he made Clay county the stage of his agricultural activities At the expiration of this period, in 1893, confident of the splendid future of western Nebraska he came to that part of Cheyenne county that now constitutes Garden county, and here he has since resided continuously on the homestead which he took up at that time and which he has developed into one of the model farms of the county He has shown much discrimination and ability in his agricultural and live-stock enterprise, through the medium of which he has gained substantial success In 1903, he added to his land holdings, under the provisions of the Kinkaid act, with the result that his estate now comprises a thousand, one hundred and sixty acres, of which three hundred are available for effective cultivation, the remainder being used for pasturage and the raising of hay and other forage crops Since 1914, Mr Curfman has given special attention to the raising of hogs, in which department of his farm enterprise he now conducts an extensive business, besides which he keeps and average of about thirty head of cattle He is one of the stockholders of the Farmers Elevator at Lisco as well as of the farmers Mercantile Company in that village and the Garden Supply Company at Oshkosh

In politics Mr Curfam is found aligned in the ranks of the Democratic party, and fraternally he is affiliated with Oshkosh Lodge, No 286, Ancient Free & Accepted Masons

December 4, 1878, recorded the marriage of Mr Curfman to Miss Hannah Askew, who was born and reared in Adams county, Illinois and the supreme loss and bereavement in his life came when his devoted wife died in 1917, at the age of fifty-nine years Of their five children brief record is here entered William F, of Oshkosh, married Miss Maude McKonkey, and they have two children, Charles F, of Lewellen, Garden county, married Miss Mamie Westcott, and they have no children, Laura is the wife of Chester Saunders, of Oshkosh; and Clarence and Pearl remain at the parental home.

PATRICK DONNELLY.— The late Patrick Donnelly, whose death occurred at his farm home in Garden county, March 23, 1918, order his life on a high plane of integrity and honor, marked the passing years with worthy achievement and his kindly and genial nature gained and retained to him a host of loyal friends, so that his untimely death at the age of fifty-three years was deeply regretted in the county in which he had established his home and in which his widow still resides.

Mr. Donnelly was born in County Monahan, Ireland, November 6, 1864, and was there reared and educated. At the age of seventeen years he severed the home ties and set forth to seek his fortune in America, where he felt assured of better opportunities for achieving independence and prosperity through his own efforts. Within a short time after his arrival he made his way, in 1881, to Nebraska and settled in Saunders county, where for a few years he was identified with farm enterprise. He then yielded to the wanderlust, and traveled somewhat widely through Nebraska and Colorado, but eventually he returned to Saunders county, where he continued to be engaged in farming, until 1907, when he came to Garden county, and took a section of land, under the provisions of the Kinkaid act. He began, with characteristic energy and judgment, the reclamation and improvement of his land, and prior to the close of his life the results of his labors were manifest in no uncertain way, for he made good improvements on his farm, brought much of the land under effective cultivation and also met with marked success in the raising of cattle, horses and hogs. The Donnelly farm, upon which Mrs. Donnelly still maintains her home, is known as one of the model places of Garden county, its improvements including a good orchard, an attractive house, equipped with individual electric-lighting system, and other farm buildings of modern type. The death of Mr. Donnelly was one of tragic order, as he was killed in an automobile accident, at North Platte, when in the very prime of his strong and useful manhood. He was a Democrat in his political allegiance, though he was affiliated with the Populist party at the time when that organization was in the zenith of its power. His religious faith was that of the Catholic church, and he exemplified the same in his righteous life and his sympathy and kindliness in his association with his fellow men. His widow and children likewise are communicants of the Catholic church, the great mother church of Christendom.

In Saunders county, September 26, 1888, was solemnized the marriage of Mr. Donnelly to Miss Bridget McCarty, who was born and reared in this state, and who is sustained and comforted by the hallowed memories of their ideal wedded life, as well as by the love and devotion of their children. In conclusion is entered brief record concerning the children of Mr. and Mrs. Donnelly: Mrs. Geneva Laudeschlager, of Wahoo, Saunder county, has three children; Francis William and his wife reside at Lincoln, this state, and they have one child; Edward A., who has the active management of the old home farm, is married but has no children; Mrs. Ida M. Richards, of Oshkosh, has four children; Mrs. Agnes Proper, of Grand Junction, Colorado, has one child; Mrs. Anna McConnell, of Oshkosh, has one child; Mrs. Theresa Hassenter, of Oshkosh, has no children; Norah and Julia D. remain with their widowed mother; and Mabel is deceased.

FREDERICK D. JAMES, who during a long and successful career, has followed various occupations in several parts of Nebraska, is now a well known resident of Potter and Cheyenne county, although his operations are by no means confined to these boundaries. During his residence in this state, he has in turn been newspaper man, postmaster, carpenter, lumberman, salesman, cashier and manager of a large commercial enterprise, and in these various fields his versatility has assisted him to well deserved prosperity.

Mr. James is a native of Iowa, born at Denmark, Lee county, August 19, 1878, the son of E. B. and Ada (Mills) James, the former also a native of Lee county, born there in 1849, and was reared on his father's farm. Mr. James grew up in the healthy environment of the country, attended the public schools and thereby gained a good practical education which was of much value to him in later life. After attaining manhood's estate he engaged in farming pursuits, as it was the business with which he was most familiar and in which he attained marked success, due to perseverity, hard work and good judgment. He was progressive in his methods, and always took an active part in all civic and communal affairs serving for many years as county commissioner, as a member of the school board and other public offices of the locality. Today he is a hale, hearty old man who is enjoying his sunset years in a quiet manner knowing that he has played his part in the development of his country as a good citizen should, and now resides with his daughter in Alberta, Kansas Mrs James was born in Lee county, Iowa, was reared, educated and met her husband there After elementary school days were over she matriculated in the Denmark Academy and upon the completion of her course graduated She passed away in 1888

Frederick James was reared in Franklin, Nebraska, where he received his educational advantages in the Franklin Academy, graduating with the class of 1899, and soon after completing his education accepted a position in the office of the *Franklin Free Press,* where he remained about a year before taking up similar work for a paper in Nelson, Nebraska, but resigned upon being appointed deputy postmaster of Nelson After his term of office was over, Mr James was employed as a carpenter for three years, learning the practical side of the contracting business, but his business ability soon was demonstrated and he was called to Sioux City, Iowa to operate a lumber yard for an Iowa concern, leaving at the end of a year to go on the road as salesman for a sash and door factory and for five years made his home in Hastings, while on the road Then the Yost Lumber Company made him an attractive offer to come to them as cashier of their head office at Lincoln, Nebraska, which he accepted, remaining with them until March, 1917, when he became established independently in business, buying the Peterson and Sons lumber yard at Potter and since then has been secretary, treasurer and manager of the F D James Lumber Company of that city Mr James is a man well and favorably known in Cheyenne county, of which he has been a resident for three years He leads an exceedingly active business life, continually coming in contact with other staple, reliable men, and at the present time he is numbered with the county's best and most dependable citizens Since locating in Potter he has taken an interested and aggressive part in all the movements for civic improvement, being a director of schools and a member of the board For some time he has been a stockholder and director of the Liberty Oil and Refining Company of Cheyenne county and has had an important influence in shaping the policies of this flourishing concern In politics Mr James is an adherent of the Republican party, while his fraternal affiliations are with the Masonic order, the Modern Woodmen of America and the United Commercial Travelers

June 23, 1904, Mr James married Miss Inez Van Valin, a native of Nuchols county, who was reared and educated there and after graduation became a school teacher, but was engaged professionally but two years before her marriage There are two children in the family Edwin and Frederick D, Jr

W O WIELAND of Mitchell, Scotts Bluff county, is assuredly a man who well merits representation in this history, for he is not only a pioneer of this section of the state but has also had a full share of the experiences marking the change of the Nebraska Panhandle from a sparsely settled cattle country into a section notable for progressive citizenship, for well developed farm properties and for thriving and attractive cities and villages He is the owner of an extensive and valuable landed estate in Scotts Bluff county and is one of its substantial citizens He knows western Nebraska thoroughly and is one of its loyal and enthusiastic "boosters," as he may well be, for here he has achieved large and worthy success through his well directed enterprise, which needed for its consummation only such opportunities as are here presented Mr Wieland owns one quarter section of land, and most of this large tract is provided with irrigation He has made good improvements on the property, and is one of the men who are doing big things on a big scale in the furtherance of agricultural and live-stock industry He is also vice-president of the State Bank of Mitchell, one of the substantial financial institutions of Scotts Bluff county

Mr Wieland takes a due amount of satisfaction in reverting to the old Keystone state as the place of his nativity He was born in Center county, Pennsylvania, August 4, 1860, a son of Daniel T and Polly (Keller) Wieland Mr Wieland acquired his early education in the public schools of his native state and was twenty-eight years of age when he came to Nebraska and initiated his experience as a cowboy in the great western section that was then known principally as a cattle country His rise to prosperity and high standing in the community has been won entirely through his own efforts, and he has the greatest confidence in the splendid future of the section that has already become known as one of the most progressive and opulent divisions of our great state In national politics he is a Democrat, but in local matters he is not bound by partisanship and votes for men and measures meeting the approval of his judgment In the time-honored Masonic fraternity he has received the Thirty-second degree of the Ancient Accepted Scottish Rite, his wife and daughter hold membership in the Eastern Star

The year 1897 recorded the marriage of Mr Wieland to Miss Mary Shull, and they have two sons Leonard and William Arthur

CHARLES HIGH — Though he has not yet reached the psalmist's span of three score years and ten and still posseses a full amount of physical and mental vigor, Mr High has the distinction of being one of the pioneer settlers of western Nebraska The memory of this man compasses almost the entire gamut that has been run in the development of this section of Nebraska from a prairie wilderness to a populous and opulent district of a great commonwealth, and it is gratifying to him that he has been able to play a part in the civic and industrial progress and upbuilding of the state and Panhandle

Charles High was born in Illinois, July 30 1852, the son of Lemuel G and Sabina (Reed) High The father was a native of the Buckeye state where he spent his boyhood and early youth receiving his excellent practical education in the public schools of his home district and after attaining manhood's estate engaged in an independent business career as a farmer in Illinois, but died while in the prime of his life, at the age of thirty-two years, in 1860. Mrs High was a Hoosier by birth, but her family removed to Illinois when she was a small child and it was there that she was reared and educated, met and married Lemuel High She was a devoted wife and mother and when left a widow with small children shouldered the responsibilities of both father and mother to them She lived to have the satisfaction of seeing her son Charles develop into a fine man, as she lived to be eighty-seven years old, passing away in 1914

Charles remained at home with his mother on the farm after his father's death and while still a young boy began to assume what farm duties he could for his age and strength, and thus at an early age was a good practical farmer He attended the district school, thus gaining a good foundation for an education which he has steadily continued through wide reading of the best and most practical literature and the varied periodicals that bear on all subjects of life and commercial interests As soon as his years permitted he assumed management of the home place which he conducted for fifteen years for his mother Mr High read much of the west and determined to take of the advantages offered by cheap land in the newer country beyond the Mississippi

Leaving Illinois, the family came to Nebraska, locating in Dodge county, about forty miles west of the Missouri river. There they established a new home and Mr High again engaged in farming He devoted much time to the study of farm conditions in this section, the best crops for the climate and soil and this study was rewarded by bountiful crops He worked hard, was thrifty and before many years became a man of means with capital to invest As the western part of the state was becoming well known as one of the most fertile sections Mr High disposed of his holdings in Dodge county and came to the Panhandle in 1913, having great faith in the future of this part of the country. Locating in Potter Mr High purchased a half section of land, where he again engaged in business as an agriculturist, raised the land to a high state of fertility, made many excellent improvements and at the end of five years sold the farm at a good advance on the original investment, having earned it all with his own hands guided by foresight, initiative and by following modern methods in his management He had specialized in alfalfa for feeding, wheat, potatoes, hogs and cattle, all of which seemed to thrive in the Lodgepole valley under his skillful guidance Having been a land holder, Mr High was not contented to be without some landed interest and, in 1917, he bought a full section of land north of Dix, where he farmed two years During this time he greatly improved the estate, so that when he disposed of it in 1919, he did so with a handsome profit After retiring from active participation in active agricultural life he became interested in the Citizens Bank of Potter and purchased a small block of stock in that flourishing institution He has branched out into commercial life as his interests include holdings in the Farmers Elevators of Potter and Dix. For many years Mr High has taken an active part in communal and civic affairs, doing his part as a prominent and progressive citizen should, as he was county supervisor of Dodge county for six years and after locating in Cheyenne county was chairman of the Potter town board for two years In politics he is an adherent of the Democratic party while his fraternal affiliations are with the Masonic order and the Modern Woodmen of America

July 2, 1885, Mr High married Miss Susan Edwards, at Abington, Illinois She was a native of the Old Dominion, but her family removed from Virginia to Illinois when she was a small child so that she was reared and educated in the latter state and there met and married her husband She was a loving wife and mother and played a gallant part in establishing the new home in the west Mrs High passed away in 1912, at the age of fifty-three years, leaving a sorrowing husband and family Mr and Mrs High had three children Mrs Sabina Catherwood, lives at North Bend Nebraska, Mrs Bertha McChihan, is a resident of Potter, and Robert, who also lives in Potter, has four children of his own

JOSEPH HERBERT FOSTER, one of the prominent business men of Dalton, who has taken an able part in the development of the town and its varied enterprises, is today regarded as one of the rising men of this section in commercial affairs

He was born near Charleston, Coles county, Illinois, April 9, 1873, the son of Josiah H and Susan F Foster The father was one of the pioneer settlers of this region as he came to Nebraska in 1888, and took up a homestead nine miles northwest of Potter He had been a soldier of the Union Army during the Civil War and made proof of his land under soldier's rights, but later moved to eastern Nebraska and after a few years there returned to the old home at Villagrove, Illinois, where he now lives at the age of eighty-five years

Herbert Foster accompanied the family when they came to Nebraska in the early days and was educated in the common schools of Illinois and Nebraska after coming to this locality Later he attended the high school at Sidney When his schooling was over Mr Foster worked for about two years on ranches in the western part of the state, spent one more winter in school and then began teaching in Cheyenne county under Miss Mattie McGee, then county superintendent He followed this profession for eight years, doing ranch work and breaking horses in between times, then for eight years devoted all his time to ranch and range work before owning and managing a ranch of his own Mr Foster was well versed in the business and was successful in his enterprise, made money and entering commercial life became one of the organizers of the local corporation known as the "Dalton Co-operative Society," at Dalton, in 1909 The company started with a paid up capital of fourteen hundred dollar, it was well managed and at once began to make money and today has a capital of fifty thousand dollars, and in 1920 did a business of nine hundred and seventeen thousand dollars The company is an independent corporation doing business in lumber, grain, coal, hardware and livestock, a general business that is of great benefit to the sur-

rounding country Mr Foster also helped to organize the Great Divide Telephone Company (incorporated), the only incorporated farmers telephone company in western Nebraska, with a paid up capital of ten thousand dollars From this it will be gathered that he has been a man who has materially assisted in the development of his community At the time of the organization of the co-operative company, Mr Foster bought a large block of its stock and later purchased more, being today one of the heaviest holders He also invested in the stock of the telephone company and in the Farmers State Bank, of Dalton, one of the promising financial institutions of the Panhandle

For two terms Mr Foster was assessor in Davison precinct, Cheyenne county, but takes no active part in politics, though he votes as a Republican At the present time he is secretary-treasurer of the Dalton Co-operative Society, secretary of the Great Divide Telephone company treasurer of the board of education of Dalton, a member of the village council and is president of the Farmers State Bank As its executive head he has initiated policies that have placed the bank on a sound financial basis, won the confidence of the people and is regarded as one of the sound and reliable bankers of the western part of the state

December 4, 1905 Mr Foster married at Sidney, Nebraska, Miss Edith M Davison, the daughter of James and Mary Davison, who settled in Cheyenne county in 1885 One daughter has been born to this union, Frances, at home

Mr Foster is a member of the Knights of Pythias, at Dalton, of the Odd Fellows, of which he is secretary, belongs to the Dalton Country Club, and attends the Presbyterian church His entire time is devoted to his various interests, and today Mr Foster is considered one of the leading financiers and bankers in this part of the Panhandle and state

HUMPHREY SMITH was born at Waverley, Canada, at the lower end of Georgian Bay, in 1847 He was the son of Charles and Mary (Le Brash) Smith He was the oldest of a family of thirteen children, only four of whom are now living The others being Jesse, who lives on the old family homestead at Waverley, Frank, living at Bakersfield, California, a carpenter by trade, and Anna, the widow of William Hagerty, living at South River, Canada

At the age of ten years Mr Smith went to Lake Superior and was reared there by different people When he was twenty-one he took out citizenship papers in the United States and came west to Scandinavia, Kansas, where he remained two years. When the Black Hills country was opened he was back at Lake Superior again, and at that time he came to Anselmo, Nebraska. The railroad was then being built, and he opened a store and hotel at that place and remained there until the drouth of the early nineties. He then went to the Black Hills, and, in 1907 came to Bayard and bought a lumber yard. He has now practically retired from active business.

While on Lake Superior he was married to Emily Hart, and to them five children were born, all of them now living. They are: Eugene, who farms and conducts a barber business at Portland, Oregon; Charles, living at Paulson, Montana; Harvey, who was a captain in the United States army and served in the Philippines in the Spanish War and was a training captain at Portland, Oregon, during the late war, and is now in the insurance business; Daisy, now Mrs. Fred Fleming, of Greeley, Colorado, and Silas, a civil engineer of Salt Lake City, Utah.

Mr. Smith was married a second time in 1901, to Ida Reed, a native of South Dakota, and they have two children, Sidney and Herbert, both of whom are at home.

Mr. Smith is a member of the Methodist church and a Republican in politics. He has had a varied life of pioneering and adventure. He has been successful and is now able to retire from the active struggle and enjoy a well earned rest in the evening of life. He is respected by his neighbors and friends as a man of honorable character and good judgment.

ROBERT OSBORNE, who is now living practically retired at Gering, judicial center of Scotts Bluff county, is consistently to be designated as one of the representative pioneers of this section of Nebraska to which this history is devoted. As a Scotchman he possesses the admirable traits of character for which the sons of bonnie old Scotland are famous, and his energy and ability have been conclusively proved by the worthy success and prosperity he has achieved through his association with farm enterprise in western Nebraska. He has accumulated and still owns one of the extensive and valuable landed estates of Banner county, and remained on this fine ranch until 1914, when he established his residence at Gering; his retirement having been prompted alike by his somewhat impaired health and specially by his desire to give to his children the excellent advantages of the Gering schools.

Mr. Osborne was born in Ayrshire, Scotland, on March 29, 1861, and is a son of Robert and Mary (Hamilton) Osborne, the former of whom died in Scotland and the latter of whom came with her children to America and located in Illinois, where she passed the remainder of her life, the father having been long identified with the coal-mining industry in Scotland.

Robert Osborne acquired his early education in the schools of his native land and was an ambitious youth when, in 1881, he came to the United States and established his residence in Illinois. Thence he later removed to Kansas, and, in 1887, he came to western Nebraska and numbered himself among the pioneers of the present Banner county, where he secured and duly perfected title to a homestead and a pre-emption claim. Here he lived up to the full tension of the pioneer days; gave his zealous labors to the development and improvement of his land and eventually accumulated in that locality his present well improved and valuable ranch of fifteen hundred and twenty acres, the same having continuously been devoted to diversified agriculure and stock-raising.

As a citizen Mr. Osborne has exemplified distinctive loyalty and public spirit, and he served four years as county assessor of Banner county, later having served for a similar period as precinct assessor in Scotts Bluff county. In politics he maintains an independent attitude and gives his support to men and measures meeting the approval of his judgment. In connection with his pioneer experiences in Banner county it may be noted that he there dug the first well on Swede Point.

The marriage of Mr. Osborne occurred in 1883, when he wedded Miss Margaret Draper, who was born and reared in Illinois and who, traveling by train, arrived in Banner county before he did, as he made the overland journey with team and wagon. For nearly two weeks after his arrival in this section of Nebraska he was unable to find trace of his wife and children, but he eventually discovered them in Banner county, where the reunion was a joyful one, as may well be imagined. In conclusion is given brief record concerning the children of Mr. and Mrs. Osborne: Robert L. has general supervision of the old homestead ranch in Banner county; Martha is the wife of F. F. Stafford, of that county; Lewis likewise resides in Banner county, where he is successfully identified with farm enterprise; Belle is the wife of Arvil Barstow, of Scotts Bluff county; Clarence and Hugh W. are associated with their older brother in the man-

agement of the old home place, and Florence, Mary and Dora, who remain at home are attending the public schools of Gering

EDWARD W SAYRE — The pioneer families of the Panhandle who played their parts in the vital drama that has turned this section of Nebraska into a paradise for the homeseeker, developing the wild rolling prairie into one of the richest farming sections of the western hemisphere, dotted with thriving communities, have reason to hold themselves responsible for much of the present day progress and prosperity, for it was their leadership and courage that opened what was for so many years known as the "Great American Desert," to settlement While many of those who experienced the actual hardships and privations incident to the early days have passed away, there still remain many, who, through sheer force of will and determination, and the necessary energy, brought out of primeval conditions what have become twentieth century actualities Among these is found Edward Sayre, who came with the homesteaders and took a pre-emption in old Cheyenne county in the year 1888

Mr Sayre was born in far off India, in 1865, being the son of Edward H and Mary C (Hulfish) Sayre, the former born in South Hampton, England The father was a Presbyterian minister who devoted his life to God's work as a missionary, being sent to India, where his son, Edward junior, was born and received his early educational training Reverend Sayre was a fine student and after taking up his residence in India devoted much study and time to the languages of that country, becoming one of the recognized authorities of the Hindustan language from the literature of which he made many and valuable translations into English After serving many years in the foreign fields as a representative of the Presbyterian church Edward Sayre retired from the ministry and is now spending the sunset years of his life at Gering

Edward W Sayre, after finishing his elementary education, took a four year's course in the high school and then spent one year in the Davenport College, at Davenport, Iowa, thus laying the foundation of the exceptionally fine education to which he has ever since been adding by his wide reading of the best English literature, a deep study of the subjects in which his business interests have led him and by keeping abreast of the present day trend of events in the periodicals

Mr Sayre was only nineteen years old when he began his mercantile career in Illinois In July 1888 Mr Sayre bought the mercantile stock of Kiefer, Hastings & Company at Gering, which he ran until thirteen years ago when he became the first merchant of Morrill Mr Sayre was the first man to build a brick store in the town as he saw it had a future and his faith has been justified

More than ten years ago he bought a quarter section of land in section 16-23 59 which he has since managed Mr Sayre was one of the first officers of the irrigation company, of which he was a promoter, he has served his county in the capacity of treasurer for two terms, and for some years was president of the village board of Gering Mr Sayre's fraternal relations are with the Masonic order, of which he has been a member for many years Politically he casts his vote with the Republican party, of which he is a staunch supporter, though his influence is given to the best man for office in all local elections

In 1889 Mr Sayre married Miss Margaret Wood, who was born and reared in Iowa, where she received her education Nine children came to bless this union Edward D, who served in the Ninety-first, the Wild Cat Division, during the war with Germany, Ruth, the wife of William S Proudfit, Kenneth Doris Faith Harry, Margaret, Elizabeth and William all of whom are still at home with their parents All of the children have been given good educational advantages, Mr and Mrs Sayre have proved themselves good friends to the schools, and are the advocates of all improvements that tend to the uplift and development of the civic and communal life of their section

JAMES O KANE, deceased, one of the pioneer residents of Cheyenne county who took a prominent part in the business and agricultural development of this section and a man to whom pioneer honors are due was a constructive builder here He not only was one of the men connected with the railroad, that opened up the county to settlement but he later became a prominent and successful business man, who took part in the civic affairs of Cheyenne county and Sidney, always assisting in all movements for the general good of his country and community

Mr O'Kane was born in Indiana, January 11, 1848, the son of Joseph C and Mary (Davis) O'Kane, to whom were born eight children Five of them are living Mr O'Kane was reared and educated in his native state, but like so many young men of that period, believed there were more opportunities for a man in the new western country, and in 1873 came to Sidney, where he was employed in the freight office of the Union Pacific Railroad for a number of years Following this position he was engaged in other railroad work for a time, but saw openings worth while in stock raising, which was at its height at the time, and invested in that business He was a man of excellent business ability, made money on his cattle in the days of the open range, became recognized as one of the prominent live stockmen of his day and was far-sighted enough to dispose of his interests when the open range days were over and thus made money by selling at an opportune time After this he bought and sold property in Sidney, he knew this county and the country of the Panhandle well and was able to give his clients the best of advice and the benefit of his knowledge and experiences The real estate business was a success and Mr O'Kane handled large land deals involving many thousands of dollars He became one of the prominent figures in the financial circles of Sidney and western Nebraska, a position he held to the time of his death

In 1878, Mr O'Kane married Miss Bridget Brown, a native of Ireland who came to the United States in 1870, and to them were born six children Mayme, who married Fred Bard, of Aberdeen, Washington, Francis J, who lives at Casper, Wyoming, James, of Great Falls, Montana, Gertrude, the wife of E S Stokes, of Sidney, and two who died in infancy

Mr O'Kane died December 21, 1902 He was a member of the Catholic church and a Democrat

CHARLES M HADLER, one of the well known and successful business men of Sidney, who is engaged in handling real estate on an extensive scale and also owns an abstract office, is a leading spirit in the development of the town and Cheyenne county

He was born in Elko Nevada, May 14, 1875, the son of James S and Mary E (Byerley) Hadley, the father being a native of Indiana and the mother of Ohio To them nine children were born William H, who lives at Fort Russell, Wyoming, Samuel A, lives at Loveland, Colorado Clarence E, resides at Steamboat Springs, Colorado, Charles M, of this review, Florence E, married Dr B W Frazey and lives in Cheyenne, Wyoming, James F, lives in Cheyenne Wyoming, Mary

E, is the wife of John Bowman of Sidney, Bertha L, married Nathan Mack, of Wymore, Nebraska, and Albert J, of Cheyenne, Wyoming

James S Hadley was a contractor who engaged in that line of business for a number of years before moving to Cheyenne county in 1886 He took up a homestead here on Pumpkin creek were he spent the remainder of his life Mr Hadley died in Sidney May 10, 1905, being survived by his wife until May 31, 1918 He raised stock after coming to the Panhandle, and was an active man to the time of his death Mr Hadley was a member of the Odd Fellows, the Methodist church and a Republican

Charles M Hadley was reared and attended the public schools of Cheyenn county, when old enough he became a cow boy and engaged in handling cattle for a few years, then entered the official life of the county as assessor in 1903 He served four years, during which time he became familiar with the country and business Upon leaving office Mr Hadley opened a real estate and abstract office where he has won a high reputation as a business man and success from a financial point of view He is popular in the business circles of Sidney, being progressive in ideas and a booster for this section of Nebraska Since locating in Sidney Mr Hadley has taken an active part in all civic and communal life and helped build up the town He is a Democrat in politics

ANDREW K GREENLEE, a pioneer settler of Cheyenne county who has played an important part in the development and settlement of this section of the Panhandle, is the principal owner of the largest department store in western Nebraska, and vice-president of the First National Bank of Sidney Mr Greenlee has taken an active part in all the changes that have taken place here and is one of the constructive men who have made the present prosperity possible He was born in Crawford county, Pennsylvania, October 25, 1860, the son of Albert Keith and Martha (Barnes) Greenlee, both natives of Pennsylvania To them were born three children Andrew, of this review, Ernest, who lives in Fostoria, Ohio, and Ralph, deceased

The father was a successful farmer who was gaining a high standing as an agriculturist as a young man He died when Andrew was seven years old and was survived by his widow who reared the family Mrs Greenlee lived until June 30 1918 The father was a Re-

MRS. JULIA SCANLON AND HUSBAND

publican in politics and a member of the Baptist church, his wife was a member of the Methodist church

Andrew Greenlee attended the elementary public schools, then entered the normal school at Edinborough, Pennsylvania, later taking some courses at Valparaiso Normal School, Valparaiso, Indiana. Soon after leaving that institution he came west to York county, Nebraska, to engage in the milling business, which he followed one year, then moved on west to Cheyenne county when this part of the state was little settled. Mr Greenlee took up a homestead here in 1885, which he still owns, engaged in farming on his land which was seven miles southwest of Sidney and remained there until 1893, when he moved into the town. For a time he worked for a merchant, learning the mercantile business thoroughly. He then started a store of his own, later entering into partnership with H P Benson. Their trade was good from the first. They enjoyed the business and weathered the hard years of the early '90s, remaining the leading merchants of Sidney from first establishing themselves. Late in 1908 the business of F L Van Gorder, a merchant of Sidney, was merged with that of Greenlee & Benson, and early in 1909 the business was incorporated under the name of The Sidney Mercantile Co. F W Vath, F L Van Gorder and A K Greenlee, incorporators; H P Benson later purchased a block of shares

The business continued to prosper and The McLernon corner which had been purchased, became too small. At this time Mr Greenlee turned in the Urbach property adjoining and in 1916 both of these locations were covered with the present modern and substantial building, than which there is none better in the western part of the state. The second story of this splendid brick structure is now leased to, and ably conducted by Voclav, Kline and wife

Following the death of Mr Benson in 1909, that of Mr Vath in 1918, and the retirement of F L Van Gorder in 1920, Mr Greenlee purchased the holdings of each so that now the entire business is owned by the Greenlees, (named above) and Glen D Van Gorder, a promising young business man

They have made a department store of the business and now carry the largest stock in western Nebraska. Mr Greenlee is a keen, far-sighted business man who has dealt squarely with all his customers, has the confidence of the people and is today rated high in the mercantile business of the state. He has not confined his energies to one line of endeavor but has entered the banking business, owns a block of the stock of the First National Bank and is vice-president of that flourishing institution. He is constructive as a banker and many of the successful policies of the bank have been adoped at his suggestion. He still is the owner of a large amount of land in the county which he believes is a valuable asset

May 20, 1888, Mr Greenlee married Miss Elizabeth McAlester, a native of Ireland, and they have had five children—one died in infancy, Martha R, Catharine, Albert D, and Roy E

Mr Greenlee is a member of the Masonic order, of the Odd Fellows, the Rebeccas, the Workmen and of the Episcopal church. In politics he is a Republican. All the years of his residence in Cheyenne county and Sidney, Mr Greenlee has taken an active part in civic affairs and given of his time and energies to assist in the development of this section. He has been a prominent figure in many good works, including the activities of the county in the World war

DENNIS JOSEPH SCANLON, deceased, one of the pioneer settlers of Cheyenne county and Sidney, was a man who did much for the opening up and development of this section of the county and the town, as he was the first druggist in all this western country and as there were few doctors here then Mr Scanlon was relied upon by the people to prescribe for them when a physician's services could not be obtained. He did not confine his energies to his drug house alone but became one of the first and ablest bankers of the Panhandle where his constructive talents, high integrity and foresight assisted many homesteaders in settling here

Mr Scanlon was born in County Tipperary, Ireland, February 9, 1849. He was reared and educated in his native land, but like so many young, ambitious Irishmen, wished to get ahead in the world and came to the United States in 1873, to seek his fortune. Soon after landing here he enlisted in the army where he served ten years as a hospital steward. It was while with the United States forces that Mr Scanlon was stationed at Sidney Post, now Sidney in 1881. He liked this western country, believed there was a great future for this section and when his last term of enlistment was finished, he settled in Sidney, opening the pioneer drug store here in 1885, which served a wide territory. People came to have great confidence in Mr Scanlon as his long term as hospital steward had given him practical knowledge of both drugs and medical practice and for years he served the people in this dual capacity and it was fortunate that the early settlers had such an able man in the drug business for there were many times when he saved lives in emergencies. The value of this work can not be estimated but was shown by the residents of this section in their love and faith in him. After Sidney began to grow and the country to settle up, Mr Scanlon saw opportunities in financial circles and was one of the men who instigated and organized the early bank which later became the First National Bank of today. He bought a large block of the stock of the bank and became its first president, an office he held to the time of his death. Mr Scanlon developed marked talents as a financier was constructive in his policies, won the confidence of the people which is so necessary in the furtherance of banking business and became one of the prominent figures in the business circles of the Panhandle, where he was always helping men to become established. It was through him that many homesteaders were able to locate in Cheyenne county, prove up and become established farmers and ranchers. The amount of good which Mr Scanlon did in this manner will never be known but he must be considered one of the great developers of his day. When he died it was a loss to the community and the county, as well as the business world.

March 21, 1880, Mr Scanlon married Miss Julia Conelly, a native of Ireland, who came to the United States in 1872. She was married in Pagosa Springs, Colorado, and in May, 1881, came to Sidney, Nebraska, and spent the remainder of her life here. Mr Scanlon died August 10, 1912, after living more than a quarter of a century in the Panhandle and Cheyenne county. He lived to see all the marvelous development here, and was a part of it, for he took an active part in all civic affairs and those which have made Sidney the city of prominence it is today. He was a member of the Ancient Order of United Workmen, belonged to the Catholic church and was a Republican.

JAMES W JOHNSON, vice-president and cashier of the American State Bank of Sidney, comes of a banking family, as his father was a financier all his life. Mr Johnson has taken an interest in Cheyenne affairs for many years, long before he came here to live, and today is one of the prominent and successful bankers of the Panhandle. He was born in Iowa, December 26, 1878, the son of Frank C and Jane (Armstrong) Johnson the former a native of Canada, while the mother was born in Ireland. There were nine children in the Johnson family, of whom three are living. Sarah, married Spencer M Brooks and lives in Omaha, Ruby also lives in Omaha, and James W, the subject of this review. Frank C. Johnson engaged in banking when a young man, became associated with other prominent men in financial affairs in Omaha, where his rise as a banker led him to become interested in other banks, and he was president of the Midland State Bank and the Citizens Bank of Omaha. He severed his connections with the banks in Omaha but continued in the banking business in Iowa, also becoming a large landholder and the owner of some fourteen hundred acres of fine farm land there. He was a member of the Odd Fellows lodge, was a Democrat in politics, and a communicant of the Methodist church. Mr Johnson died in 1896, being survived by his widow until 1913.

James W Johnson was reared in Omaha and attended the public schools, graduating from the high school in 1896. Soon after this he entered the Omaha National Bank where he remained four years, learning the practical side of banking business. Following this Mr Johnson accepted a position with the First National Bank, of Omaha. During this time he had become a proficient bank man, had studied the various branches of finance and became vice-president of the bank at Spearfish, South Dakota, where he put his excellent policies into practice, gaining a reputation as a constructive man of affairs, whose policies were progressive yet conservative. His reputation as an able banker became known and Mr Johnson was offered and accepted the office of treasurer of the Guarantee Trust Company of Chicago. For four years he held this position of responsibility and trust, and by his ability assisted materially in the management and policies of the company. For some time Mr Johnson had been identified with interests in Cheyenne county, as he became interested in banking business here in 1902. He made several trips a year to the Panhandle, oftentimes four, and in 1918, came to Cheyenne county and Sidney to live. He saw that there was a future in the banking circles of the Panhandle which were so rapidly developing. Mr Johnson bought a large block of stock of the Sidney State Bank, after coming west, at once introduced many good ideas in its management, as he was vice-president. The bank was consolidated later with the American Bank and Mr Johnson assumed the duties of vice-president and cashier. Under his able guidance the bank has grown, is

doing a fine business and today has a capital stock of $100,000 and surplus of $25,000 The American Bank is the largest bank in the state west of Kearney, which speaks well for the men who are in control and manage it The deposits in 1921 were $1 000 000

In 1912, Mr Johnson was married to Miss Helen Fowler, who was born in South Dakota, and they have two children, Virginia and Helen Jane

Mr Johnson is a member of the Benevolent and Protective Order of Elks and a 32d degree Mason and Shriner He was county chairman of the War Savings Stamps drives and took an active part in assisting the Government in its prosecution of the war He is a Republican and has been City Treasurer for three years He has severed active connection with the American Bank and is now in the insurance and farm loan business in Sidney, conducting the largest business of its kind in Cheyenne county

ALBERT N MATHERS — The business acumen that serves to make Albert N Mathers a successful banker, extends to the many lines of activity in which he has been engaged for years, some of these being of a private nature, while others are of such public importance that he is welcomed as an assistant adviser by those who have high government problems to solve Mr Mathers is a Nebraska man and no one can question his devotion to the best interests of her people His business capacity, his matured conservative judgment, his sterling honesty and high sense of personal responsibility are known and appreciated in many sections and particularly at Gering, where he is president of the Gering National Bank

Albert N Mathers was born in Otoe county, Nebraska, February 9, 1882, the only son of John C and Mary (Cowles) Mathers, the former of whom was born in Alsace Lorraine, France, the latter in the state of New York They were married in the city of Chicago Illinois and Mrs Mather's death occurred in 1911 They had but two children, Carrie and Albert N The former is the wife of John S McKibben who is in a wholesale business at Kansas City where John C Mathers now lives retired When six years old he accompanied his parents to the United States, and was reared in Illinois He enlisted in the Union army at the beginning of the Civil War, leaving good farm prospects behind, and served four years and three months as a member of the Fourth Illinois Cavalry After the war ended he came to Nebraska acquired land which included the

town site of Douglas, was one of the organizers of a bank at that place and in numerous ways became prominent in Otoe county In 1888, he moved to Douglas but for some time afterward continued interested in farming

Albert N Mathers completed his education in the University of Nebraska from which he was graduated in 1906 For one year afterward he taught school for three years following that was engaged in a mercantile business at Douglas and then went to Kansas City In 1911 he came to Scottsbluff Nebraska, as assistant cashier in the Scottsbluff National Bank, in which institution he is still interested He became interested quite early in Gering with the result that in 1913 he bought a controlling interest in the Gering National Bank, became its president and has ably directed its affairs ever since The latest bank statement shows the capital stock to be $40 000, with deposits $520,000, the deposits having increased under Mr Mather's administration to that amount from $52,000 when he took charge Mr Mathers has taken pains to encourage home enterprises and with other concerns, he is interested in the Scottsbluff Creamery Company and other corporation enterprises at both Gering and Scottsbluff He owns eight improved, irrigated farms in the county and is a heavy sheep and cattle feeder He also has much city property, including lots and business buildings and is a booster for the city and county He is associated with every enterprise for the good of the community In banking circles his standing is high and for a number of years he was a member of the board of directors of the Nebraska Bankers Association and during one year of his directorship, was president of the board He is a member of the Federal Reserve board, was one of the organizers of the Scottsbluff country club and has been director and vice president since its organization

In 1907 Mr Mathers was united in marriage to Miss Fern Johnston, who was born near Nebraska City, Nebraska Her father Albert Johnston, was a prominent stockman in Otoe county for many years, featuring Shorthorn cattle and exhibiting at many state fairs He owns three valuable farms near Gering to which city he retired in 1916 Mr and Mrs Mathers have two children Albert Lowell, who has reached his tenth year, and Elizabeth, who is four years old

In politics Mr Mathers is a Republican He was elected first mayor of Gering in 1916, the fine public school building was erected, the water works and sewer system were established during his administration For a number of years he has been chairman of the board of education He has been an active and efficient official of the Gering-Fort Laramie government ditch and worked hard to get the land signed up, making two trips to Washington, D C, to present the matters to legislators, at his own expense He was elected treasurer of the irrigation company, an office which he has filled to the entire satisfaction of the association Recognizing his fitness for such public service, Governor Neville named Mr Mathers as one of the committee of five representative men to go to Washington to confer with Food Commissioner Hoover respecting the serious condition confronting sheep and cattle feeders in the state He has taken an active part in government war work, was vice chairman of all four bond floatations, chairman of the publicity committee of the county, local chairman of the War Savings Stamps issue and a director of the Red Cross Wherever he was needed he responded to the call and was one of the most effective four-minute speakers on the Nebraska list Mr Mathers has a beautiful home at Gering He was reared in the faith of the Methodist Episcopal church

JOHN T McINTOSH, postmaster of Sidney and one of the younger business men of Cheyenne county who has made a name for himself in financial circles, was born in Potter, April 9, 1871, the son of James J and Mary (Heelan) McIntosh, the former a native of Canada while the latter was born in Ireland The father was a pioneer settler of this county, a prominent man here whose sketch appears in this volume

John McIntosh was reared in Cheyenne county and received his education in the public schools of Sidney After graduating from the high school he entered Notre Dame University, of Indiana, took a special commercial course and after receiving his degree returned to Sidney His father was one of the influential men who started the American State Bank, and the young man went into that institution of which his father was president This was just a few months after the bank was started Mr McIntosh proved to be an efficient official, holding the office of assistant cashier He learned the banking business from the ground up and soon was made cashier, holding office eleven years, and it was due to many of the policies that he and his father instituted that the bank was so successful He became well and favorably known in banking and financial

circles in the Panhandle as a sound, conservative banker Resigning from the bank, Mr. McIntosh accepted a position with the Union Pacific railroad, where his knowledge of finance was of great value, and he gained as high place with that road as he had with the bank, and only left because of a more advantageous offer from the Burlington Railroad There he remained until appointed postmaster in 1911, an office which he has since held, to the entire satisfaction of the residents of Sidney Today Mr McIntosh is regarded as one of the substantial and reliable business men of this section, and is one who always helps in the upbuilding of Cheyenne county and the Panhandle which has always been his home

October 25, 1899, Mr McIntosh married Miss Mary McFadden who came to Cheyenne county with her parents when six months old and was reared and educated here Two children have been born of this union : James Clifton and Jean Heelan, both at home

Mr McIntosh is a member of the Knights of Columbus, belongs to the Catholic church and is a Republican He is one of the men who believes in the future of the Panhandle and like his father, is a constructive builder when it comes to county and civic affairs

JAMES L McINTOSH is a native son of Nebraska He resides at Sidney, the county seat of Cheyenne county, where he is engaged in the practice of law, and also in the real estate business In his professional business alliance he is senior member of the firm of McIntosh & Martin, and formerly was associated in law practice with George W Heist, at Sidney He has proved himself one of the most aggressive and loyal supporters of progressive movements that have conserved the civic and material advancement of Sidney and Cheyenne county and had the distinction of serving as the first mayor of Sidney He represents most fully what is known by the characteristic American expression of "live wire," and is one of the leaders in community thought and action at Sidney He is a stalwart in the local camp of the Republican party, is affiliated with the Knights of Columbus and both he and his wife are communicants of the Catholic church

Mr McIntosh was born in that part of Cheyenne county, Nebraska, that is now comprised in Kimball county, near the old town of Antelope, now Kimball, and the date of his nativity was April 11, 1874 He is a son of James J and Mary Helen McIntosh, and on other pages will be found a review of the career of his father, so that further data concerning the family history are not demanded in the present connection. After due preliminary discipline in the schools of his native state James L. McIntosh entered Notre Dame University, Indiana, and after taking an academic course there he was matriculated in the law department of the University of Michigan, at Ann Arbor, where he was graduated as a member of the class of 1893, with the degree of Bachelor of Laws. Upon his return to Nebraska he was admitted to the bar of his native state and forthwith engaged in the practice of his profession at Sidney where he formed a partnership with Henry St. Rayner who is now a resident of Portland, Oregon. Later he was similarly associated with George W. Heist for some time, and his present professional and business coadjutor is Paul L. Martin. The firm has a substantial and representative law business.

In 1898 was solemnized the marriage of Mr. McIntosh to Miss Rose Pavla, a native of Iowa. Mr. and Mrs. McIntosh have three children. Mary, who won a scholarship at the time of her graduation in the Sidney high school. McKinley, who is a member of the class of 1920 in the high school, and James G., who is attending the public schools of Sidney, the family being one of marked popularity in connection with the representative social activities of the community.

MYRTLE J. LANCASTER, treasurer of Cheyenne county is one of the native daughters of Nebraska who has made a marked success in business life through her own ability and high standards. Miss Lancaster was born in York county, Nebraska, the daughter of John P. and Madge E. (Miller) Lancaster, the former a native of Illinois and the mother of Pennsylvania. They were married in Illinois and became the parents of four children Clyde W., who lives in California, William L., deceased, Myrtle, of this review, and Helen, who married Mervin E. Oliver, and lives in Sidney.

Mr. Lancaster was a farmer who came to Nebraska at an early day, locating first in York county and then coming farther west in 1886, to settle here in Cheyenne county. He took up a homestead, proved up on it and engaged in farming until 1891, when he went to Colorado and became a gold sampler at Victor, Colorado, for the Taylor and Brunton Sampling Company, later returning to Sidney he became a railroad man and is now connected with the Union Pacific Road.

Miss Lancaster received her education in the public schools of Sidney where the greater part of her life has been spent. After finishing her education she entered business life, being bookkeeper for some years for a firm in Sidney. Later she went into the county offices where she showed marked ability in handling the work, became familiar with many branches of county administration and was appointed deputy under W. R. Wood, county treasurer and S. Fishman, county treasurer. After serving under them she ran and was elected treasurer of Cheyenne county in 1918, the first woman to ever hold that office in Cheyenne county. Miss Lancaster was thus thoroughly acquainted with the administration of the county funds and her election proves in what high standing she is held by the people of the county. For nine years she has been working in some official capacity in the court house and is regarded as one of the most able and efficient officers that the county has ever had.

Miss Lancaster is a member of the English Lutheran church, of the Rebeccas and is vice-president of the State Association of County Treasurers.

CLAYTON RADCLIFF county attorney of Cheyenne county, is a native son of the Panhandle, born at Ogallala, March 16, 1889, the son of Mack and Bertie (Gasp) Radcliff, to whom were born three children. Harriet, of Sidney, Clayton, of this review and Anna, also living in Sidney. The father was a native of Ohio who ran away from home when a boy of fourteen years and went to Texas, staying there five years. When Mr. Radcliff came to Cheyenne county in 1872, with a "trail herd" from Texas, he was about nineteen years old. He found work as a cow boy which he followed for some time, then engaged in ranching business for himself, at the Ogahala ranch. Later he took up a homestead and the log cabin which he built is still standing today with the loop holes through which to fire at the Indians, when the whites were attacked. There is a picture of this cabin in the history of the Panhandle. Sidney, the nearest postoffice, was thirty-five miles distant and the nearest neighbor was twenty miles away, which shows the condition of the country at the time. Mack Radcliff still owns and operates this ranch consisting of about seven thousand acres, located on Cedar creek in Morrill county. Sidney was the great town of the Panhandle at that time the outfitting place for the northwestern section of the state and also for the Black Hills. Mr. Radcliff was foreman of the Ogallala Cattle Company for a number of years, one of the well known cattle men of this section. Clayton Radcliff's grandmother on his mother's side ran the first hotel in Ogallala and at the time of one of the Indian uprisings was the only woman left in the town. Just after the Indians had left Mack Radcliff came very near capture by the Indians when he rode to the Keystone ranch on North Platte river to warn them of the coming Indians and a few days later Mr. Radcliff acted as guide and scout for the cavalry from Fort Sidney, who pursued the Indians into the sand hills. The following winter Mr. Radcliffe was in Fort Robinson—1878—when the Indians made their last final and fatal effort to escape, which resulted in the death of most of them when they were held as prisoners at Fort Robinson. The story of this Indian fight is well told in Edgar Bronson's Cow Boy Life on the Western Plain. The following year Mr. Radcliff, as foreman for Bill Shadley, trailed the first herd of Mormon cattle from Utah to western Nebraska.

Clayton Radcliff was educated in the public schools of Sidney, graduated from the high school and then entered the State University at Lincoln where he took a course in the law school, receiving his degree in 1913. He was at once admitted to practice in Nebraska and came back to Sidney to open a law office where he has since been engaged in his professional work. Mr. Radcliff was elected county attorney in 1914 and has been re-elected to that office and is serving at the present time, which testifies to the faith and confidence which the people have in his ability.

In 1918, Mr. Radcliff married Miss Merna Swartzlander, a native of Iowa whose father is a doctor. Mr. Radcliff is one of the younger members of the Cheyenne County Bar who bids fair to go far in his chosen profession. He is a Mason and an Elk.

NELS JOHNSON, represents the progressive, sturdy type of citizens which has been furnished Cheyenne county by Sweden. Mr. Johnson had had some experience in agricultural industry in his native land, but farming is conducted in a much different manner in Europe than in America and the crops grown in that country are of another character so that he was forced to work out his own problems. He possessed willingness, inherent ability and a determination to succeed, and so well has he directed and managed his affairs that today he is one of the largest landholders in Cheyenne county; is accounted one of the substantial and progressive citizens of his community and held in high esteem by his business associates and the many friends he has made in the land of his adoption.

Nels Johnson was born in Sweden September 23, 1862, the son of Mr. and Mrs. Jons Monsen, who were natives of that land where they were reared, educated and married. The father was one of the successful small farmers of Sweden, who are known the world over for thrift, industry and the good living they are able to make of the small farms that lie between the mountain ranges, for the land of the valleys is very fertile and the husbandmen of this far northern country are able to win a comfortable living on an amount of land that is inconsiderable in our own country of broad prairies. Nels grew up on his father's farm, attending school, which is under government supervision in Sweden, until he had gained a good practical education. He assisted as a boy with such tasks as were not beyond his strength, and as he grew to manhood assumed more and harder work. He was an ambitious youth, and had heard from many of his returned countrymen of the opportunities to get ahead in America, so determined to come here for that purpose. In 1882, while not yet of age, he embarked for the New World, the "Land of Promise" which in reality has become one to him. Landing in Portland, Maine, he came west and was employed on a farm in Illinois, for five years before he accumulated enough capital to really engage in independent business as a farmer. During this time Mr. Johnson had made inquiries about the different sections of the country, where the best land was to be obtained and after long consideration chose Nebraska for his permanent home. Coming to the Panhandle he located in section 8, township 14, range 48, on a hundred and sixty acre claim, and that the location has proved satisfactory goes without saying when we learn that this is still the family home. In addition to this first tract Mr. Johnson filed on a tree claim. He proved up on all his land within the time required by the government. Mr. Johnson had some small capital, which, with his native thrift, he had been able to save from his earnings—enough at least, to make a substantial payment on the land he had selected. There he at once began permanent improvements; at first they had the usual sod house of the pioneer, but it was a warm comfortable home when the terrible winter blizzards swept across the prairies. As soon as possible he erected buildings for his stock so that they too would be protected alike from winter cold and summer heat. As Mr. John-

FRANK F. STAUFFER AND FAMILY

son says he raised little the first years and the mere question of existance in the new country proved serious, but he was young, had been reared to hard work and was not afraid of doing whatever he could to earn money. From the first he had great faith in the possibilities and future of the Panhandle, and though he suffered from drought in the early nineties, from the grasshoppers and other insect pests held on through all and his faith has been justified, for today he has a comfortable fortune and has seen this rolling prairie become a veritable garden spot of Nebraska. Mr Johnson engaged in general farming, cultivated his land with care and skill, adding improvements from time to time as his capital allowed, erected new and substantial buildings, and eventually converted what had been useless and valueless property into one of the finest cultivated farming estates in this part of Cheyenne county. As the years passed he bought land adjoining his original tract until today he owns a full section and the improvements on all of it he has the pride of knowing he placed there himself. Mr Johnson has continued to engage in diversified farming and stock-raising, managing his affairs with success so that he now has this valuable farm estate and various other investments. Among the citizens of Cheyenne county he is known as an industrious man of high principles, excellent business ability, and utmost personal probity. He has their respect for what he has achieved and the manner in which he accomplished it. Public life has never been attractive enough nor political rewards strong enough, to take his attention from the cultivation of the soil, and he has been contented to carry his career straight through as a representative of the agricultural interests of his community. Having an excellent practical education himself, Mr Johnson has taken an active part in every progressive civic movement since locating in Nebraska, he advocated good roads, good schools and all improved agricultural methods, to demonstrate this he has been a member of his school board for many years, filling it efficiently and well.

In 1888, just a year after coming to this community, Mr Johnson married Miss Elna Olson, also a native of Sweden, and this courageous woman has proved an able helpmate and devoted mother, for she stood beside her husband during all the early years, lending aid and encouragement and has played no minor part in building up the fortune which she and her husband today enjoy. There have been four children in the Johnson family. Ester M, lives on a ranch in Colorado, Sophie, resides on a farm near Sidney. Julia, now Mrs Rassmuson, also lives on a ranch in Colorado, and Mabel, who is at home with her parents. The family are members of the Lutheran church, while Mr Johnson is guided by no party lines when he casts his vote but gives his support to the man best fitted to serve the community, state or nation, as the case may be.

FRANK F STAUFFER, the vice-president and former cashier of the Banner County Bank at Harrisburg, has long been one of Banner county's representative men. He has lived here since boyhood and owns many acres of valuable land, included in which is a tract on which was sown the first alfalfa seed in Banner county.

Frank F Stauffer was born in Wayne county, Ohio, May 14, 1875. His parents were Dansel H and Mary (Frase) Stauffer, the former was born in Lancaster county, Pennsylvania and died in Nebraska, April 11, 1899. The latter was born in Ohio and died in Wayne county in January, 1881. Of their eight children, five are living and are thus distributed as to homes. Frank F lives in Banner county, Nancy, in Akron, Ohio, John at Massilon, Ohio, Cora at Warwick, Ohio, and Salamon, in Michigan. Before coming to Nebraska, in November, 1885, the father for many years had been a farmer, fruit grower and dairyman in Ohio, where he owned one hundred and sixty acres of land. After reaching this state Dansel H Stauffer pre-empted a quarter section in Banner county on which his son now lives, later filed as homesteader on the same land and also took a tree claim, Frank F now owning that land also. Mr Stauffer was greatly interested in fruit growing and set out the first orchard in the county, and was also the first successful experimenter in alfalfa. Before leaving Ohio he was married a second time, Miss Lucy Rudy becoming his wife in 1885. No children were born to this union but she survived him until 1915. Mr Stauffer was a member of the Christian church and in his political views he was a Republican. In every way he was a man of sterling character.

Frank F Stauffer was eleven years old when he accompanied his father to Banner county but in his own opinion his education had not been completed. He was quite ambitious and as there were no schools organized here for about three years, he had to study alone and applied himself so diligently to his books that he qualified himself for teaching, and later on taught one term of school in Banner county for twenty-five dollars a month, two terms in Hamilton county and one more term in Banner county, at thirty dollars a month His school district was the first to organize in the county Clara Shumway was the first county superintendent, and Minnie Shumway taught the first school in an upstairs room of Mr Stauffer's home

Beginning early to work on the farm at wages of fifty cents a day, Mr Stauffer carefully saved his money and when ready to start into the cattle business had five hundred dollars to invest He remained at home that winter and taught his last term of school in Banner county, but in the spring located at Kearney, from which place he was shortly afterward summoned home by a telegram announcing his father's serious illness His father died a week later and Mr Stauffer had to take charge and assume many responsibilities not of his own undertaking He found himself obligated to stay, when the estate was finally settled, and this has since been his home, although at that time he much preferred Hamilton county as a place of residence It was about that time that serious troubles arose between ranch owners and those who turned their hundred of cattle out on the range without paying any attention to whom the ranges belonged This trouble was not to Mr Stauffer's liking but he did the best he could where the estate's interest was concerned and received credit for acting justly and fairly

Mr Stauffer now has in farms and ranches more than twenty-two hundred acres He breeds Shire horses and Hereford cattle, raising fifty head of cattle yearly and a hundred head of horses From a field of thirty-five acres of non-irrigated land, his father's original alfalfa farm, Mr Stauffer cuts an average of a hundred tons of alfalfa a year This land had been reseeded only once in twenty-five years The orchard set out by his father consists of five acres of apples and cherries, the latter being a pretty sure crop, apples not doing so well when there have been late frosts Mr Stauffer has harvested a hundred bushels of luscious cherries in one season He and his family enjoy the comforts of a modern residence of ten rooms and bath, hot water system of heating and acetylene gas for lighting, cooking and ironing The barns Mr Stauffer constructed himself, of substantial logs and he has installed a water pressure system covering all the out-buildings and garden lots This is one of the model rural homes of the county

On June 18, 1906, Mr Stauffer was united in marriage with Miss Martha Osborne, who is a daughter of Robert and Maggie (Draper) Osborne, who came to Banner county in 1887, and now live at Gering Mr Osborne is assistant assessor Mr and Mrs Stauffer have five children Henry, Iva, Walter, Robert and Cora, for whose rearing and education Mr Stauffer has made careful provision

In 1913, Mr Stauffer became identified with the Banner County Bank at Harrisburg, of which he is the vice-president and he served the institution as cashier for four years, 1913-1916 He is also interested in the Gering National Bank He is a Republican Although he has been active in many public ways in the county, he has never accepted any political office except that of county commissioner, in which capacity he served four years

JOHN CLURE, who is one of Morrill county's most highly respected citizens and substantial farmers, came to Nebraska in those early days when many hardships and even dangers menaced the lives of the pioneers He has resided in several sections of the state and came to Morrill county in 1905, when he secured his valuable homestead of a hundred and sixty acres

John Clure was born at Aurora, Illinois, July 12, 1849 His grandfather was born in Switzerland and took part in the war of 1812 after settling in Canada, in which country the grandmother was born The parents of Mr Clure were Joseph and Mary (Burlin) Clure, and the father was born in Canada, February 10, 1819, and the mother was reared in the dominion Of their twelve children, three daughters and four sons are living

John Clure went from Illinois to Iowa and lived there for twenty-seven years before coming to Nebraska He was one of the earliest homesteaders on Pine ridge, six miles west of Belmont, in Dawes county, and lived there for ten years To the usual hardships of pioneers, the fear of Indian attacks was added and Mr Clure helped in the building of stockades for protection, but the prompt and efficient protection given the settlers by the soldiers from Fort Robinson made the stockades unnecessary In order to provide for his family before the land yielded crops, Mr Clure worked at any kind of employment he could secure for money was very scarce He has cut and hauled many a load of wood and sold the same for seventy-five cents He has put excellent improvements on his homestead in Morrill county and has practically turned over the general management to his son, Samuel R, who was honorably discharged at Fort Dodge, Iowa, having served as motor ambulance driver in France from November 13, to January 25, 1919

On March 31, 1872, in Cass county, Iowa, Mr Clure was married to Miss Sarah Parker, whose parents, Humphrey and Nancy (Cole) Parker, were natives of Pike county, Indiana The father of Mrs Clure lived to the age of eighty and the mother to the age of eighty-one years Mr and Mrs Clure have had children as follows Lula May, who died in infancy, Arthur, who lives at Mimature, Nebraska, Mrs Lucy L Wood, whose home is in Nevada, where she has a chicken ranch, Mrs Annabel Hart, who lives near McGrew, Nebraska, Mrs Myra A Shawver, who lives at Alliance Nebraska, Myrton O and Joseph H, both of whom live in Cass county, Iowa, John E, who resides at Melbeta, Nebraska, Samuel R, who lives at home, Asher E, who lives at Gering, Nebraska, and Serena A, who lives at Mc-Grew Mr Clure and his family attend the Christian church He has been a busy man all his life and never felt inclined toward political office, and at present is an independent voter, nor has he ever united with any fraternal organization He has been a good friend and neighbor, however, and in every section in which he has lived may be found those who have been helped in a friendly way He has survived many of his old neighbors of pioneer days who were closely drawn together by common misfortune and danger, but some of them remain and when they meet, their stories of early days are both interesting and instructive to the younger generation

FRANK ROBERTSON who has been a well contended resident of Nebraska since the age of sixteen years, came to the state with his parents in 1886 He has been continuously interested in the material development of the country, being a competent carpenter as well as substantial farmer

Frank Robertson was born in the state of New York, not far distant from the great metropolis, September 14, 1870 He comes of sturdy stock, both of his parents surviving and his father, though seventy-eight years old, continues work as a carpenter, with almost as sure a measuring eye as fifty years ago Frank Robertson is one of the sons of Levi and Eunice (Kenney) Robertson, both of whom were born in New York and now live in Franklin county, Nebraska There Frank Robertson grew to manhood, learned his father's trade and also engaged in farming In 1906, he came to Morrill county and homesteaded and now has a val-

uable farm of two hundred and forty acres, eighty acres of which is cattle pasture He has made many improvements and his buildings are all commodious and substantial Mr Robertson is a man of excellent business judgment and has been very successful in his undertakings

Mr Robertson was married to Miss Netta Bond, who was born June 10 1870, in Boone county, Iowa, and is a daughter of Thomas and Mary (Ford) Bond, the former of whom was born in 1848, in central Indiana, and the latter in 1852, in the state of New York They came to Nebraska and homesteaded in Franklin county in 1873, removing to Lincoln county in 1908, and now live at Bridgeport Mr and Mrs Robertson have five children Ralph, who is married, lives in Wyoming, and Florence, Olive, Seth and Eleanora, all at home They also have an adopted daughter, Esther, who is now the wife of Wallace Smith, a farmer in Morrill county Mr and Mrs Robertson are members of the Free Methodist church Mr Robertson has always been an outspoken advocate of temperance and thus identified himself with the Prohibition party, and in every other way has lent his influence to the cause of law and order wherever he has lived

WILLIAM GETTY, who is one of the enterprising and successful farmers and landowners of Morrill county, was born December 19, 1883, in Morton county, Kansas He is the oldest of seven children born to Charles and Armitta (Way) Getty, the former of whom was born in Indiana, and the latter in Morton county, Kansas

The parents of Mr Getty came to Nebraska and settled in Box Butte county in 1896, but about one year later went to Weston county, Wyoming, and there the father owns a ranch and carries on an extensive stock business William Getty remained with his father in Wyoming until manhood, then acquired ranching land for himself and still owns two hundred and twenty acres in Weston county, Wyoming In 1917, he came to Morrill county and bought eighty acres of land which he has highly improved It is all irrigated and is exceedingly valuable

In 1911, Mr Getty was united in marriage to Miss Flossie Spencer who was born in Nebraska Extended mention of the Spencer family will be found in this work Mr and Mrs Getty have two children, namely Alice and Elsie The family belongs to the Methodest Episcopal church Mr Getty is an independent voter

ARTHUR M GILBERT —In no section of Nebraska has agricultural development been more rapid than in Morrill county, aided, in recent years, by the vast irrigation projects of the Federal Government Ownership of irrigated land means certain prosperity and it has been the laudable effort of homesteaders who came and established themselves here as permanent residents to secure the advantages of irrigation as rapidly as possible Among the enterprising young men who came to Morrill county and homesteaded a hundred acres in 1908, is Arthur M Gilbert, who has ninety acres of his land ditcher It is situated on section 34 town 22, and is considered one of the best farms in the county

Arthur M Gilbert was born at Clarks, Nebraska, February 5, 1886, and is third in a family of eight children born to B M and Eliza E Gilbert His father was born in Kentucky and his mother in Wisconsin, and their marriage took place at Prairie City, Iowa The father has been a farmer all his life, first in Iowa and later in Merrick county, Nebraska, and still later in Oregon, where he still lives

In the public schools of Merrick county, Arthur M Gilbert obtained his education and his agricultural training was secured in a very practical way on his father's farm He came to Morrill county in 1907, and in the following year secured his homestead He has devoted much care to the substantial improving of his land and all his surroundings indicates thrift and plenty He carries on general farming and raises about thirty head of cattle per year, and additionally has been a moderate sheep feeder

In Morrill county, September 29, 1909, Mr Gilbert was united in marriage to Miss Lela E Brown, who was born in Iowa She is a daughter of J M and Hannah C Brown, natives of Iowa, who homesteaded in Morrill county in 1908 Mr Brown met an accidental death on his farm Mrs Brown resides at Boulder, Colorado Mr and Mrs Gilbert have three children Le Arta, Archie and Kenneth Mr and Mrs Gilbert are members of the Seventh Day Adventist church, in which they are highly esteemed, as they are in the entire neighborhood Mr Gilbert is a well read, intelligent man and votes according to his own judgment

MICHAEL M KLINE, of section 30, township 14-48 Cheyenne county, has been one of the industrious men of this locality for more than three decades, linking his name with all that is admirable in farming, and wise and progressive in individual life and bearing the distinction of being thoroughly and com-

pletely self-made He is a native of the Old Dominion of fine German stock, and has many of the admirable traits of that fine race Mr Kline was born in Virginia, December 17, 1855, the son of George and Elizabeth (Miller) Kline, both natives of that state The father was a general farmer who followed that vocation all his life, passing away at the age of sixty-seven years, while his wife survived him until her seventy-fourth year There were the following children in the Kline family Michael, the subject of this review, Elizabeth, who married Paul Kappler and lives in Iowa, Ira, a resident of Omaha, and Eva, who is married and now resides in Colorado

Michael grew up on his father's farm in Virginia, attending school during the term and helped on the farm during his vacations and after school hours, as there are many things that a sturdy boy can do that do not tax his strength and at the same time keep him out of doors When he was young boys did not have the lives of youths on a farm today, there was no using the family automobile after supper to go to town or visit friends, though a general eight hour schedule was in vogue, eight hours in the morning and another in the afternoon, so that he grew up with an excellent practical knowledge of the farming business as carried on along the Atlantic seaboard After finishing school Michael remained with his father for a period before launching himself as an independent business man in agricultural pursuits, but there was little chance for a young man in such an old and well settled state as Virginia and he decided to try his fortune in the west In the fall of 1878 he came west to Keokuk county, Iowa, where he lived nine years In 1887, he came to Nebraska, settled in Cheyenne county becoming one of the pioneers of the section He took up a homestead on which he still resides, which is ample testimony that he considered the location an excellent one and had faith in the future of the Panhandle His start, necessarily was a modest one, but as time passed and he was able to realize money from his labors and crops, he added to his equipment, made improvements and enlarged the scope of his operations From the first, Mr Kline engaged in general farming and stock raising, and as he was careful, thrifty, willing to try and adopt new methods and new farm machinery that lightened the work on the farm, was soon meeting with gratifying success He erected a comfortable home, many and substantial farm buildings and made such other improvements as highly enhanced the value of the land, so that today his is one of the finest properties in the district. He has not confined himself to farming alone as he has ever taken an active part in communal affairs and for two years was postmaster at Colton, an office he filled efficiently and well. Since first coming to Cheyenne county he has been an advocate of good schools, good roads and intensive and for more than thirty years has served as a school director in district No. 64. In politics he is a Democrat and though he never has had time or aspired to public office, takes a keen interest in the local elections. Both he and his wife are members of the Brethren church (Dunkards) of which they are liberal supporters.

Before coming to his western home, on October 21, 1880, Mr. Kline married Miss Fannie Floy, also a native of Virginia. She was born in that state May 26, 1852, the daughter of Samuel Floy who was born, reared and educated in his native state, who after attaining his majority and engaging in farming in the Old Dominion for some years, came west to Keokuk county, Iowa, where he was an early homesteader. There he spent many years as a tiller of the soil and was a minister of the Brethren church and did much church work. He passed away on his farm after living out the psalmist's span of "three score years and ten" for he was eighty-eight years old at the time of his death.

During the years he has been a resident of this locality Mr. Kline has not only gained material success and become established as one of the productive farmers of the Panhandle, but at the same time has built up a personal reputation for honesty in business, industry in the daily affairs of life and public spirit as a citizen of the community.

PATRICK O'GRADY, who undoubtedly is one of the best known men of Banner county, is serving in his third term as sheriff of the county, an office he has filled with remarkable efficiency. On many occasions he has proved his great personal courage, and the record of his public services shows that fidelity to duty has always been his aim, irrespective of danger or loss to himself. Hence Sheriff O'Grady enjoys a large measure of public esteem, all men being his friends except those who have broken the law.

Patrick O'Grady was born in County Sligo, Ireland, February 3, 1873, the only child of Owen and Mary (Casey) O'Grady. His father was a tanner by trade and this he followed all his life either in Ireland or England. Both

parents were faithful members of the Roman Catholic church.

When Patrick was thirteen years old, he came to the United States. He had already had some schooling but had additional school training in America, but the greater part of a very sound and effective education Sheriff O'Grady no doubt obtained in association with others as his life experiences have brought about. He landed on United States soil April 11, 1886. His first work was in connection with railroading and for five years

he lived at Miles City, Montana. For a short interval he worked in Banner county but went back to Montana, and it was several years later that he came to Banner to establish his permanent home. In 1893, he homesteaded and engaged in farming until called to public office.

On June 14, 1892, at Corning, Arkansas, Sheriff O'Grady was married to Miss Florence Cripp. Mrs. O'Grady died in 1896 leaving no children.

In politics the sheriff is a republican. He served one term as deputy sheriff under Sheriff Ingalls, then was elected sheriff and has served ever since. During the continuance of the World War, he was chairman of the county draft board. He belongs to the order of Royal Highlanders and also to the Knights of Pythias and in the latter organization has passed all the chairs in the local body.

JOHN B. KILGORE, who owns a fine, irrigated farm in Morrill county, successfully carries on large farm industries here, for he is thoroughly experienced, having been engaged in farm pursuits all his life. Mr. Kilgore was born at Springfield, Kansas, November 6, 1876, and is a son of James V and Sarah (Buskirk) Kilgore

James V Kilgore was born in Illinois and like others of his name still living in that state, served with honor in the Civil war, for the Kilgores were well represented in that struggle and have always been noted for their loyalty and good American citizenship In early manhood he learned the carpenter trade and later worked at the same in Kansas and in Nebraska After coming to the latter state he homesteaded in Perkins county where he engaged in farming for a time His wife is a native of Wisconsin and they both survive, living in comfortable retirement in Yamhill, Oregon

After his schooldays were over, John B Kilgore gave his father assistance, and it was not until 1911, that he homesteaded for himself, in Morrill county Like many another investor he found little encouragement at first in his efforts to raise crops on arid land and for two years a cloud of discouragement attended him, but when irrigation became a fact and the ditches brought the water, he realized that after all his judgment had not been at fault and that he owned property worth thousands of dollars in his hundred acre farm When government statistics announce an increase of yield in crops on irrigated land of twenty-eight per cent, the bountiful harvests of this section can be understood Mr Kilgore has improved his farm in every way and his attractive farm house indicates a large degree of comfort

Mr Kilgore was married to Miss Ella Davison, who was born in Sweden, April 21, 1882, a daughter of Oscar and Huldah Davison, who came to the United States from Sweden in 1887, and settled in Cass county Nebraska The father of Mrs Kilgore worked on the railroad until the time of his death, when he was forty-seven years old The mother lives in Minnesota Mr and Mrs Kilgore are members of the Methodist Episcopal church at Bayard, Nebraska In politics Mr Kilgore has always been affiliated with the Republican party

BENJAMIN LEVENSKY who is a well known and highly respected resident of Omaha was for years connected with large business enterprises in Kimball, Nebraska He was born in Russia, in 1863, and in 1882 came to the United States The first years in America, Mr Levensky spent in Connecticut, then made his way west to Iowa, lived there until the spring of 1885, when he came to western Nebraska and secured a homestead and tree claim in Sheridan county He lived on this

land five years and proved up on the claim, then sold, and under the firm name of Levensky & Litman, engaged in the mercantile business in Hay Springs, with branch stores at Bassett and Newport Because of good crops at that time the farmers bought many goods and the firm prospered, but in 1892, there was a crop failure and this was immediately felt by the merchants in their business Mr Levensky sold his interest and moved to Newport, where he carried on his business for twelve years, then traded his store for four sections of land, two of them in Kimball and two in Banner county, and now owns about five thousand acres in these counties For six years he carried on a store and also engaged in farming and stock-raising In October, 1918, he sold his store and the stock on his ranch, as his older sons entered military service and were not able to help him, and now lives retired in Omaha, at 2747 North Forty-fifth Avenue

At Leads, South Dakota, in 1891, Mr Levensky was married and he and his wife have had seven children. Dora, who is married to Earl Wolff of Nashville, Arkansas, Israel, who had the distinction to be the second young man to enlist in the World War, from Kimball county, was in an infantry unit and stationed for eighteen months at Honolulu, and received his discharge at Camp Dodge; Ephraim also enlisted and saw service with the submarine division of the navy, was a member of the crew of the Oregon and crossed the ocean carrying supplies to the troops in France, and was honorably discharged at Denver, Colorado, Sol, who is completing his high school work at St John's Military Academy, a member of the class of 1920, Mae, deceased, Jacob, who is attending the public schools, and Mier, who died in infancy

Mrs Levensky died in 1912 Mr Levensky has been a leading citizen wherever he has lived and while at Newport was a member of the school board and director of the Citizens State Bank For years he has been prominent in Masonic circles and was past master of his lodge He is also a member of the Odd Fellows in which he was noble grand On November 11, 1916, Mr Levinsky married Mrs Jennie Levinsk of New York, who came to the United States from Russia and lived in New York until her marriage to Mr Levensky

EMAL W SWANSON, who is one of Bridgeport's best known citizens, having been agent for the Burlington railroad at this point since 1909, has been identified with this system for sixteen years, during that time earning promotion and is held in high regard not only by the corporation by which he is employed, but by the traveling public generally in this section Mr Swanson was born at Aledo, Mercer county, Illinois, September 6, 1885

The parents of Mr Swanson were John and Sarah (Robinson) Swanson, the former of whom was born in Minnesota and the latter in Missouri Their marriage took place at Joy, Illinois, where the mother yet resides The father operated a restaurant at Joy, for a number of years, and his death occurred there in 1914 He was a Republican in politics and both parents belonged to the Methodist Episcopal church They had the following children Lena, who is the wife of Virgil A Love, a stockman in Mercer county, Illinois, Celesta, who is the wife of Scott Terry of Los Angeles California, a civil engineer, Emal W, who resides at Bridgeport, and George, who is employed in the government shipyard at Mobile, Alabama

Emal W Swanson obtained his education at Joy, Illinois, where he completed the high school course, then learned the art of telegraphy, a natural aptness assisting him in quickly reaching facility He was stationed first at Viola and then at Bushnell, Illinois, and in 1909, he was placed in charge of the station at Bridgeport, where he has faithfully performed his duties ever since In 1916, he established a bottling plant at Bridgeport for the manufacture of soft drinks and has prospered in this undertaking Mr Swanson has been industrious and saving and recently has made investments that have resulted in the building of a comfortable and attractive residence here Mr Swanson is a self-made man and his financial independence is the direct result of his own efforts

In 1910, Mr Swanson was united in marriage to Miss Josephine St Clair, who was born at Monmouth, Illinois, and they have one son, Robert St Clair, born October 14, 1915 They are members of the Presbyterian church In politics he is a Republican, and he belongs to the Masonic fraternity

WILLIAM RITCHIE, JR lawyer, with offices at Bridgeport and Omaha, has been identified for seventeen years with educational, professional and military affairs in Nebraska as they came within the scope of his effort On July 1, 1904, he came first to Bridgeport, and this city, as other sections, has been benefited by his vitalizing energy and by the example he has set of loyal and patriotic citizenship Mr

Ritchie was born at Ravenswood, Chicago, Illinois, July 28 1886, the son of William and Charlotte (Congdon) Ritchie, the latter of whom was born at Amboy, Illinois, and the former at Frederick Maryland, his home being across the street from the home of the heroine of Whittier's poem 'Barbara Fritchie,' his father being Barbara Fritchie's family physician For thirty five years William Ritchie, Sr, has been a lawyer of prominence in the city of Chicago, one of his distinguished clients being the late Theodore Roosevelt He has served as chancellor of the Episcopal Diocese of Chicago Of his four sons, William Jr, and his twin brother, Prescott C, were the first born, the latter of whom is connected with the Westinghouse Company in Chicago, John, who left Princeton College to enter a military training camp during the World War, and Gorton, who is a student in the University of Wisconsin

William Ritchie, Jr, was graduated from the Oak Park high school in 1904 and shortly afterward went to Morrill county, Nebraska, while Morrill county was still a part of Cheyenne county as a visitor on the Belmont ranch Finding climate and surroundings agreeable and the people friendly and congenial, Mr Ritchie decided to remain in the neighborhood of Bridgeport, accepting the offer to teach a country school and taught three terms Following this entry into the pedagogic profession he became principal of the Lodgepole schools and remained two years On the suggestion of friends he then became a candidate for county superintendent of schools of Cheyenne county, to which office he was elected in 1907, at the first primary election, having made a campaign on horseback of over three thousand miles He served until 1909, then resigned and entered the Nebraska State University as a student of law During his administration of the office of superintendent, he did a stupendous amount of business, which included the organization of twenty-five new school districts the building of forty-two new schoolhouses and passing judgment on one hundred and ninety-seven petitions for changes in school district boundaries He continued his interest in educational development after entering the university, and under appointment of Governor Morehead, did legal work on a committee to revise the school laws Also while in the university he organized the Teachers' Casualty Underwriters Association, which has become the largest insurance organization of its type in the United States

Mr Ritchie was graduated from the department of law in the university in February, 1915, and in June following embarked in the practice of his profession at Brideport and his professional future seemed fully assured On August 2, 1917, however, Mr Ritchie loyally put his personal ambitions aside and enlisted as a private soldier for service in the World War, entering Company C, Sixth Nebraska Infantry, that afterward became Company K, One Hundred and thirty-fourth U S Infantry A month later he was sent to Fort Snelling for training He was commissioned first lieutenant November 27, 1917, and assigned to Company B, Three Hundred and Ninty-fourth Infantry, Eighty-eighth Division at Camp Dodge Because of an injury he was operated on March 21, 1918, but infection set in while he was in the hospital where he was confined until July 22, 1918 He was promoted captain July 23, 1918 but his physical condition rendered it impossible for him to accompany his division abroad and he was transferred to the Tenth Division at Camp Funston, Kansas He commanded Company I, Sixty-ninth United States Infantry until October 1, 1918, when the influenza assumed an epidemic form in the ranks At the time of the signing of the armistice with the enemy, Captain Ritchie was in command of the Second battalion, Seventh colored troops He was then assigned to camp headquarters and put in charge of the discharging of enlisted men at Camp Funston, and received his honorable discharge February 4, 1919 His application for a commission with the Reserve Corps was not granted because of his physical condition at that time

On April 26, 1916, Captain Ritchie was united in marriage to Miss Eunice Arthur, daughter of Rev L A Arthur, rector of the Episcopal church at Grand Island

Mr Ritchie maintains an office at Omaha under the firm name of Ritchie, Mantz and Canaday, and is attorney for Central States Investment Company, the Skinner Packing Company, Skinner Manufacturing Company, Trans-Mississippi Life Insurance Company and other corporations At Bridgeport, Mr Ritchie is associated with Ralph O Canaday The law firm of Ritchie & Canaday handles a large proportion of the important law business at Bridgeport

Mr Ritchie has been active in political circles since early manhood He was a delegate to the National Democratic convention held in Baltimore in 1912 and voted for Woodrow Wilson, was chairman of the Democratic Congressional District in 1914, in the First District, and chairman of the Sixth Congressional District in 1916 Although but a young man he has achieved honorable success along different lines, and, experienced and levelheaded with high standards that he lives up to, may go far both professionally and politically He has the respect and confidence of his fellow citizens and the good will of hosts of friends Outside of college life he has not been active as a fraternity man, but he retains membership in the Beta Theta Pi and the Phi Delta Phi societies of university days

RALPH OLIVER CANADAY, who is one of the younger members of the Bridgeport bar, came to this city to establish himself in his profession in March, 1919, after his return from military service during the World War Lieutenant Canaday was born at Minden, Nebraska, April 4, 1891, the elder of two sons born to Joseph S and Mary Jane (Winters) Canaday His brother, Walter A Canaday is in the real estate business at Broadwater, and his sister, Golda May, is a senior in the State University

Senator Canaday, father of Ralph Oliver Canaday, was born in Indiana, lived subsequently in Illinois and now lives at Minden, Nebraska The Canadays probably settled in Kentucky contemporary with Daniel Boone and the grandfather of Senator Canaday was the only member of his family that escaped during an Indian attack on the unprotected settlement Joseph S Canaday was married in Illinois to Mary Jane Winters, who was born in Crawford county, that state, and in 1887 they came to Nebraska He bought land in Kearney county and still lives at Minden He has been very prominent in Democratic politics in that county, served in the state senate, was county superintendent of schools and also county treasurer and has frequently been suggested for other public positions of responsibility He was the organizer of the Farmers Co-operative Elevator Association found all over the state and is president of the same With his family he belongs to the Christian Science church

Ralph O Canaday was graduated from the Minden high school in 1909, and spent six years in the State University, in 1915, being graduated with the degree of A B and in 1918, with his L L B degree He was admitted to the bar in 1917, and practiced at Minden until May 17, 1918, when he entered the National army, going to the officers' training school at Camp Dodge, and was commissioned second lieutenant of Company D, Eighty-eighth Infantry on August 26, 1918 The end of hostilities came before his regi-

MARY VAN PELT

THOMAS VAN PELT

MRS. SARAH JOHNSON

WM. VAN PELT

JONITHAN VAN PELT

THOMAS VAN PELT   NANCY VAN PELT

CYRUS VAN PELT AND FAMILY

ment left Camp Dodge, and he received his discharge January 3, 1919 In march following he came to Bridgeport with William Ritchie, Jr, and has been engaged in the practice of law here ever since with encouraging success He is locally in charge of the land belonging to the Belmont Irrigating Canal & Water Power Company In politics Mr Canaday is a Democrat and fraternally a Mason, belonging to Lodge No 19, A F & A M at Lincoln He belongs to the Christian Science church

CYRUS VAN PELT — Probably no pioneers of Banner county are better or more favorably known than the Van Pelts and the records show that Mrs Nancy (Lucas) Van Pelt was the oldest living homesteader in the county when she died June 23, 1920 In association with her son Cyrus Van Pelt and her daughter, Mrs Sarah J Johnston who have been long identified with county affairs and development, they owned five thousand, five hundred acres of valuable land

Cyrus Van Pelt was born in Marion county, Iowa, April 18, 1858, a son of Thomas C and Nancy (Lucas) Van Pelt The father was born in Highland county, Ohio, and moved from there to Iowa in 1856, where he engaged in farming until the outbreak of the Civil War He enlisted in the Fortieth Iowa Infantry, later came home on furlough on account of sickness, then returned to his regiment and died in camp on July 14, 1863 He was a man of sterling character, industrious in peace and brave in war He was a member of the Methodist Episcopal church, and gave his political support to the Republican party Mr Van Pelt's mother was born in Highland county, Ohio, March 12, 1825 and resided in Banner county, esteemed and beloved by all who knew her There were six children in the family Sarah J, who is the widow of Samuel B Johnston, resided many years with her mother, Jonathan, who died in 1880, married Rose Plummer, Mary, who is the widow of A B Stanfield, lives at Greybull, Wyoming, William, who lives in Banner county, married Blanche Snyder, Cyrus of Banner county, and Thomas, who is deceased, is survived by his widow, Mrs Lottie (Brookheart) Van Pelt, who lives near Flowerfield

Cyrus Van Pelt attended school in Iowa and remained at home with his people in Marion county until fifteen years old, then lived in Page county for seven years, afterward was in Maryville, Missouri, for two and a half years, and in Des Moines, Iowa, for four and

a half years In 1887, Mr Van Pelt came to Cheyenne, now Banner county, homesteaded and bought a tree claim He now operates his own and his sister's land, raising cattle extensively For the first few years after coming here, the Van Pelts engaged in crop raising but when the long season of drought fell in the Nebraska country, they turned their attention to cattle and it was not unusual to have five hundred head a year Climatic changes have been brought about, a change in agriculture, and during the past few years Mr Van Pelt has been farming on a large scale, assisted by the latest improved machinery including tractors, threshers, trucks and modern farm implements of all kinds He raises about two hundred and thirty head of cattle since devoting more land to farming purposes

Mr Van Pelt, in recalling early days here, mentions the necessity of hauling water a distance of four miles for about four years His first well went down two hundred and fifty feet and the water was drawn in the old fashioned way with an upright shaft to which a horse was attached James Campbell had the first windmill in this neighborhood Wood was plentiful in the near by canyons and logs were cut there, from which houses and barns were built that are yet in good repair Many settlers left the country during the days of drought and Mr Van Pelt was able to buy their wire fencing for a comparatively small amount, and today has twenty-five miles of such fence on his ranch Many cattle were lost in the early blizzards that visited this section and on many occasions Mr Van Pelt had to cover a territory of twenty-six miles to get his cattle back in the home corral Among other changes in the country the growth of Gering and Scottsbluff are worthy of note In 1891, when he visited Gering the only hotel was a poor log affair where now stands a palatial structure, and Scottsbluff was only a small hamlet

When Mr Van Pelt came to Banner county his mother accompanied him, with his sister and a brother, and they homesteaded, their land cornering together, so that their log house, sixteen by thirty-two feet with an ell could rest on the corner of all three homesteads, thus all lived in the same house and at the same time on their own homesteads In all the plans and decisions, the mother was a leading factor Despite her ninety-four years she continued to enjoy good health had a very retentive memory and excellent eyesight, and enjoyed church attendance and social intercourse as if she were many years younger

Cyrus Van Pelt was married March 26, 1890, to Miss Jennie W. McKee, a daughter of David S. and Sarah A. (Savage) McKee, who came to Banner county in 1885. Mrs. Van Pelt is the youngest of their six living children, the others being: Alexander, who lives in Pennsylvania, married Sallie Patton; James S., who lives in Banner county, married Malinda Ferguson; David, who lives at Council Bluffs, Iowa, married Ida Bolton; Robert W., who lives at Council Bluffs, married Fannie M. Wheeler; and William S., who lives at Denver, married Alma Sutton.

Three children were born to Mr. and Mrs. Van Pelt: Carl P., who married Bessie Try, who died in November, 1918, of influenza, has lived at home since then; Luella, is the wife of Franklin Schumaker, of Banner county; and Edna, who died when aged fifteen years. Mrs. Cyrus Van Pelt is a member of the Presbyterian church, and Mr. Van Pelt not only was very helpful in the organization of the church in this neighborhood, but also of school district number thirty-five, of which he has been treasurer for a number of years. He has served in other positions of responsibility in the county, being the first county treasurer, in which office he served two terms. From an accident in childhood, Mr. Van Pelt has been somewhat handicapped all his life, but in his relations with others has never taken advantage of this, nor has ever claimed relief from social, neighborly duties or citizenship responsibilities. He is held in universal respect and esteem.

WALTER A. CANADAY, second son of Hon. Joseph S. and Mary Jane (Winters) Canaday, was born at Minden, Nebraska, March 22, 1893. He was graduated from the high school of Minden in 1913, after which he took a commercial course in Boyle's Business College at Omaha. He then went on his father's farm in Kearney county and remained interested there until in August, 1917, when he joined a medical corps for service in the World War, accompanied the American Expeditionary Force to France, where he served form August, 1918, until May, 1919, when he was discharged. He returned home and visited one week, then came to Bridgeport and embarked in the real estate business in partnership with R. C. Neumann. Mr. Canaday's business future looks bright. Like his brother he belongs to the Masonic lodge and Christian Science church. Both are held in the highest possible esteem.

JOHN S. EMERSON. — Many men are moulded by the circumstances of life, as lack of early opportunity oftentimes changes a man's entire career. Not every youth can or does overcome obstacles and make his way along the path to success as has Mr. Emerson, whose real estate operations and extensive farm interests make him one of the very responsible men of western Nebraska.

John S. Emerson was born near Afton, Iowa, September 25, 1875, the son of W. J. and Christian J. Emerson. The father was born in Ohio and the mother in Montreal, Canada, and they had six children, four of whom are living, but John S. is the only one in the middle west. When he was only seventeen years old, circumstnces gave Mr. Emerson the care of his mother as well as the support and education of his two sisters in addition to finishing his own education. While most boys of this age have at least a part of their time for some diversion, Mr. Emerson had to keep steadily at work and was rewarded later by seeing both his sisters graduate from school and to find himself the possessor of a good, practical education. In order to earn money for the family and himself Mr. Emerson, while very young learned the photograph business which he followed with success for several years. While working in this line at Crete, Nebraska, he also attended Doane College, taking four lines of study. This not only took all his day time but his evenings and many times a good share of the night for study. He was then twenty-five years old.

Later Mr. Emerson located at Hartington, Cedar county, Nebraska, where he had a photograph studio and before long he owned and operated two studios and two hotels. He gave personal supervision to each and made all show good returns. A good opportunity came to sell and Mr. Emerson disposed of his business after running them four years. While taking a much needed vacation in the northern part of the county he believed he saw a good opportunity and on the spot purchased the general mercantile store owned by August Krouse in the old inland town of St. James, near the Missouri river. Mr. Emerson immediately took charge of the eight thousand dollars stock and a fourth class post office. He built up a good trade by hard work and good business judgment and was making money but was not contented to live and rear his family in an inland town without the educational and other advantages. He could not sell his business or trade it for anything paying as well because St. James was so far from a railroad. He believed it was a case of moving his store to a railroad or building a road to the town and he chose the latter course. In three years this was accomplished, a thing that the citizens of St. James had been trying to accomplish for forty years, for they had made every effort to get the Northwestern Railroad to come into the town and had given it up. After the line was completed and extended and the new towns of Maskell, Obert and Wynot were established, Mr. Emerson found not only a ready sale for his store but also for his half interests in the new town sites which he had acquired in the railroad building operations. The country through which the line was built was well settled, the towns grew rapidly; Wynot being at the end of the line grew especially rapid, gaining five hundred inhabitants in a few months. After selling his store Mr. Emerson moved to the new town and also disposed of most of his real estate interests in Cedar county. He was still looking for business opportunities and while on a trip through the southwestern portion of Nebraska was impressed with the need of more railroads there and the consequent development, so he moved west and became identified with this section. He purchased a large ranch in Banner county, which he still owns and though conditions have not been favorable to railroad building he is glad he came. After spending a few years on his ranch, where he recuperated in health from the strenuous years of hard work, Mr. Emerson moved to Scottsbluff where he is extensively engaged in a real estate and farm loan business, establishing the Emerson Land Company. He has shown his faith in the Panhandle by investing heavily in farm lands until he now owns forty-five hundred acres located in Scottsbluff, Banner, Kimball and Sioux counties.

Mr. Emerson is a quiet, unassuming man who delights in his family and friends. On December 31, 1902, Mr. Emerson married Miss Florence I. Averill, the daughter of Dr. J. A. Averill, of Corning, Iowa. Three children have been born to this union: Ruth, Beatrice and Rex, all of whom are being trained in the school of Democracy, the American public school, and Mr. Emerson expects to send them to college.

The Emerson family are members of the Methodist church and Mr. Emerson is a Republican. He is a member of the Odd Fellows and is a Thirty-second degree Mason.

EDWARD D. SCHICK.— The improvements which have been made in almost every line of industry within recent years, have been kept pace with in plumbing and gas-fitting. New methods are being used and more durable parts taking the place of many of the old construction materials, while modern devises have added to comfort, safety and sanitation. The plumbing house that his skilled workers who carry on the business along modern improved lines, is the one the discriminating people of Scottsbluff prefer to deal with and this is evidenced by the success that has visited Edward D. Schick and his partner since they established ther busness here in 1918.

Edward D. Schick was born at Cleveland, Ohio, February 4, 1872. His parents were Otto and Mary (Krabach) Schick, the former of whom was born at Baden, Germany, and died aged seventy-nine years, in 1918, and the latter born in Ohio and residing at Columbus Grove. They had the following children: Fannie, who is the wife of Fred Fisher, of Toledo, Ohio; Frank, who is a steamfitter at Cleveland; Edward D., who is a respected citizen of Scottsbluff; Fred, who is a machinist at Cincinnati; and Bertha, who is the wife of Cass Clouson, a bookkeeper with Marshall Field & Company, Chicago, Illinois. The family was reared in the Roman Catholic church. By trade the father was a cooper but afterward became a farmer and for forty years lived on his own farm near Defiance, Ohio, then sold and retired to Columbus Grove.

After his school day Edward D. Schick learned the plumbing and steamfitting trade and went to Denver, at the age of seventeen years, after which he worked at his trade until 1893, when he engaged in mining in Colorado until 1901. Then he operated a shop of his own for eight years in Denver. In 1915, Mr. Schick came to Scottsbluff and immediately bought a residence on the edge of town. It now is well within the town limits and he and his family have many near neighbors, so rapid has been the growth of Scottsbluff. This growth has been profiable to Mr. Schick for his skilled work is needed in the construction of both residences and business buildings of any pretention. In 1918, with Mr. Ray, Mr. Schick established his plumbing and heating business and has prospered, giving close attention to his work and is doing well.

In 1895, Mr. Schick was united in marriage to Miss Maude Dufoe, who was born in Iowa, and is a daughter of Alexander and Emma (Doyle) Dufoe. The mother of Mrs. Schuck resides with her at Scottsbluff, but, in 1888, the family had moved to Denver and there the father died, being then aged ninety-four years Mr and Mrs Schick have the following children· Thelma, who is the wife of C L Ray, Paloma, who is the wife of Jack Brashier, a farmer in Scotts Bluff county, and Vernon and Beatrice, both of whom are in school Mr Schick and family belong to the Methodist Episcopal church In politics he in an independent voter, and fraternally he belongs to the Woodmen of the World He is one of the representative business men of the city

THOMAS KNAPP, who is one of the younger business men of Scottsbluff, in point of time, is firmly established here in the automobile repair and painting business and is proprietor of the Baxtrom Auto Paint Company and its branches Mr Knapp was born at Omaha, Nebraska, May 18, 1880

The parents of Mr Knapp were Jarred M and Ellen M (Edwards) Knapp, both of whom were born near Columbus, Ohio, in which state they were married They came west and located first at Council Bluffs, Iowa He was appointed an Indian agent and had many exciting adventures He drove the third express on the Oregon trail and made trips to Salt Lake, Utah, in the early days Three time the Mormons kidnapped his two brothers and twice he succeeded in reclaiming them, but failed in the third attempt He acquired land in Nebraska on which a part of the city of Omaha now stands and engaged in farming there He died at Omaha, where the mother of Mr Knapp yet lives, having resided on the same place for over forty-three years In politics the father of Mr Knapp was a Republican He belonged to the Masonic fraternity, was a Knight of Pythias and an Odd Fellow, and was active in the organization known as the Knights of Labor

Thomas Knapp was the only child born to his parents He grew up near and in Omaha, and attended school until he was eleven years old Since that age he has, unaided, earned his own way in the world and through industry has prospered For twelve years he worked for the Samuel Lee Orchard Carpet Company, Omaha, and then eight years for the Union Outfitting Company, making a very creditable record, in that his services were so long retained by these two representative firms During a cyclone that swept Omaha, he lost property valued at many thousands of dollars On May 7, 1918, Mr Knapp came to Scottsbluff and started his present business with two assistants, while he now requires seventeen skilled men Mr Knapp is head of the largest

auto painting, body building, and top making concern in the west, known as the L A Mann Company This firm will operate twelve places of business before the close of 1921, and have plans all made for a three story building in Scottsbluff, size forty-eight by one hundred and sixty-six feet Mr Knapp knows eighty percent of the residents of Scottsbluff and Morrill counties by name and is known as a success wherever he goes, as is shown by the many branches he has taken as a failure and made big successes with them Mr Knapp lives with his hand in his pocket for anyone he can help and is known as "Christmas basket Knapp" on account of his personal efforts to the poor at Christmas time

WILLIAM A GORDON — There is not any doubt but that the real estate man of today, succeeds by reason of the confidence placed in him by his clients; therefore the most prosperous men and firms in the real estate business are those who are able to advise wisely and willing to deal honestly A very successful real estate man at Scottsbluff, is William A Gordon, the head of the Gordon Realty Company, that is doing much work for the upbuilding of the city along lines of beauty as well as utility Mr Gordon came to Scottsbluff in 1915, but has been a resident of Nebraska since boyhood He was born in Jones county, Iowa, in 1872

The parents of Mr Gordon were William P and Mary (Lawrence) Gordon, who were born, reared and married in Iowa Of their six children, William A is the first born, the other being as follow Anna, who lives at Denver, is the widow of Lee Young, Etta, who is the wife of L W Douglass, now of Nemaha county, Bessie, who is the wife of Samuel Stone, a restaurant and bakery man of Minneapolis, Charles F, who is a promising young newspaper writer of Scottsbluff, and Ella, who resides with her widowed mother in this city By trade Mr Gordon's father was a blacksmith and worked at the same in Iowa, then in Nebraska and in Kansas; in 1874, he went back to Iowa, but returned to Nemaha county in 1879 At one time he was a heavy breeder of Hambletonian horses Mr Gordon owned a shop at Johnson, Nebraska, and did fairly well as a business man and died at that place He was a staunch Democrat in politics but never consented to hold office

William A Gordon attended school at Johnson and then learned the barber's trade which he followed until 1894, when he went into the grain business at Glenrock, Nebraska, where

John W. Spracklen and Family

he prospered for three years then bought an elevator at Cook, Nebraska, but shortly afterward sold and went to Lincoln and bought barber shop privileges in the Lincoln Hotel He disposed of his interests there eighteen months later, traded his property for a store at Havelock, which he moved to Glenrock These changes were made for excellent business reasons but Mr Gordon was too honest a man to make a financial success out of any of them at someone else's expense, and when he sold his interest at Glenrock and went to Graf, in Johnson county, his entire capital in money was just fifty dollars He had the good name, however, which gave him credit, and enabled him to again embark in the mercantile business and from the very beginning he prospered and during the next seven years he settled all his obligations and did an excellent business

About that time Mr Gordon decided to try ranch life and traded his store for a ranch on which he lived for four years and then traded the ranch for a farm of eighty acres in Scotts Bluff county, later bought a hundred and twenty acres in Morrill county and another farm in Scottsbluff county comprising eighty-eight acres, all valuable land In 1915, Mr Gordon came to Scottsbluff, where he occupies one of the most attractive residences of the city, and embarked in the real estate business with charles McElroy That partnership continued a year, when the firm name became Gordon & Osborn, and two years later, Mr Gordon, with Mr Douglass as a partner, organized the Gordon Realty Company They own an entire subdivision and are selling lots to responsible people and building houses of modern style of construction

In 1894, Mr Gordon was united in marriage to Miss Cora Belle Cook, who was born in North Carolina, a daughter of Columbus L Cook, who came to Nemaha county, Nebraska, in 1890, where the mother of Mrs Gordon died Her father then returned to North Carolina and died there The following children were born to Mr and Mrs Gordon Mable, William A Charles, Mary, Grace, Bessie, Raymond and Harold The older daughters live at home and the younger children are yet in school The two older sons, William A and Charles, have but recently been honorably discharged from military service in the World War the former with rank of orderly sergeant and the latter sergeant major William A, was a member of the One Hundred and twenty-seventh Artillery and was with the first contingent of the American

Expeditionary Force to land in France Charles A was also in the One Hundred and Twenty-seventh Field Artillery from the first, but while in training at Fort Sill was injured and sent to the hospital Both sons in their conduct and service reflected credit on their parents and country Mr Gordon and family belong to the Methodist Episcopal church In politics he is a Democrat

JOHN W SPRACKLEN — An early settler and one of the substantial and representative men of Dawes county, is found in John W Spracklen, whose business activities here have covered many years, and one whose experiences, in some ways, have been typical of a large class of western homesteaders Like others, Mr Spracklen met with unexpected difficulties, but in his case, the star of hope was never lost sight of, and resolution and common sense were constant companions There was a time when, in spite of industry and seeming exercise of good judgment, his efforts seemed futile, but that season was lived through, a change came and today the signature of few men in Dawes county carries with it more financial weight

John W Spracklen was born at Belle Plaine, Benton county, Iowa, August 23, 1862 His parents were Peter and Catherine (Russell) Spracklen, natives of Ohio and Tennessee respectively In 1852, Peter Spracklen entered government land in Iowa and engaged in farming in Benton county until 1878, when he moved with his family to Pawnee county, Nebraska During 1882, he ranged cattle near the river, then came into Dawes county to decide for himself concerning the value of the land to which he had heard settlers were beginning to come In his estimation the prospect was favorable and he immediately hastened to the filing office at Valentine, where his friend, John Danks of Long Pine joined him, and unexpectedly his son, John W Spracklen, who had also determined on homesteading in the newly opened tract Peter Spracklen had to make his way from his range camp to his camp on the Niobrara river near Valentine and after the party had secured the numbers of sections, range and town, in order to make filings, they found so many other settlers already in the entry office, each one demanding priority, that they spent an entire day, going without their meals, before they secured their papers John W. Spracklen secured the first homestead in his township, filing on section 29-32-49, and also, at the same time, April 5, 1884, filed on a timber claim on section 20, while his father filed on a homestead and timber claim adjoining his son's entries.

Peter Spracklen resided on his Dawes county land for some years or until his health failed, when he began to think of a home in another climate. In younger years he had read of the magnificent forests of Oregon and often expressed a desire to visit them and also, if opportunity presented, to kill a bear. So he sold his Dawes county land and moved to Oregon and lived in the shadow of the great trees until he had satisfied his ambition in regard to the bear, when he returned to Belle Plaine, Iowa, where he made his home for the rest of his life with a daughter, his death occurring in May, 1898, when seventy-three years old. His wife had homesteaded in Pawnee county, Nebraska, and the children completed paying for the land. She died in Pawnee county in 1913. They were members of the Methodist Episcopal church. Of their nine children six are living, John W., being the only one residing in Nebraska.

During boyhood, John W. Spracklen attended school during the winter months and also was instructed by a teacher whom he remembers with high regard, Mr. Jackson Gunn, who still resides at Belle Plaine. He was sixteen years old when he accompanied his parents to Nebraska and after that rode range for a number of years every summer. A normal, wholesome young man, as soon as he had filed on a homestead his thoughts turned to establishing a home, and on April 17, 1884, he was united in marriage to Dora M. Gillmor, a daughter of John P. and Rosanna (Howe) Gillmor, residing twelve miles north of Seneca, Kansas.

Immediately after their marriage Mr. and Mrs. Spracklen started for the homestead, he having an old covered wagon, and a steady old team. They did not own any large amount of household goods but had all that seemed actually necessary. On their way they stopped at Bradshaw to visit cousins, then went on as far as Willow Springs where other cousins lived and there Mrs. Spracklen was taken ill. During the two weeks that she was sick Mr. Spracklen plowed corn land for his cousin. They then resumed their journey and when they reached Long Pine came up with Mr. Spracklen's brother, who had become uneasy because of their delay, his father also coming in from the homestead with an ox-team. After all had finally reached the homestead locality, they spent a day and a half before they were able to identify the land, not

being familiar with the method of the early surveyors who marked section lines with pitch sticks. Only a few years ago Mr. Spracklen found one of these old sticks still remaining on one of his ranches.

Mr. Spracklen immediately broke up five acres on the homestead and the same amount on his tree claim, got in crops and watched them give promise of abundant yield, only to have them all destroyed by the range cattle. In the second year the government issued orders to the cattle rangers that led to the removal of the cattle to the north side of White river and later on still farther west. Mr. Spracklen and family received their mail at the Half Diamond-E ranch, on Chadron creek. They lived on the homestead for five years, proved up and then immediately borrowed a thousand dollars on it and bought relinquishments on adjoining land. This proved poor policy as the crops failed and he was not able to meet the interest on the loan, then surrendered his deed to a loan agent who also secured a judgment against him for eighty-five dollars. For six years he and his family then lived on a pre-emption and during that time dug two wells but neither afforded a sufficiency of water. About this time someone contested his tree claim, and while he won the court case it cost him money and anxiety and it was eighteen months before he was assured of his right by the land office.

During the first summer the Spracklens lived in their wagon, but in the fall built a log house, after which came a large amount of trouble in sinking a well. Like many other homesteaders Mr. Spracklen found it desirable to work for wages when jobs could be secured, and he remembers breaking land for five dollars an acre, when his wife had to accompany him to drive the team while he cut a road through the brush. The present highway follows very closely the original passage made by Mr. Spracklen on Dead Horse Creek. Did space permit, it would be interesting to follow these hardy and courageous settlers through those early days and note how patiently and resourcefully they met and overcame the hundred little annoyances and the losses that were important only because there were no near neighbors to help out, or supply depots where lost articles could readily be replaced.

As indicated, Mr. Spracklen lost his homestead and pre-emption. In 1896, he went to Sioux county and traded his herd of horses for the William Young farm and made money in handling horses and cattle there, but lost through unfortunate investments and then returned to Dawes county, where he bought his brother-in-law's homestead for two hundred and fifty dollars, a tract that had seventy acres of land already broken and a house that had cost five hundred dollars. He lived there two years, then bought for two hundred and fifty dollars, on a five-year time, a quarter of his father's old homestead, and additional property, giving a mortgage on his timber claim and the homestead he had bought. In the next year he bought two hundred and forty acres for eight hundred dollars, continuing to buy one tract after another, raising rye and cattle in the meantime to meet his payments, and, by 1908, was one of the largest landowners in the county. His financial good management, brought about by foresight, surprised his friends who did not have as supreme faith in the value of Dawes county land as he has had. It is a matter of satisfaction to Mr Spracklen that he has made good on every deal for the past twenty-five years. Today he values his land at a hundred dollars an acre, and has refused one hundred and twenty-five dollars for land he purchased for seventeen and a half dollars. In addition to his farms and ranches, he has a hundred and five head of registered Shorthorn cattle, two hundred head of hogs and a handsome modern residence at Chadron. He is a stockholder in the Chadron State Bank and a director and also a director in the Farmers Union Store Company.

A family of eleven children has been born to Mr and Mrs Spracklen, of whom the following survive: John P, who is a rancher on Dead Horse creek, Dawes county, married Harriet Whitehead and they have three children, Ralph L, John W and Harry; Leonora M, who is the wife of A A Vannatta, of Chadron, and they have four children, Blanche M, Lawrence A, Lester and an infant; Clement A, who is a rancher on White river; Delinda P, who is the wife of Lloyd Robbins, a rancher on Dead Horse creek, and they have one child, Evelyn K; Sadie M, who is the wife of Walter Owens, living on White river, and they have two children, Rosalie and Dora J, and Frank E, Roy W, Nellie A and Mildred L. Mrs Spracklen belongs to the Royal Neighbors, and Mr Spracklen to the Modern Woodmen of America. In politics he is a Republican but has never been willing to accept political office.

CHARLES C NELSON, who is a highly regarded citizen of Bridgeport, where he has lived retired since 1912, is a man of large capital and formerly an extensive raiser of cattle. Mr Nelson is well known in other sections of the country, for he started out a youth with but little means and sought and found work wherever he could make his industry profitable. Thus, unassisted he built up his ample fortune through his own efforts.

Charles C Nelson was born in Virginia, July 17 1849. His parents were James A and Margaret A (Trimmer) Nelson, both of whom were born in Ireland. They came to the United States in 1838 and settled in Virginia. He was a soldier in the Union army in the Civil War, served three years and was wounded in the hand at the battle of Springfield, Missouri. Mr Nelson settled in Missouri in 1861 and after the war returned to his farm and died there. The mother of Mr Nelson came later to Nebraska and died at Sidney. There were but two sons born in the family, John and Charles C, the former of whom was accidentally killed when a small boy. James A Nelson was a man of importance in Henry county, Missouri, at one time owning land there and serving on the board of county commissioners. He was a Republican in his political views and both he and wife were members of the Episcopal church. The paternal grandfather of Mr Nelson, James Nelson, brought his family to the United States in 1838, and settled in Virginia but later accompanied his son and family to Missouri and died there. The maternal grandfather, Joseph Trimmer, was born in Ireland but died in Scotland.

Charles C Nelson had district school opportunities in Missouri. He remained at home assisting his father until he was nineteen years of age, then started out for himself and finally reached Texas, where he went to work on a stock ranch and remained ten years. In 1879, he came to Nebraska and went to work on a ranch in old Cheyenne county, invested his earnings in land and in the course of time had it well stocked and continued in the cattle business for many years afterward. He still owns nine hundred and sixty acres in Nebraska. Additionally he has important oil interests in Oklahoma and is a director of the Wyoming Refining Company. He is a man of fine business judgment.

In 1874, Mr Nelson was united in marriage to Miss Alice Clark, who died in 1885, leaving two children, Fredonia, who is the wife of John Clowges, a ranchman near Bridgeport, and Margaret, who is the wife of Albert Cuddy living in Montana. Mr Nelson was married a second time in 1909, to Miss Lillian Franklin, who was born at Eureka Springs,

Arkansas Mrs Nelson is a member of the Baptist church Throughout his whole political life Mr Nelson has been a Republican and sees no reason to change his opinions at the present time He has served as school treasurer and school director but otherwise has never consented to accept public office He belongs to the Masonic lodge at Bridgeport

JAMES A GAINES — Shrewd business ability, special adaptiveness to his calling, appreciation of its many advantages and belief in his own power to succeed placed James A Gaines among the foremost and substantial merchants of Bridgeport From his small beginnings, his efforts brought forth the development or large interests, permitting his retiremen in 1917, and his consigning to younger hands the tasks that made up the sum of his existence for many years He has a modern home at Bridgeport and is regarded as one of the financially strong and morally high retired citizens of his community

Mr Gaines was born at Lexington, Illinois, March 3, 1859, a son of James and Fannie (Shotwell) Gaines, natives of old Virginia They were married in their native state, and subsequently moved to Illinois, where, in the vicinity of Lexington, James Gaines, who was the son of a Virginia planter and former slave-owner, carried on successful agricultural operations until his death in 1915 He was a self-made man in every respect, and one who won the confidence of his fellow men through a display of integrity and good citizenship, and was a Democrat in his political adherence He and Mrs Gaines, who survives him and still resides at Lexington, held membership in the United Brethren church They were the parents of four children Bert, who was for thirty-four years been in the service of the Chicago, Burlington & Quincy Railroad, and is a resident of Hastings, Nebraska, Mrs Sarah Gray, the wife of a painter of Illinois, James A, of this review, and William E, who for years was a partner of J L Miller at Bridgeport, but is now engaged in the coal business at Greeley, Colorado

James A Gaines obtained his educational training through attendance at the public schools of Lexington, Illinois, and grew up in the atmosphere of the home farm, where for some years he was engaged in assisting his father in the latter's extensive operations Desiring a career of his own, in 1899, he came to Nebraska to make his permanent home, and here embarked in the mercantile business at Bridgeport His opening venture was a modest one, but he possessed the necessary qualifications for the acquirements of success, and it was not long ere he enlarged the scope of his operations When he had his Bridgeport business operating upon a handsomely paying basis he started branches at Bayard, Morrill and Minatare, all of which he developed into successful commercial houses He also secured a concession at City Park, Denver, which he conducted for three years, and which yielded his handsome returns for his foresight and labor He established a reputation for integrity, veracity and probity, which gave him an excellent standing in business circles, and this standing has also held good in the matter of citizenship for he has always been a hearty supporter of good measures His faith in the future of Nebraska, has been evidenced in his purchase of two irrigated farms, which he still retains and the operations on which he supervises In 1917, he sold his store property and retired from active business pursuits, content with the material compensation which he had been able to lay aside during the years of his business activity

Mr Gaines was married in 1910, to Miss Olive E Millhollin, who was born in Iowa, a daughter of Hugh Millhollin, who came to Bridgeport in 1889, and for many years was engaged in the carpenter trade Three children have been born to Mr and Mrs Gaines Leland, aged nine years, Kathryn, aged seven years, and Linden, aged five years Mr Gaines is a Democrat in his political views, but not a politician His fraternal affiliaton is with the Modern Woodmen of America, and he has numerous friends in the local lodge, as he has also in business circles

RICHARD L HOFFMAN was born in Missouri on December 30, 1860, the son of Benjamin Franklin and Susan E Hoffman Both his parents were natives of Kentucky, and their family consisted of five children, of whom three are living A son, William, and a daughter, Sallie B, are deceased Another daughter, Susan E, now Mrs William Lane, lives in Kansas City, Missouri, as does the youngest son, Frank The father was a farmer by occupation He died in 1862, and the mother followed him to the grave in 1865

Richard was educated in the Missouri schools, and after finishing his schooling he took up railroading and followed that calling for a number of years As the great opportunities of the new west began to be known, and the younger spirits who are always attracted by adventure began to follow the advice

EVERETT P. WILSON

of Horace Greeley and go west to grow up with the country, Mr Hoffman came to western Nebraska and found employment as a cowboy in 1880 He followed this vocation for some years, leaving in 1902, for Montana, and spent the next ten years in the vicinity of Missoula, Montana, and Fort Collins Colorado In 1912, he returned to the North Platte valley and filed on a government homestead, and is now the owner of eighty acres of fine land and is in position to take advantages of the great increases in land values in this section of the state

Mr Hoffman was married in Gering in 1892, to Addie Ford Following his marriage he and his father-in-law conducted a hardware business in Gering for several years Two children have been born to Mr and Mrs Hoffman The older son, Richard Frank, is now connected with the United States Reclamation service, having recently been discharged from the United States army The other son, Millard J, lives at home Mr Hoffman is a Democrat in politics, is a member of the Christian church, and belongs to the Odd Fellows

EVERETT P WILSON, who is an honored member of the faculty of the State Normal School at Chadron, is widely known as an educator in Nebraska, which state has been his chosen field of effort almost since the termination of his own schooldays He has been identified with the above institution as head of the department of history and civics, since the beginning in June, 1911

Professor Wilson is a son of John W and Mary E (Magee) Wilson, natives of Ohio He is the second of four sons all of whom are living He was born September 2, 1868, near Princeton, Bureau county, Illinois In 1870 the family moved to Iroquois county, Illinois and settled on a piece of raw prairie land which they improved In 1883, they moved to southwestern Iowa, settling in Cass county, near Atlantic, where they lived for many years Mr Wilson's parents were both teachers in their younger years After their marriage they turned their attention to agriculture, in which they became unusually successful His father was a man who read widely and who possessed an unusually clear understanding of public questions His death occurred at Ames, Iowa, in 1912 His mother survives and makes her home with the oldest son at Jefferson Iowa

Mr Wilson's early education was received in a rural school in eastern Illinois and in the wholesome atmosphere of a good home and in a community environment unusually rich in that sterling pioneer stock that has given America her great place among the nations As a boy in Iowa he took an active part in a neighborhood debating and literary society At an early age he became deeply interested in current political questions After teaching three terms in rural schools he became a student at Highland Park college at Des Moines, Iowa, where he remained for several years In 1894, he entered the Lincoln Normal University located at Lincoln, Nebraska He was graduated from the classical course of this institution After his graduation he remained with the school for some time as teacher of history and civics He retired from this position in 1898, to become principal of the public school of Niobrara in Knox county, where he continued until 1900 From 1900 to 1905, he was superintendent of schools at Ponca, Nebraska, and then served in the same capacity at Wayne, Nebraska, until 1909, in which year he came to Chadron Here he served as superintendent of the city schools until 1911, when he was called to accept the position he has held ever since on the faculty of the Chadron State Normal School

This school was established by the legislature of Nebraska in the session of 1909, for the Sixth Congressional District, and its opening was in June, 1911, although the first summer session was held in the high school building as the new accommodations had not yet been completed That people were anxious to take advantage of this course of instruction was shown by the enrollment of pupils in the first year Enrollments have increased in number each year and at persent the students number more than three hundred, of both sexes

The first president of the institution was Joseph Sparks and W T Stockdale became dean, the latter still occupying that office, but the present president is Robert I Elliott The majority of the students are from Nebraska, Wyoming and South Dakota and the interest shown is very encouraging The best of educational talent is employed, and the standard is the same as is maintained in all other normal schools in the state Further state appropriations ensure the carrying out of plans for extensive expansion in the near future

Mr Wilson was married at Oakland, Nebraska, in February, 1897, to Miss Cora E Young, a daughter of Andrew and Edvinna (Brand) Young who came to Nebraska from Ohio in 1856 They stopped first in Omaha which was then a small village They settled in Burt county, which became their permanent home. They endured all the hardships of pioneer life. They were in danger from hostile Indians. Their first home was engulfed in the Missouri river, when that treacherous stream cut away a large part of the bottom land in Burt county. Their few domestic animals perished in the awful winter of 1856-57. At times they were in dire need of the necessities of life. As the country developed living conditions improved and hardships were replaced by comforts. They reared a large family of children who became successful men and women in Nebraska and in other states.

Mr. and Mrs. Wilson are the parents of five children, Mary, Eleanor, Ruth, Winifred and Evelyn. The eldest, after completing a collegiate course in the State Normal and teaching for three years in the high schools of northwest Nebraska, is now a student of piano in the American Musical Conservatory in Chicago.

In politics Mr. Wilson is a Republican. He has served the community in which he lives in various way. For many years he has been chairman of the city park board and also a member of the city library board. For a number of years he has been a member of the board of directors of the Young Men's Christian Association. During the period of the World War he delivered numerous patriotic addresses. He has been prominently identified with the development of agriculture in northwest Nebraska. He was one of the pioneer promoters of the Dawes County Farm Bureau and has served as secretary since its organization. In the Constitutional Convention of 1919-20 he represented the 74th district composed of the counties of Dawes and Sioux. He served on the committees on arrangement and phraseology and on municipal government. In the campaign for the ratification of the amendments proposed by the convention he addressed a large number of audiences in western and northwestern Nebraska. For many years before the liquor traffic was prohibited he was an ardent opponent of saloons.

In addition to his work in the class room he has been active in other lines of educational work. He is the author of a text on State and Local Civil Government in Nebraska that has been widely used in the schools. He has served as president of the North Nebraska Teachers' Association and also of the Northwest Nebraska Teachers' Association. He has been very active in Christian work, and has spoken many times in public on religious subjects.

As a man, an educator and a citizen he is held in high regard by his fellow citizens.

MICHAEL ELASS.— Here is presented a biographical sketch of a man who was one of the pioneers of western Nebraska, who has lived here to see the prairies of the past become fertile farm lands dotted with thriving communities. When he started out in life he had but few advantages to assist him along the road to success, but his diligence and judicious management have brought him ample reward in return for his labors.

Michael Elass is of German blood and like so many men of the fine German-Americans has the fine traits of his ancestors along with the progress and initiative of the native American. He is a native of the Buckeye state, born in Ohio in 1841, the son of George and Christiana Elass, both of whom were born in the German Empire. They were reared and educated in Europe but saw no chance there to get ahead in the world and decided to start life anew in the United States. The father had learned the trade of stone mason in the old country which he followed after coming to America. Like so many of his countrymen his great desire was to become a landholder and when his capital permitted he purchased a farm and in time acquired a holding of three hundred acres where he engaged in agricultural pursuits all his life as he and his wife died at the age of sixty-five years. There were eight children in the family, of whom four are living: Sophia, the wife of George Vivivel, of Nebraska, now eighty-six years of age; Christiana, who married Henry Beerline; Caroline, who is the wife of Fred Cipp, lives in Omaha, and Michael.

Michael Elass remained at home on his father's farm in Ohio during his youth and attended the schools of that section, at the same time he gave effective assistance on the home farm, early acquiring invaluable knowledge of agricultural methods that have proved of great use to him in later life. He remained in Ohio during his early manhood, as a farmer, but as he desired land of his own took advantage of the government land in Nebraska, and, in 1887, came to Cheyenne county with no other equipment than a wagon, team of horses and a cow. He drove overland to the new home and thus saw most of the country and when he located on his homestead knew just what he wanted and that he has been contented may be judged from the fact that he still resides on the original farm which he secured under government grant. Mr. Elass took up a homestead and pre-empted three hundred and twenty-four acres in section fourteen, King township, not far from the Platte river, as water was one of the important things in the early days for stock His first home was a sod house but it was warm and more of a home than we of the present generation can conceive for such houses were more comfortable during the terrible blizzards of the early days when frame houses were too cold for people to remain in them Mr Elass at once began improvements on his land, broke what he could the first year and put in a crop He endured all the trials, hardships and privations incident to a new country with few railroads and towns far away, where provisions could be obtained and produce sold, but he weathered them all, and his faith and foresight in locating in the Panhandle have been justified, for by holding out and keeping his land when so many of the other settlers sold out and went back east he has won a comfortable fortune He early determined to have thoroughbred stock and as soon as his capital permitted began to specialize in Belgian draft horses which have become the pride of this section of Nebraska He has displayed his horses at the local fairs and won two prizes in Bridgeport, but Mr Elass has not confined himself to one line of endeavor, in addition to his horses he raises a high grade of cattle and hogs, while his agricultural activities are given to diversified farm crops and forage As he looks back across the years he can visualize each improvement which has been made on the home place, all placed there by his own hands, and all looks exceedingly good even in this day of modern methods and advanced practices From his first settlement in this section Mr Elass has taken an active and interested part in all civic and communal affairs, doing his full share for the development of the district in which he made his home For more than twenty years he has been a member of the school board and when irrigation was introduced along the Platte became by unanimous choice a director of the irrigation company He is progressive in his ideas, keeps abreast of all affairs whether local, state or national and is a worthy representative of his section In politics he is bound by no party lines in either local or nation wide affairs, casting his influence ever on the side of the man who will be the most worthy and capable servant of the people and the nation

Over forty years ago Mr Elass selected a worthy helpmate for the journey through life and, in 1876, married Miss Sarah Tunson, who was born in Wisconsin and died in 1913 after a long, worthy life There were eight children in the family George, who is interested in a saw mill in Montana, Charles, a farmer in that state, Hilda, the wife of Will McBride, on a farm near Sterling, Colorado, Philip, in the United States army as a member of the Fifty-sixth Regiment, saw service in France for six months, Fannie, the wife of Fred Noie, lives in Montana, Angie, who married Horace Weaver, lives in Morrill county, and Anna, the wife of Harvey Williams who runs a lumber yard in Oregon Mr Elass appreciated what an advantage a good practical education was for life work and saw that his children had all the advantages to be obtained in the home schools and now looks with pride at the family he has reared to man and womanhood

JOHN T McCOMSEY — Among the residents of the Panhandle who came here in pioneer days as a youth and remained to assist in the development and progress that followed, a place of honor must be accorded John McComsey, who has established a record for industry, and good citizenship in the Hull district

Mr McComsey was born in Stark county, Illinois, August 6 1877, the son of Charles and Mary Elizabeth (Godfrey) McComsey, the former a native of Illinois, while the mother was born and reared in New Jersey To them nine children were born, six of whom survive Ida, a resident of Gering, Mattie, the wife of Cal Smith, lives in Torrington, Wyoming, Thomas, a resident of Eugene, Oregon, Bertha, who married George Benton, of Colorado, and John Charles McComsey, in his youth learned the trade of brick layer and plasterer, a vocation which he followed in his native state, but he desired a farm of his own, and as land in Illinois, a thickly populated state, was high he came to Nebraska in 1886, one of the hardy pioneers of this section

John received his elementary education in Illinois, as he was a boy of nine years when his parents came to Nebraska, and after the family were settled here he attended the public schools, laying the foundation of a good practical education which has been of great value to him in his business life When the school days were over he began general farming operations, but as this was the period when stock-raising was at its height, he naturally specialized in cattle and has continued in that line to the present day At first he had only high bred short horns, a specially fine beef breed but, in 1919, crossed them with Herefords and expects to obtain an exceptionally good strain, obtaining the best qualities of both. From time to time, as money came in from his varied lines of industry, Mr. McComsey branched out, purchased other tracts of land near the home place, until today he is the owner of a landed estate of two full sections of land, where he has placed excellent and permanent improvements and has a modern home. Being an advocate of progress he has the most improved and modern equipment on the farm to lighten labor and increase production, being rated as one of the substantial men of the Hull district.

In 1907, Mr. McComsey married Miss Sarah Schoemacher, a sketch of her family is to be found in this history. To Mr. and Mrs. McComsey one child has been born, Theresa Marie, at home.

Mr. McComsey is a Republican in politics, is a Mason of high standing, having taken a Thirty-second degree, while his wife belongs to the Order of the Eastern Star. They have a wide acquaintance among the pioneers of the Platte valley and can relate many interesting reminiscences of the days when conditions were still crude and primitive and can look with pride upon the country which they have seen change from wild, unbroken prairie to a rich farming district, to the development of which they have contributed liberally in work as true American citizens.

CHARLES E. ANDERSON.— In section thirty-six, township twenty-three, range fifty-five, five miles distant from the city of Scottsbluff will be found the well improved farm that is the stage of the successful activities of Mr. Anderson, who came from his native Sweden to America when a young man and whose energy and good judgment have enabled him to win independence and prosperity through his connection with farm industry in the state of his adoption. He was born in Sweden in 1870, and is a son of John and Sarah Anderson, his father having been a farmer in Sweden and he himself having thus gained practical experience that has proved of great value to him in his operations as a farmer in Nebraska.

Mr. Anderson was given the advantages of the schools of his native land and was an ambitious youth of about eighteen years when, in 1888, he came to the United States and prepared to win success through his own efforts. He came to Burt county, Nebraska, where he found employment. For five years he assisted in government surveying work in west-

ern Nebraska and, in 1904, he established his home in Scotts Bluff county and prepared to initiate independent operations as a farmer. He here took up a homestead of eighty acres and this constitutes an integral part of his present model farm property of three hundred and twenty acres, all of which has been supplied with effective irrigation, in the meantime he has made the best of improvements on the place. It should be stated that Mr. Anderson is a skilled civil engineer, as he continued his education after coming to America by completing a course in civil engineering at the University of Nebraska, where he was graduated as a member of the class of 1898. The ensuing five years he devoted to government surveying work, as previously intimated. He is a man of superabundant energy and progressiveness and his ability has been the force that has moved him forward to the goal of successful achievement in an important industrial field. In politics he is not moved by strict partisan dictates but supports men and measures meeting the approval of his judgment, irrespective of party affiliations.

In 1914, was solemnized the marriage of Mr. Anderson to Miss Esther Nyquist, of Burt county, Nebraska, she likewise being of staunch Swedish ancestry. They have two children: Melvin and Leonard.

CHRISTOPHER G. ABBOTT. — There are residents in every progressive town and city in western Nebraska, living retired from active pursuits perhaps, but by no means to be considered aged, who can, from personal experience, depict an entirely different life from the quiet, orderly, law abiding commercial activities and social enterprises of today. Through such reminiscences come the realization of the marvelous changes that the passage of thirty years have brought about. A representative of the old-time period mentioned, now one of Crawford's substantial citizens, is found in Christopher G. Abbott.

Christopher G. Abbott was born January 15, 1860, in Wabasha county, Minnesota. His parents were George and Ellen (Woods) Abbott, the former of whom was born in Ireland and the latter in England. They were married in the city of Chicago, Illinois, and in 1858, settled as farming people in Minnesota, journeying to that state by ox teams. The father bought land but had not progressed far in its development, when the Civil War came on. In 1862, he enlisted in Company K, Ninth Minnesota Volunteer Infantry, and in the same year, led by Captain Capound, the regiment took part in the battle at Mankato, where a number of white settlers were massacred by the Sioux Indians, and later thirty-eight of these savages were executed by the government. After this battle the regiment was ordered south, George Abbott re-enlisting. When the troops were permitted to land from the steamer on the Minnesota river, stopping for a brief visit at Wabasha, a great reception awaited them, and one incident lingers in Mr. Abbott's memory. He was only three years old at the time but he recalls being carried aloft on the shoulder of Duke Wellington, probably on the way to greet his father, but of the brave father he has no recollection, and never saw him again. The latter was wounded at the battle of Nashville, and although it seemed but a slight injury at first, blood poisoning resulted and his death occurred in the spring of 1864. The mother of Mr. Abbott survived until 1917 and her burial was at Minnieska, Minnesota. She was a member of the Episcopal church. There were four children in the family, Christopher G., being the youngest of the three survivors. His brothers are: John H., who lives at Arlington, South Dakota; and William J., who lives at Whitman, Grant county, Nebraska.

Christopher G. Abbott spent his boyhood on the farm which continued family property until the death of his mother. He had school advantages until he was sixteen years of age, when he accompanied his brother John on a proposed trip to California, with the idea of finding and joining an uncle who had gone to that state in 1861. In 1876, the boys progressed no farther west than St. Joseph, Missouri, and they spent the winter working at Graham. Christopher had not abandoned the search, however, for this uncle was his father's favorite brother and he had been named for him. By this time he had discovered that the west was an immensity he had never realized, and that finding his relative would be in the nature of a miracle, and this miracle actually did occur. In the spring of 1877, he went to Colorado with a cow puncher named Wiley Adams, and near Hughes they worked as cattle herders, but Mr. Abbott was not satisfied with what he was earning, therefore went to work on what was then called the Kansas Pacific Railroad, now the great Union Pacific, and worked up on that line to be fireman He was a dutiful son and during this time wrote letters to his mother and received replies and one of the letters, instead of reaching him, through some mistake in the mails, was delivered to his uncle, who was a well known man in Kan-

sas. Thus a family re-union was brought about in a rather remarkable way.

After working two years on the Kansas Pacific, Mr. Abbott was transferred to the South Park & Narrow Guage Railroad, with which line he continued until 1881, when he visited his mother for several months and then went to South Dakota until the spring of 1885, when he visited his uncle in Kansas and, in 1886, first started in the cattle business. In that year he went to Phillips county, Kansas, and with a bunch of cattle settled on a ranch on Beaver creek. At that time the building of a railroad was considered something of an interference with their business by the cattle men as they wanted a free range. A cousin of Mr. Abbott's had worked on the trails in the sand hill region of Grant county, Nebraska, and on his assurance that Grant county was so sandy that a railroad could never be built there, Mr. Abbott, on May 1, 1886, joined with the Haneys, provisioned for the long trip, and drove their cattle on toward Grant county. They camped for two days on Broadwood creek, then moved to Proctor's ranch, camped, then located twenty-five miles north for two months and branded calves. On July 3, 1886, he filed on pre-emption and tree claim in North Platte. Mr. Abbott and his uncle resided on this ranch until the spring of 1889. About one week after settling on the ranch, Mr. Abbott happened to notice, from the top of a hill, a party of men, which proved to be white men when seen through his field glasses. At first he could not make out their movements, but later recognized them to be railroad surveyors, notwithstanding the assurance he had received that no road could be laid through that sand. He was glad to see them, however, as life on the ranch was proving very lonesome. The nearest post office was North Platte, ninety-six miles distant. Among the pleasant occurrences he recalls his first Christmas dinner, when he enjoyed the hospitality of his nearest neighbor, William Proctor, who lived twenty-five miles south of him. It is probable that his hosts never realized what a great kindness they had done to their lonely neighbor. In the spring of 1887, the Circle ranch boys came up to see how the cattle had stood the winter, and he welcomed them with great heartiness.

In the spring of 1887, occurred the prairie fire near Anselmo, in which several lives were lost, Mr. Abbott being safely at home at the time, assisting his uncle to preserve their belongings. In the fall of that year the grading was finished and the steel laid for the railroad as far as Whitman, which was a tent city with stores, dance halls, saloons and other lines of frontier settlement   Fairchild & Bodine had the first store at Whitman   In the summer of 1888, the railroad was extended to Broncho lake, now Alliance, the rails being laid that fall   The Lincoln Land Company had platted a town and sold lots by auction   It was in June, 1887, that Mr Abbott killed a buffalo on Gunshot lake, southwest of Whitman   It had come into the water with range cattle, and was the last buffalo ever shot in that part of the country   The flesh Mr Abbott sold to the railroad boarding car at nine cents a pound, the hide he presented to a doctor at Anselmo, the head became the property of a conductor on the railroad work train, and the horns Mr Abbott still owns

In 1889, Mr Abbott sold his ranch to Sylvester Carothers, who is now a member of the Nebraska legislature   He then homesteaded in Cherry county and lived there until 1897, improving the property in the meanwhile, and then sold it to his brother   In the spring of 1893, he lost nine head of cattle in the worst prairie fire he ever experienced, some of his neighbors losing their lives in this fire   On July 13, 1906, Mr Abbott came to Crawford and bought a part interest in a saloon business with Ed Henderson, in May, 1908, purchasing the Henderson interest   He continued alone in the business until 1910, when he sold to the firm of Cottom & Newcomb, since which time Mr Abbott has lived retired

In the spring of 1895, in Grant county, Mr Abbott was married to Jessie Manning, who died six months later   In 1898, he was married to Bessie Chamberlain, whose widowed mother lives in North Platte   Mrs Abbott left one daughter, Hazel M, who is the wife of Omar Slayter, of Kearney, Nebraska   In the fall of 1905, Mr Abbott was married to Laura Shearer, of Sioux Falls, South Dakota   Mrs Abbott is a member of the Lutheran church, while Mrs Slayter is an Episcopalian   Mr Abbott is not a member of any body, but he helped to organize and to build the Methodist Episcopal church at Whitman   A strong Republican, as was his father, he was quite active politically while living in Grant county, which he helped to organize and of which he might have been sheriff but declined the nomination   He served on school boards and two terms as assessor of Grant county   For many years he has been identified with the Odd Fellows, has passed the chairs in the local body and belongs to the Encampment, and also is a member of the order of Elks

HORATIO G NEWCOMB —While probably no section of the United States can show more genuine culture at present, or more evidence of substantial development than western Nebraska, it must be acknowledged that at one time there was an element here that gave trouble to the authorities and menaced the peace of the quiet settler and industrious ranchman   There are men yet living who, in official capacities, had to deal with this unruly element, an example being found in Horatio G Newcomb, a retired resident of Crawford, but formerly deputy sheriff of Dawes county and later town marshal of Crawford

Horatio G Newcomb was born in Franklin county, Vermont, in November, 1851, a son of Frank Newcomb and wife, the former a native of Boston, Massachusetts, and the latter of Canada   The mother of Mr Newcomb died in 1855, before he was old enough to realize her maternal care   Of the three children in the family, Horatio G was the only one to come to the west   The father of Mr Newcomb moved to Montpelier, Vermont, in the early fifties, where he opened a meat market and continued in the same business there throughout his active life   He died there in 1895

Until he was seventeen years old, Mr Newcomb remained in his native state and attended school at Montpelier   With a boy's desire for adventure, he determined to see the great western country, finally taking passage on a coastwise vessel and in the course of time reached California, by way of the Isthmus of Panama, little dreaming of the great engineering feat that would change that region in the future   It came about in California that he secured a job to drive cattle, with which work he was not unfamiliar as he had assisted his father in handling cattle back in Vermont   He made friendly acquaintances and one of these offered him work as range rider in Wyoming   He thought the terms fair and accepted with the reservation that he should be given the opportunity of returning to Vermont on a matter of great importance, this being his marriage, which was celebrated at Pigeon Hill, Canada, April 20, 1875 when he was united with Miss Martha Holsopple, whose people were Canadians

In May, 1875, Mr and Mrs Newcomb reached Cheyenne, Wyoming, and he began his life as a cowboy riding range for Hi Kelly, who was a well known cattleman and freighter between Omaha and Laramie City, Wyoming   His work on the Cross-T ranch kept him from home much of the time, and had Mrs Newcomb realized the danger attending her, with predatory and savage Indians wandering alone and in groups through the country near her dwelling, she would not have felt safe and contented as she did at that time Fortunately no harm ever came to her from this source Late in the fall of 1877, Mr Newcomb managed to put up a two-room cabin, but the lateness of the season prevented "chinking" on the outside They spent the winter in comparative comfort but in March the worst blizzard that Mr Newcomb ever encountered in his whole western experience, came upon them, lasting for seven days and nights The furious wind drove snow to the depth of seven inches through the unchinked crevices of the back room of the cabin When the storm ceased Mr Newcomb remembered that he had loaned his shovel to his nearest neighbor, a mile distant, and in order to dig a tunnel to where he had left his horses, he had to secure the shovel In looking after his own affairs as best he could under the circumstances, he found a small herd of cattle all bunched together, only their backs showing above the snow Thousands of cattle perished on the ranges during that storm and a dreadful condition came about when the rangers, with no wells of water available, had to risk typhoid fever by drinking water from the streams infected by the dead cattle, this being particularly the case along Bear creek

During 1884-5-6, Mr and Mrs Newcomb operated the hotel at Fort Laramie, and for three years following he served as deputy sheriff These were real wild west days and Deputy Newcomb witnessed many exciting encounters and took official part in many dangerous adventures In 1889, he and wife moved to Crawford and went into business and continued until 1891, when he sold out and joined a party going to the Klondike in Alaska While in the gold fields there he found ore but not in quantity to pay for the labor and desperate hardship involved The party of four remained in the far north through June, July and August putting in most of their time in hunting and trapping caribou, moose, bear, foxes and martins In 1892, Mr Newcomb came back to Crawford and re-engaged in the saloon business in which he continued until 1917, when he retired

Mr and Mrs Newcomb have two children, a son and daughter Jay, who was born July 20, 1876, on the Cross-T ranch, resides at Grass Creek, Wyoming, and Jessie, who was born December 25, 1880, is the wife of Lieutenant Randolph, an officer in the United States army stationed at Denver Mr Newcomb has always voted with the Democratic party He has never been anxious to serve in political office but there have been times when a man of his personal courage and known resoluteness would have been deeply appreciated by his fellow citizens and he has been induced to accept responsibility During 1909, he served as town marshal of Crawford He belongs to the fraternal order of Eagles, and was treasurer of the local lodge for three years and to the Knights of Pythias, serving this organization three years as vice chancellor

EDWIN C McDOWELL. — Extensively interested in the production of cattle and growing of alfalfa, in Dawes county Edwin C McDowell devotes his three ranches situated near Crawford to these industries, and is probably one of the most progressive and enterprising agriculturists in this part of the state Mr McDowell is widely known and his acquaintances are generally also his friends

Edwin C McDowell was born in Holmes county, Ohio, November 12, 1861 one of a family of five children born to Robert and Elizabeth (Thompson) McDowell Both parents were born in Ohio, where the mother died in 1873 In 1874, the father moved to Iowa, in which state he lives retired, owning property at Ames and Des Moines Of the family, Edwin C and a half brother Earl, live at Crawford, the latter being an attorney

After completing the common school course at Ames, Iowa, Edwin C McDowell became a clerk and was in the mercantile business until he came to Nebraska On April 1, 1886, he filed on a homestead on Little Cottonwood creek, in Sioux county, which he later sold and filed on a tree claim in the same county not far distant from Ardmore, South Dakota, subsequently selling that also He came to Crawford before the first tent was raised in which the earliest merchants disposed of their wares The first general merchandise store was owned and operated by H F Cluft and was located about one mile northeast of the present city, to which it was moved when the railroad reached here Mr McDowell recalls those days well He worked for a Mr Eastman during the winter of 1886-7 His employer had a tent business house and a warehouse, the latter being a small frame affair with earthen floor Mr Eastman slept in the board building while Mr McDowell had his nightly rest in the tent Both of them barricaded themselves in at night, a sense of prudence urging them because of friction existing at the time between soldiers at the fort and cowboys that made promiscuous shooting a not unusual occurrence at night At the best of times the ground was not altogether a couch to be enjoyed peacefully, as the shelter afforded by the walls of the warehouse and tent seemed to attract rattlesnakes They were dangerous as well as unwelcome visitors and Mr McDowell tells of occasions when but for great ingenuity and quick action with his gun, he would have probably lost his life in the encounter

Before leaving Iowa Mr McDowell had learned something of the barbering trade, and while yet with Mr Eastman some of his friends prevailed upon him to open a shop to accommodate them He bought an old barber's chair that he found at Chadron, brought it to Crawford and made barbering a side line In this connection Mr McDowell tells an amusing story which has for its foundation the fact that he was busiest as a barber on Sunday Every business man worked through Sunday just as any other day On one Sunday as he was busy with a customer in one end of the tent, a preacher of his acquaintance came in and announced that by invitation of Mr Eastman, a meeting was about to be held in the tent, the church people beginning to crowd in He evidently was a man wise in his day and generation Seeing how Mr McDowell was engaged he hastened to say, "You just go ahead, I'll not interfere with you or you with me" Not at all indifferent to the religious services, the young barber found himself getting very nervous during the prayers, singing and preaching, and to such an extent, that one of the customers under the razor whispered, "Brace up, that's what Gladstone said to Bismark"

At one time Mr McDowell was in a general mercantile and hardware business with the early merchant, H F Cluff, and later, operated a branch hardware store along the railroad as partner with Ellis E Camp As the building continued west, Mr McDowell moved his store to keep up with it In those days freighting was done by team, and he recalls when "Arkansas John," a noted character, passed many times over the trail with his freighting outfit of a hundred and eighty mules, sometimes ten mule teams to a wagon In contrasting transportation facilities of that day with the present, Mr McDowell was led to speak of his first automobile ride when he was the guest of Mr Dick Richards He thinks that apart from the honor, he earned the ride as at every hill he had to climb out and assist the owner of the machine to push it up to level It may

be added that Mr McDowell travels in a very different kind of car at the present time

At Staplehurst, Nebraska, April 11, 1899, Mr McDowell was married to Miss Effie Gorton, a daughter of Edward and Jane Gorton, the former of whom is deceased The mother of Mrs McDowell resides at Crawford Four children, two sons and two daughters, were born to Mr and Mrs McDowell Robert, who went as a soldier with the American Expeditionary Forces to France and served there nine months, is now at home assisting his father, Esther, who is a college student at University Place, Nebraska, Catherine, who died when aged eleven years, and Charles, who is at home Mrs McDowell is a member of the Methodist Episcopal church

Mr McDowell is now devoting his entire attention to his three ranches, all of them situated within six miles of Crawford For some of this land, which he bought at from a dollar and a quarter to a hundred dollars an acre, he has refused to sell for a hundred and fifty dollars an acre Formerly he raised many sheep and horses, but sold all that stock some years ago and now breeds Durham cattle and raises alfalfa, in 1919 producing eight tons an acre for which he received twenty-five dollars a ton in the stack He operates according to modern methods and has introduced some unusual features, and has his own private irrigation system, which is fed by seven artificial lakes, as an example of modern agricultural progressiveness

All his life Mr McDowell has been more or less interested in Democratic politics, and for many years has served in public capacities at Crawford He served on the school board for nine years, on the city council and its chairman for a protracted period, and, in 1917, was elected mayor of the city The only fraternal organization with which he has united is the Knights of Pythias and he is one of the active members of this body, according to its foundation principles and benevolent aims, at Crawford

CHARLES A MINICK — Among the thousands of careful business men and conservative concerns that have found it profitable to maintain commercial relations with so solid and reliable a banking institution as the First National Bank at Crawford, Nebraska, there are many having a personal as well as business acquaintance with the men who serve as the bank's officials Perhaps none of these officials in Dawes county is better known or more universally esteemed than is Charles A Min-

Mr. and Mrs. Gilbert Fritcher

ick, a former mayor of Crawford, who is the bank's cashier, who has been connected with this institution since 1899

Charles A Minick was born December 21, 1871, at Orrstown, Franklin county, Pennsylvania, the eldest of three sons born to Peter D and Anna L (Hollar) Minick Both parents were born in Pennsylvania and the mother died in Iowa, in 1907 The father has been a prominent citizen of Villisca, Iowa, for a number of years, serving on the school board for a long time and for four years, during the administration of President Cleveland, as postmaster of Villisca He is a very successful dealer in real estate all through Montgomery county Of his three sons the two younger, Harry M and Austin A , are in the insurance business at Des Moines

By the time he was sixteen years old, Charles A Minick had completed his course in the Villisca High school, after which he gave his father assistance for a time, then spent some years working for the Wells-Fargo and the American express companies, the last two years in this work being at Chadron, Nebraska When Bartlett Richards organized the Bank of Crawford, in August, 1899, Charles A Minick came to Crawford to assume the duties of cashier and has been identified with the institution ever since, it now being the First National The bank has had a prosperous career Its last statement gives authority for quoting its resources as $700,000, capital and profits over $80,000, deposits $560,000 The present officers are O R Irins, president, and Charles A Minick, cashier

Mr Minick was married at Des Moines, Iowa, June 28, 1898, to Miss Emma Welsh, who is a daughter of William and Sarah (Boomer) Welsh, well known residents of Des Moines Mr and Mrs Minick have children as follows Charles A , who is attending the Nebraska State University since his return home from military service, having been in army camps and in France for eighteen months, Clifford W , who is connected with the Chicago & Northwestern Railroad at Crawford, and Robert G , John W , R Quentin and Helen A , all of whom are attending school Mr Minick and his family are members of the Methodist Episcopal church, to which he gives a generous support, being liberal minded however and equally generous to all denominations He takes a deep interest in the Young Men's Christian Association and in fact, encourages every moral enterprise that is practical in its management A thirty-second degree Mason, he has filled all the of-fices in the Blue lodge Earnest and conscientious in his political sentiments, Mr Minick has long been identified with the Republican party at Crawford, where he served ten years on the school board and fourteen as city treasurer, resigning the latter office in 1915, when elected mayor of the city

GILBERT FRITCHER, who was one of the early homesteaders and highly respected citizens of Dawes county, was born in Otsego county, New York, January 14, 1832 a son of Adam and Sallie (Lowell) Fritcher, both of whom were born near Albany, New York The father was a farmer all his life and also was a grower of hops, which he mainly sold to brewers

Gilbert Fritcher grew up on his father's farm and obtained his education in the country schools When the Civil War came on he enlisted for service in the Ninety-third New York Volunteer Infantry, in which he served two years, or until severely wounded in the arm He was a corporal in rank

At Northampton, New York, on April 3, 1866, Gilbert Fritcher was married to Miss Sarah A Wallin Her parents were John and Sarah (Howgate) Wallin, both of whom were born in England and came to the United States when young, the father when a boy of eleven years and the mother at the age of nine Their subsequent marriage took place in the city of Brooklyn Mr and Mrs Fritcher continued to live in the state of New York until the spring of 1885, when they came to Nebraska, stopping for several months at Lincoln Mr Fritcher and their son William drove over the country to Dawes county and took possession of the homestead already secured It was situated east of Whitney and Mr Fritcher also had a tree claim located six miles north of Whitney In the fall of the year Mrs Fritcher joined the family, but when cold weather came on they went to Chadron, where Mr Fritcher conducted a livery stable during the winter and the next summer The first few years on the homestead were disappointing, unusual heavy winds sweeping over the freshly cultivated fields, with such force as to blow the seed out of the ground and scatter it long drouths following, in which everything dried up In the earlier years it was almost impossible to get a sufficiency of water in the wells, its transportation from the rivers adding greatly to the drudgery of the farm In later years, from some cause probably easily explained by scientists, the wells filled with water and thereby the almost barren farm land became fertile

Mr. Fritcher and his family remained on the homestead for nineteen years, then sold and lived near Chadron for a year and a half, afterward renting a farm on Bordeaux creek, and there, several years later, on July 16, 1907, Mr. Fritcher passed away. His widow and two children survive, the latter being: William and Edith, the daughter being the wife of Walter Reed of Clay Center, Nebraska.

The family spent the year 1901, in Morrill county. Wherever they lived, Mr. Fritcher was looked on as a trustworthy, honest man. He was a Republican in his political views and always interested in public matters, in younger years in New York serving in public office. He belonged to the Grand Army of the Republic and was a credit to the organization. In public enterprise in Dawes county, such as organizing the school district, he readily donated his share and helped put up the first school building in District No. 16. It was made of slabs, with a dirt roof, and Mrs. Fritcher remembers when as many as thirty-five children attended, but attendance in later years dwindled to a half dozen. Her first home in the county was a two-room shack, in which the family lived for six years, when it was replaced by a frame house which is yet standing. Mrs. Fritcher and son William reside at Chadron, where he operated a draying line for four years and is now a railroad man.

C. L. LEITHOFF, who is president and general manager of the Midwest Monument Company, is not only prominent in business affairs in Dawes county, but, as mayor of the city of Crawford, occupies a position that fairly indicates the confidence placed in him by his fellow citizens as a leader in civic welfare.

Mr. Leithoff was born on a farm in Gearly county, Kansas, in 1871, the son of Louis and Henriette (Walter) Leithoff, being the youngest of four brothers, though he had two younger sisters. His father settled in Gearly county, Kansas, in 1863, and Charles spent his boyhood there. He attended the public school but spent more time riding as a cow boy on the range and says that is where he gained the greatest part of his education. He earned money early gathering scrap iron and bones which he sold to a Jew, and traded the junk for a violin, and his parents often had to send him to the barn when he was learning to play it. He became a musician and played for dances in the country, bought a lamb with the money and in time had quite a sheep business of thirty head, but they got into the gar-

den and he sold them to buy calves, which started him in the cattle business, in which he was engaged until 1904. He then engaged in a hardware and implement business for four years at Junction City, meeting with success. In 1908, Mr. Leithoff disposed of his interests there and came to Dawes county where he has since lived. In 1912 he built the Gate City Hotel of Crawford, became its successful manager and ran the hotel eight years, but he liked the life in the country and traded the hotel for a large ranch and a bunch of cattle which he still owns. R. N. Leithoff, his oldest son, is manager of the varied interests.

Mr. and Mrs Leithoff are the parents of five children, two boys and three girls, the oldest and youngest being deceased. The second daughter, Marie, is the wife of C. C. Cropp, of Los Angeles, California, who is connected with the Western Pacific Railroad; the youngest son, Carl, is at home.

Mr. Leithoff is a practical business man. In July, 1917, he organized the Midwest Monument Company at Crawford, purchasing one small plant at Gordon and another at Chadron. Crawford offers better business location in the way of shipping facilities and freight rates. Mr. Leithoff located at first on a side street but the business soon outgrew the quarters there and removal was made to West Main street. His trade territory covers Nebraska and reaches into Wyoming and South Dakota. Steady employment is afforded nine men and the quality of work and material is under guarantee.

MILBERNE G. EASTMAN, cashier of the Commercial State Bank of Crawford, Nebraska, is well known in the banking field in Dawes county and equally well known and esteemed in commercial life in other sections of the United States and even in the Orient. It has been his good fortune to see many parts of the world, in which he has honorably and adequately represented a department of the national government.

Milberne G. Eastman was born at Clarion, Iowa, July 3, 1868, the eldest of five children born to Oliver K. and Henrietta (Graham) Eastman. His mother was born in Michigan and now resides at Crawford. His father, a native of New York, died at Lincoln, Nebraska, in 1913. For fifty years he was in the banking business. In 1886, he engaged in the mercantile business at Crawford, in partnership with E. F. Doerr, some years later moving to Ardmore, South Dakota, where he established a bank, in later years returning to Crawford The brothers and sister of Milberne G Eastman are as follows H O who is vice-president of the Corn Exchange National Bank of Omaha, L M, who is manager of the Handcraft Furniture Company of Lincoln, G S, who is state bank examiner, lives at Crawford, and Bessie E Chapman, who is assistant cashier of the Commercial State Bank of Crawford

Mr Eastman attended the public schools of Webster City, Iowa, completing the high school course, and was nineteen years old when he came to Dawes county, Nebraska, and became manager of the post trader's store at Fort Robinson, a position he filled for several years Upon being appointed a commissary agent under the United States government, he thoroughly prepared himself for such a responsible office, which took him to Alaska, Japan, China, and other far eastern countries Before returning to the United States he spent several years in the Civil service in the Philippine Islands In 1910, Mr Eastman came back to his old home at Crawford, at the time becoming assistant cashier of the Commercial State Bank in which office he continued until February, 1919, when he became cashier of this institution, his only sister succeeding him as assistant cashier

The Commercial State Bank of Crawford, Nebraska, was established in 1886, with a capital of $15,000 The first home of the bank was in a lumber office It is now housed in a magnificent building of its own, of pressed brick construction, situated in the business center of the city, and its resources exceed $1,000,000 Its first officers were Leroy Hall, president, and Fred A MaComber, cashier Its present officers are Leroy Hall, president, Andrew Vetter, Frank L Hall and Claire E Hall, vice-presidents, M G Eastman, cashier, and Bessie E Chapman assistant cashier Practical business men control and manage this bank

On April 25, 1911, Mr Eastman was united in marriage to Miss Edith Primeaux, a daughter of Antoine Primeaux, an early merchant of Crawford Mrs Eastman died August 6, 1918 Their only child died in 1915, aged three and a half years In politics he has always been a Republican, and from 1894 to 1898, while his father was serving as county clerk of Dawes county, he served as deputy county clerk He is a Thirty-second degree Mason

DAVID S COCKRELL —There are many residents of the city of Chadron who recall David S Cockrell once well known and much esteemed here, where many substantial structures still stand testifying to his skill and thoroughness as a carpenter and builder He was an early homesteader in this section of Dawes county, and because of sterling character and sound judgment, was long looked upon as one of the most dependable and trustworthy of men

David S Cockrell was born at Charlestown, West Virginia, January 8, 1851 His ancestry can be traced to emigrants who came to the American colonies on the Mayflower His parents were David Harris and Cecelia (Miles) Cockrell, of old Virginia stock, the former of whom died in 1887 and the latter in 1892 By trade the father was a carpenter but during the war between the states, he became prominent in military life and served as captain of a company in the Confederate army He was wounded at one time through the bursting of a shell Of his seven children, David S was the only one to make his home in Nebraska An uncle of the latter crossed the plains to California in 1849

Mr Cockrell was a man of natural intelligence and always a persistent reader In youth he had some academic advantages at Charlestown, and under his competent father had thorough trade instruction Both before and after coming to Nebraska he worked as a carpenter and continued in this line as opportunity offered even up to the time of his death, on January 19, 1900 On March 17, 1884, at Charlestown, West Virginia, David S Cockrell was married to Miss Regina Hilbert, who is a daughter of John E and Elizabeth (Hilbert) Hilbert who were natives of Hesse Darmstadt Germany Three children were born to Mr and Mrs Cockrell the only survivor being Ruth Manning Cockrell, an educated and accomplished lady, who is an instructor in the schools of Malvern, Iowa

In 1885, Mr and Mrs Cockrell came to Dawes county and homesteaded east of the present city of Chadron They lived on the homestead for seven years and it was during that time that Mrs Cockrell had an experience with a prairie fire that she can never forget, not because of any real injury but because of her exposure to almost certain incineration and her alarm for the safety of Mr Cockrell At the time she happened to be in Chadron attending the wants of a sick sister She heard the fire bells ringing, according to the usual method of arousing the people, and upon inquiry found that a prairie fire was sweeping over the country south of the city, she could not subdue her anxiety over her husband on the ranch. Her brother-in-law hitched up his horses but they refused to face the dense smoke. They started on foot and one mile from town found that William Birdsall had lost his barn and with the high wind that was blowing, it seemed certain that the Cockrell ranch could not have escaped. Fortunately, however, Mr Cockrell had seen the danger in time had turned his cattle loose and when Mrs Cockrell reached him, worn out with fatigue and anxiety, having passed through lines of fire almost the whole distance, she found him safe and protecting the buildings with plenty of water at hand. It was a marvelous escape.

It was also while living on the ranch that on many occasions the family thought it wise to prepare for possible Indian attacks, and an occasion of this kind is humorously told of by Mrs Cockrell. It was during the uprising at the time of the battle of Wounded Knee and all the settlers were alert and watchful, having little confidence in the peaceful intentions of any wandering Indians. One night Mrs Cockrell felt alarmed over shadows going over the neighboring hill and communicated her fears about Indians to Mr Cockrell, and declared to him that rather than be scalped she would jump into the well if they came any closer. In his calm way he replied, "If you do, don't jump until I tell you." But her fears were so great that she had about made up her mind to jump anyhow, and told him that she would use her own judgment as to the proper minute. Fortunately the menacing shadows disappeared and on the following morning the family discovered that the intruders had been the Indians from the neighborhood going to tether their horses.

In 1892, Mr Cockrell and family moved into Chadron. He had already assisted in the building of the court house, the high school building and the Blaine hotel and was afterward engaged on many of the most important construction work in that city. Late in 1889, he was called to Lost Cabin Wyoming, to assist in the building of a fine residence for a sheep rancher, a Mr Okie. The weather was very inclement and he fell ill with pneumonia. As soon as Mrs Cockrell learned of his condition she started for Lost Cabin which entailed a stage journey of two hundred miles, in January weather. She arrived too late however to see her husband alive as he died suddenly, and after a rest of four hours she started back to Chadron. This experience she considers the worst of many since coming to Nebraska.

In 1908, Mrs Cockrell bought the store building she yet occupies, and ever since has conducted a novelty store with much success. She is prominent in other than a business way, being very active in club work, a member of the Women's Federation Club of Nebraska, which is affiliated with the National Federation, and is somewhat interested in national politics. She belongs to the Order of Rebekah and Degree of Honor. She was one of the early members of the Congregational church in this section and later of the Christian Science church, with which latter organization she is identified. In politics Mr Cockrell was a Democrat. He belonged to the A O U W and the Odd Fellows and in the latter body had been an official. He was known all over this section of the West, having helped in the building of the Pine Ridge agency and also engaged in freighting.

THOMAS J WILSON —Wealth is relative. When some individuals acknowledge possessing it, they refer to their gold, their jewels, their stocks and bonds, but when Thomas J Wilson, of Chadron, genially declares himself rich, he is not referring to his many acres of valuable ranch land in Dawes county, but to his pride in a large, intelligent, happy family, the sound health of himself and beloved wife, the great esteem in which they are universally held, and the many ways in which he has been privileged to add to the welfare of his fellow citizens during the many years he has lived among them.

Thomas J Wilson was born in Morgan county, Indiana, February 12, 1839, and was reared on a farm. When the Civil war came on, he enlisted for military service in Company C, Sixth Indiana Volunteer Infantry, and served out that enlistment of three months, was honorably discharged and returned home. On November 14, 1861, he was united in marriage to Miss Elender L Myers, who was born in Morgan county, Indiana, January 13, 1843. When Mr Wilson realized that his country had still further need of his loyal service, he re-enlisted in July, 1862, volunteering in Company B, Indiana Sixty-seventh Infantry, for three years, and was stationed at Mobile, Alabama, when the war ended. On the day of his second discharge, he saw the barber who shaved him, wipe the lather off his razor with a ten dollar Confederate bill, gladly accepting the ten cents in United States money as pay for his services. Among his many interesting relics of those and other days, Mr Wilson preserves a fifty dollar Confederate bill.

Mr Wilson returned then to his home in Indiana and shortly afterward he and his wife moved to Missouri, in which state they lived as farmers for nineteen years. In the meantime those great developing agents, the railroads, began to creep across the country, and both the Wabash and the Southern Pacific sent their representatives through the east and middle west to solicit business for their respectives lines as the emigration movement began to gather forces with the opening up of government lands. In the neighborhood in Missouri where Mr Wilson was living, much interest was aroused as to colonizing in northwest Nebraska, by a Mr Sweat, who reported favorably from his investigations concerning land and climate. The colony was formed in the winter of 1883-4, a part of the colonists embarking on the Wabash lines and coming to Valentine, Nebraska, by way of Council Bluffs, Iowa, and the other on the Southern Pacific to Sidney, Nebraska. Mr Wilson was one of the thirty travelers to come to Valentine, where the railroad ended, the train of sixteen cars being mainly loaded with household goods. Just prior to this a Mr Schmuhorn, with a body of colonists from Indiana, reached Valentine, and as this little frontier town was merely a railroad terminus, there was nothing to do but for the travelers to put up tents.

Many difficulties arose and as days lengthened into weeks before goods could be arranged for transportation across the country, a large proportion of the colonists expressed disappointment and dissatisfaction. Mr Wilson, like others did not find prospects quite as he had pictured them and like others met with unexpected hardship and loss, but he never became despondent, always looked forward with hope and in every way in those early days set an example of the celebrated "show me" spirit for which Missourians are yet noted. The Indiana colony left Valentine before the Sweat colony, setting up camp on the present site of Gordon, the latter colony passing them on their way to Dawes county two days later. They found only four houses, no public roads or bridges for a distance of a hundred and fifty miles after leaving the Minnecadoose river.

After reaching Dawes county Mr Wilson filed on a homestead on Bordeaux creek, which he sold at a later date, buying the ranch that he yet owns. In the fall after coming here Mr Wilson returned to his Missouri home.

After several starts the next fall he went on a hunting party which resulted in the killing of plenty of antelope but no buffalo. Game was very plentiful but birds were few, the latter following the settlers and the seeding of the land. In the spring of 1885, Mr Wilson came back to Dawes county with his family and was guide for the second colony of ten people and seven cars from the same neighborhood. Although the country seemed desolate there was water in the running streams and timber along the banks for building and for fuel, yet it required courage and hope to really believe in those early days, that this section could ever be developed and made as productive and valuable as some other parts of the state. To offset this, it may be stated that Mr Wilson owns four hundred and eighty acres of land that is valued at a hundred and twenty-five dollars an acre and adjacent land is yielding from a hundred and twenty-five to eighty bushels of potatoes and two to three tons of alfalfa an acre.

Mr Wilson set out his first orchard about thirty years ago and it is still bearing, later set out two others the last one, set about eighteen years ago on his home farm, produced two hundred and fifty bushels of apples in 1919. Quails are plentiful all through this section and in South Dakota. When Mr Wilson came here first he brought two pair of the birds and set them free, the present abundance being the increase from that pair. In recalling early days here, Mr Wilson refers to the lonliness and anxiety of Mrs Wilson when it was necessary for him to make the ten-day trip to Valentine for supplies. While the Indians were usually peaceable, they had the memory of one who proudly exhibited scalps of thirty white people as proof of his prowess at one time and this sight was not very reassuring when a woman and little children had to be left for a length of time practically alone and defenseless.

To Mr Wilson and his wife eleven children were born and of these the following are living: Citha Jane, who is the wife of John A Butler, of Chadron; Edward J who is a resident of Portland, Oregon, married Sadie Jones; Mary E, who is the wife of Grant Blinn, of O'Neill, Nebraska; Sarah C, who is the wife of William Jerters of Clifford, North Dakota; Martha E, who is the wife of A C Riemenschneider of Cody Nebraska; Thomas J, who lives at Spencer Wyoming, married Grace Glinn; John E who is of Edgemont, South Dakota married Margaret Van Buren; and James C, also of Edgemont, who married Laura Goble In addition, there are twenty-three grandchildren in the family and twenty great-grandchildren, wealth indeed. as Mr Wilson claims

Mr Wilson has always believed in the principles of the Republican party as the best for this country and has used his influence as a citizen to strengthen party control He was one of the first prominent men of the country to be made a justice of the peace, and was sworn in in the old town of Chadron on the White river The office in those days carried with it not only honor and responsibility but a large measure of personal danger, Judge Wilson, however, never failing in the strict performance of the law He bravely met danger in other ways, an instance being given in the following occurrence With his family he belongs to the United Brethren church and in Missouri, in addition to being superintendent of the Sunday school for fifteen years, was a local preacher For some time after the colonists settled in Dawes county, there were neither schools nor churches and the only way to nourish the needed religious spirit, was to have meetings held occasionally in the homes of settlers Upon one occasion he had promised Dr Gillespie to hold a meeting near his house A big, burly cowboy objected When the would-be worshippers had gathered, this man was seen to be present but Judge Wilson assured the crowd that there would be a meeting During one of the services the call for some one to open with prayer and the cowboy had the audacity to offer to pray Judge Wilson quietly walked to his side and placed his hand on his head, and through his calm courage so reduced the bluster of the young man that he ever afterward avoided looking Mr Wilson in the face Among his Indian treasures Mr Wilson shows a polished bowl made from a tree knot, and a spoon from an inland stream clam shell, both showing artistic skill Mr Wilson moved to Chadron, Nebraska, in 1911, and lives in a beautiful, modern home on Morehead Street He and his wife are enjoying the fruits of a well spent life revered and honored by all who know them

THEODORE R CRAWFORD — Among the younger educators of Dawes county, none have progressed more rapidly or surely in their profession than has Theodore Ray Crawford, superintendent of the entire school system at Chadron A teacher from choice, he has been thoroughly trained for the work, and nature has assisted in endowing him with those qualities that inspire confidence and arouse ambition When Professor Crawford tells his pupils that knowledge is the key wherewith they may unlock the greatest of earth's treasures, they are apt to believe him, and from that time on their progress is assured

Theodore Ray Crawford was born August 23, 1892, near Clyde, Kansas, the only child of his parents, T F and Emma D (Mickey) Crawford His father was born in Illinois, and his mother in Wisconsin His remote ancestry was Scotch and Irish His mother, who died in September, 1912, was a cousin of former Governor Mickey of Nebraska In 1880, his father came to Kansas and for some years was head miller of a large milling plant in Kansas City, later embarking in a flour manufacturing business of his own At the present time he is manager of the Farmers Elevator & Lumber Company, at Endicott, Nebraska In his political views he is a zealous Republican

Following his graduation from the high school at Blue Hill, Nebraska, Theodore R Crawford entered Hastings College, from which he was graduated with the degree of B S, and subsequently took a post graduate course in the University of Nebraska His first experience as a teacher was as principal of the high school at Edgar, Nebraska, where he continued two years, then as principal of the high school at Alliance and from that place, in 1918, he came to Chadron as superintendent At Broken Bow, Nebraska, October 11, 1913, he was married to Miss Bertha Barrett, a daughter of Daniel S Barrett, who still lives in Custer county, to which he came in pioneer days Professor and Mrs Crawford have two children Dorcas E and Theodore Ray Politically he is a Republican and fraternally is a Mason He is on the directing board of the Young Men's Christian Association, and both he and wife are members of the Methodist Episcopal church They are held in very high esteem at Chadron

EUGENE A PATTERSON, who now enjoys a life of peace and quiet in his comfortable home at Chadron, to which he retired in 1916, requires no effort of memory to transport himself back to different times when he justly was acclaimed an Indian fighter Although forty years have passed since he took part in what proved the massacre at Milk river, a tragedy that aroused the whole eastern as well as western country, he still mourns for his brave comrades who fell victims of Indian treachery For gallant conduct on that occasion, Mr Patterson was awarded a medal that he preserves among his most treasured possessions

Eugene A Patterson was born at Massilon, Ohio, March 28, 1855 His parents were William and Mary E (Warner) Patterson, the former of whom was a native of Canada and the latter of New York A cabinetmaker by trade, the father followed the same in the east until 1876, came then to Otoe county, Nebraska, but shortly afterward settled permanently in Minnesota and died there March 3, 1888 The mother of Mr Patterson died September 16, 1890 Of their thirteen children Eugene A is the only one living in Nebraska The parents were members of the Baptist church, and the father was a Republican in politics

Until he was twenty-one years old, Eugene A Patterson lived at New Lisbon, Ohio, where he attended school and followed farming for a time He came then to the west and on July 7, 1879, enlisted in the United States army for five years From Cheyenne, Fort Russell, Wyoming, as a member of Company F, Fifth United States cavalry, he went to Fort Niobrara, on the way to the Ute agency in Colorado The Indians had become troublesome and this cavalry company, in charge of Major Thornburg, volunteered to drive them back to their reservation in Colorado The Indians cut them off, corraled them for six days before help came, killed all the horses and all but twelve of the pack mules and killed eighteen out of the twenty soldiers, including Major Thornburg They held the Indians back, however, until relief came, nearly all the company coming to the rescue also being wounded, Mr Patterson suffering with the rest While stationed at Fort Niobrara, he assisted in making three surveys for the government road from Pine Ridge to Buffalo Gap, this being in 1881 It was at that time that the picture that is printed in the Indian chapters, through the courtesy of Mr Patterson, of the last Indian Sun Dance permitted by the Government was taken, which, together with the picture of the burial place of Red Cloud's daughters also loaned by Mr Patterson are exceedingly interesting bits of local history Mr Patterson served his full time of enlistment and was honorably discharged at Fort Robinson, July 6, 1884

On July 24, 1884, Mr Patterson was united in marriage to Miss Elizabeth Neu, a daughter of Frederick and Charlotte (Schwaertfager) Neu, of near Nebraska City, the former of whom was born in Prussia and the latter in Indiana From the age of five years Mrs Patterson was well acquainted with the family of Honorable Sterling Morton Three children were born to this marriage Frederick who is a prominent citizen of Dawes county, Myrtle, who is the wife of F W Clark, manager of the Patterson ranch, and they have one son, Stanley Paul, and Harry, who is a ranchman in South Dakota, married Anna Davis, and they have three children, John, Harry and Ralph All the children belong to the Presbyterian church, but Mrs Patterson was reared in the Christian church

For two years after his marriage, Mr Patterson followed farming in Otoe county, Nebraska then for nine years was located thirty miles south of Chadron In early manhood he had learned the barber's trade, and during the later years he was in the army worked at the same At one time he was promoted to corporal but resigned after a few months, finding that he could provide a better income as a barber than he received as an officer Afterward, for many years he followed his trade both at Lincoln and Dunbar In 1916, he returned to Dawes county, in the meanwhile having acquired a valuable ranch which is located between Chadron and Crawford Mr Patterson belongs to the Modern Woodmen of America and takes part in the work of the lodge at Chadron Although believing that politics has its established place in representative government, he has never been unduly active, although ever as in his early manhood, ready to do his full duty when accepting responsibility He has a wide acquaintance and is held in high regard by all

PEARL A REITZ — In marking the growth and rapid development of Chadron as a city, due credit must be given not only to its substantial older citizenship, but to the ambition and enterprise of its younger business men In this connection mention may be made of Pearl A Reitz, who is president and manager of the Reitz & Crites Lumber Company of Chadron and Wayside, Nebraska

Pearl A Reitz was born at Barneston, in Gage county, Nebraska, and is a son of C J and Mae (Beatty) Reitz, well known residents of Gage county He was educated in the public schools at Reserve, Kansas, and later had advantages at Lincoln, Nebraska In January, 1911 he came to Chadron and since then has been mainly identified with the lumber industry He was connected with Robert Hood for two years, then was manager of the Morison Lumber Company, and in March, 1919, became president and manager of the Reitz & Crites Lumber Company, an enterprise of large scope The company operates in lumber, manufactures shingles and handles coal and wood This business is one of the oldest in its line in the city, having been established by Robert Hood in 1885 and conducted by him until December, 1913, when he sold to the Schweiger Lumber Company The new owner continued the business until March 1, 1919, at which time he disposed of his interests to Reitz & Crites The present firm incorporated and elected the following officers Pearl A Reitz, president and manager, F A Crites, vice-president, and E D Crites, secretary and treasurer The company maintains yards at Chadron and also at Wayside

On June 22, 1915 Mr Reitz was united in marriage in Miss Edith Copeland, and they have one daughter, Priscilla Mr Reitz has advanced far in Masonry and is a Shriner He has not been particularly active in politics, having always been more wide awake to business opportunity, but he is thoroughly interested citizen and his influence is for law and order in every public movement

ADDISON V HARRIS, who for many useful, busy years was a man of high standing in Dawes county, came to the state of Nebraska in 1879, and to Dawes county four years later and ever afterward, until his tragic death while in the pursuit of duty, maintained his home here, acquiring a homestead and tree claim in 1884 He was one of the first men to start irrigation projects here, taking out the first water right from the White river and naming the Harris Cooper Irrigation Ditch

Addison V Harris was born in Withe county, Virginia, November 11, 1856 His parents were John L and Mary A (Eskew) Harris, and his paternal grandfather was a Presbyterian minister In 1880, the parents came to Nebraska and the mother died in Otoe county, March 11, 1892 The father followed the blacksmith trade and was variously engaged in different parts of Nebraska, but ultimately returned to his old home in Virginia and his death occurred there

Addison V Harris was a well educated, well informed young man of twenty-three years when he came to Nebraska in 1879 and located in Otoe county, where he worked as a blacksmith and also engaged in farming He remained there until the spring of 1883 when he came to Dawes county, where he homesteaded and pre-empted land four miles west of Whitney, Valentine being the filing office It was a lonely country at that time and all provisions had to be secured from Fort Robinson, then freighted to Sidney and Valentine,

from which points the settlers had to transport all supplies, the round trip often consuming six days At that time there was no town of Crawford, the mail courier leaving his packages with about the only resident, the government then calling the mail station Red Cloud, and neither had Alliance nor Chadron made efforts to rise from the prairie

At Lincoln, Nebraska, April 25, 1881, Mr Harris was united in marriage to Miss Mary M Mecham, whom he had met in Otoe county, Nebraska She was born in Lee county, Iowa, in 1846, and was an infant when she was taken by her parents to Illinois during the exodus of the Mormons They were Alonzo and Nancy (Martin) Mecham, the former of whom was born August 6, 1822, and the latter January 27, 1823 They were Mormons and when driven from Nauvoo, Illinois, went to Council Bluffs, then called Kanesville, Iowa, and from there to Washington county, Nebraska, where the father bought land in Nebraska Territory and farmed it three years before he moved to Otoe county and settled on the middle branch of the Little Nemaha river He died in Keyapaha county in 1904, when the mother of Mrs Harris came to live with her and died in 1906 Mr and Mrs Harris had three children, the eldest dying in infancy The second, Emma Leah, married David E Norman and died May 8, 1909, in Chadron, Nebraska The third, Albert Von, married Emma R Grobe, of Texas, and they live at Salt Lake, Utah

Mr and Mrs Harris lived on the homestead, Mr Harris owning three hundred and twenty acres of fine land and having an interest in several other tracts He raised thoroughbred Poled Angus and Hereford cattle Mr Harris was a good farmer and careful stockman but he had other ambitions as may be indicated by his mastering of the law through study at home and in the office of Judge Crites, and came into so much legal practice that he established a law office at Whitney and one at Crawford, hiring men to operate his farms He was very active in Democratic politics and was as early as 1885, elected county commissioner with James Patterson and D W Sperling and served three years in that office, also was a notary public and a justice of the peace Mr Harris was one of the men who assisted in the organization of Chadron and took an active part in civic affairs Fraternally he was associated with the Knights of Pythias and the Modern Woodmen He served in other official capacities and it was while operating as an officer from Judge Crites' office that he met his death on January 17, 1895 He had gone to a place in Crawford and proceeded to levy on some hay, and was killed while performing his duty as an officer The loss of a man like Addison Harris was a heavy one in Dawes county He was universally esteemed and many old settlers yet remember his early kindness and good advice given them, for he had the right kind of public spirit, desired to see the county settled, contented and happy, and made it a part of his business to seek out newcomers and make the way as easy as possible for them

After the death of her husband Mrs Harris remained on the farm for some time, greatly assisted in getting her affairs adjusted by those of whom she speaks as "good neighbors" She had a hundred and sixty acres pre-emption of her own and was not altogether unused to business transactions, but the ready kindness of her neighbors made things better for her and she has never forgotten them In the winter of 1890 occurred the Indian uprising and she had passed several nights of terror alone in her cabin Mr Harris being at Omaha, and says she had an arsenal of such weapons as knives, axes and pitchforks, with which to defend herself if necessary In 1909, Mrs Harris sold her farm land but still owns two valuable improved properties at Chadron and makes her home in this city She can recall many events of true historical interest in which she and her family have taken part and it is something of a privilege to listen first hand to these reminiscences Mrs Harris is a member of the Methodist Episcopal church, and of Ruben Pickett Lodge, D A R of Chadron She is respected and honored by all who know her

ROBERT I ELLIOTT, president of the State Normal School at Chadron, occupies a prominent position in the educational field and has proved able and resourceful in the executive office The school over which he presides was established by the legislature of Nebraska in the session of 1909, for the Sixth Congressional District Its first president was Joseph Sparks and the dean, as now was Dr W F Stockle The nominal opening of the institution was in June, 1911 although the first session was necessarily held in the high school building, but since that time the original school buildings have been made to suffice, with some few improvements, including the erecting of a girls' dormitory Under the present management, however, many plans are being formulated for the extension of facilities that Dr Elliot deems absolutely necessary

Robert I Elliott was born at Worth, Cook county, Illinois, April 18, 1883 His parents are John and Marion (Tobey) Elliott, natives of Illinois, who now live comfortably retired at University Place, Nebraska In earlier years the father of Professor Elliott was a farmer in Illinois, from which state he moved to Wayne county Nebraska, in 1884, and later engaged in banking at Winside, Nebraska He is a member of the Episcopal church, and the mother was reared in the Methodist Episcopal body They had the following children Robert I, who was an infant when brought to Nebraska, Jack, who is a resident of Scottsbluff, Nebraska, Alice, who is a teacher in the high school at Alliance, Nebraska, Olive, who is the wife of Oliver P Fulton, of Gage county, Nebraska, and Mamie, who met an accidental death

After attending the public schools, Mr Elliott entered the Wayne Normal School and subsequently the State University at Lincoln He began teaching school in 1901, in Stanton county and four years later was appointed county superintendent of the schools of Wayne county, returning then to the university, and from there came to Chadron in 1909, as superintendent of the city schools On August 1, 1916, he assumed the duties pertaining to the presidency of the Chadron Normal School

At Cambridge, Nebraska in November, 1913, Mr Elliott was married to Miss Anna Babcock and they have one son, who bears his father's name They are members of the Congregational church

President Elliott has devoted his entire life to the cause of education He is deeply interested in his present work and the three hundred students enrolled from Nebraska, Wyoming and South Dakota, find in him an inspiring leader and helpful friend It is his hope to have the legislature provide adequately for the erection of an additional school building, a gymnasium and a boys' dormitory, and also he has plans looking to the erection of a model rural schoolhouse on the eighty-acre campus His ideas are practical in the extreme and doubtless will be carried out His political opinions have always kept him within the fold of the Republican party but no proposal of political preferment has ever appealed to him otherwise than educational For three years he served as deputy superintendent of Public Instruction

ALBERT A VANNATTA — There are numerous things in life in which people may take pleasure and pride, but it is doubtful if any offer more solid satisfaction than the realization of work well done. There is justifiable pride when any goal has been reached through one's own efforts, and there are few of his fellow citizens at Chadron who will not agree that Albert A Vannatta proprietor of the Van Buren Hotel, deserves the good fortune that his own industry has brought about

Albert A Vannatta was born at Danville, Illinois, December 15, 1882, and is a son of Samuel and Clementine (Knox) Vannatta. The father was born in Rantoul, Illinois, and the mother in La Salle county. She resides at Danville, Illinois, Albert A being the only one of the five children of the family to live in Nebraska. The father died at Danville in 1908, where for twenty-five years he has been a carpenter and builder.

In his native city Albert A Vannatta had excellent school training and completed the high school course. In July, 1909, he came to Nebraska and went to work on a farm in Dawes county, farm wages at that time being twenty-five dollars a month. He had his own way to make in the world and had accepted conditions cheerfully and hopefully, proved steady and reliable in his new surroundings and soon made friends. On June 22, 1910, he was married to Miss Lenora M Spracklen, who is a daughter of John W and Dora (Gilmore) Spracklen, who have been residents of Dawes county since 1884. Mr Vannatta by this time was receiving forty dollars a month. He and young wife started housekeeping in a one-room log cabin, and through their combined industry and frugality he soon found himself able to rent land and for two years he devoted himself to raising cattle. In the third year he bought a farm but found crop raising less profitable than the growing of cattle, therefore sold his land one year later and re-entered the stock business and later purchased a ranch. He averaged four hundred and fifty head of fine cattle yearly. During all these years he has handled land to some extent and because of careful investments, has found the business quite profitable and still remains interested in land and cattle although not giving his personal attention to the same since 1919. In that year he moved into Chadron and bought the Van Buren Hotel, which he operates as one of the leading hostelries of the city. Mr Vannatta, in a comparatively short time has built up a comfortable fortune, and his example of industry, energy and enterprise, might properly be brought to the attention of other young men who lament lack of opportunity in Dawes county.

Mr and Mrs Vannatta have children as follows. Blanche M. Lawrence, Lester and an unnamed infant, the older children already being apt pupils in the Chadron public schools. In politics Mr Vannatta is a sound Republican and deeply interested in all the leading questions of the day, and undoubtedly as a citizen of Chadron his influence for law, order and economic city government will be beneficial.

BYRON L SCOVEL, whose long connection with the banking interests of Dawes county, universally trusted and highly esteemed here, has been one of the substantial upbuilders of Chadron. He came to the west from New York in 1888, and immediately found congenial duties and promise of an honorable career, and Mr Scovel has, perhaps, become as thoroughly Nebraskan as a native son. He was born at Burke, Franklin county, New York, October 24, 1856. His parents were george T and Amy C (Tower) Scovel, the former of whom was a native of Vermont and the latter of New York. His father, from the age of seven years until he retired and came to live with his son in Dawes county, Nebraska, was a farmer and resident of Franklin county, New York. He died here in 1918. His mother died in New York in 1894. Mr Scovel, senior, was largely interested in the manufacture of potatoe starch, an industry of great import before other starch bearing plants had been analyzed. He was a staunch business man, a Republican in his political opinions and a member of the Baptist church. Of his two children, Byron L alone survives.

In the country school near his father's farm and later in Malone Academy, Byron L Scovel was well instructed, and when seventeen years of age went to a commercial school in Boston. Afterwards he became interested for a time in a general store near Burlington, Vermont. In June, 1888, he came to Dawes county, and entered the employment of Bartlett Richards, who owned the controlling interest and was the financial backer of the First National Bank of Chadron. Mr Scovel worked three years as bank clerk, then ten years as assistant cashier and sixteen years as cashier. When he first entered the employ of this bank it was the largest in Dawes county, had $75,000.00 deposits and charged two per cent interest monthly. During the hard years in the early history of Dawes county, the officers of the First National Bank worked through many discouraging seasons, as for Mr Scovel, he kept at his desk without vacation or relief seven days and six nights in the week, and undoubtedly his fidelity and business sagacity did much to keep the bank on a safe and paying basis

After a continuous service of twenty-nine years Mr Scovel retired from the First National Bank in April, 1918, and in the January following was appointed State Bank Examiner, which office he held but a short time when he was elected president of the Chadron State Bank and after six months service, resigned his position on account of change in the management which made it uncongenial for him In January, 1920, he in connection with other business men and ranchmen organized the Farmers & Merchants State Bank of Chadron, with a capital of $100,000 00 and suplus $10,000 00, in August, 1920, this bank bought the Chadron State Bank and consolidated the two under the name of the Chadron State Bank with capital and surplus as above Mr Scovel is president, J H White, vice-president and W P Rooney, cashier

At Chadron on January 8, 1891, Mr Scovel was united in marriage to Anna G Campbell, who was born in Pennsylvania, where her parents died Mr and Mrs Scovel have two children, Elmira G, an accomplished musician, who teaches the science in the Chadron State Normal School, George Kenneth, who completed a law course in the Leland Stanford University in June, 1920, and was admitted to practice in the California courts and is associated with a leadng law firm at this time in Sant Ana, California During the World War he served with credit as a soldier, was a member of the American Expeditionary Force sent to France and spent sixteen months in that country, immediately resuming his interrupted studies upon his return to his native land Mr Scovel is a Republican in his political opinions He is also a York Rite and a Scottish Rite Mason and a Shriner, and has held all the offices in the Blue Lodge, Chapter and Commandery in his home city He is a member of the Elks Lodge at Norfolk, Nebraska His wife, daughter and himself are members of the Eastern Star and the ladies are members of the Woman's Club The family belongs to the Episcopal church, and there are few benevolent movements or enterprises for the public welfare that do not claim their interest and engage their assistance

HARRY B COFFEE, eldest son of Samuel B and Elizabeth (Tisdale) Coffee was born in Sioux county, Nebraska, March 16, 1890 He attended the public school at Chadron and graduated in 1909, valedictorian of his class

He then entered the University of Nebraska and graduated in 1913 While at the university Mr Coffee was honored by being elected president of his class, and also business manager of the Cornhusker, the annual college publication He was a member of the Alpha Tau Omega fraternity

After leaving the University of Nebraska he engaged in the real estate business at Chadron where he has been actively engaged ever since with the exception of one year spent in touring Canada and South America, and one year in the army, in which he holds a reserve commission in the air service as captain

During 1919, following his discharge from the army, Mr Coffee again took charge of his real estate office and sold almost a million dollars worth of real estate, specializing in large ranch sales His sales for two years, 1917 and 1919, included over forty-three thousand acres of northwestern Nebraska land — a record few in Nebraska have equalled

In addition to his real estate activities, Mr Coffee is president of the Coffee Cattle Company, a $100,000 00 corporation with holdings on the Niobrara River in Sioux county, Nebraska The company owns and leases approximately twenty thousand acres

Mr Coffee is now president of the Chadron Chamber of Commerce, and a booster for city and county development He is a member of the Methodist Episcopal church He is an Elk, a Rotarian, and Shriner Mason

SAMUEL BUFFINGTON COFFEE, for many years was an extensive ranchman well and favorably known in Nebraska He was born at Greenfield, Missouri, March 2, 1855, a son of Col John T and Harriet (Wade) Coffee, and a brother of Charles F Coffee, a prominent banker and stock raiser of Chadron Nebraska

In 1877, "Buff" Coffee, as he was familiarly known, in company with his brother, Charles F Coffee came from Texas and settled in Wyoming where together they engaged in the cattle business In 1879, "Buff" Coffee moved to Sioux county and was one of the first homesteaders in that frontier region He continued to acquire as much land as he could buy adjoining, but, in 1898, he branched out, and bought a ranch on the Niobrara river south of Harrison Practically every acre of this land acquired since 1879, is still held by the family

On March 1, 1889, Mr Coffee was married to Elizabeth Tisdale of Georgetown, Texas, and of their five children the following survive Harry B, an active young business man of Chadron, Rex T, who owns and operates the original ranch of his father, Guy H, who is vice-president and general manager of the Coffee Cattle Company which owns and operates the ranch purchased by his father in 1898, together with extensive holdings since acquired by the company, and Edna, who is the wife of John B Cook, a prominent business man of Scottsbluff, Nebraska

Mr Coffee died at Harrison. Nebraska, October 1, 1900 He was a man of sterling personal character, and great business capacity He was a member of the Methodist Episcopal church and the Masonic fraternity

WILLIAM A DANLEY — To watch a seed develop into a plant is an interesting experience that has come within the attention of almost everyone, but to watch modern cities develop from the bare prairie into power and opulence, comfort beauty and culture, all within the short space of but little more than thirty years, has not been the privilege of everyone This proof of virility and enterprise in western Nebraska, has come under the personal observation of William A Danley, who is one of Chadron's best known and highly esteemed residents In all this marvelous development Mr Danley has borne such part as opportunity afforded

William A Danley was born September 2, 1860, at Danvers, McLean county, Illinois His parents were Samuel T and Mary E (Blair) Danley, natives of Illinois, the father born December 2, 1833, and the mother December 14, 1838 Mr Danley and his father were both born in the house on the old Danley homestead which the Danleys occupied until 1875 The father carried on farming in McLean county until 1879, when he moved to Niobrara, Knox county, Nebraska In 1889, he moved to Dawes county, homesteading near the old site of Chadron, removing from there in 1899, to Colorado, where his death occurred in 1900 The mother returned then to Chadron, where she died in 1904 There were but two children William A and Margaret The latter married John Setter and her death occurred in 1898 She was the first young woman to be married in Dawes county The father was a man of political importance and active in the Republican party, and he was the first county commissioner elected in Cherry county, Nebraska Both parents were members of the Congregational church

William A Danley obtained his schooling in his native state and was nineteen years old when he accompanied his parents to Nebraska

In 1880, both father and son worked at railroad grading on what was then called the Fremont, Elkhorn & Missouri Valley line, later the Chicago and Northwestern, from O'Neill to Buffalo Gap, and they also hauled wood from the hills and sold it for five dollars a load As a summer occupation they started the first and only general merchandise store at Bordeaux, returning then to railroad labor During 1887 and 1888, Mr Danley and his father operated a dairy, and during the time they were so engaged, William A drove the milk wagon to even the most distant points and in the severest winter weather missed only three deliveries in all that time Knox county was not very closely settled at that time but great friendliness existed between the pioneers and they were willing to travel long distances to any social gathering There was a lack of public entertainment, hence, when a few public spirited men arranged to have a Fourth of July celebration in 1884, it is quite possible that every family in the county was represented The immortal Declaration of Independence was reverently listened to as it was read aloud by Mrs O'Linn, there being no doubt at that time, of the pride these lonely settlers took in their American citizenship

After leaving Princeton (Illinois) College, Mr Danley accepted the first position offered him, that of bookkeeper for the Milwaukee Beer Company, at Running Water, Nebraska, although he was no advertiser for the concern as he has never tasted liquor in his life In addition to the activities already mentioned, Mr Danley then engaged in farming for a time, and after coming to Chadron was in the hardware business for several years, and for six years was in the bakery business From 1907, to 1916, he was postmaster at Chadron, and in 1917, he embarked in the book and music business in this city

At Chadron, on December 19, 1888, Mr Danley was married to Miss Jennie Hollenbeck, a daughter of John Hollenbeck, a former well known resident of Dawes county Mr and Mrs Danley have one son, Neal, who resides with his parents They are members of the Congregational church, Mr Danley being the oldest member in point of time, of this congregation, and for ten years he served as superintendent of the Sunday School He has long been identified with the Masonic fraternity and is serving as secretary of the lodge at Chadron

A Republican through training and conviction, Mr Danley has always vigorously upheld the principles of his party On numerous occasions he has been chosen to serve in public office, the city council benefiting by his sound judgment for two terms, and, in 1911, he was appointed official register of births and deaths, an office he has faithfully filled ever since. Mr Danley in turning his thoughts backward, can remember when Chadron was but a small town and when only the long red prairie grass waved over the present sites of Atkinson, Stuart, Long Pine and Ainsworth.

FRED W PATTERSON, one of the enterprising young business men of Chadron, is junior member of the firm of Houghton & Patterson, real estate and general insurance. He has spent the greater part of his life in Nebraska, his native state, and yields to no one in loyal devotion to her best interests.

Fred W Patterson was born August 24, 1885, at Syracuse, Otoe county, Nebraska, the eldest of three children born to Eugene A and Elizabeth (Neu) Patterson, the former of whom was a native of Ohio and the latter of Indiana. They now reside in the Kenwood addition to Chadron, the father still retaining his ranch that is situated between Chadron and Crawford, in Dawes county. For five years, from 1879 to 1884, the father of Mr Patterson served as a soldier in the United States army and during those dangerous times in the Indian country, assisted in making three surveys for government road from Pine Ridge to Buffalo Gap.

Mr Patterson was two years old when his parents came to Dawes county and for seven years lived on the homestead. In the spring of 1895, he went to Dunbar in order to enjoy better educational advantages, and there completed the public school course. In 1906 he entered the university at Lincoln where he pursued his studies for the next two years, from there going to Texas, in which state he remained three years and upon his return to Nebraska, engaged to work for the Burlington Railroad, in May, 1909 and continued until April, 1919, when he came to Chadron and here entered into his present business partnership, Mr Houghton being one of the earliest real estate and insurance men of this city.

At Unadilla, Nebraska, on September 24, 1910, Mr Patterson was united in marriage to Miss Bessie M Mortimore, a daughter of George and Clara (Copes) Mortimore, who reside at Chadron. Mr and Mrs Patterson have a daughter, Lillis. They are members of the Methodist Episcopal church and take part in the city's pleasant social life. Mr Patterson being secretary of the New Community Club

of Chadron. He was reared in the Democratic party but personally prefers to be independent in his political life, at all times being an interested, honest citizen and upright business man.

MRS MARY E HAYWARD — Praiseworthy have been the efforts of innumerable young men in western Nebraska whereby they have built up their fortunes, and admiration and approval cannot be withheld by the honest historian. Unusual, however, is the equally successful example offered in Dawes county, by a member of the other sex, and even through a brief recital of the dominant facts in the life of Mrs Mary E Hayward, a prominent business woman of Chadron and a suffrage leader in the state, is revealed surpassing business ability, mental vigor and personal courage. Mrs Hayward has been a resident of Dawes county for thirty-five years and is one of the most interesting personalities of Chadron.

Mary E (Smith) Hayward was born at Liberty, in Susquehanna county, Pennsylvania, the eldest of four children born to Andrew L and Phoebe E (Law) Smith. Mrs Hayward has two sisters, Mrs Nellie Woodard, who resides at Chadron, and Mrs Sarah Ross Jacobus, who lives in Pennsylvania. The father followed agricultural pursuits in Pennsylvania during his entire life. The parents were members of the Presbyterian church and the children were reared in this religious body, good influences surrounding them from infancy.

Educational advantages were afforded Mary E Smith and she completed the high school course at Great Bend, Pennsylvania. While books were not as plentiful or as easily distributed in those days as at present, there were well patronized public libraries and long before Mrs Hayward came to Nebraska, she had read stories of the west and was particularly interested in a volume called 'Western Life.' Circumstances so adjusted themselves that there came a time when she found herself on a railroad train bound for that great western country of which she had read and thought so much, but of which she later discovered, she knew very little. Her objective point when she left home, was the Pacific coast, her intention being to locate either in Tacoma or Spokane, Washington. Of attractive personality and pleasant manner, it was not difficult for her to find agreeable traveling acquaintances and after submitting them to the test of her judgment, she found railroad travel very interesting as far west as the line then went. It ended, however, at Valentine in Cherry county, Nebraska

In pleasant reminiscence and with much humor, Mrs Hayward relates numerous incidents and adventures with which she gained first-hand information as she traveled many miles by stage coach through what was then a practically unsettled part of Nebraska. She learned how people long separated from the conventional customs of living, still managed to be happy and cheerful along the line of a ruder civilization, and what she criticized and felt resentful about at first, she later understood. It was a new and startling experience to sleep in a room next to a log saloon, with the sheriff of the county and his friends playing cards all night almost within her sight, or later to share her landlady's bed with the children of the family and the five pet dogs sleeping beneath, and it was a distinct shock when she reached Gordon, in Sheridan county, to learn that it was the home of the original "Doc" Middleton, notorious horse thief and outlaw, of whom she had read in "Western Life," far away in her peaceful eastern home

By this time Miss Smith had decided to locate permanently in Nebraska but Sheridan county did not altogether satisfy her, Rushville at that time consisting of but one log house and a tent hotel, and Hay Springs of but one house. She came then to Dawes county and reached what is now Chadron, in April, 1885, immediately consulting Benjamin Loewenthall, who had established a clothing store in a tent three miles from the present town site. The city of Chadron, it may be remarked received its name from a French trapper and squawman named Chadron, who came to this country in 1847. After necessary preliminaries, Miss Smith pre-emped land twenty-five miles west of Chadron and by September of that year had established herself in a business way, on her present location at Chadron. During the first summer she raised all the vegetables she needed on her homestead. Earlier, she went to Box Butte county and filed on a timber claim, and it was on her return trip from that mission that she stopped over night at a country boarding house that was also a grocery and incidentally the post office for that section. She discovered that occasionally the argus eyes of the government may see but blindly, for the postmistress could neither read nor write and each expectant recipient of a letter was at liberty to select what he chose. If the key to the mail bag happened to be mislaid, the bag was easily opened with a sharp knife. It seemed to be the custom at that time and Miss Smith recalls no dissatisfaction

On January 26, 1888, Mary E. Smith was united in marriage to William F Hayward, who had come to this county in 1886, and homesteaded five miles west of Chadron. Mr Hayward was a prominent man in the populist party, served one term as mayor of Chadron and one term as treasurer of Dawes county and was one of the organizers of the lodge of Odd Fellows. He died some years ago

When Mrs Hayward went into the general mercantile business, she established the firm name of M E Smith & Co, which has been maintained ever since. She began in a small way, carefully watching the tastes of her customers before laying in a heavy stock, and in order to be accommodating, kept her store open in the evenings and on Sundays, although it entailed a wearying round of toil. She succeeded. Today Mrs Hayward carries the largest stock of general merchandise in Chadron, and gives employment and pays high wages to some twenty people. She stands foremost among the business men of the county and her satisfied patrons come long distances to deal with her

Immersed in business as she has been for so many years, Mrs Hayward has not been a recluse, on the other hand, has taken an active and interested part in all that concerns the advancement of her sex, both politically and socially. She is a leading member of the Woman's Suffrage Club at Chadron, which she helped to organize, and is president of the Suffrage organization in Nebraska, working hard for legislative recognition. She is consistently charitable and has given substantial encouragement to many moral movements here. During the life of the Business Men's and the Commercial clubs, she was a working member and in that way did much to assist in the development of the city. She belongs to the order of Lady Maccabees

BENJAMIN F PITMAN, who has been one of the enterprising and representative men of Dawes county for many years, and identified with the prosperous little city of Chadron almost from its beginning, came to this section of the state in 1887, and has greatly assisted in the wonderful development that has taken place here within thirty-three years. Observant and thoughtful from youth, ambitious but always soundly practical, Mr Pitman discovered business opportunity where many others saw only a prospect of wasted effort. Time has justified his optimism that was supplemented by applied energy

Benjamin Franklin Pitman was born January 16, 1861, at Newcastle, Indiana, and his education was secured in the public schools of his native city. After completing his high school studies, he entered the Citizens State Bank in a clerical capacity, and while there, during a whirlwind political campaign, was elected city clerk, this election being considered at the time in the light of a joke played on him by his friends. Mr Pitman accepted the office and proved so efficient that he was elected without opposition for a second term, but he had no political aspirations and soon resigned in order to enter upon the duties of assistant cashier in the Rushville National Bank, at Rushville, Indiana, with a salary of $1,500 a year. To many less farsighted men this would have been quite satisfactory, but Mr Pitman in looking ahead, realized that a more active business career would be more congenial. Hence he resigned his position and gave up his comfortable salary in the bank and entered into partnership with a friend and together they established a real estate, loan and insurance business at Huron, South Dakota.

This partnership was more or less satisfactorily operated for several years and then dissolved, the partner accepting a business proposition farther west and Mr Pitman acquiring the agency of the Showalter Mortgage Company. In the meantime Mr Pitman had organized the West Coast Fire and Marine Insurance Company of Washington. After handling the agency business very successfully at Huron for some years, the Showalter company sent him to Chadron, Nebraska, in 1887, as their northwest representative. It may be stated that in that capacity he loaned homesteaders over $500,000 for the company.

Mr Pitman had by no means forgotten his training and early association with the banking business, and became financially interested in the bank at Harrison, Sioux county, became its president with C E Holmes as cashier, later selling the bank. For some years he has been a director and a member of the finance committee of the First National Bank of Chadron and the First National Bank of Hay Springs, in Sheridan county. His realty at Chadron is very valuable, consisting largely in modern buildings which reflect credit on owner and architect. In partnership with Charles F Coffee, he erected the handsome Coffee-Pitman block, the largest business structure in the city, and he built also the Rex Theater building, the old Post Office building and the Masonic Temple.

There was a time when Mr Pitman was personally acquainted with the majority of the settlers in Dawes county, knew every road in the county and could follow every trail. In 1897, he invested in land extensively in the county, acquiring twenty-five thousand acres, buying some of it for a dollar an acre. Later on he sold quarter sections of this land for two hundred and fifty dollars, this same land now commanding from seventy-five dollars to a hundred dollars an acre. Mr Pitman owns the brick-yard at Chadron but has it under rental.

After locating at Huron, South Dakota, Mr Pitman returned to Rushville, Indiana, where he was united in marriage with Miss Emma M Morgan and they came immediately to Huron, a sister of Mrs Pitman accompanying them. At that time it was indeed a desolate place for cultured young women to try to make a home. Building operations had not yet been directed to the erection of comfortable residences, and Mr Pitman had to establish his wife and sister in the Railroad Hotel. During their necessary period of residence in this pioneer hostelry, the ladies learned more facts concerning the state of western civilization and the character and personality of the Indians of whom they had read extensively, than they ever dreamed of, but they were sincere and resourceful young women and the time came when they could bravely smile over what at first, seemed unbearable conditions. Two children were born to Mr and Mrs Pitman, one of whom died in infancy. The survivor, who bears his father's name, is now completing his university studies in Washington, having served with the American Expeditionary Force for two years in France during the World War. Mrs Pitman is a member of the Episcopal church.

Mr Pitman has never been anxious to serve in any political capacity, but he has always been an ardent Republican and proud of his Americanism. For seven years he was a member of the city council of Chadron when he resigned in order to accept the position of city clerk. He has been quite active in fraternal life, is Past Master of Samaritan Lodge No 158, A F & A M, Chadron, Nebraska, and its oldest resident Past Master. Past High Priest of Occidental R A Chapter No 48, Chadron, Nebraska, and Past Grand High Priest of the state of Nebraska. Past Illustrious Master of Zerubbabel Council Royal and Select Masters, No 27 of Chadron, Nebraska, and Grand Captain of the Guard of the Grand Council of Nebraska. Past Eminent Commander of Mehta Commandery No 22,

Knights Templar of Chadron, Nebraska, Past Commander of Chadron Lodge Knights of Pythias, Tangier Temple O A O M Shrine Omaha, Nebraska, K C C H Scottish Rite Mason, Omaha, Nebraska, Consistory No 1, Omaha, Nebraska, is a trustee of the Nebraska Masonic Home, a charter member of Norfolk Lodge No 653 later transferring to Chadron Lodge No 1399 and a member of the Elks of which he is a trustee

HON ALBERT W CRITES, who for many years was accounted Chadron s leading citizen, achieved distinction at the bar, on the bench, in politics and in Freemasonry His wide learning, his professional ability, his social gifts, true manliness and distinctive personality, received generous recognition from those who knew him longest and best, while on numerous occasions he was chosen by the highest governmental authorities for positions of trust and great responsibility It was in answer to such a call that he came first to Chadron His entire life was one of honorable and satisfactory accomplishment, and in his death, his community and the county and state lost a man of unusual worth

Albert W Crites was born at Waterford, Racine county, Wisconsin, May 12, 1848, and died at Chadron, Nebraska, August 23, 1915 His parents were Joseph and Lydia (Darling) Crites The paternal genealogical line leads back to a Pennsylvania Dutch ancestor who served as a soldier under General Braddock at Fort Duquesne in 1775, and the maternal, to a Connecticut colonist who was a member of the colonial contingent sent in 1745, to assist the British in the capture of Louisburg, Cape Breton Albert Crites' boyhood was passed on his father s farm and his early education was secured in the district schools of Racine county Later he became a student in Lawrence University, Appleton, Wisconsin, and afterward taught school four years Mr Crites then entered upon the study of law in the office of Judge Hand, of Racine, was admitted to the bar on March 22, 1872, and subsequently was admitted to practice in the Circuit, District and Supreme courts of the United States

In 1886, Mr Crites was appointed by President Grover Cleveland, chief of the department of captured property claims and lands, and, in, 1887, by the same authority he was appointed the first receiver of public moneys in the United States Land office at Chadron, coming here in June of that year With the utmost efficiency he administered this office for

two years and three months, retiring when a change of administration brought Mr Harrison to the presidency From casting his first vote until the end of his life he was a loyal and ardent Democrat, serving at times as a member of the Democratic State Central Committee, and very often attending state conventions as a delegate from Dawes county In 1891, he was appointed by Governor Boyd to fill a vacancy as Judge of the Fifteenth Judicial District of Nebraska, and his career on the bench reflected credit both on himself and the judicial district He served also as county attorney of Dawes county and was a member of the Nebraska Land Transfer Commission In community affairs he assumed a natural leadership, built up the public schools as a member of the Chadron board of education, and as mayor inaugurated many substantial movements and gave encouragement to numberless worthy enterprises

On June 15, 1876, Judge Crites was united in marriage to Mary Caroline Hayt, who was born at Battle Creek, Michigan, in 1846, and still survives Two sons were born to this marriage, Edwin D and Frederick A, both of whom are prominent in the business and professional affairs of Chadron and Dawes county Judge Crites was a Mason of high rank and at the time of death was Knight Commander of the Court of Honor, the entering degree of the Thirty-third degree in Freemasonry, a distinction attained by comparatively few members of the fraternity

Edwin D Crites, son of Albert Wallace and Mary Caroline (Hayt) Crites, was born January 29, 1884, was educated in the Nebraska State University, studied law and was admitted to the Nebraska bar in 1908 Like his father, prominent in Democratic politics, he was elected county attorney of Dawes county and served five terms, from 1908 to 1919, and earlier had served as city attorney and city engineer of Chadron He is active in business circles and is a member of the Reitz & Crites Lumber Company, of Chadron and Wayside, Nebraska Mr Crites is unmarried

Frederick A Crites, son of Albert Wallace and Mary Caroline (Hayt) Crites, was born at Plattsmouth, Nebraska, July 1, 1885 He attended the Chadron high school, the Lincoln high school and the Nebraska State University at Lincoln and was admitted to the Nebraska bar In November, 1918, he was elected on the Democratic ticket, county attorney of Dawes county, and was appointed by Hon W H Munger, referee in bankruptcy for this district, and is serving most satisfac-

MR. AND MRS. ARTHUR BARTLETT

torily in both offices He has long been prominent in the lumber industry and is vice-president of the Reitz & Crites' Lumber Company of Chadron and Wayside, Nebraska

Mr Crites was married at Kansas City, April 18, 1913, to Miss Marion H Hart of that place a daughter of S E and Caroline (Smith) Hart They have two sons Albert Wallace and Sherman E Mr and Mrs Crites are members of the Congregational church He is active in civic affairs, taking an earnest and sincere interest in the welfare and progress of Chadron and is particularly interested as a member of the city board of education He belongs to the Masonic fraternity and to the Elks at Chadron

ARTHUR M BARTLETT, pioneer settler of Dawes county, president of the Dawes county Farm Bureau, one of the largest landholders in the county and a man who has taken a prominent and active part in the development of this section deserves a place in the history of the county, where he has lived close to a half century He was born in Prescott, Hampshire county, Massachusetts, December 26, 1860, the son of Alfred E and Rebecca L (Putnam) Bartlett The ancestors of this illustrious family came over in the *Mayflower* and one of the members signed the Declaration of Independence Mr Bartlett is the oldest of the six children born to his parents The father was a farmer and young Arthur was reared in the country He attended the country schools and worked on the home place in the summer time In 1867, the family moved to Audibon county, Iowa, and settled forty-five miles from a railroad, there being only three houses in the country By winter time the family had a house to move into as they were about a mile from the little village of Exira The next spring the father bought more land a half mile south of the little town which was the county seat of Audobon county, where the Bartletts lived until 1886

Arthur Bartlett continued to attend what schools were afforded in his frontier home in Iowa and when old enough helped on the farm, becoming a practical farmer at an early age December 15, 1880, Mr Bartlett married Miss Ada I Shravger, who was born at Rock Island, Illinois, the daughter of Frank and Anna (Umtsead) Shravger, the former a native of Pennsylvania, while the mother was born in Danville, Illinois Mrs Bartlett was next to the oldest in a family of four, being the only girl Her father was a carpenter

by trade, who moved to Iowa with his family in 1868 after his marriage Mr Bartlett and his bride lived on a farm for three years in Iowa, then came to Dawes county, Nebraska, in 1884, and took a homestead and timber claim twenty-six miles south of Chadron which he still owns When he came here he had but seventy-five dollars cash, a pair of mules and an Indian pony besides four head of cattle It took all the money to buy lumber for a house The first year he put in a sod crop and had thirty acres of fine oats and barley which was hailed out in July He then began to help put up hay to earn some money, working for a dollar and a quarter a ton and boarding himself as well as furnishing his own team In order to live and as he expressed it, "Make a grub stake," he obtained the mail route from Hay Springs to Hemingford and Nonpareil, which he drove for two years Mrs Bartlett joined her husband in their new home in 1885, at a time when they had to get supplies from Valentine, a hundred and fifty miles away To make the trip usually took about fourteen days The next year the railroad reached Rushville and the first settlers began to gather up the buffalo bones and those of other animals and sold them for fourteen dollars a ton Mr Bartlett made some money this way and also by breaking land for other people who were coming into the country He lived on his homestead until 1893, adding to his land from time to time He was very successful in his business and at the present time owns seven thousand acres, which makes him one of the largest landholders in the county or western Nebraska For many years he has dealt and handled in live stock as well as raising many cattle and still uses his land for this purpose

From first settling in the Panhandle Mr Bartlett has taken a prominent part in the development of the county and played a leading part in public affairs, for in 1890, he was elected county commissioner of Dawes county and served three years In 1892, he superintended the breaking of the buffalo grass on the court house grounds and had it planted to blue grass the following year Then he took up a collection in the various county offices and had twelve trees planted, elm and boxelder, the first on the grounds around the court house The county up to this time had been renting a building for a poor house and hiring a matron In 1893 Mr Bartlett persuaded the other county commissioners that it would be better to have a county poor farm and succeeded in buying a quarter section of land two and one-half miles east of Chadron on Bordeaux creek,

which is practically self supporting Mr
Bartlett only used his good judgment in this
matter as in the many others coming up during his administration so that he made many
improvements that benefited all the county,
taking care of its affairs as he would his own
business

In 1893, Mr Bartlett was elected sheriff of
Dawes county, was re-elected in 1895 In the
spring of that year the people of Crawford began to object to the canteen maintained at
Fort Robinson, and called upon Sheriff Bartlett to close it He went, capturing a large
amount of liquor which the colonel in command refused to turn over, but when Mr
Bartlett informed him that he would call the
assistance of the State Militia, the liquor was
given up Mr Bartlett has served as a member of the board of managers and general superintendent of the Dawes County Fair Association for thirteen years, he was one of the
first men to realize the value of a county farm
bureau and helped in the promotion of the
Dawes County Farm Bureau of which he is
president This is an honor for Dawes was
the third county in the state to get such farm
demonstration and shows that Mr Bartlett
is a man abreast of the times He took a
leading part in the organization of the Farmer's Union, which owns a grocery store and
elevator of twenty-four thousand bushels, also
a flour mill of a hundred barrels capacity
Though engaged in all varied endeavors, Mr
Bartlett has continued the active management
of his ranching business and at one time owned
ten thousand acres of land on which he ran
five thousand cattle, eight hundred sheep and
some horses He now is the owner of five
ranches or tracts of land comprising nearly
seven thousand acres all well improved for
the purpose for which he uses it Mr and Mrs
Bartlett have one son, Alfred T, who is associated with his father in the ranch business
and also owns a thousand acres of land of
his own He was just getting his business in
shape to enlist in the army when the armistice
was signed

Arthur Bartlett, Sr, is a Republican, has
taken an active part in public life for many
years Mrs Bartlett is a member of the
Methodist church and the family is one of
the first financially and socially in Dawes
county

BENJAMIN LOEWENTHAL, who is not
only a pioneer settler but also the pioneer merchant of Chadron, Nebraska, is also the sole
survivor here of the nineteen original town

property owners The same enterprising spirit
that led to his coming to this section thirty-five
years ago, has been a moving force ever since,
not alone in the substantial expansion of his
own business interests, but in the development
of projects for the welfare of all this section.
He has long been one of the representative
men of Dawes county, and many times has
been called upon to fill positions of trust and
responsibility

Benjamin Loewenthal was born at Brooklyn, New York, February 10, 1855, a son of
Moses and Rachel (Cohn) Loewenthal Of
their five children, four survive, Benjamin
being the only one living in Dawes county
In 1881, Moses Loewenthal came west to the
Black Hills, and in the following year his wife
and children joined him They lived in that
region until 1889, moving then to Dawes county, Nebraska, where the rest of their lives were
spent

Growing to manhood in the east, Benjamin
Loewenthal had excellent opportunities in the
way of educational and business training, and
was well equipped when he came to Dawes
county With remarkable foresight he chose
a business location on a railroad line, on April
25, 1885, setting up a tent in the village of Da-
cota Junction, three miles west of the present
city of Chadron His stock of clothing was
not extensive because all goods had to be
freighted from Sidney, a distance of about
two hundred miles, but he maintained his tent
store until the following August, when he
moved to the new town of Chadron, with which
he has been prominently identified ever since

The first town lot in the new village was
sold August 1, 1885 As an inducement to
legitimate business men instead of speculators,
the originators of the town platted the ground
and offered lots for the very conservative sum
of from one to five hundred dollars, and Mr
Loewenthal was one of the nineteen men who
took advantage of this offer He at once
started to build a store immediately opposite
his present site, but sold it to an urgent buyer
before it was completed, and moved his goods
to another building, which he occupied until
1888, when he erected and took possession of
his present commodious building The new
town a famous shipping point for cattle
and business of every kind, became brisk as
people from all over the country moved in
When Mr Loewenthal established himself at
Chadron, he enlarged the scope of his business by adding boots and shoes to his line of
clothing, and has continued to deal in these

articles to the present time, two of his sons being associated with him

At Chadron, in 1886, Mr Loewenthal was united in marriage to Miss Rose Cohn, who died March 9, 1907 She was a daughter of Julius Cohn, a resident of New York, and the mother of five children Julius, Sadie, More, Charles and George In May, 1908, Mr Loewenthal married Bertha Loeffler, of New York City

In public affairs in Dawes county, Mr Loewenthal has been prominent and useful from the first He served in 1886, as first city treasurer of Chadron, and, in 1894, was elected on the Democratic ticket a member of the board of county commissioners of Dawes county, and served in 1894-5-6, during a period of momentous importance in the affairs of the county In civic matters he has been exceedingly active, serving on the school board for over twenty-one years, and as a member of the city council during 1912 and 1913 In 1914, he was elected mayor of Chadron, of which honor no citizen could be found more capable or deserving During the life of the Chadron Business Men's and the Chadron Commercial clubs, he was an active factor, and for thirty-four years he has been a member of Chadron lodge No 36, Odd Fellows, possibly a charter member and for many years an official No man in the business life of this city or in its official administration, is held more trustworthy than Benjamin Loewenthal

ROBERT A DAY —The genius for constructive achievement has marked the career of this representative pioneer merchant and banker of Oshkosh, Garden county, and the more credit is due to him by reason of the fact that this achievement has assisted him definitely in the advancement of local interests in general, besides giving him secure vantage ground as one of the substantial and valued citizens of his county

Mr Day was born in Brown county, Ohio, March 5, 1866, and the old Buckeye state likewise figures as the birthplace of his parents, Albert M and Mary L (Brown) Day, members of sterling families that were founded in that fine old commonwealth in the pioneer era of its history Albert M Day was actively engaged in farming in Ohio at the time of the outbreak of the Civil War, and such was his physical condition that he was rejected for service when he attempted to enlist as a patriot soldier of the Union Notwithstanding his remaining in civil life, he was taken prisoner by Confederate forces under command of the famous raider, General John Morgan, whose men likewise stole the horses from the farm of Mr Day, who was not long held in captivity After the war Mr Day continued his association with agricultural industry in Ohio until 1884, when he came with his family to Nebraska and became a pioneer settler near Ulysses, Butler county, where he accumulated valuable property and where he remained until his death, which occurred when he was about seventy years of age, his wife having passed away at the age of sixty-eight years Mrs Day was a daughter of Robert and Martha (Wardlow) Brown, who passed their entire lives in Ohio Of the children of Mr and Mrs Day the eldest is Elizabeth, who is the wife of G R Pollock, of Ulysses, Nebraska, Robert A, of this review, was the next in order of birth, William D is a resident of Ulysses, Nebraska, Lillian W is the wife of Frank Palmer, of Ulysses, and Osa M, wife of J M Stephens, likewise resides at Ulysses

Robert A Day passed the period of his childhood and early youth on the old home farm in Ohio and is indebted to the public schools of his native state for his early educational discipline He was nearly nineteen years of age when he accompanied his parents to Butler county, Nebraska, where he continued to be identified with farm enterprise for two years thereafter He then came to what is now Garden county, where, in December, 1886 he took up a homestead near the present town of Oshkosh During the first five years of his residence here he was in the employ of the Rush Creek Land and Live Stock Company, and in the meantime he instituted the development of his homestead, to which he duly perfected his title After severing his connection with the company mentioned, Mr Day became manager for the first mercantile store at Oshkosh, this establishment having been conducted by George T Kendall & Company, of St Paul, this state After being thus engaged about one year Mr Day opened the first drug store in the progressive new town, and the building which he used for this purpose was the second frame structure erected in the village After conducting the drug store about three years, Mr Day removed to Chappell, county seat of Deuel county, and there he served two years as deputy county clerk He then assumed the office of county clerk, of which he continued the incumbent two successive terms—1902-6—after which he became cashier of the Deuel County Bank, at Oshkosh, which was made the judicial center of the new county of Garden While thus serving as cashier of the bank he was chosen as the first county clerk of the new county, and his previous experience fortified him most admirably for the duties of this exacting office in the formative period of Garden county history. He retained the office of county clerk from 1910 to 1915, and in the meantime he transcribed from the records of Deuel county all data requisite for the new county. Mr Day has figured continuously as one of the most loyal, liberal and progressive citizens of Garden county, where his influence and co-operation have been given in support of all measures projected for the general good of his home town and county. From 1907 to 1915, he served as United States commissioner for Nebraska, and incidental to the nation's participation in the great World War he was county chairman of the first and second Liberty Loan drives, as well as chairman of the committee having in charge the drive for the sale of war savings stamps. Upon the organization of Garden county the bank of which he was cashier changed its title to the First State Bank, and of this substantial and important institution he was elected president in January, 1919 but on the 15th of the following March he retired from this office, his resignation being prompted by his impaired health. Since that time he has lived virtually retired, save that he is associated with his son-in-law, J C Schlater, in the real estate business, his broad and exact knowledge of realty values in this section of the state making his advisory service of great value in this connection.

In politics Mr Day is found arrayed as a staunch advocate and supporter of the cause of the Democratic party, he and his wife are active members of the Christian church, and he is a charter member of Oshkosh Camp, No 4991, Modern Woodmen of America. He is a citizen who has been the true apostle of progress and civic liberality, and none commands more inviolable place in popular confidence and esteem in Garden county.

On April 19, 1896, was solemnized the marriage of Mr Day to Miss Viola E Empson, of Vallonia, Indiana. Mrs Day was born at Sumner, Missouri, and was a child of two years when she accompanied her father, after the death of her mother, to Indiana, in which state she was reared and educated, her advantages having included those of Lebanon, Ohio, College, in which she was graduated. She has been a popular figure in the representative social activities of Oshkosh and presides most graciously over a home that is known for its cordial hospitality. Mr and Mrs Day have

three children. Pearl Marguerite is the wife of J C Schlater, of Oshkosh, and Robert Stanley and William A remain at the parental home.

JUST JOHNSON, who is now living retired at Oshkosh, Garden county, established his home in this locality in 1887, when the county was still a part of Cheyenne county, and here he gained pioneer honors in connection with industrial development and progress, for he obtained land a few miles south of the present town of Oshkosh and there developed a productive farm of one hundred and sixty acres, which valuable property he still owns. He was one of the successful exponents of agricultural and live-stock industry in Garden county during the various transitions that marked its segregation from Deuel county, and previously from Cheyenne county, and he has lived up to all the possibilities that have been offered, as is demonstrated by the substantial success that has attended his efforts and the high place which he has in popular confidence and good will.

Mr Johnson was born in Sweden, on October 12, 1859, and was reared and educated in his native land. In 1880, about the time of attaining his legal majority, Mr Johnson immigrated to the United States, and his principal capital consisted of his dauntless courage and determination, his sturdy physical powers and his ambition to achieve worthy success. He first settled in Sac county, Iowa, and after having there been engaged in farm enterprise three years he came to Blair, Washington county, Nebraska, where he assisted in building the bridge across the Missouri river. He then went to Rapid City, in the Black Hills district of South Dakota, and for a period of about one year he was engaged in freighting between that locality and Chadron, Nebraska. He then, in 1887, became one of the pioneers of the present Garden county where he took up a homestead and improved the farm property, where he lived until his retirement, in 1917, since which year he has maintained his residence at Oshkosh. He is aligned as a supporter of the cause of the Democratic party and his religious faith is that of the Lutheran church, of which his wife likewise was a communicant.

In July, 1904, Mr Johnson wedded Mrs Maggie (Edwards) Adell, widow of Charlie Adell. She was born at Bull Mountain, Colorado, and was but four years old at the time of her mother's death, her father, who was a miner in Colorado, died in 1908. Mrs Johnson who proved a devoted wife and mother, died in 1912, and of the three children the eldest was Ida May, who died October 20, 1917 William and Albert remain with their father in the pleasant home which he has provided at Oshkosh

MRS SARAH E VALENTINE, needs in this publication no voucher for her popularity, not only in her home community of Oshkosh, Garden county, but also on the part of the traveling public, for she is the gracious and generous hostess of the popular Oshkosh hotel known as the Travelers' Home, her management of which has made it a place that fully justifies its title

Mrs Sarah E (Stratton) Valentine was born at New Troy, Berrien county, Michigan, and is a daughter of James and Elizabeth (Abley) Stratton, the former of whom was born in the state of New York and the latter was a girl of fourteen when she came to America from her native Switzerland, in company with her parents James Stratton was a pioneer in southern Michigan and he passed the closing period of his long and useful life at Three Oaks, that state, where he died at the patriarchal age of ninety-two years, his wife having died in the city of Chicago, at the age of sixty-four years Mrs Valentine acquired her early education in the public schools of her native place and thereafter completed a commercial course in the Savor's Institute, of Chicago She became an assistant instructor in the telegraphic department of that institution and retained this position about one year Within a short time afterwards she became the wife of James H Redding, and they established their residence on the upper peninsula of Michigan at Menominee, where their son, Emmett F Redding, was born September 5, 1886 He now lives in Chicago where he is assistant superintendent of the Metropolitan Life Insurance Company, having been with that company for ten years, working from the bottom up to his present position They remained there eighteen years and there Mr Redding died at the age of fifty-five years Mrs Redding then removed to Chicago, where she followed the profession of nursing and where her marriage to William H Valentine was solemnized

In 1913, Mrs Valentine came to Oshkosh, Nebraska, mainly to assume charge of and perfect title to the homestead that had been taken up, in Garden county, by her deceased sister, Miss Flora A Stratton On July 17 of the same year, she assumed control of the Travelers' Home, which excellent hotel she has since conducted with marked success, so that it is a specially favored resort for the traveling public She has perfected title to the land mentioned and is still the owner of the property Mrs Valentine is a popular factor in the social activities of Oshkosh, is a member of the Congregational church and is actively affiliated with the Woman's Christian Temperance Union

SAMUEL P DeLATOUR cast in his lot with the people of progressive Nebraska in the year 1880 after a previous valuable experience of official and business life in the state of Arkansas In Nebraska his career has been marked by constructive and successful enterprise and he is one of the influential pioneer citizens of the splendid Panhandle of the state, with prestige as a banker and as a prominent man of the cattle industry He is president of the Bank of Lewellen, Garden county, and the owner of valuable farm land in this county, where he has the distinction of becoming the first settler in Bear Creek precinct He has been a prominent figure in civic and material development and progress in this section of the state and is an honored citizen, meriting special recognition in this history

Samuel P DeLatour was born at Platteville, Wisconsin, September 15, 1848 and, as the name implies is able to trace his lineage back to sterling French origin, though the family has long been established in America He is a son of John J and Sarah J (Parr) DeLatour, the former of whom was born in the state of New York, in 1815, and the latter of whom was born at Greenville, Bond county, Illinois, in 1825 her parents having been numbered among the earliest of the pioneer settlers of that section of Illinois, where she was reared and educated and where her marriage was solemnized in 1844 John J DeLatour completed a course in the historic old Williams College, New York and was a man of exceptional intellectuality, his profession having been that of civil engineer About the time of attaining his legal majority he removed to the west, in 1835 and established his residence in Illinois where he became prominently identified with the real estate business as well as with pioneer farm enterprise During the climacteric period of the Civil War he gave effective service to the Union, as an official in the quartermaster's department He passed the closing period of his life in the city of Chicago, where he died at the venerable age of eighty years In the meantime he had been for a time a resident of Wisconsin, where he had various business and industrial interests, the family having returned to Illinois when the son Samuel P, of this review, was a child of two years The devoted wife and mother died when she was about eighty-four years of age, and of the five children—three sons and two daughters—Samuel P was the second in order of birth

Mr DeLatour acquired his preliminary education in the common schools of Illinois, then completed a higher course of study in the Clark Seminary, at Aurora In December, 1869, about three months after the celebration of his twenty-first birthday, Mr DeLatour established his residence at Huntsville, Arkansas, where he was appointed deputy clerk of the district court He remained at Huntsville until 1872, when he removed to Helena, Phillips county, that state, where he served not only as clerk of the district court but also clerk of the United States District court, besides where he engaged in the banking business In 1880, he resigned the office of clerk of the Federal District Court and came to Nebraska Soon after his arrival in this state, Mr DeLatour settled at Cambridge, Furnas county, where he became associated with William E Babcock in organizing the Republican Valley Bank, besides engaging also in the live-stock and mercantile business In 1883, he disposed of his various interests in Furnas county and came to the western part of the state, to establish a cattle ranch ninty miles northwest of North Platte, in that part of old Cheyenne county that is now included in Garden county He thus became the first settler in the present Blue creek precinct of Garden county, as he filed entry on a pre-emption claim of a hundred and sixty acres, in 1884, improving it into a good ranch which he devoted principally to the cattle industry, an enterprise which proved successful The ranch later was in the newly organized Deuel county, and still later was included in the newer county of Garden A man of broad experience and fine intellectual endowment, Mr DeLatour was well qualified when he was called upon to serve through appointment, as county attorney of Deuel county, an office of which he continued the incumbent one term

In 1911, Mr DeLatour organized the Garden County Bank, at Lewellen, and in this enterprise he was associated with his two younger sons—Eugene and Ben C In 1914 the father and sons purchased the business of the Bank of Lewellen and the two institutions were then consolidated, under the title of the Bank of Lewellen Mr DeLatour has since been president of this substantial and well regulated institution, which has a capital stock of $50,000 00, and deposits which have now reached an average aggregate of $260,000 00.

Well fortified in his convictions as to matters of economic and governmental, Mr DeLatour gave his support to the Republican party until 1895, when he transferred his allegiance to the Democratic party, of whose basic principles he has since continued a stalwart advocate He is affiliated with the Masonic fraternity and his religious views are in harmony with the faith of the Episcopal church, his wife having been a zealous member of this church

At Helena, Arkansas, in 1873, was solemnized the marriage of Mr DeLatour to Miss Lucy McGraw, who was reared and educated in that state, though she was born in Kentucky Her sister became the wife of Honorable Powell Clayton, who served as governor of Arkansas and later as United States senator from that state Mrs DeLatour, a woman of most gentle and gracious personality, died in January, 1897 Of the four children the eldest is John McGraw DeLatour, who served in the medical department during the World War and who is now a resident of Cheyenne Wells, Colorado, where he is engaged in a real estate business Samuel Van Allen DeLatour resides at Lewellen, Garden county, and gives his attention principally to stock-raising Eugene, whose death occurred October 23, 1918, was associated with his father in the banking business at Lewellen, and prior to the organization of Garden county he served two terms as county clerk of Deuel county, and, Ben C DeLatour, who is vice-president and general manager of the Bank of Lewellen

JOHN MEVICK—When it is stated that for five successive years this well known pioneer and live-stock man of Garden county captured the grand champion prize for car-load exhibits of hogs at the Denver Stock Show, Denver, Colorado, it becomes apparent that he has not been laggard in promoting the live stock industry and advancing stock standards in the Panhandle of Nebraska His admirably improved stock farm, which still receives his personal supervision, is situated about four miles northwest of Lewellen, in which village he maintains his residence, after having lived on the farm for nearly a quarter of a century

Mr Mevick was born at Kenosha, Wisconsin, October 3, 1860, and is a son of Peter and Mary Mevick, the former a native of Germany and the latter of Ireland their marriage having been solemnized at Kenosha, Wisconsin, from which state they removed to Illinois in the year 1861, both having been young people when they came to the United States. Peter Mevich became a substantial farmer in Henry county, Illinois, but he died in the very prime of his manhood, having been forty-seven years of age at the time of his demise, and his widow, who long survived him, having attained to the venerable age of eighty-two years, her death occurred at Henry, Illinois.

John Mevich was an infant at the time of his parents' removal to Illinois, and there his early education was acquired in the public schools at Mineral, Henry county. His initial enterprise as a farmer was prosecuted in Illinois, where he remained thus engaged for a period of three years. He then removed to Hamilton county, Iowa, where he continued in agricultural pursuits for three years, at the expiration of which he came to Nebraska and numbered himself among the pioneer settlers in that part of old Cheyenne county that now constitutes Garden county. He took up homestead and tree claim and instituted their reclamation from the prairie wilds. With the passing years he continued to make excellent improvements on the property and became a specially prominent and successful representative of live-stock industry in the present Garden county, remaining on his original farm for the long period of twenty-four years. He still owns the property to which he has added until he now has a valuable landed estate of twelve hundred acres, and though he has resided in the village of Lewellen since 1909, he continues the active management of his farm and substantial live-stock operations and stands as one of the most extensive breeders and feeders of hogs in the western part of the state. His fine ranch, known as one of the best equipped in the Nebraska Panhandle, has four miles of hog fence, and three hundred acres of his land receives effective irrigation from Blue creek, he having been prominently concerned in the building of this irrigation system. In addition to raising hogs on a large scale Mr. Mevich usually runs an average of nearly two hundred head of cattle on his ranch and about fifty to sixty head of horses, though he is gradually reducing his activities in the raising of horses. He has been a leader in movements tending to advance the agricultural and live-stock industries in Garden county and was the first president of the Garden County Fair Association, of which office he continued the incumbent from 1910 to 1917, in the meantime he wielded vital influence in forming the policies and directing the other activities that have made this organization a most successful adjunct of industrial progress in the county. In politics he is a staunch supporter of the cause of the Republican party and fraternally he is affiliated with the Modern Woodmen of America. He and his wife are members of the Methodist Episcopal church and are sterling pioneers who have a wide circle of friends in the section of Nebraska in which they have long maintained their residence.

At Lewellen, Garden county, March 31, 1890, was solemnized the marriage of Mr. Mevich to Miss Grace White, daughter of Wellington and Mary (Langton) White, the former was born and reared in Wisconsin and the latter was born in England, whence she came with her parents to America when a girl, the family home being established in Wisconsin. Mr. White served as a member of a Wisconsin volunteer regiment in the Civil War, and continued with his command during virtually the entire period of this historic conflict. He finally came with his family to Nebraska and he and his wife now maintain their home at Lewellen. Mr. and Mrs. Mevich have two children. Ruth M. is the wife of George H. Morris, of Oshkosh, Garden county, and Charlotte M. remains at home.

J. MONROE BRUNT who has secure vantage ground as one of the substantial citizens and pioneers of Garden county, where he has been successful in his operations as an agriculturist and stock-grower, claims beautiful old Union county, New Jersey, as the place of his nativity. In the village of Rahway in that historic county, he was born July 12, 1858, and is a son of Joseph and Emma Brunt, both of whom were born in England, and the latter of whom was a girl when she accompanied her parents to America, the family home being established in New Jersey. Joseph Brunt was reared and educated in England and was a young man when he came to the United States, where he became a successful contractor at Rahway, New Jersey. In that state he and his wife passed the remainder of their lives, she having been sixty-four years of age at the time of her death and he having attained to the venerable age of eighty-two years. They became the parents of seven sons and three daughters, and the subject of the review is the youngest of the number. The four older sons William, James, Harvey and George all served gallantly as soldiers of the Union during the Civil War, and all were members of the same artillery command, with assignment to the same cannon.

J. Monroe Brunt acquired his preliminary education in the public schools of his native state, where he supplemented his training by a course of higher study in Princeton University, from which historic institution he was graduated as a member of the class of 1878, with the degree of Bachelor of Law   He was well equipped for the legal profession, but his tastes and inclinations led him into other fields of endeavor, so that he never engaged actively in the practice of law   Mr Brunt became associated with his father's contracting business, of which he was superintendent about five years, and he then assumed a large contract in the supplying of natural ice to the city of New York   After the completion of this contract he conducted a general merchandise business in his native town of Rahway for a period of about ten years, and for about one and one-half years thereafter he was connected with the company operating the DeBrosse street ferry, between New York City and New Jersey   In 1881, Mr Brunt set forth for the west, and finally settled in Henry county, Illinois, where he was engaged in farm enterprise about four years   He then, in 1885, came to Nebraska and established his residence at Ogallala, but in 1887, he came to what is now Garden county and took up a pre-emption claim about three miles distant from the village of Lewellen, which was not marked by a single building at that time   Later he filed claim to a homestead adjacent to his pre-emption claim, and here he has maintained his residence during the long intervening years which have been marked by worthy achievement on his part   His landed estate now comprises five hundred acres of well improved and productive soil, and here, in connection with diversified agriculture he is giving special attention to the raising of live stock—horses, cattle and hogs   He has been notably prominent and successful in the raising of horses of fine type and has made this department of his farm enterprise one of major importance   Mr Brunt has given his influence and co-operation in the movements and enterprises that have tended to advance the general welfare of the community and in his personal activities has kept pace with the march of splendid development and progress in this section of the state   Upon the organization of the Bank of Lewellen he became the first depositor in the new institution   He has been prominently identified with irrigation development here and was for ten years superintendent of the Bratt irrigation ditch—1896-1906

In politics Mr Brunt is aligned in the ranks of the Democratic party and prior to the organization of Garden county he served, in 1888-9, as sheriff of Deuel county   Prior to coming to the west he had served as chief of police in the city of Rahway, New Jersey, for two years, 1877-8   As a young man he served about three years as a member of the New Jersey National Guard, and he was first sergeant of Company F, Third New Jersey Regiment, when it was in active service during the great strike turbulence in New York City in 1880   He is affiliated with the Modern Woodmen of America and both he and his wife hold membership in the Methodist Episcopal church at Lewellen

The marriage of Mr Brunt was solemnized at Ogallala, Nebraska, in 1887, when Miss Ella Ross became his wife   She was born in Iowa, and is a daughter of Samuel and Mary A (Newsome) Ross, who were natives of Pennsylvania and who became pioneer settlers in Iowa, the father having been seventy-six years of age at the time of his death and the mother having passed away at the age of sixty-two years   Mrs Brunt came to Ogallala, Keith county, Nebraska, in 1886, and near that place she took up a homestead, to which she perfected her title in due time and of which she eventually made a profitable sale   Of the children of Mr and Mrs Brunt the eldest is Ethel, the wife of George West, of Lewellen, and has two sons and two daughters, and Nine V, Clayton W and Edith M remain at home   Clayton W the only son, served in the coast artillery during the progress of the World War and received his honorable discharge after the signing of the historic armistice

GEORGE M COCHRAN is the owner of a well improved farm of two hundred and forty acres, three miles northwest of Lewellen, Garden county and is now giving his active supervision to the property with energy and good judgment that are giving him maximum returns in his operations as an agriculurist and stock-grower   He was long identified with railway service, and in this connection he was foreman of the extra corps of workmen that laid the track on the North Platte Valley branch of the Union Pacific Railroad, through Garden and other counties in western Nebraska

George Madison Cochran was born in Jennings county, Indiana, August 15, 1869, and is a son of Richard M and Elizabeth (May) Cochran the former a native of Pennsylvania and the latter of Indiana, to which state her parents removed from North Carolina   Richard M Cochran became a prosperous farmer in Indiana and later in Kansas   He removed to Montana, 1899, and died a year later   Mrs

George K. Cogdill and Family

Cochran died in 1907 George M Cochran gained his rudimentary education in the public schools of his native state and was about twelve years old at the time of the family removal from Indiana to Montgomery county, Kansas, where he was reared to manhood on the home farm and where he continued to attend school at intervals He continued his association with agricultural industry in Kansas until 1886, when he came to Pawnee City, Nebraska, and entered the employ of the Chicago, Rock Island & Pacific Railroad Company, with which he continued in the construction and maintenance service about eighteen months For the ensuing two years he was foreman of a section gang on the Missouri Pacific Railroad, and for the following seventeen years he was in the service of the Union Pacific Railroad—as section foreman, extra-gang foreman, and as road-master on the division between Laramie and Rawlins, Wyoming He next acted as foreman of the extra gang that laid the track on the North Platte Valley branch, in which connection he established his headquarters at Lewellen, Nebraska In the meantime he became the owner of his present farm, and in 1909 left the railway service to take charge of his place, upon which he has erected good buildings and made other modern improvements, including the providing of irrigation facilities He is one of the appreciative and loyal citizens of Garden county and has served six years as a member of the Lewellen school board His political allegiance is given to the Republican party, he and his wife are members of the Methodist Episcopal church, and since 1890 he has been affiliated with the lodge of Ancient Free & Accepted Masons at Brock, Nebraska

At Big Springs, this state, in 1892, Mr Cochran wedded Miss Jennie B Plummer, who was born in Wisconsin, where she was reared and educated Her father, Walker D Plummer, was a resident of Mount Hope, Wisconsin, at the time of his death, when he was about fifty-six years of age, and her mother, Margaret (Chisholm) Plummer, a native of Scotland, died at the venerable age of eighty-one years Mr and Mrs Cochran have two sons and two daughters George Byron, who resides at Alliance, Box Butte county, married Miss Helen Shoup, and they have one child, a daughter, Rolland Bruce, Myrtle B and Vivian M remain at home

GEORGE K COGDILL horseman rancher and groceryman, was born in Gentry county, Missouri, May 22 1866, the fifth in a family of seven children, three boys and

four girls His father, Miles Cogdill, died when George was at the age of five years Miles Cogdill was also born in the state of Missouri and was, by occupation, a farmer and blacksmith George's mother, a Miss Eliza Perkins, with her parents emigrated from Illinois to Missouri and there married Miles Cogdill and thus the beginning of our subject and surroundings that well qualified him for the life he was destined to live At the age of eight years he began to cast about and through perhaps the adventurous disposition that was his worked at the grocery store and on the farm making what spending money he could and attending the district schools until he was about twelve years old He was a lover of good books and received much knowledge from them About this time he and a cousin of the same name and about the same age were herding cattle on the Empire Prairie a tent being their only shelter One Sunday afternoon, when a dark cloud appeared in the west, the boys watched it as they rode about their cattle and before they realized their situation, they were nearly in the midst of one of the worst cyclones that ever struck Missouri, missing the boys but a short distance For the next five or six years George worked on the farm raising stock, having developed a love for stock, especially horses, and when about eighteen years of age his mother died, and the following fall, 1884, he drifted west to Valentine, Nebraska, falling in with a freight outfit and walking most of the way still farther west to Bordeaux Creek, stopping at Mr and Mrs Boner's, in Sioux county, now Dawes county, where for the next several years, it was his home when in Dawes county He stayed there about a month and at that time had the pleasant experience of being almost run over by a Black Tail deer He liked this part of the country as there were plenty of both small and big game but he returned to Missouri and in the next spring in company with a cousin by the name of Lute Russell, came back to Dawes county Chadron, at that time, was only a small village, mostly tents, so the boys, hunting adventures as much as work, walked sixty miles west to Harrison, Sioux county Nebraska, which at this early date consisted of one tent about twelve by fourteen feet, occupied by a thrifty merchant with a small stock of groceries and a large stock of forty rod whiskey They hired out there to the Fremont, Elkhorn and Missouri Valley Railroad Company to cut ties and were furnished with blankets and a grub stake and went into the hills nearby, built a shelter of poles and pine boughs, but life was too short for the boys to stay in this lonely place and still having a few dollars, they walked west thirty miles to Lusk, (then called Silver Cliff) Wyoming, stayed at Lusk one week, then walked back east fourteen miles and went to work at the Node (Flying E) Cow Ranch, where George worked on the roundup two seasons and learned the ways of handling stock in Wyoming, and in the spring of 1888, George, with a good saddle horse and Tom Lockett with his team, headed their horses west on a trip across the continental divide to Idaho Arriving at Lusk, Wyoming, they fell in with a man by the name of Joe Rogers, who had a team and wagon, so George, Joe and Tom pulled on together and when near Glenrock, Wyoming, one rainy evening they camped in an old shack at the protest of George, as they had tents, and a few days later, when at Casper, Wyoming they got word that the old shack had been damaged by fire, so not having time to fight the imputation of wrong doing, they paid $13 50 each, so leaving Tom at Casper, George and Joe headed their horses toward the setting sun and when reaching the Sweetwater country, George was taken seriously sick Joe did all he could for him and finally told him he would have to cross the high divide if he didn't get a doctor, who was at Casper, seventy-five miles away

George refused to have a doctor or even to be moved to the ranch where Joe had insisted on taking him, and after lying there in a tent for a week or ten days he finally got well, but the trip to Idaho was abandoned and George and Joe went to work at the U T Cow Ranch on Sweetwater river near Devil's Gate George rode over this part of the country on the roundup and a few days after starting north at Pine Mountain about five o'clock one evening, missed his private saddle horse that was running in the saddle horse herd, and suspected that he had been stolen He made a short circle over the country and struck the trail of the horse, discovering the horse was headed east The horse was shod and consequently more easily followed and so he followed the trail rapidly until dark, as the trail entered a steep gulch He then pulled the saddle from his horse, tied one end of the saddle rope to the front foot of the horse and the other end to the saddle horn and using the saddle for a pillow, dropped onto his blanket and caught what little sleep he could, with the sharp bark of the coyote ringing in his ears as he drifted off to dreamland The next morning when it

was light enough to see the trail and after following it for four or five hours, a large cow herd so obliterated the horse's tracks that he could not pick it up again, and after riding into Casper no clue to the horse was found The next morning, realizing his cow camp had moved, he did not attempt to go back over the same trail, but took a short cut across the rolling prairie, his cowboy training standing him in good stead, and at dusk he rode into camp, having traveled a distance of one hundred and fifty miles

On the trail George had many thrilling experiences and when at the ranch of the U M Cattle Company, branding and turning over cattle, he narrowly escaped being gored to death by a Texas cow but had the presence of mind to throw himself on the ground as the cow stumbled over him, and in the fall of 1889 George and Lew Spaulding saddled their private horses and rode to Casper As they jogged down the Sweetwater valley they were talking of the past and what she could tell, if she could only talk, and nearing Fish Creek, they could see the surroundings and talked of the hanging of Cattle Kate and a man by the name of Avery, which had happened a short time before During the following year he worked for the L O and other cattle outfits and drifted back and forth from the cattle country to his home on Bordeaux Creek, where he filed on a homestead in 1890 and in 1891 went back up the trail with the outfit from Orin Junction, Wyoming, to Red Water, Montana This trip was more clouds than sunshine — riding all day and standing two hours night guard This herd was made up of from two to four year old steers partly unbranded and when laying over a day or making an early camp, the boys would rope and brand cattle for past time It rained the first two or three days out and while branding cattle one evening, a boy had his horse jerked down on him breaking his collar bone and two ribs A few days later, while the outfit was on the move, they got the news of Tom Wagner, who had been hung to a limb, which was about ten miles away On the 14th of July, north of the Cheyenne river on the head of Lodge Pole, George and two other boys were caught in a terrific hail storm and as they drifted along together and as the hail became unbearable George reached down for his ladigo straps, slips off of his horse and in a few seconds had his saddle over his head As he peeked out from under his saddle he saw the other boys had done likewise, and then took his forty-five and fired two shots into the air and was answered back with the same report

One evening on the Little Missouri River a grey wolf got up and they gave chase, and George roped the ugly loafer They stretched him out, ear marked him and gave him his liberty

When the Yellowstone was finally reached, considerable trouble was experienced that we haven't space to describe, and after crossing and striking the Red Water range, which was a broad rolling prairie and the land of beef cattle — this was a rainy, foggy morning — here the herd was divided into three bunches and driven off in different directions to be turned loose George and Tom Berry, a man who was acquainted with this part of the country, took one bunch of the cattle and drove them northwest, turning them loose on Muddy Creek and then started for Alkali Springs, where the mess wagon had camped This was about twenty miles from the ranch and Tom got lost and they wandered around for about six hours, finally riding into camp about three o'clock in the afternoon, the outfit moving on into the ranch the next morning About the fifteenth of July they began shipping beef cattle, making four shipments of eight hundred head each, having to cross either the Yellowstone or the Big Missouri with each shipment to get to the railroad On reaching the Big Missouri on one of these drives, George took his horse and crossed the river to hold the cattle up and on reaching the opposite bank his horse kept bogging down, while the boys cheered from the other side and it was plain to see that they couldn't cross the cattle George recrossed the river and the cattle were thrown back for a week for the bank to dry off, and George had charge of the outfit as they moved back to the ranch Finally at the close of the last shipment, which they drove to Fallon, near Glendive, Montana, George bade goodbye to the boys and went to Chicago with that train load of cattle and on the way back he visited his old home in the land of the blue grass, at Stanberry, Missouri He stopped there only a few days and then come on to Chadron and his homestead on Bordeaux Creek and the following March 20, 1892, he was happily married to Miss Eva Clark at the home of the bride's parents in Antelope Valley in Dawes county

Mrs Cogdill was the third child in a family of nine children, seven girls and two boys, and at the time she was married to Mr Cogdill, was a school teacher During the time that Mr Clark and his family have lived in Dawes

county they have built up one of the finest ranches in the country and he and his good wife have retired and now live on "Easy Street" in Hay Springs, Nebarska

After his marriage, Mr Cogdill and his wife moved upon his homestead on Bordeaux Creek and for the succeeding six years, to keep the wolf from the door, he cut wood and hauled logs from Pine Ridge to the Wilson saw mill on Bordeaux Creek, giving one third of the finished lumber for having it sawed, and hauled his wood twelve miles to Chadron — receiving ten dollars per thousand for the lumber and two dollars and a quarter per cord for the wood, during those years He then changed the style of his business to pasturing and breaking horses for the neighbors and others that lacked the nerve or inclination to do it for themselves, his cowboy training standing him in good stead for this sort of work and which business he followed for several years At one time he nearly lost his life riding a wild horse that ran away with him and into a barb wire fence The wire caught on the stirrup, turning the saddle with his foot in the stirrup, and his boot pulled off, letting his foot out or he would have been dragged to death

During these years he also worked into a small herd of cattle and horses and bought adjoining deeded land until in 1919 he owned fifteen hundred and sixty acres of well improved deeded land with a very substantial herd of good stock and ranch equipment, which he and his good wife sold for a life's fair financial competence and concluded to take a rest for a year

Mr and Mrs Cogdill have seven children Denver R Cogdill, now married, living at Hat Creek, Wyoming, Hazel Munkres (nee Cogdill), married, Chadron Nebraska, Edna Hoke (nee Cogdill), married, Chadron, Nebraska, William Dale Cogdill, mechanic, single, living with parents, Raymond Cogdill, at home with parents, State Normal student, Helen G Cogdill, student at State Normal, Mary E Codgill, student at Chadron high school

The following spring after the sale of the old home ranch Mr and Mrs Codgill, with their three youngest children, Raymond, Helen and Mary, drove to the Pacific coast for a pleasure trip, shipping down the Columbia River twenty miles and on the way back from the Pacific Coast, shipped from Bremerton across the canal or bay a distance of 19 miles, to Seattle returning home by way of Yellowstone Park and in order to partially satisfy his restless spirit for industry, with his son-in-law, John Hoke, Mr Codgill purchased the Beghtol grocery stock in Chadron at 231 Main Street, and we predict a successful business career and a full share of the public patronage for them Mr Cogdill owns a modern home in Chadron and he and his estimable wife are among the best families in Dawes county Mrs Codgill takes an active part in the social functions among the ladies of the city and George is certainly finding pleasure in selling that which satisfies the inner man as he did when a small boy

JAMES A WILSON — He to whom this memoir is dedicated was numbered among the honored pioneers and successful farmers of what is now Garden county, where he established his residence in 1886, when this county was still a part of Cheyenne county and which was to become later a part of Deuel county, to which it remained attached until the organization of Garden county Mr Wilson, who died on April 15, 1906, improved and developed the fine farm property upon which his widow still resides, about five miles from the village of Lewellen He was one of the progressive farmers and honored and valued citizens of this locality at the time of his demise, when he was fifty-six years of age

Mr Wilson was born in the state of Indiana, July 6, 1850, and was a son of William and Amanda Wilson, who were natives of Greene county, Ohio and who became pioneer settlers in Warren county, Indiana where the father developed a farm and where both he and his wife passed the remainder of their lives their family consisting of three sons and seven daughters James A Wilson, the next to the youngest of the children, was reared to adult age in Warren county, Indiana, where he profited by the advantages of the public schools of the period and where he early gained fellowship with the sturdy and invigorating life of the farm In 1872, he engaged in general farm enterprise in Iriquois county Illinois, where he was actively identified with agricultural and live-stock industry until 1885, when he came to Nebraska and located in Custer county There he remained about one year and then came with his family to what is now Garden county, where he took up pre-emption and tree claims and instituted the reclamation of a farm He thus became one of the pioneer agriculturists and stock-growers of the county as now constituted, and to the improvements which he made on his land his widow has materially added since his death while she has shown marked ability and discrimination in carrying

forward successfully the farm activities which he had initiated Mr Wilson was a man of sterling integrity in all the relations of life and commanded the unqualified esteem of all with whom he came in contact His political allegiance was given to the Democratic party, but his widow, who now enjoys the franchise, is well fortified in her convictions and is found arrayed in the ranks of the Republican party Mr Wilson held membership in the Baptist church, of which Mrs Wilson likewise is a devoted adherent She is a popular figure in connection with the representative social activities of her home community, is affiliated with the Royal Neighbors and is at the time of this writing, in the the winter of 1919-20, president of the Lewellen Women's Club

On March 21, 1882, was solemnized the marriage of Mr Wilson to Miss Mollie Belle Meeker, who was born in Clark County, Illinois, on September 2, 1861 a daughter of Matthias and Elizabeth (Allstott) Meeker, the former a native of Essex county, New Jersey, and the latter of Indiana Matthias Meeker was reared and educated in Ohio and became a shipbuilder by trade and vocation, with residence in Cincinnati, Ohio He later removed to Indiana, where his marriage was solemnized, and finally he engaged in farming in Clark county, Illinois, where he died at the age of seventy-two years, his wife having been about eighty-four years of age at the time of her death Of their six sons and four daughters, three sons and two daughters are now living, three of the sons having been gallant soldiers of the Union during the Civil War, as were also three brothers-in-law of Mrs Wilson — the husbands of her sisters Mr Wilson was afforded the advantages of the public schools of Illinois and in that state her marriage occurred She is a woman of culture and gracious personality, and is the popular chatelaine of one of the attractive rural homes of Garden county In conclusion of this memoir is given brief record concerning the children of Mr and Mrs Wilson, Albert Ray, who is a resident of Billings, Montana, married Miss Myrtle Gillilard and they have two children, Addie became the wife of Alfred Fought, and by this marriage she has two sons and one daughter, after the death of her first husband she became the wife of S G Dumond, and they reside at Alliance, Nebraska Bertrand, who married Miss Edna Adams, and has one son, was in active service as a member of Company F, One Hundred and Ninth Engineers, during the period of the nation's participation in the World War, and since the close of the war he has been engaged in the drug business at Grand Island, and Earl, the youngest of the children, remains with his widowed mother on the home farm

CHARLES M EGGERS has been a resident of the state since the year 1888, and came to what is now Garden county in 1896 where he reclaimed and improved a productive farm of a hundred and sixty acres, which he bought at a cost of twelve hundred dollars He secured his homestead in Blue Creek precinct in 1902, the valuable property being still in his possession In 1911 Mr Eggers removed from his farm to the village of Lewellen, where he has since lived practically retired as one of the substantial and highly esteemed pioneer citizens of this section of the state

Charles Matthew Eggers was born in Parke county, Indiana December 23, 1867, and is a son of Enoch and Lydia (Brock) Eggers, the former of whom was born and reared in Indiana and the latter in Kentucky Enoch Eggers continued his farm operations in Indiana until 1878, when he removed with his family to Kansas, and initiated the development of a pioneer farm He passed the closing period of his life in the state of Nebraska, where he died at the venerable age of eighty-five years, his wife having passed away at the age of eighty-one years Both were earnest members of the Baptist church They became the parents of eight children, all sons and of the number four are living and all reside in Nebraska Aside from the subject of this sketch the others are, Thomas who likewise resides at Lewellen, Joseph E who also is one of the well known citizens of the Lewellen neighborhood and James M who resides at North Platte, Lincoln county

Charles M Eggers obtained his youthful education in the public schools of Indiana and Kansas to which latter state the family removed when he was ten years of age In the Sunflower state he was actively associated with his father in the developing and other work of the pioneer farm and there he eventually engaged in independent farm enterprise About one year later, however he returned to Indiana, where he was identified with farming for the ensuing two years, at the expiration of which in 1888 he came to Nebraska and turned his attention to general farming in Lincoln county There he continued operations eight years, at the expiration of which he came to that part of Deuel county which now comprises Garden county, and took up a homestead in Blue Creek precinct Here he

continued his successful activities as an agriculturist and stock-grower until 1911 when as previously noted, he retired and established his residence in the village of Lewellen He has had no ambition to enter the arena of practical politics, but has been at all times liberal and public-spirited as a citizen and he gives his support to the cause of the Democratic party, both he and his wife being members of the Baptist church

In Deuel (now Garden) county this state, January 1, 1896 Mr Eggers wedded Miss Mollie Boggs, a daughter of Hugh and Nancy (Woods) Boggs who were natives of Kentucky and who settled in Adams county, Nebraska, in 1885 Two years later Mr Boggs came with his family to what is now Garden county, where he took a homestead and engaged in farming, as a pioneer During the four years from 1891 to 1895, he served as mail carrier between Ogallala and Oshkosh, and he was one of the well known citizens of the county at the time of his death at the age of sixty-nine years his widow likewise passing away at the same age Mr and Mrs Eggers had two children Birdie, the wife of Ed McCormick, of Lewellen, and Nannie remains at the parental home

HENRY TILGNER, who has gained a competency through his able and earnest activities as an agriculturist and stock-grower in western Nebraska is now living practically retired in the pleasant village of Lewellen, Garden county, having located in this county in 1895 when it was still a part of Deuel county Mr Tilgner was born in the province of Silesia, Germany, January 13, 1850, and there his parents passed their entire lives He whose name introduces this paragraph gained his early education in the schools of his native province and, in 1869, as an ambitious and self-reliant young man who was determined to win prosperity through his own efforts, he emigrated to the United States He settled near Watertown Wisconsin, and there he eventually established himself as an independent farmer There he continued his farm activities until 1885, when he came to Nebraska and took up homestead and pre-emption claims in Frontier county where he put his energies and judgment to such effective use that he developed a valuable and productive farm property and gave his attention to both agriculture and the raising of live stock In 1895, he sold his farm and came to that part of Deuel county that is now comprised in Garden county Here he purchased a quarter-section of land, three and one-half miles east of Lewellen and made many substantial improvements, and achieved success in diversified agriculture and the raising of live stock He became one of the most extensive hog-raisers in the county and made this a prime feature of his farm enterprise until his retirement, in 1918 when he established his residence at Lewellen, where he owns one of the attractive homes of the village. A man of broad views and progressive ideas, Mr Tilgner has been liberal in the support of movements for the general good of the community and he served four years — 1915-1919 — as regent of the Garden county high school His political support is given to the Republican party, and he and his wife are members of the Lutheran church He still owns his farm, which has greatly appreciated in value in late years, and is a part owner of the Grafen ditch, which affords irrigation facilities for his land

In February, 1871, Mr Tilgner wedded Miss Minnie Schultz, who died in 1897 and was survived by eight children, all of who are still living The second marriage of Mr Tilgner was solemnized at North Platte, when Miss Betty Kirsch became his wife She was born in Germany and was a young woman when she came to America Mr and Mrs Tilgner have four children, two sons and two daughters Annie, married Norman Green and lives at Lewellen, Irving E, married Aria Higgins and lives at Lewellen, Roy C, and Alice M, live at home

IRA COPLEY is one of comparatively few citizens in Garden county who can claim the fine old Bluegrass state as the place of their nativity In Garden county he demonstrated in a marked way the possibilities in the domain of so-called dry farming, his experience having covered a period of twelve years, in none of which did he lose a crop He has, however, full appreciation of the superior claims for irrigated farm lands, and he is now conducting operations on his well improved farm of six hundred and forty acres which is irrigated from what is known as the West ditch His farm, which is given over to diversified agriculture and the raising of good types of live stock, is situated sixteen miles northwest of the village of Lewellen, which is his post-office address, and the community at large looks upon him as a sterling and progressive citizen of marked public spirit

A scion of a family that was founded in the fair old commonwealth of Virginia in the colonial era of our national history Ira Copley himself was born in Boyd county, Kentucky, February 1, 1875 His early education was received in the schools of Kentucky and West Virginia, in which latter state his father was actively identified with lumbering operations for a term of years Mr Copley is a son of James S and Martha A (Hammonds) Copley, the former of whom was born in West Virginia and the latter in Kentucky, where she was reared and educated and where her marriage was solemnized James S Copley continued to be engaged in lumbering activities in the south until 1907, when he came to Nebraska and located in that part of Deuel county that now constitutes Garden county Here he took up a homestead and by the time he had perfected his title to the property he had brought an appreciable part of the land under effective cultivation and made excellent improvements, including the erection of substantial buildings Here he continued his farm operations until 1918, when he and his wife returned to West Virginia, where they now maintain their home Of their eight children, the subject of this sketch was the fourth in order of birth Another of the sons, James S, Jr, likewise is a representative farmer in Garden county

Ira Copley early gained experience in connection with the lumbering business, through association with his father's operations in West Virginia, and in that state he was for a time engaged in farming In 1907, he came to what is now Garden county, Nebraska, and took up a homestead claim of six hundred and forty acres, near Lewellen He stocked the place well and eventually developed a productive farm that was marked by good buildings and other improvements Here he continued successful operations as an agriculturist and stock-grower during a period of twelve years, at the expiration of which he removed to his present home farm, in order to secure the facilities of irrigation, which insure to him still greater success in his farm enterprise

In politics Mr Copley is found aligned as a stalwart in the ranks of the Republican party, and he and his wife are active members of the United Baptist church

Mr Copley was twenty-three years of age when in Wayne county, West Virginia, he was united in marriage to Miss Mary Wooten, who was reared and educated in that state and who is a daughter of S P Wooten, a native of Kentucky Mr and Mrs Copley have a fine family of ten children Grace is the wife of John Williams, of Lemoyn, Keith county, Ethel is the wife of John Goodron, of Lewellen, and all of the other children are still at home at the time of this writing, in the winter of 1919-20 —Luther M , Martha A , May, Alice, Winifred, Pearl Mabel and Hazel With this interesting family, the home of Mr and Mrs Copley is known for its cheer and happiness, as well as for its gracious hospitality

WILLIAM T JACKSON, who is numbered among the successful and popular exponents of farm enterprise in Garden county, has been a resident of Nebraska since he was a lad of about twelve years , is a representative of one of the sterling pioneer families of the state and in his independent career he has exemplified most fully the self-reliant and progressive spirit of the fine commonwealth in which he was reared and in which he has found ample opportunity for productive achievement as an agriculturist and stockgrower

William Thomas Jackson was born in De-Witt county, Illinois, July 21, 1861, and is a son of John A and Rhoda Ann (Harp) Jackson, the former a native of Indiana and the latter of Illinois The father was identified with agricultural pursuits from his youth until he became one of the California argonauts of the historic year 1849, when he joined the great host of adventurous spirits who were making their way to the New Eldorado, in search of gold He was measurably successful as one of the celebrated Forty-niners in California and on his return to the east he made the trip by way of the Isthmus of Panama In 1859, he again set forth for the west, at the time of the gold excitement in Colorado, and he remained for a time in the vicinity of Pike's Peak, that state He then returned to Illinois, where he continued to be engaged in farming until 1873, when he came with his family to Nebraska and settled in Thayer county There he developed a good farm and became a successful and influential citizen of his community He was one of the honored pioneer citizens of Thayer county at the time of his death, which occurred when he was eighty-two years of age, his wife having passed away in 1895, at the age of sixty-two years

William T Jackson acquired his rudimentary education in the public schools of Illinois and after the removal of the family to Nebraska he continued his studies in the schools of Thayer county, where he was reared to manhood and where he early began to contribute his quota to the work of the home farm, so that he was well fortified in practical experience when he initiated his independent operations as a farmer He continued his

farm activities in Thayer county until 1907, when he came to that part of Deuel county that now comprises Garden county Here he entered claim to a homestead and has developed and improved one of the fine farm properties of the county his well directed enterprise including both diversified agriculture and the raising of live stock Mr Jackson has been earnest in the support of the various measures and enterprises that have conserved communal advancement and is essentially liberal and public-spirited He has had no ambition for public office but has served nearly a decade as school director, an office of which he is the incumbent at the time of this writing He is well fortified in his convictions concerning governmental policies and is aligned in the local ranks of the Democratic party

At Hebron, Thayer county, August 10, 1897, was solemnized the marriage of Mr Jackson to Miss Annie M Bell, daughter of David and Betsey (Gooding) Bell, both of whom were born in England Mrs Bell was thirty-four years of age at the time of her death and Mr Bell attained to the age of sixty-one years, he having been a representative citizen of Thayer county at the time of his death Mrs Jackson was born in Lancastershire, England, and was an infant at the time of the family immigration to the United States her parents settling in Iowa, where they remained about twelve years, at the expiration of which, in 1882, they came to Nebraska and settled in Thayer county, where the father became a successful agriculturist and stock-grower Mr and Mrs Jackson have three children Rhoda, is the wife of William B Cate, of Elizabeth, Cherry county, and they have three children; Eva G is the wife of Edward M Bredesell of Hebron Thayer county, and they have one child, George A , who has returned home after service in the army during the World War, was a member of Ambulance Company No 75 and was stationed at Camp Dodge, Iowa, at the time when the war closed

RALPH W EMERSON came to Garden county in 1897, prior to its segregation from Deuel county, and here he has shown his energy and enterprise in the developing of one of the fine farm properties of the county, and has gained precedence as one of the successful exponents of agricultural and live-stock industry in his native state His farm comprising a hundred and sixty acres, is situated six miles northwest of the village of Lewellen, where he has made good improvements, so that the place gives unmistakable evidence of thrift and prosperity

It is but natural that Mr Emerson should exemplify the progressive spirit so characteristic of western Nebraska for he claims Custer county, this state, as the place of his nativity and there breathed in his boyhood and youth the vital air, productive industry, besides gaining the practical experience that equipped him effectively for independent farm enterprise after he had attained to maturity Ralph Warren Emerson was born in Custer county, Nebraska, November 22, 1883, and is a son of Edward W and Louise Emerson, the former of whom was born at Alton, Illinois, while the latter was a native of Yorkshire, England, having been a girl when her parents came to America and settled in Missouri Mrs Emerson came to Nebraska as a young woman and she became a successful teacher in the public schools at St Paul, this state, her school work having continued until the time of her marriage She was a woman of talent and gracious personality, and was only thirty-seven years of age at the time of her death, in 1894 Edward W Emerson was reared and educated in Illinois, when he came to Nebraska in the early eighties and settled in Custer county where he took a homestead and became a pioneer farmer of the county He improved a valuable farm and still lives on the old home place, as one of the representative agriculturists and stockgrowers of Custer county

Ralph W Emerson passed the period of his childhood and early youth on the pioneer farm homestead in Custer county, and availed himself of the advantages of the public schools He continued his association with farm enterprise until 1897, when he came to what is now Garden county, and was employed on ranches and farms until he found the desired opportunity to engage independently in agricultural activities In 1908 he entered claim to a homestead upon which he proved up in due course of time He finally sold this property and purchased the quarter-section which constitutes his present attractive farm home the place having good irrigation facilities and having been developed by him into a model farm, his attention being given to diversified agriculture and the raising of excellent grades of live stock Mr Emerson is a stockholder and director of the Pasley irrigation ditch, is a Republican in politics, is affiliated with the Modern Woodmen of America and he and his wife hold membership in the Methodist Episcopal and Lutheran churches

April 5, 1907, was recorded the marriage of Mr Emerson to Miss Emma H Johnson who was born and reared in what is now Garden county, a daughter of Frederick Johnson, of Lewellen Mr and Mrs Emerson have three children Glen L, born September 25, 1911, William A, born January 9, 1916, and Jessie Mae, born May 1, 1918

ALBERT R TAYLOR — In the fine farm district of which the village of Lewellen is the normal trade center are to be found many vigorous and progressive agriculturists and stock-raisers, whose comparative youth vitalizes their activities and makes them specially successful A well known and popular member of this valued class of citizens is Albert Robert Taylor, who was born in Clay county, Nebraska, January 1, 1880, and is a son of Robert and Minerva (Reynolds) Taylor, the former a native of Pennsylvania and the latter of Ohio, where she was reared and educated Robert Taylor became one of the pioneers of Clay county, Nebraska, where he established his residence in 1872 He there took up and perfected title to a homestead, and there he continued his farm enterprise until 1885, when he initiated further pioneer experience by removing to that part of Cheyenne county that now constitutes Garden county Here he secured a tree claim and a pre-emption claim, but eventually he sold these properties and purchased a farm on the south side of the North Platte river, where he successfully continued his activities as an agriculturist and stock-grower, with secure standing as one of the honored pioneer citizens of this section of the state Albert R Taylor was reared to manhood in what is now Garden county, where his early educational advantages were those offered by the excellent public schools He assisted in the work of the home farm until he began independent operations by taking up a homestead on the south side of the river He proved up on this claim and after disposing of the property he purchased his present farm, which comprises a hundred and sixty acres situated five miles northwest of Lewellen The farm has good irrigation, is improved with excellent buildings and is the stage of very successful enterprise in the domain of agriculture and stock-raising Mr Taylor is loyal and progressive as a citizen, is a Republican in politics, and is a stockholder in the Overland Ditch Company, through the medium of which his farm gains its irrigation facilities

January 21, 1907, Mr Taylor was united in marriage to Miss Addie Luark who was born in Clay county, Nebraska, and is a daughter of Edward and Margaret (Barkman) Luark,

George C. Snow

now residents of Boulder, Colorado, to which place they removed in 1917 The parents of Mrs Taylor were born and reared in Iowa and her father became a pioneer settler in Clay county, Nebraska, where he established his home in the seventies and where he remained until his removal to Colorado, as previously noted Mr and Mrs Taylor have two children — Mina Hope and Wanda Lorraine

GEORGE C SNOW, editor and proprietor of the *Chadron Journal*, and a member of the Nebraska State Legislature, worthily occupies a position of great prominence in the state He has been the recipient of many honors, both political and personal, in his long career, and his fellow citizens have frequently testified to their sincere esteem Many have known him longest and best in the field of journalism, for Mr Snow is the oldest editor, in point of service, in this part of the country

George C Snow was born in De Kalb county, Illinois, March 5, 1874 His parents are Rev Beecher O and Stella (Lyon) Snow, natives of New York, born in 1853 and 1854 respectively They now reside at Milwaukee, Wisconsin, the father being a retired minister of the Congregational church, for twenty years having served as Home Missionary Pastor for Nebraska Of his four children, George C is the only one living at Chadron At Franklin Academy and Doane College in Nebraska, Mr Snow pursued his studies until early manhood, then accepted the superintendency of a Congregational academy at Snohomish, Washington, going from there to Eureka, Kansas, and then came to Chadron, Nebraska Mr Snow then bought the *Chadron Journal*, which is the oldest newspaper west of Valentine It was established by Edward Egan, the press and cases being set up in a wagon before any railroads had been constructed through this section It has always been Republican in political policy, and it subscription list since Mr Snow took charge has extended all through Dawes, Sioux and Box Butte counties, where not only are his editorial talents greatly appreciated but confidence is inspired as to his safe and sane leadership in questions touching upon the treasured basic principle of American independence

Mr Snow has been publishing the *Journal* for the past fifteen years He now has his own building and one of the best fitted offices and finest equipped printing plants in the state and in connection with the newspaper, operates a large and profitable job office, his printing force including a number of competent employes

Mr Snow was married at Farnam. Dawson county, Nebraska, July 24, 1901, to Miss Mary Batty, who is a daughter of Rev George and Celestine (Greswold) Batty, and they have four children, namely Clayton B, Mildred A, George B, and Mary M Mr Snow and his family are members of the Congregational church

Mr Snow has always been a consistent Republican and his work for the party has been loyally and unselfishly performed, has been called to many party councils and given yeoman service in senatorial and gubernatorial campaigns In the late legislative election of the Seventy-fourth District, that includes Dawes and Sioux counties, Mr Snow was sent to the House of Representatives for the second time, his election being widely welcomed by those who appreciate his ability and honor his sterling character In local affairs Mr Snow has never been negligent since becoming a citizen of Chadron, at all times assuming his share of citizenship responsibilities and helping bear the burden of taxation or inconvenience, with better conditions always in hopeful sight For six years he served as a member of the board of education and during three years was president of this body He now is president of the Nebraska State Press Association, being elected in February 1921, and is one of five newspaper editors appointed by the governor to attend the World Newspaper Congress in Honolulu in October, 1921

LAFAYETTE O ROBLEE first came to Nebraska more than thirty years ago, and his original stage of operations was in Custer county, where he brought into effective play the practical knowledge he had gained in connection with farm enterprise in the old Empire state, which figures as the place of his nativity From Custer county he came to the present Garden county in 1909, and here he not only reclaimed and developed a productive farm but also gained place as one of the representative business men of the village of Lewellen He finally renewed his allegiance to the basic industries of agriculture and stockgrowing and is now giving his attention to the supervision of his well improved farm of a hundred and sixty acres, situated three and one-half miles northeast of Lewellen

Mr Roblee was born in New York state, October 28, 1861, and in the same state were born his parents, Orlando and Agnes (Crandall) Roblee, who passed their entire lives there, the father having been a farmer by vocation Orlando Roblee attained to the venerable age of eighty-three years, his wife having passed away at the age of sixty-eight years

Lafayette O Roblee was afforded the advantages of the public schools of Sandusky, New York, and in his youth gained practical experience in connection with the work of the home farm Finally he engaged in independent farm enterprise in his native state, and after a period of about four years turned his attention to the operation of a saw mill and grist mill, with which enterprise he was identified for eighteen months His ambition then led him to make personal investigation of the opportunities afforded in the progressive west, and in 1887, as a young man of twenty-six years, he came to Nebraska and numbered himself among the pioneers of Custer county He took a pre-emption claim near the present village of Sargent, and was engaged in general farming and stock-raising for five years, within which time he made numerous improvements on his pioneer farm At the expiration of that period Mr Roblee went to the state of New York, but four years later returned to Custer county, where he continued his activities as a farmer for the ensuing four years He then came to that part of Deuel county that is now comprised in Garden county, and entered claim to a homestead, to the reclamation and development of which he applied himself with characteristic energy and discrimination He made good improvements on the place and was actively engaged in farming until 1909 when he removed to the village of Lewellen and established himself in the furniture and undertaking business, to which he later added a grocery department He became one of the leading business man of the village and that he gained unqualified popularity in the community is evidenced by the fact that in 1910, he became postmaster of the village, a position he held until 1917 In 1918, he sold his business at Lewellen and purchased his present farm, to the active management of which he has since given his attention, as one of the successful and representative agriculturists and stock-growers of Garden county His land is well irrigated by water from the Bratt ditch of which he is one of the owners A man of well ordered convictions in relation to political affairs, Mr Roblee is a Republican, and he is always ready to lend his support to those agencies that tend to conserve the progress and prosperity of his home county and state

August 2, 1881, Mr Roblee wedded Miss Ella Fuller, who was born in the state of Iowa, but who was reared and educated in the state of New York, having been a milliner by occupation at the time of her marriage Mrs Roblee is a daughter of Alonzo and Emily (Brady) Fuller, who were born and reared in the state of New York, the father having served as a soldier of the Union during the Civil War and died in 1865, while enroute to his home, after having received his honorable discharge, his widow passed away at the age of forty-nine years Mr and Mrs Roblee have four children Dean S, of Lewellen, married Miss Maude Sargent, and they have three children, Roy L, who likewise resides at Lewellen, married Miss Ina Durand and they have two children, Agnes is living at home, Lawrence, who is associated with his father in the management of the farm, married Vinnie Rickard and they have one child, Darline

FRED JOHNSON, known and honored as one of the sterling pioneers of Garden county, has shown the energy and good judgment that gain success in connection with ranch and farm enterprise He came to Garden county about thirty-five years ago, when it was still a part of old Cheyenne county, and in developing his farm property he encountered his full share of the burdens and trials that marked the careers of all pioneers in this section of the state Success has crowned his efforts and he is a citizen who commands the high regard of the people of the county in which he has long maintained his home

Fred Johnson was born in Germany, December 19, 1855, and was a lad of about twelve years at the time the family immigrated to America He was reared and educated in the state of Wisconsin, where his parents established their home in 1867 He is a son of Henry and Mary (Russ) Johnson, who left their native land in 1867, as above noted, and came to the United States They settled in Greenfield township, Milwaukee county, Wisconsin, where the father reclaimed and developed a good farm, and they passed the remainder of their lives there, the father died at the age of seventy-three years and the mother at the age of seventy-five years

Fred Johnson made good use of the advantages afforded in the public schools of the Badger state, and as a youth of seventeen years made his way to Michigan City, Indiana, where for seven years he was employed in that now gigantic railroad-car factory of the Haskell & Barker Company For about eighteen months thereafter Mr Johnson was employed in a gas factory at Michigan City, and later he was for five years an employe in the chair factory of Ford & Johnson, which is still one of the important furniture manufacturing concerns of the country In 1885, Mr Johnson came to that part of old Cheyenne county, Nebraska, that now comprises Garden county and took up a homestead, seven miles southwest of the present village of Lewellen He instituted the improvement of his claim, to which he proved title in due course, and with the passing years he proved successful in his agricultural and live-stock operations He has developed one of the model farms of the county, and still gives his personal supervision to the same, besides which he has done a prosperous business in the raising of cattle and horses He has been a helpful force in community advancement, is a Republican in politics, is affiliated with the Modern Woodmen of America, and he and his wife are communicants of the Lutheran church

September 16, 1880, recorded the marriage of Mr Johnson to Miss Anna Lambke, of Michigan City, Indiana, in which state she was born and reared Mrs Johnson is a daughter of Henry Lambke, who was born in Germany and was a young man when he came to America For many years he was identified with railroad operations, and for twenty-one years was in the employ of the New Albany Railroad Company of Indiana In conclusion is given brief record concerning the children of Mr and Mrs Johnson Mrs Florence Wilson resides at Lewellen and is the mother of three sons, Elmer likewise resides at Lewellen, Mrs Emma Emerson, of Lewellen, has three children, Archie has returned to Garden county after loyal service with the One Hundred and Ninth Engineers during the World War, Mrs Edmona DeLatour, of Lewellen, has three children, and Harry, Gordon and Fred, Jr, remain at home

JAMES COPLEY is one of the representative farmers of the younger generation in the Lewellen district of Garden county, where he purchased and still owns the fine homestead which was here taken by his father, James S Copley, of whom more specific mention is made on other pages, in the sketch of the career of an older son, Ira Copley, so that a repetition of the family record is not demanded in the present review

James Copley was born in Wayne county, West Virginia, and was reared and educated in his native state, the date of his birth having

been April 6, 1889 He has been a resident of Garden county since 1907, in which year he came to Nebraska from West Virginia, and for several years was in the employ of the Western Land & Cattle Company In 1916, he entered claim to the homestead upon which he now resides and upon which he has made good improvements, besides which, as previously stated he owns the old homestead of his father, who was one of the pioneers of what is now Garden county and who is still engaged in farming and live-stock enterprise The farm property of James Copley now comprises a section of valuable land, and his attractive home is situated about six miles west of the village of Lewellen, which is his postoffice address He is a Republican in his political views and he and his wife are popular factors in the social life of their community

At Oshkosh, Garden county, December 4, 1916, Mr Copley was united in marriage to Miss Mabel Nash, whose parents were pioneer settlers in the vicinity of Oshkosh, where Mrs Copley was reared and educated, she being a native of Nebraska Mr and Mrs Copley have a fine little son, Kenneth, who was born August 16, 1918, and who holds undisputed dominion in the pleasant home

HJALMAR E OLSON conducts a successful business in the buying and shipping of live stock, with residence and general headquarters in the village of Lewellen, Garden county, and he also has developed a prosperous business as a skilled and resourceful auctioneer, a capacity in which his services are much in demand He is one of the alert and progressive young men of Garden county, where he had been engaged in farm enterprise prior to his removal to Lewellen

Mr Olson was born in Kearney county, Nebraska, March 2, 1887, and is a son of Andrew and Catherine Olson, the former of whom was born in Sweden and the latter in Denmark After their marriage the parents continued their residence in Sweden until 1885, when they came to the United States and settled in Kearney county, Nebraska, where the father became a successful agriculturist and stockgrower he having continued his active association with farm enterprise in that county until 1912, when he retired and established his residence at Minden, where he and his wife have since maintained their home They have reared a fine family of twelve children—nine sons and three daughters—all of whom are living, the subject of this sketch having been the seventh in order of birth

Hjalmar E Olson passed the period of his

childhood and early youth on the home farm of his father and is indebted to the excellent schools of Kearney county for his educational training At the age of twenty-one years he engaged in independent farm enterprise in his native county, where he continued operations eight years, with special attention given to the raising of hogs At the expiration of that period he came to Garden county and purchased a tract of land, upon which he was successfully engaged in general farming and stock-raising for the ensuing two years He then removed to Lewellen and engaged in his present line of business, in which his success has been enhanced by his thorough knowledge of live-stock values His political allegiance is given to the Republican party, and while he has shown no desire for official preferment he served as treasurer of school district No 50, Garden county, from 1916 to 1919

November 10, 1908, recorded the marriage of Mr Olson to Miss Mattie Jones, who was reared and educated in Kearney county, and who is a daughter of Edmond N and Sarah Jones, natives of Kentucky Mr and Mrs Jones established their residence in Kearney county in 1880, and there he took up a homestead and instituted the development of a farm He perfected his title to the place and there his death occurred in 1897, his widow being now a resident of Minden, that county Mr and Mrs Olson have five children—Avis, Doris, Vera and Verna (twins) and Irene Mrs Olson is a member of the Baptist church

IRA PAISLEY has been prominently and worthily identified with the civic and industrial interests of Garden county since the pioneer period when it was still a part of old Cheyenne county, and after many years of earnest and active service in connection with the development of the agricultural resources and live-stock interests of this section of the state, is now living virtually retired, in a pleasant home in the village of Lewellen As one of the honored and influential pioneers of the county he is entitled to special tribute in this history

Ira Paisley a scion of the staunchest of Scotch ancestry, was born in Muskingum county Ohio, February 16, 1851 and is a son of Hugh C and Mary Anna (Haynes) Paisley the latter of whom was born and reared in Ohio her death having occurred in Louisa county, Iowa, in May, 1865 Hugh C Paisley was a member of an old and influential family of Scotland and was born in the city of Paisley, where he was reared and educated As a young man he severed the ties that bound

him to his native land and set forth to seek his fortunes in the United States He settled in Ohio, where he followed the trade of carpenter for some time, but eventually became a farmer His marriage was solemnized in the old Buckeye state where he continued to reside until 1850, when he removed with his family to Illinois About two years later he settled in Iowa, where he secured a tract of land about twenty miles north of the present city of Burlington There he continued his activities as a pioneer farmer about thirteen years, and, in 1880, came with his family to Nebraska and established his residence in Polk county, where he passed the residue of his life, having been sixty-nine years of age at the time of his death Of his nine children the subject of this sketch was the seventh in order of birth Three of the sons, Findley, Isaiah and Frank, were soldiers of the Union in the Civil War Findley and Isaiah enlisted in the Sixteenth Iowa Volunteer Infantry, and Frank became a member of Company M, Eighth Iowa Cavalry All three served during the great conflict and all were confined for a time in the historic Andersonville Prison Findley met his death at the battle of Shiloh, and Frank was severely wounded at the battle of Gregory Station

Ira Paisley was reared under the influence of the pioneer era in the state of Iowa, where he was afforded the advantages of the public schools, and, in 1870, he came to Nebraska and settled in Polk county, where he began to farm About four years later he returned to Iowa, and remained about two years He then came again to Polk county, Nebraska, to live there until 1884, when he came to the western part of the state as a pioneer settler of that part of Cheyenne county that constitutes the present Garden county He took up a pre-emption claim on the Blue river, northwest of the present village of Lewellen, and became one of the early farmers and live-stock men in this now opulent section of the state He perfected title to his claim, and also to a homestead, which likewise he developed and improved and owned until 1904 Thereafter he secured three-fourths of a section of land, under the provisions of the Kincaid law, and he continued his active association with farm industry until 1918, when he sold his land and removed to Lewellen, where he has since lived in well earned retirement, the labors and well directed enterprise of earlier years having gained to him a competency He takes great pride in the progress that has been made in this section of the state and also in the fact that he has been able to do his share in promoting this civic and industrial advancement A man of sterling character, he has ever commanded the fullest measure of popular confidence and esteem, and has been a loyal and public-spirited citizen His political allegiance is given to the Republican party, but he has never manifested any ambition for the honors of political office He is one of the stockholders of the Warner Telephone Company and has other substantial financial interests Both he and his wife are zealous members of the United Presbyterian church at Lewellen

September 20, 1875, was solemnized the marriage of Mr Paisley to Miss Mary M Wilson, who was born and reared in Iowa, where her father, John M Wilson, was a pioneer farmer, having been born in Preble county, Ohio Mr Wilson passed the closing period of his life at Pawnee City, Nebraska, where he died at the venerable age of eighty years, the maiden name of his wife was Garrison She died before her husband Mr and Mrs Paisley have nine children—Harry, of Joliet, Wyoming, Mrs Eva White, of Lewellen, Mrs Cora Clark, of Oshkosh. Garden county, Mrs Clara Sellers, of Ashton, Idaho, John M, of Joliet, Wyoming, Mrs Pearl Robinson and Mrs Irene Roberts, both residents of Lewellen, Garden county, Mrs Myrtle Robinson, of Joliet, Wyoming, and Ira Jr, of Lewellen, who served in the United States Navy during the period of the World War

JAMES W ORR is to be accorded recognition as one of the representative pioneer citizens and successful agriculturists and stockgrowers of Garden county, where he is the owner of a large and well improved farm property, to the supervision of which he gives his personal attention, though he is now living in semi-retirement, in the village of Lewellen His career has been one of varied and productive activity, and he has achieved worthy success as a result of his earnest and well directed endeavors

James William Orr was born at Rockaway, New Jersey, on September 8, 1868, and is a son of John and Mary (McCormick) Orr, the former a native of Scotland and the latter of Ireland The father was a young man when he removed from Scotland to Ireland, where he engaged in the work of his trade, that of wheelwright There his marriage was solemnized, and in 1847 he came with his family to America and settled at Rockaway, New Jersey, where he continued for many years in

the work of his trade and in that state he died at the age of seventy-four years, his wife having passed away at the age of sixty-three years Of their nine children three sons became residents of Nebraska—Calhoun, John H and James W

To the public schools of his native state James W Orr is indebted for his early educational discipline, which was received principally at Danville, Warren county At the Kislepau mines, near that place, he became operator of a stationary engine, and continued his association with the mining industry about ten years In 1892, Mr Orr came to Nebraska and took up a homestead in Keith county, where he engaged in farming and stock-raising and finally perfected his title to the land In 1898, he began to farm irrigated land, and to give special attention to the raising of hogs, but about three years later he went to the state of Washington, where he returned to the occupation of his youth, by operating the engine of a shingle mill He remained in Washington about seven months and then returned to Nebraska and resumed operations as an agriculturist and stock-grower In 1903, Mr Orr purchased four hundred acres of land in Garden county, and to provide for its irrigation effected the completion of the Bratt ditch In addition to developing the agricultural resources of his land he has here carried on successful and somewhat extensive operations in the raising and feeding of hogs, and continued to live on his well improved farm until 1918, when he removed to Lewellen, where he owns and occupies an attractive home, though he still gives a general supervision to his farming and stock interests He is part owner of the Bratt irrigation ditch and has been liberal in the support of enterprises that have tended to advance the civic and material welfare of the county His political views are in harmony with the principles of the Democratic party, he is affiliated with the Modern Woodmen of America and the Woodmen of the World, and he and his wife hold membership in the United Presbyterian church

November 12, 1904, recorded the marriage of Mr Orr to Miss Elsie Branden, who was born in England, of Scotch-Irish lineage Mrs Orr died in 1907 and is survived by two children, Mary E and John E, who remain at home On March 9, 1910, was solemnized the marriage of Mr Orr to Miss Jennie Gordon, at Julesburg, Colorado Mrs Orr was born and reared in County Tyrone, Ireland, and came to America with her brother Andrew, who settled in Garden county, Nebraska, in 1907 Mr and Mrs Orr have one son, James G

STEPHEN L BROWN — In Crawford county, Illinois, February 4, 1854, Stephen Louis Brown "ope'd wondering eyes to view a naughty world," and it may be that on this birthday of a now prominent and honored citizen of Garden county, Nebraska, the auguries foreshadowed the career that was to give to him a plethora of pioneer experience in the west He has been distinctly one of the world's productive workers, and the results of his honorable and well ordered efforts are shown in his ownership of a large and valuable farm property in this county

Mr Brown is a son of Phillip and Caroline (Dare) Brown, both of whom were born and reared in the state of Indiana, to which Mrs Brown's people came as pioneers from New Jersey Phillip Brown became a substantial farmer in the Hoosier state, whence he eventually removed to Illinois and engaged in the same fundamental vocation in Crawford county In 1878, he removed with his family to Kansas, where he became associated with construction work on the Missouri Pacific Railroad In the following year he came to Nebraska and assumed a contract for grade work on the line of the same railway, near Brownsville In 1882, he established his residence in Atchison county, Missouri, where he resumed his operations as a farmer, but eventually he returned to Illinois, where he died within a short time, at the age of eighty-two years His wife was seventy-nine years of age at the time of her death, which occurred in Boone county, Nebraska

Stephen L Brown passed the first nine years of his life in Crawford county, Illinois, where he profited by the advantages afforded in the public schools of the period, and then went to Lawrence county, that state, where he remained, and continued to attend school, until he attained his legal majority When about eighteen years of age he gained his initial experience in independent farm enterprise, in Lawrence county, Illinois, and in the autumn of 1877, he drove with team and wagon from Illinois to Mitchell county, Kansas In February of the following year he filed entry on a pre-emption claim and a tree claim on Cheyenne creek, in Lane county, that state, but, owing to ensuing difficulties with cattle men and a mistake on the part of a veteran soldier of the Civil War, who had supposed he was to have the same property, Mr Brown turned over to him the two claims On June 25 of

the same year he left Lane county and went to Cove county, from which locality he made his way on foot to Mitchell county—covering a distance of eighty-five miles in two and a half days For the ensuing three years he was engaged in grade work on the line of the Missouri Pacific Railroad, in association with his father, and he then accompanied his father to Atchison county, Missouri, where he remained about two years He then, in 1886, came to Nebraska and purchased a quarter-section of railroad land in Greeley county, where he was engaged in farming and stock-raising for fourteen years The following two years found him employed on the extensive ranch of Samuel Allerton, the prominent Chicago meat-packer, in Boone county, this state, and the next six years he was engaged in farming near Albion, that county In 1908, he established his residence in that part of Deuel county that now comprises Garden county, where he filed entry on a section of land, to which he eventually perfected his title With characteristic energy and discrimination he instituted the development and improvement of his land, which he utilized for diversified agriculture and for the raising and feeding of cattle and horses He has made this one of the valuable farm properties of the county, and although he now resides in the village of Lewellen he still gives his personal supervision, in a general way, to his well improved farm, his son Frank having the active management of the place and their live-stock operations being conducted on a somewhat extensive scale

In politics Mr Brown designates himself an independent Democrat, in a fraternal way he is affiliated with the Modern Woodmen of America, and he and his wife are zealous members of Calvary Baptist church at Lewellen, in which he formerly served as superintendent of the Sunday school

In Atchison county, Missouri, February 14, 1885, was solemnized the marriage of Mr Brown to Miss Phoebe Watts, who was born in the state of Illinois, as were also her parents, James H and Martha (Gill) Watts Mr Watts, a farmer by vocation, though a carpenter by trade, removed with his family from Illinois to Missouri in 1882, and about four years later went to Greeley county, Nebraska, where he took up a homestead and engaged in farming He eventually retired from the farm and settled at Cedar Rapids, Boone county, where he died at the age of seventy-five years, his wife having been forty years old at the time of her death, which occurred in Illinois In conclusion is given brief record concerning the children of Mr and Mrs Brown James R and his wife reside at Oshkosh, Garden county, and they have four children, Ray B, of Lewellen, is married and has two children, he served as a member of the One Hundred and Ninth Engineers Corps in the late World War, Mrs Pearl E West passed away in 1910, leaving two children, one of whom, Hazel C, resides in the home of her maternal grandparents Mr and Mrs Brown of this review, Frank P, who has the active management of his father's farm, as previously noted, is married and has two children, Louis E, who is now at home, likewise served with the One Hundred and Ninth Engineers in the late war and Millard G is the youngest member of the home circle

FRED L MELIUS is to be ascribed pioneer honors in western Nebraska, where he did his part in connection with the social and industrial development and up-building of this now opulent and progressive part of the States He resides with his brother, Jesse P, on one of the fine farms of Garden county, which is situated six and one-half miles northeast of Oshkosh, the county seat Mr Melius is still arrayed in the ranks of eligible bachelors in the county, but it may be stated that this condition of celibacy in no degree impairs his personal popularity, which is of unqualified order

Mr Melius was born in Delaware county, Iowa, September 20, 1863, and is a son of Peter F and Helen (Ingraham) Melius both of whom were born and reared in the state of New York, where the father passed his entire life, his death having occurred in 1876, at which time he was forty-eight years of age After the death of her husband, Mrs Helen Melius came with her children to the west and established her home in Nebraska, where she was a pioneer and where she reared her two sons and one daughter, to whom her devotion was unstinted and unselfish This gracious pioneer woman passed the closing period of her life in Garden county, where she died in 1916, at the age of seventy-seven years, the gentle evening of her life having been brightened by the filial love and solicitude of her children

Fred L Melius was about seventeen years of age at the time he accompanied his widowed mother to Iowa, where he was reared to manhood and received the advantages of the public schools There he turned his attention to agricultural pursuits in which he was engaged four years and by means of which he was able to provide well for his mother and the two younger children In 1880, Mr Melius came

to Nebraska and settled in Nance county, on the Pawnee Indian reservation There he continued his farm labors until 1888 when he took up a homestead in Box Butte county and instituted operations as a pioneer agriculturist and stock-grower He filed on a tree claim also, and eventually perfected title to both claims, which he effectively developed and in the activities of which his brother, Jesse P, became associated as a partner In 1916, Mr Melius sold his property in Box Butte county, and has since resided in the home of his only brother, in Garden county, a sketch of the career of the brother being given in paragraphs that immediately follow this review Mr Melius has been one of the world's productive workers, and ample success has attended his efforts during the years of his residence in Nebraska He has hewed close to the line of his chosen vocation and thus has had no desire for political activity or public office, though he is essentially loyal as a citizen and as a staunch supporter of the cause of the Democratic party

JESSE P MELIUS — In the foregoing article is given adequate review of the family record of Mr Melius, who is numbered among the vigorous exponents of agricultural and live-stock industry in Garden county, in which fields of industrial enterprise his initial experience was acquired through close association with his elder brother, Fred L, to whom the preceding biographical sketch is dedicated

Mr Melius was born in Delaware county, Iowa, June 3, 1874, and was a child when he accompanied his widowed mother to Nebraska He was about three years old at the time of the family removal to Nebraska, where he was reared to adult age and where he was afforded the advantages of the public schools that marked the pioneer days in Nance county As the preceding article indicates, he became actively associated with his brother in the live-stock business in Box Butte county, and in 1909, he took up a tract of land in Garden county, under the provisions of the Kincaid act He perfected his title to this land, and since that time he has acquired by purchase an entire section of land, as well as an additional quarter-section A portion of his land he rents for farm purposes, and his special field of activity is in the raising of Poland-China hogs, in which field he has topped the Omaha market for the past two years (1918-1919) He is developing also the admirable agricultural resources of his property, and with the assistance of his wife is proving very successful also in the raising of Columbia

Wyandotte poultry His home place is pleasantly situated, six and one-half miles northeast of Oshkosh, he has erected good buildings and made other modern improvements that denote his energy and progressiveness Mr Melius takes loyal interest in community affairs and is liberal and public-spirited as a citizen, his political allegiance being given to the Republican party

At Alliance, Box Butte county, December 20, 1905, was solemnized the marriage of Mr Melius to Miss Sadie Campbell, who was born in Kansas but reared and educated in Garden county, Nebraska, where her parents, Thomas W and Jessie (Stonehacker) Campbell were pioneer settlers, her father being still a resident of the county and her mother having died in 1907 Mr and Mrs Melius have four children—Fern, Lester, Cloyd and Vernon Mr Melius has just completed a modern nineroom home where he and his family are prepared to enjoy the fruits of their labor

WILLIAM N CAMPBELL is another of the progressive and substantial pioneer citizens of Garden county, where he is the owner of a large and well improved landed estate and is specially well known as a successful stockgrower

William Nelson Campbell was born in Mills county, Iowa, June 20, 1871, and is a son of James W and Julia (Pack) Campbell, both of whom were born and reared in Iowa, where their marriage was solemnized and whence they finally removed to Kansas, where the father became a prosperous farmer Since 1917, Mr and Mrs James W Campbell have maintained their home in California, where he is living retired William Campbell was reared to manhood in his native state, and there received the advantages of the public schools of the middle-pioneer period In 1894, at the age of twenty-two years, he came to what was then Deuel county, Nebraska, and took up a homestead near Mumper, which later was included in the present Garden county He improved this property and utilized the same as the stage of his successful activities in the raising of horses and cattle His good judgment was shown in his investing in more land from time to time as circumstances and opportunity justified, and one of his purchases was what is now known as the Lost Creek Ranch, of one thousand, three hundred and eighty acres To this he has since added until he is now the owner of about five thousand acres of the valuable land of Garden county, and he conducts large and successful operations in the raising of cattle, horses and hogs,

with special attention given to Hereford cattle, Percheron horses, and Duroc hogs which he raises and feeds for market The agricultural department of his farm enterprise likewise receives the attention that insures maximum success, and Mr Campbell is essentially a representative factor in connection with industrial activities in Garden county His political support is given to the Democratic party and he is affiliated with Oshkosh Lodge No 286, Ancient Free & Accepted Masons

At Aronoque, Kansas, February 26, 1895, was solemnized the marriage of Mr Campbell to Miss Linda J McCabe, a daughter of James F and Julia (McMullen) McCabe, who were born and reared in Missouri and who became early settlers in Kansas, where they still maintain their home, as venerable pioneer citizens of Aronoque Mr and Mrs Campbell have four children Myron Vaile, married Mayne Nash, May 25, 1918, at Scottsbluff, who was born 1901, at Oshkosh, Nebraska, the daughter of Eli F and Rhoda (Hunter) Nash of Garden county, and at present lives on the home place and is associated with his father in conducting the ranch, and John Percy, Helen and Ruth remain at home, which is known for its generous hospitality and good cheer

FREDERICK A PICKERING has been a resident of Nebraska since his boyhood and in his career has manifested in a distinct way the progressive spirit that has ever marked the history of this commonwealth He is to be designated as one of the pioneer exponents of farm enterprise in Garden county, where he still gives his active supervision to his large and well improved ranch which is devoted to diversified agriculture and the raising of live stock

Mr Pickering was born in Fulton county, Illinois, June 26, 1867, and is a son of A G and Sarah Jane (Strode) Pickering, both of whom were born in Ohio, though the latter was reared and educated in Illinois, where her marriage was solemnized A G Pickering was a young man when he engaged in farming in Illinois, and, in 1881, came with his family to Nebraska, the first year having been passed in Cass county Removal was then made to Phelps county, where he continued farm activities about five years, and he then became one of the pioneer settlers in Garden county, which was at that time still a part of Cheyenne county Here he took up and improved a homestead, to which he perfected his title, and became one of the successful agriculturists and stock-raisers of this section of the

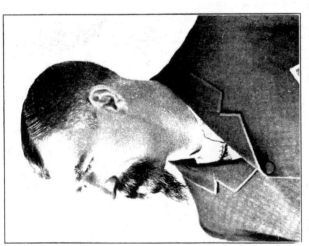

Mr. and Mrs. Albert M. Wright

state, as well as a man whose ability and sterling character made him influential in community affairs He served about twenty years as postmaster at Kowanda and was one of the venerable and honorable pioneer citizens of the county at the time of his death, in 1913, at the age of eighty-three years Mrs Sarah Jane (Strode) Pickering died in 1874, when the subject of this sketch was but seven years old, he having been the fourth in a family of six children, three of whom—James A, Charles M and Ernestine Helen—took up homesteads in Garden county and in due course proved up on the claims

Frederick A Pickering acquired his preliminary education in the public schools of Illinois and was fourteen years old at the time of the family removal to Nebraska, where he continued to attend school whenever opportunity offered, at the same time giving his share of aid in the work of his father's pioneer farm In 1893, he engaged in independent farming, in Phelps county, but two years later he came to that part of Deuel county that now comprises Garden county, where he has since given his close attention to agricultural and live-stock industry, through the medium of which he has achieved substantial and worthy success In 1903, he entered claim to a homestead of a hundred and sixty acres, and under the provisions of the Kincaid act he added to his holdings until he is now the owner of four hundred acres, his pleasant home being near Oshkosh, the county seat Mr Pickering is a bachelor, is affiliated with Oshkosh Lodge, No 286, Ancient Free & Accepted Masons, and in politics he gives his allegiance to the Republican party As a citizen he has shown his loyalty by supporting enterprises and measures that have inured to the general welfare of the community, and in Garden county he has a wide circle of friends

ALBERT M WRIGHT, United States Deputy Marshall, pioneer of Dawes county, and prominent man of affairs in the Panhandle for many years, was born in Racine county, Wisconsin, January 20, 1847, the son of Eben and Julia (Merrill) Wright, both natives of Vermont Albert was the eldest of the six children born to his parents His father was a farmer and the boy was reared in the country He was sent to the public schools and thus gained a good practical education When only eighteen years of age he enlisted in 1864, in Company H, Fifth Wisconsin Volunteer Infantry He was on the battle front March 25 1865, before Petersburg Virginia, was in the

following battles which took place in that locality when General Robert E Lee was driven from his stronghold of Petersburg and was present when Generals Ewell and Fitz Hugh Lee surrendered with six thousand troops Mr Wright was in the army that followed Lee until he too surrendered At the close of hostilities he was mustered out at Madison, Wisconsin, June 26, 1865 Almost at once he returned to his farm near Portage, Wisconsin, remaining there in farm work until 1871, when he went into the timber country of northern Michigan, but in the spring of 1872, took a position as brakeman on the Northwestern Railroad on a run from Escanaba He remained with the road until 1885, and during that time was promoted to conductor Mr Wright was married December, 1872, at Milton, Wisconsin, to Miss Sarah L Wood, who was born in Rock county, the daughter of Joel and Sarah A (Butts) Wood, both natives of New York state She was the youngest in a family of eleven children Mrs Wright was a graduate of the Milton Academy, of Milton Mr and Mrs Wright became the parents of three daughters Anna J, who married Joseph Robinson, had one daughter, who married E M Birdsall, Mary, who married G T H Babcock, an attorney of Chadron, has two children, George and Jane Mr Babcock is a Mason and also belongs to the Modern Woodmen and the Elks The third child is Gertrude S, who married E L Godsall, a passenger conductor for the Northwestern Railroad He volunteered for service during the Spanish American War, and served as lieutenant of Company H, Second Nebraska Infantry He volunteered during the World War as a member of Company H, Sixth Nebraska Infantry, then was transferred to the One Hundred and thirty-fourth Infantry, Thirty-fourth Division and went to France as captain of that company He was promoted to Major in France and commanded prisoners at Fort Rougoune, France and Is Sur Tille He returned to the United States October 20, 1919 Mr Godsall is a Knight Templar Mason and a member of the Elks

In 1885, Mr Wright came to western Nebraska, and located at Valentine, the end of the Northwestern Railroad at that time He took charge of a construction train laying track west from Chadron and held the position until August 10, when he was assigned to the operating department of the railroad and remained in the freight department until promoted to the passenger service as conductor in 1886 Mr Wright served in this capacity, and remained with the company until 1907, a long period of service. He had the honor of running the first passenger train into Deadwood, South Dakota, when the road was completed to that point, on December 29, 1890. After leaving the railroad, Mr Wright was appointed city marshall of Chadron, in 1908, served until June of the same year and then was appointed Deputy United States Marshall, a position he still holds. He has faithfully performed his duties in a most efficient manner and is regarded as a man of honor and marked ability. The Wright family have a good modern home on Bordeaux Avenue, in Chadron where they enjoy their many friends. Mr Wright is a Thirty-second degree Mason, a Shriner and also belongs to the Elks. For one year he served as mayor of Chadron, an executive office which he filled to the satisfaction of the people. Being connected with the railroad for so many years he is well known from Chadron to Escanaba, Michigan, an unusual honor.

J FRANK BLAUSEY figures as a native son of what is now Garden county, though at the time of his birth the county was still a part of Deuel county, and here he has proved his loyalty as well as his full appreciation of the advantages and attractions of this section of Nebraska, by his successful association with agricultural and live-stock industry, of which he is one of the prominent and representative exponents in Garden county.

John Frank Blausey was born at Ramsey, Deuel (now Garden) county, May 12, 1889, a date that denotes conclusively that he is a representative of one of the pioneer families of this section of the state. He received his early education in the public schools of Garden county, and prior to initiating his independent farm career he was employed about six months in connection with the construction of a government irrigation ditch north of Scottsbluff, Scotts Bluff county. Thereafter he was employed about a year by the Chicago, Burlington & Quincy Railroad Company, with headquarters in Nebraska's capital city. He is now the owner of a valuable farm of four hundred acres, four and one-half miles northeast of Oshkosh, has made good improvements on the property and here is proving most successful in his operations as an agriculturist and stock-raiser. He has had no predilection for political activity or public office, but accords loyal allegiance to the Republican party. His wife is an active communicant of the Lutheran church.

December 20, 1911, at Oshkosh, was recorded the marriage of Mr Blausey to Miss Martha Kaschke, who was reared and educated in Sedgwick county, Colorado, being a daughter of Henry Kaschke, of whom individual mention is made elsewhere in this volume. Mr and Mrs Blausey have two children—Beulah and Bernice.

JOHN C HARTMAN was a lad of twelve years at the time when his parents established themselves as pioneers in what is now Garden county, and here he has risen to secure vantage place as one of the prominent representatives of live-stock and agricultural industry in this section of the state. In Clayton county, Iowa, whose eastern borders skirt the shores of the Mississippi river, John C Hartman was born, October 26, 1875, and is a son of Sebastian R and Marie (Herman) Hartman, both natives of Austria and both young people when they came to the United States, their marriage having been solemnized in the state of Illinois. Sebastian Hartman immigrated to America in 1865, and for two years thereafter he was engaged in farm enterprise in Illinois. He then removed to Iowa and became a pioneer settler in Clayton county, where he was engaged in farming for the ensuing nine years. He then removed with his family to Kossuth county, that state, which was the stage of his successful farm operations for ten years, at the expiration of which, in the spring of 1887, he came to the part of old Cheyenne county, Nebraska, that now comprises Garden county, where he took up homestead and pre-emption claims and girded himself valiantly for the labors of a pioneer agriculturist and stock-grower in a new country. He improved his land, to which he perfected title in due time, and there he continued his activities until 1909, when he removed to Julesburg, Colorado, where he died at the age of sixty-nine years and where his widow passed away about two years later.

The rudimentary education of John C Hartman was obtained in the schools of Kossuth county, Iowa, and was rounded out by his attending the pioneer schools of what is now Garden county, Nebraska, where he was reared to manhood and early began to assist in the work of his father's farm. When about twenty-six years of age he initiated his independent activities as an agriculturist and stock-raiser, with which basic industries he has since continued his close and successful association, energy and good management having brought to him substantial prosperity, of which evidence in given in his ownership of a thousand

and forty acres of land, the major part of which is devoted to other phases of agriculture and to the raising of good grades of live stock.

Mr. Hartman has always taken loyal interest in community affairs, is a Democrat in his political proclivities and had the distinction of serving as the first assessor of Garden county, a position which he held three years.

December 7, 1904, recorded the marriage of Mr. Hartman to Miss Anna Krause, of Sedgwick, Colorado, whose mother contracted a second marriage and is now a resident of Pendleton, Oregon, her name being Mrs. Bertha Shumway. Mr. and Mrs. Hartman's cheery home circle includes their fine family of six children: Bertha A., Harman J., Clara M., Roy M., Howard E., and Herbert L.

CHARLES F. CARR, a representative agriculturist and stock-grower of Garden county, is another man who has proved that metropolitan training and vocation do not preclude the achieving of a definite success and precedence in connection with farm enterprise, for he claims the great western metropolis, Chicago, as the place of his nativity, and while he gained in his youth a measure of experience in connection with farm industry, he eventually returned to his native city, where he learned the metal-polisher's trade, to which he devoted his attention about five years, after which he was engaged in business as a painter and paperhanger for a period of about ten years. He then came to that part of Deuel county, Nebraska, that now constitutes Garden county, and in the same year, 1902, took up a homestead, ten miles northeast of Oshkosh. He has reclaimed and developed this tract into one of the excellent farms of the county, and has continued his successful activities along the lines of diversified agriculture and the raising of good types of live stock. He was a resident of Illinois at the inception of the Spanish-American War, and promptly manifested his patriotism by enlisting in the Second Illinois Volunteer Infantry, with which he continued in service until the regiment was mustered out and he received his honorable discharge. Mr. Carr has proved himself liberal in support of measures tending to foster the prosperity and advancement of the community, is independent in politics and has had no desire for public office of any kind.

Mr. Carr was born in the city of Chicago on August 4, 1860, and is a son of William F. Carr, who was born in England and who finally established his residence in Chicago, where he found employment in the McCormick agricultural implement manufactory. He eventually removed to Kansas, and his son Charles finally lost all trace of him, Charles having been an infant at the time of his mother's death. Mr. Carr continued his residence in Illinois the greater part of the time until he was about fifteen years old and he then went to Kansas, where he remained until he attained his legal majority. It was at this stage of his career that he returned to Chicago, as noted in a preceding paragraph, and concerning his activities since that time ample record has already been given.

Mr. Carr chose as his wife, Mrs. Maude (McMannen) Brigham, who was born and reared in Iowa, where her parents, both now deceased, were pioneer settlers. Mr. and Mrs. Carr have one child, by a former marriage of Mrs. Carr, Mable Brigham, who is the wife of Edward Allen, of Oshkosh, and who has two children, Charles Merle and Verona Maude.

BIRD S. RODGERS is the owner of an entire section of land in Garden county and with the energy and progressiveness that insure success, he has here carried on vigorous activities as an agriculturist, stock grower and dairy farmer, of which lines of enterprise he is one of the substantial and popular representatives in the county.

Mr. Rodgers was born in Keokuk county, Iowa, August 10, 1877, and is a son of Joshua T. and Rebecca Jane (Perry) Rodgers, both of whom were reared and educated in Iowa, in which state Mrs. Rodgers was born, her parents having been pioneer settlers in that commonwealth, as were also the parents of her husband, who was a native of Indiana, and who was a boy at the time of the family removal to the Hawkeye state. Joshua T. Rodgers continued his association with farm enterprise in Iowa until 1879, when he removed to Missouri, later coming to western Nebraska and became a pioneer settler in 1888, in what is now Morrill county, where he engaged in general farming and stock-raising. Later he was engaged in the cattle business in the sandhill district of Morrill county, and finally he resumed farm operations, south of Lisco, that county, where he continued until his retirement from active labors and has since maintained his home at Bridgeport, Morrill county, the mother of the subject of this sketch having passed away in 1888, at the age of thirty-six years.

Bird S. Rodgers was reared under the conditions and influences that marked the pioneer period of the history of the Nebraska Panhandle, and his early educational training was received principally in the rural schools of what is now Morrill county When a lad of about eleven years be began working for a cattle company, and he continued to be thus identified with the cattle business about nineteen years, within which period he was employed by various representative cattle companies In 1912, he purchased a quarter-section of land two and one-half miles north of Oshkosh and, in 1917, made an additional purchase that so enlarged his estate that he is now the owner of an entire section of the valuable land of Garden county, his progressiveness being manifest in the improvements and general condition of his farm property, which is devoted to diversified agriculture, to dairy farming and to the feeding of hogs during the winter seasons He is loyal in citizenship, is a Democrat in his political alignment and is affiliated with the Woodmen of the World

On November 15, 1905, at Sidney, Cheyenne county, Mr Rodgers was united in marriage to Miss Carrie Robinson, who was reared and educated in western Nebraska, she having been six weeks old when her parents removed to this section of the state from St Paul, Howard county On other pages the sketch of her brother, John Robinson, gives data concerning the family history Mr and Mrs Rodgers have two children Alice I, born April 12, 1910, and John T, born July 8, 1917

CYRUS L KEMPLIN is a Garden county citizen who has here proved his constructive ability through his effective enterprise as an agriculturist and stock-raiser, and he is the owner of a large and well improved landed estate, on which he lives in contentment and prosperity His career has been varied and interesting, but in his multifarious experiences in the past he reverts to none than has given him as much satisfaction as that connected with his industrial activities in Garden county

A representative of a sterling pioneer family of Iowa, Mr Kemplin was born in Story county, that state, on December 17, 1866 His father, Wilson Kemplin, was a native of West Virginia, and he was a young man when he wedded Miss Anna Simmons, who was born and reared in Ohio where their marriage was solemnized Mrs Kemplin died in Vernon county, Missouri, in 1873, when her son Cyrus subject of this review, was a lad of seven years Wilson Kemplin manifested somewhat of a nomadic spirit during the course of his long and active career, and he resided for varying intervals in different states of the west, it having been his distinction to

be a pioneer of Nebraska Prior to the admission of Nebraska to statehood he took up a pre-emption claim not far from the present capital city, and thus he became one of the earliest settlers of Lancaster county He finally returned to Iowa, but passed the closing period of his life at Lincoln, Nebraska, where he died at the venerable age of eighty-five years

Cyrus L Kemplin, gained his early education in the public schools of Iowa, and as a youth he learned the mason's trade, to which he gave his attention about seven years in Kansas City and St Louis, Missouri In 1890, he went to South Dakota, where he remained about one year, after which he returned to Kansas City His next response to the wanderlust was given when he went to Bighorn mountains, in Wyoming and Montana, in which section he found employment in saw mills, besides working at his trade for some time From that locality he came to Nebraska in 1893, making the trip with team and wagon, in true pioneer style, and finding his destination in that part of Deuel county that now constitutes Garden county He entered claim to a homestead in Antelope valley, and on this place he continued his residence thirteen years, during which time he perfected his title and made numerous improvements Success attended his efforts, and after disposing of this homestead he purchased the fine ranch of seventeen hundred and sixty acres which represents his place of abode and the stage of his vigorous activities at the time He has erected good buildings and made other excellent improvements on his extensive ranch, and here he is doing a successful business in the propagation of the various crops for which the soil and climate are best adapted and in the raising of live stock, his average run of cattle being about forty head and his place showing about forty head of horses at the opening of the year 1920

Mr Kemplin has never manifested any desire to "tinker with" practical politics and has shown his public spirit more effectively in productive industry than by seeking official preferment His allegiance is given to the Democratic party and he is steadfast in his political convictions

April 1, 1887, recorded the marriage of Mr Kemplin to Miss Minnie St Clair, who was born and reared in Indiana, where her mother still lives, at the venerable age of eighty-five years, the father, George St Clair, having been a valiant soldier of the Union during the Civil War and died shortly after the close of that great conflict Mr and Mrs

Kemplin have six children: Mrs. Mary Pickard resides at Oshkosh and is the mother of two children; Mrs. Mabel Bennett, of Gering, Scotts Bluff county; Belva remains at home; Mrs. Goldie Pratt resides at North Platte, this state, and Charles and Mary are the younger members of the home circle.

JOSEPH R. WOOLERY came to Garden county in 1890, at which time it was still a part of Deuel county, and during the intervening years he has continued as a vigorous and successful exponent of agricultural and livestock industry, and his civic loyalty and worthy achievement have contributed to the development and progress of this section of the state.

Joseph Richard Wollery was born in Pettis county, Missouri, April 1, 1864, and in the same state was born his father, Joseph Perry Woolery, a farmer by vocation who died at the age of forty-eight years, his wife, whose maiden name was Eliza Carpenter, having likewise been born and reared in Missouri and died when she was about forty years of age.

Joseph R. Woolery was reared to adult age in his native state, where he received the advantages of the public schools. After the death of his father he lived in the home of his uncle, Thomas Woolery, until he was eighteen years of age, and he then found employment in connection with the operation of a threshing machine. Thereafter he was employed a few months at farm work, for thirteen dollars a month and at the age of twenty-one years he went to Kansas, where he was employed about three years, principally at ranch work. From the Sunflower state he made his way to Denver, Colorado, where he was engaged in teaming for a period of about a year. His next venture was made in 1890, when he came to Nebraska and took up the homestead upon which he has continued to reside to the present time, having developed one of the excellent farms of Garden county and being now the owner of nine hundred and fourteen acres, with about a hundred acres under effective cultivation and the remainder used as grazing land. Mr. Woolery has been specially progressive and enterprising in his industrial activities, and has made each year mark an advancement in cumulative success. He is one of the substantial stock-growers of the county, and raises cattle, horses and hogs. He keeps an average of about fifty head of cattle and ships about a carload of hogs annually. He is one of the stockholders in the Farmers' Elevator in the village of Lisco, and also in the Farmers' Mercantile Company at that place.

In politics Mr. Woolery gives his allegiance to the Democratic party, and he has been influential in public affairs in his county, where he is serving his second term as a member of the board of county commissioners (1919-20), having been elected first in 1912. His service has been marked by earnest effort to promote the best interests of the county and its people. His wife is affiliated with the Royal Neighbors and is a popular figure in the representative social life of her home community.

May 26, 1895, recorded the marriage of Mr. Woolery to Miss Maude V. Suit, who was born at Council Bluffs, Iowa, in which state she received her earlier educational training, which was supplemented by her attending school after the removal of the family to western Nebraska, as she was a resident of Garden county at the time of her marriage, which was solemnized at Oshkosh. Mrs. Woolery is a daughter of Slathial B. and Helen (Kimble) Suit. Her father came to western Nebraska in 1887, as a pioneer farmer of this section of the state, his death occurred in 1918, when he was about seventy years of age, and his widow, who is a native of California, is now a resident of Oshkosh, Garden county. Mr. and Mrs. Woolery have three children — Joseph Percy, Mildred and Cecil Gwendolyn.

JAMES J. McCONNELL is another vigorous and progressive young man who has found in Garden county ample opportunity for successful activity in connection with the basic industries of agriculture and stock-raising, along which lines he initiated his independent career by entering into partnership with Frank O'Rouke, of St. Joseph, Missouri, with whom he continued to be associated two years, as active manager of a well improved farm and stock ranch of seventeen hundred acres, situated eighty-eight miles east of Oshkosh, the county seat. At the present time he is conducting an individual farming and live-stock enterprise, with special attention given to the raising of cattle and horses, and the base of his operations is a farm of six hundred and forty acres, located about eight miles northeast of Oshkosh.

Mr. McConnell was born in Fountain county, Indiana, October 21, 1894, and at Veedersburg, that state, he received his earlier education in the public schools. He was about nine years old at the time the family removed to Nebraska, and here his further educational discipline was received in the public schools of Lakeside, Sheridan county, and those of the Antelope valley, in Garden county. Mr. McConnell is a son of Edward and Eliza (Singleton) McConnell, the former born at Springfield, Ohio, and the latter of whom was born and reared in Fountain county, Indiana, where her marriage was solemnized When Edward McConnell was about one year old his parents returned to their old home in Ireland, and thereafter he remained on the soil of the British Isles — in Ireland, Scotland and England — until he had attained to the age of fourteen years, when he came back to his native land and became a member of the family circle of his uncle, at Springfield, Ohio He gained excellent mercantile experience along retail lines, and finally became a traveling salesman for a wholesale dry-goods house, and remained "on the road" for a period of eighteen years, during which time he traveled in eighteen different states, in the north and the south In 1903, he came with his family to Nebraska and settled at Lakeside, Sheridan county, where he conducted a hotel about one year He then took up a homestead, and also purchased an additional quarter-section of land, in the Antelope valley, in Deuel county, his land lying near the Garden county line He continued his successful activities as an agriculturist and stock-grower until 1918, when he retired, and he and his wife have since maintained their home in the thriving little city of Oshkosh

James J McConnell gained practical experience in connection with the activities of the home farm of his father, and thus was well prepared when he instituted independent enterprise as an agriculturist and stock-raiser, an industry of which he is now one of the prosperous and popular representatives in Garden county He is aligned in the ranks of the Democratic party and he and his wife are communicants of the Catholic church, in the faith of which they were reared In a fraternal way he is affiliated with the Modern Woodmen of America

January 9, 1917, at Oshkosh was solemnized the marriage of Mr McConnell to Miss Anna Donnelly, who was born and reared in Saunders county, the daughter of Patrick and Bridget (McCarty) Donnelly, the father was born in Ireland, November 6, 1864, and the mother in Illinois, December 1, 1869 They came to Nebraska in 1908 and settled in Garden county Mrs McConnell's educational advantages included those of the Nebraska State Normal School at Chadron Prior to her marriage she had been, for three terms a successful and popular teacher in the public schools of Oshkosh Mr and Mrs McConnell have a fine little son, Darrell Joseph, who was born October 7, 1918, and who is the arbiter of all affairs in the pleasant home

FREMONT G DURAND is properly to be ascribed a tribute not only as one of the representative pioneer citizens of Garden county but also as one who has furthered communal advancement and prosperity through his well ordered and unreservedly successful activities as an agriculturist and stock-raiser, his attractive home farm being situated about four miles southeast of the village of Lewellen

Mr Durand was born in Stark county, Illinois, November 24, 1857, and is a son of Mardonous and Matilda (Williams) Durand, the former a native of the state of New York and the latter of Illinois, where her parents were pioneer settlers Mardonous Durand was reared and educated in the old Empire state and was a young man when he numbered himself among the pioneers of Stark county, Illinois, where he continued to be engaged in farm enterprise until 1861, when he amplified his pioneer experience by removing with his family to Iowa There he reclaimed and improved a productive farm, and he continued his residence in the Hawkeye state until 1885, when he removed to Fort Collins, Colorado, and became again a pioneer, this time on the wide-stretching plains of Colorado There he passed the remainder of his life, and he was seventy years of age at the time of his death, his wife there passing away at the age of seventy-two years They lived up the full experiences of American pioneer life and labored to worthy ends, so that they ever commanded the confidence and high regard of all with whom they came in contact, both having been earnest members of the Methodist Episcopal church The father was first a Whig and later a Republican in his political adherency

Fremont G Durand was about three years old at the time the family removed to Iowa, where he was reared on the pioneer farm of his father, in Keokuk county and profited by the advantages afforded in the common schools of the period Finally he engaged in independent farm enterprise in Iowa, and his attention was thus engrossed for a period of six years, at the expiration of which, in 1884, he removed to Colorado and became interested in the sheep business, with headquarters in Morgan county In 1887, he drove a bunch of sheep through from Colorado to old Cheyenne county, Nebraska, his destination being the part that now constitutes Garden county Here he took up and perfected title to a homestead and a tree claim, in Ash Hollow, and continued his activities as a stock-grower, principally sheep, until 1893, when he returned to Colorado, and engaged in general farming, near Fort Collins, where he continued operations until 1900 Mr Durand then came to his land in Garden county, Nebraska, and has since resided on the tree claim which he took up in 1887 Many of the trees which he planted on this place in the pioneer days are now of stately proportions and add greatly to the attractiveness of the farm, which he has improved with good buildings and otherwise made to conform with the high standards that now obtain in this fine section of Nebraska He also purchased an additional tract of a hundred and sixty acres, which is maintained under effective cultivation, and utilizes also an entire section of school land on the South Table, where he has gained special success and prominence in the raising and feeding of cattle and hogs

Mr Durand has entered fully into the best community spirit, and has been specially liberal and progressive in his civic attitude His political allegiance is given to the Republican party and is fortified by well regulated convictions and opinions concerning matters of governmental and economic policy He is prominently affiliated with Camp No 7970, Modern Woodmen of America, at Lewellen, of which he has served for the past nineteen years as clerk, besides which he has represented the organization as a frequent delegate to the grand encampment of the state Both he and his wife are zealous members of the Methodist Episcopal church at Lewellen, and in the community their circle of friends is coincident with that of their acquaintances

At Indianola, Iowa, January 10, 1882, was solemnized the marriage of Mr Durand to Miss Mary McNaught, who was born in Illinois and was a girl at the time the family removed to Iowa, where she was reared and educated She is a sister of Mrs Louise P Clary and Mrs Maggie B Orr, both of Lewellen, and on other pages of this work the review of Mrs Clary gives the history of the McNaught family Mr and Mrs Durand have four children Frederick M and his wife reside at Cassa, Wyoming, and they have one child, Inez is the wife of Roy L Robley, of Lewellen, Nebraska, and they have two children, Gordon S, of Lewellen, married Miss Fanny Shryer and they have three children, and John H, of Lewellen, still permits his name to be enrolled on the list of eligible bachelors in Garden county

THOMAS W LALLY —Bearing a family name that has been one of eminence in Ireland for many generations, this well known citizen of Cheyenne county may well take pride in claiming the Emerald Isle as the home of his immediate ancestors, for the Irish are a people known for their versatility, marked mental ability and enthusiasm for any cause which they espouse Appreciative of the subtle attractions and undeniable opportunities offered in the field of practical journalism, Mr Lally has chosen the newspaper as his vocation, and the success which he has attained in this field of endeavor is shown by his well equipped establishment and the excellent business which he controls as the editor and publisher of the *Dalton Delegate*

Thomas Lally was born at Lewiston, Illinois, January 1, 1879, the son of Frank M and Mary A (Gibbons) Lally The father was born in County Mayo but was brought to the United States by his parents while still a young boy The family located in Illinois after reaching America and there Frank grew to manhood He received an excellent preliminary education in the public schools and later studied pharmacy, being admitted to practice in Illinois where he was engaged in the business of his profession until the time of his death at the age of sixty-four years Mary Gibbons Lally was born in Lewiston, Illinois, of Irish parents, she was reared there and given the educational advantages afforded by the public schools of that progressive commonwealth and was considered the finest Celtic scholar in Illinois and later of Nebraska She was a highly educated and cultured woman, being a graduate of the State University of Nebraska and supplemented her college course there by graduate studies in the University of Chicago and Columbia University in the city of New York At the present time she resides in Cleveland, Ohio

Thomas was raised in his parents' prosperous home in Illinois and given the best of educational advantages as both his parents were well educated and his mother an exceptionally brilliant woman and scholar He inaugurated his independent career as a school teacher in Furnas county, Nebraska, being the youngest at that time to hold a license Being a pedagogue proved rather satisfactory as an introductory business in life and Mr Lally taught for eighty-seven months in the state of Nebraska, but the Irish of his blood spurred him on to further endeavor and a vocation that would give greater scope to his talents To this end he accepted a position on a newspaper in Sterling and subsequently at Elk Creek, being engaged in professional work along the lines of his choice for three years and at the same time learning the newspaper business from the bottom up Following this he devoted a year to work as a journeyman printer before becoming the business manager of the *Daily Tribune* at Hyannis, Nebraska, and was known as the newsboy of the Forest Reserve in 1913.

As an independent newspaper publisher Mr. Lally made his first venture when he purchased the plant and established the business of the *Dalton Delegate* in 1914, at Dalton, Nebraska, a weekly newspaper that has a wide circulation in the northern part of Cheyenne county, supplying a long felt want in this section. He has made a great success of this initial venture in journalism, being ably assisted by his capable and efficient wife, who has taken an active and prominent part in building up a progressive and paying business within the short period since it was first established. The paper began its upward climb from its inception, and the advancement has continued from that time to the present, with the result that the paper has become a potent influence in public affairs in this section of the country, an effective exponent of local interests, and a vehicle through which communal progress and prosperity are furthered. The *Delegate* reached at one time a circulation of one thousand copies weekly and is to be found in the representative homes throughout the Dalton district of the county. The *Delegate* is non-partisan but nevertheless is a local political organ of no insignificant influence, the while it expresses the well fortified political views of its publisher. However, Mr. Lally endeavors to give to his readers a fair, impartial and unprejudiced view on all questions of interest, political or otherwise, and his paper merits classification among the model village publications of Nebraska, its columns having effective summary of the latest news of general character, as well as a chronical of local events and activities, and terse, well written editorials. The paper has good support from the merchants and professional men of its community and is recognized as a good advertising medium. In connection with the newspaper plant is a well equipped job-printing department, in which first class job printing of all kinds is executed. Mr. Lally is non-partisan in his political views, voting for the man he deems best fitted to serve the people in public office while he is in faith a Roman Catholic.

August 29, 1899, Mr. Lally married Miss Bertha A. Kirste, at Norton, Kansas; she was a native of Webster county, Nebraska, reared and educated in that locality and was the oldest girl scholar in Mr. Lally's first school. Her father was a native of Germany, born at Thorne, who came to the United States when a young man to take advantage of the free land offered to settlers by the government. After coming to this country he came west, locating on a homestead in Furnas county to engage in farming and still lives there, being engaged in agricultural pursuist. Mrs. Kirste, was Helen Kaforka, also of German birth, who accompanied her parents to America when a child of five years. They located in Illinois where she grew up, was educated and married and is still living on the old home farm with her husband in Furnas county.

There were eight children in the Lally family: Walter E., Inez E., Frank W., Arthur T., Helen Kareen, Orville B., Robert E., and Eugene O., all of whom are still members of the happy family circle. Both Mr. and Mrs. Lally have been ambitious for their children and have determined to give them every advantage in an educational way that is within their means, letting each child determine the line along which he or she desires to develop a taste or talent and already the two oldest have reached a high attainment in musical study, having attended the best musical schools, of Cleveland, Ohio, and Chicago.

JULIUS E. GALLOGLY, who is cashier of the Farmers State Bank at Dix, Nebraska, is a young man whose business capacity and sterling personal character have established him in the confidence of this community. He was born in 1885, in Ohio, and is the son of M. D. and Mary Gallogly, who moved to Sheridan county, Kansas in his youth. His mother still lives there and he has one sister, Mrs. Spear, who is a resident of Bushnell, Nebraska.

Julius E. Gallogly was liberally educated, after completing the high school course he attended Sheridan College, following which he took a course in the Grand Island Business and Normal College. He entered the business world in connection with a real estate firm in Kansas, remaining in that line five years, then spent three years in Wyoming, coming from there to Dix, Nebraska, in the lumber business. In 1917, he became identified with the Farmers State Bank as cashier, a position for which he is admirably fitted, being careful, conservative, and courteous.

Mr. Gallogly was married in 1914, in Wyoming, to Miss Erzinger, a lady of education and culture who was a welcome addition to Dix's pleasant social circles. Mr. Gallogly belongs to the Masons at Kimball and the Odd Fellows in Wyoming.

JAMES O BAKER —Along manifold lines has this honored early settler exerted influence during the more than thirty-six years since he came to Nebraska He is a man of broad, intellectual, keen, high ideals and gracious personality, a financier of exceptional business ability—a citizen who commands the fullest measure of popular confidence and esteem For many years he was a resourceful and progressive executive of the banks with which he was associated and wielded a strong influence in the upbuilding of the substantial financial institutions of Scottsbluff county Since disposing of his interests in the banking circles of the Panhandle, Mr Baker has become known as one of the successful and prominent ranchmen of this section, and today is one of the largest landed proprietors in the valley and Scotts Bluff county, making his home in Mitchell

James O Baker was born in Whiteside county, Illinois, June 6, 1852, the son of Captain Reuben and Elizabeth (Hubbart) Baker, the former a native of Ohio, while the mother was born in New York They both came to Illinois with their parents when children, were reared and educated in that state and later met and were married there At the outbreak of the Civil War Mr Baker responded to the president's call for volunteers, and enlisted in the Seventeenth Illinois Cavalry and became captain of his troop Most of his service was in Missouri, Arkansas and Texas After peace was declared he was sent west to fight in the Indian country of Wyoming, western Nebraska and Colorado, as the Indians went on the war path in all these states and Kansas about the time the Civil War closed Captain Baker was mustered out of the service at Fort Leavenworth, Kansas, and soon resumed his profession as a preacher of the Methodist Protestant church He had been ordained before the war and continued to serve the church for many years thereafter, passing away at the hale old age of eighty-two years He had been a member of the Abolitionist party during the troubled times of the late fifties and early sixties Mrs Baker died in Oklahoma in her eighty-eighth year

James O Baker was reared and educated at his boyhood home in Illinois He was sent to excellent public schools for his elementary education and later to the Methodist Theological Seminary, at Adrian, Michigan, but left before completing his course in that institution After leaving college the young man took up farming in Illinois and, in 1884, came west, locating at Phillips, Hamilton county, Nebraska, taking a position in the right of way department of the Burlington Railroad which was being built across the state at that time Mr Baker remained associated with the road until 1888, when he saw excellent opportunities for the establishment of a bank at Phillips, being one of the organizers of the institution He became its cashier and capably filled that position until 1901, when he disposed of his stock in the bank

From the first Mr Baker's business career was marked by courage, self reliance and progressiveness, as well as by that dynamic initiative and executive ability that brings normally in their train a full measure of success, and it was his far vision that led him in association with J W Wehn to establish banks at Bridgeport, Minatare, Bayard and Mitchell, for he foresaw that this country was to become rich and productive Under the new organization Mr Baker became cashier of the Mitchell bank, an association which continued until he sold his interests in the banks in 1903, to invest in irrigated land In this new enterprise he has given his business the same attention and energy devoted to banking and has gained secure status as one of the representative figures in agricultural circles in the upper Platte valley As a banker he showed special constructive talent, and through his effective policies furthered the success of every financial enterprise with which he became associated, since taking up agriculture Mr Baker has become recognized as one of the representative farmers and progressive and public-spirited citizens of Scotts Bluff county and as such he merits specific mention in a history of the county and the Panhandle

Since first investing in land in this valley Mr Baker has continued to add to his holdings until he is the owner of some nineteen hundred and twenty acres, most of it very valuable, worth from thirty to three hundred and fifty dollars an acre, depending upon its location with regard to the irrigation ditches More than sixteen hundred acres of his property is rented In the Ozark mountains Mr Baker has purchased extensive tracts of heavily timbered land and now holds two thousand, three hundred and sixty acres which will bring in a fortune when the trees are cut for lumber

In June, 1872, Mr Baker's marriage with Miss Emily Robinson was solemnized on his birthday Mrs Baker was a native of Illinois and lived but a few years after her marriage, her death occurred in 1878, at the age of twenty-five years On December 23, 1882, Mr Baker married Miss Minnie Brollar, a native of Iowa Mrs Baker's father, Job Brolliar, was an Iowa soldier in the Civil War Early in the war, he was discharged, came home sick and died A few years later, when she was but a little girl, her mother died She and her orphaned sister and brother attended the public school at Vinton, Iowa Her education was finished at Iowa City, after which she taught school until she was married She continued teaching two years after her marriage She was the first postmistress at Phillips, Nebraska, and held the position for six years Then she took an active part in her husband's banking business She was the bookkeeper who never closed up her books until mistakes, if any, were corrected When the bank at Minatare was started Mr Baker's partner, Mr Wehn, wanted her to take charge, insisting that she was a better banker than her husband Since the Bakers sold their banking interests, she has been prominently identified with all social interests in Mitchell For two or three years, she was president of the Woman's Club in Mitchell and was the first president of the Red Cross organized at Mitchell Mrs Baker is a Past Matron of the Eastern Star She probably did more than any one woman to make a success of the Scotts Bluff County Fair A woman of unusual business ability and entitled to a full share of credit for what success Mr Baker has had Mr and Mrs Baker have a beautiful home in Mitchell, where they dispense a gracious hospitality to their many old, warm friends

Mr Baker is known throughout the upper valley as one of the men who has played an improtant part in the opening up and development of this section His faith in the future of the irrigated land led many other men to invest along the Platte and their faith in this section has been justified, for by holding his land Mr Baker became wealthy He has taken an active and influential part in civic, county and panhandle affairs from first locating here He advocates and supports all movements for the benefit of the country and has given liberally in support of all the activities he believed worthy For many years Mr Baker has taken part in the councils of the Democratic party and though he has been urged to do so will not hold office He makes his headquarters at Mitchell, from which point he superintends the management of his farms It can well be said that Mr Baker is self-made, for when he came to Nebraska he had little of equipment in the way of worldly goods, but did have a fair education, the determination to make good and succeed He made many

friends for people soon learned that his word was as good as his bond, and today is rated one of the best known and popular men in western Nebraska Mr Baker is a member of the Masonic order, having taken his Thirty-second degree in the Scottish Rite.

WILLIAM J EWING, president of the Dalton State Bank, at Dalton, Cheyenne county, Nebraska, has been identified with commercial affairs in the state for many years and since assuming his present office February 1, 1919, has become well known and is held in high esteem by the banking circles over the Panhandle

Mr Ewing was born in Fayette county, Indiana, July 27, 1875, the son of John G and Emmiline (Shotridge) Ewing His father was also a native of Indiana, reared and educated there and when President Lincoln called for volunteers to help preserve the Union at the outbreak of the Civil War, he responded when only a boy of fifteen by running away from home and enlisting in an Indiana regiment He saw some of the hardest fighting during that memorable conflict but lived and returned to his home when peace was declared, to engage in peaceful pursuits He was a practical farmer by vocation and also an expert saw mill man, an industry in which he was engaged for several years at the same time conducting his farm During President Harrison's administration Mr Ewing was engaged by J N Huston, treasurer of the United States, to carry on agricultural business and the diversified interests on his land Being a natural farmer, having tastes that fitted him to make the most of every opportunity available in agricultural pursuits, it was but natural that Mr Ewing wanted landed property of his own and as Indiana was well settled up, the land there was high so he determined to take advantage of the government land in states farther west In 1882, he came to Nebraska with this end in view, locating in Polk county where he at once actively engaged in farm activities Now he is a retired man living at Exeter, Nebraska

William J Ewing was reared in Kansas and Nebraska, attending public schools of both states and thereby gaining an excellent practical education which has been of benefit to him in his varied and active commercial life While living at home he gained practical knowledge of farm business and when only twenty years old established himself independently in business on a farm in Jefferson county, where for nine years he raised varied farm crops and tells that he sold his corn for nine cents a bushel and managed to make money at that, so that we know he naturally had great ability in financial dealings Following this he removed to Fillmore county and again was occupied as a farmer but left the country to engage in mercantile pursuits, running a grocery store for a year, where he gained profitable experience and knowledge with regard to business methods and commercial life By 1904, Mr Ewing had gained such an excellent reputation and the confidence of his business associates that he was asked to invoice the store owned by Dan McAleese, for Harry Brown, and the following year was engaged to run the electric light plant in Sidney, but was induced to resign by Mr Brown and go to Dalton where he became the manager of a store owned by Mr Brown and several other men This was a successful undertaking and he determined to branch out in business life, first putting in a lumber yard for the Bridgeport Lumber Company The active management of the yard was placed in his efficient hands and for nearly thirteen years he was manager of the business During this time he had accumulated considerable capital by thrift, good investments and a frugal manner of living Mr Ewing looked the financial field over and decided that the banking business appealed to him most, he had gained a wide and varied circle of friends during his business life in Dalton all of whom held him in high esteem due to his careful methods, absolute honesty in all dealings with customers and they all recognized in him exceptional qualities that are necessary for finance In 1919, on February 1st, he bought an interest in the stock of the Dalton State Bank and at once assumed the management of that thriving institution as president With his varied experience in financial circles and his marked executive ability the future of the bank looks very bright under the capable guidance of such a man as Mr Ewing He has not confined all his energies to business alone, but has willingly and capably taken part in civic and communal affairs as he has been a member of the town board for eight years In politics Mr Ewing is a Republican and takes an active part in all local political affairs which tend to the benefit of the community His fraternal affiliations are with the Independent Order of Odd Fellows, Lodge No 385 and the Modern Woodmen of America

On November 14, 1908, Mr Ewing married Miss Elsie Poole, at Dalton, the daughter of Sidney and Dora Poole, who were pioneers of Cheyenne county , the former passed away in 1918, at the age of seventy years while the mother now resides near Julesburg Mrs Ewing was born in Illinois but her parents brought her to Nebraska when she was very young and thus she is nearly a native daughter of this state, having been reared and educated in this western section of the great commonwealth She spent her girlhood on the home farm and attended the local school , subsequently she graduated from the high school in Sidney and having qualified herself to teach was engaged in that profession for seven years previous to her marriage She is a woman of high attainments and a worthy partner of her successful and progressive husband They are wide readers of modern literature and the current periodicals and thus keep up with all the progressive movements of the state and nation There are two girls in the family Bessie L, at home and Clara V , to whom all the educational and social advantages of the town and state have been accorded by their parents

WILLIAM H ZILMER, the owner and manager of the Gates City Hotel of Crawford, is one of the most popular hotel men in western Nebraska where he is well known and has a wide clientele among the traveling public He is an excellent business man of marked ability, who has made a success of various lines of enterprise and is considered one of the substantial and progressive men of Crawford, where he takes an active part in all civic affairs

He was born in Stanton county, Nebraska, July 8, 1881, and is one of the typical native sons of this commonwealth, with all the energy and initiative that are credited to the Nebraskan Mr Zilmer's parents were August and Amelia (Bramer) Zilmer, both natives of Germany, the mother being born at Potsdam August Zilmer came to the United States in 1861 Within a short time Mr Zilmer went to Michigan and secured employment in an iron ore mine, working fifteen hundred feet under ground Having a sister and brother-in-law in Wisconsin, Mr Zilmer determined to go to them but as he had no money made the trip on foot from Michigan to Shell Lake The soil of that locality is poor so Mr Zilmer, his sister and her husband, Ludwig Beltz, hearing of the rich fertile lands of Nebraska, determined to come here They drove through the country in true pioneer style with an ox team and wagon and reached Omaha when that city was a town of only two hundred and fifty inhabitants The Union Pacific Railroad was laying rails on an extension at the time, and Mrs Zilmer obtained work with the road, soon becoming noted for his great strength, as he could alone lift a rail into place from the ground   For about a year he remained with the railroad and then came to Stanton, Nebraska, and took up a homestead on which he made improvements and proved up   The grant to his land was signed by President U S Grant   After living alone for four years Mr Zilmer married and eleven children were born to the union, only four of whom are now living, two boys and two girls   William was one of the youngest children and had a twin sister

William Zilmer was reared on his father's homestead in Stanton county and when a small boy he raised three pigs to make some money, using it to buy some clothes as times were hard   He attended the public schools for his education and worked on the farm in summer time   After attaining his majority Mr Zilmer remained at home and when he was twenty-seven years old his father gave up the active management of the farm and retired   Mr Zilmer then took charge and farmed the land on shares   He well remembers the grasshopper years when the pests ate everything, even making holes in the fork handles   A year after taking over the farm Mr Zilmer was married at Twin Falls, Idaho, to Miss Agnes Bohaboy, a native of Bohemia, who came to this country with her parents   Three children have been born to Mr and Mrs Zilmer, Ilo, August and Esther   Leaving the farm Mr Zilmer lived three years in Idaho, before returning to Nebraska to locate at Valentine   There he ran a hog ranch three years, meeting with success but was offered a good price for the place and sold out to come to Crawford, February 15, 1918   Soon afterward he bought the Gate City Hotel, which is one of the first class houses in western Nebraska   It is modern throughout, has a good reputation for its pleasant accommodations and Mr Zilmer has gained a fine reputation as host   He is genial, ever ready to accommodate his guests and has so gained a good reputation as an up-to-date hotel man   For some years now he has been well known over this section where he has made many warm friends not only with the usual travelers but through the large automobile trade which is growing each year   His house is a model in its way and no one leaves who has not a good word for the Gate City

Since first coming to Crawford, Mr Zilmer has taken an active part in the affairs of the town, is ready and willing to help in any movement for the upbuilding of the town and its

development and is wide awake to all business opportunities   He is a Republican

SOLOMON D HICKEL, one of the progressive and prosperous young farmers of Dawes county, who is making a success of his business due to his ability and attention paid to farming, is a native son and in his career is displaying all the initiative credited to Nebraskans

He was born in Saunders county, July 2, 1884, the son of Granville and Malinda (Woods) Hickel, being the fifth in a family of eight children consisting of five boys and three girls   His father was a pioneer settler of Saunders county, locating there in 1871   He took up a homestead near the present site of Wahoo   Granville Hickel was a successful farmer and man of affairs, taking part in his community and served one term as State Representative and one term as State Senator, being elected on the Democratic ticket

Solomon Hickel was reared on his father's farm and educated in the public schools near his home   He then entered the high school at Ashland and graduated   While still a boy he began to trap animals for their pelts, selling the muskrat skins at a profit, so that he early learned to make money   He remained at home until he was twenty years old and a year later, March 11, 1905, was married in Ashland to Miss Gertrude Sherman, the daughter of John and Mattie (Wood) Sherman, being the oldest of their five children   Mrs Hickel was educated in the public schools and at Ashland   Her father was descended from General Sherman and was one of the pioneer settlers of Lancaster county, where he farmed for many years   He lived to see the many changes come to Nebraska and sold his land for two hundred and fifty dollar an acres in 1920   Mr Sherman has now retired from active life and lives in Lincoln

Solomon Hickel came to Dawes county in 1918, bought a half section of land eight miles north of Whitney and later leased another half section nearby   He has placed good improvements on his farm, introduced modern methods of farming which he finds pay and is regarded as one of the substantial men of his locality   He is keeping abreast of all agricultural business, is energetic and a hard worker so that his farm is paying a good income to him for the capital invested   General farming is carried on at the Hickel place also some stock is raised, all with an eye to the eventual success of the ranch   Mr Hickel has made his own success and it has been through his own initiative and the determination to succeed that he has taken a place in the ranks of the prominent agriculturists of the Whitney locality where young men are coming to the front as the great producers of the day Mr Hickel takes active part in all civic affairs connected with his community and supports all the movements for the development of the county and his home district He is a member of the Methodist church

FREDERICK N SLAWSON, the efficient county clerk of Cheyenne county, who resides at Sidney is a man of diversified talents Along manifold lines has this pioneer school teacher of Nebraska exerted benign influence during more than a quarter of a century of continuous residence in Cheyenne county He is a man of broad intellectual, keen, high ideals, and gracious personality — a citizen who commands the fullest measure of popular confidence and esteem

Mr Slawson is a native of the Keystone state born at Clara, Potter county, August 14, 1873, the son of Hugh and Alice E (Brooks) Slawson, also natives of Pennsylvania who were born, reared, educated and married in that state They were ambitious people, who believed that a greater future and fortune lay before them in the west than among the mountains of their youth and with stout hearts, high courage and a determination to succeed over every obstacle came west, locating in Cheyenne county near Lodgepole in 1884 Here Hugh Slawson used what money he had in the purchase of land and also homesteaded a hundred and sixty acre tract He erected the usual sod house of this locality for the first home and began breaking the soil in order to put in his first crop Mr and Mrs Slawson worked together to make a success of their farm, they soon had good and permanent improvements for that day and when blizzard, drought and insect pests drove many of the other settlers to sell out and return to their old homes in the east, they held on, their faith in this section could not be broken and though they suffered and endured privations and hardships kept their courage high and in the end won a comfortable fortune which they enjoyed in their later years Mr Slawson became recognized as a progressive and prosperous farmer and stockman in this section, specializing in high grade stock In later years, after the country was better settled, he also engaged in the dairy business, in which line he met with marked success From time to time, as his capital permitted, he purchased land adjoining the original homestead until he

had six hundred and forty acres of the finest farming land in the Pole Creek valley A considerable estate for a man to accumulate in twenty-five years by his own unaided efforts Mr Slawson was a member of the Democratic party but never aspired to office, as his time and energies were entirely taken up by the varied agricultural pursuits in which he engaged He died at Lodgepole, Nebraska, a man of honor and years Mrs Slawson survives her husband and is still living on the old home place near Lodgepole She has been a devoted mother and loving helpmate and companion for more than a half-century — a woman whose strength has been as the number of her days and who had a remarkable share in pioneer experience in the great west and the development of this section of Nebraska

Frederick was twelve years of age when he accompanied his parents to the pioneer home in Cheyenne county He had already attended school in the east and after coming to the Lodgepole community received such instruction as was available in the frontier community His career as a representative of the pedagogic profession began at an early date as he was but seventeen years of age when he began to teach school in a sod house without floor or plaster for the munificent sum of twenty dollars a month, and boarded himself He became a well known and popular teacher in Cheyenne and Keith counties, spending several years in professional life before he determined to establish himself independently as a farmer He filed on a homestead in the vicinity of Lodgepole, made some good and permanent improvements on it, proved up and was engaged in general farming pursuits for nearly ten years, meeting with gratifying success in this business venture After this considerable period on the farm he rented the property and again taught school in Cheyenne county and Lodgepole, following this vocation until the fall of 1896, when the residents of his community elected him to the responsible office of county clerk which he filled in such an able manner that he was re-elected in 1918, and is still the incumbent of that office. Mr Slawson is a farsighted man of varied attainments who has ever had faith in this western country, he keeps abreast of the times and does not confine his interests to one line nor his activities either for he was one of the prime movers and an organizer of the Liberty State Bank of Sidney During his agricultural life and also while professionally engaged he had accumulated a large capital and with the growth of food production in the middle west saw that the financial institutions of the section were bound to prosper and so he bought a large block of stock, in fact the controlling interest in the bank at the time of its organization and became its first president This progressive banking house opened for business April 5, 1919 Mr Slawson is playing a large part in the upbuilding of the business of the bank and his influence will be potent in furthering its development along progressive but conservative lines that will win the confidence of the residents of the section which it is called upon to serve

Mr Slawson is quite active in political circles in Cheyenne county and though not a member of the Democratic party usually casts his vote with it in national affairs but is not bound by party lines in local elections, voting for the man most efficiently prepared for offices of trust in the giving of the people He is not a member of any denominational church but is interested in all church work

June 30, 1898, Mr Slawson married Miss Flora Gonson, a native of Ohio, who came to Cheyenne county with her parents in 1885, and with them shared the vicissitudes of a frontier home, developing into one of the sturdy, gracious daughters of this commonwealth Her parents were Lewis A and Elizabeth (Harper) Gonson, natives of the Buckeye state who came to this county and took up a homestead in the early eighties, where the father engaged in farming until his death The mother now resides at Kearney Mr and Mrs Slawson have one son Hugh L, who enlisted at Fort Logan when President Wilson called for volunteers when the United States entered the World War After completing his military training he was sent to France where he served for a year as a member of the quartermaster's department He received his honorable discharge at Fort Russell, Wyoming, and has returned home

GEORGE B LUFT — A man whose high ideals were crystallized into large and worthy achievement and unequivocal righteousness in all of the relations of life the late George Bowman Luft was an honored pioneer of the Nebraska Panhandle and in an unassuming way he left large and worthy impress upon the history of this section of the state He became one of the leading business men and influential citizens of Scotts Bluff county and was still actively engaged in business in the city of Scottsbluff at the time of his death which occurred on the 15th of July, 1915 His character and has achievement marked him as a man who stood "four square to every wind

that blows," and this publication exercises a consistent function when it enters a tribute to his memory

Mr Luft was born at Warsaw, Hancock county, Illinois, on the 31st of May, 1858, and was a representative of one of the well known and highly honored families of that locality. There he was reared to adult age and there he received the advantages of the public schools of the period, though he was still a boy at the time of the family removal to Nebraska, where his parents became pioneer settlers and where he continued his educational work in the schools of the day At the age of seventeen years he assumed a position as clerk in the mercantile establishment of Herman Diers, at Seward, the present judicial center of the county of the same name Later he was similarly employed in a store at Aurora, Hamilton county, and, in 1885, he became one of the pioneer settlers in that part of old Cheyenne county that now constitutes Scotts Bluff county He filed entry on a homestead three miles south of Gering, the present county seat of Scotts Bluff county, and eventual he perfected his title to this claim Upon the founding of the town of Gering he removed to that place and entered the employ of the Markham Mercantile Company, which opened one of the first business establishments in the village He was thus engaged from 1886 to 1888, and he then became associated with W H. Charlesworth in the conducting of a drug store at Gering Eventually he purchased his partner's interest and assumed sole control of the enterprise, which he continued in an individual way until 1889, when he removed to the new town of Ashford, Banner county, where he engaged in the general mercantile business and where he remained about three years He then, in 1892, returned to Gering, where he was engaged in the same line of business until 1900, when he removed to Scottsbluff and established the first dry-goods store in the present metropolis of Scottsbluff county Eventually he consolidated his business with that of the Diers Brothers, under the firm name of Luft & Diers Brothers, and a grocery department was added to the representative establishment He continued as general manager of this representative business concern until 1910, when he sold his interest to the Diers Brothers and engaged independently in the handling of men's furnishing goods, with a specially modern and attractive establishment to which his personal popularity and the effective service rendered, attracted a large and representative trade He continued this prosperous enterprise until the

Wm. Braddock   Mrs. William Braddock

time of his death, and the business still represents a part of the substantial estate which he left Mr Luft was a man of boundless energy, was liberal and progressive as a citizen, and his admirable traits of character won and retained to him the high regard of all with whom he came in contact He served about five years as vice-president of the Irrigators' Bank at Scottsbluff, his retirement from this office having occurred in 1905 The honors of public office and the activities of practical politics had no allurement for him, but he was aligned as a staunch advocate and supporter of the principles of the Democratic party He was affiliated with the Masonic fraternity and the Independent Order of Odd Fellows, and was an active member of the Methodist Episcopal church, as is also his widow, who still maintains her home at Scottsbluff

At Harrisburg, Banner county, on the 8th of December, 1890, was solemnized the marriage of Mr Luft to Miss Clara B Shumway, who was born and reared in Illinois, where she was afforded excellent educational advantages including those of Knox Academy She is a sister of G L Shumway, editor of this history of western Nebraska, and on other pages are found ample data concerning the Shumway family Mrs Luft became one of the popular and successful members of the pedagogic profession in Nebraska and in 1885 she had the distinction of becoming the first county superintendent of schools in Banner county, a position of which she continued the efficient incumbent until the time of her marriage Mrs Luft is a woman of most gracious personality and has been a leader in the literary and social circles of the city of Scottsbluff The home life of Mr and Mrs Luft was ideal in its every relation, and thus to Mrs Luft remain as consolation and compensation the gracious memories of a devoted companionship She has no children

WILLIAM BRADDOCK, deceased, pioneer and one of the most prominent figures in the development and settlement of Dawes county, was for years one of the best known ranchmen and cattle breeders in western Nebraska, where he gained a high reputation for his introduction of thoroughbred cattle, being one of the first men in this section to realize that well bred stock paid the best He won marked success with the able assistance of his wife who for years was the one whom he consulted in business matters

Mr Braddock was born December 26, 1858, near Marshalltown, Iowa, the son of Martin and Delilah (Lepley) Braddock both natives of Knox county, Ohio, the father being of English and the mother of Pennsylvania Dutch descent William was the fifth child in a family of eleven children, consisting of six boys and five girls His father moved to Marshalltown, Iowa in 1850, and homesteaded the farm where he lived the rest of his life. William was reared in the country, and worked on the farm in his younger days so that his school privileges were limited, but he came of stock that was thrifty and industrious and laid the foundation in his youth for the great accomplishments of after years The school that taught him most was that of experience and he learned well He first earned money as a young boy and early learned its value Mr Braddock remained at home most of the time until the fall of 1884 He had already heard of the great opportunities open for a young man with grit and energy in the western part of Nebraska and came that year by rail to Valentine, the end of the road at the time, then joined a company of freighters to go the rest of the way They reached Beaver valley on Thanksgiving day and Mr Braddock often spoke of that memorable day and the beautiful appearance of the valley He took a pre-emption at once which is still the property of his widow and heirs As soon as possible Mr Braddock built a dugout on his claim and prepared to pass the winter of 1884-85 In the spring he went over onto Bordeaux creek to get lumber for some building as there was a small saw mill then owned by G W Messenger The distance was only about twelve miles but on the trip a hard snow storm came up and covered all the landmarks so that Mr Braddock became lost and wandered in the white waste for two days before finding his dugout That was a hard winter as the snow was the deepest ever known in this section and it laid three feet on a level from December to spring One of the amusing experiences told by Mr Braddock was of his winter in the dugout He was lonely as neighbors were far apart and few and he did not trust the few prowling Indians as he believed they resented the settlers coming in and taking their hunting grounds, though they lived on a reservation He spent many wakeful nights and heard many queer noises and after weeks of anxiety found that some sand mice had been carrying corn from his supplies up to a human skull that he had found on the prairie and kept on a shelf After that he slept better With the spring Mr Braddock broke some of his land and farmed a little, took up a homestead and later a tree claim The tree claim was some distance from his other land and he found it impractible to handle and finally sold it, that being the only piece of land of which he disposed in his many years of ranch life As soon as he made a little money Mr Braddock would buy some cattle brand them and turn them out on the range In the fall he would join in the round up with the large cattle owners and bring his cattle back to the home range He kept doing this year after year until he had a large herd

March 28, 1899, Mr Braddock married Miss Julia Anna Jacobson, who was born near Nevada, Iowa, the daughter of John H and Dora (Tow) Jacobson, both of Scandinavian descent In the spring of 1885, the Jacobsen family came to Sheridan county, Nebraska, and took a pre-emption northwest of Rushville, seven miles, where they lived twenty years then moved near Mullen, Hooker county and continued in the ranch business Mrs Braddock went through the common schools and was teaching at the age of sixteen Later she attended the Rushville high school She received only twenty-five dollars a month but invested what money she could in calves each year and when she was married added twenty head of fine cows to her husband's large herd Of this accomplishment she was justly proud After her marriage they worked harder than ever as both she and her husband toiled early and late They attained a remarkable success however, and she feels that they were well repaid, as at one time they had twenty-five hundred head of cattle before the free range was done away with After that the owners sold many cattle and kept smaller herds Mr Braddock was one of the first men to see far ahead and realize that irrigation was to be the great thing in western Nebraska and built nine miles of ditches on his ranch, as he had the priority water rights from Beaver creek along which his land streched for fourteen miles His entire ranch is fenced with four wires and cross fenced with posts every rod so that that his improvements were some of the best in the west The Braddock ranch has five hundred acres of alfalfa which is usually cut three times a season, two hundred acres are native wheat grass meadow which usually cuts three hundred tons of hay per year, and the ranch is one of the best located and most beautiful in the state, lying in the beautiful Beaver and White river valleys It stands as an enduring monument to the man and woman who spent so many years of their life here, reclaiming the virgin prairie to productive farm purposes In 1908, Mr Braddock bought his first registered Herefords, gradually worked out of grades

and into pure bred cattle At the heighth of his career he died, January 7, 1917, a great loss to his community and mourned by all who knew him After her husband's death, Mrs Branddock with undaunted courage assumed the full control of her husband's business and has made an enviable record as a business woman in the northwestern country where she is widely and well known Her large herd of Anexiety 4th cattle, some seven hundred in number, are said to be the finest in the country by experienced cattle men who are breeders themselves Mrs Braddock had the honor of having the first show herd of this kind of cattle exhibited at a National Stock Show, from Dawes county and the county boasts that it has more pure bred white faces than in any territory of its size in the United States. Mrs Braddock exhibited at the Denver Cattle Show of January, 1921, where she won a premium on every animal exhibited This in competition with veteran breeders who have been showing cattle for many years, a rather unusual honor for a woman When asked by a friend, "Mrs Braddock, were you not surprised?" She replied, "No, this was not thought out or accomplished in a day. Many months of careful watching of the development of different individuals are necessary in the selection of a show animal, and I have made an intensive study of the various types of beef cattle for several years Right now I am planning and preparing for the 1922 shows"

Mrs Braddock is a woman of high culture and refinement as she has studied these many years in spite of the trials and hardships she endured on the ranch in the early days She has two cultured daughters, Gladys Enid, who after graduating from the Chadron State Normal School, in Chadron attended the Nebraska State University, at Lincoln and is now attending the University of Chicago, and Wilma Doris, who is in the ninth grade of the normal school at Chadron Mrs Braddock has a beautiful home in Chadron and now lives surrounded by all the luxuries and comforts that wealth and culture can afford but she says that wealth is not all in life to live for and is desirous of assisting in the farther development and improvement of Dawes county where she has played an important part in stock raising and agriculture She stands nigh in the community respected by her business associates and loved by the many old friends Few women have been able to take up such a large business enterprise and make the success that she has

J H WILHERMSDORFER, county judge of Sioux county two terms, is one of the well known business men of this locality who has taken an active part in the development of the Panhandle for many years, and is today one of the leading automobile dealers of his section.

Judge Wilhermsdorfer was born in Kirkwood, Illinois, October 16, 1866, the son of Solomon and Mary M (Kness) Wilhermsdorfer He was reared in Illinois and Iowa and received his educational advantages in the latter state, at Abingdon, where he completed his schooling The early years of his business life the judge spent in Iowa but came to Nebraska in 1901, and located in Sioux county, where he at once began to take an active part in the civic life of Harrison, as he was an aggressive business man and took a deep interest in the general welfare of the town which he made his home In 1904, he was elected county judge and took his seat on the bench of Sioux county, serving two terms, which attests to his ability as a judge and how well he satisfied the people of the county The judge's third term in office was but recent, as he was elected in 1920 After locating here the judge began to look around for good property in which to invest, and chose ranching land, which he managed himself For some time now he has owned a garage in Harrison, being the local agent of Ford cars and trucks in this part of the county He has built up an excellent business which brings in most satisfactory returns

From 1898 to 1910, the judge was a member of the National Guard of Iowa He is a Republican, a member of the Odd Fellows, the Knights of Pythias, and the Modern Woodmen of the World and A F & A M

October 25, 1892, Judge Wilhermsdorfer married Miss Zua E Bowman, at Fairfield, Iowa, the daughter of Samuel Bowman, and one son has been born to this union Moritz, who is at home He graduated from the high school and then entered the engineering department of the University of Nebraska, a member of the class of 1923 At the present time he is at home for the summer vacation

Judge Wilhermsdorfer is one of the substantial and progressive men of his community who is helping to "put the Panhandle of Nebraska on the map," as he has strong convictions about civic duty and is not afraid to voice them He is progressive in his own business and believes in having the public business run in the same manner While on the bench he served his county well and faithfully and gained a high standing as a result of his judicial duties

JOHN H NEWLIN is a pioneer settler of the Panhandle, school teacher, rancher, and newspaper man There are few people in what was old Sioux county who are not familiar with this man who has played an important part in the development of his section, and is today most highly respected and well fixed with worldly goods of his own accumulating

John Hamilton Newlin was born in Ripley, Brown county, Ohio May 22 1853, the son of Nathaniel and Melissa (Hamilton) Newlin The father was born in Brown county, Ohio, March 4, 1820, he was left motherless when less than two years of age and was reared by Samuel Pangeburn, a prominent miller at Ripley The mother was born near Cincinnati, Ohio, September 30, 1823, she was left an orphan when young and was reared by an uncle, named William Gates, of Mason county, Kentucky She met Nathaniel Newlin and they were married at Ripley, Ohio, March 15, 1843 Ten children were born to this union, of whom six still live, the oldest being seventy-seven years of age and the youngest fifty-seven years old After their marriage, Mr and Mrs Newlin moved to Peoria, Illinois, in 1855, and three years later to Mendota, the same state, and from there came west to Dallas county, Iowa, in 1868 After two years there they removed to Guthrie county, where they continued to reside many years Mrs Newlin died in 1896, and her husband in 1898 They are buried at Bayard, Iowa In early life Mr Newlin was a cooper and followed his trade until he became a farmer, an occupation he followed after coming west

John H Newlin received his education in the common schools of Illinois and later of Iowa He early became a farmer and also was a school teacher, making a success of both vocations In May, 1890, Mr Newlin came to the Panhandle, locating in Sioux county, where he filed on a government claim in Hat Creek valley, becoming one of the early settlers of that section He improved his land and was one of the homesteaders who became well known From 1897 to 1903, he lived on rented land in Wyoming Before coming to Nebraska he had lived on the frontier in Kansas, having gone there in 1877, but after spending four years in that unsettled country, returned to Iowa in February, 1881 He taught school there, also taught after taking up his land in the Hat Creek valley and later, during his residence in Wyoming becoming one of the well known and popular teachers wherever he had a school

Having a natural bent for journalism, Mr Newlin decided to engage in that profession and, on February 1, 1904, bought the Harrison *Sun*, of which he was the able owner and manager until June 1, 1921 During that long period he became one of the prominent figures among the newspaper men of western Nebraska, as he published a well printed sheet, containing all the latest news of the day, with able editorials upon important questions of the day and political matters, and supplied the territory which the *Sun* served with an up-to-date publication of which the people were proud, and one of benefit to the community It was with regret that Mr Newlin's friends learned that he was to retire from journalism From first settling in the Panhandle, Mr Newlin took an active part in all public affairs, as he served as treasurer of the Harrison school board for many years, was secretary of the Harrison Cemetery Association, and a member of the village board two years

He is a Republican in politics and says that he cast his first vote for President Hayes in 1876 He is a member of the Odd Fellows lodge, having filled all the chairs in that organization, and was delegate to the Grand Lodge of which he is also a member He belongs to the Rebekah lodge, Palestine Encampment, Chadron, Nebraska, and is a member of the Methodist church

May 29, 1890, Mr Newlin was married at Harrison Nebraska, to Miss Ella M Conner, the daughter of William W and Nancy (Carson) Conner Mrs Newlin was born near Cullom, Cass county, Nebraska, December 8, 1863, and according to the family history is a distant relative of the famous Kit Carson of frontier days and fame She was left an orphan at the age of fifteen years. but received a high school education and upon graduating taught school in Iowa, Nebraska, and Wyoming until she married Three children have been born to Mr and Mrs Newlin Jessie Eva, a graduate of the Chadron State Normal School and the holder of a life certificate in Nebraska and Wyoming, was married August 29, 1920 to Milo E Wolff, at Hot Springs, South Dakota, and they now live in Wyoming, Bessie M, also a graduate of the Chadron Normal School, who holds a life certificate in this state, has taught in Nebraska and Wyoming She is now at home, being a linotype operator in the *Sun* office, and Nellie B, who died September 26, 1903, in Harrison, aged three years

Mrs Newlin, then Miss Ella Conner, came to Sioux county from Guthrie county, Iowa, in April, 1888, and pre-empted 160 acres of government land in the Hat Creek valley, where she was living at the time of her marriage

FRED WILLIAM MEYER, the owner and editor of the Harrison *Sun*, is one of the well known newspaper men of the Panhandle, where he has taken a leading part in molding public opinion through the columns of his paper No man wields a wider influence than one who is associated with the press of the country, and Mr Meyer is no exception to this rule, for he has been enterprising and progressive in his own affairs and has reflected these qualities in the *Sun*, which supplies the Harrison district with the latest news of the day, encourages all movements for the betterment of the community and gives the people excellent service, for he is a practical man of affairs and is not merely theoretical in his views of politics and matters pertaining to public welfare The people are to be congratulated that they have such excellent newspaper service and such an able man for its manager

Fred W Meyer is a native son of Nebraska, born at Platte Center, May 20, 1886, the son of Fred and Evangeline (Rosekrans) Meyer, the former a native of Germany, while the mother was born in Pennsylvania They came to this state about fifty years ago and located on a farm near Platte Center, became prosperous farmers and today live retired, enjoying the fruits of the many years of labor There were eight children in the Meyer family Minnie, John, William, Lena, Anna, Sena, Martha, and Fred W, of this review He attended the excellent public schools of his district, then entered the high school of Platte Center, where he graduated in 1904 Being ambitious, Mr Meyer then attended the Wayne Normal College at Wayne, spent three years there and, in 1907, entered the University of Nebraska, receiving the degree of Bachelor of Science upon graduating Entering upon the duties of his profession as teacher, he now holds a life state teacher's certificate Mr Meyer became superintendent of the city schools of Platte Center, a position which he held five years, which attests to his ability as an organizer and administrative head He then accepted a position in the treasurer's office, which he filled four and a half years before resigning to become a member of the newspaper fraternity, having bought the Harrison *Sun*. In the meantime Mr Meyer had invested in a large ranch, having taken a Kinkade homestead ten miles south of Harrison which he has improved and there demonstrated that the educated man makes a good farmer. In all these enterprises he had prospered, due to his ability, foresight, and excellent management

On August 18, 1920, Mr Meyer married, at Mitchell, Nebraska, Miss Vinnie Newell, the daughter of W Newell, of that city Her father was a rancher of the district north of Mitchell, and one of the well known men of his district

Mr Meyer is a Democrat, a member of the A F & A M, Odd Fellows, and Woodmen of the World, and was raised in the Baptist church He is one of the progressive and enterprising men who are making history in the Panhandle, not alone by his own achievements, but by the wide influence he exerts through the *Sun*, which is one of the well edited, newsy sheets of western Nebraska

THOMAS EDWIN PHILLIPS, one of the popular business men of Harrison, who owns and manages the leading drug store of the town, is a man who has made a great success of his business as a druggist and, Harrison is to be congratulated that it has such a proficient, conscientious, and able man to carry on the exacting business of a druggist In addition to his prescription department, Mr Phillips also carries all the side lines that the public have come to expect and demand in a drug store, and has one of the most attractive houses in the Panhandle, where he has had a remarkable success

Thomas E Phillips was born in Truro England, the son of Thomas Edwin and Mary (Hendy) Phillips, both of whom were born in Cornwall, England Thomas attended the schools in his native town, graduating in 1897 He decided to come to the United States to make his home, and in 1909 started in business at Chadron, Nebraska, and related that when he started up there he had practically no capital but his early business training, ability to work, and the determination to success, and he has won out, for today he is one of the leading business men of Harrison In August, 1916, Mr Phillips moved to Harrison, opened his store, and from the first has met with gratifying and marked success He has taken an active part in the life of the town since settling here and during the World War was a member of the County Council of Defense also of the food administrator for the county, offices which he filled with credit to himself and the satisfaction of the people

December 24, 1907, Mr Phillips was married, at Des Moines, Iowa, to Miss Jessamine Pearl Foxwell, the daughter of William Foxwell, who was born in Wisconsin, but later moved to Columbus, Nebraska He died at Lincoln, Nebraska, in 1914, surrounded by his wife and three daughters Mrs Maude Slat-terly, of Chadron, Mrs Gertrude Richards of Lincoln, and Mrs Jessie Phillips, of Harrison One child has been born to Mr and Mrs Phillips Mary Louise, born at Chadron, October 27, 1914

Mr Phillips is associated with the Methodist church and is one of the enterprising men who takes an active part in all movements for the improvement and upbuilding of Harrison, as he is energetic and enthusiastic in his business and applies the same ideas to public welfare

JOHN ELMER MARSTELLER, deceased, was one of the early settlers of Harrison, Sioux county, and took a prominent part in the upbuilding of the city and county, as he was a man of energy, progressive in his ideas and a successful business man For years he held positions of responsibility in Harrison and Sioux county, giving of his time and money in the interests of the people and district His death was a loss not alone to his family and friends, but to the whole community where he had labored for nearly forty years

Mr Marsteller was born in Mercer county, Pennsylvania, March 19, 1863, and died in Harrison, February 10, 1921 He was the son of Mr and Mrs R E Marsteller, also natives of Mercer county, Pennsylvania The boy was reared and educated in his native county, where he later learned the carpenter's trade Hearing of the many opportunities offered young men in the new western country he came to Nebraska in the early 80's, locating near Harrison, in Sioux county, at a time when the little settlement consisted of a town of tents He at once began to work at his trade and assisted in building many of the early structures of Harrison which stand as a monument to his skill and ability Later Mr Marsteller engaged in business in Harrison as a hardware and furniture merchant When his business had grown he became a general merchant, carrying all lines demanded in such a growing community In all lines he met with gratifying success as he gained the confidence of the settlers by his fair, honest dealing He watched the great development of this section of Nebraska and took a prominent part in it, for he invested his money in land and became one of the well known and prominent ranchmen of Sioux county For some years Mr Marsteller was postmaster of Harrison, a position he was well qualified to fill, at one time and another he held most of the offices of the village board of Harrison and was county commissioner of Sioux county at the time of his death During the World War Mr Marsteller was one of the prominent figures in all war work, being an officer of the Red Cross, and was chairman of the County Council of Defense in the united war work campaign He gave freely of his time and energies in assisting the government to prosecute all war work both in county and town He was a Democrat in politics but stood for the ideas that were best for the country Mr Marsteller was for many years a member of the Masonic order, also belonged to the Woodmen of the World, and Modern Woodmen of America Reared in the Methodist faith, he early joined that church and was devoted to the religious interests of his community, willing to make great sacrifices in the interests of the church and its people For more than twenty years he was superintendent of the Sunday School

May 14, 1890. Mr Marsteller was married at Harrison to Miss Ida Smith, the daughter of Mr and Mrs W R Smith Mrs Marsteller's father was one of the leading merchants of Harrison. owning a general store Four children were born to this union Bessie Janett, the wife of L A Alexander, of Hanover, Indiana, Vernard Emmett, who married Nell Anderson, Byrdice May, and John Wesley

Mr Marsteller was one of the true Christian men whose life was above reproach, who performed many acts of charity of which no one was ever aware, ever ready to help others in distress, he took a prominent part in all civic and communal affairs and became the recognized leader of Harrison He was broad minded, charitable, yet held positive convictions as to life conduct and had the courage to express them His life was one that might well be emulated by the rising generation

SAMUEL KNORI, one of the prominent ranchmen of Sioux county, who has accumulated a large estate in Nebraska since he came here some twenty-eight years ago, through his own initiative and ability and the determination to succeed, is well and favorably known in this section where he has made a success of his vocation

Mr Knori was born in Switzerland in 1867, the son of Andrew and Anna (Zvahlen) Knori, who spent all their lives in Switzerland as farmers The mother died there in 1880, and the father in 1911

Mr Knori was reared on his father's farm and attended the public schools for his educational advantages When his schooling was finished he learned farming under his father But as there was little opportunity for a young man to get ahead in his country, he decided to come to America and learn what fortune could be gained in this land of opportunity and plenty Landing in New York in 1891, Mr Knori came west and spent two years in Wisconsin, where he learned the American methods of farming and our customs Land was high there, so he came to the Panhandle in 1893 and took up a homestead three miles north of Harrison From the first he prospered, due to his training as a farmer and willingness to work so that today he is the owner of two thousand acres of land ten miles north of Harrison on Monroe creek, for as he made money Mr Knori invested it in land, which has made him one of the largest landed proprietors of the county At the present time he runs about two hundred head of cattle annually He carries on some general farming and devotes much time to stock raising and feeding

Before leaving Switzerland, Mr Knori spent the required time of two years in the army. He is a Democrat and for twenty years has been a member of the Woodmen of the World His church affiliations are with the Lutheran organization

January 15, 1900, Mr Knori married Miss Elizabeth Noreisch, who was born in Germany in 1881 Her parents were farmers in that country and spent their lives there Four boys have been born to this union John, Samuel, Emanuel A, Gustave, and Lewis, all of whom are at home

Mr Knori has made a study of agriculture in the western part of Nebraska and is today recognized as one of the able and substantial men of his district, where he is ever ready to assist in all movements for the development of his community or the county.

GEORGE W McCORMICK, one of the well known ranchers of Sioux county, who is regarded as a substantial and progressive agriculturist of his section, was born at Green Bay, Wisconsin, June 30, 1866, the son of S W and Charlotte L (Smith) McCormick The father was a contractor and builder in Green Bay, who later moved to Neligh, Nebraska, where he followed the same business for a number of years In 1870, Mr McCormick built a boat which he loaded with lumber and went down the Mississippi river with it to the mouth of the Missouri river From there he returned up stream to St Joseph, Missouri, and within a short time of reaching that town came to Nebraska, locating at Neligh

George McCormick was reared in Green Bay and later in Nebraska, receiving his educational advantages in the public schools of his home town In 1908 he decided to come to the Panhandle and take advantages of government land, settling on a homestead in Sioux county in the Harrison district, where he has made a specialty of sheep and now has many head In addition to his interest in sheep, Mr McCormick has carried on such general farming as he found profitable along with raising cattle and sheep, and has found all lines brought good returns He has gained high standing in the county since locating here and has made a business of his farming, so that it has brought substantial income for his time and labor

February 23, 1894, Mr McCormick married Miss Jane Longeor, the daughter of Harvey Ingalside Longeor, at Lakeland, Nebraska, and they have the following children Hattie A, George W, Roy Samuel, Ralph, Nellie May, Addison, and Esther Rose, all at home

Mr McCormick is a Republican and a man who takes an interest in all the progressive affairs of his community, assisting in the upbuilding of his district and the county

SLATTERY BROTHERS

SLATTERY BROTHERS are owners of the Caledonia Ranch Among the early settlers and pioneer ranchmen of Sioux county no men are more well and favorably known than the Slattery brothers James, born in 1861, William, born in 1867, Daniel, born in 1879, are all natives of Chateaugay, New York, while Patrick, the youngest, born in 1882, is a native of Iowa They are the sons of Daniel and Alice (Ryan) Slattery, the former born in 1809, died in 1879 at Winthrop, Iowa, while the mother was born in Ireland in 1830, and died in Sioux county in 1917

Daniel Slattery and his brothers received their educational advantages in the public schools of Winthrop, Iowa, and there Daniel graduated from the high school With their mother they came west in 1887, and Daniel filed on a homestead in Sioux county in July of that year, land which was the start of their present large ranches, and which has never passed from the original owner The Slattery family prospered, they weathered all the early hard years of draught and insect pests and remained in the country when other settlers grew discouraged, gave up and left the country They believed that the land in the Panhandle could be made to pay and remained to demonstrate the fact The older boys worked hard, encouraged by their courageous mother, and as times became better made money on their cattle, bought more land from time to time until today they own some thirty-six hundred acres, known throughout this section as the Caledonia Ranch, one of the finest ranching properties in Sioux county and the northwest They run about two hundred head of cattle annually and harvest some two hundred tons of hay, which is used for feeding

The brothers have never married, but maintained their home with their mother during her life and now run a fine bachelor establishment

Known throughout this section of the state as excellent business men and successful ranchers, the Slattery brothers are always ready to boost for improvements of their section, giving of time and money to all laudable movements.

They are Republicans and members of the Catholic church Many and interesting are the events and experiences of which they can tell of the early days in the Panhandle, the cattle industry, and first attempts at farming Today they devote their time to raising and feeding cattle and the necessary farming in connection, and are large shippers to the eastern markets

WILBUR F SHEPARD

WILBUR F SHEPARD, prominent ranchman and banker of Sioux county, to whom pioneer honors should be accorded, as he came to the Panhandle in 1887, and passed through all the hardships and privations of the early days on the frontier, has gained wealth and position through his own hard work, determination to succeed in the west, and his faith in the future of this section Today Mr Shepard is one of the largest landholders in Sioux county and also owns a large tract of land in Wyoming, all of which has been under his active management and supervision for years He is regarded as one of the successful and progressive men of this section, which is known for its able men of affairs

Mr Shepard was born in Ottumwa, Iowa, being the son of John R and Mary (Swickard) Shepard, the former a native of Jefferson county, Ohio He died at Ottumwa in 1900, but the mother still lives at the old homestead in a house which was built on the land in 1876 Wilbur Shepard was sent to the public school for his elementary education and then to the schools of Ottumwa, where he received an excellent general education He learned farming under his father in Iowa and has followed that vocation all his life Coming west when the Panhandle was regarded and actually was the frontier, Mr Shepard took up land in old Sioux county when there were no towns in this section, habitations were few and far apart and the principal business was running cattle He holds the first tax receipt ever paid on land in Sioux county, which shows that he was practically the first settler perhaps not actually the first, but the first to lawfully hold land and pay taxes on it Many and interesting are the experiences and stories which Mr Shepard tells of the early days, how many of the settlers became discouraged and left, but he remained, having faith in what the country would eventually become, and this faith has been justified, for today Mr Shepard holds ten thousand acres of valuable land on which he runs two thousand head of cattle Eight thousand acres are located in Sioux county and two thousand across the line in Wyoming Starting with a small ranch, Mr Shepard made money in cattle and invested in more from time to time, so that today he is the owner of a vast estate Turning his attention to other lines of enterprise, Mr Shepard became interested in banking and bought stock in the First National Bank of Harrison, is a director of that prosperous and sound financial institution He is a Republican and a member of the Methodist church

June 25, 1895, Mr Shepard married Miss Anna E Zubst, the daughter of Frederick and Dora Zubst, both of whom are dead They came to America from Germany and were farmers all their lives Three children have been born to Mr and Mrs Shepard George H, Leroy, and Minnie May, all at home with their parents

Since 1887, when he came into Sioux county, Mr Shepard has taken a prominent part in the upbuilding and development of this section He has always been a loyal supporter of all improvements of the county and his own community, and today is one of the most substantial and prominent ranchmen of the northwest, a position he has gained by his own efforts

SAMUEL M THOMAS, prominent ranchman of the Harrison district, Sioux county, who came here in the early days and has played an important part in the opening up and development of this section of the Panhandle, where he is today recognized as a man of prominence and substantial fortune, should be accorded pioneer honors, as he came here when habitations were few, town unknown, and settlers far apart

Samuel Thomas was born near Iriquois, Illinois, in 1852, the son of Samuel and Evelyn (Courtright) Thomas The father went to California during the gold rush and died there in 1852, soon after his arrival The mother died in Iowa Samuel was reared by his mother, who was left a widow when he was an infant, and received his educational advantages in the public schools He says that he attended school only a few years, as he was compelled to earn his own living as soon as he was old enough to do so, in fact since

he was ten years of age The boy began to work for farmers and in that way learned farm business, which he has followed since his childhood Knowing that there were opportunities of securing land in the western part of Nebraska, Mr Thomas came west in 1888 and took up land in the Panhandle of Nebraska. He came across the country in true pioneer style in a prairie schooner, bringing all his worldly goods in the wagon He was one of the first men to locate in what is now Sioux county and with a smile says that he "enjoyed all the trials and pleasures of pioneer life," which consisted mostly of privations, hard work and hard knocks for many years Mr Thomas was not discouraged by all the suffering and hard work, stuck to his land when others were leaving to go back east, and has lived to see the great changes and improvements come to this section, so that today he and the other old settlers enjoy all the comforts and luxuries that were not dreamed of in the early days He has as well accumulated a comfortable fortune, and is the owner of a ranch of three thousand one hundred and sixty acres From first coming here Mr Thomas has specialized in raising stock and today is well known as a breeder of high bred Polled Hereford cattle He has gained a wide reputation for this and is one of the well known breeders of his section, owning a fine lot of stock

December 25, 1873, Mr Thomas married Miss Edeline Bouroughs of Illinois, the marriage taking place in that state Mrs Thomas's father was a prominent man of his locality, being a member of Congress at the time Black Hawk signed the treaty with the government in 1849 The following children were born to Mr and Mrs Thomas· Theressa, the wife of T C Lewis, Charles, who married Mary Turner, died in 1917, Ira, who married Christine Peterson, Henry, who married Esther Hamlin, and S E, who married Clara Larsen

Mr Thomas is a Republican and a member of the Methodist church

ALBERT L SCHNURR, president of the First National Bank of Harrison, Sioux county and vice-president of the bank of Van Tassel, Wyoming, is one of the prominent business men and financiers of the Panhandle and western Nebraska, where he has gained a prominent place in banking circles by his ability in business and progressive ideas as a banker He has gained the confidence of the people of his section by conservative methods during a period of readjustment of the world and of our own country, when the people must look to the bankers to assist so materially in tiding the country over the crisis. Young in years, Mr. Schnurr is old in sagacity and under his able guidance the banks of which he is the executive head are doing a growing business each year.

Mr. Schnurr was born in Mount Pleasant, Iowa, August 21, 1879, the son of William and Rosa Schnurr, the former a native of Ohio, while the mother was born in Virginia. Albert received his elementary education in the public schools of Mount Pleasant and, in 1896, graduated from the high school there. For a year he attended an academy and then graduated from the business department of Iowa Wesleyan College. He studied law after entering business life and was admitted to the bar in Nebraska in 1906. After leaving college Mr. Schnurr was employed in law offices in Mount Pleasant and later in Omaha; during this time he devoted all his spare time to studying law with such excellent results that he took the bar examination in Nebraska and was admitted to practice here. Mr. Schnurr came to Harrison in 1905, just a year before he became a lawyer; he was elected county judge, holding office in 1908, 1909, and 1910, proving a most efficient member on the bench. In the fall of 1910, Mr. Schnurr became interested in the First National Bank of Harrison, was elected vice-president of the institution at that time and began the business of learning banking. His professional training and business life of early years assisted him greatly; and it was due to his constructive ideas and able guidance that the First National gained in depositors and standing in the nine years before he became the actual head of the bank in 1919. Later he also became interested in the Van Tassel bank, bought a large block of stock in it and became its vice-president. Mr. Schnurr's rise as a banker has been consistent and rapid, which attests his ability as a financier.

During the World War he was chairman of Sioux county in each Liberty Bond drive; was chairman of the War Savings campaign, and a member of the executive committee of the Sioux County Red Cross Society. In every way he assisted the government in the active prosecution of the war in his community and county.

Mr. Schnurr, as has been stated, was county judge in 1908, 1909, and 1910, and is a member of the village board of Harrison, having been elected in 1920. Progressive in his business, he believes in applying the same methods to communal affairs and is an excellent man to have running the town's affairs. In addition to his banking interests, Mr. Schnurr is secretary and treasurer of the Harrison Real Estate and Loan Company, which is a growing concern.

For three years Mr. Schnurr was high secretary of the Independent Order of Forresters of Nebraska; is Past Master of Sioux Lodge No. 277, of the Masons; Past Noble Grand of Harrison Lodge of the Odd Fellows; is a member of the Brotherhood of American Yeomen, and is a Scottish Rite Mason.

At Harrison, on June 29, 1910, Mr. Schnurr married Miss Elsie M. Rohwer, the daughter of Eggert Rohwer, who settled in Sioux county in 1887, being one of the prominent ranchmen of the county. Two children have been born to this union, William Eggert, and Clarence A.

The Schnurr family is one of the prominent and well known ones of Harrison and Sioux county, where they have a host of warm friends.

RALPH B. SCHNURR. — Citizens of Sioux county owe much to the vision and sagacity, the loyalty to the county's best interests by Judge Ralph B. Schnurr. For many years he has been a staunch advocate of the possibilities of Sioux county as a combination farm and ranch community. He has practiced as well as preached it, and men have listened to his words because they have had the convincing quality of practical experience as well as the insight and philosophy of an educated and cultured man.

He was born April 15, 1886, at Mt. Pleasant, Iowa, the son of William and Rosa Schnurr, who for many years were helpful and highly esteemed residents of the charming Iowa town. The father was born at Springfield, Ohio, and the mother in Richmond, Virginia. She was the daughter of John Rukgaber, a contractor of that city. For thirty years, and up to the time of his death, in 1909, William Schnurr was a prosperous shoe merchant of Mt. Pleasant, and one of its representative citizens. He was of German descent and traced his family back to the Ruprechts.

Ralph B. Schnurr, with whom this sketch has principally to do, early gave promise of his intellectual and prosperous business career. He was a fine student in his boyhood and he was graduated from the Mt. Pleasant High school with honors. Later he attended the business college of the Iowa Wesleyan University and learned the principles of modern business, which were of incalculable benefit to him in later life.

He first engaged in different kinds of mercantile business as an employe, studying how business principles were applied in each one to make for business success. With this sort of a training he later engaged in business for himself, first in the abstract, insurance and loan business, and then in ranching operations, and as a breeder of pure Hereford cattle. ˙ At the present time his main business interests are in ranch lands and the oil industry, and he has

RANCH OF R. B. SCHNURR

the utmost faith in the future of Sioux county in both branches of these undertakings.

From January 5, 1911, to January 6, 1921, he was county judge of Sioux county, Nebraska. During this ten years of service he acted in the capacity of land commissioner and rendered valuable assistance in the settlement and upbuilding of Sioux county. The services he thus rendered were of immense benefit to the people of this county, because they came from a sagacious mind and a willing heart. Judge Schnurr's retirement at the end of his ten years of faithful service was purely voluntary.

Judge Schnurr found time, also, to devote to semi-public affairs. For several years he was secretary of the Fair Association of Sioux county, and at the present time he is secretary-treasurer of the Harrison N. F. L. Association. He has been active in various lodges, including the I. O. O. F., Patriarchs Militant, the Rebekahs, the Elks, and the Woodmen of the World.

Judge Schnurr was one of the most active and loyal of citizens during the World War, and the advice that he gave and the example he set were noted with pride and pleasure by the great circle of his friends and admirers in Sioux county. He acted as secretary of the county council of defense and chairman of the

legal advisory board for Sioux county,, and in both these capacities he rendered able and single-minded service.

Judge Schnurr was reared a Presbyterian, but he is a man of broad and liberal views on matters of religion and in politics. At the present time he is affiliated with no church but is a supporter of them all, believing that each of them has its work to do in making the world better. In politics he is a Democrat, but here again he is liberal in his views, able to discern the weaknesses as well as extol the virtues of his own party and being generous enough to accord just commendation to men in other parties whom he believes to be able and upright.

Mrs. Schnurr's maiden name was Margaret M. Bixler. She was married to the judge, November 13, 1920. She is the daughter of Mr. and Mrs. James M. Bixler, both of them now deceased. They were early settlers in Illinois, later pioneers of Indian Territory and, after the death of her husband, Mrs. Bixler moved with her adult sons and daughters to Sioux

RESIDENCE OF R. B. SCHNURR

county, where they acquired valuable ranch holdings.

An active, upright, educated man, Judge Schnurr is a fitting type of the men who have made the barren wilderness fruitful, who have transformed its wildness and monotony into prosperous ranches, comfortable, stately homes, and towns where the hum of industry never ceases. These are the men who are the bone and the sineu and the brains of America and from them and the areas they inhabit this nation gets its virility and courage which enables it successfully to pass every crisis and to face every foe, both without and within.

D. R. WILLIAMS

J. W. KINNAMON

DAVID R. WILLIAMS is a representative of one of the prominent and well known pioneer families of western Nebraska and here he has won for himself individual prestige in connection with farm industry and as a dealer in real estate, in which latter domain of enterprise he has developed a substantial business and incidentally aided in furthering the march of development in this section of the state. He is the owner of a large tract of land in Garden and Morrill counties, as well as his attractive home property in the village of Lisco, where he is a stockholder in the Farmers' Elevator Company and where also he maintains his general business headquarters.

David R. Williams was born in Harrison county, Missouri, June 22, 1878, and in the same state were born his parents, George D. and Martha E. (Johnson) Williams, whose marriage was solemnized while the groom was home on a brief furlough from service as a Union soldier in the Civil War. George D. Williams was born in Harrison county, Missouri, a member of one of the old and honored families of that section, and when the Civil War was precipitated on the nation he manifested his patriotism by enlisting in the Thirteenth Missouri Cavalry, with which he served during the entire course of the great conflict between the states of the north and the south. After the surrender that brought the war to an end, it was found necessary to maintain troops for some time in repelling the uprisings of western Indians, who had taken advantage of war conditions, and thus Mr. Williams continued in service about a year, at Fort Sedgwick and at Moore's ranch, Colorado, where he aided in fighting the Indians. He then resumed the occupations of peace, and he continued as a representative of farm industry in Missouri until 1884, when he came to western Nebraska and numbered himself among the pioneers of that part of Cheyenne county that now comprises Deuel county. There he took up and eventually proved title to homestead, preëmption and tree claims, and he became a successful agriculturist, besides raising cattle and horses upon a substantial scale. He served four years as deputy sheriff of old Cheyenne county, and for three years he conducted a livery stable and business at Chappell, the present judicial center of Deuel county. In 1896 he removed to Kearney, where he engaged in the livery business, and in 1901 he removed to Julesburg, Colorado, where he continued to reside until his death, in December, 1917. His widow still resides at Julesburg, and in 1919 she celebrated the seventy-sixth anniversary of her birth. She

has gained wide experience in connection with life in the west, and both she and her husband gained a host of friends during their years of residence in western Nebraska.

David R. Williams has satisfaction in reverting to the fact that he attended the first school established at Chappell, the present county seat of Deuel county, and that his teacher in this pioneer school was Mrs. Onie Neil. Thereafter he continued his studies in the Kearney Military Academy, at Kearney, this state, and he was about eighteen years of age when he accompanied his parents on their removal to Julesburg, Colorado. In that state he remained variously engaged from the time he attained to his legal majority until he was twenty-eight years old, when he obtained 640 acres of land in what is now Morrill county, Nebraska, the tract being situated about nine miles northwest of the village of Lisco, having been at that time in Deuel county. He perfected his title to this land, which he still owns and upon which he has made good improvements, in consonance with the progressive spirit so definitely in evidence in this section of Nebraska. Mr. Williams now owns practically 3,000 acres of land in Garden and Morrill counties, and at Lisco he has the office headquarters of his well ordered and very successful real-estate business.

Mr. Williams is found arrayed as a stanch advocate and supporter of the principles of the Republican party. He is affiliated with Logan Lodge, No. 70, Ancient Free & Accepted Masons, at Julesburg, Colorado, and his wife holds active membership in the Presbyterian church.

At Kearney, Nebraska, on the 26th of April, 1905, was solemnized the marriage of Mr. Williams to Miss Esther J. Ewey, of Amherst, Buffalo county, where her parents were pioneer settlers; she having been born and reared in Nebraska and having received the advantages of the schools in the city of Kearney. Mr. and Mrs. Williams have three children, whose names and respective dates of birth are here noted: Helen S., April 18, 1906; John Kenneth, June 18, 1911; Earl Palmer, February 18, 1912.

JOSEPH W. KINNAMON is another of the progressive citizens who came in an early day to what is now Scotts Bluff county, then a part of Cheyenne county, and here his cumulative success has been on a parity with the remarkable development and advancement of the county during the intervening years. In 1888 he here entered claim to a homestead, near Gering, and his good judgment likewise led him to file also on preemption and tree claims, the whole comprising a tract which he reclaimed from the wilds and developed into a productive and valuable farm estate He disposed of a portion of his land, but still retains a model farm of 130 acres, worth $400 00 per acre, which he utilizes for diversified agriculture and stock-growing and which is situated one-half mile northwest of Gering Mr Kinnamon has been essentially one of the influential men of this locality and has achieved prosperity of a substantial order, the while he has at all times commanded unqualified popular confidence and good will His farm has the best of irrigation facilities and is well improved in all respects He is chairman of the company controlling the Gering irrigation ditch, and has aided in the construction of other ditches in the county He was one of the organizers of the Gering National Bank and continued as a stockholder and director of this representative institution In former years Mr Kinnamon conducted for a time a meat market at Gering, as did he also a feed and implement store His political allegiance is given to the Democratic party He is affiliated with the Masonic fraternity and his religious faith is that of the Methodist Episcopal church, of which his wife likewise was a devout member, she having been the daughter of a clergyman of that denomination

Mr Kinnamon was born in Clinton county, Ohio on the 16th of March, 1861, and is a son of James and Louisa (Wherry) Kinnamon, both likewise natives of the old Buckeye state, where the father became the owner of a good farm and was an honored citizen of Clinton county Both he and his wife passed the closing years of their lives in the state of Ohio Of the ten children, four are living James A is a farmer in Illinois, Thos W likewise follows agricultural pursuits in that county, Harry, a half-brother, resides in Ohio, and Joseph W is the immediate subject of this review The father was a Democrat in politics and both he and his wife held membership in the Christian church

Joseph W Kinnamon gained his youthful education in the public schools of Ohio and was sixteen years of age when he became a resident of Illinois When he was twenty-four years old he came to Nebraska and established his residence in Gage county, where he remained three years He then, in 1888, came to what is now Scotts Bluff county, where he has since maintained his home and where his success has kept pace with the growth and development of the county

In 1902 was solemnized the marriage of Mr

Kinnamon to Miss Loura E Mann, who was born and reared in this state and whose death occurred in 1917 Of the three children of this union two died in infancy, and the survivor, Joseph Carl, remains on the home farm and is attending school

ALSON J SHUMWAY, who started in the abstract business when he came to Scottsbluff in 1905, continued in the same until he entered the National army, October 16, 1918, and went to France as a member of an ammunition train, 77th division sector Y M C A

Mr Shumway was born at Oxford, Illinois, May 1, 1869, and is a son of G L Shumway, extended mention of whom will be found in this work He first attended the country schools and later Knox College, beginning business life in the newspaper business and prior to coming to Scottsbluff was editor of a journal published at Harrisburg, Nebraska

On September 1, 1896, Mr Shumway was united in marriage to Mrs Jennette (McKinnon) Rosenfeldt, who was born at Muskegon, Michigan, the fifth in a family of ten children born to Hugh and Elizabeth (Mickel) McKinnon, the other survivors being Mrs John R Kelley, of Harrisburg, Nebraska, Edward J, who is a farmer near Flowerfield, Nebraska, and M M and H O, both of whom are residents of Scottsbluff Mr and Mrs Shumway have two sons Burgess McKinnon and Hugh S, the latter of whom, born June 19, 1906, is yet in school The elder son was born June 19, 1898, was well educated and entered the National army for military training July 5, 1918 He remained in the training camp at Mare Island, California, until his honorable discharge, February 19, 1919 Mr and Mrs Shumway are members of the Christian Science church He belongs to the Knights of Pythias and the Royal Neighbors, is a Scottish Rite Mason and both he and wife are members of the Eastern Star Mr Shumway is a Republican in politics During his absence, Mrs Shumway carried on the abstract business very efficiently This office has the only set of abstract books in the county that have been photographed from the original records Mrs Shumway is of Scotch ancestry Her people came to Muskegon, Michigan in 1870, moved from there to Chicago, where her father was a machinist in the railway shops, and came to Nebraska in 1889 and homesteaded Both parents died in this state

MARK SPANOGLE, financier, and lawyer, is in years of service here one of the old-

Yours Truly
A.J. Shumway

Mark Spanogle

est active bankers in the north Platte valley and one of the best known men in Morrill county, having been cashier of the Bridgeport bank for twenty years. Ever since his arrival in Bridgeport he has taken an active part in the development of this section of the Panhandle; has been an indefatigable worker for the opening up of the valley; stood behind all movements for progress in all the varied industries of the county; and has helped turn the wheels of enterprise for the good of his community and Morrill county.

Though practically a native of Nebraska, as he has spent the greatest part of his life in this state, Mr. Spanogle was born in Lewiston, Mifflin county, Pennsylvania, April 27, 1868, the son of Andrew and Margaret (Rice) Spanogle, who came west in 1879, reaching Hamilton county, Nebraska, on March 4th of that year. As a young boy Mr. Spanogle had attended school in his native state and after accompanying his parents to the new home in the west, he attended the frontier schools maintained in Hamilton county at that time, thus completing his elementary education. When only nineteen years of age he began his business career as bookkeeper in a bank at Phillips, Nebraska, in 1887. Having a natural aptitude for finance, he was rapidly advanced and, in 1892, was elected manager of the bank of Phillips, though very young to hold such a responsible position. From first entering the financial circles of the state Mr. Spanogle's career has been marked by executive ability, initiative, self-reliance, and progressiveness. His high standing has gained him popular confidence and esteem, which has furthered the success of every bank with which he has been associated.

Matriculating in the law department of Drake University, Des Moines, Iowa, in 1892, Mr. Spanogle took this course with the idea of fitting himself still further for his business. He received his degree in 1894 and located in Clay county to practice law. Three years later he was elected county attorney of Clay county. serving in that office until the expiration of his term. In 1901, Mr. Spanogle came to the Panhandle, as he had long believed that this section of the state had a great future. Locating in Bridgeport, in August of that year, he was elected cashier of the Bridgeport bank on September 1st, a position which he has since continuously filled. Just two years after settling in the valley, Mr. Spanogle was elected attorney of Cheyenne county in 1903. Within a short period he became one of the organizers of the Union State bank of Broadwater, Nebraska, and, in 1917, was elected its president.

Mr. Spanogle is the pioneer banker of the north Platte valley and his influence here has been wide and of great value to the people; for his policies have ever been constructive and of benefit in the development of every community where he has banking interests.

When the question of dividing Cheyenne county came up, Mr. Spanogle was made chairman of the campaign committee which worked for the erection of Morrill county as a separate unit of the state. This was a long, hard fight and it was due to the able work and administrative ability of the chairman that in the end it was settled in an amicable manner. The new county was erected and no hard feeling remained between the people of old Cheyenne county and new Morrill county, and their relations have been cordial and happy ever since. When the selection of the seat of justice came up for Morrill county, Bayard made a hard fight for it and again Mr. Spanogle was called on to head the campaign to secure it for Bridgeport. It was due to his able management that Bridgeport became the county seat, and an easy winner in the contest.

In 1904 Mr. Spanogle, Robert Willis, L. R. North, and Charles F. Clawgs became convinced that the rich alluvial soil of the valley was excellent for raising sugar beets; others were skeptical, and in order to convince people these men planted forty acres in beets. They knew little or nothing of sugar beet culture, were rank amateurs as farmers, but Colonel Atkins helped them, hired a crowd of high school boys to hoe the beets during the summer and in the fall when the crop had matured the beets were harvested, and shipped to Grand Island to the sugar factory. The test proved a great success and fully demonstrated that this section was a fine location for raising beets. From the date of shipping this first crop inquiries began to come in with regard to locating in the valley for beet growing. More settlers arrived, new farms were opened up, with the result that the country around Bridgeport is today one of the most highly productive, not only in western Nebraska but the whole country. Thousands of acres of land are now under cultivation that were virgin soil; more are and will be broken and sowed to beets. Four sugar factories are already in operation, with the prospect of others in the near future to care for the increasing crops; and much of this is due to the small group of men who had the enterprise and courage to go ahead and show what could be done in this district.

The early irrigation projects in the valley had been largely by independent corporations, operating individually, all with different water rights and rates. As the new land was opened up and entered new methods were needed, and it soon became apparent that some general unified irrigation system was needed, as the old companies were all working on different plans and basis. It was necessary to enact new laws that would tend to unify the general system. When this was known, Mr Spanogle was asked to assist in getting the different old companies to join with the new projects for a unified district and federal system, which was accomplished with little friction and for the benefit of all who bought water for irrigation.

During the war Mr Spanogle devoted much of his time to assisting the government in its prosecution. He was drafted as chairman of the Liberty Loan committee and was placed in charge of the western six counties of the state. His work was continuous and he prosecuted a vigorous campaign with the result that the Panhandle did its full share in buying bonds, due largely to his personal activities.

April 22, 1899, Mr Spanogle married Miss Gertrude L Hurd, of Harvard, Nebraska, and since coming to Bridgeport they have been recognized as leaders of every communal movement for the development of the town. Mr Spanogle is a member of the Masonic order, while his wife belongs to the Episcopal church.

A hard worker, Mr Spanogle is naturally a busy man, but never too much so to give a word of personal advice or listen to a proposition of benefit to Bridgeport or Morrill county. He has a host of warm friends and is known best to them as "Mark." A man of great constructive ability he has builded soundly and well for the upper valley, ever placing its welfare before his own personal affairs and interests.

INDEX OF PORTRAITS

ANDERSON, VICTOR M D 17
Atkins, Auburn W 21

BAILEY, MR AND MRS ARTHUR 141
Baker James O 121
Barkhoff, Mr and Mrs William and Family, 545
Bartlett, Arthur M, 689
Beard, Andrew Broaddus, 341
Beard, Mr and Mrs George 521
Beck Mr and Mrs 197
Bellows, Mr and Mrs Frank J, 533
Braddock, Mr and Mrs Wm, 727
Brown, Hope and Family 73
Bushee, Berton Kenyon 5

CAMPBELL, MR AND MRS JOHN H, 565
Chambers, R O 213
Chew, Fred and Family 561
Coffee Charles F, 29
Cozdill George K and Family 697
Colbert, Frederick J MD 425
Crawford Andrew T, 41
Cromer, Edward P 169
Crume Harvey Wallace and Family, 245

DeBILY, MR AND MRS WILLIAM, 269
DeLaMatter, Enos S 349
DeLaMatter Mr and Mrs Robert, 165
Bean Richard F, James Emmett, Remsberg Asa 601
Dickinson, Mr and Mrs Seymour, 285

LEHRMAN, GEORGE, 173
Lhrman, George, Farm Residence of 173
Ehrman Mr and Mrs Fred and Residence, 369
Engstrom John and Family 93
Engstrom John Ranch of, 93
Everett Fred F 465
Ewbank, Mr and Mrs John 529

FALLK ANDREW J, MD, 105
Filer, John I, 525
Fisher, Aaron P and Five Generations, 237
Fisher, Frank F, 453
Flower, Charles H and Family, 229

French John L and Family 161
Fincher Mr and Mrs Gilbert 673
Foreman Frank L and Family 85

GARRARD ROBERT 549
Goos Adolf L 325
Green Mr and Mrs J N 209
Green Farm Residence of Mr and Mrs 209
Grewell C T and Family 585
Gumner Alfred W 413
Gurnsey, Horace W, 317

HALL EUGENE A 478
Howard Mrs and Mrs Thomas M 461
Howard Mr and Mrs James 277
Hiersche Wenzel 449
Halley Tullius 473
Hampton Rodolphus M 9
Hodder Ernest C 593
Holliday Mr and Mrs and Son 553
Hotchkiss Mr and Mrs Charles B 389
Hunt Geo J 493

JENKINS LEMUR Z AND WIFE 113
Jennings Mr and Mrs Walter 81
Jones John N and Family 97

KAMANS GOTTFRIED 129
Kennedy Mr and Mrs Anthony 481
Kinnamon Joseph W, 737
Kronberg Chris and Family 77

LARSON S N AND FAMILY 319
Lancomer George 485
Linden William D 117
Linn Gus 509
Lockwood Charles Elmer 33
Lovelace Oscar R 629
Lyda Curtis O 401

McCOSKEY ALLEN B, 469
McCormick Mr and Mrs Chris McCormick Edd W McCormick James McCormick Jack McCormick Jennie McCormick Robert N 597
McKinnon Mr and Mrs Ed J 581
McKinnon, Ranch of Ed J, 581

McClendhin Mrs Elizabeth, 65

MAGINNIS, PATRICK 25
Manning Charles I 501
Marcott Miles J 443
Mason George I 49
Morris Mr and Mrs 177
Moss Dr A F 509
Mathers A N 645
Maupin, William and Family, 357

NEIGHBORS MR AND MRS JOSEPH 111
Nelson Mr and Mrs Peter 557
Newell, John and Family 573
Nichols, Yorick, 149

O'BANNON, MR AND MRS OSCAR AND SON (0)
O'Harra, Thomas L 365
Orr John H, 589
Orr John A and Wife, 457

PARMENTIER MR AND MRS A M 427
Peterson Mr and Mrs 577
Peterson Petrus and Family 541
Petite Captain Albert M and Wife 37
Pickett Mr and Mrs 61
Pickering Mr and Mrs Jesse, 221
Plummer Deaver N and Wife 125

RANDALL, WILL N AND FAMILY, 189
Raymond Lewis L 13
Roush, Isaac 538
Reeves Ruben Thomas, 45
Russell James R and Family 137
Reynolds Edward M and Family 605

SAMS AND McCATER 417
Sayre Edward W and Family, 641
Scanlon Mrs Jula and Husband 643
Schooley Wm H and Wife 193
Schooley Elmer 193
Scott Ambrose I 445
Seger Bert J 441
Sherwood Alson I 738
Shumway Mr and Mrs S B 418
Simmons Charles H and Family 469

Smith, Alva A and Family, 89
Smith, H Leslie, 413
Snow, George T, 705
Spanogle Mark, 738
Spracklen, John W and
  Family, 661
Stauffer, Frank F and
  Family, 649

THOELICKE, MR AND
  MRS J T, 53
Thomas, Gustav Adoph,
  Thomas, Gottfried, Thomas
  Carl, Thomas, Christian
  Henry, 57
Thornton, Henry M, 381

Troy, Francis M, 373

VAN PELT THOMAS, CYRUS
  and Family, Nancy, Mary,
  Wm, Thomas, Mrs Sarah
  Johnson, Jonathan, 657
Vogler, Mr and
  Mrs Henry, 513
Vonburg, Peter, 133

WADE, J B AND FAMILY, 360
Wallace Mr and
  Mrs D E, 293
Wallace, Mr and
  Mrs William L, 157
Walters, Mr and

Mrs William T, 261
Westervelt, Eugene T, 405
White, W W, 477
Whitehead, James Theaker,
  633
Wilcox, George L, 429
Willis, Robert H, 489
Wilson, Everett P, 665
Wisner, Mr and
  Mrs I O 537
Wood, Asa B, 333
Wright, Mr and Mrs
  Albert M, 713
Wright, William H, 433
William, David R, 737

ZLINER, ERNEST, 301

GENERAL INDEX

ABBOTT, CHRISTOPHER C 668
Abbott, Frank W., 290
Abegg, Frank 367
Adams, George M 495
Adams, John H, 420
Ahlstrom, Andrew F, 108
Allen, Anson B, 632
Allen, George F, 448
Allen, Silas G, M D, 44
Allison, Frank 223
Alexander Frederick, 540
Alvis, William E, 132
Alumbaugh, Rolla W, 325
Amos, Horace C, 153
Anderson, Andrew, 176
Anderson, Charles E, 668
Anderson, Charles S, 547
Anderson, James T, 76
Anderson, John M, 510
Anderson, Victor, M D, 17
Ansen, Fred, 86
Arnold, Richard H, 476
Atkins, Auburn W, 21
Atkins, William D, 148

BABCOCK, HARVEY I, 604
Babcock, Roy Allen 546
Bailey, Arthur J, 141
Baird, Harry I, 247
Baker, Francis O, 152
Baker, James O, 721
Baker, Walter E, 250
Bald, Frederick A, 351
Bald, Harvey K, 505
Banks, James H, 490
Barbour, W M, 516
Barkell, Philip R, 315
Barkell, Thomas C, 611
Barkhoff, William, 545
Barlow, Robert A, 523
Barnes, William 462
Barnwell, William 450
Barrett, Ewing F, 612
Barrett, Robert F, 111
Bartlett, Arthur M, 689
Barton, Samuel 258
Basye, G Lee, 327
Baur, John G, 178
Baxter, James, 98
Baxter, James A, 190
Bean, Richard E, 601
Beatty, Wallace Dwight 554
Beatty, Warren, 465
Beard, Andrew Broaddus 341
Beard, Edwin A, M D, 243
Beard, George W, 521
Beck, William H, 197
Becker, Frank R, 68
Beckwith, Roy, 371
Beebe, Harvey, 220
Beerline, George W, 422
Beerline, Mike 118

Bellows, Frank J, 533
Bennett, George N, 321
Bennett, Sewell F, 451
Berry, Lyman A, 346
Bigelow, Ira 189
Bigler, Mitchell I, 526
Bigsby, Everett 616
Bigsby, Rolland B, 306
Birt, Clarence F, 172
Black, Frank L, 482
Blak, Emar, M D, 378
Blausey, J Frank 714
Blood, Henry C, 196
Bly, John W, 100
Boatsman, John, 241
Boggs, Clarence E, 59
Bogle, Charles L, 157
Bogle, James W, 181
Bogle, John F, 163
Bogle, Thomas L, 166
Bowen, Harry T, 542
Bower, Walter W, 440
Bowers, Thomas F, 170
Bowman, Daniel 187
Bowman, Leroy 187
Brackman, Charles, 191
Brady, John, 196
Braddock, William, 727
Brashear, H C, 213
Braziel, Perry 153
Brennan, Frank I, 309
Brewster, Benjamin A, 490
Brever, Fred 568
Bristol, Martin, 560
Brunt, J Monroe 695
Brodhead, John A, 408
Brodrick, Mrs William 576
Broshar, John W, 92
Brown, Horace E, 47
Brown, Harlin I, M D, 77
Brown, Hope 73
Brown, John J, 193
Brown, Olmsted Busher 408
Brown, Stephen L, 710
Brown, William G, D D S, 22
Bryan, Fred M, 74
Bryan, Paul F, 75
Buckner, George R, 583
Buettner, W H, 524
Burk, C H, 199
Burlew, Charles A, 377
Burnett, Arthur I, 291
Burns, Fred L, 197
Bushee, Burton Kenyon, 5
Buske, August 448

CADWELL, CHARLES J, 528
Caldwell, James A, 144
Calhoun, William F, 82
Canaday, Ralph O, 115
Canaday, Walter A, 116
Campbell, Frank T, 156

Campbell, Jessie 550
Campbell, John H, 565
Campbell, Runes C, 264
Campbell, William N, 712
Capper, Ralph F, 566
Card, James A, 425
Cargill, Vert B, 147
Carlson, Guy, 52
Carlson, Oscar A, 187
Carlson, Theodore, 186
Carr, Charles E, 715
Carr, James M, 60
Casper, Charles D, 109
Catton, Isaac, 502
Chadron State Bank 490
Chambers, Charles P, 625
Chambers, Clarence S, 456
Chambers, Robert O, 213
Chambers, Thomas L, 272
Chapen, Orin Delbert 284
Chew, Frederick C, 561
Chiles, Asa E, 72
Christensen, Joseph C, 531
Christensen, Michael 486
Christenson, Ola 592
Churchill, Charles R, 535
Clark, Richard 559
Clark, Walter, 522
Clary, Morse P, 557
Clausen, John Jr, 171
Clawges, Charles F, 121
Cleveland, Herbert L, 192
Cleveland, Ziba Valette, 147
Cline, William S, 87
Clough, Charles E, 348
Cluck, Curtis M, 236
Cluck, Millard F, 99
Clure, John, 650
Clure, John E, 194
Cochran, George M, 696
Cockrell, David S, 675
Coffee, Charles F, 29
Coffee, Harry B, 683
Coffee, Samuel Buffington 683
Cogdill, George K, 697
Colbert, Frederick J, M D, 125
Collins, Earl W, 202
Conklin, G R, 200
Cook, John B, 34
Cook, Orla F, 235
Coomes, Jesse C, 78
Cooper, Charles C, 469
Cooper, Cyrus D, 259
Copley, Ira 702
Copley, James 707
Copsey, Herbert A, M D, 245
Cory, William M, 584
Cosper, Robert J, 80
Coursey, Harry P, 347
Cowen, Elisha M, 104
Cowen, Frank E, 103
Cox, Erastus W, 308

Cox, Thomas W G , 307
Crawford, Andrew, M D , 41
Crawford Andrew T , 41
Crawford, Theodore R , 678
Crites, Albert W , Hon 688
Cromer, Edward P 169
Cromer, George C , 260
Cronkleton George G 126
Cronn, Wilham J , 151
Crosby, Mark H , 619
Cross, Edd S , 296
Cross, Seward E 299
Cross, Vance J , 406
Crume, Harvey Wallace, 245
Curfman, George O , 635
Currie, Edwin A , 101
Curry, Robert, 247
Curtis, Albert Ernest, 207

Dailey, Bernard F 544
Dailey, Robert F , 553
Daniels, S W , 431
Danley, William A , 684
Darnall, Arthur, 287
Darnall, Ralph 287
Davis, A C , 214
Davis, Daniel D , 96
Davis, Evan G 18
Davis, Sydney J , 261
Day, Robert A , 691
Deal. Frank, 528
DeBely, William, 269
DeConly, Frank B , 64
DeGraw, George, 428
DeLaMatter, Enos S , 349
DeLaMatter, Robert M , 165
Delatour, Samuel P , 693
Denslow, Lloyd 15
Denton, George B , 194
Deutsch Theodore D , 39
Dickinson, Seymour S 285
Diehl, William N , 511
Donnelly, Patrick, 636
Donovan Millard F , 375
Doran, Ruby P , 79
Dormann, August, 91
Douglas John B 208
Dow, Clare A , 343
Downer, Alfred G , 403
Downer, Amon R , 7
Dubbs, Ralph E , 320
Dubuque, Harry A , 356
Dunham, Edward C , 87
Dunham, Andrew J 146
Dunn, Patrick J 363
Dunn, William E 297
Durnal, R F , 530
Dyson, William I 454

Eastman, Milburne, 674
Eber, William Christian, 572
Eberhardt Henry, 102
Eggers, Charles M , 701
Ehrman, Fred 309
Ehrman George, 173
Eliss Michael 666
Elliott, Robert I , 681
Elmquist John, 624
Elquist, Amos, 217
Engstrom John 93
Enlow, Jesse Franklin 204
Emerson, John S 658
Emerson Ralph W , 703
Emick, William W , 88

Ericson, Walter J , 125
Ernst, George A 164
Evans, William T , 205
Evans, Winfield, 48
Everett, Fred F , 465
Everett, Thomas R , 466
Ewbank, John E , 529
Ewing, William J , 722

Faden Frank Sydney, 571
Fairbairn, Robert H Jr , 533
Faught, Arthur M , M D , 27
Faught, William W , 628
Faulk, Andrew J , M D , 105
Ferguson, J H , 459
Fickes David F 435
Filer, John I , 525
Fink, John, 552
Finn, James, 463
Fischer, Frank F , 453
Fisher, Aaron P , 237
Fisher, Albert E , 113
Fleisbach, Chester, 42
Fleisbach Harry S , 40
Flower, Charles Henry, 229
Flower Lorenzo, 531
Folmsbee S S , 193
Ford, Riley, 507
Ford, George, 268
Foreman, Frank L , 85
Forsling, Oscar E , 148
Fornander, S August, 598
Foster, Joseph Herbert, 639
Foster, Charles B , 249
Foster, James C , 559
Franklin Robert I , 192
French, John E , 161
French William F , 17
Fricke, Lawrence A 138
Fritcher, Gilbert, 673
Fugate, Jefferson Davis, 460
Fuller, Herbert R , 519

Garr William H , 100
Gaines James A , 664
Gallogly, Julius E , 720
Garrard George H , 238
Garrard Robert P , 549
Garrett, Elton 257
Garrett, Vincent A 248
Garvey John T , 388
Garvey, Samuel S , 455
Gatliff Charles H 255
Gault Caleb W , 576
Gebauer Julius 422
Gentry, Benjamin F , 6
Gering Edson, 397
Getty William 651
Gilbert Arthur M , 652
Gilbert, Parvin E , 271
Gilbert Robert I 78
Gingrich Charles V 265
Gillette Robert E 78
Glenn John H , 487
Godhey Cyrus H , 209
Gooling Finest Eugene 154
Golden Theodore F 192
Gompert, Gerhard 227
Gompert Jacob 227
Goodwin Zadock 121
Goos Adolf F 325
Gordon William A 600
Graham Robert 339
Grimrell John W 598

Grassmuck, Gotleib, 189
Graves, Johnson H , 50
Green, Harry G , 536
Green, Milton M , 299
Greene, Thomas A , 380
Greenlee, Andrew K , 642
Grewell, Charles T , 585
Grimes, Sidney A D , 382
Grimm Joseph L , 12
Grisham, Claude E , 525
Grubbs, William A , 411
Gunmer, Alfred W , 613
Gumaer, H G , 617
Gumaer, William F , 444
Gummere, Sheridan, 214
Gunderson, Hans, 156
Gurnsey, Horace W , 317
Guthrie, John W 336
Guthrie, William E 112

Haas, G F , 183
Hadler, Charles M , 642
Hagerty, Charles F , 595
Hain, Corie J , 183
Haldeman, Ada M , 10
Hale, William A , 518
Hall, Eugene A 478
Hall, John H , 278
Halley, Tullus C , 473
Hamilton, Luther F , 23
Hampton, A J , 470
Hand, George J , M D , 355
Hanks Robert M , 18
Hanna, George Max, 526
Hanna, Samuel B , 150
Hannawald, Martin, 114
Hampton Rodolphus M , 9
Hanson, James Christian, 191
Hashman, Calvin L , 329
Harding, William Henry, 11
Hargraves M S 348
Harness, Elmer J , 446
Harpole, Charles H , 128
Harris, Addison V , 680
Harris, Milton E 95
Harrison, Clyde L , 63
Harshman Robert J , 273
Hartman, John C , 714
Harvey, Earl, 305
Harvey, George W , 151
Harvey, Mrs Nellie, 354
Harward, Harvey, 198
Hatterman, Henry C , 505
Hatterman, William G , 500
Hayward, Mrs Mary E , 685
Haxby, Robert, 98
Heinz, John, 472
Hendrikson Lars J , 318
Hermann, Arthur H , 291
Hershman, Chas E M D , 357
Hewett, James H H , 330
Hewitt, Mrs Louisa, 622
Hersche, Anton, 282
Hersche Wenzel, 449
Hickel, Solomon D , 724
High, Charles, 638
Higgins, Mathew J , 68
Hill, David W , 103
Hill Miss Myrtle, 195
Hills, Charles, 268
Hills, George H , 458
Hjalmar E Olson, 707
Hobart Ralph W , 28
Hodnett, William P , M D , 67

Hodder, Ernest C., 59ა
Hoffmann, Richard L., 664
Hohnstein, John, 190
Holloway, Frank, 569
Holladay, John B., 553
Honnold, Arthur R., L.L.B., 32
Hooker, Mark R., 571
Hopkins, Lemuel M., 305
Horrum, Emory, 160
Hotchkiss, Charles B., 389
Houghton, Fred J., 498
Howard, James N., 277
Howard, Roy L., 406
Howard, Thomas M., 461
Howe, Emory C., 158
Hubbard, Albert, 181
Hudson, John N., 239
Huffman, Walter D., 252
Hunt, Cole, 428
Hunt, Frank N., 116
Hunt, Judge J., 493
Hutchinson, Cecil Fay, 262

Ingles, George W., 612
Ireland, George, 228
Ireland, Ted L., 22
Ireland, Wilbur J., 79
Irion, Charles H., 57

Jackson, William T., 703
Jacoby Charles E., 180
James, Emmett, 602
James, Frederick B., 636
Janssen, Peter, 230
Jeffers, Charles W., 366
Jeffords, Arthur A., 196
Jenkins, Elmer Z., 113
Jennings, Walter E., 81
Jenny, John L., 398
Jensen, Christian, 608
Jensen, Gotfred, 575
Jensen, Jens, 574
Jessup, James, 292
Johns, William, 199
Johnson, Charles O., 323
Johnson, Conrad A., 572
Johnson, Emil, 309
Johnson, Esther M., 587
Johnson, Frank E., 501
Johnson, Fred, 706
Johnson, James W., 644
Johnson, John, 600
Johnson, Julius J., 160
Johnson, Just, 692
Johnson, Mabel J., 107
Johnson, Nels, 648
Johnson, Syver, 623
Johnston, William, 423
Jones, Earl F., 317
Jones, George A., 610
Jones, Howard O., D.D.S., 20
Jones, John A., 97
Jones, Woodford G., 158
Jones, Woodford R., 164
Jones, Z. Harold, 108
Jordan, Willis B., 512
Jurberg, Ernest, 168

Kamann, Gottfried, 129
Karpf, Henry G., 242
Kaschka, Henry, 426
Keebaugh, John E., 218
Keller, Joseph E., 280
Kelley, E. Frank, 106

Kellums, John H., 254
Kelly, John R., 69
Kelly, Samuel, 618
Kendall, William I., 283
Kelly, Ted., 411
Kemplin, Cyrus L., 716
Kennedy, Anthony, 481
Kennedy, James M., D.D.S., 361
Kennedy, Thomas F., 74
Kent, William E., 82
Kerns, Albert B., D.D.S., 73
Kibble, Ephriam T., 339
Kiesel, Michael L., 203
Kilgore, John B., 653
Kimbrough, George F., 64
King, Martin J., 140
Kinnamon, Joseph W., 737
Kipp, J. J., 201
Kirkham, Valle B., 7
Kline, Michael M., 652
Klingman, Ernest H., 70
Knapp, Richard S., 515
Knapp, Thomas, 660
Knori Samuel, 732
Konkle, John, 279
Kronberg, Chris., 77
Kuehne, Herman, 622

Lackey, William M., 190
Lally, Thomas W., 719
Lamm, William H., 8
Lamm, William, Sr., 19
Lancaster, Myrtle J., 647
Lane, Barton E., 224
Lane, Guy., 46
Lane, J. Ray, 45
Lane, Jesse B., 44
Lane, Thomas B., Jr., 534
Langmaid, Owen W., 302
Larson, Hans C. L., 174
Larson, John A., 251
Larson, Swen N., 319
Laucomer, George, 485
Law, William L., 444
Lawton, Gus W., 75
Lawyer, George W., 262
Leafdale, George W., 316
Leavitt, Heyward G., 539
Leithoff, C. L., 674
Lester, Edward E., 171
Levensky, Benjamin, 654
Lewis, Lee E., 51
Liggett Furniture Co., 94
Lindberg, Fred R., 538
Linden, William D., 117
Linderman, Conrad, 630
Linn, Gus, 509
Lisco, Reuben, 436
Lockwood, Charles Elmer, 33
Loewenthal, Benjamin, 690
Loewenstein, John L., 127
Logan, Frank Lincoln, 206
Lovelace, Oscar R., 629
Luft, George B., 726
Lyda, Curtis O., 401
Lyman, William H., 14
Lynholm, Arthur & Harold, 549

McCain, William A., 54
McCaffree, Floyd S., 421
McComber, William, 301
McClenahan, Elijah, 65
McComsey, John T., 667
McConnell, James J., 717

McCorkle, John C., 350
McCormick, Christopher, 597
McCormack, Jack G., 603
McCormick, George W., 732
McCorick, Philip, 590
McCoskey, Allen B., 409
McCreary, Frank A., 35
McCreary, James C., 42
McCue, Charles J., 515
McDaniel James W., 520
McDonald, Kenneth W., 442
McDowell, Edwin C., 671
McElroy, Charles C., 45
McFeron, Charles W., 568
McHenry, Llyn O., 145
McHenry, Matthew H., 12
McIntosh, James L., 646
McIntosh, John T., 646
McKelvey, William T., 123
McKinnon, Edward J., 581
McNett, John, 294
McProud, Guy C., 186
McRae, Clarence V., 430
McSween, Philo J., 104

Maginnis, Patrick, 25
Magruder, Frank C., 52
Mahaffy, W. J., 363
Mainard, Jonnie S., 233
Manning, Charles F., 501
Manning, John R., 163
Manser, Gotleib C., 507
Mantor, Hugh E., 582
Mark, George E., 135
Mark, Hartson A., 432
Marlin, Jesse H., 266
Marlin, William, 136
Marshall, John F., 192
Marsteller, John Elmer, 731
Martin, John M., 70
Maryott, Miles J., 443
Mason, George E., 49
Matheny, Charles M., 60
Mathers, Albert N., 645
Mathson, John, 269
Maupin, Will M., 357
Maxwell, Hiram W., 453
Meek, William Baston, 188
Meglemre, Clyde E., 538
Melick, C. Russell, 389
Melick, Frederick W., 385
Melick, Grant G., 372
Melius, Fred L., 711
Melius, Jesse P., 711
Melton, William G., 510
Meredith, Lawson E., 258
Meston, Alexander, 56
Mevick, John, 694
Meyer, Fred William, 730
Meyer Louis, M., 452
Michael, Philip J., 387
Middaugh, Asa F., 55·
Miles, William P., 586
Miller, Adelbert A., 101
Miller, Harry I., 312
Miller, Henry, 139
Miller, James W., 328
Miller, Jesse M., 331
Miller, Melvin, 275
Miller, Robert C., 386
Miller, Robert G., 20
Miller, S. A., 368
Miller, True, 341
Miller, William P., 311

Minick, Charles A , 672
Minshall, Jess R 420
Mintle, Thomas C , 456
Mitchell, Wilson, 304
Montz, John W 63
Montz, Labarnah A , 276
Montz, Martie R , 64
Moomaw, Austin, 195
Moomaw, Leon A , 143
Moon, Alonzo L., 221
Moore, Clyde N . M D , 46
Moore, George W , 459
Morby, Fred, 167
Morgan, Frank B , 90
Morgan, Fred Reynard, 517
Morris, George II , M D , 434
Morris, John W , 177
Morrison, Amos C , 224
Morrison, Charles O , 123
Morrow, John C , 334
Morrow, Thomas M , 28
Morrow, William, 32
Moss, Albert E , D O , 569
Moyer, Augustus L , 494
Mueller John L 233
Muhr, John L , 288
Muhr, William C 288
Muhr, Walter A , 288
Muirhead Alexander, 390
Munroe, George A , 255
Murphy, James R , 52

Nash, Eli F , 551
Neely, Calvin, 149
Neeley, Charles E 200
Neeley, Franklin E , 36
Neeley, Robert G , 8
Neighbors, Joseph G , 253
Neighbors, Melville, 195
Neighbors, Thomas F , 111
Neff, Henry W , 50
Nelson, Charles C , 663
Nelson, Charles G 179
Nelson, Hans P , 174
Nelson, Jacob M 176
Nelson, Peter, 557
Nelson, Philip, 154
Neumann, Raymond C , 116
Newberry, Chema A , 578
Newcomb, Horatio G , 670
Newell, John, 573
Newell, William W 225
Newlin John H 729
Newman, August G 511
Newsum, I W , 136
Nichols, Yorick, 149
Niehus, Henry, 532
Nolan, Michael F , 337
Noyes, George, 303
Nusz, Christian, 140

O'Bannon Oscar O , 609
O Grady, Patrick, 653
O Hair Thomas L , 365
O Kane James, 641
O'Keefe, E B , D D S 364
O Keefe, John and
    William I , 352
O Keefe, Thomas J , 580
Oldaker Charles J , 180
Olson, Lars, 401
Olson Hjalmar F 707
Orr, James W 709
Orr, John A , 457

Orr, John H , 589
O'Shea, Peter, 38
Osborne, Dale B , 429
Osborne, Robert, 640
Osborne, Robert L , Jr , 611
Osborne, Rev Thomas C , 142
Otte, Henry J , 267
Otte, William, 205

Paisley, Ira, 708
Palm, F John, 315
Parmenter, Albert M , 427
Patterson, Fred W , 685
Patterson, Eugene A , 678
Pattison, Harry B , 263
Pattison, Joe N , 263
Patton, James, 310
Paxton, Albert, 93
Pearson, Gust, 309
Pebley, Joseph, 447
Peckham, John S , 16
Peckham, George B , 16
Pedersen, Jens C , 543
Pedrett, Jacob, 159
Peterson, Albert W 92
Peterson, Claus E , 577
Peterson, Frank, 322
Peterson, Henry C , 606
Peterson, Petrus, 541
Peterman, Shadrach, 492
Petite, Captain Albert M , 37
Pfeifer, Christian, 615
Pfeifer, John Harvey, 201
Phelps, Edwin A , Sr , 599
Phinney, Zina, 245
Phillips, Joseph H , 166
Phillips, Thomas Edwin, 731
Pickering, Frederick, 712
Pickering, Jesse, 221
Pidgeon, Carl, 503
Pieper, Hugo, 211
Pieper, T H , 212
Pickett, Dick, 61
Pickett, Oxley D , 153
Pindell, James D , 503
Pitman, Benjamin F , 686
Plummer, Denver Newton, 125
Plummer, John W , 228
Pollock, D J 67
Porter, Edgar C , 122
Potts, William A , 484
Potmesil James V 383
Powell, Frank E , 253
Price, L Frank, 155
Prohs, Otto J , 19
Pullen, William F , 568
Putnam, Frank H , 110

Quint, Robert, 439
Quinn, Edward M , 530
Quivey, William, 562

Ratcliff, Clayton, 647
Randall, Dean E , 567
Randall, Henry E , 124
Randall Will N 189
Raymond, John F , 84
Raymond, Lewis L , 13
Rasmussen, John N , 175
Rasmussen Nelson H , M D , 27
Read David R 182
Reddish, Robert O , 361
Reddish Frank L , 362
Redheld, Franklin A , 198

Redfield, William C , 219
Reed, Monroe J , 188
Reel, Joseph L , 544
Reeves, William, 45
Reid, Edward H , 219
Reitz, Pearl A , 679
Remsberg, Asa, 602
Reynolds, Edward M , 605
Reynolds, William H , 488
Rhoades, W H , 530
Rice, Claude D , 395
Rice, J S , 266
Richardson, George E , 508
Rider, Cyrus W , 324
Ridge, Hugh, 608
Rinne, Herman, 528
Ripley, Samuel Willard, 71
Ritchie, William, Jr , 655
Roberts, B F , 514
Roberts, Clarence E , 531
Roberts, Frederick H , 541
Robert, H Willis, 489
Roberts, Wilbern, 231
Robertson, Frank, 651
Robertson, John, 144
Robinson, John, 438
Robinson, John W 163
Roblee, Lafayette, 705
Rockey, Nathan A , 579
Rodgers, Allen D , 369
Rodgers, Bird S , 715
Rolph, Edward L , M D , 180
Romine, Mrs Freda, 495
Romine, Mrs Gertrude H , 496
Rosebrough Benajah A , 109
Rosenfelt, James S , 277
Ross, Ida M , 378
Ross, Gilbert, 142
Roudehush, Jacob H 446
Rouse, Earnest G , 467
Roush, Isaac, 537
Rowland Patrick 424
Rumer, George P , 449
Russell, Henry E , 222
Russell, Herbert G , 464
Russell, James R , 137
Rutledge, Fred D , 170
Ryburn, Agnew R , 506

Sams, Harvey L , 417
Sands, Frank M , 566
Sayre, Edward W , 641
Chambers, Charles P , 625
Scanlon, Dennis Joseph, 643
Schaffer, Roy, 185
Schenck, Daniel R , 48
Schick, Edward, 659
Schindler, George, 620
Schlater, Jacob C , 439
Schlosser, Charles T , 451
Schmode, Henry A , 86
Schneider, Charles, 532
Schnurr, Albert L , 734
Schnurr, Ralph B , 735
Schooley, Elmer, 193
Schooley, James R , 491
Schooley, William H , 193
Schroeder, Max, 184
Schuemaker, Franklin W , 416
Schumacher, John, 90
Schumacher, William Jos , 210
Schwarner, Lou, 72
Scott, Ambrose F , 445
Scott, Elbert, 119

Scott, Fremont, 24
Seger, Bert J., 441
Scovel, Byron L., 682
Selzer, Arthur L., 54
Selzer, Michael, 55
Seybolt, Albert T., 120
Shafto, Milton E., 402
Shaul, William D., 408
Shaw, James D., 59
Shepard, Wilbur F., 733
Shumway, Alson J., 738
Shumway, Stephen B., 418
Sickels, George B., 527
Simmons, Charles H., 469
Simmons, Charles S., 91
Simmons, Otis W., 95
Simmons, Robert G., 14
Simmons, William L., 102
Simonian, Armenag, 274
Skinner, Richard, 413
Slattery Brothers, 733
Slawson, Frederick N., 725
Smice, Henry H., 394
Smith, Alva A., 89
Smith, Charles H., 534
Smith, Fenner A., 395
Smith, Frank H., 392
Smith, H. Leslie, 413
Smith, Jerome H., 43
Smith, Humphrey, 639
Smith, Orval, 427
Smith, Stephen, 232
Smith, William T., 208
Snook, Uzell T., 410
Snow, George C., 705
Snyder, Mervin, 324
Snyder, Nathaniel M., 56
Soderquist, Peter, 627
Sonday, David H., 167
Sorensen, Severin, 36
Sowerwine, George, 274
Spahr, William C., 298
Spanogle, Mark, 738
Spanogle, Clyde, 110
Spear, Levi B., 300
Spencer, Elwin M., 431
Spillman, Homer J., 592
Spohn, William, 631
Spracklen, John W., 661
Springer, Henry M., 215
Stalnaker, Sarah Rosella, 588
Stark, Sanford, 83
Stauffer, Frank F., 649
Stearns, Frank, 565
Steen, Clarence G., D.D.S., 76
Steuteville, Charles E., 114
Steuteville, John H., 107
Stockwell, George W., 59
Stockdale, William Tobert, 483
Stone, Pearl M., 138
Stone, Fred W., 555
Streeks, George E., 615
St. Agnes Academy, 353
Stewart, Alfred J., M.D., 134
Stewart, Allison E., 263
Stewart, Edward, 470
Stewart, Herman G., 131
Steward, John C., 508

Sudman, Angust, 441
Swanson, Emal W., 654
Swanson, Charles E., 246
Swanson, Paul, 600
Swanson, Roy E., 426
Swindell, William B., 270

Tanner, Frank G., 182
Taplin, George F., 237
Taylor, Albert R., 704
Tennis, William T., 191
Thoelecke, Julius Theo., 53
Thomas, Carl, 260
Thomas, Elmer E., 313
Thomas, Harold S., 474
Thomas, Lloyd C., 358
Thomas, Samuel M., 734
Thomas, Valentine, 57
Thompson, James N., 430
Thompson, Peter, 281
Thornton, Henry M., 381
Thostesen, Ova N., 455
Thurman, Alonzo, 185
Tilgner, Henry, 701
Trout, Edward W., 230
Trowbridge, Arthur J., 326
Trowbridge, Stephen W., 326
Troy, Francis M., 373
Turnbull, George H., 162

Uglow, Charles H., 267
Uhrig, Anton, 332
Utter, Lewis Emerson, 473

Vanderberg, Edward F., 270
Valentine, Sarah E., Mrs., 693
Vannatta, Albert A., 681
Van Pelt, Cyrus, 657
Van Pelt, Thomas U., 415
Van Pelt, William, 314
Vaughn, Joseph H., 344
Vivian, Emmons C., 146
Vogel, John, 374
Vogler, George Lester, 574
Vogler, Henry, 513
Vonburg, Peter, 133

Wade, Jason B., 360
Wagoner, Carl A., 416
Waitman, Price P., 527
Waitman, Vernon, 532
Walker, Adam, 184
Wallace, Mrs. Lydia, 548
Wallace, William B., 552
Wallace, William L., 157
Wallage, Elmer, 293
Walters, William T., 261
Walrath, Andrew J., 502
Walsh, Henry, 563
Walsh, Robert G., 256
Warner, Arthur G., 313
Warner, Bert, 296
Warner, Daniel W., 424
Warner, William W., 314
Warren, Joseph E., 397
Watkins, Thomas F., 126
Watson, Claude R., M.D., 226
Walsh, Henry, 563

Waterman, W. W., 596
Watson, John T., 475
Webber, Charles V., 608
Weber, Bert R., 464
Weber, William, 128
Webster, James, 233
Wehn, Martin L., 419
Wells, L. P., 471
Wertz, John R., 603
West, Charles N., 240
Westervelt, Eugene T., 405
Westervelt, James P., 11
Willis, Robert H., 489
Willis, William H., 497
Whipple, Edward A., 285
Whitaker, George Keiper, 200
White, William Allen, 485
White, William W., 477
Whitehead, James Theaker, 633
Whitcomb Bros., 160
Whitman, George L., 232
Wieland, W. O., 637
Wiggins, Lloyd, 119
Wikston, Gustave, 535
Wilcox, George L., 429
Wiles, Isaac, 570
Wilhermsdorfer, J. H., 729
Williams, David R., 737
Williams, Joseph C., 130
Wilson, Thomas J., 676
Wilson, Everett P., 665
Wilson, James A., 700
Wilson, Jesse W., 396
Wilson, Manuel G., 186
Wilson, Roy D., 400
Wilson, William G., 391
Winterbotham, William H., 591
Wisner, Ray A., 537
Wisner, William M., 289
Wood, Asa B., 333
Wood, Edward S., 435
Wood, John T., 621
Wood, Joseph P., 198
Woodman, Joseph G., 231
Woolery, Joseph R., 717
Woten, Rev. Frank A., 84
Wynne, Clarence, 295
Wynne, Johnnie T., 295
Wright, Albert M., 713
Wright, Fred A., 437
Wright, S. B., 365
Wright, William H., 433
Wyatt, Clyde O., 286
Wyatt, Harvey L., 404

Young, Frank B., M.D., 23
Young, Fred L., 261
Young, Thomas H., 216
Youngheim, John W., 262

Zalman, Godfrey M., 624
Zehrer, Ernest, 301
Zehr, Nicholas E., 588
Zilmer, William H., 723
Zimmerman, Adam H., 504
Zoellner, Jonas, 43
Zorn, Elmer S., 403
Zorn, William H., 407

1481

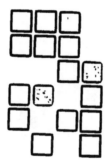

Printed in the USA
CPSIA information can be obtained
at www.ICGtesting.com
LVHW011929251123
764906LV00005B/190